2007 International Conference on Power Electronics and Drive Systems

Bangkok, Thailand
27-30 November 2007

Pages 1423-1897

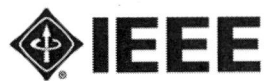

IEEE Catalog Number:	CFP07PEL-PRT
ISBN 10:	1-4244-0644-7
ISBN 13:	978-1-4244-0644-9

Copyright © 2007 by The Institute of Electrical and Electronics Engineers, Inc.
All Rights Reserved

Copyright and Reprint Permissions: Abstracting is permitted with credit to the source. Libraries are permitted to photocopy beyond the limit of U.S. copyright law for private use of patrons those articles in this volume that carry a code at the bottom of the first page, provided the per-copy fee indicated in the code is paid through Copyright Clearance Center, 222 Rosewood Drive, Danvers, MA 01923.

For other copying, reprint or republications permission, write to IEEE Copyrights Manager, IEEE Operations Center, 445 Hoes Lane, Piscataway, New Jersey USA 08854. All rights reserved.

IEEE Catalog Number:	CFP07PEL-PRT
ISBN 10:	1-4244-0644-7
ISBN 13:	978-1-4244-0644-9
LOC:	2006933010

Additional Copies of This Publication Are Available from:

IEEE Service Center
445 Hoes Lane
Piscataway, NJ 08854
Phone: (800) 678-IEEE
              (732) 981-1393
Fax: (732) 981-9667
E-mail: customer-service@ieee.org

Organizers/Committees

Organizers:

Chulalongkorn Univ.
Center of Excellence in Electrical Power Tech., Chulalongkorn Univ.
King Mongkut's Univ. of Tech. Thonburi
King Mongkut's Inst. of Tech. Ladkrabang
King Mongkut's Inst. of Tech. North Bangkok
IEEE Thailand Section
IEEE IAS/PELS Joint Chapter, Singapore Section

Technical Co-Sponsors:

IEEE Power Electronics Society
IEEE Industry Applications Society
IEEE Industrial Electronics Society
IEEE IAS/PELS/IES Joint Chapter, Thailand Section

Organizing Committees

General Chairman	Doncker, R. D.	RWTH-Aachen Univ.
Advisory Board	Lavansiri, D.	Chulalongkorn Univ.
	Jaovisidha, V.	Chulalongkorn Univ.
	Tandhavatana, S.	IEEE Thailand Section
	Pungprasert, V.	IEEE Thailand Section
	Leelarasmee, E.	Chulalongkorn Univ.
	Pichetchumroen, V.	King Mongkut's Inst. of Tech. Ladkrabang
	Yingwatana, A.	King Mongkut's Inst. of Tech. North Bangkok
	Liang, Y. C.	National Univ. of Singapore
	Panda, S. K.	National Univ. of Singapore
General Co-Chairmen	Karnasuta, K.	IEEE Thailand Section
	Vilathgamuwa, D. M.	Nanyang Tech. Univ.
Organizing Committee Chairman	Phoomvuthisarn, S.	Center of Excellence in Electrical Power Tech. Chulalongkorn Univ.
Co-Chairman	Kulvitit, Y.	Chulalongkorn Univ.
	Yungyuen, U.	King Mongkut's Univ. of Tech. Thonburi
	Khan-ngern, W.	King Mongkut's Inst. of Tech. Ladkrabang
	Chunkag, V.	King Mongkut's Inst. of Tech. North Bangkok
Technical Program Committee	Khan-ngern, W.	King Mongkut's Inst. of Tech. Ladkrabang
	King Jet, T.	Nanyang Tech. Univ.
	Sangwongwanich, S.	Chulalongkorn Univ.
	Chunkag, V.	King Mongkut's Inst. of Tech. North Bangkok
	Sirisukprasert, S.	Kasetsart Univ.
	Boonyaroonate, I.	King Mongkut's Univ. of Tech. Thonburi
Treasurers	Bunnagulrote, B.	Center of Excellence in Electrical Power Tech. Chulalongkorn Univ.
	Battul, D.	School of Electrical & Electronic Engineering Singapore Polytechnic
Publications	Tarateeraseth, V.	Srinakharinwirot Univ.
	Jangwanitlert, A.	King Mongkut's Inst. of Tech. Ladkrabang
Tutorials	Chunkag, V.	King Mongkut's Inst. of Tech. North Bangkok
	Liutanakul, P.	King Mongkut's Inst. of Tech. North Bangkok
Local Arrangements	Kinnares, V.	King Mongkut's Inst. of Tech. Ladkrabang
	Polmai, S.	King Mongkut's Inst. of Tech. Ladkrabang
	Yutthagowith, P.	King Mongkut's Inst. of Tech. Ladkrabang
	Kittiratsatcha, S.	King Mongkut's Inst. of Tech. Ladkrabang
	Fuengwarodsakul, N.	King Mongkut's Inst. of Tech. North Bangkok
Publicity	Suwankawin, S.	Chulalongkorn Univ
Exhibition	Jangwanitlert, A.	King Mongkut's Inst. of Tech. Ladkrabang
Secretariat	Suwankawin, S.	Chulalongkorn Univ.

iii

International Steering Committee

Acarnley, P.	Univ. of Newcastle	Kennel, R.	Univ. of Wuppertal
Akagi, H.	Tokyo Inst. of Tech.	Kolar, J. W.	Swiss Federal Inst. of Tech. (ETH) Zürich
Alex, Q. H.	North Carolina State Univ.	Lai, J. S.	Virginia Polytechnic Inst. and State Univ.
Amaratunga, G. A. J.	Univ. of Cambridge	Longya, X.	Ohio State Univ.
Bhat, A. K. S.	Univ. of Victoria	Lorenz, R. D.	Univ. of Wisconsin-Madison
Boroyevich, D.	Virginia Polytechnic Inst. and State Univ.	Matsuse, K.	Meiji Univ.
Bose, B. K.	Univ. of Tennessee	Mohan, N.	Univ. of Minnesota
Chan, C.C.	Univ. of Hong Kong	Nakaoka, M.	Yamaguchi Univ.
Clare, J. C.	Univ. of Nottingham	Ninomiya, T.	Kyushu Univ.
Dehong, X.	Zhejiang Univ.	Okuma, S.	Nagoya Univ.
Divan, D.	Georgia Inst. of Tech.	Qian, Z.	Zhejiang Univ.
Elbuluk, M. E.	Univ. of Akron	Rahman, M. A.	Memorial Univ. of Newfoundland
Enjeti, P.	Texas A&M Univ.	Schroeder, D.	Univ. of Munich
Ertan, B.	Middle East Technical Univ.	Sekiya, H.	Chiba Univ.
Forsyth, A.J.	Univ. of Manchester	Sen, P.C.	Queen's Univ.
Green, T. C.	Imperial College	Shoyama, M.	Kyushu Univ.
Guo, X. D.	Harbin Inst. of Tech.	Suetsugu, T.	Fukuoka Univ.
Holtz, J.	Univ. of Wuppertal	Teck, O. B.	McGill Univ.
Hui, R. S. Y.	City Univ. of Hong Kong	Tenti, P.	Univ. of Padova
Husain, I.	Univ. of Akron	Tsai, P. C.	National Tsing Hua Univ.
Jahns, T. M.	Univ. of Wisconsin–Madison	Undeland, T. M.	Norwegian Univ. of Science and Tech.
Jain, P.	Queen's Univ.	Wu, B.	Ryerson Univ.
Jezernik, K.	Univ. of Maribor	Wyk, J. D. V.	Virginia Polytechnic Inst. and State Univ.
Kazimierczuk, M. K.	Wright State Univ.	Zhengming, Z.	Tsing Hua Univ.

History of PEDS Conference

International Conference on Power Electronics and Drive Systems, PEDS Conference, originated in Singapore and the first PEDS conference was held in Singapore in 1995. The aim of the PEDS Conference is to provide a forum for participants from the industry and academia in the area of power electronics and drives to exchange ideas and have interactions. The conference is biennial and since 1995 the PEDS Central Committee, Singapore in collaboration with overseas organizing committees, have organized PEDS Conference series held in various Asia Pacific (IEEE Region 10) countries. All the PEDS conferences are being held in technical co-sponsorship with the IEEE Power Electronics Society and IEEE Industry Applications Society.

PEDS Conference	Venue
PEDS 1995	Singapore
PEDS 1997	Singapore
PEDS 1999	Hong Kong
PEDS 2001	Bali, Indonesia
PEDS 2003	Singapore
PEDS 2005	Kuala Lumpur, Malaysia
PEDS 2007	Bangkok, Thailand

List of Reviewers

Abbaszadeh, K.
Abe, S.
Acarnley, P. P.
Adnani, M. E.
Afjei, E.
Ahmad, G.
Ahmed, M.
Al-Haddad, K.
Amirifar, R.
Ang, S.
Apte, A. A.
Attaviriyanupap, P.
Awad, H.
Azli, N. A.
Baiju, M. R.
Bakan, A. F.
Beig, A. R.
Bhadra, S. N.
Bharanikumar, R.
Bhat, A. K. S.
Bina, M. T.
Biswas, S. K.
Boonyaroonate, I.
Bunlaksananusorn, C.
Chang, K.-T.
Chen, H.
Chen, J.-J.
Cheng, M.-Y.
Chengfeng, Y.
Cheung, N. C.
Chiba, A.
Chien, F. T.
Chiu, H.-J.
Choi, B.
Chou, J.-H.
Chunkag, V.
Clare, J. C.
Colli, V. D.
Corzine, K. A.
Covic, G. A.
Cruden, A.
Dahono, P. A.
Daming, Z.
Dianguo, X.
Doki, S.
Dong-Hee, L.
Duffy, M.
Dzung, P. Q.
Elbuluk, M. E.
Ertugrul, N.
Eskander, M.
Farhangi, S.
Filho, E. R.
Forsyth, A.

Fujiwara, O
Fukuda, S.
Garvey, S. D.
Grabner, H.
Griva, G.
Gueldner, H.
Guo, Y.
Hagh, M. T.
Hakimie, H.
Hamzah, M. K.
Hamzah, N.
Hanamoto, T.
Hava, A. M.
Hayashi, Y.
Hennen, M. D.
Higuchi, T.
Ho, S.-T.
Hofmann, W.
Hori, Y.
Howe, D.
Hsieh, G.-C.
Hua, S.
Huang, L.
Huang, S.-J.
Hung, J. Y.
Hur, J.
Hussien, Z. F.
Idris, N. R. N.
Jain, P. K.
Janakiraman, P. A.
Jangwanitlert, A.
Jerome, J.
Jianxin, S.
Khan, P. K. S.
Khan-Ngern, W.
Kim, I.-S.
Kim, Y.-H.
Kinnares, K.
Kobayashi, S.
Kolar, J. W.
Komurcugil, H.
Kubota, H.
Kulvitit, Y.
Kurokawa, F.
Lafoz, M.
Lai, C.-K.
Lai, J.-S.
Lecci, A.
Lee, D.-C.
Lee, E.-W.
Lee, S. C.
Lee, Y.-S.
Li, D. D.
Li, G.

Li, H.
Li, J.
Li, W.
Liang, Y. C.
Liaw, C.-M.
Lin, C.-H.
Lin, R.-L.
Liserre, C.
Lo, Y.-K.
Loh, A.
Lorenz, L.
Low, K.-S.
Manmek, T
Markadeh, G. A.
Marques, G. D.
Martins, J. F. A.
Matsui, M.
Matsuo, K.
Mekhilef, S.
Morales-Castorena, A.
Morimoto, M.
Morimoto, S.
Mukerjee, R.
Muni, B. P.
Murthy, S. S.
Mutoh, A.
Muyeen, S. M.
Nagaraju, J.
Narayanan, G.
Nho,N. V.
Noguchi, T.
Nussbaumer, T.
Okou, A. F.
Omar, A. M.
Pai, F.-S.
Palandurkar, M. V.
Pan, C.-T.
Panda, S. K.
Patel, H. K.
Phuong, L. M.
Pichetjamroen, V.
Ping, H. W.
Pires, A.
Pires, V.
Polmai, S.
Ponce, M.
Qian, Z.
Rafael, S.
Rahman, M. A.
Ramasamy, A. K.
Rashad, E. E. M.
Ratanapanachote, S.
Rizk, J.
Ruan, X.

Saied, B. M.
Sangwongwanich, S.
Saudemont, C.
Saxena, T. K.
See, K. Y.
Senthilkumar, R.
Shaojun, X.
Sharma, V. K.
Shieh, H.-J.
Shimizu, T.
Shin, G.-H.
Shing, C. S.
Shinnaka, S.
Shuhua, F.
Singh, B.
Sirisukprasert, S.
Soltani, J.
Sopavanit, C.
Staines, C. S.

Sumedha
Sun, K.
Suwankawin, S.
Tahami, F.
Takahashi, R.
Tanaka, T.
Tarnekar, S. G.
Tenti, P.
Thounthong, P.
Tomita, H.
Tseng, K.-J.
Tsui, M.
Vaclavek, P.
Vaez-Zadeh, S.
Veszpremi, K.
Vijayarajan, K.
Vilathgamuwa, D. M.
Villasenor, A. G.
Wang, C.-M.

Wang, H.-P.
Wang, L.
Weiming, M.
Wen, F.-L.
Wolbank, T. M.
Wu, L.
Wu, T.-F.
Xu, D. (David)
Xu, D. (Dehong)
Xu, L.
You, K.
Yousfi, D.
Zhang, X.
Zhengyu, L.
Zhong , Q.-C.
Zhu, J.
Zirn, O.
Zolghadri, M. R.

This page intentionally left blank.

Table of Contents

Fuel cell systems and applications 1
Bernard Davat

Recent Trends iin Power Qualliity Improvements Techniiques 58
Bhim Singh

Power Electronics for Future Utility Applications 213
Rik W. De Doncker, Christoph Meyer, Robert U. Lenke, Florian Mura

Digital Control Generations -- Digital Controls for Power Electronics through the Third Generation 221
Philip T. Krein

Power Electronics and Control of Renewable Energy Systems 226
F. Iov, M. Ciobotaru, D. Sera, R. Teodorescu, F. Blaabjerg

Design and Evaluation of a 60 000 rpm Permanent Magnet Bearingless High Speed Motor 249
T. Schneider, A. Binder

Performance Investigation of Two-, Three- and Four-Phase Bearingless Slice Motor Configurations 257
M.T. Bartholet, S. Silber, T. Nussbaumer, J.W. Kolar

Compensation of Pole Position Estimation Error for Sensor-less IPMSM Drives with DC Link Current Detection 265
Hisao Kubota, Yusuke Shibano, Takayuki Kobayashi

A Novel Dual-Stator Hybrid Excited Synchronous Wind Generator 270
Liu Xiping, Lin Heyun, Yang Chengfeng, Fang Shuhua, Guo Jian

Application of Multi-level Multi-domain Modeling in the Design and Analysis of a PM Transverse Flux Motor with SMC Core 275
Youguang Guo, Jianguo Zhu, Dikai Liu, Haiyan Lu, Shuhong Wang

New Approximate 2DOF Digital Controller for DC-DC Converter with Second-Order Differential Characteristics 280
Eiji Takegami, Kohji Higuchi, Kazushi Nakano, Satoshi Tomioka, Kazushi Watanabe, K.K. Densei-Lambda

High Accuracy CMOS Current Sensing Circuit for Current Mode Control Buck Converter 286
Yuang-Shung Lee, Chih-Jen Hsu

Small Signal Analysis of a dual-switch forward Converter with non-ideal transformer in Current-Programmed Control 291
Weiping Zhang, Yuzhou Lei, Xiaoqiang Zhang, Yuanchao Liu

High Frequency Transformer Designs for Improving Cross Regulation in Multiple-Output Flyback Converters 295
Kusumal Chalermyanont, Pairote Sangampai, Anuwat Prasertsit, Surapon Theinmontri

Operation of a wye Connected Three- Level Active Power Filter under Nonideal Conditions 299
H.B. Zhang, A.M. Massoud, S.J. Finney, B.W.Williams, T.C. Lim, H. Hotait

Application of GPRS Techniques for Wide-Area Power Quality Monitoring 305
Shun-Yu Chan, Jen-Hao Teng, David Chang, Li-Yuan Chin

Design and Development of Autotransformer Based 24-Pulse AC-DC Converter fed Induction Motor Drive 310
Bhim Singh, Vipin Garg, G.Bhuvaneswari

Power Quality Monitoring System Using Real-Time Operating System 318
Krisda Yingkayun, Suttichai Premrudeepreechacharn, Kosol Oranpiroj

Technology Performance Comparison of Triacs Subjected to Fast Transient Voltages 322
L. Gonthier, A. Passal

Table of Contents

On-line Junction Temperature Measurement of CoolMOS Devices 327
Andreas Koenig, Thomas Plum, Peter Fidler, Rik W. De Doncker

Analytical Design of High-Power MTO Thyristors 333
Thomas Plum, Rik W. De Doncker

A Novel Gate Driver with Output Voltage Having Double Source Voltage 338
K. I. Hwu, Y. T. Yau

Effects of Internal Feedback and Gate-Drive Signal on the Turn-off Loss of MOSFET ZVS 342
Youthana Kulvitit, Puckapon Opanuruk, Tanvaa Tansatit

A Novel Bridge Type FCL Based on Single Controllable Switch 350
Wanmin Fei, Yanli Zhang, Qi Wang

A Novel Isolation Power Supply for Gating Multiple Devices in FACTS Equipment 354
Yanli Zhang, Wanmin Fei, Zhengyu Lu

Voltage and Frequency Controller for Parallel Operated Isolated Asynchronous Generators 357
Bhim Singh, Gaurav Kumar Kasal

DSP controlled Semiconductor based High-Voltage Source 363
F. Martin, T. Leibfried, O. Kerz, K. Mossner

Open Switch Fault Diagnosis for a Doubly-Fed Induction Generator 368
W. Sae-Kok, D M Grant

Rapid Analysis & Design Methodologies of High- Frequency LCLC Resonant Inverter as Electrodeless Fluorescent Lamp Ballast 376
Yong-Ann Ang, David Stone, Chris Bingham, Martin Foster

Analysis and Control of Dual-Output LCLC Resonant Converters, and the Impact of Leakage Inductance 382
Y. Ang, C. M. Bingham, M. P. Foster, D. A. Stone

A Novel QR ZCS Switched-Capacitor Bidirectional Converter 388
Yuang-Shung Lee, Yi-Pin Ko, Chien-An Chi

Analysis of a Half - Bridge Inverter for a Small- Size Induction Cooker Using Positive-Negative Phase-Shift Control under ZVS and NON-ZVS Operation 394
P. Achara, P. Viriya, K. Matsuse

Adaptive Phase Control Method for Load Variation of Resonant Converter with Piezoelectric Transformer 401
S. T. Yun, J. M. Sim, J. H. Park, S. J. Choi, B. H. Cho

Adaptation of Motor Parameters in Sensorless PMSM Drives 406
Antti Piippo, Marko Hinkkanen, Jorma Luomi

Development of 150000 r/min, 1.5 kW Permanent- Magnet Motor for Automotive Supercharger 414
Toshihiko Noguchi, Masaru Kano

Analysis and Performance Evaluation of Radial Flux Air-Cored Permanent Magnet Machines with Concentrated Coils 420
P.J. Randewijk, M.J. Kamper, R-J. Wang

Analysis and Experimental Investigation for Field-Control Capability of a Novel Hybrid Excitation Claw-Pole Synchronous Machine 427
Yang Chengfeng, Lin Heyun, Liu Xiping, Fang Shuhua, Guo Jian

A single-Capacitor Turn-off Snubber for Interleaved Boost Converter with Coupled Inductor 433
S.-Y. Tseng, J. Z. Shiang, Y.-H. Su

Buck-Boost Converter Associated with Active Clamp Forward Converter for PV Power System 440
S. Y. Tseng, W. C. Chen, Y. J. Li, J. S. Kuo

Table of Contents

Comparison of Three-Phase DC-DC Converters vs. Single-Phase DC-DC Converters 448
Christian P. Dick, Andreas Konig, Rik W. De Doncker

Applying Modified One-Comparator Counter-Based PWM Control Strategy to Flyback Converter 456
K. I. Hwu,, Y. H. Chen

Analysis of Conducted EMI Reduction on a Boost Converter Using Progressive Inductor Winding Technique 460
Kritsada Saritsiri, Werachet Khan-Ngern

Practical Issues Concerned with Zero sequence component and Harmonic Compensation in Four-Wire systems 465
E. Pashajavid, K. Kanzi, M. Tavakoli Bina

Automated Design and Implementation of Resonant Controllers for Current Control of Shunt Active Filters 470
W. Lenwari, M. Sumner, P. Zanchetta

A Modular Structured Multilevel Inverter Active Power Filter with Unified Constant-Frequency Integration Control for Nonlinear AC Loads 475
P. Y. Lim, N. A. Azli

HCC PWM Control of the Single-Phase Bi- Directional Buck Converter giving IEEE 519 Compliance at any Power Factor 480
A. N. Arvindan, V. K. Sharma

Passive EMI Filter Performance Improvements with Common Mode Voltage Cancellation Technique for PWM Inverter 488
C. Khun, W. Khan-Ngern, M. Kando

Novel Auxiliary Diagnosis Method for State-of-Health of Lead-Acid Battery 493
Yu-Hua Sun, Hurng-Liahng Jou, Jinn-Chang Wu

Electromechanical Model of a Longitudinal Mode Piezoelectric Transformer 498
Shine-Tzong Ho

Latest Development of Transformer Parasitic Inductive Components and Lossless Inductive Snubber-Assisted Series Resonant High-Frequency ZCS-PFM DC-DC Converter for RF Generator 504
Hisayuki Sugimura, Manabu Ishitobi, Bishwajit Saha, Sang Pil Mun, Soon Kurl Kwon, Mutsuo Nakaoka

A General Method for Deciding the Input Filter Capacitance of Flyback Switching AC-DC Converter with Peak Current-Controlled Mode 510
Jiaxin Chen, Jianguo Zhu, Youguang Guo

Design of High Performance and Low Cost Line Impedance Stabilization Network for University Power Electronics and EMC Laboratories 515
D. Sakulhirirak, V. Tarateeraseth, W. Khan-Ngern, N. Yoothanom

A Robust Output Current Control Method with Disturbance Observer for Matrix Converter under Unbalanced Input Voltage 521
Kazuo Oka, Kouki Matsuse

FPGA Design of Single-phase Matrix Converter Operating as Cycloconverter 527
Z. Idris, M.K. Hamzah, A. Saparon, N.R.Hamzah, N.Y. Dahlan

Input and Output Ripple Analysis of AC Chopper 534
Arwindra Rizqiawan, Dessy Amirudin, Deni, Pekik Argo Dahono

A Three-level 4 × 3 Conventional Matrix Converter 541
Runjie Rong, Poh Chiang Loh, Peng Wang, Frede Blaabjerg

A novel primary-side controlled contactless battery charger 546
Yi-Hwa Liu, Shun-Chung Wang, Rong Ceng Leou

Table of Contents

Research on Digital Soft-switch Welding/Cutting Inverter Power Source 551
G.R. Zhu, Z. Liu, X. Li, B.Y. Liu, S.X. Duan, Y. Kang

Design of an Adjustable High Output Voltage Asymmetrically Switched Class D Converter 556
M. Rentzsch, H. Guldner, C. Ditmanson

New Direct High Frequency Soft-Switching Inverter-Fed AC-DC Converter with Voltage Doubler for Consumer Magnetron Drive 563
Hisayuki Sugimura, Bishwajit Saha, Hidekazu Muraoka, Sang Pil Mun, Tomokazu Mishima, Hideki Omori, Mutsuo Nakaoka

Complete loading Characteristics Modeling of an Axial Flux Permanent Magnet Synchronous Machine Using Ck Spline Functions 569
Z. Lakhdari, F. Amrane, L. Adélaide, Ph. Makany

The Bearingless 2-Level Motor 574
P. Karutz, T. Nussbaumer, W. Gruber , J.W. Kolar

Analysis and Design of a Sliding Mode Controller for Buck Converters Operating in DCM with Adaptive Hysteresis Band Control Scheme 581
Hung-Chih Lin, Tsin-Yuan Chang

Buck Converter Simulation Technique Based on the Fourier Transform 587
Acacio M. R. Amaral, A. J. Marques Cardoso

ANALYSIS OF HOPF BIFURCATION IN DC-DC LUO CONVERTER USING CONTINUOUS TIME MODEL 595
A.Kavitha, G.Uma

Analysis of a Mixed-Signal Control for DC-DC Converters based on Hysteresis Modulation And Estimated Inductor Current 600
D. Trevisan, S. Saggini, P. Mattavelli, L. Corradini, P. Tenti

Power Quality Study in Macao 607
Sio-Un Tai, Man-Chung Wong, Ming-Chui Dong, Ying-Duo Han

Some Findings on Harmonic Measurement in Macao 614
Sio-Un Tai, Man-Chung Wong, Ming-Chui Dong, Ying-Duo Han

Coordinated design of PSS and TCSC dynamics model for power system network oscillations 620
M. Tarafdar Haque, A. Roshan Milani, A. Lafzi

An Analytic Approach To Harmonic Analysis of 48-Pulse Voltage Source Inverter 626
B. Geethalakshmi, P. Dananjayan

Detailed losses Analysis of High-Frequency Planar Power Transformer 632
Yu Ma, Peipei Meng, Junming Zhang, Zhaoming Qian

Design of a Nuclear Magnetic Resonance Fast Field Cycling Air Cored Magnet 636
Duarte M. Sousa, Gil D. Marques, Pedro J. Sebastiao,, Antonio C. Ribeiro

Using DFT to Obtain the Equivalent Circuit of Aluminum Electrolytic Capacitors 643
Acácio M. R. Amaral, Gustavo M. Buatti, Hugo Ribeiro, A.J. Marques Cardoso

A Mathematical Analysis on Vector Inversion Generators 648
D. J. Thrimawithana, U. K. Madawala

Novel Multi-Level High Voltage Pulsed Power Generator 654
D. J. Thrimawithana, U. K. Madawala

Potential and Electric Field Distribution Analysis of Field Limiting Ring and Field Plate by Device Simulator 660
C.N. Liao, F.T. Chien, Y.T. Tsai

xii

Table of Contents

Wire and Wireless Linked Remote Control for the Group Lighting System Using Induction Lamps 665
Kyu Min Cho, Jae Eul Yeon, Ma Xian Chao, Hee Jun Kim

Induction Heating with Traveling Magnetic Field for Uniform Heating to Flat Metal 671
T. Sekine, H. Tomita, Y. Saito, S. Obata, S. Yoshimura

Three-Phase (LC)(L)-Type Series-Resonant Converter with Capacitive Output Filter 677
M. Almardy, A.K.S. Bhat

Analysis of a Full-Bridge Inverter for Induction Heating Using Asymmetrical Phase-Shift Control under ZVS and NON-ZVS Operation 685
N. Yongyuth, P. Viriya, K. Matsuse

FPGA-Based Phase-Shift ZVS Full-Bridge DC-DC Converter Using One-Comparator Counter-Based PWM Control Strategy 692
K. I. Hwu, Y. T. Yau

A Simplified Power Control Scheme for Resonant Inverter with Purely Resistive Load 697
Pramoch Dorkmai, Youthana Kulvitit, Tanvaa Tansatit

Voltage Injection Based Initial Rotor Position Estimation Method for Three-Phase Star- Connected Switched Reluctance Machines 703
P. Somsiri, P. Champa, P. Wipasuramonton, K. Tungpimonrut, P. Aree

Control Scheme for Switched Reluctance Drives with Minimized DC-Link Capacitance 710
Christoph R. Neuhaus, Rik W. De Doncker, Nisai H. Fuengwarodsakul

Multiphase Torque-Sharing Concepts of Predictive PWM-DITC for SRM 716
Helge J. Brauer, Martin D. Hennen, Rik W. De Doncker

A New Two Phase Configuration for Switched Reluctance Motor with High Starting Torque 722
E. Afjei, K. Navi, S. Ataei

Application of Power Electronics for Damping of Torsional Vibrations 726
T. Zoller, T. Leibfried, A. M. Miri

Application of Battery Energy Operated System to Isolated Power Distribution Systems 731
Bhim Singh, A. Adya, A.P. Mittal, J.R.P Gupta

Pulse Doubling in 18-Pulse AC-DC Converters 738
Bhim Singh, Sanjay Gairola

Magnetic Field Analysis and Control Strategy of Permanent Magnet Actuator for Low Voltage Vacuum Circuit Breaker 745
Fang Shuhua, Lin Heyun, Yang Chenfeng, Liu Xiping, Guo Jian

Analysis of Transformer Inrush Current under Harmonic Source 749
Chien-Lung Cheng, Jim-Chwen Yeh, Shyi-Ching Chern, Yi-Hung Lan

Voltage Sag Compensation Performance by DSTATCOM with Series Inductor and Energy Storage 755
Sumate Naetiladdanon

Cooperative Operation of Active Power Filters by Instantaneous Complex Power Control 760
Elisabetta Tedeschi, Paolo Tenti, Paolo Mattavelli

Impact of Adjustable Speed PWM drives on Operation and Harmonic Losses of Nonlinear Three Phase Transformers 767
M.A.S. Masoum, Paul S. Moses, Amir S. Masoum

Real-Time Implementation of Voltage Dip Mitigation using D-STATCOM with Fast Extraction of Instantaneous Symmetrical Components 773
Thip Manmek, Chathura P. Mudannayake

Table of Contents

Combined System of Static Synchronous Series Compensation and Passive Filter applied to Wind Energy Conversion System.. 781
A. Singer, W. Hofmann

Control of active injector for multi-pulse rectifiers operating on variable frequency supplies 788
Ismael Araujo-Vargas, Andrew J. Forsyth, F. Javier Chivite-Zabalza

36-pulse hybrid ripple injection for high performance aerospace rectifiers 796
F. Javier Chivite-Zabalza, Andrew J. Forsyth, Ismael Araujo-Vargas

A 48-pulse converter using dc-ripple injection.. 804
F. Javier Chivite-Zabalza, Andrew J. Forsyth

A Study of Different Possible Switched Mode Chopper Circuits for Multi-Magnet Based DC Electromagnetic Levitation System.. 812
Subrata Banerjee, Dinkar Prasad, Jayanta Pal

Power Supply with Potential Use in Magnetic Stimulation.. 817
Duarte M. Sousa, Antonio Ferraz

A Novel Maximum Power Point Tracking Method for the Photovoltaic System.................................. 824
Hurng-Liahng Jou, Wen-Jung Chiang, Jinn-Chang Wu

Maximum Power Point Algorithm in PV Generation: An Overview.. 829
Hardik P. Desai, H. K. Patel

A DC-Module-Based Power Configuration for Residential Photovoltaic Power Application 836
Bangyin Liu, Shanxu Duan, Yong Kang

Analysis and Improvement of Maximum Power Point Tracking Algorithm Based on Incremental Conductance Method for Photovoltaic Array 842
Bangyin Liu, Shanxu Duan, Fei Liu, Pengwei Xu

Application of Maximum Power Point Tracker with Self-organizing Fuzzy Logic Controller for Solar-powered Traffic Lights.. 847
Noppadol Khaehintung, Phaophak Sirisuk

Supply-side Current Harmonics Control of Three Phase PWM Boost Rectifiers Under Distorted and Unbalanced Supply Voltage Conditions 852
Xinhui Wu,, Sanjib K. Panda, Jianxin Xu

A Two-stage Converter with a Coupled-Inductor 858
Hirotaka Nakanishi, Yoshihiro Tomihisa, Terukazu Sato, Takashi Nabeshima, Kimihiro Nishijima, Tadao Nakano

Three-Phase AC to DC Converter with Minimized DC Bus Capacitor and Fast Dynamic Response 863
U. Kamnarn, Y. Kanthaphayao, V. Chunkag

A Simple Effective Duty Cycle Controller for High Power Factor Boost Rectifier 869
Hussain S. Athab, P. K. Shadhu Khan

A Cost Effective Method of Reducing Total Harmonic Distortion (THD) in Single-Phase Boost Rectifier 874
Hussain S. Athab, P. K. Shadhu Khan

Comparison of Different Methods to Detect Static Air Gap Asymmetry in Inverter Fed Induction Machines.. 880
T.M. Wolbank, P. Macheiner

Analysis of the Synchronous Torques in a Split Phase Induction Motor.. 886
P. Scavenius Andersen, D. G. Dorrell, N. C. Weihrauch, P. E. Hansen

On-Line Diagnosis of Three-Phase Closed Loop Induction Motor Drives Using an Eigenvalue aß-Vector Approach 894
J. F. Martins, V. Fernao Pires, A. J. Pires

xiv

Table of Contents

Design and Development of a 36-Pulse AC-DC Converter for Vector Controlled Induction Motor Drive................899
Bhim Singh, Sanjay Gairola

Comparison of Outer- and Inner-Rotor Switched Reluctance Machines........................907
Martin D. Hennen, Rik W. De Doncker

Optimization of Predesign of Switched Reluctance Machines Cross Section Using Genetic Algorithms................912
Satit Owatchaiphong, Christian Carstensen, Rik W. De Doncker

Shaft Position for an 8/6 Switched Reluctance Machine: Theoretical concept, FEM analysis and Experimental results........................917
Silviano Rafael, P.J. Costa Branco, A.J. Pires

Sensorless Control of Brushless Doubly-Fed Reluctance Machines using an Angular Velocity Observer................922
Milutin G Jovanovic, David G Dorrell

A Half-Bridge PV System with Bi-direction Power Flow Controlling and Power Quality Improvement................930
C.L. Shen, S.T. Peng

Response of DSTATCOM under Voltage Flicker In Farm Wind........................937
K. Aodsup, P. N. Boonchiam, A. Sode-Yome, P. Kongsuk, N. Mithulananthan

A Comparative Study of Fixed Speed and Variable Speed Wind Energy Conversion Systems Feeding the Grid........................941
S.S. Murthy, Bhim Sing, P.K. Goel, S.K. Tiwari

Prediction of Wind Power Generation based on Chaotic Phase Space Reconstruction Models........................949
Dong Lei, Wang Lijie, Hu Shi, Gao Shuang, Liao Xiaozhong

Power Flow Control for Efficiency Improvement in a Forward-Flyback Mixed Converter........................954
Yoshito Kusuhara, Asahi Nakayama, Tamotsu Ninomiya, Shin Nakagawa

Hammerstein Model-Based Robust Control of DC/DC Converters........................959
F. Alonge, F. D'ippolito, T. Cangemi

A New Model Control DC-DC Converter to Improve Dynamic Characteristics........................968
F. Kurokawa, S. Sukita

Fuzzy Incremental Controller for the 3rd Order Buck Converter........................973
M. Veerachary, Deepen Sharma

Design of a Single-Stage Single-Switch Power- Factor-Corrected (S4-PFC) AC/DC Converter........................977
P. Kongthawornwattana, C. Bunlaksananusorn, S. Kittiratsatcha

A DSP-Based Unified Three Phase/Switch/Level Unity Power Factor Rectifier Using Feedback Linearization for DC-Bus Voltage Control........................983
Ali Moallem, Hesameddin Mirzaee Teshnizi, Mohammadreza Zolghadri

A Soft-Switched AC-DC Symmetrical Boost Converter with Power Factor Correction........................989
A. Jangwanitlert, J. Songboonkaew

Education Reforming for Power Electronics........................994
Weiping Zhang, Xiaohan Guan, Dongyan Zhang

A Novel Current Control System for PMSM Considering Effects from Inverter in Overmodulation Range........................999
Smith Lerdudomsak, Shinji Doki, Shigeru Okuma

Modelling of the Feeding Network of a Linear Synchronous Machine and Estimation of Model Parameters........................1006
J. Rost, H. Gueldner, R. Hellinger, A. Weller

Analysis of Losses in Inverter Fed Large Scale Synchronous Machines using 2D FEM Software........................1012
Samer Shisha, Chandur Sadarangani

Table of Contents

Position sensorless control of the Reluctance Synchronous Machine considering High Frequency inductances 1017
H.W. De Kock, M.J. Kamper, O.C. Ferreira, R.M. Kennel

Carrier PWM algorithm in overmodulation range for Multileg Multilevel Inverter 1027
Nguyen Van Nho, Hong Hee Lee

Carrier Based Single-state PWM Technique In multilevel Inverter 1033
Nguyen Van Nho, Quach Thanh Hai, Hong Hee Lee

Implementation of a Single-carrier Multilevel PWM Technique Using Field Programmable Gate Array (FPGA) 1041
N. A. Azli, L. Y. Teng, P. Y. Lim

SPACE VECTOR PWM FOR MULTILEVEL INVERTERS - A FRACTAL APPROACH 1047
Anish Gopinath, M.R. Baiju

Elimination of Harmonics in a Five-Level Diode-Clamped Multilevel Inverter Using Fundamental Modulation 1055
Sule Ozdemir, Engin Ozdemir, Leon M. Tolbert, Surin Khomfoi

Compensation of DC-Link Oscillations of Cascaded H-Bridge Converters 1060
M. Tavakoli Bina, B. Eskandari

Combined DC-Filter and optimized Modulation to Absorb DC-Link Oscillations of Cascaded H-Bridge Converters 1065
M. Tavakoli Bina, B. Eskandari

Control Strategies of a Hybrid Multilevel Converter for Expanding Adjustable Output Voltage Range 1070
Shoji Fukuda, Takatsugu Yoshida, Shigeta Ueda

High Efficiency Single Phase Multi-level Inverter by New Controlled Switch Signal 1078
Ruthapong Kumchaiyo, Itsda Boonyaroonate

FPGA Implementation of Quasi-BLDC Drive 1082
C.S. Soh, C. Bi, K.K. Teo

A Practical Method to Eliminate the Conduction Torque Ripple in BLDCM Using Cascade Topology 1088
Xiaofeng Zhang, Zhengyu Lu, Yu Ma, Zhaoming Qian

Program Architecture for Realizing Design Optimization of a BLDC Motor 1092
Dong-Hun Kim, Giwoo Jeung, Heung-Geun Kim, In Dong Kim

Stable Operation of the Brushless Doubly-Fed Machine (BDFM) 1096
Shiyi Shao, Ehsan Abdi, Richard Mcmahon

Sail Generator Feasibility Study 1102
Ha Pham Ngoc, Yasuaki Matsui, Pathom Attaviriyanupap, Osamu Iso

Braking Circuit of Small Wind Turbine Using NTC Thermistor under Natural Wind Condition 1109
Y. Matsui, A. Sugawara, S. Sato, T. Takeda, K.Ogura

Flywheel Energy Storage Drive for Wind Turbines 1115
K. Veszpremi, I. Schmidt

Theory, Simulation and Experimental Verification of a New Integral Cycle Robust Control Strategy for Self Excited Induction Generators 1123
S.S. Murthy, A.J.P. Pinto

Performance Comparison of DC Link Voltage Controllers in Vector Controlled Boost Type PWM Converter for Wind Turbine System 1129
W. Sudmee, B. Neammanee

Analysis and Design of Class DE Amplifier with Nonlinear Shunt Capacitance 1136
Hiroo Sekiya, Takayuki Watanabe, Tadashi Suetsugu, Marian K. Kazimierczuk

xvi

Table of Contents

A Novel Control Strategy of the Class-D Stereo Audio Amplifier...1142
Kyu Min Cho, Won Seok Oh, Hai Xu, Hee Jun Kim

Robust H_infinity Control Design for PFC Rectifiers..1147
F. Tahami, H. Molla Ahmadian, A. Moallem

Parallel Operation of Power Factor Corrected AC-DC Converter Modules With Two Power Stages........................1152
Aravind Pothana, Krishna Vasudevan

Noise Radiation of Switched Reluctance Drives..1160
K. A. Kasper, M. Bosing, R. W. De Doncker, S. Fingerhuth, M. Vorlander

Iron Losses in Electrical Machines Due to Non Sinusoidal Alternating Fluxes......................................1167
J. A. Walker, D. G. Dorrell, E. Ritchie

Design Requirements for Doubly-Fed Reluctance Generators...1174
D. G. Dorrell

A Magnetic Gear Box for application with a Contra-rotating Tidal Turbine..1182
Laxman Shah, A. Cruden, Barry W. Williams

Mechatronic . Advanced Computational Intelligence...1187
D. Schroder, H. Schuster, C. Westermaier

New Space Vector Control Approach for Four Switch Three Phase Inverter (FSTPI)..................................1195
Phan Quoc Dzung, Le Minh Phuong, Pham Quang Vinh, Nguyen Minh Hoang, Tran Cong Binh

**The Development of Artificial Neural Network Space Vector PWM for Four-Switch Three- Phase
Inverter**..1202
Phan Quoc Dzung, Le Minh Phuong, Pham Quang Vinh

**Voltage Losses Compensation Using Artificial Neural Network for Estimation Nonlinear Characteristic of
Switches**..1208
N. Pothi, S. Premrudeepreechacharn, C. Rakpenthai

**A Simple Carrier-Based PWM Method For Three-Phase Four-Leg Inverters Considering All Four Pole
Voltages Simultaneously**...1213
Nakharet Chudoung, Somboon Sangwongwanich

Inverted Sine Carrier Pulse Width Modulation for Fundamental Fortification in DC-AC Converters.................1221
R.Nandhakumar, S.Jeevananthan

**Fault Detection and Reconfiguration Technique for Cascaded H-bridge 11-level Inverter Drives
Operating under Faulty Condition**..1228
Surin Khomfoi, Leon M. Tolbert

**Investigation into Harmonic Losses in a PWM Multilevel cascaded H-Bridge Inverter Fed Induction
Motor**...1236
Prasopchok Hothongkham, Vijit Kinnares

**Extend the Use of Auxiliary Circuit to Start up, Shut down, and Balance of the Modified Diode Clamped
Multilevel Inverter**...1242
Ahmed Ali Ashaibi, S.J. Finney, B.W. Williams, Ahmed Massoud

Five-Level Z-Source Neutral-Point-Clamped Inverter..1247
F. Gao, P. C. Loh, F. Blaabjerg, R. Teodorescu, D. M. Vilathgamuwa

Capacitor Voltage Balancing Using Redundant States for Five-Level Multilevel Inverter..........................1255
Hadi A Hotait, Ahmed M Massoud, Steve J. Finney, Barry W. Williams

Sliding Mode Repetitive Control of PWM Voltage Source Inverter...1262
Sufen Chen, Y. M. Lai, Siew-Chong Tan, Chi K. Tse

Output Current Ripple Analysis of Five-Phase PWM Inverters...1267
Deni, E. G. Supriatna, P. A. Dahono

xvii

Table of Contents

An Improved 'DC-DC Type' High Frequency Transformer-Link Inverter by Employing Regenerative Snubber Circuit 1274
Z. Salam, S. M. Ayob, M. Z. Ramli, N. A. Azli

A Novel Dimming Technique for Cold Cathode Fluorescent Lamp 1278
K. I. Hwu, Y. H. Chen

Time Delay Compensation For A DSP-Based Current-Source Converter Using Observer-Predictor Controller 1284
Huu-Phuc To, Muhammed Fazlur Rahman, Colin Grantham

Implementation of Hysteresis Current Control for Single-Phase Grid Connected Inverter 1290
Krismadinata, Nasrudin Abd Rahim, Jeyraj Selvaraj

Use of Air-Cored Axial Flux Permanent Magnet Generator in Direct Battery Charging Wind Energy Systems 1295
F.G. Rossouw, M.J. Kamper

Transverse Flux Machines for Sustainable Development - Road Transportation and Power Generation 1301
D. Svechkarenko, A. Cosic, J. Soulard, C. Sadarangani

Low Voltage Ride-Through Capability for Wind Turbines based on Current Source Inverter Topologies............... 1308
Pierluigi Tenca, Andrew A. Rockhill, Thomas A. Lipo

Optimal Control of Direct Driven Feed Axes with Flexible Structural Components 1316
Ekkehard Batzies, Tobias Scholler, Volkmar Welker, Oliver Zirn

Leakage Energy Recovered Narrow Pulsed Voltage Generator Associated with Ultrasound Generator for Liquid Food Sterilization 1321
S. Y. Tseng, Y. D. Chang, P. L. Huang, T. F Wu, Y. M. Chen

Energy Harvesting from Exercise Bicycle 1327
Suchart Janjornmanit, Samart Yachiangkam, Aswin Kaewsingha

Modeling and Analysis of Igniter for HID Lamps 1330
Weiping Zhang, Qiang Cheng

Design of a Single Bi-directional DC-DC Converter for Onboard Energy Improving of Zero Emission Electric Vehicles 1335
Werachet Khan-Ngern

Speed Sensorless Control with Neuron MRAS Estimator of an Induction Machine.. 1340
Dong Lei, Yang Dong, Liao Xiaozhong

Adaptive Flux model for commissioning of signal injection based zero speed sensorless flux control of induction machines 1346
T.M. Wolbank, M.A. Vogelsberger, R.H. Stumberger

Design and Performance of a Single Stator, Dual Rotor Induction Motor.. 1352
S. Sinha, N. K. Deb, N. Mondal, S. K. Biswas

Investigation of skew effect on the Performance of Self - Excited Induction Generators 1356
B. Sawetsakulanond, V. Kinnares

Analysis of Double Loops Discrete Single Input PI Fuzzy for Single phase Inverter.. 1363
S.M. Ayob, Z. Salam, N.A. Azli

A new three-phase varying-band hysteresis current controller for voltage-source inverters.. 1368
Vinciane Chereau, Francois Auger, Luc Loron

Diode-Assisted Buck-Boost Current Source Inverters 1376
F. Gao, C. Liang, P. C. Loh, F. Blaabjerg

Table of Contents

Single-Stage Fluorescent Lamps Electronic Ballast Using Class-DE Low dv/dt Rectifier for Power-Factor Correction...1383
Chainarin Ekkaravarodome, Adisak Nathakaranakule, Itsda Boonyaroonate

Output Impedance Design Consideration of Three Control Schemes for Bus Converter in On-Board Distributed Power System...1388
Seiya Abe, Masahiko Hirokawa, Tamotsu Ninomiya

Optimal Generation Rescheduling for Security Operation of Power Systems Using Optimal Control Theory...1394
J. Q. Sun, K. W. Chan, D. Z. Fang

Improvement of Transient Response of Thermal Power Plant Using VVVF Inverter...............1398
N. Matsui, F. Kurokawa

A Novel Circuit Topology for Three-Phase Four-Wire Distribution Electronic Power Transformer.......1404
H.Mirmousa, M.R.Zolghadri

A Half-Bridge DC/DC Converter for Plasma Cutting Machine...1412
N. Sanajit, A. Jangwanitlert

Ripple Estimation for Paralleled Converter System with Automatic Interleaving Function.............1417
Teruhiko Kohama, Ryota Tsunesada, Tamotsu Ninomiya

Design of a New Hysteretic PWM Controller for All Types of DC-to-DC Converters..................1423
Min Lin, Takashi Nabeshima, Terukazu Sato, Kimihiro Nishijima

Implementation of Fuzzy Logic Controller with Bifurcation Control of a Current-mode Boost Converter...............1429
Noppadol Khaehintung, Phaophak Sirisuk, Anantawat Kunakorn

Phase Advance Approach to Expand the Speed Range of Brushless DC Motor.........................1434
Binhminh Nguyen, Minh C. Ta

Nonlinear Decoupled Control for a Six-Phase Series-Connected Two Induction Motor Drive Using the Sliding-Mode Technique...1442
J. Soltani, N. R. Abjadi, Gh. R. Arab Markadeh

The Decoupled Stator Flux and Torque Sliding-Mode Control of Induction Motor Drive Taking the Iron Losses into Account...1449
M.Hajian, J.Soltani, S.Hosein Nia, G.R.Arab

AN EFFICIENT DIRECT TORQUE CONTROL SCHEME FOR SPLIT PHASE INDUCTION MOTOR..........1455
A. Khajeh, J. S. Moghani, M. Shahbazi

A Method of Speed Sensorless Vector Control Parallel -Connected Dual Induction Motors Fed by One Inverter in a Rotor Flux Feedback Control..1460
Jun Nishimura, Kazuo Oka, Kouki Matsuse

A Combined Model Flux Observer for Vector Control of Traction Asynchronous Motors...............1465
F. Tahami, S. Chini Foroosh

Torque Ripple Elimination for Doubly-Fed Induction Motors under Unbalanced Source Voltage...............1471
Hong-Geuk Park, Ahmed G. Abo-Khalil, Dong-Choon Lee, Kwang-Myoung Son

Online H8 Speed Control of Sensorless Induction Motors with Rotor Resistance Estimation...............1477
Peda V Medagam, Farzad Pourboghrat

Analysis and Comparative Study on the Performance between Standard and High Efficiency Induction Machines operating as Self - Excited Induction Generators..1483
B. Sawetsakulanond, V. Kinnaraes

A simple Approach to Capacitance Determination of Self - Exited Induction Generators for Terminal Voltage Regulation...1489
B. Sawetsakulanond, V. Kinnares

xix

Table of Contents

Symmetrical Components-Based Control Technique of Doubly Fed Induction Generators under Unbalanced Voltages for Reduction of Torque and Reactive Power Pulsations........................1495
S. Wangsathitwong, S. Sirisumrannukul, S. Chatratana, W. Deleroi

A New Switching Technique for Direct Torque Control of Induction Motor using Four-Switch Three-Phase Inverter...........................1501
Phan Quoc Dzung, Le Minh Phuong, Pham Quang Vinh, Nguyen Minh Hoang, Nguyen Xuan Bac

Detection of Some Parameters of Induction Motors a Proposal and Its Verification........................1507
H. Bulent Ertan, Volkan Sezgin, Baris Colak

Comparison of Basic Direct Torque Control Designs for Permanent Magnet Synchronous Motor...........................1514
M. N. Abdul Kadir, S. Mekhilef, W.P. Hew

Improved DSVM-DTC Based Current Sensorless Permanent Magnet Synchronous Motor Drive........................1520
Bhim Singh, Devendra Goyal

A High Performance Direct Torque Control Scheme of Permanent Magnet Synchronous Motor...........................1527
Dong-Hee Lee, Young-Joo An, Eui-Chel Nho

Low Cost Position Sensor for Permanent Magnet Linear Drive........................1533
Ralf Wegener, Florian Senicar, Christian Junge, Stefan Soter

Design of One Rotary-linear Permanent Magnet Motor with Two Independently Energized Three Phase Windings........................1538
L. Chen, W. Hofmann

Position Estimation of Permanent Magnet Synchronous Motor Using Un-known Input Observer...........................1543
Masaru Hasegawa, Satoshi Yoshioka, Keiju Matsui

Switched Reluctance Motor Drive for Electric Motorcycle Using HFNN Controller........................1549
Chih-Hong Lin

STATE - SPACE AVERAGING, SIMULATION, STABILITY STUDIES FOR STEP UP POSITIVE OUTPUT SWITCHED CAPACITOR DC-DC CONVERTER........................1555
E. Jayashree, G. Uma, M. Vaigundamoorthi

Active Clamp Interleaved Boost Converter with Coupled Inductor for High Step-up Ratio Application.................1560
S. Y. Tseng, J. Z. Shiang, W. S. Jwo, C. M. Yang

Active Clamp Interleaved Flyback Converter with Single-Capacitor Turn-off Snubber for Stunning Poultry Applications........................1567
S. Y. Tseng, C. T. Hsieh, H. C. Lin

Novel Current Feedforward Average Current Mode Control Technique to Improve Output Dynamic Performance of DC-DC Converters........................1575
P. Chrin, C. Bunlaksananusorn

Stability Analysis of Cascaded DC-DC Power Electronic System........................1581
M. Veerachary, S. Bala Sudhakar

Averaged Switch Modeling of DC/DC Converters using New Switch Network........................1586
Chien-Min Lee, Yen-Shin Lai

Soft Transition Operation of UPS in High- Power-Factor Mode of Three-Phase Front- End Rectifier........................1590
G. A. Dhomane, H. M. Suryawanshi

Specific Harmonic Power Suppression of Direct- Power-Controlled Current-Source PWM Rectifier........................1595
Toshihiko Noguchi, Kohji Sano

Frequency-Controlled LCC Resonant Converter with Synchronous Rectifier........................1601
Yu Ma, Xiaogao Xie, Zhaoming Qian

Selection of the Filter Capacitor for Power Supplies using 1-Phase Diode Rectifier........................1605
N. Mondal, S. K. Biswas, S. Sinha, N. K. Deb

Table of Contents

High Performance Single-Phase Voltage Regulator with a Simple Circuit Topology 1610
Chien-Ming Wang, Ching-Hung Su, Chang-Hua Lin, Maw-Yang Liu, Kuo-Lun Fang

Small-Signal Modeling of Series Resonant Converter 1615
Weiping Zhang, Peng Mao, Yuanchao Liu

Modelling of Three phase Z-Source Boost Buck Rectifiers 1620
D M Vilathgamuwa, P C Loh, K Karunakar

A NEW SINGLE-PHASE CONTROLLED RECTIFIER USING SINGLE-PHASE MATRIX CONVERTER WITH REGENERATIVE CAPABILITIES 1626
R. Baharom, M.K. Hamzah, A. Saparon, S.Z. Mohammad Noor, N.R.Hamzah

Implementation of Space Vector Modulated 3. to 3 . Matrix Converter Fed Induction Motor 1632
S. Ganesh Kumar, S. Siva Sankar, S. Krishna Kumar, G. Uma

A Single-Phase High-Power-Factor Neutral-pointer Clamped Multilevel Rectifier 1636
Yun Xu, Yunping Zou, Chengzhi Wang, Wei Chen, Bangyin Liu

Two Phase Inverter Drive of Three Phase Motor 1641
Saksit Jangjaempradit, Masayuki Morimoto

Predictive Current Controller for Inverter Fed Medium Voltage Drives with LC Filter 1645
T. Laczynski, A. Mertens

Novel Control Strategy of Instantaneous Power Based CVCF Inverter 1651
Akira Sato, Toshihiko Noguchi

An Improved Parallel Processing UPS Using a Voltage-Controlled Voltage Source Inverter 1657
S.W. Lee, H. Dehbonei, S.H. Ko, S.R. Lee, B.H. Jang, Y.H. Moon, T.K. Ko

A PEMFC/Battery Hybrid UPS System for Backup and Emergency Power Applications 1662
Yuedong Zhan, Jianguo Zhu, Youguang Guo, Hua Wang

Design of the Two Parallel Inverter Modules by Circular Chain Control Technique 1667
K. Piboonwattanakit, W. Khan-Ngern

Investigation of Topologies of Low Voltage Multilevel Inverters 1672
Yanli Zhang, Wanmin Fei, Shoufang Wang

Solution for PWM converter switching for Voltage Source Inverter using Non- Traditional Method 1677
V. Jegathesan, Jovitha Jerome

Piecewise Linear Control Surface for Single Input Nonlinear PI-Fuzzy Controller 1682
S. M. Ayob, Z. Salam, N. A. Azli

Open-Loop Control of a Stepping Motor through IP Network 1686
K. Matsuo, T. Miura, T. Taniguchi

Fuzzy Logic Controller for Electric Vehicle Braking Strategy.....**Fig 4. adjusted due to text re-flow 1691
Xixi. Wang, K.W.Eric Cheng, Xiaozhong Liao, Norbert C. Cheung, Lei Dong

Skid Steering in 4-Wheel-Drive Electric Vehicle 1697
Gao Shuang, Norbert C. Cheung, K. W. Eric Cheng, Dong Lei, Liao Xiaozhong

A Flexible Multi-Pulse Control Strategy for Universal Nail Collator 1703
Chien-Lung Cheng, Shyi-Ching Chern, Jim-Chwen Yeh, Ming-Yi Wu

Cycloconverter Based Three Phase Induction Motor to Replace Flywheel of the Process Machine 1708
M.V. Palandurkar, M. A. Chaudhari, J. P. Modak, S. G. Tarnekar

A Novel Zero-Voltage-Switching Single-Stage High-Power-Factor Electronic Ballast 1712
Chien-Ming Wang, Ching-Hung Su, Chang-Hua Lin, Maw-Yang Liu, Kuo-Lun Fang

Opto-Mechatronic System Design of the LED Projector by Using Brushless DC Motor 1717
Jian-Long Kuo, Tzu-Hsuan Fang

xxi

Table of Contents

The Color Measurement System of PWM-Controlled LCD by Using Back-Propagation Neural Network................1722
Jian-Long Kuo, Xian-Lin Liu

Gapped Air-cored Power Converter for Intelligent Clothing Power Transfer......................................1727
Y. Lu, K.W.E.Cheng, Y. L. Kwok, K. W. Kwok, K.W. Chan, N.C.Cheung

Simulation Program for Switching Converters Using Numerical Fourier Transform........................1734
Yoshihiro Tomihisa, Hirotaka Nakanishi, Terukazu Sato, Takashi Nabeshima, Kimihiro Nishijima, Tadao Nakano

The Most Suitable Application of SiC Diode..1740
Tomoaki Makino, Atsushi Hirota, Satoshi Nagai

Multi-Domain System Simulation and Rapid Prototyping of Digital Control Algorithms using VHDL-AMS 1744
P.J. Randewijk

Reforming Power Electronics Laboratory 1752
Xiaohan Guan, Weiping Zhang, Xusen Zhao, Yuanchao Liu

Online performance monitoring and testing of electrical equipment using Virtual Instrumentation 1757
S.S. Murthy, Raghu K. Mittal, Avneesh Dwivedi, G. Pavitra, Sonika Choudhary

A Balancing Strategy and Implementation of Current Equalizer for High Power LED Backlighting........1762
Chang-Hua Lin, Tsung-You Hung, Chien-Ming Wang, Kai-Jun Pai

Modeling of the Parasitical Capacitance Effect in LCD Panel and Corresponding Elimination Strategy................1767
Chang-Hua Lin, Tsung-You Hung, Chien-Ming Wang, Kai-Jun Pai

On-line SOC Estimation of Battery for Wireless Tram Car..1773
Hiroyuki Miyamoto, Masayuki Morimoto, Katsuaki Morita

Narrow- control-bandwidth Operation of Piezoelectric-transformer Converter 1777
Weiping Zhang, Xiaoqiang Zhang, Yuzhou Lei, Yuanchao Liu

Modified Map of Variable Active Passive Reactance for Stability Evaluation with Consideration of Capacitor Mode 1782
S. Mohammad Shariatmadar, Jalal Nazarzadeh

Design of the Longitudinal Mode Piezoelectric Transformer...1788
Shine-Tzong Ho

The Comparison of Conducted EMI Emission and Electrical Performances of Lamps 1794
C. Uyaisom, W. Khan-Ngern

Neural Identification of Average Model of STATCOM using DNN and MLP 1799
M. Tavakoli Bina, S. Rahimzadeh

Hybrid Simulation of Power Systems with Dynamic Phasor SVC Transient Model........................1804
E. Zhijun, K. W. Chan, D. Z. Fang

CONTROL OF CURRENT- SOURCE ACTIVE POWER FILTER USING UNIT VECTOR TEMPLATE IN THREE PHASE FOUR WIRE UNBALNCED SYSTEM..1810
K. Vadirajacharya, Pramod Agarwal, H.O. Gupta

Improved Control of Three Phase Active Filters Using Genetic Algorithms 1816
Bhim Singh, Varun Singhal

A Fuzzy Adaptive Detecting Approach of Harmonic Currents for Active Power Filter....................1822
Yilong Qu, Weipu Tan, Yihan Yang

Comparative Evaluation of Harmonic Extraction Techniques for Three-Phase Three-Wire Active Power Filter..1827
R. Chudamani, Krishna Vasudevan, C.S. Ramalingam

xxii

Table of Contents

Hybrid Passive Filter Design for Distribution Systems with Adjustable Speed Drives 1834
M.A.S. Masoum, A. Ulinuha, S. Islam, K. Tan

A Graphic User Interface-based Program for Voltage Sag Calculation 1840
T. Tayjasanant, K. Yossombut, P. Sawatpipat

Operational Characteristics of Fault Current Limiting Reactor Combined with Multi- Functional Inverter 1846
S. H. Ko, S. H. Lim, S. R. Lee, S. W. Lee, I. C. Kim, S. H. Ko, H. S. Kim

Low Cost AC Solid State Circuit Breaker 1851
W. Pusorn, W. Srisongkram, W. Subsingha, S. Deng-Em, P. N. Boonchiam

A Variable Gain Control Scheme of Digital Automatic Voltage Regulator for AC Generator 1857
Dong-Hee Lee, Jin-Woo Ahn, Tae-Won Chun

A Graphic User Interface-based Program for Harmonic Impedance Calculation 1862
T. Tayjasanant

The analysis and simulation of power circuits for AC high-voltage converters 1868
Y.Y. Skorokhod, S.I. Volskiy

A Single Stage Flyback PFC Converter for Testing Distance Relay Systems 1875
V. Fernao Pires, J. F. Martins, J. Fernando Silva

H-Infinity Control Theory Apply to New Type Arc-suppression Coil System 1880
Yilong Qu, Weipu Tan, Yihan Yang

Characteristics of a novel topology of a DC-AC Converter for Fuel Cells 1885
K. Fukushima, T. Ninomiya, I. Norigoe, Y. Harada, K. Tsukakoshi, Z. Dai

A Comparative Study of PWM Schemes for Grid Connected PV Cell 1891
Vineeta Agarwal, Alok Vishwakarma

This page intentionally left blank.

Design of a New Hysteretic PWM Controller for All Types of DC-to-DC Converters

Min Lin, Takashi Nabeshima, Terukazu Sato, and Kimihiro Nishijima

Oita University

Department of Electrical and Electronic Engineering

700 Dannoharu, Oita 870-1192, Japan

E-mail : nabesima@eee.oita-u.ac.jp

Abstract— A new control method using a hysteretic PWM controller for all types of converters and its proper design method are presented. The triangular voltage obtained from a simple RC network connected between comparator output and converter output is superimposed to the output voltage and as a feedback signal to a hysteretic comparator. Since the hysteretic PWM controller essentially has derivative characteristics and has no error amplifier, the presented method provides no steady-state error voltage on the output and excellent dynamic performances for the load current transient by choosing proper values of time constants in the RC network. Performances of the proposed controller are experimentally verified for the buck, buck-boost and boost converters

Index Terms-- DC-DC converter, Hysteretic PWM control, Transfer function, Transient response

I. INTRODUCTION

Many of recent digital applications using LSIs, such as DSP, FPGA and memory chips, demand high current and low voltage for their power supply. In addition, the supply current may change very often from low to maximum with high slew rate and a tight tolerance for the transient voltage is required even in such a case. Most effective approach to meet small transient voltage and fast response is to use smaller inductance in the LC filter circuit. Multi phase converters are especially effective to reduce the ripple current of the capacitor for using small inductance.

Most of the feedback controllers for dc-to-dc converters generally employ a high gain operational amplifier as an error amplifier in order to regulate the output voltage. Phase compensation circuits are generally used not only to improve the dynamic performance but also to compensate the effect of phase lag in the operational amplifier itself. However, it has become very difficult to realize fast transient response only by the voltage feedback control due to the higher bandwidth of the LC filter.

It is known that a current mode control together with the voltage feedback is one of solution to improve the stability and the dynamic performance [1-3]. In this case, a current-sensing resistor which may cause the increase of power loss and a number of components in the control circuit are necessary. In voltage mode, on the other hand, a hysteretic PWM control for buck converters using output ripple voltage shows inherently stable

performances since the feedback signal includes inductor current information [4,5]. This control method, it is generally called "Bang bang control" is essentially stable and robust for many control objects. However, in the converter applications, the switching frequency depends on the value of ESR in the output capacitor. In order to avoid this effect, a triangular voltage generated by a simple RC integral circuit connected in parallel to the inductor winding is introduced instead of using the output ripple voltage. The application of this method for the buck converter and its control characteristics have been reported [6]. The transfer function of the controller shows derivative characteristics and excellent transient behaviors are obtained by this method. For the boost and the buck-boost converters, a similar control method by introducing an auxiliary winding of the inductor has been proposed and it also showed excellent control characteristics [7].

This paper proposes a new hysteretic PWM control method applicable to any type of PWM converters without using additional winding or active devices and provides its design consideration. The RC integral circuit is employed to obtain the triangular voltage similar to the inductor current and it is connected between the output of the converter and the hysteretic comparator output. Furthermore, two more elements, a resistor and a capacitor are introduced into the integral circuit to feed the output dc voltage into the comparator input. The steady state characteristics and transient responses of three types of the converter are experimentally examined and key transfer functions are analyzed.

II. CIRCUIT CONFIGURATION AND OPERATION

Fig.1 shows a new scheme of the hysteretic PWM control technique with RC network connected between the converter output and the output of a hysteretic comparator having a window voltage V_{hys}. The triangular voltage similar to the inductor current is generated by C_1 and R_1 according to the comparator output V_{cp} and its ac component is fed to the comparator input through a capacitor C_2 together with the output voltage through R_2.

Consider the output voltage rapidly changes to lower level form the steady state for instance. The variation of the output voltage appears on the feedback signal V_{fb}

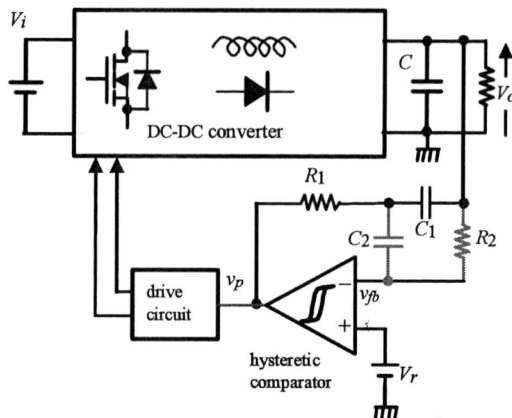

Fig. 1. A new hysteretic PWM controller

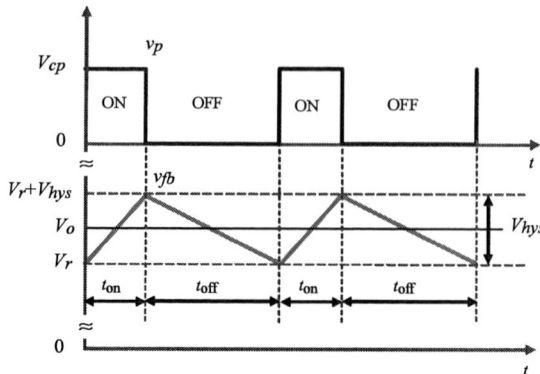

Fig. 2. Voltage waveforms of v_{fb} and comparator output signal v_p

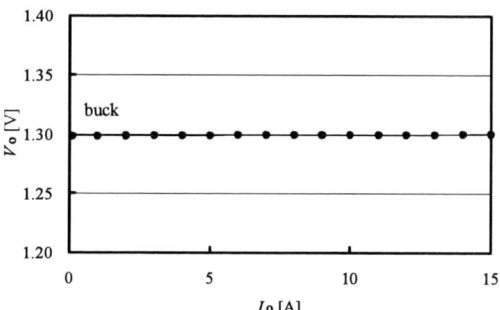

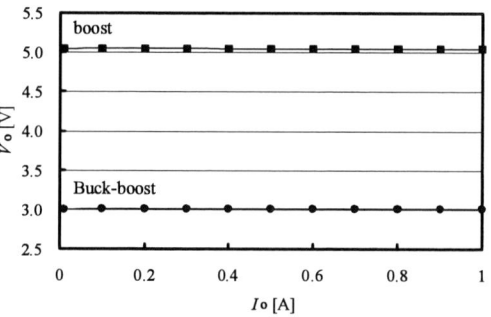

Fig. 3. Load characteristics

through the capacitor C_1. When it happens in the on state of the switch, the duration of the on state is extended without delay. When it happens during the off state, on the other hand, the off state is shorten or changed to the on state. For slow change of the output voltage, the voltage deviation is transferred through the resistor R_2. The transmission of this voltage has a certain time constant mainly determined by R_2 and C_2.

III. STEADY STATE AND DYNAMIC PERFORMANCES

The duty cycle of the switches in the steady state is automatically adjusted to a certain value to set the output voltage V_o between the threshold voltages V_r and $V_r + V_{hys}$ as show in Fig. 2. When the wave shape of the input signal feeding to the comparator is nearly triangular, the steady-state output voltage V_o is expressed as

$$V_o = V_r + \frac{1}{2}V_{hys} \tag{1}$$

and is independent of the input voltage and the load current. Therefore the presented method has no steady-state error on the output such as a ripple regulator or V^2 control methods.

Fig. 3 shows the experimental results of the load characteristics of the buck, buck-boost and boost converters. As can be seen from these results that all three types of converters have excellent load characteristics. It is evident that the presented method shows no steady-state error on the output. For the variation of the input voltage, similar results are also obtained.

Transient responses of the output voltages for the step load change are shown in Fig. 4 taking a resistance R_2 as a parameter. The load current in (a) changes 1A to 15A with a slew of 30/μs for the buck converter and 0.2A to 1A for buck-boost and boost converters. It is seen from the figure that no oscillation is observed after undershoot or overshoot. This means that the proposed method provides very stable dynamic performances even to the boost and buck-boost converters. It is also observed that the settling time is well improved by decreasing R_2. Common parameters used in the experiments are as follows:

[Buck converter]

V_i =5V, L=0.4μH, C=200μF, r_L=2mΩ,
r_c (ESR of C)= 1mΩ, on resistance of switches=5mΩ

[Buck-boost and boost converter]

V_i =3.6V, L=7μH, C=100μF, r_L=12mΩ,
r_c (ESR of C)= 2mΩ, on resistance of switches=8mΩ

IV. SWITCHING FREQUENCY

As can be seen form the circuit of Fig. 1, the proposed control circuit is self oscillating type and the switching frequency is not set by an external clock signal. The factors that define the switching frequency the most are comparator output voltage V_{cp}, the hysteretic window voltage V_{hys} and the time constant $T_1 = R_1 C_1$ of RC network when neglecting the output ripple voltage and propagation delays in the control circuit. In the ideal case,

1424

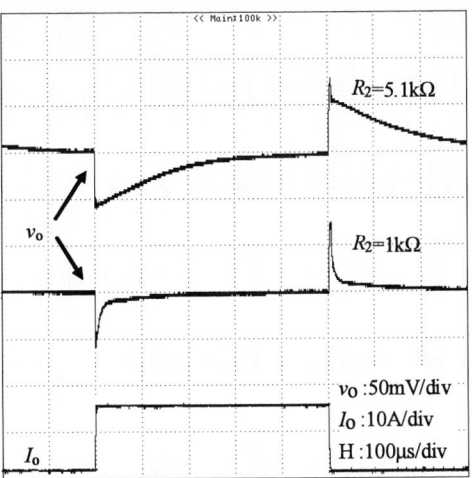

(a)　buck converter

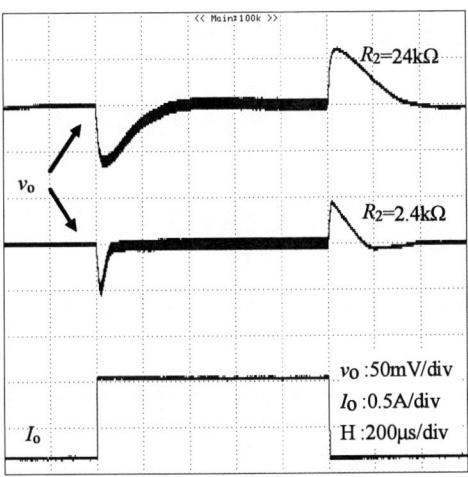

(b)　buck-boost converter

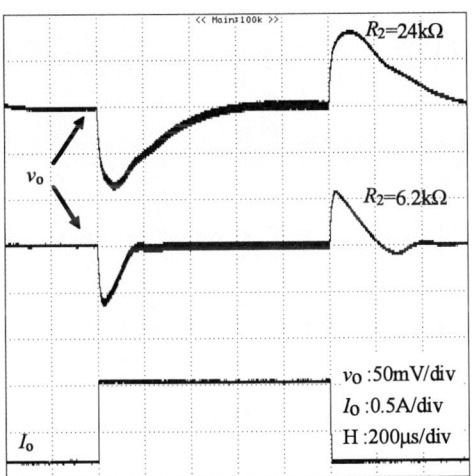

(c)　boost converter

Fig. 4. Transient responses of the output voltage for step change of the load
(R_1=2.7kΩ, C_1=0.047μF, C_2=0.01μF)

the switching frequency f_s is expressed as

$$f_s = \frac{(V_{cp} - V_o)V_o}{V_{hys}T_1 V_{cp}} \qquad (2)$$

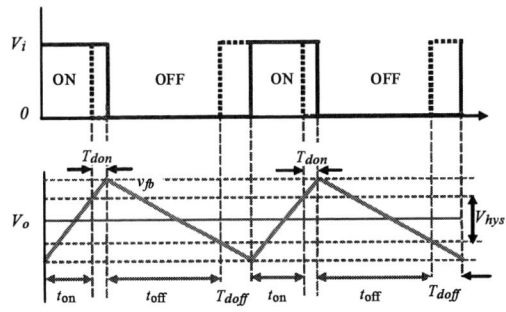

Fig. 5. Effect of delay propagation

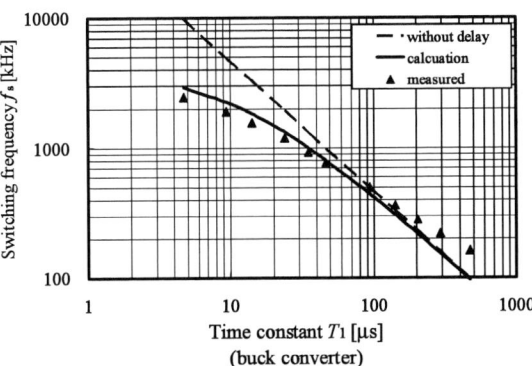

Fig. 6. Switching frequency f_s for time constant T_1

It is preferable to make the product of V_{hys} and T_1 as small as possible to realize high switching frequency operation.

In actual case, however, the switching frequency differs from (2) depending on the propagation delay time existing in the hysteretic comparator and the drive circuit. As shown in Fig. 5, the timing of the switching is delayed and the switching frequency becomes lower than (2).

Taking the delay in the control circuit into account, the switching frequency is expressed as

$$f_s = \left[\frac{V_{cp}V_{hys}}{(V_{cp} - V_o)V_o}T_1 + \frac{V_{cp}}{V_{cp} - V_o}T_{don} + \frac{V_{cp}}{V_o}T_{doff} \right]^{-1} \qquad (3)$$

where symbols T_{don} and T_{doff} designate the delay time at turn on and turn off, respectively. As is evident from (3), the switching frequency is limited by T_{don} and T_{doff}. When the hysteresis window voltage V_{hys} is chosen to obtain high switching frequency, the output ripple voltage also affects it for smaller capacitance or larger ESR of the smoothing capacitor of the converter. By employing parallel connected ceramic capacitors for the buck converter, these effects can be negligible. Fig. 6 shows the switching frequency for the time constant T_1 when ceramic capacitors having small ESR were used as parallel connection. The dotted line in this figure indicates the case without delay time. On the other hand, for the boost and buck-boost converters, the effect of the ripple voltage should be taken into account since it strongly depends on the load current [7].

V. TRANSFER CHARACTERISTICS

Before starting analysis of the proposed controller, the following transfer function $H_{cp}(s)$ of the hysteretic comparator between the positive input and the output duty cycle is introduced here when a triangular voltage is used as a negative input signal [8].

$$H_{cp}(s) = \frac{1+sT_1}{V_{cp}} \qquad (4)$$

It is obvious that the sign of (4) changes to negative for the feedback voltage v_{fb}. The transfer function (4) is available as long as the voltage across the capacitor C_1 varies linearly.

An ac equivalent circuit of the RC network is shown in Fig. 7. The transfer function $H_{fv}(s)$ between the output voltage Δv_o and the feedback voltage ΔV_{fb} is expressed as

$$H_{fv}(s) = \frac{\Delta V_{fb}(s)}{\Delta V_o(s)} = \frac{1+s(T_1+T_3)+s^2 T_1 T_2}{1+s(T_1+T_2+T_3)+s^2 T_1 T_2} \qquad (5)$$

where time constants $T_1 = R_1 C_1$, $T_2 = R_2 C_2$ and $T_3 = R_1 C_2$. The transfer function (5) is similar to that of bridged-T network and the gain characteristics has minimum value at ω_p.

$$\left| H_{fv}(j\omega_p) \right| = \frac{T_1+T_3}{T_1+T_2+T_3}$$
$$\omega_p = \frac{1}{\sqrt{T_1 T_2}} \qquad (6)$$

The transfer function $H_c(s)$ between the output voltage and the duty cycle is then

$$H_c(s) = H_{fv}(s) H_{cp}(s) \qquad (7)$$

Fig. 8 shows the frequency responses of the gain calculated by (7) taking the time constant T_2 as a parameter. The gain in the middle frequency region, the settling time of the step response might be almost determined by it, is mainly determined by the time constant T_2. On the other hand, the gain higher than the mid frequency linearly increases with the frequency. This means the controller has derivative characteristics in the high frequency and therefore the stability and the transient performance are improved by T_1. The block diagram of the regulator system is illustrated in Fig. 9.

VI. DESIGN CONSIDERATION

Employing the above analyses, a design procedure of RC network in the control circuit is discussed here. Key parameters to be discussed are the time constant T_1 and T_2. As mentioned the above sections, the switching frequency and the transient peak voltage of the output for the variations of the load current and the reference voltage are strongly influenced by T_1. Therefore the design starts on finding optimum value of T_1, and then finding T_2 as the next step.

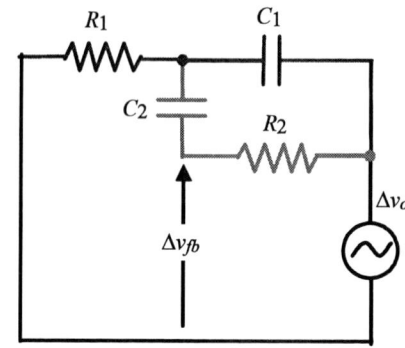

Fig. 7. Equivalent circuit Transfer function from output voltage to feedback voltage

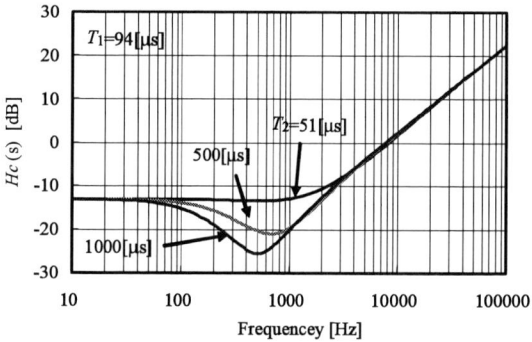

Fig. 8. Frequency responses of the $H_c(s)$

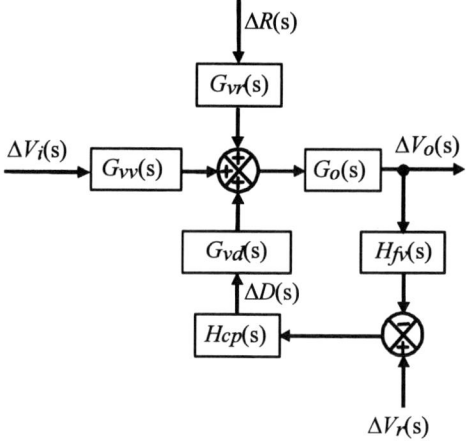

Fig. 9. Block diagram of the regulator system

(1) Time constant T_1

From the view point of the dynamic performance of the converter, the time constant T_1 should be chosen as large as possible to obtain higher loop gain in the high frequency. When an allowable minimum switching frequency f_{smin} and a possible minimum hysteresis window voltage V_{hysmin} are given by the specifications, the maximum value T_{1max} is expressed by

$$T_{1max} = \frac{1 - f_{s\min}\left(\dfrac{V_{cp}}{V_{cp} - V_o}T_{don} + \dfrac{V_{cp}}{V_o}T_{doff} \right)}{f_{s\min} \cdot \dfrac{V_{cp}V_{hys\min}}{(V_{cp} - V_o)V_o}} \qquad (8)$$

(2) Time constant T_2

The capacitor C_2 is used as an ac coupling one between the negative comparator input and C_1 since the voltage across the capacitor C_1 contains a certain level of dc component. On the contrary, the resistor R_2 is used to superimpose dc or lower frequency component of the output voltage on the triangular voltage passed through C_2. In this case, R_2 should be much larger than the impedance of C_1 at the switching frequency in order to keep the linearity of the integrated voltage of C_1. Therefore R_2 must be

$$R_2 \gg \frac{1}{2\pi f_s C_1}. \qquad (9)$$

As can be seen from (6), T_2 should be smaller than $T_1 + T_3$ to reduce the large dip of the control gain lower than mid frequency region. For instance, when the pole frequency ω_p is chosen as same with the zero frequency of (4), this means $T_1 = T_2$, and the gain dip of (6) is smaller than -3 dB, the following conditions must be satisfied to make the gain dip smaller than -3 dB.

$$T_3 \geq \sqrt{2}\,T_1 \qquad (10)$$

With above conditions, C_2 and R_2 are determined as

$$C_2 = \frac{T_3}{R_1}, \quad R_2 = \frac{T_2}{C_2} \qquad (11)$$

(3) Design example

As an example, the buck converter controlled by proposed method is examined with specifications listed in Table 1.

The maximum time constant T_{1max} is directly calculated from (8) as 82 μs. The resistance R_1 is connected to the comparator output and its value should not be too small to prevent overload for the comparator. Then 2 kΩ of R_1 is chosen first and 41nF of C_1 is then determined.

There is no special rule to specify the time constant T_2. If short settling time during the transient is required, this value is to be small. In addition, it is preferable to locate ω_p near or higher than the zero frequency of $H_{cp}(s)$ to have small dip of the loop gain as discussed above. Then two times of $1/T_1$ is used as ω_p here, which provides $T_2 = T_1/4$. Furthermore, from (6), the time constant 102 μs of T_3 is obtained when -1 dB of the gain dip is assumed. All values of RC network are summarized in Table 2 from T_1, T_2, T_3 and (11).

Fig. 10 shows the transient output voltage for the step load change with using parameters listed in Table 2. The measured switching frequency is 480 kHz in this

Table 1 Specifications

Parameter	Value
Input voltage V_i	5.0 V
Output voltage V_o	1.3 V
Switching frequency f_{smin}	500 kHz
Filter inductance L	0.4 μH
Filter capacitance C	300 μF
Delay time T_{don}, T_{doff}	50 ns
Hysteretic window voltage V_{hysmin}	20 mV
Comparator output voltage V_{cp}	4.5 V

Table 2 Parameters of RC network

Parameter	Calculated	Experiment
R_1	2.0 kΩ	2.0 kΩ
C_1	41 nF	47 nF
R_2	488 Ω	470 Ω
C_2	42 nF	47 nF

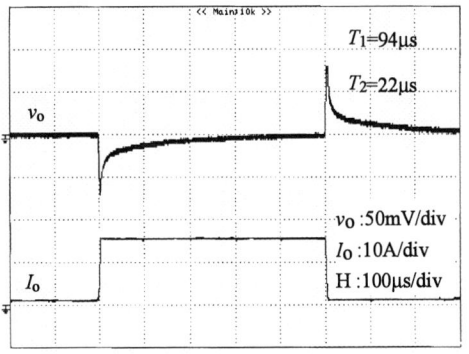

Fig. 10. Transient responses of the output voltage for step change of the load

experiment. It can be seen from the response that the settling time is well shorten. For more small transient voltage, the larger time constant T_1 is required. However it makes the switching frequency lower when the hysteretic window voltage V_{hys} can not be made small.

VII. CONCLUSIONS

This paper has presented a new hysteretic PWM method for all type of PWM converters and its design considerations. Experimental and analytical investigations of the steady state and dynamic characteristics of the presented control method for the buck, boost and buck-boost converters show excellent performances. The most interesting properties of the presented controller are two time constants of the RC network, which improve the response time and the settling time, respectively. Design procedure of RC network in the controller has been discussed in detail.

REFERENCES

[1] R. Redl,and N. O. Socal, "Near-optimum dynamic regulation of DC-DC converters using feed-forward of output current and input voltage with current-mode control," IEEE Tans. On Power Electronics, Vol. PE-1, No.3, pp. 181-192, 1986

[2] G. K. Schoneman and D. M. Mitchell, "Output Impedance Considerations for Switching Regulators with Current-Injected Control" Proc. of Applied Power Electronics Conference, pp.423-335, 1987

[3] G. Garcera,M. Pascual and E. Figueres. "Robust average current mode control of DC-DC PWM converters based on a three controller scheme" Proc. of the IEEE international symposium on industrial electronics, Volt 2, pp.596-600, 1999

[4] R. Miftakhutdinov, "Analysis and optimization of synchronous buck converter at high slew-rate load current transients," proceedings of power electronics specialists conference, pp.714-720,2000

[5] Jason, "Characterization and Performance Comparison of ripple-based control methods for voltage regulator modules," proceedings of power electronics specialists conference, pp.3713-3720,2004

[6] T. Nabeshima,T,Sato,S.Yoshida,S.Chiba and Condi, "Analysis and design consideration of a buck converter with a hysteretic PWM controller," proceedings of power electronics specialists conference, pp.1711-1716,2004

[7] T. Nabeshima,T,Sato,K. Nishijima and S. Yoshida, "A novel control method of boost and buck-boost converters with a hysteretic PWM controller," EPE'05,CD-ROM,2005

[8] T. Nabeshima T. Sato S. Yoshida and K. Onda, "Control Characteristics of a Buck Converter Controlled by Hysteretic PWM Control Method with CR Integral Circuit" THE IEICE TRANS. ON COMMUNICATIONS (JAPANESE EDITION) VOL.J89-B NO.5 2006

Implementation of Fuzzy Logic Controller with Bifurcation Control of a Current-mode Boost Converter

Noppadol Khaehintung[*], Phaophak Sirisuk[**], and Anantawat Kunakorn[***]

[*]Dept. of Control and Instrumentation Engineering, [**]Dept. of Computer Engineering, Faculty of Engineering,
Mahanakorn University of Technology, Bangkok, Thailand 10530
[***]Dept. of Electrical Engineering, Faculty of Engineering, King Mongkut's Institute of Technology Ladkrabang,
Bangkok, Thailand 10520
E-mail: *noppadol@mut.ac.th, phaophak@mut.ac.th* and *kkananta@kmitl.ac.th*

Abstract-- **This paper presents the design of a basic fuzzy logic controller for switching current-mode DC/DC boost converters. The proposed simple fuzzy logic controllers used only nine rules to regulate output voltage. Moreover, they provide an optimal slope compensation to keep the system adequately remote from the first bifurcation point by means of stabilizing around the reference signal. In spite of nonlinear characteristics and instabilities of the converter, the performance of the closed-loop control system can be considerably improved to avoid bifurcation phenomena. The experimental results found that these techniques introduced in this paper give satisfactory results with the regulating and tracking modes under changes in operating conditions.**

Index Terms--- **Current-mode Boost Converter, Fuzzy Logic Controller, Bifurcation Control, Slope Compensation.**

I. INTRODUCTION

Nowadays, DC/DC converter is widely used, especially as a key part of a switching power supply. Generally, this is achieved by chopping and filtering the input voltage through an appropriate switching action, mostly implemented via Pulse Width Modulation (PWM) circuits. A current-programmed converter becomes the regulating scheme of choice in DC/DC converter and offers significant improvement over a direct duty ratio control such as automatic overload protection and design flexibility in improving small-signal dynamics [1]. However, due to nonlinear characteristics of switching devices employed in the converter, the design of a control scheme for such a converter may be difficult with a rich variety of bifurcation [2].

Recently, the main antecedent in the study of bifurcations and chaos has been observed and analyzed for various kinds of power electronic circuits [3]. For a system that exhibits bifurcation when a certain parameter is changed, the key design problem is the control of bifurcation or chaos control [4]. These circuits are designed and guaranteed for stable operations. In most practical situations, the required stable operation is a period-1. Several researches have focused on the control of bifurcation for instance, the work presented in [5], with uses of the discrete time map to analyze and control

the bifurcation for switching control converters. The proposed border bifurcation curves have been presented in [6] to provide useful information for circuit design and control. Although, a variable-ramp compensation in a DC/DC boost converter has been presented and implemented in [7], the design for automatic closed-loop operation has not been mentioned. In addition, applications of chaos anti-control to switch-mode power supplies to reduce spectral peaks have been shown in [8]. Although the gain feedback to control of bifurcation was proposed in [9], however, this system was not guarantee the over compensation, leads system to the discontinuous-conduction mode (DCM).

In the past decades, Fuzzy Logic Control (FLC) has become a popular candidate for applications in power electronic circuits, nonlinear control or optimal search problems [10]. The outmost performance of FLC comparing with a conventional controller has been illustrated in [11]. The applied FLC for a current-mode DC/DC boost converter (CMBC) to optimize a ramp compensation signal has been presented in [12]. However, its structure is generally more complicated and requires relatively high performance processor.

In this paper, the circuit configuration of the proposed a CMBC with bifurcation control is introduced. A simple FLC is employed to search an optimal ramp compensation for CMBC. Apart from simplicity, the fuzzy rules, used in this research, occupy a very small memory space in a computation process, and are easy to implement. The performances of the control system are realized using SAB C167CR microcontroller [13] with various operating conditions of the system.

The paper is organized as follows. In Section II, the circuit configuration of CMBC is introduced. The proposed FLC for control bifurcations is then expressed in Section III. Simulation results from the control system are given in Section IV. Finally, conclusions are drawn in Section V.

II. CIRCUIT CONFIGURATION

A. The proposed CMBC

The proposed FLCs for a CMBC with bifurcation control are depicted in Fig. 1. The system consists of a constant DC voltage source, a CMBC and a load with the parameters given in Table I. One FLC is used to regulate the output voltage (v_{out}), accordingly to the desired output voltage (V_{ref}), in spite of the fluctuations in the input voltage (v_{in}) and the changes in load levels (R_{Load}). The other FLC is used to control of bifurcation which provides an optimal slop compensation m_c depended on an output v_{com}. In stead of using Fourier transform (FFT) to measure current spectral peaks [12], the proposed system uses a simple frequency feedback to capture switching frequency operation. Fig. 2 depicts the implementation of the proposed control algorithm on SAB C167CR microcontroller [13]

B. Nonlinear dynamics of CMBC

In CMBC system, the inductor current, i_L, is chosen as the programming variable which, by comparing with reference current, I_{ref}, from current controller, generates the turning-output signal for switch S. The switch S is turned on by a clock pulse signal at time equal to nT and turned off when i_L climbs up to the denoted value I_{ref} and remains in off position until the begin of the next cycle as shown in Fig. 3.

Typically, the system state equation can be represented by [7]

$$\begin{bmatrix} \dot{v}_{out} \\ \dot{i}_L \end{bmatrix} = \begin{bmatrix} -1/R_{Load}C & q/C \\ -q/L & 0 \end{bmatrix} \begin{bmatrix} v_{out} \\ i_L \end{bmatrix} + \begin{bmatrix} 0 \\ 1/L \end{bmatrix} v_{in}, \quad (1)$$

where q is a switching function which is one during the switch off interval while zero during on interval.

Let us define $i_L(n)$ and $i_L(n+1)$ as the inductor current at a $t = nT$ and $t=(n+1)T$, respectively. In steady-state condition, the inductor current can be derived from [7] as:

$$i_L(n+1) = I_{ref} + \frac{(v_{in} - v_{out})T}{L} t'_n \quad (2)$$

and

$$t'_n = T - T\left[\left(\frac{t_n}{T}\right) \bmod 1\right], \quad (3)$$

where, T is a clock pulse period and mod(.) is the modulus function. The slope of inductor current can be expressed by inspecting as:

$$\frac{I_{ref} - i_L(n+1)}{(1-D)T} = \frac{v_{out} - v_{in}}{L} \quad (4)$$

and

$$\frac{I_{ref} - i_L(n)}{DT} = \frac{v_{in}}{L} \quad (5)$$

The inductor current shown as solid line in Fig. 3 under period-1 operation, which frequency of i_L equal f, can be expressed as:

$$i_L(n+1) = \left(1 - \frac{v_{out}}{v_{in}}\right) i_L(n) + \frac{I_{ref}v_{out}}{v_{in}}\left(\frac{v_{out}-v_{in}}{L}\right)T \quad (6)$$

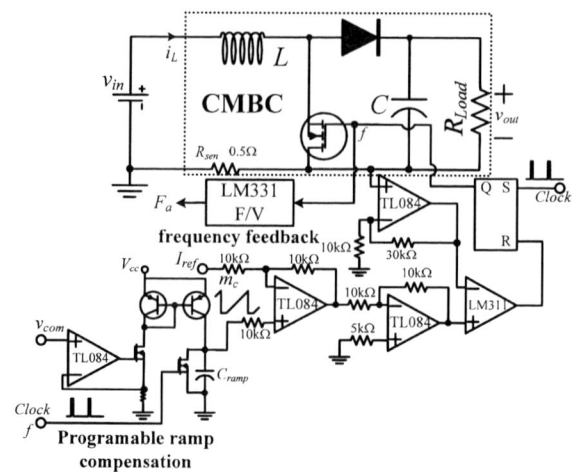

Fig. 1. The proposed CMBC circuit.

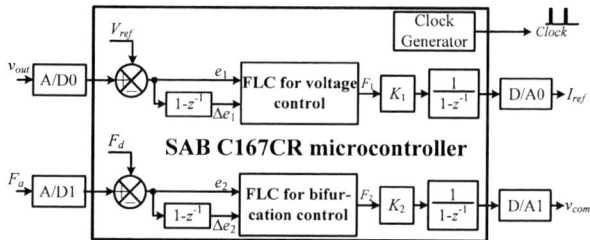

Fig. 2. The proposed control algorithm.

TABLE I
THE CMBC PARAMETERS

Variable	Definitions
Inductor (L)	$160\,\mu$H
Capacitor (C)	$47\,\mu$F
Switching Freq ($1/T$).	10kHz
Output voltage (v_{out})	12-15V
Input voltage (v_{in})	5V

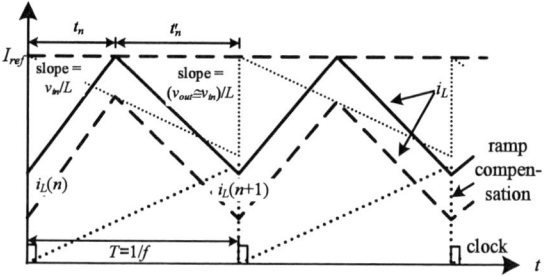

Fig. 3. The inductor current i_L of CMBC.

III. THE PROPOSED CONTROLLERS

A. Compensation for CMBC

In order to maintain a stable period-1 operation [1,7], a CMBC must operate with duty ratio D set below 0.5, so that the criterion of no bifurcation can be achieved. For the CMBC, the equivalent criterion of no bifurcation in terms of I_{ref} using the steady-state equation relating R_{Load} and D is given by:

1430

$$i_{ref} < \frac{v_{in}}{R_{Load}} \left[\frac{DR_{Load}T}{2L} + \frac{1}{(1-D)^2} \right]_{D=0.5} \quad (7)$$

To prevent the operation in period-doubling when I_{ref} exceeds the above-stated limit, a variable compensated ramp is used to raise the upper bound of I_{ref}. Thereby widening the operation range shows in Fig. 3, with dot and dash lines described in [7]. The variable-ramp compensation for a CMBC needs only be controlled according to:

$$\frac{m_c L}{v_{in}} = M_c(v_{in}) \geq \frac{v_{out}}{2v_{in}} - 1, \quad (8)$$

where, m_c is a slope of compensating ramp which is appropriate for system environments and $M_c = m_c L / v_{in}$.

B. The proposed FLC for Voltage Control Loop

In this closed-loop control system, I_{ref} is controlled via feedback loop which attempts to keep v_{out} fixed by adjusting I_{ref}. The I_{ref} for a CMBC can be found as:

$$I_{ref}(k) = I_{ref}(k-1) + K_1 F_1, \quad (9)$$

where, K_1 and F_1 are the gain factor and the output of FLC for voltage control loop, respectively. Let us define the error $(e_1(k))$ and the associated change of error $(\Delta e_1(k))$ as:

$$e_1(k) = V_{ref}(k) - v_{out}(k) \quad (10)$$

$$\Delta e_1(k) = e_1(k) - e_1(k-1), \quad (11)$$

It is noted that, the time instant k is corresponding to the sampling time of an A/D and is intentionally chosen to be different from n, which is corresponding to the switching time of the switch S. The controller's function is to force the error, which is the difference of the current and desired output voltage to zero.

C. The proposed FLC for Control of Bifurcation

Referring to Fig. 1, the FLC can be established to search an optimal value of m_c from (8) by using frequency feedback. In particular, this optimal value of m_c can be achieved by the fuzzy algorithm as an FLC for voltage control loop. Let us define the error $(e_2(k))$ and the associated change of error $(\Delta e_2(k))$ as:

$$e_2(k) = F_d(k) - F_a(k) \quad (12)$$

$$\Delta e_2(k) = e_2(k) - e_2(k-1), \quad (13)$$

where, $F_d(k)$ and $F_a(k)$ are the desired and actual frequency, respectively. Therefore, forcing the error

function in (12) and its associated change of error in (13) to zero, the stable operation at period-1 can be achieved. The optimal slope m_c can be found as:

$$m_c(k) = m_c(k-1) + K_2 F_2, \quad (14)$$

where, K_2 and F_2 are the gain factor and the output of FLC for control of bifurcation.

D. FLC rules

To control the DC/DC converter, the criterions of the linguistic rules in the controller are given in [11]. This can be translated meta-rules into following fuzzy control rule:

Rule (i) : if $e_x(k)$ is A_i and $\Delta e_x(k)$ is B_j
        then $F_x(k)$ is C^l,   (15)

where, A_i and B_j, are the fuzzy subsets in their universe of discourse including positive (PO), zero (ZE) and negative (NG), C^l, is the output fuzzy subsets, or fuzzy singleton for Sugeno fuzzy model [11] and $x=1,2$.

Note that the number of fuzzy function should be optimally determined with considering the interrelation of control accuracy and calculation capacity. The membership function of each fuzzy set is selected based on trial-and-error such that the region of interest is covered appropriately. To avoid complicated calculation process the triangular shape membership will be used. The membership functions in each universe of discourse are shown in Fig. 4, and the fuzzy rule base is shown in Table II. For any given input pair of $(e_x(k), \Delta e_x(k))$, the crisp value of controller output $F_x(k)$ is calculated by fuzzy inference system. Firstly, the *product* operation is adopted to obtain the output fuzzy set μ_{F_x} for each rule. The *sum* operation is then adopted to combine each output fuzzy set μ_{F_x} into a single fuzzy set. Finally, through the defuzzification process and using the *centroid* method, a crisp value for $F_x(k)$ is obtained by [14]

$$F_x(k) = \frac{\sum_{l=1}^{M} F_x^l w_l}{\sum_{l=1}^{M} w_l}, \quad (16)$$

where, $w_l = min[\mu_{e_x}(e_x(k)), \mu_{\Delta e_x}(\Delta e_x(k))]$ is the compatibility (weighting factor) and F_x^l is a value corresponding to the membership function of $F_x(k)$.

IV. EXPERIMENTAL RESULTS

To evaluate performances of the proposed system, the FLC algorithms were developed on SAB C167CR microcontroller equipped with AD5333 D/A 12 bit resolution. The CMBC model described in the previous section with the parameters given in Table II was invoked. Fig. 5 depicts the hardware experimentation for this research work. The FLCs were implemented and set sampling time at 1ms to compare the performance of CMBC without and with FLC for bifurcation control.

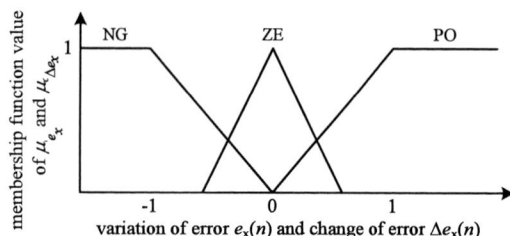

Fig. 4. The membership functions of the proposed FLC.

TABLE II
A FUZZY RULE BASE

$\Delta e(n)$ \ $e(n)$	NG	ZE	PO
NG	C^1 (-1.0)	C^2 (-0.5)	C^3 (-0.2)
ZE	C^4 (-0.2)	C^5 (0)	C^6 (0.2)
PO	C^7 (0.2)	C^8 (0.5)	C^9 (1.0)

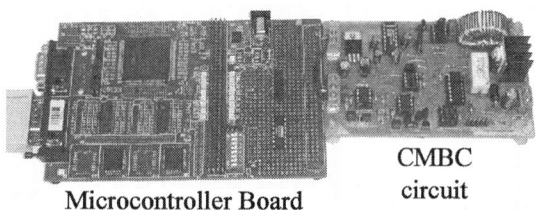

Microcontroller Board   CMBC circuit

Fig. 5. Hardware Experimentation.

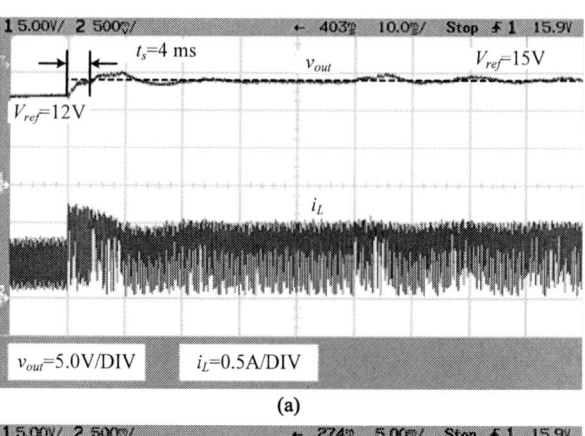

(a)

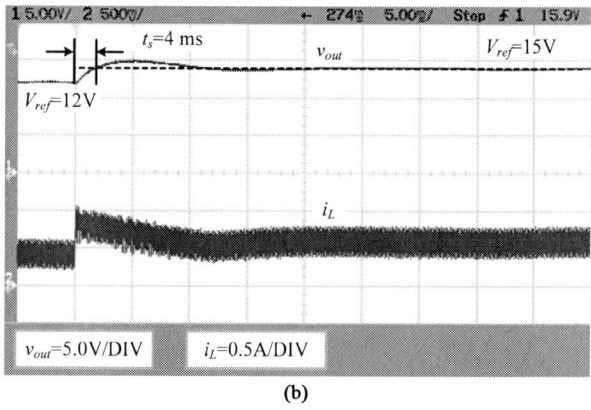

(b)

Fig. 6. Output response of v_{out} and i_L at 50Ω of R_{Load} while V_{ref} change from 12V to 15V (a) without FLC and (b) with FLC for bifurcation control.

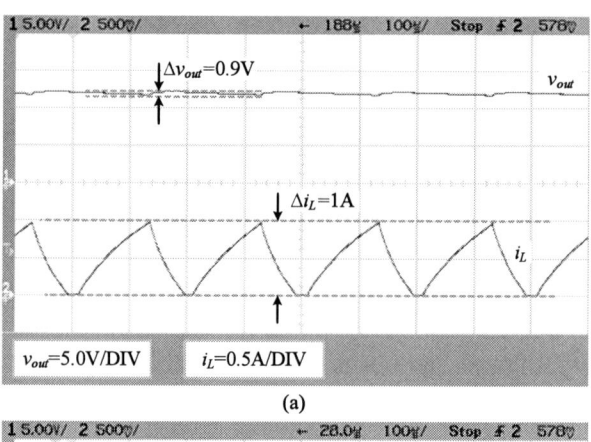

(a)

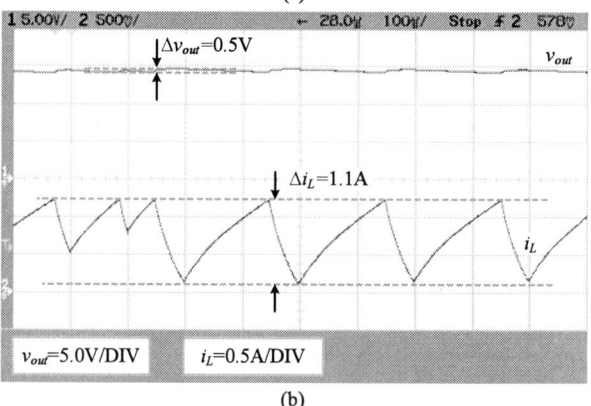

(b)

Fig. 7. Steady state of v_{out} and i_L at 50Ω of R_{Load} without FLC for bifurcation control. (a) 12V of V_{ref} and (b) 15V of V_{ref}.

In the experiments, the input supply voltage (v_{in}) is fixed at 5V. The values of load resistance (R_{Load}) were 25Ω and 50Ω. The output voltage (v_{out}) converges from the starting point to the steady point at V_{ref} = 12V and 15V. The comparisons for output voltage of CMBC without and with FLC for bifurcation control, when V_{ref} changed from 12V to 15V, are depicted in Fig. 6 (a) and (b), respectively. It can be seen clearly that, the steady-state response of the CMBC without FLC for bifurcation control was irresolution as shown in Fig. 6 (a).

For 50Ω of R_{Load}, the output voltage and inductor current for the non-compensated CMBC in steady-state situation are depicted in Fig. 7 (a) and (b) when V_{ref} = 12V and V_{ref} = 15V, respectively. It is noticeable that without FLC for bifurcation control, the system can not operates at 10kHz, period-1 operation, with the limitation for the ripple of i_L (Δi_L) at 1.0A and 1.1A, respectively. These results are quite different when applied FLC for bifurcation control to CMBC with Δi_L are 0.6A and 0.65A for 12V and 15V of V_{ref} as depicted in Fig. 8 (a) and (b).

For different load condition at 25Ω of R_{Load}, the system, without FLC compensation, operates at 12V of V_{ref} is submitted to chaotic, with Δi_L at 1.4A and Δv_{out} at 1V approximately as shown in Fig. 9 (a). With the proposed FLC for bifurcation control, the output remains stable and decreases Δi_L and Δv_{out} fluctuation as depicted in Fig. 9 (b). It is noticeable that, the output voltage and current response between with and without FLC compensation are clearly different.

1432

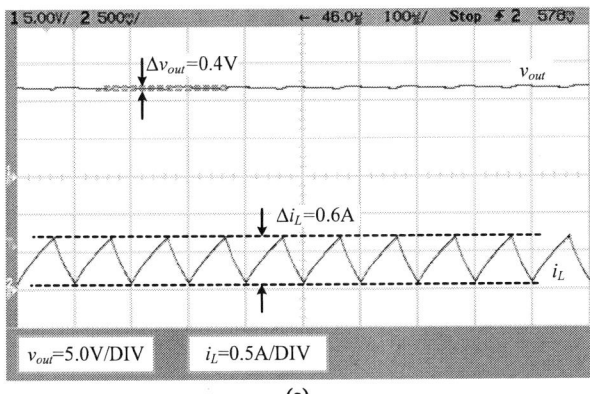

(a)

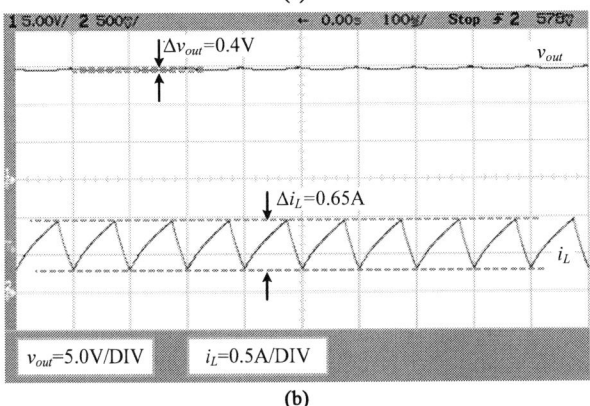

(b)

Fig. 8. Steady state of v_{out} and i_L at 50Ω of R_{Load} with FLC for bifurcation control. (a) 12V of V_{ref} and (b) 15V of V_{ref}.

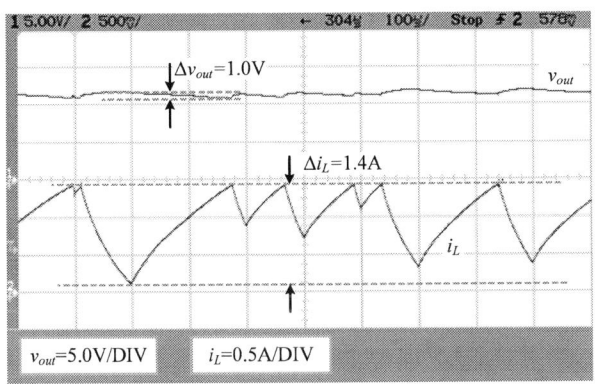

(a)

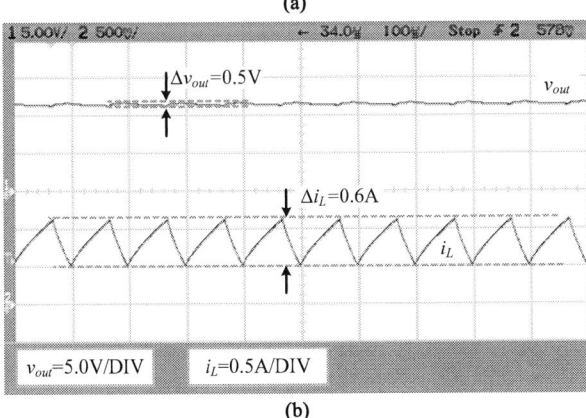

(b)

Fig. 9. Steady state of v_{out} and i_L at 25Ω of R_{Load} and 12V of V_{ref} (a) without FLC and (b) with FLC for bifurcation control.

V. CONCLUSIONS

In this paper, the advantages of a fuzzy logic controller for a current-mode boost converter with control of bifurcation have been demonstrated and implemented for stable system operations. The experimental results guarantee that the proposed FLC controller can give the desired stable period-1 under changes in operating conditions. The technique introduced in this paper can be efficiently realized by means of a simple FLC, which can be easily implemented in a real-time control scheme.

REFERENCES

[1] F.D. Tan and R.D. Middlebrook, "A unified model for current-programmed converters," *Trans. on IEEE Power Electronics*, Vol. 10, No. 4 , July, 1995, pp. 397-408.

[2] S. Banerjee and G.C. Verghese, *Nonlinear Phenomena in Power Electronics*, IEEE press, New York, 2001.

[3] M.D. Bernardo, F. Garofalo, L. Glielmo and F. Vasca, "Switchings, Bifurcations, and Chaos in DC-DC Converters," *Trans. on IEEE CAS-I*, Vol. 45, No. 2, 1998, pp. 133–141.

[4] C. Morel, M. Bourcerie and F. Chapeau-Blondeau, "Extension of Chaos Anticontrol Applied to the Improvement of Switch-Mode Power Supply Electromagnetic Compatibility," *Proc. of the IEEE ISIE 04 Conf*, pp. 447–452, 2004.

[5] M. D. Bernardo and F. Vasca, "Discrete-Time Maps for the Analysis of Bifurcations and Chaos in DC-DC Converters," *IEEE Trans. on CAS-I*, Vol. 47, No. 2, 2000, pp. 130–143.

[6] M. Yue and H. Kawakami, "Control of bifurcation in DC/DC PWM switching converters," *Proc. of the 8th ICARCV 2004 Conf.*, Vol. 2, pp. 1421 – 1426, 2004.

[7] C.K. Tse and Y.M. Lai, "Control Bifurcation in Current-Programmed DC/DC Converters: A reexamination of Slope Compensation," *Proc. of IEEE ISCAS 2000 Conf.*, pp. I-671-674, 2000.

[8] W.C.Y. Chan and C.K. Tse, "Study of Bifurcations in Current-Programmed DC/DC Boost Converters: From Quasiperiodicty to Period-Doubling," *Trans. on IEEE CAS-I*, Vol. 43, No. 12, 1997, pp. 1129-1142.

[9] N. Khaehintung, P. Sirisuk and T. Wiangtong, "Control of Bifurcation for a Current-mode DC/DC Boost Converter by Self-inductor Current Feedback," *Proc. of IEEE ISCIT 2006 Conf.*, pp. W3D-2, 2006.

[10] N. Khaehintung, P. Sirisuk and W. Kurutach, "A novel ANFIS controller for maximum power point tracking in photovoltaic systems," *Proc. of the IEEE PEDS 2003 Conf.*, Vol. 2, pp. 833-836, 2003.

[11] N. Khaehintung, A. Kunakorn, M.Aorpimai1 and P. Sirisuk, "An Adaptive Fuzzy Logic Controller by Sliding Mode Control Method for DC/DC Converter," *Proc. of the IEEE PEDS 2005 Conf.*, Vol. 1, pp. 833-836, 2005.

[12] N. Khaehintung, P. Sirisuk and A. Kunakorn, "Control of Bifurcation by Fuzzy Logic Controller for Current-mode Boost Converters," *Proc. of IEEE IPEMC 2006 Conf.*, Vol. 1, pp. 368-372, 2006.

[13] http://www.infineon.com/

[14] N. Khaehintung and P. Sirisuk "Implementation of maximum power point tracking using fuzzy logic controller for solar-powered light-flasher applications," *Proc. of IEEE MWCAS 2004 Conf.*, Vol. 3, pp. iii - 171-4, 2004.

Phase Advance Approach to Expand the Speed Range of Brushless DC Motor

BinhMinh Nguyen, Minh C. Ta

Department of Industrial Automation, Faculty of Electrical Engineering
Hanoi University of Technology, 1 Dai Co Viet street, Hanoi, Vietnam
e-mail: ngbminh20021846@yahoo.com, minhtc-auto@mail.hut.edu.vn

Abstract--In order to extend the speed range of a brushless DC motor drive, it is common to control the phase current to lead the phase back EMF by an "advance angle". It is based on the "phase advance approach", which was proposed for brushless DC motor drive in 1995. This paper completes the "phase advance approach" theory in several aspects: a) the stator resistance is taken into account in the analysis and calculation of phase reference current; b) the situation where the back EMF is smaller than the applied voltage is also taken into consideration; and c) an analytical equation to calculate the advance angle is also derived. The obtained equation are therefore more general than those in the prior art. Extended simulation results demonstrate the validity of the approach.

Index Terms--Trapezoidal-type PM synchronous motors, Brushless DC Motor Drive, Phase Advance Angle, Wide Speed-range Operation.

NOMENCLATURE

θ_0	Phase advance angle
u	Phase applied voltage
e	Phase back electromagnetic force (BEMF)
i	Phase current
V	Amplitude of phase applied voltage
E	Amplitude of phase back electromotive force
K_e	Electromotive force constant
K_t	Torque constant
p	Number of pole pairs
R	Resistance of phase winding
L	Inductance of phase winding
ω	Electrical speed
ω_m	Motor speed
τ	Time constant

I. INTRODUCTION

It is well known that in the trapezoidal-type PM synchronous motors (often called Brushless DC Motor) the BEMF wave-form is trapezoidal and the current shape is rectangular as shown in Fig. 1. Brushless DC Motor (BLDCM) has many advantages: it requires less maintenance, operates more quietly than a brushed DC motors. BLDCM also produces more output power per frame size than other kinds of motors. Due to the wave-forms of BEMFs and currents, the BLDCM has however two main disadvantages: it produces more torque ripple and it's difficult to control at high speed. This paper focuses on overcoming the second disadvantage. In section 2, the basic problems of controlling BLDCM in high-speed region including the "phase advance approach" in literature [1] are analyzed. The general mathematical derivation of phase advance approach is developed in details in section 3. In section 4, the calculation of phase advance angle is described. Computer simulation results are given in section 5. Finally, the conclusion of this paper is given in section 6.

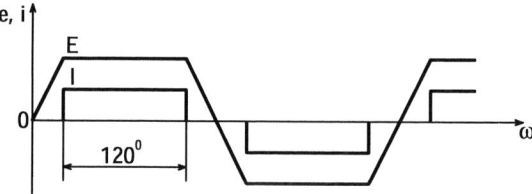

Fig. 1. Typical phase BEMF and phase current of BLDCM.

II. BASIC PROBLEMS OF CONTROLLING BLDCM

A. Limit of Operation Region

Due to the fact that the BEMF and current of BLDCM is not sinusoidal, the vector control principle can not be readily applied. To control a BLDCM, it is common to use the conventional phase-current-control configuration as shown in Fig. 2 with two close loops for current and speed control (R_ω is the speed regulator and R_i is the current regulator). With this drive system, BLDCM can operate with rated torque at any speed under the rated speed.

In many applications such as electric vehicles, tools machines, etc, wide speed range (over the rated speed) is required. Above the rated speed, because the power of a BLDCM is a limit determined value, the torque decreases inversely proportional to the motor speed. Fig. 3 shows the typical characteristics of a motor where the *bd* is often called the power limit curve, on which the speed is maximal for a given load torque. Conventional drive system cannot however operate on the power limit curve due to voltage saturation. The lower curve *bef* illustrated in Fig. 3 is the locus of operating points of conventional drive system without any additional special control technique. Although Fig. 3 is general for any kind of motors, the phenomena of

978-1-4244-0644-9/07/$25.00 ©2007 IEEE

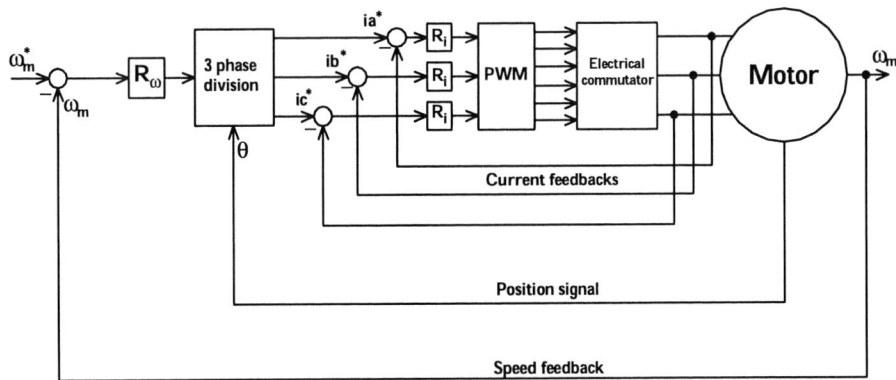

Fig. 2. Conventional BLDCM drive system with current loop and speed loop.

saturation can be explained for the BLDCM as follows.

PWM technique is used to control the applied voltage of BLDCM. Operating at rated speed, the power switches are turned on 120 electrical degrees each half cycle, so the applied voltage attaints its rated value. This is the maximal applied voltage for a BLDCM.

The voltage equation for each phase of the motor can be expressed as

$$u = Ri + L\frac{di}{dt} + e \qquad (1)$$

where u is the phase applied voltage, R is the resistance of phase winding, L is the inductance of phase winding, i is the phase current and e is the phase BEMF.

Because the applied voltage is limited, if the motor current is large enough to satisfy the load torque (for example T_1), the BEMF cannot increase to a higher value. BEMF is directly proportional to the motor speed, so the motor speed cannot be developed ($\omega_{m2} < \omega_{m1}$) as shown by point e in Fig. 3. Similarly, if we want the motor to operate at a higher speed (for example ω_{m1}), the output torque of the motor cannot be developed as required ($T_2 < T_1$ point f in Fig. 3). As the operating points e, f are under the power limit curve, the total power of the motor cannot be utilized.

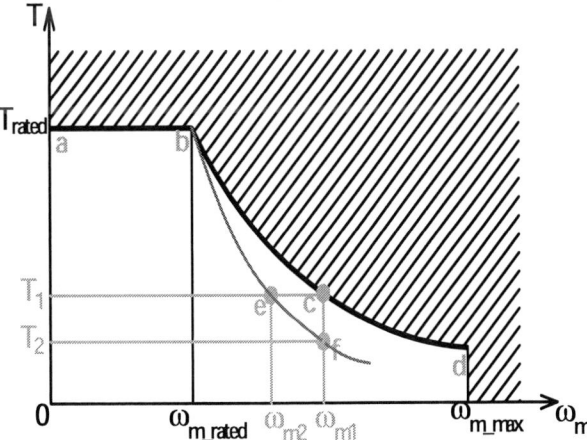

Fig. 3. Operating points of BLDCM.

B. Phase Advance Control of BLDCM

In order to expand the speed range of a drive, it is common to weaken the existing field. For the separately excited DC motor, as the motor has separate windings for the flux-producing and torque-producing currents, it is very easy to control this motor above rated speed by decreasing the flux in keeping the applied voltage at its rated value. For a vector-control IM drive, flux-weakening operation can be realized by decreasing the d-axis (flux-producing) current. For a vector-control PM drive, the field of PM can be weakened by introducing a negative d-axis current ($Id < 0$). For a BLDCM, flux-weakening operation is not a matter of acting on d-axis current as compared to other of AC motors, because the vector control is not applicable to this kind of motor as mentioned previously.

In 1995, C. C. Chan *et al.* proposed "phase advance approach" to solve this problem for BLDCM [1]. The main idea of this approach can be explained as follows. There are two types of electromotive forces in the phase winding of a BLDCM. One is called Back electromotive force (BEMF), e, which is induced by the magnet field of the rotating of the permanent magnet. Another one is called Transformer electromotive force (TEMF), $L\frac{di}{dt}$, which is induced by the transformer action of the time-varying stator current in the phase windings. Below the rated speed, BLDCM is controlled conventionally, the phase current is in phase with the phase EMF. When operating above the rated speed, the phase-current leads the phase BEMF. Thus, the TEMF is utilized to counteract the BEMF. This is equivalent to the flux-weakening for the DC motor drive.

Chan *et al.* examined the phase current when the amplitude of BEMF is higher than the amplitude of phase applied voltage (E >V) with the assumption that the phase winding resistance is negligible. But the amplitude of BEMF can be lower than the amplitude of phase applied voltage (E < V), and the phase winding resistance is not always negligible.

There is a corresponding advance angle for every given reference speed. The locus of these points is the "phase advance curve". In the paper in 1995, Chan *et al.* proposed that the value of "advance angle" is governed by an approximate linear relationship with the motor speed. In reality, the "phase advance curve" is not a straight line, but a complicated curve. In the latter paper in 1998, Chan *et al.* proposed an algorithm which is called "Adaptive searching trajectory" to determine the corresponding advance angle [2]. This algorithm increases the complicatedness of the BLDCM drive. Beside the works of Chan *et al.*, other researches in phase advance approach have been investigated. J. S. Lawler *et al.* identified several limitations of the phase advance approach. The phase advance approach is especially sensitive to the motor inductance that must be larger than a threshold value to maintain motor current within rated value when operating at rated power and high speed [4]. If the motor inductance is low, additional cooling will be necessary for the motor and inverter components and the current rating of the inverter will have to be increased. J. S. Lawler *et al.* also proposed an inverter topology and control scheme which is called dual-mode inverter control [5]. With this control scheme, the range of motor inductance is widened.

For the purpose of redounding to the encompassment of the theory of "phase advance approach", the authors of this paper examine two cases: E < V and E > V and the phase winding resistance is taken into account.

III. PRINCIPLE AND MATHEMATICAL DERIVATION OF PHASE ADVANCE APPROACH

The BLDCM which is used to examine this approach has three phases, two pairs of poles and *wye* stator connection. In the following mathematical analysis, the assumption that the BEMF is trapezoidal is used. Fig. 4 illustrates the per-phase equivalent circuit diagram of BLDCM, Q1 and Q2 are the power switchers, D1 and D2 are feedback diodes. When operating above the rated speed, the conduction period of Q1 leads the BEMF by a spatial angle, which is called the "advance angle" (θ_0). Thus, Q1 is fired θ_0 ahead of the instant that the phase BEMF reaches its positive maximum.

A. Amplitude of Phase BEMF Lower than Amplitude of Phase Applied Voltage E < V

Starting from the turn-on of Q1, the first half operating cycle of the equivalent circuit (Fig. 4) is divided into four intervals as illustrated in Fig. 5. In the interval 1-2, the phase BEMF is increasing and the phase applied voltage is positive so that the phase current increases rapidly. In the interval 2-3, both the phase BEMF and phase applied voltage reaches their positive amplitudes. Since the phase applied voltage is larger than the BEMF, the phase current increases gradually. At the end of the interval 2-3, Q1 is turned off. After that, the phase current flows through D2 in the interval 3-4. Since the phase applied voltage is negative

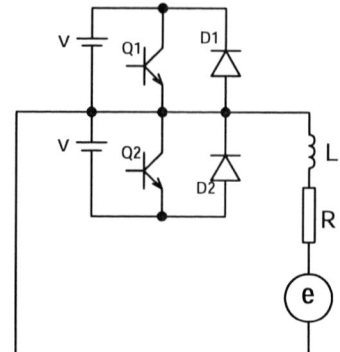

Fig. 4. Per-phase equivalent circuit diagram of BLDCM.

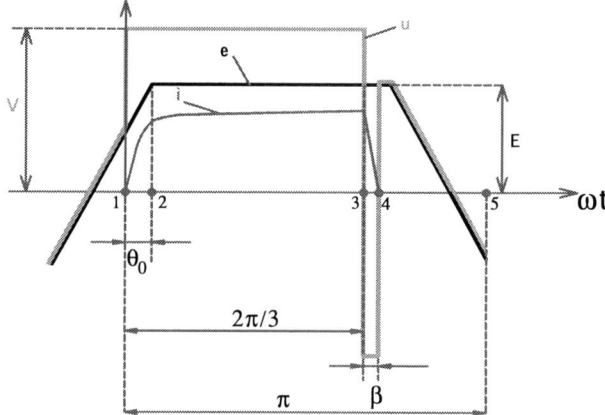

Fig. 5. Wave form of phase current when E < V.

and the BEMF is positive, the phase current drops rapidly to zero. In the interval 4-5, since the circuit is opened, the phase applied voltage equals the BEMF while the phase current is always zero. Therefore, in both intervals 1-2 and 2-3, the TEMF $L\dfrac{di}{dt}$ is positive but decreases. In the interval 3-4, TEMF is negatives. The second half cycle is analyzed similarly.

The phase current is divided into four stages over the first half cycle.

a) Stage I, [$0 \le \omega t \le \theta_0$]

This stage covers the interval 1-2. The phase applied voltage and the BEMF are given by

$$\begin{cases} u = V \\ e = \dfrac{6E\omega}{\pi}t + \dfrac{\pi - 6\theta_0}{\pi}E \end{cases} \qquad (2)$$

Substituting (2) into (1), the phase current is obtained as

$$i_1(t) = A + Bt - Ae^{\frac{-t}{\tau}} \qquad (3)$$

where $\tau = \dfrac{L}{R}$ is the time constant and

1436

$$\begin{cases} A = \dfrac{V-E}{R} + \dfrac{6E\omega L}{\pi R^2} + \dfrac{6E}{\pi R}\theta_0 \\[3mm] B = \dfrac{-6E\omega}{\pi R} \end{cases} \qquad (4)$$

At the end of this stage, the value of the phase current is expressed as

$$I_2 = i_1\left(\frac{\theta_0}{\omega}\right) = A + B\frac{\theta_0}{\omega} - Ae^{\frac{-\theta_0}{\omega\tau}} \qquad (5)$$

b) Stage II, $[\,\theta_0 \le \omega t \le \dfrac{2\pi}{3}\,]$

This stage covers the interval 2-3. The phase applied voltage and the BEMF are given by

$$\begin{cases} u = V \\ e = E \end{cases} \qquad (6)$$

Substituting (6) into (1), the phase current is obtained as

$$i_2(t) = M + I_2 - Me^{-\left(t - \frac{\theta_0}{\omega}\right)\!\big/\tau} \qquad (7)$$

where $M = \dfrac{V-E}{R}$

At the end of this stage, the value of the phase current is expressed as

$$I_3 = i_2\left(\frac{2\pi}{3\omega}\right) = M + I_2 - Me^{-\left(\frac{2\pi}{3\omega} - \frac{\theta_0}{\omega}\right)\!\big/\tau} \qquad (8)$$

c) Stage III, $[\,\dfrac{2\pi}{3} \le \omega t \le \dfrac{2\pi}{3} + \beta\,]$

This stage covers the interval 3-4. The phase applied voltage and the BEMF are given by

$$\begin{cases} u = -V \\ e = E \end{cases} \qquad (9)$$

Substituting (9) into (1), the phase current is obtained as

$$i_3(t) = P + I_3 - Pe^{-\left(t - \frac{2\pi}{3\omega}\right)\!\big/\tau} \qquad (10)$$

where $P = \dfrac{-V-E}{R}$

At the end of this stage, the value of the phase current is zero, so

$$i_3\left(\frac{2\pi}{3\omega} + \frac{\beta}{\omega}\right) = P + I_3 - Pe^{-\left(\frac{2\pi}{3\omega} + \frac{\beta}{\omega} - \frac{2\pi}{3\omega}\right)\!\big/\tau} = 0. \quad \text{Solve this}$$

equation, the duration of this stage can be found as:

$$\beta = \omega\tau\ln\left(\frac{P}{P+I_3}\right) \qquad (11)$$

d) Stage IV, $[\,\dfrac{2\pi}{3} + \beta \le \omega t \le \pi\,]$

This stage covers the interval 4-5. Because the circuit is opened, the phase current is always zero. The derivation of the second half cycle can be obtained similarly.

B. Amplitude of Phase BEMF Higher than Amplitude of Phase Applied Voltage $E > V$

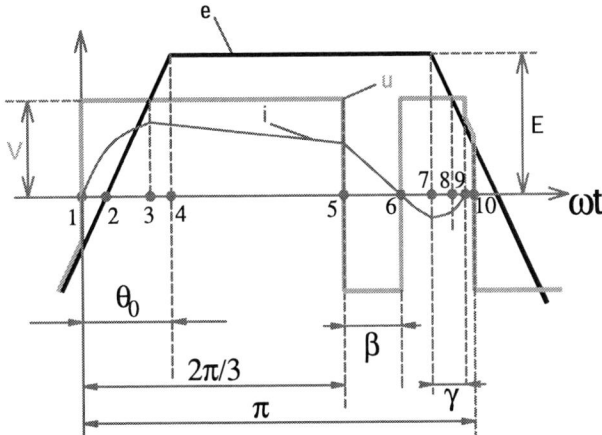

Fig. 6. Wave form of phase current when E > V.

Starting from the turn-on of Q1, the first half operating cycle of the equivalent circuit (Fig. 4) is divided into nine intervals as illustrated in Fig. 6. In the interval 1-2, the negative phase BEMF acts to strengthen the phase applied voltage so that the phase current increases rapidly. In the interval 2-3, both the phase applied voltage and the phase BEMF are positive. Since the applied voltage is still larger than the phase BEMF, the phase current increases gradually. In the interval 3-4 and 4-5, the phase BEMF is larger than the phase applied voltage. Thus, the phase current decreases gradually. Since the phase TEMF $L\dfrac{di}{dt}$ is negative, it acts to counteract the phase BEMF. This phenomenon is the key point of the approach. At the end of the interval 4-5, Q1 is turned off. After that, the phase current flows through D2 in the interval 5-6. Since the phase applied voltage is negative while the BEMF is positive, the current drop rapidly to zero. Then, in the interval 6-7 and 7-8, the phase BEMF is larger than the phase applied voltage, the phase current begins to flows negatively through D1. When the phase BEMF becomes smaller than phase applied voltage in the interval 8-9, the phase current goes back to zero. Since the circuit is opened in the interval 9-10, the phase applied voltage equals the phase BEMF and the phase current is always zero. The second half cycle is analyzed similarly. The phase current can be divided into six stages over the first half cycle.

a) Stage I, $[\,0 \le \omega t \le \theta_0\,]$

This stage covers the intervals 1-2, 2-3 and 3-4. The phase current can be expressed as

1437

$$i_1(t) = A + Bt - Ae^{\frac{-t}{\tau}} \tag{12}$$

At the end of this stage, the value of the phase current is expressed as

$$I_4 = i_1\left(\frac{\theta_0}{\omega}\right) = A + B\frac{\theta_0}{\omega} - Ae^{\frac{-\theta_0}{\omega\tau}} \tag{13}$$

b) Stage II, [$\theta_0 \leq \omega t \leq \dfrac{2\pi}{3}$]

This stage covers the interval 4-5. The phase current can be expressed as:

$$i_2(t) = M + I_4 - Me^{-\left(t - \frac{\theta_0}{\omega}\right)/\tau} \tag{14}$$

At the end of this stage, the value of the phase current is expressed as:

$$I_5 = i_2\left(\frac{2\pi}{3\omega}\right) = M + I_4 - Me^{-\left(\frac{2\pi}{3\omega} - \frac{\theta_0}{\omega}\right)/\tau} \tag{15}$$

c) Stage III, [$\dfrac{2\pi}{3} \leq \omega t \leq \dfrac{2\pi}{3} + \beta$]

This stage covers the interval 5-6. The phase current can be expressed as

$$i_3(t) = P + I_5 - Pe^{-\left(t - \frac{2\pi}{3\omega}\right)/\tau} \tag{16}$$

At the end of this stage, the value of the phase current is zero, so the duration of this stage can be found as

$$\beta = \omega\tau \ln\left(\frac{P}{P + I_5}\right) \tag{17}$$

d) Stage IV, [$\dfrac{2\pi}{3} + \beta \leq \omega t \leq \dfrac{2\pi}{3} + \theta_0$]

This stage covers the interval 6-7. The phase current can be expressed as

$$i_4(t) = M - Me^{-\left(t - \frac{2\pi}{3\omega} - \frac{\beta}{\omega}\right)/\tau} \tag{18}$$

At the end of this stage, the value of the phase current is expressed as

$$I_7 = i_4\left(\frac{2\pi}{3\omega} + \frac{\theta_0}{\omega}\right) = M - Me^{-\left(\frac{\theta_0}{\omega} - \frac{\beta}{\omega}\right)/\tau} \tag{19}$$

e) Stage V, [$\dfrac{2\pi}{3} + \theta_0 \leq \omega t \leq \dfrac{2\pi}{3} + \theta_0 + \gamma$]

This stage covers the intervals 7-8, 8-9. The phase current can be expressed as

$$i_5(t) = A' + I_7 + B'\left(t - \frac{2\pi}{3\omega} - \frac{\theta_0}{\omega}\right) - A'e^{-\left(t - \frac{2\pi}{3\omega} - \frac{\theta_0}{\omega}\right)/\tau} \tag{20}$$

where

$$\begin{cases} A' = \dfrac{V - E}{R} - \dfrac{6E\omega L}{\pi R^2} \\[2mm] B' = \dfrac{6E\omega}{\pi R} \end{cases} \tag{21}$$

At the end of this stage, the value of the phase current is zero, so it's possible to calculate the duration of this stage.

f) Stage VI, [$\dfrac{2\pi}{3} + \theta_0 + \gamma \leq \omega t \leq \pi$]

This stage covers the interval 9-10. Because the circuit is opened, the phase current is always zero. The derivation of the second half cycle can be obtained similarly. All the factors (A, B, M, P) in this second case are calculated similarly in the first case.

IV. CALCULATION OF PHASE ADVANCE ANGLE

For each given reference speed there is a corresponding advance angle. The trajectory of advance angles in function of speed is a complicated curve which is called "phase advance curve". Calculating the phase advance corresponding to the reference speed is an important problem in the approach of phase advance.

Below the rated speed, the output torque of the motor is directly proportional to the amplitude of the phase current:

$$T = K_t . I \tag{22}$$

The current flows through the phase winding 120 degrees each half cycle, so the rms value of the current can be expressed as

$$I_{rms} = \sqrt{\frac{1}{T}\int_0^T i^2 dt} = \sqrt{\frac{1}{\pi}\int_{\frac{\pi}{6}}^{\frac{5\pi}{6}} I^2 d(\omega t)} = \frac{\sqrt{6}}{3}I \tag{23}$$

From (22) and (23):

$$T = K_t^* . I_{rms} \tag{24}$$

where

$$K_t^* = \frac{3}{\sqrt{6}}K_t \tag{25}$$

The problem of calculating the advance angle can be expressed as follows. For a given speed ω_m and torque

$T = \dfrac{\omega_{m\_rated}}{\omega_m}.T_{rated}$, find out the phase advance which is

satisfies equation (24). That is with this advance angle, the current is large enough to satisfy the load torque:

$$I_{rms} = f(\omega, \theta_0) = \frac{T}{K_t^*} \tag{26}$$

In these expressions, T_{rated} is the rated torque, ω_m the mechanical speed, ω_{m_rated} the rated mechanical speed of the motor, ω the electrical speed and $\omega_m = \dfrac{\omega}{p} = \dfrac{\omega}{2}$ in our case. The rms value of the phase current is a function of ω, θ_0, other motor parameters and the supplied voltage V.

V. COMPUTER SIMULATION RESULTS

Table I shows the technical data of the BLDCM which is used to examine in this paper.

TABLE I
TECHNICAL DATA OF THE MOTOR

Rated applied voltage	$U_{rated} = 12V$
Maximum no-load speed	$n_0 = 2826$ rpm
Rated speed	$n_{rated} = 685$ rpm
Rated current	$I_{rated} = 95A$
Rated torque	$T_{rated} = 3{,}75$ Nm
Electromotive force constant	$K_e = 0{,}0202$ V/rad
Torque constant	$K_t = 0{,}0405$ N.m/A
Number of pole pairs	$p = 2$

The maximum amplitude of BEMF which is correspondent to the maximum speed is

$$E = K_e.\omega_{m\_max} = 0.0202 * \left(\frac{2826}{60}.2\pi\right)$$

$$= 5.97(V) < V = \frac{U_{rated}}{2} = \frac{12}{2} = 6(V)$$

Since the amplitude of BEMF is always smaller than half of the rated applied voltage, only the first case (E < V) is examined. The expressions of phase currents contain exponents, so the Taylor expansion must be utilized.

$$e^x = 1 + x + \frac{x^2}{2!} + \frac{x^3}{3!} + \dots \tag{27}$$

Using the Taylor expansion, the approximate expressions of phase currents can be expressed as:

$$i_1(t) = A + Bt - Ae^{\frac{-t}{\tau}} = Bt - A\left[\left(\frac{-t}{\tau}\right) + \frac{1}{2}\left(\frac{-t}{\tau}\right)^2\right] \tag{28}$$

$$i_2(t) = M + I_2 - Me^{\frac{-\left(t - \frac{\theta_0}{\omega}\right)}{\tau}}$$

$$= I_2 - M\left[\left(\frac{-\left(t - \frac{\theta_0}{\omega}\right)}{\tau}\right) + \frac{1}{2}\left(\frac{-\left(t - \frac{\theta_0}{\omega}\right)}{\tau}\right)^2\right] \tag{29}$$

$$i_3(t) = P + I_3 - Pe^{\frac{-\left(t - \frac{2\pi}{3\omega}\right)}{\tau}}$$

$$= I_3 - P\left[\left(\frac{-\left(t - \frac{2\pi}{3\omega}\right)}{\tau}\right) + \frac{1}{2}\left(\frac{-\left(t - \frac{2\pi}{3\omega}\right)}{\tau}\right)^2\right] \tag{30}$$

The expression of the rms value of the phase current is

$$I_{rms} = \sqrt{\frac{1}{\frac{\pi}{\omega}}\left(\int_0^{\frac{\theta_0}{\omega}} i_1^2\,dt + \int_{\frac{\theta_0}{\omega}}^{\frac{2\pi}{3\omega}} i_2^2\,dt + \int_{\frac{2\pi}{3\omega}}^{\frac{2\pi}{3\omega}+\frac{\beta}{\omega}} i_3^2\,dt\right)} \tag{31}$$

For a given reference speed above the rated speed, it is possible to found an advance angle by using computer to solve equation (31). The results are summarized in Table II. Fig. 7 shows the "phase advance curve".

Fig. 8 illustrates the closed loop control system of BLDCM drive, which is suitable for both constant-power operation and constant torque operation. When the reference speed ω_m^* is set, the BLDCM starts up. From the output of the encoder, the speed feedback signal ω_m and position signal can be obtained. When the speed feedback is lower than or equal to the rated speed, the BLDCM operates in the constant torque region and the advance angle is set to zero. When the speed feedback is above rated speed, the BLDCM operates in the constant power speed with a corresponding advance angle. This advance angle is calculated by using a "lookup table" which is obtained from table II.

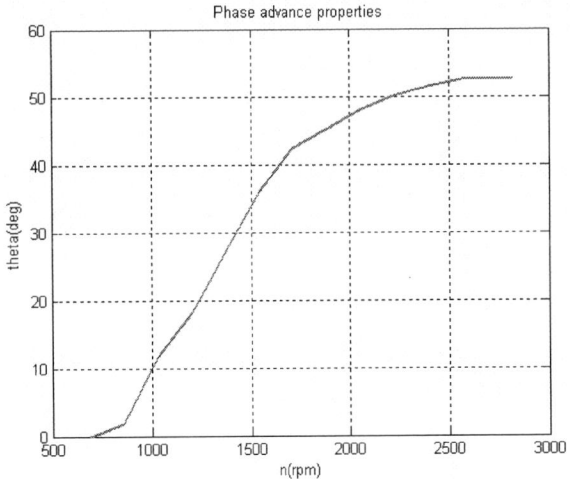

Fig. 7. Phase advance curve.

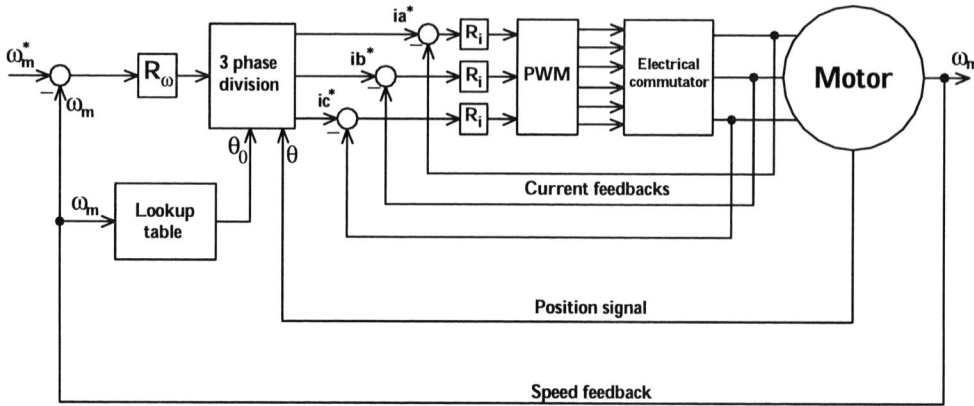

Fig. 8. Block diagram of phase advance drive system.

Both the BLDCM drive system in Fig. 8 and the conventional drive system in Fig. 2 were examined in using Matlab/Simulink. The simulation was performed with reference speed $n^* = 2500$ rpm and the load torque

$$T = \frac{n_{rated}}{n^*} \cdot T_{rated} = \frac{685}{2500} * 3.75 = 1.03 (Nm).$$ If the

conventional drive system is used, the speed of the motor can increase up only to 1800 rpm and the motor torque ripple is considerable as shown in Fig. 9. This speed is much lower than the reference speed. If the configuration of Fig. 8 is used, the motor speed can reach the reference speed of 2500 rpm with less torque ripple (Fig. 10).

The higher speed, the higher switching frequency of the power switches which are often MOSFETs. This

phenomenon affects to the torque-producing. So it is evident that operating at high speed, the BLDCM produces more torque ripple than in low speed. Reducing torque ripple, or smoothing the output torque of BLDCM is also an important problem. It is beyond the scope of this paper and therefore is not described here.

Many computer simulations were performed and the results are shown in Fig. 11 in form of the limit-operating locus with the conventional drive system (the dot curve) and with the phase advance drive system (the solid curve). It is clear that the speed range is wider with phase advance drive system than that of the conventional drive. For example, if the load torque is $T = 1.03\ N.m$, with use of the phase advance, the drive can reach the maximum speed of 2500 rpm, while the conventional drive can reach the maximal speed of only 1800 rpm.

VI. CONCLUSIONS

Phase advance approach to widen the speed range of BLDCM is presented. The principle of this approach is analyzed in more details in this paper compared to other works by Chan et al. The wave form of phase current is examined in both cases: E > V and E < V. The stator resistance is taken into consideration in the analysis and calculation of phase current and advance angle. The obtained equations are therefore more general than those in literature. The calculation of advance angle is proposed for the first time in this paper. The advance angle is calculated off-line by computer simulation and can be then implemented by a "lookup table". Extended simulation results have shown that the speed range of the drive can be considerably widened. Although each motor needs its own "lookup table" or "phase advance curve", this approach is particularly suitable for mass productions.

TABLE II
REFERENCE SPEED AND CORRESPONDING ADVANCE ANGLE

Mechanical speed (rpm)	Electrical speed (rad/s)	Advance angle (degree)
685	143.46	0
1.25×685 = 856.25	179.32	1.9
1.5×685 = 1027.5	215.2	11.75
1.75×685 = 1198.75	251.06	18.2
2×685 = 1370	286.94	27.5
2.25×685 = 1541.25	322.78	36.1
2.5×685 = 1712.5	358.66	42.3
2.75×685 = 1883.75	394.54	45.1
3×685 = 2055	430.40	47.9
3.25×685 = 2226.25	466.26	50.1
3.5×685 = 2397.5	502.14	51.5
3.75×685 = 2568.75	538.00	52.5
2826	591.88	52.7

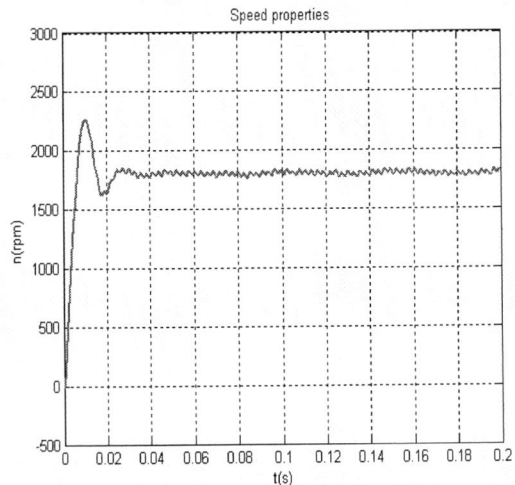 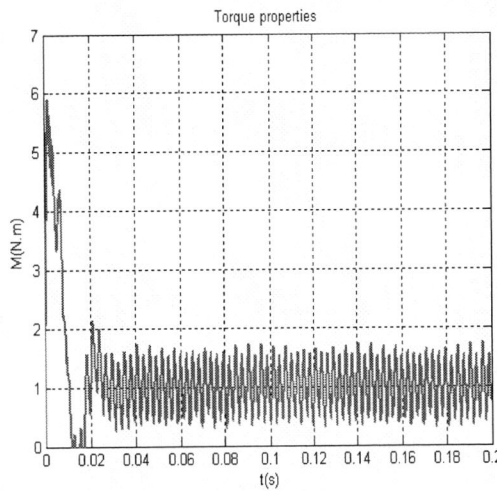

Fig. 9. Speed and output torque of the motor with conventional drive system.

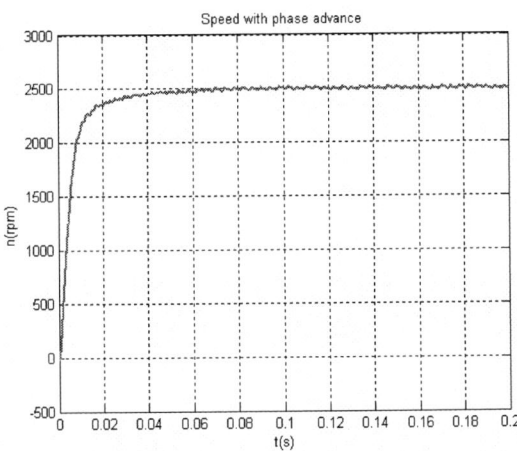 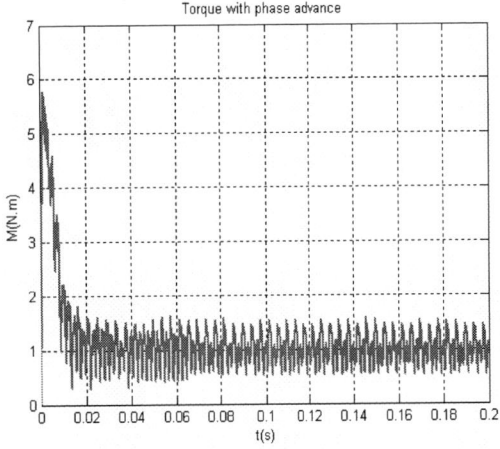

Fig. 10. Speed and output torque of the motor with phase advance drive system.

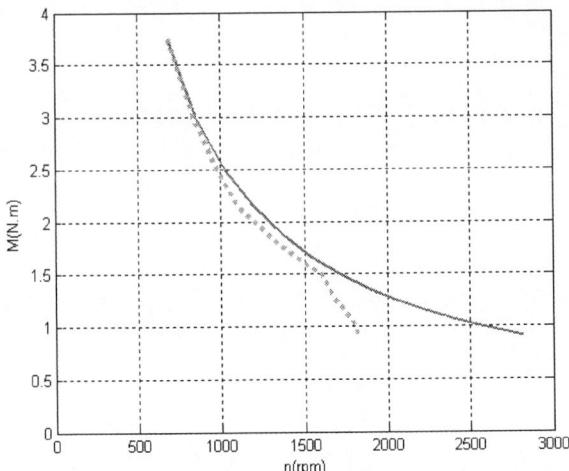

Fig. 11. Limit-operating locus of the conventional drive (dot curve) and the phase advance drive (solid curve).

REFERENCES

[1] C. C. Chan, J. Z. Jiang, W. Xia and K. T. Chau, "Novel wide range speed control of permanent magnet brushless motor drives", *IEEE Trans. on Power Electronics*, vol. 10, Sept. 1995, pp. 539 - 546.

[2] C. C. Chan, W. Xia, J. Z. Jiang, K. T. Chau and M. L. Zhu, "Permanent magnet brushless drives", *IEEE Industry Application Magazine*, Nov./Dec. 1998, pp. 16-22.

[3] Bimal K. Bose, *Modern Power Electronics and AC Drives*, Prentice Hall PTR, 2002.

[4] J. S. Lawler, J. M. Bailey, J. W. Mc Keever, Ioao Pinto, "Limitation of the conventional phase advance method for constant power operation of the brushless DC motor", *Proc. IEEE Southeast Conf.*, Apr. 2002, pp. 174–180.

[5] J. S. Lawler, J. M. Bailey, J. W. McKeever, Joao Pinto, "Extending the constant power speed range of the brushless DC motor through dual-mode inverter control", *IEEE Transactions on Power Electronics*, vol. 19, Issue 3, May 2004, pp. 783-793.

Nonlinear Decoupled Control for a Six-Phase Series-Connected Two Induction Motor Drive Using the Sliding-Mode Technique

J. Soltani*, N. R. Abjadi**, Gh. R. Arab Markadeh***

* Faculty of Engineering, University of Malaya, Kuala-Lumpur, Malaysia
** Department of Electrical and Computer Engineering, Isfahan University of Technology, Isfahan, Iran
*** Department of Engineering, Shahrekord University, Shahrekord, Iran
e-mail: soltani@um.edu.my, navidabjadi@yahoo.com

Abstract — **In this paper a nonlinear decoupled control is designed for a series connected two six-phase squirrel induction motor (IM) drive which is supplied by a six-phase space vector pulse width modulation voltage source inverter (SVPWM VSI). By using a well known phase transposition in the series connection, the independent control of two machines can be realized completely. The nonlinear controller is developed in a stationary (α, β) reference frame with rotor fluxes ($\lambda_{\alpha r}$, $\lambda_{\beta r}$) and stator currents ($i_{\alpha s}$, $i_{\beta s}$) as state variables. At the first, an ideal feedback linearization control (IFLC) system is adopted in order to decouple the torque and rotor flux amplitude of each IM. Then to enhance the performance of the drive system against uncertainties of the plant, such as electrical and mechanical parameter variation, external torque disturbance and unmodelled system dynamics, a sliding-mode feedback linearization control (SMFLC) system is applied, that comprises a SM flux controller and a SM speed controller. Moreover a two level SVPWM is used to supply the two motors drive system. Finally, the effectiveness and capability of the proposed control strategy is verified by computer simulation.**

Index Terms-- **Multimotor drives, multiphase machines, sliding-mode control, adaptive input-output feedback linearization.**

I. INTRODUCTION

Recently using the multiphase machines in various industrial applications especially ones require more than one electric motor drive such as electric vehicles, textile, web processing and paper mills is increasing,.

The advantage of multiphase machines are higher torque density, higher efficiency, reduced torque pulsations, greater fault tolerant, improvement of the drive noise characteristic, and reduction in the required rating per inverter leg (and therefore simpler and more reliable power conditioning equipment) [1, 2].

These advantages stem from the fact that control of machine's flux and torque, produced by the interaction of fundamental field component and the fundamental stator current component, requires only two stator current components [1, 2].

Six-phase ac motor drives are often considered as viable solutions when reduction of the inverter per-phase rating is required due to the high motor power. Although the basic concept is old [3], there has been an upsurge in the interest in this type of ac motor drive in recent time [2, 4,

5]. The standard choice is a six-phase induction or synchronous machine with two three-phase windings on the stator. The spatial displacement between the two three-phase windings is 30 (so called quasi-six-phase machine) and neutral points of the two windings can be isolated or connected. The main reason for selecting the asymmetrical six-phase winding instead of the true six-phase winding (60 displacement between any two consecutive phases), elimination of the sixth harmonic from the torque [3], was important in the pre-pulse width-modulation (PWM) era of voltage-source inverter (VSI) control. A quasi six-phase induction motor is used in the work described here, with the underlying idea of realizing a two-motor drive system with independent control, while utilizing a single six-phase SVPWM VSI as the supply.

In the past decade, the variable structure control strategy using the SM has been the focus of much research on the control of the AC servo drive system [6, 7]. The feature of a SM control system is that the controller is switched between two distinct control structures. In general, the design of the SM controller can be divided into two phases; the hitting phase and the sliding phase. Before the system reaches the switching surface (hitting phase), there is a control directed towards a switching surface, and when all the states of the controlled systems are constrained to lie within a switching hyper plane, the SM occurs (sliding phase). Once the states of the controlled system enter the SM, the dynamics of the system are determined by the choice of sliding hyper plane and are independent of uncertainties and external disturbances [8].

In [6, 7], a robust nonlinear controller has been proposed for a primary type Linear IM (LIM) which is based on combination of SM control and input-output feedback linearization. The system control laws are obtained based on Lyapunov theory, such that the asymptotic stability of the control system can be guaranteed under the occurrence of system uncertainties. Using the proposed surface of [6, 7], there is no reaching phase as in the traditional SM control.

Selection of the upper bound of lumped uncertainty has a significant effect on the SM control performance [6]. If the selected bound is too large, the sign function will result in serious chattering phenomena in the control efforts. The undesired chattering control efforts will wear the bearing mechanism and might excite unstable system dynamics. On the other hand, if the selected bounds are too small, the stability conditions may be not satisfied, and this will cause the controlled system to be unstable. In [6, 7] two adaptive algorithms are utilized to adjust the upper bounds on the lumped uncertainties in real time for the SMFLC system.

Reference [7] has used the proposed method of [6] but generate the voltage references for a PWM VSI. The main aim of this paper is to use the control method of [7] for a two six-phase series-connected IM drive which is fed by a SVPWM VSI, using a special configuration described in [1]. It will be shown that the proposed control strategy is capable of tracking the speed and rotor flux reference signals in spite of motor parameters variation and uncertainty.

A two level SVPWM scheme has been used for an IM with 30° double stator winding in [4]. This inverter has been designed to generate approximately pure sinusoidal voltage outputs. Using the same idea described in [4], in this paper a SVPWM voltage source is developed which is capable of supplying the two motor drive system with different rotor speeds. The SVPWM scheme is designed to generate an arbitrary reference voltage space vector which constitutes the motors main frequencies ω_1 and ω_2.

II. DESCRIPTION AND MODELING OF THE DRIVE SYSTEM

The stator windings of a quasi six-phase machine consists of two three-phase windings are mutually displaced in space by 30°, as illustrated in Fig. 1.

Referring to Fig. 2 the Six-phase stator windings of the two machines are connected in series, with an appropriate phase transposition [1]. The two-motor drive system is supplied from a two level SVPWM VSI.

From Fig. 2 the relationship between voltages and currents can be obtained as:

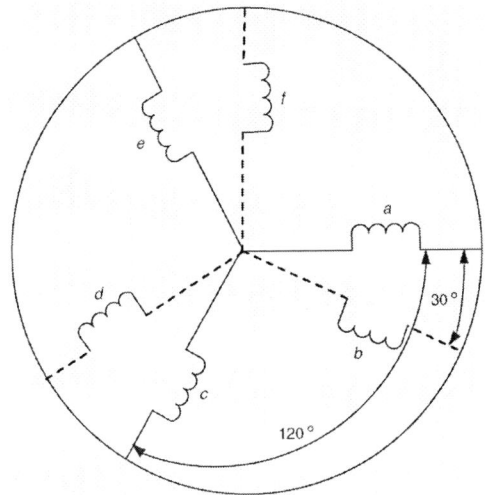

Fig. 1. Schematic representation of a quasi six-phase (split phase, dual three-phase) machine stator winding

$$
\begin{aligned}
v_A &= v_{as1} + v_{as2} & i_A &= i_{as1} = i_{as2} \\
v_B &= v_{bs1} + v_{ds2} & i_B &= i_{bs1} = i_{ds2} \\
v_C &= v_{cs1} + v_{es2} & i_C &= i_{cs1} = i_{es2} \\
v_D &= v_{ds1} + v_{bs2} & i_D &= i_{ds1} = i_{bs2} \\
v_E &= v_{es1} + v_{cs2} & i_E &= i_{es1} = i_{cs2} \\
v_F &= v_{fs1} + v_{fs2} & i_F &= i_{fs1} = i_{fs2}
\end{aligned}
\tag{1}
$$

Each machine is supposed to have its own parameters [1]. Based on Clark's transformation described by matrix C, the $(\alpha - \beta)$, $(x - y)$ and (zero sequence) equivalent circuits of each motor in the stationary reference frame are depicted in Fig. 3.

$$
C = \sqrt{\frac{2}{6}}
\begin{array}{c}
\alpha \\ \beta \\ x \\ y \\ 0+ \\ 0-
\end{array}
\begin{bmatrix}
1 & \cos\phi & \cos 4\phi & \cos 5\phi & \cos 8\phi & \cos 9\phi \\
0 & \sin\phi & \sin 4\phi & \sin 5\phi & \sin 8\phi & \sin 9\phi \\
1 & \cos 5\phi & \cos 8\phi & \cos\phi & \cos 4\phi & \cos 9\phi \\
0 & \sin 5\phi & \sin 8\phi & \sin\phi & \sin 4\phi & \sin 9\phi \\
1 & 0 & 1 & 0 & 1 & 0 \\
0 & 1 & 0 & 1 & 0 & 1
\end{bmatrix}
\tag{2}
$$

where $\phi = \pi / 6$ [4].

From Figs. (1) and (3) the (α, β) and (x, y) voltages and currents of the six-phase SVPWM VSI are obtained as:

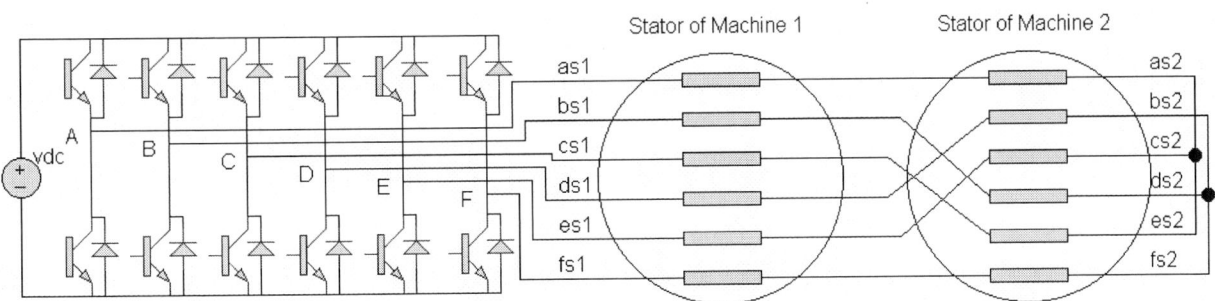

Fig. 2. Two six-phase machine drive supplied by a single inverter

1443

$$
\begin{bmatrix}
v_\alpha^{INV} \\
v_\beta^{INV} \\
v_x^{INV} \\
v_y^{INV} \\
v_{0+}^{INV} \\
v_{0-}^{INV}
\end{bmatrix}
= C
\begin{bmatrix}
v_A \\
v_B \\
v_C \\
v_D \\
v_E \\
v_F
\end{bmatrix}
= C
\begin{bmatrix}
v_{as1}+v_{as2} \\
v_{bs1}+v_{ds2} \\
v_{cs1}+v_{es2} \\
v_{ds1}+v_{bs2} \\
v_{es1}+v_{cs2} \\
v_{fs1}+v_{fs2}
\end{bmatrix}
=
\begin{bmatrix}
v_{\alpha s1}+v_{xs2} \\
v_{\beta s1}+v_{ys2} \\
v_{xs1}+v_{\alpha s2} \\
v_{ys1}+v_{\beta s2} \\
0 \\
0
\end{bmatrix}
\tag{3}
$$

$$
i_\alpha^{INV} = i_{\alpha s1} = i_{xs2}
$$
$$
i_\beta^{INV} = i_{\beta s1} = i_{ys2}
$$
$$
i_x^{INV} = i_{xs1} = i_{\alpha s2}
$$
$$
i_y^{INV} = i_{ys1} = i_{\beta s2}
$$
$$\tag{4}$$

From Fig. 3, the stator voltage equations of each machine are

$$
v_{\alpha sk} = R_{sk} i_{\alpha sk} + \frac{d}{dt}(L_{sk} i_{\alpha sk} + L_{mk} i_{\alpha rk})
$$

$$
v_{\beta sk} = R_{sk} i_{\beta sk} + \frac{d}{dt}(L_{sk} i_{\beta sk} + L_{mk} i_{\beta rk})
$$

$$
v_{xsk} = R_{sk} i_{xsk} + \frac{d}{dt}(L_{lsk} i_{xsk})
$$

$$
v_{ysk} = R_{sk} i_{ysk} + \frac{d}{dt}(L_{lsk} i_{ysk})
$$

$$\tag{5}$$

where k=1, 2.

The rotor voltage equations of each machine are

$$
0 = R_{rk} i_{\alpha rk} + \omega_{r1}(L_{rk} i_{\beta rk} + L_{mk} i_{\beta sk}) +
$$
$$
\frac{d}{dt}(L_{rk} i_{\alpha rk} + L_{mk} i_{\alpha sk})
$$

$$
0 = R_{rk} i_{\beta rk} - \omega_{rk}(L_{rk} i_{\alpha rk} + L_{mk} i_{\alpha sk}) +
$$
$$
\frac{d}{dt}(L_{rk} i_{\beta rk} + L_{mk} i_{\beta sk})
$$

$$\tag{6}$$

where k=1, 2.

The torque equations of the two series-connected machines are given as follows

$$
T_{ek} = P_k L_{mk}(i_{\alpha rk} i_{\beta sk} - i_{\beta rk} i_{\alpha sk}) \tag{7}
$$

Here k=1, 2 and P_k are pole pairs.

From (4) and (7), one can seen that the torque production currents of the first motor ($i_{\alpha s1}, i_{\beta s1}$) are equal to none producing torque currents of the second motor (i_{xs2}, i_{ys2}). As a result, the two motors can be controlled independently.

III. IDEAL FEEDBACK LINEARIZATION CONTROL SYSTEM (IFLC)

The state-coordinate transformed model of each machine is expressed by

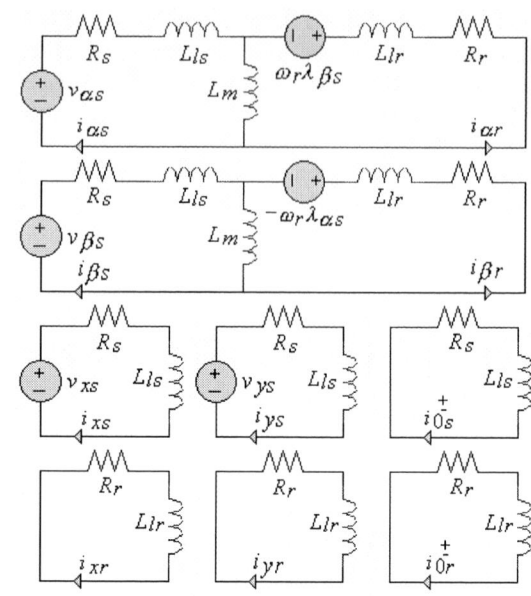

Fig. 3. Six-phase IM equivalent circuits of one machine

$$
\dot{i}_{ds} = -\frac{1}{\sigma L_s}\left(R_s + \frac{R_r L_m^2}{L_r^2}\right)i_{ds} + \frac{R_r L_m}{\sigma L_s L_r^2}\lambda_{dr} +
$$
$$
\frac{\omega_r L_m}{\sigma L_s L_r}\lambda_{qr} + \frac{1}{\sigma L_s}v_{ds}
$$

$$
\dot{i}_{qs} = -\frac{1}{\sigma L_s}\left(R_s + \frac{R_r L_m^2}{L_r^2}\right)i_{qs} - \frac{\omega_r L_m}{\sigma L_s L_r}\lambda_{dr} +
$$
$$
\frac{R_r L_m}{\sigma L_s L_r^2}\lambda_{qr} + \frac{1}{\sigma L_s}v_{ds}
$$

$$
\dot{\lambda}_{dr} = \frac{R_r L_m}{L_r}i_{ds} - \frac{R_r}{L_r}\lambda_{dr} - \omega_r \lambda_{qr}
$$

$$
\dot{\lambda}_{qr} = \frac{R_r L_m}{L_r}i_{qs} + \omega_r \lambda_{dr} - \frac{R_r}{L_r}\lambda_{qr}
$$

$$
T_e = \frac{J}{P}\dot{\omega}_r + \frac{B}{P}\omega_r + T_L
$$

$$\tag{8}$$

where $\sigma = 1 - \dfrac{L_m^2}{L_s L_r}$.

The rotor flux amplitude is defined as follows:

$$
\lambda_r = \sqrt{\lambda_{dr}^2 + \lambda_{qr}^2} \tag{9}
$$

and using the two rotor equation voltage in (8), the time derivative of λ_r can be derived as follows:

$$
\dot{\lambda}_r = \frac{-R_r}{L_r}\lambda_r + \frac{R_r L_m}{L_r \lambda_r}(\lambda_{dr} i_{ds} + \lambda_{qr} i_{qs}) \tag{10}
$$

From (10) and the last equation of (8), the IM motion dynamics can be expressed as

$$\begin{bmatrix} \dot{\lambda}_r \\ \dot{\omega}_r \end{bmatrix} = \begin{bmatrix} \dfrac{-R_r}{L_r}\lambda_r \\ \dfrac{-B}{J}\omega_r - \dfrac{T_L P}{2J} \end{bmatrix} +$$

$$\begin{bmatrix} \dfrac{R_r L_m}{L_r \lambda_r}\lambda_{dr} & \dfrac{R_r L_m}{L_r \lambda_r}\lambda_{qr} \\ \dfrac{-P^2 L_m}{J L_r}\lambda_{qr} & \dfrac{P^2 L_m}{J L_r}\lambda_{dr} \end{bmatrix}\begin{bmatrix} i_{ds} \\ i_{qs} \end{bmatrix}$$

$$= b + A\begin{bmatrix} i_{ds} \\ i_{qs} \end{bmatrix} \tag{11}$$

If the system parameters are available and all system states are measurable, an IFLC system can be designed as follows:

$$\begin{bmatrix} i_{ds} \\ i_{qs} \end{bmatrix} = A^{-1}(-b + \begin{bmatrix} I_\lambda \\ I_\omega \end{bmatrix}) \tag{12}$$

where I_λ and I_ω are the new control inputs, which should be designed [6].

IV. SLIDING-MODE FEEDBACK LINEARIZATION CONTROL SYSTEM (SMFLC)

To ensure the stability of the controlled system, despite the existence of the uncertain system dynamics, a newly designed SMFLC system, that comprises a sliding-mode flux controller and a sliding-mode position controller, is used as follows [6]:

$$\begin{bmatrix} i_{ds} \\ i_{qs} \end{bmatrix} = \overline{A}^{-1}(-\overline{b} + \begin{bmatrix} I_\lambda^a \\ I_\omega^a \end{bmatrix}) \tag{13}$$

where I_λ^a and I_ω^a are the new actual control inputs; $\overline{A}$ and $\overline{b}$ can be obtained from (11) using the nominal parameter values without external force disturbance. Consider the system parameter variations and external force disturbance, the relationship between the inputs and outputs of the dynamic model can be rewritten as follows:

$$\begin{bmatrix} \dot{\lambda}_r \\ \dot{\omega}_r \end{bmatrix} = \overline{b} + \overline{A}\begin{bmatrix} i_{ds} \\ i_{qs} \end{bmatrix} + \begin{bmatrix} \psi_\lambda \\ \psi_\omega \end{bmatrix} = \begin{bmatrix} I_\lambda^a + \psi_\lambda \\ I_\omega^a + \psi_\omega \end{bmatrix} \tag{14}$$

where the lumped uncertainty vector $[\psi_\lambda \quad \psi_\omega]^T = \Delta b + \Delta A[i_{ds} \quad i_{qs}]^T$.

The switching surface are chosen as

$$S_\lambda = (\frac{d}{dt} + \gamma)\int_0^t e_\lambda(\tau)d\tau < S_\omega = (\frac{d}{dt} + \rho)\int_0^t e_\omega(\tau)d\tau \tag{15}$$

where γ and ρ are positive constants.

Using the follow sliding mode controllers the errors converge to zero [6]:

$$I_\lambda^a = -\gamma e_\lambda(t) + \gamma e_\lambda(0) + \dot{\lambda}_r^* - \hat{\eta}_\lambda \operatorname{sgn} S_\lambda(t) \qquad ,$$

$$\dot{\hat{\eta}}_\lambda = \alpha |S_\lambda(t)|$$

$$I_\omega^a = -\rho e_\omega(t) + \rho e_\omega(0) + \dot{\omega}_r^* - \hat{\eta}_\omega \operatorname{sgn} S_\omega(t) ,$$

$$\dot{\hat{\eta}}_\omega = \beta |S_\omega(t)|$$

where α and β are positive constants.

V. SPACE VECTOR PWM (SVPWM)

In [4], a technique of vector space decomposition control of voltage source inverter fed dual three-phase induction machines was presented. The goal of that space vector PWM control was to synthesize the d-q voltage vectors to satisfy the machine's torque control requirements, and, at the same time, to maintain the average volt-seconds on the x-y and zero planes to be zero during every sampling interval.

Using the same idea described in [4], in this paper a SVPWM voltage source is developed which is capable of supplying the two motor drive system with different rotor speeds. The SVPWM scheme is designed to generate an arbitrary reference voltage space vector which constitutes the motors main frequencies ω_1 and ω_2.

From Fig. 4 in each plane, the number of 64 space vectors can be recognized, from which 62 vectors are active and the rest are zero vectors at origin. The decimal numbers in Fig. 4 denote the switching modes. When each decimal number is converted to a six digit binary number, the 1's in this number indicate that the upper switches in the corresponding switching arms are "on", while the 0's indicate the "on" state of the lower switches.

The MSB (most significant bit) of the number represents the switching state of phase a, the second MSB for phase b, and so on [4].

From the average vector concept during one sampling period, the reference voltage vectors on the d-q and x-y plane can be realized by adjusting the applying times of 4 active vectors and 2 zero vector.

The nonzero vectors can be classified into four set with different length. These sets are illustrated in Fig. 5 with four dodecagons in each plane.

If the d-q reference vector is greater than the reference vector in x-y plane, d-q reference vector will determine the switching vectors and vice versa, in order to enable a better utilization of the available DC bus voltage. For example if such reference vector is between vectors 48 and 56 in d-q plane (sector 1), the active switching vectors are 48, 56, 57 and 52, as shown in Fig. 5.

This strategy can be repeated in all of the sectors and shown in flowchart of Fig. 6.

In order to prevent output voltage from producing high dv/dt, the changes of output voltage should be minimized,

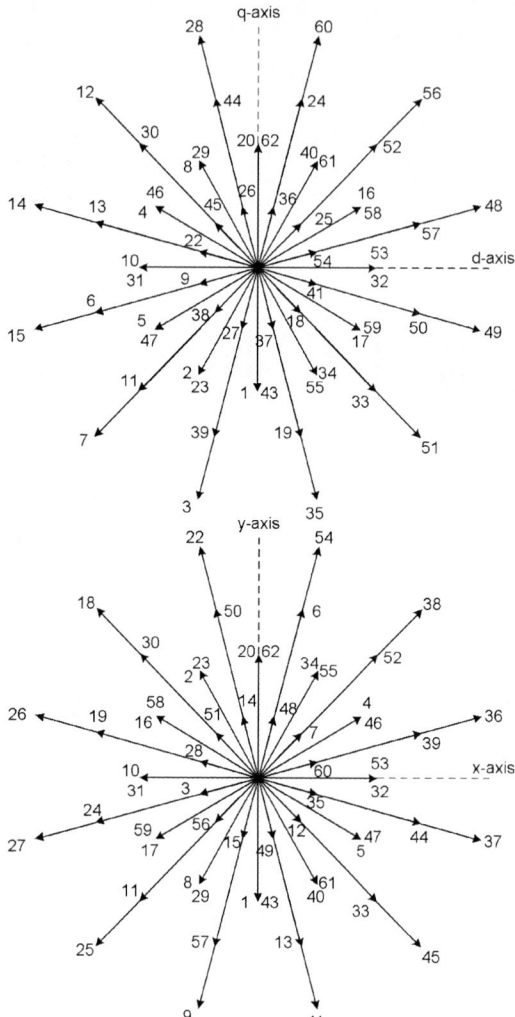

Fig. 4. The location of switching vectors on dq and xy planes

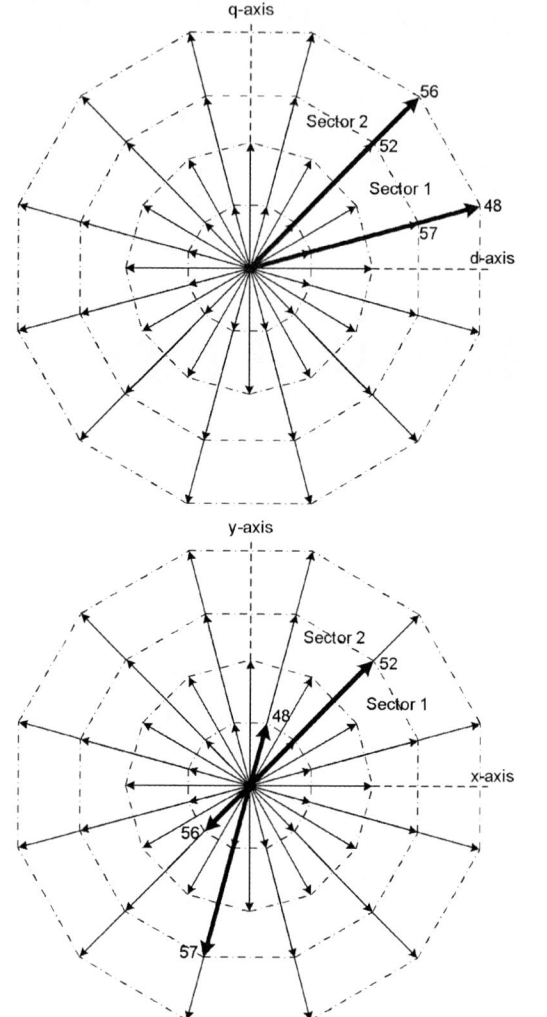

Fig. 5. Realization of a reference voltage vector located in sector 1 on dq plane

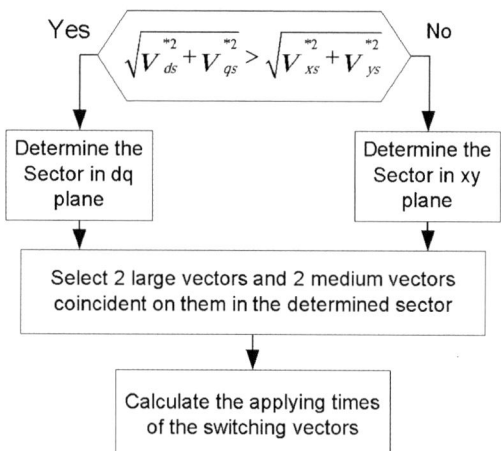

Fig. 6. Flowchart to realize a reference voltage vector

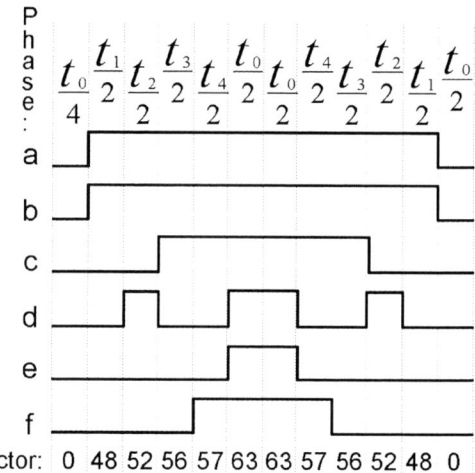

Fig. 7. Switching signals in sector 1 of dq plane

for example the switching signals in sector 1 of dq plane should be as shown in Fig. 7. In addition the operating sequence of voltage vectors in all sections of dq plane is shown in Table I. In this table the switching vectors in half the period are shown. A similar table can be obtained for xy plane.

The space vector PWM strategy is accomplished by the following equation:

1446

$$
\begin{bmatrix} t_1 \\ t_2 \\ t_3 \\ t_4 \\ t_0 \end{bmatrix} = T_s \begin{bmatrix} v_d^1 & v_d^2 & v_d^3 & v_d^4 & 0 \\ v_q^1 & v_q^2 & v_q^3 & v_q^4 & 0 \\ v_x^1 & v_x^2 & v_x^3 & v_x^4 & 0 \\ v_y^1 & v_y^2 & v_y^3 & v_y^4 & 0 \\ 1 & 1 & 1 & 1 & 1 \end{bmatrix}^{-1} \begin{bmatrix} v_d^* \\ v_q^* \\ v_x^* \\ v_y^* \\ 1 \end{bmatrix} \quad (16)
$$

where v_j^k is the projection of kth voltage vector on the j-axis and T_s is the period time of one sequence [4].

TABLE I. Operating sequence of voltage vectors on dq plane for half period

Times Sectors	$\dfrac{t_0}{4}$	$\dfrac{t_1}{2}$	$\dfrac{t_2}{2}$	$\dfrac{t_3}{2}$	$\dfrac{t_4}{2}$	$\dfrac{t_0}{4}$
1	0	48	52	56	57	63
2	0	24	56	52	60	63
3	0	24	28	44	60	63
4	0	12	44	28	30	63
5	0	12	13	14	30	63
6	0	6	14	13	15	63
7	0	6	7	11	15	63
8	0	3	11	7	39	63
9	0	3	19	35	39	63
10	0	33	35	19	51	63
11	0	33	49	50	51	63
12	0	48	50	49	57	63

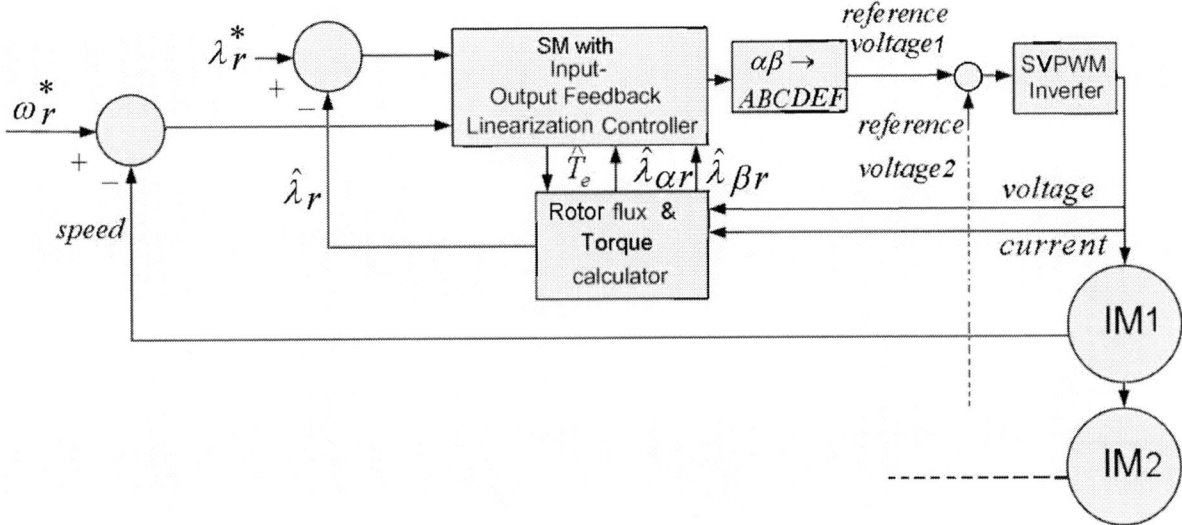

Fig. 8. Drive system block diagram

VI. SIMULATION RESULTS

The proposed control scheme is implemented in a block diagram shown in Fig. 8.
Simulation results are obtained for two six-phase squirrel cage IM with parameters shown in Table II [9].

Assuming the initial errors of +100% and +40% respectively in R_s, R_r. The IMs speeds control results are obtained and shown in Fig. 9. In addition, the motors fluxes control results are shown in Fig. 10. Notice that the SM controller gains are obtained by trial and errors.

TABLE II. IMs parameters

Poles	6	R_s	1.5 Ω
R_r	0.56 Ω	L_s	128.5 mH
L_r	128.5 mH	L_m	120 mH
P_n	2 hp	f_n	50 Hz

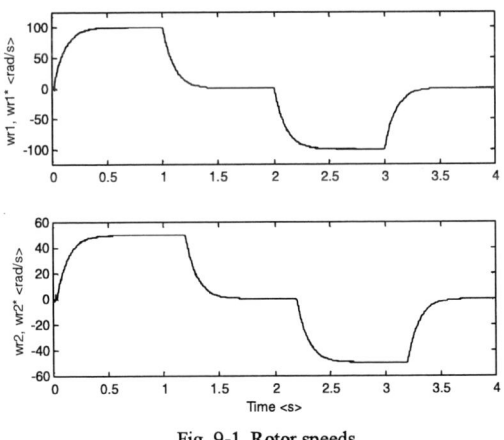

Fig. 9-1. Rotor speeds

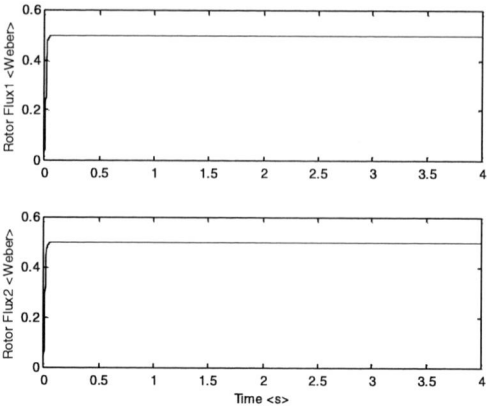

Fig. 9-2. Rotor fluxes

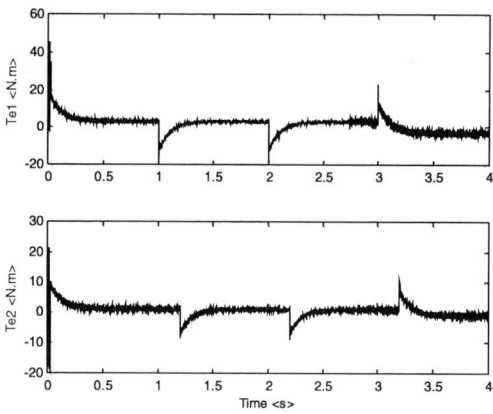

Fig. 9-3. Motors phase a currents

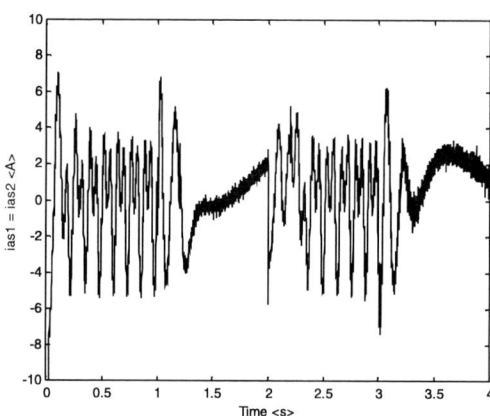

Fig. 9-4. Motors torques

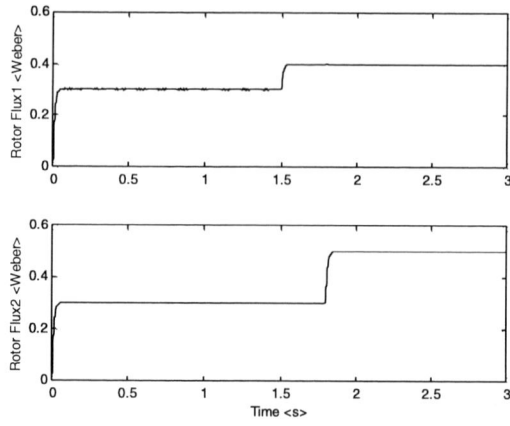

Fig. 10. Rotor fluxes

VII. CONCLUSION

This paper has discussed a series-connected six-phase two motor drive which is supplied by a six-phase SVPWM VSI.

The robust nonlinear tracking controllers have been designed capable of controlling the rotor flux and rotor speed of each motor separately. The proposed controllers are able to track the desired rotor speeds and rotor fluxes for each motor perfectly in spite of motors resistances mismatching. The effectiveness and validity of the proposed control method is shown by simulation results.

REFERENCES

[1] E. Levi, S. N. Vukosavic and M. Jones, "Vector control schemes for series-connected six-phase two-motor drive systems", *IEE Proc.-Electr. Power Appl.*, Vol. 152, No. 2, March 2005.

[2] M. Jones, S. N. Vukosavic, E. Levi and A. Iqbal, "A Six-Phase Series-Connected Two-Motor Drive With Decoupled Dynamic Control", *IEEE Trans. on Ind. Appl.*, Vol. 41, No. 4, July/August 2005.

[3] R. H. Nelson and P. C. Krause, "Induction machine analysis for arbitrary displacement between multiple winding sets," *IEEE Trans. Power App. Syst.*, vol. PAS-93, no. 3, pp. 841–848, may/Jun. 1974.

[4] Y. Zhao and T. A. Lipo, "Space Vector PWM Control of Dual Three-phase Induction Machine Using Vector Space Decomposition", *IEEE Trans. on Ind. Appl.*, Vol. 31, No. 5, Sep./Oct. 1995, pp. 1100-1109.

[5] R. Bojoi, M. Lazzari, F. Profumo and A. Tenconi, "Digital field oriented control for dual three-phase induction motor drives," in *Proc. IEEE-IAS Annu. Meeting*, Pittsburgh, PA, 2002, CD-ROM, Paper 22P1.

[6] R.-J.Wai and W.-K.Liu, "Nonlinear decoupled control for linear induction motor servo-drive using the sliding-mode technique", *IEE Proc.-Control Theory AppL*, Vol. 148, No. 3, May 2001.

[7] J. Soltani and M.A. Abbasian, " Robust Nonlinear Control of Linear Induction Motor taking into account the Primary End Effects", International Power Electronics and Motion Control Conf. (IPEMC), 2006.

[8] Y. Wang, Z. Wang, J. Yang and R. Pei, "Speed Regulation of Induction Motor Using Sliding Mode Control Scheme", pp. 72-76, IAS 2005.

[9] G. K. Singh, K. Nam, and S. K. Lim, "A Simple Indirect Field-Oriented Control Scheme for Multiphase Induction Machine", *IEEE Trans. Ind. Elect.*, Vol. 52, No. 4, pp. 1177-1184, August 2005.

The Decoupled Stator Flux and Torque Sliding-Mode Control of Induction Motor Drive Taking the Iron Losses into Account

M.Hajian[*], J.Soltani[**], S.Hosein nia[*], G.R.Arab[***]

[*]Faculty of Electrical and Computer Engineering, Isfahan University of Technology, Isfahan, Iran
[**]Engineering Department, University of Malaya, 50603 Kuala Lumpur
[***]Faculty of Engineering, Shahrekord University, Shahrekord, Iran

Abstract- **This paper presents a novel decoupled Sliding-Mode (SM) stator flux and torque controller for three-phase Induction motor drive which is supplied by a two level SVM voltage source inverter (VSI). The nonlinear controllers are developed in a stationary two axis (α, β) IM reference frame with stator fluxes ($\varphi_{s\alpha}, \varphi_{s\beta}$) and stator currents ($i_{s\alpha}, i_{s\beta}$) as state variables. In this model the motor iron losses is modeled by a shunt rotor speed dependent core resistance (R_c) in IM two axis equivalent circuits. Using the proportional integral type SM switching surfaces, the Lyapunov based control efforts are designed so that the errors between stator flux (φ_s^*) and motor reference torque (T_e^*) and the actual values of these signals (φ_s, T_e) asymptotically converge to zero in spite of machine parameters mismatching and uncertainty. In this control strategy, since the control of stator flux and torque are decoupled hence, for a given motor load torque and a given motor speed, the drive system efficiency can be easily minimized by adjusting the magnitude of stator reference flux. To do so, both the model based and on-line search methods are adopted. Finally some simulation results are presented to show the capability and validity of the proposed control scheme.**

Index Terms- **Induction Motor, Sliding-Mode, Decoupled Control, Stator Flux, Torque**

I. INTRODUCTION

Among nonlinear control strategies, Sliding-Mode(SM) theory is one of the prospective control methodologies for IM flux and speed control because of its order reduction, disturbance rejection, strong robustness in comparison with other control schemes and particularly its simplicity of practical implementation by power converters [1-3]. Upon these advantages so far, a few research notes have been reported for IM speed and flux control or estimation, using sliding-mode technique.

Induction drive control techniques are well treated in the literature. Field orientation control (FOC) and Direct Torque Control (DTC) method appear to be very convenient for good dynamic performance operation. DTC has a faster dynamic response due to the absence of the PI current controllers. However DTC control method based on bang-bang controllers has some serious drawbacks such as a high amount of torque and flux pulsations and variable switching frequency.

To come up with above drawbacks, in the last two decades many researchers have tried to develop the different DTC Space Vector Modulation(DTC-SVM) schemes with constant switching frequency [4].The presented methods are almost parameter dependent. In these control methods, the parametric deviation will significantly affect the dynamic performance and the stability for practical implementation. Therefore, recently, for the nonlinear feed back control of induction motors, many studies have presented compensators for the influence of the variation of motor parameters [1-4].

The electrical drives efficiency is greatly reduced at light loads, where the flux magnitude is held on its rated value. However, most of the time these drives operate at some loads lower than their rated values. In these cases, the significant energy saving is a major gain. Hence so far various methods have been investigated to achieve such a goal [5-6].

The main contribution of this paper is to introduce a novel DTC-SVM SM scheme for decoupled control of the stator flux and torque of three-phase IM drives even if parameters uncertainty and mismatching exists. It is well worth to mention that, in [1], such a controller has been developed in the IM stator flux field oriented reference frame, using the proportional derivative type SM switching surfaces.

Considering the motor core losses, the SM controllers are derived in an IM stationary reference frame with stator fluxes ($\varphi_{s\alpha}, \varphi_{s\beta}$) and stator currents ($i_{s\alpha}, i_{s\beta}$) as state variables, using the proportional integral type SM switching surfaces. As a result compared to SM control of [1], our control method does not need to any frame transformation which effectively results in less computation time and less computation errors. Moreover, the overall stability of the drive system is guaranteed since the IM full order model is applied to design the nonlinear flux and torque controllers. In addition, the stator flux observer described in [1] is used to detect this quantity. One may note that reference [1], combines the IM model in stationary and FOC reference frames in order to detect the stator and rotor fluxes simultaneously.

Our second contribution is on-line minimizing of the IM total losses for a desired load torque and a desired rotor speed. That simply can be achieved by adjusting the magnitude of the stator reference flux signal while the motor real input power becomes minimum. The main

978-1-4244-0644-9/07/$25.00 ©2007 IEEE

reason is that in our proposed approach, the stator flux and torque controls are absolutely decoupled. It is also possible to obtain the stator reference flux from IM FOC model in steady state condition, However this method is almost parameter dependent[2]. According to the model-based method described in [2], it is proved that for a given load torque and a given rotor speed, the IM total losses is minimized when in the FOC steady-state condition the (d,q) axis motor losses (P_{lossd}, P_{lossq}) becomes equal. In [2], a conventional PI controller is used to get this aim. One may note that P_{lossd}, P_{lossq} are time variant functions both in the transient and steady-state condition and as a result it is extremely difficult to tune this regulator for the requested aim.

In this paper for a desired load torque and rotor speed, we use the IM FOC equivalent circuit in steady state condition as described in [2] to obtain the slip speed and the magnitude of stator reference flux. The proposed control scheme described in this paper is verified by some simulation results.

II.   IM DTC-SVM SCHEME

Considering the IM core losses through a speed dependent shunt resistance (R_c) in the stationary (α, β) axis equivalent circuits of this motor as depicted in Fig.1, the IM fifth order model with stator fluxes ($\varphi_{s\alpha}, \varphi_{s\beta}$), stator currents ($i_{s\alpha}, i_{s\beta}$) and rotor speed (ω_r) as state variables is derived as:

$$\frac{d}{dt}\begin{bmatrix} \psi_s \\ i_s \end{bmatrix} = \begin{bmatrix} A_{11} & A_{12} \\ A_{21} & A_{22} \end{bmatrix}\begin{bmatrix} \psi_s \\ i_s \end{bmatrix} + \begin{bmatrix} B_1 \\ B_2 \end{bmatrix} v_s$$

$$\frac{d\omega_r}{dt} = \frac{n_P}{J}(T_e - T_l) - \frac{B}{J}\omega_r$$

With :

$$\psi_s = \begin{bmatrix} \varphi_{s\alpha} \\ \varphi_{s\beta} \end{bmatrix}, i_s = \begin{bmatrix} i_{s\alpha} \\ i_{s\beta} \end{bmatrix}, v_s = (\frac{R_c}{R_c + R_s})\begin{bmatrix} v_{s\alpha} \\ v_{s\beta} \end{bmatrix}$$

$$A_{11} = 0 , A_{12} = (R_s \| R_c)I$$

$$A_{21} = -\frac{R_r}{\sigma L_m^2}I + \frac{\omega_r L_r}{\sigma L_m^2}J , A_{22} = \frac{R_r L_s + (R_s \| R_c)L_r}{\sigma L_m^2}I + \omega_r J$$

$$B_1 = \begin{bmatrix} 1 \\ 1 \end{bmatrix}, B_2 = \begin{bmatrix} -\frac{L_r}{\sigma L_m^2} \\ -\frac{L_r}{\sigma L_m^2} \end{bmatrix}, I = \begin{bmatrix} 1 & 0 \\ 0 & 1 \end{bmatrix}, J = \begin{bmatrix} 0 & -1 \\ 1 & 0 \end{bmatrix} \quad (1)$$

Where n_p is number of pole pairs, L_r, L_s are stator and rotor inductances, L_m is magnetizing inductance, σ is leakage factor and R_s, R_r are respectively stator and rotor resistances.

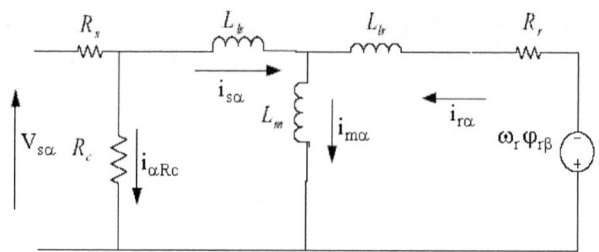

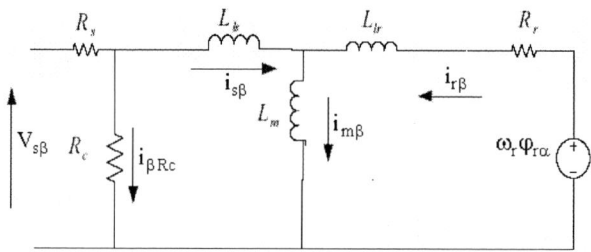

Fig. 1. The IM two axis equivalent circuits in the stationary reference frame

Using the following Sliding-Mode switching surfaces:

$$S_{te} = (\mu_1 \frac{d}{dt} + 1)\int(T_e - T_{er})dt \quad (2)$$

$$S_{\lambda s} = (\mu_2 \frac{d}{dt} +)\int(|\lambda_s^2| - |\lambda_{sr}^2|)dt \quad (3)$$

Where, μ_1, μ_2 are positive design parameters, λ_s, λ_s^* are stator flux real and desired values and T_e, T_e^* are real and reference values of motor electromagnetic torque.

Using the IM model described above, it is proved that for a candidate Lyapunov function $V = \frac{1}{2}(S_t^2 + S_{\lambda s}^2)$, the

SM reaching condition ($S.\dot{S} < 0$) is satisfied if:

$$\frac{dS_t}{dt} = -k_1 S_t - k_2 \, sgn(S_t)$$

$$\frac{dS_{\lambda s}}{dt} = -k_3 S_{\lambda s} - k_4 \, sgn(S_{\lambda s})$$

$\quad (4)$

and:

$$\begin{bmatrix} v_{\alpha m} \\ v_{\beta m} \end{bmatrix} =$$

$$\begin{bmatrix} \frac{3}{2}n_P(i_{s\beta} + \frac{L_r}{\sigma L_m^2}\lambda_{s\beta}) & -\frac{3}{2}(i_{s\alpha} + \frac{L_r}{\sigma L_m^2}\lambda_{s\beta}) \\ 2\lambda_{s\alpha} & 2\lambda_{s\beta} \end{bmatrix}^{-1}\begin{bmatrix} U_1 \\ U_2 \end{bmatrix}$$

$\quad (5)$

With:

$$U_1 = [-k_1 S_t - k_2 \, sgn(S_t) + T_{er} - T_e]/\mu_1 - F_1$$

$$U_2 = [-k_3 S_{\lambda s} - k_4 \, sgn(S_{\lambda s}) - |\lambda_s|^2 + |\lambda_{sr}|^2]/\mu_2 - F_2 \quad (6)$$

$$F_1 = \frac{3}{2} n_P [\frac{\omega_r L_r}{\sigma L_m^2}(\lambda_{s\alpha}^2 + \lambda_{s\beta}^2) +$$

$$\frac{R_r L_s + (R_s \| R_c)L_r}{\sigma L_m^2}(\lambda_{s\alpha}i_{s\beta} - \lambda_{s\beta}i_{s\alpha}) +$$

$$\omega_r (\lambda_{s\alpha}i_{s\alpha} + \lambda_{s\beta}i_{s\beta})] - \frac{dT_{er}}{dt}$$

$$F_2 = -2(R_s \| R_c)(i_{s\alpha} + i_{s\beta}) - 2\lambda_{sr}\frac{d\lambda_{sr}}{dt} \qquad (7)$$

where,

$$v_{\alpha m} = \frac{R_c}{R_c + R_s}v_{s\alpha}$$

$$v_{\beta m} = \frac{R_c}{R_c + R_s}v_{s\beta}$$

Based on above control theory, the overall block diagram of the IM drive control is shown in Fig. 2.

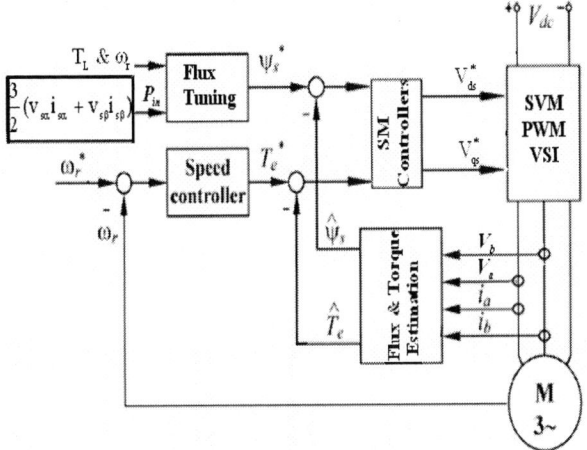

Fig.2. The overall IM DTC-SVM control

III. SIMULATION RESULTS

A C++ computer program was developed to model the control system shown in Fig.2. In this program, the system equations are solved to obtain the system state variables, using the static fourth order Runge-Kutta numerical method. Simulation results are obtained for a three-phase squirrel cage induction motor with characteristics given in Tab. I.

Assuming a +50% initial errors in stator and rotor resistances and -50% initial errors in motor inductances, and for a moment inertia of $J = 1.5 J_n$ where n denotes the nominal value, simulation results are obtained for IM speed, torque and stator flux controls as shown in Figs. 1-3.

In addition, for a given load torque and rotor speed, the magnitude of stator reference flux is on-line adjusted until the motor total losses becomes minimum. This procedure is also repeated by using the IM steady-state equivalent circuit in FOC as described in the introduction part of this paper. However this method is almost parameter dependent. The simulated results of this test are shown in Figs. 4-5.

TABLE I. INDUCTION MOTOR SPECIFICATION

$n_P = 2$	$\text{Cos}(\varphi) = 0.81$
$P_n = 2.2\text{kW}$	$N_r = 1430\text{r/min}$
$V_s = 400\text{v}$	$T_L = 14.7\text{N.m}$
$I_s = 4.9\text{A}$	$L_m = 0.328\text{H}$
$R_s = 2.89\ \Omega$	$R_r = 1.88\ \Omega$
$L_{ls} = 0.016\text{H}$	$L_{lr} = 0.013\text{H}$

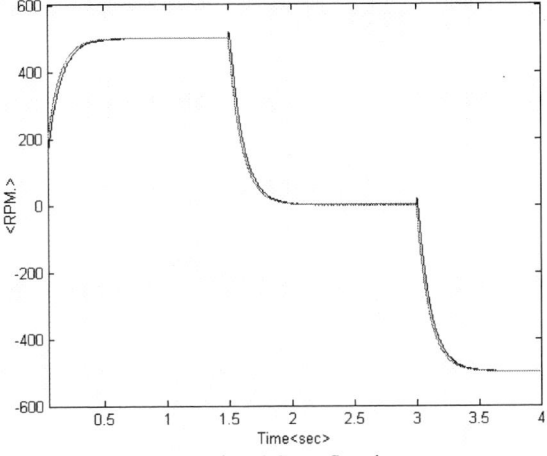

Fig.1-1. Rotor Speed

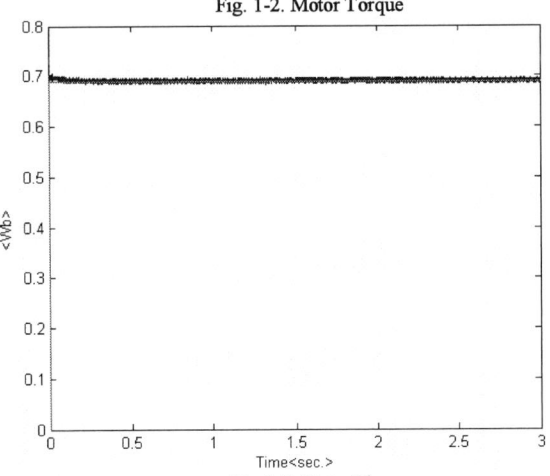

Fig. 1-2. Motor Torque

Fig 1-3. Stator Flux

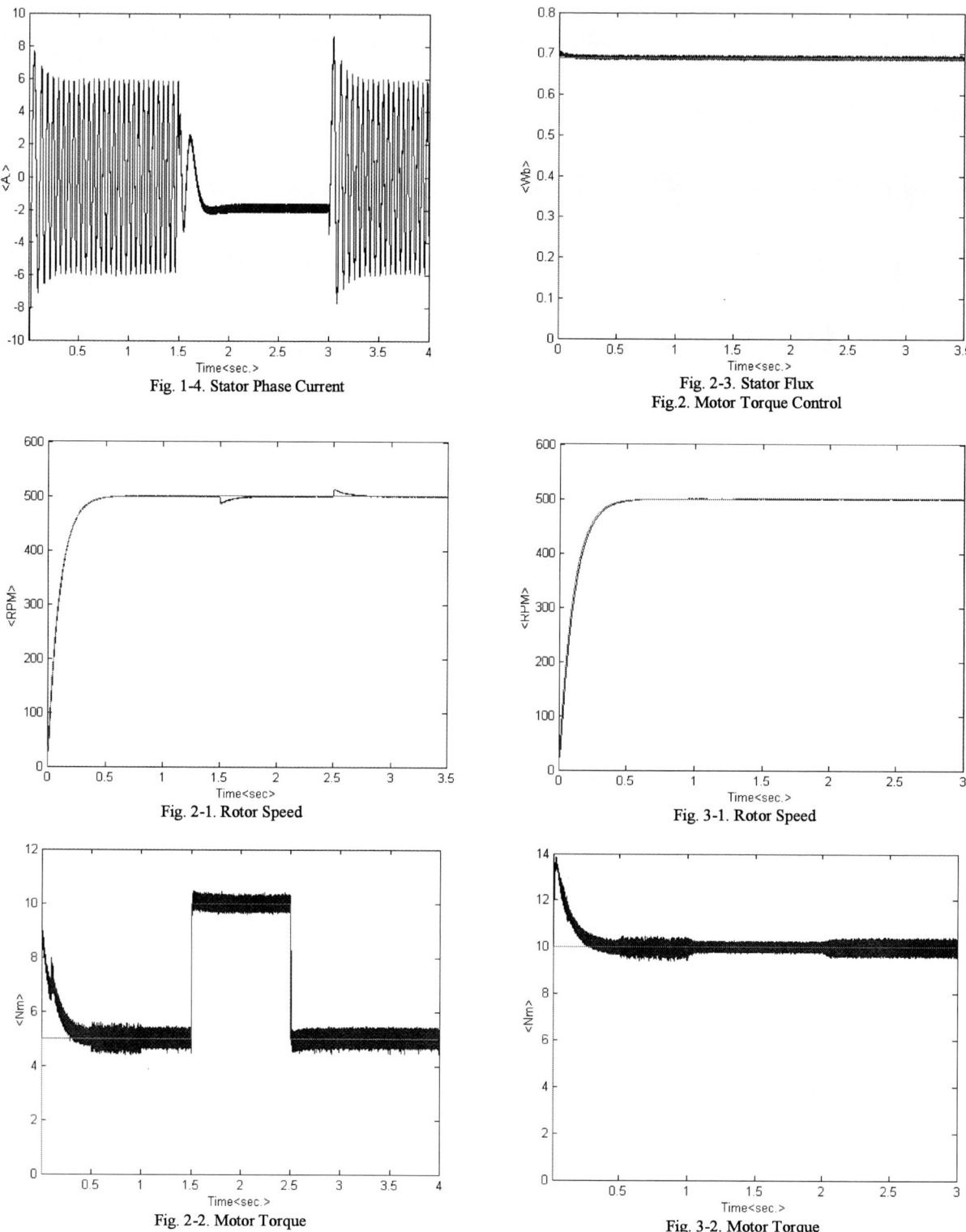

Fig. 1-4. Stator Phase Current

Fig. 2-3. Stator Flux
Fig.2. Motor Torque Control

Fig. 2-1. Rotor Speed

Fig. 3-1. Rotor Speed

Fig. 2-2. Motor Torque

Fig. 3-2. Motor Torque

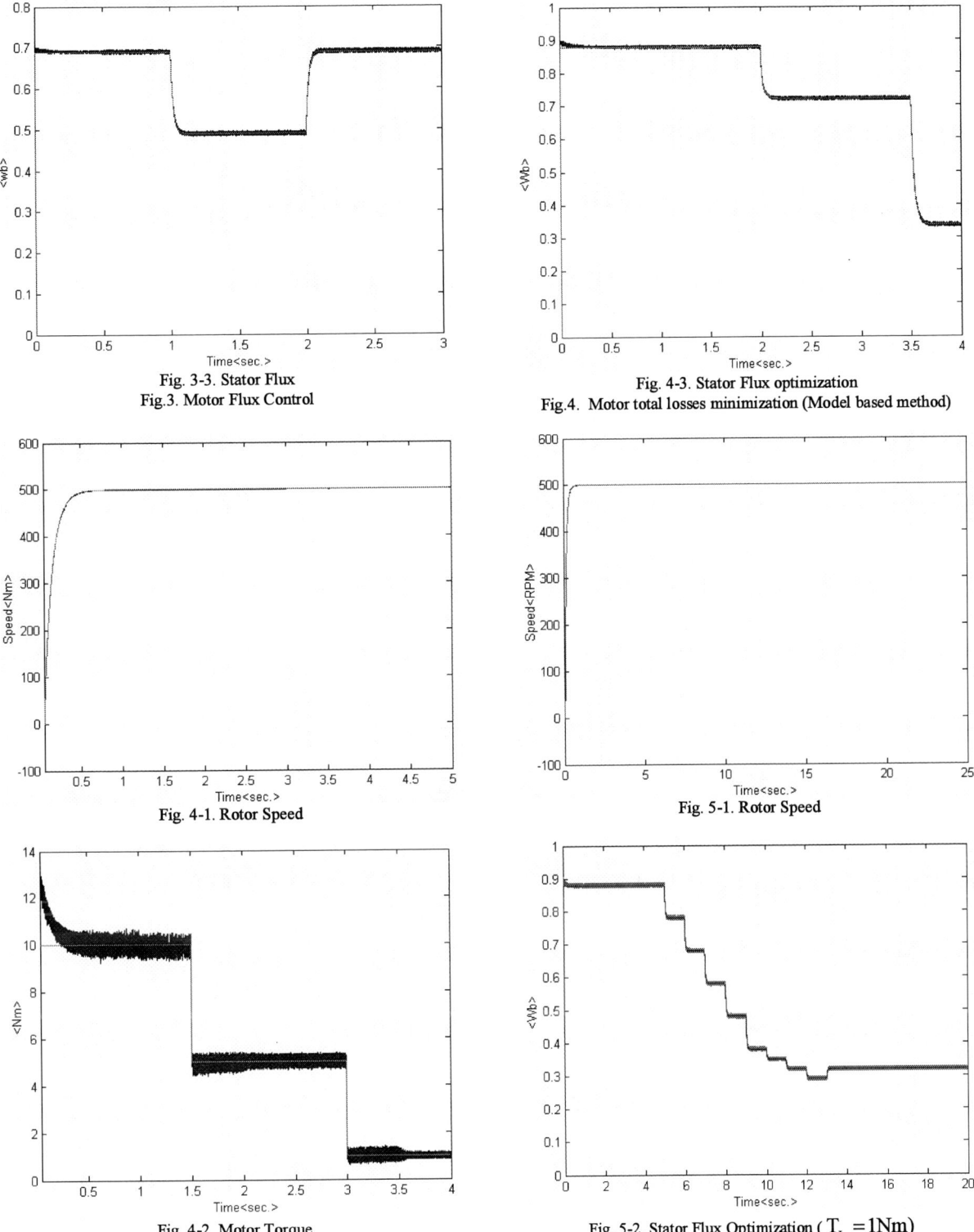

Fig. 3-3. Stator Flux
Fig.3. Motor Flux Control

Fig. 4-3. Stator Flux optimization
Fig.4. Motor total losses minimization (Model based method)

Fig. 4-1. Rotor Speed

Fig. 5-1. Rotor Speed

Fig. 4-2. Motor Torque

Fig. 5-2. Stator Flux Optimization ($T_L = 1Nm$)

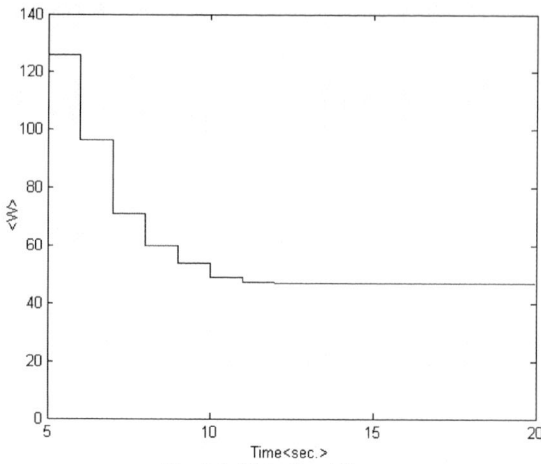

Fig. 5-3. Motor Input Power

Fig5. Motor total losses minimization (search method)

IV. CONCLUSION

A DTC-SVM scheme has been described for three-phase IM drive which is designed in the stationary (α, β) axis reference frame with stator fluxes and currents and rotor speed assumed as state variables taking into account the iron loss resistance. The proposed control scheme is developed based on Sliding-Mode control so that the decoupled control of stator flux and torque is achieved in spite of parameter uncertainty and variations. The capability of nonlinear control approach has been tested by simulation. The simulated results obtained, confirm the validity and effectiveness of the introduced control method. Although the simulated results have been presented in this paper, our experimental setup is in process and practical results will be presented in the next paper.

REFERENCES

[1] Cristian Lascu, Ion Boldea , Ferede Blaabjerg, "Direct torque control of sensorless induction motors : A sliding mode approach ", IEEE Transactons on Industry Applications, Vol.40, No.2 ,2004.

[2] Flemming Abrahamsen , et al , " On the Energy Optimized Control of Standard and High-Efficiency Induction Motors in CT and HVAC Applications", IEEE Trans ON Industry Application, Vol. 34 , NO 4 , 98.

[3] Mehmet Dal, "Sensorless Sliding Mode Direct Torque Control of Induction Motor", IEEE ISIE 2005 , Crotia.

[4] G.R.Arab , J.Soltani, "Robust Direct Torque Control of Adjustable Speed Sensorless Induction Machine drive based on space vector modulation using a PI predictive controller", pp.: 485-496,Springer 2006.

[5] Kouki Matsuse, et al , "A Speed-Sensorless Vector Control of Induction Motor Operating at High Efficiency Taking Core Loss into Account", IEEE Transacton ON Industry Application , Vol. 37, NO2, 2001.

[6] Gan Dong, et al , "Efficiency Optimizing Control of Induction Motor Using Natural Variables", IEEE Trans ON Industrial Electronics , Vol. 53 , NO 6, 2006.

AN EFFICIENT DIRECT TORQUE CONTROL SCHEME FOR SPLIT PHASE INDUCTION MOTOR

A. Khajeh*, J. S. Moghani*, and M. Shahbazi*

* Department of Electrical Engineering, Amirkabir University of Technology, 424 Hafez Ave. Tehran 15914, Iran
(ahmad_khajeh79@yahoo.com , moghani@aut.ac.ir)

Abstract– in this paper, a predictive direct torque control (DTC) scheme for split phase induction machine (SPIM) is established. The induction motor has two sets of stator three-phase windings spatially shifted by 30 electrical degrees. The major drawback of SPIMs is occurrence of extra harmonic currents. Thus in the DTC of SPIMs in addition to control of torque and flux we should consider simultaneously minimizing harmonic components of stator current. Predictive DTC along with optimized SVPWM is used in this paper. Simulation results show that in addition to a good dynamic response, current harmonics in this scheme is significantly reduced.

Index Terms-- direct torque control, predictive DTC, split phase induction machine.

I. INTRODUCTION

In the early 1980s, current-source inverters (CSIs) were used to drive induction machines. Due to lower switching frequency of operation, sixth harmonic torque pulsations were predominant. Sixth harmonic torque pulsations are produced mainly due to the interaction between the fundamental flux and the fifth and seventh harmonic rotor currents. To eliminate sixth harmonic torque pulsations, the split-phase induction motor (SPIM) structure was proposed. An SPIM has two sets of three-phase windings, each of them separated by 30 in space (Fig. 1). In the split-phase motor configuration, sixth harmonic torque pulsations produced by the two sets of windings are in phase opposition. Therefore, sixth harmonic torque pulsations are completely absent in the SPIM [1]. Other potential advantages of SPIM over three-phase ones are such as reduction of the rotor harmonic losses, the rated current of the power electronic switches is halved respect to the three-phase inverter of same power, the possibility to increase the torque per ampere ratio for the same machine volume, and improved reliability.

Direct torque control (DTC) of induction motors, which has been developed in the recent decades, is a powerful control method for motor drives. Featuring direct control of stator flux and electromagnetic torque, stator current and voltage are controlled indirectly. Excellent dynamic performance of both speed and torque are obtained.

In principle, DTC method is based on instantaneous space vector theory. Supplied by a voltage source inverter

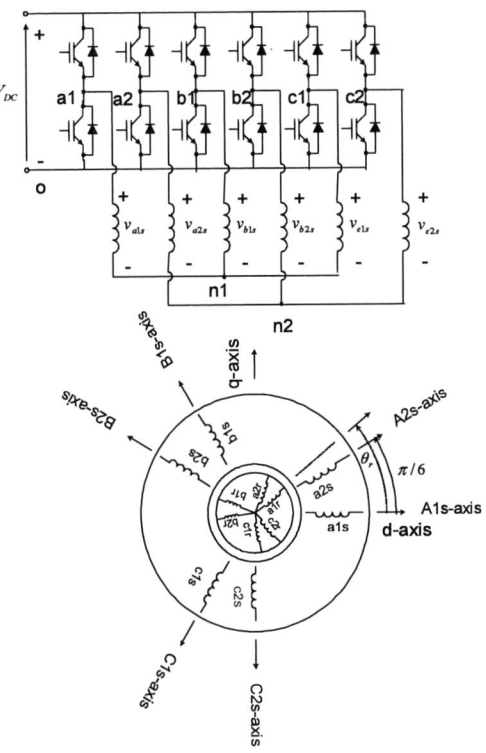

Fig. 1. Schematic of SPIM drive

(VSI), it is possible to control directly the stator flux linkage and the torque by selection of optimum inverter switching modes. The selection is made to restrict the flux and torque errors within respective flux and torque hysteresis bands, to obtain fast torque response. The major drawback of SPIM is occurrence of extra harmonic currents. Thus in the DTC of SPIM in addition to control of torque and flux we should consider simultaneously minimizing harmonic components of stator current. Among all of the PWM techniques proposed for SPIMs, in minimizing the current harmonics, the best one is SVPWM based on vector space decomposition (VSD) technique(VSD-SVPWM) [2, 3]. But the execution time is the problem of VSD technique that reported in [2]. An optimized implementation of this SVPWM based on maximum offline calculation is used in this paper.

II. MOTOR MODEL

According to conventional generalized electrical

978-1-4244-0644-9/07/$25.00 ©2007 IEEE

machine theory, the following assumptions have been adopted in deriving the model of SPIM:

1. The machine air gap is uniform.
2. Machine windings are sinusoidal distributed.
3. Flux path is linear.
4. Iron loss is neglected.

The phase variable mathematical model for SPIM in matrix form is

$$\begin{bmatrix} V_s \\ V_r \end{bmatrix} = \begin{bmatrix} R_{ss} + pL_{ss} & pL_{sr} \\ pL_{rs} & R_{rr} + pL_{rr} \end{bmatrix} \begin{bmatrix} I_s \\ I_r \end{bmatrix} \quad (1)$$

Where, V_s, I_s stator phase voltage and current vectors, 6×1 matrix,

V_r, I_r rotor phase voltage and current vectors, 6×1 matrix,

R_{ss}, R_{rr} stator and rotor resistance matrix, 6×6 diagonal resistance matrix,

L_{ss}, L_{rr} stator and rotor inductance matrix, 6×6 matrix,

$L_{sr} = (L_{rs})^t$, stator to rotor 6×6 mutual inductance matrix.

For analysis and control purposes, the original six dimensional machine system can be decomposed into three two-dimensional orthogonal subspaces $(d,q), (z_1,z_2), (o_1,o_2)$ and by using the transformation matrix [2]:

$$[T] = \frac{1}{\sqrt{3}} \begin{bmatrix} 1 & \cos(\theta) & \cos(4\theta) & \cos(5\theta) & \cos(8\theta) & \cos(9\theta) \\ 0 & \sin(\theta) & \sin(4\theta) & \sin(5\theta) & \sin(8\theta) & \sin(9\theta) \\ 1 & \cos(5\theta) & \cos(8\theta) & \cos(\theta) & \cos(4\theta) & \cos(9\theta) \\ 0 & \sin(5\theta) & \sin(8\theta) & \sin(\theta) & \sin(4\theta) & \sin(9\theta) \\ 1 & 0 & 1 & 0 & 1 & 0 \\ 0 & 1 & 0 & 1 & 0 & 1 \end{bmatrix} \quad (2)$$

Where $\theta = \dfrac{\pi}{6}$.

The transformation matrix has the following properties:

1) The fundamental components of the machine variables and the harmonics of order $k = 12m \pm 1$ ($m = 1,2,3,...$) are mapped in the (d,q) subspace. These components will contribute to the air-gap flux.

2) The harmonics of order $k = 6m \pm 1$ ($m = 1,3,5,...$) are transformed in the (z_1,z_2) subspace. These harmonics (the 5th 7th, 17th, 19th ...) will not contribute to the air-gap flux because the (d,q) and (z_1,z_2) subspaces are orthogonal.

3) The zero-sequence components are mapped in the (o_1,o_2) subspace.

If the two stator sets have isolated neutral points, no current components flow in the (z_1,z_2) Subspace consequently, the machine model referred to the stationary reference frame can be reduced to two sets of decoupled equations corresponding to the machine (d,q) and (z_1,z_2) subspaces.

Using complex vector notation $\left(\overline{f} = f_d + jf_q\right)$, the machine model in the (d,q) subspace is:

$$\begin{cases} \overline{v_s} = R_s \cdot \overline{i_s} + p \cdot \overline{\psi_s} \\ 0 = R_r \cdot \overline{i_r} + p \cdot \overline{\psi_r} - j \cdot w_r \cdot \overline{\psi_r} \\ \overline{\psi_s} = L_s \cdot \overline{i_s} + M \cdot \overline{i_r} \\ \overline{\psi_r} = M \cdot \overline{i_s} + L_r \cdot \overline{i_r} \\ T_e = \dfrac{P}{2} \left(\psi_{sd} i_{sq} - \psi_{sq} i_{sd} \right) \end{cases} \quad (3)$$

Where w_r is the rotor speed, P is the number of poles and p is the derivative operator.

As shown by (3), the torque production involves only quantities in the (d,q) subspace. And consequently the machine control is simplified and in (d,q) subspace is similar to three phase counterpart. Then the synchronous reference frame oriented either on the machine rotor flux, stator flux or air gap flux is obtained as for a three phase machine.

The machine model in the (z_1,z_2) subspace is:

$$\begin{bmatrix} v_{sz_1} \\ v_{sz_2} \end{bmatrix} = \begin{bmatrix} R_s + L_{ls} \cdot p & 0 \\ 0 & R_s + L_{ls} \cdot p \end{bmatrix} \begin{bmatrix} i_{sz_1} \\ i_{sz_2} \end{bmatrix} \quad (4)$$

As can be seen from (4) only stator resistance and leakage inductance are associated with (z_1,z_2) plane variables. It is important to note that the nonelectromechanical energy conversion related variables on the (z_1,z_2) plane should be controlled to be as small as possible to reduce the extra losses in the machine.

III. SWITCHING TABLE DTC

In the conventional three-phase DTC scheme all control parameters are measured in the stationary reference frame [4, 5]. According to the equation (3), stator flux linkage can be written as follow:

$$\begin{cases} \psi_{sd} = \int (v_{sd} - R_s i_{sd}) dt \\ \psi_{sq} = \int (v_{sq} - R_s i_{sq}) dt \end{cases} \quad (5)$$

According to space vector theory, the instantaneous electromagnetic torque is proportional to the cross-vectorial product of the stator flux-linkage space vector $\overline{\psi_s}$ and rotor flux-linkage space vector $\overline{\psi_r}$. That is torque equation can be expressed in the stationary reference frame as follows:

$$T_e = c|\psi_r||\psi_s|\sin(\gamma) \tag{6}$$

In the equation (6) γ is the angle between the stator and rotor flux-linkage space vectors, c is a constant. If the stator flux-linkage is assumed to be constant, the torque can be rapidly changed by appropriate space voltage vector.

A six-phase inverter provides 48 independent nonzero vectors and four zero vectors to form a 12-sided four-layer polygon in each machine subspace. As shown in [2] the outermost polygon vectors in the (d, q) subspace have the minimum absolute in the (z_1, z_2) subspace. Thus for minimizing content of harmonic currents and maximum utilization of machine flux capability the outermost polygon vectors is selected. Even with a larger number of available voltage vectors compared with the case of three-phase inverters, the switching table DTC (STDTC) has the relatively high harmonic content of the phase currents due to the current harmonics generated in the (z_1, z_2)-subspace. Thus, the overall drive Efficiency will be reduced.

A STDTC scheme has been applied for the SPIM prototype in this paper using the schematic block diagram in Fig. 2. Simulation results are shown in Fig. 3. system is drived under no load condition, followed by a step change in load.

Simulation results for the STDTC show distorted phase currents. For this reason, sampling period in this strategy is important. According to simulation results in this work, current harmonics for sampling period larger than 100 µs are too large. Sampling period in simulation results that is presented in Fig. 3. is 80 µs.

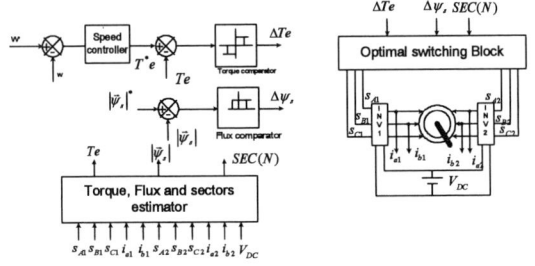

Fig. 2. Block diagram of simulated STDTC

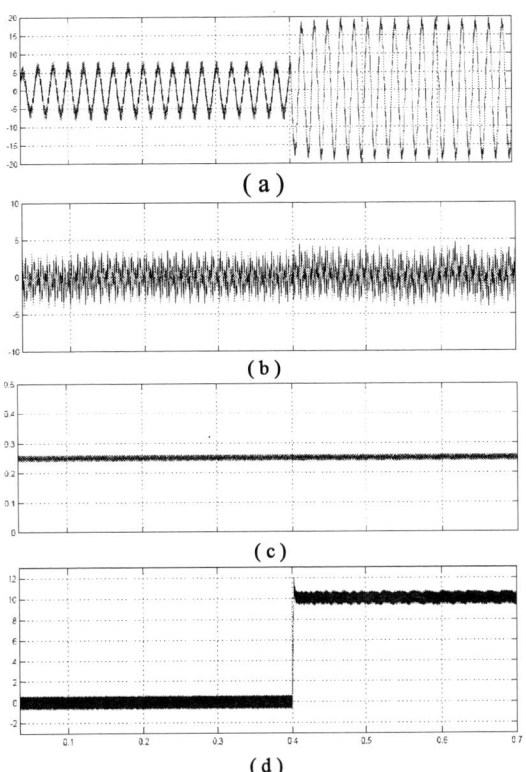

Fig. 3. Simulation results for STDTC: (a) d-current,id(A). (b) z1-current, iz1s (A). (c) Stator flux, fs(Wb). Electromagnetic Torque, Te (N.m)

IV. PREDICTIVE DTC

In the previous section it is shown that STDTC drive has a good dynamic responses but with the large amplitude of (z_1, z_2) subspace current harmonics. For achieving the good dynamic responses of DTC in controlling of stator flux and electromagnetic torque associated with the minimum harmonic currents, predictive DTC presented in this paper. Because only the parameters of (d, q) subspace are related to the electromagnetic torque produced by the motor, we can utilize this analogy for using the predictive DTC that used in three-phase machines [6, 7, and 8]. The basic principle of the scheme consists in finding a relation between the quantities to be controlled (the stator flux magnitude and the torque) and the stator (d, q) voltage components. This done using the simplified model of the induction machine (Fig. 4). In this model, the back EMF, $\overline{e}$, has to be estimated from the measured stator (d, q) voltage components and from the stator (d, q) current components, in the stationary reference frame.

After generating the (d, q) voltage components by the predictive DTC, it is needed a PWM block to impose this voltages to the machine. Among all of the PWM techniques proposed for SPIMs, in minimizing the current harmonics, the best one is SVPWM based on

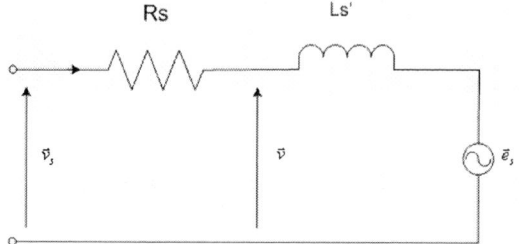

Fig. 4. Simplified model of induction machine in the d-q stationary reference frame

vector space decomposition (VSD) technique [2, 8]. But the execution time is the problem of VSD technique that reported in [2]. VSD-SVPWM is used in this paper. In the transient conditions the voltage that obtained from the predictive scheme is large and can't synthesize with VSD-SVPWM, so the control scheme switch to switching table DTC.

The torque equation at (3) can be rewritten as:

$$T_e = \frac{P}{2}\left(\overline{\psi_s} \times \overline{i_s}\right) \qquad (7)$$

From Fig. 4, the change in stator current vector over a constant period Ts is given by:

$$\Delta \overline{i_s} = \frac{\overline{v} - \overline{e}}{L_s'} Ts \qquad (8)$$

The period Ts is constant in the proposed scheme in order to maintain the constant switching frequency. At speeds above a few hertz, the stator IR drop can be neglected, in which case $\overline{v}$ is equal to $\overline{v_s}$. It is assumed that the stator electrical time constant is much longer than Ts, and therefore, the change in current over the period Ts is linear. In this paper we don't consider speed above nominal speed then we can neglect the flux change term in the differential equation of electromagnetic torque. The corresponding change in electromagnetic torque over the period Ts is:

$$\Delta T_e = T_e^* - T_e = \frac{P}{2}\left(\overline{\psi_s} \times \Delta \overline{i_s}\right) = \frac{P}{2}\left(\overline{\psi_s} \times \frac{\overline{v_s} - \overline{e}}{L_s'} Ts\right) \qquad (9)$$

From Fig. 4, the voltage behind the transient reactance is:

$$\overline{e} = \overline{v_s} - R_s \overline{i_s} - \frac{d}{dt}\left(L_s' \overline{i_s}\right) = \frac{d}{dt}\left(\overline{\psi_s} - L_s'\right) \qquad (10)$$

If it is assumed that $\overline{e}$ is sinusoidal, then:

$$\overline{e} = jw_e\left(\overline{\psi_s} - L_s' \overline{i_s}\right) \qquad (11)$$

As shown in [6], solving of this equation along with flux equation in the stationary reference frame is time consuming. The goal is to obtain the (d,q) voltage components from these equations. From the machine equation it is appeared that in stator flux reference when neglect the IR drops the d-voltage component is proportional to differential of flux as:

$$v_d^s = \frac{\Delta \psi_s}{Ts} = \frac{\psi_s^* - \psi_s}{Ts} \qquad (12)$$

With substitution of the d-voltage component in the (9) readily q- voltage component obtained as:

$$v_q^s = \frac{2\Delta T_e L_s'}{P\psi_s Ts} + e_q^s \qquad (13)$$

After obtaining the reference voltage vector in the stator reference frame it is needed to transform this voltage vector to the stationary reference frame. Then the stator IR drop is added to reference voltage vector in the stationary reference frame.

Simulation results with the same condition at STDTC are shown in Fig. 5. From simulation results it is appear that in addition of the good dynamic response, in predictive DTC current harmonics are significantly reduced.

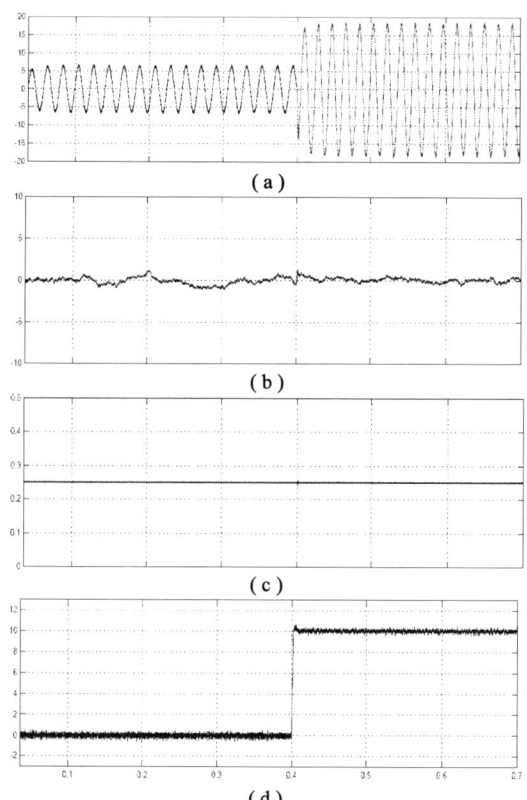

Fig. 5. Simulation result for predictive DTC: (a) d-current,id(A). (b) z1-current, iz1s (A). (c) Stator flux, fs(Wb). Electromagnetic Torque, Te (N.m)

1458

Table I show comparison of important parameters in two schemes in the same condition of simulation.

TABLE I

COMPARISON OF IMPORTANT PARAMETERS IN TWO SCHEMES

Parameter Scheme	5th harmonic in phase current	7th harmonic in phase current	Stator Flux Ripple	Torque Ripple
STDTC	9.2%	6.02%	5%	10%
Predictive DTC	3.21%	2.15%	1.2%	2%

V. CONCLUSION

The major drawback of SPIMs is occurrence of extra harmonic currents. Thus in the DTC of SPIMs in addition to control of torque and flux we should consider simultaneously minimizing harmonic components of stator current. In this paper the conventional DTC scheme first investigated based on switching table of the optimal vectors. Simulation results presented show that the machine has the large harmonic current especially in (z_1, z_2) subspace. Then Predictive DTC along with optimized SVPWM is used in this paper. Simulation results shown that in addition to good dynamic response, current harmonics in this scheme is significantly reduced.

References:

[1] K. Gopakumar, S. Sathiakumar, S. K. Biswas, and J. Vithayathil, "Modified current source inverter fed induction motor drive with reduced torque pulsations," *Proc. Inst. Elect. Eng.*, pt. B, vol. 131, no. 4, pp. 159–164, Jul. 1984.

[2] Y. Zhao and T. A. Lipo : "Space vector PWM control of dual three-phase induction machine using vector space decomposition", *IEEE Transactions on Industry Applications.*, vol. 31, no. 5, pp. 1100-1109, 1995.

[3] R. Bojoi, A. Tenconi, F. Profumo, G. Griva and D. Martinello, "Complete Analysis and Comparative Study of Digital Modulation Techniques for Dual Three-Phase AC Motor Drives", *Conf. Rec. IEEE PESC*, Vol.2, pp.851-857, 2002.

[4] I. Takahashi and T. Noguchi, "A new quick response and high effiency control strategy of an induction motor," *IEEE Trans. Ind. Appl.*, vol. 25, no. 5, pp. 820–827, Sep./Oct. 1986.

[5] I. Takahashi and T. Noguchi, "High-performance direct torque control of an induction motor," *IEEE Trans. Ind. Appl.*, vol. 25, no. 2, pp. 257–264, Mar./Apr. 1989.

[6] C. Lascu, I. Boldea and F. Blaabjerg, "A Modified Direct Torque Control for Induction Motor Sensorless Drive," *IEEE Trans. Ind. Appl.*, vol. 36, no. 1, pp. 122–130, Jan./Feb. 2000.

[7] T. G. Habetler, F. Profumo, M. Pastorelli, and L. M. Tolbert, "Direct torque control of induction machines using space vector modulation," *IEEE Trans. Ind. Appl.*, vol. 28, no. 5, pp. 1045–1053, Sep./Oct. 1992.

[8] G. Griva, F. Profumo, M. Abrate, A. Tenconi, and D. Berruti, "Wide speed range DTC drive performance with new flux weakening control," in *Conf. Rec. IEEE Power Electronics Specialists Conf. (PESC)*, Fukuoka, Japan, 1998, vol. 2, pp. 1599–1604.

[9] K. Hatua and V.Y. Ranganathan, "Direct Torque Control Schemes for Split-phase Induction Machines", Conf. Rec. IEEE IAS, vol.1, pp.615-622, 2004.

A Method of Speed Sensorless Vector Control Parallel -Connected Dual Induction Motors Fed by One Inverter in a Rotor Flux Feedback Control

Jun Nishimura, Kazuo Oka and Kouki Matsuse

Department of Electrical Engineering, Meiji University, Japan
1-1-1, Higashimita, Tama-ku, Kawasaki, JAPAN

Abstract– **A purpose of this paper is to present the speed-sensorless vector method of parallel-connected dual induction motors fed by one inverter. This system may be unstable when the ratings of each induction motor are different and an extremely unbalanced load for both motors is added. Then we suggest a control method that the system can be stable in this condition. According to this method, the rotor flux is directly controlled to be constant by a rotor flux feedback control though it isn't controlled in an ordinary system.**

Index Terms—**Adaptive Rotor Flux Observer, Directed-Field-Oriented Control and Rotor Flux Feedback Control**

I. INTRODUCTION

At present, parallel-connected dual induction motors by one inverter are employed to save a cost, space and weight.

We use the direct-field-oriented control with an adaptive rotor flux observer for this system to cut the speed-sensor [1-2].

For this system, it may be unstable when the ratings of each induction motor (IM1 and IM2) are different and an extremely unbalanced load is added. Fig.1 shows a current vector diagram of both motors in unbalanced load.

We assume that the load added to IM2 is heavier than that of IM1. In this case, since the same command is only given, the phase of stator current for IM2 is lag compared with IM1. A torque control of IM2 is realized but the rotor flux decreases whereas that of the IM1 increases.

Therefore, a phase and amplitude difference of rotor flux in the rotor reference frame for each motor are caused (Fig.1). Finally, a change of the rotor flux is made the system unstable. Where, we suggest a vector control method with the rotor flux feedback control. In employing proposed method, a mean value of rotor flux in the rotor reference frame for each motor can be kept at constants for controlling that directly. As a result, the system is controlled to be stable.

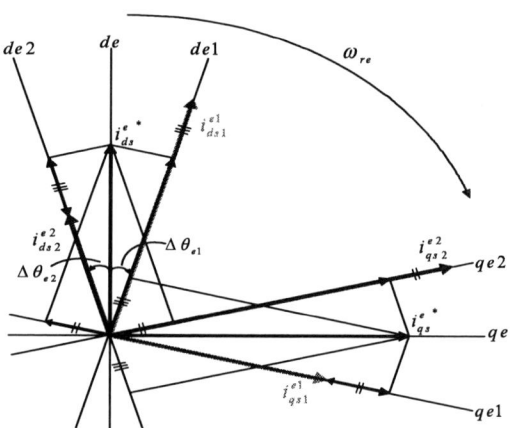

Fig. 1. Current Vector Diagram of Two Induction Motors in Unbalanced Load

II. DUAL INDUCTION MOTORS

A. Current Model of Dual Induction Motor

The stator currents i_{s1}, i_{s2} of the parallel connected dual induction motors are shown in Fig.2.

The stator currents i_{s1}, i_{s2} can be represented by $\overline{i_s}$, the average of i_{s1} and i_{s2}. An equation (1) shows the relation of these currents. As $\overline{i_s}$ is half the input current i_s, it is possible to control directly it with an inverter.

$$\overline{i_s} = \frac{i_{s1} + i_{s2}}{2} \qquad (1)$$

B. Flux Model of Dual Induction Motors

The formula of an induction motor for rotor flux in a flux in a rotating frame of reference is given in Eq.(2).

$$\frac{d}{dt}\mathbf{\Phi_r^e} = U\mathbf{i_s^e} - \{S_r\mathbf{I} + \mathbf{j}(\omega - \omega_r)\}\mathbf{\Phi_r^e} \qquad (2)$$

ω_r : Excitation frequency

ω_r : Motor angular velocity.

Each of $\mathbf{\Phi_{r1}^e}$ and $\mathbf{\Phi_{r2}^e}$ is defined as the rotor flux of each induction motor. And $\overline{\mathbf{\Phi}}_r^e$ is defined as the average of each rotor flux.

The average of each induction motor for Eq.(2) take the following formula.

$$\frac{d}{dt}\overline{\mathbf{\Phi}}_r^e + \{\overline{S}_r\mathbf{I} + \mathbf{j}(\omega - \overline{\omega}_r)\}\overline{\mathbf{\Phi}}_r^e = \overline{U}\overline{\mathbf{i}}_s^e \qquad (3)$$

where

$$\overline{\mathbf{\Phi}}_r^e = \frac{\mathbf{\Phi}_{r1}^e + \mathbf{\Phi}_{r2}^e}{2} \qquad \overline{\mathbf{i}}_s^e = \frac{\mathbf{i}_{s1}^e + \mathbf{i}_{s2}^e}{2}$$

$$\overline{S}_r = \frac{S_{r1} + S_{r2}}{2} \qquad \overline{U} = \frac{U_1 + U_2}{2}$$

$$\overline{\omega}_r = \frac{\omega_{r1} + \omega_{r2}}{2}$$

C. Torque of Dual Induction Motors

Equation (4) for the average torque of two motor induction motors. This Equation can be also represented by machine parameters.

$$\overline{T}' = p\overline{M}'(\overline{\mathbf{i}}_s^e \times \overline{\mathbf{\Phi}}_r^e) = \overline{T}_e \qquad (4)$$

where

$$\overline{T}_e = \frac{T_{e1} + T_{e2}}{2} \qquad \overline{M}' = \frac{1}{2}(\frac{M_1}{L_{r1}} + \frac{M_2}{L_{r2}})$$

L_{r1}, L_{r2} : Rotor self-inductance.
p : Pole pairs

D. Calculation of Reference Current

Equation (3) and (4) are separated into **dq** components, where the positive d-axis correspond with the average rotor flux in order to calculate the reference current to control the average torque of two induction motors.

The q-axis component of Eq.(4) for the average torque is represented by the following equation. The average torque can be controlled with this equation.

$$\overline{i}_{qs}^{e^*} = \frac{\dfrac{\overline{T}'^*}{p\overline{M}'}}{\overline{\phi}_{dr}^e} \qquad (5)$$

The d-axis component of Eq.(3) for the average rotor flux is represented by the following equation. The average rotor flux can be controlled with this equation.

$$\overline{i}_{ds}^{e^*} = \frac{\overline{\phi}_{dr}^e}{M} \qquad (6)$$

where
*: indicates reference values.

III. SYSTEM CONFIGURATION

Fig.2 shows a system configuration of proposed method. For this system, the directed-field-oriented control with the adaptive rotor flux observer is used and the rotor flux feedback control is introduced. But according to conventional method, it couldn't be kept the rotor flux at constants not to have used the rotor flux feedback control.

So now we substitute an estimated rotor flux with the adaptive rotor flux observer for a detected rotor flux as an actual value of that. It is more difficult for that to keep the system to be stable. That is because this system only detects the sum of current flowing into the motor.

Therefore the rotor flux feedback control is introduced in the system for keeping the mean value of rotor flux for each motor in the rotor reference frame to be constants.

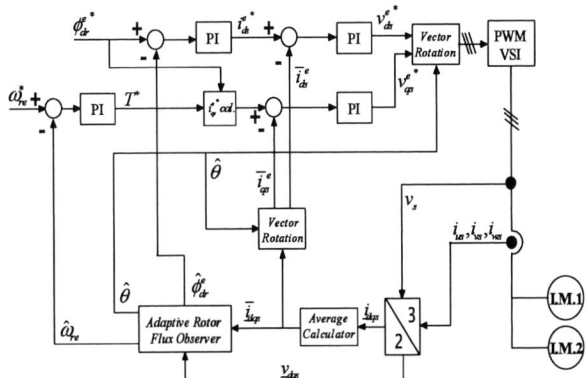

Fig.2 Configuration of Proposed System

IV. ROTOR FLUX FEEDBACK CONTROL SYSTEM

Fig.3 expresses the block diagram of rotor flux feedback control system. The plant model of this system is a first-order lag system with the time constant $1/S_r$.

Where, $1/S_r$ expresses L_r/R_r, which is a rotor time constant. We substitute an estimated rotor flux with the adaptive rotor flux observer for a detected rotor flux as an actual value of that.

For the system, a flux current command is calculated by Proportional-Integral (PI) controller.

In this system, the mean value of the rotor flux is controlled. Consequently, the rotor flux of both motors can be kept at constants.

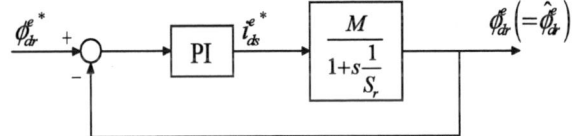

Fig.3 Block Diagram of Rotor Flux Feedback Control System Diagram

V. SIMULATION RESULT

The rating and parameters of two induction motors employ in simulation are shown in Table1.

Table.2 shows the simulation conditions.

TABLE I
RATING AND PARAMETERS OF TWO INDUCTION MOTORS

Output	2.2kW	Stator Resistance	0.7185Ω
Pole Pairs	2	Rotor Resistance	0.5965Ω
Voltage	180V	Stator Inductance	63.38mH
Current	10A	Rotor Inductance	63.38mH
Motor Speed	1750rpm	Mutual Inductance	61.28mH
Inertia	0.024kg.m²	Torque	12N.m

TABLE II
SIMURATION CONDITIONS

time[s]	1	2	3	4	5	6
Speed Command [rpm]	500					
Load of motor 1[N.m]	18			24		
Load of motor 2[N.m]	18	24				

In the conventional method, it is impossible for the system to be stable because the rotor flux isn't controlled (Fig.4). Otherwise, as can be seen from Fig.5, we can keep the mean value of rotor flux for each motor at constant in the unbalanced load. Moreover, the mean value of rotor speed for both motors is appropriately estimated.

By the way, Fig.6 shows the torque response waveforms. Fig.6 (a), (b) are that of conventional proposed method respectively. Then a torque control can't be realized in the conventional method but that of proposed method is realized.

According to proposed method, a stable speed control can be realized though the rotor speed difference is caused.

Consequently, the rotor speed of them can be stable too.

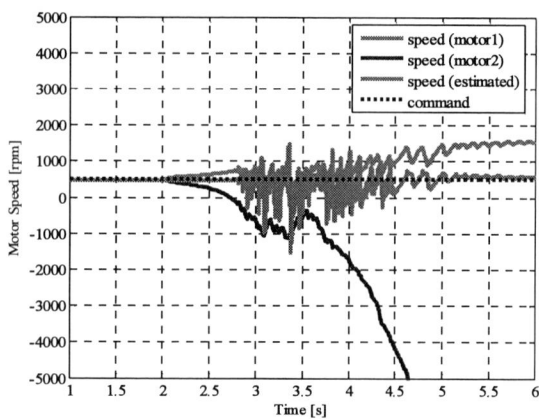

(a) Motor Speed

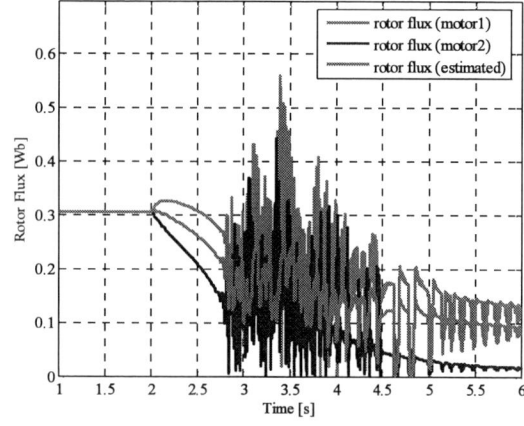

(b) Rotor Flux

Fig. 4 Conventional Method

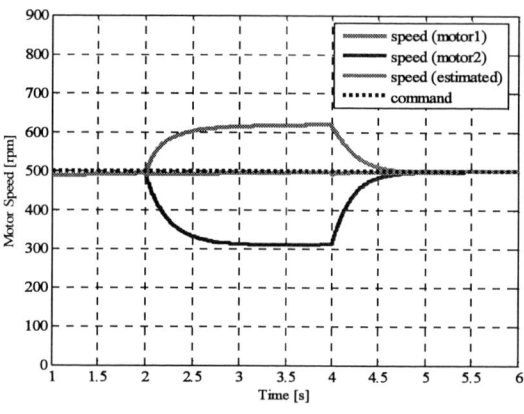

(a) Motor Speed

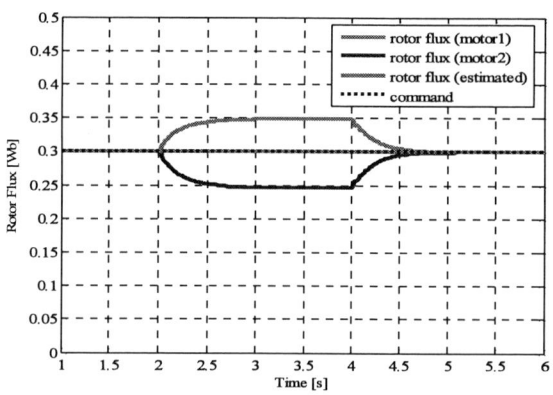

(b) Rotor Flux

Fig.5 Proposed Method

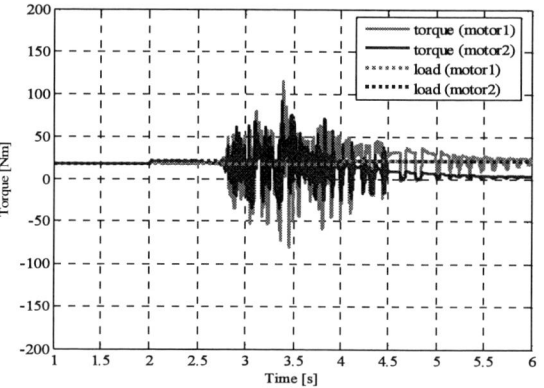

Fig. 6 (a) Conventional Method

Fig.6 Torque Response

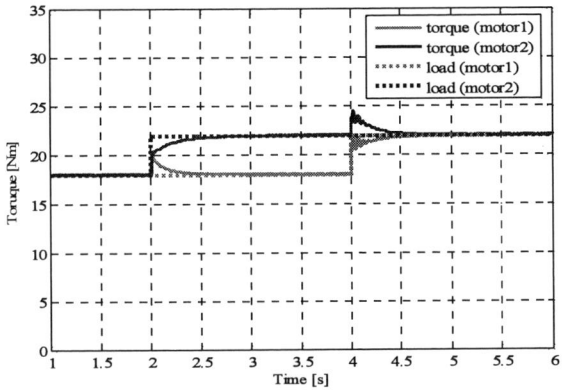

Fig.6 (b) Proposed Method

Fig.6 Torque Response

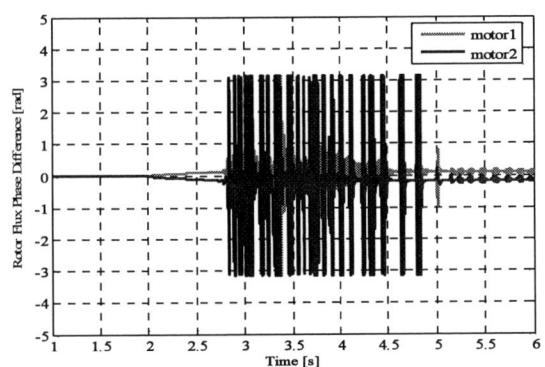

Fig. 7 (a) Conventional Method

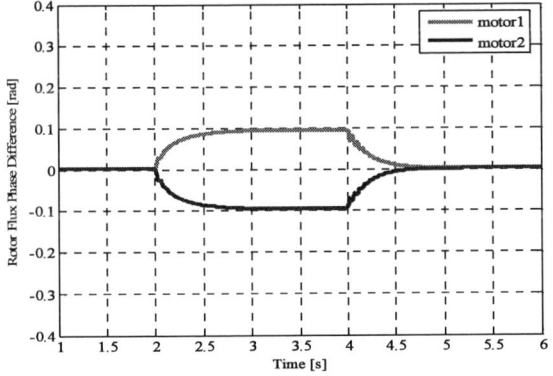

Fig.7 (b) Proposed Method

Fig.7 Rotor Flux Phase Difference

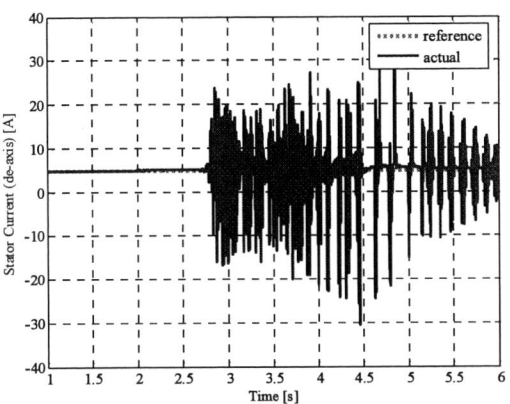

Fig. 8 (a) Conventional Method

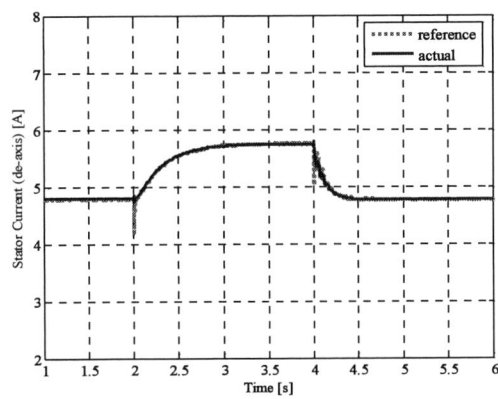

Fig.8 (b) Proposed Method

Fig.8 Stator Current (de-axis)

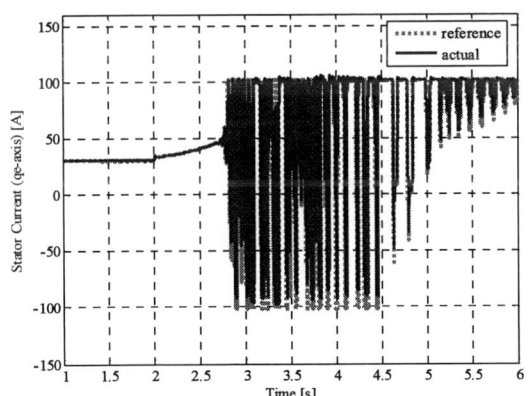

Fig. 9 (a) Conventional Method

Fig.9 Stator Current (qe-axis)

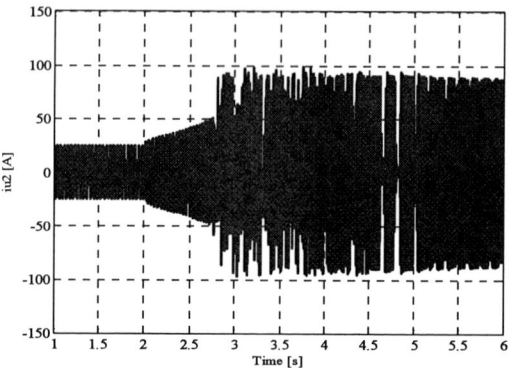

Fig. 11 (a) Conventional Method (Motor2)

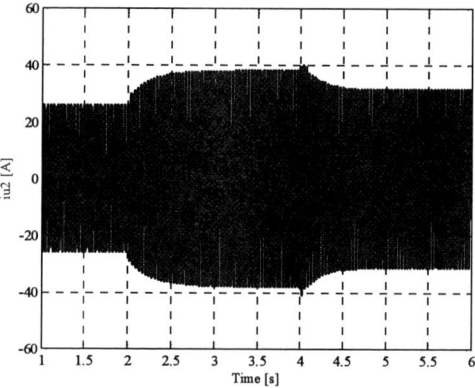

Fig.11 (b) Proposed Method (Motor2)

Fig.11 Stator Current (u-phase)

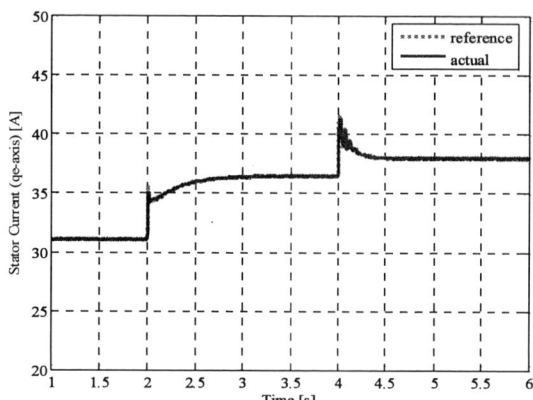

Fig.9 (b) Proposed Method

Fig.9 Stator Current (qe-axis)

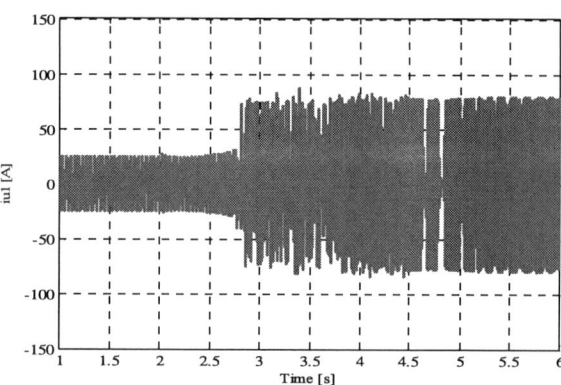

Fig. 10 (a) Conventional Method (Motor1)

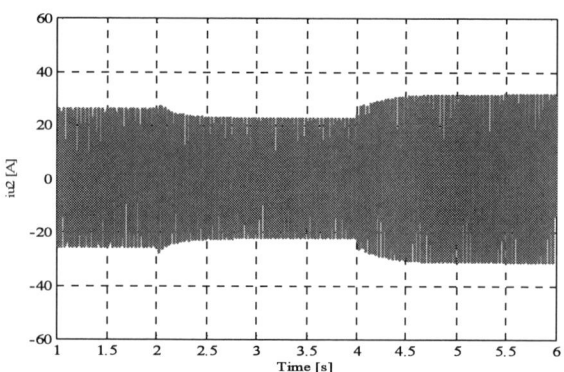

Fig.10 (b) Proposed Method (Motor1)

Fig.10 Stator Current (u-phase)

VII. CONCLUSION

This paper presents a rotor flux control method for the system of parallel-connected dual induction motors by one inverter. And the system uses a method of direct-field-oriented control of parallel-connected two induction motors.

According to the proposed method, it can be kept the rotor flux.

REFERENCES

[1] M.Taniguchi, T.Yoshinaga, and K.Matsuse,"A Speed-Sensorless Vector Control of Parallel-Connected Multiple Induction Motor Drives with Adaptive Rotor Flux Observer" IEEE PESC'06, 2006, CD-ROM.

[2] M. Taniguchi, J. Nishimura, and K. Matsuse, "Characteristics of Speed-Sensorless Vector Control of Parallel-Connected Multiple Induction Motor", IWSEM, 2006, 2-p-2, pp.2-57-60.

[3] T. Fushima, H. Natsume and K.Matsuse," A Method of Speed Sensorless Vector Control of Parallel Connected Induction Motors",pp1-2.

[4] J. Nishimura, K. Oka and K. Matsuse, '' A Method of Speed Sensorless Vector Control Of Parallel Connected Dual Induction Motors By a Single Inverter with a Rotor Flux Control," ICEMS, 2007, pp1, pp3-5.

A Combined Model Flux Observer for Vector Control of Traction Asynchronous Motors

F. Tahami, S. Chini Foroosh

Department of Electrical Engineering, Sharif University of Technology, Tehran, Iran

Abstract– **Vector control induction motor drives are particularly sensitive to the variation of motor parameters. As a consequence, the field orientation is missed, leading to substantial degradation of performance. The performance of traction motors can be even more degraded by the substantial core loss due to their high supply frequency. In this paper, a flux observer using combined voltage and current model is presented. The influence of the core loss resistance is considered in motor models. This control system is robust against simultaneous variations of the stator resistance, rotor resistances, as well as the mismatch of the equivalent stator core-loss resistance, yielding a globally stable observer with highly robust performance.**

Index Terms-- **Field Oriented Control, Induction motor, Iron loss, flux observer.**

I. INTRODUCTION

The number of adjustable speed drives with induction machines is increasing in the industry. New industrial and appliance motor drives have become more advanced in performance and demand for sophisticated motor control systems that ensure high levels of efficiency and robust operation.

In rotor flux oriented control state observers are frequently used to estimate the rotor flux. Stability of indirect field-oriented control of induction motor drives is greatly influenced by motor parameters. Variations of motor parameters deteriorate the drive performance by introducing errors in the estimated magnitude and position of the flux vector.

The parameter sensitivity problem of an induction motor vector control is well stated and several solutions have been considered to improve the flux vector control accuracy [1-5]. Stability of indirect field-oriented control (IFOC) of induction motor drives is greatly influenced by estimated value of rotor time constant. Due to the requirement for an accurate rotor time constant to give high-performance control, estimation techniques of rotor resistance have been proposed by some authors [6-11]. An online tuning scheme for stator and rotor resistances is presented in [12] to overcome rotor resistance variation problem.

The voltage model flux observer is an alternative approach to overcome the problem of rotor resistance variation. The main drawback of the voltage model is that it requires an open loop integrator, which fails in low speed operation. The current model on the other hand fails at high speed due to speed dependent eigen-values which become less damped at high speed [13]. It is

proposed to use voltage model during operation at high speeds and current model for low speeds. Therefore hybrid solutions have been proposed [14, 15].

Nevertheless, analysis of parameter variations in conjugation with iron loss has hardly received enough attention so far. Iron loss is a possible source of performance deterioration. The impact of iron loss is significant in motors used for traction applications. Traction motors are characterized by their high speed and operation deep into the field weakening region. Due to their high supply frequency, traction motors make a significant amount of eddy-current and hysteresis loss. If we simply neglect the iron loss, then it detunes the overall vector controller and results in an error in the torque control. In this case, the core-loss resistance mismatch degrades the performance of the torque control accuracy and becomes a serious problem.

In this paper a combination of motor models are used to mitigate the effects of parameter variations. The influence of the mismatch of the equivalent stator core-loss resistance is shown in order to emphasize a need for robustness against the core-loss resistance mismatch as well as the stator and the rotor resistance variations. This control system is robust against simultaneous variations of the stator and rotor resistances, and the mismatch of the equivalent core-loss resistance, yielding a dependable observer.

II. MODELING OF IRON LOSS IN INDUCTION MOTORS

Generally, iron loss in induction motors is modeled by means of insertion of the iron loss resistor connected in parallel with the magnetizing branch. Fig.1 shows the vector model for induction motor in an arbitrary reference frame considering iron loss. This model represents a sixth order system. Standard state space equations of the system can be written in the following form [16, 17]:

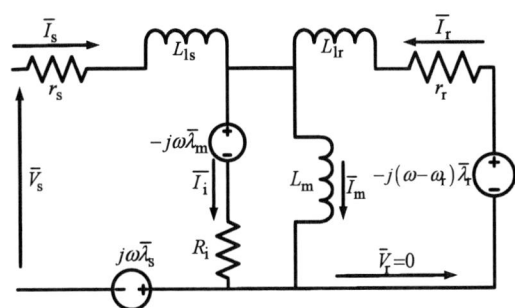

Fig. 1. Dynamic model of induction machine considering iron loss

$$p\bar{\lambda}_r = \frac{r_r}{L_{lr}} L_m \bar{I}_m - \left(\frac{r_r}{L_{lr}} + j(\omega - \omega_r) \right) \bar{\lambda}_r \qquad (1)$$

$$p\bar{I}_m = \frac{R_i}{L_m} \bar{I}_s - \left(\frac{R_i L_r}{L_m L_{lr}} + j\omega \right) \bar{I}_m + \frac{R_i}{L_m L_{lr}} \bar{\lambda}_r \qquad (2)$$

$$p\bar{I}_s = -\frac{1}{L_{ls}} (r_s + R_i + j\omega L_{ls}) \bar{I}_s + \frac{R_i L_r}{L_{ls} L_{lr}} \bar{I}_m - \frac{R_i}{L_{ls} L_{lr}} \bar{\lambda}_r + \frac{1}{L_{ls}} \bar{V}_s \quad (3)$$

Where $\bar{\lambda}_r$, $\bar{I}_m$, and $\bar{I}_s$ are rotor flux linkage, magnetizing current, and stator current vectors respectively and have been chosen as the state variables of the system. Parameters r_s and r_r are stator and rotor resistances and L_{ls} and L_{lr} are the corresponding leakage inductances. R_i is the equivalent iron loss resistor. ω and ω_r are the reference frame angular frequency and rotor velocity respectively and p is the differential operator. L_s and L_r are defined as:

$$L_s = L_m + L_{ls} \qquad (4)$$

$$L_r = L_m + L_{lr} \qquad (5)$$

Where L_m is the magnetizing inductance.

Conventional flux estimators use the simplified vector model for the induction motor which neglects the iron loss. In this paper modified flux estimators are used to establish a combined model flux estimator. These modified estimators are derived from the above vector model which accounts for iron loss.

The modified current model in arbitrary reference is derived directly from the equations (1) and (2). First, the magnetizing current is estimated form (2). Rotor flux is then derived from (1). Fig. 2 shows the block diagram of this estimator. In this figure, superscript ^ denotes the estimated quantities which are used in the control system. Rotor speed and stator currents are assumed to be measured by appropriate sensors.

To set up the modified voltage model, the stator flux in the stationary reference is first calculated as:

$$p\hat{\bar{\lambda}}_s = \bar{V}_s - \hat{r}_s \bar{I}_s \qquad (6)$$

Rotor Flux should now be calculated from stator flux. Stator and rotor fluxes are given by:

$$\bar{\lambda}_s = L_{ls} \bar{I}_s + L_m \bar{I}_m \qquad (7)$$

$$\bar{\lambda}_r = L_{lr} \bar{I}_r + L_m \bar{I}_m \qquad (8)$$

On the other hand, in Fig. 1, KVL and KCL can be written in the magnetizing branch as follows:

$$R_i \bar{I}_i = (p + j\omega) \bar{\lambda}_m \qquad (9)$$

$$\bar{I}_s + \bar{I}_r = \bar{I}_m + \bar{I}_i \qquad (10)$$

The currents in the magnetizing and the iron loss branches are calculated from (7) and (9) respectively:

$$\bar{I}_m = \frac{\bar{\lambda}_s - L_{ls} \bar{I}_s}{L_m} \qquad (11)$$

$$\bar{I}_i = \frac{(p + j\omega) \bar{\lambda}_m}{R_i} \qquad (12)$$

Substituting (12) in (10), the rotor current is given by:

$$\bar{I}_r = \left(\frac{L_m p}{R_i} + 1 \right) \bar{I}_m - \bar{I}_s \qquad (13)$$

Substituting (11) and (13) in (8), the rotor flux is then obtained from the stator flux in the stationary reference frame by:

$$\bar{\lambda}_r = \left(\frac{\hat{L}_{lr} p}{\hat{R}_i} + \frac{\hat{L}_r}{\hat{L}_m} \right) \bar{\lambda}_s - \left(\frac{\hat{L}_{lr} \hat{L}_{ls} p}{\hat{R}_i} + \frac{\hat{\sigma}}{1 - \hat{\sigma}} \hat{L}_m \right) \bar{I}_s \quad (14)$$

Where $\hat{\sigma}$ is the total leakage coefficient:

$$\hat{\sigma} = 1 - \frac{\hat{L}_m^2}{\hat{L}_s \hat{L}_r} \qquad (15)$$

Block diagram of this estimator is shown in Fig. 3.

III. PARAMETER SENSITIVITY ANALYSIS OF THE REDUCED ORDER FLUX OBSERVERS CONCERNING CORE LOSS

The accuracy of the flux estimate used for FOC is influenced by motor parameters. Thus, analysis of the observer accuracy under different operating conditions is very appropriate. To evaluate the parameter dependency of flux observers, the estimated flux angle given by modified current and voltage models including the iron loss resistance with mismatched parameters, are compared with those of the exact motor model. The parameters of a 12kw traction motor as tabulated in table-I are used for this purpose.

The simulation results showing the estimation error in different loads are illustrated in Fig. 4. The influence of parameter errors on the accuracy of the estimated flux as a function of speed is illustrated in Fig 5.

At light loads, the current model is much sensitive to the rotor resistance, whilst at heavy loads the flux estimate is less sensitive to the rotor resistance.

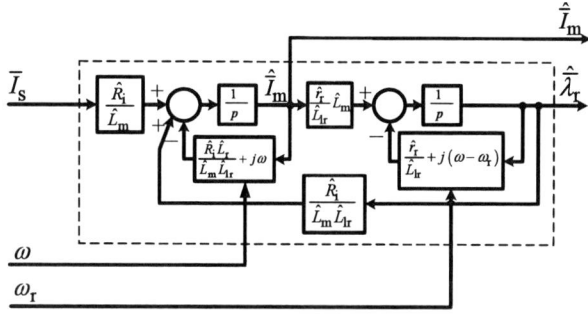

Fig. 2. Modified current model based flux estimator

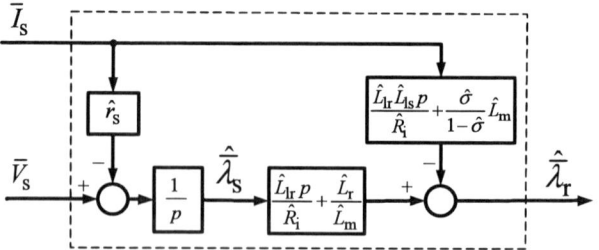

Fig. 3. Modified voltage model based flux estimator

TABLE I
MOTOR PARAMETERS

SPECIFICATION	QUANTITY
NOMINAL OUTPUT POWER	12 KW
NOMINAL VOLTAGE	100VΔ
NOMINAL CURRENT	86.3A
NOMINAL FREQUENCY	200 Hz
NOMINAL SPEED	5920 RPM
MAX. SPEED	12000 RPM
POWER FACTOR	0.857
EFFICIENCY	93.6%
$R_1, R'_2, X_1, X'_2 (\Omega)$	0.0305, 0.029, 0.198, 0.25
X_M (FLUX= 120%, 110%, 100%, 80%, 70%) (Ω)	2.45, 2.58, 2.70, 2.88, 2.93
R_{FE} (F=50, 100, 200, 300, 400 Hz) (Ω)	41, 57.2, 67, 79.4, 83

Unlike the current model based observer, the accuracy of voltage model observer is completely insensitive to rotor resistance but is most sensitive to stator resistance at low rotor velocities. It is evident from Fig. 5 that near the rated speed, the stator resistance error is less significant. However, the low speed sensitivity is a well acknowledged limitation of this observer [14]. In the field weakening region, both models show to be sensitive to mismatch in rotor resistance, while the voltage model is very sensitive to stator resistance variations as well.

IV. THE PROPOSED COMBINED FLUX OBSERVER

The rotor flux vector can be estimated by any of voltage and current models. Because the attributes of both open-loop observers are in many ways complementary, we can create a better observer by directly merging the two models. Several techniques have been successfully used to aggregate different estimates. The most widely used idea is to take an arithmetic average. Aggregating data using techniques such as averaging is an operation common to many information fusion processes. In the arithmetic average, we combine all the estimates with equal weights. In this particular problem, it makes sense to give more weight to consistent estimates and less weight to estimates that are far away from the consensus of the majority.

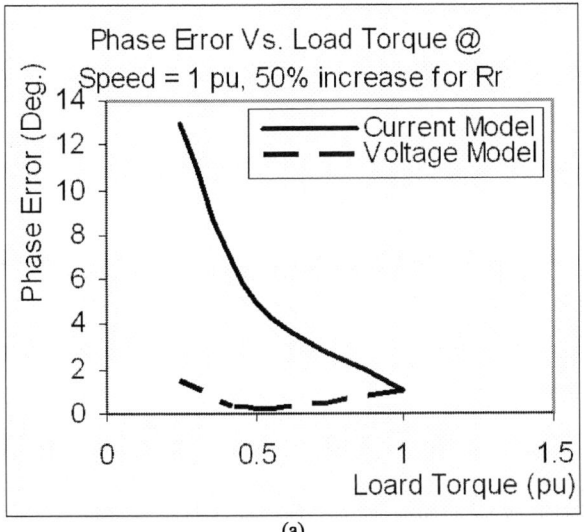

(a)

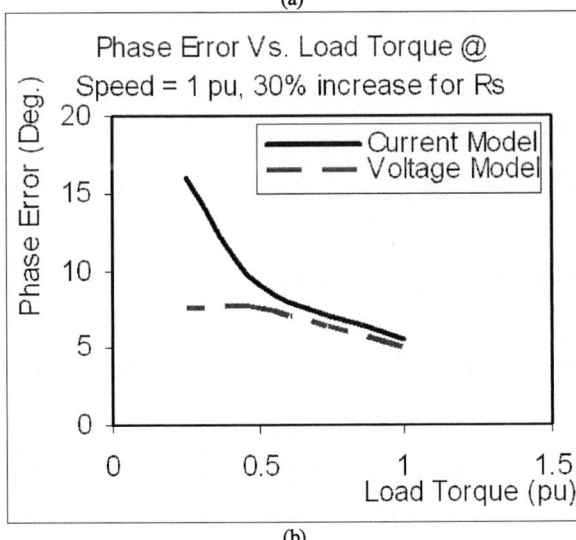

(b)

Fig. 4. Error in estimated flux angle using modified current and voltage models vs. load torque considering iron loss, at rated speed operation, a) 50% increase in rotor resistance, b) 30% increase in stator resistance

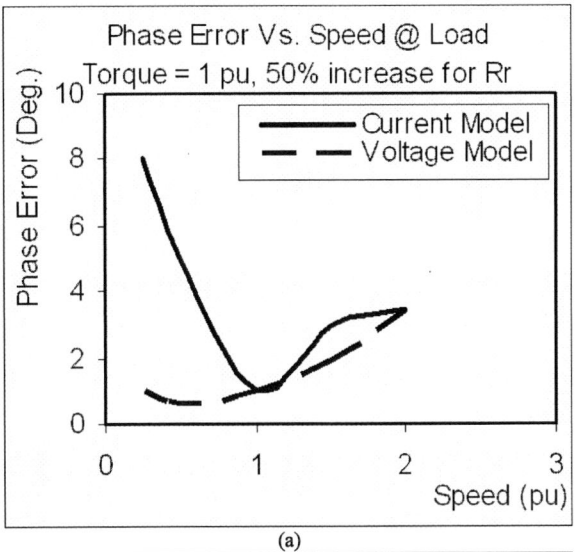

(a)

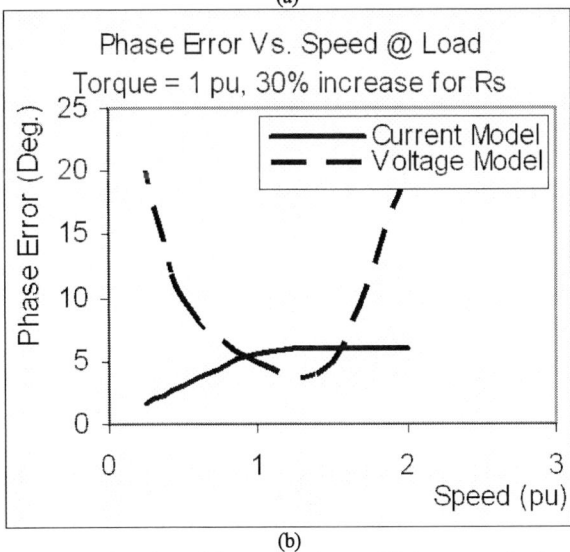

(b)

Fig. 5. Error in estimated flux angle using modified current and voltage models vs. speed considering iron loss, at rated load torque operation, a) 50% increase in rotor resistance, b) 30% increase in stator resistance

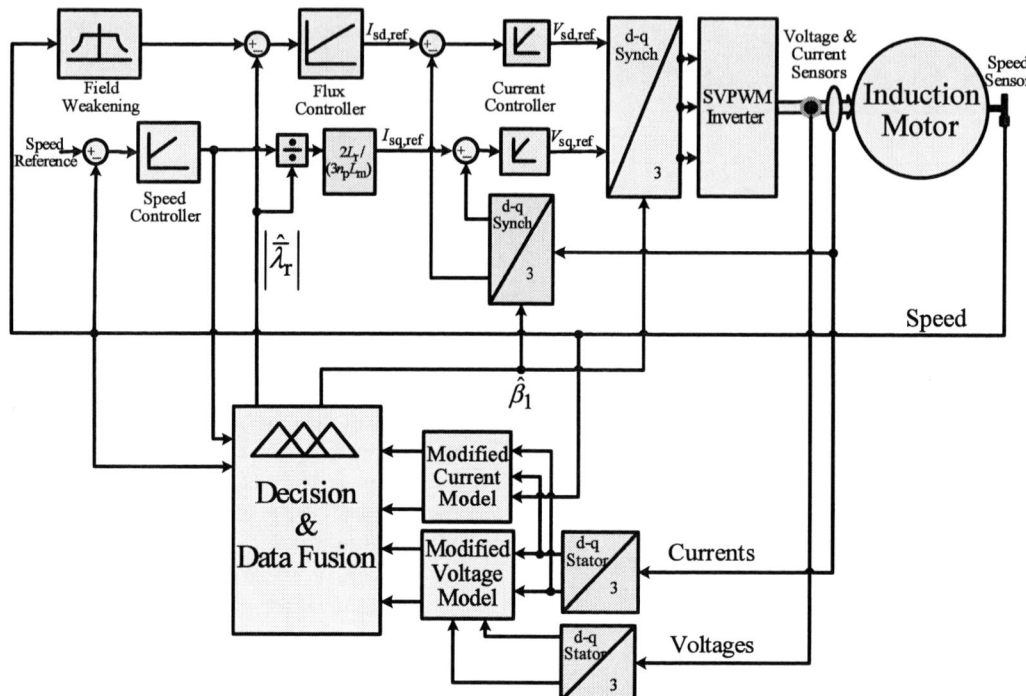

Fig. 6. The proposed vector control scheme utilizing a combined model flux estimator

The flux angle (β_1) then can be computed by averaging the weighted estimates:

$$\beta_1 = \frac{w_c \beta_{1c} + w_v \beta_{1v}}{w_c + w_v} \qquad (16)$$

Where w_c and w_v are the weights that are devoted to the estimated flux given by the current and voltage models (β_{1c} and β_{1v}) respectively. Fig. 6 depicts the proposed control scheme.

The weights are given by a Fuzzy aggregating technique. Operation in the field weakening region, where the speed is above the rated value has also been considered. The Fuzzy rules have been established on the basis of linguistic terms such as:

- In low speed operations, the current model weight is set to high.
- In high speed and light load operation, the current model weight is set to low.
- In high speed and mid load operation the current model weight is set to medium.
- In high speed and heavy load operation the current model weight is set to high.
- In medium speed and light load operation the current model weight is set to medium.
- In medium speed and mid load operation, the current model weight is set to low.
- In medium speed and heavy load operation, the current model weight is set to high.
- In the field weakening region and light load operation the current model weight is set to low.
- In the field weakening region and mid load operation the current model weight is set to medium.

- In the field weakening region and heavy load operation the current model weight is set to low.

V. SIMULATION RESULTS

To evaluate the proposed observer, a simulated program is conducted for the aforesaid traction motor. First, a 50% increase in the value of rotor resistance at the rated speed and torque is introduced. Fig. 7 shows the actual (solid lines) and estimated (dotted lines) rotor fluxes for the current model, the voltage model, and the proposed model observers. The results show that the proposed model operates satisfactory in these operating conditions.

Fig. 8 shows the actual (solid lines) and estimated (dotted lines) rotor flux for the three observers at 1.5pu speed and the corresponding reduced torque, as the motor operates in the constant power region. A 50% increase in each of the rotor and stator resistances is assumed as error in parameters. The magnetizing inductance is also identified by the data provided by the manufacturer as shown in Table.1. It is obvious from the results that the system is also capable of producing appropriate estimates for the rotor flux in this region.

In general, the proposed method can provide a precise flux estimate, in sense of magnitude and angle, in a wide range of operation including the field weakening region.

VI. CONCLUSIONS

The influences of variation in rotor and stator resistances on current and voltage flux observers in association of the equivalent core-loss resistance investigated by numerical simulations. A robust combined model observer against the stator and rotor resistance mismatch showed. The performance of the controller is examined by extensive simulation studies over a traction motor. The simulation results show that the proposed method is very effective in flux estimation of traction motors working with high supply frequency and in deep flux weakening.

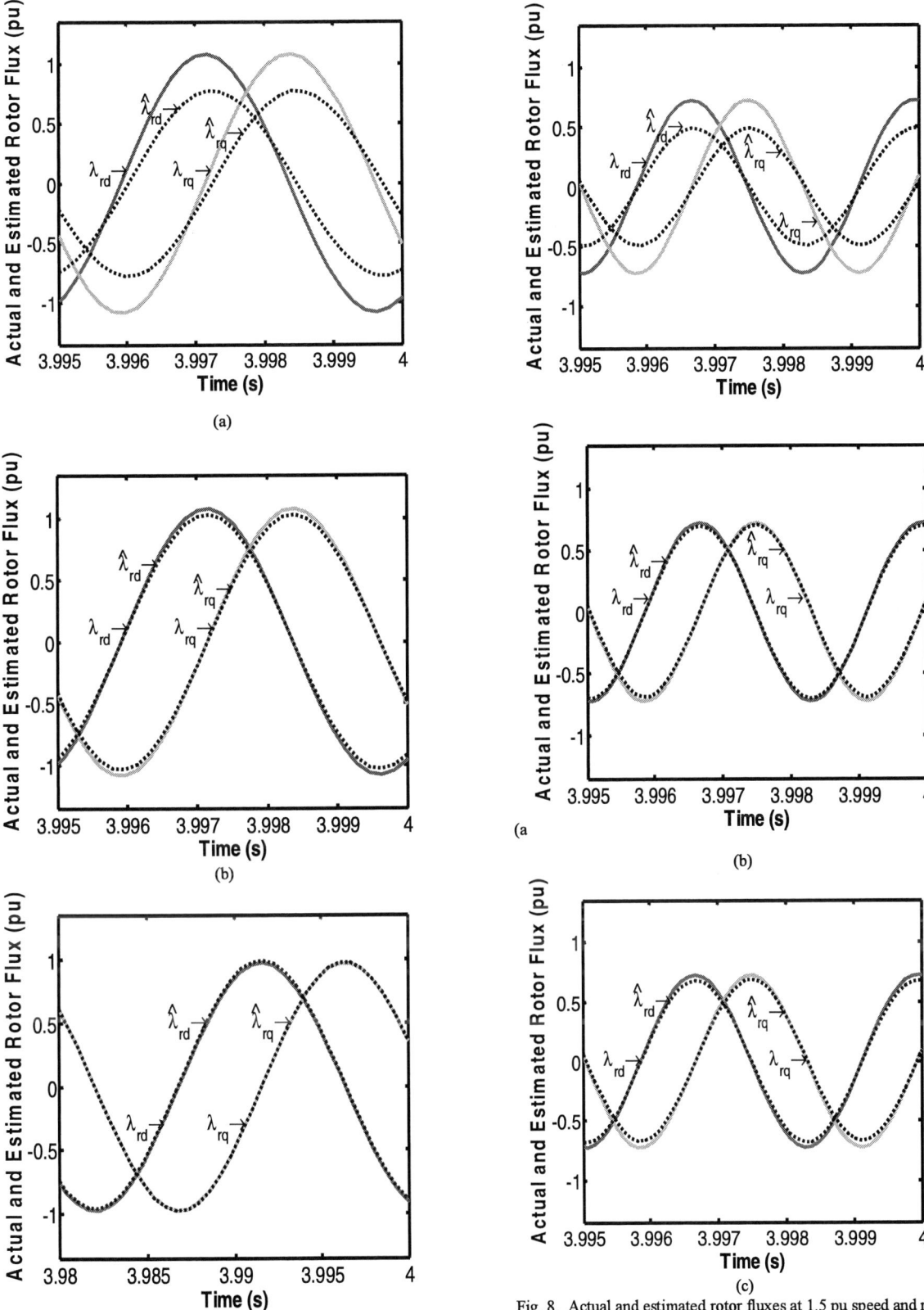

Fig. 7. Actual and estimated rotor fluxes at nominal speed and load with 50% increase in the value of rotor resistance; a) current model, b) voltage model, c) the proposed model

Fig. 8. Actual and estimated rotor fluxes at 1.5 pu speed and reduced torque with 50% increase in the values of rotor and stator resistances; a) current model, b) voltage model, c) the proposed model

REFERENCES

[1] L. Harnefors, "Design and analysis of general rotor–flux-oriented vector control systems," *IEEE Trans. Ind. Electron.*, vol. 48, pp. 383–390, Apr. 2001.

[2] R. Marino, S. Peresada, and P. Tomei, "Output feedback control of current-fed induction motors with unknown rotor resistance," *IEEE Trans. Contr. Syst. Technol.*, vol. 4, pp. 336–347, July 1996.

[3] X. Roboam, C. Andrieux, B. de Fornel, and J. C. Hapiot, "Rotor flux observation and control in squirrel-cage induction motor: Reliability with respect to parameters variations," *Proc. Inst. Elect. Eng. D.*, vol. 139, pp. 363–370, 1992.

[4] T. Matsuo and T. A. Lipo, "A rotor parameter identification scheme for vector controlled induction motor drives," *IEEE Trans. Ind. Applicat.*, vol. 21, pp. 624–632, May/June 1985.

[5] R. Marino, S. Peresada, and P. Tomei, "Global adaptive output feedback control of induction motors with uncertain rotor resistance," *IEEE Trans. Automat. Contr.*, vol. 44, pp. 967–983, May 1999.

[6] J.C. Moreira, K.T. Hung, T. A. Lipo, and R.D. Lorenz "A simple and robust adaptive controller for detuning correction in field-oriented induction machines," *IEEE Trans. Ind. Appl.*, 1992, 28, (6), pp. 1359–1366.

[7] J.C. Moreira, and T. A. Lipo 'A new method for rotor time constant tuning in indirect field oriented control', *IEEE Trans. Power Electron.*, 1993, 8, (4), pp. 626–631

[8] D. J. Atkinson, J.W. Finch, and P. P. Acarnley "Estimation of rotor resistance in induction motors," *IEE Proc. Electr. Power Appl.*, 1996, 143, (1), pp. 87–94.

[9] L.C. Zai, C.L. Demarco, and T.A. Lipo "An extended Kalman filter approacch to rotor time constant measurement in PWM induction motor drives," *IEEE Trans. Ind. Appl.*, 1992, 28, (1), pp. 96–104

[10] R. Marino, A. S. Peresada, and P. Tomei "Exponentially convergent rotor resistance estimation for induction motors," *IEEE Trans. Ind. Electron.*, 1995, 42, (5), pp. 508–515

[11] K. Tungpimolrut, F. Z., Peng and T. Fukao "Robust vector control of induction motor without using stator and rotor circuit time constants," *IEEE Trans. Ind. Appl.*, 1994, 30, (5), pp. 1241–1246

[12] S. H. Joen, K. K. Oh, and J. Y. Choi "Flux observer with online tuning of stator and rotor resistances for induction motors," *IEEE Trans. Ind. Electron.*, vol. 49, no. 3, pp. 653–664, Jun. 2002.

[13] H. Rehman, A. Derdiyok, M. K. Guven, and L. Xu, "A new current model flux observer for wide speed range sensorless control of an induction machine," *IEEE Trans. Power Electron.*, vol. 17, no. 6, pp. 1041–1048, Nov. 2002.

[14] P.L. Jansen and R.D. Lorenz, "A Physically Insightful Approach to the Design and Accuracy Assessment of Flux Observers for Field Oriented Induction Machine Drives", in IEEE Trans. on Ind. Appl., Jan/Feb 1994, pp. 101-110.

[15] P.L. Jansen, C.O. Thompson, R.D. Lorenz, and D.W. Novotny, "Observer-Based Direct Field Orientation: Analysis and Comparison of Alternative Methods", in IEEE Trans. on Ind. Appl., vol. 30, Jul/Aug 1994, pp. 945-953.

[16] E. Levi, "Impact of iron loss on behavior of vector controlled induction machines," *IEEE Trans. Ind. Applicat.*, vol. 31, no. 6, pp. 1287-1296, 1995.

[17] Jong-Woo Choi, Dae-Woong Chung, Seung-Ki Sul, "Implementation of Field Oriented Induction Machine Considering Iron Losses," *Proc. APEC*, vol. 1, pp. 375–379, 1996

Torque Ripple Elimination for Doubly-Fed Induction Motors under Unbalanced Source Voltage

Hong-Geuk Park *, Ahmed G. Abo-Khalil *, Dong-Choon Lee*, Kwang-Myoung Son**

* Dept. of Electrical Eng. Yeungnam University, 214-1, Daedong,Gyeongsan,Gyeongbuk ,Korea
** Dept. of Electrical Eng. Dongeui University , 995 Eomgwangno Busan_Jin Gu, Busan, Korea

Abstract — This paper proposes a control scheme which can eliminate the torque ripple of the DFIM at unbalanced source voltage. The machine torque is expressed using the positive and negative sequence components of the voltage and current. For suppression of the torque ripple, the stator reactive power ripple is controlled to be zero. The proposed torque elimination method is verified by simulation results using PSCAD/EMTDC.

Index Terms—DFIM, torque ripple elimination, unbalanced source voltage.

I. INTRODUCTION

As the industry is growing up more and more, about 50% of the electric power has been consumed for the pumps and ventilation systems [1]. For this, high power induction machines have been used, especially, DFIM (doubly-fed induction motor) is used for the system higher than 1 [MW] [2]-[4].

Conventional control methods for the DFIM is the static Scherbius or Kramer methods. Thereafter, the cyclo-converter has been used to control the DFIM drives, however the control performance is limited [5].

Recently, the back-to-back type PWM converters have been used to control the doubly-fed induction generators for wind power generation [6]. The stator of the DFIM is connected to the source directly and the rotor can be controlled using the back-to-back converters. Since the converter handles only the slip power, the converter power rating is determined by the speed control range. For full speed control ranges, the power rating of the converter is as large as the rated power. However, the converter power rating can be reduced if the operating speed range is narrow near synchronous speed, which is different from the cage-type induction motors. On the other hand, the stator reactive power of the DFIM can be controlled by controlling the rotor d-axis current, so power factor can be controlled.

Fig. 1 shows the DFIM drive system fed by the back-to-back PWM converters. As expected from Fig. 1, if there exists the source voltage unbalance, it influences the

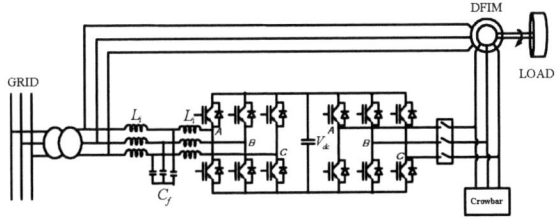

Fig. 1. Configuration of DFIM systems

DFIM operation directly, which causes the torque pulsation. According to the NEMA and IEEE standard, the system is required to keep operation if the variation of the phase voltage is within ±10% or the voltage unbalance factor is less than 3% [7][8]. Even though the source voltage condition is not exceeded, the DFIM causes significant torque ripples due to the source voltage unbalance. This torque pulsation results in mechanical stresses of the drive train and acoustic noise [9].

Recently, the operation of DFIG (doubly-fed induction generator) under unbalanced source voltage has been investigated. In [6], the torque ripple can be reduced, however, it is difficult to increase the bandwidth of the control system since the controlled quantity has double the synchronous frequency. Also, another control method was presented which can suppress the torque pulsation of the DFIG by eliminating the stator reactive power ripple [10] [11].

In this paper, a torque ripple elimination method under unbalanced source voltage is investigated in view of the motor drives. The torque ripple can be suppressed by eliminating the reactive power ripple components. Also, a smooth starting of the DFIM is shown with help of the crowbar. The proposed control method is verified by simulation results for 2[MW] DFIM using PSCAD / EMTDC.

II. MODELING AND CONTROL OF DFIM

For high-performance drives of the DFIM, a stator-flux oriented vector control is usually employed [12]. In this section, the modeling and control method of the DFIM will be described.

978-1-4244-0644-9/07/$25.00 ©2007 IEEE

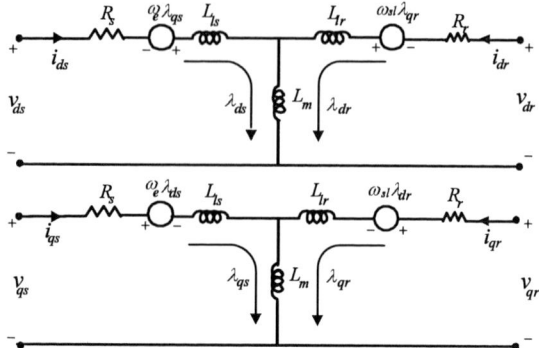

Fig. 2. d-q equivalent circuits of DFIM.

A. DFIM modeling

Fig. 2 shows the d-q equivalent circuits of the DFIM. Under stator flux-oriented vector control, the fluxes, currents and voltages can be expressed as [13]

$$\lambda_{dqs} = L_s i_{dqs} + L_m i_{dqr} \tag{1}$$

$$\lambda_{dqr} = L_r i_{dqr} + L_m i_{dqs} \tag{2}$$

$$v_{dqs} = R_s i_{dqs} + p\lambda_{dqs} + j\omega_e \lambda_{dqs} \tag{3}$$

$$v_{dqr} = R_r i_{dqr} + p\lambda_{dqr} + j\omega_{sl} \lambda_{dqr} \tag{4}$$

where

L_m : Magnetizing inductance
L_s : Stator self-inductance
L_r : Rotor self-inductance
λ_{dqs} : Stator d-q axis flux linkage
λ_{dqr} : Rotor d-q axis flux linkage
i_{dqs}, i_{dqr} : Stator and rotor d-q axis currents
ω_e, ω_{sl} : Source and slip angular frequencies
p : differential operator

where the stator related terms are transformed in a synchronous reference frame using the stator flux angle, which is calculated as

$$\lambda_{dqs}^s = \int (v_{dqs}^s - R_s i_{dqs}^s) dt \tag{5}$$

$$\theta_e = \tan^{-1} \frac{\lambda_{qs}^s}{\lambda_{ds}^s} \tag{6}$$

where the superscript 's' indicates quantities in the stationary reference frame.

On the contrary, the rotor related terms are transformed using the slip angle θ_{sl} as

$$\theta_{sl} = \theta_e - \theta_r \tag{7}$$

where θ_r is the rotor position angle.

The developed torque is given by

$$T_e = \frac{3}{2}\frac{P}{2}\frac{L_m}{L_s}(\lambda_{ds} i_{qr} - \lambda_{qs} i_{dr}) \tag{8}$$

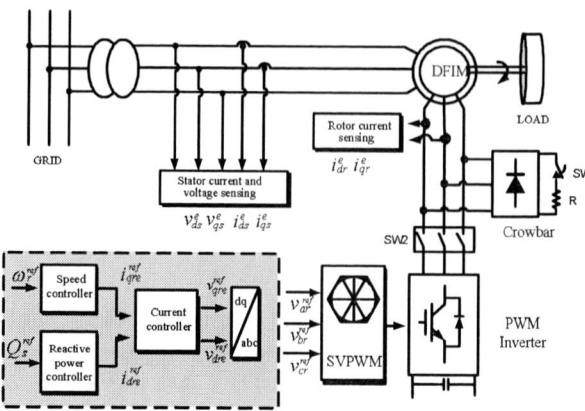

Fig. 3. Control block diagram of DFIM.

B. Control of DFIM

From (1)-(6), the stator reactive power can be expressed as [12]

$$Q_s = \frac{3}{2}\frac{L_m}{L_s} v_{qs}(i_{ms} - i_{dr}) \tag{9}$$

where i_{ms} is the magnetizing current.

The excitation current of the DFIM can be supplied from the stator and/or the rotor. So, the stator side can be operated at unity power factor if the excitation is supplied completely through the d-axis rotor current ($Q_s = 0$). The torque of the DFIM is controlled by the q-axis rotor current of which reference comes from the speed controller output. Fig. 3 shows the control block diagram of the DFIM.

III. STARTING OF DFIM

In the DFIM, the rotor induced voltage is proportional to the slip. When the rotor is connected to the source through the PWM converter for starting, the rotor induced voltage equal to the rated value is applied to the converter. If the DFIM operates at a limited speed range around the synchronous speed, for example, ±0.3 [p.u.], the voltage rating of the converter can be reduced to 0.3 [p.u.]. In case of the limited-speed drive application, the cost of the converter is much decreased. For this reason, a special starting circuit is required to accelerate the DFIM speed higher than 0.7[p.u.] of the synchronous speed.

Conventionally, starting is done by inserting the discrete resistances in the rotor circuit to achieve the same purpose. Nevertheless, this method suffers from several disadvantages such as (i) inclusion of mechanical switches with their associated problems, (ii) discontinuity of starting torque, and (iii) abrupt variation of the supply current [14]. Other methods have been employed to start-up the wound-rotor motors using reactors or saturators. The cost of these methods is higher than the conventional discrete resistance starting which is also

1472

quite expensive due to its numerous components and switching control circuit [15].

The liquid rheostat is the preferred method for starting large loads, primarily because it offers step-less transition through the acceleration period. This is an advantage, particularly when torque pulsations are a concern. A liquid rheostat may be used to start more than one motor. Starting can be designed for manual or automatic operation. A separate shorting contactor should be used with each motor [16]

In Fig. 3, for starting, the SW1 is turned on to make the rotor connected to the crowbar circuit while the SW2 is turned off. At higher speed than 0.7[p.u.], the SW1 is turned off and the SW2 is turned on, so the back-to-back PWM converter starts to operate normally. If the motor speed decreases or increases beyond 30% of the rated speed, the crowbar is turned on, then the rotor winding is short-circuited through the crowbar and the rotor current flows through this. There are several advantages for this method (i) no mechanical switches are used, and (ii) the crow bar can be used for protection in case of faults.

III. CONTROL OF DFIM UNDER UNBALANCED SOURCE VOLTAGE

The control scheme of doubly-fed induction generators at unbalanced source voltage has been presented in [10]. The same derivation is developed below.

A. DFIM torque

Under unbalance source voltage, dynamic equations of the DFIM in (1)-(4) are expressed with the superscripts of "p" for the positive sequence component. For the negative sequence component, they are modified with the negative source angular frequency ($-\omega_e$) and the superscripts of "n" such as

$$\lambda_{dqs}^n = L_s I_{dqs}^n + L_m I_{dqr}^n \tag{10}$$

$$\lambda_{dqr}^n = L_r I_{dqr}^n + L_m I_{dqs}^n \tag{11}$$

$$V_{dqs}^n = R_s I_{dqs}^n + \frac{d}{dt}\lambda_{dqs}^n + j(-\omega_e)\lambda_{dqs}^n \tag{12}$$

$$V_{dqr}^n = R_r I_{dqr}^n + \frac{d}{dt}\lambda_{dqr}^n + j(-\omega_e - \omega_r)\lambda_{dqr}^n \tag{13}$$

The total apparent power flowing into the DFIM can be expressed as

$$S_T = 1.5(V_{dqs}^s I_{dqs}^{s*} + V_{dqr}^s I_{dqr}^{s*}) \tag{14}$$

where

$$V_{dqr}^s = e^{j(\omega_e - \omega_r)t}V_{dqr}^p + e^{j(-\omega_e - \omega_r)t}V_{dqr}^n$$

$$I_{dqr}^s = e^{j(\omega_e - \omega_r)t}I_{dqr}^p + e^{j(-\omega_e - \omega_r)t}I_{dqr}^n$$

where ω_r is the rotor speed.

Taking the real part of (14) and dividing it by the mechanical speed, the instantaneous torque is obtained as [13]

$$T_e(t) = T_{e0} + T_{ec2}\cos(2\omega_e t) + T_{es2}\sin(2\omega_e t) \tag{15}$$

Where

$$T_{e0} = 1.5L_m(i_{qs}^p i_{dr}^p + i_{qs}^n i_{dr}^n) \tag{16}$$

$$T_{ec2} = 1.5L_m(i_{qs}^p i_{dr}^n + i_{qs}^n i_{dr}^p) \tag{17}$$

$$T_{es2} = 1.5L_m(i_{qs}^p i_{qr}^n - i_{qs}^n i_{qr}^p) \tag{18}$$

where T_{e0} is the average torque, T_{ec2} and T_{es2} are the magnitude of the torque ripples.

These torque ripple components can be suppressed by controlling the negative sequence components of the d-q axis rotor currents. However, they can also be eliminated through the reactive power control, which will be described below.

B. DFIM power

Under unbalanced source voltage, the stator apparent power of the DFIM can be expressed in terms of the positive and negative sequence components as [10], [17]

$$S_s = 1.5(V_{dqs}^s I_{dqs}^{s*}) \tag{19}$$

where

$$v_{dqs}^s = e^{j\omega_e t}v_{dqs}^p + e^{j(-\omega_e)t}v_{dqs}^n$$

$$i_{dqs}^s = e^{j\omega_e t}i_{dqs}^p + e^{j(-\omega_e)t}i_{dqs}^n .$$

From (19), the instantaneous active power $p_s(t)$ and reactive power $q_s(t)$ can be expressed as

$$p_s(t) = P_{s0} + P_{sc2}\cos(2\omega_e t) + P_{ss2}\sin(2\omega_e t) \tag{20}$$

$$q_s(t) = Q_{s0} + Q_{sc2}\cos(2\omega_e t) + Q_{ss2}\sin(2\omega_e t) \tag{21}$$

where

$$P_{s0} = 1.5(v_{ds}^p i_{ds}^p + v_{qs}^p i_{qs}^p + v_{ds}^n i_{ds}^n + v_{qs}^n i_{qs}^n)$$

$$P_{sc2} = 1.5(v_{ds}^p i_{ds}^n + v_{qs}^p i_{qs}^n + v_{ds}^n i_{ds}^p + v_{qs}^n i_{qs}^p)$$

$$P_{ss2} = 1.5(v_{qs}^n i_{ds}^p - v_{ds}^n i_{qs}^p - v_{qs}^p i_{ds}^n + v_{ds}^p i_{qs}^n)$$

$$Q_{s0} = 1.5(v_{qs}^p i_{ds}^p - v_{ds}^p i_{qs}^p + v_{qs}^n i_{ds}^n - v_{ds}^n i_{qs}^n)$$

$$Q_{sc2} = 1.5(v_{qs}^p i_{ds}^n - v_{ds}^p i_{qs}^n + v_{qs}^n i_{ds}^p - v_{ds}^n i_{qs}^p)$$

$$Q_{ss2} = 1.5(v_{ds}^p i_{ds}^n + v_{qs}^p i_{qs}^n - v_{ds}^n i_{ds}^p - v_{qs}^n i_{qs}^p)$$

In stator flux-oriented vector control, the positive and negative sequence components of the stator d-axis voltage are zeros. Hence, the coefficients of the active and reactive power ripple components are reduced to

$$P_{sc2} = 1.5(v_{qs}^p i_{qs}^n + v_{qs}^n i_{qs}^p) \tag{22}$$

$$P_{ss2} = 1.5(v_{qs}^n i_{ds}^p - v_{qs}^p i_{ds}^n) \tag{23}$$

$$Q_{sc2} = 1.5(v_{qs}^p i_{ds}^n + v_{qs}^n i_{ds}^p) \tag{24}$$

$$Q_{ss2} = 1.5(v_{qs}^p i_{qs}^n - v_{qs}^n i_{qs}^p) \tag{25}$$

On the other hand, substituting (1)-(4) with the superscripts of "p" and (10)-(13) in (22)-(25) and neglecting the differential terms of the current in steady state,

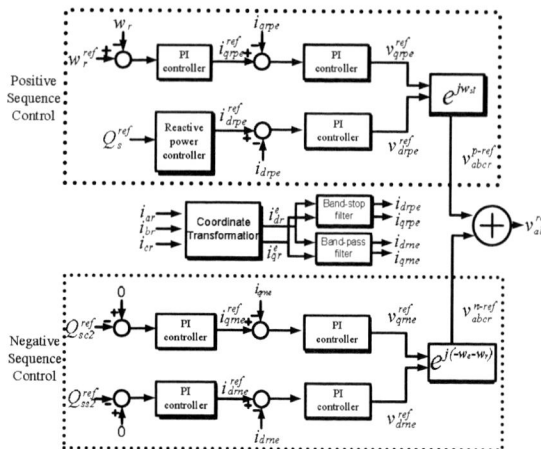

Fig. 4. Control block diagram of DFIM under unbalanced source voltage.

$$P_{sc2} = 1.5\left\{2(R_s + \frac{dL_s}{dt})i_{qs}^p i_{qs}^n - \omega_e L_m(i_{qs}^p i_{dr}^n - i_{qs}^n i_{dr}^p)\right\} \quad (26)$$

$$P_{ss2} = -1.5\{2\omega_e L_s i_{qs}^p i_{qs}^n + \omega_e L_m(i_{qs}^p i_{qr}^n + i_{qs}^n i_{qr}^p)\} \quad (27)$$

$$Q_{sc2} = -1.5\omega_e L_m(i_{qs}^p i_{qr}^n - i_{qs}^n i_{qr}^p) \quad (28)$$

$$Q_{ss2} = 1.5\omega_e L_m(i_{qs}^p i_{dr}^n + i_{qs}^n i_{dr}^p) \quad (29)$$

It is important to note that the reactive power ripples in (28) and (29) have the same components as in the torque ripples (18) and (17), respectively. Therefore, if the reactive power ripples can be controlled to be zero, the machine torque ripples will be eliminated.

Fig. 4 shows the control block diagram of the DFIM under source voltage unbalance. For the reactive power ripple elimination, Q_{cs2} and Q_{ss2} controllers are added with inner control loops of the negative sequence components of the rotor current.

IV. SIMULATION RESULTS

Simulation is carried out to show the validity of the proposed control algorithm, where PSCAD software is used. The parameters of the DFIM used for the simulation is given in Table I in Appendix.

The switching frequency of the power converter is 2[kHz], DC-link voltage is 1,200[V], and the drop of a phase voltage is −10%.

Fig. 5 shows the starting performance of the DFIM. The connection of the crowbar is controlled for starting and disconnected at 0.5[sec] and the converter is connected to the DFIM rotor. From the top in the figure, the motor speed, the motor torque, three-phase rotor voltages, the dq-axis rotor currents, the dq-axis stator currents, and the stator active and reactive power are shown. At 0.5[sec], the stator reactive power control is activated, so the d-axis rotor current is supplied to excite the motor instead of the stator.

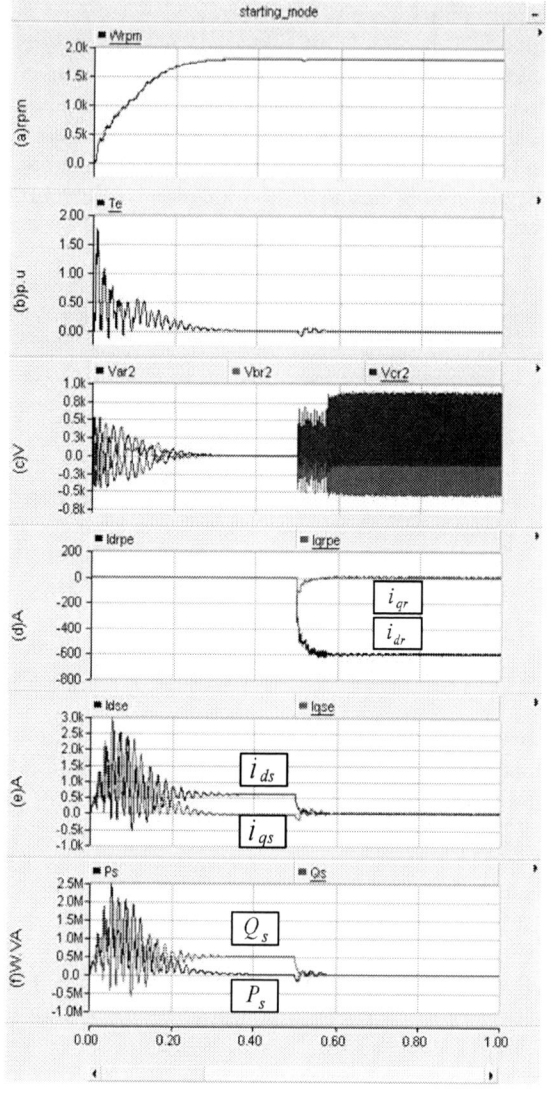

Fig. 5. DFIM performance during starting.
(a) rotating speed   (b) torque
(c) rotor-phase voltage   (d) dq-axis rotor currents
(e) dq-axis stator currents
(f) stator active and reactive power

Fig. 6 shows the DFIM performance at source voltage unbalance. (a) the high torque pulsation appears during the period of the source voltage unbalance between 2.5[sec] and 3.2[sec], of which magnitude is 3% of the rated torque. (b) the dq-axis rotor positive currents, (b) the dq-axis rotor negative currents, (c) the motor speed, (d) the stator active power, and (e) the stator reactive power are shown. Due to the unbalance, the rotor negative currents occur, which cause the power and torque pulsations.

Fig. 7 shows the DFIM performance with the torque ripple elimination control, which corresponds to Fig. 6. the torque in which the ripple components have been almost eliminated by the control. The reactive power ripples have been eliminated in (e), however, the active power ripples are remained.

1474

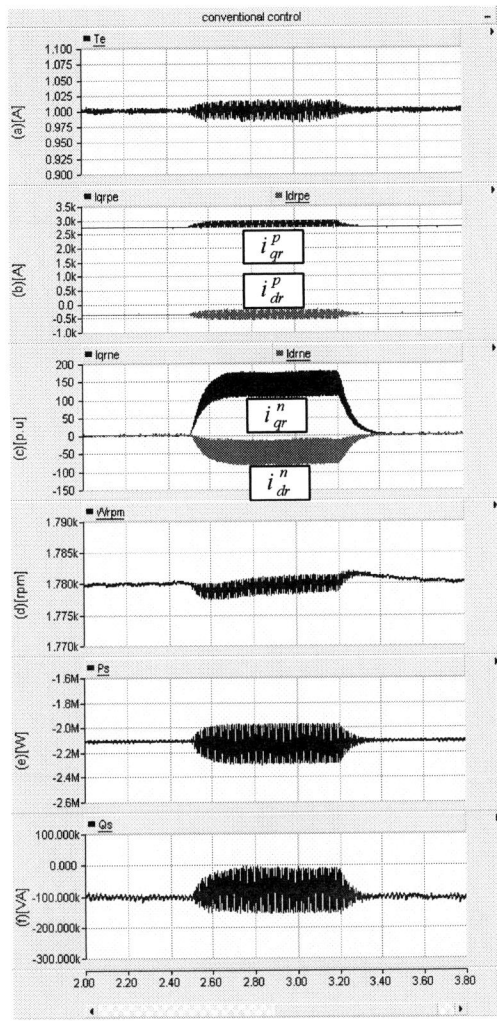

Fig. 6. DFIM performance under unbalanced source voltage.
  (a)   dq-axis rotor positive currents
  (b)   dq-axis rotor negative currents
  (c)   rotating speed
  (d)   stator active power
  (e)   stator reactive power

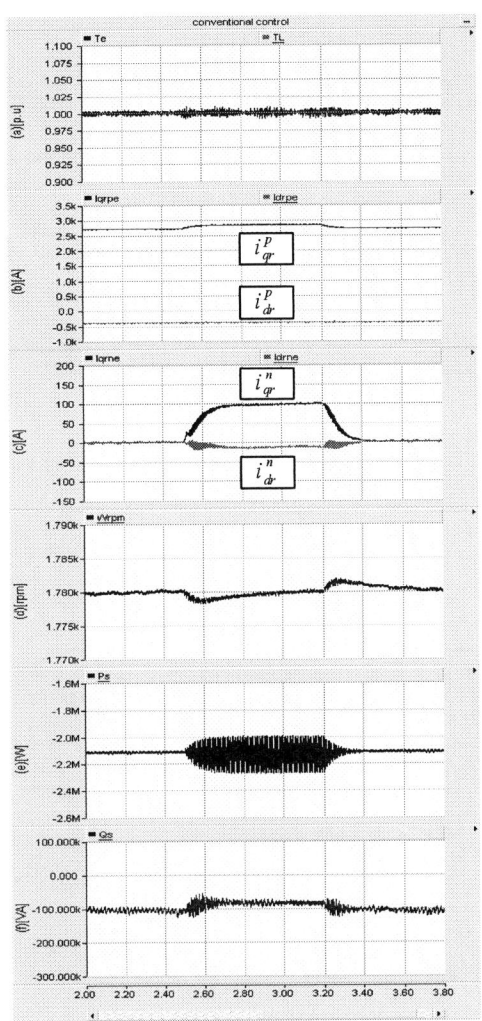

Fig. 7. DFIM performance under unbalanced source voltage.
  (a)   dq-axis rotor positive currents
  (b)   dq-axis rotor negative currents
  (c)   rotating speed
  (d)   stator active power
  (e)   stator reactive power

VI. Conclusions.

In this paper, a control scheme of the DFIM has been proposed at source voltage unbalance. To eliminate the torque ripples of the DFIM under source voltage unbalance, the reactive power components at double the source voltage frequency are cancelled by controlling the negative sequence current component. For applications of narrow speed operating range of high power DFIM drives, a starting method using the crowbar system has been used to reduce the power rating of the converter. The proposed control algorithm has been validated with simulation results for 2[MW] DFIM systems using PSCAD/EMTDC.

APPENDIX

The parameters of the 2[MW] DFIM used for the simulation are listed in Table I.

Table I.   Parameters of DFIM

Parameters	Value
Rated power	2[MW]
Rated line voltage	690[V]
Stator resistance	0.00488 [p.u.]
Rotor resistance	0.00549 [p.u.]
Equivalent iron loss resistance	0.018 [p.u.]
Stator leakage inductance	0.09241 [p.u.]
Rotor leakage inductance	0.09955 [p.u.]
Mutual inductance	3.95279 [p.u.]
Number of poles	4
Rated frequency	60[Hz]
Moment of inertia	4800 kg.m^2

References

[1] W. Leonhard, *Control Electrical Drives*, Springer Verlag, 1997

[2] A. Bocquel and J. Janning,"Analysis of a 300MW variable speed drive for pump-storage plant applications," *Power Electronics and Applications, 2005 European Conf. on*, pp. 11-14. Sept. 2005.

[3] S. Drid, M. S. Nait-Said, and M. Tadjine, "Double flux oriented control for the doubly fed induction motor," *Electric Power Components and Systems*, vol. 33, no. 10, Oct. 2005.

[4] C. Batlle, A. Doria-Cerezo, and R. Ortega, "Power flow control of a doubly-fed induction machine coupled to a flywheel," *IEEE Int. Conf. on Control Applications*, vol. 2, pp. 1645-1650, Sept. 2004.

[5] B. K. Bose, *Power Electronics and AC drives*, Prentice-Hall,1986

[6] T. Brekken and N. Mohan, "A novel doubly-fed induction wind generator control scheme for reactive power control and torque pulsation compensation under unbalanced grid voltage conditions," *IEEE PESC Conf. Proc.*, vol. 2, pp. 760 - 764, 2003.

[7] NEMA Standards, "Application guide for AC adjustable speed drive systems,"

[8] IEEE Standard, "IEEE guide for AC motor protection," IEEE Standard C 37.96-2000

[9] E. Muljadi, T. Batan, D. Yildirim, and C.P. Butterfield, "Understanding the unbalanced-voltage problem in wind turbine generation," *IEEE IAS Conf. Proc.*, pp. 1359-1365, 1999.

[10] J.-I. Jang, Y.-S. Kim, and D.-C. Lee, "Active and reactive power control of DFIG for wind energy conversion under unbalanced grid voltage," *IPEMC 2006proc., Shanghai*, vol. 3, pp. 1487-1491, Aug. 2006.

[11] Lie Xu and Yi Wang,"Dynamic modeling and control of DFIG based wind turbines under unbalanced network conditions," *IEEE Trans on Power Systems*, vol. 22, no. 1, Feb. 2007.

[12] Longya Xu, Li Zhen, and Eel-Hwan Kim, "Field-orientation control of a doubly excited brushless reluctance machine," *IEEE Trans on Industry Applications*, vol. 34, no. 1, pp. 148-154, Jan-Feb.1998.

[13] C. M. Ong, *Dynamic simulation of electric machinery*, Prentice Hall, Inc, pp.167–258, 1998.

[14] M. A. Abodel-Halim, M. A. Badr, and A. I. Alolah, "Smooth starting of slip ring induction motors," *IEEE Trans on Energy Conversion*, vol. 12, no. 4, pp. 317-322, Dec. 1997.

[15] L. Yu, "Constant starting torque control of woundrotor induction motors"*, IEEE Trans on Power Apparatus and Systems*, vol. 89, no. 4, pp. 646-651, April. 1970.

[16] Nathan Schachter, "Experience with synchronous and slip ring induction motors driving cement mills," *Cement Industry Technical Conference proc., IEEE/PCA*, pp. 75-96, May. 1998.

[17] H. S. Song and K. Nam, "Dual current control scheme for PWM converter under unbalanced input voltage conditions," *IEEE Trans. on Ind. Electron.*, vol. 46, no. 5, pp. 953-959, Oct., 1999.

Online H_∞ Speed Control of Sensorless Induction Motors with Rotor Resistance Estimation

Peda V Medagam
Embedded Control Systems Laboratory
Southern Illinois University, Carbondale
medagam@siu.edu

Farzad Pourboghrat
Electrical & Computer Engineering Department
Southern Illinois University, Carbondale
pour@siu.edu

Abstract — **This paper proposes an online technique for the H_∞ control of speed and flux norm of current-fed induction motors (IM). Integrals of the speed and flux norm tracking errors are considered as elements of the state variables. A robust H_∞ optimal control strategy for this problem is determined by solving the algebraic Riccati equation (ARE), using multilayer recurrent neural networks. The proposed controller allows for the simultaneous and independent control of both speed and flux norms of the induction motors. The control implementation involves estimating the rotor flux, rotor resistance, and speed of the induction motor, using continuous-time extended Kalman filter (EKF). The simulation results show the effectiveness of the proposed controller even in the presence of rotor resistance and load torque disturbances.**

Index Terms — **Induction motor; Online H_∞ control; Neural networks; Extended Kalman filter; Speed and rotor resistance estimation**

Nomenclature

$i_{s\alpha}, i_{s\beta}$	-	α-β components of stator current
$u_{s\alpha}, u_{s\beta}$	-	α-β components of stator voltage
$\Phi_{r\alpha}, \Phi_{r\beta}$	-	α-β components of rotor flux
ω_m	-	Mechanical speed of the motor
R_r, R_s	-	Rotor resistance and stator resistance
L_r, L_s	-	Rotor and stator inductances
M	-	Mutual inductance
P	-	Number of pole pairs
J	-	Moment of inertia
t_l	-	Load torque
k_T	-	Electromagnetic torque constant
$\Phi = \sqrt{\phi_{r\alpha}^2 + \phi_{r\beta}^2}$	-	Flux norm

I. Introduction

Induction motors (IM) are very popular in the industry due to their efficiency, robustness, low cost, and ability to operate in wide torque and velocity ranges. But their models are relatively more complicated and difficult to control as compared to those of DC motors. In recent years many advanced techniques like nonlinear control, adaptive control and sliding mode control techniques have been proposed to control the induction motors. With the advent of fast digital signal processors (DSP), micro controllers and power electronics devices, the implementation of these advanced control techniques has become much easier.

Optimal control is an aspect of control theory, which is widely used because of guaranteed closed loop performance [1]. Optimal control techniques have been successfully applied to induction motors [2], [3], to minimize the stored magnetic energy and coil losses, while satisfying torque tracking control objective. Marino *et al.* [4] developed an offline linear quadratic control for induction motors based on the extended state model to track flux and torque. For general parameter perturbations, the LQG optimal control loses its robust properties and stability margins. H_∞ optimal control, however, is one of the robust design approaches that provides the optimal control with disturbance rejection. Chaiverini *et al.* [5] proposed an offline H_∞ control for induction motors by solving the algebraic Riccati equation (ARE) for the linearized system model. The controller is implemented by estimating the rotor flux components using sliding mode observer and speed measurement. However, speed sensors require a careful mounting on the shaft of the motor, which reduces reliability and increases the cost of the drive.

Kwan *et al.* [6] developed a neural network robust controller based on back-stepping techniques. However, this controller is not easy to implement because it requires the measurement of the states of the induction motors. To circumvent this problem, Kenne *et al.* [9] proposed a robust flux and speed estimator for induction motors, along with an online estimator for rotor resistance. Similarly, in [10], a robust vector control technique was reported for induction motors, with a full order observer, using both the gain scheduled control and linear matrix inequality approach.

This paper presents the design of a sensorless field-oriented robust H_∞ control for the three-phase induction motors with unknown rotor resistance and varying load torque. To achieve our control objectives, the induction motor model is decomposed into two linear decoupled systems using feedback linearization technique [7], where rotor speed and flux norm are considered as control inputs. Moreover, by introducing reference values for the rotor speed and flux norm, the induction motor model is expressed in terms of tracking errors dynamics. Thus the control problem is formulated as a robust optimal regulation problem. The robust control can be designed by solving the algebraic Riccati equation (ARE) using multilayer recurrent neural networks proposed in [8]. The implementation of this control needs the knowledge of the rotor flux, rotor speed, and rotor resistance, which are, in general, difficult to measure. To overcome this problem, extended Kalman filters are designed for the estimation of states and parameters of the induction motor.

The paper is organized as follows. Section II presents IM modeling, while section III presents the robust control strategy. Extended Kalman filter estimation method and simulation results are given in sections IV and V, respectively. Finally, conclusions are summarized in section VI.

II. IM MODELING AND STATEMENT OF PROBLEM

The dynamic model of a three-phase induction motor in a stationary reference frame, by choosing the stator current components, angular velocity and rotor flux components as state variables, is in the following form

$$\dot{i}_{s\alpha} = -\left(\frac{R_s}{\sigma L_s} + \frac{R_r M^2}{\sigma L_s L_r^2}\right)i_{s\alpha} + \frac{MR_r}{\sigma L_s L_r^2}\phi_{r\alpha} + \frac{pM}{\sigma L_s L_r}\omega_m\phi_{r\beta} + \frac{u_{s\alpha}}{\sigma L_s}$$

$$\dot{i}_{s\beta} = -\left(\frac{R_s}{\sigma L_s} + \frac{R_r M^2}{\sigma L_s L_r^2}\right)i_{s\beta} + \frac{MR_r}{\sigma L_s L_r^2}\phi_{r\beta} - \frac{pM}{\sigma L_s L_r}\omega_m\phi_{r\alpha} + \frac{u_{s\beta}}{\sigma L_s}$$

$$\dot{\phi}_{r\alpha} = -\frac{Rr}{Lr}\phi_{r\alpha} - p\omega_m\phi_{r\beta} + M\frac{R_r}{L_r}i_{s\alpha} \qquad (2.1)$$

$$\dot{\phi}_{r\beta} = p\omega_m\phi_{r\alpha} - \frac{Rr}{Lr}\phi_{r\beta} + M\frac{R_r}{L_r}i_{s\beta}$$

$$\dot{\omega}_m = \frac{k_T}{J}\left(\phi_{r\alpha}i_{s\beta} - \phi_{r\beta}i_{s\alpha}\right) - \frac{t_L}{J}$$

where $\sigma = 1 - \dfrac{M^2}{L_s L_r}$ and $k_T = \dfrac{pM}{L_r}$. For the current source inverter (CSI) control design, only the last three equations of (2.1) are required for control derivation. The aim of our control design is to track the speed ω and flux norm $\Phi = \sqrt{\phi_{r\alpha}^2 + \phi_{r\beta}^2}$ to their reference values of ω_r and Φ_r, respectively

Define the flux norm tracking error ε_Φ and its integral ξ, as

$$\varepsilon_\Phi = \Phi_r - \Phi \qquad (2.2)$$

$$\frac{d\xi}{dt} = \varepsilon_\Phi \qquad (2.3)$$

Also, define the speed tracking error ε_ω and its integral η, as

$$\varepsilon_\omega = \omega_r - \omega_m \qquad (2.4)$$

$$\frac{d\eta}{dt} = \varepsilon_\omega \qquad (2.5)$$

The steady state flux norm tracking error equals zero when $\dfrac{d\xi}{\partial t} \to 0$ and it can be represented in terms of induction motor parameters using feedback linearization. To achieve this, define the decoupling controls as

$$u_\Phi = \frac{1}{\Phi}\left(\phi_{r\alpha}i_{s\alpha} + \phi_{r\beta}i_{s\beta}\right) \qquad (2.6)$$

$$u_\omega = \left(\phi_{r\alpha}i_{s\beta} - \phi_{r\beta}i_{s\alpha}\right) \qquad (2.7)$$

Now, from equation (2.1) and (2.2), we can write

$$\begin{aligned}\frac{d\varepsilon_\Phi}{dt} &= \frac{d\Phi_r}{dt} - \frac{1}{\Phi}\left(\phi_{r\alpha}\frac{d\phi_{r\alpha}}{dt} + \phi_{r\beta}\frac{d\phi_{r\beta}}{dt}\right) \\ &= \frac{d\Phi_r}{dt} - \left(-\frac{R_r}{L_r}\Phi + \frac{MR_r}{L_r}\left(\frac{\phi_{r\alpha}i_{s\alpha} + \phi_{r\beta}i_{s\beta}}{\Phi}\right)\right) \\ &= -\frac{R_r}{L_r}\varepsilon_\Phi + \frac{d\Phi_r}{dt} + \frac{R_r}{L_r}\Phi_r - \frac{MR_r}{L_r}u_\Phi \end{aligned} \qquad (2.8)$$

where $u_\Phi = \dfrac{1}{\Phi}\left(\phi_{r\alpha}i_{s\alpha} + \phi_{r\beta}i_{s\beta}\right)$, from (2.6), has been used as a new control input, [1].

Similarly the steady state error of speed is defined in terms of induction motor parameters using equations (2.1) and (2.4), as

$$\begin{aligned}\frac{d\varepsilon_\omega}{dt} &= \frac{d\omega_r}{dt} - \frac{d\omega_m}{dt} \\ &= \frac{d\omega_r}{dt} - \frac{k_T}{J}\left(\phi_{r\alpha}i_{s\beta} - \phi_{r\beta}i_{s\alpha}\right) + \frac{t_L}{J} \\ &= \frac{d\omega_r}{dt} + \frac{t_L}{J} - \frac{k_T}{J}u_\omega \end{aligned} \qquad (2.9)$$

where $u_\omega = \left(\phi_{r\alpha}i_{s\beta} - \phi_{r\beta}i_{s\alpha}\right)$, from (2.7), has been used as a new control input, [1].

Equations (2.8) and (2.3) represent the electrical subsystem model and for robust control design it can be written as

$$\dot{x}_e = A_e x_e + B_e u_\Phi + K_e w_e \qquad (2.10)$$

where $x_e = \begin{bmatrix}\varepsilon_\Phi \\ \xi\end{bmatrix}, A_e = \begin{bmatrix}-\dfrac{R_r}{L_r} & 0 \\ 1 & 0\end{bmatrix}, B_e = \begin{bmatrix}-\dfrac{MR_r}{L_r} \\ 0\end{bmatrix}, K_e = \begin{bmatrix}1 \\ 0\end{bmatrix}$

and $w_e = \dfrac{d\Phi_r}{dt} + \dfrac{R_r}{L_r}\Phi_r$.

Equations (2.9) and (2.5) represent the mechanical subsystem model and for robust control design it can be rewritten as

$$\dot{x}_m = A_m x_m + B_m u_\omega + K_m w_m \qquad (2.11)$$

where $x_m = \begin{bmatrix}\varepsilon_\omega \\ \eta\end{bmatrix}, A_m = \begin{bmatrix}0 & 0 \\ 1 & 0\end{bmatrix}, B_m = \begin{bmatrix}-\dfrac{k_T}{J} \\ 0\end{bmatrix}, K_m = \begin{bmatrix}1 & \dfrac{1}{J} \\ 0 & 0\end{bmatrix}$ and

$$w_m = \begin{bmatrix}\dfrac{d\omega_r}{dt} \\ t_L\end{bmatrix}$$

Equations (2.10) and (2.11) are two decoupled systems with the control inputs u_Φ, u_ω and disturbances w_e, w_m. The control design for these two decoupled systems is explained in the following section.

III. H_∞ CONTROL

This section shortly presents the proposed online robust H_∞ optimal control approach for the sensorless induction motor. The corresponding neural solution to algebraic Riccati equation is investigated in detail in Wang *et al.* [8].

Consider the following controllable linear system

$$\dot{x} = Ax(t) + Bu(t) + B_w w(t) \qquad (3.1)$$
$$y = Cx(t)$$

where $x \in \Re^n$ is the system state, $u \in \Re^m$ is the control input, $w \in \Re^r$ is the disturbance signal and $y \in \Re^p$ is the system output. Also, consider a performance measuring output

$$z = C_z x + D_z u \qquad (3.2)$$

where $R = D_z^T D_z$ and $Q = C_z^T C_z$ are positive definite and positive semi-definite matrices, respectively.

The aim is to find an H_∞ control for system (3.1), that is a state feedback control law

$$u(t) = -Kx(t) \qquad (3.3)$$

such that the closed loop system is asymptotically stable and its transfer matrix from w to z has an L_2 gain less than or equal to a small positive number $\gamma > 0$. The control exists if and only if there exists a positive definite and symmetric matrix solution $P > 0$ to the generalized algebraic Riccati equation

$$A^T P + PA + Q + P \left(\frac{1}{\gamma^2} B_w B_w^T - BR^{-1}B^T \right) P = 0 \qquad (3.4)$$

The H_∞ control gain, required in (3.3), is then given by $K = R^{-1}B^T P$. It is well known that the solution of the Riccati equation in (3.4) could not be found analytically, except for a very limited number of systems. Therefore, for real-time implementation, one should require an online computation method for Riccati equation. Here, we consider a fast, simple and efficient real-time method for numerical solution of Riccati equation using gradient-type neural networks [8].

As it is well known, the solution of P must be positive definite and symmetric. However, (3.4) has a unique positive–definite and symmetic solution if it has a Cholesky factorization. Therefore, a constraint for positive–definiteness and symmetricity of P is added by requiring,

$$G_1(P,L) = [g_{1,jk}] = LL^T - P = 0, \quad j,k = 1,...,n \qquad (3.5)$$

where $g_{1,jk}$ is the jk^{th} element of the objective function G_1, and L is a Cholesky factor. Also to satisfy (3.4), define

$$G_2(P) = [g_{2,jk}]$$
$$= A^T P + PA + Q + P \left(\frac{1}{\gamma^2} KK^T - BR^{-1}B^T \right) P = 0, \qquad (3.6)$$
$$j,k = 1,...,n$$

where $g_{2,jk}$ is the jk^{th} element of the objective function G_2. To solve P from (3.5) and (3.6), the following Lyapunov energy function is defined [11],

$$E[G_1(P,L), G_2(P)] = \frac{1}{2} \sum_{j=1}^{n} \sum_{k=1}^{n} \left[g^2_{1,jk} + g^2_{2,jk} \right] \qquad (3.7)$$

Then, a matrix-oriented gradient algorithm is developed to find the update rule for P by changing the variables in the direction of the negative gradient of the energy function E to minimize (3.7), as

$$\frac{dP}{dt} = -n_P \frac{\partial E}{\partial P} \qquad (3.8)$$

$$\frac{dL}{dt} = -n_L \frac{\partial E}{\partial L} \qquad (3.9)$$

Therefore, the update law can be given as [8],

$$\dot{P}(t) = -\eta_p \left[P(t)SU(t) + U(t)SP(t) - AU(t) - U(t)A^T - Y(t) \right] \qquad (3.10)$$

$$\dot{L}(t) = -\eta_V Y(t)L(t) \qquad (3.11)$$

$$U(t) = f \left(P(t)SP(t) - A^T P(t) - P(t)A - Q \right) \qquad (3.12)$$

$$Y(t) = f \left(L(t)L(t)^T - P(t) \right) \qquad (3.13)$$

where $S = BR^{-1}B^T - \frac{1}{\gamma^2} B_w B_w^T$, η_p and η_V are positive scalar learning factors, f is activation function, $U(t)$ is input layer, $P(t)$ is output layer, and $L(t)$ and $Y(t)$ are hidden layers of the network.

The online control strategy found by the above method can be implemented with the knowledge of the system states (flux components and speed) and unknown motor parameters (rotor resistance). The only measurable states, however, are current components. Hence, we need to design an observer that estimates the flux components and speed. In addition, we need to design a parameter estimator to estimate the rotor resistance. The unknown load torque, however, is considered as a disturbance and is taken care of using H_∞ control.

An extended Kalman filter is used to estimate the unknown state and parameters of IM, which is explained in the following section.

IV. EXTENDED KALMAN FILTER ESTIMATION

In this section, the continuous time extended Kalman filter design is briefly reviewed and is applied to the induction motor model to estimate the flux components, speed and rotor resistance of the motor. The filter is described as

$$\dot{x} = f(x,u) + w \qquad (4.1)$$
$$y = h(x) + v$$

where $f(x,u)$ is a nonlinear function of states and inputs; $h(x)$ is a function of states; w and v are white noise vectors effecting the state and output equations with covariance matrices Ξ and Θ, respectively. The EKF algorithm is to estimate the unknown states of the system and is given as

$$\dot{\hat{x}} = F\hat{x} + Gu + K_f\left(y - C\hat{x}\right) \tag{4.2}$$

where $F = \left.\dfrac{\partial f(x,u)}{\partial x}\right|_{x=\hat{x}}$, $G = \left.\dfrac{\partial f(x,u)}{\partial u}\right|_{x=\hat{x}}$, $C = \left.\dfrac{\partial h(x)}{\partial x}\right|_{x=\hat{x}}$

and the filter gain matrix K_f is given by

$$K_f = NC^T\Theta^{-1} \tag{4.3}$$

where $N > 0$ is a symmetric positive definite matrix, which satisfies the filter Riccati equation

$$\dot{N} = FN + NF^T + \Xi - NC^T\Theta^{-1}CN$$
$$N(t_0) = N_0 \tag{4.4}$$

where Ξ and Θ are assumed to be, respectively, positive semi-definite and positive definite matrices with appropriate dimensions.

The dynamic model of induction motor described in section II and that of the unknown parameters are augmented first. Then continuous time extended Kalman filter is used for estimating the states of the augmented system. The augmented state variables are selected as $z^T = \left[x_1, x_2, x_3, x_4, x_5, x_6\right] = \left[i_{s\alpha}, i_{s\beta}, \phi_{r\alpha}, \phi_{r\beta}, \omega_m, R_r\right]$. The dynamic model of induction motor augmented with that of rotor resistance is given as

$$\dot{z} = f(z,u) + GW$$
$$Y = Cz + v \tag{4.5}$$

where $f(z,u) = \begin{bmatrix} f_m(x,u) \\ 0 \end{bmatrix}$ such that

$$f = \begin{bmatrix} -\left(\dfrac{R_s}{\sigma L_s} + \dfrac{x_6 M^2}{\sigma L_s L_r^2}\right)x_1 + \dfrac{Mx_6}{\sigma L_s L_r^2}x_3 + \dfrac{pM}{\sigma L_s L_r}x_5 x_4 + \dfrac{u_{s\alpha}}{\sigma L_s} \\[2mm] -\left(\dfrac{R_s}{\sigma L_s} + \dfrac{x_6 M^2}{\sigma L_s L_r^2}\right)x_2 + \dfrac{Mx_6}{\sigma L_s L_r^2}x_4 - \dfrac{pM}{\sigma L_s L_r}x_5 x_3 + \dfrac{u_{s\beta}}{\sigma L_s} \\[2mm] -\dfrac{x_6}{Lr}x_3 - px_5 x_4 + M\dfrac{x_6}{L_r}x_1 \\[2mm] px_5 x_3 - \dfrac{x_6}{Lr}x_4 + M\dfrac{x_6}{L_r}x_2 \\[2mm] \dfrac{k_T}{J}\left(x_3 x_2 - x_4 x_1\right) - \dfrac{t_L}{J} \\[2mm] 0 \end{bmatrix},$$

$$C = \begin{bmatrix} 1 & 0 & 0 & 0 & 0 \\ 0 & 1 & 0 & 0 & 0 \end{bmatrix}, \quad G = I_{6\times6}, \text{ and that } W \in \Re^6 \text{ and}$$

$v \in \Re^2$ are white noise vectors with covariance matrices Ξ and Θ.

The EKF estimator dynamics is then given as

$$\dot{\hat{z}} = F\hat{z} + Gu + K_f\left(Y - C\hat{z}\right) \tag{4.6}$$

where $F = \left.\dfrac{\partial f(z,u)}{\partial z}\right|_{z=\hat{z}}$, $G = \left.\dfrac{\partial f(z,u)}{\partial u}\right|_{z=\hat{z}}$ and these values

are given in the appendix.

Selecting symmetric positive semi-definite and positive definite matrices $\Xi \geq 0$ and $\Theta > 0$, and assuming the system

is observable, the Riccati equation (4.4) can be solved for the covariance matrix $N > 0$ and the time varying gain K_f can be determined from equation (4.3). The motor states and parameters are then estimated using equation (4.6) with the knowledge of K_f and measurement of motor currents. The H_∞ control (3.3) is then calculated, online, for the two decoupled systems (2.10) and (2.11) to achieve the robust control objective for the induction motor.

V. SIMULATION RESULTS

The proposed online neural network based H_∞ control method was applied to the computer simulated induction motor, for the time interval of $t \in [0,10]$ seconds. The specifications of induction motor are given in Table I. The ARE is solved using recurrent neural network explained in section III. The design parameters of the neural networks for both the electrical and mechanical subsystems are selected as $\eta_p=100$, $\eta_V=1000$, and hyperbolic tangent function is used as activation function of the neural network in equations (3.10) and (3.11). The positive definite matrices Q, R of equation (3.4) are selected as $Q=diag[1, 1]$, $R=1$ for both electrical and mechanical decoupled subsystems. The positive scalar γ is selected as 0.13 for electrical and 0.8 for mechanical subsystems. The estimation of flux components, speed and rotor resistance is done using continuous time extended kalman filter presented in section IV. Fig. 1 shows the schematic diagram of the proposed control. The inverter used in this simulation is current control voltage source inverter (CCVSI). All the simulations are carried out using MATLAB/SIMULINK. The simulation results are shown in Fig. 2- Fig.5.

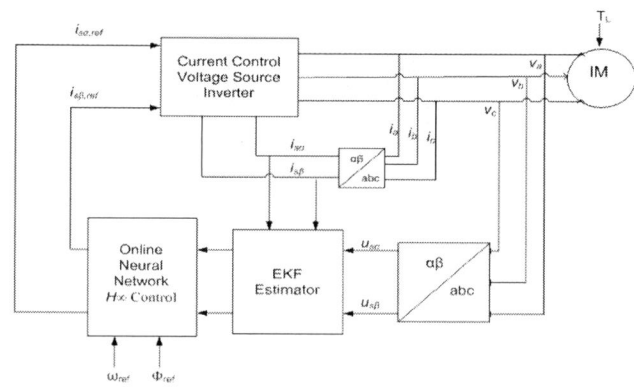

Fig. 1. The overall simulated block diagram of the proposed H_∞ control

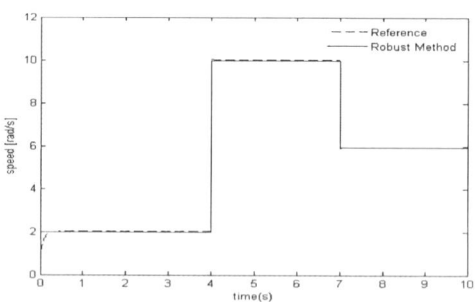

Fig.2- Rotor speed without any disturbances

1480

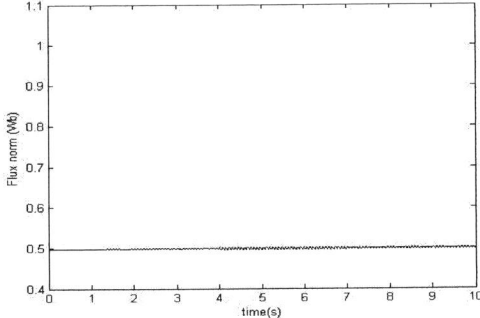

Fig.3. Flux norm without any disturbances

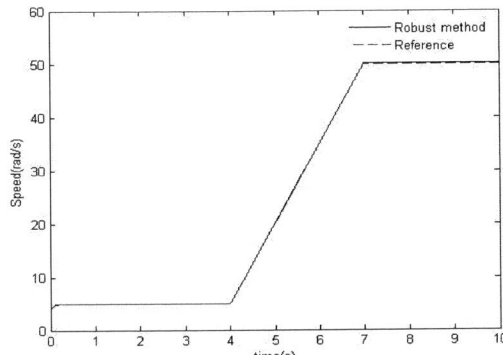

Fig.4. Rotor speed with 50% variation of rotor resistance

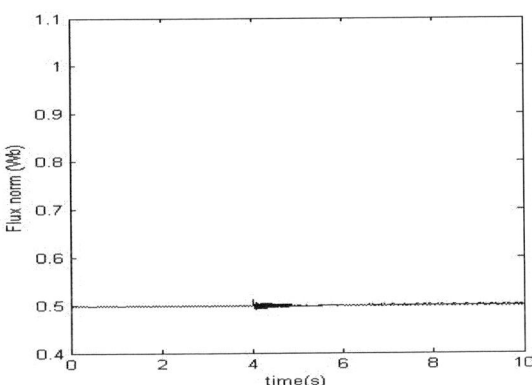

Fig.5. Flux norm with 50% variation of rotor resistance

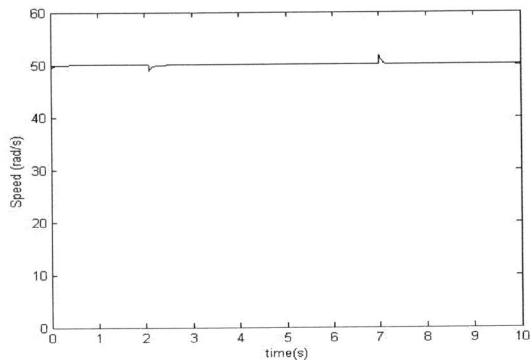

Fig.6. Rotor speed with 50% variation of load torque

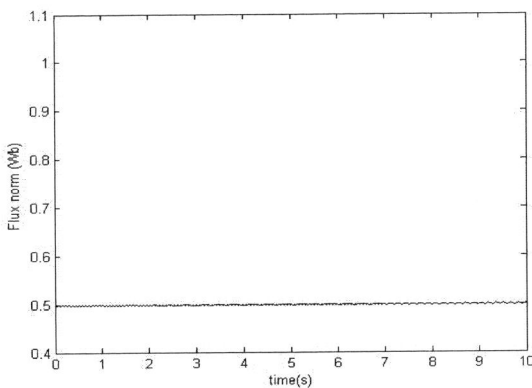

Fig.7. Flux norm with 50% variation of load torque

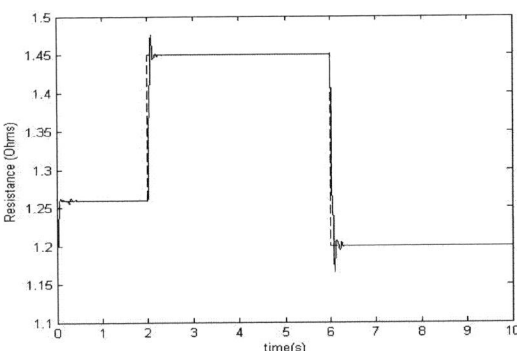

Fig.8. Rotor resistance (dashed line) and estimated resistance (solid line)

Fig.2 shows the motor speed for a step change reference value. It is obvious that the motor speed can track its desired value in a short time for rotor resistance of 1.26Ω and load torque of 8.7 Nm and also when these values change. The motor speed follows the reference trajectory successfully. Fig. 3 shows the flux norm for the reference value of 0.5 Wb. Fig. 4 and Fig. 5 show the speed and flux of the motor respectively with rotor resistance change. Fig 6 and Fig. 7 show the speed and flux of the motor respectively for 50% load torque change. Clearly, the load torque change did not affect the speed and flux.

The continuous time extended Kalman filter (EKF) is used to estimate the flux components, motor speed and rotor resistance. The positive definite matrices Ξ and Θ are selected as $\Xi=diag[1e-6,\ 1e-6,\ 1e-6,\ 1e-2,\ 1e-5]$ and $\Theta=diag[0.09,\ 0.09]$. The initial value of covariance matrix N_0 is selected as identity matrix. Fig.8 shows the actual and estimated values of the rotor resistance.

TABLE I
MODEL PARAMETERS

Model Symbols	Values
R_s	1.05 Ω
R_r	1.26 Ω
L_d	0.149 H
L_q	0.149 H
P	1
M	0.144 H
J	0.00448 Nms^2/rad

It is obvious that EKF successfully estimates the unknown rotor resistance despite its large variations.

VI. CONCLUSION

This paper presents an online robust control law for the current fed induction motor to control the speed and flux norm. The flux and speed errors are considered as new state variables and converted to robust control problem using transformed induction motor model. H_∞ control is found by solving algebraic Riccati equation numerically using recurrent neural networks. The rotor flux components, motor speed and rotor resistance are estimated using continuous time extended Kalman filter. Simulation results demonstrate the effectiveness of the proposed method in spite of rotor resistance and load torque variations.

REFERENCES

[1] Peter Dorato "Linear Quadratic Control, An Introduction", *Krieger publishing company, Florida,* 2000
[2] M. Negam, M. salem, O. Ebrahim "Experimental investigation of high performance induction motor control based on optimal preview controller" *Proc. IEEE Conf. electronics* 2004.
[3] Georges, D, & Canudas de "Nonlinear H_2 and H_∞ optimal controllers for current fed induction motors". *IEEE Trans. Automatic control,* vol. 44, pp. 1430-1435, 1999.

[4] P. Marino, M. Milano, F. Vasca "Linear quadratic state feedback and robust neural network estimator for field oriented controlled induction motor" *IEEE Trans. Industrial Electronics,* Vol. 46, No.1, 1999
[5] S. Chiaverini, Giuseppe Fusco "Bandwidth vs. gains design of H_∞ tracking controllers for current fed induction motors" *Automatica* 2002, pp. 1575-1581.
[6] C.M. Kwan, F.L. Lewis "Robust back stepping control of induction motor using neural networks" *IEEE Trans. Neural networks,* vol. 11, no.5, 2000
[7] W. J. Wang, C. C. Wang " A new composite adaptive speed controller for induction motor based on feedback linearization" *IEEE Trans. Energy Conversion,* Vol. 13, No.1 , 1998
[8] J. Wang, G. Wu "A multilayer recurrent neural network for solving continuous-time algebraic Riccati equations" Neural *Networks,* pp. 939-950, 1998
[9] Kenne, G. Ahmed-Ali, T. Nkwawo, H. L. Lagarrigue, F. "Robust Rotor Flux and Speed Control of Induction Motors Using On-Line Time-Varying Rotor Resistance Adaptation" *Proc. IEEE Conf. Decision and Control ECC,* 2005
[10] M. Hasegawa, S. Furutani, S. Doki, and S. Okuma "Robust Vector Control of Induction Motors Using Full-Order Observer in Consideration of Core Loss" *IEEE Trans. Industrial Electronics,* Vol. 50, pp. 912-919, 2003.
[11] C. L. Lin, C. C. Lai, T. H. Huang, "A Neural Network for Linear Matrix Inequality Problems", *IEEE Trans. Neural Networks,* Vol.11, No.5, pp.1078-1092, 2000.

APPENDIX

$$
F = \begin{bmatrix}
-\left(\dfrac{R_s}{\sigma L_s} + \dfrac{x_6 M^2}{\sigma L_s L_r^2}\right) & 0 & \dfrac{M x_6}{\sigma L_s L_r^2} & \dfrac{pM}{\sigma L_s L_r} x_5 & \dfrac{pM}{\sigma L_s L_r} x_4 & -\dfrac{x_6 M^2}{\sigma L_s L_r^2} \\[2ex]
0 & -\left(\dfrac{R_s}{\sigma L_s} + \dfrac{x_6 M^2}{\sigma L_s L_r^2}\right) & -\dfrac{pM}{\sigma L_s L_r} x_5 & -\dfrac{pM}{\sigma L_s L_r} x_3 & \dfrac{M x_6}{\sigma L_s L_r^2} x_4 & -\dfrac{pM}{\sigma L_s L_r} x_5 x_3 \\[2ex]
M\dfrac{x_6}{L_r} & 0 & -\dfrac{x_6}{Lr} & -p x_5 & -x_4 & -\dfrac{1}{L_r} x_3 + \dfrac{M}{L_r} x_1 \\[2ex]
0 & M\dfrac{x_6}{L_r} & p x_5 & \dfrac{x_6}{Lr} & x_3 & -\dfrac{1}{L_r} x_4 + \dfrac{M}{L_r} x_2 \\[2ex]
\dfrac{k_T}{J} x_3 & -\dfrac{k_T}{J} x_4 & \dfrac{k_T}{J} x_1 & \dfrac{k_T}{J} x_2 & 0 & 0 \\[2ex]
0 & 0 & 0 & 0 & 0 & 0
\end{bmatrix}
$$

$$
G = \begin{bmatrix}
\dfrac{1}{\sigma L_s} & 0 \\[2ex]
0 & \dfrac{1}{\sigma L_s} \\[2ex]
0 & 0 \\
0 & 0 \\
0 & 0 \\
0 & 0
\end{bmatrix}
$$

Analysis and Comparative Study on the Performance between Standard and High Efficiency Induction Machines operating as Self – Excited Induction Generators

B. Sawetsakulanond and V. Kinnaraes

Dept. of Electrical Engineering, King Mongkut's Institute of Technology Ladkrabang, Bangkok 10520, Thailand
Fax 662-3269902 E-mail:Budhapon@hotmail.com, E-mail: kkwijit@kmitl.ac.th

Abstract--**This paper proposes the performance analysis and comparative study between standard and high efficiency squirrel cage self – excited induction generators (SEIG). Consideration of capacitance of SEIG based on equivalent circuit model is given. Testing and performance comparison under dynamic and steady state conditions for both pure resistive and resistive – inductive load conditions have been made. The interpretation of experimental results can be guidelines for effective design and development of SEIG for wind energy applications.**

Index Terms--self - excited induction generator, induction machines

I. INTRODUCTION

Owing to the continuous increase in energy need, it is difficult to meet the growing demand by exploiting energy from the limit conventional source, such as coal, oil, gas, and so on. As a consequence, a greater emphasis is now being given to harness energy from non - conventional sources such as wind, biogas and small hydro heads[1,3]. A three phase induction machine can be made to work as a self - excited generator (SEIG) when its rotor is driven at suitable speed by wind energy and its excitation is provided by connecting a three phase capacitor bank at the stator terminals in order to build-up voltage and regulate terminal voltage. It offers various advantages over other machines such as reduced unit cost, brushless rotor (squirrel cage construction), absence of separate DC source and ease of maintenance. Numerous papers have attempted to analyze the SEIG using equivalent circuit approach [1-5]. Generally two different (but related) methods of capacitance value solution for voltage build-up and terminal voltage regulation have been employed, namely, the loop impedance method and nodal admittance method. Steady state analysis of such generators seems to be more interested than dynamic analysis. Excitation capacitors affects stator current. Therefore careful selection of this capacitor is required. Controlled static var compensator in conjunction with the ac load voltage regulator based on dynamic model can be found in [6]. These techniques are based on power electronic applications which provides good performance under a wide range of operation. However there are a few research works on investigation of various types of squirrel cage induction machines with suitable capacitance subject to load conditions based on separation of capacitors into built-up capacitor and compensating capacitor. Therefore this paper present the analysis of capacitance with the proposed approach depending on load conditions and comparative study on the performance between standard and high efficiency machines operating as self – excited induction generators. The results will be graphically presented and discussed in order to provide guidelines for the designers.

II. EQUIVALENT CIRCUIT ANALYSIS

For proposed capacitance consideration of the SEIG, the system can be shown in Fig 1. Capacitors are divided into two parts namely, a built-up or excitation capacitor (C_b) and a compensating capacitor (C_c) for terminal voltage regulation. The C_b is responsible for no-load operation whilst both C_b and C_c are responsible for on-load operation. For no-load operation, the per phase equivalent circuit is shown in Fig2. neglecting harmonic effect and core loss[3].

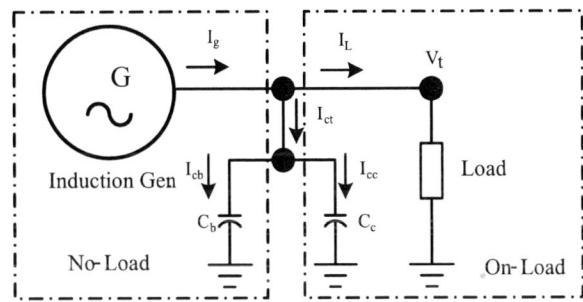

Fig.1. Single line diagram of the SEIG

From Fig.2, impedance, Z_{CD} can be written as equations (1) - (3)

$$Z_{CD} = R_{CD} + jX_{CD} \qquad (1)$$

$$R_{CD} = \frac{(a-b)R_2 X_m^2}{R_2^2 + (a-b)^2 (X_m + X_2)^2} \qquad (2)$$

$$X_{CD} = \frac{R_2^2 X_m + (a-b)^2 X_m X_2 (X_m + X_2)}{R_2^2 + (a-b)^2 (X_m + X_2)^2} \qquad (3)$$

Where a is the per unit frequency
  b is the per unit speed

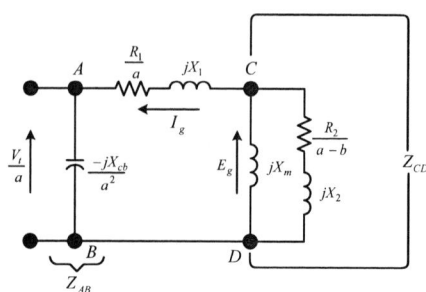

Fig.2. Per phase equivalent circuit of the SEIG at no - load

Using KVL results in

$$I_g\left(\frac{-jX_{cb}}{a^2}+\left(\frac{R_1}{a}+jX_1\right)+Z_{CD}\right)= 0 \qquad (4)$$

Since I_g is not definitely zero during voltage build-up, as a consequence, equation (4) can be rewritten as

$$\left(\frac{-JX_{cb}}{a^2}+\left(\frac{R_1}{a}+JX_1\right)+Z_{CD}\right)= 0 \qquad (5)$$

From equation (5), the real part is zero. Then

$$\frac{R_1}{a}+\frac{(a-b)R_2 X_m^{~2}}{R_2^{~2}+(a-b)^2\left(X_m+X_2\right)^2}= 0 \qquad (6)$$

According to equation (6), we can obtain $a_{\max}$ as

$$a_{\max}=b-\frac{b}{2}\left[\frac{1-\sqrt{1-\left(\dfrac{b_c}{b}\right)^2}}{1+\left(\dfrac{R_1}{R_2}\right)\left(1+\dfrac{X_2}{X_m}\right)^2}\right] \qquad (7)$$

Also, b_c can be determined as

$$b_c=\frac{2R_1}{X_m}\sqrt{\frac{R_2}{R_1}+\left(1+\frac{X_2}{X_m}\right)^2} \qquad (8)$$

Where b_c is the critical speed

Therefore capacitance for built-up voltage during no-load can be determined as

$$C_b=\frac{1}{\left\{2\pi f_b Z_b a_{\max}^2\left(X_1+X_{CD}\right)\right\}} \qquad (9)$$

Where C_b is the per phase built – up capacitor value
Z_b is the base impedance
f_b is the base frequency

For analysis of the on - load operation, C_c is introduced. According to Fig.3, when supplying the resistive – inductive load, the terminal voltage level will be reduced since the load draws reactive power from the system. In order to maintain the terminal voltage constant, C_c can be determined as follows. Total reactive power of overall capacitors (i.e. C_b and C_c) is

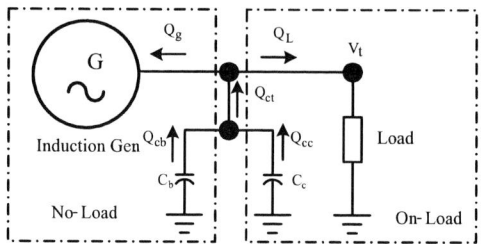

Fig.3. Power flow diagram

$$Q_{ct} = Q_g + Q_L \qquad (10)$$

When

$$Q_g = \sqrt{S_g^{~2} - P_g^{~2}} \qquad (11)$$

$$Q_L = \frac{\left(\dfrac{V_t}{a}\right)^2}{X_L} \qquad (12)$$

Where S_G is the per phase apparent power of the SEIG
P_G is the per phase active power of the SEIG
Q_G is the per phase reactive power of the SEIG
Q_L is the per phase reactive power of load
Q_{ct} is the per phase reactive power of total capacitor

Compensated reactive power is

$$Q_{cc} = Q_{ct} - Q_{cb} \qquad (13)$$

When

$$Q_{cb} = 2\pi\left(\frac{V_t}{a}\right)^2\left(f C_b\right) \qquad (14)$$

Where Q_{cb} is the per phase reactive power of the built - up capacitor
Q_{cc} is the per phase reactive power of the compensating capacitor

Per phase current of the compensating capacitor (C_c) is determined from

$$I_{cc}=\frac{Q_{cc}}{V_t\big/a} \qquad (15)$$

1484

Per phase compensated capacitance is calculated as following equations

$$X_{cc} = \frac{V_t/a}{I_{cc}} \qquad (16)$$

$$C_c = \frac{1}{2\pi f (jX_{cc})} \qquad (17)$$

Thus, total capacitor value for the SEIG is

$$C_t = C_b + C_c \qquad (18)$$

Where C_c is the per phase compensating capacitor

C_b is the per phase built – up capacitor

C_t is the per phase total capacitor

The procedure for determining capacitor values for the SEIG under on – load conditions when supplying the resistive load can be performed as same as for the resistive – inductive load with $Q_L = 0$

III. EXPERIMENTAL TESTS AND DISCUSSION

The tests have been made for finding parameters and for investigation performance

A. Parameters Test

Three phase, 3kVA, 2.2 kW, 220/380 volt, 4 poles induction machines with standard and high efficiency types are used. Table 1 shows parameters of the under test machines obtained from the test complied with IEEE std. 112-1996 testing [7].

Table 1
Parameters of the Machines

Type	R1	R2	Rc	X1	X2	Xm
Standard Machine	3.17 Ω	2.56 Ω	627.37 Ω	3.48 Ω	3.48 Ω	93.43 Ω
High Efficiency Machine	1.40 Ω	1.81 Ω	568.50 Ω	3.31 Ω	3.31 Ω	75.76 Ω

B. Operating Test

The capacitance analysis uses Maple program for determining capacitance under various load conditions and 0.85 lagging power factor. Tables 2-3 give results with constant speed of 1500 rpm and regulated terminal voltage of 220 V, Y connected. Figs. 4-16 show the SEIG performance in terms of voltage build – up, power quality, frequency variation and efficiency.

From Fig.4, it can be seen that characteristics between E_g and X_m for both SEIGs are different. X_m of the high efficiency SEIG is smaller than that of the standard SEIG.

Table 2
Capacitance for resistive load

Type	No-Load C_b (µF)	ON-Load ; C_c (µF)					C_T (µF)
		192 (W/ph)	384 (W/ph)	576 (W/ph)	768 (W/ph)	Total C_c (µF)	
Standard	35	5	7	9	12	33	68
Hi-efficiency	40	3	4	5	6	18	58

Table 3
Capacitance for resistive - inductive load

Type	No-Load C_b (µF)	ON-Load ; C_c (µF)					C_T (µF)
		213 (W/ph)	424 (W/ph)	648 (W/ph)	859 (W/ph)	Total C_c (µF)	
Standard	35	14	17	20	24	75	110
Hi-efficiency	40	10	13	16	20	59	99

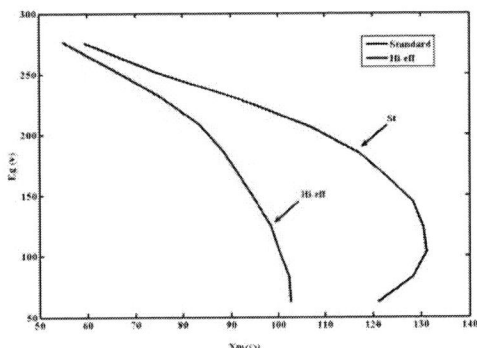

Fig.4. Characteristic between E_g and X_m

It is true that in the design of the high efficiency machine, maximum flux density level is low in order to reduce core loss. As a result, X_m is low. For SEIG applications of the high efficiency induction machine, C_b for the high efficiency SEIG is higher than for the standard SEIG.

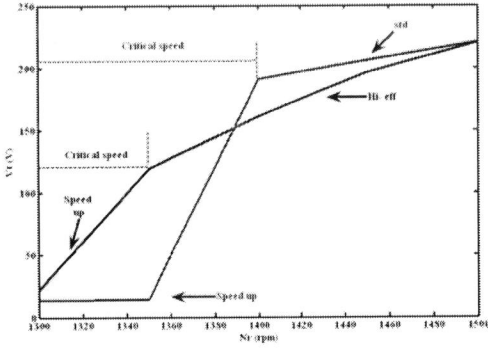

Fig.5. Variation of speed with terminal voltage for SEIG

Figs.5-6 demonstrate the impact of the speed change on built- up voltage. When increasing the speed of the prime mover, the high efficiency SEIG has a critical speed

1485

point. which is lower than the standard SEIG. When decreasing the speed of the prime mover, built – up voltage of the high efficiency SEIG is decreased lower than the standard SEIG. The reason is that in the design of high efficiency machines, magnetic material type with narrow hysteresis loop is always selected in order to reduce the hysteresis losses. As a consequence, the capability of voltage regulation is influenced with this view point, the high efficiency SEIG requires improvement to overcome this problem.

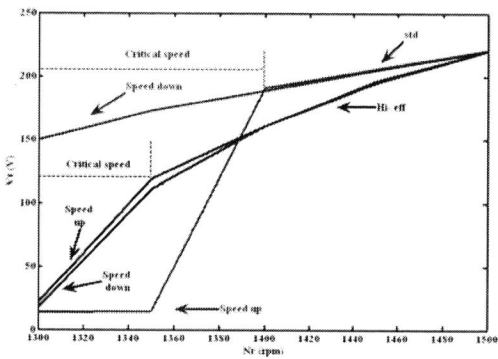

Fig.6. Variation of hysteresis voltage for SEIG

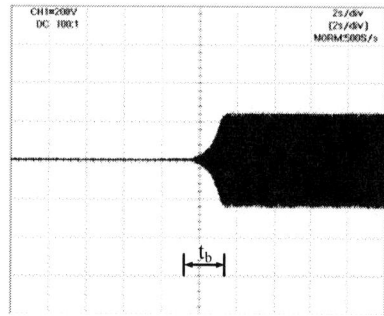

Fig.7. Built – up voltage of the standard SEIG

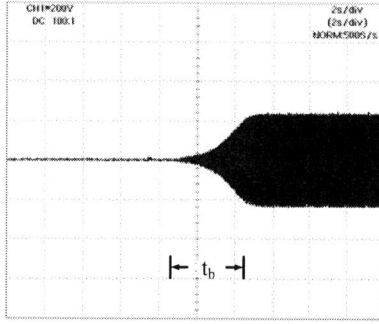

Fig.8. Built – up voltage of the high – efficiency SEIG

Figs.7-8 illustrate to duration (built – up voltage time, t_b) of voltage build- up of the SEIG. The duration for the

high efficiency SEIG is longer than for the standard SEIG since the high efficiency SEIG is designed for low magnetic flux density.

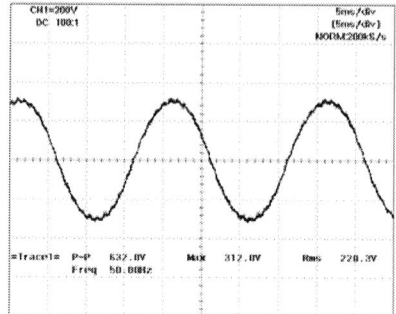

Fig.9. Terminal voltage waveform of the standard SEIG under steady – state , % THDv = 4.7 %

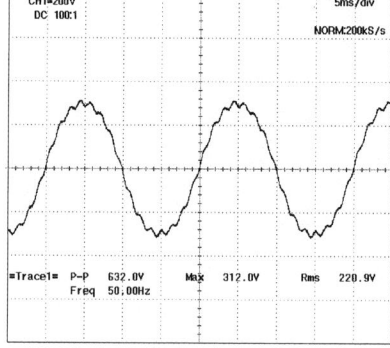

Fig.10. Terminal voltage waveform of the high efficiency SEIG under steady – state , % THDv = 6 %

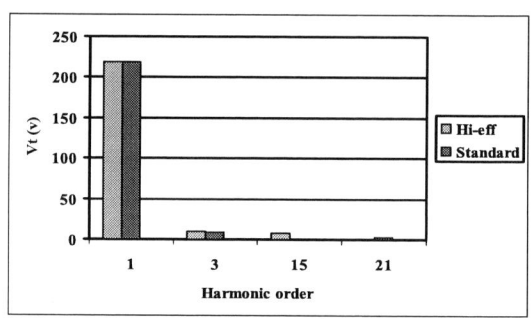

Fig.11. Harmonic Spectra of the SEIG terminal voltage

Figs. 9-11 show terminal voltage waveforms. The standard SEIG offers more nearly sinusoidal waveform than the high efficiency SEIG. Fig.11. confirms this point. According to this disadvantage the high efficiency SEIG should be improved. Figs. 12-13 show variation of reactive power with output power under various load conditions. Q_{cb} of the standard and high efficiency is decreased due to the impact of a reduction in frequency of induced voltage. Q_{cb} of the high efficiency SEIG is higher than that of the standard SEIG due to lower X_m.

1486

As a result, it needs higher capacitance. Q_{cc} is increased with increasing loads. Q_{cc} of the high efficiency SEIG under pure resistive load and resistive – inductive loads. As a consequence, it needs lower compensated capacitance since the high efficiency SEIG has voltage drop due to low stator impedance and low Q_G. The sum of reactive power of total capacitors is equal to the sum of reactive power of the SEIG and loads.

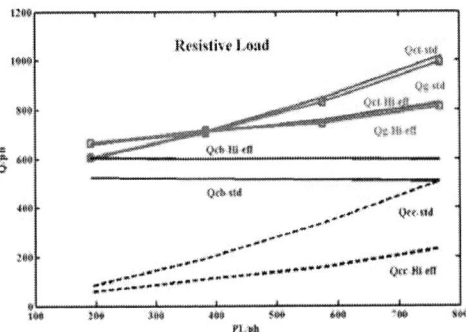

Fig.12. Variation of reactive power with output power for resistive load

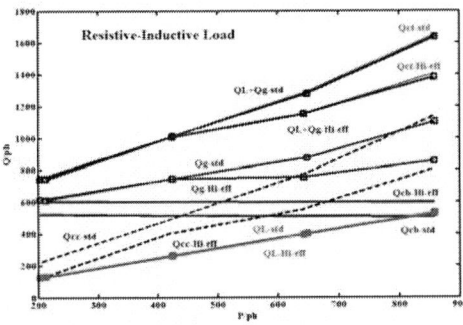

Fig.13. Variation of reactive power with output power for resistive - inductive loads

Figs.14-15 show a change in frequency of the SEIG. The high efficiency SEIG gives smaller change than the standard SEIG under various types of loads since the rotor resistance of the high efficiency SEIG is less than that of the standard SEIG.

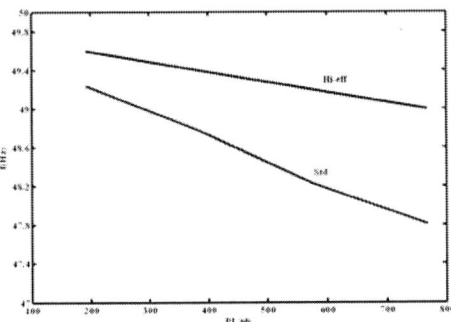

Fig.14. Variation of frequency with output power for resistive load

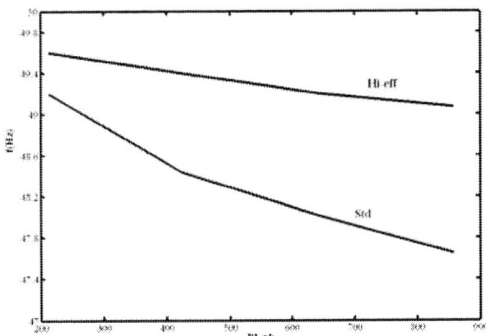

Fig.15. Variation of frequency with output power for resistive - inductive loads

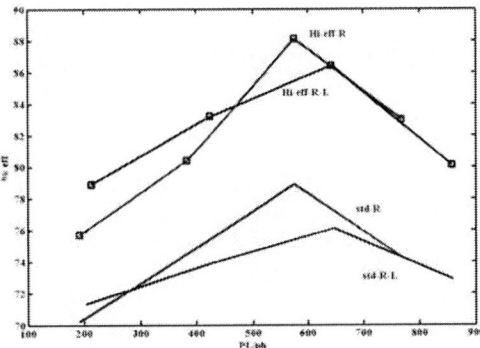

Fig.16. Variation of efficiency with output power for both SEIG

From Fig.16, it can be seen that the efficiency of the high efficiency SEIG is higher than that of the standard SEIG under various types of loads. The fact is that in the design of the high efficiency machine, a reduction in losses is the main objective. As a result, the high efficiency machine still produce low losses when operating as the SEIG.

IV. CONCLUSION

From the tests and analysis of the performance of the high efficiency and standard SEIG, it found that the high efficiency SEIG is suitable for generating electric power from wind energy. However, improvement and development are still required in term of capability of voltage regulation under a speed change, harmonic reduction of terminal voltage and start – up duration

APPENDIX

V_b: 220 V, I_b: 5.2 A (standard machine) and 4.7 (high efficiency machine), P_b: 2.2 kW, Z_b: 42.30 Ω (standard machine) and 46.80 Ω (high efficiency machine) , Z_L: per phase resistive load 242 Ω and resistive- inductive load 166.97+j103.45 Ω .

REFFERENCES

[1] S.S. Murthy , O.P. Malik , and A.K. Tandon , " Analysis of self – excited induction generators," *Proceedings of the IEE* , Vol.129, No.6, pp.260-265, November 1982.

[2] S.S. Murthy , B.P. Singh , C. Nagmani and K.V.V. Satyanarayanna , " Study on the use of conventional induction motors as self excited induction generators," *IEEE Transactions on Energy Conversion*, Vol.3, No.4, pp.842-848, December 1988.

[3] T.F. Chan , "Capacitance Requirements of Self-exited Induction Generators," *IEEE Transactions on Energy Conversion*, Vol.8, No.2, pp.304-310, June 1993.

[4] S.P Singh , B.S Singh and M.P. Jain , "Comparative study on the performance of a commercially designed induction generator with induction motors operating as self excited induction generators," *Proceedings of the IEE* , Vol.140, No.5, pp.374-380, September 1993.

[5] L. Shridhar , Bhim Singh , C.S. Jha , "A step towards improvements in the characteristics of Self excited Induction Generator ," *IEEE Transactions on Energy Conversion*, Vol.8, No.1, pp.40-46, March 1993.

[6] Tarek Ahmed, Katsumi Nisshida, Mutsou Nakaoka and Hyun Woo Lee, "Self - Excited Induction generator with Simple Voltage Regulation Scheme for Wind Energy," *Proceedings of the 30 th Annual Conference of the IEEE*, Vol.140, No.5, pp.86-91, November 2004.

[7] IEEE std 112 – 1996 , *"IEEE Standard Test Procedure for Polyphase Induction Motors and Generators ,"* pp.28-53, 1996.

A simple Approach to Capacitance Determination of Self - Exited Induction Generators for Terminal Voltage Regulation

B. Sawetsakulanond* and V. Kinnares**

*Dept. of Electrical Engineering, Mahanakorn University of Technology
Nongjok, Bangkok 10530, Thailand
Fax 662-9884040 E-mail:Budhapon@hotmail.com
**Dept. of Electrical Engineering, King Mongkut's Institute of Technology Ladkrabang,
Bangkok 10520, Thailand
Fax 662-3269902 E-mail: kkwijit@kmitl.ac.th

Abstract--This paper proposes a simple method for determining of capacitance requirements of three phase self-exited induction generators(SEIG). Consideration of capacitance of the SEIG based on equivalent circuit model under steady - state load conditions and a power flow diagram is given. Performance testing has been performed under a wide range of operating conditions such as no – load and on – load with pure resistive and resistive – inductive load conditions. Experimental results can be guidelines for simply suitable capacitance determination method for SEIGs with reasonable terminal voltage regulation under a wide range of operation.

Index Terms--Induction generator, self - excited

I. INTRODUCTION

A three phase induction machine can be made to work as a self - excited generator (SEIG) when its rotor is driven at suitable speed by wind energy and its excitation is provided by connecting a three phase capacitor bank at the stator terminals. Numerous papers in the researchers have attempted to analyze the SEIG using equivalent circuit approach. Two different methods of solution have been employed, namely, the loop impedance method and nodal admittance method [1-5]. Many research works have been done to determine the minimum capacitance for SEIG [1-5]. Most of these works used loop equations in the analysis of steady - state equivalent circuit model. By using these techniques, it has much difficulty and it needs numerical iterative techniques to obtain the minimum capacitance required. Steady – state analysis of such generators is of interest, both from the design and operational point of view.

This paper presents the simple method analysis of capacitance requirements of self-excited induction generators with terminal voltage regulation for a wide range of operating conditions.

II. EQUIVALENT CIRCUIT ANALYSIS

For the proposed capacitance consideration of the SEIG, the system can be shown in Fig.1. Capacitors are divided into two parts such as a built-up or excitation capacitor (C_b) and a compensating capacitor (C_c) for terminal voltage regulation. The C_b is responsible for no-load operation whilst both C_b and C_c are responsible for on-load operation. For no-load operation, the per

phase equivalent circuit is shown in Fig.2 neglecting harmonic effect and core loss[3].

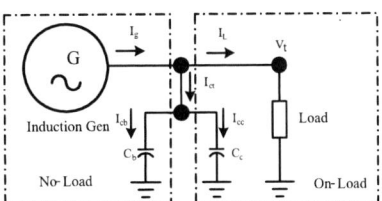

Fig.1. Single line diagram of the SEIG

From Fig.2, impedance, Z_{CD} can be written as equations (1) - (3)

$$Z_{CD} = R_{CD} + jX_{CD} \qquad (1)$$

$$R_{CD} = \frac{(a-b)R_2 X_m{}^2}{R_2{}^2 + (a-b)^2 (X_m + X_2)^2} \qquad (2)$$

$$X_{CD} = \frac{R_2{}^2 X_m + (a-b)^2 X_m X_2 (X_m + X_2)}{R_2{}^2 + (a-b)^2 (X_m + X_2)^2} \qquad (3)$$

Where a is the per unit frequency
b is the per unit speed

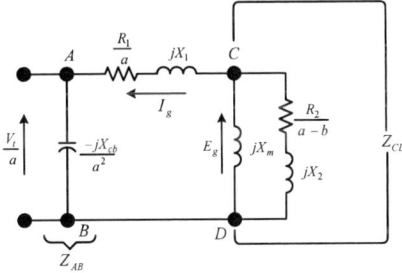

Fig.2. Per phase equivalent circuit of the SEIG at no - load

Using KVL results in

$$I_g \left(\frac{-jX_{cb}}{a^2} + \left(\frac{R_1}{a} + jX_1 \right) + Z_{CD} \right) = 0 \qquad (4)$$

Since I_g is not definitely zero during voltage build-up, as a consequence, equation (5) can be rewritten as

$$\left(\frac{-JX_{cb}}{a^2} + \left(\frac{R_1}{a} + JX_1 \right) + Z_{CD} \right) = 0 \qquad (5)$$

From equation (5), the real part is zero. Then

$$\frac{R_1}{a} + \frac{(a-b)R_2 X_m^2}{R_2^2 + (a-b)^2(X_m + X_2)^2} = 0 \qquad (6)$$

According to equation (6), we can obtain a_{max} as

$$a_{max} = b - \frac{b}{2}\left[\frac{1 - \sqrt{1 - \left(\frac{b_c}{b}\right)^2}}{1 + \left(\frac{R_1}{R_2}\right)\left(1 + \frac{X_2}{X_m}\right)^2}\right] \qquad (7)$$

Also, b_c can be determined as

$$b_c = \frac{2R_1}{X_m}\sqrt{\frac{R_2}{R_1} + \left(1 + \frac{X_2}{X_m}\right)^2} \qquad (8)$$

Where b_c is the critical speed

Therefore capacitance for built-up voltage during no-load can be determined as

$$C_b = \frac{1}{\left\{2\pi f_b Z_b a_{max}^2 \left(X_1 + X_{CD}\right)\right\}} \qquad (9)$$

Where C_b is the per phase built – up capacitor value

Z_b is the base impedance

f_b is the base frequency

The proposed simple method determines capacitance by including the determination of the power flow diagram for regulating terminal voltage under a wide range of operating conditions. Reactive power flow can be shown in Fig 3.

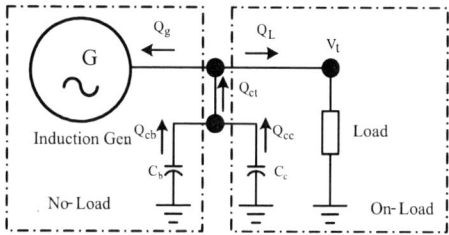

Fig.3. Power flow diagram of the SEIG

For the analysis of the on - load operation, C_c is included. According to Fig.3, when supplying the resistive – inductive load, the terminal voltage will be reduced since the load draws reactive power from the system. In order to maintain the terminal voltage constant, C_c can be determined as follows. Total reactive power of overall capacitors (i.e. C_b and C_c) is

$$Q_{ct} = Q_G + Q_L \qquad (10)$$

When

$$Q_G = \sqrt{S_G^2 - P_G^2} \qquad (11)$$

$$Q_L = \frac{\left(\frac{V_t}{a}\right)^2}{X_L} \qquad (12)$$

Where S_G is the per phase apparent power of the SEIG

P_G is the per phase active power of the SEIG

Q_G is the per phase reactive power of the SEIG

Q_L is the per phase reactive power of load

Q_{ct} is the per phase total reactive power of the capacitors

Compensated reactive power is

$$Q_{cc} = Q_{ct} - Q_{cb} \qquad (13)$$

When

$$Q_{cb} = 2\pi V^2 (f C_b) \qquad (14)$$

Where Q_{cb} is the per phase reactive power of the built - up capacitor

Q_{cc} is the per phase reactive power of the compensating capacitor

Per phase current of the compensating capacitor (C_c) is determined from

$$I_{cc} = \frac{Q_{cc}}{V_t / a} \qquad (15)$$

Per phase compensated capacitance is calculated as following equations

$$X_{cc} = \frac{V_t / a}{I_{cc}} \qquad (16)$$

$$C_c = \frac{1}{2\pi f \left(j X_{cc}\right)} \qquad (17)$$

Thus, total capacitor value for the SEIG is

$$C_t = C_b + C_c \qquad (18)$$

Where C_c is the per phase value of the compensating capacitor

C_b is the per phase value of the built – up capacitor

C_t is the per phase value of the total capacitor

The procedure for determining capacitor values for the SEIG under on – load conditions when supplying the resistive load can be performed as same as for the resistive – inductive load with $Q_L = 0$

III. EXPERIMENTAL TESTS AND DISCUSSIONS

Two induction machines with 0.75 kW and 2.2 kW, 220/380 volt, 4 poles are tested as the SEIG. The tests have been performed for two parts, namely a parameter test and an operation testing

A. Parameters Test

The parameter test has been made with no - load test , blocked rotor test and V-I method test complied with IEEE std 112 – 1996   Method F-F1 [6]. The results are illustrated in table 1. The objective of this test is to determine the built – up capacitance for the SEIG

Table 1
Parameters of the Machines

Rated	R1	R2	Rc	X1	X2	Xm
0.75 kW	10.47 Ω	13.8 Ω	1308. 3 Ω	14.27 Ω	14.27 Ω	189.76 Ω
2.2 kW	3.17 Ω	2.56 Ω	627.3 7 Ω	3.48 Ω	3.48 Ω	93.43 Ω

B. Operating Test

Maple program is used for determining the capacitance under resistive load and resistive – inductive load for both SEIGs with star connected. Speed is kept at 1500 rpm and terminal voltage is regulated at 220 V.

Table 2
Capacitance for 0.75 kW SEIG
under pure resistive – load

No-Load C_b (µF)	ON-Load ; C_c (µF)					C_T (µF)
	63 (W/ph)	123 (W/ph)	190 (W/ph)	252 (W/ph)	Total C_c (µF)	
16.5	1.5	2	3	4	10.5	27

Table 3
Capacitance for 0.75 kW SEIG under
resistive – inductive loads,0.53 lagging power factor

No-Load C_b (µF)	ON-Load ; C_c (µF)					C_T (µF)
	84 (W/ph)	172 (W/ph)	274 (W/ph)	359 (W/ph)	Total C_c (µF)	
16.5	12	14	18	22	66	82.5

Table 4
Capacitance for 2.2 kW SEIG
under pure resistive – load

No-Load C_b (µF)	ON-Load ; C_c (µF)					C_T (µF)
	192 (W/ph)	384 (W/ph)	576 (W/ph)	768 (W/ph)	Total C_c (µF)	
35	5	7	9	12	33	68

Table 5
Capacitance for 2.2 kW SEIG under
resistive – inductive loads,0.85 lagging power factor

No-Load C_b (µF)	ON-Load ; C_c (µF)					C_T (µF)
	213 (W/ph)	424 (W/ph)	648 (W/ph)	859 (W/ph)	Total C_c (µF)	
35	14	17	20	24	75	110

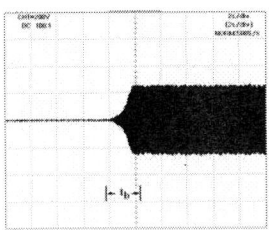

Fig.4. Built – up voltage waveform of the 0.75 kW SEIG with C_b of 16.5 µF during start - up

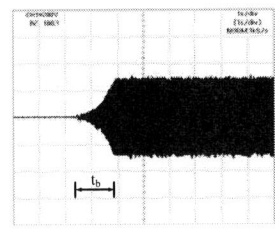

Fig.5. Built – up voltage waveform of the 2.2 kW SEIG with C_b of 35 µF during start - up

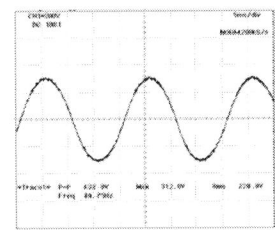

Fig.6. Terminal voltage waveform of the 0.75 kW SEIG with C_b of 16.5 µF at no - load

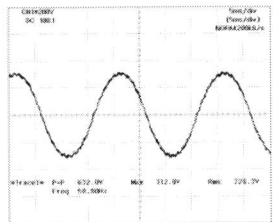

Fig.7. Terminal voltage waveform of the 2.2 kW SEIG with C_b of 35 µF at no - load

Figs.4-7 show performance during start – up and steady state of the 0.75 kW and 2.2 kW SEIG with C_b of 16.5 µF and 35 µF at no – load. Terminal voltage wave forms of the SEIGs are nearly sinusoidal.

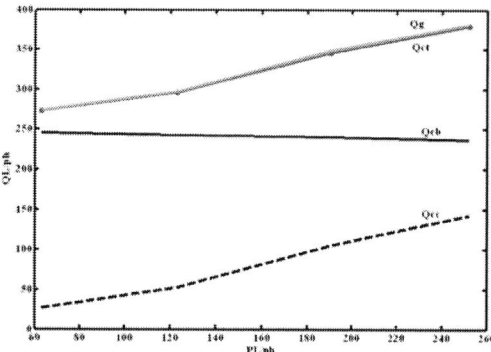

Fig.8. Variation of reactive power with output power of the 0.75 kW SEIG under pure resistive – load

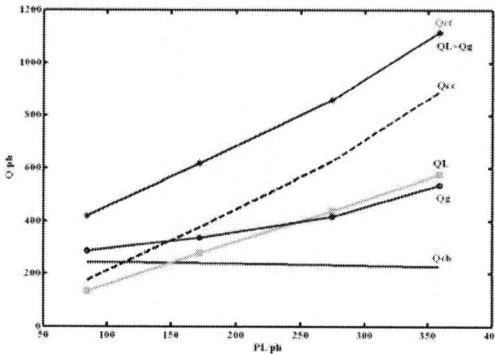

Fig.9. Variation of reactive power with output power of the 0.75 kW SEIG under resistive – Inductive loads

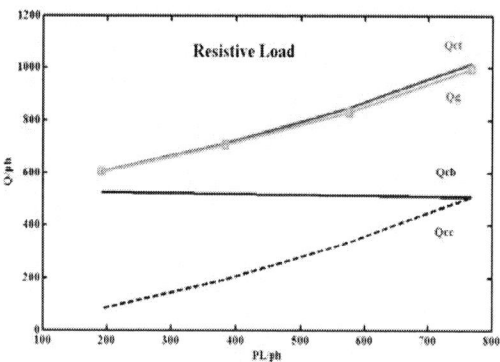

Fig.10. Variation of reactive power with output power of the 2.2 kW SEIG under pure resistive – load

Figs.8-11. show reactive power with increasing load power of the 0.75 kW and 2.2 kW SEIG under pure resistive load and resistive – inductive loads. Q_{cb} is reduced with increasing load power due to the impact of the decreased frequency of the induced voltage. Q_{cc} is increased with increasing load power. Q_{cc} for the resistive load is less than for the resistive – inductive loads resulting in less C_c requirement. The reason is that

resistive – inductive loads draw the reactive power form the system. As a consequence, C_c is higher in order to keep terminal voltage constant. The sum of reactive power supplied from total capacitors is equal to the sum of the reactive power of the SEIG and loads.

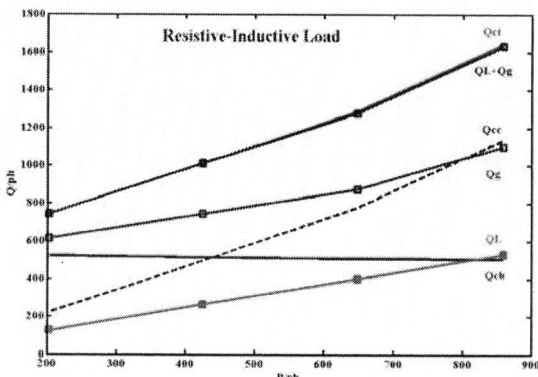

Fig.11. Variation of reactive power with output power of the 2.2 kW SEIG under resistive – inductive loads

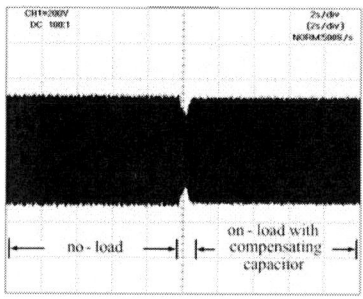

Fig.12. Terminal voltage waveform of the 0.75 kW SEIG during connecting 3.5μF compensated capacitor for voltage regulation under no – load (before) and resistive load of 123 W/ph (after)

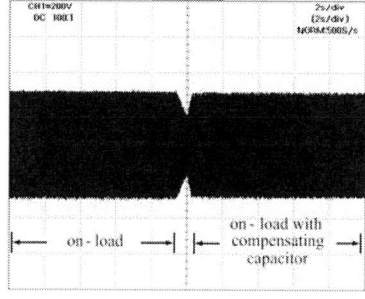

Fig.13. Terminal voltage waveform of the 0.75 kW SEIG during connecting 7 μF compensated capacitor for voltage regulation under resistive load of 123 W/ph (before) and resistive load of 252 W/ph (after)

Figs.12-15 show voltage regulation performance for both SEIGs under resistive load. Capacitance for regulating terminal voltage is non – linearly increased with increased loads

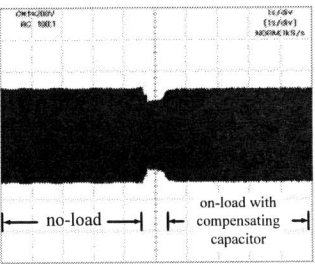

Fig.14. Terminal voltage waveform of the 2.2 kW SEIG during connecting 12 μF compensated capacitor for voltage regulation under no – load (before) and resistive load of 384 W/ph (after)

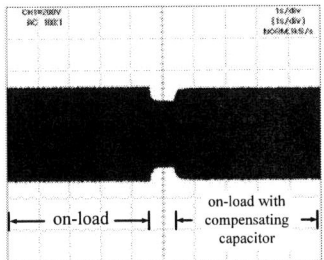

Fig.15. Terminal voltage waveform of the 2.2 kW SEIG during connecting 21 μF compensated capacitor for voltage regulation under resistive load of 384 W/ph (before) and resistive load of 768 W/ph (after)

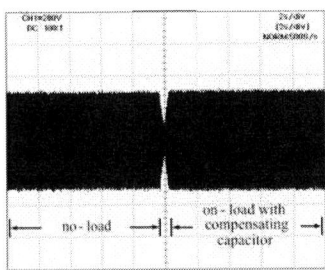

Fig.16. Terminal voltage waveform of the 0.75 kW SEIG during connecting 26 μF compensated capacitor for voltage regulation under no – load (before) and resistive – inductive loads of 172 W/ph (after)

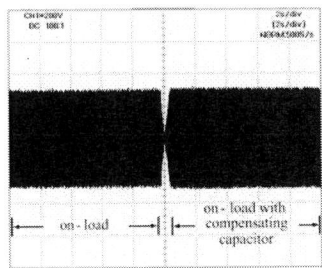

Fig.17. Terminal voltage waveform of the 0.75 kW SEIG during connecting 40 μF compensated capacitor for voltage regulation under resistive - inductive loads of 172 W/ph (before) and resistive – inductive loads of 359 W/ph (after)

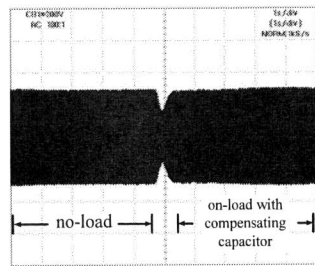

Fig.18. Terminal voltage waveform of the 2.2 kW SEIG during connecting 31 μF compensated capacitor for voltage regulation under no – load (before) and resistive – inductive loads of 424 W/ph (after)

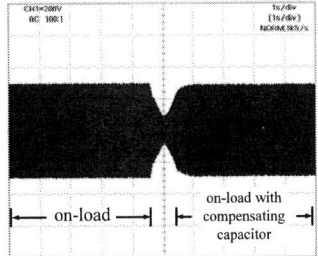

Fig.19. Terminal voltage waveform of the 2.2 kW SEIG during connecting 44 μF compensated capacitor for voltage regulation under resistive - inductive loads of 424 W/ph (before) and resistive – inductive loads of 859 W/ph (after)

Figs. 16-19 show voltage regulation performance of both SEIG under resistive – inductive loads. Capacitance for regulating terminal voltage is non – linearly increased with increased loads due to the impact of drawing the reactive power from the system. As a result, the capacitor value is higher than for the resistive load.

V. CONCLUSION

This paper has proposed the simple approach to capacitance determination for the SEIG with different rating under various types of loads. The proposed analysis is easy and is ale to analyze suitable capacitance for voltage regulation under various operating conditions and different rating

APPENDIX

V_b: 220 V, I_b: 2.0 A (0.75 kW) and 5.2 A (2.2 kW), P_b: 0.75 kW and 2.2 kW, Z_b: 110 Ω (0.75 kW) and 42.30 Ω (2.2 kW), Z_L: per phase resistive load 765 Ω and resistive- inductive load 163.7+j206.04 Ω (0.75 kW), per phase resistive load 242 Ω and resistive- inductive load 166.97+j103.45 Ω (2.2 kW).

REFFERENCES

[1] S.S. Murthy , O.P. Malik , and A.K. Tandon , "Analysis of self – excited induction generators," *Proceedings of the IEE* , Vol.129, No.6, pp.260-265, November 1982.

[2] S.S. Murthy , B.P. Singh , C. Nagmani and K.V.V. Satyanarayanna , "Study on the use of conventional induction motors as self excited induction generators," *IEEE Transactions on Energy Conversion*, Vol.3, No.4, pp.842-848, December 1988.

[3] T.F. Chan , "Capacitance Requirements of Self-exited Induction Generators,"   *IEEE Transactions on Energy Conversion*, Vol.8, No.2, pp.304-310, June 1993.

[4] S.P Singh , B.S Singh and M.P. Jain , "Comparative study on the performance of a commercially designed induction generator with induction motors operating as self excited induction generators," *Proceedings of the IEE* , Vol.140, No.5, pp.374-380, September 1993.

[5] L. Shridhar , Bhim Singh , C.S. Jha , "A step towards improvements in the characteristics of Self excited Induction Generator ," *IEEE Transactions on Energy Conversion*, Vol.8, No.1, pp.40-46, March 1993.

[6] IEEE std 112 – 1996 , "*IEEE Standard Test Procedure for Polyphase Induction Motors and Generators ,*" pp.28-53, 1996.

Symmetrical Components-Based Control Technique of Doubly Fed Induction Generators under Unbalanced Voltages for Reduction of Torque and Reactive Power Pulsations

S. Wangsathitwong, S. Sirisumrannukul, S. Chatratana* and W. Deleroi**

King Mongkut's Institute of Technology North Bangkok
1518 Pibulsongkram Rd., Bang sue, Bangkok, Thailand
*Technology Management Center, National Science and Technology Development Agency
111 Thailand Science Park, Paholyothin Rd. Klong 1, Klong Luang, Phatumthani 12120, Thailand
**B-4728 Corsostraße 3, Hergenrath, Belgium
E-mail: ssww@kmitnb.ac.th, spss@kmitnb.ac.th, somchaich@nstda.or.th, werner.deleroi@skynet.be

Abstract--This paper presents a control technique to reduce the effect of unbalanced voltages in the doubly fed induction generator (DFIG) in wind energy conversion systems. The unbalanced voltages cause negative effects to the DFIG such as torque pulsation and increased stator current. Based on the symmetrical component theory, the negative sequence of the stator voltage can be extracted from the three-phase stator voltage and fed to compensate the rotor voltage to reduce the unbalanced effect. The magnitude of the compensating negative sequence is calculated from the negative sequence of the stator voltage and rotor speed. The proposed method shows a promising result for reduction of torque and reactive power pulsations.

Keywords- Doubly fed induction generator, Unbalanced voltages, Symmetrical components

NOMENCLATURE

i_{R+}, i_{R-}	Positive and negative sequence components of RMS rotor current phasors
i_{S+}, i_{S-}	Positive and negative sequence components of RMS stator current phasors
$i_{Rd\pm}, i_{Rq\pm}$	d and q axis rotor current symmetrical components in synchronous reference frame
$i_{Sd\pm}, i_{Sq\pm}$	d and q axis stator current symmetrical components in synchronous reference frame
L_0, X_0	Magnetizing inductance and reactance
L_R, X_R	Rotor inductance and reactance
L_S, X_S	Stator inductance and reactance
$X_{\sigma S}, X_{\sigma R}$	Stator and rotor leakage reactance
P	Magnetic poles
R_S, R_R	Stator and rotor winding resistances
s	Slip
T_e	Electrical torque
V_{a+-}	Positive and negative sequence components of phase a voltage
V_a, V_b, V_c	Three phase RMS voltage
V_R, V_S	Rotor and stator voltages
$v_R(t), i_R(t)$	Time varying rotor voltage and current vectors
$v_{Rd\pm}, v_{Rq\pm}$	d and q axis rotor voltage symmetrical components in synchronous reference frame
$v_s(t), i_s(t)$	Time varying stator voltage and current vectors
v_{S+}, v_{S-}	Positive and negative sequence components of RMS stator phase voltage phasors
$v_{Sd\pm}, v_{Sq\pm}$	d and q axis stator voltage symmetrical components in synchronous reference frame
ε	Rotor angle
ω_0	Stator line voltage frequency

I. INTRODUCTION

Wind energy has been recognized as a sustainable source of energy and experienced rapid growth due to the technological development of wind turbines. Such development has lead to a considerable decrease in cost and therefore allowed wind energy source to compete with conventional energy for electricity production. Many large wind farms employ doubly fed induction generator (DFIG) with variable speed wind turbines [1]. The DFIG-based wind turbines have several advantages such as variable speed control, four-quadrant active and reactive power operation. Moreover, converter rating is about 20 – 30 % of generator rating and therefore less power loss. For a DFIG-based wind turbine, its stator side is directly connected to the grid whereas its rotor side is connected through a four-quadrant converter, as shown in Fig. 1.

Wind turbines are usually located in rural area or

This work was supported by college of industrial technology, King Mongkut's Institute of Technology North Bangkok.

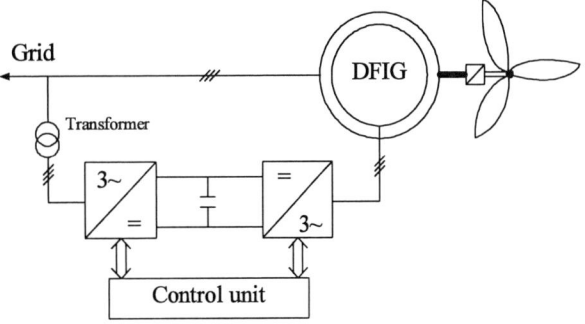

Fig. 1. Schematic diagram of DFIG for wind turbine system.

connected to a weak grid system. In that system, the unbalance of the grid voltage may arise even during normal operation. The unbalanced voltage is due to, for example, unbalanced loads, large single-phase loads, untransposed lines. Severe voltage unbalances can also be caused by unbalanced faults. The unbalanced voltage condition introduces many negative effects to the performance of three-phase DFIGs. This condition will cause unbalanced currents in the stator and rotor circuits whereas a small unbalanced voltage can cause a large unbalanced current. These effects create a pulsating torque with a twice frequency of that of the grid supply and consequently, speed fluctuation, acoustic noise, increased losses and reduced insulation life. This may be acceptable in small machines but not in larger ones. The pulsating torque can fatigue the rotor shaft, gearbox or wind turbine mechanical system.

Although a conventional PI control system can reduce unbalanced voltage to some extent, its significant effects remain to be seen. As a result, a number of methods to reduce the unbalanced effect have been proposed. Of these is an unbalanced operation of three-phase machine based on the symmetrical component theory, which states that the total voltage of any phase is equal to the sum of the corresponding components of the different sequences in that phase [2]. The set of symmetrical components are classified as positive, negative and zero-sequence networks. Because the neutral point of a DFIG is, in general, not grounded, its zero sequence network does not exist. With the symmetrical components, the positive-sequence network is the same as a normal balanced three-phase voltage system while an unbalanced-system is a combination of the positive and negative-sequence networks. Therefore, the negative-sequence currents are the main cause for uncommon operation of a DFIG.

Many methods to reduce pulsating torque and to balance stator current have been introduced in literatures. Reference [3] employs compensated rotor currents added to the rotor currents to eliminate perturbation terms in magnetizing current. In [4], a supplementary control loop is added to a conventional controller. A separate control of positive and negative of rotor current is presented in [5], with an additional control loop of negative sequence to reduce pulsating torque and reactive power.

This paper proposes a control method for DFIG under unbalanced source voltage. The main control loops are same as those of conventional controller but compensated rotor voltages are directly added to minimize the negative sequence component of the stator and rotor currents. The compensation terms are determined from the unbalanced stator voltage and rotor speed.

This paper is organized as follows. The definition of unbalanced voltage and current are described in Section II. The mathematical model of DFIG and the proposed method are detailed in Section III. The verification results are shown in Section IV for steady state model while dynamic model are illustrated in Section V. Conclusion is given in Section VI.

II. DEFINITION OF UNBALANCED VOLTAGE AND CURRENT

Three-phase voltages or currents are balanced or symmetrical if all the three phases have the same amplitude and a phase shift of 120^0 respect to each other. If either or both of these conditions are not met, the system is called unbalanced or asymmetrical.

There are two general definitions for measuring the voltage unbalanced, NEMA and IEC. Based on the IEC standard [6], unbalanced voltage is defined by the voltage unbalanced factor given by

$$\text{Voltage unbalanced factor, VUF (\%)} = \frac{V_2}{V_1} \times 100 \qquad (1)$$

Similarly, the unbalanced current factor is defined as

$$\text{Current unbalanced factor, CUF (\%)} = \frac{I_2}{I_1} \times 100 \qquad (2)$$

where V_1 and I_1 are positive sequence components of voltage and current; V_2 and I_2 are negative sequence components of voltage and current.

III. MATHEMATICAL MODEL OF DFIG

Based on the symmetrical components, time-varying stator voltage and current vectors in the stationary reference frame can be represented as the superposition of positive and negative sequence components [7], as shown in (3) and (4). The rotor current vector in stator coordinate can be expressed by (5).

$$\underline{v}_S(t) = \frac{3\sqrt{2}}{2}\left[\underline{v}_{S+}e^{j\omega_0 t} + \underline{v}_{S-}^*e^{-j\omega_0 t}\right] \qquad (3)$$

$$\underline{i}_S(t) = \frac{3\sqrt{2}}{2}\left[\underline{i}_{S+}e^{j\omega_0 t} + \underline{i}_{S-}^*e^{-j\omega_0 t}\right] \qquad (4)$$

$$\underline{i}_R(t)e^{j\varepsilon} = \frac{3\sqrt{2}}{2}\left[\underline{i}_{R+}e^{j\omega_0 t} + \underline{i}_{R-}^*e^{-j\omega_0 t}\right] \qquad (5)$$

As can be seen from (3) - (5), the positive and negative component vectors rotate with equal angular velocity in opposite direction. In steady state, if the synchronously rotating positive sequence is considered as a reference frame, the d-q components of the positive sequence terms are constant and the negative sequence terms in this frame are alternating components with a frequency of $2\omega_0$. The torque equation is calculated from

$$T_e = \frac{2}{3}\frac{P}{2}L_0\,\text{Im}\left[\underline{i}_S(\underline{i}_R e^{j\varepsilon})^*\right] \qquad (6)$$

Substituting (4) and (5) in (6),

$$T_e = \frac{2}{3}\frac{P}{2}L_0[(i_{Sq+}i_{Rd+} - i_{Sd+}i_{Rq+}) + (i_{Sd-}i_{Rq-} - i_{Sq-}i_{Rd-})$$
$$+ ((i_{Sd+}i_{Rq-} + i_{Sq+}i_{Rd-}) - (i_{Sd-}i_{Rq+} + i_{Sq-}i_{Rd+}))\cos 2\omega_0 t$$
$$+ ((i_{Sd+}i_{Rd-} - i_{Sq+}i_{Rq-}) - (i_{Sd-}i_{Rd+} - i_{Sq-}i_{Rq+}))\sin 2\omega_0 t]$$
$$(7)$$

Obviously from (7), the interaction between the stator and rotor currents rotating in the same sequence produces a constant torque during steady state operation while the pulsating torque occurs due to the interaction between the stator and rotor currents in difference sequence, the frequency of pulsating is $2\omega_0$. Consequently, the average torque is an algebraic sum of the two constant torques produced by the stator and rotor currents in the same sequence.

The DFIG stator and rotor voltage vectors in stationary reference frame are mathematically expressed by (8) and (9).

$$v_S(t) = R_S i_S + L_S \frac{d}{dt} i_S + L_0 \frac{d}{dt} i_R e^{j\varepsilon} \qquad (8)$$

$$v_R(t) = R_R i_R + L_R \frac{d}{dt} i_R + L_0 \frac{d}{dt} i_S e^{-j\varepsilon} \qquad (9)$$

In synchronously rotating reference frame, stator and rotor voltage equations for positive and negative sequence reference frames are

$$v_{Sd\pm} = (R_S + L_S \frac{d}{dt})i_{Sd\pm} - \omega_0 L_S i_{Sq\pm} + L_0 \frac{d}{dt} i_{Rd\pm} - \omega_0 L_0 i_{Rq\pm} \qquad (10)$$

$$v_{Sq\pm} = \omega_0 L_S i_{Sd\pm} + (R_S + L_S \frac{d}{dt})i_{Sq\pm} + \omega_0 L_0 i_{Rd\pm} + L_0 \frac{d}{dt} i_{Rq\pm} \qquad (11)$$

$$v_{Rd\pm} = L_0 \frac{d}{dt} i_{Sd\pm} - (\omega_0 \mp \omega_m) L_0 i_{Sq\pm} + (R_R + L_R) \frac{d}{dt} i_{Rd\pm} - (\omega_0 \mp \omega_m) L_R i_{Rq\pm} \qquad (12)$$

$$v_{Rq\pm} = (\omega_0 \mp \omega_m) L_0 i_{Sd\pm} + L_0 \frac{d}{dt} i_{Sq\pm} + (\omega_0 \mp \omega_m) L_R i_{Rd\pm} + (R_R + L_R) \frac{d}{dt} i_{Rq\pm} \qquad (13)$$

In steady state, the stator and rotor voltage equations of the positive and negative sequence system in (10) - (13) are written in forms of phasor equations by setting the differential term in (10) – (13) to be zero.

$$V_{S+} = (R_S + jX_S)I_{S+} + jX_0 I_{R+} \qquad (14)$$

$$V_{R+} = jX_0 s I_{S+} + (R_R + jX_R s)I_{R+} \qquad (15)$$

$$V_{S-} = (R_S + jX_S)I_{S-} + jX_0 I_{R-} \qquad (16)$$

$$V_{R-} = jX_0(2 - s)I_{S-} + (R_R + jX_R(2 - s))I_{R-} \qquad (17)$$

From (14) – (17), the steady state equivalent circuits of the positive and negative sequence components are illustrated in Fig. 2 (a) and (b), respectively.

Therefore, positive and negative torque equations are

$$T_{e+} = \frac{2}{3} \frac{P}{2} \frac{X_0}{\omega_0} \text{Im}(I_{S+} I_{R+}^*) \qquad (18)$$

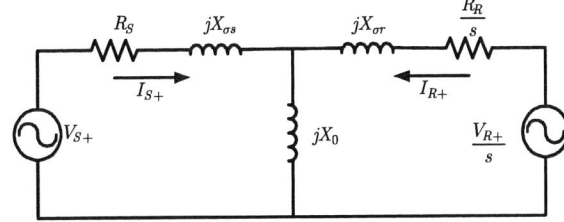

Positive sequence.

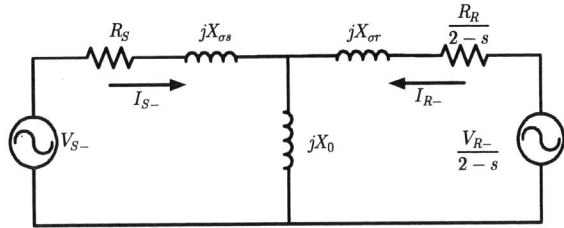

Negative sequence.

Fig. 2. Equivalent circuits of DFIG.

$$T_{e-} = \frac{2}{3} \frac{P}{2} \frac{X_0}{\omega_0} \text{Im}(I_{S-}^* I_{R-}) \qquad (19)$$

The total torque is the summation of both positive and negative torque. Clearly, the negative torque and also pulsating torque can be minimized if the negative sequence component of both stator and rotor current are zero.

From (16) and (17), the negative stator and rotor currents can be determined by

$$I_{S-} = \frac{1}{D}\left[\left(\frac{R_R}{2-s} + jX_R\right)V_{S-} - jX_0 \frac{V_{R-}}{2-s}\right] \qquad (20)$$

$$I_{R-} = \frac{1}{D}\left[(R_S + jX_S)\frac{V_{R-}}{2-s} - jX_0 V_{S-}\right] \qquad (21)$$

where $D = (R_S + jX_S)\left(\dfrac{R_R}{2-S} + jX_R\right) + X_0^2$

In (20), the negative rotor voltage that minimizes the negative stator current can be calculated by (22).

$$V_{R-} = \frac{X_R}{X_0}(2 - s)V_{S-} - j\frac{R_R}{X_0}V_{S-} \qquad (22)$$

Likewise, the negative rotor voltage to minimize the negative rotor current in (21) is

$$V_{R-} = \frac{X_R X_0}{R_S^2 + X_S^2}(2 - s)V_{S-} + j\frac{X_0 R_S}{R_S^2 + X_S^2}(2 - s)V_{S-} \qquad (23)$$

Because R_S and R_R are small, the imaginary part in (22) and (23) are nearly equal to zero and the real part can be approximated by

$$V_{R-} = (2 - s)V_{S-} \qquad (24)$$

Therefore, if the rotor voltage is fed with the negative sequence component given in (24), the negative sequence stator and rotor currents can be reduced and also pulsating torque.

The three phase negative sequence rotor voltages are determined from the instantaneous negative sequence component of stator voltage and added to the rotor voltage command from a conventional controller as illustrated in Fig. 3. The detail of instantaneous symmetrical components is given in Appendix.

IV. VERIFICATION OF STEADY STATE MODEL

In steady state, the typical torque-slip characteristic of a doubly fed induction machine under unbalanced voltage with a VUF of 100% and without a compensation for the negative sequence rotor voltage is illustrated in Fig. 4, The solid line in the figure represents the positive torque curve, the dot line is the negative torque curve, and the dash line is a total torque curve. In Fig. 4, the net torque is increased in generator mode while in motor mode the net torque is reduced. When the negative sequence rotor voltage calculated from (24)) is fed, the resulting torque is illustrated in Fig. 5. Obviously, the negative torque is minimized nearly equal to zero in the whole range. The average torque is closed to the positive torque as in normal operation.

V. VERIFICATION OF DYNAMIC MODEL

Initially, a DFIG operates under balanced voltage and thereafter its voltage in phase A is decreased by 27% or VUF = 10% at t = 15 s. When the system voltages are unbalanced, the torque response of the DFIG in Fig. 6 from t = 15 s to t =18 s pulsates with a conventional PI control and the oscillation frequency is equal $2\omega_0$. Later at t = 18 s, the compensating negative rotor voltage obtained from (24) is applied as illustrated in Fig. 3. Apparently, the pulsating torque is significantly reduced from 0.25 p.u. to 0.02 p.u. The stator reactive power

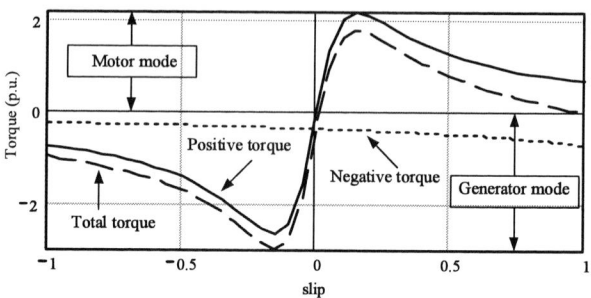

Fig. 4. Torque-slip curve of DFIG under unbalanced voltage.

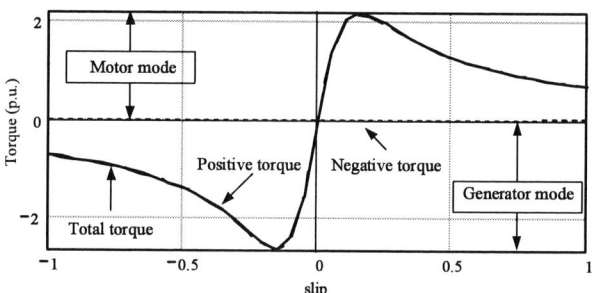

Fig. 5 Torque-slip curve of DFIG with negative voltage compensation.

shown in Fig. 7 indicates that the reactive pulsation is reduced from 0.35 p.u. to 0.026 p.u.. Moreover, the unbalanced of three phases stator current is also decreased as can be seen from the rms current with and without a compensation for the negative sequence rotor voltage in Fig. 8.

If VUF is varied, the magnitudes of the pulsating torque and the pulsating stator reactive power as a function of the unbalanced voltage are shown in Figs. 9 and 10, respectively. The results show that both magnitudes are reduced compared with those of the conventional method by a factor of 10. For the current unbalanced factor as a function of the unbalanced voltage

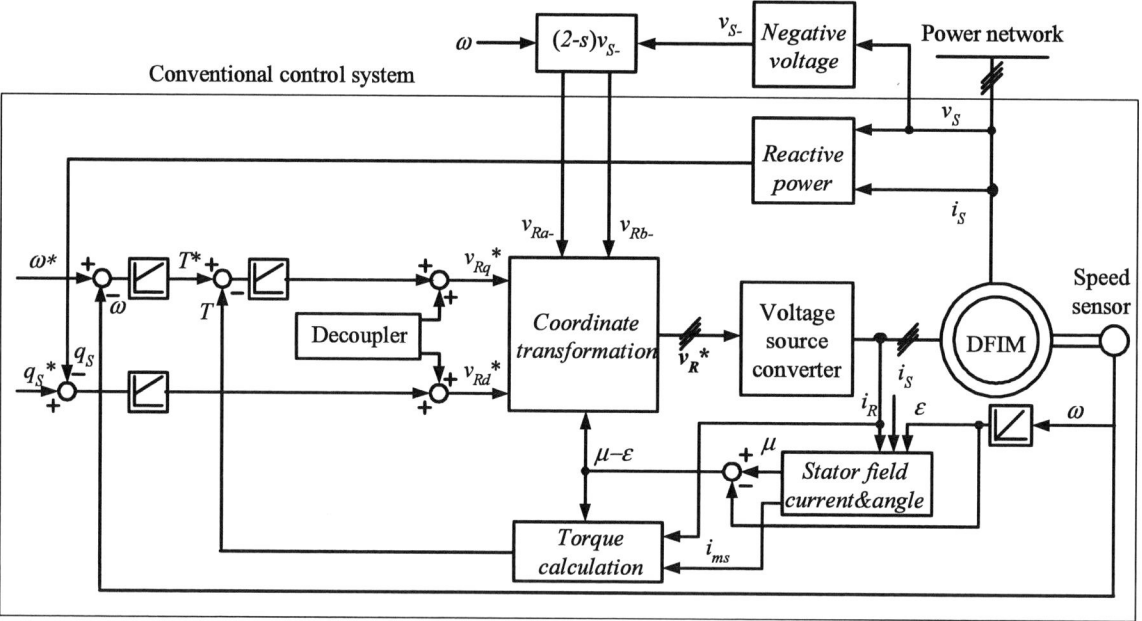

Fig. 3. Control system of DFIG on unbalanced voltage.

1498

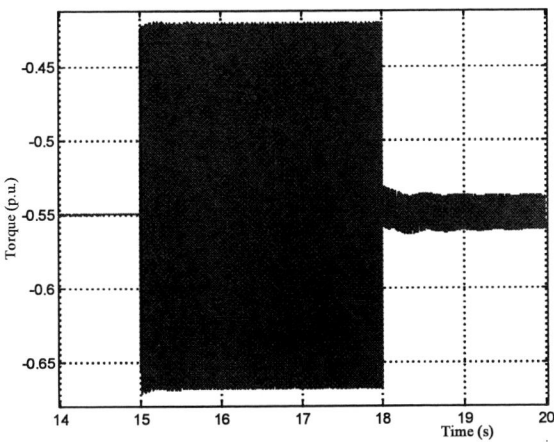

Fig. 6. Electrical torque response with VUF = 10%.

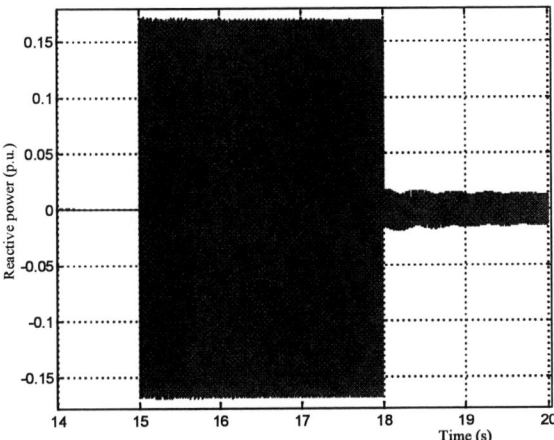

Fig. 7. Stator reactive power response with VUF = 10%.

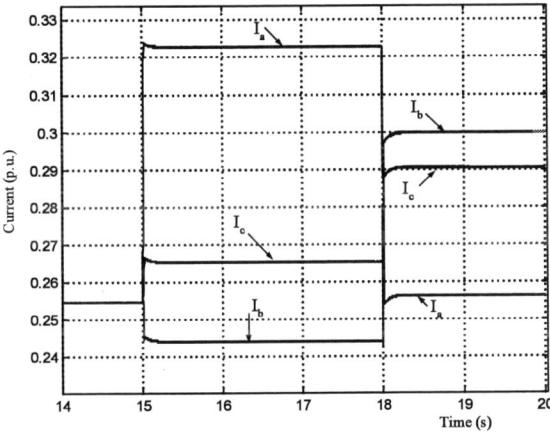

Fig. 8. Three phases stator rms current response.

factor, the result is depicted in Fig. 11.

If the voltages of phases B and C are dropped, the result is shown in Fig. 12. It is seen that the pulsating torque and pulsating reactive power with the compensation method are about ten times as low as those with the conventional control system.

VI. CONCLUSIONS

The study of unbalanced stator voltage effect on the DFIG is presented in this paper. Electrical torque and

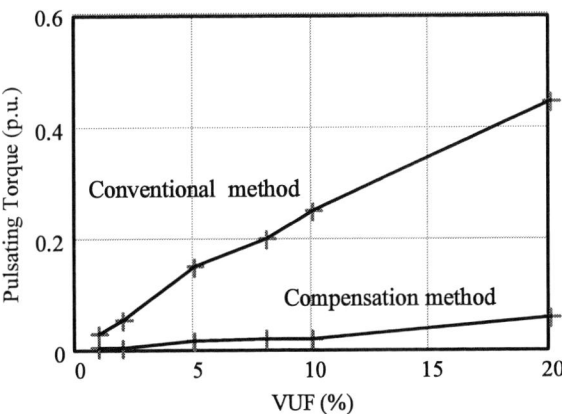

Fig. 9. Magnitude of pulsating torque as function of voltage unbalanced factor.

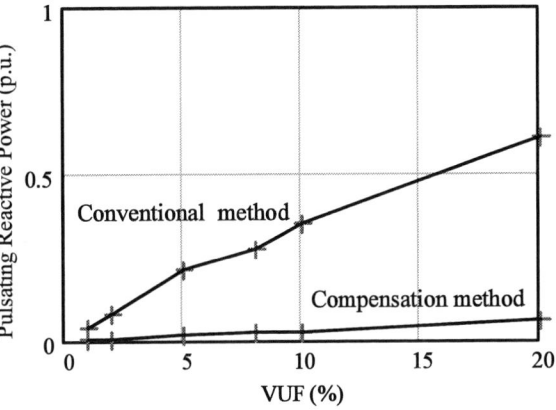

Fig. 10. Magnitude of pulsating stator reactive power as function of voltage unbalanced factor.

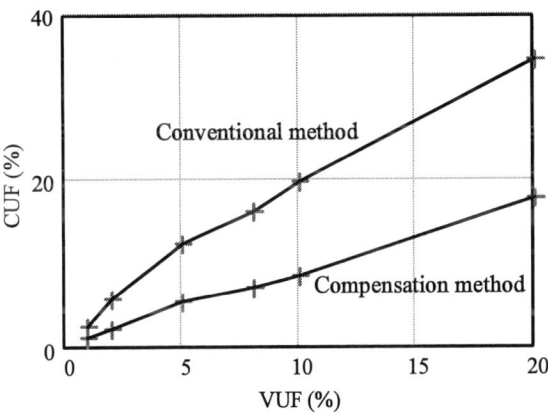

Fig. 11. Current unbalanced factor as function of voltage unbalanced factor.

reactive power pulsation due to this unbalance introduce several negative effects to the DFIG. These pulsating are occurred caused by the interaction between the stator and rotor negative sequence component currents. The proposed method is based on the symmetrical component theory and combines with the conventional control method. The negative sequence used to compensate the rotor voltage is determined from the negative sequence of the stator voltage and rotor speed. This compensating

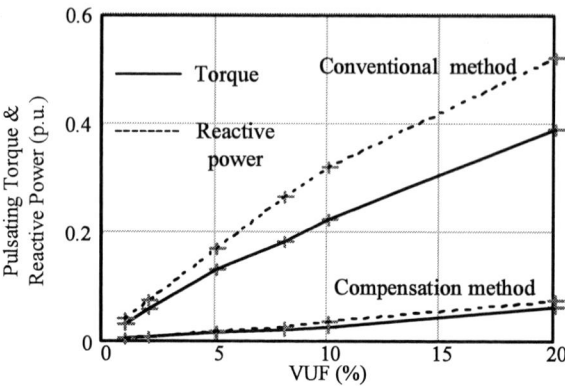

Fig. 12. Magnitude of pulsating torque and stator reactive power as function of voltage unbalanced factor

voltage is directly fed to the rotor voltage to reduce the negative sequence of the stator and rotor currents. It can be verified from the case study performed that the proposed method can significantly reduce torque and reactive power pulsations, thus improving the overall operational performance of the DFIG.

APPENDIX

The method of symmetrical components was introduced in [9] in complex steady state phasors. The set of sequence components are positive, negative and zero and can be determined from

$$
\begin{bmatrix} V_{a0} \\ V_{a+} \\ V_{a-} \end{bmatrix} = \frac{1}{3} \begin{bmatrix} 1 & 1 & 1 \\ 1 & a & a^2 \\ 1 & a^2 & a \end{bmatrix} \begin{bmatrix} V_a \\ V_b \\ V_c \end{bmatrix} \tag{20}
$$

where $a = e^{j\frac{2\pi}{3}}$

For instantaneous symmetrical components, (20) can be modified by replacing complex phasor a with a 120^0 phase shift operator in time domain as in (21)

$$
\begin{bmatrix} v_{a0} \\ v_{a+} \\ v_{a-} \end{bmatrix} = \frac{1}{3} \begin{bmatrix} v_a + v_b + v_c \\ v_a + S_{120}v_b + S_{240}v_c \\ v_a + S_{240}v_b + S_{120}v_c \end{bmatrix} \tag{21}
$$

Another transform method of symmetrical components based on 90^0 phase shift operator is written as in (22) [10].

$$
\begin{bmatrix} v_{a0} \\ v_{a+} \\ v_{a-} \end{bmatrix} = \frac{1}{3} \begin{bmatrix} v_a + v_b + v_c \\ v_a - \frac{1}{2}(v_b + v_c) + \frac{\sqrt{3}}{2}S_{90}(v_b - v_c) \\ v_a - \frac{1}{2}(v_b + v_c) - \frac{\sqrt{3}}{2}S_{90}(v_b - v_c) \end{bmatrix} \tag{22}
$$

REFERENCES

[1] T. Ackerman, *Wind Power in Power System,* John Wiley & Sons, Chichester, 2005.

[2] C. F. Wagner, R. D. Evans, *Symmetrical Components as Applied to the Analysis of Unbalanced Electrical Circuits,* Mcgraw-Hill, 1933, New York.

[3] T. Brekken, N. Mohan, "A Novel Doubly-fed Induction Wind Generator Control Scheme for Reactive Power Control and Torque Pulsation Compensation Under Unbalanced Grid Voltage," in Power Electronics Specialist Conference, vol. 2, pp. 760 – 764, 2003.

[4] T. Brekken, N. Mohan and T. Undeland, "Control of a Doubly Fed Induction Wind Generator Under Unbalanced Grid Voltage Conditions," *IEEE Trans. Energy Convers.,* vol. 22, no. 1, pp. 129-135, Mar 2007.

[5] J. I. Jang, Y. S. Kim and D. C. Lee, "Active and Reactive Power of DFIG for Wind Energy Conversion under Unbalanced Grid Voltage," in Power Electronics and Motion Control Conference, pp. 1 – 5, 2006.

[6] European standard EN-50160, Voltage Characteristics of Electricity Supplied by Public Distribution Systems, CENELEC, Brussels, Belgium, 2000.

[7] W.Leonhard, *Control of Electrical Drive,* Berlin German, Springer, 1996.

[8] P. C. Krause, *Analysis of Electric Machinery,* Mcgraw-Hill, NY, 1986.

[9] C. L. Fortescue, "Method of symmetrical Coordinates Applied to the Solution of Polyphase Networks," Trans. *AIEE,* pt. II, vol. 37, pp.1027 – 1140, 1918.

[10] M. R. Iravani, M.Karimi-Ghartemani, "Online estimation of steady state and instantaneous symmetrical components," in *Porc. IEEE* Gener. Transm. Distrib., vol. 150, pp.616 – 622, 2003.

A New Switching Technique for Direct Torque Control of Induction Motor using Four-Switch Three-Phase Inverter

Phan Quoc Dzung*, Le Minh Phuong*, Pham Quang Vinh**,

Nguyen Minh Hoang***, Nguyen Xuan Bac*

* Faculty of Electrical & Electronic Engineering, HCMC University of Technology, Ho Chi Minh City, Vietnam
** Siemens AG Presentation, Ho Chi Minh City, Vietnam
*** NARC, Ulsan University, Korea

Abstract–This paper presents a new switching technique for Direct Torque Control of Induction Motor using Four-Switch Three Phase Inverter (DTC-FSTPI-IM) for low power applications. The modified switching table in this method is based on the principle of similarity between FSTPI and SSTPI (Six-Switch Three Phase Inverter), where the αβ plan is divided into 6 sectors and the formation of the voltage space vector is done in the same way as for SSTPI by using effective (mean) vectors. This approach allows using the well-knowing established switching table of SSTPI for FSTPI, in order to reduce torque ripples in comparison with the conventional DTC method for FSTPI. The validity of new proposed technique is verified by simulation results using Matlab/Simulink and also experimentally by using DSpace 1104 Card.

Index Terms-- Direct Torque Control, Four-Switch Three-Phase Inverter (FSTPI), Optimal Switching Table, Pulse-Width- Modulation, Six-Switch Three-Phase Inverter (SSTPI), Space Vector.

I. INTRODUCTION

Recently, several scientific researches have been done for Four-Switch Three-Phase Inverters (FSTPI) with the target for reducing the cost of electric drives. Several inverter schemes with reduced number of switches have been proposed [1-5].

To obtain the simple, effective performances, the fast control of torque and flux; a DTC system for FSTPI-IM has been proposed [6]. In this paper, the optimal switching look-up table is established with four basic space vectors of FSTPI and in according with four main sectors in the αβ plan. Comparison with DTC of induction motor fed by conventional SSTPI confirm that FSTPI topology can be alternative to the conventional topology for low power low cost induction motor drives.

DTC method for SSTPI-IM has been improved in some researches [7-13], while the torque and speed ripples are reduced. The inverter switching frequency can be increased by mixing high-frequency dither signals with the error signals of torque and flux [7]. In order to reduce the speed (torque) ripple, the space vector modulation (SVM) modulator has been used as shown in [8-12].

However, up till now no attentions are paid to apply these improvements for DTC method for FSTPI-IM, because of the difference between the basic space vectors of FSTPI and SSTPI.

The proposed switching technique for DTC-FSTPI-IM in this paper has been done by using the new approach based on the principle of similarity between FSTPI and SSTPI, where the αβ plan is divided into 6 sectors and the formation of the required reference voltage space vector is done in the same way as for SSTPI by using effective (mean) vectors. This approach allows the possibilities in using these well-know established improvements of DTC technique for SSTPI [13] in the DTC-FSTPI-IM system (Fig.1).

II. VOLTAGE SPACE- VECTOR ANALYSIS FOR FSTPI AND THE PRINCIPLE OF SIMILARITY

According to the scheme in Fig.2 the switching status is represented by binary variables S_1 to S_4, which are set to "1" when the switch is closed and "0" when open. In addition the switches in one inverter branch are controlled complementary (1 on, 1 off), therefore:

$$S_1 + S_2 = 1$$
$$S_3 + S_4 = 1 \qquad (1)$$

Phase to common point voltage depends on the turning off signal for the switch:

$$V_{a0} = (2S_1 - 1) \cdot \frac{V_{dc}}{2}; V_{b0} = (2S_3 - 1) \cdot \frac{V_{dc}}{2}; V_{c0} = 0 \qquad (2)$$

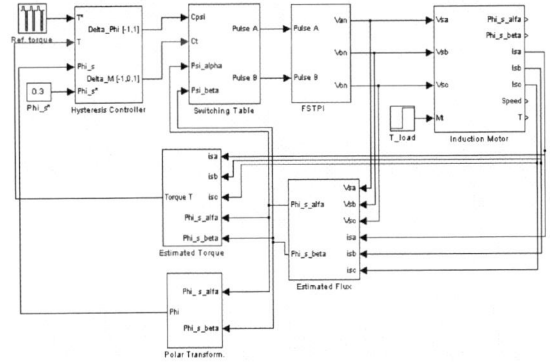

Fig. 1. DTC schema for FSTPI-Induction Motor.

This work was supported by Vietnamese Ministry of Science and Technology (MOST).

978-1-4244-0644-9/07/$25.00 ©2007 IEEE

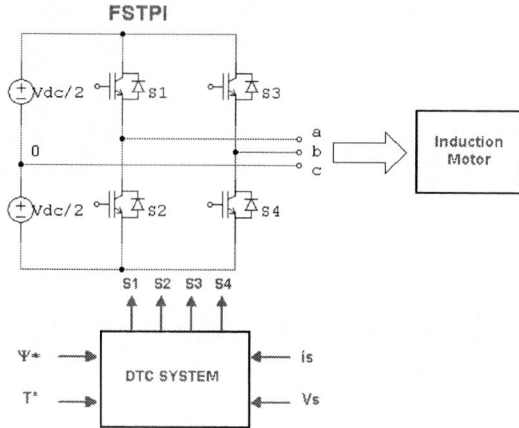

Fig. 2. Power circuit of FSTPI.

Combinations of switching S_1-S_4 result in 4 general space vectors $\vec{V}_1 \rightarrow \vec{V}_4$ (Fig.3, Table 1), components αβ of the voltage vectors are gained from abc voltages by using Clark's transformation:

$$\begin{bmatrix} V_\alpha \\ V_\beta \end{bmatrix} = \frac{2}{3}\begin{bmatrix} 1 & -\frac{1}{2} & -\frac{1}{2} \\ 0 & \frac{\sqrt{3}}{2} & -\frac{\sqrt{3}}{2} \end{bmatrix}\begin{bmatrix} V_a \\ V_b \\ V_c \end{bmatrix} \qquad (3)$$

Where V_a, V_b, V_c: output phase voltages on the load (Y connection), defined by:

$$V_a = \frac{1}{3}\left(2V_{a0} - V_{b0}\right) \ ;$$
$$V_b = \frac{1}{3}\left(2V_{b0} - V_{a0}\right); \qquad (4)$$
$$V_c = -\frac{1}{3}\left(V_{a0} + V_{b0}\right)$$

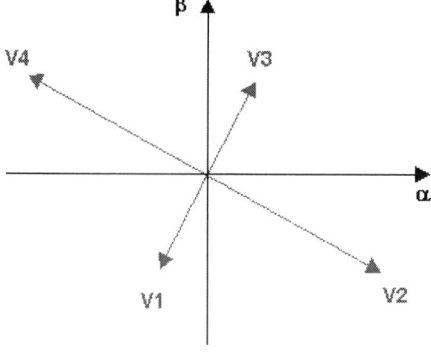

Fig. 3. Voltage space vectors in the plan αβ (*)

TABLE 1
COMBINATIONS OF SWITCHINGS AND VOLTAGE SPACE VECTORS

S1	S3	$\vec{V} = V_\alpha + jV_\beta$
0	0	$\vec{V}_1 = \frac{V_{dc}}{3}e^{-j\frac{2\pi}{3}}$
1	0	$\vec{V}_2 = \frac{2V_{dc}}{3}e^{-j\frac{\pi}{6}}$
1	1	$\vec{V}_3 = \frac{V_{dc}}{3}e^{j\frac{\pi}{3}}$
0	1	$\vec{V}_4 = \frac{2V_{dc}}{3}e^{j\frac{5\pi}{6}}$

To simulate six non-zero vectors in SSTPI, beside the two V_1 and V_3, it can be used the effective vectors V_{23M}, V_{43M}, V_{14M} and V_{12M}. These vectors are formed as follows:

$$\vec{V}_{23M} = \frac{1}{2}\left(\vec{V}_2 + \vec{V}_3\right) = \frac{V_{dc}}{3}e^{j0};$$

$$\vec{V}_{43M} = \frac{1}{2}\left(\vec{V}_4 + \vec{V}_3\right) = \frac{V_{dc}}{3}e^{j\frac{2\pi}{3}}; \qquad (5)$$

$$\vec{V}_{14M} = \frac{1}{2}\left(\vec{V}_1 + \vec{V}_4\right) = \frac{V_{dc}}{3}e^{j\pi};$$

$$\vec{V}_{12M} = \frac{1}{2}\left(\vec{V}_1 + \vec{V}_2\right) = \frac{V_{dc}}{3}e^{-j\frac{\pi}{3}};$$

To simulate zero vectors of SSTPI, we use the effective V_{0M}:

$$\vec{V}_{0M} = \frac{1}{2}\left(\vec{V}_1 + \vec{V}_3\right) \qquad (6)$$

The similarity between space vectors of FSTPI (Fig.4) and SSTPI (Fig.5) is presented in Table 2.

III. NOVEL SWITCHING TECHNIQUE FOR DTC

The objective of the DTC is to maintain the motor torque and stator flux within a defined band of tolerance by selecting the most convenient voltage space vector from the look-up table (switching table). In the case of the conventional switching table of DTC for FSTPI-IM, one of four active vectors is chosen (Table 3) [6].

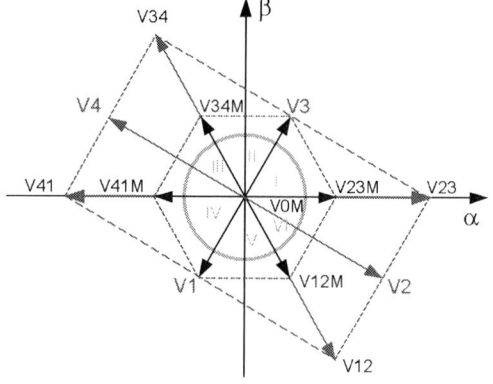

Fig. 4. Voltage Space-Vectors for FSTPI on the principle of similarity.

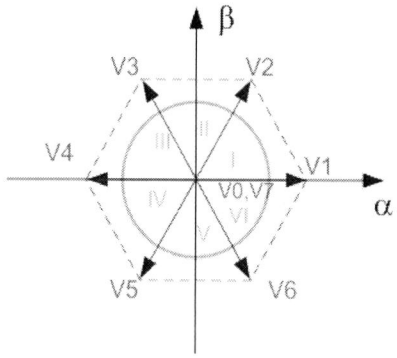

Fig. 5. Base space vectors in SSTPI.

TABLE 2
SIMILARITY BETWEEN SPACE VECTORS OF FSTPI AND SSTPI

Used voltage space vectors for SSTPI	Used voltage space vectors for FSTPI
V1	V23M
V2	V3
V3	V43M
V4	V14M
V5	V1
V6	V12M
V0, V7	V0M

TABLE 3
CONVENTIONAL SWITCHING TABLE FOR DTC CONTROL METHOD

Δψ	ΔT	Sector 1 240°÷330°	Sector 2 -30°÷60°	Sector 3 60°÷150°	Sector 4 150°÷240°
1	1	V2	V3	V4	V1
1	-1	V1	V2	V3	V4
0	1	V3	V4	V1	V2
0	-1	V4	V1	V2	V3

TABLE 4
PROPOSED SWITCHING TABLE FOR DTC CONTROL METHOD

Δψ	ΔT	SECTOR					
		I -30° ÷30°	II 30° ÷90°	III 90° ÷150°	IV 150° ÷210°	V 210° ÷270°	VI 270° ÷330°
1	1	V3	V43M	V14M	V1	V12M	V23M
	-1	V12M	V23M	V3	V43M	V14M	V1
	0	V13M	V13M	V13M	V13M	V13M	V13M
-1	1	V43M	V14M	V1	V12M	V23M	V3
	-1	V1	V12M	V23M	V3	V43M	V14M
	0	V13M	V13M	V13M	V13M	V13M	V13M

Table. 4. Proposed switching table for DTC method

The simulated DTC system driven by FSTPI uses a switching table (Table 4). The torque controller has 3 levels: -1, 0, 1; the flux controller has 2 levels: -1, 1.

The parameters used in simulation are given below:
- Vdc = 300V
- Torque hysteresis band = 10%
- Flux hysteresis band = 1%
- Reference flux ψ_s* = 0.3 Wb.
- Reference torque: T^* = 5 Nm when $0 \text{ s} \leq t \leq 0.1$ s; T^* = 8 Nm when $0.1 \text{ s} \leq t \leq 0.2$s; T^* = 5 Nm when $0.2 \text{ s} \leq t \leq 0.3$ s; T^* = -10 Nm when $0.3 \text{ s} \leq t \leq 0.4$s; T^* = 5 Nm when $0.4 \text{ s} \leq t \leq 0.5$s.
- Sample time : T_s = 5e-5
- Load torque T_L = 5Nm at t = 0.1s.
- Time of simulation t = 0.5s.

Furthermore, the comparison between the conventional DTC method [6] and the proposed one has been done.

In order to reduce the torque and speed ripples by using the principle of similarity for voltage space vectors, optimum switching table in the proposed method is established similarly for the SSTPI switching table. The αβ plan is divided in to six sectors, and for each sector, the optimal space vector is chosen accordingly to the required torque and flux by using the effective vectors (equations 5, 6). These vectors are synthesized using the basic space vectors with the duty cycle of 50% (switching period is Ts). The same way is done for effective zero space vector (Table 4).

The flux and torque calculations remain the same. The stator flux is estimated as follows:

$$\psi_{s\alpha} = \psi_{s\alpha0} + (v_{s\alpha} - i_{s\alpha} \cdot R_s) \cdot T_s$$
$$\psi_{s\beta} = \psi_{s\beta0} + (v_{s\beta} - i_{s\beta} \cdot R_s) \cdot T_s \quad (7)$$

The estimated flux $\widetilde{\Psi}$ and flux angle sector are defined as follows:

$$\widetilde{\Psi}_s = \sqrt{\psi_{s\alpha}^2 + \psi_{s\beta}^2} ; \theta_i = \arctan\left(\frac{\psi_{s\beta}}{\psi_{s\alpha}}\right) \quad (8)$$

The torque is estimated by the following formula:

$$\widetilde{T} = \frac{3P}{2}(\psi_{s\alpha} i_{s\beta} - \psi_{s\beta} i_{s\alpha}) \quad (9)$$

Where: $\mathbf{v_s}$, $\mathbf{i_s}$ stator voltage and current vectors
R_s stator resistance
P number of pole pair
T electromagnetic torque
$\mathbf{\psi_s}$ stator flux vector
T_s sampling time

IV. SIMULATION OF THE PROPOSED DTC METHOD FOR FSTPI-IM

A Simulink/Matlab is used to validate the proposed DTC method for FSTPI-IM.

The induction motor model for the simulation studies has the follows parameters:

Type: Three-phase, squirrel-cage induction motor. 220V, 1HP, 1680r/min, R_s = 3.2 (Ω), R_r = 2.336 (Ω), L_s = 0.2965 (H), L_r = 0.2965 (H), L_m = 0.2931 (H), P = 2, J_m = 0.0034 (kgm²).

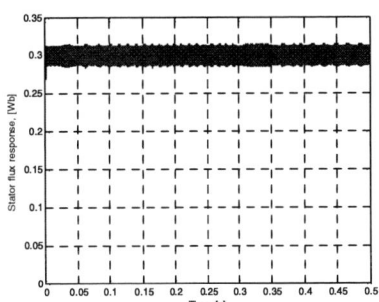

a) Conventional method

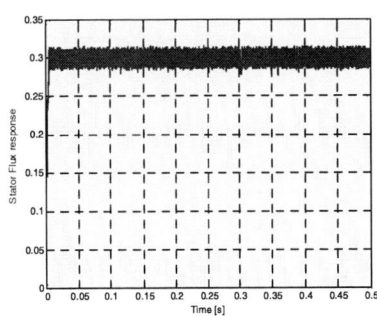

b) Proposed method

Fig. 6. Stator Flux responses.

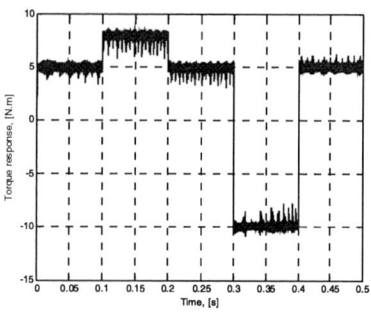

a) Conventional method

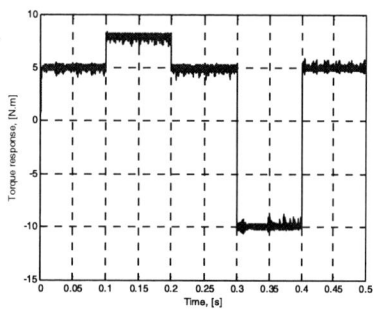

b) Proposed method

Fig. 7. Torque responses.

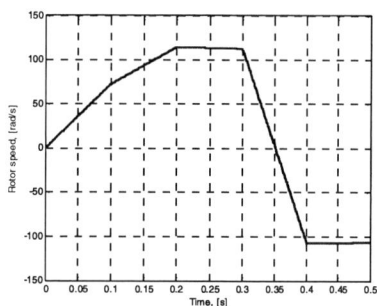

a) Conventional method

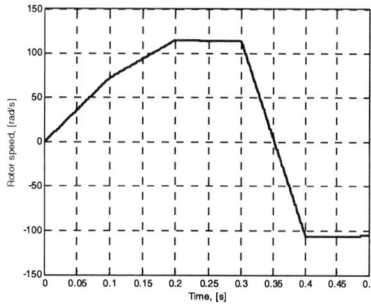

b) Proposed method

Fig. 8. Rotor speeds.

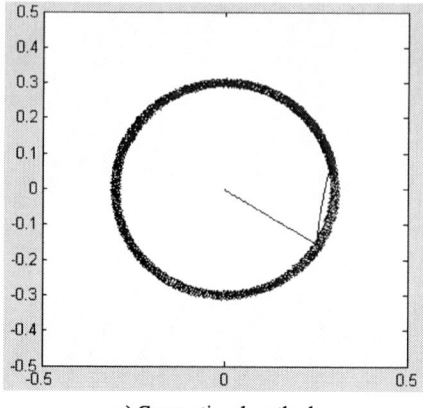

a) Conventional method

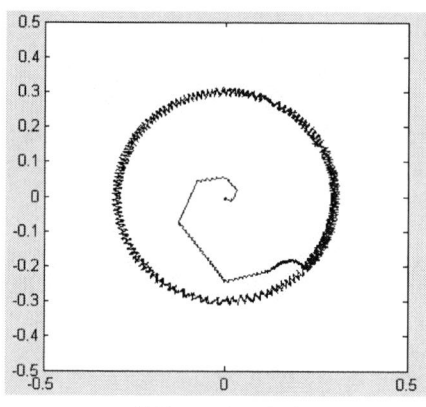

b) Proposed method

Fig. 9. Trajectories of flux space vector.

Simulation results demonstrate the performance of the proposed DTC for FSTPI-IM, while the good responses of the flux, torque, and speed are obtained (fig.6 - 9) with reduced torque ripples (fig. 7).

V. EXPERIMENTAL RESULTS

The experimental setup is carried out by a DSpace 1104 system with I/O card for real time control (Sample time: T_s = 5e-5). An interface board was build to receive the gate-drive signal, isolated them and connected to the four switches which were implemented using integrated IGBT 60A. The output from FSTPI was connected to a three phase induction motor (Fig.10).

The induction motor has the follows parameters: Three-phase, squirrel-cage induction motor 380V, Y connected, ½ HP, 1420r/min., R_s = 5.4972 (Ω), R_r = 7.466 (Ω), L_s = 1.3040 (H), L_r = 1.3040 (H), L_m = 1.2278 (H), P = 2, J_m = 0.0008 (kgm^2).

The DC link voltage was adjusted at 100V, and the split capacitors are rated at 6800μF.

The Hall-effect current sensors ($i_{sa,b,c}$) and voltage sensors ($V_{DC1,2}$), which have been used to receive feedback signals, are LEM LA55-P and LV25-P respectively.

The DTC system driven by FSTPI uses a proposed switching table (Table 4). The control algorithm is executed by Matlab/Simulink program (Fig.11) and it provides the duty cycles d_{Aon} and d_{Bon} for generating control signals. The parameters of the torque controller, the flux controller, torque hysteresis band, flux hysteresis band are the same as the simulated DTC system. Reference flux: $\psi_s^* = 0.2$ Wb. Reference torque: $T^* = 5$ Nm in starting time; $T^* = 2.5$ Nm in steady- state. Load torque: $T_L = 2.5$Nm.

The experimental responses including steady-state stator line voltages, stator currents, and transient response of stator flux are shown in Fig.12-14 correspondingly.

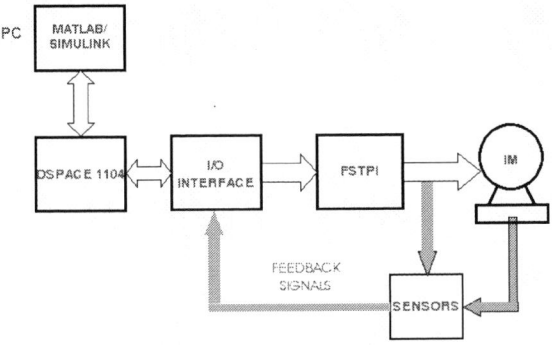

Fig. 10. Interface between Matlab/Simulink and Dspace 1104 Card for FSTPI-IM .

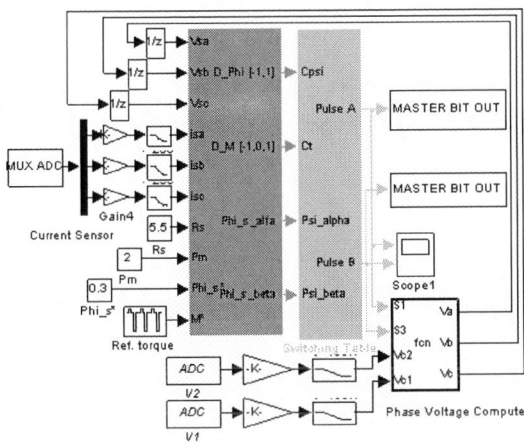

Fig. 11. DTC Control Algorithm using Matlab Simulink with DSpace Card DS1104.

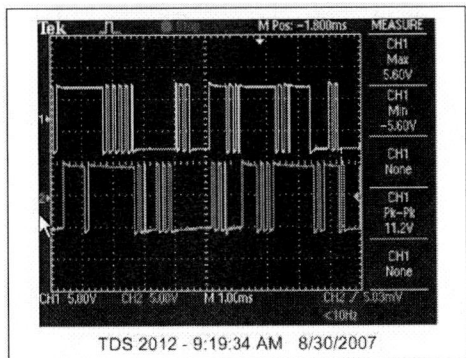

Fig. 12. Stator line voltage waveforms V_{ac}, V_{bc} (K=10) for FSTPI (From Oscilloscope Tektronix TDS 2012).

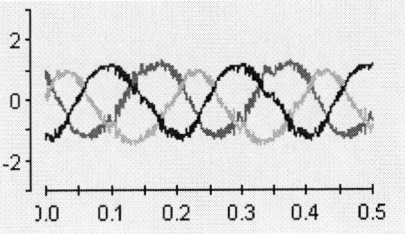

Fig. 13. Stator current waveforms for FSTPI.

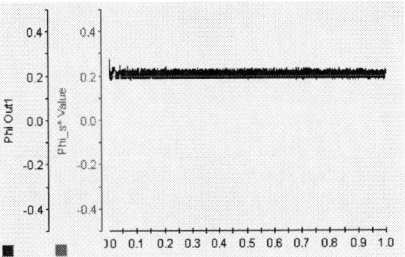

Fig. 14. Reference and estimated stator flux.

VI. CONCLUSION

A new switching technique for Direct Torque Control of Induction Motor using Four-Switch Three Phase Inverter (DTC-FSTPI-IM) for low power applications has been presented. The modified switching table in this method is based on the principle of similarity between FSTPI and SSTPI (Six-Switch Three Phase Inverter), where the $\alpha\beta$ plan is divided into 6 sectors and the formation of the voltage space vector is done in the same way as for SSTPI by using effective (mean) vectors. This approach allows using the well-knowing established switching table of SSTPI for FSTPI, in order to reduce torque ripples in comparison with the conventional DTC method for FSTPI. The validity of new proposed technique is verified by simulation and experimental results using Matlab/Simulink and DSpace 1104. Simulation and experimental results demonstrate the good performance of the proposed DTC for FSTPI-IM, while the good responses of the flux, torque, and speed are obtained.

REFERENCES

[1] Frede Blaabjerg,, Sigurdur Freysson, Hans-Henrik Hansen, and S. Hansen "A New Optimized Space-Vector Modulation Strategy for a Component-Minimized Voltage Source Inverter " *IEEE Trans. on Power Electronics*, Vol. 12, No. 4, July 1997,pp 704-710

[2] F. Blaabjerg, S. Freysson, H. H. Hansen, and S. Hariseri. "Comparison of a space-vector modulation strategy for a three phase standard and a component minimized voltage source inverter". *In Conf. Rec. EPE*, pages 1806-1813, Sevilha - Spain, September 1995.

[3] M. B. R. Correa, C. B. Jacobina, E. R. C. Da Silva, and A. M. N. Lima. "A General PWM Strategy for Four-Switch Three-Phase Inverters" *IEEE Trans. on Power Electronics*, Vol. 21, No. 6, Nov. 2006, pp 1618-1627.

[4] G. A. Covic, G. L. Peters, and J. T. Boys, "An improved single phase to three phase converter for low cost ac motor drives," *in Proc. PEDS '95*, Singapore, vol. 1, pp. 549–554.

[5] G.-T. Kim and T. Lipo, "VSI-PWM rectifier/inyerter system with a reduced switch count," *in Conf. Rec. IAS*, 1995, pp. 2327 - 2332.

[6] Mohamed Azab and A.L. Orille, "Novel Flux and Torque Control of IM Drive using FSTPI", *in IECON'01*, 2001,pages 1268 -1273.

[7] T. Noguchi, M. Yamamoto, S. Kondo, and I. Takashi, "High frequency switching operation of PWM inverter for direct torque control of induction motor," *in Conf. Rec. IEEE-IAS Annu. Meeting*, 1997, pp. 775–780.

[8] Y. S. Lai, T. Y. Shihn, Y. S. Kuan, and H. C. Huang, "A novel inverter control technique for direct torque control drives" (in Chinese), *J. Power Electron. Technol.*, vol. 39, pp. 71–77, 1997.

[9] C. Lascu, I. Boldea, and F. Blaabjerg, "A modified direct torque control (DTC) for induction motor sensorless drive," in Conf. Rec. IEEE-IAS Annu. Meeting, 1998, pp. 1887–1894.

[10] Y. S. Lai and J. H. Chen, "A new approach to direct torque control of induction motor drives for constant inverter switching frequency and torque ripple reduction," *IEEE Trans. Energy Conversion*, vol. 16, pp. 220–227, Sept. 2001.

[11] T. G. Habetler, F. Profumo, M. Pastorelli, and L. M. Tolbert, "Direct torque control of induction machines using space vector modulation," *IEEE Trans. Ind. Applicat.*, vol. 28, pp. 1045–1053, Sept./Oct. 1992.

[12] G. Buja, D. Casadei, and G. Serra, "Direct stator flux and torque control of an induction motor: Theoretical analysis and experimental results," *in Proc. IEEE IECON'98*, vol. 1, 1998, pp. T50–T64.

[13] Yen-Shin Lai, Wen-Ke Wang, and Yen-Chang Chen. "Novel switching techniques for reducing the speed ripple of AC Drives with DTC" *IEEE Trans. on Ind. Electronics*, Vol. 51, No. 4, 2004, pp 768-775.

Detection of Some Parameters of Induction Motors a Proposal and Its Verification

H. Bülent Ertan and Volkan Sezgin

Middle East Technical University
Electrical & Electronics Eng. Department
06531 Ankara, TURKEY

Barış Çolak

TUBITAK – Space Technologies Research Institute,
Middle East Technical University Campus
Ankara, TURKEY

Abstract- This paper describes a novel approach for offline stator leakage inductance estimation method that can be used for self-tuning of induction motor drives. The paper briefly describes the theory behind the approach. The proposed method is based on applying a DC voltage to the motor windings and measuring the DC current that flows in the circuit over a short interval of time. The measured variables are DC quantities and relatively noise free as compared to some other available methods. The proposed approach is tested on three industrial induction motors. It is shown that the prediction accuracy is very good. However, it is noted that some means of filtering the prediction fluctuations in the measurement is desirable. An on-line stator resistance prediction method is also introduced in the paper. This method requires very little computation effort as most of the variables requires for the calculations are already done within a sensorless vector control implementation. When this approach is tested on three industrial motors its estimation results are found to be very accurate.

Index terms- Self-commissioning, parameter estimation, online parameter identification

I. INTRODUCTION

To obtain a better performance from an induction motor drive it is essential to identify the circuit parameters of the motor driven. Most modern drives house a micro-processor inside. Even in simple scalar drives the user enters the rated values of the motor. The processor estimates the parameters of the motor [1]. The estimated parameters are used to obtain a better performance.

High performance drives employ, field orientation control (FOC) or direct torque control (DTC). These techniques require motor parameters (stator resistance, inductance, etc) to estimate the stator and rotor field vector positions and employ a control strategy, which tries to keep their relative position so that the desired torque output is obtained at the desired flux level. Such drives therefore employ a measurement process, before the actual torque control operation, for determining the parameters of the motor connected to their terminals. This procedure is called "self-commissioning" or "self-tuning". Once the motor starts running, some of motor parameters vary due to changing temperature of the motor or due to other operating conditions, such as operating the motor at various levels of mutual flux. Hence, an "intelligent" motor driver is required to measure the motor parameters also during the operation [2] of the motor drive.

The theory and application of FOC or DTC techniques are discussed in many books and papers. Therefore this issue shall not be discussed here.

Such motor drives commonly measure the terminal voltage and phase currents of the motor they drive. This paper shall discuss parameter estimation in the field oriented control context. There are two basic models for field-oriented control of induction motors [3]:

- Current model: This model requires an accurate measurement of the rotor time constant and accurate measurement of mechanical speed of the motor [4]. This type of drive can be controlled down to zero speed;
- Voltage model: This model does not require speed measurement; however its accuracy depends on precise measurement of stator resistance and leakage inductance. In this approach; it is not possible to maintain field orientation at very low speeds.

The first approach is advantageous if the additional cost of the speed measurement is not a problem and if a speed measurement device can be implemented to the motor shaft. The second approach is called "sensorless" in the sense that a shaft speed measurement is not required.

This paper presents the application results of an approach for determination of the stator leakage inductance of an induction motor for self-commissioning purposes. Also, a novel on-line stator resistance measurement technique and its application results are presented. Before proceeding with the introduction of the methods and their application results it is appropriate to discuss their merits in comparison with existing methods for the same purposes. Many proposals exist in the literature for determination of the said parameters. It is virtually impossible to discuss them all within the available space. So the discussion here will be limited to methods investigated by the authors.

II. OFF-LINE STATOR LEAKAGE INDUCTANCE ESTIMATION

In this part of the paper, we shall concentrate on methods for determination of the leakage inductance of an induction motor for self-commissioning. In the literature quite a number of studies exist which address the self-tuning problem [2-5]. One of the solutions for the offline estimation of the stator leakage inductance is applying a series of voltage pulses to the machine and stator transient time constant is measured [4]. The highest level of current in this method is at most 30% of the rated current; however rated flux or current should be present during estimation process for more realistic results. In [10], a complex voltage waveform is applied to the machine during the estimation of the stator leakage inductance; hence this method requires a powerful microcontroller and complex hardware. There are other methods that require high performance hardware for the calculation of the stator leakage inductance [6,7] The method in this paper is much simpler and efficient than the methods mentioned above for stator leakage inductance estimation. In short its advantages are as follows:

- Estimation can be performed at rated flux or current, hence its results are realistic [4];
- The region where the estimation is performed is much longer then the software control cycle to take enough number of current samples, thus enabling the automated estimation [6];
- This method creates little computational burden for the microprocessor; extra hardware is not needed [6,7];
- Current measurement, used for leakage inductance measurement, is a noise free signal [4,5] (the phase is short circuited during the measurement and the current measured is DC);
- This method works even for the motors with low stator phase voltages, for which parameter estimation is particularly difficult [13].

In the following section, first the theory behind the proposed algorithm shall be briefly introduced as it is already described elsewhere [13]. Following this, the off-line leakage inductance predictions using the proposed approach on three 3-phase induction motors will be given and compared with the conventional measurement findings. The basic label data of the test motors are given in Table 1.

TABLE 1
THE TEST MOTOR DATA

Rating Test Motor	Power rating (kW)	poles	Rated current (A)	Voltage (V)	Rotor time constant (ms)
1	1.0	2	2.7	380 Y	66.1
2	1.1	6	3.1	380 Y	44.2
3	2.2	2	4.8	380 Y	129.7

A. The theory

Stator transient inductance term (σL_S) refers to the inductance value measured from stator winding terminals. This term is defined by the following equation [5]:

$$\sigma L_S = \frac{L_S L_r + L_m (L_S + L_r)}{L_m + L_r} \quad (1)$$

Where, L_S and L_r are stator and rotor leakage inductances, respectively; and L_m is the magnetizing (mutual) inductance. With the assumption of $L_m \gg L_r$, usually, stator transient inductance (σL_S) is considered as the sum of stator and rotor leakage inductances. When the machine is not energized and the initial flux is zero at standstill, the motor equivalent circuit may be considered as a series connected resistive-inductive load.

The proposed stator transient inductance measurement method relies on forcing a user-defined dc current through the stator phase windings for short time duration. This current establishes the flux in the motor. Note that; applying a controlled current assures that the motor and the inverter is not subjected to over currents [3].

After the rotor flux is built up to the desired value; the stator phase windings are short-circuited with the aid of the inverter switches, As soon as the stator phases are short-circuited, stator phase voltages decrease to zero and the stator phase currents start to decrease. At this condition, the stator voltage (U_S) and rotor magnetizing current (I_{mr}) are given by the following equations [4]:

$$I_{mr} = \psi_r / L_m \quad (2)$$

$$U_S = R_S I_S + \sigma L_S \frac{dI_S}{dt} + (1 - \sigma) L_S \frac{dI_{mr}}{dt} \quad (3)$$

Where, R_S is the stator resistance, I_S is the stator phase current, and ψ_r is the rotor flux.

A typical variation of the phase current after the short circuit is shown in Fig. 1. A short time interval right after short-circuiting a phase (this interval is much shorter than the rotor time constant, see Table 1) is called the "sub-transient region". Since the rotor flux is a slow changing variable, in the sub-transient region, the change in the rotor flux may be neglected and it may be assumed that the flux stays at the value set by the drive. Therefore, Equation (2) may be simplified and take the form shown in (4). σL_S (stator transient inductance) may be calculated using this equation, provided that the stator resistance (R_S) is known and the stator current is measured.

$$U_S = R_S I_S + \sigma L_S \frac{dI_S}{dt} \quad . \quad (4)$$

Suppose that the stator current is measured at every Δt seconds. In this case, dI_S may be approximated by the difference of the two consecutive measurements of current, namely between 'n'th and 'n-1'th measurements. Therefore, Equation (4) may be written as in (5) and each variable is measured or calculated at every Δt seconds.

$$U_S(n) = R_S I_S(n) + \sigma L_S(n) \frac{I_S(n) - I_S(n-1)}{\Delta t} \quad (5)$$

$$\sigma L_S(n) = \frac{(U_S(n) - R_S I_S(n))\Delta t}{I_S(n) - I_S(n-1)} \qquad (6)$$

In the application here, σL_s is calculated "n" times (10 times here), with 0.1 ms intervals using (6). Since the stator resistance is known (can be measured via some well-known method), L_s is easy to calculate.

B. Test Results

Fig. 1 displays the voltage and current recordings on motor 2 during off-line inductance measurement. Note that the measurement process lasts 1 ms while the rotor time constant of this motor is 66.1 ms, therefore the rotor flux remains virtually constant during the measurement.

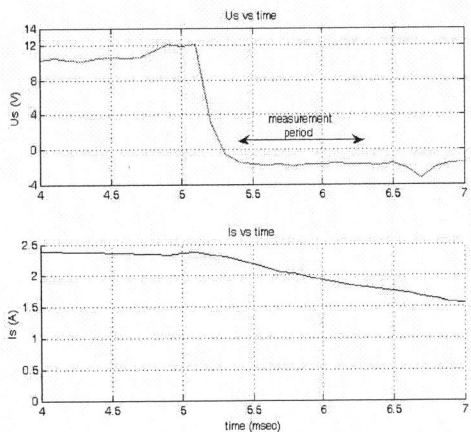

Figure 1 Measured data for leakage inductance prediction

TABLE 2
COMPARISON OF CONVENTIONAL AND PROPOSED STATOR RESISTANCE
AND LEAKAGE INDUCTANCE MEASUREMENT RESULTS

Test motor	Stator Resistance measurement (ohm)		Stator leakage inductance (mH)	
	Proposed on-line measurement algorithm	Conventional measurement	Proposed off-line measurement algorithm	Conventional measurement
1	4.74 (%5.5)	4.48	11.98 (%2.3)	11.70
2	6.22 (%6.2)	6.47	31.28 (%5.1)	29.75
3	2.66 (%2.7)	2.74	14.85 (%13.0)	13.14

During the measurement first few data points are not used in the calculation to avoid parasitic effects of the switching that could lead the sensor reading errors. The calculations are done for the following 10 measurements. Fig. 2 displays the calculated leakage inductance values for the data points recorded. The average of the calculated leakage inductance values is assumed to be the predicted value. The scattering in the data points is due to the difference term in Eq. 6. The predicted leakage inductance value 31.28 mH, and the measured value using conventional methods is 29.75 mH. The error is only about 5%. The predicted and measured values for the other two test motors are given also in Table 2. It can be observed that the largest prediction error is about 13%.

It is obvious that the prediction error can be reduced if the scattering in the calculations can be reduced by filtering the data in some way. This issue needs to be investigated further. However, even with the raw processing applied here the prediction accuracy is acceptable.

III. ON-LINE STATOR RESISTANCE ESTIMATION METHOD

For both the rotor and stator-flux oriented vector control methods; which are based on the "voltage model" accurate estimation of feedback signals (e.g. flux vector, speed, and frequency), is very important. The variation of the stator resistance with temperature therefore is especially very important in the prediction process. The estimation error becomes even more important near zero speed, where the stator resistance voltage drop tends to be comparable with machine counter-electromotive force

(EMF). Therefore, some means of accounting for the resistance variation due to temperature is essential.

One method for correctly estimating the resistance value during operation is placing temperature-sensing thermistors in the stator winding in a distributed manner and measuring the average stator winding temperature. Once the temperature of the winding is known, estimating the stator resistance becomes a simple matter, given its value at a known temperature [7]. However, the use of such sensors in a sensorless drive is not acceptable.

Another method for online estimation of the stator resistance, based on the zero sequence model is presented in [8]. The proposed method provides an estimate of the stator resistance that does not require the knowledge of any other parameter and is independent of the drive control strategy. The method requires the neutral wire of the machine be available and the measurement of the phase currents and voltages. However, in some cases neutral wire is not available.

Ha and Lee proposed a novel online identification method that can be used for estimating both R_s and R_r [9]. In this approach a small magnitude and low frequency AC component is injected to the flux command. Their identification algorithm for R_s is motivated from the steady state power flow between stator and rotor through the air gap. This method can be applied only under steady state conditions. Furthermore the injected low frequency signal may cause problems at low speed.

The estimation scheme reported in [14] uses a model

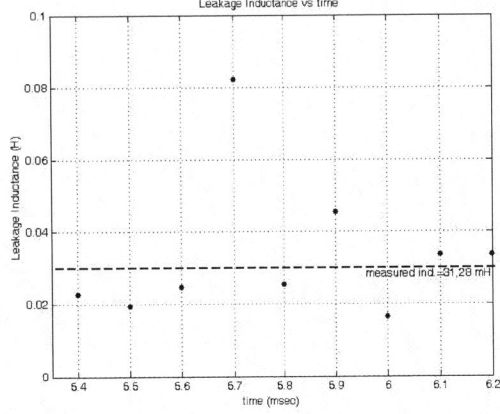

Figure 2 Predicted leakage inductance in the measurement period

1509

reference adaptive approach with proportional-integral controllers. The difference between the measured and observed current is an input for the estimation system.

[15] describes also a stator resistance estimation method developed in conjunction with the rotor flux-based MRAS speed estimator and it operates in the stationary reference frame. Moreover, the error quantity is formed on the basis of differences in rotor flux component values, obtained at the output of the reference and the adjustable model.

Another method for stator resistance estimation is utilizing artificial neural networks [16]. The error between the measured stator current and the estimated stator current using neural network is back propagated to adjust the weights of the neural network.

The method proposed here is simple and promises to eliminate the disadvantages of available resistance estimation methods.

A. The Field Orientation Control Method

The stator resistance estimation algorithm in this paper is developed for a drive using "voltage model" for vector control. As it is well known the "voltage model" approach requires the stator resistance and leakage inductance to be estimated accurately. This section briefly describes the vector control method used by the authors

In an induction machine the torque is a function of both the stator and rotor currents, also when stator current is controlled both the rotor flux and torque values are affected. Hence, a direct control of torque with stator currents becomes impossible [1]. However, by transforming the three phase system (a-b-c) into stationary (α-β) and synchronously rotating (d-q) reference frames and using Clarke and Park transformations respectively and coinciding the d-axis and rotor flux (Ψ_r); the electrical torque equation becomes as follows [11]:

$$T_e = \frac{p}{2}\frac{3}{2}\frac{L_m}{L_r}\Psi_{rd}I_{sq} \tag{7}$$

where p is the number of poles, Ψ_{rd} is the d-axis

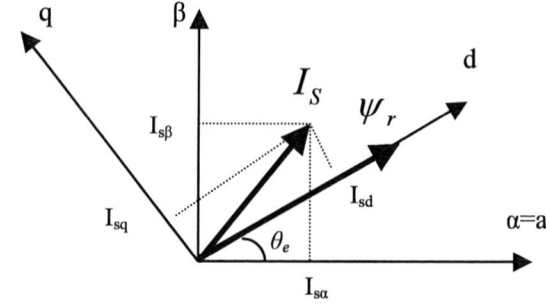

Figure 3 Vector diagram for field orientation of induction motor

component of the rotor flux and I_{sq} is the q-axis component of the stator current.

Fig. 3 shows the vector diagram of the reference frames, the projections of stator current and rotor flux, and the rotor flux position with respect to α-axis (θ_e). In the steady state, namely if the derivative of Ψ_{rd} is zero, Ψ_{rd} can be expressed as follows:

$$\psi_{rd} = L_m I_{sd} \tag{8}$$

Therefore, Equation (7) can be rewritten as in (9); which means that when the rotor flux is maintained constant, the torque of the motor can be controlled through I_{sd} and I_{sq}.

$$T_e = \frac{p}{2}\frac{3}{2}\frac{L_m^2}{L_r}I_{sd}I_{sq} \tag{9}$$

Fig. 4 shows the Field Orientation Control (FOC) scheme used by the authors. The drive utilizes the voltage control method without speed sensors; hence the observed parameters in the drive are the stator voltages (U_S) and the stator currents (I_S). Since it is assumed that the system is balanced, it is enough to measure the two of the currents in the three-phase motor with sensors. After two stator currents and voltages are measured and transformed into d-q components (I_{sd} and I_{sq}), these values are compared with the reference values, which are calculated from the reference torque and flux values using (8) and (9). The error is inserted into a PI regulator in order to generate the voltage references for the PWM

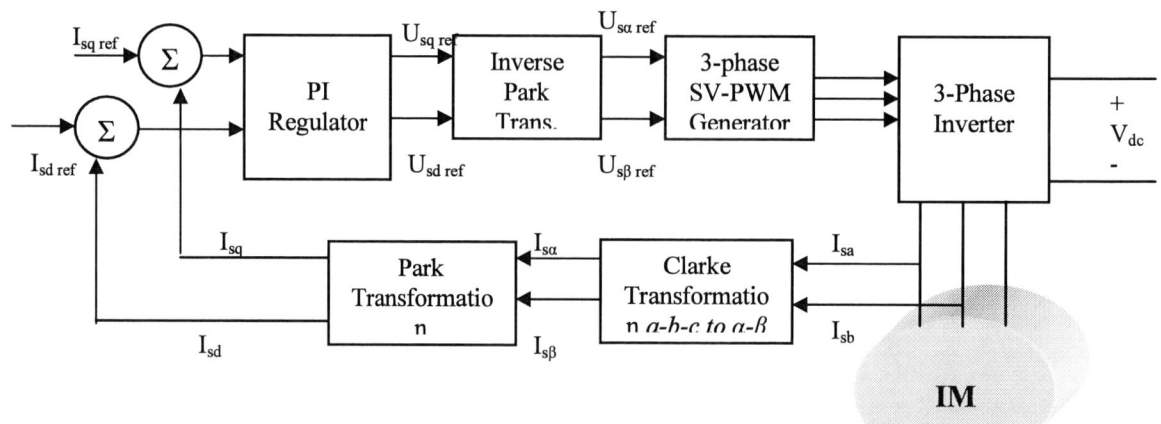

Figure 4   Field Orientation Control (FOC) scheme for the motor drive under test

generator, which utilizes Space Vector Pulse Width Modulation (SVPWM) technique.

The actual stator (Ψ_S) and rotor flux (Ψ_r) are given by the following equations:

$$\psi_S = \int (U_S - R_S I_S)dt \qquad (10)$$

$$\psi_r = \frac{L_r}{L_m}(\psi_S - \sigma L_S I_S) \qquad (11)$$

The synchronous speed (ω_S), which is the derivative of the flux position with respect to α-axis (θ_e), is calculated with the following equation:

$$\psi_{r\alpha} = \frac{L_r}{L_m}(\psi_{s\alpha} - \sigma L_s i_{s\alpha}) \quad \psi_{r\beta} = \frac{L_r}{L_m}(\psi_{s\beta} - \sigma L_s i_{s\beta})$$

$$\psi_r = \sqrt{\psi_{r\alpha}^2 + \psi_{r\beta}^2}$$

$$\theta = \tan^{-1}(\frac{\psi_{r\beta}}{\psi_{r\alpha}}) \qquad \cos\theta = \frac{\psi_{r\alpha}}{\psi_r} \qquad \sin\theta = \frac{\psi_{r\beta}}{\psi_r}$$

$$\omega_s = \frac{d\theta}{dt} = \dot{\theta} = \frac{(\psi_{r\alpha}\cdot\dot{\psi}_{r\beta} - \psi_{r\beta}\cdot\dot{\psi}_{r\alpha})}{\psi_r^2} \qquad (12)$$

The dotted terms indicate the derivative of the variables involved.

B. Online Resistance Estimation Method

The stator voltage vector magnitude at the α-β reference frame ($U_{s\alpha}$ and $U_{s\beta}$) is given by [1]:

$$U = \sqrt{U_{s\alpha}^2 + U_{s\beta}^2} \qquad (13)$$

$$U_{s\alpha} = R_s I_{s\alpha} - \omega_s \psi_{s\beta} \qquad (14.a)$$

$$U_{s\beta} = R_s I_{s\beta} + \omega_s \psi_{s\alpha} \qquad (14.b)$$

From the above equations the stator resistance is given by:

$$R_s = \frac{U_{s\beta} - \omega_s \psi_{s\alpha}}{I_{s\beta}} \qquad (15)$$

In (15) $U_{s\alpha}$, $I_{s\beta}$ are calculated from measurements and $\Psi_{s\alpha}$, can be determined as explained below. The algorithm introduced here relies on measurements of stator voltage and current at the zero crossing instant of I_{sa}; $I_{sa} = I_{sa} = 0$. At this condition;

$$U_{s\alpha} = -\omega_s \psi_{s\beta} \qquad \psi_{s\beta} = -(\frac{U_{s\alpha}}{\omega_s}) \qquad (16)$$

where ω_s can be calculated from (12). The total flux linkage (Ψ_s) is the vector summation of the d and q components of the flux linkage:

$$\text{where} \quad \psi_s = \sqrt{\psi_{s\alpha}^2 + \psi_{s\beta}^2} = \sqrt{\psi_{sd}^2 + \psi_{sq}^2} \qquad (17)$$

I_{sq} is found from measured current and $\psi_{sd} = \Psi_{sd(ref)}$

$$\psi_{sq} = \sigma L_s I_{sq} \qquad (18)$$

In (18) $\Psi_{sd(ref)}$ is the reference value for the d-component of the stator flux linkage. Substituting (17) and (18) into (16), Ψ_s can be found. Therefore from (16) $\psi_{s\alpha}$ can be found and R_s can be calculated form (15). Note that a sensorless vector controlled drive algorithm calculates $U_{s\alpha}$, $I_{s\beta}$, ω_s and $\psi_{s\alpha}$ therefore the computational burden of the stator resistance prediction method, introduced here, is minimal.

C. Experimental Results

For verifying the online stator resistance estimation algorithm, the method proposed is applied to the three test motors described in table 1. During the test the motors are driven with the vector control algorithm presented in the section III.A. The test arrangement is such that right after the estimation process is completed the motor terminals are connected to a resistance measurement instrument via a switch and the stator resistance of the test motor is recorded. In this manner it is assured that the predicted value can be compared with a conventional measurement and the temperature rise effect on the results can be minimized.

It is of interest to find out how the algorithm performs while the test motors are operating at various speeds (frequencies). It is also interesting to know whether the value of the phase resistance plays any role in the prediction accuracy. However the results presented here are limited to the findings at 10 Hz. The results found while the measurement is performed at other frequencies are in fact quite similar. The measurement results presented here are made under steady-state conditions. Since the focus here is on the performance of the algorithm the information provided here is limited to measurements and calculations made at zero crossing of phase current "a" for one cycle.

When phase current "a" is reduced below a certain value the drive starts estimating the stator resistance. The estimation is done once in every 100 μs interval. Fig. 5 displays $U_{s\beta}$, $I_{s\beta}$, $\psi_{s\alpha}$ and $I_{s\alpha}$ calculated from the information recorded for two cycles for test motor 1. The voltage and current recordings in the figure are obtained after filtering the actual waveforms with a filter whose corner frequency is 2 kHz. A slight distortion can be observed on the voltage waveform as well as on the current waveform. However, the flux waveform is very smooth.

In Fig. 6 the variables used in the resistance calculation are displayed in the vicinity of the zero crossing region of phase "a" current The figure covers a duration of 1 ms in which 10 resistance calculations are made The last graph displays the calculated resistance values at measurement points. It can be observed that the variation of the value of the measured resistance remains within a narrow band. The average value of the measurements taken is 4.74 ohm. The value found from the conventional measurement is 4.48 ohm. The error in predicted value is only 5.5%. The predicted resistance value for the other two motors is given in Table 2. It can

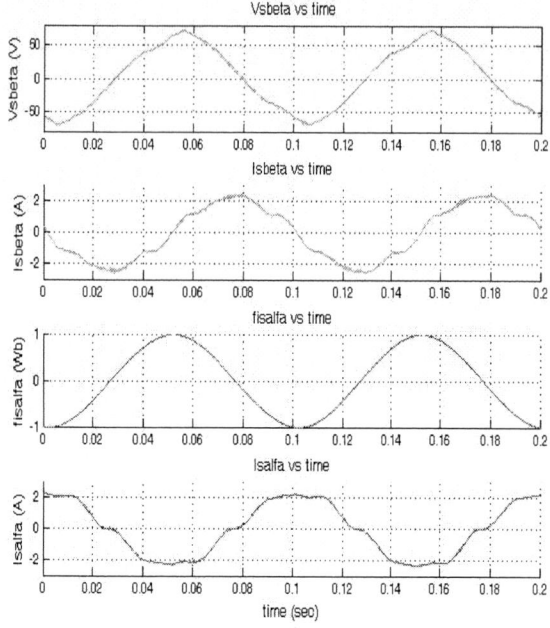

Figure 5 Time variation of the variables used for stator resistance prediction (Test motor 1)

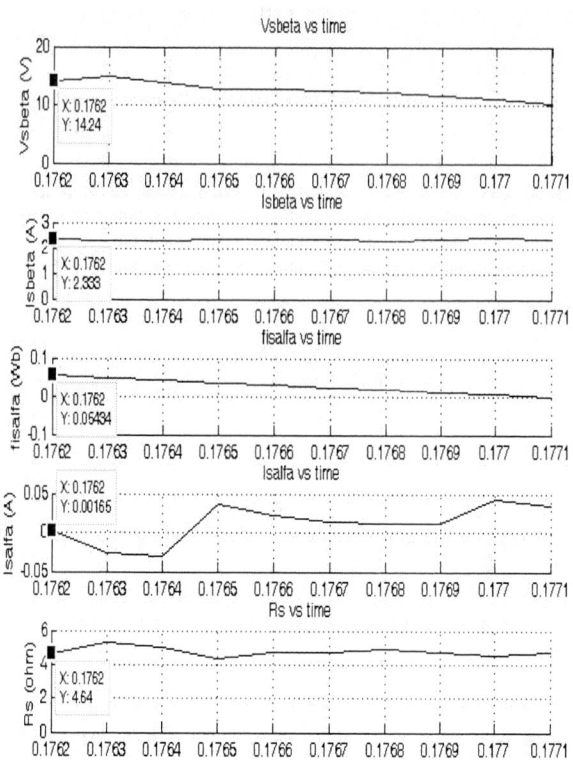

Figure 6 The variables used for resistance prediction during resistance prediction (test motor 1). The last graph displays the predicted resistance

be observed that the prediction error is even smaller for these two motors.

The proposed approach is tested also while the test motors are operated under load. The test motor resistance is predicted every 15 minutes and also measured with a resistance meter. The findings for test motor 1 are presented in Fig. 7. It can be observed that the predicted resistance is within 3% of the conventional measurement during the test period of 60 minutes. Note that if the stator resistance of the motor is known at a specified temperature, the resistance prediction method can be used not only for a more accurate sensorless control of the motor but also for prediction of the motor temperature.

The results presented in this section are very promising and leads to the conclusion that the proposed algorithm can be used for the estimation of the stator resistance in an induction motor controller.

IV. CONCLUSION

This paper first describes a novel approach for measurement of the stator leakage inductance of an induction motor during self-tuning. The approach is tested on three different motors of different pole numbers and power ratings. The prediction results are found to be acceptably accurate. Although the measurements needed for the proposed measurement method are relatively noise free, the difference equation involved in the calculations is found to lead considerable scattering in the predictions. Some means of filtering the data needs to be developed.

In section II an on-line stator resistance method is introduced. This method is suitable for measurements both at steady state and dynamic conditions. The calculation burden of the approach is very small for a

Rs vs time

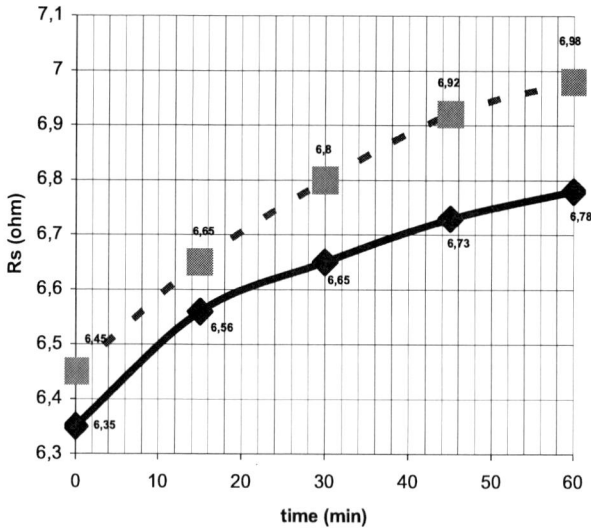

Figure 8 The variation of stator resistance of test motor 1. Solid line is from conventional resistance measurement. The dashed line presents the predicted values.

sensorless vector controlled drive as the measurements and most of the calculations needed are already done within the vector control algorithm. The proposed approach for stator resistance prediction is tested on three different test motors, operating at steady-state conditions and is found to lead very good results. Although it is not

1512

reported here the tests are also performed at different supply frequencies with the driving inverter. These tests indicate that the proposed approach leads to good results irrespective of the supply frequency.

REFERENCES

[1]. H. B. Ertan, Ç. H. Özyurt "Prediction of Induction motor Parameters From Manufacturer's Data" SIELA 2005, Plovdiv, Bulgaria, 2-3 June 2005

[2]. P. Vas, *Sensorless Vector and Direct Torque Control*, Oxford University Press, 1998.

[3]. H. Can, "Implementation of Vector Control for Induction Motors", *MSc. Thesis,* Middle East Technical University, Ankara, Turkey, 1999

[4]. A.M. Khambadkone, J. Holtz, "Vector Controlled Induction Motor Drive With a Self Commissioning Scheme", *IEEE Transactions on Industrial Electronics,* Vol 38 No 5, pp. 322-327, Oct. 1991.

[5]. P. Vas, *Parameter Estimation, Condition Monitoring, and Diagnosis of Electrical Machines*, Clarendon Press, Oxford 1993.

[6]. J.L. Silvino, B.C. Rabelo, "An Improved Estimation of The Inducion Machine Leakage Inductances", *IEEE Transactions on Industrial Electronics*, Letter to the Editor, Vol 46, No.5, pp. 1040-1042, Oct. 1999.

[7]. Y. Lai, J. Lin, J. Wang, "Direct Torque Control Induction Motor Drives with Self Commissioning Based on Taguchi Methodology", *IEEE Transactions on Power Electronics*, Vol 15, No 6, pp. 1065-1071, Nov. 2000.

[8]. Y. Lin, C. Chen, "Automatic IM Parameter Measurement under Sensorless Field-Oriented Control", *IEEE Transactions on Industrial Electronics,* Vol. 46, No 1, pp. 111-127, Feb. 1999.

[9]. B. Bose, N. Patel, "Quasi-Fuzzy Estimation of Stator Resistance Estimation of Induction Motor", *IEEE Transactions On Power Electronics*, Vol. 13, No. 3, pp. 401-409, May 1998.

[10]. C.B. Jacobina, J.E.C. Filho, A.M.N. Lima, "Online Estimation of the Stator Resistance of Induction Machines Based on Zero Sequence Model", *IEEE Transactions on Power Electronics*, Vol. 15, No.2, pp. 346-353, March 2000.

[11]. I. Ha, S. Lee, "An Online Identification Method for Both Stator and Rotor Resistances of Induction Motors without Rotational Transducers", *IEEE Transactions on Industrial Electronics*, Vol.47, No. 4, pp. 842-853, Aug. 2000

[12]. E. Akin, H.B. Ertan, M.Y. Uctug, "Vector Control of Induction Motor through Rotor Flux Orientation with Stator Flux Components as Reference", *International Conference on Electrical Machines 92*, Manchester, pp.853-857, 1992.

[13]. H. B. Ertan, E. Murat, B. Çolak, "A Novel Approach to Detection of Some Parameters of Induction Motors" , IEEE - IEMDC'07 May 3-5, 2007, Antalya Turkey, AF 012327.

[14]. G. Guidi, H. Umida, "A novel stator resistance estimation method for speed-sensorless induction motor drives," *IEEE Trans. Ind. Appl.*, vol. 36, pp. 1619–1627, Nov./Dec. 2000.

[15]. V. Vasic, S. N. Vukosavic, E. Levi, "A Stator Resistance Estimation Scheme for Speed Sensorless Rotor Flux Oriented Induction Motor Drives", *IEEE Trans On Energy Conversion*, Vol.18, No.4, pp. 476-483, Dec 2003.

[16]. B. Karanayil, M.F.Rahman, C. Grantham, "Online Stator and Rotor Resistance Estimation Scheme Using Artificial Neural Networks for Vector Controlled Speed Sensorless Induction Motor Drive", *IEEE Transactions on Industrial Electronics*, Vol. 54, No. 1, pp. 167-146, Feb 2007.

Comparison of Basic Direct Torque Control Designs for Permanent Magnet Synchronous Motor

M. N. Abdul Kadir, S. Mekhilef, and W.P. Hew

Malaya University, 50603 Kuala Lumpur, Malaysia

Abstract- **Basic PMSM-DTC drive controller consists of flux and torque hysteresis comparators and switching table. In the literature, three different designs have been reported. This paper investigates the three designs and compares their performance. The study aims to trace the reasons behind the recommendation of avoiding zero voltage states for PMSM-DTC reported in some studies. The system which avoids the zero states is compared with two systems include the zero states in their switching tables. The three systems are compared in their response time, switching losses, torque and flux ripples with three different input commands. The results as well as the theoretical discussion, show that the zero states should not be excluded from the switching table. The study concludes that the two level torque comparator, eight-state table design is suitable for two quadrant drives, while the three-level comparator, eight-state table is the most suitable for the four-quadrant drives.**

Index Terms—**DTC, permanent magnet synchronous motor, torque ripple, torque control.**

I. INTRODUCTION

The concept of the basic direct torque control (DTC) drive is to determine the required changes in the torque and flux of the motor under control, and operate the inverter supplying the motor in the switching state of voltage vector that best produce the required changes. The concept has been applied successfully for the induction motors and extended to the permanent magnet synchronous motor (PMSM) drives [1]. It has been shown that the PMSM-DTC inherits the main advantages of the induction motor-DTC drive such as the absence of current control, omitting the axis transformation, and less dependence on the motor parameters [1].

The basic PMSM-DTC drive is shown in Fig.1, in this drive the switching table is formed from the predetermined selections of the next state aiming to apply the above concept. Switching table designs vary according to the number of output levels of the torque hysteresis controller and the operation quadrants. The main purpose of this study is to compare these designs. The comparison includes the speed of response, the ripple contents and the switching losses. The three designs considered can be categorized into two classes, the first omitting the zero states [1]-[3], or 6-state class and the next includes the zero states or 8-state class [4] and [5].

Besides comparing the three DTC schemes the study aims to examine and discuss the recommendation to avoid the zero state selection reported in [1]–[3] as other studies [4]-[7] have not acknowledged this practice. The justifications for excluding the zero state from the operation states are:

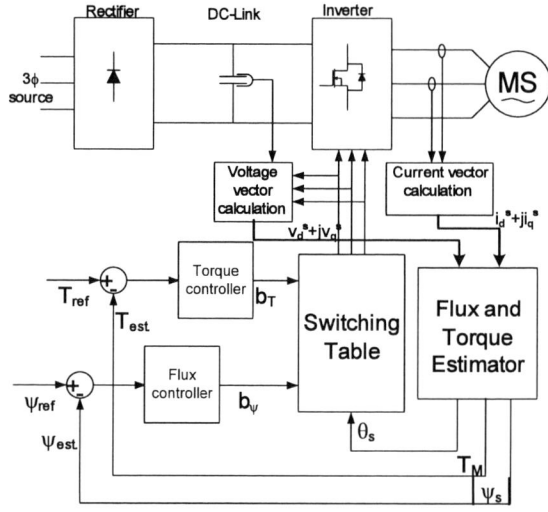

Fig. 1 basic PMSM-DTC drive

1. It is suggested that the stator flux vector stays at its position when zero voltage vector is applied, for induction motor. As for the PMSM the stator flux will change in the case of applying zero stator voltage due to the rotor flux, and for this reason the zero voltages should be avoided [1]. Where, the induction motor DTC drive includes the zero state in its switching table.

2. Since the torque is proportional to the torque angle, δ, not to the slip frequency as in induction motor, for controlling the amplitude of the stator flux and for changing the torque or δ quickly, the zero voltage vectors are not used in PMSM [2].

3. At low speed the ability to change δ, and hence the torque, is mainly determined by the motion of the rotor when zero voltage vectors are selected. To control torque at low speed without this dependency to the motion of the rotor, quick change of δ can be forced by avoiding the zero voltage vectors and applying voltage vectors which move the stator flux linkage vector as quickly as possible [3]. The study suggests that this may not be necessary at high speed where the rotor may move sufficiently quickly to produce the required change in δ and hence the torque. Accordingly, the zero state must be avoided in low speed. Experimental results consistent with the simulation results are presented in this reference to support this argument. Where, δ is the angle between the rotor and stator fluxes denoted by the torque angle as the motor torque is proportional to its sine.

The above three explanations are questionable either because of some inconsistency in the comparison between the induction motor and the PMSM or due to the inappropriate reasoning of the drive behavior as discussed in the following paragraphs.

With regarding to justification (1) above, the stator flux of the PMSM, as well as the induction motor, is the time integration of the stator voltage minus the stator resistance voltage drop, indeed both motors have similar stator flux expression from the stator terminals side. And the assumption of constant stator flux with zero voltage application for both motors is the approximation of neglecting the stator voltage drop agreed to be acceptable for short interval.

As for justification (2), the instantaneous motor torque is proportional to the torque angle δ in PMSM as well as the induction motor. On the other hand, the slip concept in the induction motor is related to the steady state model and its effect on the torque is indirect. Therefore the slip of the induction motor it is not comparable to δ of the PMSM, but for both motors the torque is proportional to sine δ.

Justification (3) suggests that at low speed, the rotor ability to change δ dominates, where with zero voltage, the stator flux will be stationary-unlike the argument of justifications (1) and (2)- but δ will change due to the rotor motion. This agrees with the flux stator expression but it is not indicated why this effect dominants only at low speed operation. Indeed, the rate of change of δ where the stator flux is constant is the negative of the motor speed, and therefore at higher speed this effect should be dominant if the explanation is correct.

Due to the above arguments and to the fact that the practice of omitting the zero switching state has been abandoned when the DTC drive upgraded to operate with SVM modulator [6], [7], this study investigates this matter and discusses the results demonstrated in the previous studies.

In this paper the basic motor equations are presented in section II. In section III the three basic PMSM-DTC schemes have been presented. Section IV describes the simulated system. The simulation results and discussion are presented in section V.

II. BASIC DTC CONCEPT AND PMSM EQUATIONS

In PMSM, the stator flux, φ_s, vector equals to the time integration of the stator back emf vector:

$$\vec{\psi_s} = \int (\vec{V_s} - \vec{i_s}R_s) \tag{1}$$

Where v_s, i_s and R_s are the stator voltage and current vectors and the stator winding resistance respectively.

Neglecting the stator resistance voltage drop, the stator flux can be described as the time integration of the stator voltage. As the motor is supplied by a three phase inverter, at any time the stator voltage vector will be any of the six non-zero vectors shown in Fig 2 or zero. By operating the inverter is a certain switching state, the motor flux will build up in the direction of the voltage vector corresponding to this state.

The motor torque is given by

$$T_M = \frac{3p_p\psi_s}{4L_dL_q}\left[2\psi_fL_q sin\delta + \psi_s\left(L_d - L_q\right)sin2\delta\right] \tag{2}$$

Where ψ_s and ψ_f are the stator and rotor fluxes, L_d and L_q are the stator winding direct and quadrature inductances.

For the motoring mode with positive rotational direction, the phase relationship between the stator and rotor fluxes is as shown in Fig. 3. By selecting one of the non-zero switching states, we can direct the stator flux vector to move at any of the six directions indicated at the tip of the vector as shown Fig. 3. The required change in the flux amplitude and torque can be achieved be selecting the proper inverter state. Clearly, the stator flux vector position determines the type of change a voltage vector causes. Therefore, the 360° space has been divided into six 60°-sectors as indicated in Fig. 2. The stator flux position has been abbreviated to the sector at which the vector is located. Selecting the zero state has negligible effect on the flux amplitude and δ, providing that the switching time is fast enough.

A. The Problem Description

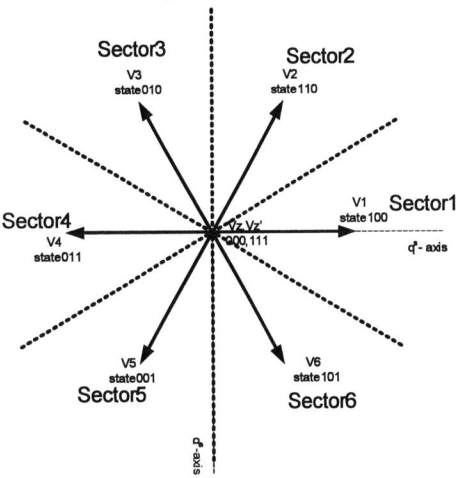

Fig. 2 Voltage vectors of the 3-phase VSI and the six stator flux position sectors

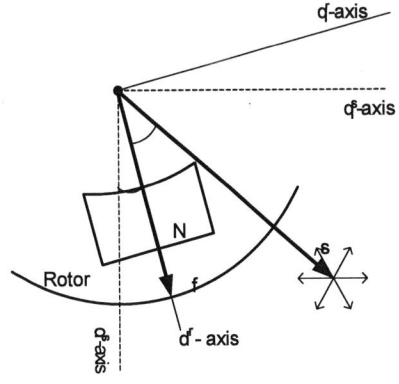

Fig. 3 The stator and rotor flux vectors for positive motoring mode

III. THE THREE DTC SCHEMES

The structure of the basic PMSM-DTC drive shown in Fig. 1 is general for all three schemes considered in this study. They, however, differ in their switching tables design as indicated below:

1) Two-level hysteresis controller/six-state table (switching table-1)

This scheme has been adopted in studies [1]-[3]. The switching table used in these studies is given in Table I, where (b_T and b_ψ) are the discrete torque and flux error signals respectively and the resultant vector follows the notation given in Fig. 2. As indicated in section I, this design has slashed the zero states.

2) Positive torque /eight-state table (switching table-2)

The switching table submitted in [3] (for comparison with table 1 above) and adopted in [4] is shown in Table II. This table provides zero inverter state whenever the torque error signal (b_T) is low.

3) Bipolar Torque/ eight-state table (switching table-3)

In the study presented in [5] not only the switching table has been modified, but also the hysteresis controller has been replaced by a three-level controller. The flux controller is, as in other designs, a two-level controller. The switching table is shown in Table III. This table produces a zero state only when the torque controller output is in the middle level (b_T=0), which can be interpreted as: *no significant torque change is required.* The characteristics of the 2- and 3-level hysteresis controllers are shown in Fig.4.

TABLE I
SWITCHING TABLE 1

b_ψ	b_T	Sector 1	Sector 2	Sector 3	Sector 4	Sector 5	Sector 6
1	1	V2	V3	V4	V5	V6	V1
1	0	V6	V1	V2	V3	V4	V5
0	1	V3	V4	V5	V6	V1	V2
0	0	V5	V6	V1	V2	V3	V4

TABLE II
SWITCHING TABLE 2

b_ψ	b_T	Sector 1	Sector 2	Sector 3	Sector 4	Sector 5	Sector 6
1	1	V2	V3	V4	V5	V6	V1
1	0	V7	V0	V7	V0	V7	V0
0	1	V3	V4	V5	V6	V1	V2
0	0	V0	sV7	V0	V7	V0	V7

TABLE III
SWITCHING TABLE 3

b_ψ	b_T	Sector 1	Sector 2	Sector 3	Sector 4	Sector 5	Sector 6
1	1	V2	V3	V4	V5	V6	V1
1	0	V7	V0	V7	V0	V7	V0
1	-1	V6	V1	V2	V3	V4	V5
0	1	V3	V4	V5	V6	V1	V2
0	0	V0	V7	V0	V7	V0	V7
0	-1	V5	V6	V1	V2	V3	V4

IV. SIMULATION MODEL DESCRIPTION

The motor parameters are given in Table IV. The digital controller is operated at two values of sampling time; 100 and 50µsec. The hysteresis bands are set to ±0.06 Nm for the torque hysteresis comparator and ±0.07Wb for the flux comparator. The reference stator flux is fixed to 0.7 Wb. The load torque is composed of a constant torque of 0N.m plus a friction torque with friction coefficient of 0.001147N.m.sec.

V. SIMULATION RESULTS

The three systems are compared at three different operating conditions:

1) Starting for high speed operation with a reference torque of $T_{M,ref}$=1.2N.m.

2) Starting for low speed operation with a reference torque of $T_{M,ref}$=1.05N.m.

3) Pulsed reference torque as shown in figure-5.

A. High Speed Operation:

Speed variation during starting for the three designs is shown in Fig. 6. The variation in the steady state speed is due to the fact that the drive speed loop is open and implies a different average torque. Averaging block has been used to show the torque average in the simulation model. The corresponding torque response is seen in Fig. 7, where (+Tm) indicates the unipolar torque or the two quadrant design of Table II, while (±Tm) indicates the bipolar torque or the four quadrant design of Table III. This figure shows a considerable torque ripple variation between the three designs.

The flux response of the three designs is shown in Fig.8. This figure shows that the flux average and ripple is basically determined by amplitude of the input reference and the hysteresis band. However, a slight variation in the time response speed can be noted, where the six-state switching table is the shortest.

To compare the switching losses the switching pulses of the three switching signals have been counted and plotted against time in Fig. 9, from this data the average switching frequency has been determined. The performance parameters are summarized in Table-V.

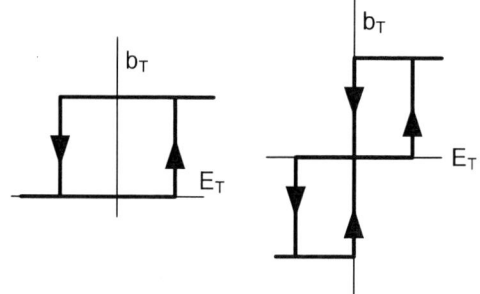

Fig.4 Two and three level torque hysteresis controllers characteristics

TABLE IV
PARAMETERS OF THE IPMSM USED IN THE COMPUTER SIMULATION

Rated power , W	1000	Maximum speed, rpm	1500
Rated voltage, V	138.56	Stator resistance, Ω	5.8
Magnetic flux , Wb	0.533	d-axis inductance, mH	44.8
Number of poles	4	q-axis inductance, mH	102.7
Rated torque, N.m	6	moment of inertia, kg-m²	0.001

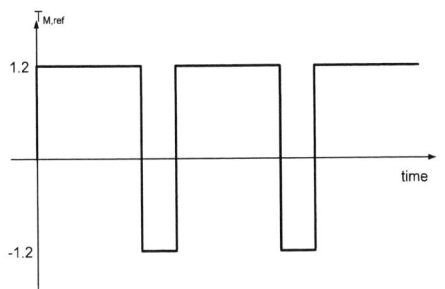

Fig.5 Pulsed reference torque for the third operation condition

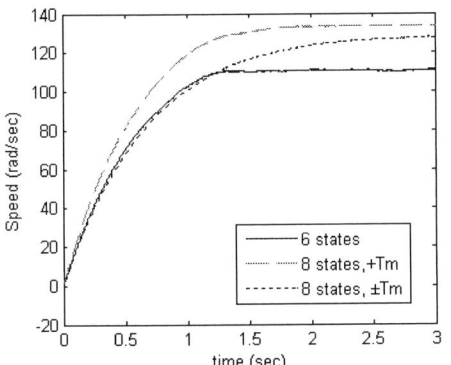

Fig.6 Acceleration with reference torque 1.2 N.m.

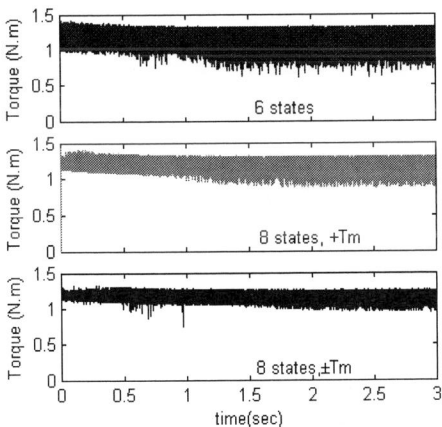

Fig.7 Motor torque when the reference torque =1.2 N.m. for, from the top table I, table II and tableIII

B. Low Speed Operation

The simulation has been repeated with a reference torque of 1.05N.m. The simulation results are shown in Fig.10 and Fig. 11 for the speed and torque variation with the time

respectively. The performance indicators are summarized in Table VI. The results show that for the six-state switching table, the staring time is about 30% longer that of switching table-2. Also the torque ripple is considerably higher than that of the other two methods. And the switching frequency is about three times that of the other two methods. These results in short show that switching table-2, the eight-state unipolar torque, is the most suitable one for low speed operation.

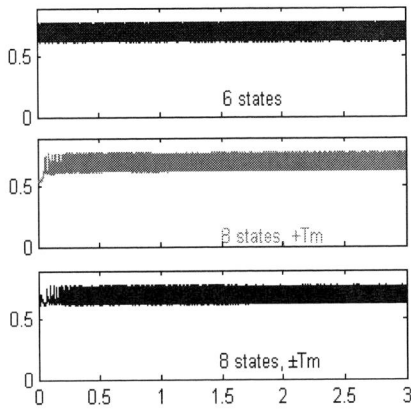

Fig. 8 Flux response during starting for, from the top table1, table 2 and table3

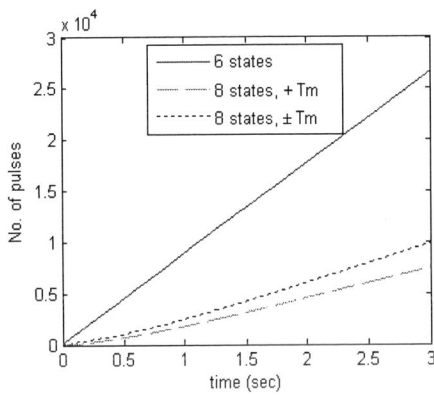

Fig. 9 The number of switching pulses.

TABLE V
THE PERFORMANCE OF THE THREE DTC DRIVES WITH T_{REF}=1.2 N.M

Comparison parameter	Switching table I	Switching table II	Switching table III
Starting time, sec	0.98	1.4	1.9
Torque average, N.m.	1.128	1.153	1.148
Torque ripple (p-p), N.m	0.53	0.29	0.19
Flux average, Wb.	0.7007	0.6768	0.6872
Flux ripple (p-p), Wb.	0.14	0.14	0.14
Switching frequency (Hz)	1964	813	1505

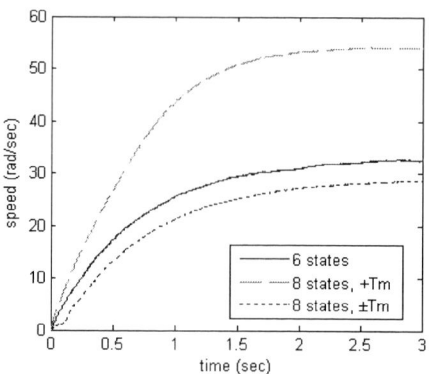

Fig.10 Acceleration with reference torque =1.05 N.m

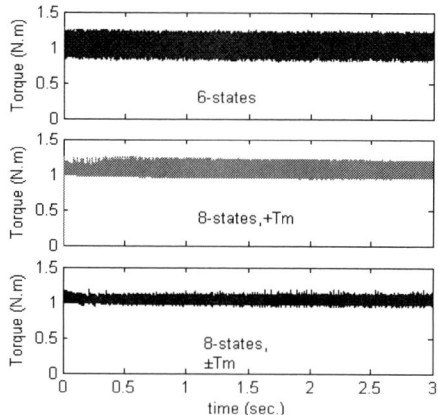

Fig.11 Motor torque when the reference torque =1.05 N.m for, from the top table I, table II and table III

TABLE VI
THE PERFORMANCE OF THE THREE DTC DRIVES WITH Tref=1.2 N.m

Comparison parameter	Switching table I	Switching table II	Switching table III
Starting time, sec	2.0	1.55	2.0
Torque average, N.m.	1.038	1.062	1.034
Torque ripple (p-p), N.m	0.45	0.26	0.19
Flux average, Wb.	0.7007	0.6768	0.6872
Flux ripple (p-p), Wb.	0.152	0.152	0.156
Switching frequency (Hz)	2957	833	1090

C. Pulsed Reference Torque

The simulation has been repeated when the reference torque is pulsating with the pattern in shown Fig. 5. The load torque model has been modified to pure friction torque with friction coefficient of 0.011147N.m.s.

The Simulation results are shown in Fig.12 and Fig.13 for the speed and torque responses respectively. From these results we notice that the unipolar torque drive of Table II is not applicable for four quadrant operation as the drive is not operational in the third quadrant. This is due to the design of the switching table which has not

considered the production of negative torque. However, the negative torque can be produced transiently if the speed is positive and the stator flux vector lags the rotor (δ negative). The performance parameters have been summarized for comparison in Table VII. These results show that the eight-state bipolar design of Table III have almost identical speed response with less torque ripple and switching losses compared to the six-state scheme of Table-I.

The results shown in Fig. 12 and Fig 13 are consistent with those obtained in [3] where the two schemes of Table-I and Table-II have been compared. However, the reason for the failing of scheme II to follow the reference torque is not due to the quick change of δ a low speed, as indicated in section I, but it is due to the fact that the switching table is not designed for negative torque production. With positive rotational speed the negative torque is produced due to the increased zero state intervals leads to negative δ. This can be seen by examining the change in the motor sector and inverter switching state during the negative reference torque interval for the scheme of Table II as shown in Fig. 14. This figure shows that, when the reference torque becomes negative the inverter goes to the zero states (0 and 7) for longer intervals compared to the positive torque production state. However the inverter does take nonzero states and this occurs when the negative torque produced is less than the reference torque, and therefore the motor torque will increase. The negative torque production capability lasts as long as δ is negative as in (2). And this occurs when the rotor flux leads the stator flux. When the rotor stops, and this occurs as shown in Fig 14 when the operation sector holds its value, applying zero voltage will not lead to δ decrease. Shortly, after the rotor stops the motor current follows the motor voltage to zero and motor torque becomes zero.

Unlike the recommendation given in [3], the zero state inclusion indeed is more important for low speed operation, as it effectively reduces the motor voltage and this is done indirectly in the SVM control.

VI. Conclusion

The Study has compared three designs of the switching table. It has been shown that eliminating the zero states from the switching tables provides faster flux response, and negligible effect on the speed response. On the other hand eliminations the zero states increases the switching frequency and the torque ripple.

It has been also shown that the two-level torque controller is more suitable for the two-quadrant drive. For four-quadrant drive the three-level torque controller with eight switching state table drive is the most suitable

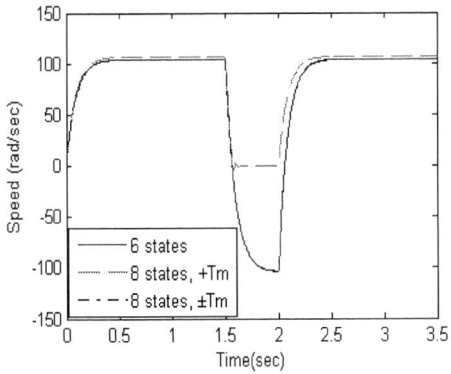

Fig. 12 Speed variation with time of the DTC drive with pulsating torque

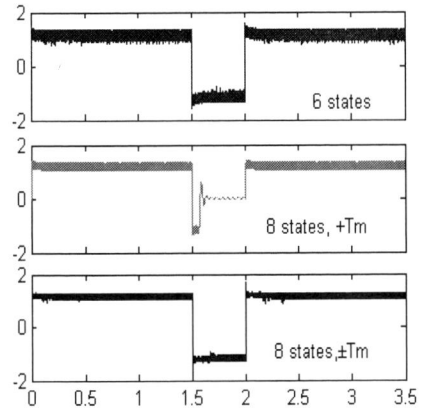

Fig.13 Motor torque variation with time for, from the top table I , table II and table III

TABLE VII
THE PERFORMANCE OF THE THREE DTC DRIVES WITH Tref=1.2 N.m

Comparison parameter	Switching table I	Switching table II	Switching table III
Starting time, sec	2.0	1.55	2.0
Torque average, N.m.	1.038	1.062	1.034
Torque ripple (p-p), N.m	0.45	0.26	0.19
Flux average, Wb.	0.7007	0.6768	0.6872
Flux ripple (p-p), Wb.	0.152	0.152	0.156
Switching frequency (Hz)	2957	833	1090

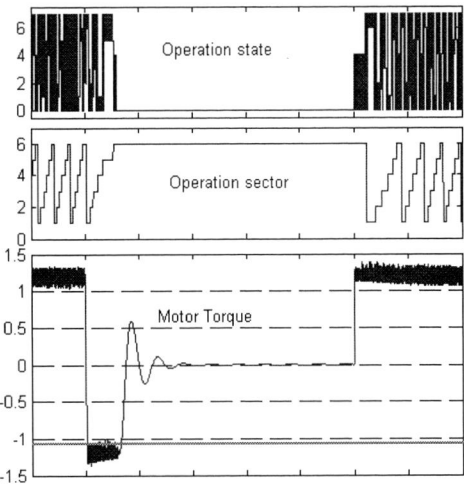

Fig.14 The variation of inverter's operation state, flux operation sector and the motor torque with time for table II DTC scheme during the application of negative reference toqque

REFERENCES

[1]. L. Zhong, and M.F. Rahman; "Analysis of direct torque control in permanent magnet synchronous motor drives", *IEEE Trans. on Power Electronics,* vol. 12, Issue 3, Pages: 528-536, May 1997.

[2]. L. Zhong, M.F. Rahman, W.Y. Hu, and K.W. Lim, "A direct torque controller for permanent magnet synchronous motor drives", *IEEE Trans. on Energy Conversion,* vol 14, Issue 3, Pages: 637-642, Sept. 1999.

[3]. M.F. Rahman, and L. Zhong, "Voltage switching tables for DTC controlled interior permanent magnet motor", *in IECON eds* , San Jose, California, USA, pp. 1445 - 1451 Nov.29-Dec.3, 1999.

[4]. M. F. Zolghadri, E.M. Olasagasti, and D. Roye, "Steady state torque correction of a direct torque controlled PM synchronous machine", *Proceedings of IEEE International Electrical Machines and Drives Conference,* IEMDC'97, Wisconsin, USA, Pages: MC3/4.1-MC3/4.3, May 18-21 1997.

[5]. G. Diamantis, and J.M. Prousalidis, "Simulation of ship propulsion system with DTC driving scheme", *PEMD 2004 Second International Conference on Power Electronics, Machines and Drives* (con. Publication No. 498), 2004.

[6]. L. Tang, L. Zhong, M.F. Rahman, and Y. Hu, "A novel direct torque controlled interior permanent magnet synchronous machine drive with low ripple in flux and torque and fixed switching frequency", *IEEE Trans. On Power Electronics,* Vol. 19, No. 2, pp. 346-354, March 2004.

[7]. L. Tang, L. Zhong, M.F. Rahman, and Y. Hu; "A novel direct torque control for interior permanent magnet synchronous machine drive with low ripple in torque and flux-A speed-sensorless approach", *IEEE Trans. On Industry Applications,* Vol. 39, No. 6, pp. 1748-1756, November 2003.

Improved DSVM-DTC Based Current Sensorless Permanent Magnet Synchronous Motor Drive

Bhim Singh, *Senior Member, IEEE*, and Devendra Goyal

Department of Electrical Engineering, Indian Institute of Technology, Delhi, New Delhi-110016, India
Emails: bhimsinghr@gmail.com and goyal.devin@gmail.com

Abstract--**This paper presents the current sensorless control of a direct torque controlled (DTC) permanent magnet synchronous motor (PMSM) drive using an improved discrete space vector modulation (DSVM) technique. After considering the primary ideas of the DSVM-DTC technique, a new set of voltage vectors are synthesized in optimized sequence by applying three standard voltage space vectors for three equal time intervals at each sampling period for the realization of the DSVM-DTC methodology of PMSM drive which further reduces torque ripple. However, it addresses the problem by introducing a higher number of predefined voltage space vectors. The current estimation is based on discrete electrical motor model in rotor reference frame. Such sensor elimination results in reduction in drive cost, size, noise immunity, increases drive efficiency and reliability. The effectiveness of the proposed drive system is demonstrated and clarified by several drive tests in MATLAB/SIMULINK platform under loads perturbation and speed reversal dynamics incorporating constant power operation.**

Index Terms-- **Direct torque control (DTC), permanent magnet synchronous motor (PMSM), discrete space vector modulation (DSVM), modeling equations, current estimation.**

I. INTRODUCTION

Direct torque control is one of the high performance control strategies for AC machines put forward by German scholars Takahashi et. al. [1] and Depenbrock [2] in the mid 1980's, mainly developed for induction motors. It has now been used to implement DTC of permanent magnet brushless motors [3]. The permanent magnet synchronous motor (PMSM) is widely used in high performance variable speed drive because it has many advantages such as maintenance free, high controllability, robustness against environment, high power density, overall efficiency, reliability and high power factor operation.

Conventional DTC is implemented during bang-bang control in two hysteresis controllers to regulate flux and torque directly and only one discrete voltage space vector (VSV) is applied during the whole sampling interval which creates large torque and flux ripples in drive system. Therefore, the decoupling of the nonlinear ac motor structure is obtained by the use of on-off control, which can be related to the on-off operation of the inverter power switches. Although DTC has many advantages i.e. fast response, elimination of PWM controllers and complex co-ordinate transformation, it still has some drawbacks as reported in recent literature [4].

Recently, many researchers have tried to reduce the torque ripple and to fix the switching frequency of the DTC scheme [5-9] which are classified into three categories. The first category is hardware related, by employing multi-level

inverter, there are more voltage space vectors available to control the flux and torque [6-7]. An attempt to improve the switching table has been made in second category [8]. In third category, predictive algorithms are developed to calculate the most appropriate voltage space vectors to minimize torque and flux linkage ripples is called space vector modulation (SVM) [9]. However, few disadvantages of SVM are the calculation of complicated equations online and sensitivity of PI controllers to the changes of motor parameters, speed and load.

Another alternative to increase the number of available vectors is an on-line modulation between active and null vectors, in order to obtain a theoretically infinite number of applicable vectors in each of the six spatial directions [10]. Simple switching table is replaced by several switching tables, obtaining a combination of three voltage vectors into the prefixed sampling period which is called discrete space vector modulation (DSVM) [11-12]. In general, the determination of the switching tables is carried out on the basis of physical considerations concerning the effects determined by radial and tangential variations of the stator flux vector on torque and flux values. The current sensorless drive system uses the comparatively high-resolution position sensor in order to achieve a vector transformation and a sinusoidal current drive [13].

This paper deals with the realization of the improved discrete space vector modulation based direct torque controlled permanent magnet synchronous motor drive without current sensors in order to simplify the high performance drive system. Direct sensing of the DC voltage rather than the AC voltages of the inverter output is adopted here since the inverter output voltage normally contains higher harmonics which require more computational time to eliminate. Modeling of improved DSVM-DTC based drive system is presented and MATLAB/SIMULINK models are developed to study the performance of the drive system under steady state and dynamic conditions during starting, and speed reversal and load perturbations. The four quadrant and field weakening operation of the proposed drive is also studied and results show that the ripples in flux and torque are greatly reduced both below and above base speed with improved estimation of winding currents.

II. OPERATING PRINCIPLE OF DIRECT TORQUE CONTROL

In direct torque controlled PMSM drive, while the amplitude of the stator flux linkage is kept invariant, the electromagnetic torque of a PMSM is directly proportional to the sine value of electrical angle between the stator flux linkage and rotor flux linkage vector. The stator flux linkage (D-Q, i.e. stator frame) and rotor flux linkage (d-q, i.e. rotor frame) are represented in Fig. 1 where d_s-q_s is synchronous rotating frame. Therefore the precise torque control can be achieved by controlling the instantaneous speed of the stator flux linkage with its amplitude keeping constant below base speed and flux weakening above base speed. It is fulfilled through the application of proper selecting the voltage space vector generated by an inverter and therefore fast torque response can be obtained by increasing the rotating speed of the stator flux linkage as fast as possible.

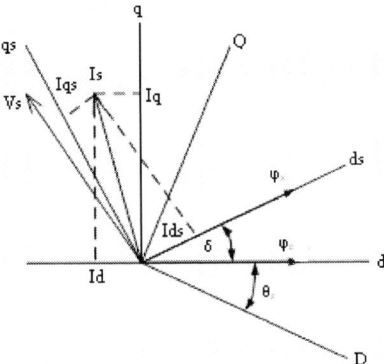

Fig. 1. Stator and rotor flux linkages in different reference frames

III. DSVM-DTC SYSTEM FOR PMSM

A. System Configuration

The schematic of current sensorless DSVM-DTC based PMSM drive system is shown in Fig. 2. The drive consists of a PI speed controller, a two level flux hysteresis comparator, five level torque hysteresis comparator, optimum sequence synthesized voltage vector switching table, current estimation block, the motor and IGBT based voltage source inverter (VSI). Estimated three phase winding currents are used to calculate the stator flux linkage and electromagnetic torque in torque and flux estimation block.

B. Operating Principle of DSVM-DTC Technique

The main idea of the DSVM-DTC control strategy is to force the torque and stator flux to approach their reference values by applying in one sampling period three voltage vectors instead of only one voltage vector as in classical DTC. With DSVM-DTC strategy, 19 voltage vectors can be available for each sector, based on the rotor speed, the flux and the torque errors range as is represented in Fig. 3 and Table I. The black dots represent the ends of the synthesized voltage vectors. As an example, the label 23Z denotes the voltage vector which is synthesized by using the voltage space vectors V_2, V_3 and V_0 or V_7 each one applied for one third of the cycle period. The advantage of using the DSVM technique is that one can choose among 19 voltage vectors instead of the five of classical DTC.

C. Improved DSVM-DTC Technique

An improved switching table for optimized sequence of synthesized voltage vectors is shown in Table I. Synthesized voltage vectors are selected by a two-level hysteresis band for the flux, and a five-level hysteresis band for the torque. The five level torque comparator used in improved DSVM implementation is shown in Fig. 4. In improved switching

table, changing the sequence of the 3 voltage vectors applied to the three equal time intervals of one sampling period does not change the final synthesized voltage vector. However, the sequence can greatly affect the torque ripple. The basic concept of algorithm is given below.

It is assumed that the rotor speed is in high speed range, and stator flux vector is in sector 1+. In Fig. 5(a), at time t_k, the beginning of a sampling period, $d\varphi_s$ is +1 and dT_e is -1, which indicates that the actual torque value is greater than the reference torque value so the torque should be decreased. In this case, voltage vector '3ZZ' is applied in classical DSVM-DTC. If V_3 is firstly applied, the torque error will be enlarged as shown in Fig. 5(a) where T_s is a sampling period time. In contrast, 'ZZ3' is selected to apply in improved DSV-DTC technique which reduces the torque ripple as shown in Fig. 5(b).

When the motor is in the low-speed region the induced voltage is close to zero and the voltage vectors are selected symmetrically around zero. If torque error is +1 or -1, a moderate increase or decrease torque respectively is wanted which leads to application of short active vectors i.e. 2ZZ, 3ZZ, 5ZZ or 6ZZ. When the torque error $dT_e= \pm2$, DSVM-DTC chooses voltage vectors as the classical DTC. In the medium-speed region, the induced emf voltage starts to introduce asymmetry of the voltage vectors on torque behavior. For $dT_e=-1$ a moderate decrease torque is wanted which leads to application of ZZZ and for $dT_e= +1$, voltage vector 22Z is selected when flux should is to be raised and 33Z when lowered. During the high-speed region, two switching tables are defined corresponding to sector 1+ and sector 1-. When the motor is rotating in counter-clockwise and the torque error $dT_e= +1$, four voltage vectors ('333', '332', '223', '222') are selected among which '333' and '332' can reduce the flux and '223' and '222' can increase the flux. In order to reduce the flux, in sector 1+, '333' can be chosen and in sector 1-, '332' can be chosen.

IV. MODELING OF DSVM-DTC BASED PMSM DRIVE SYSTEM

A. PI Speed Controller

The desired reference value of the speed signal is compared with the feedback speed signal and the speed error is processed in the speed controller. The output of the speed controller is the torque value. This value of the torque is fed to the limiter and the final reference torque is obtained from the limiter.

The speed error at the mth instant of time is given as

$$dw_e(m) = \omega_e*(m) - \omega_e(m) \qquad (1)$$

The output of the speed controller at the mth instant is given as

$$T(m)^* = T_e(m-1)+K_p[d\omega_e(m) - d\omega_e(m-1)]+K_I d\omega_e(m) \qquad (2)$$

where $\omega_e(m)$ and $T_e(m)$ are the electrical speed and reference torque at m^{th} instant.

B. Flux Weakening Controller

Field-weakening controller sets the reference value of the flux depending upon the base speed and actual speed of the motor. If the speed of the motor is below the base or the are to nominal speed then the flux reference is maintained at

1521

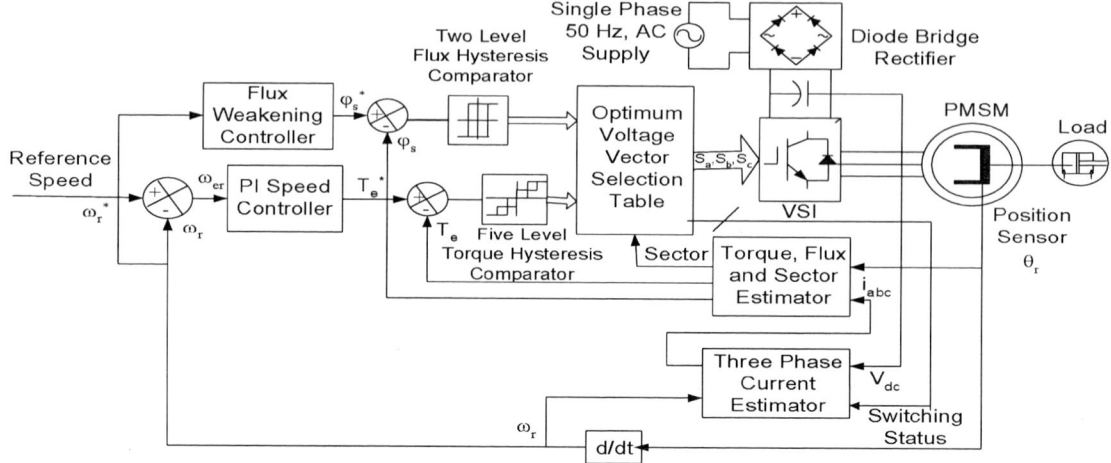

Fig. 2. Block Diagram for DSVM-DTC based current sensorless PMSM drive

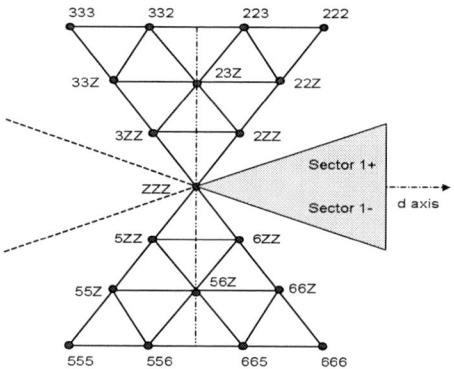

Fig. 3. Implementation of DSVM-DTC technique

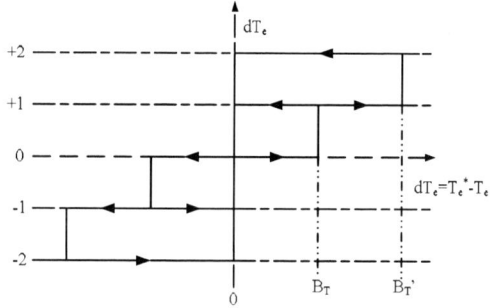

Fig.. 4. Proposed five level torque comparator

constant at the nominal value. If speeds above the base speed be attained then the flux of the motor is weakened. The mathematical relations are given below

$$\begin{cases} \varphi_s = |\varphi_{sref}| & \text{for} (|\omega_r| < \omega_b) \\ |\varphi_{sref}| = \varphi_s(\omega_b/|\omega_r|) & \text{for } (|\omega_r| > \omega_b) \end{cases} \tag{3}$$

where ω_r, ω_b are the mechanical and base angular speeds respectively.

C. Flux and Torque Comparator

Similar to the conventional DTC, the amplitude and position angle of the actual flux linkage and torque in the improved DSVM DTC system are also estimated from the

sensed values of voltage and current using following equations

$$\varphi_D(m) = \varphi_D(m\text{-}1) + \left\{ v_D(m\text{-}1) - R_s \overline{i_D(m)} \right\} T_s \tag{4}$$

$$\varphi_Q(m) = \varphi_Q(m\text{-}1) + \left\{ v_Q(m\text{-}1) - R_s \overline{i_Q(m)} \right\} T_s \tag{5}$$

where

$$\overline{i_D(m)} = \left\{ i_D(m\text{-}1) + i_D(m) \right\}/2 \tag{6}$$

$$\overline{i_Q(m)} = \left\{ i_Q(m\text{-}1) + i_Q(m) \right\}/2 \tag{7}$$

$$\varphi_S(m) = \sqrt{\varphi_D(m)^2 + \varphi_Q(m)^2} \angle \tan^{-1}\left(\frac{\varphi_Q(m)}{\varphi_D(m)} \right) \tag{8}$$

$$T_{est}(m) = 3P/4 \left\{ \varphi_D(m) i_Q(m) - \varphi_Q(m) i_D(m) \right\} \tag{9}$$

where $\varphi_D(m)$ and $\varphi_Q(m)$ are D- and Q- axis components of the stator flux linkage of the motor at m^{th} sample instant respectively, R_s is the stator resistance and T_s is the sampling time. Finally the sector, in which flux vector is present, is computed from the stator flux angle.

D. Mathematical Model of PMSM

The well-known stator flux linkage and voltage equations in the rotor d-q reference frame as shown are as follows [10]

$$\begin{bmatrix} v_d \\ v_q \end{bmatrix} = \begin{bmatrix} R_s + pL_d & -\omega_e L_q \\ \omega_e L_q & R_s + pL_q \end{bmatrix} \begin{bmatrix} i_d \\ i_q \end{bmatrix} + \begin{bmatrix} 0 \\ \omega_e \varphi_f \end{bmatrix} \tag{10}$$

where ω_e, is the electrical angular velocity, R_s is the stator resistance, φ_f is the magnet flux linkage, L_d and L_q are the d-q axis inductances respectively, i_d and i_q are d-q axis currents, and $p = d/dt$ is the time derivative operator.

The developed electromagnetic torque is given as [11]

$$T_e = 3P/4 \left(\varphi_f i_q + (L_d - L_q) i_d i_q \right) \tag{11}$$

where P is the number of magnet poles.

Considering the dynamics of the motor [12] as

$$T_e - T_m = J\frac{dw_r}{dt} + B\omega_r \tag{12}$$

where ω_r is the mechanical angular velocity, T_m is the motor load torque, J is the motor moment inertia and B is friction coefficient.

TABLE I

IMPROVED SWITCHING TABLE FOR DSVM-DTC IMPLEMENTATION (STATOR FLUX IN SECTOR 1, COUNTER CLOCKWISE ROTATION)

For low speed range (sector 1)					
$\dfrac{dT_e}{d\varphi_s}$	-2	-1	0	+1	+2
+1	555	5ZZ	ZZZ	3ZZ	333
-1	666	6ZZ	ZZZ	2ZZ	222
For medium speed range (sector 1)					
$\dfrac{dT_e}{d\varphi_s}$	-2	-1	0	+1	+2
+1	555	ZZZ	Z3Z	33Z	333
-1	666	ZZZ	Z2Z	22Z	222
For high speed range (sector 1+)					
$\dfrac{dT_e}{d\varphi_s}$	-2	-1	0	+1	+2
+1	555	ZZ3	3Z3	333	333
-1	666	ZZ2	2Z3	223	222
For high speed range (sector 1-)					
$\dfrac{dT_e}{d\varphi_s}$	-2	-1	0	+1	+2
+1	555	ZZ3	3Z2	332	333
-1	666	ZZ2	2Z2	222	222

The above equations represent the model of salient pole PMSM where d-q axis inductances are not equal as in the case of surface mounted (non-salient pole) PMBL motors. When $L_q > L_d$, the motor utilizes the reluctance torque and results in high efficiency of the drive.

The phase currents are computed using inverse Park's transformation as

$$i_a = i_d \cos\theta_r - i_q \sin\theta_r \qquad (13)$$

$$i_b = i_d \cos(\theta_r - 2\pi/3) - i_q \sin(\theta_r - 2\pi/3) \qquad (14)$$

$$i_c = i_d \cos(\theta_r - 4\pi/3) - i_q \sin(\theta_r - 4\pi/3) \qquad (15)$$

where, θ_r is the position angle of the rotor.

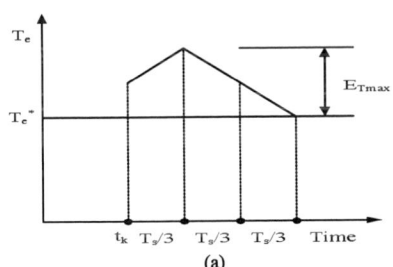

(a)

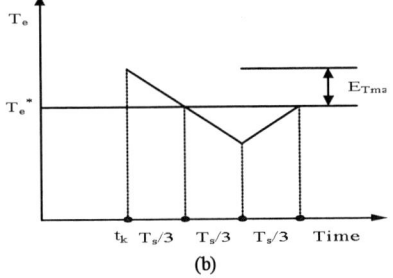

(b)

Fig. 5. Comparison of torque waveform when dT_e=-1 between (a) classical DSVM-DTC and (b) improved DSVM-DTC

V. CURRENT SENSORLESS CONTROL OF PMSM DRIVE

A. Current Estimation

Rotor reference d-q currents are estimated using motor model in discrete equation based on eq. (16)-(17) in winding current estimation block with sampling time T_s which is feedback to torque and flux estimator block as shown in Fig. 1. Estimated rotor reference d- and q- axes currents (i_{dest}, i_{qest}) are calculated as follows

$$i_{dest}(m+1) = i_{dest}(m) + \frac{T_s}{L_d}\{v_d(m) + \omega_e(m)L_q i_{qest}(m) \\ - R_s i_{dest}(m)\} \qquad (16)$$

$$i_{qest}(m+1) = i_{qest}(m) + \frac{T_s}{L_q}\{v_q(m) - \omega_e(m)\varphi_f - \\ \omega_e(m)L_d i_{dest}(m) - R_s i_{qest}(m)\} \qquad (17)$$

where voltages $v_d(m)$ and $v_q(m)$ are d- and q- axis components of the stator voltage supplied to the motor at m^{th} sample instant respectively.

B. Stator Voltage Estimation

The applied three phase stator winding voltages are reconstructed using a voltage sensor mounted at the dc-link terminal of the VSI and the applied switching pulses to the inverter legs. The three phase stator voltages are estimated from the mathematical relation as

$$v_a = \frac{V_{dc}}{3}(2S_a - S_b - S_c) \qquad (18)$$

$$v_b = \frac{V_{dc}}{3}(2S_b - S_a - S_c) \qquad (19)$$

$$v_c = \frac{V_{dc}}{3}(2S_c - S_a - S_b) \qquad (20)$$

where S_a, S_b and S_c are switching signals of upper switches for a, b and c phases respectively and V_{dc} is the dc-link voltage. These three phase stator voltages can be represented in d-q rotor reference frame as

$$
\begin{bmatrix} v_d \\ v_q \\ 0 \end{bmatrix} = \frac{2}{3} \begin{bmatrix} \cos\theta_e & \cos(\theta_e - 2\pi/3) & \cos(\theta_e + 2\pi/3) \\ \sin\theta_e & \sin(\theta_e - 2\pi/3) & \sin(\theta_e + 2\pi/3) \\ 1/2 & 1/2 & 1/2 \end{bmatrix} \begin{bmatrix} v_a \\ v_b \\ v_c \end{bmatrix} \quad (21)
$$

C. Performance of Current Estimation Control

In order to evaluate the accuracy of the current estimation and performance of the current sensorless drive system, the drive is tested under steady state and dynamic condition with speed reversal and load perturbation. Since current estimation includes resistance and d-q axis inductances so the accuracy of current estimator depends on the accuracy of the motor parameters.

VI. SIMULATION STUDY OF CURRENT SENSORLESS IMPROVED DSVM-DTC BASED PMSM DRIVE SYSTEM

To validate the model of theoretical analysis and system modeling, a simulation study of current sensorless DSVM-DTC based PMSM drive system has been carried out with improved switching table. To realize the optimum performance control, it is necessary to synthesize both the requirement of flux linkage amplitude and torque. Since the torque is obtained from a PI speed controller and the flux linkage error is controlled by another PI controller so the PI controller parameters, K_P and K_I, should be tuned to achieve high system performance of the system. A large K_P accelerates the system dynamic response while the K_I mainly influences the system static error. In this way, the amplitude, rotating velocity and direction of the stator flux linkage vector can be controlled and, of course, the quick torque response can be reached. Simultaneously the current estimator block gives the accurate tracking of three phase winding currents to meet the desired drive performance. The current estimation is based on eq. (16) and (17) which is also represented as MATLAB model shown in Fig. 6. The rated speed of the motor is 300 rad/s and rated load torque applied is 5.5 Nm. The sampling interval is taken 100 μs for the torque control loop. A three-phase insulated gate bipolar transistor (IGBT) intelligent power module is used for the inverter, which is supplied at a dc-link voltage of 340 V which is derived from output capacitor connected at the output of single phase rectifier.

VII. RESULTS AND DISCUSSION

The current sensorless improved DSVM-DTC based PMSM drive is simulated in MATLAB environment along with Simulink and Power System Blockset toolboxes. The motor specifications are given in Appendix.

A. Performance of Proposed PMSM Drive

The steady state performance of the current sensorless direct torque controlled PMSM drive has been tested at full load and rated speed. Fig. 7 shows the steady state response of the drive at full load torque of 5.5 Nm and

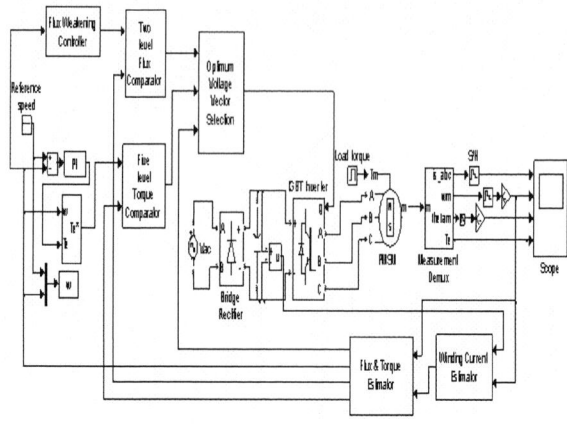

Fig. 6. MATLAB model of current sensorless improved DSVM-DTC based PMSM drive

rated speed of 300 rad/s in terms of reference and actual motor speed ω_r, developed electromagnetic torque T_e, three phase motor currents i_{act}, three phase estimated motor currents i_{est} and D-Q axis stator flux.

Current estimation also operates satisfactorily under load perturbation from 25% - 100% - 25% of rated load torque under constant rated speed condition as shown in Fig. 8. It is also seen from these figures that the response is fast enough and current estimation is validated. Fig. 9 shows the four quadrant operation characteristics of the proposed current sensorless drive system. This is examined by applying initially +400 rad/s speed step reference at no-load. The motor accelerates to rated speed (+300 rad/s) under constant torque operation and achieve +400 rad/s under flux weakening operation where stator flux is decreased to 0.1575 wb and stays for some time in first quadrant and then reverse rated speed is applied which leads to the motor in another quadrant. The drive operation shows the proper tracking of winding currents in all four quadrant operation and during transition between two quadrants. The corresponding stator currents have constant magnitude and speed varying frequency during transitions but the constant magnitude and frequency in steady-state. Figs. 10(a)-(c) show that the torque ripple improvement from 28.4%-14.7%-10.8% of rated load respectively in improved DSVM-DTC as compared to classical DSVM-DTC and conventional DTC at rated condition.

VIII. CONCLUSION

High performance DSVM-STC based PMSM current sensorless drive with optimized switching sequence has been investigated through analysis and simulation. The proposed drive has offered the advantage of reducing flux linkage and torque ripple, constant switching frequency operation of an inverter and it maintains good dynamic response as of classical DTC without any current sensors. The implementation of the DSVM technique requires only a small increase (20–25%) of the computational time required by basic DTC scheme. The current sensorless algorithm is deduced, improved DSVM-DTC operating principle has been explained and simulation has been

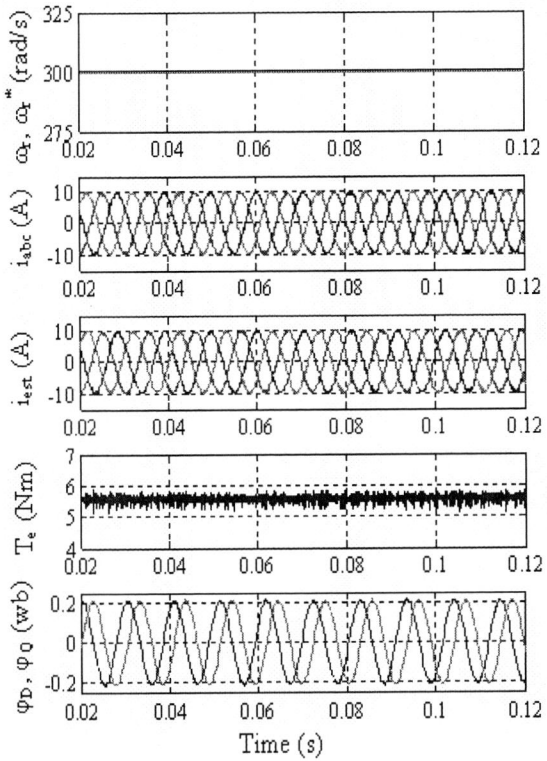

Fig. 7. Steady state response of the current sensorless improved DSVM-DTC based PMSM drive at rated condition.

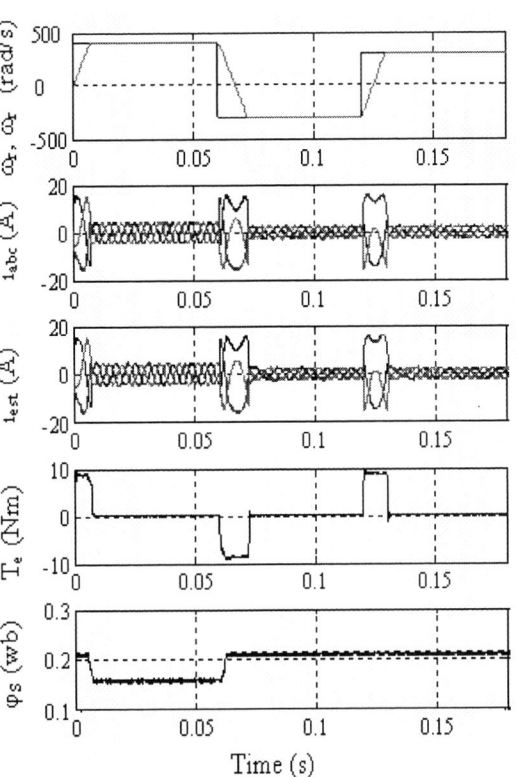

Fig. 9. Four quadrant operation of the current sensorless improved DSVM-DTC based PMSM drive incorporating wide speed operation.

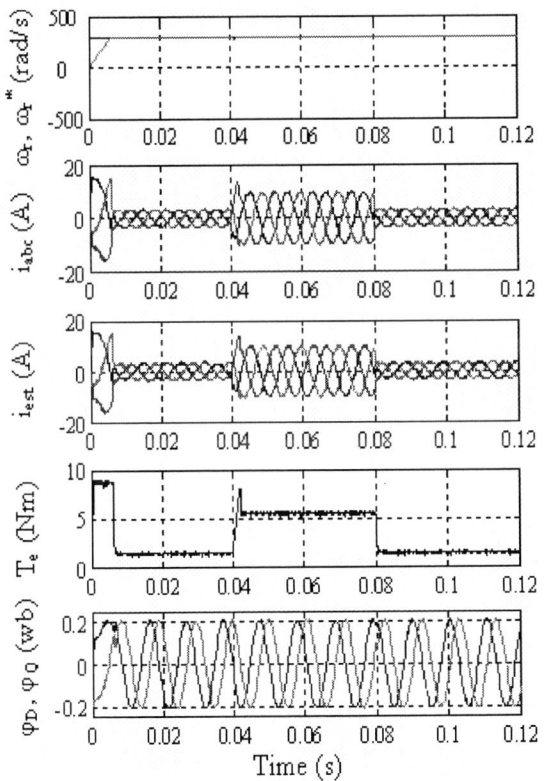

Fig. 8. Load perturbation response of the current sensorless improved DSVM-DTC based PMSM drive at rated speed.

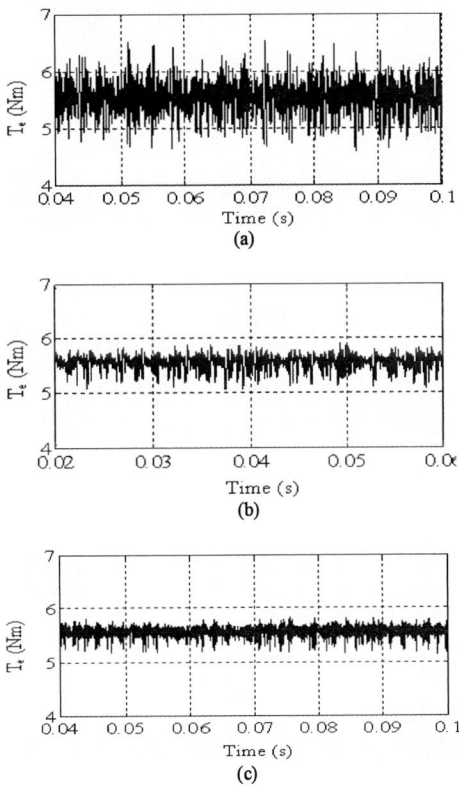

Fig. 10. Torque ripple performance (a) conventional DTC (b) classical DSVM-DTC and (c) improved DSVM-DTC.

carried out to validate the theoretical analysis. The results have shown the satisfactorily control of the drive under steady state and dynamic conditions.

APPENDIX

Motor parameters:

Number of poles: P=4; Stator resistance: R=0.61 Ω; Magnet flux linkage φ_f= 0.181 wb; d- and q- axis inductance: L_d= 9.1 mH and L_q= 11.5 mH; DC link voltage: V_{dc}= 340V; Rated phase current = 8A, Base speed: w_r= 300 rad/s.

REFERENCES

[1] Takahashi and T. Noguchi, "A new quick-response and high-efficiency control strategy of an induction motor," *IEEE Trans. Ind. Applicat.*, vol. 22, no. 5, pp. 820–827, Sept./Oct. 1986.

[2] M. Depenbrock, "Direct self-controlled (DSC) of inverter fed induction machine," *IEEE Trans. Power Electron.*, vol. 3, pp. 420–429, Oct. 1988.

[3] M. F. Rahman, L. Zhong, and K.W. Lim, "A direct torque controlled interior magnet synchronous motor drive incorporating field weakening," in *Proc. 1997 IEEE Ind Applicat. Soc. Annu. Meeting*, vol. 1, New Orleans, LA, Oct. 5–9, 1997, pp. 67–74.

[4] D. Casadei, G. Serra, and A. Tani, "Implementation of a direct torque control algorithm for induction motors based on discrete space vector modulation," *IEEE Trans. Power Electron.*, vol. 15, pp. 769–777, July 2000.

[5] Z. Tan, Y. Li, and M. Li, "A direct torque control of induction motor based on three level inverter," in *Proc. IEEE-PESC'01 Conf.*, vol. 2, 2001, pp. 1435–1439.

[6] C. Martins, X. Roboam, T. A. Meynard, and A. S. Caryalho, "Switching frequency imposition and ripple reduction in DTC drives by using a multilevel converter," *IEEE Trans. Power Electron.*, vol. 17, pp. 286–297, Mar. 2002.

[7] C. G. Mei, S. K. Panda, J. X. Xu, and K. W. Lim, "Direct torque control of induction motor-variable switching sectors," in *Proc. IEEE-PEDS Conf.*, Hong Kong, July 1999, pp. 80–85.

[8] A. Tripathi, A. M. Khambadkone, and S. K. Panda, "Space-vector based, constant frequency, direct torque control and dead beat stator flux control of AC machines," in *Proc. IEEE-IECON'01 Conf.*, Nov. 2001, pp. 1219–1224.

[9] D. Swierczynski, M. Kazmierkowski, and F. Blaabjerg, "DSP based direct torque control of permanent magnet synchronous motor (PMSM) using space vector modulation (DTC-SVM)," in *Proc. IEEE Int. Symp. Ind. Electron.*, vol. 3, May 2002, pp. 723–727.

[10] D. Casadei, G. Serra, A. Tani, "Constant frequency operation of a DTC induction motor drive for electric vehicle," in *Proc. of ICEM '96*, Vigo, Spain, September 10-12, 1996, Vol. 111, pp. 224- 229.

[11] D. Casadei, G. Serra, and A. Tani, "Stator flux vector control for high performance induction motor drives using space vector modulation," in *Proc. Electromotion*, vol. 2, no. 2, pp. 79–86.

[12] C. Lochot, X. Roboam, and P. Maussion, "A new direct torque control strategy for an induction motor with constant switching frequency operation," in *Proc. EPE'95*, Sevilla, Spain, pp. 2.431–2.436.

[13] K. Ohishi and Y. Nakamura, "High performance current sensorless speed servo system of PM motor based on

current estimation," in *Proc. 2001 IEEE IAS Ann. Meet.*, 2001, pp. 1240-1246.

BIOGRAPHIES

Bhim Singh was born in Rahamapur, U. P., India in 1956. He received B. E. (Electrical) degree from University of Roorkee, India in 1977 and M. Tech. and Ph. D. degrees from Indian Institute of technology (IIT), New Delhi, in 1979 and 1983, respectively. In 1983, he joined as a Lecturer and in 1988 became a Reader in the Department of Electrical Engineering, University of Roorkee. In December1990, he joined as an Assistant Professor, became an Associate Professor in 1994 and Professor in 1997 at the Department of Electrical Engineering, IIT Delhi. His field of interest includes power electronics, electrical machines and drives, active filters, static VAR compensator, analysis and digital control of electrical machines. Prof. Singh is a Fellow of Indian National Academy of Engineering (INAE), Institution of Engineers (India) (IE (I)) and Institution of Electronics and Telecommunication Engineers (IETE), a Life Member of Indian Society for Technical Education (ISTE), System Society of India (SSI) and National Institution of Quality and Reliability (NIQR) and Senior Member of IEEE (Institute of Electrical and Electronics Engineers).

Devendra Goyal was born in Bharatpur, Rajasthan, India in 1982. He received B. Tech. (Electrical) degree from National Institute of Technology, Kurukshetra, India in 2004. In 2005, he worked as a GET in Automobile Industry, Subros Limited, Noida. Presently he is a pursuing M.Tech degree in the Department of Electrical Engineering, IIT Delhi. His field of interest includes power electronics, electrical machines and drives.

A High Performance Direct Torque Control Scheme of Permanent Magnet Synchronous Motor

Dong-Hee Lee, Young-Joo An*, and Eui-Chel Nho*

Dept. of EE & Mechatronics, Kyungsung University, Busan, Korea
* Pukyong National University, Busan, Korea

Abstract-- This paper presents an advanced DTC (Direct Torque Control) of PMSM. The proposed DTC method uses a conventional torque estimator and torque error. But the switching signal is generated by PWM method according to the switching rules and torque error. A simple calculation of PWM without any complex determination of space vector can assure the constant switching frequency with an excellent control performance.

The proposed torque control scheme for PMSM is verified by computer simulations.

***Index Terms**—PMSM (Permanent Magnet Synchronous Motor), DTC(Direct Torque Control), DTC-PWM*

I. INTRODUCTION

PMSM(Permanent Magnet Synchronous Motor) is much used in industrial applications, home appliances and robot system due to the high efficiency, high torque and high control performance[1-4].

In order to get an excellent torque control performance, many techniques are investigated[3-8]. Among the various control schemes, SVPWM (Space Vector Pulse Width Modulation) with PI current controller is widely used for torque control of PMSM due to a high control performance.

Although SVPWM with PI current controller has an excellent control performance, the calculation of space vector is somewhat complex and the practical performance is dependent on the PI control gains. To overcome these difficulties, DTC (Direct Torque Control) is introduced to PMSM drive.

DTC is very simpler than SVPWM with PI current controller, and it is not required any control gains exception of torque error bandwidth. Since the output torque is controlled within the error bandwidth, control performance and maximum switching frequency are dependent on the designed error band width of DTC. In order to get a more excellent control performance, shorter sampling time for torque estimation and higher switching frequency are essential. Furthermore, the switching frequency is changed according to load torque and operating speed. So, Input line filter design for EMI/EMC is very difficult.

This paper presents an advanced DTC of PMSM. The proposed DTC method uses a conventional torque estimator and torque error. But the switching signal is generated by PWM method according to the switching rules and torque error. A simple calculation of PWM without any complex determination of space vector can assure the constant switching frequency with an excellent control performance.

The proposed torque control scheme uses a conventional control rule of DTC, but the switching pulse width is controller by the torque error. Therefore, a simple implementation but a high control performance can be obtained.

The proposed torque control scheme for PMSM is verified by some computer simulation.

II. DTC OF PMSM

Fig. 1 shows the conventional DTC method for PMSM. **Torque Estimator** and **Flux Estimator** calculate output torque Tm and stator flux linkage vector from current and voltage information, respectively.

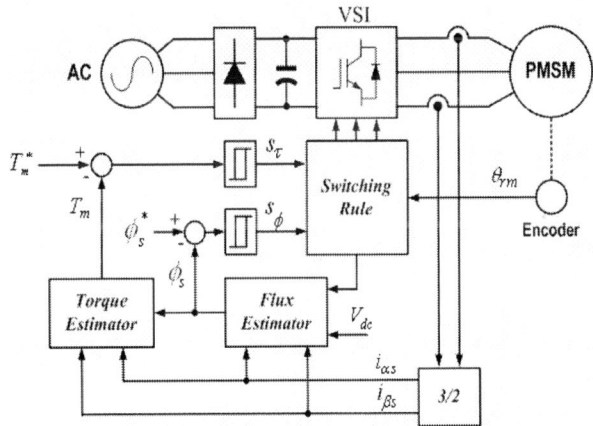

Fig. 1. Conventional DITC scheme of PMSM

In order to generate suitable switching signals for VSI(Voltage Source Inverter), voltage vector should be properly selected according to torque and flux linkage error.

The stator flux linkage of a PMSM can be expressed in a stationary reference frame is

$$\phi_{\alpha s} = \int \left(v_{\alpha s} - R_s \cdot i_{\alpha s} \right) \cdot dt \tag{1}$$

$$\phi_{\beta s} = \int \left(v_{\beta s} - R_s \cdot i_{\beta s} \right) \cdot dt \tag{2}$$

$$\phi_s = \sqrt{\phi_{\alpha s}^2 + \phi_{\beta s}^2} \tag{3}$$

where, $\phi_{\alpha s}$, $\phi_{\beta s}$ and ϕ_s are the α- and β-axes stator flux linkage and flux linkage, respectively. $v_{\alpha s}$, $v_{\beta s}$ and $i_{\alpha s}$, $i_{\beta s}$ are the winding voltages and currents of PMSM respectively at stationary reference frame. R_s is the resistance of the PMSM.

And the output torque of PMSM can be obtained as follow.

$$T_m = \frac{3}{2} \cdot \frac{p}{2} \cdot \left(\phi_{\alpha s} \cdot i_{\beta s} - \phi_{\beta s} \cdot i_{\alpha s} \right) \tag{4}$$

where, p is the number of pole pairs.

s_τ and s_ϕ are the sign of toque error and flux linkage error, respectively which are calculated as 1 and -1 in the control block diagram.

To select the voltage vectors for controlling the amplitude of the stator flux linkage, the voltage vector plane is divided into six regions as shown in Fig. 2.

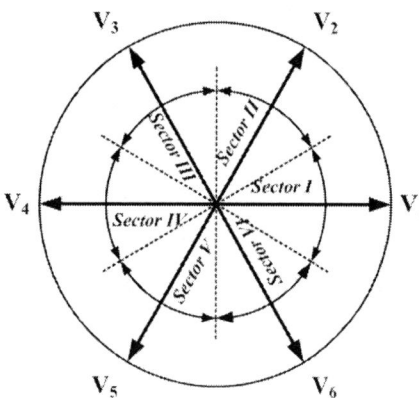

Fig.2. The control vector of stator flux linkage

In each region, two adjacent voltage vectors are selected to control the flux linkage.

The switching rule is determined by the torque, flux error and sector of stator flux linkage. Table 1 shows the switching rule for a DTC of PMSM.

TABLE 1. SWITCHING RULE FOR DTC OF PMSM

s_ϕ	s_τ	Sector of flux linkage					
		I	II	III	IV	V	VI
1	1	V_2	V_3	V_4	V_5	V_6	V_1
	-1	V_6	V_1	V_2	V_3	V_4	V_5
-1	1	V_3	V_4	V_5	V_6	V_1	V_2
	-1	V_5	V_6	V_1	V_2	V_3	V_4

DTC is very simple and the torque ripple is dependent on selected torque error bandwidth and sampling period. But, the switching frequency is changed according to operating condition, and a short sampling period is required for the high performance. In practical application, a random switching frequency make difficult to design of EMI/EMC filter, and the maximum switching frequency is limited by power devices.

III. THE PROPOSED DTC-PWM METHOD

A. The problem of conventional DTC

Fig. 3 shows the torque reference and actual torque waveform in a conventional DTC. In the case of Fig. 3(a), the switching signal is changed according to torque error band T_H. Generally, the actual torque is in the band of torque error. However, the actual torque can be exceeded over the band due to the hysterisis effect shown in Fig. 3(a). In this case, the torque error ΔT_m is increased according to motor speed, switching voltage and sampling period T_S. If the sampling time is short, the torque error can be reduced, but the maximum switching frequency is increased and the switching loss is increased, consequently.

In the case of Fig. 3(b), the switching signal is determined by the signature of torque error. The torque error can be reduced than the case of Fig. 3(a), but the average of actual torque can has some error. This average torque error can be oscillated, and the control performance is decreased due to this average torque error.

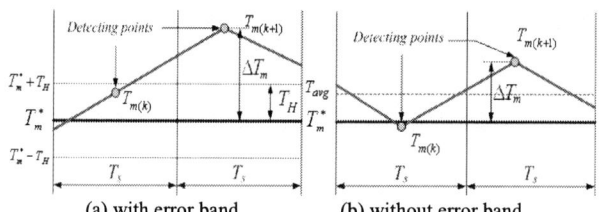

(a) with error band (b) without error band

Fig. 3 Torque error in a conventional DTC

B. The proposed DTC-PWM method

In order to overcome the problem of conventional DTC method, a novel DTC-PWM method is proposed. The proposed DTC-PWM method is combined a conventional DTC and a PWM technique.

Fig. 4 shows the proposed torque control scheme. The proposed method uses the same switching rule shown as Fig. 2 and Table I, but has additional PWM block with duty ration calculation. And the calculation of stator flux $\phi_{\alpha s}$, $\phi_{\beta s}$ are not required. In the calculation of stator flux, the information of $v_{\alpha s}$, $v_{\beta s}$ are essential. But, it is very difficult to direct detect the stator voltage in practical application, so they are estimated from the switching pattern and link voltage. Furthermore, the integral term can increase calculation error.

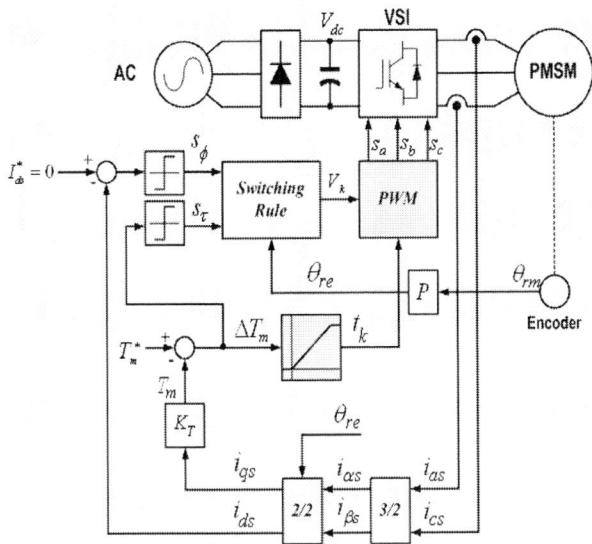

Fig. 4. Proposed torque control scheme

Exceptional case of field-weakening control, the linkage flux of PMSM is produced by permanent magnet of the motor. So, the d-axis current(flux component) is controlled as zero for the effective control of PMSM. And the torque is controlled by q-axis(torque component) current.

In the proposed control scheme, the accuracy stator linkage flux is not required, but just the signature of linkage flux can be easily obtained by the d-axis current of the PMSM as follows[9] :

$$
\begin{bmatrix} i_{\alpha s} \\ i_{\beta s} \end{bmatrix} = \sqrt{\frac{2}{3}} \cdot \begin{bmatrix} 1 & -\frac{1}{2} & -\frac{1}{2} \\ 0 & \frac{\sqrt{3}}{2} & -\frac{\sqrt{3}}{2} \end{bmatrix} \cdot \begin{bmatrix} i_{as} \\ i_{bs} \\ i_{cs} \end{bmatrix} \tag{5}
$$

$$
\begin{bmatrix} i_{ds} \\ i_{qs} \end{bmatrix} = \begin{bmatrix} \cos\theta_{re} & \sin\theta_{re} \\ -\sin\theta_{re} & \cos\theta_{re} \end{bmatrix} \cdot \begin{bmatrix} i_{\alpha s} \\ i_{\beta s} \end{bmatrix} \tag{6}
$$

$$
T_m = K_T \cdot i_{qs} \tag{7}
$$

where, K_T is torque constant of PMSM.

From the simple transformation, s_ϕ can be easily obtained by i_{ds}, and the signature of torque error s_τ is calculated as follow.

$$
\begin{aligned}
s_\phi &= sign \left| I_{ds}^* - i_{ds} \right| \\
s_\tau &= sign \left| T_m^* - T_m \right|
\end{aligned} \tag{8}
$$

From (8) and Table I, the switching rule block just determines one voltage vector V_k. And the final switching signals are determined by the voltage vector

and switching time t_k as shown in Fig. 4. The switching time t_k is proportional to torque error ΔT_m.

$$
\Delta T_m = T_m^* - T_m \tag{9}
$$

The switching time t_k and zero vector time t_0 are calculated as follows.

$$
t_k = \begin{cases} \dfrac{\Delta T_m}{T_H} \cdot T_s \Big| & when \ \ \Delta T_m < T_H \\ T_s & \Big| \ \ when \ \ \Delta T_m \geq T_H \end{cases} \tag{10}
$$

$$
t_0 = T_s - t_k \tag{11}
$$

where, T_H and T_s denote torque error bandwidth and sampling period, respectively. Actually, the maximum torque ripple of DTC is determined by sampling period T_s (switching frequency), load current and motor speed. The maximum torque ripple can be reduced in high speed and short sampling period. So, the torque ripple can be changed by selection of T_H.

C. The Effect of Torque Bandwidth Selection

In this paper, the switching time of effective voltage vector is calculated by the torque error and torque bandwidth T_H. In the same torque error, switching time is changed by the torque bandwidth. If we select small torque bandwidth, actual torque ripple is increased. Otherwise, output torque during sampling time can not be satisfied.

The maximum torque variation during sampling time is determined by the motor speed, load torque, sampling time and motor parameters. The instantaneous voltage equation of PMSM becomes as follows in d-q axis.

$$
\begin{aligned}
v_{ds} &= R_s \cdot i_{ds} + L_s \cdot \frac{d}{dt} i_{ds} - L_s \cdot \omega_r \cdot i_{qs} \\
v_{qs} &= L_s \cdot \omega_r \cdot i_{ds} + R_s \cdot i_{qs} + L_s \cdot \frac{d}{dt} i_{qs} + K_e \cdot \omega_r
\end{aligned} \tag{12}
$$

where, v_{ds}, v_{qs} and i_{ds}, i_{qs} are the voltage and current in d-q axis. R_s and L_s denote the winding resistance and inductance, respectively. K_e and ω_r are the back EMF constant and electrical angular speed of motor.

In the ideal case, we can assume d-axis current is zero. The applied voltage during sampling period can be obtained by the switching vector and DC link voltage. Since the output torque of PMSM can be derived by the q-axis current, the maximum torque variation according to the sampling period can be derived as follows.

$$\Delta T_m = \frac{K_T}{L_s} \cdot \left(v_{qr} - R_s \cdot \frac{T_m}{K_T} - K_e \cdot \omega_r \right) \cdot T_s \quad (13)$$

Fig. 5 shows the maximum torque variation according to sampling period, motor speed and torque. Here, torque variation ΔT_m is the percent value to rated torque. Fig. 5(a) shows the torque variation according to the sampling period and motor speed in rated torque. Fig. 5(b) shows the torque variation according to the sampling period and torque in rated speed. As shown in Fig. 5, the torque variation in the fixed sampling period is increased according to speed and load torque decreasing.

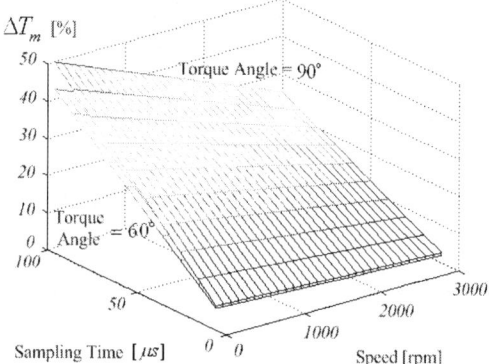

(a) Torque variation in rated torque

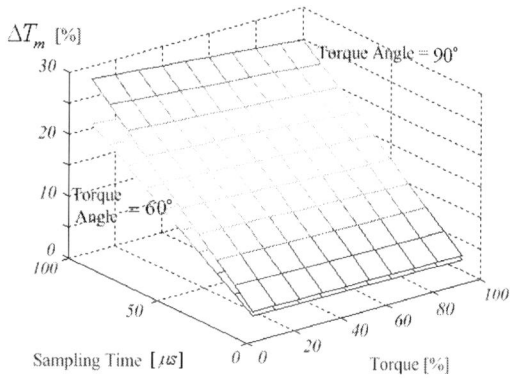

(b) Torque variation in rated speed
Fig. 5 Torque variation according sampling time, speed and load torque

According to the speed increasing, back EMF is increased, maximum torque variation is dramatically decreasing in the same sampling period. In this reason, a constant torque bandwidth can not be satisfied the torque ripple reduction and dynamic characteristics. For the high control performance, torque bandwidth in the calculation of effective voltage switching time should be changed by the motor speed and torque.

If $s_\phi = 1$ and $s_\tau = 1$, voltage vector is selected as V_2. And the switching time of V_2 and zero-vector V_0 are calculated from (11) and (12).

Fig. 6 shows the switching signal comparison between the conventional method and the proposed one. The conventional method uses only one switching vector at every sampling period. But the proposed method uses effective voltage vector and zero-vector as shown in Fig. 6(b). Fig. 6(c) shows the torque waveform in the proposed control scheme. Comparing the conventional DTC, effective voltage vector and zero-vector are supplied in a sampling period.

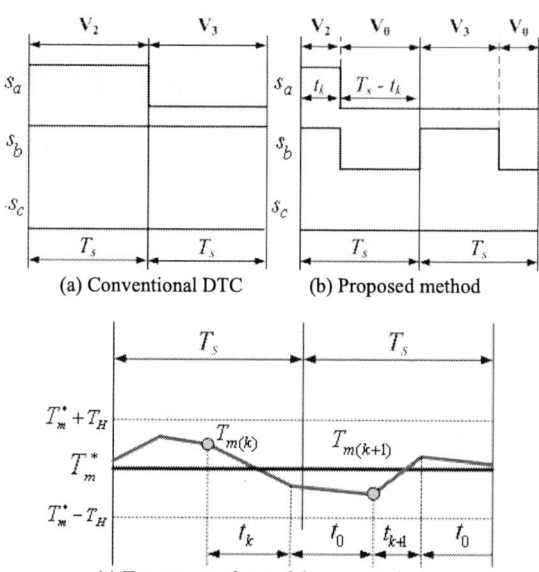

Fig. 6 The comparisons of switching signals and torque waveform

IV. SIMULATIONS RESULTS

In order to verify the proposed torque control scheme, computer simulation is implemented.

A 3-phase 1.2[kW] PMSM is used for computer simulation. Table 2 shows the specifications of the tested PMSM.

TABLE 2. THE SPECIFICATIONS OF TESTED PMSM

Rated power	1.2[kW]	Rated speed	3750[rpm]
Rated torque	1.7[Nm]	Max. speed	5000[rpm]
L_s	14[mH]	R_s	4.765[Ω]
Rotor inertia	1.051×10^{-4} [kgm²]	Flux density	0.1848[Wb]

Fig. 7 and Fig. 8 show simulation results with a conventional DTC. The sampling period of torque estimator is selected as 25[us] and 50[us], respectively. As shown in Fig. 7 and Fig. 8, the torque ripple is in-proportional to sampling period. And the switching frequency is not constant.

Fig. 9 and Fig. 10 show simulation results with the proposed control scheme. The sampling period is selected same as the conventional one. But the torque ripple can be reduced than that of the conventional method with a constant switching frequency.

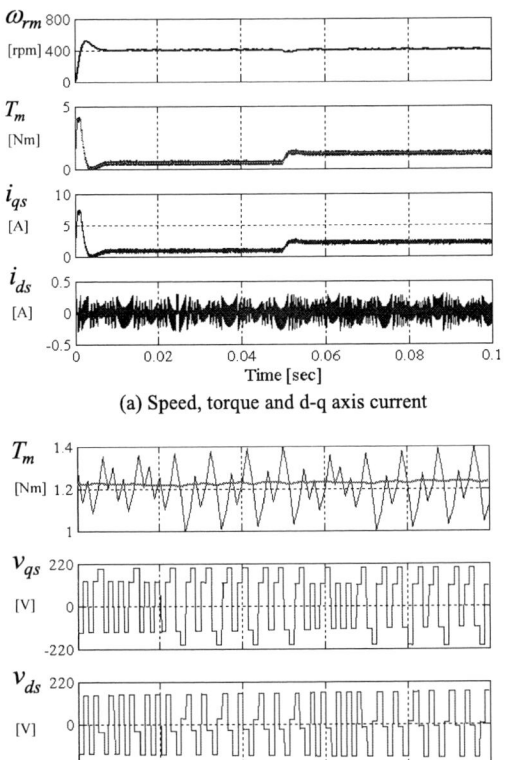

(a) Speed, torque and d-q axis current

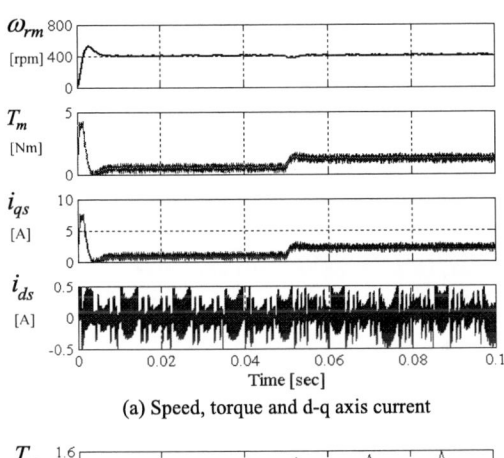

(b) Torque and d-q axis voltage

Fig. 7 Simulation results of conventional DTC($T_S = 25[\mu s]$)

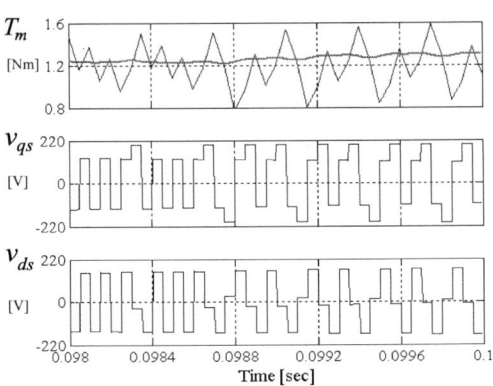

(a) Speed, torque and d-q axis current

(b) Torque and d-q axis voltage

Fig. 8 Simulation results of conventional DTC($T_S = 50[\mu s]$)

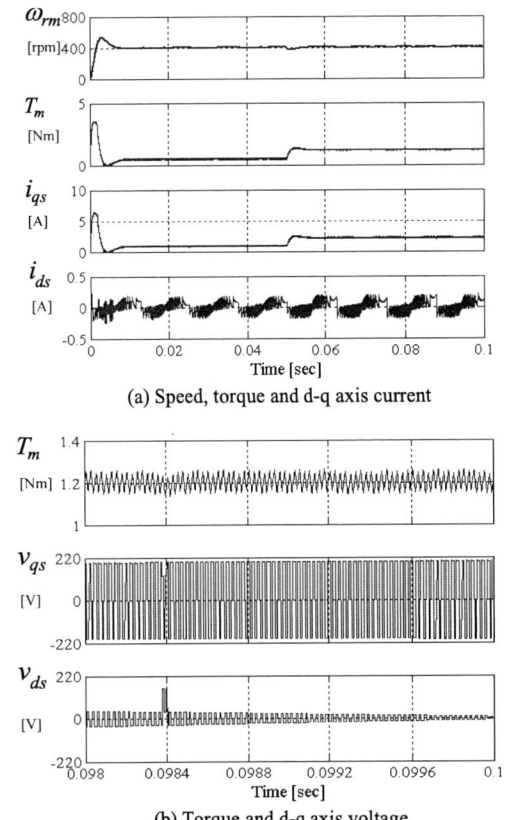

(a) Speed, torque and d-q axis current

(b) Torque and d-q axis voltage

Fig. 9 Simulation results of the proposed control scheme($T_S = 25[\mu s]$)

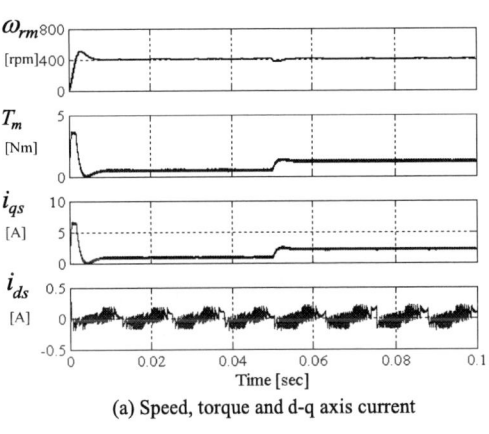

(a) Speed, torque and d-q axis current

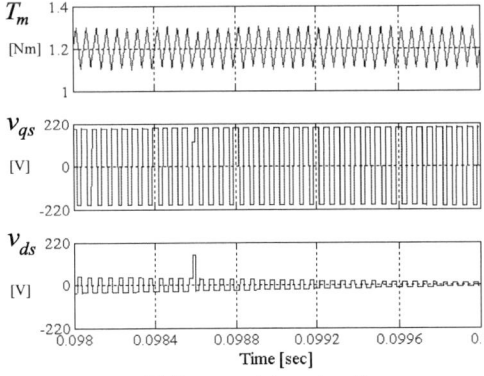

(b) Torque and d-q axis voltage

Fig. 10 Simulation results of the proposed control scheme($T_S = 50[\mu s]$)

1531

V. CONCLUSIONS

This paper presents a novel DTC-PWM control method for high performance torque control of PMSM. The proposed torque control scheme is combined by a conventional DTC and PWM technique. In order to reduce torque ripple with a same sampling period, adjustable switching vector is supplied according to torque error and switching vector. The torque of PMSM is calculated as q-axis current and the stator flux calculation is not required. Just a signature of d-axis current and torque error are used for selection of the switching vector.

Computer simulation results show a better control performance than that of a conventional DTC method.

ACKNOWLEDGEMENT

This work has been supported by KESRI(R-2005-B-109), which is funded by MOCIE(Ministry of Commerce, Industry and Energy). Korea

REFERENCES

[1] M. Depenbrock, "Direct self-control of inverter-fed machine," IEEE Trans. Power Electron., vol. 3, pp. 420–429, Oct. 1988.

[2] I. Takahashi and T. Naguchi, "A new quick-response and high-efficiency control strategy of an induction motor," IEEE Trans. Ind. Applicat., vol. IA-22, pp. 820–827, Sept./Oct. 1986.

[3] C. French and P. Acarnley, "Direct torque control of permanent magnet drives," IEEE Trans. Ind. Applicat., vol. IA-32, pp. 1080–1088, Sept./Oct. 1996.

[4] L. Zhong, M. F. Rahman, W. Y. Hu, and K. W. Lim, "Analysis of direct torque control in permanent magnet synchronous motor drives," IEEE Trans. Power Electron., vol. 12, pp. 528–536, May 1997.

[5] M. F. Rahman, L. Zhong, and K. W. Lim, "A direct torque controlled interior permanent magnet synchronous motor drive incorporating field weakening," IEEE Trans. Ind. Applicat., vol. 34, pp. 1246–1253, Nov./Dec. 1998.

[6] I. Takahashi and T. Noguchi, "Take a look back upon the past decade of direct torque control," in Proc. IEEE-IECON'97 23rd Int. Conf., vol. 2, 1997, pp. 546–551.

[7] D. Casadei, G. Serra, and A. Tani, "Implementation of a direct torque control algorithm for induction motors based on discrete space vector modulation," IEEE Trans. Power Electron., vol. 15, pp. 769–777, July, 2000.

[8] C. G. Mei, S. K. Panda, J. X. Xu, and K. W. Lim, "Direct torque control of induction motor-variable switching sectors," in Proc. IEEE-PEDS Conf., Hong Kong, July 1999, pp. 80–85.

[9] A. Tripathi, A. M. Khambadkone, and S. K. Panda, "Space-vector based, constant frequency, direct torque control and dead beat stator flux control of AC machines," in Proc. IEEE-IECON'01 Conf., Nov. 2001, pp. 1219–1224.

Low Cost Position Sensor for Permanent Magnet Linear Drive

Ralf Wegener[1], *Student Member, IEEE*, Florian Senicar[2], Christian Junge[2], Stefan Soter[1], *Member, IEEE*

[1]Institute of Electrical Drives and Mechatronics, University of Dortmund, Germany
http://eam.e-technik.uni-dortmund.de; Fax: +49 231 755 7374;
ralf.wegener@uni-dortmund.de; stefan.soter@ieee.org

[2]Electrical Machines and Drives Group, University of Wuppertal, Germany
http://www.ema.uni-wuppertal.de

Abstract— **This paper deals with a custom made low cost sensor for measuring the position of a permanent magnet linear motor. The principle how to measure position and movement direction with two analog hall sensor elements is described. The following simulated and detailed error and failure treatment is very important to know exactly the performance and the possibilities of this low cost sensor element. Afterwards this position sensor is build and some measurements with a linear machine is done. After filtering, the accuracy of the two signals is high enough to be an input of a converter control to determine the correct current which has to be injected. If there is another higher ranking closed-loop control, e.g. pressure, flow or force, in the control system this low cost sensor is sufficient and works very well. It is possible to implement the very small sensor in the housing of the linear drive. This sensor costs less than 15 dollar and can not be compared to a very precise working linear senor for some hundred dollar in order to position the linear drive very exact but the accuracy is high enough to build a lower ranking closed-loop control and to stabilize a complex control system of converter, linear drive and load.**

Keywords— **Position Sensor, Permanent Magnet, Linear Drive**

Topic— **Motor drives and motion control**

I. MOTIVATION

In some application fields linear drives are not used for precise positioning but to regulate e.g. pressure, flow or force of an industrial process. If two parts have to be pressed together the force is the higher ranking closed-loop control and the position is on-

ly needed for lower control loops. In this cases it is not possible to control the linear drive sensorless, without any linear positioning sensor but the standard types with accuracies from 50μm up to 10μm are very expensive. In this application field a linear sensor with an accuracy of 200μm to 1mm is sufficient. New analog hall sensors are available on the market for less than two dollar. To build a positioning sensor with these attributes two low cost hall sensor elements, some surface mounted devices and a small pcb is enough. A requirement is, that the permanent magnets provides a sinusoidal field outside the winding area of the linear drive, which is fulfilled in most cases in an optimal distance. This small linear positioning sensor can be mounted nearby the magnets and the windings and needs no extra space like a conventional linear sensor.

II. MEASUREMENT OF THE MAGNETIC FIELD

The magnetic rotor field of a permanent magnet linear motor outside the stator windings has different shapes depending on the distance of measurement. At the surface of the magnets the field has a nearly rectangular characteristic. When the distance between sensor and magnets increases the shape becomes softer and in the optimal position the curve can be described as a sinu-

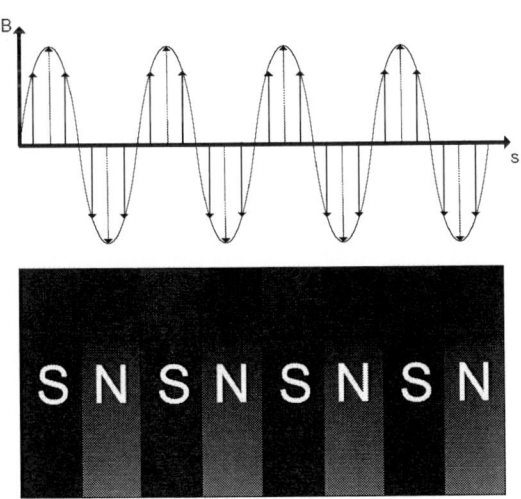

Fig. 1. Characteristic of the magnetic field based on the permanent magnets on the rotor, measured in the optimal distance.

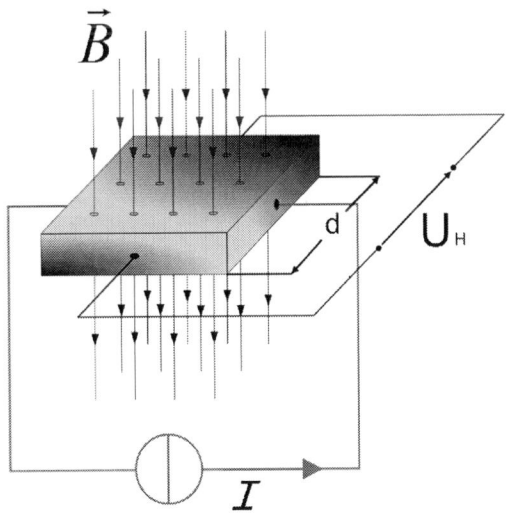

Fig. 2. Sensor element for measuring the magnetic field using the hall effect.

soidal wave along the rotor of the linear machine. It is based of the permanent magnets stringed together in opposite magnetic directions shown in figure 1.

The magnetic field can be measured using the well known Hall-Effect with a sensor shown in figure 2. The magnetic field B is penetrating the sensor element which is fed with a constant measurement current I. The Hall effect diverts the electrons which results in a voltage U_H measurable in orthogonal direction of the sensor element. The voltage U_H is proportional to the absolute value of the magnetic field like in equation (1).

$$U_H = \frac{1}{n \cdot q} \cdot \frac{I \cdot B_\perp}{d} \qquad (1)$$

The principal function of the sensor is proven by the simple

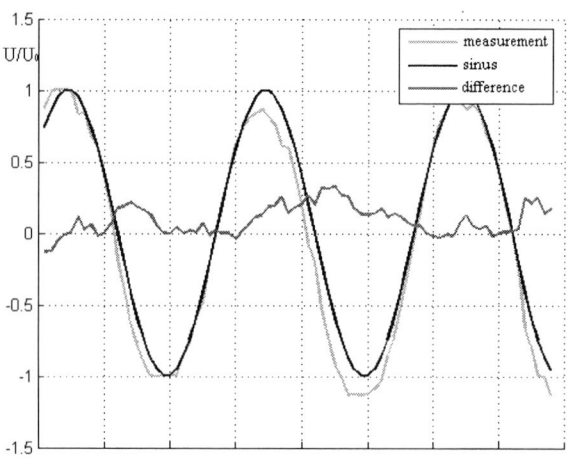

Fig. 3. Magnetic field of the permanent magnets of the armature

measurement of the magnetic field relative to the position of the armature shown in figure 3. The measured signal, printed in green, is approximated wich a sinusoidal signal with the error printed in red.

III. CONSTRUCTED SENSOR

To give the possibility to determine the direction of the movement it is necessary to measure the magnetic field with two sensors in a distance of 90° of the sinusoidal approximation. This equals the half length of the magnet.

The measured signals have to be adapted to the inputs of the converter. This is done by an operational amplifier with a differential output and a second operational amplifier to correct the offset of the sensor. The circuit is shown in figure 4.

IV. POSSIBLE ERRORS

In the following chapter the possible errors of the position sensor is researched. This is done by simulations of the incorrect sensor signals with determination of the position error.

A. Reduced Amplitude

The first possible error is an unbalanced gain of the operational amplifier or the sensor in one direction as shown in figure 5. The same error can be caused by an unequal magnetization of

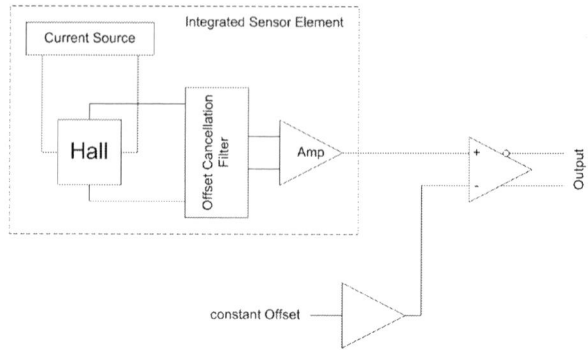

Fig. 4. Functional description of the built sensor.

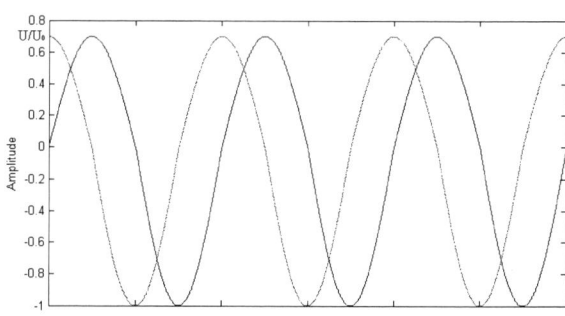

Fig. 5. Unbalanced signal amplitude in positive and negative direction.

the positive and negative magnet which is much more likely. In the simulation the magnetic field of the magnets with the north pole in front of the sensors is scaled by 0.7. As shown in figu-

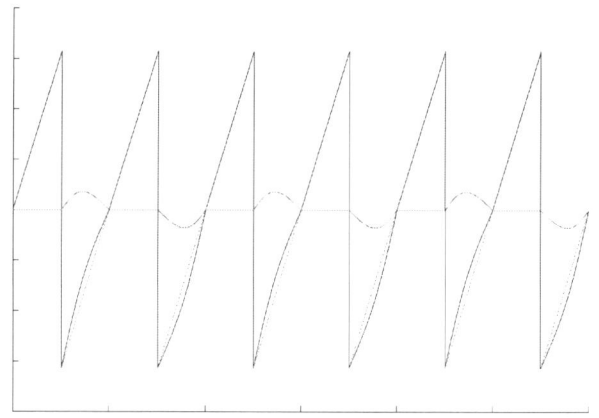

Fig. 6. Position error caused by the unbalanced signal amplitude.

re 6 the maximum position error is around 10°. In a real linear machine, which is also used for measurements, this equals a distance of 1.7mm because the magnets are arranged with a period of 6omm. This is an error of approximately 2.8%.

B. Phase Error

As described in chapter III the position sensor consists of two separate hall sensors which are positioned with an angle of 90° relative to the magnet period. The exact distance of the sensors can be varied because of factory tolerances. In the simulation the two sensors are positioned with an extra distance of 1mm which equals an angle of 96°. The results are shown in figure 7. This

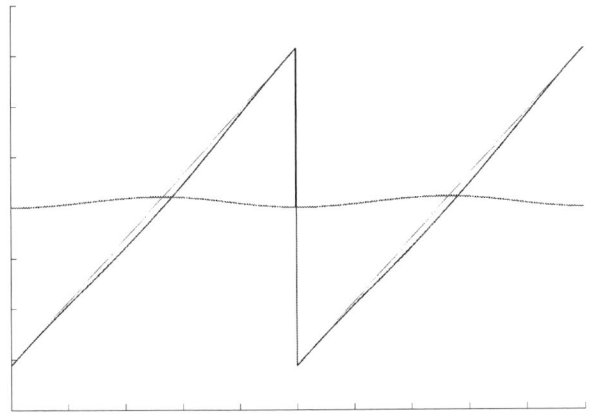

Fig. 7. Measured angle with a sensor-position error of 1mm.

error results in a permanent offset of the measured value related to the calculated position. This is irrelevant for the position control, because the position is measured in a relative way and the offset is neglected. In addition the position signal is deformed in a nonlinear way, caused an absolute position failure in the area of 1mm.

C. Range Error

The magnetic field of the used permanent magnets is too high for the used hall sensors. Therefore the sensors distance to the magnets have to be increased till the sinusoidal signal is not cut any longer at high amplitudes. In case of an overload of the two sensors the measured values are shown in figure 8. The calcu-

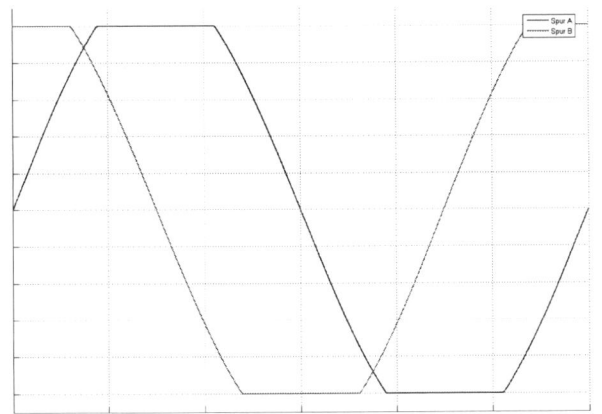

Fig. 8. Sensor signal in case of an overload because of too large magnetic fields.

lated position results in an error but this failure can be easily detected (see figure 9). Because of that the overload condition is not relevant to the motor control.

V. MEASURED RESULTS

In order to test the built sensor it is mounted at a distance of approximately 20mm in the above mentioned reference linear drive in orthogonal direction from the permanent magnets of the

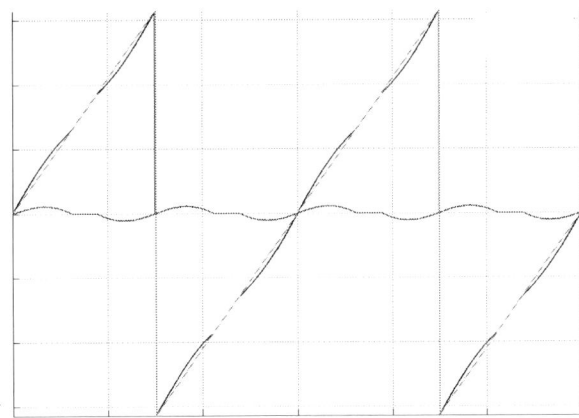

Fig. 9. Calculated position in an overload condition.

armature. This is now moved with a fixed speed and the measured signals are recorded. The results are shown in figure 10. The two sinusoidal signals are faulty at two positions. This is

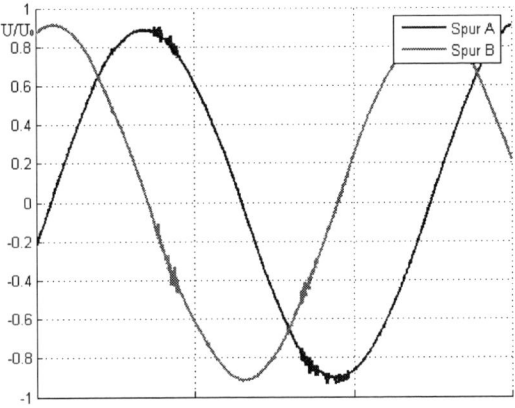

Fig. 10. Measured sensor signals with fixed an armature speed

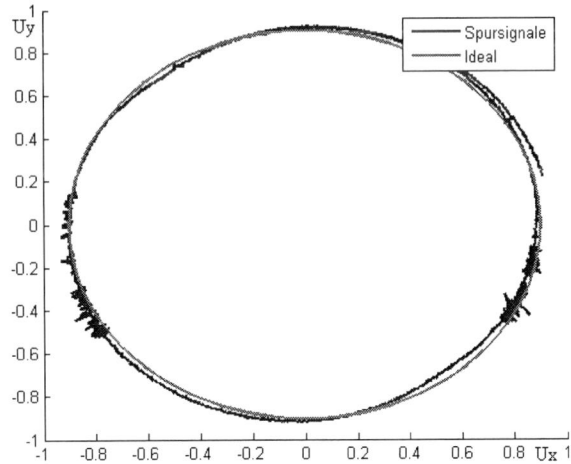

Fig. 11. x-y diagram of the measured signals

explainable with the near placement of the sensor to the coils which currents interferes with the signals.

The measurement can be visualized in x-y diagram (figure 11) where the sine and cosine signals are plotted in both axis. With an optimal sensor the resulting diagram has to be a perfect circle.

VI. BUILT FILTER

The results of the built filter structure are shown in figures 12 to 15.

The position signals, presented in figure 12, are measured with

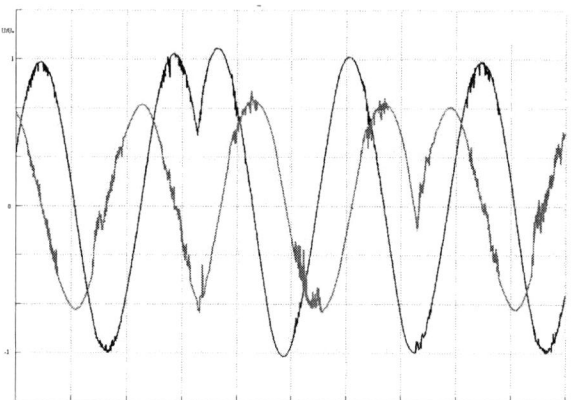

Fig. 12. Unfiltered measurement signals for a movement of 10cm

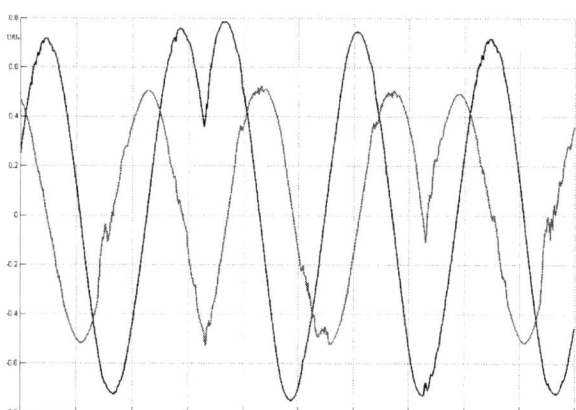

Fig. 13. Filtered measurement signals for a movement of 10cm

a movement of the armature in both direction for approximately 10cm. These signals are filtered in order to produce a suitable position signal shown in figure 13.

The interference of the coil current is magnified in figure 14. The noise is pulsating because the current in only one coil nearest to the hall sensors interferes the measurement.

The resulted position signal is compared to the reference signal measured with a 10μm optical linear sensor in figure 15. The absolute error of the position is very low during the whole measurement range.

VII. CONCLUSION

The simulation and measurement results, shown and described in this paper, have demonstrated that is it possible to build a low cost linear sensor with elements available at the market. The dimensions are small enough to implement the sensor pcb in the

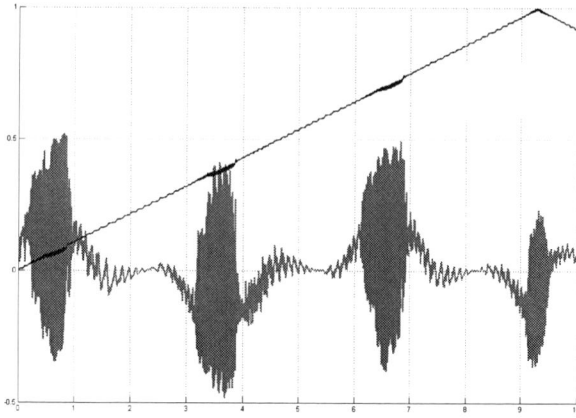

Fig. 14. Interference of the position signals caused to the coil current.

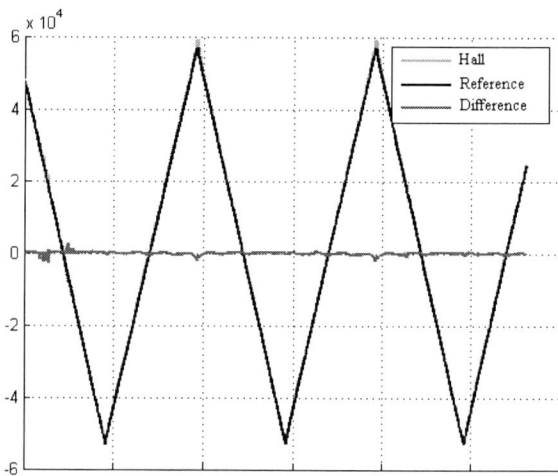

Fig. 15. Position signals of the system with the hall sensor and a reference sensor

linear drive itself in order to need no extra space and to get a compact linear drive system. If the highest ranking closed-loop control is not the position, the accuracy of the custom made sensor is sufficient to be a part of the lower ranking converter control system. With very low costs of less than 15 dollar this linear sensor can be used in spite of the very expensive standard solution. Further detailed measurements have to be done to improve accuracy and interference resistance.

REFERENCES

[1] Senicar, F. *Entwicklung einer Hall-Sensor basierten Lageerfassung für einen Linearmotor mit permanenterregtem Sekundärteil*, Studienarbeit am LS EAM, University of Dortmund, 2006

[2] Morimoto, S.; Sanada, M.; Takeda, Y. *High-performance current-sensorless drive for PMSM and SynRM with only low-resolution position sensor* Industry Applications, IEEE Transactions on; Volume 39, Issue 3, May-June 2003 Pages: 792 - 801

[3] Guang Liu; Kurnia, A.; De Larminat, R.; Rotter, S.J. *Position sensor error analysis for EPS motor drive* Electric Machines and Drives Conference, 2003. IEMDC03. IEEE International Volume 1, Issue , 1-4 June 2003 Page(s): 249 - 254 vol.1

[4] Lidozzi, A.; Solero, L.; Crescimbini, F.; Di Napoli, A. *SVM PMSM Drive With Low Resolution Hall-Effect Sensors* Power Electronics, IEEE Transactions on; Volume 22, Issue 1, Jan. 2007 Page(s):282 - 290

[5] Yong Ho Yoon; Mu Sun Woo; Seung Jun Lee; Chung Yuen Won; You

Young Choe; *Speed control system of slotless PM brushless DC motor using 2Hall-ICs* Industrial Electronics Society, 2004. IECON 2004. 30th Annual Conference of IEEE Volume 2, 2-6 Nov. 2004 Page(s):1374 - 1379 Vol. 2

[6] Xu Zheng; Li Tiecai; Lu Yongping; Guo Bingyi *Position-measuring error analysis and solution of hall sensor in pseudo-sensorless PMSM driving system* Industrial Electronics Society, 2003. IECON03. The 29th Annual Conference of the IEEE Volume 2, Issue , 2-6 Nov. 2003 Page(s): 1337 - 1342 Vol.2

[7] Baris Ozturk, S.; Akin, B.; Toliyat, H.A.; Ashrafzadeh, F. *Low-cost direct torque control of permanent magnet synchronous motor using Hall-effect sensors* Applied Power Electronics Conference and Exposition, 2006. APEC06. 19-23 March 2006

Design of One Rotary-linear Permanent Magnet Motor with Two Independently Energized Three Phase Windings

L. Chen*, W. Hofmann**

* Chemnitz University of Technology, Electrical Machines and Drives, Germany
** Chemnitz University of Technology, Electrical Machines and Drives, Germany

Abstract-- Electrical drives employing motors able to directly manage multiple degrees of freedom may result in relevant interest in several application fields, e.g. robotics and tooling machines. In this paper, a new type of rotary-linear permanent magnet motor with one motion of two degrees of freedom has been proposed. Its main performance advantage is its characteristics of decoupling linear and rotary motions comparing another types of rotary-linear motors. The proposed motor consists of two stator three–phase winding systems separately distributed along radial and axial axes. By using independently energized three–phase windings, two space rotating magnetical fields along axial axis and radial axis are achieved, which are utilised to product rotary and linear motion. One special rotor permanent magnet structure capable of generating one radial and one axial magnetical field has been put forward and analyzed to produce enough torque, drag force and decouple rotary and linear motions. Based on the theoretical analysis and calculation, an experimental motor has been designed and its design process has been detailed introduced in the paper. The validity of analysis and design technique has been confirmed by one three-dimension finite element method (FEM) calculation.

Index Terms-- permanent magnet, rotary-linear, motor, three-dimension FEM

I. INTRODUCTION

Recent advances in robotics, office automation, and intelligent flexible manufacturing and assembly systems have necessitated the further development of programmable, servo-controlled, high speed actuators with multiple degrees-of-freedom. In most cases, such movements are obtained independently by mechanically combining linear or rotary elementary motions. Nevertheless, this requires a suitable mechanical structure able to perform such combination, equipped with as many standard single-DoF drives as the total number of elementary motions. Usually such solutions result in expensive, heavy and cumbersome apparatuses [1]. Till now in the technical literature a small number of configurations of multi-DoF electric machines has been proposed. The most diffused one is the Sawxyer step planar motor [2]. Furthermore, for what concerns instead rotary-linear machines, the conceptually simplest solutions consist in mechanically coupling a standard induction rotary machine with a linear machine, having a

prolonged bulk mover, such as a voice-coil actuator [3] or a linear induction motor [4].

Nevertheless, the poor integration level of such solutions leads to modest overall performances. A more integrated machine is proposed in [5], featuring 4 linear poly-phase induction stators evenly surrounding a cylindrical bulk prolonged mover. Here the rotary torque is obtained by suitably time-shifting the supply current terns relative to each stator. Anyway, four 3-phase inverters are required to supply the windings, while a true decoupled regulation of force and torque appears problematic. In [6] a "screwthread" cylindrical linear reluctance motor is allowed to also rotate and is supplied with the superimposition of a low frequency tern and a high frequency tern of currents. In fact, the linear motion is generated primarily due to the low-frequency component, while the high-frequency component is mainly responsible of eddy currents induced in the bulk mover, giving rise to a net torque by induction effect. Nevertheless a true decoupled regulation of force and torque appears problematic too.

In this paper, we are concerned with a new rotary-linear permanent magnet synchronal motor. Its main performance advantage is its decoupling characteristics of linear and rotary motions. The proposed motor consists of two three–phase stator windings separately distributed along radial and axial axes in order to achieve two rotating magnetical fields, which are utilised to product one rotatary and one linear motion separately. To be compatible with the structure of two stator windings, one special permanent rotor magnet structure has been put forward and analyzed in order to not only produce enough torque and drag force but also decouple rotary and linear motions in the paper. Motor structure is shown in Fig.1. Based on theoretical analysis, an experimental motor will be designed and its design process is detailed introduced. The validity of analysis and design technique is confirmed by FEM calculation.

978-1-4244-0644-9/07/$25.00 ©2007 IEEE

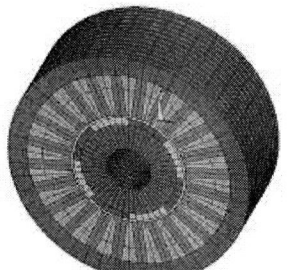

a.) Motor structure

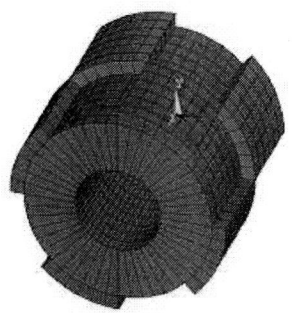

b.) Rotor structure with permanent magnets

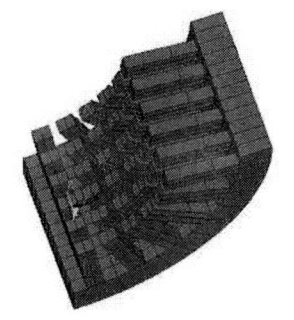

c.) Stator structure

Fig. 1. Motor structure

II. PRINCIPLE AND STRUCTURE

Figure 2 shows structure principle of the proposed rotary-linear PM motor. In Fig. 2, two sets of three phase windings are equipped in radial and axial stator slots. This rotor is supported by a rotary-linear bearing system or an air-pressure bearing. By supplying two three-phase currents in each primary winding, one rotary and one linear magnetic field are generated.

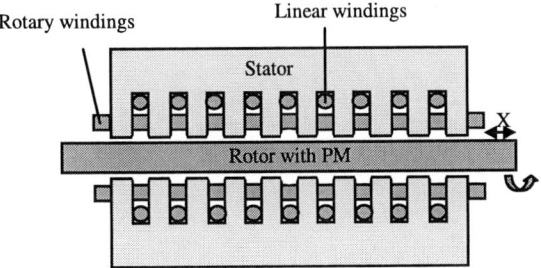

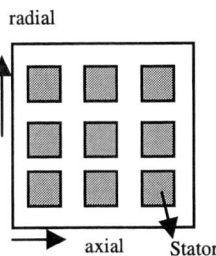

Fig. 2. Structure principle of one rotary-linear PM motor and stator structure

Fig.3 shows the structure of rotor permanent magnets. One pair of permanent magnets are intersected distributed on rotor surface along axial direction. It is confirmed by theoretical analysis, the permanent magnet structure is available of creating force, torque and decoupling rotary and linear motions by generating two magnetical fields along radial and axial directions. It means, that it is possible to carry out any one of six motions in linear and rotary plane, as shown in Fig. 4.

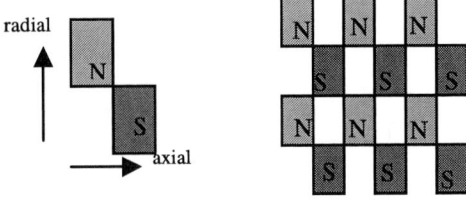

Fig. 3. Structure of rotor permanent magnets

The generating principle of force and torque is depicted so in Fig. 5. Two magnetical fields along radial and axial directions are produced from permanent magnets, which interacts with stator axial and radial alternative current magnetical field and produces decoupled force and torque.

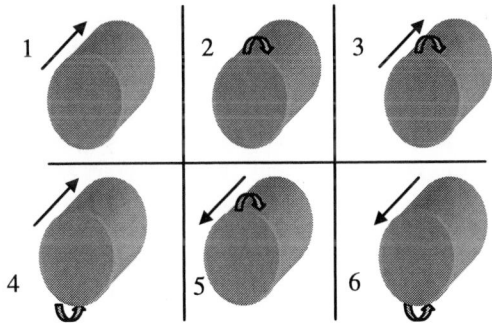

Fig. 4. Six motions in linear and rotary plane

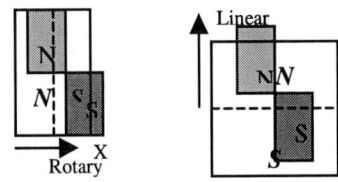

Fig. 5. Generating principle of force and torque

The torque of a PM motor can be expressed so:

1539

$$T = 2\pi r^2 l B_1 K_{1s} \sin \beta , \qquad (1)$$

where r is rotor radius and l is motor length. The air gap fundamental flux density due to permanent magnet is B_1 and the linear current density along stator periphery is K_{1s}. β is the angular displacement between two fields produced by permanent magnet and stator current. B_1 is considered as one radial variant used to generate torque. Based on one simplified model of linear motor with surface mounted PMs, the equation of generating thrust can be derived and expressed as

$$\vec{F}_2 = \vec{J} \times \vec{B} = p J_0 B_2 a_x , \qquad (2)$$

where J_0 is applied current density, B_2 is the flux density of axial direction. a_x is the pole pitch.

III. DESIGN OF ONE ROTARY-LINEAR MOTOR

One eight pole motor ($p_1 = 4$ and $p_2 = 4$) with a rated output power of $P_{1N} = 250$ W at rated speed of $n_{N1} = 3000$ rpm and a rated output thrust of $F_{N2} = 60$ N at rated speed of $V_{N2} = 1$ m/s has been designed. The sheets with outer diameter $d_a = 100$ mm, inner diameter $d_i = 47.6$ mm and $Q = 24$ slots are chosen, based on the parameters of one screwthread rotary-linear motor, which has been made in our institute. The design flow chart is shown in Fig. 6.

The maximum flux density B_t in the stator teeth is usually limited to about 1.6 T to avoid excessive saturation. The desired value of air gap flux density above the magnet is then given by $B_\delta = B_t \cdot w_t / A$ T for a tooth width w_t, slot width w and a slot pitch A= $w_t + w$. For most machines, the slot width of the stator is equal to tooth width. Thus, a typical value for air gap density is about 0.8 T. The mechanical airgap of 1 mm has been chosen. By chooosing NdFeB magnets with the characteristics as shown in Table 1, a fundamental air gap density of o.8 T is achieved as design aim. The magnet length can been analytically determined in (4), where $B_{\delta,1}(r=r_m)$ is the air gap fundamental flux density on the permanent magnet surface, μ_m is the relative magnet permeability of magnets, τ_p is the pole pitch, k_s and k_{fr} represent the influence of magnet curve and air gap leakage flux ("fringing"), k_p and k_l represent influence

$$B_m = \frac{\pi}{4} \cdot B_{\delta,1}(r = r_m) \qquad (3)$$

$$h_m = \frac{\mu_m \cdot \delta}{\dfrac{B_R}{B_m} - 1} \cdot k_s \cdot k_{fr} \cdot k_p , k_l \qquad (4)$$

here: $k_s = \dfrac{r_m}{r_m - h_m}$, $\qquad k_{fr} = 1 + \dfrac{2\delta}{\tau_p (r = r_m)}$,

$k_p = 1 + \dfrac{2(h_m + \delta)}{\tau_i (r = r_{in})}$, $\qquad k_l = 2$

of air gap leakage flux between magnets and considering two magnetic flux variants along axial and radial directions. Based on practical motor parameters, the above factors are approximately calculated and equal $k_s = 1.1$, $k_{fr} = 1.11$, $k_p = 1.367$. The length of magnet is calculated as 3.34 mm. Considering the end leakage flux, one magnet length of 4 mm is chosen. The rotor data are summarised in Table 1.

TABLE 1
ROTOR DIMENSIONS AND PARAMETERS

Remanence flux density	B_R (80°C)	1.233 T
Coercive force	H_{CB} (80°C)	956 A/m
Number of rotor pole pairs	P_1	4
Stator bore diameter	$d_i = 2 \cdot r_s$	47.6 mm
Mechanical air gap	δ_{mech}	1 mm
Magnet height	h_m	4 mm
Fundamental air gap flux density	$B_{\delta,1}$	0.8 T

Before designing axial windings, which is placed together with radial windings in the same stator slots, the iron length l_{Fe} of the machine is first determined. A current loading $A_1 = 100$ A/cm for the axial winding is choosen in order to provide enough thermal margin for the additional radial windings. Here, one type of single layer windings with one slot per pole and phase is chosen and winding factor is $k_{w,1} = 1$. For a desired output power of $P_N = 0.25$ kW and rated efficiency of $\eta = 0.9$, power factor of $\cos\varphi = 0.9$, the iron length is determined in (5).

$$l_{Fe} = \frac{P_N}{\dfrac{\pi^2}{\sqrt{2}} \cdot k_w \cdot B_1 \cdot l_s \cdot n_s \cdot D_i^2 \cdot \cos\varphi} = 0.0532 \, \text{m} \qquad (5)$$

Considering the especial rotor PM structure, the length of stator should be chosen to be two time as much as the calculated length. It is choosen to be 150 mm and the length of rotor is choosen to be 300 mm as the moving part. Furthermore, the stator winding parameters are calculated, which are summarised in Table 2. The drive windings take an area of $A_{slot,1} = 20$ mm² from the stator slot and slot filling factor is about $k_{slot} = 0.4$.

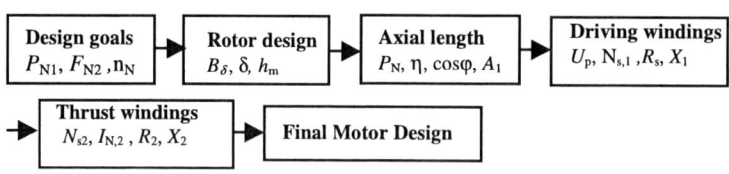

Fig. 6. Flow chart of a rotary-linear motor design

1540

TABLE 2
KEY DRIVING WINDING PARAMETERS

Number of pole pairs	p_1	4
Number of slots per pole and phase	$q_1=Q/2mp_1$	1
Winding factor (single layer)	$k_{w,1}$	1
Number of turns per phase	$N_1 \cong 0.9 \cdot U_{s1} /(\sqrt{2}\pi \cdot f \cdot k_{w,1} \cdot 2/\pi \cdot \tau_{p,1} \cdot l_{Fe} \cdot B_{\delta,1})$	166
Number of parallel branches	a_1	1
Number of turns per coil	$N_{c,1} = a_1 \cdot N_1 /(p_1 \cdot q_1)$	42
Back e.m.f. (r.m.s.)	$U_{p1} = \sqrt{2}\pi \cdot f \cdot k_{w,1} \cdot 2/\pi \cdot \tau_{p,1} \cdot l_{Fe} \cdot B_{\delta,1} \cdot N_1$	202 V
Rated current (r.m.s.)	$I_{N,1} = P_{N1} /(3 \cdot U_p)$	0.46 A
Current loading	$A_1 = 2 \cdot m \cdot N_{s,1} \cdot I_{s,1} /(2p_1 \cdot \tau_{p,1})$	32 A/cm
Current density	J_1	2.5 A/mm²

To generate a steady state thrust force of $F_2 = 60$ N, the related current loading A_2 is determined. Key linear winding parameters are shown in Table 3.

TABLE 3
KEY LINEAR WINDING PARAMETERS

Rated linear current (r.m.s.)	$I_{N,2}$	1 A
Number of turns per phase	$N_{s,2}$	70
Number of turns per coil	$N_{c,2}$	18
Current density	J_2	0.7 A/mm²

In order to verify the design, one three-dimension FEM calculations with program code ANSYS are carried out. For no-load and load investigations, calculating model has be created due to its symmetry of motor structure. The comparison of design goals and FEM results in Table 4 verifies correction of the design.

TABLE 4
COMPARISION OF ANALYTICAL CALCULATIONS AND FEM RESULTS

Quantity	Analytical design value	FE verification result	Deviation
No load air gap flux density $B_{\delta,1}$	0.8 T	0.7 T	12.5 %
Rated shaft torque M_N	0.79 Nm	0.706 Nm	10 %
Rated lateral force F_N	60 N	60 N	0 %

The following figures show the flux distribution without current load, with driving current and with torque current.

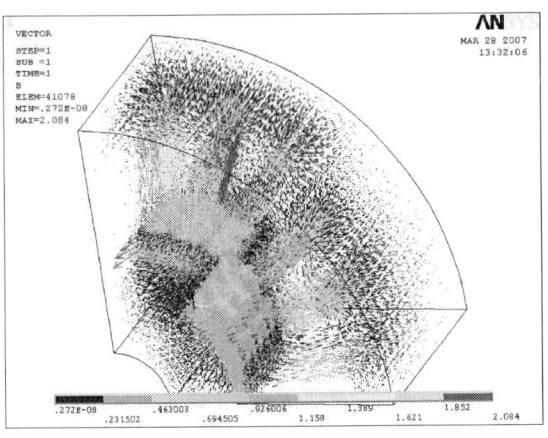

Fig. 7. Flux vector distribution without current load

Flux vector distribution without load is shown in Fig.7 and flux vector generated from rotor permanent magnet develops separately along axial and radial axes. Therefore it is possible to provide two flux separately to generate torque and thrust force. Fig. 8 shows the flux distribution with current load of thrust force of 1A and a thrust force of 60 N is generated. Fig. 9 shows the flux distribution with current load of torque of 0.46 A and a torque of 0.79 N is producted.

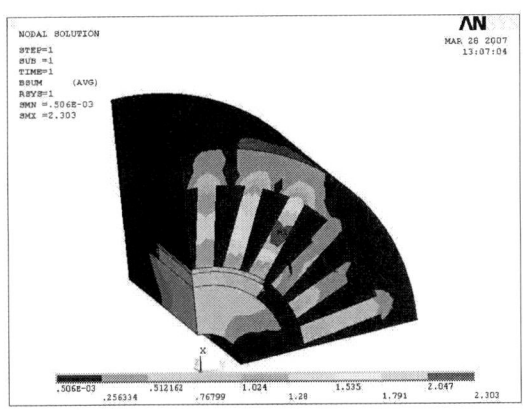

Fig. 8. Flux distribution with current load of thrust force

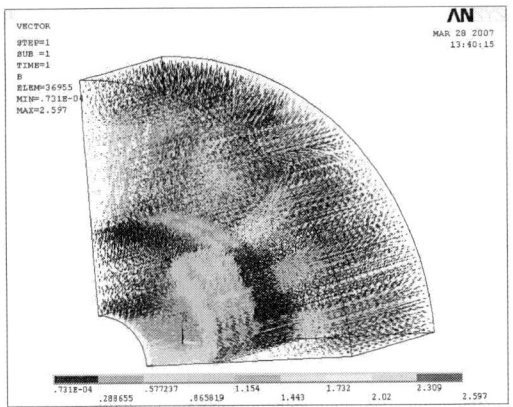

Fig. 9. Flux vector distribution with current load of torque

IV. CONCLUSIONS

In this paper, a new type of rotary-linear permanent magnet motor with one motion of two degrees of freedom has been proposed. Theoretical analysis and FEM calculation show that the motor with the especial permanent magnet structure is able to successfully carry out two-dimension motion. Future work will include production and testing of a prototype, so that the above calculation can also be verified by measurements.

REFERENCES

[1] J. Wang, K. Mitchell, G.W. Jewell and D. Howe, "Multi-Degree-of-Freedom Spherical Permanent Magnet Motors," *proc. of IEEE ICRA '2001 Conference*, Seoul Korea 2001, pp. 1798-1805.

[2] J. Ish-Shalom, "Modeling of Sawyer Planar Sensor and Motor Dependence on Planar Yaw Angle Rotation," *proc. of IEEE ICRA '97 Conference*, Albuquerque 1997, v. 4. pp. 3499-3504.

[3] G.P. Widdowson, T.H. Kuah, L. YouYong, C.J. Vath, "Design. Analysis and Testing of a Linear-Theta Motor for Semiconductor Applications," *proc. of IEEE PEDS '97 Conference*, Singapore 1997, v. 1. pp. 357-364.

[4] P. de Nwit, J. van Dujck, T. Biomer, P. Rutgers, "Mechatronic Design of an Induction actuator," *proc. of IEE EMD '97 Conference*, Cambridge 1997, pp. 279-283.

[5] W.J. Jeon, M. Tanabiki, T. Onuki, J.Y. Yoo, "Rotary-Linear Induction Motor Composed of 4 Primaries with independently Energized Ring-Windings," *proC. of IEEE-LAS Annual Meeting*, News Orleans 1997, v. 1. pp. 365-372.

[6] L. Goebel, W. Hoffman, "Control of a Rotational-Thrust Drive with Helical N\loto," *proc. of IEEE-IES IECON '97 Conference*, News Orleans 1997, v. 3. pp. 1343-1348.

Position Estimation of Permanent Magnet Synchronous Motor Using Un-known Input Observer

Masaru Hasegawa, Satoshi Yoshioka, Keiju Matsui

Dept. of Electrical Eng., Chubu University
Matsumoto-cho, Kasugai, Aichi, 487-8501 JAPAN
TEL:+81-568-51-1111; FAX:+81-568-51-1141
E-mail: mhasega@isc.chubu.ac.jp

Abstract— This paper proposes a new position estimation method for Interior Permanent Magnet Synchronous Motor(IPMSM) drives fed by over-modulation mode PWM voltage source inverter. It is being attractive to expand drive region of IPMSM to high speed range for application to electric vehicles, electrical household appliance and so on. Although over-modulation mode PWM voltage source inverter makes it possible to solve above problems, position sensorless drive based on stator voltage knowledge cannot directly be utilized because voltage reference is not equal to inverter output voltage. In this paper, un-known input observer is proposed to estimate rotor position without stator voltage knowledge.

Index Terms— Over Modulation, Permanent Magnet Synchronous Motor(PMSM), Position Estimation, Un-known Input Observer

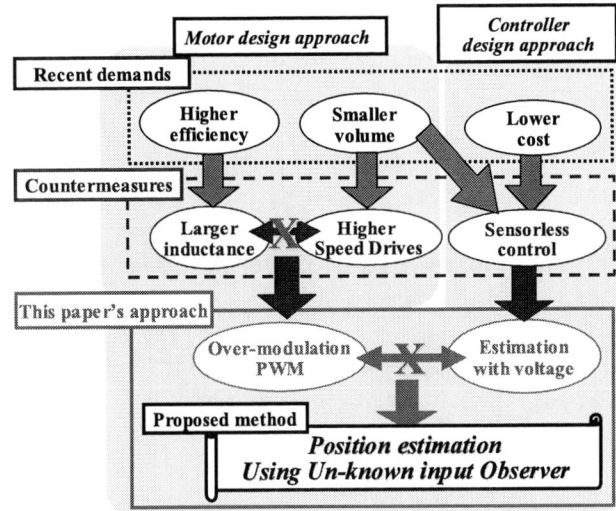

Fig. 1. Background and Approach of This Paper

I. INTRODUCTION

This paper proposes a new position estimation method for Interior Permanent Magnet Synchronous Motor(IPMSM) drives fed by over-modulation mode PWM voltage source inverter.

Recent demands of motor drive system are summarized as follows:

1. to improve efficiency,
2. to drive high-speed mode,
3. to realize high torque response.
4. to achieve low cost.

IPMSM has widely been utilized from 1). Next, much attention about high speed motor drives 2) has been attracted because the high speed operation makes the motor volume so small. In addition, the stator inductances recently tend to enlarge so that the stator current can be reduced while maintaining MMF, yielding higher efficiency. From the viewpoint of 2), however, high speed drives cannot be realized because back EMF reaches DC bus voltage of PWM inverter in middle-speed range due to large stator inductances. In this situation, 2) is not consistent with 1). Hence,

this implies that high torque response cannot be realized in middle and high speed range 3). In order to solve this problem, over-modulation mode of PWM inverter is often employed to enlarge inverter output voltage while keeping stator current waveform[1].

On the other hand, position sensorless control system has been much attractive from 4). It should be noted that much many strategies of position estimation require stator voltage to estimate rotor position and speed based on IPMSM mathematical model. This describes that the fundamental component of the stator voltage needs to be measured or must be equivalent to its reference for PWM inverter. These facts mean that much many known strategies for position estimation are inapplicable to drive system under over-modulation PWM mode. Fig.1 depicts aforementioned background and this paper's approach.

Hence, rotor position needs to be estimated without stator voltage (motor input quantities). This paper proposes a new position estimation method using un-known input observer which is constructed on multirate discrete time system.

978-1-4244-0644-9/07/$25.00 ©2007 IEEE

This observer estimates rotor position without stator voltages, so that can be applied to over-modulation PWM mode. This paper describes the construction of un-known input state observer and robustness to magnetic saturation. Feasibility and effectiveness of the proposed system are shown by MATLAB simulation and some experiments.

II. UN-KNOWN INPUT FLUX OBSERVER

A. Time-Continuous Voltage Equation of IPMSM based on flux model

This subsection derives a novel mathematical expressions of IPMSMs which is a linear flux model of IPMSMs on a stationary reference frame, which plays an important role to construct the proposed un-known input observer.

Let $v_d(t)$, $v_q(t)$, $i_d(t)$ and $i_q(t)$ be voltages on the rotating coordinate aligned with rotor position($d - q$ axis) and currents on $d - q$ axis, respectively. A time-continuous mathematical model of IPMSMs on the rotating coordinate is well-known as

$$
\begin{bmatrix} v_d(t) \\ v_q(t) \end{bmatrix} = \begin{bmatrix} R + pL_d & -\omega_{re}L_q \\ \omega_{re}L_d & R + pL_q \end{bmatrix} \begin{bmatrix} i_d(t) \\ i_q(t) \end{bmatrix} + \begin{bmatrix} 0 \\ \omega_{re}K_E \end{bmatrix},
\tag{1}
$$

where L_d and L_q mean d-axis inductance and q-axis one, respectively. Also, R and ω_{re} stand for winding resistance and rotor speed, respectively. p means the differential operator.

This paper considers the adaptive flux observer on $\alpha - \beta$ coordinate for position estimation and speed adaptation. Hence, a linear flux model of IPMSMs, which takes current and a kind of flux as state variables, needs to be constructed. Thereafter, this paper discusses the linear flux model of IPMSMs from the viewpoint of an appropriate flux definition.

Consider torque generation mechanism of IPMSMs. Output torque τ of IPMSMs is written by

$$
\begin{aligned}
\tau(t) &= (L_d - L_q)i_d(t)i_q(t) + K_E i_q(t) \\
&= \{(L_d - L_q)i_d(t) + K_E\}i_q(t) = |\boldsymbol{\lambda}(t)|i_q(t) ,
\end{aligned}
\tag{2}
$$

where,

$$
|\boldsymbol{\lambda}(t)| = (L_d - L_q)i_d(t) + K_E .
\tag{3}
$$

This equation means that a flux which really contributes to torque generation is $\boldsymbol{\lambda}(t)$, and that the direction of $\boldsymbol{\lambda}(t)$ aligns with d-axis. Therefore, the rotor position θ_{re} can be obtained by estimating flux

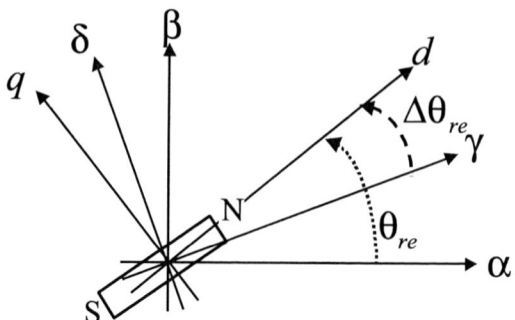

Fig. 2. Coordinates for IPMSM drives

vector $\boldsymbol{\lambda}$ on stationary reference frame($\alpha - \beta$ coordinate) as shown in Fig.2.

From the above definition, $\boldsymbol{\lambda}$ is expressed on $\alpha - \beta$ coordinate by

$$
\boldsymbol{\lambda}(t) = \begin{bmatrix} \lambda_\alpha(t) \\ \lambda_\beta(t) \end{bmatrix} = \{(L_d - L_q)i_d(t) + K_E\} \begin{bmatrix} \cos\theta_{re} \\ \sin\theta_{re} \end{bmatrix},
\tag{4}
$$

and its derivative is shown by

$$
\begin{aligned}
p \begin{bmatrix} \lambda_\alpha(t) \\ \lambda_\beta(t) \end{bmatrix} &= (L_d - L_q)\dot{i}_d(t) \begin{bmatrix} \cos\theta_{re} \\ \sin\theta_{re} \end{bmatrix} \\
&+ \omega_{re}\{(L_d - L_q)i_d(t) + K_E\} \begin{bmatrix} -\sin\theta_{re} \\ \cos\theta_{re} \end{bmatrix} .
\end{aligned}
\tag{5}
$$

On the other hand, (1) can be manipulated by the follow:

$$
\begin{aligned}
\begin{bmatrix} v_d(t) \\ v_q(t) \end{bmatrix} &= \begin{bmatrix} R + pL_q & -\omega_{re}L_q \\ \omega_{re}L_q & R + pL_q \end{bmatrix} \begin{bmatrix} i_d(t) \\ i_q(t) \end{bmatrix} \\
&+ \begin{bmatrix} (L_d - L_q)\dot{i}_d(t) \\ 0 \end{bmatrix} \\
&+ \begin{bmatrix} 0 \\ \omega_{re}\{(L_d - L_q)i_d(t) + K_E\} \end{bmatrix} .
\end{aligned}
\tag{6}
$$

In addition, transforming this equation onto $\alpha - \beta$ coordinate, the following equation can be obtained as

$$
\begin{aligned}
\begin{bmatrix} v_\alpha(t) \\ v_\beta(t) \end{bmatrix} &= \begin{bmatrix} R + pL_q & 0 \\ 0 & R + pL_q \end{bmatrix} \begin{bmatrix} i_\alpha(t) \\ i_\beta(t) \end{bmatrix} \\
&+ (L_d - L_q)\dot{i}_d(t) \begin{bmatrix} \cos\theta_{re} \\ \sin\theta_{re} \end{bmatrix} \\
&+ \omega_{re}\{(L_d - L_q)i_d(t) + K_E\} \begin{bmatrix} -\sin\theta_{re} \\ \cos\theta_{re} \end{bmatrix} .
\end{aligned}
\tag{7}
$$

The second term and the third one of (7) equal to (5). Therefore, the linear flux model can be obtained as follow:

$$
\begin{bmatrix} v_\alpha(t) \\ v_\beta(t) \end{bmatrix} = \begin{bmatrix} R + pL_q & 0 \\ 0 & R + pL_q \end{bmatrix} \begin{bmatrix} i_\alpha(t) \\ i_\beta(t) \end{bmatrix}
$$
$$
+ p \begin{bmatrix} \lambda_\alpha(t) \\ \lambda_\beta(t) \end{bmatrix} . \tag{8}
$$

It can be seen that the non-diagonal terms do not appear in impedance matrix and that this voltage equation of IPMSMs is fully identical to that of SPMSMs without any approximation. It means that IPMSMs can be regarded as SPMSMs, which makes it possible to construct the adaptive observer for IPMSMs sensorless drives on $\alpha - \beta$ coordinate.

To realize position sensorless control system, the rotor position θ_re can be obtained by estimating flux vector $\boldsymbol{\lambda}$:

$$
\theta_{re} = \tan^{-1}\left(\frac{\lambda_\beta}{\lambda_\alpha}\right) .
$$

Next subsection discusses flux observer for position estimation.

B. State Equation of IPMSM on Discrete Time System

Next, this subsection describes state equation of IPMSM on discrete time system, which is based on un-known input flux observer.

Let $\boldsymbol{i}(t)$, $\boldsymbol{\lambda}(t)$ be stator current, flux (which contribute to torque production), respectively. State equation of IPMSM on continuous time system based on complex notation is expressed as follow [2]:

$$
\frac{d}{dt}\begin{bmatrix} \boldsymbol{i}(t) \\ \boldsymbol{\lambda}(t) \end{bmatrix} = \begin{bmatrix} -\frac{R}{L_q} & -\frac{\boldsymbol{j}\omega_{re}}{L_q} \\ 0 & \boldsymbol{j}\omega_{re} \end{bmatrix} \begin{bmatrix} \boldsymbol{i}(t) \\ \boldsymbol{\lambda}(t) \end{bmatrix}
$$
$$
+ \begin{bmatrix} \frac{1}{L_q} \\ 0 \end{bmatrix} \boldsymbol{v}(t) , \tag{9}
$$

$$
\boldsymbol{i}(t) = \begin{bmatrix} 1 & 0 \end{bmatrix} \begin{bmatrix} \boldsymbol{i}(t) \\ \boldsymbol{\lambda}(t) \end{bmatrix} = C \begin{bmatrix} \boldsymbol{i}(t) \\ \boldsymbol{\lambda}(t) \end{bmatrix} \tag{10}
$$

Following equation can be obtained by discretizing above equation with control period τ,

$$
\begin{bmatrix} \boldsymbol{i}[k+1] \\ \boldsymbol{\lambda}[k+1] \end{bmatrix} = \begin{bmatrix} \exp(-\frac{R}{L_q}\tau) & A_{12} \\ 0 & \exp(\boldsymbol{j}\omega_{re}\tau) \end{bmatrix} \begin{bmatrix} \boldsymbol{i}[k] \\ \boldsymbol{\lambda}[k] \end{bmatrix}
$$
$$
+ \begin{bmatrix} \frac{1}{R}\left(1 - \exp(-\frac{R}{L_q}\tau)\right) \\ 0 \end{bmatrix} \boldsymbol{v}[k]
$$
$$
= A_d \begin{bmatrix} \boldsymbol{i}[k] \\ \boldsymbol{\lambda}[k] \end{bmatrix} + B_d \boldsymbol{v}[k] , \tag{11}
$$

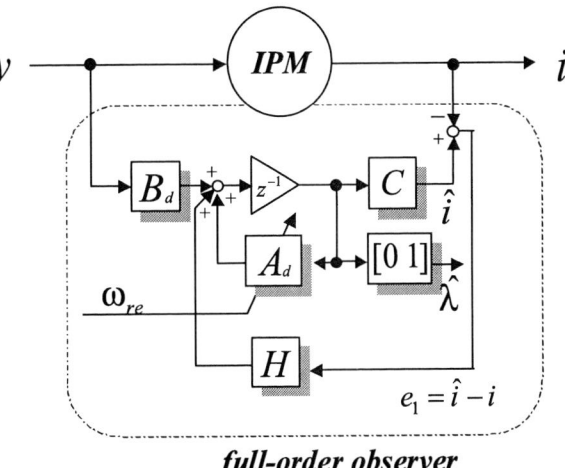

full-order observer

Fig. 3. Configuration of full-order observer

in which

$$
A_{12} = \frac{\omega_{re}}{L_q} \frac{\omega_{re}\exp(-\frac{R}{L_q}\tau) - \omega_{re}\cos(\omega_{re}\tau) + \frac{R}{L_q}\sin(\omega_{re}\tau)}{\frac{R^2}{L_q^2} + \omega_{re}^2}
$$
$$
+ \boldsymbol{j}\frac{\omega_{re}}{L_q} \frac{\frac{R}{L_q}\exp(-\frac{R}{L_q}\tau) - \frac{R}{L_q}\cos(\omega_{re}\tau) - \omega_{re}\sin(\omega_{re}\tau)}{\frac{R^2}{L_q^2} + \omega_{re}^2} .
$$

C. Construction of Un-known input observer

In general, stator voltage is necessary to estimate rotor position of PMSM. A full-order observer, for instance, is constructed as the follows:

$$
\begin{bmatrix} \hat{\boldsymbol{i}}[k+1] \\ \hat{\boldsymbol{\lambda}}[k+1] \end{bmatrix} = A_d \begin{bmatrix} \hat{\boldsymbol{i}}[k] \\ \hat{\boldsymbol{\lambda}}[k] \end{bmatrix} + B_d \boldsymbol{v}[k] + \boldsymbol{H}\left(\hat{\boldsymbol{i}}[k] - \boldsymbol{i}[k]\right) , \tag{12}
$$

in which $\boldsymbol{H}$ stands for observer gain. Fig.3 shows this observer diagram.

Consider the observer without voltage quantities in this paper. Fig.4 illustrates the voltage and the current characteristics in control period. To find the un-known input observer[3], the stator current $\boldsymbol{\xi} = \boldsymbol{i}[k+1/2]$ is inter-sampled. Hence, $\boldsymbol{\xi}$ can be expressed by

$$
\begin{bmatrix} \boldsymbol{i}[k+1/2] \\ \boldsymbol{\lambda}[k+1/2] \end{bmatrix} = \widetilde{A_d} \begin{bmatrix} \boldsymbol{i}[k] \\ \boldsymbol{\lambda}[k] \end{bmatrix} + \widetilde{B_d}\boldsymbol{v}[k] \tag{13}
$$

$$
\boldsymbol{\xi} = C \begin{bmatrix} \boldsymbol{i}[k+1/2] \\ \boldsymbol{\lambda}[k+1/2] \end{bmatrix} , \tag{14}
$$

where $\widetilde{A_d}$ and $\widetilde{B_d}$ are system matrix and input matrix discretized with $\tau_o = \tau/2$, respectively.

Once the inter-sampled current $\boldsymbol{z}$ is measured, the

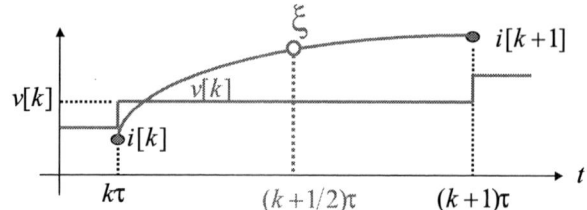

Fig. 4. Voltage and Current Characteristics in Control Period

stator voltage $v[k]$ is given by (13) and (14) as follows:

$$v[k] = \left(C\widetilde{B}_d\right)^{-1}\left(\xi - C\widetilde{A}_d\begin{bmatrix}\hat{i}[k]\\\hat{\lambda}[k]\end{bmatrix}\right), \quad (15)$$

which the flux $\boldsymbol{\lambda}$ is substituted by its estimate $\hat{\boldsymbol{\lambda}}$ as well as stator current $\boldsymbol{i}$. This equation and the full-order observer (12) result in the un-known input observer (16) as

$$\begin{bmatrix}\hat{i}[k+1]\\\hat{\lambda}[k+1]\end{bmatrix} = \left(A_d - B_d\left(C\widetilde{B}_d\right)^{-1}C\widetilde{A}_d\right)\begin{bmatrix}\hat{i}[k]\\\hat{\lambda}[k]\end{bmatrix}$$
$$+ B_d\left(C\widetilde{B}_d\right)^{-1}\xi$$
$$+ H\left(\hat{i}[k] - i[k]\right). \quad (16)$$

Finally, estimated rotor position $\hat{\boldsymbol{\theta}}_{re}$ can be obtained by

$$\hat{\boldsymbol{\theta}}_{re}[k] = \angle\hat{\boldsymbol{\lambda}}[k]. \quad (17)$$

Fig.5 demonstrates the block diagram of the unknown input observer. This system becomes observable if $\omega_{re} \neq 0$, H can be designed so that state estimation might be stabilized. In other words, the proposed method cannot be applied to drive at standstill.

III. EVALUATION OF ROBUSTNESS TO MAGNETIC SATURATION

A. Evaluation Model

In recent years, some problems caused by magnetic saturation has been focused due to motor volume miniaturization of IPMSM. From the viewpoint of mathematical model, magnetic saturation means L_q variation(decrease). Hence, much attention should be paid to L_q variation due to magnetic saturation, where L_q usually tend to be decreased by magnetic saturation. This section derive mathematical model for robustness evaluation to L_q variation.

From analogy to (16), the mathematical model of PMSM can be written as follow:

$$\begin{aligned}x[k+1] &= A_u x[k] + B_u \xi\\ &= (\hat{A}_u - \Delta A_u)x[k] + (\hat{B}_u - \Delta B_u)\xi,\end{aligned}$$
$$(18)$$

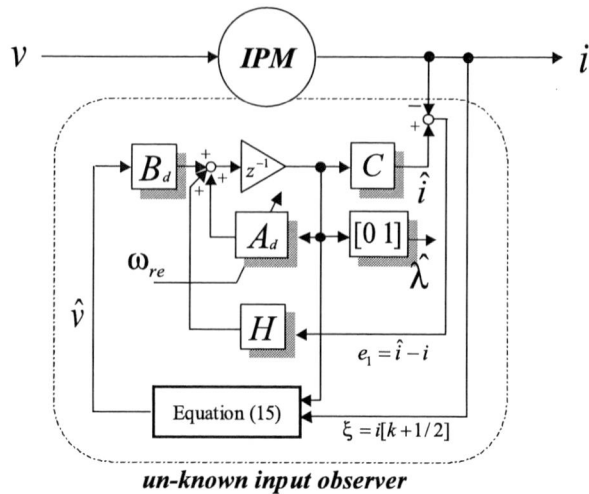

un-known input observer

Fig. 5. Configuration of un-known input observer

in which, motor parameter matrix A_u and B_u are express by controller parameter $\hat{A}_u$ and $\hat{B}_u$, and their variation ΔA_u and ΔB_u, respectively.

Subtracting (18) from (16), error system can be obtained by

$$\begin{aligned}e[k+1] &= \hat{x}[k+1] - x[k+1]\\ &= \left(\hat{A}_u + HC_1\right)e[k] + \Delta A_u x[k] + \Delta B_u \xi\end{aligned}$$
$$(19)$$

Hence, flux estimation error can be given as follows:

$$\begin{aligned}e_2[k] &= \hat{\lambda}[k] - \lambda[k] = [0 \quad 1]e[k]\\ &= [0 \quad 1]\left(zI - \hat{A}_u - HC\right)^{-1} *\\ &\quad * \left(\Delta A_u x[k] + z^{1/2}\Delta B_u i[k]\right). (20)\end{aligned}$$

Therefore, estimated flux can be written by

$$\hat{\lambda}[k] = \lambda[k] + e_2[k] = \left(1 + \frac{e_2[k]}{\lambda[k]}\right)\lambda[k] \quad (21)$$

in which $\lambda[k]$ is given by (3).

From the above equations, position estimation error $\Delta\theta_{re}$ can be calculated as following:

1. Parameter matrix variation ΔA_u due to L_q variation is given by $\hat{A}_u - A_u$. ΔB_u is also given as same manner.

2. Operation point $x[k]$ is defined by

$$x[k] = [\ i_d[k] + ji_q[k]\quad |\lambda| + j0\ ]^T.$$

3. $\hat{\lambda}[k]$ is obtained by (20) and (21), where

$$z = \varepsilon^{j\omega_{re}\tau} \qquad z^{1/2} = \varepsilon^{j\omega_{re}\tau/2}$$

in the case of considering steady state error.

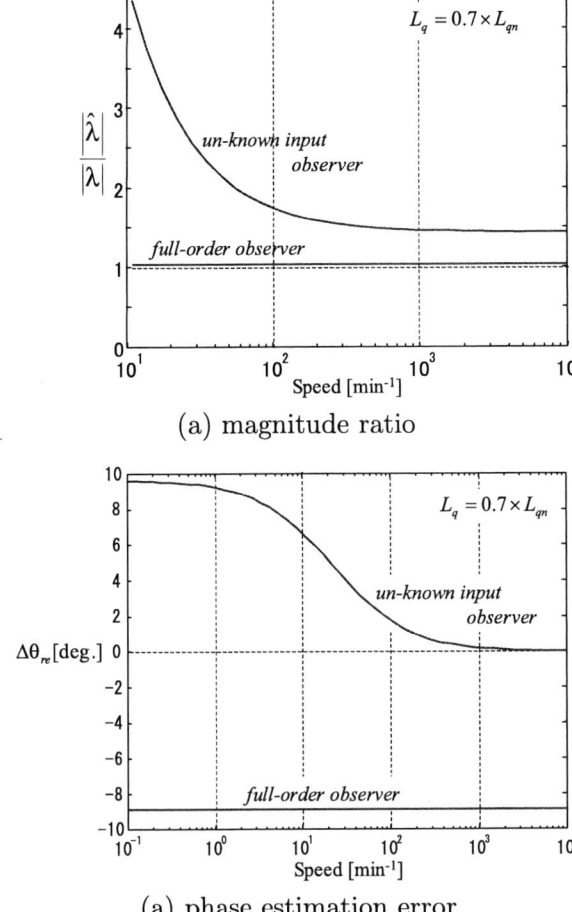

(a) magnitude ratio

(a) phase estimation error

Fig. 6. Robustness to L_q variation

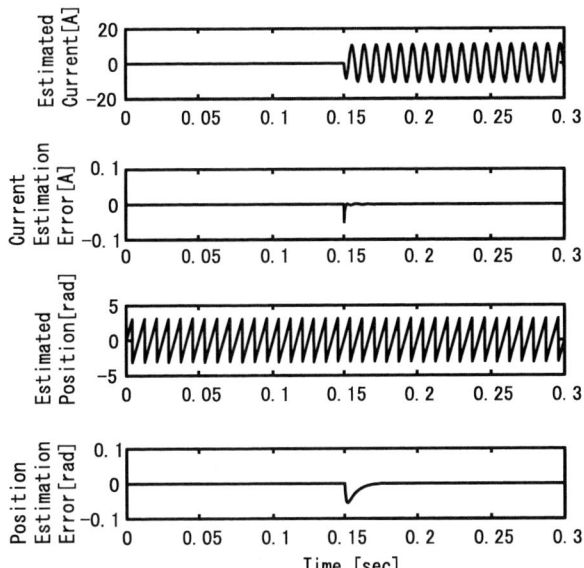

Fig. 7. Simulation results

IV. SIMULATION AND EXPERIMENTAL RESULTS

Simulation and experiments were carried out to show feasibility of the proposed strategy, in which IPMSM was controlled by vector control with actual rotor position θ_{re}. Un-known input observer just estimates current $\hat{i}[k]$ and flux $\hat{\lambda}[k]$ to confirm the feasibility, firstly.

Fig.7 shows MATLAB simulation results of un-known input flux observer (16). In addition, the rated load torque was applied at $t = 0.15$sec. It can be seen from this figure the current estimation error and the flux estimation error converge to zero even after load torque was applied. It should be noted that rotor position can be estimated successfully with no voltage knowledge. Hence, it turns out that this proposed approach can be applied to over-modulation PWM mode inverter fed IPMSM.

A. Steady Characteristic

Fig.8 shows experimental results at in $\omega_{rm}^* = 1000\text{min}^{-1}$. θ_{re}, $\hat{\theta}_{re}$, $\Delta\theta_{re}$ are actual rotor position, estimated rotor position by the un-known input observer and position estimation error, respectively. Poles of the un-known input state observer were assigned at 0.94 and 0.9. It can be seen from Fig.8 that position estimation error is averagely $5°$ and position estimation precisely can be realized.

B. Speed Step Response

Fig.9 shows experimental results of speed step response. In (a), the speed reference value ω_{rm}^* was stepwise given from 1000min^{-1} to 1500min^{-1}, ω_{rm}^*

B. Robustness Evaluation to Magnetic Saturation

In this subsection, robustness of flux estimation to L_q variation (70% to nominal value) is surveyed.

Fig.6 shows characteristics of robustness of flux estimation under 70% L_q and 100% load condition, in which figures (a) and (b) stand for magnitude ratio and phase estimation error, respectively. These characteristics of un-input known state observer can be obtained using previous defined evaluation model. That of time-continuous full-order observer is referred to appendix.

It is obvious from these figures that magnitude of estimated flux $\hat{\lambda}$ is considerably varied due to L_q variation. It should be noted, however, phase estimation error can extremely be suppressed, compared with time-continuous full-order observer. It is well-known that phase estimation characteristics are more important than other, as a result, these results can conclude that the proposed observer structure is effective in a sense of robustness to magnetic saturation.

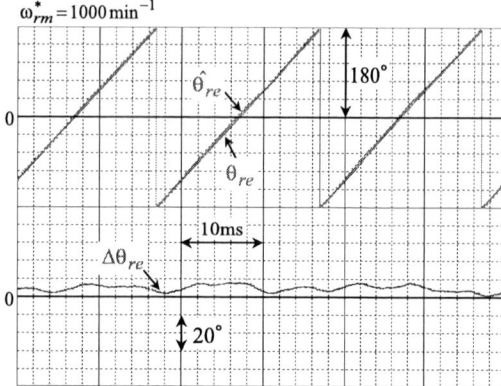

Fig. 8. Steady state characteristics

was stepwise changed from 1500min^{-1} to 1000min^{-1} in the case of (b). Poles of the un-known input state observer were assigned at 0.9 and 0.85. Although, transitional position estimation error was caused at speed reference change, estimation error both (a) and (b) demonstrate that converges immediately.

V. Conclusions

This paper has proposed PMSM rotor position estimation method using un-known input observer while considering over-modulation PWM mode operation.

Recent motor development is toward higher speed drive, smaller volume and higher efficiency, so novel control systems as well as motor design need to be developed while considering specialized applications. This paper approach might be useful for such applications as one of the high performance motor control system.

We are working toward the adaptive observer development without voltage, gain design of the observer gain(pole assignment) and further experimental validation.

Appendix

Evaluation Model of Full-Order Observer

A full-order observer is generally constructed based on time-continuous flux model (9) as follows:

$$
\begin{aligned}
\frac{d}{dt}\hat{x} &= \hat{A}\hat{x} + \hat{B}v + HC(\hat{x} - x) \\
&= \left(\hat{A} + HC\right)\hat{x} + \hat{B}v - HCx . \quad (22)
\end{aligned}
$$

Hence, estimated flux is obtained by

$$
\begin{aligned}
\hat{\lambda} &= [1 \ 0]\hat{x} \\
&= [1 \ 0]\left(sI - \hat{A} - HC\right)^{-1}\left(\hat{B}v - HCx\right) . \quad (23)
\end{aligned}
$$

Considering robustness to parameter variation at steady state, s is replaced by $j\omega_{re}$. In addition,

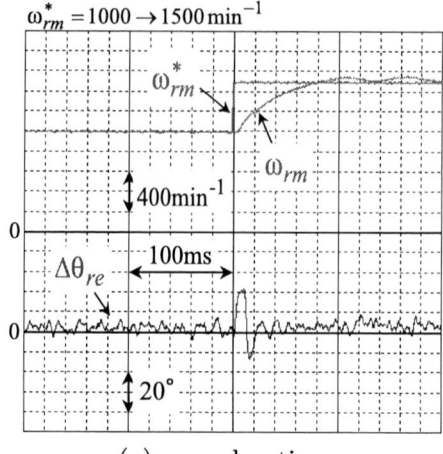

(a)   acceleration

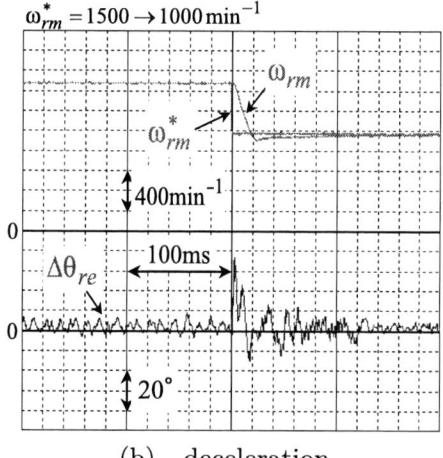

(b)   deceleration

Fig. 9. Speed step response

$i = i_d + ji_q$ is defined as load condition. From the above, λ and v are given by

$$
\begin{aligned}
\lambda &= \{(L_d - L_q)i_d + K_E\} + j0 \\
v &= v_d + jv_q \\
&= (R + j\omega_{re}L_q)(i_d + ji_q) + j\omega_{re}|\lambda|
\end{aligned}
$$

Therefore, $\hat{\lambda}$ is evaluated by (23).

References

[1] M.Yabe, et al: "A Sensor-less Drive of IPMSM Motor with Over-modulation PWM", The Papers of Technical Meeting, IEE of Japan, RM-01-160, pp.8-12, 2001(in Japanese)

[2] M.Hasegawa, et al: "Robust Adaptive Full-Order Observer Design Based on γ-Positive Real Problem for Speed Sensorless Vector Controlled Induction Motors", Proc. of IEEE-IECON, pp.60-65, 2003

[3] T.Mita, et al: "2-Delay Digital Feedback Control and Its Applications — Avoiding the Problems on Unstable Zeros —", Trans. of the Society of Instrument and Control Engineers(SICE), Vol. 24, No.5, pp.467-474, 1988 (in Japanese)

Switched Reluctance Motor Drive for Electric Motorcycle Using HFNN Controller

Chih-Hong Lin
Department of Electrical Engineering
National United University, Miao Li, Taiwan
No. 1, Lien-Da, Kung-Chieng Li, Miao Li 360
E-mail: jhlin@nuu.edu.tw

Abstract—The switched reluctance motor (SRM) drive system using hybrid fuzzy neural network (HFNN) controller is developed to control electric motorcycle in this paper. First, the dynamic models of a SRM drive system and electric motorcycle are builted though experimental tests and parameters measurements. Then, a HFNN speed control system that combined supervisor control, FNN control and compensated control is developed to control SRM drive system in order to drive electric motorcycle. In the proposed HFNN control scheme, an optimum phase advancing is achieved by continuous adaptation of the optimum conduction position for the FNN. The electric motorcycle is operated to provide constant disturbance torque. Finally, the effectiveness of the proposed control schemes is demonstrated by experimental results.

I. INTRODUCTION

In recent years, for the purpose of reducing air pollution and enhancing environmental protection, quite a few countries require their automotive industries to develop electric vehicles in place of gasoline-powered automobiles gradually. Here we put our attention on the development and research of electric motorcycles since motorcycles are much more widespread than automobiles for individual transportation in Asia. The switched reluctance motor (SRM) has emerged as an attractive alternative to DC and AC motors for general purpose industrial drives [1-3]. Some of these methods use off-line data collected from the machine during the static conditions, which change during the motor operation due to changes in the motor parameters. Some methods use a linear model of the machine, which may not be suitable for high-performance applications.

Recently much research has been done on the applications of fuzzy neural network (FNN) systems, which have the advantages of both fuzzy systems and neural networks, in the control fields to deal with nonlinearities and uncertainties of the control systems [4-5]. Moreover, the FNN's are universal approximators [6-7], which can approximate any dynamics to a prespecified accuracy by the learning process [6]. The fuzzy-neural approximator presented in Leu, Lee and Wang [8] tuned on line to approximate the unknown nonlinear dynamic systems for adaptive control. The supervisory FNN controller proposed in Lin, Hwang and Wai [9] comprised a supervisory controller which is designed to stabilize the system states around a defined bound region, and an FNN sliding-mode controller which combines the advantages of the sliding-mode control with robust characteristics and the FNN with on-line learning ability. The adaptive control schemes of nonlinear systems that incorporate the

techniques of FNNs have also grown rapidly [10-11]. Therefore, this paper presents a hybrid fuzzy neural network (HFNN) control SRM drive for electric motorcycle. The method is not dependent upon the predetermined characteristics of the motor and can adapt to any change in the motor characteristics. Finally, some experimental results are provided to demonstrate the effectiveness of the proposed control schemes.

II. CONFIGURATION OF SRM DRIVE

The machine model of a switched reluctance motor (SRM) can be described in matrix as follows [12-14]

$$p\lambda_{abcd} = \mathbf{v}_{abcd} - \mathbf{r}_s \mathbf{i}_{abcd} \qquad (1)$$

where $p = \dfrac{d}{dt}$ is the differential operator, $\lambda_{abcd} = [\lambda_a \ \lambda_b \ \lambda_c \ \lambda_d]^T$ is the vector of the stator flux linkage, $\mathbf{v}_{abcd} = [v_a \ v_b \ v_c \ v_d]^T$ is the vector of the stator voltages, $\mathbf{r}_s = diag[r_s \ r_s \ r_s]^T$ is the vector of the stator resistances, $\mathbf{i}_{abcd} = [i_a \ i_b \ i_c \ i_d]^T$ is the vector of the stator currents.

The vector of the stator flux linkage can be represented as

$$\lambda_{abcd} = \mathbf{L}_{abcd} \mathbf{i}_{abcd} \qquad (2)$$

in which

$$\mathbf{L}_{abcd} = \begin{bmatrix} L_{aa} & L_{ab} & L_{ac} & L_{ad} \\ L_{ab} & L_{bb} & L_{bc} & L_{bd} \\ L_{ac} & L_{bc} & L_{cc} & L_{cd} \\ L_{ad} & L_{bd} & L_{cd} & L_{dd} \end{bmatrix} \qquad (3)$$

where $\mathbf{L}_{abcd}$ is called stator inductance matrix, and L_{aa}、L_{bb}、L_{cc}、L_{dd} are self inductance of per phase, L_{ab}、L_{ac}、L_{ad}、L_{bc}、L_{bd}、L_{cd} are mutual inductance of phase to phase. Due to the stator windings of the SRM are exciting using one-phase exciting method, i.e., mutual inductances are zero, therefore, the stator inductance matrix can be simplified as

$$\begin{aligned} \mathbf{L}_{abcd} &= diag[L_{aa} \ \ L_{bb} \ \ L_{cc} \ \ L_{cd}] \\ &= diag[L_s \ \ L_s \ \ L_s \ \ L_s] \end{aligned} \qquad (4)$$

where L_s is the self inductance.

The electromagnetic torque can be expressed as

978-1-4244-0644-9/07/$25.00 ©2007 IEEE   1549

$$T_e = \frac{1}{2}\frac{\partial L_s}{\partial \theta_e} i_s^2 = \frac{1}{2}(\frac{\partial L_s}{\partial \theta_e} i_s) i_s \underline{\Delta} K_t(\theta_e, i_s) i_s \qquad (5)$$

where $\dfrac{\partial L_s}{\partial \theta_e}$ is derivative of self inductance corresponding

with electric angular θ_e, $K_t(\theta_e, i_s) = (\dfrac{\partial L_s}{\partial \theta_e} i_s)$ is the

torque coefficient, i_s is the exciting current of stator, T_e is the electromagnetic torque.

The equation of the motor dynamics is

$$\frac{d\theta_r}{dt} = \omega_r = \frac{\omega_e}{P_r} \qquad (6)$$

$$T_e = T_L + B\omega_r + J\dot{\omega}_r \qquad (7)$$

In (6) and (7), θ_r is the rotor position, ω_r is the rotor speed, P_r is the number of poles, T_L stands for the load torque (external load disturbance), B represents the viscous frictional coefficient and J is the moment of inertia.

The basic principle in controlling a SRM drive is based on constant torque coefficient $K_t(\theta_e, i_s)$, so the electromagnetic torque T_e is then proportional to i_s, which is determined by closed-loop control. Since the generated torque is linearly proportional to the stator current in (5), the torque per ampere can be achieved. The block diagram of a switched reluctance motor drive system for electric motorcycle is shown in Fig. 1. This consists of five major parts: a motor and wheels of electric motorcycle, a current-regulated converter, commutation mechanism, a controller, and sensors. First the control algorithms are executed by digital signal processor (DSP) control board. Then, the output current command to the current-regulated converter is determined by the DSP controller. The current-regulated converter controls the four phase currents to follow the current commands. The sensors consist of the current and position sensors. For the speed control system, the electric motorcycle is operated to provide constant disturbance torque. The SRM used in this drive system is a four-phase six-pole 48V, 1.13Kw, 35A, 6000rpm type.

With the implementation of digital signal processor (DSP) control, the SRM drive can be simplified to a control system, in which

$$H_p(s) = \frac{1}{Js + B} \qquad (8)$$

The parameters of the SRM are given as follows: $R = 1.21\Omega$, $L_{max} = 32.25\,mH$, $L_{min} = 124.82\,mH$. For the convenience of controller design, position and speed signals in the control loop are set at 1V=50rad/sec, respectively. The results turn out to be

$$\bar{J} = 1.04 \times 10^{-3} Nms^2 = 0.052 Nmsrad / V$$
$$\bar{B} = 6.18 \times 10^{-3} Nms / rad = 0.309 Nm / V \qquad (9)$$

The overbar symbol represents the system parameter under the nominal condition.

Due to the function $\lambda(\theta_r, i)$ of the flux linkage is relating to current i and rotor position θ_r. The function

$\lambda(\theta_r, i)$ is periodic in θ_r, and monotonically increasing in i. It can be expressed as $\lambda(\theta_r, i) = L(\theta_r, i)i$, in terms of the inductance function L. Models assuming linear magnetic ignore the current dependence of L, i.e., $L = L(\theta_r)$, while in more realistic, nonlinear magnetic model $L(\theta_r, i)$ is monotonically decreasing in i (due to saturation). The analysis and control design approach that are presented here are based on linear magnetic in this paper.

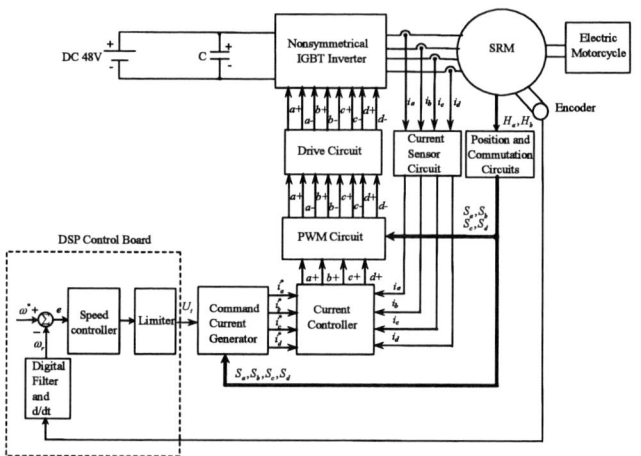

Fig. 1. Configuration of a SRM drive.

III. Hybrid Fuzzy Neural Network Controller Design for Electric Motorcycle

The dynamic equation of the SRM servo drive system can be formulated by rewriting (7) as follows:

$$\dot{\omega}_r = -\frac{B}{J}\omega + \frac{K_t}{J} i_s^* - \frac{1}{J}T_L = A_P\omega_r + B_PU_t + C_Pw \qquad (10)$$

where ω_r is the rotor speed of the SRM; i_s^* is the command current of the stator of the SRM; $A_P = -B/J$, $B_P = K_t/J > 0$, $C_P = -1/J$ and $T_L = w$. Consider the variation of system parameters and external force disturbance, the parameters in (10) are assumed to be bounded, i.e., $|A_P\omega_r| \le F^U(\omega_r)$, $|C_PT_L| \le G^U$, $B_L \le B_P$, where $F^U(\omega_r)$ is a known continuous function, G^U and B_L are know constants. Define the tracking error as follows:

$$e = \omega_m - \omega_r \qquad (11)$$

where ω_m represents the desired rotor speed; e is the tracking error of rotor speed. If the parameters of the SRM servo drive system and w are well known, the idea control law can be defined as follows [9]:

$$U^* = [-A_P\omega - C_Pw + \dot{\omega}_m + k_1e] / B_P \qquad (12)$$

where k_1 is positive constants. Substituting (12) into (11), the following equation can be obtained:

$$\dot{e} + k_1e = 0 \qquad (13)$$

1550

which implies that $\lim_{t \to \infty} e = 0$, i.e., the system state can track the desired trajectories asymptotically. However, the parameter variations of the system are difficult to measure, and the exact value of the external load disturbance is also difficult to know in advance for practical applications. In order to control the rotor speed of the SRM effectively, a hybrid control system is proposed in this section. The configuration of the proposed hybrid control system, which combines a supervisory control system, FNN control system and compensated control system, is depicted in Fig. 2. The reference model shown in Fig. 2 is designed to generate the desired reference trajectories. The control law is assumed to take the following form [9]:

$$U_t = U_S + U_F + U_C \tag{14}$$

where U_S is the supervisory control, U_F is the FNN control and U_C is a compensated control. The supervisory control U_S is designed so that the states of the controlled system are stabilized around a redetermined bound region. Due to the excessive and chattering control effort induced by the supervisory control law, the FNN control and compensated control are introduced to reduce and smooth the control effort when the system states are inside the predefined bound region. The FNN control U_R is the main tracking controller used to mimic an idea control law, and the compensated control U_C is designed to compensate the difference between the idea control law and the FNN control. The supervisory control law fires only when the FNN approximation properties can not be guaranteed.

The design of a supervisory hybrid control system is necessary for the condition of divergence of states to pull the states back to the predetermined bound region and guarantee the stability of the system. Inside the bound region the FNN control system approximates uniformly the idea control law. From (10), (12), and (14), an error equation is then obtained as follows:

$$\dot{e} = k_1 e + B_P \left[U^* - U_F - U_C - U_A \right] \tag{15}$$

In order to achieve a zero steady-state error, an integral equation can be defined as

$$e_2 = e = \dot{\eta} = \dot{e}_1 \tag{16}$$

Then an error equation can be redefined as follows:

$$\dot{\mathbf{E}} = \Lambda \mathbf{E} + \mathbf{B_m} \left[U^* - U_F - U_C - U_A \right] \tag{17}$$

where $\mathbf{E} = \begin{bmatrix} e_1 & e_2 \end{bmatrix}^T$ is tracking error vector, $\Lambda = \begin{bmatrix} 0 & 1 \\ 0 & -k_1 \end{bmatrix}$ is a stable matrix and $\mathbf{B_m} = \begin{bmatrix} 0 & B_P \end{bmatrix}^T$.

Then, the Lyapunov function candidate is chosen as follows:

$$V_A = \frac{1}{2} \mathbf{E}^T \mathbf{P} \mathbf{E} \tag{18}$$

where $\mathbf{P}$ is a symmetric positive definite matrix which satisfies the following Lyapunov equation [10]:

$$\Lambda^T \mathbf{P} + \mathbf{P} \Lambda = -\mathbf{Q} \tag{19}$$

and $\mathbf{Q} > 0$ is selected by the designer. Take the derivative of the Lyapunov function and use (17) and (19), then

$$\dot{V}_B = \frac{1}{2} \mathbf{E}^T \mathbf{P} \dot{\mathbf{E}} + \frac{1}{2} \dot{\mathbf{E}}^T \mathbf{P} \mathbf{E}$$
$$= \frac{1}{2} \mathbf{E}^T \mathbf{P} \Lambda \mathbf{E} + \frac{1}{2} \mathbf{E}^T \Lambda^T \mathbf{P} \mathbf{E} + \mathbf{E}^T \mathbf{P} \mathbf{B_m} \left[U^* - U_F - U_C - U_A \right]$$
$$\leq -\frac{1}{2} \mathbf{E}^T \mathbf{Q} \mathbf{E} + \left| \mathbf{E}^T \mathbf{P} \mathbf{B_m} \right| \left(\left| U^* \right| + \left| U_F - U_C \right| \right) - \mathbf{E}^T \mathbf{P} \mathbf{B_m} U_A \tag{20}$$

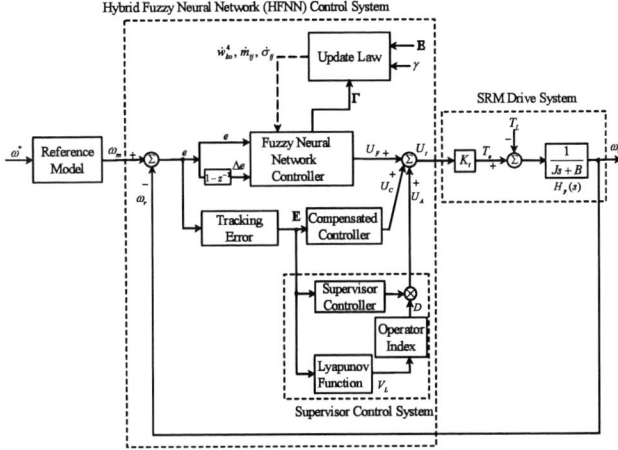

Fig. 2. HFNN control system.

To satisfy $\dot{V}_B \leq 0$, the supervisory control V_A is designed as follows [10]:

$$U_A = D \mathrm{sgn} \left(\mathbf{E}^T \mathbf{P} \mathbf{B_m} \right) \left[\left| U_F - U_C \right| + \frac{1}{B_L} \left(F^U(\omega) + G^U \right) \right]$$
$$+ D \mathrm{sgn} \left(\mathbf{E}^T \mathbf{P} \mathbf{B_m} \right) \left[\frac{1}{B_L} \left(\left| \omega_m \right| + \left| k_1 e \right| \right) \right] \tag{21}$$

where $\mathrm{sgn}(\cdot)$ is a sign function, and the operator index

$$D = \begin{cases} D = 1, & \text{if } V_A \geq \overline{V} \\ D = 0, & \text{if } V_A < \overline{V} \end{cases} \tag{22}$$

in which $\overline{V}$ is positive constant. Substitute (12) and (19) into (18) and consider the case $D = 1$, then

$$\dot{V}_B \leq -\frac{1}{2} \mathbf{E}^T \mathbf{Q} \mathbf{E} + \left| \mathbf{E}^T \mathbf{P} \mathbf{B_m} \right| \left(\left| U^* \right| + \left| U_F - U_C \right| \right) - \mathbf{E}^T \mathbf{P} \mathbf{B_m} U_A$$
$$= -\frac{1}{2} \mathbf{E}^T \mathbf{Q} \mathbf{E} + \left| \mathbf{E}^T \mathbf{P} \mathbf{B_m} \right| \left[\frac{1}{B_P} \left(\left| A_P \omega \right| + \left| C_P w \right| + \left| \dot{\omega}_m \right| + \left| k_1 e \right| \right) \right.$$
$$\left. + \left| U_F - U_C \right| - \left| U_F - U_C \right| - \frac{1}{B_L} \left(F^U + G^U + \left| \dot{\omega}_m \right| + \left| k_1 e \right| \right) \right]$$
$$\leq -\frac{1}{2} \mathbf{E}^T \mathbf{Q} \mathbf{E} \leq 0 \tag{23}$$

Using the designed supervisory control U_A as shown in (21), the inequality $\dot{V}_B < 0$ can be obtained for nonzero value of the tracking error vector E when $V_B > \overline{V}$. As the results from (23), the supervisory control system is capable to drive the tracking error to zero without using the FNN control and compensated control system. However, owing to the selection of the bounds $F^U(\omega)$, G^U, B_L and sign function, the supervisory control law can result in excessive and chattering control effort. Therefore, a FNN control and compensated control are designed to overcome the mentioned drawback. The FNN control is proposed to mimic the idea control law U^*, and a compensated control is proposed to compensate the difference between U^* and the FNN control instead of increasing the inference rules of the FNN.

A four-layer FNN as shown in Fig. 3, which comprises the input (the i layer), membership (the j layer), rule (the k layer) and output layer (the o layer), is adopted to implement the FNN controller in this study. The signal propagation and the basic function in each layer of the FNN are introduced as follows:

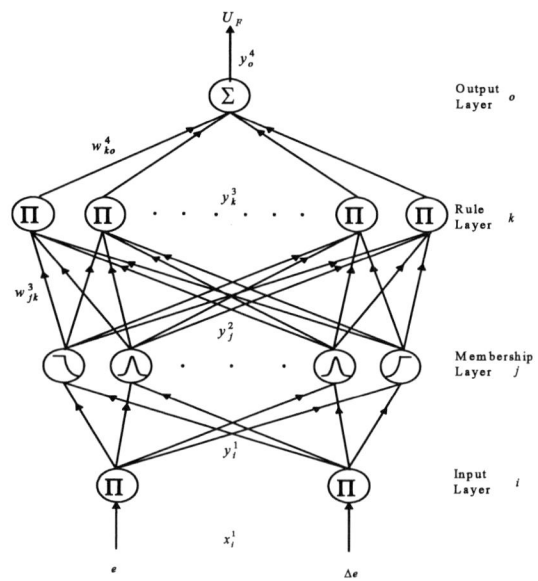

Fig. 3. Structure of FNN.

Layer 1: Input Layer
For every node i in this layer, the net input and the net output are represented as

$$net_i^1(N) = \prod_o x_i^1(N) w_{oi} y_o^4(N-1),$$

$$y_i^1(N) = f_i^1\big(net_i^1(N)\big) = net_i^1(N), \quad i = 1, 2$$
(24)

where $x_1^1 = \omega_m - \omega_r = e$ is the command error between the desired command, ω_m, and the rotor speed of the SRM, ω_r; $x_2^1 = e(1 - z^{-1}) \equiv \Delta e$ is the command error change, in which z^{-1} represents a time delay; N denotes the number of iterations; y_o^4 is the output of the FNN.

Layer 2: Membership Layer
In this layer, each node performs a membership function. The Gaussian function, is adopted as the membership function. For the jth node

$$net_j^2(N) = -\frac{\big(x_i^2 - m_{ij}\big)^2}{\big(\sigma_{ij}\big)^2},$$

$$y_j^2(N) = f_j^2\big(net_j^2(N)\big) = \exp\big(net_j^2(N)\big), \quad j = 1, \cdots, n$$
(25)

where m_{ij} and σ_{ij} are, respectively, the mean and the standard deviation of the Gaussian function in the jth term

of the ith input linguistic variable x_i^2 to the node of layer 2, and n is the total number of the linguistic variables with respect to the input nodes.

Layer 3: Rule Layer
Each node k in this layer is denoted by $\prod$, which multiplies the input signals and outputs the result of product. For the kth rule node

$$net_k^3(N) = \prod_j w_{jk}^3 x_j^3(N),$$

$$y_k^3(N) = f_k^3\big(net_k^3(N)\big) = net_k^3(N), \quad k = 1, \cdots, l$$
(26)

where x_j^3 represents the jth input to the node of layer 3; w_{jk}^3, the weights between the membership layer and the rule layer, are assumed to be unity; $l = (n/i)^i$ is the number of rules with complete rule connection if each input node has the same linguistic variables.

Layer 4: Output Layer
The single node o in this layer is labeled with $\sum$, which computes the overall output as the summation of all input signals

$$net_o^4(N) = \sum_k w_{ko}^4 x_k^4(N),$$

$$y_o^4(N) = f_o^4\big(net_o^4(N)\big) = net_o^4(N), \quad o = 1$$
(27)

where the connecting weight w_{ko}^4 is the output action strength of the oth output associated with the kth rule; x_k^4 represents the kth input to the node of layer 4, $y_o^4 = U_F$. Moreover, U_F can be rewritten as follows:

$$U_F = U_F(\mathbf{E}|\mathbf{O}) = \mathbf{O}^T \mathbf{\Gamma}$$
(28)

where the error signal is the input of the FNN; $\mathbf{O} = \big[w_{11}^4 \ w_{21}^4 \ \cdots \cdots \ w_{l1}^4\big]^T$ is the adjustable parameter vector of the FNN; $\mathbf{\Gamma} = [x_1^4 \ x_2^4 \ \cdots \cdots \ x_l^4]^T$, in which x_k^4 is determined by the selected membership function and $0 \leq x_k^4 \leq 1$.
To develop the compensated control law U_C, first, a minimum approximation error ρ is defined as follows:

$$\rho = U^* - U_F\big(\mathbf{E}|\mathbf{O}^*\big)$$
(29)

where $\mathbf{O}^*$ is an optimal weight vector achieves the minimum approximation error, and the absolute value of ρ is assumed to be less than a small positive constant, η (i.e., $|\rho| < \eta$). The error (17) can be rewritten as follows:

$$\dot{\mathbf{E}} = \Lambda \mathbf{E} + \mathbf{B}_m \Big\{ \big[U^* - U_F\big(\mathbf{E}|\mathbf{O}\big)\big] - U_C - U_A \Big\}$$

$$= \Lambda \mathbf{E} + \mathbf{B}_m \Big\{ \big[U^* - U_F\big(\mathbf{E}|\mathbf{O}^*\big)\big]$$
$$+ \big[U_F\big(\mathbf{E}|\mathbf{O}^*\big) - U_F\big(\mathbf{E}|\mathbf{O}\big)\big] - U_C - U_A \Big\}$$

$$= \Lambda \mathbf{E} + \mathbf{B}_m \Big[\rho + \big(\mathbf{O}^* - \mathbf{O}\big)^T \mathbf{\Gamma} - U_C - U_A \Big]$$
(30)

Then, the Lyapunov function is defined as

$$V(t) = \frac{1}{2}\mathbf{E^T PE} + \frac{1}{2\gamma}\left(\mathbf{O^*} - \mathbf{O}\right)^T\left(\mathbf{O^*} - \mathbf{O}\right) \tag{31}$$

Take the derivative of the Lyapunov function and use (30), then

$$\dot{V}(t) = \frac{1}{2}\mathbf{E}^T\mathbf{P}\dot{\mathbf{E}} + \frac{1}{2}\dot{\mathbf{E}}^T\mathbf{PE} - \frac{1}{\gamma}\left(\mathbf{O^*} - \mathbf{O}\right)^T\dot{\mathbf{O}}$$

$$= -\frac{1}{2}\mathbf{E}^T\mathbf{QE} + \mathbf{E}^T\mathbf{PB}_m\left[\rho - U_C - U_A\right]$$

$$+ \mathbf{E}^T\mathbf{PB}_m\left(\mathbf{O^*} - \mathbf{O}\right)^T\Gamma - \frac{1}{\gamma}\left(\mathbf{O^*} - \mathbf{O}\right)^T\dot{\mathbf{O}} \tag{32}$$

To satisfy $\dot{V}(t) \leq 0$, the adaptation laws $\dot{\mathbf{O}}$ and the compensated controller U_C are designed as follows [9]:

$$\dot{\mathbf{O}} = \gamma\mathbf{E}^T\mathbf{PB}_m\Gamma \tag{33}$$

$$U_C = \eta\,\mathrm{sgn}\left(\mathbf{E}^T\mathbf{PB}_m\right) \tag{34}$$

where $\gamma > 0$ is denoted as adaptation gains. Substitute (33) into (32) and use (21), then

$$\dot{V}(t) = -\frac{1}{2}\mathbf{E}^T\mathbf{QE} + \mathbf{E}^T\mathbf{PB}_m\rho - \mathbf{E}^T\mathbf{PB}_m U_C - \mathbf{E}^T\mathbf{PB}_m U_A$$

$$\leq -\frac{1}{2}\mathbf{E}^T\mathbf{QE} + \left|\mathbf{E}^T\mathbf{PB}_m\right|\left|\rho\right| - \mathbf{E}^T\mathbf{PB}_m U_C \tag{35}$$

use (34), thus

$$\dot{V}(t) \leq -\frac{1}{2}\mathbf{E}^T\mathbf{QE} \leq 0 \tag{36}$$

Since $\dot{V}(t) \leq 0$, $\dot{V}(t)$ is negative semidefinite (i.e., $V(t) \leq V(0)$), which implies $\mathbf{E}$ and $(\mathbf{O^*} - \mathbf{O})$ are bounded using (31). Let function $\Xi(t) = -\dot{V}(t) = \mathbf{E}^T\mathbf{QE}/2$, and integrate function $\Xi(t)$ with respect to time

$$\int_0^t \Xi(\varsigma)\,d\tau = V(0) - V(t) \tag{37}$$

Because $V(0)$ is bounded, and $V(t)$ is nonincreasing and bounded, then $\lim_{t\to\infty}\int_0^t\Xi(\varsigma)\,d\varsigma < \infty$. Differentiate $\Xi(t)$ with respect to time, then $\dot{\Xi}(t) = \mathbf{E}^T\mathbf{Q}\dot{\mathbf{E}}$. Since all the variables in the right side of (30) are bounded, it implies $\dot{\mathbf{E}}$ is also bounded. Then $\Xi(t)$ is uniformly continuous [15]. By using Barbalat's lemma [15-16], it can be shown that $\lim_{t\to\infty}\Xi(t) = 0$. Therefore, $\mathbf{E}(t) \to 0$ as $t \to \infty$.

In order to train the FNN effectively, an on-line parameter training methodology [9], which is derived using the Lyapunov stability theorem and the gradient descent method, is omitted. Moreover, the update law [9] is omitted here.

IV. EXPERIMENTAL RESULTS

A block diagram of the DSP-based computer control system for a SRM servo drive is depicted in Fig. 1. The current-controlled PWM VSI is implemented by the IGBT power modules with a switching frequency of 15kHz. The control algorithm was implemented by a TMS320C32 DSP-based control system. A TMS320C32 DSP-based control board includes multi-channels of D/A,

eight-channels programmable PWM and encoder interface circuits.

The control gains of the proposed HFNN control system are given in the following:

$$\mathbf{P} = \begin{bmatrix} 1 & 0 \\ 0 & 1 \end{bmatrix},\quad k_1 = 220,\quad \overline{V} = 1,\quad \gamma = 0.9,\quad \eta = 0.5 \tag{38}$$

First, a 2nd-order transfer function of the following form with rise time 0.3sec is chosen as the reference model for the periodic step command:

$$\frac{\omega_n^2}{s^2 + 2\zeta\omega_n s + \omega_n^2} = \frac{1512.43}{s^2 + 77.8s + 1512.43} \tag{39}$$

Some experimental results are provided to demonstrate the control performance of SRM drive system using HFNN control for electric motorcycle. Two conditions are provided in the experimentation here, one being the 1500 rpm and 3000rpm case without mounted electric motorcycle, another being the 1500 rpm and 3000rpm case with mounted electric motorcycle. The experimental results of the HFNN control system without mounted electric motorcycle and with mounted electric motorcycle due to periodic step command at 1500 rpm and 3000rpm case are shown in Fig.4 and Fig. 5. The speed responses of the SRM without and with mounted electric motorcycle at 1500 rpm and 3000rpm case are shown in Figs. 4(a), 4(c), 5(a) and 5(c); the associated control efforts are shown in Figs. 4(b), 4(d), 5(b) and 5(d), respectively. However, owing to the on-line adaptive mechanism of FNN and the compensated controller, accurate tracking control performance of the SRM can be obtained. From the experimental results, the control performance of the proposed the HFNN control system is suitable for SRM drive for electric motorcycle.

V. CONCLUSIONS

The purpose of this paper is to develop a switched reluctance machines (SRM) in order to control electric motorcycle using a HFNN control. Then, a HFNN speed control system that combined supervisor control, FNN control and compensated control is developed to control SRM drive system in order to drive electric motorcycle. In the proposed HFNN control scheme, an optimum phase advancing is achieved by continuous adaptation of the optimum conduction position for the FNN. The electric motorcycle is operated to provide constant disturbance torque. Finally, the effectiveness of the proposed control schemes is demonstrated by experimental results.

ACKNOWLEDGMENTS

The author would like to acknowledge the financial support of the National Science Council in Taiwan, R.O.C. through its grant NSC 96-2221-E-239-035.

REFERENCES

[1] P. N. Materu, R. Krishnan, "Steady-state analysis of the variable-speed switched reluctance motor drive," *IEEE Transactions on Industrial Electronics*, vol. 36, no. 4, pp. 523-529, Nov. 1989.

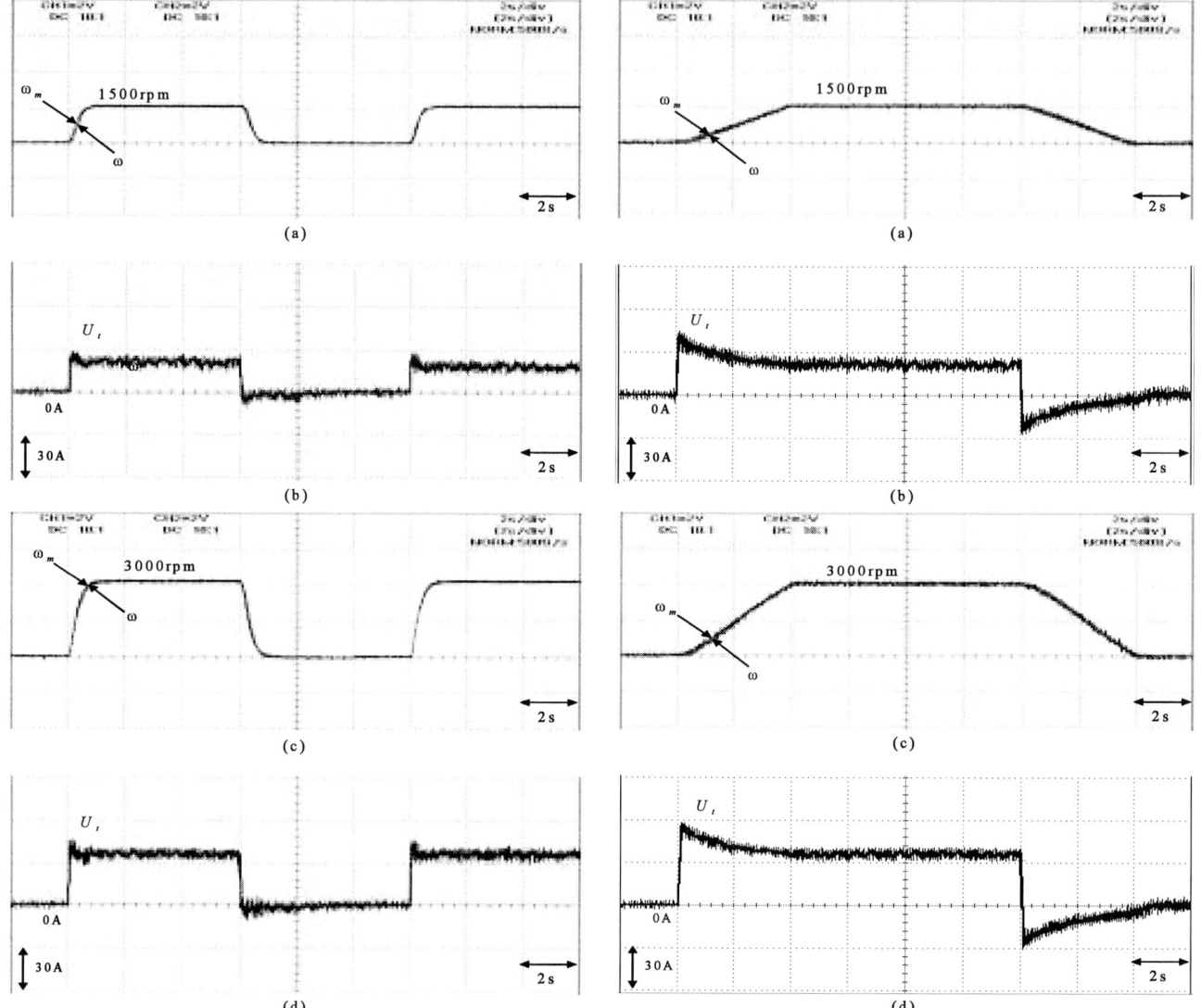

Fig. 4. Experimental results of HFNN control system without mounted electric motorcycle due to periodical step command: (a) speed at 1500 rpm case; (b) control effort at 1500 rpm case; (c) speed at 3000 rpm case; (d) control effort at 3000 rpm case

Fig. 5. Experimental results of HFNN control system with mounted electric motorcycle due to periodical step command: (a) speed at 1500 rpm case; (b) control effort at 1500 rpm case; (c) speed at 3000 rpm case; (d) control effort at 3000 rpm case.

[2] A. V. Randun, "High-power density switched reluctance motor drive for aerospace applications," *IEEE Transactions on Industry Applications*, vol. 28, pp. 113-119, Jan./Feb. 1992.

[3] K. M. Rahman, B. Fahini, G. Suresh, A. V. Rajarathnam, M. Ehsani, "Advantages of switched reluctance motor applications to EV and HEV: design and control issues," *IEEE Transactions on Industry Applications,* Vol. 36, no. 1, pp. 111-121, Jan. /Feb. 2000.

[4] C. T. Lin and C. S. G. Lee, "Neural-network-based fuzzy logic controland decision system," *IEEE Trans. Comput.*, vol. 40, pp. 1320–1336, Dec. 1991.

[5] Y. C. Chen and C. C. Teng, "A model reference control structure using a fuzzy neural network," *Fuzzy Sets Syst.*, vol. 73, no. 3, pp. 291–312, Aug. 1995.

[6] J. Zhang and A. J. Morris, "Fuzzy neural networks for nonlinear systems modelling," *Proc. IEE—Contr. Theory Applicat.*, vol. 142, no. 6, pp. 551–556, 1995.

[7] T. S. R. Jang and C. T. Sun, "Neural-fuzzy modeling and control," *Proc. IEEE*, vol. 83, pp. 378–405, Mar. 1995.

[8] Y. G. Leu, T. T. Lee and W. Y. Wang, "On-line turning of fuzzy-neural network for adaptive control of nonlinear dynamical systems," *IEEE Trans. Syst. Man, Cybern.*, vol. 27, pp. 1034-1043, 1997.

[9] F. J. Lin, W. J. Hwang, and R. J. Wai, "A supervisory fuzzy neural network control system for tracking periodic inputs," *IEEE Trans. Fuzzy Syst.,*, vol. 7, no.1, pp. 41-52, 1999.

[10] L. X. Wang, *A Course in Fuzzy Systems and Control.* Englewood Cliffs, NJ: Prentice-Hall, 1997.

[11] F. J. Lin, R. J. Wai, and H. P. Chen, "A PM synchronous servo motor drive with an on-line trained fuzzy neural network controller," *IEEE Transactions on Energy Conversion*, vol. 13, no. 4, pp. 319-325, Dec. 1998.

[12] T. J. E. Miller, *Switched reluctance motors and their control,* Clarendon Press, Oxford, 1993.

[13] A. V. Radun, C. A. Ferreira, E. Richter, "Two channel switched reluctance starter/generator results," *IEEE Transactions on Industry Applications,* vol. 34, no. 5, pp. 1026-1034, Sep./ Oct. 1998.

[14] J. G. J. O'Donovan, P. J. Roche, R. C. Kavanagh, M. G. Egan, and J. M. D. Murphy, "Neural network based torque ripple minimization in a switched reluctance motor," in *Conf. Rec. IEEE-IAS Annu. Meeting,* 1994, pp. 1226–1231.

[15] J. J. E. Slotine and W. Li, *Applied Nonlinear Control.* Englewood Cliffs, NJ: Prentice-Hall, 1991.

[16] K. J. Astrom, and B. Wittenmark, *Adaptive Control.* New York, Addison-Wesley, 1995.

STATE - SPACE AVERAGING, SIMULATION, STABILITY STUDIES FOR STEP UP POSITIVE OUTPUT SWITCHED CAPACITOR DC-DC CONVERTER

E.Jayashree, Dr.G.Uma and M.Vaigundamoorthi.

College of Engineering Guindy, Anna University, Chennai-25, India.

Abstract- **This paper presents state-space non-linear approach to analyze and design a switched capacitor dc-dc converter for obtaining controllable linear model for the converter circuit. This work also presents the stability analysis for the converter circuit using root locus method.**

Keywords: **Push-pull state, Switched capacitor, voltage -lift technique, voltage transfer gain, PID control**

I. INTRODUCTION

Electronic power converter must be suitably controlled in ordered to supply the voltages, currents, or frequency ranges needed for the load and to guarantee the requested dynamics. Furthermore, they can be designed to serve as "clean" interface between most loads and electrical utility system. Powerful modeling methodologies and sophisticated control processes must be used to obtain stable controlled power converters not only with satisfactory static and dynamic performance, but also with low sensitivity against load or line disturbances or, preferably, robustness.

Non linear state –space model provides a more consistent way of handling the control problem of power converters .This approach requires measurement of state variables, but eliminates conventional modulators and linear feedback compensators, enabling better performance and robustness.

Switched capacitor positive output dc –dc converters can sort into several subs –series [10]:
- Main series
- Additional series

A representative switched capacitor dc-dc converter topology [10] is used to demonstrate the modeling principles and to bring linear model for the elementary circuit. The main switch is S in all circuits and other switches are slaves.

The main switch S is on and slaves off during the switch-on period, and switch S is off and slaves on during the switch- off period. Use Vs and Vd to present voltage drop of switches and diodes, respectively. The load is resistive load R. The input voltage and current are Vin and Iin, and the output voltage and current are Vo and Io

II. SYSTEM DESCRIPTION AND OPERATION

Main series -2 lift circuit

The circuit of 2 lift elementary circuit is given below,

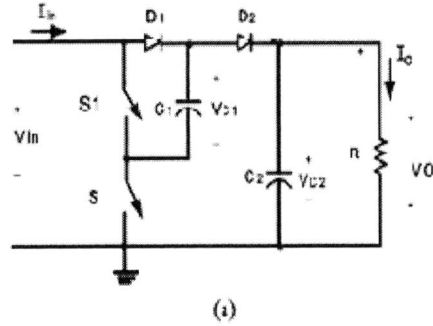

(i)

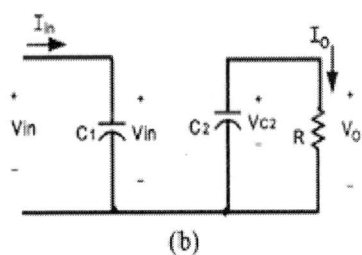

(b)

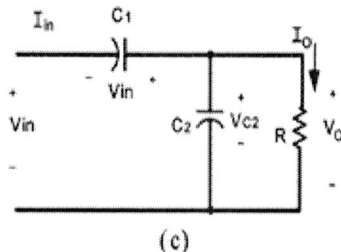

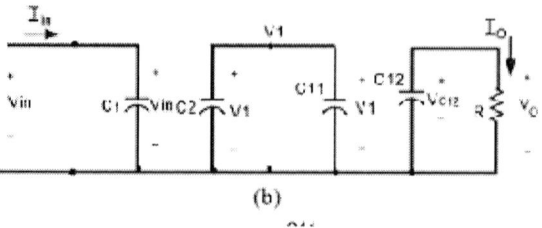

(b)

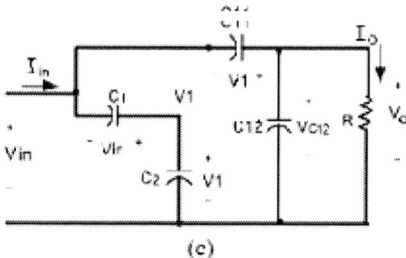

(c)

(c)

Fig1. Elementary circuit-2lift circuit
    (a) Circuit Diagram
    (b). Equivalent circuit during switching-on (S on).
    (c).Equivalent circuit during switching -off (S1 on).

The elementary circuit and its equivalent circuits during switching -on and off in Fig.1. Two switches S and S1 operate in push-pull mode. The voltage across capacitor C_1 is charged to V in during the switch S is on .The voltage across capacitor C_2 is charged to $V_0 = 2$ Vin during switch S is off (S1 is on). Therefore, the output voltage is

$$V_o = 2 V_{in}.$$

Considering the voltage across the diodes and switches, the real output voltage is

$$V_o = 2 V_{in} - \Delta V_i$$
$$= 2 V_{in} - (VD_1 + V_S + VD_2 + V_{S1}).$$

Where,

$$\Delta V_i = VD_1 + V_S + VD_2 + V_{S1}$$

VD_1, VD_2, V_{S1} and V_S are drop across the diodes and switch.For higher order lift circuit, we can analogously design the higher order lift circuit like 4, 8, 16, 32, 64 and so on. This can be achieved by multiple repeating of the parts (switches, capacitors and diodes). If the slave switches number is m, the n [th] lift circuit is countered below

$$n = 2^m$$

Additional series: - 3 lift circuit

The circuit of 3 lift additional series is given below,

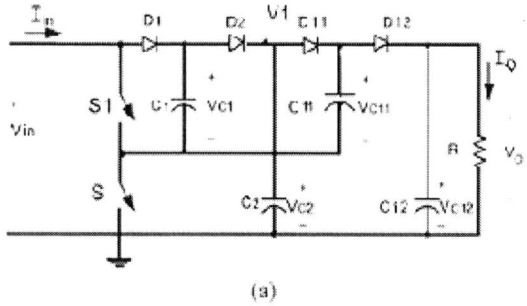

(a)

Fig2.   (a). 3 lift circuit
    (b). Equivalent circuit when the main switch is on
    (c).equivalent circuit when the slave Switch is on

The 3 lift circuit is derived from the elementary (2 lift circuit) by adding two switched capacitor and two diodes which are called additional enhanced series (AES). Two switches S and S_1 operate in push pull state .The voltage across the capacitor C1 is charged to V_{in} and C_{11} is charged to V_1 during on. The voltage across capacitor C_2 is charged to V_1 during switch off. Considering the voltage across the diodes and switches

$$V_1 = 2 V_{in} - \Delta V_1$$
$$= 2 V_{in} - (V_{D1} + V_S + V_{S1} + V_{D2})$$
$$= 2 V_{in} - (2 V_S + 2 V_d)$$

Therefore the net output voltage Vo is

$$V_o = V_1 + V_{in} - \Delta V_o$$
$$= 3 V_{in} - (4V_s + 4 V_d)$$

For higher order lift circuit, we can analogously design the higher order lift circuit like 3, 6 ,24, 48,96 and so on. This can be achieved by multiple repeating of the parts (switches, capacitors and diodes) if the slave switches number is m, the n [th] lift circuit is countered below.

$$n = 1.5 * 2^m$$

For the n th -order lift circuit, the final output voltage is,

$$V_o = 1.5 * \left(2^m V_{in} - \sum_{i=1}^{m-1} 2i \, \Delta V_{m-i} \right) - \Delta V_m - \Delta V_o$$

III. STATE VARIABLE DESCRIPTION FOR EACH CIRCUIT STATE

State space models provide a general and strong basis for dynamic modeling of various systems including power converters. State-space models are useful to design the needed linear control loops and can also be designed to computer simulate the steady state, as well as the dynamic behavior of the power converter, fitted with the designed feedback control loops and subjected to external perturbations. Furthermore state space averaging and linearization provides an elegant solution for the application of widely known linear control techniques to most power converters.

Supposing the power semiconductors as controlled ideal switches, the time (t) behavior of the circuit, over period T, can be represented by the general form of the state space model

$\overline{X} = Ax + Bu$
$Y = Cx + Du$

Where x is the state vector, u the input or control vector, and A, B, C, D are respectively the dynamics (or state), the input, the output and direct transmission (or feed-forward) matrices.

In positive multiple output push pull dc –dc converter, there are two circuit states: one state corresponds to when the switch is on and the other one is off. And the other state is vice versa. During each circuit state, the linear circuit is described by means of the state variable vector X consisting of a capacitor voltages Vc_1 and Vc_2. V_d is the input voltage. Therefore each circuit state, one can write the following state space equations.

$\overline{X} = A_1 X + B_1 V_{in}.$ During DT sec

$\overline{X} = A_2 X + B_2 V_{in}.$ During (1-D) T sec

Where x -State vector, A_1, A_2, B_1, B_2 are state-space matrices of appropriate dimensions. V_{in} is the input voltage and D is the duty ratio.

State Variable Description For On State
(Switch S is on and S_1 is off):

For the fig 1(b),
$I_{in} = Vin - X_1 / R_1$ ------- (1)
and
$I_{in} = C_1 dX_1 / dt$ ------- (2)

Sub (2) in(1)

$\overline{X}_1 = -1/R_1C_1 * X_1 + 1/R_1C_1 * V_{in}$

Also,
$C_2 dV_2/dt + X_2/R = 0$ -------(3)
$X_2 = -X_2 / RC_2$ --------- (4)
I_{in} -Input current

R- Load resistance; R_1 – Small supply resistance
X_1- state variable for capacitive voltage Vc_1
X_2- state variable for the capacitive voltage Vc_2

The average state –space equations are obtained by solving the above equations.

The state equation is,

$$\begin{bmatrix} \bar{x}_1 \\ \bar{x}_2 \end{bmatrix} = \begin{bmatrix} -1/R_1C_1 & 0 \\ 0 & -1/R_1C_2 \end{bmatrix} \begin{bmatrix} X_1 \\ X_2 \end{bmatrix} + \begin{bmatrix} \dfrac{1}{R_1C_1} \\ 0 \end{bmatrix} V_{in}$$

     ------(5)

The output equation is,
$Y = X_2$ ---------(6)
Where Y is the output vector

State Variable Description for Off state

For the fig 1.c,
$V_{in} / R_2 = V_1 / R_2 + C \, d/dt \, (V_1 - V_2)$

$V_{in} / R_2 = V_1 / R_2 + C (\overline{X}_1 - \overline{X}_2)$ ------- (7)
Also,
$C \, d/dt \, (V_2 - V_1) + C_2 \, d V_2 / dt + V_2 / R = 0$

$C \, d/dt \, (X_2 - X_1) + C_2 \, d V_2 / dt + V_2 / R = 0$ ------- (8)

Assume R_2 is a small resistance, Solving (7) and(8),
The state equation is,

$$\begin{bmatrix} \bar{x}_1 \\ \bar{x}_2 \end{bmatrix} = \begin{bmatrix} -2/R_2C & -1/RC \\ -1/R_2C & -1/RC \end{bmatrix} \begin{bmatrix} X_1 \\ X_2 \end{bmatrix} + \begin{bmatrix} 2/R_2C \\ 1/R_2C \end{bmatrix} V_{in}$$

     -------(9)

The output equation is,

$Y = X_2$ ------ (10)

IV. CONVERTER TRANSFER FUNCTION

The converter transfer function is,
$$V_O / V_{in} = -C * A^{-1} * B$$

Where,
$A = A_1 D + A_2 (1-D)$

$B = B_1 D + B_2 (1-D)$

$C = C_1 D + C_2 (1-D)$ and

$D = 0$
Assume the duty ratio D=0.5

1557

$$A = \begin{bmatrix} 0.5/C & 0.5/RC \\ 0.5/C & 1/RC \end{bmatrix}$$

$$B = \begin{bmatrix} 1.5/C \\ 0.5/C \end{bmatrix}$$

$$C = \begin{bmatrix} 0 & 1 \end{bmatrix}$$

$$D = \begin{bmatrix} 0 \end{bmatrix}$$

$$A^{-1} = R\,C^2/(1-D)^2 \;*\; \begin{bmatrix} 1/RC & -0.5/C \\ -0.5/C & 0.5/C \end{bmatrix}$$

For the small value of load resistance, the converter voltage transfer function is,

$$V_0 / V_{in} = D /(1-D)^2 \qquad\text{-------(11)}$$

If $D = 0.5$, $V_0 = 2\,V_{in}$. Where, V_0 is output voltage., V_{in} is input voltage.
The variation of duty ratio with voltage transfer gain is shown in the below fig 3.

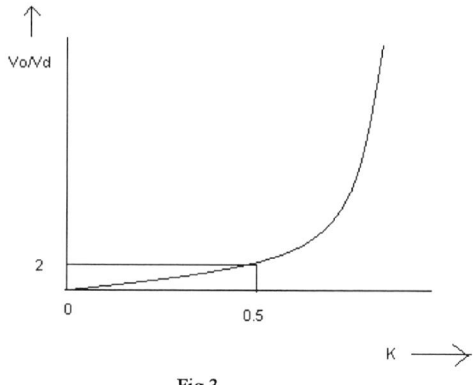

Fig 3.

For an nth -order lift circuit, the final output voltage is ,

$$V_0 = 2^m\,V_{in} - \sum_{i=0}^{m-1} 2^i\,\Delta V_{m-i}$$

V. SMALL SIGNAL DUTY RATIO TO OUTPUT TRANSFER FUNCTION

The small signal duty ratio to the output transfer function is,

$$V(S)/\,\overline{d}(s) = C\,[SI - A]^{-1} * [(A_1 - A_2)\,X + (B_1 - B_2)\,U] + [(C_1 - C_2)\,X + (D_1 - D_2)\,U] \qquad\text{------ (12)}$$

Where I is the identity matrix. A1 and A2 are the state matrices and $C_1 = C_2$; $D_1 = D_2 = 0$. Assume the input voltage $U = 10V$ (input); $D = 0.5$
At steady state, the space vector value will be

$$X = \begin{pmatrix} -10 \\ 20 \end{pmatrix}\;;$$

The voltage across the capacitor C_1 at steady state will be -10V and the voltage across the capacitor C_2 will be 20V &assume
$C_1 = C_2 = 2\mu F$.
The small signal duty ratio to output transfer function is,

$$V_0 /\,\overline{d}(s) = \frac{40\,SC + 10}{S^2 C^2 + 1.5\,SC + .25} \qquad\text{---- (13)}$$

From the above equation (13), we have two poles which may be placed in the left of the S plane.

VI. STABILITY STUDIES

From the above expression (8), it is concluded that the converter circuit is stable for the system parameters variations. The following fig 2 clearly shows the poles and zero movement for the system parameters.

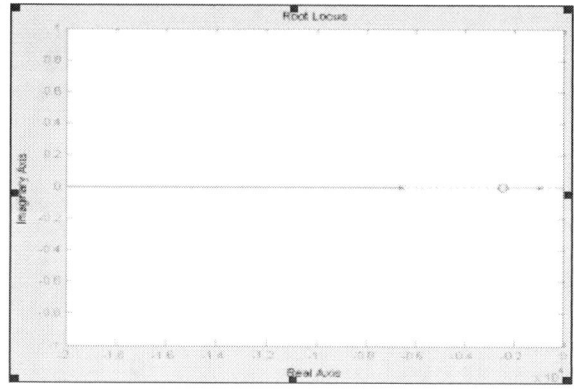

Fig 4 Root locus

VII. SIMULATION RESULTS

To verify the design and calculation results, the PSpice simulation package was applied to these circuits. Choosing =12v, all capacitors C= 2μF, and f=100 kHz, we can obtain the output voltage value which is shown below,

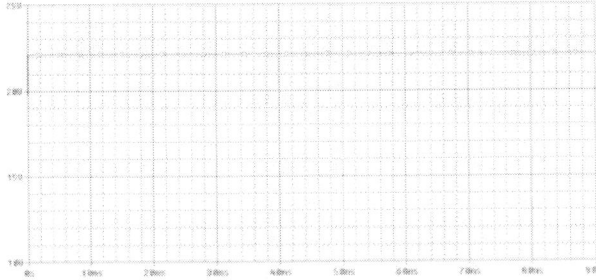

Fig 6 Simulated output voltage wave form.

VII. SUMMARY OF THE TECHNIQUE

Using this technique we can easily design a higher order lift circuit to obtain high output voltage. all these may be classified in to two categories: main series (4-,8-,16-,32- 64-,) and additional series(such as 3-,6-,12-,24-,48-, lift circuits). If the slave switches number is m, the nth –lift circuit is countered below.

$$n = 2^m \quad \text{for main series.}$$

$$= 1.5 * 2^m \quad \text{for additional series.}$$

From the above formulas, we can obtain that output voltage is doubled by adding one stage.

Equation (13) clearly tells that the converter is stable for all the circuit parameters variations. Since we have one zero and two poles which are placed in the left of the S plane These converters are very attractive because they use no magnetic components, are small in size, and are amenable to monolithic integration. This converter is suitable for industrial applications.

VIII. CONCLUSION

A new series of step up dc-dc converter- positive output multiple lift push- pull switched –capacitor converter has been correctly analyzed using State-space averaging method and stability also checked for system parameter variations. Since these converters are constructed with switched capacitor, their size is very small and power density is very high. Using this method largely increases the voltage transfer gain, high output voltage is easily obtained. Simulation and experimental results verified the design and calculations. This series dc-dc converter is suitable and convenient to be applied industrial applications with high output voltage

REFERENCES

[1] Power electronics hand book, M.H Rashid, d Academic, San Diego, CA, 2001.

[2] Luo, F.L. Luo converter- Voltage lift technique IEEE Trans Power Electronics, PP1783 -1789, and May1998.

[3] Advanced DC – DC converter by F.L. Luo

[4] Luo, F.L. Re lift converter design, test, simulation and stability analysis. IEEE Trans power Electron Vol 145, PP 315 -425, and App 1998.

[5] I.F.L. Luo and H.Ye positive output super lift converters. IEEE Trans power Electron Vol18, PP 105-113, and Jan 2003.

[6] I.F.L. Luo and H.Ye negative output super lift converters. IEEE Trans power Electron Vol18, PP 1113-1121, 2003.

[7] Luo, F.L Positive output luo converter. Voltage lift technique IEEE Trans power Electron Vol 146, PP 415 -432, and App 1999.

[8] Luo, F.L Negative output luo converter. Voltage lifts technique IEEE Trans power Electron Vol 146, PP 208 -224, and App 1999.

[9] Luo, F.L Double output luo converter. Advanced voltage lift technique IEEE Trans power Electron Vol147, PP 469 -485.App 2000.

[10] I.F.L. Luo and H.Ye positive output multiple lift push switched capacitor luo converters IEEE trans.power electron, pg 331-336 app 2002

Active Clamp Interleaved Boost Converter with Coupled Inductor for High Step-up Ratio Application

S. -Y. Tseng, #J. -Z. Shiang, #W.-S. Jwo and C.-M. Yang

GreenPower Evolution Application Research Lab.
(GPEARL)
Department of Electrical Engineering
Chang-Gung University
E-mail: sytseng@mail.ccu.edu.tw
TEL : +886-3-2118800
FAX :+886-3-2118026
Kwei-Shan Tao-Yuan , Taiwan, R.O.C

#Department of Electrical Engineering
Chien Kuo Technology University
Changhua, Changhua City, Taiwan, R.O.C

Abstract--This paper presents a soft-switching interleaved boost converter with coupled inductor for high step-up ratio applications. In the proposed converter, an interleaved boost converter associated with a coupled inductor to raise its powering capability and to increase its step-up ratio. To achieve high conversion efficiency, an active clamp circuit is introduced into the proposed one for achieving zero-voltage switching at turn-on transition. In addition, switches in the boost converter and active clamp circuit are integrated with a synchronous switch technique to reduce circuit complexity and component counts, resulting in a lower cost, smaller volume and higher efficiency. Finally, a prototype with step-up ratio of 8, and output voltage and current of 350 V/ 2 A have been verified the feasibility of the proposed converter.

Keywords: Interleaved boost converter, active clamp, synchronous switch technique

I. INTRODUCTION

Limited fossil energy and increased air pollution have spurred researchers to develop clean energy sources. One of these sources is photovoltaic (PV) power generation system, which is clean, quiet and an efficient method for generating electricity. In practical applications, the PV power generation system is usually used to supply power to utility line, as shown in Fig. 1. To increase the reliability of PV, power supply of PV power system is usually connected with battery source in parallel. Thus, DC/DC converter needs high step-up voltage for supplying energy to utility line of AC 220 V system.

To achieve a high step-up voltage ratio, a transformer or coupled inductor is usually used in converters [1]-[2]. Compared with an isolation transformer, a coupled inductor has a simpler winding structure, lower conduction loss and continuous conducted current in the primary winding, resulting in a smaller primary winding current ripple and lower input filtering capacitance. Therefore, a coupled inductor used in converter is relatively attractive. However, since the energy is trapped in the leakage inductor of transformer, it will not only increase voltage stress of the switch but induces significant loss.

To solve these problems, several methods have been proposed [2]-[5]. In [3], a resistor-capacitor-diode (R-C-D) snubber can alleviate voltage stress of the switch, but the energy trapped in the leakage inductor is dissipated. A passive lossless clamped circuit [2], [4] can recover the energy and reduce voltage spike, but the active switch is still in hard switching. In [5], an active-clamp circuit is added to the converter for recovering leakage energy and limiting voltage spike, and the clamping circuit can also achieve zero-voltage switching (ZVS), which makes the converter more viable.

To further increase the powering capability of converters, an interleaved boost converter is proposed [6]-[7], as shown in Fig. 2. As mentioned above, a coupled inductor is used to achieve high step-up voltage ratio applications. Thus, an interleaved boost converter associated with a coupled inductor is proposed in the paper, as shown in Fig. 3. To simplify the proposed converter, switches between the active clamp circuit and the interleaved boost converter are integrated with the synchronous switch technique [8], as shown in Fig. 4(c). From the performance comparison between the conventional interleaved boost converter with coupled inductor and the proposed one, the proposed one can yield higher efficiency, reduce weight, size and volume significantly.

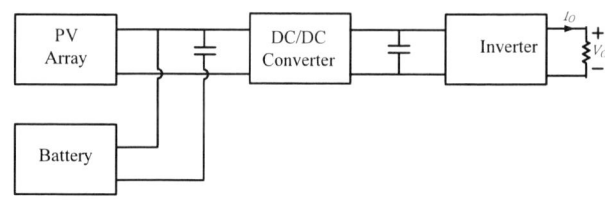

Fig. 1. Block diagram of PV power generation system.

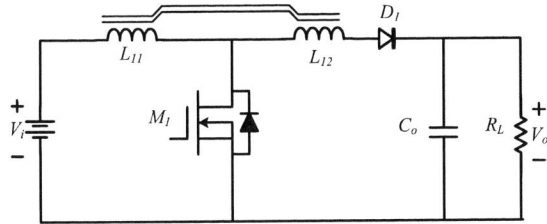

Fig. 2. Schematic diagram of boost converter with coupled inductor.

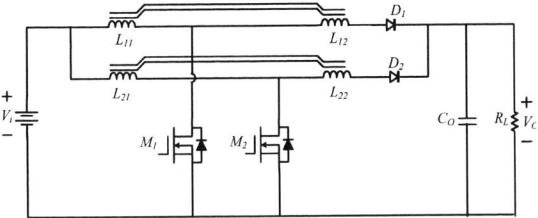

Fig. 3. Schematic diagram of an interleaved boost converter with coupled inductor

II. DERIVATION OF THE PROPOSED CONVERTER

To increase powering capability and step-up ratio, an interleaved boost converter is adopted and a coupled inductor is inserted into a conventional interleaved boost converter, as shown in Fig. 3. In high step-up ratio applications, a leakage inductor existed in the coupled inductor of boost converter is very high. Therefore, an active clamp circuit is used in the interleaved boost converter with coupled inductor to recover the energy trapped in leakage inductor.

Two sets of active clamp circuit are used in an interleaved boost converter with coupled inductor to reduce switching loss, as shown in Fig. 4 (a). Since two sets of switch pairs (M_1, M_4) and (M_2, M_3) respectively share a common mode, they can meet the requirements of synchronous switch integration technique [8] if each switch pair can be operated in synchronous. To describe each switch pair operated in synchronous, concept switch voltage and current waveforms are shown in Figs. 5 and 6. Fig. 5 shows those waveforms operated in the boundary soft-switching features. While, Fig. 6 shows those waveform operated at the same load condition. Additionally, Fig. 5(a) and Fig. 6(a) shows those waveforms with the switch integration. Fig. 5(b) and Fig. 6(b) shows those waveforms with the conventional complementary method to control the active clamp circuit. From Fig. 5, it can be seen that the proposed one operated at the boundary condition using switch integration needs more large current to achieve ZVS features. That is, operation range of one is less than that of the conventional complementary method. However, at the same load condition shown in Fig. 6, loss duty of one is also less than that of the conventional complementary method. Namely, effective duty of one is greater than that of the conventional complementary method. According to transfer function M of boost converter with coupled inductor, M is proportional to duty ratio D. Therefore, the interleaved coupled-inductor boost converter with switch integration can increase its step-up ratio. It is suitable for high step-up ratio applications. As

mentioned above, the interleaved coupled-inductor boost converter can use synchronous switch technique to simplify its circuit structure. Based on the principle of the synchronous switch technique, when switches M_1 and M_4 (or M_2 and M_3) are operated in synchronous and they have a common node, the converter shown in Fig. 4 (a) can be degenerated, as shown in Fig. 4 (b). In Fig. 4(b), switch M_{23} and M_{14} are respectively replaced with M_1 and M_2 to simplify switch sign. Similarly, diodes D_{231}、D_{232}、D_{141} and D_{142} are also separately substituted for diodes $D_3 \sim D_6$, as shown in Fig. 4(c). Since switch M_1 or M_2 needs a bidirectional switch, diodes $D_3 \sim D_6$ must be reserved. Note that when switches of the interleaved boost converter and the active clamp circuit are integrated, duty ratios of switches M_1 and M_2 in the proposed one must be limited within 0.5.

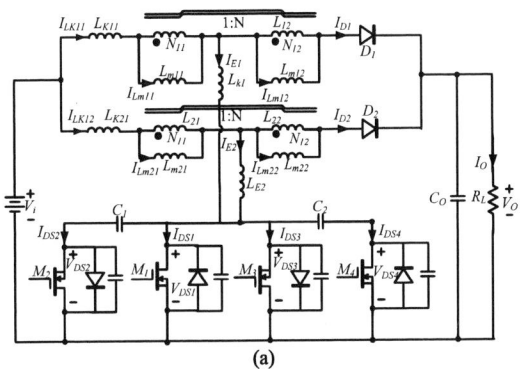

(a)

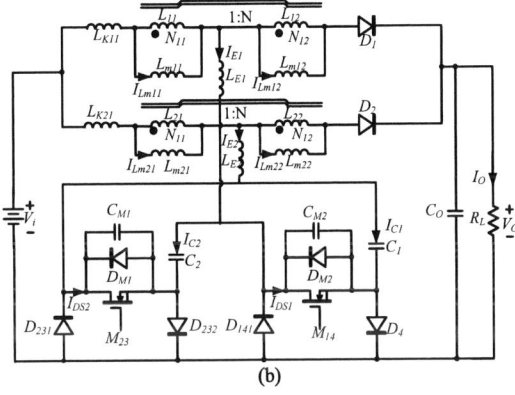

(b)

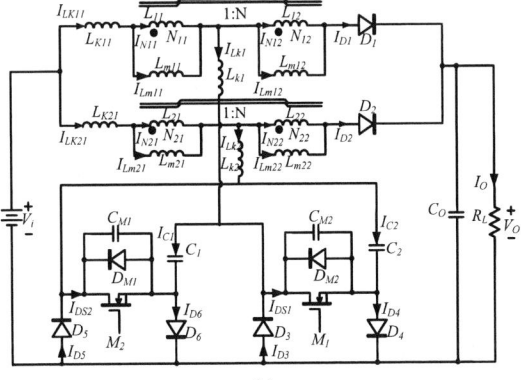

(c)

Fig. 4. Derivation of the proposed interleaved boost converter with coupled inductor.

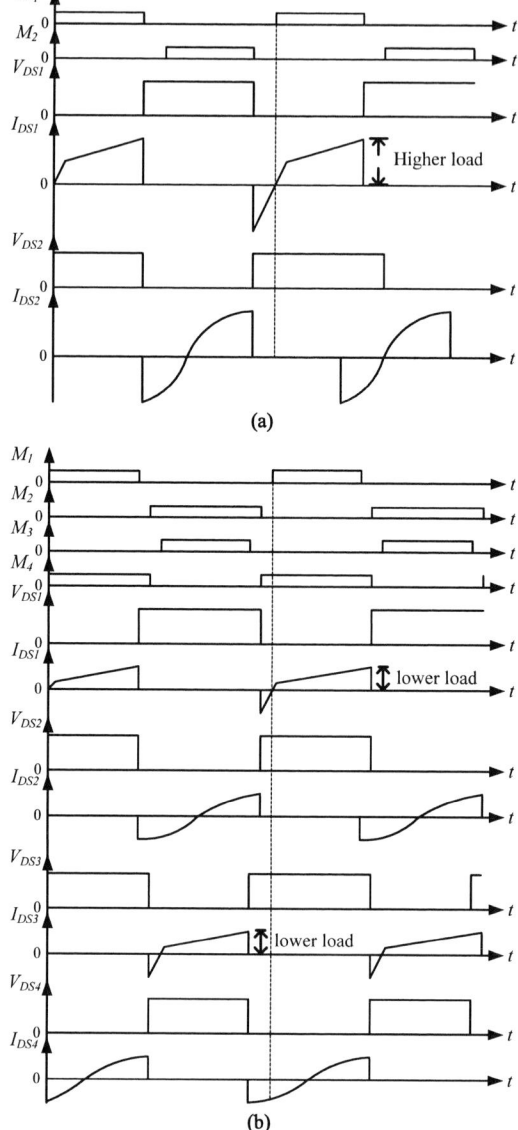

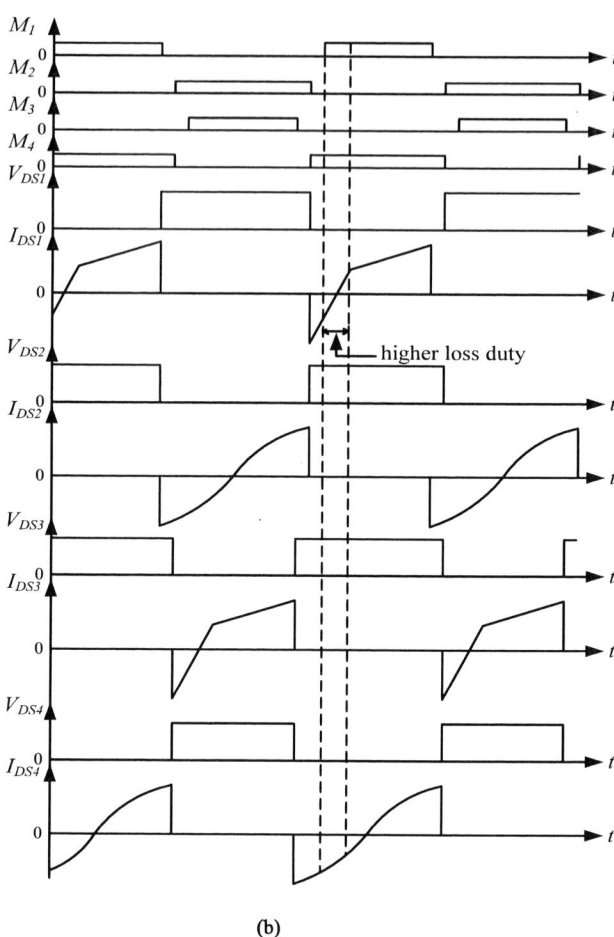

(b)

Fig.6. concept voltage and current waveforms of switches M_1 and M_2 in (a) the proposed active clamp boost converter, and (b) the conventional one with coupled inductor operated at the same load condition.

III. OPERATIONAL PRINCIPLE OF THE PROPOSED CONVERTER

The proposed converter consists of two boost converters with coupled inductor and two active clamp circuits, as shown in Fig. 4(c). According to circuit operational principle , operation of overall converter is divided into *12* modes, as shown in Fig. 7 , and their key waveform are illustrated in Fig. 8. Since operation modes between $t_0 \sim t_5$ are similar to those between $t_5 \sim t_{10}$ except that the operation of switch changes from M_1 to M_2. Thus, each operational mode during half one switching cycle is briefly described in the following.

Mode 1 [Fig. 7 (a); $t_0 \le t < t_1$]: Before t_o, diodes D_1 and D_2 are in freewheeling through coupled inductors (L_{m11} and L_{m12}) and (L_{m21} and L_{m22}), respectively. While, body diode D_{M1} is forwardly bias. When $t = t_0$, switch M_1 is turned on. Since body diode D_{M1} is forwardly biased before switch M_1 is turned on, switch M_1 is operated with zero-voltage switching (ZVS) at turn-on transition. Within this time interval, leakage inductors L_{K11} and L_{K21}、external inductors L_{E1} and L_{E2}, and capacitor C_2 form a resonant network and they are in resonant manner. Additionally, switch current I_{DS1} varies from a negative value to *0*.

Fig.5. Concept voltage and current waveforms of switches M_1 and M_2 in (a) the proposed active clamp boost converter and (b) the conventional one with coupled inductor operated in soft-switching boundary.

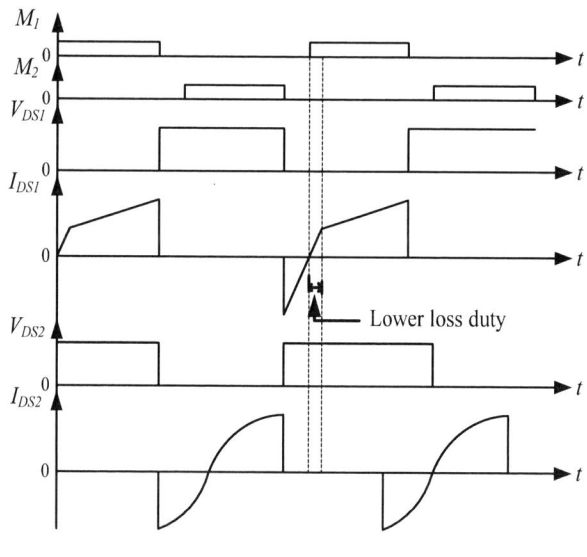

(a)

Mode 2 [Fig. 7 (b); $t_1 \leq t < t_2$]: At t_1, switch current I_{DS1} is equal to 0. At the moment, diode D_{M1} is reversely biased. During this time interval, switch current I_{DS1} equals inductor current I_{E1}. While, inductor current I_{LK11} is the sum of current I_{E1} and I_{D1}. In addition, diode current I_{D4} is the sum of switch current I_{DS1} and snubber current I_{C2}. The energies stored in coupled inductors (L_{m11} and L_{m12}) and (L_{m21} and L_{m22}) are transferred to load, and inductor currents I_{Lm12} and I_{Lm22} decrease linearly.

Mode 3 [Fig. 7 (c); $t_2 \leq t < t_3$]: At t_2, inductor current I_{E1} is equal to current I_{LK11}. Diode D_1 is reversely biased. Therefore, the energy stored in inductor L_{m12} is transferred to primary winding of coupled inductors (L_{m11} and L_{m12}) through winding N_{11} and N_{12}. During this time interval, inductor Lm_{11} is in the storing energy state, while inductors L_{m21} and L_{m22} is in the releasing energy state. In addition, inductor L_{E2} and L_{K21}, and capacitor C_2 still stay in the resonant manner, while inductors L_{E2} and L_{K11} are in the storing energy state through switch M_1 and diode D_4, respectively.

Mode 4 [Fig. 7(d); $t_3 \leq t < t_4$]: At t_3, switch M_1 is turned off. The energies stored in inductors L_{E1} and L_{E2} are released to junction capacitors C_{M1}, C_{D3} and C_{D4} for charging, and to junction capacitors C_{M2}, C_{D5} and C_{D6} for discharging, respectively. Within this time interval, diode D_2 is in freewheeling through inductors L_{m21} and L_{m22}.

Mode 5 [Fig. 7(e); $t_4 \leq t < t_5$]: When $t = t_4$, inductor current I_{E1} is equal to inductor current I_{LK11} and its current varies from a positive maximum value to a negative maximum value with the resonant manner. At the moment, diode D_1 is in forwardly bias, while the body diode D_{M2} is also in forwardly bias. In ideal condition, inductor current I_{E1} equals current I_{E2}. Within this time interval, diodes D_1 and D_2 are in freewheeling through coupled inductors (L_{m11} and L_{m12}) and (L_{m21} and L_{m22}), respectively. In addition, their current decrease linearly. When switch M_2 is turned on at the end of mode 5, the other half one switching cycle will start.

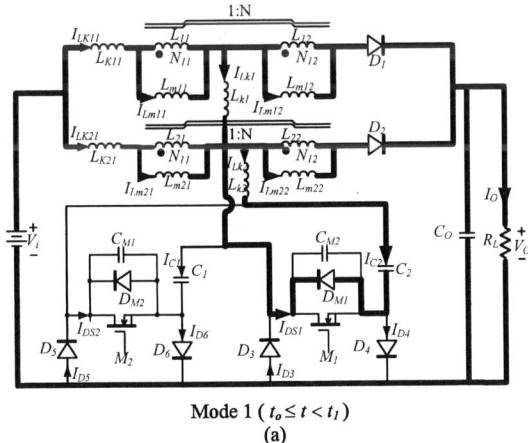

Mode 1 ($t_o \leq t < t_1$)
(a)

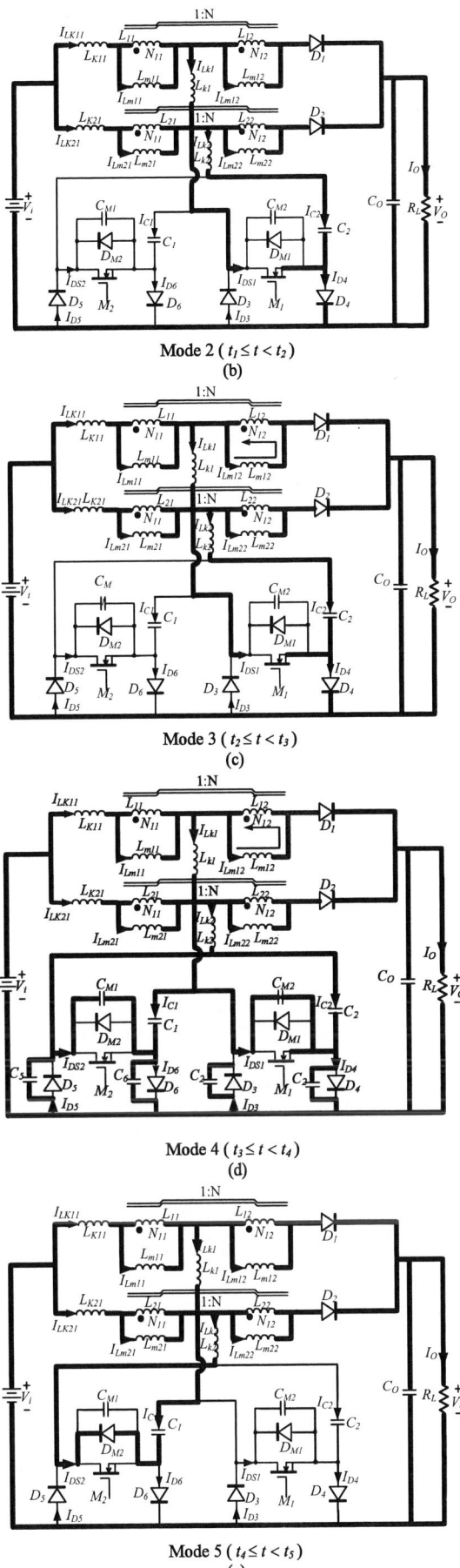

Fig. 7. Derivation of the proposed interleaved boost converter with coupled inductor.

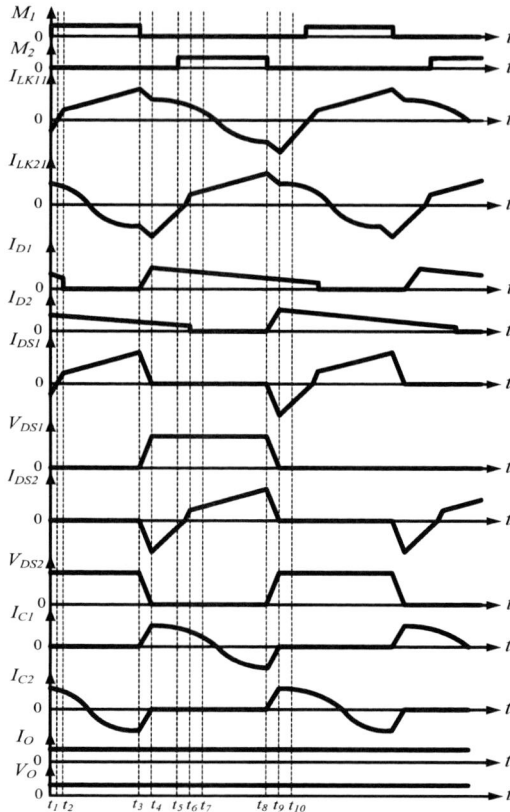

Fig. 8. Key waveforms of the proposed converter operating over one switching cycle.

IV. DESIGN OF THE PROPOSED CONVERTER

The proposed converter is composed of two active clamp boost converters with coupled inductor operated in an interleaved fashion to recover the energy trapped in leakage inductor of coupled inductor. Since the switches in each boost converter are integrated with the synchronous switch technique, duty ratios of ones are limited within 0.5. In design of the interleaved active-clamp boost converter, determination of duty ratio D_1, coupled inductances (L_{m11} and L_{m12}) and (L_{m21} and L_{m22}), active clamp capacitors C_1, and C_2, and output filter are important. In the following, their designs are analyzed briefly.

A. Duty Ratio D

To determine duty ratio, we must first attain input to output voltage transfer ratio M. Since the active clamp circuit only helps switch M_1 or M_2 to achieve soft-switching feature, it does not affect transfer ratio M of the proposed boost converter. That is, transfer ratio M will be the same as the conventional one. According to volt-second balance principle, the following equation can be obtained:

$$V_i DT_s + \left[\frac{-(V_o - V_i)}{N+1}(1-D)T_s \right] = 0, \quad (1)$$

where N is turns ratio of coupled inductor (L_{m11} and L_{m12}) or (L_{m21} and L_{m22}). From (1), transfer ratio M can be expressed by

$$M = \frac{V_o}{V_i} = \frac{(1+ND)}{1-D}. \quad (2)$$

Based on the operational condition of the proposed converter with switch integration, duty ratio of switch M_1 or M_2 is limited within 0.5, and they are operated in complementary. According to (2), a large duty ratio D corresponds to a smaller turns ratio N of coupled inductor, which results in a lower current stress imposed on switches M_1 and M_2, as well as voltage stress on diodes D_1 and D_2. However, in order to accommodate variations of load, line voltage, component value and duty loss, it is better to select an operating range as D = 0.35 ~ 0.4.

B. Coupled Inductors (L_{m11} and L_{m12}) or (L_{m21} and L_{m22})

Once the duty is selected, the turns ratio of coupled inductors L_{m11} and L_{m12} can be determined using (2), which yields

$$N = \frac{(1-D)V_o - DV_i}{DV_i}. \quad (3)$$

By applying the Faraday's law, N_{11} of the coupled inductor can be given as

$$N_{11} = \frac{DV_i T_s}{A_C \Delta B}, \quad (4)$$

where A_C is the effective cross-section area of the coupled inductor core and ΔB is the working flux density. According to (3) and (4), N_{12} can be therefore determined.

To achieve a ZVS feature, the energy stored in leakage inductor L_{K111} (or L_{K21}) and external inductor L_{E1} (or L_{E2}) must satisfy the following inequality:

$$\frac{1}{2}(L_{K11} + L_{E1})(I_{LE1(tv9)} - I_{E1(tv8)})^2 \geq \frac{1}{2}C_T V^2_{DS1(max)}, \quad (5)$$

where L_{E1} is the external inductor to achieve soft-switching feature, $I_{LE1(tv9)}$ is the current of L_{E1} at time t_9, $I_{E1(tv8)}$ is that at time t_8, C_T is the total capacitors which are the sum of C_{m1}, C_{D5} and C_{D6}, and $V_{DS1(max)}$ is the voltage across switch M_1 and its value is equal to $(V_i + (V_o - V_i) / (N + 1))$. According to circuit operational principle, the voltage V_{C1} can be approximately expressed by

$$V_{C1} = \frac{NV_i + V_o}{N+1} \quad (6)$$

Once C_T, and $I_{E1(tv9)}$ and $I_{E1(tv8)}$ and specified, the inequality of the inductor L_T ($= L_{E1} + L_{K11}$) can be determined as

$$L_T \geq \frac{C_T(NV_i + V_o)^2}{(N+1)^2(I_{LE1(tv9)} - I_{LE1(tv8)})^2} \quad (7)$$

Since the proposed converter is operated in continuous conduction mode (CCM), the inductance L_{m11} and L_{m12} must be greater than L_{m11B} and L_{m12B}, respectively, which are the inductance at the boundary of CCM and discontinuous conduction mode (DCM). Its boundary current waveforms is shown in Fig. 9. From Fig. 4(c), it can be seen that when switch M_1 is turned on, inductor current I_{LK11} is the sum of current I_{Lm11} and I_{N11}, which is the equivalent current from secondary winding N_{12} to primary winding N_{11}. Thus, current I_{LK11} can be expressed by

$$I_{LK11} = I_{Lm11} + I_{N11}, \quad (8)$$

1564

where I_{N11} is equal to $NI_{N12}(=NI_{Lm12})$. Thus, $I_{LK11(1)}$ can be determined as

$$I_{LK11(1)} = \frac{V_i}{L_{m11}}DT_S + \frac{N^2 V_i}{L_{m12}}DT_S, \quad (9)$$

where L_{m11} is the magnetizing inductor of primary winding of coupled inductor and $L_{m12}(=N^2 L_{m11})$ is that of secondary winding of coupled inductor.
According to (9), $I_{LK11(1)}$ can be rewritten by

$$I_{LK11(1)} = \frac{2V_i}{L_{m11}}DT_S. \quad (10)$$

While, $I_{LK11(2)}$ can be given by

$$I_{LK11(2)} = \frac{I_{LK11(1)}}{(1+N)} = \frac{2V_i}{(1+N)L_{m11}}DT_S \quad (11)$$

Since inductor current $I_{D1(1)}$ is equal to $I_{LK11(2)}$, and the average current $I_{D1(av)}$ equals half of output current I_O, the average current $I_{D1(av)}$ can be expressed as follows:

$$I_{D1(av)} = \frac{I_O}{2} = \frac{V_i}{(N+1)L_{m12B}}D(1-D)T_S \quad (12)$$

According to (12), the boundary inductance L_{m12B} can be determined as

$$L_{m12B} = \frac{2V_i}{(N+1)I_O}D(1-D)T_S \quad (13)$$

Based on the operational principle of the coupled inductor, the relationship between inductances L_{m11B} and L_{m12B} can be expressed as follows:

$$L_{m11B} = \frac{1}{N^2}L_{m12B} \quad (14)$$

Substituting (13) in (14), inductor L_{m11B} can be determined as

$$L_{m11B} = \frac{2V_i}{N^2(N+1)I_O}D(1-D)T_S \quad (15)$$

According to operational requirement of the proposed converter which is operated in CCM, inductors L_{m11} and L_{m12} must be greater than L_{m11B} and L_{m12B}, respectively. Similarly, inductances L_{m21} and L_{m22} can be separately determined by (15) and (13).

C. Active Clamp Capacitor C_1 or C_2

The active clamp capacitors C_1 and C_2 are used to achieve soft-switching feature. To achieve ZVS feature, one half of the resonant period formed by L_T and C_1 or L_T and C_2 should be equal to or greater than the maximum off time of switch M_1 or M_3. Thus, capacitor C_1 (or C_2) must satisfy the following inequality:

$$\pi\sqrt{L_T C_1} \geq t_{off} = (1-D)T_S \quad (16)$$

From (7) and (16), when L_T is specified, the capacitance range of the capacitor C_1 (or C_2) can be determined as

$$C_1 \geq \frac{(1-D)^2 T_S^2}{\pi^2 L_T} \quad (17)$$

D. Output capacitor C_O

The output capacitor C_O is primarily designed for reducing ripple voltage. The ripple voltage across output capacitor C_O is determined by

$$\Delta V_{rco} = \frac{\Delta Q_{CO}}{2C_O}$$
$$= \frac{1}{2C_O}(I_{O(max)} \times DT_S)$$
$$= \frac{2DI_{O(max)}T_S}{2C_O} \quad (18)$$

where $I_{O(max)}$ is the maximum output current.

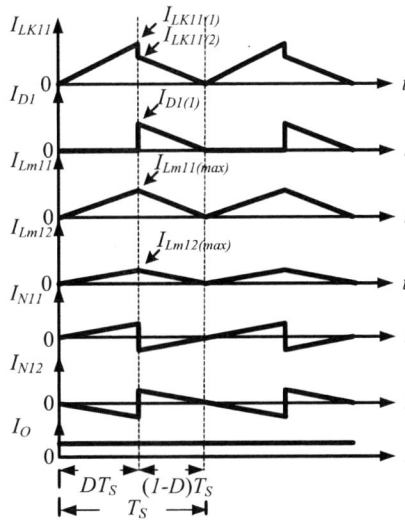

Fig. 9. Conceptual current waveforms of inductor currents and output current in the proposed converter operated in the boundary of CCM and DCM.

V. MEASURED RESULTS

To verify the proposed soft-switching converter, a prototype which is a regulator for supplying an inverter of utility line system, and uses a PV as the source with the following specifications was implemented.

- input voltage V_i: 39 ~ 47 V_{dc} (P_V arrays),
- switching frequency f_S: 100 kHz,
- output voltage V_o: 350 V_{dc},
- maximum output current $I_{o(max)}$: 2 A, and
- maximum output power $P_{o(max)}$: 700 W.

Measured voltage V_{DS} and current I_{DS} waveforms of switches M_1 and M_2 are respectively shown in Figs.10 and 11. Fig. 10 shows those under 20 % of full load, while Fig. 11 shows those under full load. From Figs. 10 and 11, it can be seen that switches M_1 and M_2 can be operated with ZVS at turn-on transition. Comparison of conversion efficiency between the interleaved boost converter with hard-switching circuit and the proposed one, illustrating that efficiency of the proposed converter is higher than the hard-switching one. Its efficiency is 91 % under full load. Fig.13 shows the step-load change between 20 % and 100 % of the full load, from which it can be observed that the voltage regulation of output voltage V_O has been limited within ±1 % to prove a good dynamic response.

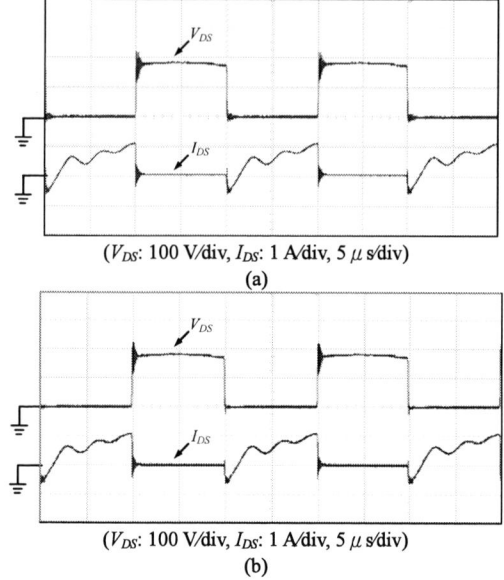

(V_{DS}: 100 V/div, I_{DS}: 1 A/div, 5 μ s/div)

(a)

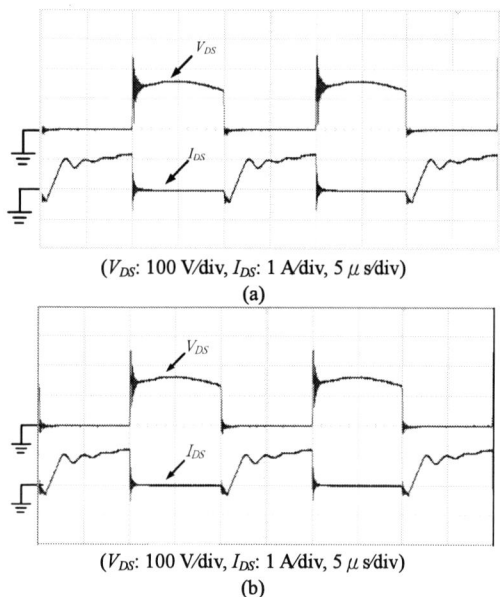

(V_{DS}: 100 V/div, I_{DS}: 1 A/div, 5 μ s/div)

(b)

Fig. 10. Measured voltage V_{DS} and current I_{DS} waveforms of (a) switch M_1 and (b) switch M_2 of the proposed converter under 20% of full load.

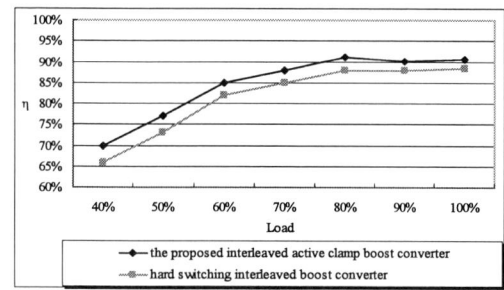

Fig. 12. Comparison conversion efficiency between the interleaved active clamp boost converter with hard switching and the proposed one.

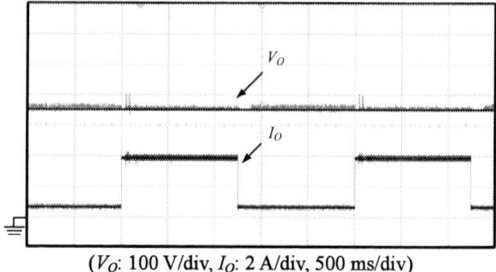

(V_O: 100 V/div, I_O: 2 A/div, 500 ms/div)

Fig. 13. Output voltage V_O and output current I_O under step-load changes between 20 % and 100 % of the full load of the active clamp interleaved boost converter.

VI. CONCLUSION

In this paper, derivation of the proposed converter with the synchronous switch technique has been briefly described. Operational principle, steady-state analysis and design of the proposed interleaved active-clamp boost converter with coupled inductor have been proposed. A prototype, in which its output voltage 350 V and its maximum output current 2 A, has been implemented. Additionally, the proposed converter with switch integration can achieve the efficiency around 91 % under full load condition. From experimental results, it can be also found that the proposed system can attain a good voltage regulation.

REFERENCES

[1] N. Mohan, T. M. Undeland and W. P. Robbins, Power Electronics, third edition, John Wiley & Sons, Inc, 2003, pp. 195.

[2] F. A. Himmelstoss and P. H. Wurm, "Low-loss converters with step-up conversion ration working at the border between continuous and discontinuous mode," Proceedings of Electronics, Circuits and System, 2000, Vol. 2, pp. 734–737.

[3] Z. Hossain, K. J. Olejniczak, K. C. Burgers and J. C balda, "Design of RCD Snubber Based Upon Approximations to the Switching Characteristics: Part I. Theoretical Development," Proceedings of Electric Machines and Drives Conference, 1997, pp. TA2/6.1-TA2/6.3.

[4] Z. Qun and F. C. Lee, "High-efficiency, high step-up DC-DC converters," IEEE Trans. on Power Electronics, 2003, Vol. 18, pp. 65–73.

[5] R.-J. Wai, P.-H. Yao and L.-W. Liu, "Soft-Switching Converter with Active-Clamp Technique," Proceedings of Taiwan Power Electronics conference, 2004, Vol. 1, pp. 528–533.

[6] M. Veerachary, T. enjyu and K. Uezato, "Maximum Power Point Tracking of Coupled Inductor Interleaved Boost Converter Supplied PV System," Proceeings of Electric Power Applications, 2003, pp. 71–80.

[7] P.-W. Lee, et al., "Steady-state Analysis of An Interleaved Boost Converter with Coupled Inductors," IEEE Trans. on ndustrial Electronics, 2000, pp. 787–795.

[8] Tsai-Fu Wu, et al., "Unified Approach to Developing Single Stage Power Converters," IEEE Trans. on Aerospace and Electronic Systems, 1998, pp. 211–223.

Fig. 11. Measured voltage V_{DS} and current I_{DS} waveforms of (a) switch M_1 and (b) switch M_2 of the proposed converter under full load.

Active Clamp Interleaved Flyback Converter with Single-Capacitor Turn-off Snubber for Stunning Poultry Applications

S. -Y. Tseng, C. -T. Hsieh and H.-C. Lin

GreenPower Evolution Application Research Lab.
(GPEARL)
Department of Electrical Engineering
Chang-Gung University
Kwei-Shan Tao-Yuan , Taiwan, R.O.C
E-mail: sytseng@mail.cgu.edu.tw
TEL: +886-3-2118800
FAX: +886-3-2118026

Abstract--The paper proposes an active clamp interleaved flyback converter with a single-capacitor turn-off snubber associated with a full-bridge inverter for stunning poultry applications. The proposed converter can use active clamp circuit to recover the energy of transformer and to achieve zero-voltage switching at turn-on transition. In addition, the proposed one adopts a single-capacitor snubber not only to smooth out switch turn-off transient for reducing turn-off loss but to reduce the ringing voltage of diodes in secondary winding of transformer when switch of the proposed one is operated at turn-on transition. Compared with the counterparts of the conventional converter topologies, the proposed converter has the merits of less component counts, higher efficiency, smaller size, and they are easier to implement. Performance measurements from a prototype have verified feasibility of the overall system design. This designed system has contributed a lot of humane slaughter and has attracted much attention.

Keywords: active clamp interleaved flyback converter, single-capacitor snubber, humane slaughter

I. INTRODUCTION

In recent years, the issue related to welfare of animals has attracted much attention. In particular, livestock and poultry must be rendered unconscious and insensible to pain before they are exsanguinated with humane slaughter methods. In many countries, carbon dioxide (CO_2) and manual electrical stunning are the two methods always used to stun poultry, such as chicken, before slaughtering [1]-[10]. Since the CO_2 method is subject to many limitations and requires higher cost, the manual electrical stunning method has been used more popular.

Poultry stunning is to cause unconsciousness by generating an epileptiform seizure, which includes two phases: a tonic phase and a clonic phase [1]. Degrees of an epileptiform seizure are dependent upon the amount of current passing the brain. The minimum current and voltage required for stunning a chicken is about 40 mA and 60 V, respectively, and they must sustain at least 3s. Conventionally, to generate the specified electrical

waveforms, rectified line voltage or battery voltage is chopped into square waveforms with power switches and they are boosted through a low-frequency step-up transformer. As mentioned above, the stunner system is in large volume, size and heavy weight, and there exist bone fractures and ecchymosis in the carcasses of chicken, resulting in low meat quality. To solve above problems, a DC/DC converter with PWM control is adopted for manual electrical stunning applications, which can properly limit current level, regulate output voltage and reduce chicken stress during stunning time.

Since the poultry stunner system belongs to low power level applications, a DC/DC converter with flyback circuit is usually adopted due to its simple circuit structure [11]. To recover energy trapped in leakage inductance of transformer in the flyback converter and achieve zero-voltage switching (ZVS) features at turn-on transition, an active clamp flyback converter is used. Although active clamp flyback can supply an enough power to single-channel poultry stunner system, it doesn't supply an enough power to multi-channel one to speed up poultry slaughter . To relieve this problem, an active clamp flyback with an interleaved manner is adopted, as shown in Fig. 1. Since the active clamp interleaved flyback converter can only solve turn-on switching loss, a single-capacitor turn-off snubber can be used to reduce turn-off switching loss [12], as shown in Fig. 2. Furthermore, when switch M_1 or M_3 is turned on, snubber capacitor C_S and diode D_1 or D_2 form a snubber to avoid the spike voltage across switch M_1 or M_3 due to the resonant network which consists of junction capacitor of diode D_1 or D_2 and leakage inductor of secondary winding in transformer T_{r1} or T_{r2}. As mentioned above, the proposed converter can recover the energy trapped in leakage inductor, achieve zero-voltage switching in switches, and reduce turn–off switching losses significantly.

II. MECHANISM OF POULTRY STUNNING

Carcass quality is highly dependent on the parameters of electrical stunning. For determining the desired

978-1-4244-0644-9/07/$25.00 ©2007 IEEE

electrical parameters, mechanism of poultry stunning is briefly described. When an enough stunning voltage is applied to the skin of poultry, it will cause an epileptic seizure in the poultry, resulting in a loss consciousness and sensibility. In poultry stunning, degrees of an epileptic seizure are dependent upon the amount of current passing the brain. To generate enough stunning current, a voltage source applied to poultry skin must overcome its impedance.

Electrical properties of skin have been studied over a century and often characterized with impedance spectra a [13], as shown in Fig. 3. An equivalent circuit composed of resistor R_S series with the parallel combination of resistor R_{SC} and capacitor C_{SC} have been used by many investigators to represent the electrical properties of skin, as shown in Fig. 4. Both skin capacitance and resistance have been shown to be proportional to contact area [14]. The skin impedance versus voltage is shown in Fig. 5, illustrating that the skin impedance is reversley proportional to the applied voltage [15].

In general, there are three main parts to form neurons: soma, axon and dendrite. A stimulation signal is sensed by sensory receptors which exist in dermis or subcutaneous layer. When sensory receptors receive into nerve impulse. Its propagation direction is, in turn, through one neuron to the other neurons, in which they are connected in series by synapses, until the nerve impulse propagation is transmitted to a receiver of brain. That is, synapses can play a transducer role which is to transfer electrical stimulation signals to chemical signals, and vice versa, as illustrated in Fig. 6.

To explain the relationship between electrical stunning for poultry and a suppression of nerve impulse propagation, an equivalent circuit for describing the impulse signal propagation between neurons is shown in Fig. 7. In Fig. 7, since synapses play a role of transducer, it can be considered as a switch Q_1. When a voltage E_i is applied then a current I_1 will pass the body of the poultry. In this stunning duration, if current I_1 is large enough, it will induce a high potential V_1 to turn on switch Q_1. When switch Q_1 is turned on, nerve impulse can reach the sensory receivers of the brain through propagation impedance R_i of neuron. As a result, poultry can sense a stimulation signal. As described previously, it can be observed that electrical stunning can inhibit nerve impulse propagation.

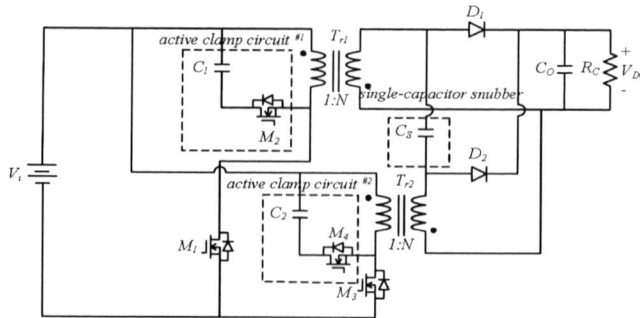

Fig. 2. Schematic diagram of an active clamp flyback converter with a single-capacitor snubber.

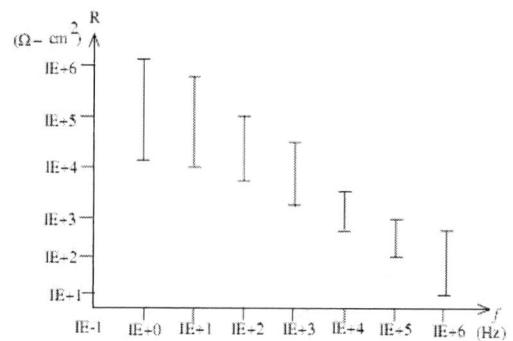

Fig. 3. Illustration of the ranges of skin impedance versus frequency

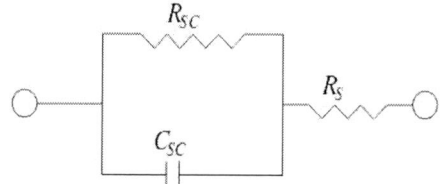

Fig. 4. An equivalent impedance of skin.

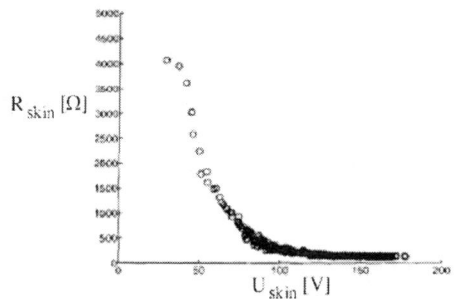

Fig. 5. Plot of the skin impedance of poultry versus the applied voltage.

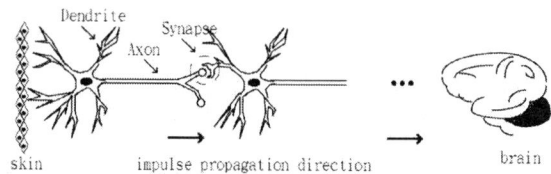

Fig.6. Illustration of impulse propagation between neurons.

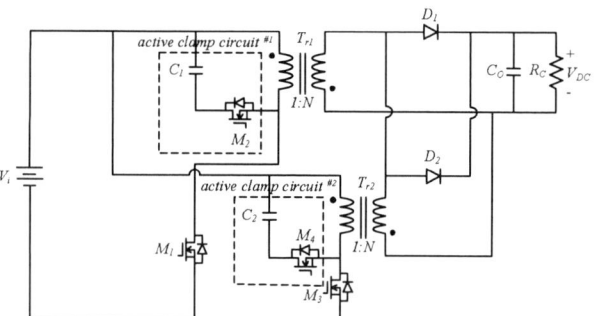

Fig. 1. Schematic diagram of an active clamp flyback converter

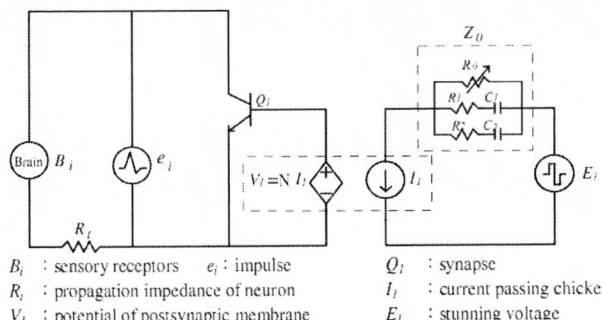

B_i : sensory receptors	e_i : impulse
R_i : propagation impedance of neuron	Q_i : synapse
V_i : potential of postsynaptic membrane	I_i : current passing chicken
Z_0 : equivalent impedance of chicken	E_i : stunning voltage

Fig. 7. An equivalent circuit for describing the impulse signal propagation between neurons.

III. OPERATIONAL PRINCIPLE OF THE PROPOSED CONVERTER

The proposed poultry stunning system, an interleaved flyback converter associated with a full-bridge inverter are proposed, as shown in Fig. 8. The interleaved flyback converter can generate a DC voltage to supply the full-bridge inverter and use an active clamp circuit and a single-capacitor turn-off snubber to reduce switching losses at turn-on and turn-off transitions. While, the full-bridge one can chop a DC voltage into a low frequency square waveform, in which its frequency varies from tens of Hz to hundreds of Hz. In the following, operational principle of the proposed one is briefly described.

Operation of the overall converter is divided into 14 modes. Its each operational mode is shown in Fig. 9 within half one switching cycle. While, their key waveforms are illustrated in Fig. 10. Since operation modes between $t_0 \sim t_7$ are similar to those between $t_7 \sim t_{14}$ except that the operation of switch changes from (M_1 and M_2) to (M_3 and M_4). Thus, each operational mode during half one switching cycle is briefly described as follows.

Mode 1 [Fig. 9(a); $t_0 \leq t < t_1$]: Before t_0, diodes D_{11} and D_{21} are in freewheeling, while diode D_{M1} is in forwardly bias. Additionally, leakage inductor L_{K21} and capacitor C_2 are in the resonant manner through switch M_4. Voltage V_{CS} across snubber capacitor C_S is equal to 0. When $t = t_0$, switch M_1 is turned on. Since body diode D_{M1} is in forwardly bias, switch M_1 is operated with zero-voltage switching (ZVS) at turn on. During this time interval, current I_{N22} is the sum of snubber current I_{CS} and diode current I_{D11}. When switch M_1 is turned on, snubber capacitor C_S and secondary winding N_{12} form a low impedance path. Thus, current I_{N22} can be abruptly replaced by $-I_{CS}$. Additionally, switch current I_{DS1} increases from a negative value to 0.

Mode 2[Fig. 9(b); $t_1 \leq t < t_2$]: At t_1, capacitor current I_{CS} is equal to $-I_{N22}$, and I_{CS} also equals current I_{N12}. Therefore, diodes D_{11} and D_{21} are reversely biased. Within this time interval, snubber capacitor C_S and

magnetizing inductor L_{m21} form a resonant network through transformer T_{r2} and they starts to resonate. While, switch current I_{DS1} is the sum of inductor current I_{Lm11} and I_{N11} which is equal to $-NI_{CS}$. Additionally, leakage inductor L_{K21} and capacitor C_2 still stays in the resonant manner. The energy stored in output capacitor C_O is released to load, and the voltage V_{CS} across snubber capacitor C_s varies from 0 to ($NV_i + V_{DC}$).

Mode 3 [Fig. 9(c); $t_2 \leq t < t_3$]: When $t = t_2$, voltage V_{CS} across snubber capacitor C_S is equal to ($NV_i + V_{DC}$) and is clamped at ($NV_i + V_{DC}$). At the same time, diode D_{21} is forwardly biased. During this time interval, magnetizing inductor L_{m11} is in the storing energy state. While, capacitor C_2 and leakage inductor L_{K21} are still kept in the resonant manner. Therefore, inductor current I_{Lm11} increase linearly, while current I_{Lm21} decreases linearly through transformer T_{r2} to output load.

Mode 4 [Fig. 9(d); $t_3 \leq t < t_4$]: At t_3, switch M_1 is turned off. At the same time, inductor current I_{Lm11} is sustained in continuous through transformer T_{r1}, snubber capacitor C_s and diode D_{21}. Within this time interval, since the energies trapped in magnetizing inductor L_{m11} and leakage inductor L_{K11} are released to capacitor C_S, C_{M1} and C_{M2}, the voltage V_{DS1} across switch M_1 is smoothed to reduce spike voltage. Therefore, switch M_1 can be operated with zero-voltage transition (ZVT). Additionally, voltage V_{CS} across sunbber capacitor C_S is released to load and its value from ($NV_i + V_o$) to 0.

Mode 5 [Fig. 9(e); $t_4 \leq t < t_5$]: At t_4, voltage V_{DS1} reaches ($V_i + V_{DC}/N$) and voltage V_{CS} across snubber capacitor C_S is nearly equal to 0. At the moment, diodes D_{M2} and D_{11} are in forwardly bias. Thus, leakage inductor L_{K11} and capacitor C_1 form a resonant network and they starts to resonate. During this time interval, the energies stored in inductors L_{m11} and L_{m21} are released to load through transformers T_{r1} and T_{r2}, and diode D_{11} and D_{21}, respectively .Their currents I_{Lm11} and I_{Lm12} decrease linearly.

Mode 6 [Fig. 9(f); $t_5 \leq t < t_6$]: At t_5, switch M_2 is turned on, while switch M_4 is turned off. Since body diode D_{M2} is forwardly biased before switch M_2 is turned on, switch M_2 is operated with ZVS at turn on. During this time interval, leakage inductor L_{K11} and capacitor C_1 are still kept in the resonant manner. Additionally, the energy trapped in leakage inductor L_{K21} is transferred to capacitors C_{M2} and C_{M4}. While, diodes D_{11} and D_{21} are in freewheeling through inductors L_{m11} and L_{m21}, respectively.

Mode 7 [Fig. 9(g); $t_6 \leq t < t_7$]: At t_6, voltage V_{DS2} is clamped to 0, while voltage V_{DS4} is equal to ($V_{C2} + V_{DC}/N$). During this time interval, diodes D_{11} and D_{21} still stays in freewheeling. Current I_{LK11} abruptly increases from a negative value to 0. When switch M_3 is turned on at end of mode 7, the other half one switching cycle will start.

1569

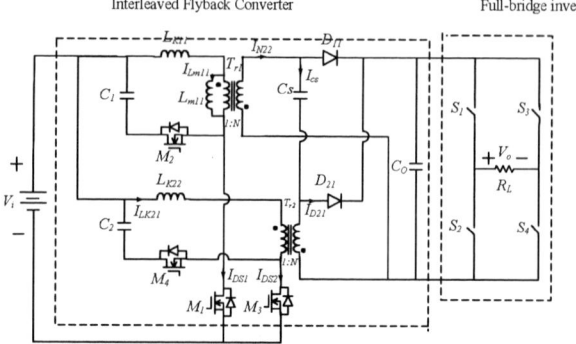

Fig. 8. Schematic diagram of a complete stunners system for stunning poultry applications.

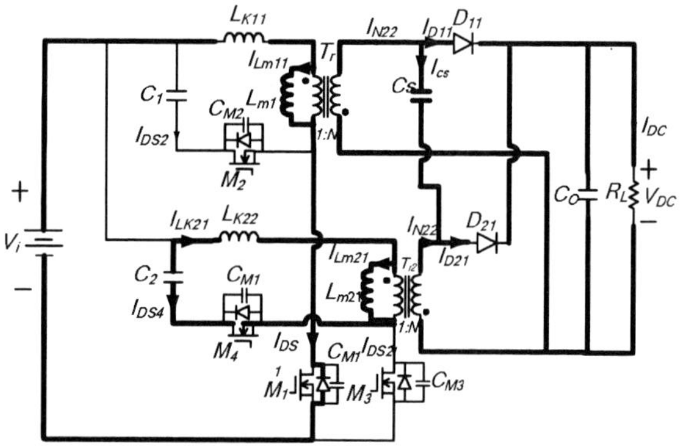

Mode 1 ($t_0 \le t < t_1$)
(a)

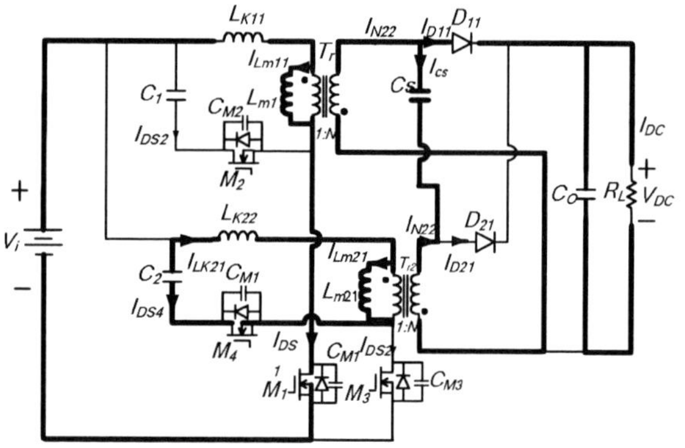

Mode 2 ($t_1 \le t < t_2$)
(b)

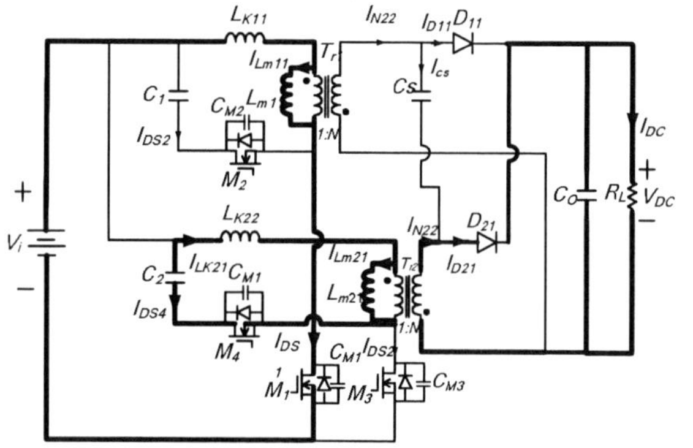

Mode 3 ($t_2 \le t < t_3$)
(c)

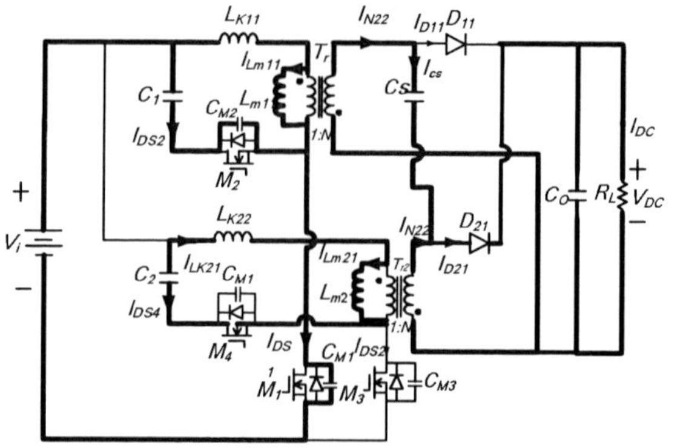

Mode 4 ($t_3 \le t < t_4$)
(d)

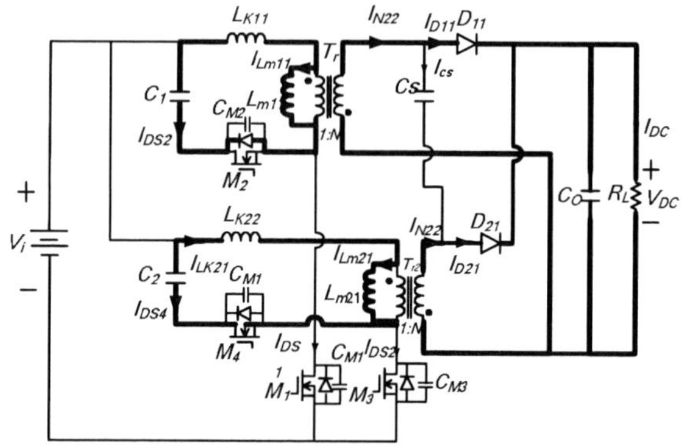

Mode 5 ($t_4 \le t < t_5$)
(e)

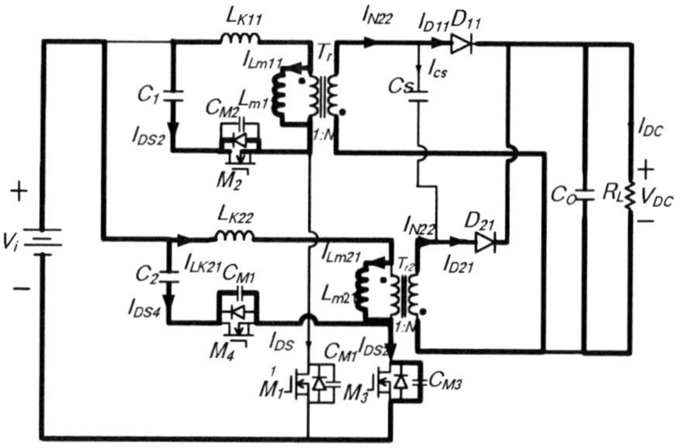

Mode 6 ($t_5 \le t < t_6$)
(f)

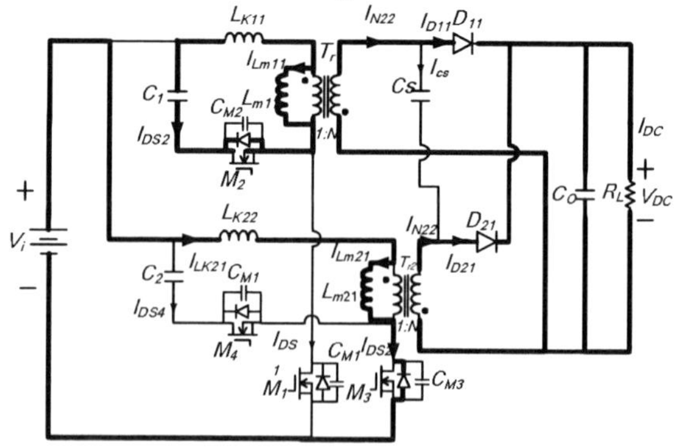

Mode 17 ($t_6 \le t < t_7$)
(g)

Fig. 9. Operational modes of the proposed converter during half one switch cycle.

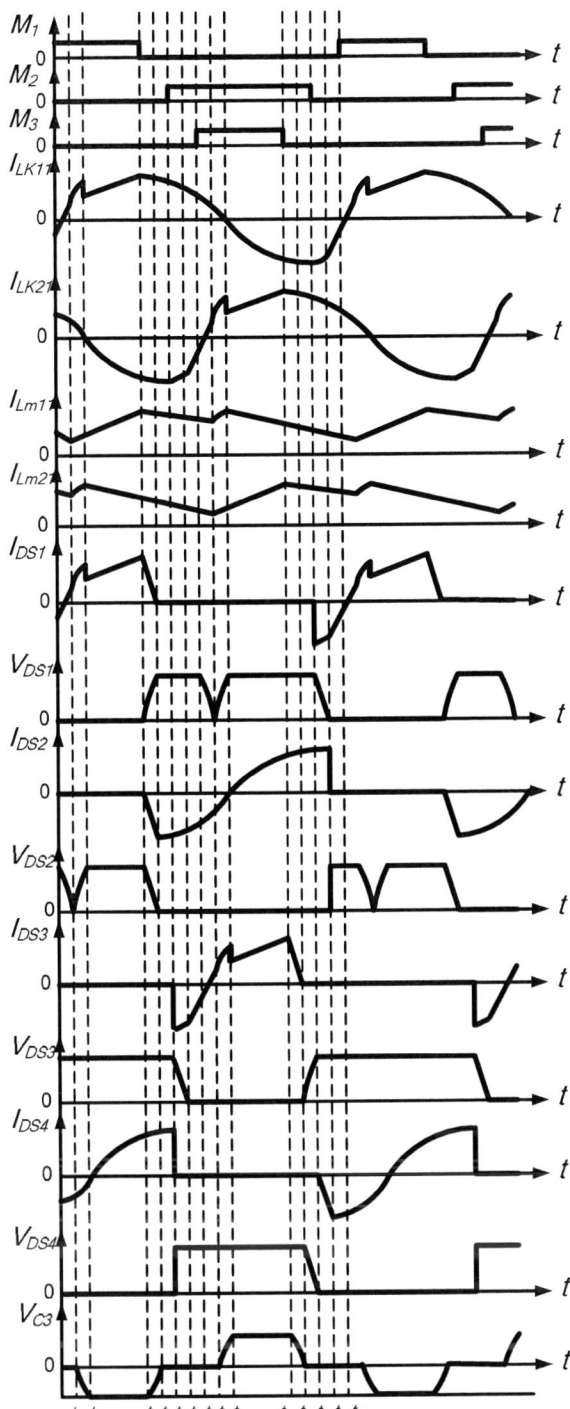

Fig. 10. key waveforms of the proposed active clamp flyback converter with single-capacitor sunbber over one switching cycle.

IV. DESIGN OF THE PROPOSED CONVERTER

The proposed converter is composed of an active clamp flyback converter and a single-capacitor snubber. To design the proposed converter systematically, determination of duty ratio D, transformer T_r, active clamp capacitor C_1 (or C_2), snubber capacitor C_S and output capacitor C_O are presented.

A. Duty Ratio D

To determine duty ratio, it needs to first obtain input to output voltage transfer ratio M. Since the active clamp circuit and a single-capacitor circuit only help switch M_1 or M_2 to achieve soft-switching feature, they do not affect transfer ratio M of the proposed flyback converter. That is, transfer ratio M will be the same as the conventional one. According to volt-second balance principle, the following equation can be obtained:

$$V_i D T_S + \left(-\frac{V_{DC}}{N} \right)(1-D)T_s = 0, \qquad (1)$$

where N is the turns ratio of transformer T_{r1} (or T_{r2}) and V_{DS} is the output dc-link voltage. From (1), it can be found that transfer ratio M can be expressed as

$$M = \frac{V_{DC}}{V_i} = \frac{ND}{1-D}. \qquad (2)$$

According to (2), once transfer ratio M and turns ratio N are specified, duty ratio D can be determined. Additionally, based on the operational condition of the single-capacitor snubber, duty ratio of switch M_1 or M_2 is limited within 0.5, and switches M_1 and M_2 are operated in complementary. To determine a proper duty ratio, the relationship between duty ratio D and component stress in the proposed converter must be considered. According to (2), a larger duty ratio D corresponds to a smaller transformer turns ratio N, which results in a lower current stress imposed on switches M_1 and M_2, as well as lower voltage stress on freewheeling diodes D_{11} and D_{21}. However, in order to accommodate variations of load, line voltage, component value, and a limited and loss duty, it had better select an operating range between $D = 0.3 \sim 0.35$.

B. Transformer Tr

Once the duty is selected, the turns ratio of transformer T_r can be determined from (2), which yields

$$N = \frac{(1-D)V_{DC}}{DV_i}. \qquad (3)$$

In the interleaved flyback converter, design values of transformer T_{r1} is the same as those of transformer T_{r2}. By applying the Faraday's law, the number N_{11} of turns at the primary winding can be determined as

$$N_{11} = \frac{DV_i T_S}{A_C \Delta B}, \qquad (4)$$

where A_C is the effectively cross-section area of the transformer core and ΔB is the working flux density. According to (3) and (4), N_{12} can be therefore determined. Similarly, primary winding N_{21} and secondary winding N_{22} of transformer T_{r2} are also determined by (3) and (4).

For the flyback converter, magnetizing inductor L_{m11} of transformer T_{r1} is determined by taking into account the current down slope, which corresponds to the off-time of switch M_1, and the inductance must be large enough to maintain continuous mode (CCM) operation. The inductance of L_{m11} must satisfy the following inequality:

$$L_{m11} \geq \frac{V_{DC}(1-D)T_S}{N^2 \Delta I_{D11(\max)}}, \qquad (5)$$

where $\Delta I_{DII(max)}$ is the maximum ripple of the secondary winding current of transformer T_{rI}, and it is equal to $\Delta I_{Lm11(max)}/N$. When the maximum current ripple is specified, the minimum magnetizing inductance can be determined. Similarly, inductor L_{m21} are also determined by (5).

C. Active Clamp Capacitor C_1 and C_2

The active clamp capacitor C_1 (or C_2) is used to achieve soft-switching feature. To achieve a ZVS feature, the energy stored in inductor L_{K11} (or L_{K21}) must satisfy the following inequality:

$$\frac{1}{2}L_{K11}(I_{LK(t12)} - I_{LK(t13)})^2 \geq \frac{1}{2}(C_{M1} + C_{M2})V^2_{DS(max)} \quad (6)$$

where $L_{K11(tv12)}$ is the leakage inductor current at time t_{12}, $I_{LK(tv13)}$ is that at time t_{13}, C_{M1} and C_{M2} are respectively the junction capacitors of switches M_1 and M_2, and $V_{DS(max)}$ is the voltage across switch M_1 and its value is equal to $(V_i + V_{DC}/N)$. Once C_{M1}、C_{M2}、$I_{LK11(tv12)}$ and $I_{LK11(tv13)}$ are specified, leakage inductor L_{K11} can be determined as

$$L_{K11} \geq \frac{(C_{M1} + C_{M2})V^2_{DS1(max)}}{(I_{LK11(tv12)} - I_{LK21(tv13)})}. \quad (7)$$

To achieve ZVS feature using active clamp circuit, one half of the resonant period formed by L_{K11} and C_1 should be equal to or greater than the maximum off time of switch M_1. Thus, capacitor C_1 must satisfy the following inequality:

$$\pi\sqrt{L_{K11}C_1} \geq t_{off} = (1-D)T_S. \quad (8)$$

From (8), when L_{k11} is specified, the capacitance range of the clamp capacitor C_1 can be determined as

$$C_1 \geq \frac{(1-D)^2 T_S^2}{\pi^2 L_{K11}}. \quad (9)$$

Similarly, inductor L_{K21} and capacitor C_2 are also determined by (7) and (9), respectively.

D. Snubber capacitor C_S

In the proposed converter, capacitor C_S resonates with inductor L_{m11} or L_{m21} to smooth out switch turn-off transition. The energy stored in C_S can be determined as

$$W_{CS} = \frac{1}{2}C_S(NV_i + V_{DC})^2 \quad (10)$$

To completely eliminate the switch turn-off loss, the energy stored in capacitor C_S must be at least equal to the turn-off w_{soff}, which is expressed by

$$W_{soff} = \frac{t_{soff}}{2}(V_i + V_{DC}/N)I_{DP}, \quad (11)$$

where t_{soff} is the turn-off fall time of switch M_1 and I_{DP} is the current passing the switch. Therefore, capacitor C_S can be determined as

$$C_S \geq \frac{t_{soff}I_{DP}}{N(NV_i + V_{DC})} \quad (12)$$

E. Output capacitor C_O

The output capacitor C_O is primarily designed for reducing ripple voltage. The ripple voltage across output capacitor C_O is determined as follows:

$$\begin{aligned}
\Delta V_{rco} &= \frac{\Delta Q_{CO}}{C_O} \\
&= \frac{1}{C_O}(I_{O(max)} \times \frac{DT_S}{2}) \\
&= \frac{DI_{O(max)}T_S}{2C_O},
\end{aligned} \quad (13)$$

where $I_{O(max)}$ is the maximum output current.

V. Measure Results

To verify the performance of the proposed intelligent stunner, as shown in Fig. 8, a prototype with the following specifications was implemented.

A. Active Clamp Flyback Converter
- input voltage V_i: 150 V_{DC},
- switching frequency f_{S1}: 50 kHz,
- output voltage V_{DC}: 120 V_{DC}, and
- maximum output current I_{DC}: 2 A.

B. Full-Bridge Inverter
- input voltage V_{DC}: 120 V_{DC},
- maximum output power: 240 W,
- output voltage V_O: 肇120 V,
- maximum output current I_O: 肇2 A, and
- switching frequency f_{S2}: 400 kHz.

According to the specifications, components of the active clamp flyback converter associated with turn-off snubber are determined as follows:

- turns ratio of transformers T_{r1} and T_{r2}: 1,
- magnetizing inductors L_{m11} and L_{m21}: 627 μH
- transformer core: EI-42,
- leakage inductors L_{K11} and L_{K21}: 7.32 μH,
- switches M_1 and M_2: IRF840, and
- diodes D_{11} and D_{21}: UF304.

To generate ac voltage waveforms, the switches of the full-bridge inverter are also determined as $S_1 \sim S_4$: IRF840.

Measured waveforms of V_{DS} and I_{DS} of switch M_1 is shown in Fig. 11 Fig. 11(a) shows those waveforms of the interleaved active-clamp flyback converter, while Fig. 11(b) shows those waveforms of the proposed converter with single-capacitor snubber. From Fig. 11, it can be seen that the proposed one can achieve ZVS feature at turn-on transition and ZVT feature at turn-on transition. Fig. 12 shows measured waveforms of V_{DS} of switch M_2, illustration that switch M_2 can be operated with ZVS at turn-on transition. Efficiency comparison between the proposed converter and conventional interleaved one with hard-switching circuit is depicted in Fig. 13. From Fig. 13, it can be seen that the proposed converter can yield higher efficiency over the conventional one, and its efficiency is 92% under full load condition. In addition, the efficiency of the overall stunner system is 85% under full load condition. Fig.14 shows measured waveforms of DC-link voltage V_{DC} and output voltage V_O of the stunner system under a load of ±2 A, in which the waveforms are with frequency of 400 Hz and ratio of 50%. From Fig. 14, it can be observed that the output voltage has been regulate within 1%, improving meat quality significantly. Measured waveforms of output voltage V_O and output current I_O during goose stunning interval are shown in Fig. 15, from which it can be found that the poultry stunner with the proposed stunner system is first controlled with the regulation voltage manner. Then, when output current I_O is greater that a set value of 350 mA, the proposed one will enter the regulation current manner to sustain constant output current. Form Fig. 15, it can be seen that the poultry stunner with the proposed stunner system can reduce poultry stress and increase meat quality. Additionally, from practical experimental results for stunning a goose, it can be observed that the coma time of goose under voltage of 120 V, current of 350 mA, and stunning time of 6s can sustain about 30s, and it is long enough for bleeding.

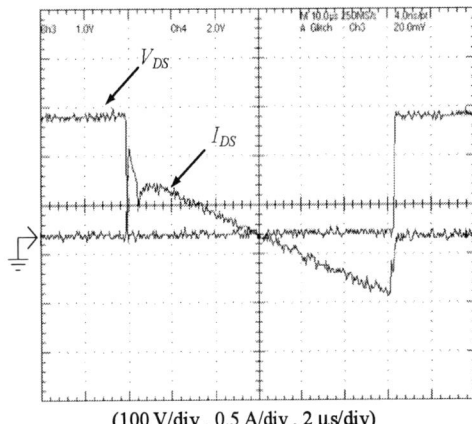

(100 V/div , 0.5 A/div , 2 μs/div)

Fig. 12. Measured voltage V_{DS} and current I_{DS} waveforms of switch M_2, under 50 % of full load in the proposed converter, illustrating ZVS at turn on.

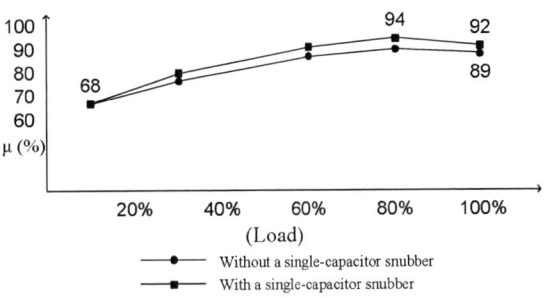

Fig. 13. Comparison conversion efficiency between the active clamp interleaved flyback converter without a single-capacitor snubber and the proposed one.

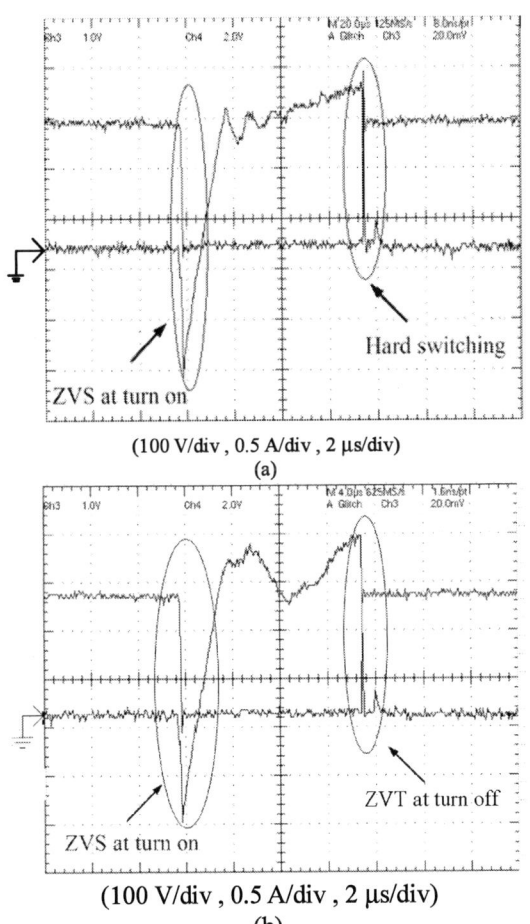

(100 V/div , 0.5 A/div , 2 μs/div)

(a)

(100 V/div , 0.5 A/div , 2 μs/div)

(b)

Fig. 11. Measured voltage V_{DS} and current I_{DS} waveforms of switch M_1 of (a) the interleaved active-clamp flyback converter and (b) the proposed converter with single-capacitor snubber, illustrating ZVS at turn on and ZVT at turn off under 50% of the full load.

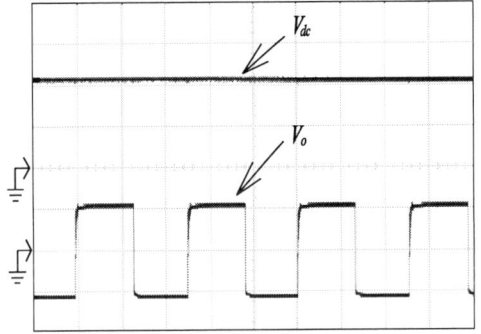

(V_{DC} : 50 V/div, V_o : 100 V/div, 1 ms/div)

Fig. 14. Measured waveforms of dc output voltage V_{DC} and output voltage V_o under a load of ±2 A.

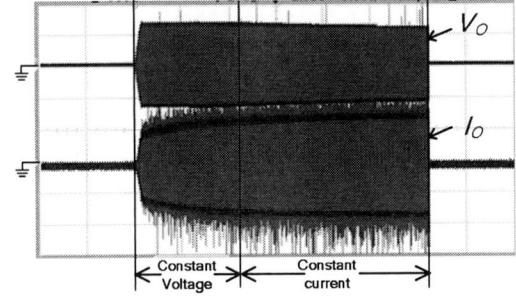

(V_O : 50 V/div, I_o : 50 mA/div, 1 s/div)

Fig. 15. Measured waveforms of output voltage V_o and output current I_o during chicken stunning interval.

VI. CONCLUSION

In this paper, coma mechanism of poultry with electrical stunning has been briefly reviewed. Operational principle, steady-state analysis and design of the proposed active clamp converter with a single-capacitor snubber associated with full-bridge inverter has been implemented to generate stunning electrical parameters, in which its current amplitude is 350 mA and its voltage amplitude is 120 V. In addition, the proposed interleaved active-clamp flyback converter with the single-capacitor snubber can achieve the efficiency around 92 % under full load condition, and the efficiency of overall stunner system is about is about 85 %. From experimental results, it can also found that proposed one can attain a good meat quality, which meet the regulation of animal welfare.

REFERENCE

[1] H. A. Channon, A. M. Payne and R. D. Warner, "Comparison of CO2 Stunninng with Manual Electrical Stunning (50Hz) of Pig on Carcass and Meat Quality," *Trans. on Meat Science*, 2002, pp.63–68.

[2] S. B. Wotton and M. O. Callaghan, "Electrical Stunning of Pigs: the Effect of Applied Voltage on Impedance to Current Flow and the Operation of a Fail-Safe Device," *Trans. on Meat Science*, 2002, pp. 203–208.

[3] A. Velarde, *et al.*, "Effect of Electrical Stunning on Meat and Carcass Quality in Lambs," *Trans. on Meat Science*, 2003, pp. 35–38.

[4] H. A. Channon, A. M. Payne and R. D. Warner, "Effect of Stun Duration and Current Level Applied During Head to Back and Head only Electrical Stunning of Pigs on Pork Quality Compared with Pigs Stunned with CO2, " *Trans. on Meat Science*, 2001, pp. 1325–1333.

[5] E. Lambooij, *et al.*, "Some neural and behavioural aspects of electrical and mechanical stunning in ostriches," *Trans. on Meat Science*, 1999, pp. 339–345.

[6] E. Lambooij, *et al* l., "The Effects of Captive Bolt and Electrical Stunnin, and Restraining Methods on Broiler Meat Quality," *Trans. on Poultry Scien ce*, 1999, pp. 600–607.

[7] B. Savenije, *et al.*, "Electrical Stunning and Exsanguination Decrease the Extracellular Volume in the Broiler Brain as Studied with Brain Impedance Recordings," *Trans. on Poultry Science*, 2000, pp. 1062–1066.

[8] V. Sante, *et al.*, "Effect of Stunning Current Frequency on Carcass Downgrading and Meat Quality of Turkey," *Trans. on Poultry Science*, 2000, pp. 1208–1214.

[9] S.F. Bilgili, "Recent Advances in Electrical Stunning," *Trans. on Poultry Science*, 1999, pp. 282–286.

[10] W. D. McNeal, *et al.*, "Effects of Stunning and Decapitation on Broiler Activity During Bleeding, Blood Loss, Carcass, and Breast Meat Quality," *Trans. on Poultry Science*, 2003, pp. 163–168.

[11] R. Watson, F. C. Lee and G. C. Hua, "Utilization of an Active-clamp Circuit to Achieve Soft Switching in Flyback Converters," *IEEE Trans. on Power Electronics*, Vol. 11, Jan. 1996, pp.162 – 169.

[12] J. Du and H. Ohsaki, "Numerical analysis of eddy current in the EMS-maglev system." Proc. of 6th Int. Conf. on Electrical Machines and Systems (ICEMS 2003), Beijing (China), Nov. 2003, pp.761-764.

[13] M. R. Prausnitz, "The Effects of Electric Current Applied to Skin: A Review for Transdermal Drug Delivery," *Proceedings of the Advanced Drug Delivery Reviews*, Vol. 18, 1996, pp. 395-425.

[14] U. Pliquett, R. Langer and J. C. Weaver, "Changes in the Passive Electrical Properties of Human Stratum Corneum Due to Electroporation,"Trans. On Biophysica Acta, 1995, pp.111–121.

[15] M. R. Prausnitz, "A Practical Assessment of Transdermal Drug Delivery by Skin Electroporation," Proceedings of the Advanced Drug Delivery Reviews, Vol. 35, 1999, pp. 61–7.

Novel Current Feedforward Average Current Mode Control Technique to Improve Output Dynamic Performance of DC-DC Converters

P. Chrin* and C. Bunlaksananusorn*

* Faculty Engineering, King Mongkut's Institute of Technology Ladkrabang (KMITL), Bangkok 10520, Thailand

Abstract–This paper proposes a novel Current Feedforward Average Current Mode Control (CFACMC) technique to improve output dynamic performance of DC-DC converters. In this new control scheme, besides its usual role as a feedback variable to a current controller, the sensed inductor current is fed forward, through a low-pass filter circuit, to sum with an output signal from a voltage controller to form the control signal. In the paper, the operating principle of CFACMC and its small signal model are described. From the small-signal model, the expression for selecting the gain of the low-pass filter to yield the improved output voltage response is derived. Both simulated and experimental results are provided to show that significant improvement in output dynamic response is achieved with CFACMC.

Index Terms—Average current mode control, DC-DC converters.

I. INTRODUCTION

Average Current Mode Control (ACMC) is a control technique commonly used in modern DC-DC to regulate the output voltage. The control structure comprises of an output voltage and inductor current loops. In ACMC, a current amplifier is employed in the current loop to force the inductor current to closely track the reference current, instead of directly comparing the two quantities using a comparator as in Peak Current Mode Control (PCMC). Because of this, ACMC has a better noise immunity, higher current loop gain, no stability problem when the duty cycle is above 0.5. These appealing features have contributed to its popularity in today's DC-DC converters. In [1-3], operating principle, modeling and control design of ACMC have been described. The small-signal model of ACMC [2,3], together with the small-signal model of the converter's power stage [4], form a complete converter system, from which transfer functions for the control design can be derived. Based on these transfer functions, the controllers in the voltage and current loops are designed so that the converter exhibits the desired output performance, i.e. having good output regulation and fast dynamic response. However, even with the well designed controllers, the converter's performance still may not meet the stringent requirements of modern electronic loads, which demands the tightly regulated voltage and wide current variations from the

converter. To enhance the performance of the conventional ACMC, the novel three-controller ACMC has recently been proposed [5]. The authors modified the voltage loop to include an auxiliary controller. The transfer function of the auxiliary controller was carefully selected and designed to increase the system robustness against variations in the input voltage, load current and *L-C* output filter. It was demonstrated experimentally that the three controller technique yields a faster output voltage response than the conventional ACMC.

This paper presents a novel Current Feedforward Average Current Mode (CFACMC) technique to improve output dynamic performance of DC-DC converters. The sensed inductor current signal, which is readily available in the conventional ACMC, is used as a feedforward signal to enhance a corrective action of the voltage loop. The feedforward path consists only of a low-pass filter circuit. Hence, the implementation of CFACMC is simple, which requires just a minor modification to the standard ACMC circuit.

II. CURRENT FEEDFORWARD AVERAGE CURRENT MODE CONTROL

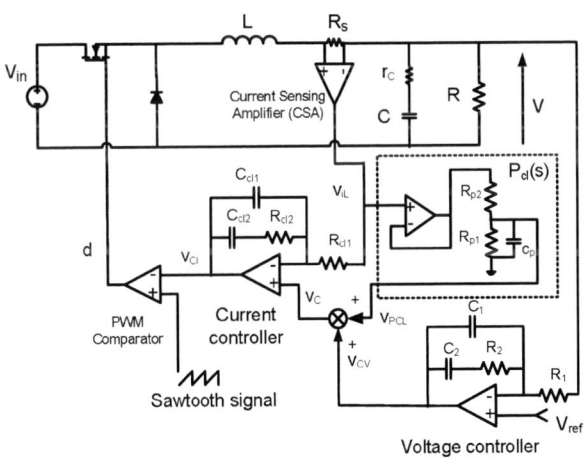

Fig. 1. Buck converter with CFACMC

A buck converter with CFACMC is shown in Fig. 1. To implement CFACMC, a low-pass filter circuit, $P_{cl}(s)$, is added to the conventional ACMC circuit to provide a feedforward path to the voltage loop. The inductor current is sensed by a resistor R_S and amplified by a current sensing amplifier (CSA) to give the sensed

---

This work is supported by JICA under the ANU/SEED-Net Project.

978-1-4244-0644-9/07/$25.00 ©2007 IEEE   1575

current signal V_{iL}. Besides being fed back to a current controller, V_{iL} is also fed forward through $P_{cl}(s)$ to sum with the output signal from a voltage controller, V_{CV}, to form the control signal, V_C. The difference between V_C, which represents the desired current, and V_{iL}, which represents the actual inductor current, is amplified by the current controller. The resulting current error signal, V_{CI}, is then compared with the sawtooth voltage, V_p, to generate the duty cycle signal, d, to drive the MOSFET.

The proposed CFACMC can improve the output voltage response of the converter as follows. Assume that the output voltage drop is occurred due to a step load change, the voltage controller will respond by increasing V_{CV}. The increase in V_{CV} will trigger the increase in V_C, V_{CI}, d, i_L, V_{iL}, and V_{PCL}, respectively. With the contribution from V_{PCL}, V_C, which is equal to $V_{CV} + V_{PCL}$, will increase more than that in the conventional ACMC, which relies on V_{CV} alone to act. In effect, CFACMC produces the stronger corrective action than the conventional ACMC, and enables the possibility to further improve the converter's dynamic response. To achieve the improved performance, the amplitude and phase of the feedforward signal, V_{PCL}, must be appropriate as the wrong signal can lead to a poor response or even instability. The amplitude and phase of V_{PCL} are dependent on the low-pass filter $P_{cl}(s)$. Therefore, the gain and cut-off frequency of $P_{cl}(s)$ must be chosen such that the improved dynamic response of the converter is attained, while their impacts on system stability are kept to a minimum.

III. MODELING OF BUCK CONVERTER WITH CFACMC

A. Modeling of Power Stage

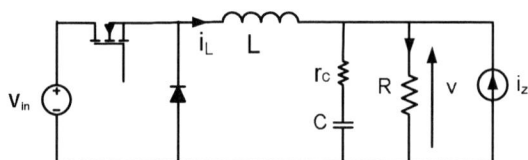

Fig. 2. Buck DC-DC Converter

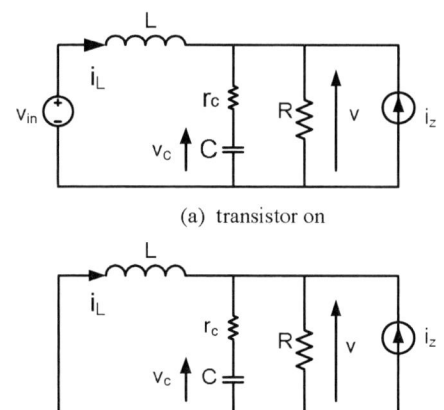

(a) transistor on

(b) transistor off

Fig. 3. Circuit configurations of the CCM buck converter in one switching period

A buck converter's power stage is shown in Fig. 2. In the figure, r_c is the equivalent series resistance (ESR) of the capacitor C, R is the standing load, and the current source i_z models the load current. In Continuous Conduction Mode (CCM) where the inductor current flows continuously over one switching period, the converter exhibits two circuit states as shown in Fig. 3. During the MOSFET turn-on interval (Fig. 3(a)) the inductor is charged and its current increases linearly. During the MOSFET turn-off interval (Fig. 3(b)), the inductor is discharged and its current decreases linearly.

From Fig. 3, the state-space equation for each circuit configuration can be written as given by (1) and (2).

$$\begin{cases} \begin{bmatrix} \dot{\mathbf{i}}_L \\ \dot{\mathbf{v}}_c \end{bmatrix} = \begin{bmatrix} \dfrac{-R.r_c}{L(R+r_e)} & \dfrac{-R}{L(R+r_e)} \\ \dfrac{R}{C(R+r_e)} & \dfrac{-1}{C(R+r_e)} \end{bmatrix} \begin{bmatrix} \mathbf{i}_L \\ \mathbf{v}_c \end{bmatrix} + \begin{bmatrix} \dfrac{1}{L} & \dfrac{-R.r_c}{L(R+r_e)} \\ 0 & \dfrac{R}{C(R+r_e)} \end{bmatrix} \begin{bmatrix} \mathbf{v}_{in} \\ \mathbf{i}_z \end{bmatrix} \\ \mathbf{v} = \begin{bmatrix} \dfrac{R.r_c}{R+r_e} & \dfrac{R}{R+r_e} \end{bmatrix} \begin{bmatrix} \mathbf{i}_L \\ \mathbf{v}_c \end{bmatrix} + \begin{bmatrix} 0 & \dfrac{R.r_c}{R+r_e} \end{bmatrix} \begin{bmatrix} \mathbf{v}_{in} \\ \mathbf{i}_z \end{bmatrix} \end{cases} \quad (1)$$

$$\begin{cases} \begin{bmatrix} \dot{\mathbf{i}}_L \\ \dot{\mathbf{v}}_c \end{bmatrix} = \begin{bmatrix} \dfrac{-R.r_c}{L(R+r_e)} & \dfrac{-R}{L(R+r_e)} \\ \dfrac{R}{C(R+r_e)} & \dfrac{-1}{C(R+r_e)} \end{bmatrix} \begin{bmatrix} \mathbf{i}_L \\ \mathbf{v}_c \end{bmatrix} + \begin{bmatrix} 0 & \dfrac{-R.r_c}{L(R+r_e)} \\ 0 & \dfrac{R}{C(R+r_e)} \end{bmatrix} \begin{bmatrix} \mathbf{v}_{in} \\ \mathbf{i}_z \end{bmatrix} \\ \mathbf{v} = \begin{bmatrix} \dfrac{R.r_e}{R+r_e} & \dfrac{R}{R+r_e} \end{bmatrix} \begin{bmatrix} \mathbf{i}_L \\ \mathbf{v}_c \end{bmatrix} + \begin{bmatrix} 0 & \dfrac{R.r_e}{R+r_e} \end{bmatrix} \begin{bmatrix} \mathbf{v}_{in} \\ \mathbf{i}_z \end{bmatrix} \end{cases} \quad (2)$$

The averaged state-space equation of the buck converter is obtained by weight average of (1) and (2) in accordance with [4], which gives:

$$\begin{cases} \begin{bmatrix} \dot{\mathbf{i}}_L \\ \dot{\mathbf{v}}_c \end{bmatrix} = \begin{bmatrix} \dfrac{-R.r_e}{L(R+r_e)} & \dfrac{-R}{L(R+r_e)} \\ \dfrac{R}{C(R+r_e)} & \dfrac{-1}{C(R+r_e)} \end{bmatrix} \begin{bmatrix} \mathbf{i}_L \\ \mathbf{v}_c \end{bmatrix} + \begin{bmatrix} \dfrac{d}{L} & \dfrac{-R.r_e}{L(R+r_e)} \\ 0 & \dfrac{R}{C(R+r_e)} \end{bmatrix} \begin{bmatrix} \mathbf{v}_{in} \\ \mathbf{i}_z \end{bmatrix} \\ \mathbf{v} = \begin{bmatrix} \dfrac{R.r_e}{R+r_e} & \dfrac{R}{R+r_e} \end{bmatrix} \begin{bmatrix} \mathbf{i}_L \\ \mathbf{v}_c \end{bmatrix} + \begin{bmatrix} 0 & \dfrac{R.r_e}{R+r_e} \end{bmatrix} \begin{bmatrix} \mathbf{v}_{in} \\ \mathbf{i}_z \end{bmatrix} \end{cases} \quad (3)$$

Linearization of (3) yields the small-signal state-space equation in (4).

$$\begin{cases} \begin{bmatrix} \dot{\hat{\mathbf{i}}}_L \\ \dot{\hat{\mathbf{v}}}_c \end{bmatrix} = \begin{bmatrix} \dfrac{-R.r_e}{L(R+r_e)} & \dfrac{-R}{L(R+r_e)} \\ \dfrac{R}{C(R+r_e)} & \dfrac{-1}{C(R+r_e)} \end{bmatrix} \begin{bmatrix} \hat{\mathbf{i}}_L \\ \hat{\mathbf{v}}_c \end{bmatrix} + \begin{bmatrix} \dfrac{D}{L} & \dfrac{-R.r_e}{L(R+r_e)} & \dfrac{V_{in}}{L} \\ 0 & \dfrac{R}{C(R+r_e)} & 0 \end{bmatrix} \begin{bmatrix} \hat{\mathbf{v}}_{in} \\ \hat{\mathbf{i}}_z \\ \hat{\mathbf{d}} \end{bmatrix} \\ \hat{\mathbf{v}} = \begin{bmatrix} r_e & 1 \end{bmatrix} \begin{bmatrix} \hat{\mathbf{i}}_L \\ \hat{\mathbf{v}}_c \end{bmatrix} + r_e\, \hat{\mathbf{i}}_z \end{cases} \quad (4)$$

By applying the Laplace transform to (4), the converter's transfer functions can be found in matrix form:

$$\begin{bmatrix} \hat{\mathbf{i}}_L(s) \\ \hat{\mathbf{v}}(s) \end{bmatrix} = \begin{bmatrix} G_{vi}(s) & G_{zi}(s) & G_{di}(s) \\ G_{vv}(s) & G_{zv}(s) & G_{dv}(s) \end{bmatrix} \begin{bmatrix} \hat{\mathbf{v}}_{in}(s) \\ \hat{\mathbf{i}}_z(s) \\ \hat{\mathbf{d}}(s) \end{bmatrix} \quad (5)$$

Since only the duty cycle-to-inductor current and the duty cycle-to-output voltage transfer functions, $G_{di}(s)$ and

1576

$G_{dv}(s)$, are required in the control design, they are determined as expressed by (6) and (7) respectively.

$$G_{di}(s) = \frac{V_{in}}{LCR} \frac{1+s(R+r_c)C}{s^2+s(\frac{1}{RC}+\frac{r_c}{L})+\frac{1}{LC}} \quad (6)$$

$$G_{dv}(s) = \frac{V_{in}}{LC} \frac{1+sr_cC}{s^2+s(\frac{1}{RC}+\frac{r_c}{L})+\frac{1}{LC}} \quad (7)$$

B. Modeling of Control Stage

Transfer functions of the current controller, $G_{cl}(s)$, and voltage controller, $G_c(s)$, in Fig. 1 are expressed by (8) and (9), respectively.

$$G_{cl}(s) = \frac{w_{cl}}{s} \frac{1+s/w_{clz}}{1+s/w_{clp}} \quad (8)$$

where $w_{cl}=1/R_{cl1}(C_{cl1}+C_{cl2})$, $w_{clp}=(C_{cl1}+C_{cl2})/R_{cl2}C_{cl1}C_{cl2}$ and $w_{clz}=1/R_{cl2}C_{cl2}$.

$$G_c(s) = \frac{w_c}{s} \frac{1+s/w_{cz}}{1+s/w_{cp}} \quad (9)$$

where $w_c=1/R_1(C_1+C_2)$, $w_{cz}=1/R_2C_2$, and $w_{cp}=(C_1+C_2)/R_2C_1C_2$.

A transfer function of the PWM comparator is well known and given by:

$$F_m = 1/V_p \quad (10)$$

where V_p is peak to peak sawtooth voltage.

The inductor current is sensed by R_S and then multiplied by A_{cl}, which is the gain of the current sensing amplifier (CSA). Hence, the sensed inductor current signal is:

$$V_{iL} = R_i i_{iL} \quad (11)$$

where $R_i=A_{cl}.R_S$

A transfer function of $P_{cl}(s)$ is given by:

$$P_{cl}(s) = \frac{k_p}{1+s/w_p} \quad (12)$$

where $k_p=R_{p1}/(R_{p1}+R_{p2})$ and $w_p=(R_{p1}+R_{p2})/R_{p1}R_{p2}C_p$.

C. Small-signal Model of Buck Converter with CFACMC

The small-signal model of the buck converter with CFACMC in Fig. 1 is shown in Fig. 4(a). It is manipulated into Fig. 4(b) which is then simplified into Fig. 4(c). The transfer functions $H_{cl}(s)$, $K_{cl}(s)$ and $G_{rb}(s)$ are given by:

$$H_{cl}(s) = \frac{F_m(G_{cl}(s)+1)}{1+F_mR_iG_{cl}(s)G_{di}(s)} \quad (13)$$

$$K_{cl}(s) = R_i P_{cl}(s) G_{di}(s) \quad (14)$$

$$G_{rb}(s) = \frac{H_{cl}(s)}{1-K_{cl}(s)H_{cl}(s)} \quad (15)$$

If $P_{cl}(s) = 0$, then $K_{cl}(s) = 0$, $G_{rb}(s) = H_{cl}(s)$, and the small-signal model in Fig. 4 will beome the small-signal model of the conventional ACMC [3]. It can be observed from (15) that CFACMC has altered the transfer function of the conventional ACMC by a factor of $1/(1-K_{cl}(s)H_{cl}(s))$. Note that from (14) $K_{cl}(s)$ is dependent on $P_{cl}(s)$. At low frequencies, $K_{cl}(s)$ should be high so that the product $K_{cl}(s)H_{cl}(s)$ approaches one. As a result, $G_{rb}(s)$ in (15) is increased as well as the loop gain $G_c(s)G_{rb}(s)G_{dv}(s)$ in Fig. 4(c). The increased loop gain will lead to the improved dynamic response of the converter. At high frequencies, $K_{cl}(s)$ should be low so that the product $K_{cl}(s)H_{cl}(s)$ approaches zero. As a result, $G_{rb}(s)$ in (15) is converged to $H_{cl}(s)$. Therefore, if the cut-off frequency of $P_{cl}(s)$ is set equal to the crossover frequency of the voltage loop gain of the conventional ACMC, $G_c(s)H_{cl}(s)G_{dv}(s)$, then $G_{rb}(s)$ will be increased only up to the crossover frequency and converged to $H_{cl}(s)$ above the crossover frequency. In this way, the inclusion of the circuit block $P_{cl}(s)$ in Fig. 1 will only serve to enhance the converter's dynamic response and not affect the system stability.

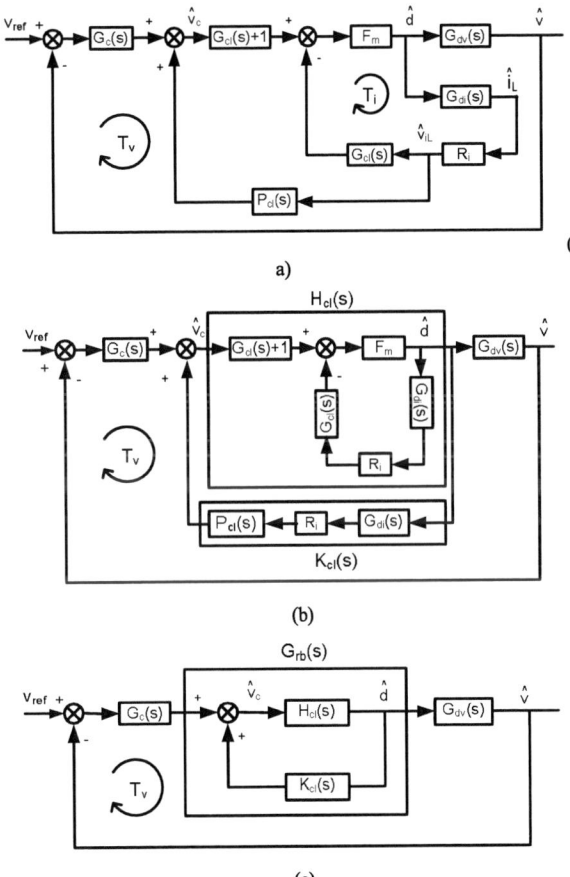

Fig. 4. Small-signal model of CFACMC

At DC frequency ($s\approx0$), $H_{cl}(0)$, $K_{cl}(0)$ and $G_{rb}(0)$ can be expressed as:

$$H_{cl}(0) = \frac{R}{V_{in} R_i} \tag{16}$$

$$K_{cl}(0) = k_p \frac{V_{in} R_i}{R} \tag{17}$$

$$G_{rb}(0) = \frac{H_{cl}(0)}{1 - k_p} \tag{18}$$

Assume that the input voltage V_{in} is constant. From (16) $H_{cl}(0)$ is proportional to the load resistor R and from (18) $G_{rb}(0)$ is proportional to $H_{cl}(0)$. Thus, $G_{rb}(0)$ is proportional to R. Fig. 5 shows the plot of $G_{rb}(0)$ in (18) as a function of k_p at the two load resistor values, R_{max} and R_{min}. When the load current is changed from minimum to maximum (i.e. the load resistor changed from R_{max} to R_{min}), $G_{rb}(0)$ will be reduced (Fig. 5). To prevent the effect of load change on $G_{rb}(0)$, from (18) k_p has to be increased, in this case, by the amount of k_1, which can be found by solving

$$G_{rb}(0)\Big|_{R_{min}} = \frac{H_{cl}(0)\Big|_{R_{min}}}{1 - k_1} = H_{cl}(0)\Big|_{R_{max}} \tag{19}$$

From (19), the gain of the low-pass filter can be found:

$$k_p = k_1 = \frac{R_{max} - R_{min}}{R_{max}} \tag{20}$$

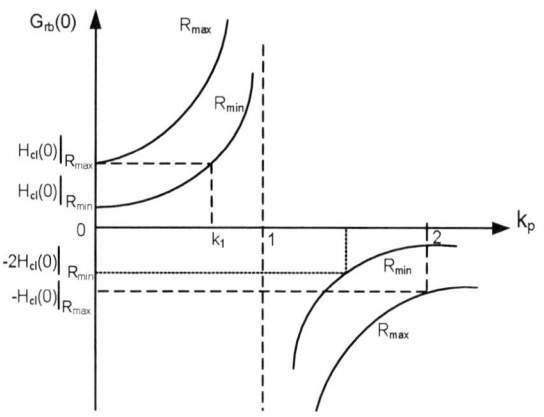

Fig. 5. Graph of $G_{rb}(0)$ versus k_p

IV. DESIGN EXAMPLE

TABLE I
PARAMETERS OF PROTOTYPE BUCK CONVERTER

V_{in}	V	L	C	r_C	R
5V	2V	45.2uH	1230uF	0.015Ω	0.4-2Ω

Circuit parameters of a buck converter are listed in Table I. The peak-to-peak voltage of the sawtooth signal is V_p=1.8V. The value of a sensing resistor is R_S=0.01Ω and the current sensing amplifier's gain is A_{cI}=7.5. The switching frequency is f_s=100 kHz.

From the parameters listed in Table I, the transfer

functions in (6), (7), (10) and (11) can be calculated:

$$G_{di}(s) = 4.5 \times 10^7 \frac{2.5 \times 10^{-3} s + 1}{s^2 + 738.6s + 1.8 \times 10^7} \tag{21}$$

$$G_{dv}(s) = 8.99 \times 10^7 \frac{1.84 \times 10^{-5} s + 1}{s^2 + 738.6s + 1.8 \times 10^7} \tag{22}$$

$$F_m = 1/V_p = 0.55 \tag{23}$$

$$R_i = R_S A_{cI} = 0.075 \tag{24}$$

A. Frequency Responses

The current controller and voltage controller are designed in the same way as do in the conventional ACMC. Hence, the design procedure outlined in [2,3] can be adopted. For the current controller design, the crossover frequency of the current loop is selected at 10kHz. The first pole of $G_{cI}(s)$ is at the origin, the zero is placed at $w_{clz} = 4.482 \times 10^3$rad/sec, and the second pole is placed at one-third of the switching frequency or w_{clp}= 188×10^3rad/sec. The design yields C_{clI}=500pF, C_{cl2}=22nF R_{clI}=560Ω, and R_{cl2}=10kΩ. Substitution of these component values into (8) results in

$$G_{cl}(s) = \frac{7.93 \times 10^4}{s} \frac{1 + 2.2 \times 10^{-4} s}{1 + 4.89 \times 10^{-6} s} \tag{25}$$

For the voltage controller design, the crossover frequency of the voltage controller is selected at 5kHz. The first pole of $G_c(s)$ is at the origin, the zero is place at w_{cz}=4.3$\times$10³rad/sec, and the second pole is placed at one-third of the switching frequency or w_{cp}=188$\times$10³rad/sec. The design yields C_1=500pF, C_2=22nF, R_1=3.9kΩ, and R_2=10kΩ. Substitution of these component values into (9) results in

$$G_c(s) = \frac{1.12 \times 10^4}{s} \frac{1 + 2.2 \times 10^{-4} s}{1 + 4.89 \times 10^{-6} s} \tag{26}$$

As shown in the Table I, the maximum and minimum load resistances are 2Ω and 0.4Ω, respectively. From (20), k_p can be calculated:

$$k_p = 0.8 \tag{27}$$

The cut-off frequency of $P_{cl}(s)$ is selected to be equal to the crossover frequency of the voltage loop gain of the conventional ACMC, which was set at 5kHz. Thus, the cut-off frequency of $P_{cl}(s)$ is w_p=31.4$\times$10³rad/s. From the values of k_p and w_p, the low-pass filter's component are calculated, yielding R_{p1}=4kΩ, R_{p2}=1kΩ, and C_p=38nF. Substitution of these component values into (12) results in

$$P_{cl}(s) = \frac{0.8}{1 + 3.04 \times 10^{-5} s} \tag{28}$$

The open loop transfer function of the CFACMC in Fig. 4(c) is given by:

1578

$$T_v(s) = G_c(s)G_{rb}(s)G_{dv}(s) \qquad (29)$$

Bode plot of (29) is shown by the solid lines in Fig. 6. In Fig 4(b), if $P_{cl}(s) = 0$, the model will become that of conventional ACMC, whose open loop transfer function is given by:

$$T_v(s) = G_c(s)H_{cl}(s)G_{dv}(s) \qquad (30)$$

Bode plot of (30) is shown by the dashed lines in Fig. 6. It can be seen that the loop gain of CFACMC is greater than the loop gain of the conventional ACMC up to the crossover frequency. Above the crossover frequency, the two plots are converged. This result confirms that the designed $P_{cl}(s)$ only serves to improve the dynamic response of the converter, while keeping its impact on system stability at the minimum.

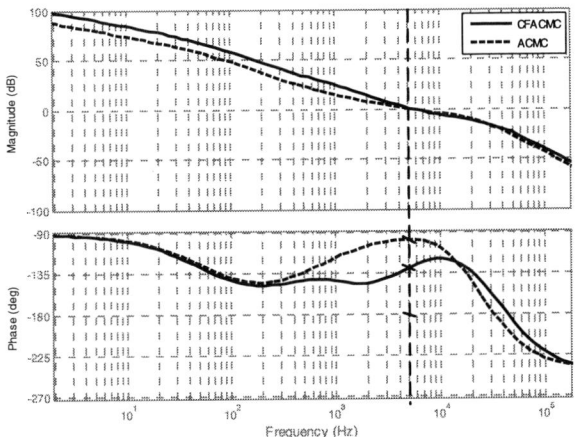

Fig. 6. Open loop frequency responses of CFACMC versus ACMC

B. Time Response

The prototype converter circuit was built in order to verify the performance of CFACMC against the conventional ACMC. In the prototype circuit board, selection between the conventional ACMC (i.e. $P_{cl}(s) = 0$) and CFACMC was made by a selecting switch. The converter was subjected to a step load current change and an output voltage response of the converter recorded. Figs. 7 and 8 show the simulated and measured output voltage responses of the converter with the conventional ACMC and CFACMC respectively. In Fig. 7, the maximum voltage drop/raise is about 65mV and the settling time is about 750μs. In Fig. 8, they are about 60mV and 200μs respectively. It can be seen that while the maximum voltage drop/raise given by the two control methods is roughly the same, the settling time given by CFACMC is faster than the conventional ACMC by 450μs. This represents a significant improvement in the dynamic response time of the converter. Fig. 9 compares the results in Fig. 7 and 8 on the same scale, where the improvement yielded by CFACMC is evident.

In Figs. 7-9, the simulated results are seen to be in good agreement with their experimental counterparts. These simulated results were produced by the SIMULINK models developed in [6]. Good correspondence between the simulated and experimental

results indicates the accuracy and validity of the models in predicting the converter's dynamic response.

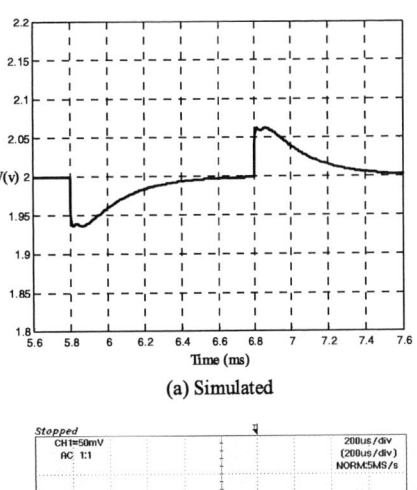

(a) Simulated

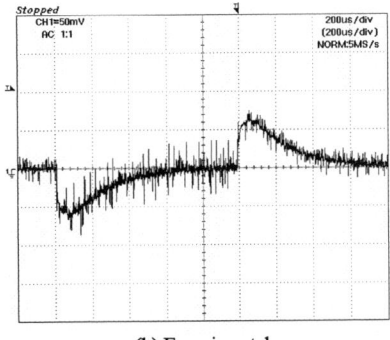

(b) Experimental

Fig. 7. Output voltage response of the buck converter with ACMC as the load current is stepped back and forth between 1A and 4A.

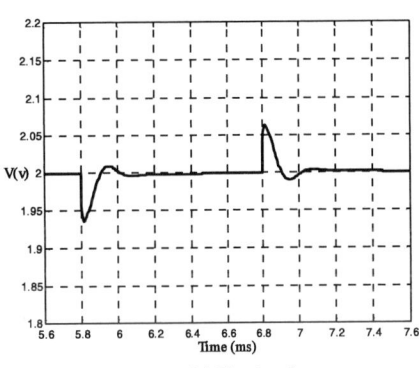

(a) Simulated

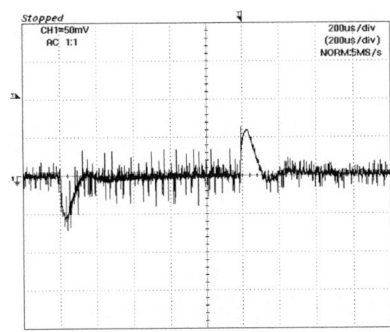

(b) Experimental

Fig. 8. Output voltage response of the buck converter with CFACMC subjected to the same loading condition as in Fig. 7

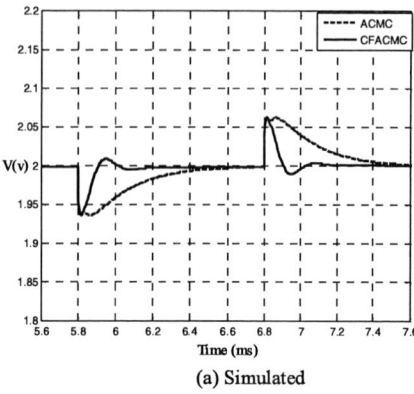

(a) Simulated

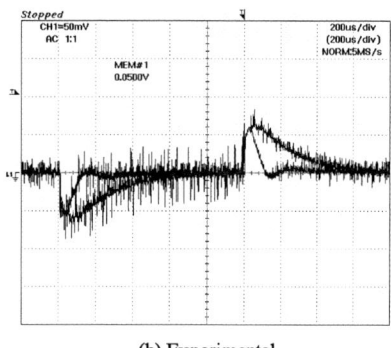

(b) Experimental

Fig. 9. Comparison between the output voltage responses of the buck converter with ACMC and CFACMC

V. CONCLUSIONS

The novel Current Feedforward Average Current Mode (CFACMC) technique to improve output dynamic performance of DC-DC converters has been proposed in this paper. It is based on feedforward of the sensed inductor current, through the low-pass filter circuit, $P_{cl}(s)$, to sum with the output signal from the voltage controller to produce the control signal. The feedforward signal helps strengthen the corrective action of the voltage loop, hence making it possible for the converter's dynamic response to be further improved from the conventional ACMC. To achieve the improved output dynamic performance, the gain and cut-off frequency of $P_{cl}(s)$ must be appropriately chosen. The equation for selecting the low-pass filter gain was derived in (20) and the low pass filter's cut-off frequency must be set equal to the crossover frequency of the voltage loop of the conventional ACMC. The capability of CFACMC in enhancing the converter's output dynamic performance was confirmed by the simulated and experimental results. Due to its simplicity and capability to significantly improve the converter's output dynamic response, the proposed CFACMC technique has a good potential for practical usage.

REFERENCES

[1] L. H. Dixon, "Average current mode control of switching power supplies," *Unitrode power supply design seminar*, SEM-700, 1990.

[2] J. Sun and R. Bass, "Modeling and practical design issues for average current mode control," *IEEE Applied Power Electronic Conference*, pp. 980-986, 1999.

[3] P. Cooke, "Modeling average current mode control," *IEEE Applied Power Electronics Conference and Exposition*, pp. 256-262, 2000.

[4] R. D. Middlebrook and Slobodan Cuk, "A general unified approach to modelling switching-converter power stages," *IEEE Power Electronics Specialists Conference*, Record, pp. 36-57, 1976.

[5] G. Garcera, E. Figueres, and A. Mocholi, "Novel three-controller average current mode control for DC-DC PWM converters with improved robustness and dynamic response," *IEEE transactions on power electronics*, Vol. 15, No. 3, pp. 516-528, 2000.

[6] P. Chrin and C. Bunlaksananusorn, "Large-signal average modeling and simulation of DC-DC converters with SIMULINK," *Power Conversion Conference*, pp. 27-32, 2007.

Stability Analysis of Cascaded DC-DC Power Electronic System

M. Veerachary, S. Bala Sudhakar

Department of Electrical Engineering
Indian Institute of Technology Delhi
New Delhi, India

Abstract-- **Stability analysis of the cascaded dc-dc power electronic system is analyzed in this paper. For demonstration boost converter supplying the hybrid switched capacitor converter considered as an example. The boost converter is acting as the bus converter, 42 V bus, while the switched capacitor converter is the point of load converter. The two converters are provided with voltage-mode and peak-current-mode controllers. Converter two-port network models are developed and then stability of the cascaded system has been analyzed. Cascaded system interaction effects, (i) source converter power handling capability with switching load, and (ii) load converter interfacing capability with bus converter, are analyzed. Simulation and experimental results are provided for verification purpose.**

Index Terms— **Boost converter, Two-port network model, Peak current-mode control, Switched capacitor converter.**

I. INTRODUCTION

The power demand in telecom and automotive applications is continuously increasing day-by-day. In order to cater these increased power demand the recent dc power supply distribution system (DCDS) is becoming a complex network and ensuring the stability of such system is an important task of the power supply designer. The DCDS typically consists of several smaller subsystems, and each subsystem is provided with its own controller. The subsystems are then integrated to form a complete DCDS [1]. The most important issue involved in the system integration is the stability and interactions among the individual subsystems. Although the individual subsystems are stable in stand-alone mode, but the integrated system may become unstable. These stability issues are mainly due to (i) lack enough power supplying capability of the bus converter at a predefined bus voltage, (ii) interface mismatch of load converter with its bus converter, (iii) reflected interactions of downstream converters, etc. To avoid all these problems, involved in the DCDS, the design engineer is required to verify the impedance criterion. Although this verification is not the problematic in the DCDS, but identifying their induced interactions between them is a complex task. The easiest method to predetermine the induced interactions of each sub-system is by using two-port modeling [ ]

methodology. In this paper a non-isolated cascaded DCDS is considered for stability analysis.

II. STABILITY ANALYSIS OF CASCADED DC DISTRIBUTED POWER ELECTRONIC SYSTEM

Fig. 1 shows a typical two-stage distributed power supply (DPS) consisting of a front-end converter, source subsystem, and several other downstream load converters in parallel, load subsystem. One possible design procedure for the DCDS, Fig. 1, is to design the line conditioner separately using an ac un-terminated modeling approach. Then, based on the output impedance characteristics of the line conditioner, a specification can be set for the input impedance of the load subsystem to ensure the system stability. One sufficient condition to guarantee the system stability is to force the magnitude of the input impedance of the load subsystem to be greater than the magnitude of the output impedance of the source subsystem.

A. Two Port Network Modeling of Cascaded System

The dynamics of a regulated converter is represented using of a set of transfer functions known as the G-parameters [2]-[3] given by Eqn.(1) for open-loop system and Eqn. (2) for a closed-loop system. G parameters based analysis has advantageous as it is easy to measure these parameters and almost all electrical circuits can be replaced with G-parameter based equivalent circuits. In this modeling the input port of the network is Norton's equivalent circuit, while the output port is Thevenin's equivalent circuit. In terms of matrix notation the input port is defined in the first row, while second row defines the output port. For analysis of the dc-dc converters, used in DCDS, a set of transfer functions, called un-terminated transfer functions, are formulated by eliminating source and load. The reason for this assumption is that, in DCDS the nature of load and source is not know explicitly just like in stand-alone converters. From the two-port network model we can easily define the following matrix relationships.

$$
\begin{bmatrix} \hat{i}_g \\ \hat{v}_o \end{bmatrix} = \begin{bmatrix} Y_{in-os} & T_{oi-os} & G_{ci-s} \\ G_{io-os} & -Z_{o-os} & G_{co-s} \end{bmatrix} \begin{bmatrix} \hat{v}_g \\ \hat{i}_o \\ \hat{d} \end{bmatrix} \quad (1)
$$

This work was supported by MHRD, Govt. of India under the R & D Project: Design and Development of Fuzzy Controllers for High Frequency DC-DC Conversion Systems.

$$\begin{bmatrix} \hat{i}_g \\ \hat{v}_o \end{bmatrix} = \begin{bmatrix} Y_N & T_{oi-cs} \\ G_{io-cs} & -Z_T \end{bmatrix} \begin{bmatrix} \hat{v}_g \\ \hat{i}_o \end{bmatrix} \quad (2)$$

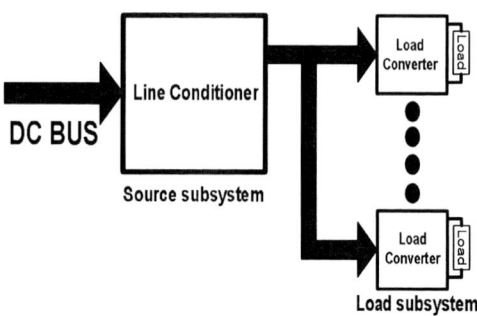

Fig. 1a. Block diagram of the two-stage distributed power supply system.

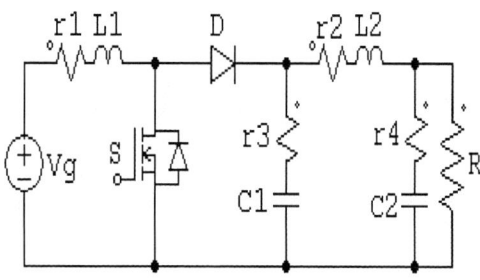

Fig. 1b. Boost Converter with output Filter (source sub-system)

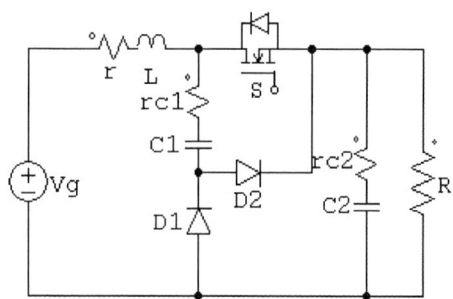

Fig. 1c. Hybrid switched capacitor converter (load sub-systems)

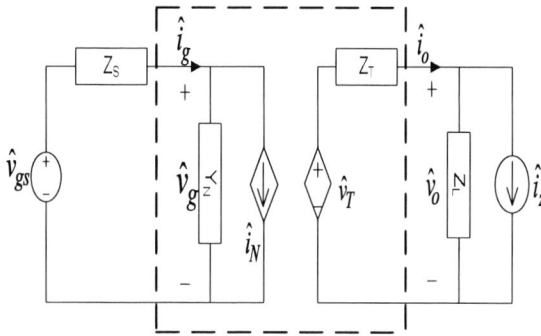

Fig. 2. The two-port network model of the converter.

B. Load interactions

By including the actual load the load induced effects can be found, from Fig. 2, by computing load terminal quantities, $\hat{v}_o$ and $\hat{i}_o$, and then substituting in Eqns. 1 and 2. The resulting model equations are defined by (3) and (4).

$$\begin{bmatrix} \hat{i}_g \\ \hat{v}_o \end{bmatrix} = \begin{bmatrix} Y_{in-os} + \dfrac{G_{io-os}T_{oi-os}}{Z_L + Z_{o-os}} & \dfrac{Z_L T_{oi-os}}{Z_L + Z_{o-os}} & G_{oi-s} + \dfrac{G_{oo-s}T_{oi-os}}{Z_L + Z_{o-os}} \\[3ex] \dfrac{Z_L G_{io-os}}{Z_L + Z_{o-os}} & \dfrac{Z_L Z_{o-os}}{Z_L + Z_{o-os}} & \dfrac{Z_L G_{oo-s}}{Z_L + Z_{o-os}} \end{bmatrix} \begin{bmatrix} \hat{v}_g \\ \hat{i}_z \\ \hat{d} \end{bmatrix} \quad (3)$$

$$\begin{bmatrix} \hat{i}_g \\ \hat{v}_o \end{bmatrix} = \begin{bmatrix} Y_{in-cs} + \dfrac{G_{io-cs}T_{oi-cs}}{Z_L + Z_{o-cs}} & \dfrac{Z_L T_{oi-cs}}{Z_L + Z_{o-cs}} \\[3ex] \dfrac{Z_L G_{io-cs}}{Z_L + Z_{o-cs}} & -\dfrac{Z_L Z_{o-cs}}{Z_L + Z_{o-cs}} \end{bmatrix} \begin{bmatrix} \hat{v}_g \\ \hat{i}_z \end{bmatrix} \quad (4)$$

C. Supply interactions

By including the actual source subsystem the source induced effects can be found, from Fig. 2, by computing source terminal quantities, $\hat{v}_g$ and $\hat{i}_g$, and then substituting in Eqns. 1 and 2. The resulting model equations are defined by (5) and (6).

$$\begin{bmatrix} \hat{i}_g \\ \hat{v}_o \end{bmatrix} = \begin{bmatrix} \dfrac{Y_{in-d}}{1+Z_S Y_{in-d}} & \dfrac{T_{oi-d}}{1+Z_S Y_{in-d}} & \dfrac{G_{oi-l}}{1+Z_S Y_{in-d}} \\[3ex] \dfrac{G_{io-d}}{1+Z_S Y_{in-d}} & \dfrac{1+Z_S Y_{in-sc}}{1+Z_S Y_{in-d}}Z_{o-d} & \dfrac{1+Z_S Y_{in-\infty}}{1+Z_S Y_{in-d}}G_{oo-l} \end{bmatrix} \begin{bmatrix} \hat{v}_{gs} \\ \hat{i}_o \\ \hat{d} \end{bmatrix} \quad (5)$$

$$\begin{bmatrix} \hat{i}_g \\ \hat{v}_o \end{bmatrix} = \begin{bmatrix} \dfrac{Y_{in-cl}}{1+Z_S Y_{in-cl}} & \dfrac{T_{oi-cl}}{1+Z_S Y_{in-cl}} \\[3ex] \dfrac{G_{io-cl}}{1+Z_S Y_{in-cl}} & -\dfrac{1+Z_S Y_{in-sc}}{1+Z_S Y_{in-cl}}Z_{o-cl} \end{bmatrix} \begin{bmatrix} \hat{v}_{gs} \\ \hat{i}_o \end{bmatrix} \quad (6)$$

$$Y_{in-\infty} = Y_{in-ol} - \dfrac{G_{io-ol}G_{ci-l}}{G_{co-l}} \quad (7)$$

$$Y_{in-sc} = Y_{in-ol} + \dfrac{G_{io-ol}T_{oi-ol}}{Z_{o-ol}} \quad (8)$$

where $Y_{in-\infty}$: ideal or infinite-bandwidth input admittance, Y_{in-sc} : short-circuit input admittance

III. CRITERION FOR SYSTEM STABILITY

Fig. 3 shows two dc-dc converter subsystems connected in series. Let the source subsystem has an input-to-output transfer function of F_S, and the load subsystem has an input-to-output transfer function of F_L

1582

then the overall input-to-output transfer function of the cascaded system is given by

$$F_{SL} = \frac{V_o}{V_g} = \frac{F_S F_L}{1 + T_m} ; \qquad (9)$$

where $T_m = Z_o / Z_{in}$; Z_o is the output impedance of the source subsystem, and Z_{in} is the input impedance of the load subsystem. The impedance ratio at the interface port, defined as $T_m = Z_o / Z_{in}$, can be considered as the loop gain of the integrated system and it can be used to determine the stability as well as the loading effects.

If $|Z_{in}| \gg |Z_o|$ for all frequencies, then the loading effect is negligible and the system stability will depend only upon the stability of the individual subsystems. However, in many cascaded systems, it is often impossible for the system to have $|Z_{in}| \gg |Z_o|$ at all interconnections, while still meeting all other system specifications. When $|Z_o|$ is larger than $|Z_{in}|$, a considerable loading effect exists. However, a loading effect does not necessarily imply a stability problem.

IV. SIMULATION AND EXPERIMENTAL RESULTS

To demonstrate the proposed modeling and stability analysis, a two stage distributed power supply system, cascading of two converters, is considered here. For the source subsystem boost converter with output filter has been used while hybrid switched capacitor converter is used for load subsystem. Here, the source subsystem is controlled with voltage-mode control technique, while the load subsystem is controlled with peak current mode control technique. State-space models for the individual models have been derived and then input and output impedances have been determined, converter parameters are listed in Table 1, using MATLAB. The controller parameters for each individual subsystem are designed by using MATLAB program. Type-III compensator has been used with voltage-mode control technique and Type-II compensator has been used for with peak-current mode controller. The final design parameters of the controllers for this operating condition are given in Table 1. Various frequency response characteristics are plotted in Figs. 4 and 5.

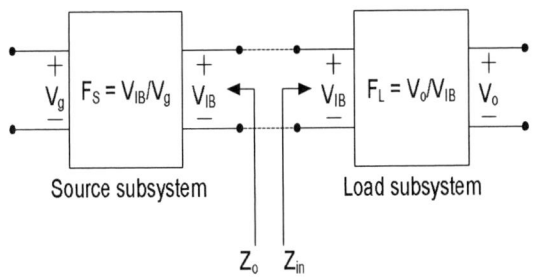

Fig. 3. Block diagram of the two stage cascaded DPS.

TABLE I
CONVERTER PARAMETERS

Hybrid switched capacitor converter parameters	
Power stage parameters	Compensator parameters
L=400 μH	R₁=10 kΩ
C₁=10 μF	R₂=5.3 kΩ
C₂=220 μF	CC₁=100 pF
Rₗ=100 Ω	CC₂=50 pF
Boost converter with output filter parameters	
Power stage parameters	Compensator parameters
L1=100 μH	R₁=10 kΩ
L2=25 μH	R₂=18.6 kΩ
C₁=20 μF	R₂=1.8 kΩ
C₂=220 μF	CC₁=30 nF
Rₗ=20 Ω	CC₂=5 nF
	CC₂=10 nF

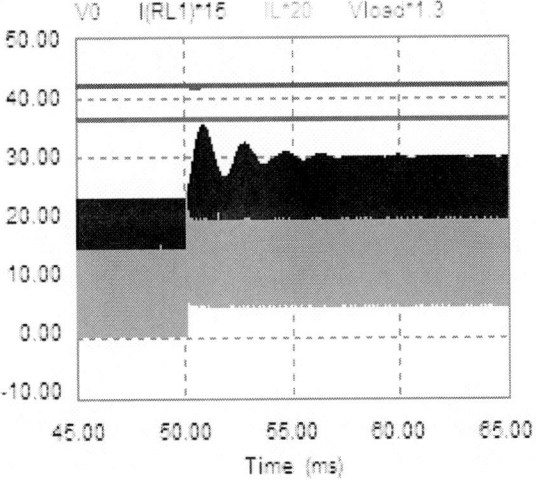

(a)   Load disturbance on the downstream converter.

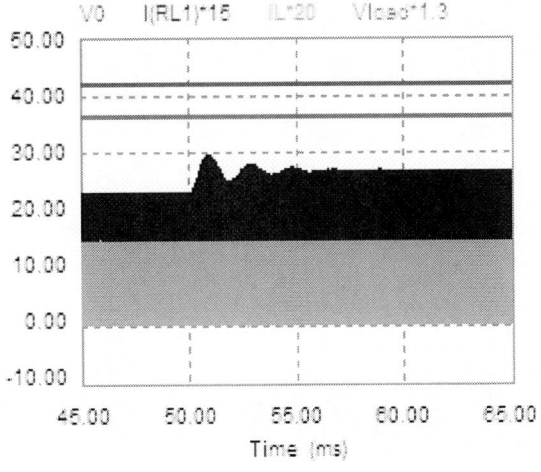

(b) Load disturbance at the intermediate stage.

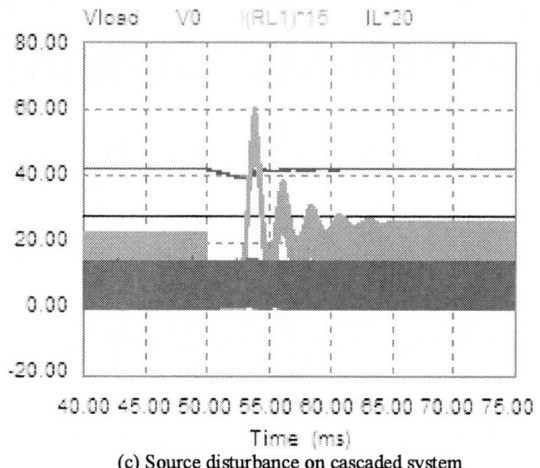

(c) Source disturbance on cascaded system

Fig. 6. Simulated dynamic responses of the cascaded DCDS.

For the case where $|Z_o|$ exceeds $|Z_{in}|$, for some frequencies, further analysis is needed to determine the system exact stability information. Consider an example case of two impedances overlapping is shown in Fig. 5. In the frequency range where $|Z_{in}| < |Z_o|$ the magnitude of loop gain T_m is greater than 0 dB, as shown in Fig. 6, and crosses the 0 dB line at the two frequencies, i.e $|Z_{in}| = |Z_o|$. Since the $|T_m|$ crosses the 0 dB line twice, two different phase margins can be defined, assuming the subsystems are individually stable; one at the initial point of the overlap and the other at the final point of the overlap. From the Fig. 6 we can observe that, in closed loop system the minor loop gain plot is not crossing the 0 db axis. That means the output impedance of the source subsystem is less than the input impedance of the load subsystem and hence the cascaded system is stable.

To verify the theoretical analysis simulation studies have been made using PSIM simulator, parameter values listed in Table 1. The voltage levels used at different stages are: (i) supply voltage to the source subsystem is 24 V, (ii) intermediate bus voltage is 42 V and, (iii) load subsystem voltage is 28 V. DCDS regulation capability is tested for the following cases: (i) supply voltage change 24 → 20 V at the input of the source subsystem, (ii) load change from 50 → 30 Ω, and (iii) at intermediate stage the load is changed from 180 → 120 Ω. The simulation results for the above cases are shown in Fig. 6. Notice that under all these operating conditions/ disturbances voltage levels at various points are regulating and also there are no current-mode induced unstable oscillations. Further, the dynamic response time is also very small.

To verify the theoretical analysis and simulation results an experimental prototype cascaded system, parameter values same as in simulation studies, has been built. MOSFET: IRF540, Diode: MUR820, Driver: IR2110, Optocoupler: 6N136 were used in the prototype model. UC3825 is used to realize the peak current-mode control and TL494 is used for realizing the voltage-mode control. It is observed that by using only the normal boost converter, the cascaded system is operating in unstable mode even through each converter is stable in stand-alone

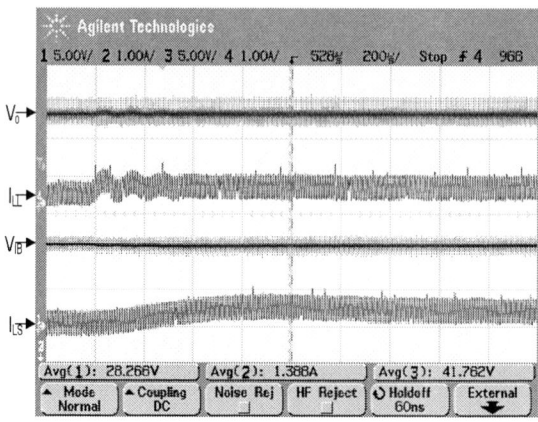

(a)　Load disturbance on the downstream converter

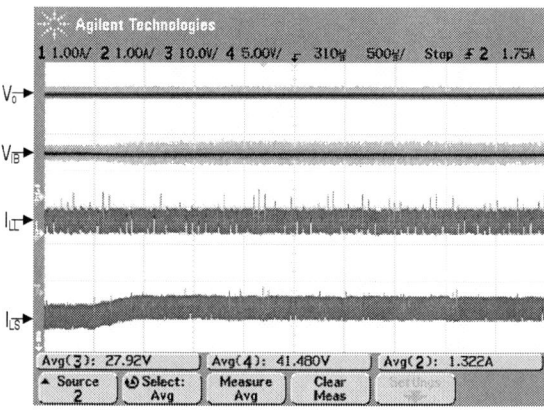

(b)　Load disturbance at the intermediate stage

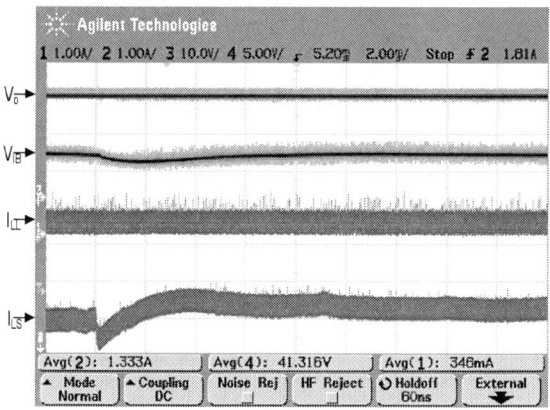

(c) Source disturbance on cascaded system

Fig. 7. Experimental dynamic responses of the cascaded DCDS.

mode. In order to avoid this instability problem a filter has been added in between the two cascaded sub-systems so that the individual sub-systems fulfill the impedance criterion. Experimental investigations have been made for all these cases, as was done in simulations, and these are shown in Fig. 7. These results are closely matching with the simulation results shown in Fig. 6. Slight discrepancy between the simulation and experimental values are attributed to the following factors: (i) voltage drops in the switching devices, (ii) simulation modeling methods, (iii) error's in measuring instruments, etc.

V. Conclusions

Stability analysis of the cascaded dc-dc power electronic system was analyzed in this paper. For demonstration boost converter supplying the hybrid switched capacitor converter considered as an example. The two converters were provided with voltage-mode and peak-current-mode controllers. Converter two-port network models are developed and then system interaction effects, like source and load converter, were analyzed. Simulation and experimental results were in close agreement each other.

Acknowledgement

The authors would like to thank the MHRD, Govt. of India for supporting this research through R & D Project entitled Design and Development of Fuzzy Controllers for High Frequency DC-DC Conversion Systems.

References

[1] Wildrick, C.M, Lee, F.C, Cho, B.H, and Choi, B, "A method of defining the load impedance specification for a stable distributed power system", IEEE Trans. Power Electron., 1995, Vol. 10, pp. 280–285.

[2] Hankaniemi,M, Suntio, T, and Sippola, M, "Load-impedance based interactions in regulated converters", in Proc. 27th IEEE Int. Telecommunications Energy Conf., 2005, pp. 569–573.

[3] Suntio, T., Hankaniemi, M. and Karppanen, M, "Analysing the dynamics of regulated converters", IEE Proc.-Electr. Power Appl., 2006, Vol. 153, pp. 905–910.

[4] Lee, F.C, Barbosa, P, Xu, P., Zhang, J, Yang, B, and Canales, F, "Topologies and design considerations for distributed power system applications", Proc. IEEE, 2001, Vol. 89, pp. 939–950.

[5] Brush, L, "Distributed power architecture demand characteristics", in Proc. IEEE Applied Power Electronics Conf., 2004, pp. 342–345.

[6] Lam, E., Bell, R, and Ashley, D, "Revolutionary advances in distributed power systems", in Proc. IEEE Applied Power Electronics Conf., 2003, pp. 30–36.

[7] Veerachary M, and Singamaneni Bala Sudhakar, "Peak Current-Mode Control of Hybrid Switched Capacitor Converter", International Conference on Power Electronics, Drives and Energy Systems (PEDES), 2006, pp. 1-6.

[8] Suntio, T., "Unified average and small-signal modeling of direct-on time control", IEEE Trans. Ind. Electron., 2006, Vol. 53, pp. 287–295.

[9] Suntio, T, Gadoura, A., and Zenger, K., "Input filter interactions in peak-current-mode controlled buck converter operating in CICM", IEEE Trans. Ind. Electron, 2002, Vol. 49, pp. 76–86.

[10] Li, P, and Lehman, B, "Performance prediction of DC-DC converters with impedances as loads", IEEE Trans. Power Electron., 2004, Vol. 19, pp. 201–209.

[11] Hankaniemi, M, Suntio, T, and Sippola, M, "Characterization of regulated converters to ensure stability and performance", in Proc. 27th IEEE Int. Telecommunications Energy Conf., 2005, pp. 533–538.

Averaged Switch Modeling of DC/DC Converters using New Switch Network

Chien-Min Lee   Yen-Shin Lai, *Senior Member, IEEE*

Center for Power Electronics Technology, National Taipei University of Technology

Abstract - **This paper presents a switch network based upon averaged switch model to represent the power devices including MOSFET and diode to give a unified approach to modeling of various kinds of converters. The non-ideal features of power devices and circuit components, including turn-on resistor of power devices, DC resistor (DCR) of inductor and equivalent series resistor (ESR) of capacitor, are taken into consideration. As compared to previous approach, the presented switching network can be applied to modeling various kinds of DC/DC converters, including buck, boost, buck-boost, Cuk, Zeta and Sepic, by simply changing two parameters. Simulation results derived from the presented methods are consistent with those derived from SPICE software and thereby validating the presented unified modeling method.**

I. INTRODUCTION

It is well known that the circuit model of converters is very essential to simulation and controller design. This becomes more and more relevant recently, especially for the development of digital-controlled DC/DC converters [1-3]. There are several methods for deriving accurate and less calculation model [4-10]. In [4], circuit averaging method is applied to the AC-modeling of PWM converters. Furthermore, state-space averaging technique [5-6] based upon the modern control theory is used to describe the dynamics of converters. The circuit is represented by simultaneous state space equations and the related solutions are derived by the theory of linear algebra. Similar matrix form for buck, boost, buck-boost, Cuk, Zeta and Sepic converters can be derived by two-port method as shown in [7].

Although the dynamics can be derived from the matrix formulation approach, the related physical features of the converters may not be fully explored [8]. Averaged switch model [8-10] has been used for the modeling of PWM converters based upon three kinds of switch networks, including buck switch network, boost switch network and two-switch network. More physical interpretation to the circuit of converters can be given by this approach.

The main theme of this paper is to further simplify the averaged switch modeling using *only one* switch network rather than *three* for DC/DC converters. The presented new switching network based upon averaged switch model is used to represent the power devices including MOSFET and diode. The non-ideal features of power devices and circuit components, including turn-on resistor of power devices, DC resistor of inductor and equivalent series resistor of capacitor, are included.

As compared to previous approach, the presented

switching network can be applied to modeling various kinds of DC/DC converters, including buck, boost, buck-boost, Cuk, Zeta and Sepic. Simulation results derived from the presented methods are consistent with those derived from SPICE software and thereby validating the presented unified modeling method.

II. AVERAGED SWITCHING MODEL

A switching-mode converter can be regarded as a switch network connected with a time-invariant network as shown in Fig. 1 [8]. There are three basic switching networks for the modeling of various kinds of converters. These basic switch networks are: buck, boost and buck-boost switch networks as shown in Fig. 2, Fig. 3 and Fig. 4 respectively. As shown in Fig. 2 to Fig. 4, the power devices, including MOSFET and diode can be replaced by the related switch networks. Moreover, as shown in Fig. 4 (A), there is **_no_** direct connection between the power devices for buck-boost switch network. Based upon this buck-boost switch network, a new switch model is presented to model various kinds of converters by sampling changing two parameters.

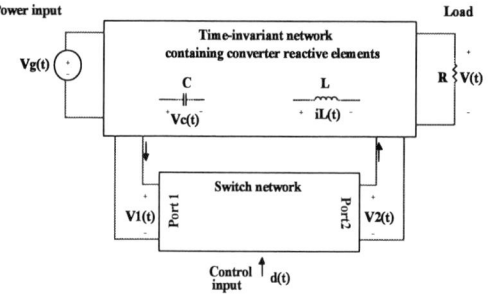

Fig. 1 Equivalent representation of switching-mode converter [8]

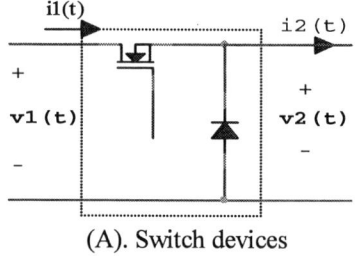

(A). Switch devices

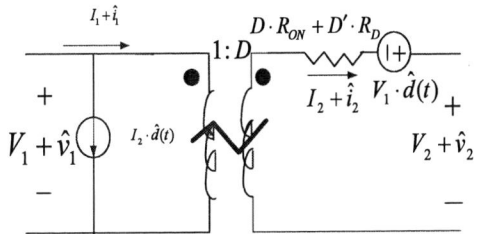

(B). Switch network

Fig. 2 Buck switch network

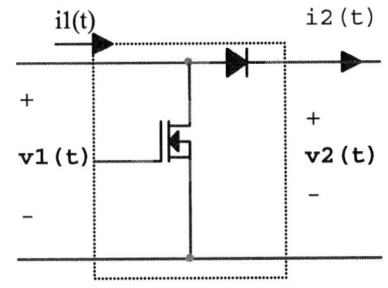

(A). Switch devices

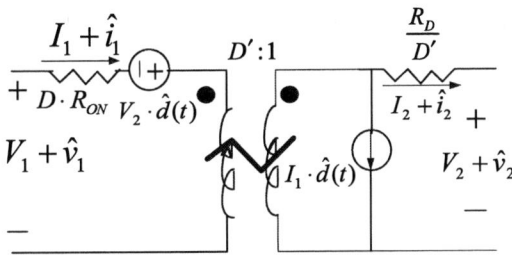

(B). Switch network

Fig. 3 Boost switch network

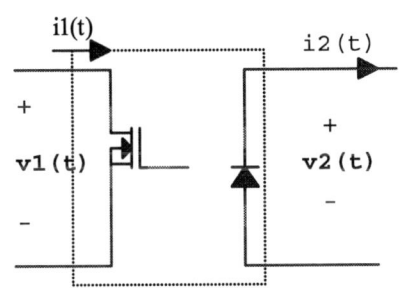

(A). Switch devices

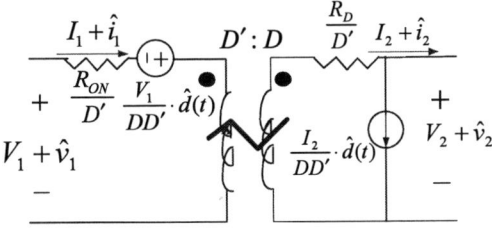

(B). Switch network

Fig. 4 Buck-boost switch network

III. PROPOSED SWITCH NETWORK AND PARAMETERS

The proposed switch network is shown in Fig. 5.

As shown in Fig. 5, the network contains two parameters, K_1 and K_2. We can apply this switch network to modeling various kinds of DC/DC converters, including buck, boost, buck-boost, Cuk, Zeta and Sepic, simply by changing the parameters as summarized in Table. 1. More details of the deduction of this Table will be illustrated in the coming section using buck and boost converters as examples.

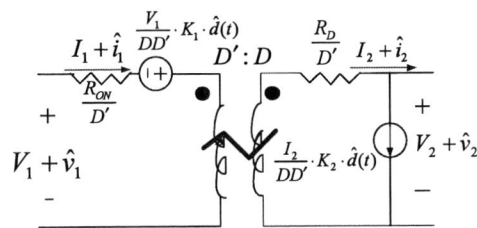

Fig. 5 Proposed switch network, basic type

By Fig. 5 and the relationship:

$$I_2 + \hat{i}_2 \gg \frac{I_2}{DD'} \cdot K_2 \cdot \hat{d}(t) \qquad (1)$$

Fig. 6 can be derived from Fig. 5. This new switch network will be used to derive the unified models of various kinds of converters.

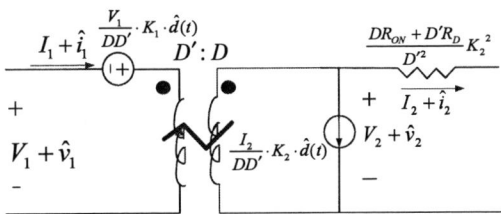

Fig. 6 Proposed switch network

Table 1 Parameters of proposed switch network

Converter Type	K_1	K_2
Buck	1	D'
Boost	D	1
Buck-Boost	1	1
Cuk	1	1
Zeta	1	1
Sepic	1	1

IV. MODELING OF CONVERTERS USING THE PROPOSED SWITCH NETWORK

A. Buck Converter

The model of buck converter based upon conventional averaged switching model is shown in Fig. 7. In contrast, Fig. 8 shows its counter part for the presented switch network. For the buck converter shown in Fig. 7, the following equation can be derived:

$$V_1 = V_g \qquad (2)$$

Similar relationship for Fig. 8 can be determined as:

1587

$$V_1 = D'V_g \qquad (3)$$

By the relationship of turn ratio between Fig. 7 and Fig. 8:

$$\frac{V_1}{D} = \frac{V_1}{DD'}K_1 \qquad (4)$$

Substituting (2) and (3) into (4) yields:

$$\frac{V_g}{D} = \frac{D'V_g}{DD'}K_1 \qquad (5)$$

Therefore,

$$\therefore K_1 = 1 \qquad (6)$$

For the secondary side of transformer shown in Fig. 7, the conversion ratio of dependent current source is D. Similarly, for the secondary side of transformer shown in Fig. 8, the conversion ratio of dependent current source is D. Since the corresponding turn ratio should be the same for these two approaches and therefore:

$$\frac{I_2}{D} = \frac{I_2}{DD'}K_2 \qquad (7)$$

The value of K_2 is:

$$\therefore K_2 = D' \qquad (8)$$

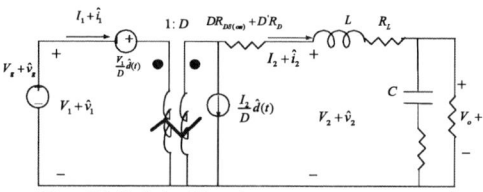

Fig. 7 Model of buck converter, averaged switching model

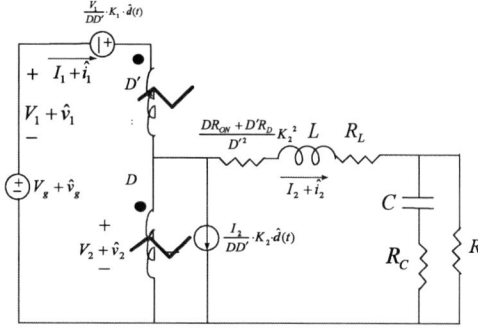

Fig. 8 Model of buck converter, proposed switch network

B. Boost Converter

The model of boost converter based upon conventional averaged switching model is shown in Fig. 9. In contrast, Fig. 10 shows its counter part for the presented switch network. For the boost converter shown in Fig. 9, the following equation can be derived:

$$V_1 = V_g \qquad (9)$$

Similar relationship for Fig. 10 can be determined as:

$$V_1 = V_g \qquad (10)$$

By the relationship of turn ratio between Fig. 9 and Fig. 10:

$$V_2 = \frac{V_1}{DD'}K_1 \qquad (11)$$

Substituting (9) and (10) into (11) yields:

$$\frac{V_2}{V_1} = \frac{V_o}{V_g} = \frac{1}{D'} \qquad (12)$$

Therefore,

$$\therefore K_1 = D \qquad (13)$$

For the secondary side of transformer shown in Fig. 9, the conversion ratio of dependent current source is 1. Similarly, for the secondary side of transformer shown in Fig. 10, the conversion ratio of dependent current source is D. Since the turn ratios should be proportional inversally for these two approaches and therefore:

$$I_1 = \frac{I_2}{DD'}K_2 D \qquad (14)$$

$$\frac{I_2}{I_1} = D' \qquad (15)$$

The value of K_2 is:

$$\therefore K_2 = D' \qquad (16)$$

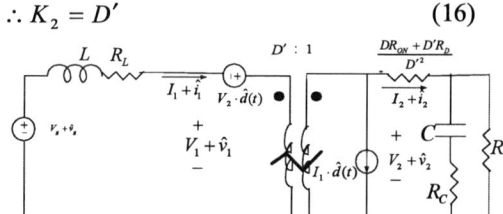

Fig. 9 Model of boost converter, averaged switching model

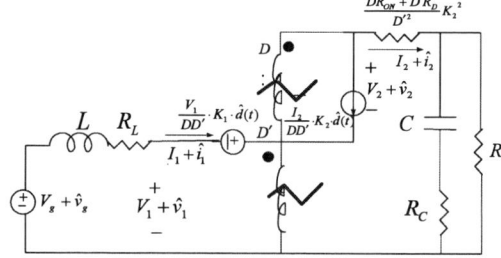

Fig. 10 Model of boost converter, proposed switch network

V. SIMULATION RESULTS

The specifications and parameters for the simulation of the converter are as follows.
V_{IN}: 3.3 V, V_o:8.23 V, I_o:150 mA, R_L:0.095Ω, R_{ON}: 0.21Ω, R_c:37 mΩ, R_D:0.35 Ω, L:6.1477 μH, C_{out}: 8.3177 μF.

1588

Fig. 11 (B) shows the Bode plot of the boost converter based upon the SPICE software as shown in Fig. 11 (A). Fig. 12 shows similar results derived from the presented model based upon the proposed switch model. Comparing Fig. 11 (B) and Fig. 12 (B), the Bode plot of transfer functions for these two methods are the same and thereby confirming the presented model and proposed switch network.

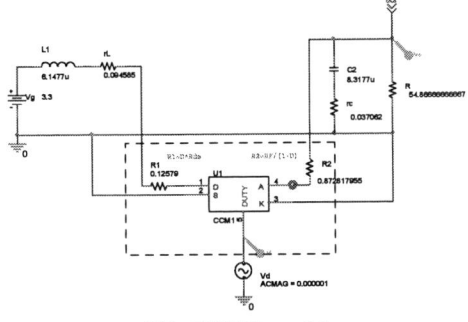

(A). SPICE model

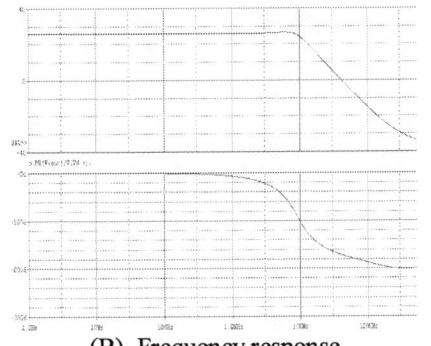

(B). Frequency response

Fig. 11 Simulation results, averaged switch model

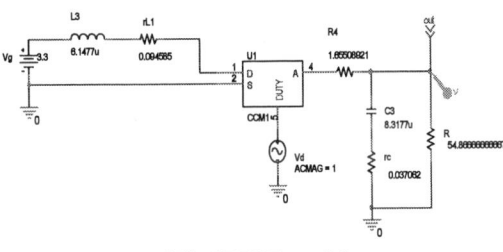

(A). SPICE model

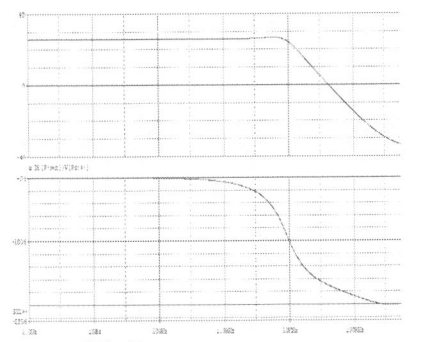

(B). Frequency response

Fig. 12 Simulation results, unified model using proposed switch network

VI. CONCLUSION

This paper presents a new switch network which can be used to replace the power devices of DC/DC converters to give a unified model. The non-ideal features of power devices and circuit components, including turn-on resistor of power devices, DC resistor (DCR) of inductor and equivalent series resistor (ESR) of capacitor, are taken into consideration.

As compared to previous approach, the presented switching network can be applied to modeling various kinds of DC/DC converters, including buck, boost, buck-boost, Cuk, Zeta and Sepic, by simply changing two parameters.

A unified model for various types of DC/DC converters based upon the presented switch network is developed and confirmed by simulation results.

REFERENCES

[1]. J. Chen, A. Prodic, R.W. Erickson and D. Maksimovic, "Predictive digital current programmed control," *IEEE Trans. on Power Electronics*, Vol.18, No.1, pp.41 1-419, Jan., 2003.

[2]. A. R. Oliva, S. S. Ang and G. E. Bortolotto, "Digital control of a voltage-mode synchronous buck converter," *IEEE Trans. on Power Electronics*, Vol.21, pp.157-163, Jan., 2006.

[3]. Y. S. Lai, C. A. Yeh and K. Y. Lee, "Novel self-commissioning digital power converter control using low sampling frequency A/D converter," *Proc. of the 4th Power Conversion Conference*, pp. 1644-1650, Nagoya, 2007.

[4]. G. W. Wester and R. D. Middlebrook, "Low-frequency characterization of switched DC-DC converters," *IEEE Trans. on Aerospace and Electronic Systems*, Vol. AES-9, pp. 376-385, May 1973.

[5]. R. D. Middlebrook, and S. Cuk, "A general unified approach to modeling switching converter power stage" Proc. of *IEEE PESC*, pp.18-34, 1976.

[6]. J. Sun, D. M.Mitchell, M. F. Greuel, P. T. Krein, and R. M. Bass, "Averaged modeling of PWM converters operating in discontinuous conduction mode", *IEEE Trans. on Power Electronics*, Vol. 16, No. 4, pp.482-492, 2001.

[7]. T. F. Wu and Y. K. Chen," Modeling PWM DC/DC Converters out of basic converter units," *IEEE Transactions on Power Electronics*, vol. 45, pp. 870-881, February 1998.

[8]. R. W. Erickson, Fundamentals of Power Electronics, 2nd Ed., Kluwer Academic Publishers, pp.226-247, 2001.

[9]. A. F. Witulski and R. W. Erickson, "Extension of state-space averaging to resonant switches and Beyond, "*IEEE Trans. on Power Electronics*, Vol.5, no.1, pp.98-109, January 1990.

[10]. O. Al-Naseem, R.W. Erickson and P. Carlin, "Prediction of switching loss variations by averaged switch modeling," Proc. of the *IEEE APEC*, pp. 242-248, February 2000.

Soft Transition Operation of UPS in High-Power-Factor Mode of Three-Phase Front-End Rectifier

G. A. Dhomane*, *Student Member, IEEE,* H. M. Suryawanshi**, *Member, IEEE*
* Research Scholar, Elect. Engg. Dept., VNIT, Nagpur, India
**Assistant Professor, Elect. Engg. Dept., VNIT, Nagpur, India

Abstract– A three-phase to single-phase ac-to-ac converter, with soft-transition of switches is presented. The converter has THD less than 4% and operates at almost unity-power-factor for wide variations in load, due to high frequency current injection, at the input of front-end-rectifier. A small filter is required at the output for filtering the high-frequency content. Sinusoidal PWM technique is used for controlling the output voltage. DSP is used for generating the desired gate pulses. The converter has high efficiency, low EMI emissions, high power packing density and suitable for UPS system. A Simulation and experimentation is carried out on a 3 kW converter and experimental results are in good agreement with simulation results.

Index Terms—High-frequency-current-injection, High-power-factor, Soft-transition, Power-factor-correction circuit.

I. INTRODUCTION

The increased use of power electronic equipments in the power system has a profound impact on power quality. The high power non linear loads (such as static power converter, arc furnace, adjustable speed drives etc) and low power loads (such as fax machine, computer, etc) produce voltage fluctuations, harmonic currents and an imbalance in network system which results into low power factor operation of the power system. There is a need of improved power factor and reduced harmonics content in input line currents as well as voltage regulation during power line over-voltage and under-voltage conditions. The uninterruptible power supplies (UPSs) have been extensively used for critical loads such as computers used for controlling important processes, some medical equipment, etc. The traditional UPS draws harmonic currents [1]. The uncontrolled diode bridge rectifier with capacitive filter is used as the basic block in many power electronic converters. Due to its non-linear nature, non-sinusoidal current is drawn from the utility and harmonics are injected into the utility lines. The total harmonic distortion (THD) factor increases to 70% [2]. The harmonics cause the malfunction of the equipments connected to the point of common coupling (PCC). They are not only responsible for increased losses but also cause excessive heating in the system [3]. Therefore regulations on line current harmonics have made power factor control, a basic requirement for power electronic

equipments [4]. Several active power ac-to-dc converters are presented in [5-7]. Resonant converter based and high-frequency current injection methods for power-factor control are presented in [8-11]. Several soft-switching converters are presented in [12-16] In this paper, high power factor operation of ac-to-ac converter with zero voltage transition (ZVT) and zero current transition (ZCT) is presented. The ZCT reduces the switching losses in the system. The ZCT operation is accomplished by taking away the main device current prior to the switching transitions, by the resonant circuit. The proposed ac-to-ac converter is shown in Fig.1.It consists of three-phase input line bridge rectifier(D_1-D_6) with power factor correction (PFC) circuit, a half-bridge inverter with two main switches (S_{m1}-S_{m2}) and two auxiliary switches(S_{a1}-S_{a2}) and L_R-C_R resonant circuit. The PFC consists of three-phase bridge inverter (S_1-S_6) with feed back capacitors (C_{f1}-C_{f3}) and inductors (L_{f1}-L_{f3}). The L_{S1}- L_{S3} are the source inductors. The diodes of the rectifier, main and auxiliary switches of half-bridge inverter operate at ZVT and ZCT. The switches of three-phase inverter show ZVT, reducing switching losses considerably. Digital Signal Processor (DSP) TMS320F2812 is used for gating the inverters. The sinusoidal PWM is used for the output voltage control. Small low pass filter is used at the output to filter the high-frequency content in the voltage. Computer simulation and experimentation is carried out for 3 kW, operating at a switching frequency of 50 kHz.

II. PRINCIPLE OF OPERATION

The proposed ac-to-ac converter is shown in Fig.1. It mainly consists of PFC circuit and a soft-switched inverter for zero voltage and zero current transitions.

A. Operation of PFC Circuit

The PFC circuit consists of three phase inverter, capacitors C_f and switched inductors L_f The inverter is switched with high frequency The high-frequency (HF) current is injected at the input of three-phase diode bridge rectifier through capacitor C_f causing modulation of input voltage of the diode bridge rectifier. This forces the diodes of the three-phase bridge rectifier to turn-on and turn-off at the switching frequency over the complete cycle of the input supply voltage. In a switching cycle, the input current is the sum of average values of injected current i_{Cf1} and $i_{Lf1,}$ as shown in Fig. 2. Average value of

978-1-4244-0644-9/07/$25.00 ©2007 IEEE   1590

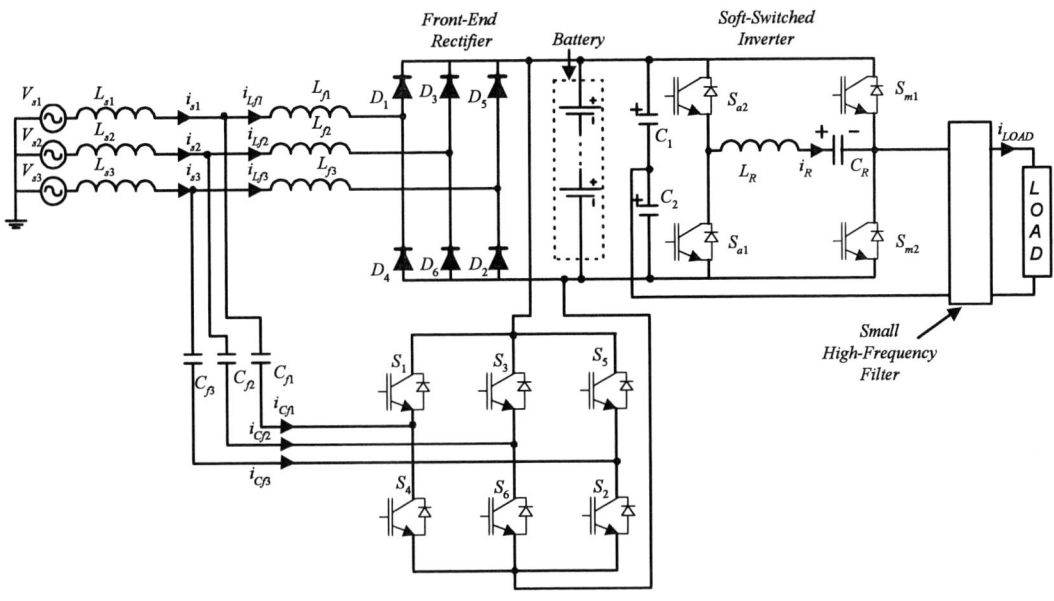

Fig. 1. A proposed ac-to-ac converter

i_{Cf1} over a switching cycle is zero and peak value of i_{Lf1} follows an envelope of the input supply phase voltage. In each switching cycle this current is reset to zero. Therefore average value of i_{Lf1} also follows the envelope of input voltage. When none of the diodes conducts then supply current flows through C_{f1}. Thus L_S operates in continuous conduction mode (CCM).therefore the input current is always in phase with the input supply phase voltage, v_{S1}. Hence the converter operates at high-power-factor. For CCM the output voltage of the rectifier should be twice the peak value of input phase voltage (2).

turned on. The L-C resonant circuit starts resonating and resonating current i_R starts to build up and the current in S_{m1} starts to decrease and i_R reaches I_{Load} at t_1. Thus the current in S_{m1} falls to zero and the body diode across S_{m1} starts to conduct surplus current. The gate driver signal can be removed at the zero current condition without causing turn off loss. The same concept is applicable for turn on transition also. As shown in Fig. 3(b), I_{Load} initially flows through body diode of S_{m2}. During turn on topological

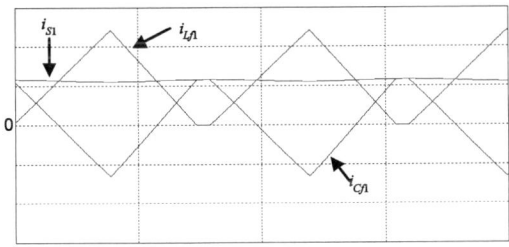

Fig. 2. Input current, capacitor current and inductor current.
Scale: 5A/div, 20µs/div

B. ZV and ZC Transitions

A zero current transition (ZCT) and zero voltage transition (ZVT) are accomplished by a circuit consisting of a half-bridge inverter (S_{m1}-S_{m2}), two auxiliary switches and a resonant network (L_R-C_R) [16].The basic concept is explained by a simplified circuit shown in Fig. 3a and 3b., the auxiliary switches, (S_{a1}-S_{a2}) are switched alternately in a definite pattern (Fig. 4).To assist the top main switch S_{m1} for turn-off, an auxiliary switch S_{a2} is

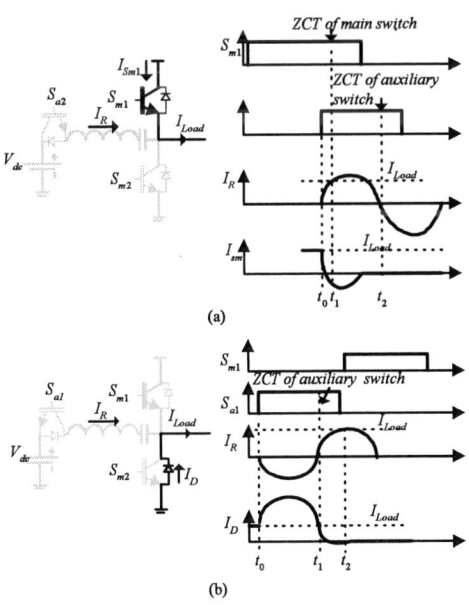

Fig. 3. Simplified circuit. a) Turn-off transition of S_{m1}
b) Turn-on transition of S_{m1}

1591

stage, the direction of S_{a1} is equivalently changed. Prior to turning on S_{m1}, S_{a1} is turned on for short duration. The current i_R starts to build up in negative direction and reverses its direction at t_1. The current through body diode of S_{m2} decreases due to increasing i_R in positive direction and surplus current passes through body diode of S_{m1} and it can be turned on at t_1. If S_{m1} is gated at this moment then zero voltage switching can be achieved. Moreover i_R flows through body diode of S_{a1}, at this moment the auxiliary switch S_{a1} can be turned off at zero-current. The same principle is also applicable to turn on and turn off of S_{m2}. Prior to turn on off S_{m2}, auxiliary switch S_{a2} is gated for short duration. Required gating pattern is generated using digital signal processor (DSP) TMS320F2812.The battery is charged from dc link voltage. Digital sinusoidal PWM technique is used for output voltage wave shaping and magnitude control. A small output filter is used to filter HF content in the output voltage.

III. DESIGN PROCEDURE
A. PFC circuit
The total three phase input power is given by,

$$P_i = 3 \cdot \frac{1}{2\pi} \int_0^{2\pi} V_s \cdot I_s \cdot d(\omega t) \qquad (1)$$

The value of the switched inductor is given by [2]

$$L_f = \frac{3}{4} \cdot \frac{V_m^2 \cdot d}{f_s \cdot P_i} \qquad (2)$$

If, P_0 is the output power of the converter, then

$$L_f = \frac{3}{4} \cdot \frac{V_m^2 \cdot d \cdot \eta}{f_s \cdot P_0} \qquad (3)$$

Where,

V_s -supply phase voltage; V_m -peak value of phase voltage; d-duty cycle of the inverter, η –efficiency of the converter, f_s -switching frequency.

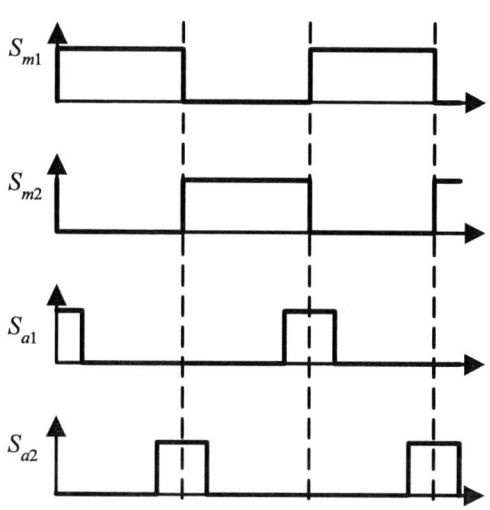

Fig. 4. Pulse pattern for turn-on and turn-off transitions

B. Soft –Transition Circuit
If V_{dc} is dc link voltage, I_{Load} is load current, Z_0 is resonant tank impedance, then

$$I_{Rp} = \frac{V_{dc}}{Z_0}, \quad Z_0 = \sqrt{L_R \big/ C_R} \qquad (4)$$

For the design of resonant components, a ratio M is defined as

$$M = \frac{I_R}{I_{Load}}$$

Therefore from (4)

$$Z_0 = \frac{V_{dc}}{(M \cdot I_{Load})} \qquad (5)$$

M should be at least 1.1.
The resonant time period T_o is given by

$$T_0 = 2\pi \sqrt{L_R C_R} \qquad (6)$$

From equations (5) and (6), the resonant tank elements are estimated as

$$L_R = Z_0 \frac{T_0}{2\pi}, \quad C_R = \frac{L_R}{Z_0^2} \qquad (7)$$

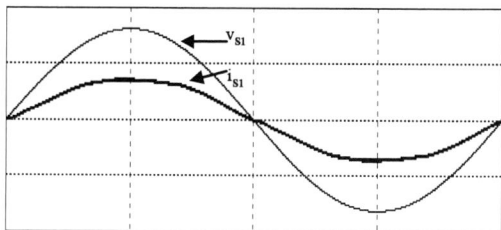

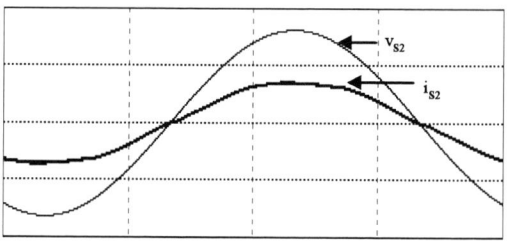

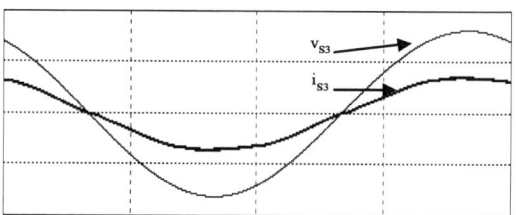

Fig. 5a. Simulated waveforms of supply voltage and current.
Scale: 200 V/div, 10 A/div, 5 ms/div.

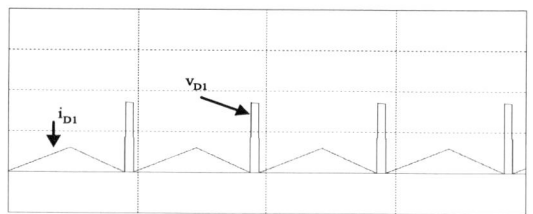

Fig. 5b. Simulated waveforms: Rectifier voltage and current
Scale: 200 V/div, 20 A/div, 20µs/div

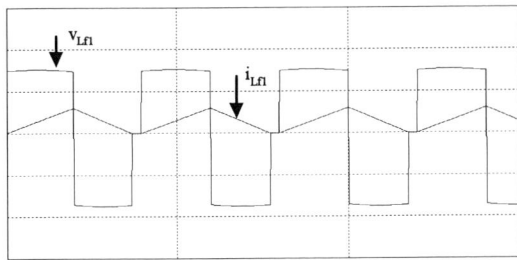

Fig. 5c. Simulated waveforms: Current through and voltage across inductor L_f. Scale: 20 A/div., 200 V/div.

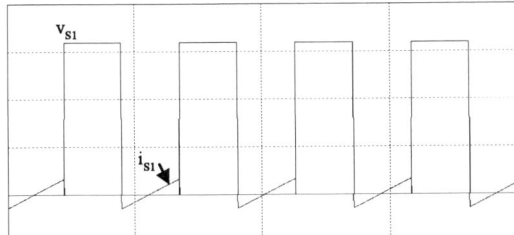

Fig. 5d. Simulated waveforms: Current through and voltage across inverter switch S_1. Scale: 20 A/div., 200 V/div.

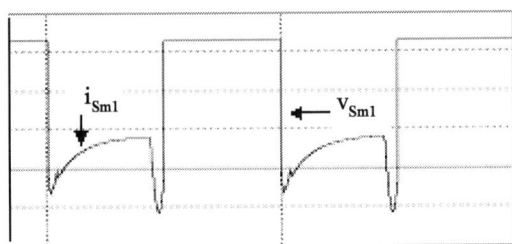

Fig. 5e. Simulated waveforms: Current through and voltage across main switch S_{m1} of a half-bridge.
Scale: 20 A/div., 200 V/div.

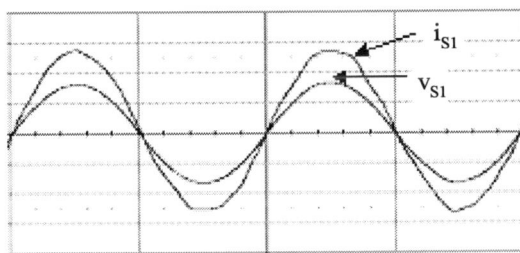

Fig. 5f. Experimental waveforms of supply voltage and current. Scale: 200 V/div, 2 A/div, 10 ms/div.

Using procedure for design outlined above, the converter specifications and components values are as follows:

Input: Three-phase, 400 V, 50 Hz
Output: Single-phase, 220 V, 50 Hz, 3kW
Inverter switching frequency, f_S=50 kHz, Source inductors, L_S=5mH, Feedback inductors L_f=250 µH, Feedback capacitors, C_f=2µF, Split capacitors, C_1=C_2=1000 µF, Resonant components, L_R=20 µH, C_R=10 ηF.

IV. RESULTS

The computer simulation of proposed converter designed as per the procedure in section III is carried out and simulation waveforms are shown in Fig. 5 (a)-(e). A laboratory prototype of 3 kW is designed and tested with a switching frequency of 50 kHz. Experimental waveform is depicted in Fig. 5 (f). The THD of supply current is found to be greatly improved and it is less than 4%.

V. CONCLUSION

A high-power-factor operation with soft-switching transition, three-phase ac-to-ac converter is proposed. The soft-switching of main and auxiliary switches are achieved thereby greatly reducing the switching losses and EMI emissions. The switches have lower stresses and can be used with high switching frequency. The proposed converter has many advantages such as high packing density, small filter requirement, high efficiency and high power factor. Better output voltage control is obtained with the flexible programming using DSP.

REFERENCES

[1] D. David Shipp, and W.S. Vilcheck, "Power quality and line considerations for variable speed drives." *IEEE Trans. Ind. Applicat.*, Mar. /Apr. 1996.

[2] M. A. Chaudhary, H. M. Suryawanshi, "High–power–factor operation of three–phase ac-to-dc resonant converter," *IEE-PA*, vol. 153, nO. 6, Nov 2006, pp 873-882.

[3] H.M.Suryawanshi, K.L.Thakre, S.G.Tarnekar, G.Kothari, and D.P.Kothari, "Power factor and closed loop control of an AC-to-DC Converter," *IEE proc.-Electr. Power Apppl.*, vol.149.

[4] J.S. Lai, and T.S. Key, "Harmonic standards: Impact of power electronics equipment design," *Power electronic Technology*, vol. 58, Aug. 2000, pp.1-13.

[5] A. R. Prasad, P.D. Ziogas, and S. Manias, "An active power factor correction technique for three-phase diode rectifiers," *IEEE- PESC Record*, pp. 58-66, 1989.

[6] D. Y. Qiu, S. C.Henry, S. H. Chung, and S. Y. Ron Hui, "Single current sensor control for single-phase active power factor correction," *IEEE transactions on Power Electronics*, vol. 17, no. 5, September 2002, pp. 623-631.

[7] Q. Huang, and F. C. Lee, "Harmonic reduction in a single-switch, three-phase boost rectifier with higher order harmonics injected PWM," *IEEE- PESC 1996 Rec.*, pp. 1266- 1271.

[8] A. I. Maswood, and F. Liu, "A unity power factor front end rectifier with hysteresis current control," *IEEE transactions on Energy Conversion*, vol. 21, no. 1, March 2006, pp. 69-76.

[9] F.S. Hamdad, and A.K.S. Bhat, "Three-phase single stage ac/dc boost integrated parallel resonant converter," *IEEE transactions on Aerospace and Electronic Systems*, vol. 40, no. 4, Oct. 2004, pp. 1311-1323.

[10] Belaguli, and A. K. S. Bhat, "High power factor operation of DCM series-parallel resonant converter," *IEEE Transactions on Aerospace and Electronic Systems*, vol. 35, no. 2, April 1999, pp. 602-613.

[11] M. A Cross, and A. J. Forsyth, "A high-power-factor, three-phase isolated ac–dc converter using high-frequency current injection," *IEEE transactions on Power Electronics*, vol. 18, no. 4, July 2003, pp. 1012-1019.

[12] M. Bellar, T. Wu, A. Tchamdjou, J. Mahdavi, and M. Ehsani, "A review of soft-switched dc-ac converters," *IEEE transactions on Industrial Appl.*, vol. 34, July 1998, pp. 847-860.

[13] D. M. Divan, "The resonant dc-link converter- A new concept in static power conversion," in *proc. IEEE-IAS Annu. Meeting,* 1986, pp. 648-656.

[14] V. Vlatkovic, D. Boroyevich,F. C. Lee, C. Caudros, and S. Gatatric, "A new zero-voltage-transition three-phase PWM rectifier/inverter circuit," in *proc. IEEE-PESC Conf.,* 1993, pp. 868-873.

[15] P. Tomasin, "A novel topology of zero-current switching voltage-source PWM inverter for high power applications," in *proc. IEEE-PESC Conf.,* 1995, pp. 1245-1251.

[16] Y. Li, F. C. Lee, D. Boroyevich, "A three-phase soft-transition inverter with a novel control strategy for zero-current and near zero-voltage switching," IEEE transactions on Power Electronics, vol.16, no. 5, Sept. 2001, pp.710–723.

Specific Harmonic Power Suppression of Direct-Power-Controlled Current-Source PWM Rectifier

Toshihiko Noguchi [*], *IEEE Senior Member*, and Kohji Sano [*]

[*] Nagaoka University of Technology
Address: 1603-1 Kamitomioka, Nagaoka, Niigata 940-2188, Japan
Phone: +81-258-47-9510, Fax: +81-258-47-9500
e-mail: tnoguchi@vos.nagaokaut.ac.jp

Abstract— This paper focuses on specific harmonic active power suppression of a direct-power-controlled (DPC) PWM current-source rectifier (CSR) in order to achieve low distortion of the input line currents. Total input power factor of the DPC-based PWMCSR becomes worse as the load gets lower due to the low-order harmonics in the line currents, especially the fifth and the seventh. Since the dominant low-order harmonic currents cause an oscillation in the active power at frequency of sixth, suppression of the sixth-order harmonic active power is essential to improve the total power factor particularly in the low-load range. The paper describes a theoretical aspect and a suppression technique of the harmonic active power, followed by basic configuration and operation of the DPC-based PWMCSR. Effectiveness of the proposed technique is confirmed through computer simulations and experimental tests, using a 2-kW prototype. As a result, the total harmonic distortion of the line currents is effectively reduced by 10 %, which results in approximately 20-% improvement of the total input power factor at a 350-W load condition

Index Terms— current-source PWM rectifier, direct power control, relay control, instantaneous power, and harmonic power suppression.

I. INTRODUCTION

In general, PWM rectifiers are extensively used as an AC/DC power converter in order to improve the total input power factor. There are two classes of the PWM rectifiers, i.e., a voltage-source rectifier and a current-source rectifier. The two rectifiers are dual with each other from the viewpoint of the circuit topology, and the former has been intensively investigated and widely been applied to various industry applications, compared with the latter. In both cases, conventional control strategies of the PWM rectifiers achieve a unity power factor operation by forcing the input line currents in phase with the power-source voltages. Therefore, most of the conventional systems have a current minor control loop with a rotational coordinate transformation like AC motor drive systems. However, this approach has some drawbacks as pointed out below.

(1) It inherently suffers from a slow response due to the current minor loop, especially due to PI regulators.
(2) It is difficult to reduce its capacitive or inductive value in the DC-bus energy buffering devices.
(3) Its control algorithm is inherently complicated due to the rotational coordinate transformation.
(4) It requires a pulse width modulator to generate switching signals for the power devices.

(5) In order to make the unity power factor operation possible, high-resolution in detecting relative phase of the power-source voltage is indispensable.
(6) Under some conditions where waveform distortion and/or unbalance in the power-source occur, it requires additional circuits to cope with such cases.

The authors have been investigating a direct power control (DPC) strategy of various power converters to overcome the drawbacks mentioned above. The key of this strategy is a direct selection of optimum switching states to perform high-speed relay (bang-bang) control of the instantaneous active and reactive power of the converters. This eliminates the pulse width modulator and the PI regulators in the current minor loop, which leads to extremely quick response and high controllability of the power. By controlling the active power at a constant value and the reactive power to be zero, sinusoidal line currents in phase with the power-source voltages are resultantly obtained with neither the current minor loop nor the rotational coordinate transformation. Furthermore, the DPC system does not need any auxiliary compensator against the waveform distortion and/or the unbalance of the power-source, which is a unique feature of this approach.

However, the DPC still has some problems as follows:
(1) Switching frequency of the converter varies as the load and/or operating condition change, which makes input filter design difficult.
(2) Waveform distortion of the line currents apt to be worse in the low-load range.
(3) The system performance is rather sensitive to delay time in the power feedback paths.

This paper discusses a compensation technique for the waveform distortion of the line currents in order to improve the total power factor in the low-load range, focusing on the DPC based PWM current-source rectifier (CSR). Through several computer simulations and experimental tests, the proposed technique is found to be effective to reduce the low-order harmonic currents and to improve the total input power factor without sacrificing inherent advantages of the DPC strategy.

II. BASIC CONFIGURATION AND OPERATION

A. System Configuration

Fig. 1 shows a schematic diagram of the DPC-based PWMCSR. As shown in the figure, relay (bang-bang) control of the instantaneous active power P and the

978-1-4244-0644-9/07/$25.00 ©2007 IEEE

instantaneous reactive power Q is performed with use of their feedback values. Both of P and Q are calculated as expressed in the following equations:

$$\begin{bmatrix} v_\alpha \\ v_\beta \end{bmatrix} = \sqrt{\frac{2}{3}} \begin{bmatrix} 1 & -1/2 & -1/2 \\ 0 & \sqrt{3}/2 & -\sqrt{3}/2 \end{bmatrix} \begin{bmatrix} v_u \\ v_v \\ v_w \end{bmatrix}, \tag{1}$$

$$\begin{bmatrix} i_\alpha \\ i_\beta \end{bmatrix} = \sqrt{\frac{2}{3}} \begin{bmatrix} 1 & -1/2 & -1/2 \\ 0 & \sqrt{3}/2 & -\sqrt{3}/2 \end{bmatrix} \begin{bmatrix} i_u \\ i_v \\ i_w \end{bmatrix}, \text{ and} \tag{2}$$

$$\begin{bmatrix} P \\ Q \end{bmatrix} = \begin{bmatrix} v_\alpha & v_\beta \\ v_\beta & -v_\alpha \end{bmatrix} \begin{bmatrix} i_\alpha \\ i_\beta \end{bmatrix}. \tag{3}$$

As expressed in (3), P and Q can simply be calculated with an inner product and an outer product between the power-source voltage vector and the line current vector, respectively. On the other hand, the instantaneous active power command P^* is provided from a DC bus current control block, while the instantaneous reactive power command Q^* is directly given from the outside of the controller, according to the desired input total power factor. Zero command value for Q^* is normally given to the controller to achieve a unity input power factor operation. Control errors of the active and the reactive power, i.e., $\Delta P = P^* - P$ and, $\Delta Q = Q^* - Q$ are quantized with hysteresis comparators, of which outputs are digital signals S_p and S_q, respectively. In addition, a relative phase of the power-source voltage vector is quantized to six sectors Θ_n by using several comparators as follows:

$$(n-2)\frac{\pi}{3} \leq \Theta_n < (n-1)\frac{\pi}{3} \quad \because n = 1, 2, \cdots, 6. \tag{4}$$

A combination of these quantized signals S_p, S_q and Θ_n is used to select uniquely the most appropriate switching state of the PWMCSR. In other words, every time the PWMCSR changes its switching state, S_p, S_q and Θ_n determine the next unique and optimum switching state to restrict ΔP and ΔQ within the corresponding hysteresis bands. In order to achieve this function, a switching state table is composed as shown in Fig. 2, of which contents are predefined so that both of the active and the reactive power follow their commands with small control errors. The uniquely selected switching state turns on or off every switching device in the PWMCSR. In the switching state table shown in Fig. 2, "P" stands for a state of $S_{u, v, w}$ = "ON" and $S_{x, y, z}$ = "OFF", "O" is a state of $S_{u, v, w}$ = "OFF" and $S_{x, y, z}$ = "OFF", "N" is a state of $S_{u, v, w}$ = "OFF" and $S_{x, y, z}$ = "ON", and "S" is a state of $S_{u, v, w}$ = "ON" and $S_{x, y, z}$ = "ON".

B. Optimum Switching State Selection

Since the proposed DPC system is, in principle, based on relay control, it is absolutely significant to investigate relationship between the switching states of the PWMCSR and polarities of time derivatives of the active and reactive power dP/dt and dQ/dt. Their polarities correspond to the quantized signals S_p and S_q; hence, the time derivatives of dP/dt and dQ/dt need be solved with respect to the switching states and the phase information

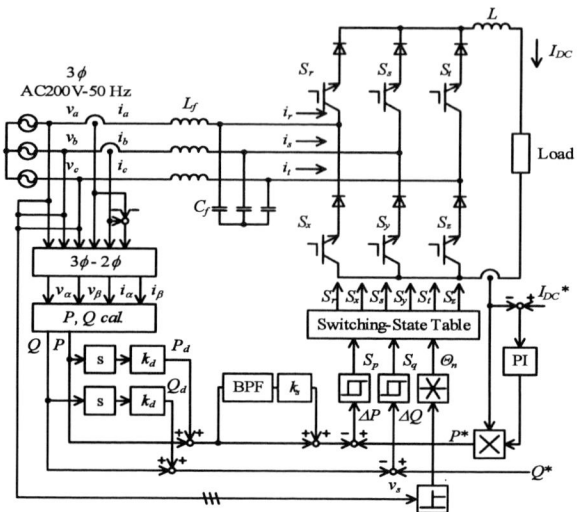

Fig. 1. System configuration of direct power control (DPC) based current-source PWM rectifier with specific harmonic power suppression

S_p	S_q	$S_r \uparrow$ Θ_1	$S_x \uparrow$ Θ_2	$S_s \uparrow$ Θ_3	$S_y \uparrow$ Θ_4	$S_t \uparrow$ Θ_5	$S_z \uparrow$ Θ_6
0	0	SOO	OOS	OSO	SOO	OOS	OSO
0	1	ONP	PNO	PON	OPN	NPO	NOP
1	0	PON	OPN	NPO	NOP	ONP	PNO
1	1	PNO	PON	OPN	NPO	NOP	ONP

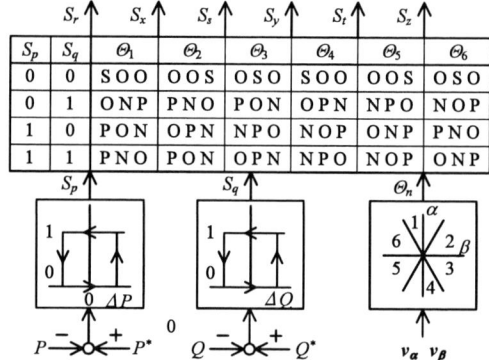

Fig. 2. Switching state table and power regulators.

of the power-source voltage vector in order to compose the switching state table appropriately.

From a mathematical model of the PWMCSR connected to the power grid shown in Fig. 3, the following current equation is established:

$$i_s + i_c = i_s + C_f \frac{dv_c}{dt} = i_s', \tag{5}$$

where the power-source current vector is defined by

$$i_s = i_\alpha + ji_\beta = \sqrt{\frac{2}{3}}\left(i_u + i_v e^{j\frac{2\pi}{3}} + i_w e^{j\frac{4\pi}{3}} \right). \tag{6}$$

In the case of three-phase balanced power source, the above current is represented as a rotating vector with constant amplitude of I_{rms} expressed as

$$i_s = \sqrt{3}I_{rms}e^{j\omega t}. \tag{7}$$

In addition, the input current vector i_s' drawn by the PWMCSR is a function of the switching state as shown below, where I_{DC} is a DC bus current:

$$i_s' = \sqrt{\frac{2}{3}}I_{DC}\left(S_u + S_v e^{j\frac{2}{3}\pi} + S_w e^{j\frac{4}{3}\pi} \right). \tag{8}$$

On the other hand, the time derivatives of dP/dt and

1596

dQ/dt are derived from (3) and are approximated as indicated by the following equations because variation of the power-source current i_s can be regarded as negligibly small during switching intervals of the PWMCSR and capacitor voltage v_c is almost equal to the power-source voltage v_s:

$$\frac{dP}{dt} = i_s \bullet \frac{dv_s}{dt} + \frac{di_s}{dt} \bullet v_s \approx i_s \bullet \frac{dv_c}{dt}, \text{ and} \quad (9)$$

$$\frac{dQ}{dt} = i_s \times \frac{dv_s}{dt} + \frac{di_s}{dt} \times v_s \approx i_s \times \frac{dv_c}{dt}. \quad (10)$$

Substituting (5), (7) and (8) into the above equations, dP/dt and dQ/dt can be solved as follow:

$$\frac{dP}{dt} = -\frac{3I_{rms}^2}{C_f} + \frac{\sqrt{2}I_{rms}I_{DC}}{C_f}\left\{(S_u - \frac{S_v}{2} - \frac{S_w}{2})\cos\theta \right.$$
$$\left. + \frac{\sqrt{3}}{2}(S_v - S_w)\sin\theta\right\} \quad , \text{ and}$$

$$(11)$$

$$\frac{dQ}{dt} = \frac{\sqrt{2}I_{rms}I_{DC}}{C_f}\left\{(S_u - \frac{S_v}{2} - \frac{S_w}{2})\sin\theta \right.$$
$$\left. - \frac{\sqrt{3}}{2}(S_v - S_w)\cos\theta\right\} \quad , \quad (12)$$

where θ is an argument of the power-source voltage vector. According to the polarities of dP/dt and dQ/dt solved as (11) and (12), one of the switching states of the PWMCSR can uniquely be determined to restrict the control errors ΔP and ΔQ within the hysteresis bands, which leads to determination of the whole contents of the optimum switching state table. Fig. 4 is an example of behaviors of the active and the reactive power when the power-source voltage vector is in the sector Θ_1, where the time derivatives dP/dt and dQ/dt are symbolized with tilted arrows.

C. LC filter Resonance Suppression

The PWMCSR requires a LC filter at input terminals, which possibly causes current oscillation at the resonant frequency. Therefore, the DPC based PWMCSR has a compensator to damp the oscillation. A Laplace-transformed circuit equation of the PWMCSR with the LC filter shown in Fig. 3 can be described on a synchronous rotating reference frame as follows:

$$V_s = sL_f I_s + \frac{1}{sC_f}(I_s - I_s'). \quad (13)$$

Assuming that the power-source voltage and the input line current have a relative phase of φ, they can be expressed as

$$V_s = \sqrt{3}V_{rms}, \text{ and} \quad (14)$$

$$I' = \sqrt{3}I_{rms}'e^{-j\varphi}. \quad (15)$$

Substituting the above equations into (13), the power-source current I_s can be derived in an approximated expression as follows:

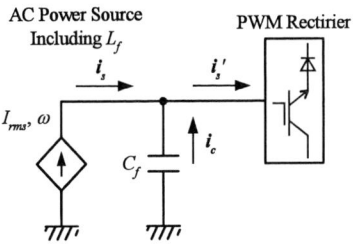

Fig. 3. Simplified mathematical model of PWMCSR.

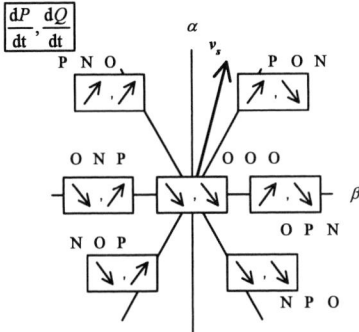

Fig. 4. Symbolized time derivatives of instantaneous active and reactive power in case of sector Θ_1.

$$I_s = \frac{\sqrt{3}V_{rms}C_f}{L_fC_fs^2+1} + \frac{\sqrt{3}I_{rms}'e^{-j\varphi}}{s(L_fC_fs^2+1)} \approx \frac{\sqrt{3}I_{rms}'e^{-j\varphi}}{s(L_fC_fs^2+1)}. $$

$$(16)$$

Applying an inverse Laplace-transform to (16), the following instantaneous apparent power S is calculated in the time domain:

$$S = v_s\overline{i_s}$$
$$= 3V_{rms}I_{rms}'e^{j\varphi} - 3V_{rms}I_{rms}'e^{j\varphi}\cos\frac{1}{\sqrt{L_fC_f}}t. \quad (17)$$

This equation shows that the first term corresponds to the fundamental frequency components of the active and the reactive power, while the second term is the resonant frequency components of them. Since the resonant components have no damping factor as indicated in (17), differential elements are added to the power feedback in order to damp the oscillation. Applying this feedback compensation converts the transfer function of the power-source current I_s to the following form:

$$I_s \approx \frac{\sqrt{3}I_{rms}'e^{-j\varphi}}{s(L_fC_fs^2+k_ds+1)}. \quad (18)$$

Therefore, S can be damped by the derivative compensation elements, which are inserted in the feedback paths of the active and the reactive power. When the derivative gain is set at $k_d = 2\sqrt{L_fC_f}$, a critically damped response of S is achieved as

$$S = 3V_{rms}I_{rms}'e^{j\varphi}$$
$$- 3V_{rms}I_{rms}'e^{j\varphi}\left(1 + \frac{1}{\sqrt{L_fC_f}}t\right)e^{-\frac{1}{\sqrt{L_fC_f}}t}. \quad (19)$$

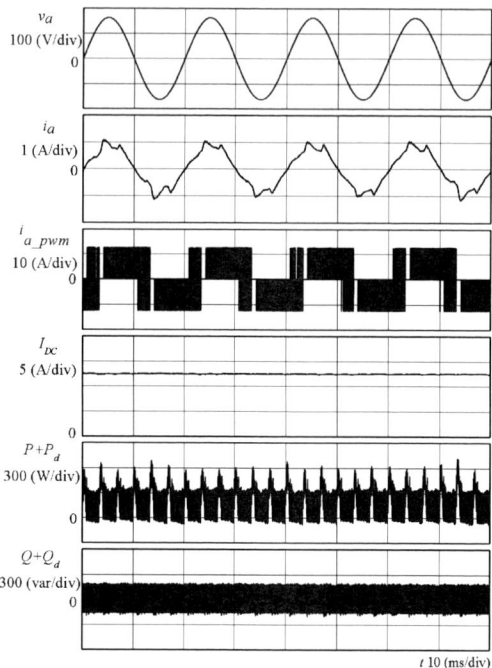

(a) Waveforms of power-source line-to-neutral voltage, input line current, PWM current, DC bus current, and active and reactive power.

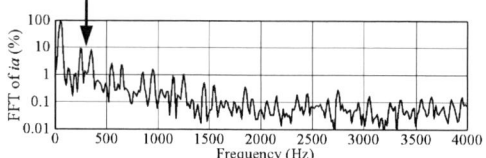

(b) Frequency spectra of input line current.

Fig. 5. Operating waveforms without compensation at low load condition (simulation result).

Table 1. Electric parameters of power circuit and test conditions.

Power source	AC 3ϕ 200 V, 50 Hz
Input filter	$L_f = 2.7$ mH, $C_f = 40$ μF
DC bus reactor	$L = 40$ mH
DC bus current command	$I_{DC}^* = 12.5$ A
Load power	220 W
Hysteresis bandwidth	300 W for P, 300 var for Q
Central frequency of BPF	300 Hz (six times of 50 Hz)
Quality factor of BPF	20

III. SPECIFIC HARMONIC POWER SUPPRESSION

A. Requirements of Harmonic Power Suppression

As described earlier in this paper, although the control errors ΔP and ΔQ as well as the relative phase of the power source voltage vector θ are quite roughly quantized, sinusoidal waveforms can resultantly be obtained in the input line currents without using a current minor loop, which is one of the significant features of the DPC based system. However, when the load power is in a relatively low range, low-order harmonics appear in the input line currents and detrimentally affects not only the total harmonic distortion (THD) level of the currents but also the total input power factor and the total efficiency. This phenomenon is particularly remarkable in as low-

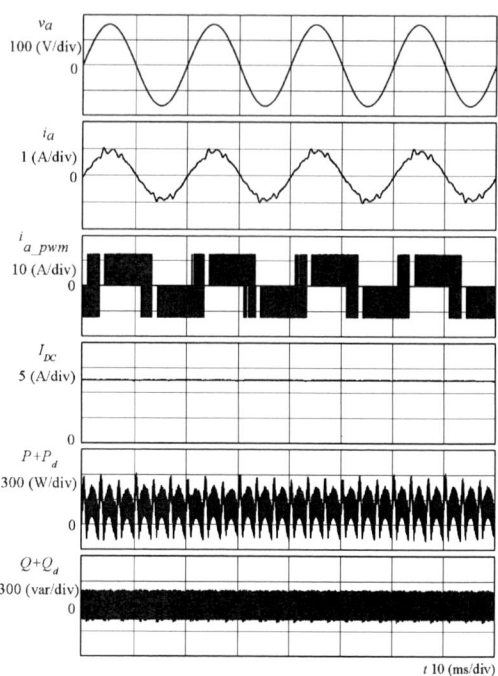

(a) Waveforms of power-source line-to-neutral voltage, input line current, PWM current, DC bus current, and active and reactive power.

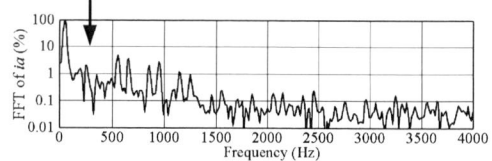

(b) Frequency spectra of input line current.

Fig. 6. Operating waveforms with sixth-order harmonic active power suppression at low load condition (simulation result).

load range as the hysteresis bandwidths ΔP and ΔQ. If the hysteresis bandwidths can be narrowed, this problem may be avoided. Due to the delay time in detection and calculation of the instantaneous active and reactive power, however, it is not impossible to restrict ΔP and ΔQ within the specified hysteresis bandwidths even though the bandwidths are reduced to nearly zero.

B. Theoretical Analysis of Harmonic Power

Assuming that the power-source supplies ideally balanced sinusoidal three-phase voltages to the DPC based PWMCSR, the power-source voltage vector can be expressed as

$$v_s = \sqrt{3} V_{rms}^{\omega} e^{j\omega t}, \qquad (20)$$

where a superscript of the root-mean-square value denotes an angular frequency ω of the voltage vector. As described previously, the input line current vector has low-order harmonics; hence, its mathematical expression can be as follows:

$$i_s = \sqrt{3} I_{rms}^{\omega} e^{j(\omega t - \varphi^{\omega})}$$
$$+ \sqrt{3} I_{rms}^{5\omega} e^{-j(5\omega t - \varphi^{5\omega})} + \sqrt{3} I_{rms}^{7\omega} e^{j(7\omega t - \varphi^{7\omega})} + \cdots \qquad (21)$$

In this expression, the input line current vector is composed of a fundamental component and the harmonic

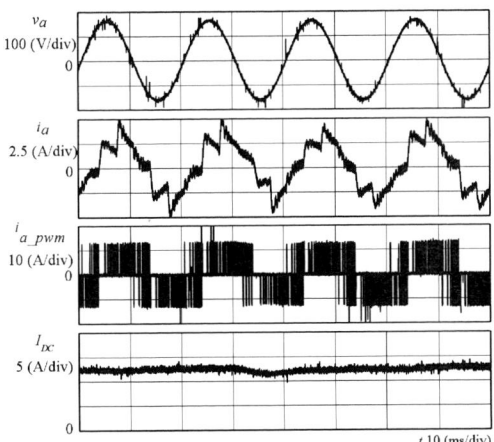

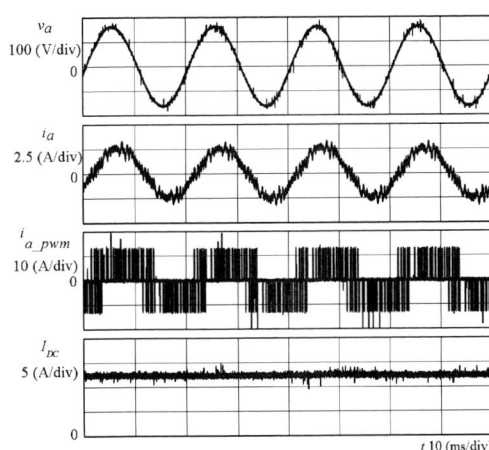

(a) Waveforms of power-source line-to-neutral voltage, input line current, PWM current, and DC bus current.

(a) Waveforms of power-source line-to-neutral voltage, input line current, PWM current, and DC bus current.

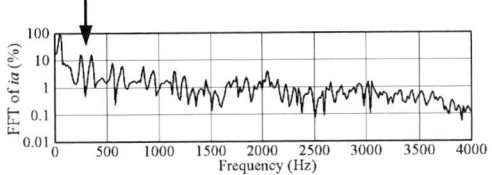

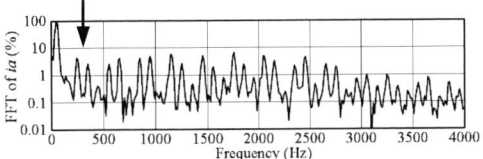

(b) Frequency spectra of input line current.

Fig. 7. Operating waveforms without compensation at low load condition (experimental result).

(b) Frequency spectra of input line current.

Fig. 8. Operating waveforms with sixth-order harmonic active power suppression at low load condition (experimental result).

components except for multiples of the third-order harmonics because of a three-phase and three-line system. As can be seen in the second term of the above equation, the fifth-order harmonic component causes a negative sequence, whereas the fundamental and the seventh-order components constitute a positive sequence. Using these equations, the apparent power S is derived as

$$
\begin{aligned}
\boldsymbol{S} = & 3V_{rms}^{\omega} I_{rms}^{\omega} e^{j\varphi^{\omega}} \\
& + 3V_{rms}^{\omega} I_{rms}^{5\omega} e^{j(6\omega t - \varphi^{5\omega})} + 3V_{rms}^{\omega} I_{rms}^{7\omega} e^{-j(6\omega t - \varphi^{7\omega})} + \cdots
\end{aligned}
$$
(22)

As indicated in (22), the apparent power S has a static DC component brought by the fundamental components of the voltage and the current and the multiples of the sixth-order harmonic components, which result in the dominant oscillations of the active and the reactive power. Particularly, the real part of S, i.e., the active power, is a major component of the sixth-order frequency. Therefore, suppression of the sixth-order harmonic active power is effective and essential to reduce the input line current distortion, especially low-order harmonic components such as the fifth and the seventh.

C. Specific Harmonic Power Suppression

The fifth and the seventh harmonic distortion of the input line currents can simultaneously be suppressed by restricting specifically the sixth-order harmonic active power; thus minimization of the hysteresis bandwidth only for the sixth-order harmonic active power is required in the DPC system. In order to achieve this goal, a feedback signal of only the sixth-order harmonic active power is selectively magnified by extracting the sixth-

order harmonic component with a band-pass filter (BPF) as shown in Fig 1. The central frequency of the BPF is adjusted at a frequency of the sixth, and the quality factor is tuned to have an appropriate damping characteristic. The proposed technique can effectively diminish the fifth and the seventh harmonic currents at the same time without sacrificing inherent simple configuration of the DPC based PWMCSR.

IV. COMPUTER SIMULATION RESULTS

In order to examine basic operation characteristics of the proposed technique, some computer simulations were conducted with PSIM, where electric parameters of the power circuit and test conditions are listed in Table 1.

Fig. 5 shows operating waveforms and frequency spectra of the input line current at as low load as 220 W with no compensation. As shown in Fig. 5 (a), a large waveform distortion can be seen in the line current due to the low-order harmonics caused by active power deviation out of the specified hysteresis band. The ripples of the active power appear six times per cycle, i.e., 300 Hz, whereas the reactive power is properly restricted within the predetermined hysteresis band. This sixth-order harmonic active power brings the fifth and the seventh harmonic currents as indicated by a downward arrow in Fig. 5 (b), which are more than 10 % of the fundamental component. On the other hand, Fig. 6 depicts operating characteristics at 220-W load power where the proposed compensation technique is applied to the system. As shown in the line current waveform, undesirable large ripples are effectively reduced around peaks of the current, resulting in the sinusoidal waveform. This effect can be confirmed in the frequency spectra of

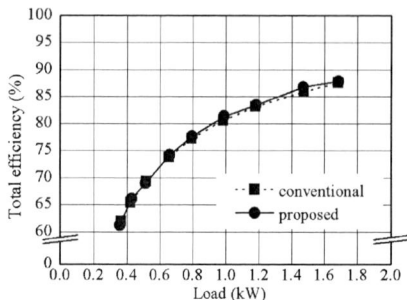

Fig. 9. Total efficiency characteristics with/without sixth-order harmonic active power suppression (experimental result).

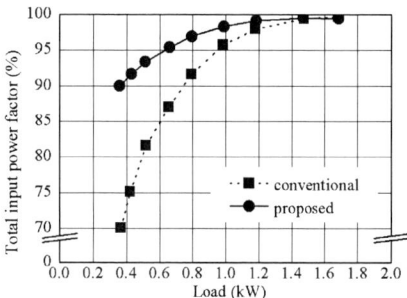

Fig. 10. Total input power factor characteristics with/without sixth-order harmonic active power suppression (experimental result).

the line currents shown in Fig. 6 (b), where the fifth and the seventh harmonics are diminished down to approximately 1 %.

V. EXPERIMENTAL RESULTS

The proposed compensation technique was examined with a 2-kW prototype that is composed of analog and digital mixed signal hardware. The prototype has the similar specifications and parameters as used in the computer simulations.

Fig. 7 shows an experimental result at low-load condition in the case of no compensation. It is found that the input line current includes acute ripples around its peaks, of which main components are the fifth and the seventh as indicated in the FFT analysis result of Fig. 7 (b). A relative amount of these current harmonics with respect to the fundamental component is almost same as that of the simulation result. Fig. 8 demonstrates an experimental result with the proposed sixth-order harmonic active power suppression. As can be seen in the figure, the current ripples caused by the low-order harmonics are dramatically rejected and an appropriate sinusoidal current waveform is drawn by the PWMCSR.

Figs. 9 and 10 depict the total efficiency and the total input power factor, respectively. Striking difference in the total efficiency is not seen between with and without proposed technique, and the maximum efficiency is 88.5 % in both cases. However, the total input power factor is remarkably improved owing to the sixth-order harmonic active power suppression, especially in the lower-load range. This improvement is made mainly by the low-order harmonics rejection in the input line currents, which can be confirmed by the THD characteristics shown in Fig. 11. It is confirmed

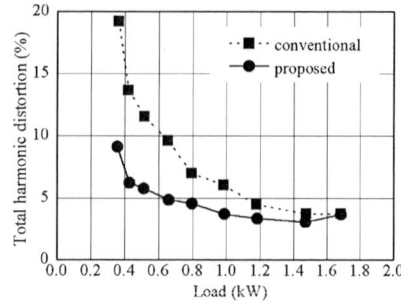

Fig. 11. Total harmonic distortion characteristics with/without sixth-order harmonic active power suppression (experimental result).

throughout the experimental tests that the average switching frequency of the PWMCSR is almost constant at 2 kHz over the full load range regardless of implementation of the compensation.

VI. CONCLUSIONS

This paper described a technique to suppress a specific harmonic power in the DPC based PWMCSR. The proposed technique selectively reduces the specified order of the harmonic active power, which resultantly improves not only the input line current waveforms but also the total input power factor, especially in a low-load range. The performance of the proposed compensation technique was examined through computer simulations and experimental tests, and overall effectiveness was consequently confirmed by the operating waveforms, the total input power factor and the THD characteristics.

REFERENCES

[1] T. Ohnishi, "Three-Phase PWM Converter/Inverter by Means of Instantaneous Active and Reactive Power Control," *IEEE IECON Proc.*, vol. 1, 1991, p.p. 819-824.

[2] T. Noguchi, H. Tomiki, S. Kondo, I. Takahashi and J. Katsumata, "Instantaneous Active and Reactive Power Control of PWM Converter by Using Switching Table," *IEE-Japan Trans. Ind. App.*, vol.116-D, no. 2, p.p. 222-223, 1996.

[3] T. Noguchi, H. Tomiki, S. Kondo, and I. Takahashi, "Direct Power Control of PWM Converter Without Power-Source Voltage Sensors", *IEEE Trans. Ind. App.*, vol. 34, no. 3, 1998, p.p. 473-479.

[4] M. Malinowski, M. Jesinski, and M. P. Kazmierkowski, "Simple Direct Power Control of Three-Phase PWM Rectifier Using Space-Vector Modulation (DPC-SVM)," *IEEE Trans. Ind. App.*, vol.51, no.2, 2004, p.p.447-454.

[5] K. Toyama, O. Mizuno, T. Takeshita, and N. Matsui, "Suppression for Transient Oscillation of Input Voltage and Current-Source Three-Phase AC/DC PWM Converter", *IEEJ Trans. Ind. App.*, vol. 117-D, no. 4, 1997, p.p.420-426.

[6] Y. Sato, T. Kataoka, "An Investigation of Waveform Distortion and Transient Oscillation of Input Current in Current Type PWM Rectifiers", *IEEJ Trans. Ind. App.*, vol. 114-D, no. 12, 1994, p.p.1249-1256.

[7] Toshihiko Noguchi, Daisuke Takeuchi, Somei Nakatomi, and Akira Sato, "Novel Direct-Power-Control Strategy of Current-Source PWM Rectifier," *IEEE PEDS Proc.*, CDROM, 2005, p.p. 860-865.

Frequency-Controlled LCC Resonant Converter with Synchronous Rectifier

Yu Ma, Xiaogao Xie, and Zhaoming Qian

College of Electrical Engineering, Zhejiang University, Hangzhou, China

Abstract– **This paper presents the theoretical analysis and experimental results for a frequency controlled LCC resonant converter. A prototype based on LCC resonant converter employing resonant gate drive synchronous rectifier is constructed to validate the theoretical analysis. The experimental results are in good agreement with the theoretical predictions.**

Index Terms—**LCC Resonant Converter, Synchronous Rectifier, Frequency-Control, DC-DC Conversion**

I. INTRODUCTION

Resonant converters enjoy many advantages over pulse-width modulation (PWM) converters, namely low device stress[1-3](low electromagnetic interference), and high efficiency at high frequency operations. For reasons such as these they have developed an increasing interest in designing high frequency dc-to-dc power supplies.

Basically, there are two kinds of resonant converters. One is known as the Series Resonant Converter(SRC), in which the load is connected to the resonant circuit in series. The other is the Parallel Resonant Converter(PRC), in which the load is usually connected to the resonant circuit in parallel, and the output voltage is obtained from the voltage across the resonant capacitor. The main characteristics of the SRC include: 1) inherent overload protection, 2) load sensitivity, and 3) poor operation at no load, whereas those of the PRC are 1) load insensitivity, 2) no load operation, and 3) need of overload protection.

Among these resonant converter topologies, LCC resonant converter is very similar to the parallel resonant converter, expect that a series capacitor is added to the resonant tank[4-5]. Therefore, LCC resonant converter combines the advantages of series resonant converters (SRCs) and PRCs: Firstly, the series capacitor Cs makes the equivalent tank capacitance smaller; this results in an increase of the characteristic impedance of the resonant tank, and is helpful to limit the circulating current. Secondly, the voltage conversion characteristics allow the converter to operate in a wide load range. The LCC resonant converter behaves more like a PRC under light load, and a SRC under full load. Thirdly, the LCC converter has an inherent short circuit protection. Due to the above characteristics, LCC resonant converter is preferable for voltage regulator applications having wide load variation, such as in switching power supplies.

This paper presents the theoretical analysis and experimental results of the frequency-controlled LCC resonant converter with a current-doubler synchronous rectifier that is suitable for applications with high-output current and low-output voltage. The fundamental frequency approximation of the resonant circuit is used in the analysis. In section II, the analysis of LCC resonant converter is carried out. Section III contains the design procedure and experimental results of the resonant converter. Finally, the conclusions are given in section IV.

II. ANALYSIS OF THE LCC RESONANT CONVERTER

Fig.1 shows the LCC resonant converter with the resonant gate drive synchronous rectifier. The LCC resonant tank comprises of a resonant inductor Ls, a series-resonant capacitor Cs, and a parallel-resonant capacitor Cp. ZVS operation can be accomplished when the switching frequency is above the resonant frequency and the switches are alternately turned on and off with a duty cycle of 50%. The half-bridge LCC inverter at the primary side operates with a frequency modulation mode and Fig.2 shows the ideal voltage and current waveforms. If the loaded quality factor is high, the current through the resonant tank circuit is nearly a sine wave and the converter operates in the continuous mode. The output voltage of the half-bridge LCC inverter, which is the input voltage of the rectifier, VR is sinusoidal. When VR is positive, the driving signal of the transformer auxiliary winding NR1 is applied to the synchronous switch SR1.

The switch SR1 is on and SR2 is off, and $v_{ds(SR2)} \approx \frac{v_R}{n}$.

When VR is negative, the driving signal of the winding NR2 is applied to the synchronous switch SR2. The switch

SR2 is on and SR1 is off, and $v_{ds(SR1)} \approx \frac{v_R}{n}$.

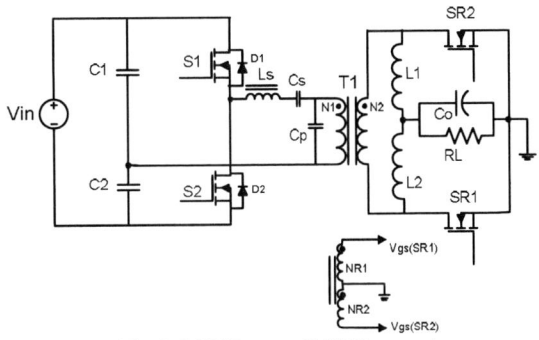

Fig. 1. LCC Resonant DC-DC converter

This work was supported by Astec Corporation and China National Science Fund (No.50237030).

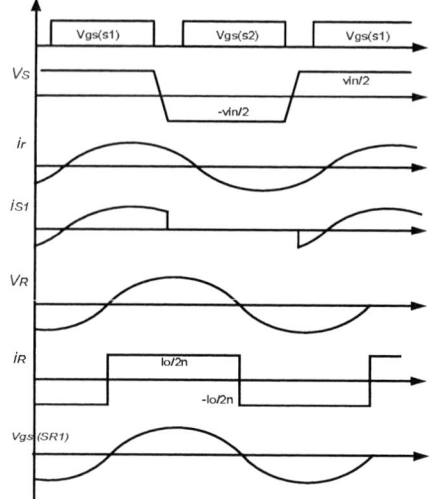

Fig. 2. The ideal key waveforms of the LCC resonant converter

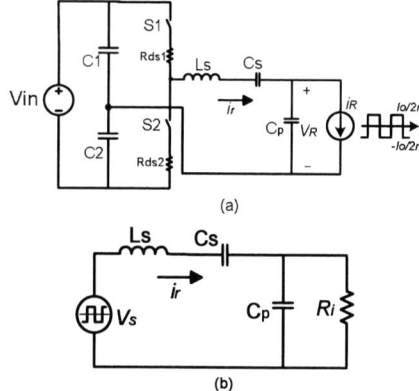

Fig.3.The equivalent circuits of LCC resonant converter

The two inductances L1 and L2 are assumed to be high enough so that each inductor conducts a dc current, equal to $\frac{I_o}{2}$. The current through the transformer is a square wave with peak values $-\frac{I_o}{2n}$ and $\frac{I_o}{2n}$.

Fig.3 shows the equivalent circuits of the LCC resonant converter with the resonant gate drive synchronous rectifier. The synchronous rectifier can be replaced by a square-wave current sink, as shown in Fig.3(a). The peak values –Io/(2n) and Io/(2n) of the current waveform i_R are valid for the current-doubler rectifier. The fundamental frequency approximation of the current through the resonant circuit is used in the analysis of the converter. The synchronous rectifier is replaced by its input resistance Ri, as shown in Fig.3(b). The dc voltage source and the switches S1 and S2 are replaced by a square-wave voltage source, which is given by

$$v_s = \begin{cases} \dfrac{v_{in}}{2} & for\ 0 < \omega t \le \pi \\[2mm] \dfrac{v_{in}}{-2} & for\ \pi < \omega t \le 2\pi \end{cases} \qquad (1)$$

Its fundamental component is $v_{i1} = v_m \sin \omega t$, where

$v_m = \dfrac{2v_{in}}{\pi}$. The rms value of v_{i1} is $v_{rms} = \dfrac{v_m}{\sqrt{2}} = \dfrac{\sqrt{2}v_{in}}{\pi}$.
The voltage transfer function of the resonant circuit is

$$|M| = \frac{V_{Ri}}{V_{rms}} = \frac{1}{\sqrt{(1+A)^2 \left[1 - \left(\dfrac{\omega}{\omega_o}\right)^2\right]^2 + \left[\dfrac{1}{Q_L}\left(\dfrac{\omega}{\omega_o} - \dfrac{\omega_o}{\omega}\dfrac{A}{1+A}\right)\right]^2}} \qquad (2)$$

where V_{Ri} is the rms value of the fundamental component of the rectifier input voltage, V_{rms} is the rms value of the fundamental component of the voltage Vs at the input of the resonant circuit, A=Cp/Cs, C=CpCs/(Cp+Cs), $\omega_o = 1/\sqrt{LsC}$ is the corner frequency, $Z_o = \sqrt{\dfrac{Ls}{C}}$, $Q_L = \dfrac{R_i}{Z_o}$ is the normalized load of the inverter. The magnitude of the ac-to-ac voltage transfer function is plotted as a function of $\dfrac{f}{f_o}$ at different values of Q_L for A=1, 2.2, 3.2 and 4.4 in Fig.4.

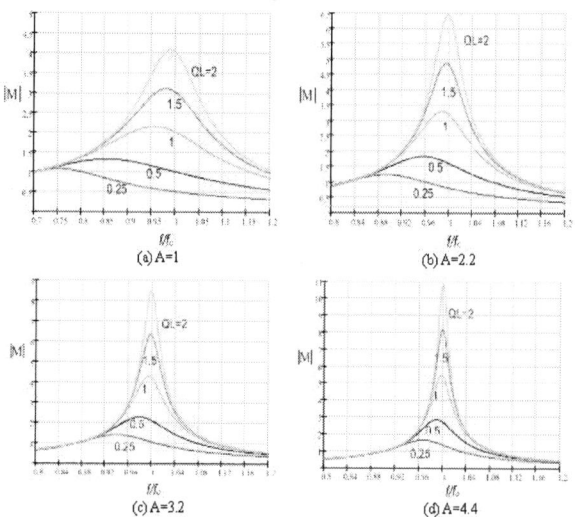

Fig.4.Voltage transfer gain in functions of f/fo with different A

III. DESIGN PROCEDURE AND EXPERIMENTAL RESULTS

For the high frequency LCC resonant converter, the following specifications are given:

Input voltage: Vin=48v   Output voltage: Vo=4v
Maximum output current: Io(max)=20A

The selection of the operating region, aimed at optimizing the power stage design needs to be carefully considered. To achieve the zero voltage switching for the main switches, the operating region should be higher than the resonant frequency. To reduce the reactive power and current stress of the resonant circuit, an operating point should be right on the peak conversion ratio curve. In this case, the current through the resonant inductor and the voltage Vs will be in phase, and most of the power transferred to the resonant tank is delivered to the load.

From the theoretical analysis in Section II, the equivalent input resistance of the rectifier Ri should be firstly obtained to design the resonant tank voltage transfer gain and the operation frequency ranges. The value of the Ri is derived as below:

The input current of the rectifier is a square wave and given by

$$i_R = \begin{cases} \dfrac{I_o}{2n} & \text{for } 0 < \omega t \le \pi \\[2mm] \dfrac{I_o}{-2n} & \text{for } \pi < \omega t \le 2\pi \end{cases} \qquad (3)$$

where n is the transformer turns ratio. The fundamental component of the input current i_R is $i_1 = I_{1m}\sin\omega t$, where $I_{1m} = \dfrac{2I_o}{\pi n}$. The input power contains only the power of the fundamental component and can be calculated as

$$P_i = \frac{I_m^2 Ri}{2} = \frac{2I_o^2 Ri}{\pi^2 n^2} \qquad (4)$$

Assuming that the efficiency of the rectifier is η_R, $\eta_R = \dfrac{P_o}{P_i} = \dfrac{I_o^2 R_L}{P_i}$. Substituting (3) into (4), one obtains the input resistance of the rectifier:

$$Ri = \frac{\pi^2 n^2 R_L}{2\eta_R} \qquad (5)$$

The equivalent input resistance of the rectifier changes as the load changes. The equivalent input resistance of the rectifier at full load condition is

$$R_{i(full\,load)} = \frac{\pi^2 n^2 R_{L(\min)}}{2\eta_R} \qquad (6)$$

From the plots of the voltage transfer gain with different values of A in Fig.4, a higher value of A results in a reduced switching frequency range and higher circulating current. The design trade-off should be carefully considered. The chosen operating region is shown in the Fig.5, according to the theoretical analysis mentioned above, where Md is the desired voltage gain.

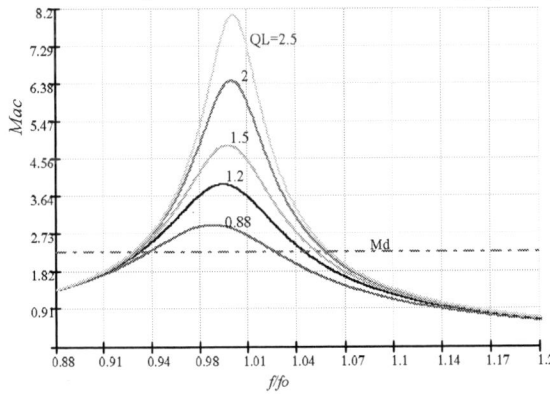

Fig.5.Voltage transfer gain in functions of *f/fo*

The LCC resonant converter is built up and tested for a dc input voltage of 48V, a dc output voltage of 4V, and a load resistance from 0.2ohm to 2ohm. The main parameters of the power stage is chosen as below:

Ls=6.8uH;　Cs=10nF;　Cp=22nF;

Resonant frequency: *fo*=736kHz;

Turns ratio: N1:N2:NR1(NR2) =5:1:1.

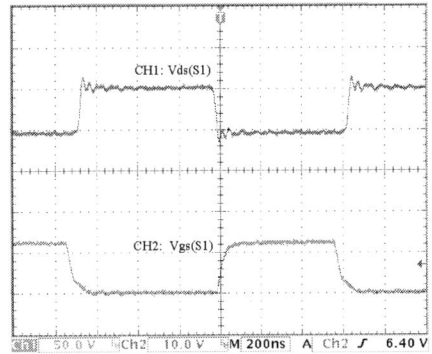

Fig.6. Waveforms of S1at full load, Vin=48v, *fs*=755KHz

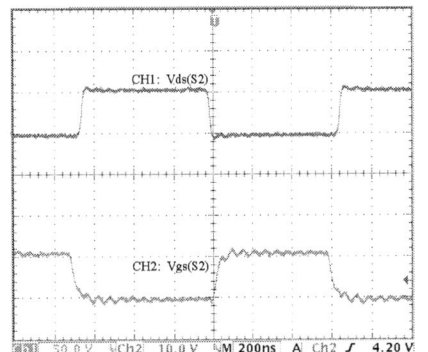

Fig.7. Waveforms of S2at full load, Vin=48v, *fs*=755KHz

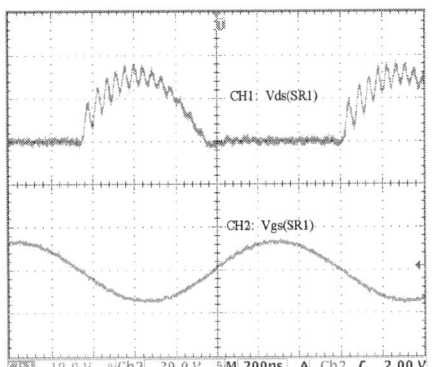

Fig.8. Voltage waveforms of SR1 at light load condition

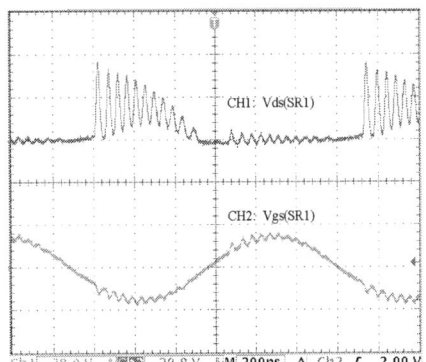

Fig.9. Voltage waveforms of SR1 at full load condition

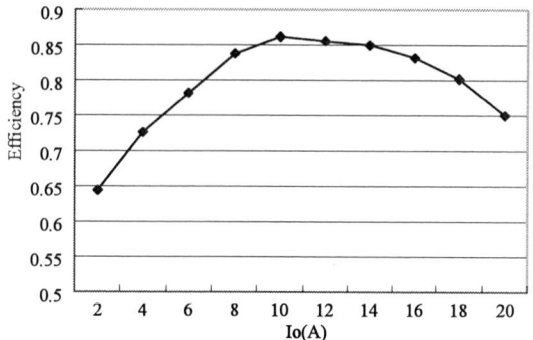

Fig.10. Measured Efficiency of LCC Resonant Converter

As the load current increases from 2A to 20A, the switching frequency is decreased from 810KHz to 755KHz. Fig.6 and Fig.7 show the waveforms of the primary side switches S1 and S2 at full load respectively. It can be seen that both switches achieve zero voltage switching. Fig.8 and Fig.9 show the voltage waveforms of the synchronous switch SR1 at light load and at full load, respectively. As the load current increases, the SR body diode conducts for a short time after the gate drive voltage becomes negative. Therefore, the overall efficiency of the resonant converter will be deteriorated. The measured efficiency of the resonant converter is shown in Fig.10. The experimental results are in good agreement with the theoretical predictions of the LCC resonant converter.

IV. CONCLUSIONS

This paper presents the theoretical analysis and experimental results of the frequency-controlled LCC resonant converter with a current-doubler synchronous rectifier. The fundamental frequency approximation of the resonant circuit and complete dc analysis are given. A prototype based on LCC resonant converter employing the resonant gate drive synchronous rectifier has also been constructed to validate the theoretical analysis. The experimental results are in good agreement with the theoretical analysis.

REFERENCES

[1] A.K.S.Bhat, "Analysis and Design of a series-parallel resonant power supply," *IEEE Transactions on Aerospace and Electronic Systems*, vol.28, No.1, pp.249-258, January 1992.

[2] R.L.Steigerwald, "A comparison of half-bridge resonant converter topologies," *IEEE Transactions on Power Electronics*, vol.3, No.2, pp.174-182, Apr 1998.

[3] Young-Goo Kang; Upadhyay, A.K.; "Analysis and design of a half-bridge parallel resonant converter," *IEEE Transactions on Power Electronics*, vol.3, Issue 3, pp.254-265, July 1988

[4] Tsai, M.-C.; "Analysis and implementation of a full-bridge constant-frequency LCC-type parallel resonant converter," *IEE Proceedings of Electric Power Applications*, vol.141, Issue.3, May 1994, pp.121-128

[5] K.Kazimierczuk, Nandak, "Analysis of Series-Parallel Resonant Converter", *IEEE Transaction on. Aerospace and Electronic Systems*, vol.29, Issue.1, Jan 1993, pp.88-98

[6] Marian K.Kazimierczuk, Shan Wang, "Frequency-Domain Analysis of Series Resonant Converter for Continuous Conduction mode", *IEEE Transaction on. Power Electronics*, Vol.7, Issue 2, Apr,1992 ,pp.270-279

[7] Marian K.Kazimierczuk, Wojciech Szaraniec, "Analysis and Design of Parallel Resonant Converter at high QL", *IEEE Transaction on. Aerospace and Electronic Systems*, vol.28, Issue 1, Jan 1992, pp.35-48

[8] Yang, E.X.; Lee, F.C.; Jovanovic, M.M.; "Small-signal modeling of LCC resonant converter ," *The 23rd Annual IEEE Power Electronics Specialists Conference*, July, 1992, vol.2, pp.941 – 948.

Selection of the Filter Capacitor for Power Supplies using 1-Phase Diode Rectifier

N. Mondal**, S. K. Biswas (Sr. Member, IEEE)**, S. Sinha* and N. K. Deb**

* Dept. of Electrical Engg. Central Calcutta Polytechnic, Kolkata 700014, India
** Dept. of Electrical Engg. Jadavpur University, Kolkata 700032, India

Abstract–Diode bridge rectifiers with capacitor-input filters are commonly used at the input stage for power supplies operating from the single phase mains input, as they provide a cost effective solution in the lower power range. However, as there is no simple design rule existing in published literature that gives a reasonable prediction based on real-life conditions, the design of the rectifier-capacitor stage is still a trial and error or based upon experience. This paper presents an approximate (but simple) solution based upon empirical relations, for the prediction of the currents involved and specifies the capacitor, so that it can be selected.

Index Terms—Capacitor-input Filter, Diode Rectifier, Ripple Current in Capacitor, SMPS.

I. INTRODUCTION

Single-phase diode bridge rectifiers with capacitor filters are commonly used at the input stage for power supplies as they provide a cost effective solution in the lower power range. Low power switch mode power supplies (SMPS) of less than about 300 Watts are today commonly used for various portable power supplies or power adapters. However, the design of the rectifier-capacitor stage is still a trial and error or based upon experience, as there is no simple design rule existing in published literature [1-4] that gives a reasonable prediction of the currents and capacitor selection based on real-life conditions. The best values was given by Schade [1] in 1943 using a set of curves obtained experimentally, but it requires a prior knowledge of the ratio of source resistance to load resistance (R_S/R_L) – a rather difficult prediction. This paper presents an approximate (but simple) solution based upon empirical relations, to the prediction of the currents involved and specifies the capacitor, so that it can be selected.

The circuit of such a 1-phase rectifier input stage with capacitor filter, interfaced between the load and the alternating supply source, is shown in Fig 1. In real life, the source V_s includes its series equivalent inductance L_s and series resistance R_s. The dc filter capacitor C stores energy during its charging time and feeds the load during its discharge time. The load for the capacitor, which is a power supply, usually has a constant voltage and constant current output. Therefore, for a high efficiency switching converter power supply operating at constant output voltage and current, the reflected load to the capacitor is a constant power load. For a linear power supply operating at constant output voltage and current, the reflected load

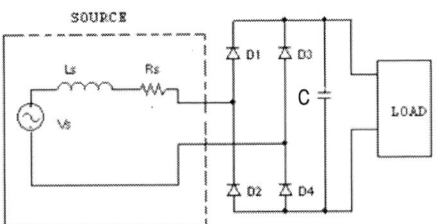

Fig. 1. The load interfaced with source through rectifier & filter

to the capacitor is a variable resistance. The usual practical waveform [2-3] is shown in Fig 2, where it will be noticed that the capacitor starts charging to a point beyond the instant when the supply voltage peak occurs, due to the energy stored in the source inductance. After that, the capacitor discharges till its voltage falls below the increasing sine wave of the next half-cycle of input. Thus, the actual duration of charging is higher than the ideal case when source impedance is zero (when capacitor charging stops at the peak of the input sinewave). All the energy stored in the capacitor during the charging interval is supplied to the load during the discharge interval such that the average current through the capacitor is zero (since its total charge remains constant after a cycle). The nature of discharge is exponential for constant resistance load as described in existing literature, but is different for the case of a constant power load like SMPS, as will be discussed in this paper.

Further, load like SMPS will have a ripple current drawn from the same capacitor due to its own switching stage, operating at high frequency. In this case also, no average current will be supplied from the capacitor. Thus, In low power SMPS, there is a common capacitor that

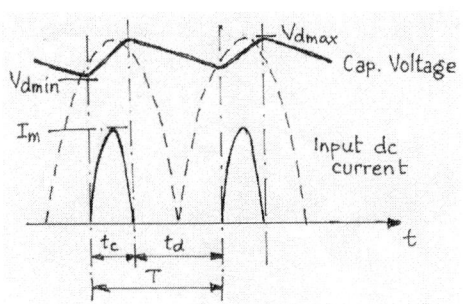

Fig. 2. Voltage & current waveforms

978-1-4244-0644-9/07/$25.00 ©2007 IEEE

serves the dual function of filtering the input dc on one hand, as well as the function of filtering the ripple current input of the switching converter on the other. Thus, its ripple current will have contribution from both sides at different frequency components.

The method of analysis proposed in this paper is based on the following assumptions:

1. The ripple on the dc bus voltage is small, such that the power supply input current can be approximated by its average value I_d.
2. The input ac current waveform in each half cycle can be approximated by a sinusoidal half wave of half time period t_c equal to the charging interval and a peak of I_m.
3. The actual charging interval is higher than the ideal condition charging period by a factor α, which is dependant on source inductance, but not much influenced by the values of capacitor C or power supply current, for a specific dc bus voltage ripple measured by the factor β.
4. The dc bus voltage ripple is measured by a factor β, which is the ratio of the minimum instantaneous voltage V_{dmin} by the maximum dc bus voltage V_{dmax}. It is customary to select its value close to 0.866 in a 1-phase rectifier with capacitor filter, since this value is 0.866 in a 3-phase 6-pulse rectifier without any capacitor on the dc bus. Thus β has usual values from 0.8 to 0.9. Lower values result in too large voltage ripple while higher values will result in too small charging intervals and hence too large charging current peaks (to deliver same charge).
5. The actual discharging interval of the capacitor is t_d, during which the power flow to the power supply is constant for switch mode power supplies

II. ANALYSIS.

A. Calculation of Capacitor Charging Current

Considering the input dc current (i_{in}) waveform to be parts of sinusoidal half-waves, the expression for duration t_c is (with $\omega = 2\pi f$ and $f = 1/2t_c$)

$$I_{in} = I_m \sin \omega t = I_m \sin 2\pi f t$$
$$= I_m \sin (\pi/t_c)t \qquad (1)$$

Consider the repetition time period of input dc current pulses to be T, where

$$T = 1/2f_s \qquad (2)$$

where f_s = ac supply frequency, eg 50Hz.

The rms value of input dc current (pulsating) to the capacitor and load combined, is:

$$I_{in}(rms) = \sqrt{\frac{1}{T}\int_0^{t_c} i_{in}^2 dt} = \sqrt{\frac{I_m^2}{T}\int_0^{t_c} (\sin^2 \omega t)dt}$$

$$= I_m \sqrt{\frac{t_c}{2T}} \quad \text{Amps} \qquad (3)$$

The dc bus capacitor cannot carry dc current component since there is no net build-up of charge in it. Thus, it carries only the ac component of input dc current i_{in}. The dc component of load current is designated as I_d.

Thus the rms ripple current due to charging, through dc bus capacitor is

$$I_{ccap} = \sqrt{I_{in(rms)}^2 - I_d^2} = \sqrt{I_m^2 \frac{t_c}{2T} - I_d^2}$$

$$= I_d \sqrt{\left[\frac{I_m^2}{I_d^2}\right]\frac{t_c}{2T} - 1} \quad \text{Amps} \qquad (4)$$

The charge input to the capacitor during the charging period is the total charge input minus the average charge to the load during the same period. This will be equal to the average charge withdrawn by load during the discharge period.

Upon integration, total charge input to capacitor during one pulse duration of input dc current (i_{in}) is

$$Q_{in} = \int_0^{t_c} I_m \sin \omega t = (2I_m/\pi).t_c \quad \text{Coulombs} \qquad (5)$$

Thus, equating the charges,

$$\frac{2I_m}{\pi}t_c - I_d t_c = I_d t_d \qquad (6)$$

while,

$$t_d = T - t_c = \frac{1}{2f_s} - t_c \qquad (7)$$

Thus, substituting (7) in (6), yields the peak current as

$$I_m = \frac{\pi}{4f_s}.\frac{I_d}{t_c} \quad \text{Amps} \qquad (8)$$

Substituting eq (8) in (3)

$$I_{in}(rms) = \frac{\pi}{4f_s}\frac{I_d}{t_c}\sqrt{\frac{t_c}{2T}} = \frac{\pi}{4f_s}\frac{I_d}{t_c}\sqrt{t_c f_s}$$

$$= \frac{\pi}{4\sqrt{t_c f_s}}I_d \quad \text{Amps} \qquad (9)$$

Similarly, substituting eq (8) in (4),

$$I_{ccap} = I_d \sqrt{\frac{\pi^2 I_d^2}{16 f_s^2 t_c^2}\frac{1}{I_d^2}\frac{t_c}{2T} - 1}$$

$$= I_d \sqrt{\frac{\pi^2}{16 t_c f_s} - 1} \quad \text{Amps} \qquad (10)$$

The equations (8), (9) & (10) indicate that if I_d, t_c and f_s are known, then the values of peak input ac current I_m, the rms input ac current $I_{in}(rms)$ and the capacitor charging ripple current (rms) I_{ccap} can be predicted.

If the capacitor is connected to load through a switching power converter operating at constant output voltage and current, then during the discharge interval t_d of the dc bus capacitor, power is constant, ie.,

$$P = i(t)^2 R \quad \text{Watts} = \text{constant} \qquad (11)$$

Note that the equivalent load resistance R as seen by the capacitor C varies with current to maintain constant power.

At the instant when discharge starts, energy in capacitor is

$$E_1 = 0.5CV_{dmax}^2 \quad \text{Joules} \qquad (12)$$

1606

After discharge time t_d, energy in the same is

$$E_2 = 0.5CV_{dmin}^2 \quad \text{Joules} \tag{13}$$

The energy has been dissipated in the equivalent resistor R with constant power for duration t_d, ie.,

$$E_1 - E_2 = 0.5CV_{dmax}^2 - 0.5CV_{dmin}^2 = P \cdot t_d \tag{14}$$

Re-arranging eq. (14),

$$V_{dmin} = \sqrt{V_{dmax}^2 - \frac{2Pt_d}{C}} \quad \text{Volts} \tag{15}$$

Hence from (15), the required capacitance is :

$$
\begin{aligned}
C &= \frac{2Pt_d}{V_{dmax}^2 - V_{dmin}^2} \\
&= \left(\frac{1}{2f_s} - t_c\right)\frac{2P}{\left(V_{dmax}^2 - V_{dmin}^2\right)} \quad \text{Farads} \tag{16}
\end{aligned}
$$

Thus once again, t_c holds the key to solution of the equations.

Now, if source inductance is neglected, then the capacitor charges up sinusoidally from V_{dmin} as

$$v = V_{dmax}\sin\omega t \tag{17}$$

The charging duration is t_c', ending when $v = V_{dmax}$, Hence ,

$$\left|V_{dmin}\right| = \left|V_{dmax}\cos\omega.t_c'\right| \tag{18}$$

Re-arranging eq. (18),

$$t_c' = \frac{\cos^{-1}(V_{dmin}/V_{dmax})}{2\pi.f_s} = \frac{\cos^{-1}\beta}{2\pi.f_s} \tag{19}$$

In actual condition, it is observed that due to the presence of source inductance, the value of t_c' increases from that given by eq. (19) to t_c depending on the value of β predominantly, irrespective of actual values of R & C individually. Thus, it is convenient to express as:

$$t_c = \alpha t_c' = \frac{\alpha\cos^{-1}\beta}{2\pi.f_s} \quad \text{sec} \tag{20}$$

Experiments are carried out to plot the typical value of α versus β for typical source impedances at 230 volts, 50 Hz ac supply. An experimental 1-phase rectifier with capacitor input filter and a switching power stage, loaded with resistors, was used to obtain data through an oscilloscope. The value of load, dc bus capacitance, as well as the input ac voltage was varied. Data was

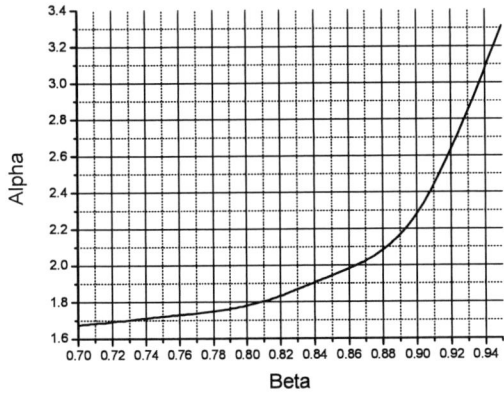

Fig. 3. Experimental plot of α versus β

collected at different sites and an average set of values was obtained. This value of α is plotted versus β as depicted in Fig. 3.

During design, the value of β acceptable to the designer must be decided first, corresponding to I_d. Then, find corresponding typical value of α. Now, the approximate values of the various currents can be found from the expressions (8), (9) and (10) modified below. These current values will form the basis of the selection of the diodes, the input fuse and the dc capacitor.

$$I_m = \frac{\pi}{4\alpha.t_c'f_s}I_d = \frac{\pi^2}{2\alpha.\cos^{-1}\beta}I_d \quad \text{Amps} \tag{21}$$

The rms value of input ac current (I_{ac}), has the same numerical value as that of the input dc current I_{in}. Hence,

$$
\begin{aligned}
I_{ac} &= \frac{\pi}{4\sqrt{\alpha.t_c'f_s}}I_d = \frac{\pi\sqrt{2\pi}}{4\sqrt{\alpha.\cos^{-1}\beta}}.I_d \\
&= I_d\sqrt{\frac{\pi^3}{8\alpha.\cos^{-1}\beta}} \quad \text{Amps} \tag{22}
\end{aligned}
$$

$$
\begin{aligned}
I_{ccap} &= I_d\sqrt{\frac{\pi^2}{16\alpha.t_c'f_s} - 1} \\
&= I_d\sqrt{\frac{\pi^3}{8\alpha.\cos^{-1}\beta} - 1} \quad \text{Amps} \tag{23}
\end{aligned}
$$

$$
\begin{aligned}
C &= \left(\frac{1}{2f_s} - \frac{\alpha\cos^{-1}\beta}{2\pi f_s}\right)\frac{2P}{V_{dmax}^2\left(1-\beta^2\right)} \\
&= \frac{P\left(\pi - \alpha\cos^{-1}\beta\right)}{\pi f_s V_{dmax}^2\left(1-\beta^2\right)} \quad \text{Farads} \tag{24}
\end{aligned}
$$

The average values of dc bus voltage and discharge current are :

$$V_d = \frac{V_{dmin} + V_{dmax}}{2} = \frac{(1+\beta)V_{dmax}}{2} \quad \text{Volts} \tag{25}$$

$$I_d = \frac{P}{V_d} = \frac{2P}{(1+\beta)V_{dmax}} \quad \text{Amps} \tag{26}$$

B. Calculation of Capacitor Discharge Current

The discharge current for the capacitor has to be calculated individually depending on the specific type of switching power stage used (with switching frequency f), as given in the following section.

CASE I : Flyback Converter load operating in DCM

In the case of a flyback converter being used as the switching stage (being most common in low power SMPS), in Discontinuous Conduction Mode (DCM), its input current waveform is as shown in Fig. 4(a). It is a saw-tooth wave-shape during the on-time (t_{on}) of the switch with a peak magnitude of I_p.

Average input current of the converter is given as:

$$I_d = \frac{I_p.t_{on}}{2T} = \frac{I_p D}{2} \quad \text{Amps} \tag{27}$$

1607

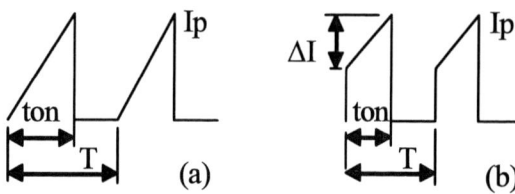

Fig. 4. Input current of switching stage
(a) Flyback converter in DCM
(b) Flyback converter in CCM
or Forward converter

RMS input current of converter is :

$$I_{rms} = I_p\sqrt{\frac{t_{on}}{3T}} = I_p\sqrt{\frac{D}{3}} \quad \text{Amps} \qquad (28)$$

Power input to converter is :

$$\begin{aligned}
P &= 0.5LI_p^2 f = V_d I_d \\
&= 0.5V_d I_p D \quad \text{Watts} \qquad (29)
\end{aligned}$$

Since average current cannot flow through the capacitor (whose voltage is constant at V_d),
RMS discharge current through capacitor C is :

$$\begin{aligned}
I_{dcap} &= \sqrt{I_{rms}^2 - I_d^2} = I_p\sqrt{\frac{D}{3} - \frac{D^2}{4}} \\
&= \frac{2P}{V_d D}\sqrt{\frac{D}{3} - \frac{D^2}{4}} \quad \text{Amps} \qquad (30)
\end{aligned}$$

From eq. (29), for constant power output from a given converter, I_p is constant. Hence, D is inversely proportional to V_d. Hence, for constant power, I_{dcap} is maximum when D = 2/3

CASE II : Flyback Converter load operating in CCM

In the case of a flyback converter being used as the switching stage in Continuous Conduction Mode (CCM), its input current waveform is as shown in Fig. 4(b). It is a trapezoidal wave-shape during the on-time (t_{on}) of the switch with a peak magnitude of I_p and a slope of ΔI at the top. Thus, the current rises from a value of I_p - ΔI to a value of I_p during the on-time.

Average input current of the converter is given as:

$$I_d = \frac{(2I_p - \Delta I)}{2}D \quad \text{Amps} \qquad (31)$$

RMS input current of converter is :

$$I_{rms} = \sqrt{\left(I_p^2 - I_p\Delta I + \frac{\Delta I^2}{3}\right)D} \quad \text{Amps} \qquad (32)$$

Power input to converter is :

$$\begin{aligned}
P &= 0.5L\Delta I(2I_p - \Delta I)f = V_d I_d \\
&= 0.5V_d D(2I_p - \Delta I) \\
&= 0.5V_d DI_p(2 - \Delta I/I_p) \quad \text{Watts} \qquad (33)
\end{aligned}$$

Since average current cannot flow through the capacitor (whose voltage is constant at V_d), the rms discharge current through Capacitor C is :

$$\begin{aligned}
I_{dcap} &= \sqrt{I_{rms}^2 - I_d^2} \\
&= \sqrt{\left(I_p^2 - I_p\Delta I + \frac{\Delta I^2}{3}\right)D - \left(\frac{(2I_p - \Delta I)D}{2}\right)^2} \\
&= I_p\sqrt{\left(1 - \left(\frac{\Delta I}{I_p}\right) + \frac{1}{3}\left(\frac{\Delta I}{I_p}\right)^2\right)D - \left(1 - \left(\frac{\Delta I}{I_p}\right) + \frac{1}{4}\left(\frac{\Delta I}{I_p}\right)^2\right)D^2} \\
&\qquad\qquad\qquad\qquad\qquad \text{Amps} \qquad (34)
\end{aligned}$$

The value of peak current I_p is obtained from (33) as :

$$I_p = \frac{2P}{V_d D[2 - (\Delta I/I_p)]} \qquad (35)$$

Thus, the ratio of $\Delta I/I_p$ is vital in deciding the value of capacitor discharge current and the duty cycle at which its maximum value occurs.

CASE III : Forward Converter load :

In the case of a forward converter being used as the switching stage, its input current waveform will be similar to that of a Flyback converter operating in CCM as shown in Fig. 4(b). It is a trapezoidal wave-shape during the on-time (t_{on}) of the switch with a peak magnitude of I_p and a slope of ΔI at the top. Thus, the current rises from a value of I_p - ΔI to a value of I_p during the on-time.

Thus, all expressions for I_d, I_{rms}, P and I_{dcap} will be similar to that of a Flyback converter operating in CCM (except that the first expression of eq. (33) involving L is not valid here).

C. Selection of Capacitor

The total rms ripple current through the capacitor is the combined effect of its charging and discharge ripple currents, ie., I_{ccap} and I_{dcap}, obtained respectively from eq. (23) and eq. (30) or (34), as below :

$$I_{cap} = \sqrt{I_{ccap}^2 + I_{dcap}^2} \qquad (36)$$

The minimum required value of capacitor is given by eq. (24). The selected capacitor must simultaneously have the capability to carry the above ripple current at the specified operating temperature, as well as have a value just above the minimum specified value.

III. DESIGN PROCESS

The design or selection process for the rectifier and capacitor starts with the noting of the input ac supply voltage range. At the lowest supply voltage, the value of β will be maximum. Now fix the value of β and note down possible value of α from Fig. 3. Hence, V_{dmax} is $\sqrt 2$ times the ac minimum rms voltage and V_{dmin} is β times V_{dmax}. Average dc voltage is known from V_{dmax} and V_{dmin} using eq. (25). From the power requirement at full load, average input current to the power supply is calculated using eq. (26). Hence, the input current to rectifier and the charging ripple current through capacitor are known along with the minimum capacitor value.

Now design the switching converter and estimate its duty cycle working range to facilitate calculation of its input discharge ripple current depending on its specific topology. For the case of a flyback converter being used as the switching stage in DCM, the discharge ripple current is calculated easily from eq. (30) where D is the value closest to 2/3 (where the ripple current is maximum). For the case of a flyback converter being used as the switching stage in CCM or a forward converter, the discharge ripple current is calculated from eq. (34) and (34) but it will be rather difficult to estimate the maximum value without a few trials. In these conditions, the designer has to fix the maximum value of $\Delta I/I_p$ and use this along with the suitable value of D to find out the maximum capacitor discharge ripple current.

Having obtained the charging and discharge ripple currents, calculate the total rms ripple current through the capacitor from eq. (36). The maximum voltage rating of the capacitor is however to be higher than $\sqrt{2}$ times the ac maximum rms voltage.

A design example is given as follows:

A rectifier system is to operate from an AC supply varying between 165V – 275V, 50 Hz to supply a constant power switching dc-dc converter that gives an output of 275W with efficiency of 85.5%.

Select β of 0.8.

Input to converter = P = 275/0.855 = 321.6W

V_{dmax} = 165*$\sqrt{2}$ = 233.3V

V_{dmin} = 0.8*233.3 = 186.6V

V_{av} = 209.95V

I_{dc} = 321.6/209.95 = 1.53A

Corresponding to β of 0.8, from Fig. 3, α=1.78. Hence, from given equations,

I_m = 6.59A

I_{ac} = 2.81A

I_{ccap} = 2.36A

C = 209μF (minimum required)

The switching converter is selected to be a flyback topology working in DCM with a maximum duty cycle of 0.5. Thus, maximum ripple current is at D = 0.5.

I_{dcap} = [(2*321.6)/(209.95*0.5)]*$\sqrt{[(0.5/3)-(0.25/4)]}$
= 1.98A

I_{cap} = $\sqrt{[2.36^2 + 1.98^2]}$ = 3.08A

IV. EXPERIMENTAL VERIFICATION

The actual values measured experimentally in a system using such a converter with 220μF capacitor and 165V ac supply are:

V_{dmax} = 230V

V_{dmin} = 186V

I_m = 7.0A

I_{ac} = 2.65A

I_{cap} = 2.85A

The results are very close to their predicted values.

The oscillogram of capacitor voltage and input ac current are given in Fig. 5. Note that the peak ac voltage is lower than expected due to distortion created in the ac voltage by the peak current of the rectifier. The same effect has extended the base width of the current from a near-sinusoidal waveform. The reason is attributed to the increase of source reactance by the autotransformer used to reduce and control the ac voltage during experiment.

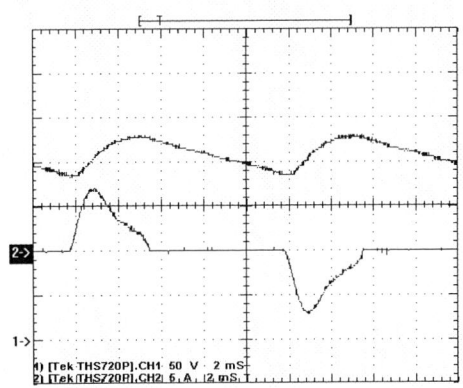

Fig. 5. Oscillogram of actual voltage & current of designed system

V. CONCLUSIONS

An approximate method of expressing the currents in a 1-phase diode bridge rectifier with capacitor filter has been presented and the results indicate reasonable accuracy. Unlike available approximate solutions, this method takes into account the effect of the source impedance. Most important is that it can predict the peak current through the diode, the input rms ac current and the ripple current through the capacitor, enabling the proper selection of the rectifier diodes, the input fuse and the dc capacitor. This process was verified experimentally. Although the results are not very accurate, the process is simple enough to generate some idea regarding the magnitude of the currents, a feat that was not possible earlier without detailed analysis.

REFERENCES

[1] S. C. Schade, "Analysis of Rectifier Operation", *Proc. Of IRE*, vol 31, no. 7, July 1943.

[2] J. Schaefer, *Rectifier Circuits*, J. Wiley & Sons, N. York, 1965, pp221-245.

[3] N. Mohan et al, *Power Electronics*, J. Wiley & Sons, N. York, 1995, pp95-100.

[4] *Rectifier Applications Handbook*, ON-Semiconductor, Denver, 2001, pp116-119

High Performance Single-Phase Voltage Regulator with a Simple Circuit Topology

Chien-Ming Wang*, Ching-Hung Su**, Chang-Hua Lin***, Maw-Yang Liu*, and Kuo-Lun Fang*

* Department of Electrical Engineering, National Ilan University, I-Lan, Taiwan
** Department of Electronic Engineering, Lunghwa University of Science and Technology, Taoyuan, Taiwan
*** Department of Electrical Engineering, Tatung University, Taipei, Taiwan

Abstract-- **This paper proposes a novel high performance single-stage voltage regulator to give high input power factor and low current distortion on the rectifier side and provide clean and stable ac voltage on the inverter side. The proposed voltage regulator only employs four switches to perform the power factor correction in input side and provide regulation of the output ac voltage magnitude. Thus, its cost is lower than the others. It has a common arm between the rectifier and inverter, and adopts an appropriate switching strategy. A significant reduction in the conduction losses is achieved, since the rectifier in the proposed converter uses a single converter instead of the conventional configuration composed of a four-diode front-end rectifier followed by a boost converter. An average-current-mode control is employed in the rectifier side of proposed converter to synthesize a suitable low harmonics sinusoidal waveform for the input current. The sinusoidal pulse-width modulation (SPWM) control strategy is employed in the inverter of proposed converter to achieve well dynamic regulation. A design example of 1000W single-phase voltage regulator is examined to assess its performance.**

Index Terms—**Voltage regulator, rectifier, inverter, SPWM**

I. INTRODUCTION

Matrix converters have received considerable attention [1]-[3] due to their potentiality to provide direct AC/AC conversion without energy storage. However, they turned into wide application due to severe requirements: four-quadrant switches, critical timing, sensing of switch voltage and current, snubber circuits needed to absorb overvoltages coming from the inductive commutation. As a result, circuit efficiency and reliability are affected. More popular is the indirect ac/dc/ac conversion by means of PWM rectifier-inverter systems with dc link. As compared to matrix converters, these systems show improved reliability and allow a greater output voltage. The one of conventional single-phase ac/dc/ac converters is shown in Fig. 1(a) and (b). In these system, a big tank capacitor in the dc link provides decoupling between the rectifier and the inverter, so that the two converters can be driven independently according to usual PWM techniques [4]-[5], providing excellent input and output performances. In fact, this system is the combination of the boost rectifier and the buck inverter. The boost rectifier performs the functions of power factor correction and boost ac/dc conversion. The buck inverter performs the function of buck dc/ac conversion with output voltage of variable amplitude. Therefore, these ac/dc/ac systems have been widely used in industrial application such as uninterruptible power supplies (UPS),

static frequency changes and variable speed drives. However, for the application in UPS, it suffers from the large number of switches. It also requires an isolation transformer at the backend, which is bulky, heavy, and expensive. Thus, from the practical and commercial points of view, they have several disadvantages such as high cost, large size, and control complexity due to the large number of switches. For improve these drawbacks, a three-leg six switches type voltage regulator is proposed in [6]-[7] shown in Fig. 1(c). It has the capability of delivering sinusoidal input current with unity power factor and good output voltage regulation. But it still needs six switches and more complex controller.

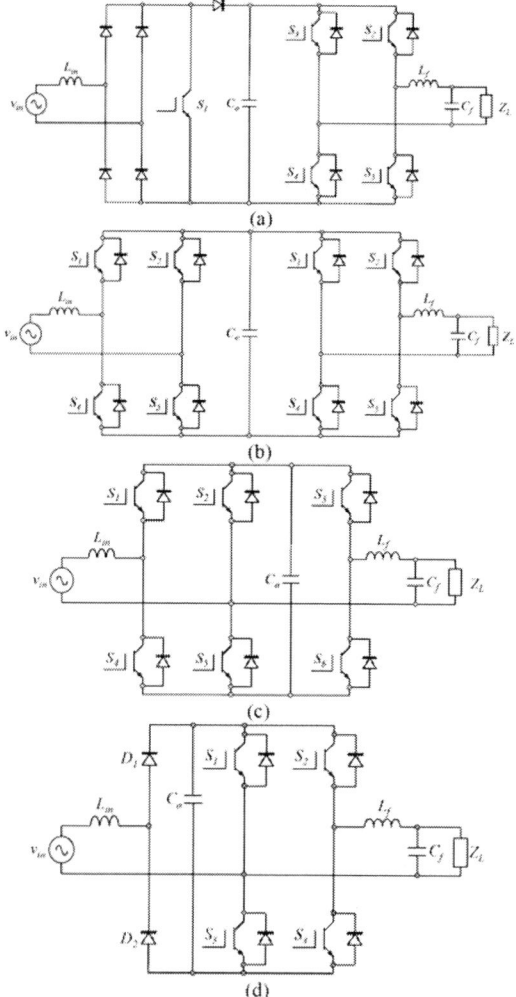

Fig. 1 (a) Conventional circuit configuration of discontinuous input current AC/DC/DC converter. (b) Conventional circuit configuration of discontinuous input current AC/DC/DC converter. (c) Three-leg six switches type AC/DC/DC converter (d) Proposed single-phase voltage regulator.

For more reducing the number of switches and simplifying the controller, a novel common-neutral single-phase voltage regulator with high input power factor and new control strategy is proposed in this paper. In the proposed voltage regulator, focuses on the practical and commercial feasibility of the ac/dc/ac converter shown in Fig. 1(c) by reducing the number of the main switches, eliminating the isolation transformer, designing a simple and low-cost controller, and decreasing the cost of the filter. Therefore, the proposed voltage regulator has a common arm between the rectifier and inverter. The proposed voltage regulator only employs four switches to perform the power factor correction in input side and provide regulation of the output ac voltage magnitude. Thus, its cost is lower than the others. Therefore, it can more reduce the number of switches, cost and size under maintaining the capabilities of the circuit configuration in Fig. 1(c). An average-current-mode control is employed in the rectifier side of proposed converter to synthesize a suitable low harmonics sinusoidal waveform for the input current. The sinusoidal pulse-width modulation (SPWM) control strategy is employed in the inverter of proposed converter to achieve well dynamic regulation. System analysis for predicting and evaluating the voltage regulator performance are conducted.

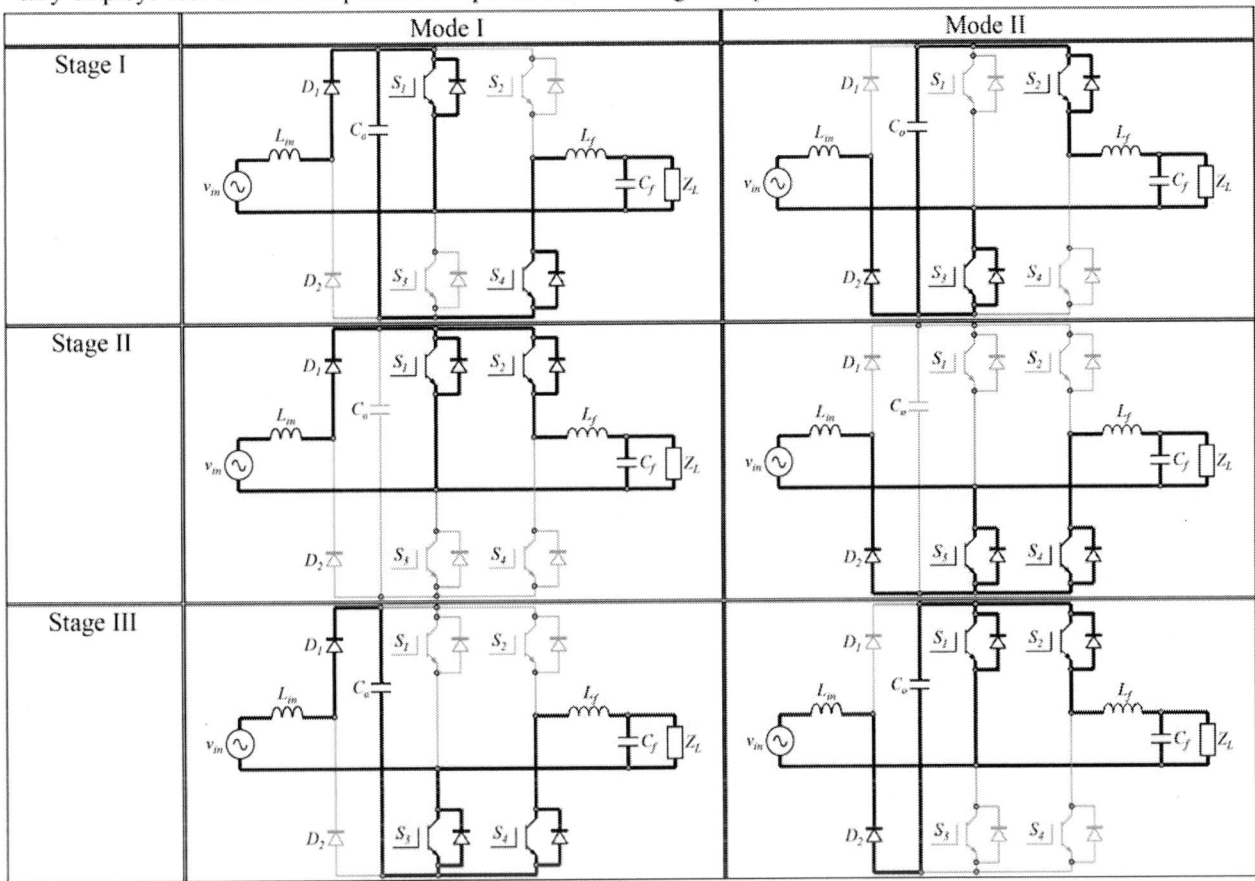

Fig. 2 The topology stages of the proposed common-neutral single-phase voltage regulator.

II. OPERATION PRINCIPLE

A. Main Circuit and Operation

The power stage diagram of the novel common-neutral single-phase voltage regulator is shown in Fig. 1(d). The circuit can be divided in two sections. The first section is a conventional PWM buck inverter with unipolar voltage switching. It is composed of the switches S_1, S_2, S_3, S_4, and output filter L_f, C_f. This section performs the function of buck dc/ac conversion with output voltage of variable amplitude. The second section is the pulse-width modulation continuous conduction mode step-up ac/dc converter, composed of L_{in}, S_1, S_3, D_1, D_2, and C_o. This section performs the functions of power factor correction and boost ac/dc conversion at fixed frequency. The switches operate on a PWM pattern to shape both the input current and output voltage to follow the reference commands. The inductor L_{in} provides voltage boost operation, and L_f and C_f also provide filter operation of the output voltage. The dc-link capacitor C_o acts as a dc voltage source and provides filtering function. For satisfying the step up/down function in the proposed converter, the dc-link capacitor voltage must be sufficiently higher than the peak voltage of the ac main source. Because the frequencies of input voltage and the desired output voltage are the same, we can divide the circuit operation into two modes during one line voltage period. The dynamic equivalent circuits during one switching period for each mode are shown in Fig. 2. To simplify the analysis, it is assumed that the proposed common-neutral single-phase voltage regulator is operating in steady-state, all devices are ideal and the losses in L_{in}, L_f, C_o, and C_f are all neglected. The operational principle of the proposed converter can be described as follow.

Mode I: (positive half cycle of line voltage)

Stage I: Before this stage, the switches S_1 and S_2 operate at turn-off state, the switches S_3 and S_4 operate at turn-on state. The energy stored in inductor L_{in} is delivered to capacitor C_o through D_1 and the opposite diode of switch S_3 while the output loop of the inverter is in a freewheeling state and the freewheeling loop is formed by S_3, S_4, the opposite diodes of switch S_3, S_4, and output filter loop. This stage begins when S_1 turns on and S_3 turns off. The input inductor L_{in} is charged from input voltage v_{in} through the switch S_1 and the diode D_1. The energy stored in capacitor C_o supplies inverter stage through S_1 and S_4.

Stage II: During this stage, the input inductor L_{in} is continuously charged from input voltage v_{in} through the switch S_1 and the diode D_1. The energy stored in capacitor C_o does not supply inverter stage and the output loop of the inverter returns a freewheeling state. The freewheeling loop is formed by S_1, S_2, the opposite diodes of switch S_1, S_2, and output filter loop.

Stage III: In this stage, the output loop of the inverter is still in a freewheeling state. But the freewheeling loop in this stage is not the same one in stage II. It is formed by S_3, S_4, the opposite diodes of switch S_3, S_4, and output filter loop. The energy stored in inductor L_{in} is delivered to the filter capacitor C_o through D_1.

Mode II: (negative half cycle of line voltage)

Stage I: Before this stage, the switches S_1 and S_2 operate at turn-on state, the switches S_3 and S_4 operate at turn-off state. The energy stored in inductor L_{in} is delivered to capacitor C_o through D_2 and the opposite diode of switch S_1 while the output loop of the inverter is in a freewheeling state and the freewheeling loop is formed by S_1, S_2, the opposite diodes of switch S_1, S_2, and output filter loop. This stage begins when S_3 turns on and S_1 turns off. The input inductor L_{in} is charged from input voltage v_{in} through the switch S_3 and the diode D_2. The energy stored in capacitor C_o supplies inverter stage through S_2 and S_3.

Stage II: During this stage, the input inductor L_{in} is continuously charged from input voltage v_{in} through the switch S_3 and the diode D_2. The energy stored in capacitor C_o does not supply inverter stage and the output loop of the inverter returns a freewheeling state. The freewheeling loop is formed by S_3, S_4, the opposite diodes of switch S_3, S_4, and output filter loop.

Stage III: In this stage, the output loop of the inverter is still in a freewheeling state. But the freewheeling loop in this stage is not the same one in stage II. It is formed by

S_1, S_2, the opposite diodes of switch S_1, S_2, and output filter loop. The energy stored in inductor L_{in} is delivered to the filter capacitor C_o through D_2.

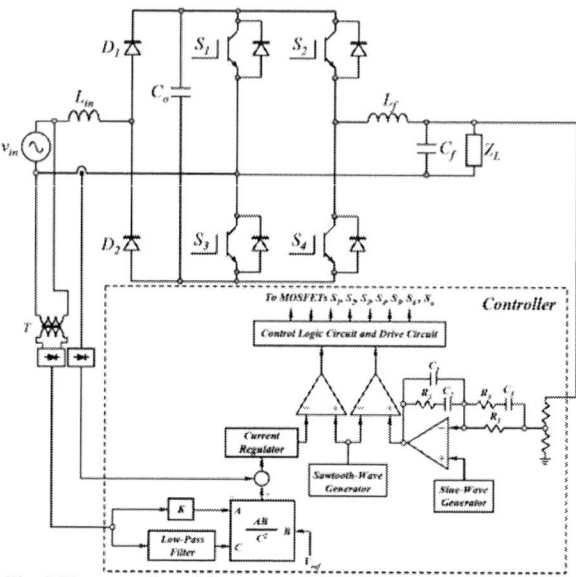

Fig. 3 The controller of the proposed single-phase voltage regulator.

B. Power Factor Correction

Traditionally, conversion of the ac line voltage from the utilities has been dominated by phase-controlled or diode rectifiers. The nonideal character of the input current drawn by these rectifiers creates a number of problems for the power distribution network and for other electrical systems in the vicinity of the rectifier, such as high input current harmonics, low input power factor, lower rectifier efficiency, ac source voltage distortion, and high reactive-component size. Therefore, the optimal rectifier would be one in which the input would draw a pure sinusoidal current at unity power factor. The topology usually employed in power factor correction single-phase power supplies in composed by a front end rectifier followed by a boost converter. In this topology, the boost converter in continuous conduction mode (CCM) with the average current control and pulse-width modulation (PWM) technique has been the most popular circuit [1]-[3]. But the significant conduction loss in the PFC circuit always includes two diode losses from the front-end bridge rectifier and one (or two) power switch loss, the conduction losses is larger. For improve this problem, the rectifier circuit has been revised. The presented rectifier circuit has only two power semiconductor conduction drops in the power flow path. Therefore, the conduction loss is considerably reduced.

C. Inverter

The PWM inverter is required to synthesize a sinusoidal waveform at its output port under different types of loads. Since the PWM inverter plays such an important role in converting a dc voltage to an ac voltage, the performance of an ac power conditioning system in highly dependent on the built-in controller of the PWM inverter. To minimize the harmonic distortion of the output waveform of a PWM inverter, many methods based on modulation strategies have been proposed. Any PWM technique can be used with this inverter. A general sinusoidal PWM technique is employed in this paper for simplicity. In this case, because the duty ratio D_{lk} is directly proportional to the amplitude of output voltage in the kth switching period, it can be designed with ease according to the desired load voltage.

$$\left|V_{ok}\right| = D_{lk} \bullet V_C \tag{1}$$

Therefore, the magnitude of the output voltage can be simply obtained using (1). Because the frequencies of ac line voltage and desired output voltage are the same in proposed voltage regulator, the reference signal of controller must be synchronization with the ac line voltage.

D. Control Strategy

The controller of the proposed high performance single-stage voltage regulator is shown in Fig. 3. In this controller, the average current mode is used as the control reference in the boost power factor pre-regulator. The boost power factor pre-regulator is designed to operate in continuous-conduction mode (CCM). This average current controller can prescribe the shape and the frequency of the input current due to its inherently synchronous feedback loop. In order to obtain almost unity power factor, the synchronous signal is sensed from a rectified sinusoidal waveform. Thus, the signal of bridge rectifier is necessary to obtain the desired synchronous signal and the *rms* input voltage for the control IC. Hall Effect sensor for detecting the input current is installed for the average current mode control. The reference current is then generated by a multiplier/divider combination of the synchronous feedback loop and input voltage feed-forward loop. In the buck inverter, a sinusoidal PWM (SPWM) technique is used to regulate the system dynamics. The feedback circuit includes sinusoidal generator, error amplifier and compensator network. The sawtooth-wave generator is common for eliminating noise interference each other.

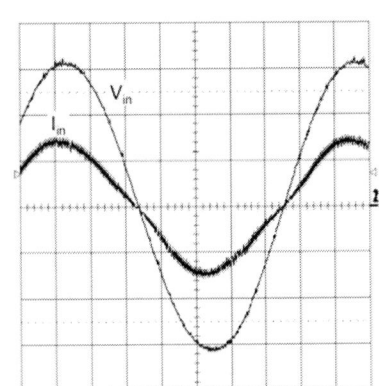

Fig. 4 Input voltage (V_{in}) and input current (I_{in}). V_{in}: 100V/div; I_{in}:5A/div, time:2ms.

III. EXPERIMENTAL RESULTS

An example of a 1kW high performance single-stage voltage regulator is designed and realized. The implemented power stage circuit is shown in Fig. 1. The boost inductance value L_{in} and the filter capacitances C_o to minimize the ripple voltage of voltage V_C are calculated as L_{in}=650μH, C_o =940μF. The output filter inductor L_f and capacitor C_f to minimize the undesired harmonics of the output ac voltage are specified as L_f=1mH, C_f=4.7μF.

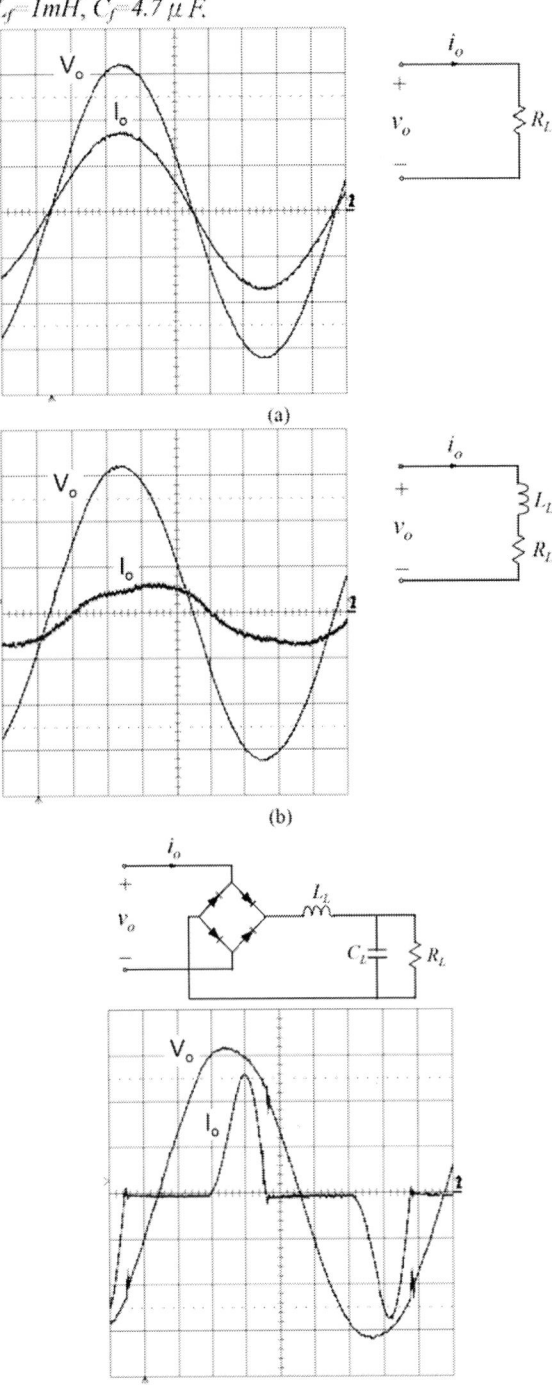

Fig. 5 Experimental results of $v_{in}(t)$, $i_{in}(t)$, $v_o(t)$ and $i_o(t)$ with

(a) resistive load (R_L=48Ω), (V_{in}:200V/div; I_{in}:2A/div; V_o: 100V/div; I_o:5A/div, time:2ms/div)

(b) inductive load (R_L=100Ω, L_L=0.26H), (V_{in}:200V/div; I_{in}:1A/div; V_o: 100V/div; I_o:5A/div, time:2ms/div)

(c) rectifier with RLC load (R_L=48Ω, L_L=1mH, and C_L=1740μF), (V_{in}:200V/div; I_{in}:5A/div; V_o: 100V/div; I_o:10A/div, time:2ms/div)

In hardware realization, we use IGBT's IRG4PC50UD as the power switches and diodes. The waveforms of the input voltage and current of the proposed high performance single-stage voltage regulator at the related 1000W are shown in Fig. 4, in which the waveforms of the input voltage and current are almost in phase and the measured power factor is over 0.99. In order to assess the dynamic performance of the presented inverter, three kinds of loads (resistive R, inductive RL, and rectifier with RLC) are examined in Fig. 5, in which the measured total harmonic distortions (THDs) for the mentioned loads are given as 1.91% for the R load, 2.63% for the RL load, and 5.25% for the rectifier with the RLC load.

IV. CONCLUSION

A novel high performance single-stage voltage regulator with high input power factor and clean ac output voltage is presented. Its configuration is simple and compact. Thus, the proposed voltage regulator is applicable in UPS and ac source design. An average-current-mode control is employed in the rectifier side of proposed converter to detect the transition time and synthesize a suitable low harmonics sinusoidal waveform for the input current. The sinusoidal pulse-width modulation (SPWM) control strategy is employed in the inverter of proposed converter to achieve well dynamic regulation. The circuit operation has been described and discussed. The design procedure and example of the novel high performance single-stage voltage regulator is described. Some experiment results prove the truth of the theoretical prediction.

ACKNOWLEDGMENT

The authors gratefully acknowledge the National Science Council Sponsored the work, Project no. NSC96-2628 - E-197-001-MY2.

REFERENCES

[1] M. Venturini, "A new sine wave in, sine wave out, conversion technique eliminates reactive elements," in *Powercon* 7, San Diego, CA, 1980, pp. E3-El5.

[2] J. Oyama and T. Higuchi *er al.*, "Novel control strategy for matrix converter," pp. 360-367.

[3] P. Tenti, L. Malesani, and L. Rossetto, "Optimum control of N-input K-output matrix converters," in *IEEE Trans. Power Electron.* vol. 7, no. 4, pp. 707-713, Oct. 1992.

[4] D. Divan, T. Habetler, and T. Lipo, "PWM techniques for voltage source inverters," in *IEEE PESC Conf. Rec.* 1990, Tutorial Notes.

[5] R. Wu, S. Dewan, and G. Slemon, "Analysis of an ac to dc voltage source converter using PWM with phase and amplitude control," in *IEEE Ind. Applicat. Soc. Annu. Meefing,* San Diego, CA, Oct. 1989, pp. 1156-1163.

[6] H. W. Park, S. J. Park, J. G. Park, and C. U. Kim, "A novel high-performance voltage regulator for single-phase ac sources," *IEEE Trans. Ind. Electron.*, vol. 48. pp. 554-562, June 2001.

[7] J. H. Choi, J. M. Kwon, J. H. Jung and B. H. Kwon, "High-

performance online UPS using three-leg-type converter," *IEEE Trans. Ind. Electron.*, vol. 52, no. 3, pp. 889-897, June 2005.

Small-Signal Modeling of Series Resonant Converter

Weiping Zhang, Peng Mao, and Yuanchao Liu

College of Information Engineering, North China Univ. of Tech., Beijing, P.R. China

Fax: 86-10-88802880 E-mail:zwp@ncut.edu.cn

Abstract—**Based on the basic modeling approach, a new small-signal modeling technique applied to the series resonant converters is presented and the models are in good agreement with the simulation and measurement results.**

Index Terms –**modeling, resonant converter model, series resonant converter, small-signal model**

I. INTRODUCTION

The main difficulties in the analysis and control of switching regulators are due to the highly non-linear nature of the dc-dc converters. To solve this problem, different linearizing modeling techniques were proposed in the past. However, they were either complicated or obtained numerically.

In this paper, a new modeling technique is presented based on the steady-state model of series resonant converter, which avoids the above mentioned problems. The series resonant converter circuit and steady-state model are shown in Fig.1 and Fig.2.

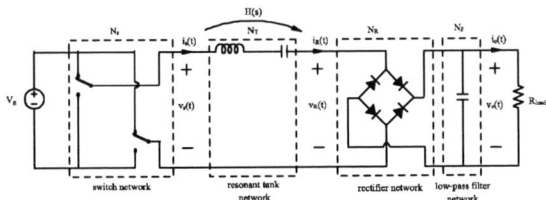

Fig. 1 Series resonant converter circuit..

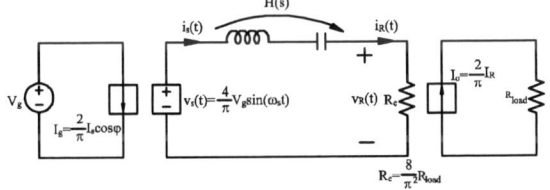

Fig. 2 Steady-state model of series resonant converter

II. THE SMALL-SIGNAL MODEL FROM INPUT TO OUTPUT

Suppose that there is small-signal perturbation $\hat{v}_g(t)$ on the input voltage V_g and $\hat{v}_g(t)$ is equal to $\Delta v_g \cos\Omega t$, illustrated in the Fig.3.

Project supported by Natural Science foundation of China (No.50477054)

Project supported by Beijing Natural Science foundation of China (No.4052011)

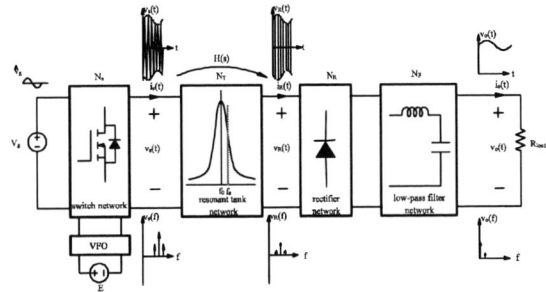

Fig. 3 Circuit block when the perturbation operates on the input

Hence the input voltage can be expressed that

$$v_g(t) = V_g + \hat{v}_g(t) = V_g + \Delta v_g \cos\Omega t = V_g(1 + m\cos\Omega t), m = \frac{\Delta v_g}{V_g} \quad (1)$$

Assume that $\Omega \ll \omega_s$, $\Delta v_g \ll V_g$, and resonant tank has high Q-factor, hence the input voltage of the resonant tank network $v_s(t)$ is

$$v_s(t) = \frac{4}{\pi}V_g(1 + m\cos\Omega t)\sin\omega_s t$$
$$= v_s \sin\omega_s t + \frac{2\Delta v_g m}{\pi}\sin(\omega_s + \Omega)t + \frac{2\Delta v_g m}{\pi}\sin(\omega_s - \Omega)t \quad (2)$$

ω_s is the frequency of the switch. It can be assumed that the voltage transfer functions of the resonant tank network are as followings:

$$\begin{cases} H_1(j\omega_s) = A_1 e^{-j\varphi_1} \\ H_2[j(\omega_s + \Omega)] = A_2 e^{-j\varphi_2} \\ H_3[j(\omega_s - \Omega)] = A_3 e^{-j\varphi_3} \end{cases} \quad (3)$$

So that the output voltage of the resonant tank is

$$v_R(t) =$$
$$v_s A_1 \sin(\omega_s t - \varphi_1) +$$
$$\frac{2\Delta v_g A_2}{\pi}\sin[(\omega_s + \Omega)t - \varphi_2] +$$
$$\frac{2A_3 \Delta v_g}{\pi}\sin[(\omega_s - \Omega)t - \varphi_3] \quad (4)$$

Assume that

$$\begin{cases} a = \frac{A_1 v_s}{2} = \frac{2A_1 V_g}{\pi} \\ b = \frac{2A_2 \Delta V_g}{\pi} \\ p = \frac{2A_3 \Delta V_g}{\pi} \end{cases} \quad (5)$$

$$v_R(t) = \{a\sin(\omega_s t - \varphi_1) + b\sin[(\omega_s + \Omega)t - \varphi_2]\}$$
$$+\{a\sin(\omega_s t - \varphi_1) + p\sin[(\omega_s - \Omega)t - \varphi_3]\} = x + y$$

$$x = a\sin(\omega_s t - \varphi_1) + b\sin[(\omega_s + \Omega)t - \varphi_2]$$
$$= [a + b\cos(\Omega t + \varphi_1 - \varphi_2)]\sin(\omega_s t - \varphi_1) +$$
$$b\sin(\Omega t + \varphi_1 - \varphi_2)\cos(\omega_s t - \varphi_1) = u_x\sin(\omega_s t - \varphi_1 + \alpha)$$
$$u_x = \sqrt{a^2 + b^2 + 2ab\cos(\Omega t + \varphi_1 - \varphi_2)}$$

Because $a \gg b$ and $\sqrt{1+k} \approx 1 + \dfrac{k}{2}$, $\quad u_x \approx a + b\cos(\Omega t + \varphi_1 - \varphi_2)$

$$\alpha = \arctan\frac{b\sin(\Omega t + \varphi_1 - \varphi_2)}{a + b\cos(\Omega t + \varphi_1 - \varphi_2)}$$

$$y = u_y\sin(\omega_s t - \varphi_1 + \beta)$$
$$u_y \approx a + p\cos(\Omega t + \varphi_3 - \varphi_1)$$

$$\beta = \arctan\frac{-p\sin(\Omega t + \varphi_3 - \varphi_1)}{a + p\cos(\Omega t + \varphi_3 - \varphi_1)}$$

$$v_R(t) = [a + b\cos(\Omega t + \varphi_1 - \varphi_2)]\sin(\omega_s t - \varphi_1 + \alpha)$$
$$+ [a + p\cos(\Omega t + \varphi_3 - \varphi_1)]\sin(\omega_s t - \varphi_1 + \beta)$$

Because $a \gg b$ and $a \gg p$, $\quad \alpha \approx \beta = 0$

$$v_R(t) = x + y = [2a + b\cos(\Omega t + \varphi_1 - \varphi_2) + p\cos(\Omega t + \varphi_3 - \varphi_1)]\sin(\omega_s t - \varphi_1) \quad (6)$$

The output voltage of rectifier network is

$$v_o(t) = \frac{\pi}{4}[2a + b\cos(\Omega t + \varphi_1 - \varphi_2) + p\cos(\Omega t + \varphi_3 - \varphi_1)] \quad (7)$$

Through the above analysis, the small-signal model from input to output is obtained, illustrated in Fig.4.

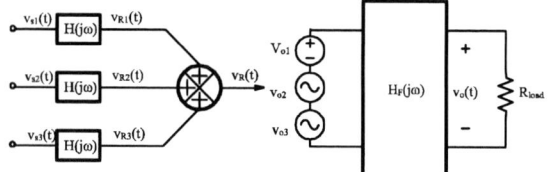

Fig. 4. Small-signal model from resonant tank input to output

$$\begin{cases} v_{s1}(t) = v_s \sin\omega_s t \\ v_{s2}(t) = \dfrac{2\Delta v_g}{\pi}\sin(\omega_s + \Omega)t \\ v_{s3}(t) = \dfrac{2\Delta v_g}{\pi}\sin(\omega_s - \Omega)t \end{cases} \quad (8)$$

$$\begin{cases} v_{R1}(t) = v_s A_1 \sin(\omega_s t - \varphi_1) \\ v_{R2}(t) = \dfrac{2\Delta v_g}{\pi} A_2 \sin[(\omega_s + \Omega)t - \varphi_2] \\ v_{R3}(t) = \dfrac{2\Delta v_g}{\pi} A_3 \sin[(\omega_s - \Omega)t - \varphi_3] \end{cases} \quad (9)$$

$$\begin{cases} V_{o1} = \dfrac{\pi}{4} v_s A_1 \\ v_{o2}(t) = \dfrac{\Delta v_g A_2}{2}\cos(\Omega t + \varphi_1 - \varphi_2) \\ v_{o3}(t) = \dfrac{\Delta v_g A_3}{2}\cos(\Omega t + \varphi_3 - \varphi_1) \end{cases} \quad (10)$$

Because of the amplitude and phase frequency characteristic of series resonant tank shown in Fig.5,

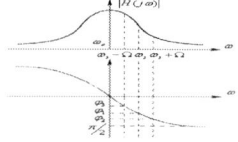

Fig. 5. Frequency characteristic of series resonant tank

when $(\omega_s - \Omega)$ and $(\omega_s + \Omega)$ are close to ω_s, φ is almost linear along the changes of the ω, hence $(\varphi_1 - \varphi_2) = \varphi_3 - \varphi_1 = \Delta\varphi$

$$v_o(t) = \frac{\pi}{4}[2a + (b + p)\cos(\Omega t + \Delta\varphi)] \quad (11)$$

$$\hat{v}_o(t) = \frac{\pi}{4}(b + p)\cos(\Omega t + \Delta\varphi)$$
$$= \frac{\Delta V_g}{2}\{|H[j(\omega_s + \Omega)]| + |H[j(\omega_s - \Omega)]|\}\cos(\Omega t + \Delta\varphi) \quad (12)$$

Assume that the low pass filter (LPF) current transfer function is

$$H_F(j\omega) = |H_F(j\omega)|e^{j\varphi_F} \quad (13)$$

The small-signal transfer function from input to output is

$$G_{vg} = \frac{\hat{v}_o(j\Omega)}{\hat{v}_g(j\Omega)}$$
$$= \frac{|H[j(\omega_s + \Omega)]| + |H[j(\omega_s - \Omega)]|}{2}|H_F(j\Omega)|e^{j(\Delta\varphi + \varphi_F)} \quad (14)$$

It also can be approximated as following:

$$G_{vg} = \frac{\hat{v}_o(j\Omega)}{\hat{v}_g(j\Omega)}$$
$$= \frac{2|H[j(\omega_s)]| + \dfrac{d|H[j(\omega)]|}{d\omega}\Big|_{\omega=\omega_s}\Omega - \dfrac{d|H[j(\omega)]|}{d\omega}\Big|_{\omega=\omega_s}\Omega}{2}|H_F(j\Omega)|e^{j(\Delta\varphi + \varphi_F)}$$
$$G_{vg} \approx |Hj(\omega_s)||H_F(j\Omega)|e^{j(\Delta\varphi + \varphi_F)} \quad (15)$$

Because

$$v_R(t) = [2a + (b + p)\cos(\Omega t + \Delta\varphi)]\sin(\omega_s t - \varphi_1) \quad (16)$$

the current of resonant tank $i_s(t)$ is

$$i_s(t) = \frac{[2a + (b + p)\cos(\Omega t + \Delta\varphi)]\sin(\omega_s t - \varphi_1)}{R_e}$$

$$i_s(t) = (i_s + \hat{i}_s(t))\sin(\omega_s t - \varphi_1) \quad (17)$$

$$\langle i_g(t)\rangle_{T_s} = (i_s + \hat{i}_s(t))\cos\varphi_1 = I_g + \hat{i}_g(t) \quad (18)$$

Based on the above analysis, the small-signal model of

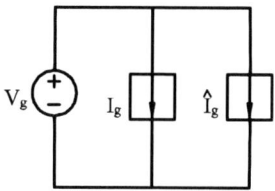

Fig. 6. Small-signal model of input port

the input port is obtained, illustrated in Fig.6.

III. The Simulation and Experiment Results When Perturbation Operate on Input

Pspice has been employed to simulate, illustrated in Fig. 7.

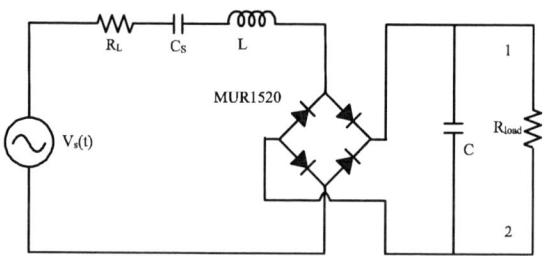

Fig. 7. Schematic circuit of simulation

1616

The parameters are as the followings:
$V_S(t) = (39.3 + 3.14 \times \sin(2\pi 100)) \sin(2\pi 50.4264k)$,
C_s=7.4n, L=1.45m, C=2.24n, R_{load}=27.36, R_L= 4.3721(the equivalent resistance of inductance).

The output waveforms in time domain and frequency domain are shown in Fig 8 and Fig 9 respectively.

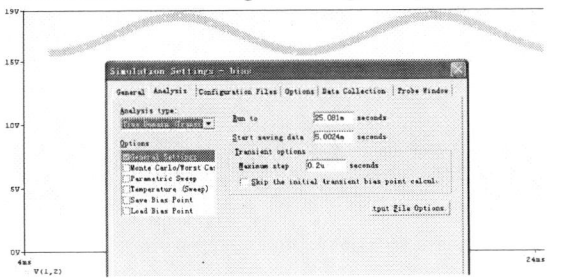

Fig. 8. The waveform of output voltage V(1,2) in time domain

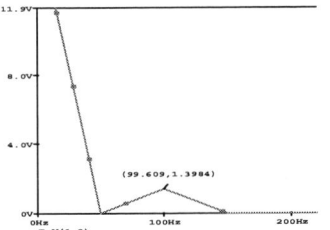

Fig. 9. FFT of the output voltage V(1,2)

The waveform in time domain and spectral analysis of output voltage V(1,2) in the experiment using oscilloscope TDS 5000 are shown in Fig.10 and Fig 11 respectively.

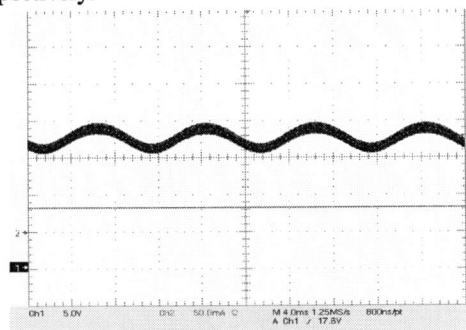

Fig. 10. The waveform of the output in time domain

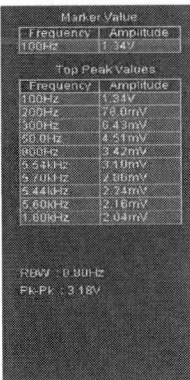

Fig. 11. Spectral analysis of output voltage

Compared three results:

experimental result: $\dfrac{\Delta V_o}{\Delta V_g} = 0.5434$

simulatiion result: $\dfrac{\Delta V_o}{\Delta V_g} = 0.56703$

the error is 4.35%

theoretical result: $\dfrac{\Delta V_o}{\Delta V_g} \approx |Hj(\omega_s)| \, | H_F(j\Omega)| = 0.5087$

the error is -6.39%.

IV. the SMALL-SIGNAL MODEL from CONTROL to OUTPUT

As is shown in Fig 12, VFO is controlled by DC source E and perturbation $\hat{E}(t)$.

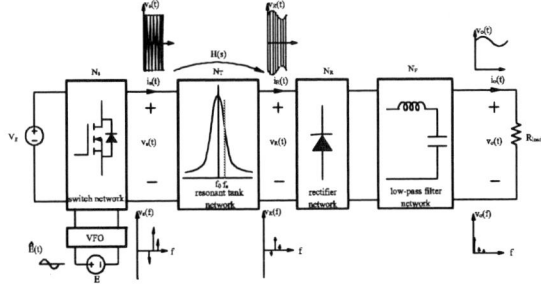

Fig. 12. Circuit block when the perturbation operate on the control

Process of frequency regulation was illustrated in Fig.13

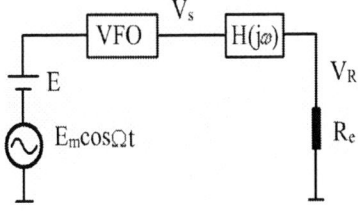

Fig. 13. Process of frequency regulation..

Assume that the perturbation is $E_m \cos \Omega t$, $\Omega \ll \omega_s$, $E_m \ll E$ and as shown in Fig.14, when close to the quiescent operating point(E, ω_s), the relation of V-ω is approximate linear, and the slop is K_o.

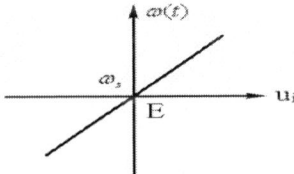

Fig. 14. Control characteristic of VFO.

When the input voltage is $u_i = E + E_m \cos \Omega t$, the output frequency of VFO is

$$\omega(t) = \omega_s + K_o E_m \cos \Omega t = \omega_s + \Delta\omega \cos \Omega t \qquad (19)$$

Hence the total phase is

$$\theta(t) = \int \omega(t)dt = \omega_s t + k_o E_m \int \cos \Omega t dt = \omega_s t + \frac{\Delta\omega}{\Omega} \sin \Omega t \qquad (20)$$

$$\theta(t) = \omega_s t + m_f \sin \Omega t \qquad (21)$$

The coefficient of frequency modulation is

$$m_f = \frac{\Delta\omega}{\Omega}. \qquad (22)$$

So that the frequency modulation signal $v_s(t)$ is

$$v_s(t) = \frac{4V_g}{\pi}\sin(\omega_s t + m_f \sin \Omega t) = v_s \sin(\omega_s t + m_f \sin \Omega t) \quad (23)$$

Assume that $m_f \ll 1$,

$$\cos(m_f \sin \Omega t) \approx 1, \sin(m_f \sin \Omega t) \approx m_f \sin \Omega t$$

$$v_s(t) = v_s \sin \omega_s t + \frac{v_s m_f}{2}[\sin(\omega_s + \Omega)t - \sin(\omega_s - \Omega)t] \quad (24)$$

$$v_s(t) = v_{s1} + v_{s2} + v_{s3}$$

$$\begin{cases} v_{s1}(t) = v_s \sin \omega_s t \\ v_{s2}(t) = \dfrac{v_s m_f}{2}\sin(\omega_s + \Omega)t \\ v_{s3}(t) = -\dfrac{v_s m_f}{2}\sin(\omega_s - \Omega)t \end{cases} \quad (25)$$

It can be assumed that the voltage transfer functions of the resonant tank network are as followings:

$$H_1(j\omega_s) = A_1 e^{-j\varphi_1}$$
$$H_2[j(\omega_s + \Omega)] = A_2 e^{-j\varphi_2}$$
$$H_3[j(\omega_s - \Omega)] = A_3 e^{-j\varphi_3}$$

So that the output voltage of the resonant tank is

$$v_R(t) = A_1 v_s \sin(\omega_s t - \varphi_1) + \frac{A_2 v_s m_f}{2}\sin[(\omega_s + \Omega)t - \varphi_2]$$

$$-\frac{A_3 v_s m_f}{2}\sin[(\omega_s - \Omega)t - \varphi_3]$$

$$v_R(t) = v_{R1}(t) + v_{R2}(t) + v_{R3}(t)$$

$$\begin{cases} v_{R1}(t) = A_1 v_s \sin(\omega_s t - \varphi_1) \\ v_{R2}(t) = \dfrac{A_2 v_s m_f}{2}\sin[(\omega_s + \Omega)t - \varphi_2] \\ v_{R3}(t) = -\dfrac{A_3 v_s m_f}{2}\sin[(\omega_s - \Omega)t - \varphi_3] \end{cases} \quad (26)$$

Assume that

$$\begin{cases} a_1 = \dfrac{A_1 v_s}{2} \\ b_1 = \dfrac{A_2 v_s m_f}{2} \\ p_1 = \dfrac{A_3 v_s m_f}{2} \end{cases} \quad (27)$$

$$v_R(t) = x + y = \{a_1 \sin(\omega_s t - \varphi_1) + b_1 \sin[(\omega_s + \Omega)t - \varphi_2]\}$$
$$+ \{a_1 \sin(\omega_s t - \varphi_1) - p_1 \sin[(\omega_s - \Omega)t - \varphi_3]\}$$

$$x = u_x \sin(\omega_s t - \varphi_1 + \alpha_1)$$

$$u_x \approx a_1 + b_1 \cos(\Omega t + \varphi_1 - \varphi_2), \alpha_1 = \arctan \frac{b_1 \sin(\Omega t + \varphi_1 - \varphi_2)}{a_1 + b_1 \cos(\Omega t + \varphi_1 - \varphi_2)}$$

$$y = u_y \sin(\omega_s t - \varphi_1 + \beta_1)$$

$$u_y \approx a_1 - p_1 \cos(\Omega t + \varphi_3 - \varphi_1), \beta_1 = \arctan \frac{p_1 \sin(\Omega t + \varphi_3 - \varphi_1)}{a - p_1 \cos(\Omega t + \varphi_3 - \varphi_1)}$$

$$v_R(t) \approx [2a_1 + b_1 \cos(\Omega t + \varphi_1 - \varphi_2) - p_1 \cos(\Omega t + \varphi_3 - \varphi_1)]\sin(\omega_s t - \varphi_1) \quad (28)$$

The output voltage of rectifier network is

$$v_o(t) = \frac{\pi}{4}[2a_1 + b_1 \cos(\Omega t + \varphi_1 - \varphi_2) - p_1 \cos(\Omega t + \varphi_3 - \varphi_1)] \quad (29)$$

$$v_o(t) = v_{o1}(t) + v_{o2}(t) + v_{o3}(t)$$

$$\begin{cases} V_{o1}(t) = \dfrac{\pi}{2}a_1 = \dfrac{\pi}{4}A_1 v_s \\ v_{o2}(t) = \dfrac{\pi}{4}b_1 = \dfrac{\pi}{8}A_2 v_s m_f \\ v_{o3}(t) = -\dfrac{\pi}{4}p_1 = -\dfrac{\pi}{8}A_3 v_s m_f \end{cases} \quad (30)$$

Through the above analysis, the small-signal model from control to output is obtained, illustrated in Fig.15.

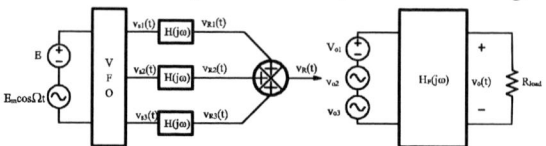

Fig. 15. Small-signal model from control to output.

When $(\omega_s - \Omega)$ and $(\omega_s + \Omega)$ are close to ω_s, φ is almost linear along the changes of the ω, hence $\Delta\varphi = \varphi_1 - \varphi_2 = \varphi_3 - \varphi_1$

$$\hat{v}_o(t) = v_{o2} + v_{o3}$$
$$= \frac{\pi}{8}v_s m_f (A_2 - A_3)\cos(\Omega t + \Delta\varphi) \quad (31)$$

Assume that the low pass filter (LPF) current transfer function is

$$H_F(j\omega) = |H_F(j\omega)|e^{j\varphi_F}$$

$$G_{V\omega} = \frac{\hat{v}_o(j\Omega)}{\Delta\omega} = $$
$$v_s \frac{\pi(|H[j(\omega_s + \Omega)]| - |H[j(\omega_s - \Omega)]|)}{8\Omega}|H_F(j\omega)|e^{j(\varphi_F + \Delta\varphi)} \quad (32)$$

It also can be approximated as following:

$$G_{V\omega} = \frac{\pi}{4}v_s \frac{d|H[j(\omega)]|}{d\omega}\bigg|_{\omega=\omega_s}|H_F(j\omega)|e^{j(\varphi_F + \Delta\varphi)} \quad (33)$$

Because $\omega_s \gg \Omega$, the frequency of $v_s(t)$ is changed slowly. Therefore, during a little time, it is seemed that the frequency of $v_s(t)$ is constant, but different time has different frequency. Assume that the admittance of resonant tank is $Y(j\omega) = |Y(j\omega)|e^{j\varphi}$.

Hence, $i_s = Y(j\omega)v_s, \omega = \omega_s + \Delta\omega \cos \Omega t \quad (34)$

The amplitude of $i_s(t)$ is $i_s = |Y(j\omega)|v_s$.

Linearizing the above equation,

$$i_s = |Y(j\omega_s)|v_s + \frac{d|Y(j\omega)|}{d\omega}\bigg|_{\omega=\omega_s}\Delta\omega \cos(\Omega t)v_s = i_{ss} + \hat{i}_s \quad (35)$$

The phase of $i_s(t)$ is

$$\varphi(\omega) = \varphi(\omega_s + \Delta\omega)$$
$$= \varphi_y(\omega_s) + \frac{d\varphi(\omega)}{d\omega}\bigg|_{\omega=\omega_s}\Delta\omega \cos(\Omega t) = \varphi_s + \hat{\varphi} \quad (36)$$

Therefore,

$$i_g(t) = (i_{ss} + \hat{i}_s)\cos(\varphi_s + \hat{\varphi})$$

$$i_g(t) \approx i_{ss}\cos \varphi_s + \hat{i}_s \cos \varphi_s - i_{ss} \sin \varphi_s \sin \hat{\varphi} \quad (37)$$

$$i_g(t) = I_g + \hat{I}_{g1} - \hat{I}_{g2} \quad (38)$$

When

$$\frac{d|Y(j\omega)|}{d\omega}\bigg|_{\omega=\omega_s} = |Y(j\omega_s)|\frac{d\varphi(\omega)}{d\omega}\bigg|_{\omega=\omega_s}\tan \varphi_s, \quad (39)$$

$$\hat{I}_{g1} = \hat{I}_{g2}, i_g(t) \equiv I_g.$$

Under this condition, there will be no influence on the average of the input current while ω is perturbed.

Based on the above analysis, the small-signal model of the input port was obtained, illustrated in Fig.16.

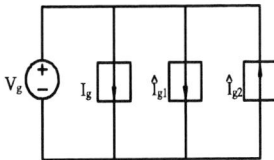

Fig. 16. Small-signal model of the input port.

V. The SIMULATION and EXPERIMENT RESULTS WHEN PERTURBATION OPERATE ON CONTRL

Pspice has been employed to simulate, illustrated in Fig.17.

The parameters are as the followings:

$V_S(t) = 35.1\sin(2\pi 50.4264kt + m_f\cos(2\pi \times 1kt))$, m_f=0.1,

$\Delta f = 100HZ$, C_s=7.4n,L=1.45m,C=2.24n,R_{load}=27.36,

R_L= 4.3721 (the equivalent resistance of inductance),

R_{in} = 0.5, L_{in}= 1.6u. (R_{in} and L_{in} constitute the internal impedance of $V_s(t)$)

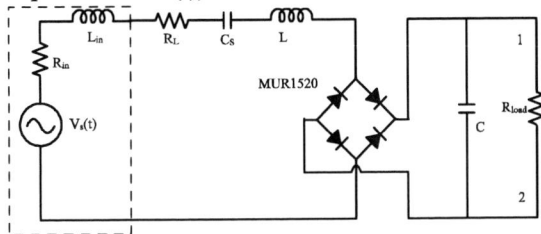

Fig. 17. Schematic circuit of PSPICE

The output waveforms in time domain and frequency domain are shown in Fig 18 and Fig 19 respectively.

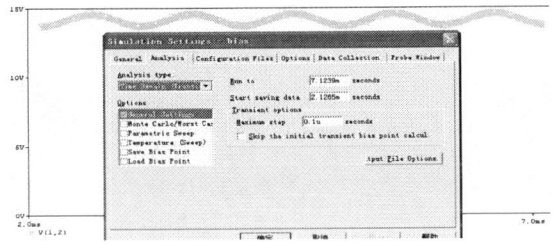

Fig. 18. The waveform of the output voltage V(1,2) in time domain.

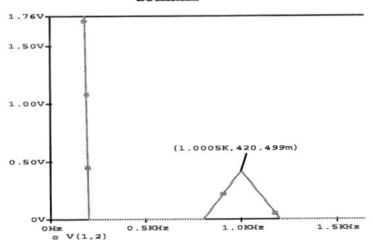

Fig. 19. FFT of the output voltage V(1,2)

The waveform in time domain and spectral analysis of output voltage V(1,2) in the experiment using oscilloscope TDS 5000 are shown in Fig.20 and Fig 21 respectively.

Compared three results:

experimental result: $\dfrac{\Delta V_o}{\Delta f} = 4$m

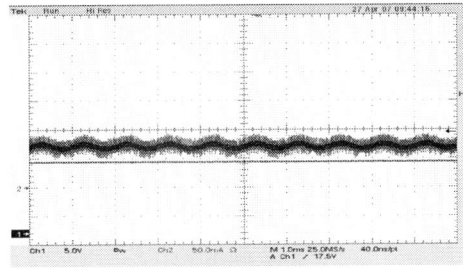

Fig. 20. The waveform of the output in time domain

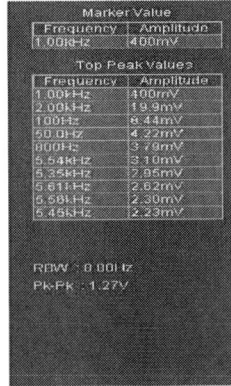

Fig. 21. Spectral analysis of output voltage.

simulation result: $\dfrac{\Delta V_o}{\Delta f} = 4.20499$m

the error is 5.12%.
theoretical result:

$\dfrac{\Delta V_o}{\Delta f} = \dfrac{\pi}{4}v_s \dfrac{d|H[j(f)]|}{df}|_{\omega=\omega_k} |H_F(jf)| = 4.5431$m

the error is 13.58%.
(If the dynamic resistance of rectification bridge is considered, the error would be reduced to 6.975%)

VI. CONCLUSIONS

This paper presents a novel approach to obtain the small-signal model of resonant converter. The experimental results are also agreed with the theoretic results, simulation results and it is also shown that the new model has many merits such as simpleness, practicality and so on.

REFERENCES

[1] Weiping Zhang, The model and control of switch converter. Beijing: Electric Power Publishing House,2005.pp.285-300

[2] Weiping Zhang, "The model of DC-DC resonant converter" unpublished.

[3] Middlebrook R D and Cuk S, "A general unified approach to modelling switching converter power stages [A]." Proceeding of IEEE PESC [C] .Los Angeles: IEEE PE Society, 1976.

[4] Divan D M, "The resonant DC link converters a new concept in static power conversion." IEEE Trans Ind Applicat,1989,25(2):317～325

Modelling of Three phase Z-Source Boost Buck Rectifiers

D M Vilathgamuwa, P C Loh and K Karunakar

Nanyang Technological University, School of Electrical and Electronic Engineering, Nanyang Avenue, Singapore 639798

Abstract-- The Z-source rectifier is a recently proposed converter topology which has a unique X-shaped impedance network on its DC side. In the process of designing the controller circuits for the Z-source rectifier, knowledge of transfer functions relating the dynamics of various variables is essential. This paper deals with the modeling of Z-source rectifier with the intent of developing a robust controller. Modelling is carried out using a set of state equations obtained by state space averaging the circuit equations in shoot through and non shoot through states of the rectifier. The derived model is verified with both simulation and experimental results.

Index Terms—Z-source converters, buck-boost rectifiers

I. INTRODUCTION

Power electronic converters are employed in almost all electronic equipment and in industrial power conversion systems. In most applications it is required to convert power from AC to DC or DC to AC. Traditionally switched-mode AC to DC conversion is carried out using voltage source rectifier (VSR) topology as shown in Fig. 1. In some applications current source rectifier (CSR) topology is also used. Both have their own limitations in terms of their working principles, i.e. they can either boost or buck the output voltage with reference to the AC input voltage. In view of reliability, any short circuiting of switches in any phase leg of VSR could damage the rectifier.

Power rectifiers are increasingly used in energy conversion systems including uninterruptible power supplies and, motor drives. Most of these applications would require the rectifiers to have both buck and boost capabilities for increased load current and input voltage variations. In general, this can be achieved by two stage power conversions with the connection of an additional dc-dc converter stage. However, this leads to a less efficient system due to increased switching losses and also this causes an increase in system complexity. Therefore, other alternative single stage buck-boost rectifier topologies are desired. The Z-source rectifier has been reported recently as such an alternative topology with great advantages [1]. It differs from conventional converters like VSR and CSR due to the presence of unique X-shaped impedance network on its DC side, interfacing H-bridge and load as shown in Fig. 2. This rectifier can overcome limitations of the conventional VSR. It can produce a DC output voltage greater or smaller than the input AC voltage while operating at unity or at any desired power factor without any extra circuitry.

The unique feature of the Z source rectifier is its impedance network. Study of Z-network is helpful in analyzing the DC side quantities. Unlike a conventional rectifier, the three phase Z–source rectifier has one extra switching state i.e. shoot through state. It has totally nine permissible switching states consisting six active vectors and three zero vectors including one with shoot-through. The extra (shoot-through) state can be treated as short circuiting of both the upper and lower switches of any one phase leg, two phase legs or all three phase legs. During shoot through state, the switch SW7 must be kept open. This shoot-through zero state provides unique feature of buck-boost operation. Equivalent circuits of the Z-source rectifier when the bridge is in any one of eight non-shoot-through states and shoot-through zero state are shown in Figs. 3(a) and (b) respectively.

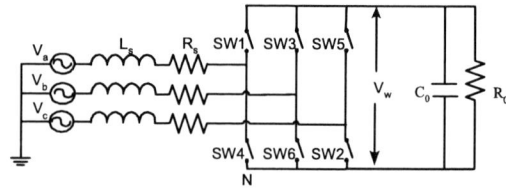

Fig.1. Conventional voltage source rectifier

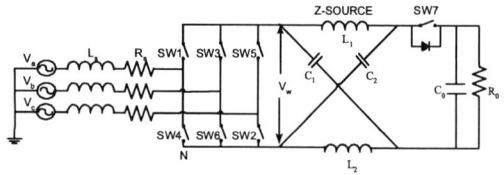

Fig.2. Voltage type Z-source rectifier

II. PULSE WIDTH MODULATION

All traditional pulse width modulation schemes can be used to control the Z source rectifier [2]. Fig. 4(top) shows the traditional PWM scheme based on the triangular carrier method. When the AC supply voltage decreases, the DC output voltage can be maintained with the adjustment of PWM modulation index. However when the AC supply voltage increases, the bucking action of the Z-source can be used with the application of the shoot-through states. The modified PWM scheme with the insertion of shoot-through state is shown in Fig.

978-1-4244-0644-9/07/$25.00 ©2007 IEEE   1620

4(bottom) [3]-[5]. The detailed analysis of buck-boost operation will be carried out in the next section. Due to high switching frequencies of the rectifier viewed from Z-source network, the inductance of the Z-source network can be reduced significantly. The available shoot through period is limited only by the zero state that is determined by the PWM modulation index.

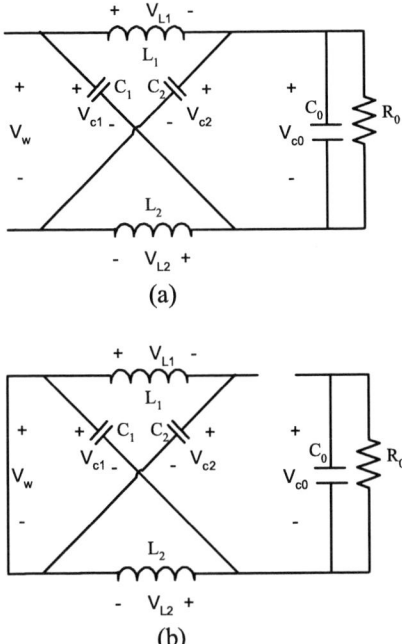

(a)

(b)

Fig. 3. Z-source network in (a) non-shoot through state and (b) shoot-through state

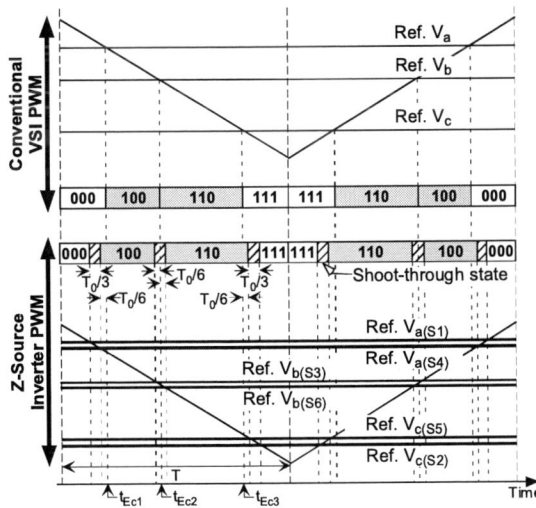

Fig. 4. PWM schemes

III. MODELLING

In this Section, the modeling and analysis of boost-buck AC to DC rectifier is discussed. The modeling is carried out with obtaining a set of differential equations, applying state space averaging and transforming the resultant state-space equations into a synchronous d-q reference frame. The converter is analyzed from the basic power circuit shown in Fig. 2 for rectifier operation. Selecting the inductor currents and capacitor voltages as state variables, the following set of differential equations are obtained:

Three phase three-wire system implies $i_a + i_b + i_c = 0$, while for a balanced three phase voltage supply, $V_a + V_b + V_c = 0$. The summation of equations from (1) to (3) and application of the above conditions would result in $V_{NO} = -V_w\left(\dfrac{d_a + d_b + d_c}{3}\right)$.

By considering the equivalent circuits in Figs. 3(a) and (b) in shoot-through and non-shoot-through states and state space averaging relevant circuit equations in the two states, the following equations can be obtained.

$$V_w = (2V_C - V_{co}) \qquad (4)$$

$$\frac{di_L}{dt} = \frac{V_C(1-2d_z)}{L} - \frac{V_{co}(1-d_z)}{L} \quad (5)$$

$$I_C = I_i(1-d_z) - I_L(1-2d_z) \quad (6) \quad \text{where}$$

$$I_i = d_a i_a + d_b i_b + d_c i_c$$

$$\frac{dV_c}{dt} = \frac{(1-d_z)d_a i_a + (1-d_z)d_b i_b + (1-d_z)d_c i_c}{C}$$

$$- \frac{I_L(1-2d_z)}{C} \quad (7)$$

$$I_{co} = (I_L - I_c)(1-d_z) - I_o \quad (8)$$

$$I_o = \frac{V_{co}}{R_o} \quad (9)$$

$$\frac{dv_{co}}{dt} = \frac{2(1-d_z)^2 I_L}{C_o}$$

$$- \frac{(1-d_z)^2(d_a i_a + d_b i_b + d_c i_c)}{C_o} - \frac{V_{co}}{R_o C_o} \quad (10)$$

Substituting V_w from (4) into equations (1) through (3),

1621

$$V_a - R_s i_a - L_s \frac{di_a}{dt} - 2d_a V_C + d_a V_{co} = 0 \quad (11)$$

$$V_b - R_s i_b - L_s \frac{di_b}{dt} - 2d_b V_C + d_b V_{co} = 0 \quad (12)$$

$$V_c - R_s i_c - L_s \frac{di_c}{dt} - 2d_c V_C + d_c V_{co} = 0 \quad (13)$$

Three phase *abc* quantities in the stationary reference frame can be converted into a rotating synchronous reference frame *dq0* quantities using the standard Park's transformations. To simplify the analysis in the AC side, it is assumed that the rotating q-axis is in-phase with the supply voltage vector. For phase and amplitude control, duty ratio *d* is defined using phase shift δ and modulation index *m*. It is also assumed that the switching frequency is much higher than the modulating frequency. Three phase duty ratios can therefore be expressed as,

$$d_a = \frac{1}{2} m \cos\left(\omega_e t - \delta\right) \quad (14)$$

$$d_b = \frac{1}{2} m \cos\left(\omega_e t - 120^0 - \delta\right) \quad (15)$$

$$d_c = \frac{1}{2} m \cos\left(\omega_e t + 120^0 - \delta\right) \quad (16)$$

The function *d* is time variant. Therefore it would be convenient to transform *d* into the rotating frame of reference. The equivalent duty ratios in the *d-q* domain can be obtained from the Park's transformations as given below:

$$d_q = \frac{1}{2} m \cos\delta \quad (17)$$

$$d_d = \frac{1}{2} m \sin\delta \quad (18)$$

The order of the rectifier state-space equations can be reduced from six to five because of balanced three phase three wire system where $V_d=0$. The simplified matrix is given in (19) at the bottom of this page.

The obtained model is a non-linear time invariant system even after transformed into the rotating reference frame. Small signal linearization around its DC operating point can be applied for solution. Linearization can be done by assuming the following small-signal disturbance:

For state variables, $i_q = I_q + \hat{i}_q$, $i_d = I_d + \hat{i}_d$, $v_c = V_c + \hat{v}_c$, $v_{co} = V_{co} + \hat{v}_{co}$, $i_L = I_L + \hat{i}_L$.

For control inputs, $d_z = D_z + \hat{d}_z$, $d_q = D_q + \hat{d}_q$ and $d_d = D_d + \hat{d}_d$, and for the voltage supply, $v_q = V_q + \hat{v}_q$.

By substituting these equations into (19) and neglecting second order terms and separating steady-state components from dynamic variables, the steady state DC model and small signal AC model can be obtained separately as explained below.

The small signal AC model of the Z source converter has been derived in (20) at the bottom of the next page. By applying Laplace transformation we can find the transfer functions between the state variables and controls. But this involves mathematical calculations with 5X5 matrix inversion.

IV. STEADY STATE ANALYSIS

Steady state solution gives the relationship between state variables and system parameters. The steady state analysis is important because the dynamic response is dependent on the steady state operating point. The steady state values for any operating point can be obtained from (19) and are further simplified as shown below.

$$V_{co} = \frac{N}{D} \quad (21)$$

where $N = 1.5(1 - D_z)^2(1 - 2D_z)V_q R_o(R_s D_q + \omega_e L_s D_d)$

and $D = 1.5(1 - D_z)^2 R_o R_s(D_q^2 - D_d^2) + (1 - 2D_z)^2(R_s^2 + \omega_e^2 L_s^2)$

$$\frac{d}{dt}\begin{bmatrix} i_q \\ i_d \\ V_C \\ V_{co} \\ I_L \end{bmatrix} = \begin{bmatrix} -\dfrac{R_s}{L_s} & -\omega_e & -\dfrac{2d_q}{L_s} & \dfrac{d_q}{L_s} & 0 \\ \omega_e & -\dfrac{R_s}{L_s} & -\dfrac{2d_d}{L_s} & \dfrac{d_d}{L_s} & 0 \\ \dfrac{3}{2}\dfrac{(1-d_z)d_q}{C} & \dfrac{3}{2}\dfrac{(1-d_z)d_d}{C} & 0 & 0 & \dfrac{-(1-2d_z)}{C} \\ \dfrac{-3}{2}\dfrac{(1-d_z)^2 d_q}{C_o} & \dfrac{-3}{2}\dfrac{(1-d_z)^2 d_d}{C_o} & 0 & \dfrac{-1}{R_o C_o} & \dfrac{2(1-d_z)^2}{C_o} \\ 0 & 0 & \dfrac{1-2d_z}{L} & \dfrac{-(1-d_z)}{L} & 0 \end{bmatrix} \begin{bmatrix} i_q \\ i_d \\ V_C \\ V_{co} \\ I_L \end{bmatrix} + \begin{bmatrix} \dfrac{V_q}{L_s} \\ 0 \\ 0 \\ 0 \\ 0 \end{bmatrix} \quad (19)$$

1622

$$I_q = \frac{(1-2D_z)R_sV_q - V_{co}(R_sD_q - \omega_e L_s D_d)}{(1-2D_z)(R_s^2 + \omega_e^2 L_s^2)} \quad (22)$$

$$I_d = \frac{(1-2D_z)\omega_e L_s V_q - V_{co}(R_sD_d + \omega_e L_s D_q)}{(1-2D_z)(R_s^2 + \omega_e^2 L_s^2)} \quad (23)$$

$$I_L = \frac{V_{co}}{R_o(1-2D_z)} \quad (24)$$

$$V_c = \frac{(1-D_z)V_{co}}{(1-2D_z)} \quad (25)$$

For a given operating point all the variables can be found from the (21) through (25). Neglecting dissipative components, the following expressions can be derived from the first and fifth rows of (19) in the steady state.

$$V_q - \omega_e L_s I_d - 2D_q V_c + D_q V_{co} = 0 \quad (26)$$
$$(1-2D_z)V_c - (1-D_z)V_{co} = 0 \quad (27)$$

Assuming unity power factor operation i.e. $I_d=0$ and from (26) and (27), the following output to input voltage ratio is obtained.

$$\frac{V_{co}}{V_q} = \frac{1-2D_z}{D_q} \quad (28)$$

This equation is similar to the expression derived in [1], i.e. $V_o = \frac{b}{M}\frac{2V_m}{\cos\delta}$, where b is the buck factor, and $D_q = \frac{1}{2}M\cos\delta$.

The limit of voltage step down operation, i.e. lowest output voltage, is determined by the maximum value of D_q. The maximum D_q is 0.5. Therefore the minimum output to input voltage conversion ratio for a given shoot through is given as,

$$\left(\frac{V_{co}}{V_q}\right)_{min} = 2(1-2D_z) \quad (29)$$

For example, for a 188 V phase voltage supply, the minimum output DC voltage for a maximum shoot through of $D_z = 0.3$ is 75.2 V. Furthermore, using (22) and (28), the following expression for I_q, can be obtained.

$$I_q = \frac{V_{co}D_d}{(1-2D_z)\omega_e L_s} = \frac{V_q}{\omega_e L_s}\frac{D_d}{D_q} \quad (30)$$

Therefore, it is clear that I_q is only dependent on D_d,

$$\frac{d}{dt}\begin{bmatrix} \hat{i}_q \\ \hat{i}_d \\ \hat{V}_C \\ V_{co} \\ \hat{i}_L \end{bmatrix} = \begin{bmatrix} \dfrac{R_s}{L_s} & -\omega_e & \dfrac{2D_q}{L_s} & \dfrac{D_q}{L_s} & 0 \\[2mm] \omega_e & \dfrac{R_s}{L_s} & \dfrac{2D_d}{L_s} & \dfrac{D_d}{L_s} & 0 \\[2mm] \dfrac{3}{2}\dfrac{(1-D_z)D_q}{C} & \dfrac{3}{2}\dfrac{(1-D_z)D_d}{C} & 0 & 0 & \dfrac{-(1-2D_z)}{C} \\[2mm] \dfrac{-3}{2}\dfrac{(1-D_z)^2 D_q}{C_o} & \dfrac{-3}{2}\dfrac{(1-D_z)^2 D_d}{C_o} & 0 & \dfrac{-1}{R_oC_o} & \dfrac{2(1-D_z)^2}{C_o} \\[2mm] 0 & 0 & \dfrac{1-2D_z}{L} & \dfrac{-(1-D_z)}{L} & 0 \end{bmatrix}\begin{bmatrix} \hat{i}_q \\ \hat{i}_d \\ \hat{V}_C \\ \hat{V}_{co} \\ \hat{I}_L \end{bmatrix} + \begin{bmatrix} \dfrac{\hat{V}_q}{L_s} \\ 0 \\ 0 \\ 0 \\ 0 \end{bmatrix} +$$

$$\begin{bmatrix} 0 \\ 0 \\ \dfrac{2I_L - \dfrac{3}{2}D_q I_q - \dfrac{3}{2}D_d I_d}{C} \\ \dfrac{(1-D_z)(-4I_L + 3D_q I_q + 3D_d I_d)}{C_o} \\ \dfrac{-2V_C + V_{co}}{L} \end{bmatrix}\left[\hat{D}_z\right] + \begin{bmatrix} \dfrac{-2V_C + V_{co}}{L_s} & 0 \\[2mm] 0 & \dfrac{-2V_C + V_{co}}{L_s} \\[2mm] \dfrac{3}{2}\dfrac{(1-D_z)I_q}{C} & \dfrac{3}{2}\dfrac{(1-D_z)I_d}{C} \\[2mm] \dfrac{-3}{2}\dfrac{(1-D_z)^2 I_q}{C_o} & \dfrac{-3}{2}\dfrac{(1-D_z)^2 I_d}{C_o} \\[2mm] 0 & 0 \end{bmatrix}\begin{bmatrix} \hat{D}_q \\ \hat{D}_d \end{bmatrix} \quad (20)$$

D_q and the supply voltage V_q but not on the shoot-through duty ratio D_z.

Small-signal frequency analysis has been carried out to determine how certain rectifier state variables vary with input variables within a given frequency range. The following are the transfer functions between V_{co} and D_z and V_{co} and i_q respectively. Their frequency responses are shown in Fig. 5.

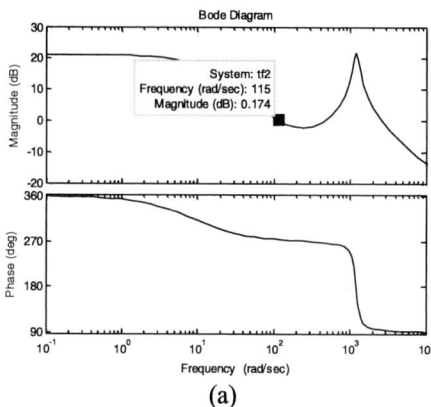

(a)

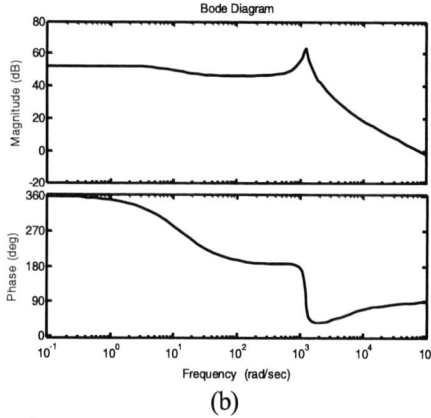

(b)

Fig. 5. Frequency responses of (a) $\hat{V}_{co}/\hat{D}_z$
and (b) $\hat{V}_{co}/\hat{i}_q$

Therefore, it is clear that I_q is only dependent on D_d, D_q and the supply voltage V_q but not on the shoot-through duty ratio D_z. Small-signal frequency analysis has been carried out to determine how certain rectifier state variables vary with input variables within a given frequency range. The following are the transfer functions between V_{co} and D_z and V_{co} and i_q respectively. Their frequency responses are shown in Fig. 5.

$$\frac{\hat{V}_{co}}{\hat{D}_z} = \frac{-6.435e^{-6}\, s^2 + 0.45}{3.146e^{-9}\, s^3 + 5.72e^{-7}\, s^2 + 0.00462\, s + 0.04}$$

$$\frac{\hat{V}_{co}}{\hat{i}_q} = \frac{-0.0002375\, s^2 - 0.9249\, s + 16.61}{3.146e^{-9}\, s^3 + 5.72e^{-7}\, s^2 + 0.00462\, s + 0.04}$$

V. RESULTS

A detailed simulation of the Z-source rectifier system using MATLAB/SIMULINK software program and an experimental study were carried out in order to verify the transient modeling. In the simulation, the rectifier has been realistically modeled to represent the switching nature of the process. The parameters for the rectifier simulation are R_s=1.6 Ω, L_s=6 mH, R_o=551Ω, C_o=220μF, L =6.5mH, C =2200 μF while the input supply (phase voltage) is 10 V at 50 Hz.

Fig. 6 shows the hardware configuration of the laboratory prototype rectifier. A 3-phase low voltage programmable power source supplies the rectifier bridge through a series line inductance. The voltage source rectifier consists of six IGBT switches with anti-paralleled diodes connected across each switch. The PWM switching signals for the voltage source rectifier are generated by a DSP controller board. The sampling frequency of the control system is set at 10 kHz.

According to simulations and experiments when the modulation index m=0.6, the phase shift $\delta = 10^0$ and the shoot-through $D_z = 0$, the output voltage is boosted to 38.05 V and input currents i_q and i_d are found to be 0.305A and 0.85A which coincide with theoretically calculated values using the steady state model equations developed. The experimental results are shown in Figs. 7(a) and (b).

When the modulation index m=0.6, phase shift $\delta = 10^0$ and shoot through D_z =0.3, simulation and experimental results reveal that both output voltage and Z network capacitor voltage were boosted to 27 .05 V and 15.75 V which are consistent with the theoretical values. Simulated and experiment input currents i_q and i_d are 0.2281A and 0.995A respectively. The theoretical calculation results obtained from the steady state model developed in (21) to (25) for the aforementioned operating conditions are given in Table I.

VI. CONCLUSION

This paper presents modeling of unique Z source rectifier, which can boost or buck the voltage using the application of shoot through state. From the analysis of the Z- source rectifier, average large signal model and small signal AC model are derived. The theoretical modeling results are compared with the simulation and experimental results.

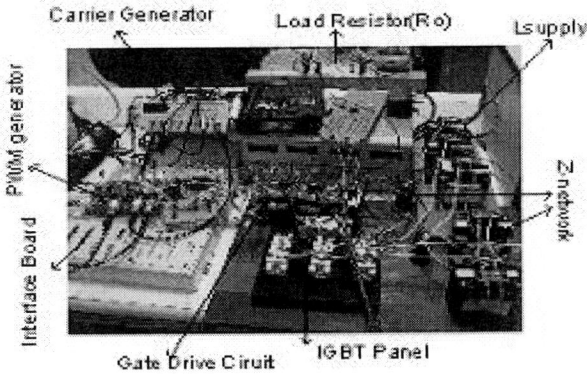

Fig. 6. Experimental setup

Table I. Theoretical values from mathematical modeling

Parameter	m=0.6 and $\delta=10^0$	
	$D_z=0$	$D_z=0.3$
$I_q(A)$	0.3056	0.2281
$I_d(A)$	0.8943	0.9945
$V_c(V)$	38.05	27.6374
$V_{co}(V)$	38.05	15.7928
$I_L(A)$	0.0691	0.0409

REFERENCES

[1] Xinping Ding and Zhaoming Quian, "Three Phase Z source rectifier," in proc. of *IEEE Power Electronics Specialists Conference*, June 2005.

[2] F.Z.Peng, "Z-Source inverter," *IEEE Transaction on Industry Applications*, vol, 39, No.2, pp. 504-510, March/April 2003.

[3] P.C.Loh, D.M Vilathgamuwa, Y.S.Lai, "Pulse width modulation of Z –source inverters," *IEEE Transactions on Power Electronics,* Volume 20, Issue 6, Nov. 2005, pp. 1346 – 1355.

[4] C J Gajanayake, D M Vilathgamuwa and P C Loh, " Small signal and signal-flow-graph modeling of switched z-impedance network", *IEEE Power Electronics Letters*, Volume 3, Issue 3, Sept. 2005, pp. 111 – 116.

[5] C J Gajanayake, D M Vilathgamuwa, P C Loh, "Development of comprehensive model and multi-loop controller for Z-source inverter*", IEEE Industrial Electronics Transactions,* Volume 54, No. 4. Aug. 2007, pp. 2352 - 2359.

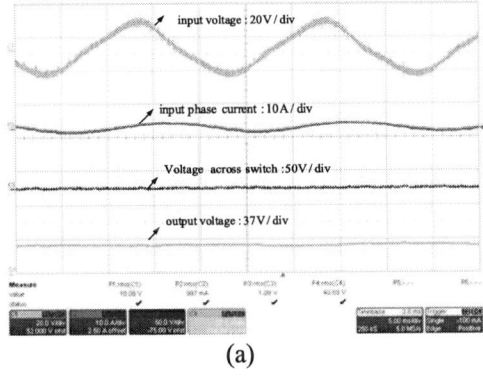

(a)

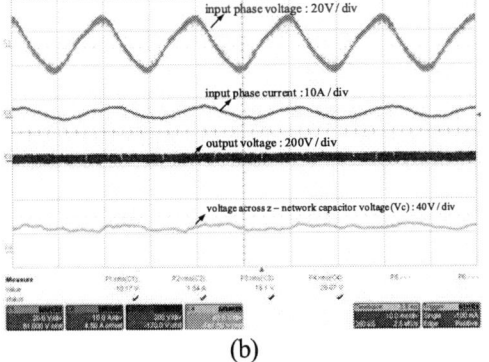

(b)

Fig. 7. Experimental results. (a) for modulation index $m= 0.6$, $D_z = 0$ and $\delta = 10^0$, (b) for modulation index $m= 0.6$, $D_z = 0.3$ and $\delta = 10^0$

A NEW SINGLE-PHASE CONTROLLED RECTIFIER USING SINGLE-PHASE MATRIX CONVERTER WITH REGENERATIVE CAPABILITIES

R. Baharom, M.K. Hamzah, A. Saparon, S.Z. Mohammad Noor & N.R.Hamzah

Faculty of Electrical Engineering, Universiti Teknologi MARA, 40450 Shah Alam, Malaysia. mustafar@ieee.org (Tel/Fax : +603-5543 6093)

Abstract :- A new single-phase controlled rectifier using single-phase matrix converter (SPMC) with fully controllable regenerative capabilities is presented. PWM technique was used to calculate the switch duty ratio to synthesize the output. Safe commutation strategy was implemented to avoid voltage spikes due to inductive load. Selected simulation and experimental results are presented to verify proposed operation.

Keywords : Single-Phase Matrix Converter (SPMC), Controlled Rectifier, Regenerative operation.

1. INTRODUCTION

Since the last millennium, development of semiconductor manufacture brought power devices such as power diode, thyristor (or silicon controlled rectifier, SCR), gate turn-off (GTO), Triac, bipolar transistor (BT), isolated gate bipolar transistor (IGBT) and metal oxide semiconductor field effected transistor (MOSFET) and so on into the DC power supply which make it becomes a common place within modern commercial and industrial environment particularly in applications for AC-DC conversions. [1]. Conventional diode bridge rectifiers normally uses bridge-diode as in Fig. 1(a) without affording any control function, are unidirectional in nature [2] and without regenerative capabilities [3]. Bidirectional operation is also possible with the inclusion of anti-parallel switch in H-bridge topology of Fig. 1(b) but is not fully controllable.

In this work the SPMC topology are used to operate as a controlled rectifier by suitable switching schemes. IGBTs are used for the main power switching device. Simple resistive load is initially used, followed by simple inductive and capacitive loads. Commutation problems that lead to switching spikes are also discussed with the necessary safe commutation algorithm. Due to the symmetrical nature of SPMC, fully controllable regenerative operation is possible together with an associated bidirectional power flow providing attractive future solutions.

2. CONTROLLED RECTIFIER USING SPMC

Matrix Converter (MC) offers an "all silicon" solution potential for AC-AC conversion, removing the need for reactive energy storage components used in conventional rectifier-inverter based system [4-6]. Previous works are based on three-phase circuit topologies; with single-phase matrix converter (SPMC) emerging [7] with works on AC-AC [8] and DC-DC conversion [9]. The single-phase matrix converter (SPMC) used in this work is as shown in Fig. 2 with common emitter anti-parallel IGBT, diode pair [8,9,10 & 11] which is capable of conduct current in both directions, whilst at the same time it is capable for blocking voltages. The four switching state used, can be manipulated to controlled AC-DC converter in this work (Fig. 3 to 6). For controlled rectifier operation, only State 1 and 4 [12] will be used, making State 2 and 3 redundant. However, these redundant switches could be used to add features to the controlled rectifier operation that may include, amongst others; safe-commutation when using RL load and unity power factor operation particularly when RC loads are used.

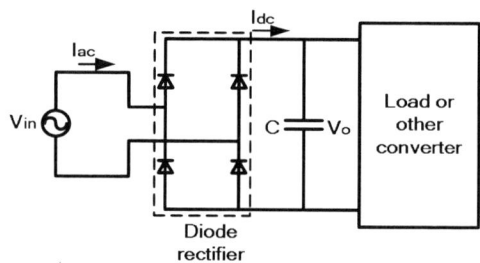

Fig 1: (a) Single-phase diode bridge rectifier with capacitive output filter

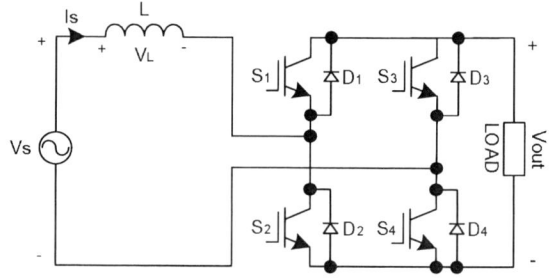

Fig 1: (b) H-Bridge Topology

978-1-4244-0644-9/07/$25.00 ©2007 IEEE

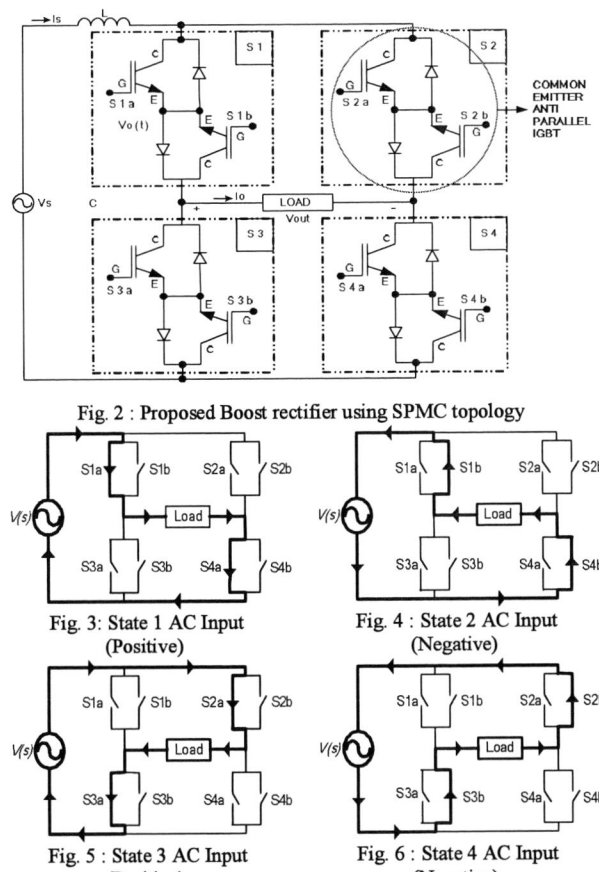

Fig. 2 : Proposed Boost rectifier using SPMC topology

Fig. 3 : State 1 AC Input
(Positive)

Fig. 4 : State 2 AC Input
(Negative)

Fig. 5 : State 3 AC Input
(Positive)

Fig. 6 : State 4 AC Input
(Negative)

3. COMMUTATION PROBLEM

One common problem in the use of matrix converter is the commutation spikes, which has limited the initial development for almost 25 years. This relates to the use of inductive loads. The commutation occurs when switches are turned 'off' and hence the existence of possible reversal current. Theoretically the switching in the SPMC must be instantaneous and simultaneous; unfortunately it is impossible for realization in practical systems due to turn-on/off IGBT characteristics, where the turn-off delay time is 420ns whilst the turn-on delay time of 75ns as discovered in this work; illustrated as shown in Fig. 7. This may create a short circuit. Further commutation problems also include; voltage spikes based on (Ldi/dt). To solve these difficulties the switching arrangements is modified to Fig. 8 & 9 with further details in [12].

Table 1: switching state for safe commutation strategy

switch	s1a	s1b	s2a	s2b	s3a	s3b	s4a	s4b
positive cycle	pwm	off	off	off	off	on	on	off
negative cycle	on	off	off	on	off	pwm	off	off

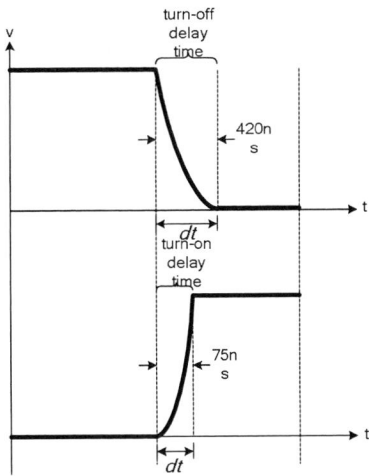

Fig 7: Time on/off proposed switch

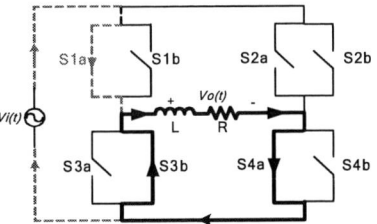

Fig. 8: Safe commutation strategy (positive state)

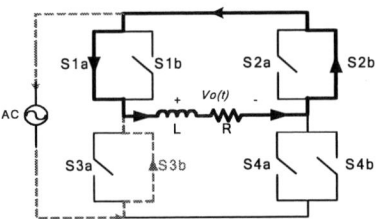

Fig. 9: Safe commutation strategy (positive state)

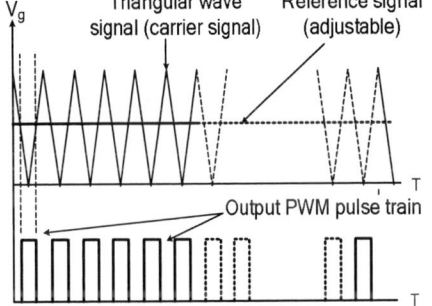

Fig. 10 : PWM waveform

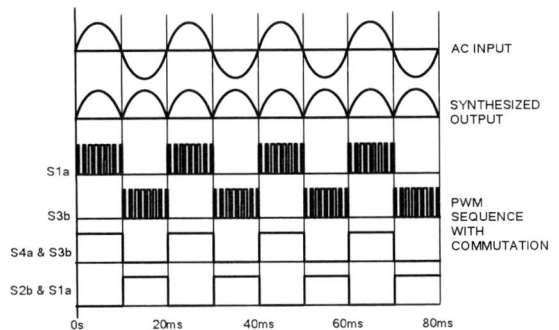

Fig. 11 : Switching Pattern for Commutation Strategy

4. PROPOSED CONTROLLED RECTIFIER USING SPMC WITH REGENERATIVE OPERATION

The use of SPMC allows regenerative energy of loads, such as motor to be recovered back to the supply so that it can absorb the energy instead of being stored in a large capacitor bank or dissipated in a braking resistor. The regenerative operation can be implemented according to the switching state in table 2; illustrated as in Fig. 12-14. Incorporating safe-commutation strategy for regenerative operation table 3 is developed complemented by illustrations as in Fig. 15-16.

Table 2: Switching state for regenerative operation

switch	s1a	s1b	s2a	s2b	s3a	s3b	s4a	s4b
positive cycle	off	pwm	off	off	off	off	off	pwm
negative cycle	off	off	pwm	off	pwm	off	off	off

Table 3: Switching state for regenerative operation with safe-commutation problem

switch	s1a	s1b	s2a	s2b	s3a	s3b	s4a	s4b
positive cycle	off	pwm	off	on	off	off	off	on
negative cycle	off	off	on	off	pwm	off	on	off

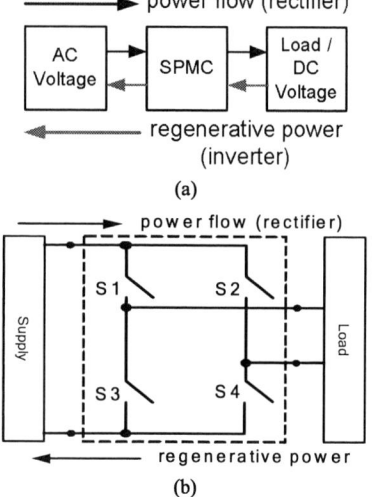

(a)

(b)

Fig.12 : Proposed regenerative operation (a) Block diagram (b) Circuit arrangement

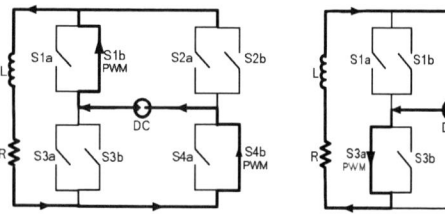

Fig. 13 : Positive cycle of inverter without safe-commutation

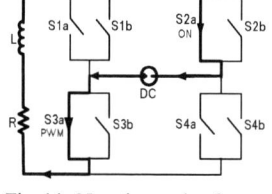

Fig. 14 : Negative cycle of inverter without safe-commutation

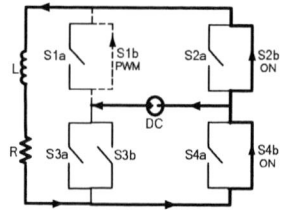

Fig. 15 : Positive cycle of inverter with safe-commutation

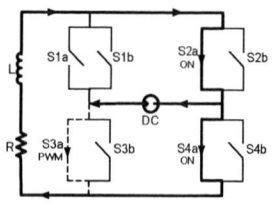

Fig. 16 : Negative cycle of inverter with safe-commutation

5. SIMULATION AND EXPERIMENTAL VERIFICATIONS

Proposed behaviour of operation is initially investigated using Matlab/Simulink (MLS) with SimPowerSystems. Comparisons are also made with PSpice to ensure accuracies in simulation. Subsequently an experimental test rig is developed to verify the operation. The single-phase matrix converter (SPMC) is built with four bidirectional switches as shown in Fig 2. For this purpose two units of IGBT switch types BUP314D are arranged in series to form one switch cell. Investigations on changes to modulation index with variation in frequencies from 1kHz up to 5kHz are recorded. The output rms voltage was also determined in this work. Switch control implementations uses PIC16F84A to generate PWM for controlled rectifier and regenerative operations. The AC-DC rectifier was supplied by 50 Volt AC voltage source; the load takes the form of a pure resistive 50Ω and 4mH inductor. Fig. 17 & 18 shows the various simulation models developed in MLS and PSpice whilst Fig. 19 shows the experimental test rig of the proposed work.

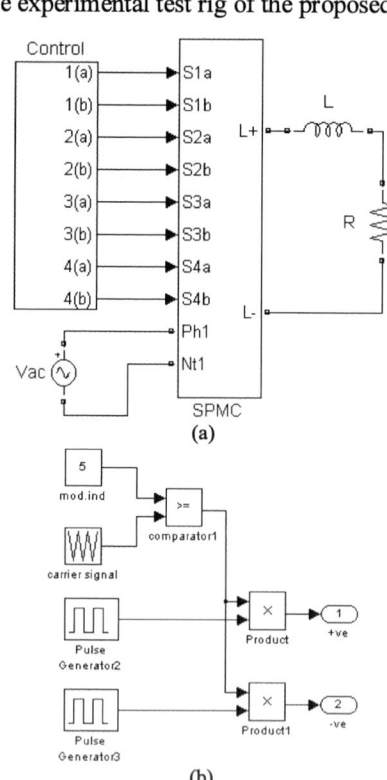

(a)

(b)

1628

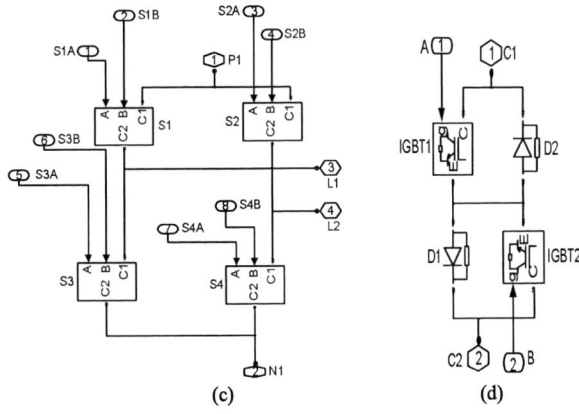

(c)   (d)

Fig. 17 : Simulation circuit using MLS (a) SPMC with regenerative operation (b) PWM generator (c) SPMC Arrangement and (d) Bidirectional Switch

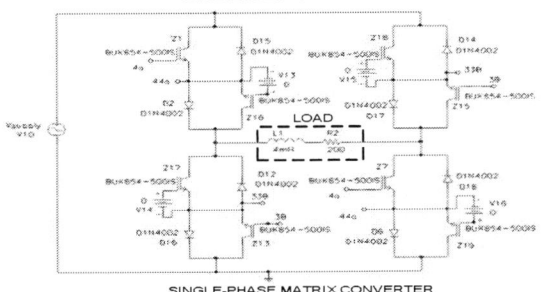

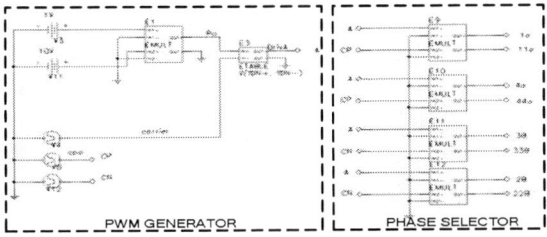

Fig. 18 : Simulation circuit using PSpice

Fig. 19 : Experimental test rig

6. RESULTS AND DISCUSSION

The result presented are arranged in accordance to the following; Fig. 20 to 22 presents the various results that include; effect of inductance and safe-commutation implementations obtained from MLS whilst Fig. 23 to 27 are those obtained using PSpice. Observe; a similar behaviour were produced. Fig. 28 to 31 shows the experimental results obtained in investigations. With pure resistive load, no spikes are apparent (Fig. 20, 23, 24 &

28). Upon the introduction of inductive load, damaging spikes was observed (Fig. 21, 25 & 29). Implementation of the commutation strategy as proposed has resulted in spikes being eliminated (Fig. 22, 26, 27 & 30). The regenerative operation represent by Fig. 33 which is inverted from the output DC voltage is as shown in Fig. 32.

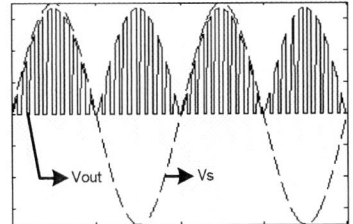

Figure 20 : Output voltage of R load from MLS

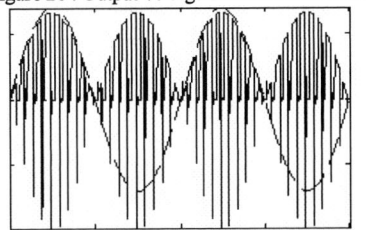

Figure 21 : Output voltage of RL load without commutation from MLS

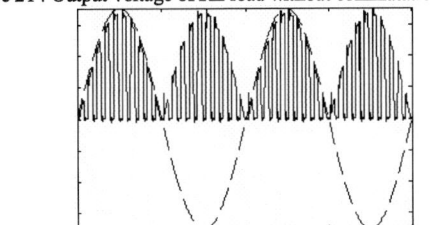

Figure 22 : Output voltage for RL load with safe commutation from MLS

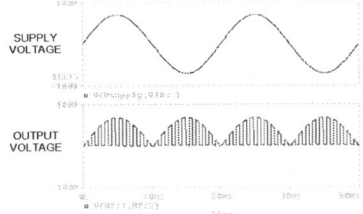

Fig. 23: Input and Output voltage for R load from PSpice

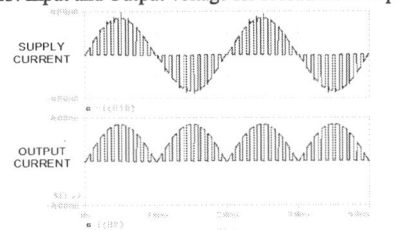

Fig. 24: Input and Output Current for R load from PSpice

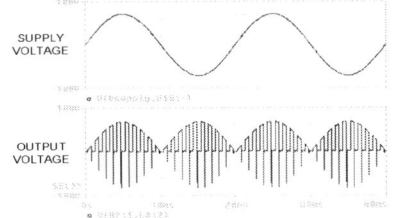

Fig. 25: Input and Output voltage for RL load Without safe-commutation Strategy from PSpice

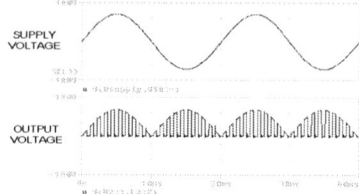

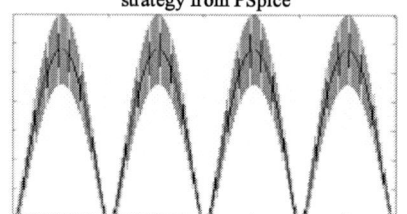

Fig. 26: Input and Output voltage for RL load With safe-commutation strategy from PSpice

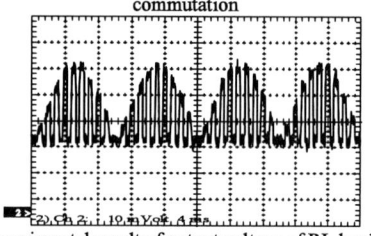

Fig. 27: Output current of rectifier with commutation

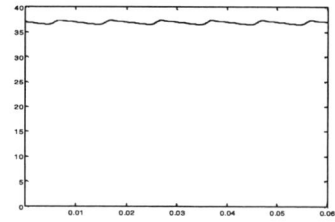

Fig. 28: Experimental result of output Voltage of R load

Fig. 29: Experimental result of output voltage of RL load without commutation

Fig. 30: Experimental result of output voltage of RL load with safe commutation strategy

Output voltage (rms) versus Modulation index

Fig. 31: Output voltage versus modulation index from MLS

Fig. 32: Output DC voltage

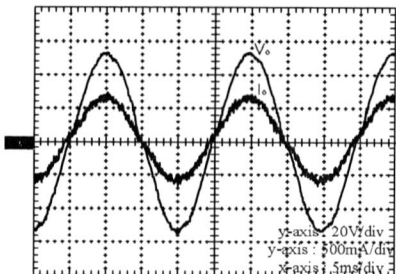

Fig. 33: Experimental Result of Regenerative Operation with filter

7. CONCLUSION

This paper illustrates with details on the operation of SPMC topology operating as a controlled rectifier operation incorporated with commutation strategy to remove damaging voltage spikes. The switching arrangements are then modified to provide regenerative capabilities. It shows that the SPMC is a very versatile topology with future potential, extending beyond the direct AC-AC converter, DC chopper operations and basic control rectifier operation. The simulation and experimental results was observed to confirm the predicted behaviour on the operation of the SPMC. Further advancement could be developed with redundant switches available but are subject to future research. This paper is one such advancement that has been explored.

7. ACKNOWLEDGEMENT

Financial support from Ministry of Science Technology and Innovation (MOSTI) Malaysia EScience Grant No : 03-01-01-SF0146 is gratefully acknowledged for implementation of this project.

8. REFERENCES

[1]   Bhim Singh, Brij N. Singh, Ambrish Chandra, Kamal Al-Haddad, Ashish Pandey, Dwarka P. Kothari., "A Review of Single-Phase Improved Power Quality AC-DC Converters," IEEE Transactions on industrial electronics, Vol. 50, No.5, October 2003.

[2]   Guichao Hua, Ching-Shan Leu, Yimin Jiang and Lee, F.C.Y.; "Novel zero-voltage-transition PWM converters", IEEE Transactions on Power Electronics, Volume 9, Issue 2, March 1994 Page(s):213 – 219

[3]   Po-Tai Cheng, Chung-Chuan Hou and Jian-Shen Li; "Design of an Auxiliary Converter for the Diode Rectifier and the Analysis of the Circulating Current", 37th IEEE Power Electronics Specialists Conference, 2006. Page(s): 1- 7

[4]   Gyugyi,L and Pelly,B.R, "Static Power Chargers, Theory, Performance and Application," John Wiley & Son Inc, 1976

[5]   Sobczyk, T., "Numerical Study of Control Strategies for Frequency Conversion with a Matrix Converter," Proceedings of Conference on Power Electronics and Motion Control, Warsaw, Poland, 1994, pp. 497-502

[6]   Cho, J.G., and Cho, G.H, "Soft-switched Matrix Converter for High Frequency direct AC-to-AC Power Conversion," Int. J. Electron., 1992, 72, (4), pp. 669-680

1630

[7] Zuckerberger, A., Weinstock, D., Alexandrovitz A., "Single-phase Matrix Converter," IEE Proc. Electric Power App, Vol.144(4), Jul 1997 pp. 235-240

[8] Zahiruddin Idris, Mustafar Kamal Hamzah & Ahmad Maliki Omar "Implementation of Single-Phase Matrix Converter as a Direct AC-AC Converter Synthesized Using Sinusoidal Pulse Width Modulation with Passive Load Condition", IEEE Sixth International Conference PEDS 2005, Kuala Lumpur, Malaysia

[9] Siti Zaliha Mohammad Noor, Mustafar Kamal Hamzah & Ahmad Farid Abidin, "Modelling and Simulation of a DC Chopper Using Single Phase Matrix Converter Topology" IEEE Sixth International Conference PEDS 2005, Kuala Lumpur, Malaysia

[10] Lixiang Wei and Thomas. A Lipo, "A Novel Matrix Converter Topology With Simple Commutation", Thirty-Sixth IAS Annual Meeting. Conference Record

of the 2001 IEEE Industry Applications Conference, 2001. Volume 3, 30 Sept.-4 Oct. 2001 Page(s):1749 – 1754

[11] Kwon, B.-H.; Min, B.-D.; Kim, J.-H.; "Novel Commutation Technique of AC-AC Converters", Electric Power Application, IEE Proceedings-, Volume: 145, Issue: 4, July 1998 Pages: 295-300

[12] Baharom R.,A.S.A. Hasim, M.K.Hamzah and A.F Omar; "A New Single-Phase Controlled Rectifier Using Single-Phase Matrix Converter Topology". Proceeding for IEEE First International Power and Energy Conference, 2006.

[13] Nastran, J.; Cajhen, R.; Seliger, M.; Jereb, P; "Active Power Filter for Nonlinear AC loads", IEEE Transaction on Power Electronics, Vol. 9, No. 1, January 1994.

Implementation of Space Vector Modulated 3Φ to 3 Φ Matrix Converter Fed Induction Motor

S. GANESH KUMAR[*], S.SIVA SANKAR[**], S.KRISHNA KUMAR[**], G. UMA[$]

[*] Research Scholar, [**] PG Student, College of Engg., Guindy, Anna University, India.
[$]Assistant Professor/ Power Electronics and Drives division, College of Engg., Guindy, Anna University, India.

Abstract– **Matrix Converters (MC) are compact voltage source converters capable of providing variable voltage and variable frequency at the output. Compared with traditional topologies the MC does not require an intermediate dc link and provides sinusoidal output waveform with minimum higher order harmonics. To yield higher rms O/P voltage, it is proposed to use Space Vector Modulation (SVM) algorithm for the voltage control of converter. The proposed modulation technique is derived from Indirect Transfer Approach (ITF).The time arc information of the supply voltages and the desired output frequency are used to identify the sectors of the SVM technique, to generate the pulses of desired duration. The digital implementation of Space Vector Modulated (SVM) switching scheme for a three phase to three phase MC is done through Digital Signal Processor (DSP) TMS 320C 2407A. To verify the validity of the algorithm a 15 KVA MC is built. SVM pulses are generated and given to the appropriate MC switches through a decoder circuit and the experimental results are presented. The proposed modulation algorithm can be used for v/f control of Induction Motor**

Keywords:-**Matrix Converter, Space Vector Modulation (SVM), Indirect Transfer function approach.**

I.INTRODUCTION

In many ac drive applications it is desirable to use a compact voltage-source converter, which can provide sinusoidal output voltages of varying amplitude and frequency, while drawing sinusoidal input currents with unity power factor from the ac source. Matrix Converters (MCs) are increasingly becoming popular because they don't have any intermediate energy storage devices except small ac filters for the elimination of switching ripples. Using MCs the v/f method of speed control of induction motor can be achieved smoothly with an improvement in power factor and reduction in harmonics. Laszel Huber et.al (1989) developed a method of voltage space vector based PWM control of MC [3].The above authors (1991) developed a method of space vector modulation with unity input power factor for MCs [4].Yanhui Xie et.al (2004) implemented DSP based three phase to three phase MC [6].Klumpner et.al [16], reduce the input unbalanced problem by taking the input/output power balance equation. A new ride-through strategy was implemented by the same author [16]. The strategy takes control of the drive when a power outage occurs, disconnecting the motor from the grid and recovering the energy stored in the inertia into dc capacitor in the clamp circuit.

II. THREE PHASE TO THREE PHASE MC

A. Matrix Converter Principle of Operation

The circuit diagram for three phase to three phase MC is shown in fig.1.It consists of nine Bi-Directional Switches

(BDS) arranged in the form of matrix. V_1, V_2, V_3 are the supply Phase voltages and Z is the load impedance. Since the MC is supplied by the voltage source the input phases must never be shorted and due to the inductive nature of the load, the output phases must not be left open. If the switching function of a switch, s_{jk}, is defined as

$$S_{jk}(t) = \begin{cases} 1, S_{jk} \text{ closed} \\ 0, S_{jk} \text{ open}, \end{cases} j \in \{A,B,C\}, k \in \{a,b,c\}$$

(1)

The constraints can be expressed as

$$S_{ja} + S_{jb} + S_{jc} = 1, j \in \{A,B,C\},$$ (2)

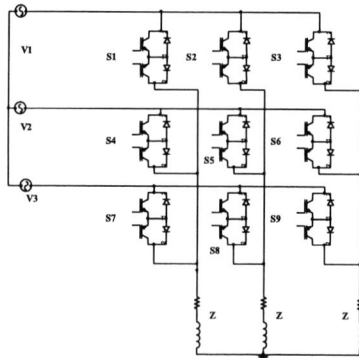

Fig 1   Circuit Diagram of 3Φ to 3 Φ MC

With these constraints, only 27 switching combinations are possible and it is classified into three groups. The first group includes six combinations where each output phase is connected to a different input phase. In the second group, there are 3 x 6 = 18 combinations with only two output phases shorted. The third group includes three combinations with all three-output phases shorted. Considering the above constraint and to improve the voltage transfer ratio SVM based gate pulse generation method is adopted in the present work.

III.SWITCHING STRATEGY FOR MC

A. Space Vector Pulse Width Modulation

In order to apply the standard inverter SVM technique to the MCs, Indirect Transfer Function (ITF) approach is used. In this approach, the modulation process is fictitiously divided into rectification and inversion process. For low harmonic distortion of the MC in the output voltage, SVM technique is used in the inversion step. For maximal voltage gain the 3 φ full-wave rectification is applied. The 3 φ balanced input source voltages and the arcs of the 3 φ full-wave rectified Input Voltage Envelope (IVE), V_{ienv} (t) are defined. As the same converter has to perform both, the rectification and

978-1-4244-0644-9/07/$25.00 ©2007 IEEE   1632

inversion, detailed analysis of the allowed switching combinations are to be done, respecting the constraints imposed by source and load. MC Space Vector is synthesized with second and third group vectors as shown in fig.3 which can assume only seven discrete positions in the complex plane, called the switching state vectors. The six non-zero switching state vectors form the MC hexagon. The desired output voltage space vector, called the reference vector V_r can be approximated by two adjacent switching state vectors, V_α and V_β, and the zero voltage vector V_O, using SVPWM. In order to enable the synthesis of the maximum possible reference vector amplitude, it is necessary to use the instantaneously largest available input voltage. This is the positive or the negative envelope of the 3 phase full-wave rectified input line voltages. At a given time instant, one and only one of the six available switching combinations produces the switching state vector magnitude, which corresponds to the Input Voltage Envelope (IVE).

The switching states of the MC are based on three factors. The position of the reference vector in the MC hexagon at a given time instant, determines the switching state vectors to be used for the output voltage synthesis. Then based on the IVE arc, at that instant and the selected switching state vectors, a particular switch combination can be selected from Table 1. Then the general expression for the conduction time of the switches in α- β coordinates is given by

$$
\left.
\begin{aligned}
T_\alpha &= \frac{2}{3} \frac{v_r}{v_{ienv}(t_k)} T_c \sin(60° - \theta_s) \\
T_\beta &= \frac{2}{3} \frac{v_r}{v_{ienv}(t_k)} T_c \sin(\theta_s)
\end{aligned}
\right\} \quad (3)
$$

$$
T_o = T_c - T_\alpha - T_{\beta,}
$$

Where

T_α—Time duration of active state in α- β coordinates

T_β---Time duration of active state in α- β co ordinates,

T_c—switching time

T_0---Time duration of zero state,

v_r—Reference voltage,

v_{ienv}—Input Voltage Envelope

As the position of the reference vector sweeps through the sectors of the MC hexagon, different switching state vectors have to be selected for each sector. In Table 1 the switching sequences within the first half of the switching cycle are shown for each IVE arc, which can also be used for second half when reading from right to left.

The general expression for the conduction time of the switches shows that the on-times of the switching state vectors are inversely proportional to the input voltage values sampled in each switching cycle. This compensates for the input voltage magnitude variations, so that the mean value of the MC output line voltages, averaged in each switching cycle, depends only on the reference vector. Vectors T_α and T_β, should be halved and distributed symmetrically around the centered zero voltage interval

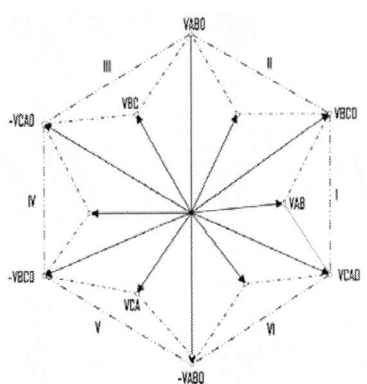

Fig 3 Vector Diagram

IV. HARDWARE IMPLEMENTATION OF MATRIX CONVERTER SYSTEM

To realize the SVM algorithm with improved voltage transfer ratio in real time MC fed R-L load is considered as shown in fig.1. The block diagram of the MC system is shown in fig 2, which consists of a three-phase supply, nine BDS arranged in the form of matrix, and a load. The input three-phase supply is also given to ZCD to identify the time arc information. DSP is used for generating the pulses. Experimental setup for MC is shown in fig.7 and it consists of nine BDS (IGBT-IR IR 4PC 40 KD) with its own driver circuit.

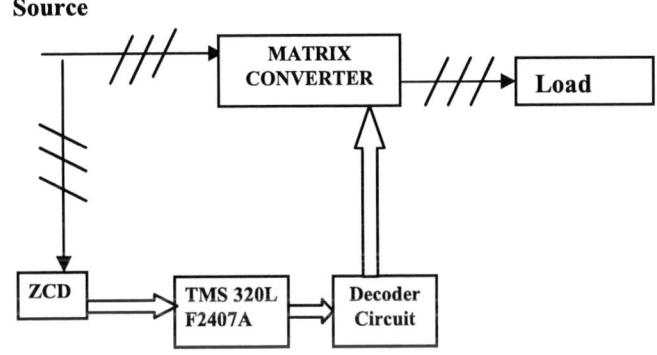

Fig.2 Block Diagram

A. Input Time Arc Identification &Gate Pulse Generation

The circuit diagram for the input time arc identification circuit is shown in fig.6. The input to the ZCDs are given through three 230/5 V step down transformers. The output of the ZCDs are given to suitable logic circuit which is used for identifying the time arc information I_1 to I_6 The space vector algorithm is implemented with TMS 320C 2407A system. Only three pulses P_1, P_2 and P_3 are generated from the DSP processor for the control of the nine BDS. The pulses for the switches are derived with the help of decoder circuits. The SVM pulse generation circuit is shown in fig4. The pulse pattern from the DSP processor and SVM pulse generation circuit are shown in figs 8 and 9. The pulses P_1, P_2 and P_3 are converted to P_α, P_β and P_o using the logic equation as follows

$$P_\alpha = \overline{x}\,yz + xyz \qquad (4)$$

$$P_\beta = \overline{xy}z + x\,\overline{yz} \qquad (5)$$

$$P_0 = \overline{x}\,yz + xy\overline{z} \qquad (6)$$

Where x,y,z are the Pulses obtained from DSP

B. Decoder Design

The decoder circuits are used to deliver the PWM pulses from the DSP processor to the four quadrant switches and realize the output from MC. The input data of the decoder are the PWM pulses from the SVM pulse generation circuit. The outputs of the decoder circuit are used to determine the state of the switches one to nine. By using the decoder circuit the pulses are distributed to the appropriate switches depending on the input time arc and output sector codes. The state of the switches S_1 is given by

$$S_1 = [\,P_\alpha\,\{(\,I_1 + I_6)(\,O_1 + O_2 + O_3) + (I_3 + I_4)(O_4 + O_5 + O_6)\}\,] + [\,P_\beta(I_1 + O_6) + (I_3 + I_4)(O_3 + O_5 + O_6 + O_6)] + P_0\,[(I_1 + I_6)(O_2 + O_6) + (I_3 + I_4)\,(O_1 + O_3 + O_5) + I_1 O_4)] \qquad (7)$$

Where I_x-Input time arc information
O_y- Sector information
Similarly the expressions can be derived for other switches. The decoder circuit for the switches S_2 is shown in fig 6

V. RESULTS

The pulses for the four quadrant switches for improved voltage transfer ratio are obtained with the help of the DSP Processor to generate an output voltage of frequency 30Hz. The second step carried out is the redistribution of the pulses obtained from the processor by using SVM pulse distribution circuit. From the SVM circuit the pulses to the four quadrant switches are given with the help of a decoder circuit. The pulse pattern obtained from the DSP processor with a switching frequency of 20 KHz for pulses P1 and P2 is shown in fig8

TABLE I
SWITCHING PATTERN

Time	$T_\alpha/2$			$T_\beta/2$			$T_0/2$		
Arc	A	B	C	A	B	C	A	B	C
1	a	c	a	a	c	c	c	c	c
2	b	c	b	b	c	c	c	c	c
3	b	a	b	b	a	a	a	a	a
4	c	a	c	c	a	a	a	a	a
5	c	b	c	c	b	b	b	b	b
6	a	b	a	a	b	b	b	b	b

Simulation is carried out with load parameters R=1Ω and L= 10mH and Induction motor, with supply voltage 220V rms /phase , and with a switching frequency 1000Hz. MATLAB Simulink model of MC fed Induction motor(220V,3HP) . Simulation results for Induction motor is shown in fig.13 The output frequencies obtained are 25Hz, 50Hz, 100Hz. (See fig.10,11 12).

VI. CONCLUSION

The proposed SVM algorithm work satisfactorily and produces pulses of desired frequency. The v/f control of induction motor using MC is in progress.

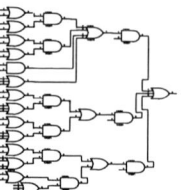

Fig.4 SVM Pulse
Realization Circuit

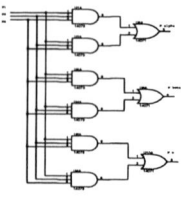

Fig 5 I/P time arc identification circuit

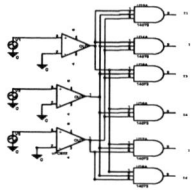

Fig 6 Decoder circuit
for switch 2

Fig.7 Hardware Set Up

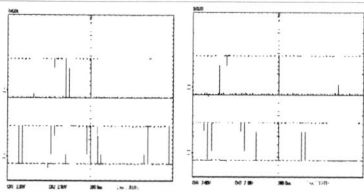

Fig .8 Pulses from DSP

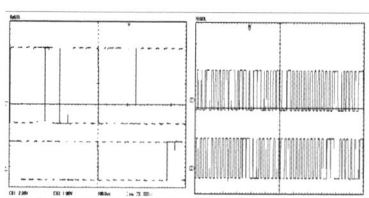

Fig.9 Pulse output from logic gates

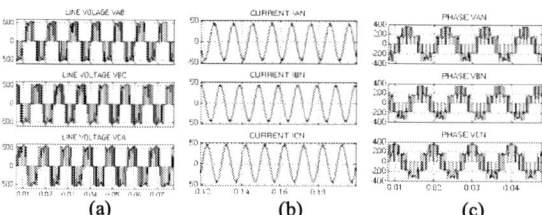

(a)   (b)   (c)

Load Voltages & Load currents of MC supplying RL load with O/P frequency 100 Hz

Fig. 10(a) Line voltage (b) Phase voltage (c) Phase Current for 100Hz

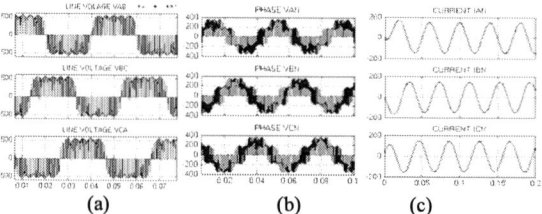

(a)   (b)   (c)

Load Voltages & Load currents of MC supplying RL load with O/P frequency 25Hz

Fig.11 (a) Line voltage (b) Phase voltage (c) Phase Current for 25 Hz

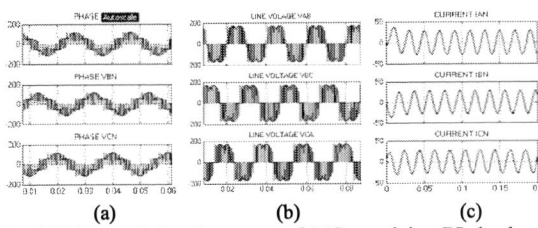

(a)   (b)   (c)

Load Voltages & Load currents of MC supplying RL load with O/P frequency 50 Hz

Fig. 12(a) Line voltage (b) Phase voltage (c) Phase Current for 50 Hz

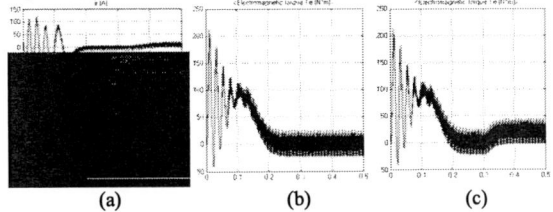

(a)   (b)   (c)

Fig. 13(a) Rotor& Stator Currents after applying load,
(b) Torque at No load(c) Torque at Load

VII.REFERENCES

[1]. B.K. Bose.,2006 "Power Electronics and AC drives ", PHI, pp.207-208

[2]. Laszel Huber, DuSan Borojevic, Nandor Burany "Voltage Space Vector based PWM control of Forced Commutated Cyclo converters" IEEE Trans pp. 871-876 1989.

[3]. Laszel Huber, DuSan Borojevic (1991) "Space Vector Modulation with unity input power factor for MCs " IEEE Trans pp. 1032-1041 year 1991.

[4]. Lazel Huber and Borojevic "Design and implementation of the three phase to three phase Matrix Converter with input power factor correction" IEEE Trans pp. 860-864 year 1993.

[5]. Yanhui Xie and Yogde Ren "Implementation of DSP based three phase to three phase AC-AC Matrix Converter" IEEE Trans pp. 843-847 year 2004.

[6]. A,Alsenia, M.G.B,Venturini (1981) "Solid State Power Conversion a Fourier Analysis Approach to Generalized Transformer Synthesis" IEEE Transaction on Circuit & System Volcas 28 ,4 April

[7]. A.Alsenia, M.Venturini (1989) "Analysis and Design of Optimum Amplitude Nine- Switch Direct Ac-Ac Converters", IEEE Transaction on Power Electronics, vol.4, No.1 January, pp 101-112.

[8]. M.Imayavaramban,K.Latha and G.Uma,(2004)"Matlab Simulink Implementation For Reducing the Motor Derating and Torque pulsation of Induction Motor using Matrix Converter with maximum voltage conversion ratio"in proceedings of the IEEE-PSCE.

[9]. C.L.Neft and Scauder (1988) "Theory and Design of a 30Hp Matrix Converter ", IEEE –IAS annual meeting, PP 931-939.

[10].Patrick W.Wheeler Jose Rodriguez, Lee Empringham (2002),"Matrix Converter Technology Review" IEEE Transaction on Industrial Electronics, vol.49, No.2, PP 276- 289.

[11].Somndia Ratanapanachote, Han ju cha, Prasad N.Enjeti(2006) "A Digitally Controlled Switch Mode Power Supply Based On Matrix Converter "IEEE Transactions on Power Electronics Vol.21, No.1, January, pp124-130.

[12].Toliyat "DSP based electro-mechanical energy conversion" Page no307-325

[13].L.Zhang, C. Watthanasaran, W.Shepherd (1995) "A Novel Switch Sequencer Circuit for Safe Commutation of A Matrix Converter", Electron Lett, 31, (18), pp 1530-1532.

[14]. F.Blaajjerg, D.Casadei, C.Klumpner and M.Matteini," Comparison of two current modulation strategies for matrix converters under unbalanced input voltage conditions," IEEE Trans. on Industrial Electronics, vol.49, pp.289-296, April 2002.

[15].J.Kang, H.Hara, A.M.Hava, E.Yamamoto, E.Watanabe, and T.Kume,"The Matrix Converter drive performance under abnormal input voltage conditions," IEEE Trans.on Power Electronics, vol.17, pp.721-730, Sept.2002.

[16].C.Klumpner, I.Boldea, and F.Blaabjerg,"Limited ride through capabilities for direct frequency converters," IEEE Trans .on Power Electronics, vol.16, pp.837-845, Nov.2001.

VIII.BIOGRAPHIES

S.Ganesh Kumar completed B.E from Madras University (1998) and M.E (PED) from Anna University (2005).He is pursuing PhD at Anna University. Presently he is working as Senior Lecturer in the EEE department of Rajalakshmi Engineering College, Chennai. His areas of interests are Matrix Converter, Multilevel Inverters.

S.Krishna Kumar completed B.E from Madras University. He completed M.E by 2007 from Anna University. Presently he is working as Assistant Professor in Prathyusha Inst.of Technologys and Management, Tiruvallore .His areas of interests are converters& Drives.

G.Uma Completed B.E from Annamalai University and M.E(C&I) from Anna University.She completed Ph.D degree in DC-DC Converters from Anna University. Presently she is working as an Assistant Professor, EEE department, College of Engineering, Guindy, Anna University, Chennai, India. Her areas of interest are Power Quality, Resonant Converters, and Matrix Converters.

A Single-Phase High-Power-Factor Neutral-pointer Clamped Multilevel Rectifier

Yun Xu, Yunping Zou, Chengzhi WANG, Wei Chen, Bangyin Liu

Power Electronic Research Center, College of Electrical and Electronic Engineering,
Huazhong University of Science and Technology, Hubei Wuhan 430074, China
Phone: 86-27-87558054 Fax: 86-27-87543658-815
E-mail:goodtom@163.com

Abstract-**A single-phase rectifier based on neutral-clamped converter (NPC) is proposed to achieve high power factor, lower current distortion and low voltage stress of power semiconductors. The principle of single-phase three-level PWM rectifier is introduced. Neutral-point balance problem is an inherent problem of three-level converter, so a bangbang control method is schemed to control the capacities' voltage. A dual close-loop control is used to make the rectifier have favorable dynamic and steady state performance, sinusoidal input current and unity power factor. Finally, the feasibility and validity of the proposed control scheme and neutral-point voltage control scheme are verified through the simulation using MATLAB.**

***Index Terms*-- Multilevel Rectifier; Neutral-pointer Clamped; High-Power-Factor; Voltage balance Control.**

I. INTRODUCTION

With the strictness of the harmonic standard of the power system, using the PWM rectifier which is reduced harmonics and have high power factor to replace the traditional diode bridge rectifier and phase-controlled rectifier is a doubtless trend. Compare with the traditional rectifier, the PWM rectifier can control current to have lower distortion and achieve high power factor, at the same time can gain the bidirectional power flow ability. Furthermore, the PWM rectifier can markedly cut down the devices' bulk and weight, and get dynamic response faster remarkably [1] [2].

In the high-voltage and high-power field, the Neutral-pointer Clamped Multilevel Rectifier can be employed. The topology of a single-phase Neutral-pointer Clamped Multilevel Rectifier is shown in Fig.1. There are four high frequency switches in one of its bridge leg that can generate there level -Udc/2, 0, Udc/2 on one bridge leg. The converter offer the advantages of low voltage stresses on switches, reduced losses at reduced switching frequency for the same level of performance in terms of reduced harmonics and high power factor at the input ac mains. The converter has bidirectional power flow and is used for even high-power applications such as BESS, metros, traction, etc. The topology can be developed for a higher number of levels for high-voltage and high-power applications. It has been reported that the ac supply current THD can be reduced below 1% without using PWM control [3].

In this paper, a composite control strategy based on a **dual close-loop control** that draws a sinusoidal input current, low current harmonics, and high power factor in the single-phase three-level rectifier is proposed. The NPC converter is used in the adopted single-phase rectifier to drive the mains

current which is following the reference current. The dc-link voltage unbalance problem across two capacitors can be solved by the proposed bangbang control based on capacitors' voltage comparison. This paper is organized in the following manner. The circulate principle and math model of the proposed single-phase three-level rectifier is discussed in Section II. The control method of the capacities' output voltage and the dual-loop composite control strategy are described in Sections III and IV. The simulation results are discussed in Section V, followed by a conclusion in Section VI.

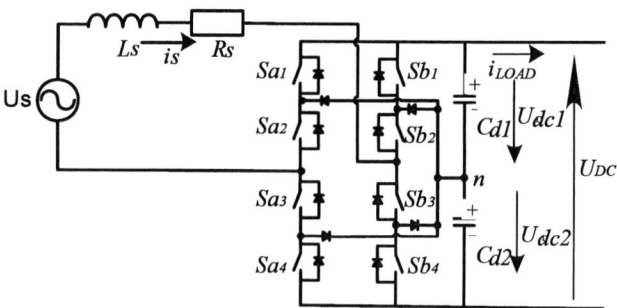

Fig.1. 3-level NPC PWM rectifier equivalent circuit

II. PRINCIPLE AND MODULATION METHOD

A. Principle of the Rectifier

The principle of single-phase NPC voltage-source PWM rectifier will be expounded. The Subharmonic Modulation method and related waveforms are systematically analyzed. In the light of high-frequency PWM rectifiers' control scheme nowadays, a dual-loop controller will be designed.

Fig.1 shows the main circuit of the single-phase NPC PWM rectifier. The Us is the input AC voltage; inductance L_S is the equivalent inductance which can deliver the energy, depress the high order harmonic and balance the voltage between the input and the output; R_S is the equivalent resistance which is very small generally; the DC capacity C_{d1} and C_{d2} are the filter capacities which can offer the route to the high order current harmonic, and reduce the ripple of the output DC voltage.

Assuming the original phase θ=0, and the magnitude is U_m, and the AC side current is I_s, and then the output voltage is U_{out}, which is the sum of the two capacity voltage U_{dc1} and U_{dc2}, the following equations can be obtain:

$$u_S = u_m \sin(\omega t) \qquad (1)$$

978-1-4244-0644-9/07/$25.00 ©2007 IEEE

$$L_S \frac{di_S}{dt} = u_S - u_{out} - R_S \cdot i_S \qquad (2)$$

By the equation (2), the vector figure which is consisted of the AC side vectors is obtained, as shown in the Fig.2.

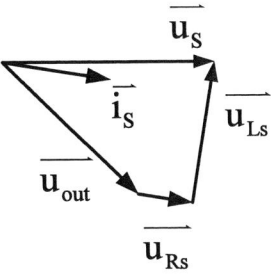

Fig.2. the vector diagram of the AC side

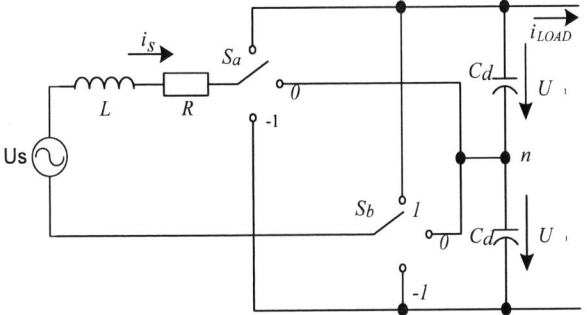

Fig.3. 3-level NPC PWM rectifier equivalent circuit

Fig.3 is the equivalent model of a single-phase NPC PWM rectifier. Herein, every bridge leg is equivalent to ternary switches, and their connection and break can be described by the logic function Sa and Sb. For the convenient purpose, the logic function can be sequentially decompounded: when Sa=1, we define S1a=1, S2a=0, S3a=0; when Sa=01, we define S1a=0, S2a=0, S3a=1; when Sa=-1, we define S1a=0, S2a=1, S3a=0. Sb will be transformed by the same way. Omitting the intricate formula computation, the math model can be gained.

B. The Subharmonic Modulation method (SHPWM)

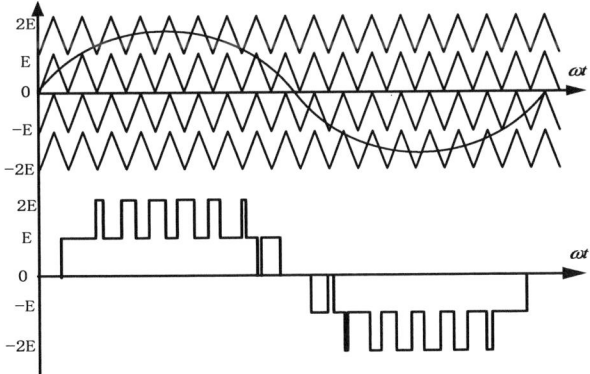

Fig.4. the sketch map of the SHPWM modulation method

SHPWM is a conventional control method fit for multilevel converter [4]. The control principle of the SHPWM method is to use several triangular carrier signals with only one modulation wave per phase. For the 5-level IGBT inverter, 4 triangular carriers of the same frequency and the same peak-to-peak amplitude are engaged so that the bands they occupy are contiguous. The zero reference is placed in the middle of the carrier set. The modulation wave is a sinusoid waveform. On every band each carrier is compared with the modulation waveform.

III. SOLVE OF THE NEUTRAL-POINT BALANCE PROBLEM

Neutral-point balance problem is an inherent problem of three-level inverter [5] [6]. Without neutral-point control, the harmonic components of output voltage will greatly increase, and the DC-link capacitors and the switching devices may probably be destroyed. Table.1 shows the voltage and the switch states of the NPC 5-level inverter. There are some redundant states, so these states can be used to control the neutral-pointer voltage balance. But by the analyses of the influence of the capacity's voltage, the ④⑤⑥ states have no affect of the balance of the capacity voltage; the ③⑦states make the C1 voltage higher; and the ②⑧states affect the C2 voltage higher. Herein, a principle that keeps the neutral-pointer voltage balance and reduces the switches loss is proposed. When the output state sequentially changes, only one leg output voltage change, and only a pair of switches trigger signals change. Hereby there are just two group states we can choose. One is ①②④⑧⑨ that is called STATE1 and another is ①③④⑦⑨that is called STATE2.

Consequently, a bang-bang control method of the neutral-pointer voltage is designed. The switch frequency is 10 KHz, then the sampling time is enacted to 10us.After the AD module samples, the voltages of the capacity C1 and C2 are compared. If the C1 voltage is lower, the inverter will choose the STATE2, otherwise the inverter will choose the STATE1. The result in Fig. will prove the feasibility of the proposed control method.

STATE	V$_{IGBT}$	Switch State							
		Sa1	Sa2	Sa3	Sa4	Sa'1	Sa'2	Sa'3	Sa'4
①	+2E	1	1	0	0	0	0	1	1
②	+E	0	1	1	0	0	0	1	1
③	+E	1	1	0	0	0	1	1	0
④	0	0	1	1	0	0	1	1	0
⑤	0	1	1	0	0	0	1	1	0
⑥	0	0	0	1	1	0	0	1	1
⑦	-E	0	1	1	0	1	1	0	0
⑧	-E	0	0	1	1	0	1	1	0
⑨	-2E	0	0	1	1	1	1	0	0

Table.1 the voltage and the switch state of the NPC 5-level rectifier

IV. THE DUAL-LOOP CONTROL DESIGN

In the various kinds of direct current control strategies in the PWM rectifier, the series structure of the dual loop consisted of the voltage outer-loop and current inner-loop is prevalent [7]. The main advantages of the dual loop control are: it has clear physical meaning, and the structure is simple, the performance is excellent, and as a result of the current loop, if the magnitude of the current reference is limited, the rectifier can work on the constant current state, and when the PWM rectifier starts up softly, the current is limited to the tolerance region of the semiconductors switch, so it's advantageous to the protection of the switches. So in the light of high-frequency PWM rectifiers' control scheme nowadays, current inner-loop controller and voltage outer-loop controller are designed with typical I type and II type system. Compound control with voltage forward-feedback is designed and verified the feasibility by simulation result.

A. The design of the current inner-loop controller

The PI controller is used considering the veracity of the system parameters and the expectation of no static error of the current control. In view of the sampling delay and the small signal character of the PWM control, the structure of the current inner-loop controller is gained in the Fig.5.

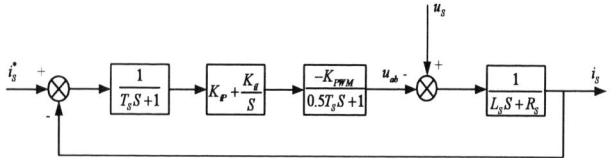

Fig.5 The block diagram of the current-loop control system

For the sake of simplify the analysis, the PI function is changed to zero-pole form, and the little inertial part can be emerged, as shown in Fig.4.

$$K_{iP} + \frac{K_{iI}}{S} = K_{iP}\frac{\tau_i S + 1}{\tau_i S} \qquad K_{iI} = \frac{K_{iP}}{\tau_i} \qquad (3)$$

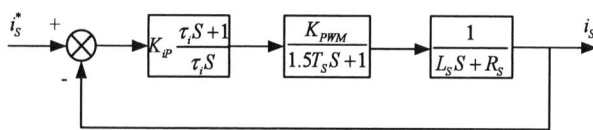

Fig.6 The compactable block diagram of the current-loop control system

The current regulator can design to I type system to gain a faster respondence of the track of the current. First, the zero point of the PI controller is designed to countervail the object function's pole point, so $\tau_i = L_s/R_s$, and the open-loop transfer function turn to a typical 1-order system.

$$W_{oi}(S) = \frac{K_{iP}K_{PWM}}{R_S\tau_i S(1.5T_S S + 1)} = \frac{K}{S(1.5T_S S + 1)} \qquad (4)$$

in the equation, $K = \dfrac{K_{iP}K_{PWM}}{R_S\tau_i}$ (5)

Base on the setting relation of the typical 1-order system, when the $\varepsilon = 0.707$,

$$\frac{1.5T_S K_{iP} K_{PWM}}{R\tau_i} = \frac{1}{2} \qquad (6)$$

So the expressions of the PI controller can be got.

$$K_{iP} = \frac{R\tau_i}{3T_S K_{PWM}} \qquad K_{iI} = \frac{K_{iP}}{\tau_i} = \frac{R}{3T_S K_{PWM}} \qquad (7)$$

And now the current close-loop transfer function Wci(S) is

$$W_{ci}(S) = \frac{W_{oi}(S)}{1 + W_{oi}(S)} = \frac{1}{1 + \dfrac{R\tau_i}{K_{iP}K_{PWM}}S + \dfrac{1.5T_S R\tau_i}{K_{iP}K_{PWM}}S^2} \qquad (8)$$

In virtue of the high switch frequency, the TS is small enough to ignore the S2 term. So the transfer function can be predigested to

$$W_{ci}(S) \approx \frac{1}{1 + 3T_S S}. \qquad (9)$$

Obviously, the current loop can be equivalent to an inertial loop; and the inertial time constant is $3T_S$. Accordingly, the current inner-loop has a fast dynamic response.

B. The design of the voltage outer-loop controller

From the rectifier's mathematic model, the equivalent black diagram of the voltage-loop control system.

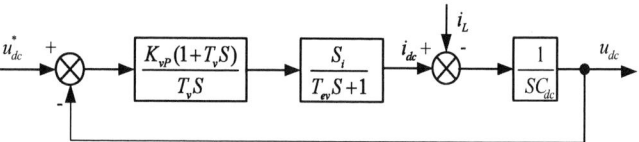

Fig.7 The compactable block diagram of the voltage-loop control system

To voltage close-loop, transfer function Wov(S) is

$$W_{ov}(S) = \frac{K_{vP}S_i(1 + T_v S)}{T_v C_{dc}S^2(1 + T_{ev}S)} = K\frac{(1 + T_v S)}{S^2(1 + T_{ev}S)} \qquad (10)$$

in the equation $K = \dfrac{K_{vP}S_i}{T_v C_{dc}}$ (11)

The compacted voltage-loop is a typical II-order system. The close-loop transfer function is

$$W_{cv}(S) = \frac{W_{ov}(S)}{1 + W_{ov}(S)} = \frac{KT_v S + K}{T_{ev}S^3 + S^2 + KT_v S + K}. \qquad (12)$$

The voltage PI controller can be designed by the 3-order best setting relation, and can get:

$$T_v = 4T_{ev} \qquad (13)$$

$$K = \frac{1}{8T_{ev}^2} \qquad (14)$$

The control system based on 3-order best setting relation has quick dynamic response, and can depress the disturbance. And the load current is feed forward, and the rectifier system can have a better dynamic and steady characteristic, and lighten the controller's burden.

It is well known that there is a ripple voltage at the

frequency of twice supply frequency appearing at the output voltage, which results in distortion of supply current [8]. So a low frequency filter has to add in the voltage-loop. Fig.8 shows the whole block diagram of the dual-loop control.

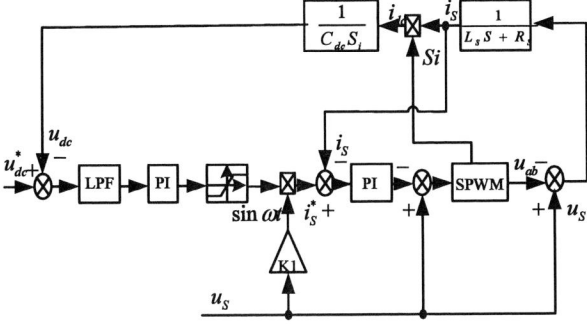

Fig.8 the block diagram of the dual-loop control

V. THE SIMULATION AND EXPERIMENT RESULT

Based on the above control scheme, the basic simulation parameters are: Usm=1555V, f= 50Hz, L_s= 4mH, R_S= 0.5Ω, C_{d1}= C_{d2}= 4700μF, Udc= 2000V, the frequency of IGBT and sample is 10kHz.

Fig.9 is the input voltage and the input current wave, and it's easy to see the high factor and low current distortion. Fig.10 is the NPC AC side output voltage, and the output wave is 5-level. Fig.11 is output DC voltage when the load abruptly changes. Fig.12 is the subtracting of the two capacities' voltage value, and the value is small. The neutral-point balance control strategy is available. Fig.13 is the experiment wave of the input voltage and the input current. In the experiment, the smaller Us is employed, the value is 80V. The experiment improved the feasibility of the rectifier.

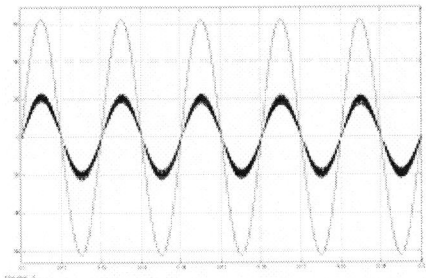

Fig.9 the input voltage and the input current wave

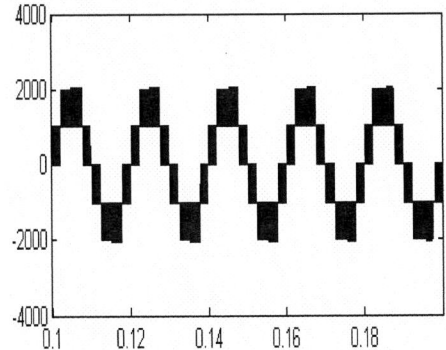

Fig.10 the NPC AC side output voltage

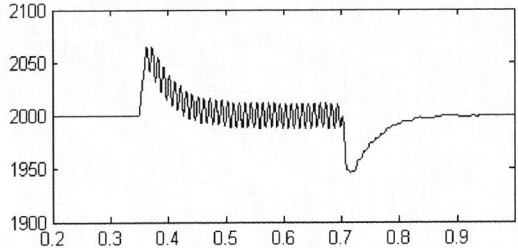

Fig.11 the output DC voltage when the load abruptly changes

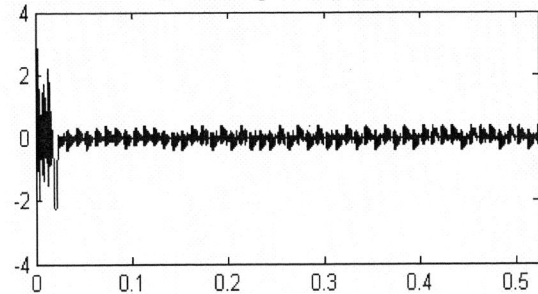

Fig.12 the result of subtracting the two capacities' voltage value

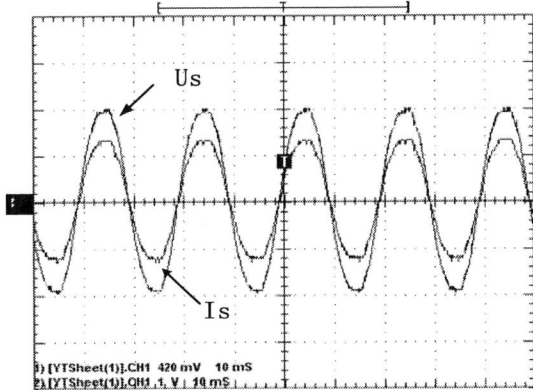

Fig.9 the input voltage and the input current wave in experiment (U$_S$: 40V/div, 10ms/div; I$_S$: 2A/div, 10ms/div)

VI. CONCLUSION

This paper presents a neutral-clamped converter (NPC) single-phase rectifier to achieve high power factor, lower current distortion and low voltage stress of power semi-conductors. A dual-loop control scheme considering the neutral-point balance problem is employed. Dual-loop control is used in the system to obtain a constant DC bus voltage and to track the phase current command. The SHPWM method is adopted to modulate the waveform to generate the output on the AC terminal of rectifier PWM. The feasibility is validated by the simulation with MATLAB and the experiment.

REFERENCES

[1] H. Mao, et al, "Review of high-performance three-phase power-factor correction circuits," *IEEE Trans. Ind. Electron.*, vol. 44, no. 4, pp. 437-446. Aug. 1997.

[2] M. Tou, et al, "Analysis and design of single-controlled switch Three-phase rectifier with unity power factor and sinusoidal input current," *IEEE Trans. Power Electron.*, vol. 12, no. 4, pp. 608-614, July 1997.

[3] J. S. Lai, F.Z. Peng, "Multilevel Converter—a new breed of power converters," *IEEE Trans. on Industry Applications*, vol. 32, no. 3, May/June 1996.

[4] G. Carrara, S. Gardella, M. Marchesoni, *et al,.* "A new multilevel PWM method: a theoretical analysis," *IEEE Trans. on Power Electron.*, 1992, 7(3), pp. 497- 505.

[5] Nabae, I. Takahashi, H. Akayi, "A New Neutral-point-clamped PWM Inverter," *IEEE Trans. Ind. Applicat.*, 1981, vol. 17, no. 5, 518~523

[6] N. Celanovic, "Space vector modulation and control of multilevel converters," Ph. D. Dissertation, Virginia Power Electronics Center (VPEC). Virginia Polytechnic Institute & State University, Blacksburg, VA (USA), 2000

[7] J. R. Rodriguez, J. W. Dixon, J. R. Espinoza, J. Pontt and P. Lezana, "PWM regenerative rectifiers: state of the art," *IEEE Trans. on Industrial Electronics*, Feb. 2005, vol. 52, pp. 5-22.

[8] J. Sebastian, M Jureguizar and J. Uceda, "An overview of power factor correction in single phase off-line power supply systems," *in Proc. 1994 IEEE Industrial Electronics Control and Instrumentation Conf*, pp. 1688-1693.

Two Phase Inverter Drive of Three Phase Motor

Saksit Jangjaempradit, Masayuki Morimoto

Department of Electrical and Electronic Engineering, Tokai University,
1117 Kitakaname Hiratsuka-shi Kanagawa 259-1292, Japan

Abstract–**The number of three phase motors drive system by inverter are increasing. Two phase inverter which decreases the number of switches in the inverter is one method to realize low cost application. This paper shows how to use 4 switches of two phase inverter in order to drive three phase motor. The results show that the mean value and fundamental component of output 3 phase voltage are balanced. The output voltage can be controlled by modulation factor. The experiment result of three phase induction motor fed by two phase pulse width modulation inverter shows that three phase current of motor is balanced.**

Index Terms--**Pulse Width Modulation, Fast Fourier Transform, Total Harmonic Distortion**

I. INTRODUCTION

Recently, three phase induction motor is the most popular electric machine in the world. The speed of three phase induction motor is varied by three phase inverter which controls ac voltages as the variable frequency and variable voltage. Three phase inverters for three phase motor use minimum of 6 switches. However, the idea of using four switch inverter to drive three phase motor has been proposed[1]-[7]. Four switch inverter can be called as two phase inverter. Two phase inverter is one of possible method to realize the low cost, small size and high reliability. Recently, the application of two phase inverter is proposed as the fault tolerant inverter when one leg of three phase inverter faults. This idea can be extend to the use of the strategy in the inverter of electric vehicle. Two phase inverter can be used like a spare-tire. This paper shows the principle of PWM control method of two phase inverter. Simulation results and experimental results about induction motor fed by two phase PWM inverter will be described.

II. PRINCIPLE OF TWO PHASE INVERTER

Principle of two phase inverter can be explained by sinusoidal circuit which consists of two sinusoidal voltage sources V_1 and V_2 and three phase balanced load R_1, R_2 and R_3, shown in Fig.1. Three phase alternating current can be synthesized by two sinusoidal voltage sources. When the phase of two AC voltage sources V_1 and V_2 is $\pi/3$. The resulted phase voltage at three phase load Van, Vbn and Vcn is three phase balanced voltage. In addition, the rotation of three phase voltage can be controlled by phase order of two phase inverter.

Synthesis of three phase voltage by two phase inverter is assured by the simulation. The simulation model is shown in Fig.2. Table I gives the two phase inverter parameters for this simulation. The PWM waveform of two phase voltage source, SW_1 leads $\pi/3$ from SW_2. The

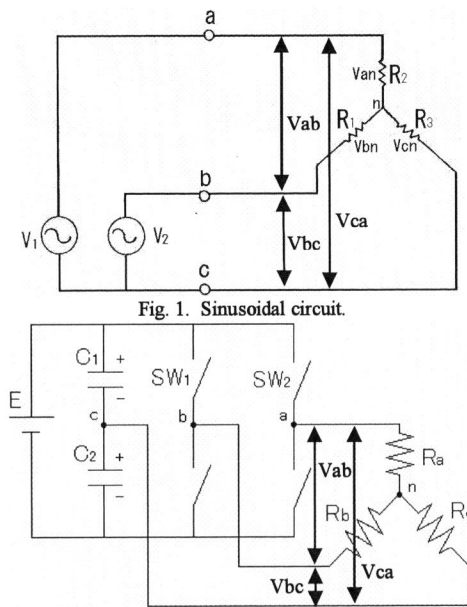

Fig. 1. Sinusoidal circuit.

Fig. 2. Principle of two phase inverter circuit.

TABLE I
TWO PHASE INVERTER PARAMETERS FOR PRINCIPLE.

Sign	Name	Value
E	DC voltage	100V.
F	Frequency of sine wave	50Hz.
F_c	Carrier frequency of triangle wave	600Hz.
M	Modulation factor	0.99
SW_1, SW_2	Ideal switch	
C_1, C_2	Large ideal balanced capacitor	
R_a, R_b, R_c	Ideal balanced resistor	
Vab, Vbc, Vca	Line to line voltage	

fullwave sinwave-triangular wave PWM method is used. Switching sequences of SW_1 and SW_2 are shown in Fig.3. The resulted three phase line to line voltage waveforms Vab, Vbc and Vca are shown in Fig.4. Waveform of Vab differ from Waveform of Vbc, Vca. Peak voltage of Vab is E. But, peak voltage of Vbc, Vca is E/2. Calculated mean value voltage and RMS voltage are shown in table II. Mean values voltage are balanced in three phase. RMS values are unbalanced. This results show that two phase inverter can produce balanced mean values of PWM waveform of three phase line to line voltage. The line to line voltage waveform is evaluated by Fast Fourier Transformer shown in Fig.5. The results are shown in table II. Fundamental components of each phase are the same value. But, harmonics are unbalanced. The source of motor torque is a fundamental component of PWM waveform. It is possible to generate the torque of three phase motor by this PWM waveform. In addition, calculated mean values corresponds to RMS expression of the fundamental component of the spectrum.

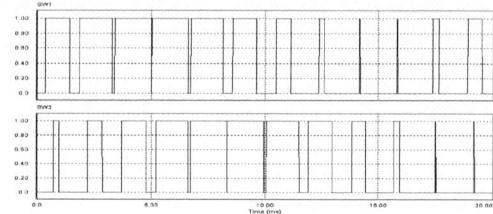

Fig. 3. Two phase voltage source waveforms

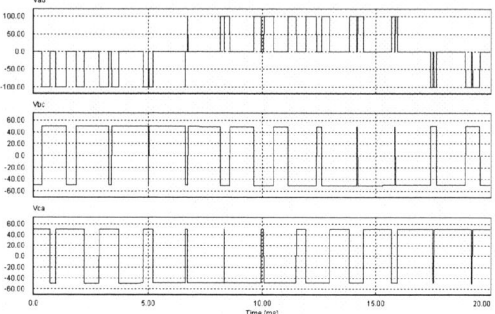

Fig. 4. Three phase line to line voltage waveforms.

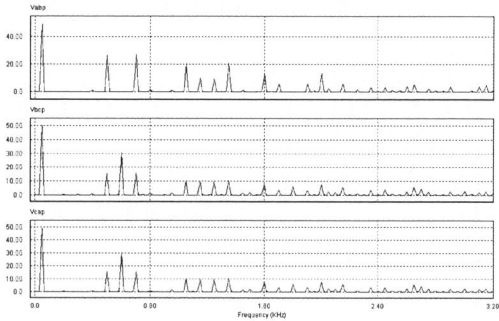

Fig. 5. Three phase line to line voltage spectrums.

TABLE II
RMS VALUE, MEAN VALUE AND FUNDAMENTAL COMPONENT
INSTANTANEOUS VALUE RESULTS.

	RMS value [V]	Mean value [V]	Fund.*[V_p]
Vab	54.8	30.1	48.1
Vbc	50.2	30.0	48.6
Vca	49.8	30.0	47.7

*Fund. : Fundamental component Instantaneous value

III. SIMULATION

The simulation circuit is shown in Fig.6. Table III gives the parameters for simulation. This simulation uses large capacitors C_1, C_2 in order to balance voltage of each capacitor. In the simulation, IGBT switches are treated as the ideal switch. The resulted three phase line to line voltage waveforms Vab, Vbc and Vca are shown in Fig.7. In Fig.7, calculated of mean values voltage and RMS values are shown in table IV. Mean values are balanced. RMS values are unbalanced. Fig.8 shows spectrum of line to line voltage waveform. The fundamental component of three phase line to line voltage is balanced. Numerrical results are shown in table IV. The simulation results are the same as to that of the principle.

IV. VOLTAGE CONTROL

The voltage control of two phase inverter is simulated. The simulation circuit is shown in Fig.6. Table III gives

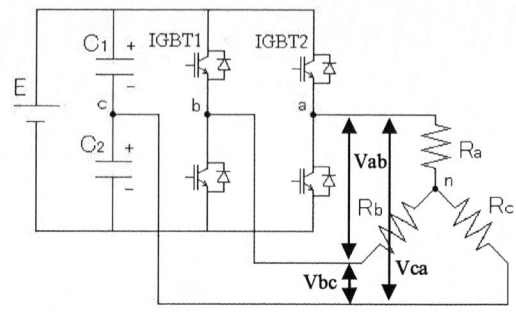

Fig. 6. Simulator circuit.

TABLE III
TWO PHASE INVERTER PARAMETERS FOR SIMULATION.

Sign	Name	Value
E	DC voltage	100V.
C_1, C_2	Capacitors	8200µF.
F	Frequency of sine wave	50Hz.
F_c	Carrier frequency of triangle wave	5kHz.
M	Modulation factor	0.99
R_a, R_b, R_c	Resistors	10Ω.
IGBT1, IGBT2	Ideal IGBT switch	
Vab, Vbc, Vca	Line to line voltage	

the two phase inverter parameters for simulation. Voltage modulation factor M is varied from 0.1 to 0.99. Figure 9 shows the result of simulation. The mean values of line to line voltage are proportional to modulation factor and three phase mean values of each phase are balanced. The result expresses that VVVF operation of 3 phase motor can be realized by two phase PWM inverter. However, the RMS values are unbalanced.

The total harmonic distortion THD is shown in Fig.10. THD is calculated by Eq.(1).

$$THD = \frac{\sqrt{\sum_{n=2}^{\infty} V_n^2}}{V_1} \tag{1}$$

From Fig.10, THD of two phase inverter is inverse proportional to modulation factor. And Vbc and Vca of harmonic component are greater than Vab of harmonic component at low M. Loss of motor is high at low M, therefore, two phase inverter is low loss when modulation factor M is high.

V. EXPERIMENT

The experimental system is shown in Fig.11. It consists of computer, DSP board, OPTO isolator, inverter, 3 phase induction motor, power meter, oscilloscope and FFT analyzer. PWM control software is implemented DSP board TMS320C6713. IGBT switches PM75RSD060 of inverter are controlled by DSP board via OPTO isolator.

1642

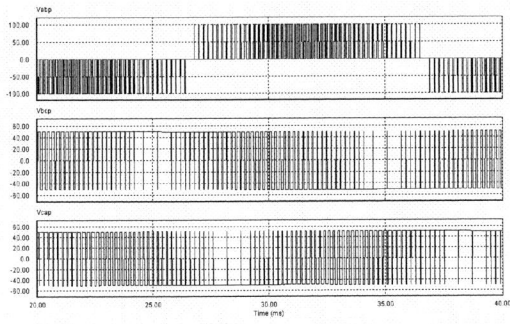

Fig. 7. Three phase line to line voltage waveforms.

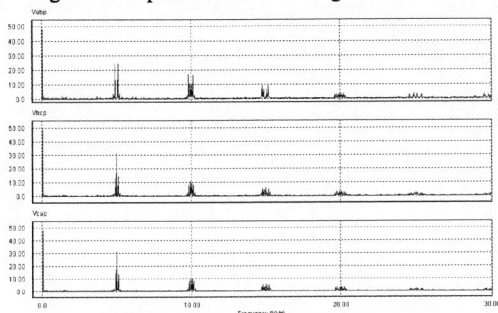

Fig. 8. Three phase line to line voltage spectrums.

TABLE IV
RMS VALUE , MEAN VALUE AND FUNDAMENTAL COMPONENT
INSTANTANEOUS VALUE RESULTS.

	RMS value [V]	Mean value [V]	Fund.*[V_p]
Vab	54.8224	30.0550	48.1481
Vbc	50.2216	30.4302	48.6396
Vca	49.7593	29.9605	47.6602

*Fund. : Fundamental component Instantaneous value

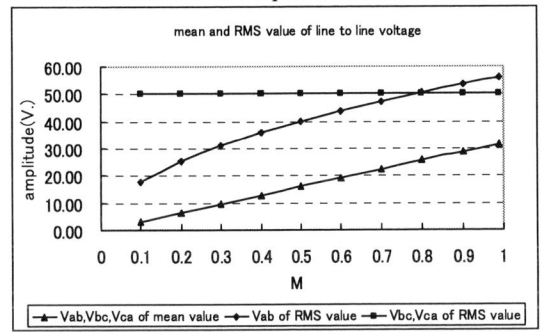

Fig. 9. Voltage control.

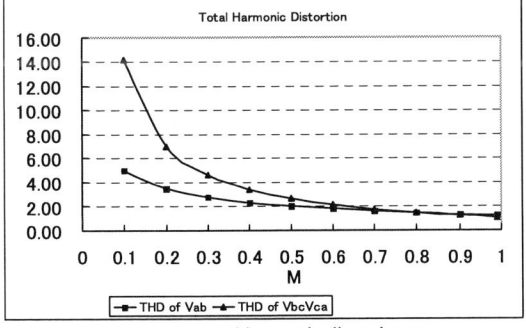

Fig. 10. Total harmonic distortion

Two phase inverter circuit is shown in Fig.6. The load of inverter is no load induction motor. Experimental condition is shown in table V. Motor parameters used in the experiment are shown in table VI. Voltage waveforms of Vab, Vbc and Vca are shown in Fig.12 by oscilloscope

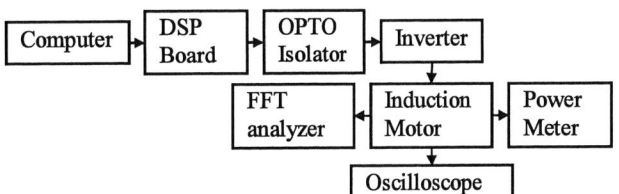

Fig. 11. Experiment circuit.

TABLE V
TWO PHASE INVERTER PARAMETERS FOR EXPERIMENT.

Sign	Name	Value
E	DC voltage	96.7V.
C_1, C_2	Capacitors	8200μF.
F	Frequency of sine wave	50Hz.
F_c	Carrier frequency of triangle wave	5kHz.
M	Modulation factor	0.99
DT	Dead time	4μs
IM	Induction motor	No load

TABLE VI
INDUCTION MOTOR PARAMETERS FOR EXPERIMENT.

Sign	Name	Value
	Power rating	3 phase, 200W.
	Voltage rating	200V.
	Current rating	1.1A.
	Number of poles	4 poles
	Frequency rating	50Hz.
R_1	Stator resistance	14.95Ω.
L_1	Stator inductance	33.4mH.
R_2	Rotor resistance	9.99Ω.
L_2	Rotor inductance	33.4mH.
R_m	Magnetizing resistance	12.25Ω.
L_m	Magnetizing inductance	555.64mH.

YOKOGAWA DL1740E. RMS value is measured by power meter HIOKI 3192, shown in table VII. Fig.12 and table VII show that the experimental result is the same as to that of the simulation.

Three phase current waveforms Ia, Ib and Ic are shown in Fig.13. RMS values of Ia, Ib and Ic are shown in table VII. Fig.13 and table VII show that Ia, Ib and Ic is balanced in the condition of unbalanced RMS voltage. From Fig.13, Ia, Ib and Ic waveforms become sine curve.

Three phase line to line voltage spectrums Vab, Vbc and Vca and three phase current spectrum Ia, Ib and Ic of experimented results are shown in Fig.14 by FFT analyzer ONO SOKKI CF-4220Z. Fig.14(a) shows Vab, Vbc and Vca of spectrum of Experimental results. From Fig.14(a), fundamental component of three phase line to line voltage is balanced. However, harmonics of three phase line to line voltage are unbalanced. Fig.14(b) shows Ia, Ib and Ic of spectrum of experimental results. From Fig.14(b), fundamental component of three phase current is balanced. The results show few harmonics component of three phase current. Vab, Vbc, Vca, Ia, Ib and Ic shows measured fundamental component Instantaneous value in table VII.

Experimented results are the same as to that of simulated results. Experimented result of motor shows that number of revolutions is 1170 RPM by no load test. Experimented results of Vab, Vbc and Vca are smaller than simulation results because experiment condition

1643

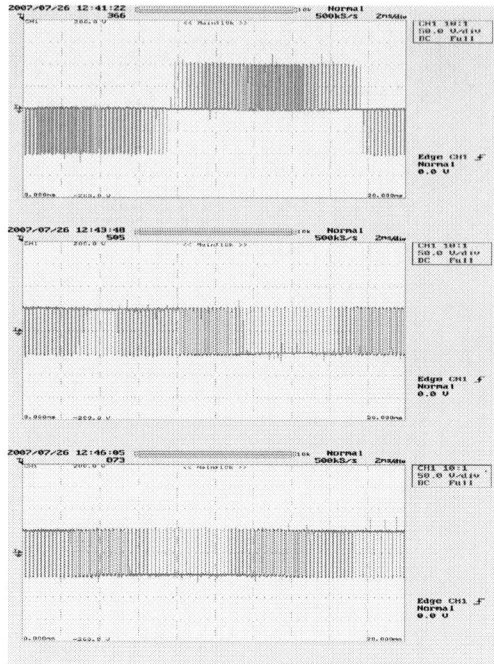

Fig. 12. Three phase line to line voltage waveforms.

Fig. 13. Phase current waveforms

TABLE VII
RMS VALUE AND FUNDAMENTAL COMPONENT INSTANTANEOUS
VALUE RESULTS.

	RMS value [V]	Fund.* [peak]
Vab[V]	51.43	43.00
Vbc[V]	48.36	44.80
Vca[V]	48.53	43.90
Ia[A]	0.264	0.370
Ib[A]	0.284	0.309
Ic[A]	0.293	0.362

*Fund. : Fundamental component Instantaneous value

includes dead time DT. Two phase inverter is possible to drive three phase Motor.

VI. CONCLUSION

In this paper, we proposed PWM control of two phase inverter drive of three phase motor. The simulation results show that two phase inverter can generate motor torque, and VVVF control can be possible by two phase inverter. The experiment results show fundamental component of three phase line to line voltage and three phase current is balanced, therefore, two phase inverter is possible to drive 3 phase Motor. In addition to harmonic component of three phase current is low. However, harmonic component of three phase line to line voltage is

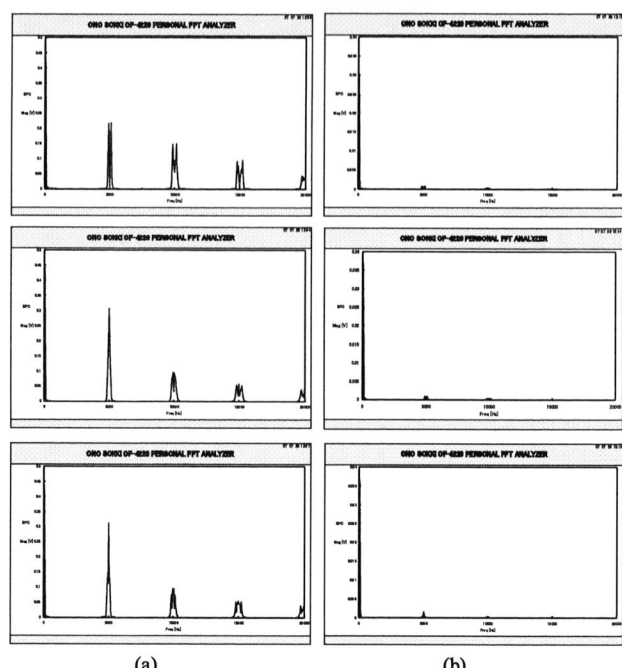

(a)   (b)
Fig. 14. Line to line voltage and phase current spectrums.

unbalanced.

Future study will show efficiency with motor load and improvement of the loss from phase unbalance.

REFERENCES

[1] Heinz W. Van Der Broeck and Hans-Christoph Skudelny, "Analytical analysis of the harmonic effects of a PWM AC drive," *IEEE. Trans. Power Electronic*, vol. 3, no.2, pp. 216-223, April 1988.

[2] Gi-taek Kim and Thomas A Lipo, "VSI-PWM Rectifier/Inverter system with a reduced switch count," *IEEE. Trans. Ind. Applicat.*, vol. 32, no.6, pp. 1331-1337, Nov./Dec. 1996.

[3] Frede Blaabjerg, Dorin O. Neacsu and John K. Pedersen, "Adaptive SVM to compensate DC-link voltage ripple for four-switch three-phase voltage-source inverters," *IEEE. Trans. Power Electronic*, vol. 14, no.4, pp. 743-752, July 1999.

[4] Olorunfemi Ojo, Zhiqiao Wu, Gan Dong and Sheetal K. Asuri, "High-performance speed-sensorless control of an induction motor drive using a minimalist single-phase PWM converters," *IEEE. Trans. Ind. Applicat.*, vol. 41, no.4, pp. 996-1004, Jul./Aug. 2005.

[5] Saksit Jangjaempradit,Yukio Nabeshima,Yusuke Hayase and Masayuki Morimoto, "2 phase inverter drive of 3 phase induction motor," *The 2007 Annual Meeting Record I.E.E. Japan*, no. 4-161, pp. 269-270, 15-17 March 2007.

[6] Do-Hyun Jang, "PWM methods for two-phase inverters," *IEEE Industry Applications Magazine*, pp. 50-61, Mar./Apr. 2007.

[7] Andre M. S. Mendes, Xose M. Lopez-Fernandez and A.J.Marques Cardoso, "Thermal Behavior of a three-phase Induction Motor Fed by a Fault-Tolerant Voltage Source Inverter," *IEEE. Trans. Ind. Applicat.*, vol. 43, no.3, pp. 724-730, May/June 2007.

Predictive Current Controller for Inverter Fed Medium Voltage Drives with *LC* Filter

T. Laczynski, A. Mertens

University of Hannover, Institute for Drive Systems and Power Electronics
Welfengarten 1, 30167 Hannover, Germany

Abstract--The switching frequency of power semiconductor devices in high power medium voltage drives is limited because of high switching losses. These drives often comprise an *LC* output filter. The filter introduce a resonant circuit that may be excited for instance by fast control transients. One option to avoid such oscillations is applying damping control methods, which on the other hand usually require switching frequencies well above the filter resonance. This paper presents a new predictive stator current controller that avoids the excitation of the filter resonance, realizes active damping and enables fast current control while maintaining low switching frequency. Feasibility and good dynamic performance of the proposed control method is demonstrated by simulation results of a 2.4 kV induction motor drive and by experimental results obtained from a 55 kW prototype a.c. drive.

Index Terms--*LC* filter, medium voltage drive, predictive current control, three level voltage source inverter.

I. INTRODUCTION

In order to minimize the switching losses of the power semiconductors high power medium voltage inverters typically operate at a switching frequency lower than 500 Hz [1-5]. They are frequently applied to the retrofit of existing fixed speed induction motors with medium voltage variable speed drives, i.e. to achieve energy savings. Because existing motors usually are not designed for inverter supply, the use of *LC* output filters becomes necessary in order to avoid isolation problems and bearing currents [6]. The filter creates sinusoidal output voltages and thereupon conditions for the machine similar to operation from the grid. The inverter switching frequency that is acceptable for thermal reasons lies only slightly above the resonance frequency of the *LC* filter. At the same time, a suitable motor current control is necessary that does not excite the filter resonance. Otherwise, sudden changes of the control reference would create considerable weakly damped transient currents [7].

This paper proposes a new predictive stator current control scheme which takes the nonlinear nature of the inverter into account. The controller computes the future stator current trajectories for every possible inverter switching state and chooses the one which minimizes a performance index. It turns out that the approach makes fast motor current control possible while maintaining switching frequency of maximal 570 Hz. By means of the controller, excitation of filter resonance is avoided and active damping of filter oscillations is achieved.

This paper presents simulation results of the proposed predictive current control method for a 2.4 kV medium voltage drive fed by a Neutral Point Clamped (NPC) three level inverter. In addition, measurement results of a prototype 55 kW induction motor drive system fed by a two level inverter are presented.

II. CURRENT CONTROL

A. System Description

The signal flow graph of the proposed controlling system for the inverter fed induction machine with *LC* filter is shown in Fig. 1.

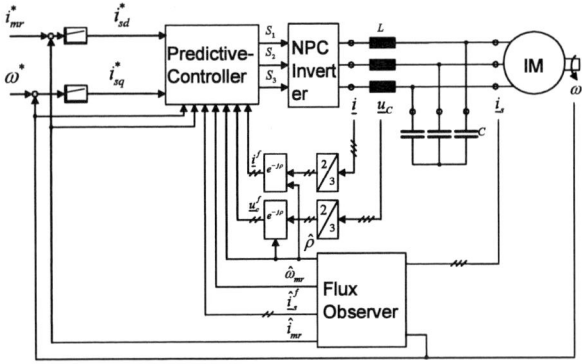

Fig. 1: Signal flow graph of the proposed control structure

The predictive current controller receives its reference value of the complex stator current vector from the flux and speed PI controllers, and computes the switching functions for the inverter using the measured or estimated instantaneous state variables of the motor and filter. The controller makes use of the values of the stator current, the choke current, the capacitor voltage and the rotational speed. Additionally, a flux observer provides values for the magnetizing current, the angle and the angular velocity of the rotor flux. An NPC three level inverter generates 19 different voltage space vectors (Fig. 2). The machine stator currents are the controlled variables. The controller predicts stator current trajectories which come with each of the voltage space vectors by solving a discrete-time state space model of the filter and motor. By means of a weighting function taking into consideration the deviation of the current trajectory from the stator reference current, a performance index is

978-1-4244-0644-9/07/$25.00 ©2007 IEEE

computed. For every sample interval, the switching state having the minimal performance index is identified, and is then generated by the inverter.

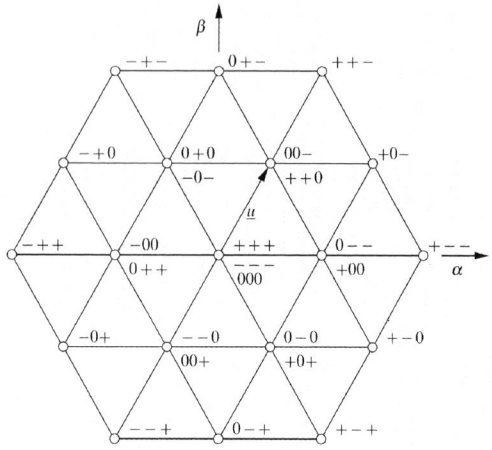

Fig. 2: Voltage space vectors of the NPC three level inverter

B. System Model

The dynamic behavior of an induction machine can be described in field coordinates by the following equations [8]:

$$u_{sd} = R_s i_{sd} + \sigma L_s \frac{di_{sd}}{dt} - \sigma L_s \omega_{mr} i_{sq} + (1-\sigma) L_s \frac{di_{mr}}{dt}, \tag{1}$$

$$u_{sq} = R_s i_{sq} + \sigma L_s \frac{di_{sq}}{dt} + \sigma L_s \omega_{mr} i_{sd} + (1-\sigma) L_s i_{mr} \omega_{mr}, \tag{2}$$

$$i_{sd} = T_r \frac{di_{mr}}{dt} + i_{mr}, \tag{3}$$

$$\omega_{mr} = \frac{i_{sq}}{T_r i_{mr}} + p\omega. \tag{4}$$

The nomenclature used is shown in Tab. I. The equation of motion is:

$$J \frac{d\omega}{dt} = \frac{3}{2} p (1-\sigma) L_s i_{mr} i_{sq} - m_l. \tag{5}$$

In the above equations an amplitude invariant space vector transformation is used:

$$x_\alpha + x_\beta = \frac{2}{3} \begin{bmatrix} 1 & e^{j\frac{2}{3}\pi} & e^{j\frac{4}{3}\pi} \end{bmatrix} \begin{bmatrix} x_1 \\ x_2 \\ x_3 \end{bmatrix}. \tag{6}$$

The differential equations of the LC filter inductor current space vector are:

$$\frac{di_d}{dt} = \frac{1}{L} u_d - \frac{1}{L} u_{cd} + \omega_{mr} i_q - \frac{R_l + R_c}{L} i_d + \frac{R_c}{L} i_{sd} \tag{7}$$

and

$$\frac{di_q}{dt} = \frac{1}{L} u_q - \frac{1}{L} u_{cq} - \omega_{mr} i_d - \frac{R_l + R_c}{L} i_q + \frac{R_c}{L} i_{sq}. \tag{8}$$

The capacitor voltages are:

$$\frac{du_{cd}}{dt} = \omega_{mr} u_{cq} + \frac{1}{C} i_d - \frac{1}{C} i_{sd} \tag{9}$$

and

$$\frac{du_{cq}}{dt} = -\omega_{mr} u_{cd} + \frac{1}{C} i_q - \frac{1}{C} i_{sq}. \tag{10}$$

Using the above equations a continuous state space system can be formed (11), which allows the computation of the stator current trajectories. The inputs of the model are the inverter output voltages u_d, u_q and the magnetizing current i_{mr}. The prediction is carried out based on the assumption that the motor speed ω, the rotor flux velocity ω_{mr} and magnetising current i_{mr} are constant during the prediction period.

$$\frac{d}{dt} \begin{bmatrix} i_{sd} \\ i_{sq} \\ i_d \\ i_q \\ u_{cd} \\ u_{cq} \end{bmatrix} = \begin{bmatrix} a_{11} & a_{12} & a_{13} & 0 & a_{15} & 0 \\ -a_{12} & a_{11} & 0 & a_{13} & 0 & a_{15} \\ a_{31} & 0 & a_{33} & a_{12} & a_{35} & 0 \\ 0 & a_{31} & -a_{12} & a_{33} & 0 & -a_{35} \\ a_{51} & 0 & -a_{51} & 0 & 0 & a_{12} \\ 0 & a_{51} & 0 & -a_{51} & -a_{12} & 0 \end{bmatrix} \begin{bmatrix} i_{sd} \\ i_{sq} \\ i_d \\ i_q \\ u_{cd} \\ u_{cq} \end{bmatrix}$$

$$+ \begin{bmatrix} 0 & 0 & b_{13} \\ 0 & 0 & b_{23} \\ b_{31} & 0 & 0 \\ 0 & b_{31} & 0 \\ 0 & 0 & 0 \\ 0 & 0 & 0 \end{bmatrix} \begin{bmatrix} u_d \\ u_q \\ i_{mr} \end{bmatrix} = \underline{A}\underline{x} + \underline{B}\underline{u} \tag{11}$$

The matrix elements of the above state space model are:

$$a_{11} = -R_c/(\sigma L_s) - 1/(T_s) - (1-\sigma)/(\sigma T_r)$$
$$a_{12} = \omega_{mr}$$
$$a_{13} = -R_c/(\sigma L_s)$$
$$a_{15} = -1/(\sigma L_s)$$
$$a_{31} = -R_c/L$$
$$a_{33} = -(R_l + R_c)/L$$
$$a_{35} = -1/L$$
$$a_{51} = -1/C$$
$$b_{13} = (1-\sigma)/(\sigma T_r)$$
$$b_{23} = -(1-\sigma)/\sigma \, p\omega$$
$$b_{31} = -a_{35}$$

TABLE I
NOMENCLATURE

R_s, R_r	stator and rotor resistance
L_s, L_r	stator and rotor self inductance
L_h	magnetizing inductance
T_s	stator time constant, $T_s = L_s/R_s$
T_r	rotor time constant, $T_r = L_r/R_r$
σ	leakage coefficient, $\sigma = 1 - L_h^2/(L_s L_r)$
ω_{mr}	angular velocity of the rotor field
ω	rotational speed
p	number of pole pairs
J	total moment of inertia
m_l	load torque
R_l, R_c	filter inductor and filter capacitance resistance
L, C	filter inductance and capacitance

In order to reduce the computational demand, the current trajectories are predicted by means of a time discrete state space model:

$$\underline{x}(k+1) = \underline{\Phi}\,\underline{x}(k) + \underline{\Gamma}\,\underline{u}(k). \tag{12}$$

The discrete state space model is obtained through discretization of the state space model (11) using the following series:

$$\underline{\Phi}(T) = \sum_{i=0}^{\infty} \frac{\underline{A}^i T_{Step}^i}{i!}, \tag{13}$$

$$\underline{\Gamma}(T) = \sum_{i=1}^{\infty} \frac{\underline{A}^{i-1} T_{Step}^i}{i!} \underline{B}. \tag{14}$$

T_{Step} is the sample period of the discrete prediction model.

C. Predictive stator current controller

An NPC three level inverter generates 19 different voltage space vectors ($z = 1,...,19$). Every voltage space vector z causes a specific stator current trajectory. At the beginning of every sample interval $t(i) = T*i$ ($i = 0,1,2,3,...$) possible stator current trajectories are predicted by solving (12). Measured instant values of the stator current, the filter inductor current and the filter capacitor voltage provide the initial values. Computation of possible current trajectories $i_{sd}(z,k)$ and $i_{sq}(z,k)$ is carried out at discrete instants $t_{step}(k) = T_{step}*k$ ($k = 0,1,2,3,...,k_{max}$). The parameter k_{max} defines the prediction length $T_{pred} = T_{step}*k_{max}$ which has a significant effect on the controller performance.

The resulting trajectory errors are defined as:

$$e(z,k+1) = \sqrt{\left(f_d\left(i_{sd}^*(k) - i_{sd}(z,k+1)\right)\right)^2 + \left(i_{sq}^*(k) - i_{sq}(z,k+1)\right)^2} \tag{15}$$

and are computed for the entire prediction length. The parameter f_d is a weighting factor. The average area between the error function and the abscissa results from:

$$e_A(z,k+1) = \frac{T_{step}}{2}\left(e(z,i+k) + e(z,i+k-1)\right). \tag{16}$$

Summation of the above values yields the performance index:

$$e_{A,tot}(z) = \sum_{k=0}^{k_{max}} e_A(z,i+k). \tag{17}$$

Following the inverter realizes a switching state z with the minimal performance index $e_{A,tot}(z)$ (17).

The measurement of the instant values and the calculation of the control algorithm demand a finite time period and create that way a dead time T_d. Therefore the measured instant values at the end of the calculation period differ from the actual one. This has to be taken into account by one sample ahead prediction of the measured instant values using (12-14) with the dead time T_d as the sample period T_{Step}.

III. SIMULATION RESULTS

Simulation results for a 2.4 kV medium voltage drive system with an LC filter were obtained using MATLAB/Simulink software and the PLECS toolbox. The a.c. network, the 12-pulse diodes rectifier, the dc-link, the three level NPC inverter and the LC filter were modelled using the PLECS toolbox. The toolbox models the power semiconductors as ideal switches in order to speed up the simulation. The induction motor and the time time-discrete controller were modelled in MATLAB/Simulink. The parameters of the simulated model are given in the appendix in Table II.

A sampling frequency of $f = 1/T = 5$ kHz was used and the prediction length T_{pred} chosen to 800 µs. The dead time was $T_d = 1/(2T)$. Fig. 3 shows the motor speed, motor torque, direct and quadrature components of stator current during start-up and under load with the rated torque.

Fig. 4 shows the step response of the current controller to a step command in speed, when the motor was operating at 20 % of rated speed and 10 % rated torque with the rated magnetising current. The quadrature current reaches its maximal value in 4 ms from the step command. This explains the dynamic performance of the current controller.

The total harmonic distortion (THD) of the stator current k_{is} is about 5 % in the entire speed range of the motor at rated load torque (Fig. 5). This corresponds to a modulation index range of 0.11,..., 1.1. The modulation index is defined as $M = u_1/(U_{dc}/2)$ where u_1 is the fundamental motor phase voltage and U_{dc} the dc-link voltage. The THD of the motor phase voltage k_{us} is shown in Fig. 6.

In the entire speed range at rated torque load the maximum average switching frequency of all twelve switches is

1647

approximately $f_s = 570$ Hz (Fig. 7). The average switching frequency by operation at nominal values is about $f_s = 370$ Hz and lies below the resonance frequency of 460 Hz. A further reduction of the average switching frequency in the entire speed region is possible by reduction of the controller sampling frequency. However, with the reduction of the sampling frequency comes a degradation of the stationary performance, especially an increasing of the THD values.

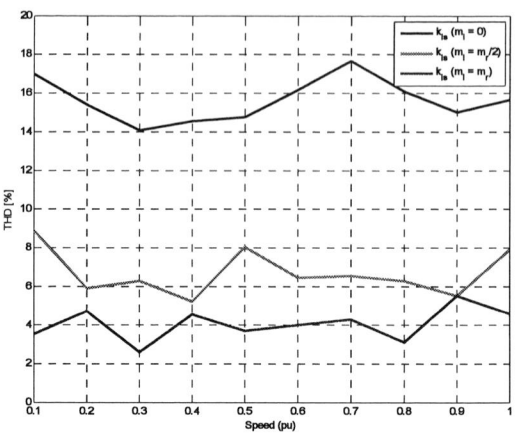

Fig. 5: Total harmonic distortion of the stator currents (simulation)

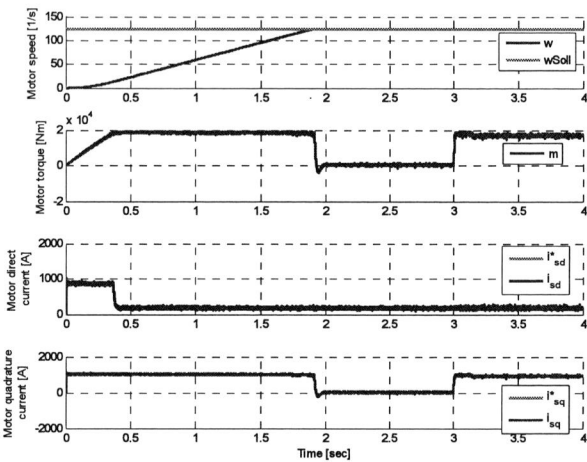

Fig. 3: Motor speed, torque, direct and quadrature components of stator current (simulation)

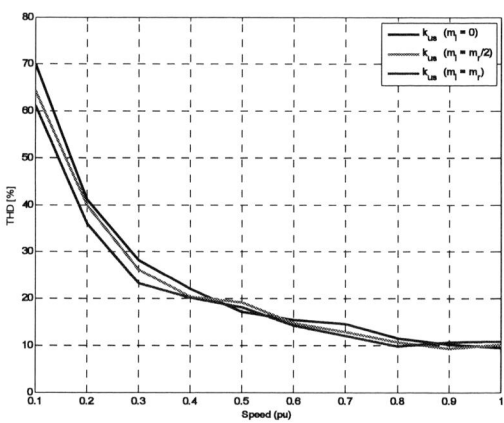

Fig. 6: Total harmonic distortion of the motor phase voltages (simulation)

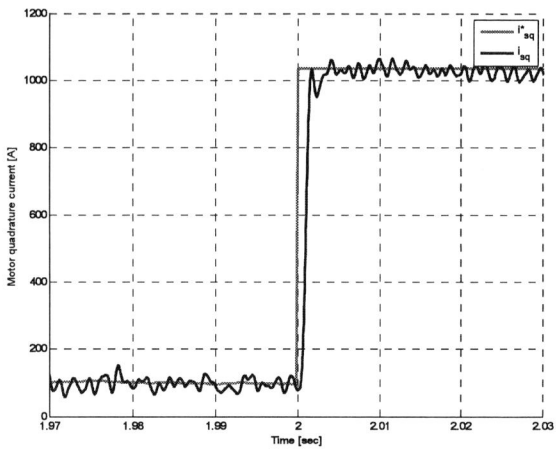

Fig. 4: Quadrature current i_{sq} (reference speed step, simulation)

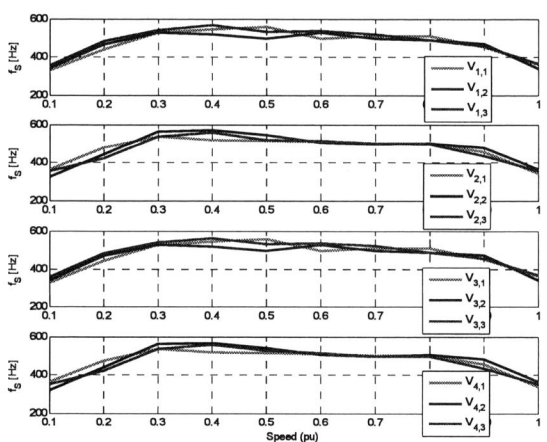

Fig. 7: Average switching frequency f_s of the IGBTs (simulation). The first index denotes IGBT number (from top to bottom) and the second index the inverter half bridge (from left to right).

1648

IV. EXPERIMENTAL RESULTS

In order to experimentally verify the performance of the proposed approach, the controller was implemented in a dSPACE DS1103 PPC controller board. The sampling frequency was set at $f = 5$ kHz. The prediction length was $T_{pred} = 1$ ms and the prediction sampling period $T_{step} = 1/(2f) = 100$ μs. The dead time was $T_d = 1/(2T)$. A 55 kW induction machine drive with LC filter was fed by a two level inverter with a 540V dc-link voltage. The drive is speed controlled with a d.c. motor delivering the mechanical load. The parameters of the experimental setup are given in the appendix in Table III.

The dynamic performance of the drive was tested using a reference speed step. Before applying the step, the drive was operated at 20% speed, rated magnetizing current and 10 % load torque. The computed motor torque (5) is shown Fig. 8. After applying the reference speed step at 1.496 s, the motor generates a maximal torque of 190 Nm within 1.5 ms. The limitation of the maximum torque to 53 % of the rated torque results from using an inverter with a 540 V dc-link voltage (400 V mains supply) instead of a 932 V dc-link voltage (690 V mains supply).

The quadrature current i_{sq} and its reference value during the speed reference step are shown in Fig. 9. The corresponding stator current is depicted in Fig. 10. The results show that a fast current control is possible without excitation of the LC filter resonance.

The stationary motor current and motor torque during operation at rated speed are presented in Fig. 11 and Fig. 12. The average value of the switching frequency was $f_s = 552$ Hz.

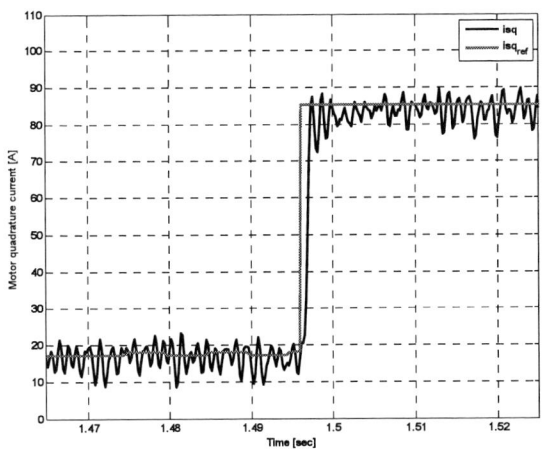

Fig. 9: Quadrature current i_{sq} (reference speed step)

V. CONCLUSIONS

The presented results show, that the proposed predictive current controller enables fast dynamic operation of drive systems with an LC output filter. At the same time the average switching frequency is kept at a maximal value of 570 Hz. At nominal values of the drive the average switching frequency is about $f_s = 370$ Hz and lies below the resonance frequency. Fast stator control is possible, filter resonance is not excited and any filter oscillations due to disturbances are dampened. The proper operation principle of the controller was verified by simulation of a 2.4 kV medium voltage drive and by measurements on a prototype 55 kW drive system.

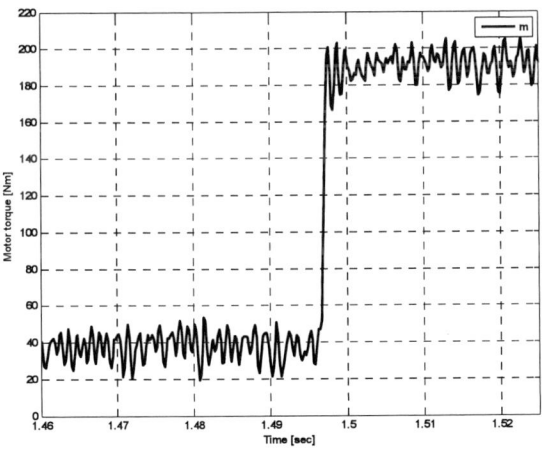

Fig. 8: Motor torque (reference speed step)

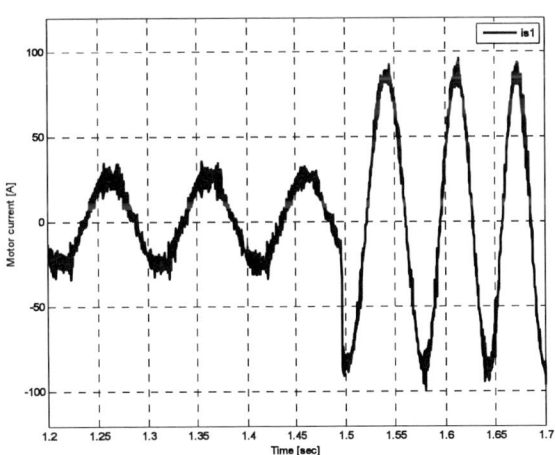

Fig. 10: Stator current (reference speed step)

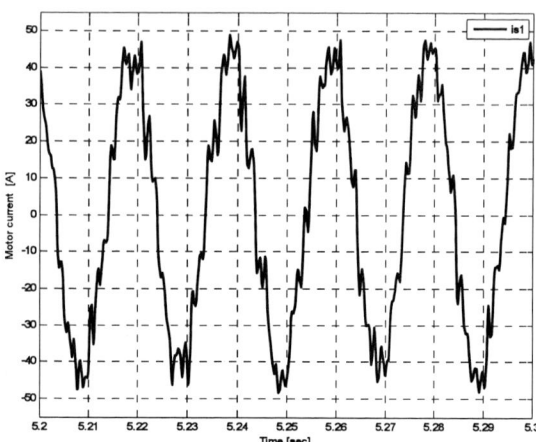

Fig. 11: Stator current (rated speed, maximal torque)

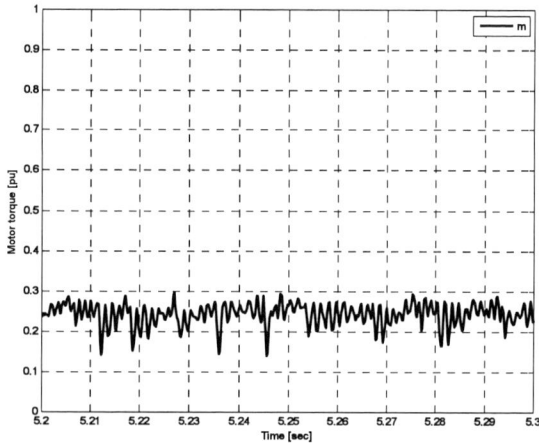

Fig. 12: Motor torque (pu)

APPENDIX

The parameters of the used drive systems are listed in Table II and Table III.

TABLE II
PARAMETERS OF THE 2.4 KV DRIVE

U	2.4 kV
I	600 A
n	1190 min^{-1}
f	60 Hz

TABLE III
PARAMETERS OF THE MOTOR AND THE LC FILTER (55 KW DRIVE)

U	660 V
I	62 A
n	1475 min^{-1}
f	50 Hz
m	357 Nm
J	0.55 kg m^2
P	55 kW
$cos\ \varphi$	0.84
p	2
R_s	114 mΩ
R_r	100 mΩ
L_s	46.932 mH
L_r	56.036 mH
L_h	46 mH
L	3.13 mH
C	39.8 µF
R_l	60 mΩ
U_{dc}	540 V

REFERENCES

[1] A. Nabae, I. Takahashi, H, Akagi, "A New Neutral Point Clamped PWM Inverter", IEEE Trans. Ind. Appl., vol. IA-17, pp. 518-522, 1981.

[2] R. Sommer, A. Mertens, M. Griggs, H.-J. Conraths, M. Bruckmann, T. Greif, "New Medium Voltage Drive Systems Using Three-Level Neutral Point Clamped Inverter With High Voltage IGBT", IEEE IAS 1999, Phoenix, AZ.

[3] P.N. Enjeti, R. Jakkli, " Optimal Power Control Strategies for Neutral Point Clamped (NPC) Inverter Topology", IEEE Trans. Ind. Appl., vol. 28, no. 3, pp. 558-566, 1992.

[4] J. Holtz, "Pulsewidth Modulation for Electronic Power Converters", IEEE Trans. Ind. Appl., vol. 31, no. 5, pp. 1110-1120, 1995.

[5] A. Sapin, P.K. Steimer, J.-J. Simond, " Modeling, Simulation and Test of a Three-Level Voltage-Source Inverter With Output LC Filter and Direct Torque Control", IEEE Trans. Ind. Appl., vol. 43, no. 2, pp. 469-475, 2007.

[6] B.P. Schmitt, R. Sommer, "Retrofit of Fixed Speed Induction Motors With Medium Voltage Drive Converters Using NPC Three-Level Inverter High-Voltage IGBT Based Topology", IEEE ISIE 2001, Pusan, Korea.

[7] J. Holtz, B. Beyer, "Fast Current Trajectory Control Based on Synchronous Optimal Pulsewidth Modulation", IEEE Trans. Ind. Appl., vol. 31, no. 5, pp. 1110-1120, 1995.

[8] W. Leonhard, "Control of Electrical Drives", Springer-Verlag Berlin, 1996.

Novel Control Strategy of Instantaneous Power Based CVCF Inverter

Akira Sato[*], *Member*, and Toshihiko Noguchi[*], *IEEE Senior Member*

[*] Nagaoka University of Technology
Address: 1603-1 Kamitomioka, Nagaoka, Niigata 940-2188, Japan
Phone: +81-258-47-9510, Fax: +81-258-47-9500
e-mail: asato@stn.nagaokaut.ac.jp, and tnoguchi@vos.nagaokaut.ac.jp

Abstract—This paper describes a novel control strategy of an instantaneous power based CVCF inverter. The most important feature of this strategy is a direct selection of switching states of the CVCF inverter to restrict errors between feedback values and command values of the active and reactive power. A theoretical analysis on relationship between the instantaneous power and the switching states is investigated, which is essential to compose a switching state table in the controller. Effectiveness of the proposed technique was examined through several computer simulations and experimental tests, using a 2.5-kW prototype. As a result, the output voltages were confirmed to be stable sinusoidal waveforms with total harmonic distortion of only 1.13 % under a nonlinear load condition. Excellent voltage balance is also achieved by the proposed system even under unbalanced load conditions.

Index Terms— direct power control, instantaneous active and reactive power CVCF inverter, relay control, switching state table, and hysteresis element.

I. INTRODUCTION

Information and telecommunication systems are indispensable infrastructure in modern societies, and are a symbol of civilization to enrich human communities. On the other hand, reliability relevant to power supply to the systems as well as data handling in the systems is a matter of vital importance. Therefore, these systems normally have redundant protection mechanisms against voltage sag or failure in the power grid to prevent losses of significant data and fatal system down in the worst case. Uninterruptible power supplies (UPS) are a powerful solution to overcome such possible troubles in the power grid, and are extensively used to improve quality and reliability of power feeding not only to the information and telecommunication systems but also medical electronic instruments, etc.

In general, a sinusoidal output voltage with stable frequency and low harmonic distortion is demanded by many of the load electric appliances, which can be attained by a voltage-source PWM inverter with an LC filter (constant-voltage and constant-frequency inverter: CVCF inverter). In order to make such operation characteristics possible even in nonlinear load cases, a variety of control algorithms have been proposed in the past works. One of the most well known approaches is an output voltage control method, where its major loop uses a proportional-integral-derivative (PID) regulator and a repetitive controller (voltage oriented control:

VOC) [1], [2]. The VOC demonstrates an excellent performance both in dynamic and static operations by means of adopting a current minor control loop of the filter capacitor [3]-[5]. Therefore, performance of the VOC method dominantly depends on controllability of the applied current control scheme.

The authors have intensively been investigating a direct power control (DPC) strategy of power converters and its applications to various power conversion systems such as a voltage-source pulse width modulated (PWM) rectifier, a current-source PWM rectifier, a multi-level converter, matrix converter, etc. This strategy features direct relay (bang-bang) control of the instantaneous active and reactive power by means of switching operation in the power converters. Owing to its own inherent unique features, the DPC achieves a high-speed response in controlling the instantaneous power without sacrificing simple configuration of the system because it requires neither current minor loops nor rotating coordinate transformations [6], [7].

This paper presents a novel application of the DPC strategy, which enhances performance of the CVCF inverter [8]. Control algorithm of a DPC based CVCF inverter is described in the paper, followed by theoretical investigation on relationship between behavior of the instantaneous active and reactive power and the switching states of the inverter. In addition, effectiveness of the proposed technique is examined through various computer simulations and experimental tests, using a 2.5-kW prototype. Consequently, excellent operation characteristics were confirmed both in a steady state and in a transient state under a nonlinear load condition as well as under unbalanced load conditions, which proves overall feasibility of the proposed strategy.

II. SYSTEM CONFIGURATION AND CONTROL ALGORITHM

A. Control Principle

Fig. 1 depicts a mathematical model of the CVCF inverter. Assuming that the output voltage v_o is a stable sinusoidal waveform, it can be expressed as a rotating vector with an amplitude of $\sqrt{3}\,V_{rms}$ and a frequency of ω, i.e.,

$$v_o = \sqrt{3}V_{rms}e^{j\omega t}, \qquad (1)$$

where V_{rms} is a root-mean-square value of the line voltage.

978-1-4244-0644-9/07/$25.00 ©2007 IEEE

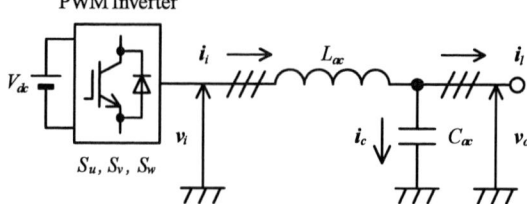

Fig. 1. Mathematical model of CVCF inverter.

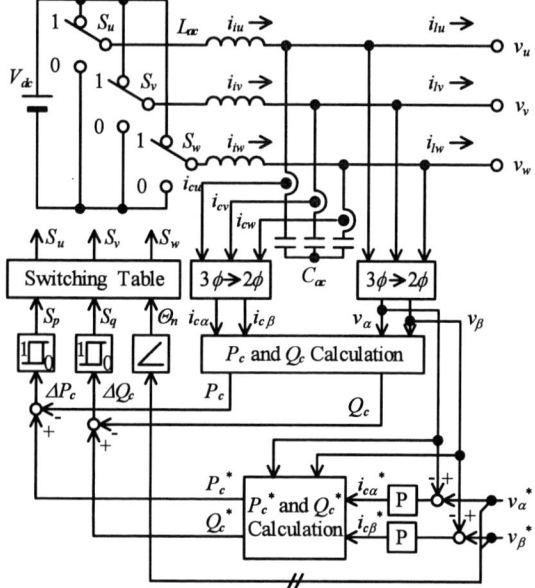

Fig. 2. System configuration of direct power control (DPC) based CVCF inverter.

From this model, a filter-capacitor current vector i_c is given as

$$i_c = C_{ac}\frac{dv_o}{dt} = j\sqrt{3}\omega C_{ac}V_{rms}e^{j\omega t}. \tag{2}$$

Using (1) and (2), the instantaneous active power P_c and the instantaneous reactive power Q_c of the capacitor C_{ac} in the LC filter are calculated with the following equations:

$$P_c = v_o \bullet i_c = 0, \text{ and} \tag{3}$$

$$Q_c = v_o \times i_c = -3\omega C_{ac}V_{rms}^2, \tag{4}$$

where $\bullet$ and $\times$ denote a scalar product and a vector product, respectively.

Therefore, charging and discharging energy of the filter capacitor allow controlling the output voltages, and stable sinusoidal output voltage waveforms can be obtained by appropriate power management of the capacitor.

B. System Configuration

Fig. 2 shows a system configuration of the direct power control based CVCF inverter, where the instantaneous active power P_c and the instantaneous

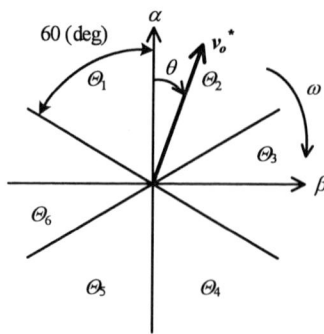

Fig. 3. Quantized phase of output voltage command vector.

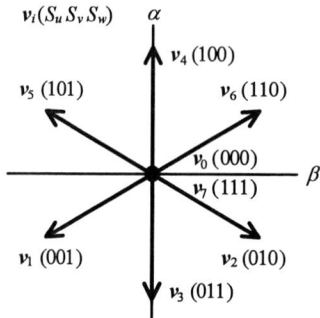

Fig. 4. Output voltage vectors of CVCF inverter.

reactive power Q_c are directly controlled by switching actions of the CVCF inverter.

The inverter system detects its output voltages v_u, v_v and v_w, and filter-capacitor currents i_{cu}, i_{cv} and i_{cw} to calculate P_c and Q_c. These detected three-phase quantities are transformed to two-phase quantities according to the following equations:

$$\begin{bmatrix} v_\alpha \\ v_\beta \end{bmatrix} = \sqrt{\frac{2}{3}}\begin{bmatrix} 1 & -1/2 & -1/2 \\ 0 & \sqrt{3}/2 & -\sqrt{3}/2 \end{bmatrix}\begin{bmatrix} v_u \\ v_v \\ v_w \end{bmatrix}, \text{ and} \tag{5}$$

$$\begin{bmatrix} i_{c\alpha} \\ i_{c\beta} \end{bmatrix} = \sqrt{\frac{2}{3}}\begin{bmatrix} 1 & -1/2 & -1/2 \\ 0 & \sqrt{3}/2 & -\sqrt{3}/2 \end{bmatrix}\begin{bmatrix} i_{cu} \\ i_{cv} \\ i_{cw} \end{bmatrix}. \tag{6}$$

The instantaneous active power P_c and the instantaneous reactive power Q_c of the filter capacitor C_{ac} can easily be obtained with (7) by using the above v_α, v_β, $i_{c\alpha}$ and $i_{c\beta}$, respectively.

$$\begin{bmatrix} P_c \\ Q_c \end{bmatrix} = \begin{bmatrix} v_\alpha & v_\beta \\ v_\beta & -v_\alpha \end{bmatrix}\begin{bmatrix} i_{c\alpha} \\ i_{c\beta} \end{bmatrix}. \tag{7}$$

On the other hand, the instantaneous active power command P_c^* and the instantaneous reactive power command Q_c^* are provided from the output voltage control block. The sinusoidal two-phase output voltage commands v_α^* and v_β^* are compared with their calculated values v_α and v_β, respectively. The control errors between them are delivered to proportional regulators and their output, i.e., the filter-capacitor current commands

1652

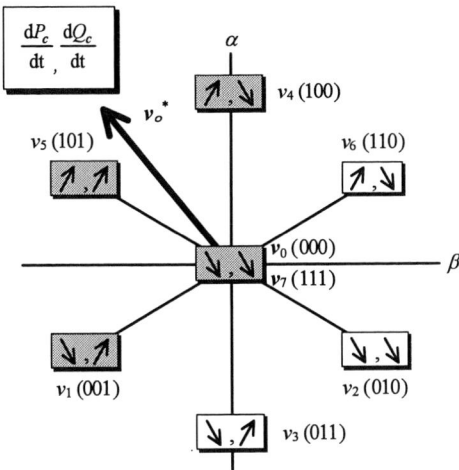

Fig. 5. Calculation results of dP_c/dt, dQ_c/dt in $\Box_1$.

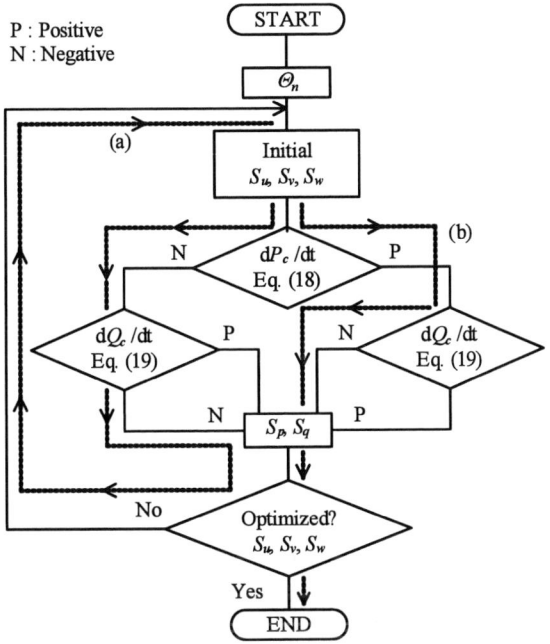

P : Positive
N : Negative

Fig. 6. Algorithm of switching pattern selection.

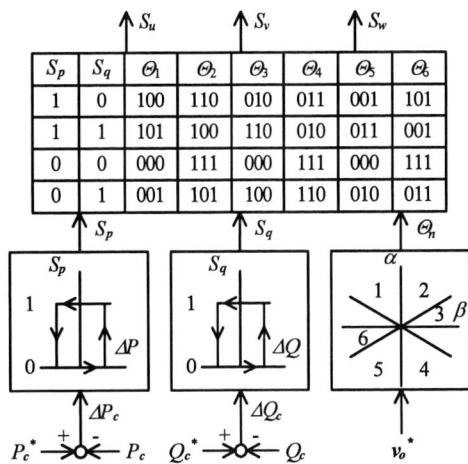

Fig. 7. Optimum switching state table and regulators.

Table 1. Electrical parameters of power circuit.

DC bus voltage V_{dc}	340 (V)
Output filter inductor L_{ac}	0.4 (mH)
Output filter capacitor C_{ac}	20 (μF)
Output voltage command v_o^*	200 (V_{rms}), 50 (Hz)

$$(n-2)\frac{\pi}{3} \leq \Theta_n < (n-1)\frac{\pi}{3} \quad \because n = 1, 2, \cdots, 6. \quad (9)$$

These quantized signals S_p, S_q and Θ_n are used to select uniquely the most appropriate switching states of the CVCF inverter, and control ON/OFF of each leg. In other words, every time the CVCF inverter changes its switching state, a combination of S_p, S_q and Θ_n determines the next unique and optimum switching state. As shown in Fig. 4, the inverter can generate six non-zero-voltage vectors v_1-v_6 and two zero-voltage vectors v_0 and v_7, according to its switching state S_u, S_v, and S_w.

C. Composing Switching State Table

Since the proposed system is, in principle, based on relay (bang-bang) control of the instantaneous active and reactive power, it is significant to investigated relationship between the switching states of the CVCF inverter and polarities of time derivatives of the active and reactive power dP_c/dt and dQ_c/dt in a specified Θ_n. The polarities of dP_c/dt and dQ_c/dt correspond to "1" or "0" of the quantized signals S_p and S_q. Therefore, it is necessary to compose the switching state table so that the active and reactive power control is achieved with small control errors restricted within the predetermined hysteresis bandwidths.

In the following analysis, the time derivative of the instantaneous active power and the instantaneous reactive power dP_c/dt and dQ_c/dt in an exceedingly short duration, which corresponds to switching operation of the CVCF inverter. The time derivatives dP_c/dt and dQ_c/dt are theoretically solved as discussed below.

From the mathematical model of the CVCF inverter illustrated in Fig. 1, the current equation can be written as

$i_{c\alpha}^*$ and $i_{c\beta}^*$. P_c^* and Q_c^* are calculated by the following equation:

$$\begin{bmatrix} P_c^* \\ Q_c^* \end{bmatrix} = \begin{bmatrix} v_\alpha & v_\beta \\ v_\beta & -v_\alpha \end{bmatrix} \begin{bmatrix} i_{c\alpha}^* \\ i_{c\beta}^* \end{bmatrix}. \quad (8)$$

Control errors of the active power and the reactive power, i.e., $\Delta P_c = P_c^* - P_c$ and $\Delta Q_c = Q_c^* - Q_c$, are quantized to generate digital signals S_p and S_q with hysteresis comparators. S_p and S_q correspond to either an ascent mode or a descent mode of the instantaneous power, according to the digital values "1" or "0." In addition, a phase angle of the output voltage vector command v_o^* is spatially quantized to six sectors Θ_n by 60 degrees by using several different comparators as shown in Fig. 3, where α-axis and β-axis denote real and imaginary parts of the vector, respectively. The quantized sectors Θ_n is mathematically expressed as

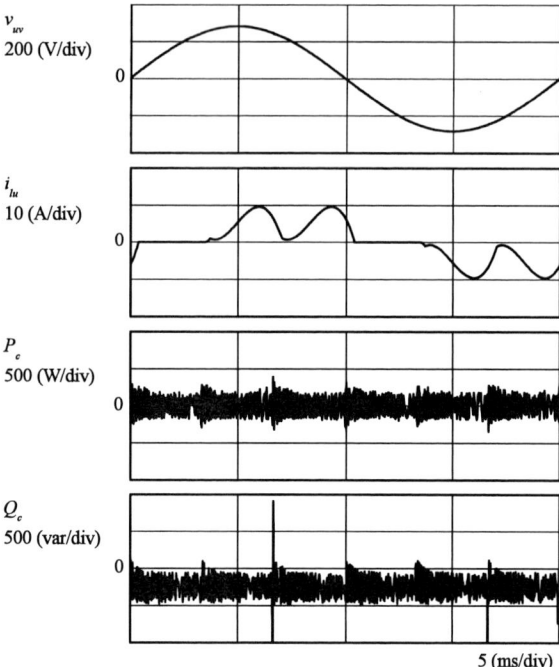

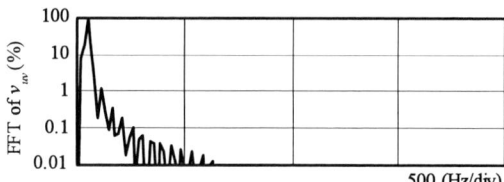

(a) Output voltage, line current, active and reactive power.

(b) Frequency spectra of output voltage.

Fig. 8. Operation waveforms in computer simulation.

$$v_i = L_{ac}\frac{di_i}{dt} + v_o = L_{ac}\frac{d}{dt}(i_{load} + i_c) + v_o. \qquad (10)$$

The output voltage vector v_i of the CVCF inverter is defined by the following equation, where V_{dc} is an inverter DC bus voltage:

$$v_i = \sqrt{\frac{2}{3}}V_{dc}\left(S_u + S_v e^{j2\pi/3} + S_w e^{j4\pi/3}\right). \qquad (11)$$

The output voltage vector v_o is defined by the following equation:

$$v_o = v_\alpha + jv_\beta = \sqrt{\frac{2}{3}}\left(v_u + v_v e^{j2\pi/3} + v_w e^{j4\pi/3}\right), \text{ and}$$

$$\qquad (12)$$

$$\because \quad \begin{cases} v_u = \sqrt{2}\,V_{rms}\cos\theta \\ v_v = \sqrt{2}\,V_{rms}\cos(\theta - 2\pi/3). \\ v_w = \sqrt{2}\,V_{rms}\cos(\theta - 4\pi/3) \end{cases} \qquad (13)$$

Substituting (13) into (12), the output voltage vector v_o can be expressed as a rotating vector as shown below:

$$v_o = \sqrt{3}V_{rms}e^{j\theta}. \qquad (14)$$

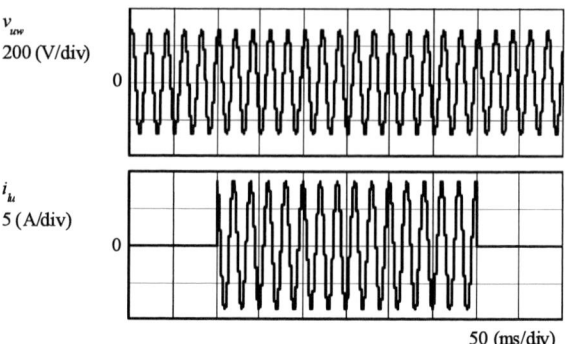

Fig. 9. Disturbance step response (simulation result).

In an exceedingly short time like a switching interval of the CVCF inverter, a variation of the load current vector i_l is negligibly small; hence, i_l can approximately be considered to be a constant value. From what is mentioned above, the time derivative of the filter-capacitor current di_c/dt is simply derived from (10) as follows:

$$\frac{di_c}{dt} \approx \frac{v_i - v_o}{L_{ac}}. \qquad (15)$$

Similarly, the output voltage vector v_o in this situation is also negligibly small, and the time derivative of the active and reactive power dP_c/dt and dQ_c/dt can be approximated as

$$\frac{dP_c}{dt} = \frac{dv_o}{dt} \bullet i_c + v_o \bullet \frac{di_c}{dt} \approx v_o \bullet \frac{di_c}{dt}, \text{ and} \qquad (16)$$

$$\frac{dQ_c}{dt} = \frac{dv_o}{dt} \times i_c + v_o \times \frac{di_c}{dt} \approx v_o \times \frac{di_c}{dt}. \qquad (17)$$

Substituting (11), (14), and (15) into the above equations, dP_c/dt and dQ_c/dt are solved as follows:

$$\frac{dP_c}{dt} = \frac{V_{rms}}{L_{ac}}\left\{-3V_{rms} + \sqrt{2}V_{dc}(S_1\cos\theta + S_2\sin\theta)\right\}, \text{ and}$$

$$\qquad (18)$$

$$\frac{dQ_c}{dt} = \frac{\sqrt{2}V_{rms}V_{dc}}{L_{ac}}(S_1\sin\theta - S_2\cos\theta), \qquad (19)$$

$$\because \quad S_1 = S_u - \frac{1}{2}S_v - \frac{1}{2}S_w, \; S_2 = \frac{\sqrt{3}}{2}S_v - \frac{\sqrt{3}}{2}S_w,$$

where θ is an argument of the output voltage vector command v_o^* as shown in Fig. 3. Equations (18) and (19) allow evaluating dP_c/dt and dQ_c/dt that depends on the switching states of the CVCF inverter and the output voltage vector phase. Fig. 5 illustrates an example of increasing and decreasing behaviors of the instantaneous active and reactive when the output voltage vector is in the sector Θ_1. The optimum output voltage vector from the CVCF inverter can be selected directly from this illustration, which is a manipulated value to regulate the active and reactive power. Suppose that one of the

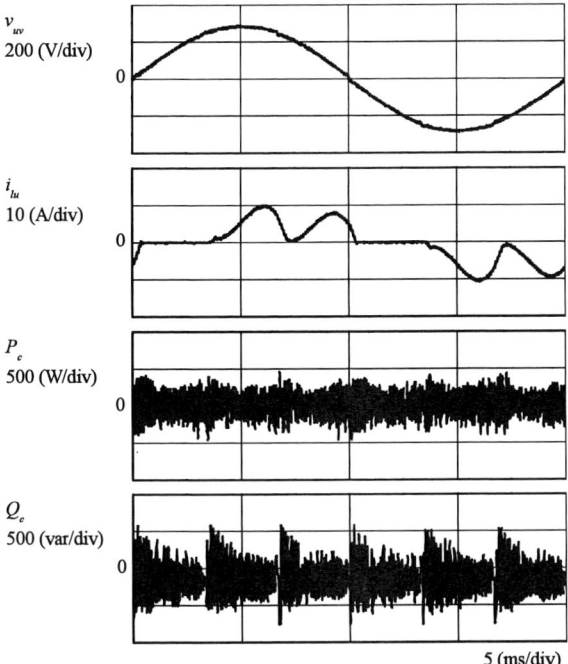

(a) Output voltage, load current, active and reactive power.

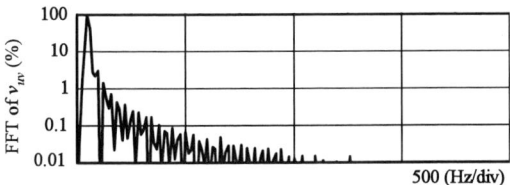

(b) Frequency spectra of Output voltage.

Fig. 10. Operation waveforms in experimental test.

voltage vectors is selected to increase the active power (S_p = "1") and to increase the reactive power (S_q = "1"), for example. In this case, the following process is made on the basis of a flow chart shown in Fig. 6 until the optimum switching state is finally determined. If a switching state S_u S_v S_w = "100" is taken up as an initial value, dP_c/dt is found to be positive, while dQ_c/dt is negative, substituting S_u S_v S_w = "100" into (18) and (19). This does not satisfy the requirement of S_p = "1" and S_q = "1"; thus, another switching state S_u S_v S_w = "101" is tried one by one in a similar manner. The switching state S_u S_v S_w = "101" lets both of dP_c/dt and dQ_c/dt positive, so this can be a candidate of the appropriate switching state of the inverter. If there should appear two appropriate switching state that equally satisfy the requirement of S_p = "1" and S_q = "1", one of the two, which is closer to the output voltage command vector v_o^* than the other, is chosen. Fig. 7 shows an optimum switching state table and hysteresis regulators (relay elements) derived by the above-mentioned process.

III. COMPUTER SIMULATION RESULTS

In order to examine basic operation characteristics of the proposed system, computer simulations were conducted under C++ programming environment. Electric parameters of the power circuit used in the

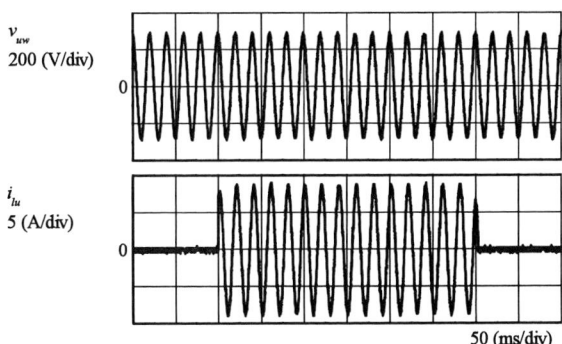

Fig. 11. Disturbance step response (Experimental result).

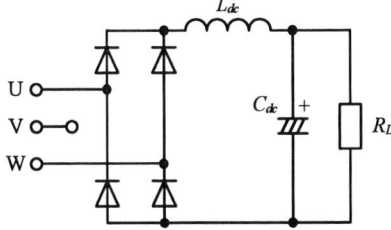

Fig. 12. Single-phase capacitor input-rectifier load.

simulations are listed in Table 1.

Fig. 8 shows steady state waveforms and frequency spectra of the output voltage when the capacitor input rectifier load of 1.7 kVA and 0.85 load power factor was connected to the CVCF inverter. As indicated in the figure, the output voltage is stably controlled to be a sinusoidal waveform with amplitude and frequency of the output voltage command. Furthermore, the FFT analysis result indicates that the output voltage waveform has low total harmonics distortion.

Fig. 9 shows waveforms of a disturbance step response, when a linear resistive load is changed stepwise between 0 and 2 kW. The output voltage is kept almost constant at the command value even though the sudden load change is applied.

IV. EXPERIMENTAL RESULTS

A 2.5-kW prototype was developed to confirm experimentally feasibility and effectiveness of the proposed DPC based control of the CVCF inverter. A controller of the prototype was constituted with analog and digital mixed signal hardware. The prototype has the same specifications and parameters used in the computer simulations.

Fig. 10 shows experimental results obtained under the identical test condition as those of the simulations. As can be seen in the figure, the output voltage waveform is controlled to be a stable sinusoidal waveform with no large distortion, where the total harmonic distortion (THD) is only 1.13 %.

Fig. 11 shows waveforms of a dynamic disturbance step response, where the disturbance is applied to the CVCF inverter by changing the load between 0 and 2 kW. Even under this severe test condition, the output voltage hardly deviate at the moment of load step change, and the stable output voltage is provided to the load.

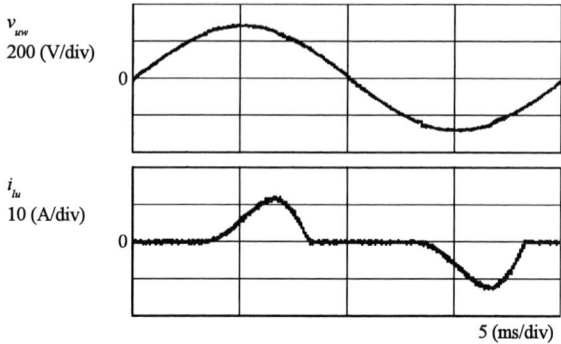

(a) Output line-to-line voltage and load current.

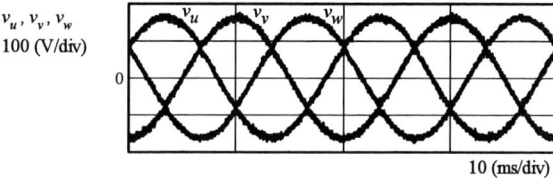

(b) Output phase voltages.

Fig. 13. Operation waveforms under unbalanced load condition.

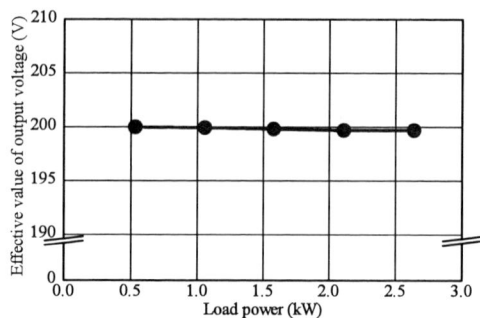

Fig. 14. Deviation characteristic of output voltage.

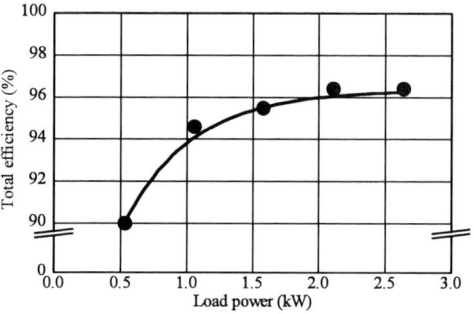

Fig. 15. Total efficiency characteristic.

In order to verify effectiveness of the proposed system under unbalanced load conditions, a single-phase capacitor input rectifier was used as a load of the CVCF inverter as indicated in Fig.12. Fig. 13 shows the output line-to-line voltage and the load current when only the U and the W terminals of the rectifier are connected to the inverter. In the unbalanced load situation, the output voltage waveform is also controlled to be low distortion and balanced.

Figs. 14 and 15 show the output voltage and the total efficiency characteristics with respect to the load power. As can be seen in the figures, the output voltage deviates from its command by 0.5 % only, and the maximum total efficiency is 96.4 % around the rated load. As confirmed in the above experimental results, the proposed strategy is quite effective to enhance performances of the CVCF inverter in various possible situations.

V. CONCLUSIONS

This paper described a novel control strategy of the CVCF inverter, which features direct control of the instantaneous active and reactive power. The key of this strategy is a relay (bang-bang) control algorithm of the instantaneous power by means of an optimum switching operation of the inverter, which makes dynamic response of the power control extremely high. In a theoretical analysis of the proposed method, a relationship between the instantaneous active and reactive power and the switching state of the inverter was clarified, which leaded to composition of the optimum switching state table in the controller. In addition, feasibility of the proposed system was examined through computer simulations and experimental tests. As a result of the tests, the low-distorted output voltage was obtained in the steady state, and excellent stability of the output voltage was confirmed in a disturbance step response. Furthermore, operation characteristics under an unbalanced nonlinear load were investigated, where the excellent voltage balance was achieved without sacrificing the THD of the output voltage waveforms.

REFERENCES

[1] S. Sato, H. Ohshima, and K. Nakajima, "Improvement of Output Waveform for CVCF Inverter Using IGBT," *IEEJ Tech. Meet. Rec.,* EDD-88-60, SPC-88-58, 1998, p.p. 11-20 (in Japanese).

[2] I. Takahashi, I. Ando, Y. Ito, and K. Amei, "Development of a Long Life Three Phase Flywheel UPS using an Electrolytic Capacitor-Less Converter/Inverter," *IEEJ Trans. on IAS,* vol.118-D, no. 2, 1998, p.p. 173-178 (in Japanese).

[3] Y. Ito, and S. Kawauchi, "Microprocessor-Based Robust Digital Control for UPS with Three –Phase PWM Inverter," *IEEE Trans. on PELS,* vol. 10, no. 2, 1995, p.p. 196-204.

[4] G. Escobar, A. M. Stankovic, and P. Mattavelli "Dissipativity - Based Adaptive and Robust Control of UPS in Unbalanced Operation," *IEEE Trans. on PELS,* vol. 18, no. 4, 2003, p.p. 1056 -1062.

[5] P. Mattavelli, "An Improved Deadbeat Control for UPS using Disturbance Observers," *IEEE Trans. on IES,* vol. 52, no. 1, 2005, p.p. 206-212.

[6] T. Noguchi, H. Tomiki, S. Kondo, and I. Takahashi, "Direct Power Control of PWM Converter Without Power-Source Voltage Sensors," *IEEE Trans. on IAS,* vol. 34, no. 3, 1998, p.p. 473-479.

[7] T. Noguchi, D. Takeuchi, S. Nakatomi, and A. Sato, "Novel Control Strategy of Instantaneous-Power Based Current-Source PWM Converter and Experimental Results," *IEEJ Trans. on IAS,* vol. 126-D, no. 4, 2006, p.p. 1486-1492 (in Japanese).

[8] A. Sato, and T. Noguchi, "Novel Control Strategy of Instantaneous Power Based CVCF Inverter," *IEEJ IAS Conf Proc.,* vol. 1, 2007, p.p. 433-434 (in Japanese).

An Improved Parallel Processing UPS Using a Voltage-Controlled Voltage Source Inverter

S.W. Lee*, H. Dehbonei**, S.H. Ko***, S.R. Lee****, B.H. Jang*****, Y.H. Moon******, and T.K. Ko******

* Institute of TMS Information Tech., Yonsei Uni., Seoul, Korea
** School of Electrical & Computer Eng., Curtin University of Tech., Perth, Australia
*** Advanced Graduate Education Center of Jeonbuk for Electronics & Information Tech., Chonbuk National Uni., Korea
**** School of Electronic & Information Eng., Kunsan National Uni., Kunsan, Korea
***** EHWA Technologies Information, Seoul. Korea
****** School of Electrical & Electronic Eng., Yonsei Uni., Seoul, Korea

Abstract--This paper presents an improved parallel processing uninterruptible power supply (UPS) for a strong grid using a bi-directional voltage-controlled voltage source inverter (VCVSI). To maintain the load voltage at the desired value and to control the active power flow between the VCVSI and grid, the amplitude and phase angle of the inverter output voltage (power angle) must be controlled. Selecting the power angle operating range is an important factor which has a direct effect on various parameters, such as the size of a decoupling inductor, grid power factor and the power ratings of VCVSI components. It is shown how the optimum power angles can be chosen and by restricting the operation of this, the power factor can be maintained above 0.9 at different loads and operating conditions. The paper examines the steady state modeling and analysis of a single phase parallel processing UPS while maintaining a high system power factor under different conditions. Experimental and simulation results of a prototyped 1KVA VCVSI confirms the validity of the proposed method.

Index Terms—Parallel processing UPS, power angle, VCVSI, power factor.

I. INTRODUCTION

Uninterruptible power supplies (UPSs) have been used to supply clean and uninterrupted power to critical loads. The number of connected sensitive loads to strong grids are increasing every day, for instance, in progressive demand for extension of information systems, internet data centers (IDCs), on-line banking systems, etc., all of which require constant uninterrupted supply. Moreover, today's modern controls which rely on real-time monitoring cannot tolerate the loss of even a small portion of information (not acceptable) and can cause huge financial burdens and/or safety hazards for different service providers. Hence, UPSs have become an increasingly important system element. Over the last several years, a considerable number of studies have been conducted to improve grid power quality and load voltage stabilization using various UPS [1]-[7].

The advantages and limitations of current-controlled and voltage-controlled inverters (CCVSI and VCVSI) have been recently explained [8]. This study shows that while a VCVSI can provide good load voltage stabilization, the grid power factor is changing by altering the inverter power angle. A parallel processing UPS has many advantages over an online UPS. For instance, in a parallel processing UPS the power is not taken via a separate rectifier to a DC bus in the same way as an online UPS. The power is fed directly via a decoupling inductor (lossless impedance) to the battery by controlling the VCVSI power angle. This means that a parallel processing UPS can offer many advantages over an online UPS, such as higher efficiency, no delay or break in supply load in changing from rectifier to inverting modes, etc. However, the parallel processing UPS performs at low power factor at a small inverter's power angles. This is due to the decoupling inductor between the CCVSI and the grid. To address this problem, a new UPS scheme using a combined CCVSI and VCVSI is proposed [9].

This paper deals with a parallel processing UPS using a single phase bi-directional VCVSI, connected to a strong grid where the grid voltage may fluctuate ±6% (220V±13V). It is proposed to limit the operating power angle of the VCVSI to an optimum limit where it does not affect the inverter performance and guarantees high grid power factor operation,

II. PARALLEL PROCESSING UPS SYSTEM

The power angle is an important parameter since it controls not only the power flow, but also dictates the grid power factor considering the system parameters (ie., size of the decoupling inductor, etc). The higher power angle results in higher active power flow to and from the grid at a superior power factor. On one hand, to operate the VCVSI at lower power angles requires a smaller decoupling inductor and relevant reactive power compensation by the inverter or grid (depending on the inverter and grid voltage). On the other hand, operating the VCVSI at a higher power angle, while requiring a bigger decoupling inductor, also requires more reactive power to flow between the VCVSI and the grid, just to compensate the decoupling inductor's reactive power requirement. Hence, it is important to properly size the decoupling inductor and the maximum power angle

---

This work has been supported by Yonsei University Institute of TMS Information Technology, a Brain Korea 21 program, Korea.

respectively. Moreover, the VCVSI operates at a low power factor while the power angle is small.

A. System Description

Fig. 1(a) shows the simplified schematic diagram of the parallel processing UPS using a VCVSI and Fig. 1(b) shows the equivalent circuit diagram of the system.

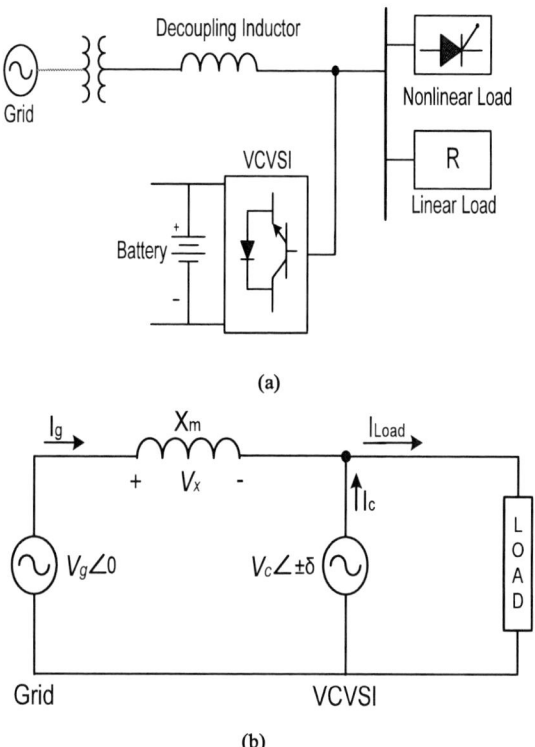

(a)

(b)

Fig. 1. Parallel processing UPS. (a) the schematic. (b) the equivalent circuit diagram.

The VCVSI is synchronized and connected to the grid through the decoupling inductor (X_m) to limit the amount of power flow to or from the grid. The VCVSI is connected to a battery and power flow is controlled by changing the VCVSI power angle (both rectification (charging) and inversion (discharging)).

B. Steady State Analysis

Using Fig. 1(b), the grid current (I_g) can be expressed as fellow.

$$I_g = \frac{V_g\angle 0 - V_c\angle\delta}{jX_m} = -\frac{V_c\sin\delta}{X_m} - j\frac{V_g - V_c\cos\delta}{X_m} \quad (1)$$

Where the power angle (δ) is a phase between the VCVSI output voltage (V_c) and the grid voltage (V_g). Assuming that the maximum permitted voltage fluctuation of the grid is limited to ±6% (220V±13V), the grid voltage can be higher, lower or equal to VCVSI voltage.

Using per unit values ($S_{base}= V^2_{base}/Z_{base}$, $V_{base}=V_c$ and $Z_{base}= X_m$) where V_{base}, Z_{base} and S_{base} are the base voltage, impedance and apparent power values respectively, the

apparent power of the grid, inverter and the decoupling inductor are given by [8]:

$$S_{gpu} = P_{gpu} + jQ_{gpu} = -V_{gpu}\sin\delta + j(V^2_{gpu} - V_{gpu}\cos\delta) \quad (2)$$

$$S_{cpu} = P_{cpu} + jQ_{cpu} = -V_{gpu}\sin\delta + j(V_{gpu}\cos\delta - 1) \quad (3)$$

$$S_{xpu} = jQ_{xpu} = j(V^2_{gpu} - 2V_{gpu}\cos\delta + 1) \quad (4)$$

where S_{gpu}, S_{cpu} and S_{xpu} are per unit values of the grid, inverter and decoupling inductor's apparent power respectively, and V_{gpu} is the per unit value of the grid voltage.

Since the load voltage must remain constant (load voltage stabilization), the only controllable parameter in the system is the power angle (δ). Here we allow for a strong grid where the grid voltage (V_g) may fluctuate by ±6%(220V±13V). Using (2) to (4) the theoretical power variation of the system for the three possible operational modes can be computed as shown in Fig. 2.

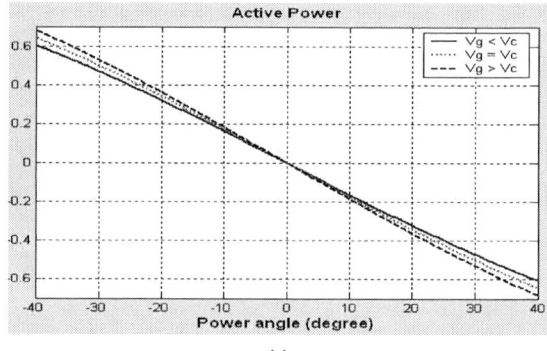

(a)

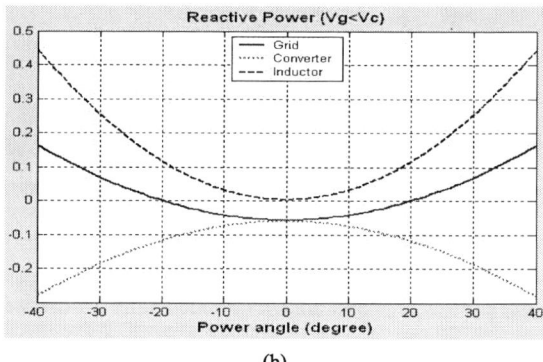

(b)

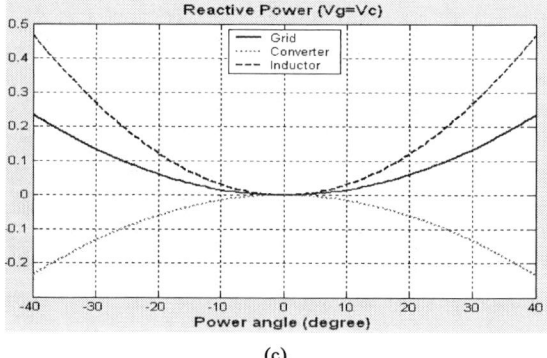

(c)

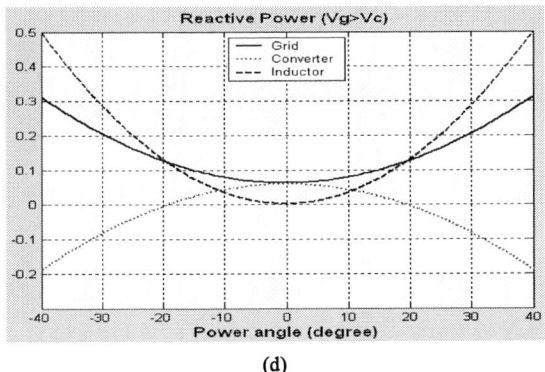

(d)

Fig. 2. System active and reactive power flow at different power angles (δ). (a) Active power flow. (b) Reactive power flow ($V_g < V_c$), (c) ($V_g = V_c$), and (d) ($V_g > V_c$).

Fig. 2(a) shows that the active power varies approximately linear to power angles. When the power angle is negative, the active power flows from the grid to the inverter and vice versa, regardless of the grid voltage. Fig. 2(b), (c) and (d) illustrate the system's reactive power flow at different grid voltage and power angles. These figures reveal the fact that the required decoupling inductor reactive power has to supply either the grid or inverter depends on the grid voltage magnitude.

C. Design Consideration of the System

To define the δ_{max}, it is necessary to consider various parameters, including the size of the decoupling inductor, cost and the amount of reactive power required by the decoupling inductor. Note that the greater the δ_{max}, the larger the size and cost of the decoupling inductor. This is because the reactive power required by the decoupling inductor must be compensated by either the grid or VCVSI and it has a direct effect on the system's power factor, especially at lower power angles.

From the Fig. 2(a), where if δ is limited to a maximum of $\delta_{max}=20°$, the active power flow of the system is merely $P_g =0.32pu$ when the grid voltage is 0.94pu ($V_g < V_c$). Therefore, in order to raise the active power flow to $P_g=1.0pu$, we need to multiply P_g by a scale factor (D). Using per unit values as mentioned in section B, this relationship can be expressed as:

$$S_{basepu} = P_{basepu} =1pu = D \times P_{gpu} \tag{5}$$

Where, S_{basepu} and P_{basepu} is a per unit value of the apparent power (S_{base}) and active power (P_{base}) respectively. In equation (9), the scale factor has a maximum value when $V_g=0.94$ pu. Using (2) and (5), the scale factor can rewritten as fellows:

$$D = \frac{1}{\left(-0.94 \times \sin \delta_{max}\right)} \tag{6}$$

In order to scale up the active power flow, we need to recalculate $X'_m = X_m/D$. Assuming that the output voltage of the VCVSI (load voltage) must be maintained at 220V (V_c= 220V) and S_{base}= 1KVA, X_m can be calculated as 48.4[Ω] for δ_{max}.=40°, $X_m=22.72Ω$ for $\delta_{max}=30°$ and

$X_m=15.56Ω$ for $\delta_{max}=20°$. While a bigger decoupling inductor provides a smaller step in changing power flow, it will require higher reactive power depending on the voltage difference of the grid and VCVSI.

Using (2), the variation of the grid power factor at various power angles for different possible operational conditions can be calculated. Fig. 3 illustrates that the power factor of the grid deteriorates when the grid voltage differs from the VCVSI voltage.

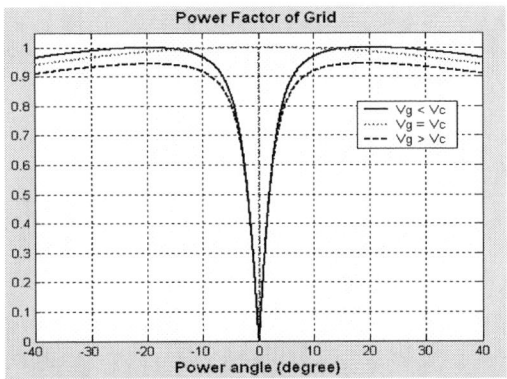

Fig. 3. Power factor versus power angle at different grid voltages.

Fig. 3 reveals that power factor is worst when the grid voltage is higher than the VCVSI's voltage and the VCVSI operates at a low power angle (± 8°). Considering the above, it seems suitable to limit the power angle to 8° and the $\delta_{max}=20°$ to minimize the size of the decoupling inductor and maximize the power factor. This figure illustrates that while the VCVSI operates at a higher power angle (more than ±8°), the power factor can maintain more than 0.9 under operating conditions.

On the hand, the required maximum apparent power of the grid (S_g), decoupling inductor (S_x), and the VCVSI (S_c) at different δ_{max} is shown in Fig. 4.

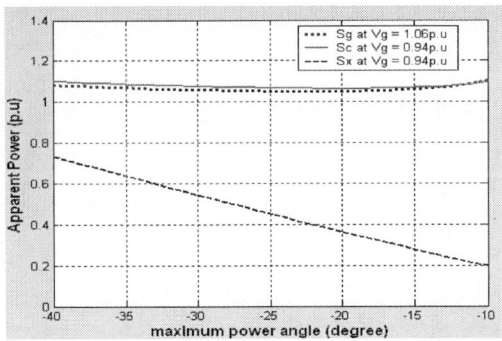

Fig. 4. The maximum power demand of each parameter in parallel processing UPS according to the δ_{max} for handling the full active power flow.

It can be seen that S_x is proportional to δ_{max} while S_g and S_c not sensitive as δ_{max} increases. In a strong grid system, improving the grid power factor and reducing the reactive power S_x across the inductor are more important than S_g and S_c . Hence, it is proposed to restrict the minimum power angle to (± 8°) and the maximum power angle to $\delta_{max}=20°$.

III. EXPERIMENTAL RESULTS

To examine the validity of the proposed design considerations, 1kVA VCVSI has been prototyped. The system's operation is divided into two modes (grid-connected (normal) and stand-alone (UPS)). Details of system specifications and parameters are given in Table I. And a photograph of the experimental set up is shown in Fig. 5.

TABLE I
SYSTEM PARAMETERS AND SPECIFICATIONS

Parameter	Value	Parameter	Value
Vac	220±13V	Vdc	200V
Frequency	60Hz	Fsw	10kHz
Full load	1KVA	Lm	42mH
δ_{max}	20°	Transformer	1:2

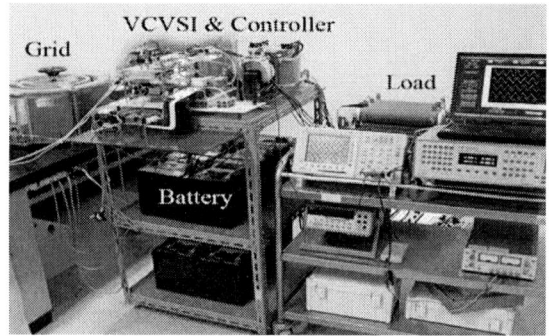

Fig. 5. A photograph of the prototyped system.

In this paper, a digital system based on the TMS320 C33 DSP board has been used to implement the proposed system. The DSP320C33 provide an internal Timer Interruption Service Routine (TISR), which is caused by TSTAT bit of the timer control register. The frequency of TISR depends on whether the timer is set up. The input data can be updated every period of TISR, instantaneously. In this paper, the one-period of TISR is decided at 46.3[us] and it is performed the iteration of 360 during one period for the grid voltage (frequency 60[Hz]).

Fig. 6 shows the voltage and current waveforms of the grid and load when supplying the full resistive load of 1p.u (R=50Ω) at V_g =0.94pu.

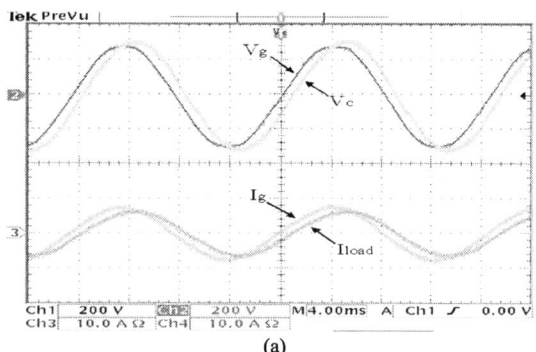

(a)

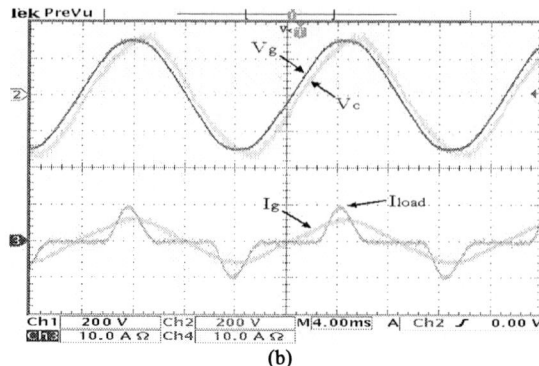

(b)

Fig. 6. Experimental waveforms at full loads (δ_{max}=20°), (a) Resistive load condition. (b) Nonlinear load condition.

As can be seen in Fig. 6, the grid supplies the load active power, and the required reactive power of the system can be supplied by the VCVSI. This result illustrates the capability of the VCVSI to operate at a high power factor in the presence of a nonlinear load.

Fig. 7 shows the voltage and current waveforms of the grid and load when the power angle is limited at 8°. In this figure, while the grid current is limited to 2.6A the load current can change from 1.9A to 0.95A. This means that the extra current will be used to charge or discharge the battery while this guarantees that the system operates above 0.9 power factor.

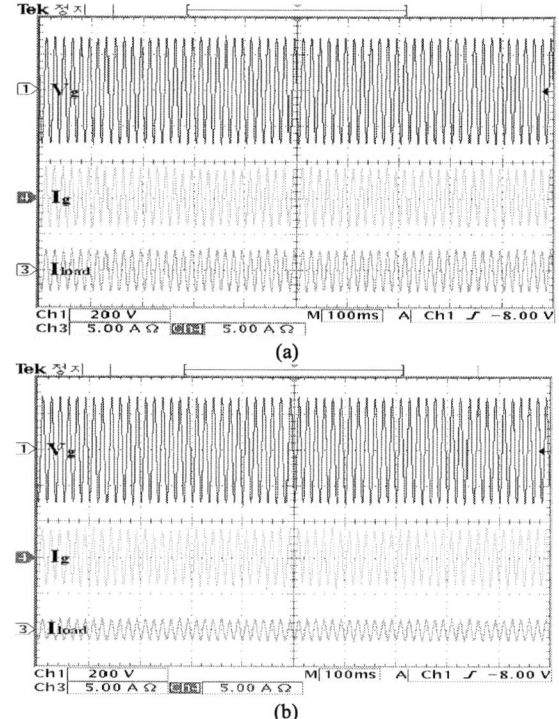

Fig. 7. Experimental waveforms of power angle limitation. (a) R=120Ω (load power: 0.4 p.u). (b) R=240Ω(load power: 0.2 p.u).

Fig. 8 shows the system operation with restricted power angle when the grid voltage is changing from 1.06p.u to 0.94p.u and vice versa. This figure shows that the VCVSI with restricted power angle can respond quickly to grid voltage changes, while not affecting in system operation and performance (the load voltage is maintained at 1.0p.u without being affected by the grid voltage fluctuations).

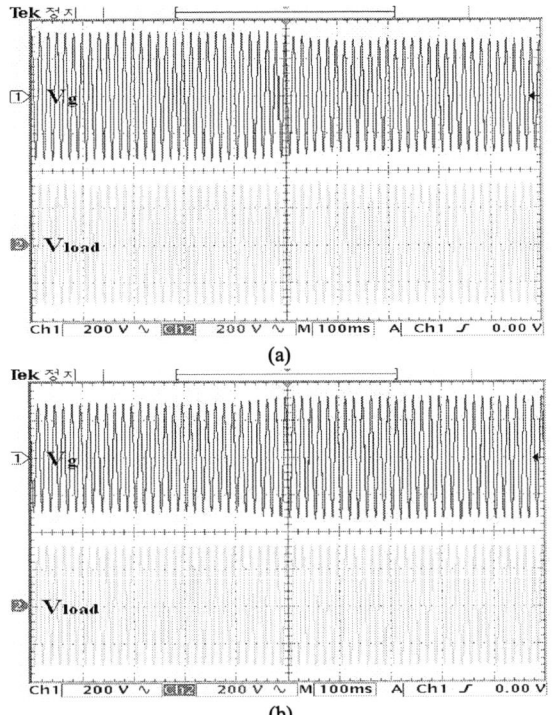

(a)

(b)

Fig. 8. Experimental waveforms of voltage stabilization with resistive load. (a) Grid voltage changes from 1.06pu to 0.96pu. (b) Grid voltage change from 0.96pu to 1.06pu

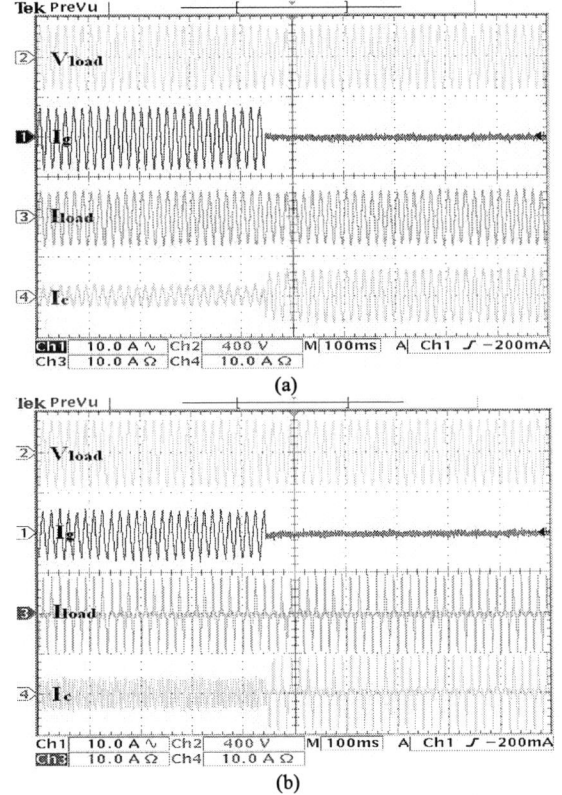

(a)

(b)

Fig. 9. Experimental waveforms of power failure performance. (a) Resistive load condition. (b) Nonlinear load condition.

Fig. 9 shows the proposed system when its mode of operation changes from normal to UPS in the presence of a nonlinear load. It is shown that the VCVSI, while providing a fast response to grid failure, can provide good active filtering support to the grid at high power factor (> 0.9) during normal mode

IV. CONCLUSIONS

This paper presents an improved parallel processing uninterruptible power supply (UPS) using a bi-directional voltage-controlled voltage source inverter (VCVSI) while connecting to a strong grid. The selection criteria for a suitable power angle has been provided. It is shown that by restricting the operation of the power angle, the grid power factor can be maintained above 0.9 under various operating conditions and in the presence of different loads. Experimental results of a prototyped 1KVA system confirm the validity of the proposed method.

REFERENCES

[1] H. Deng, R. Oruganti, and D. Srinivasan, "Modeling and Control of Single-Phase UPS Inverter: A Survey", *at presented 2005 PEDS Conf. Power electronics and Drives Systems*, vol. 2, pp. 848-853, Nov. 2005.

[2] S. Karve, "Three of a Kind[UPS topologies, IEC standard]," *IEE Review*, vol. 48, no. 2, pp. 27-31, March 2000.

[3] B.H. Kwon, J.H. Choi, and T.W. Kim, "Improved Single-Phase Line-Interactive UPS," *IEEE Trans. Ind. Electron.*, vol. 48, no. 4, Aug. 2001.

[4] W.J. Ho, J.B. Lio and W.S. Feng, "Economic UPS structure with phase-controlled battery charger and input-power-factor improvement," *IEE Proc. Eletr. Power Appl.*, vol. 144, no. 4, pp. 221-226, July 1997.

[5] J.C. Wu and H.K. Jou, "A New UPS Scheme Provides Harmonics Suppression and Input Power Factor Correction", *IEEE Trans. Ind. Electr.*, Vol. 42, No. 6, pp. 629-636, Dec. 1995.

[6] Soren Rathmann, Henry A. Warner, "New generation ups technology, The delta conversion principle", *IAS 96, Conference Record of the 1996 IEEE*, Vol. 4, pp. 2389-2395, 1996.

[7] P. Jakobsen, "Revolution within UPS system," *at the present 15th INTELEC conf. Telecommunications Energy Conf.*, vol. 1, pp. 413-419, Sept. 1993.

[8] S.H. Ko, S.R. Lee, H. Dehbonei, and C.V. Nayar, "Application of Voltage and Current Controlled Voltage Source Inverters for Distributed Generation Systems," *IEEE. Trans. Energy Conver.*, vol. 21, no. 3, pp. 782-792, Sep. 2006.

[9] H. Dehbonei, C.V. Nayar, and L. Borle, "A Multifunctional power processing unit for an off-grid PV diesel hybrid power system," *presented at the IEEE 35th Ann. Power Electronics Specialists Conf.*, vol. 3, pp. 1969-1975, June 2004.

A PEMFC/Battery Hybrid UPS System for Backup and Emergency Power Applications

Yuedong Zhan*,**, Jianguo Zhu**, Youguang Guo**, and Hua Wang*

* Department of Automation, Kunming University of Science and Technology, Kunming, 650093, China
** Faculty of Engineering, University of Technology, Sydney (UTS), NSW 2007, Australia

Abstract–This paper presents the development of a proton exchange membrane fuel cell (PEMFC) and battery hybrid uninterruptible power system (UPS) for backup and emergency power applications. A sixty-cell PEMFC stack is employed as the main power source at normal load and a 3-cell lead-acid battery is employed as the auxiliary power of the UPS at overload or during the PEMFC startup. The PEMFC consists of two valves for the hydrogen input and output respectively, a mass flow controller to adjust the hydrogen mass flow, and a pressure sensor to control the hydrogen pressure, and their control units for the management of the whole system. The design procedures of the UPS hybrid system are discussed. Experimental setups are presented and the experimental results verify the performances of the PEMFC/battery hybrid power source and the UPS system under the condition of computer load. The developed UPS system proves to be a cost-effective solution for backup and emergency power applications.

Index Terms--Backup and emergency power applications, fuel cell/battery hybrid power source, proton exchange membrane fuel cell (PEMFC), uninterruptible power supply (UPS).

I. INTRODUCTION

To provide backup and emergency power for critical loads, such as computers, medical and life support systems, communication systems, and industrial controllers, in case of power failures, the uninterruptible power supply (UPS) systems have played an important role. A high performance UPS system should have a clean output voltage with low total harmonic distortion (THD) for the unbalanced linear and nonlinear loads, high efficiency, great reliability and fast transient response for sudden load changes [1]. Besides these, an ideal UPS should have the following features: regulated sinusoidal output voltage with low THD and independent from the changes in the input voltage or in the load, on-line operation that means zero switching time from normal to backup mode and vice versa, low THD sinusoidal input current and unity power factor, low electromagnetic interference (EMI) and acoustic noise, electric isolation, low maintenance, and low cost, weight, and size. Therefore, UPS systems that can keep the information and data from being destroyed have become very popular. Particularly, with the popularization of personal computers and Internet, low capacity UPS

products will take an increscent part in the industrial and domestic markets.

As to the backup and emergency power applications, conventional approaches using batteries have problems with life, pollution, maintenance and weight. Similarly, diesel generators have problems with startup, maintenance, noise and emissions. Although the metal hydride fuel cells offer a fundamentally new approach to fuel cells that results in a practical, low cost technology with unique performance advantages, including intrinsic energy storage, instant start capability, good low-temperature performance, and fuel "hot swap" capabilities [2], the liquid-fed direct methanol fuel cells (DMFC) and proton exchange membrane fuel cells (PEMFC) using hydrogen as fuel are still considered as the well-developed and the most promising fuel cell technologies [3]. In this paper, a PEMFC stack is selected because it features higher power densities than the DMFC; the latter has additional problems due to crossover and slow anode electro-catalysis of methanol.

Compared with the fuel cells, lead-acid rechargeable batteries have a rapid transient response without any warm up or start up time, and their specific power capability is also much higher than that of fuel cells [4]. Combining the fuel cells and small capacity battery, the PEMFC/battery hybrid UPS system make the best use of the advantages of each individual device, and may meet the requirements for the above mentioned applications regarding both high power and high energy density.

This paper mainly presents the experimental study of an actively and passively controlled PEMFC/battery hybrid UPS system for backup and emergency power applications. It addresses the design of the UPS system with PEMFC/battery hybrid power source, and the experimental results obtained during the normal and backup modes of the UPS system, as well as the current-voltage and power performances of the PEMFC stack. Additionally, the dynamic characteristics are tested as a function of the PEMFC and battery load sharing in the backup hybrid power source. The PEMFC is applied as the main power source for the UPS system, the battery is connected as an auxiliary power source, and the characteristics of the change of current and power depending on the load in the PEMFC/battery hybrid power source are analyzed. The development of this PEMFC/battery hybrid power system would provide a sound basis for further study on portable backup and emergency power supply applications.

This work was partly supported by the UTS Pro Vice Chancellor R&D Seed Fund.

II. SYSTEM SETUP

Fig. 1 illustrates schematically the structure of a single-phase high frequency 300 W UPS with a backup 300 W PEMFC generating unit and battery, which is composed of AC/DC rectifier, AC/DC charger, DC/AC inverter and DC/DC converter and their data acquisition and control devices. The UPS system supplies the load with the required and uninterruptible AC power. The PEMFC stack operates on hydrogen and air. Because of the sudden changes of the UPS load and the slow start-up of the PEMFC stack, which may take a few seconds before its output voltage reaches the rated voltage, a small capacity battery or ultra-capacitor is required for UPS applications.

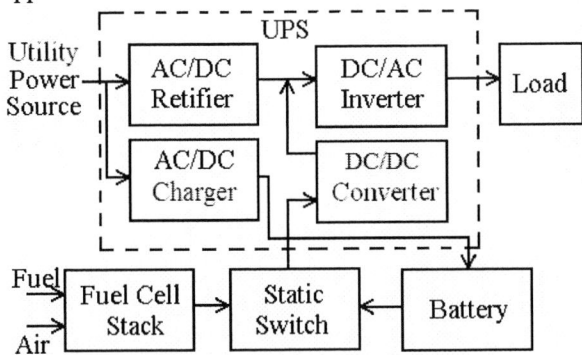

Fig. 1. UPS system with backup PEMFC and battery

A. PEMFC Stack

The PEMFC stack is a self-humidified, air-breathing, 60-cell stack with an overall size of 22.0 cm x 10.5 cm x 7.0 cm, as shown in Fig. 2. In the presented PEMFC system, there is an option to humidify the hydrogen gas and to use oxygen instead of air. To improve the cooling of the stack, three fans are used. The stack has a maximum operating temperature of 65 °C and an operating pressure of 4.55-5.5 psi for hydrogen.

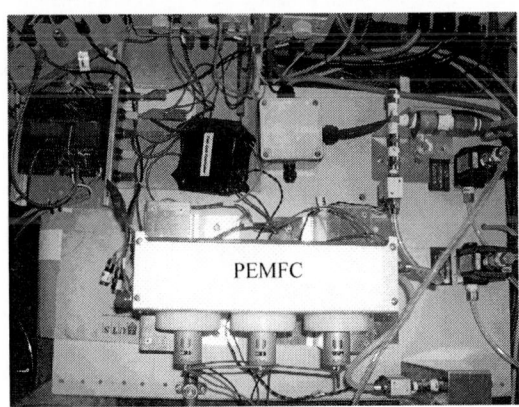

Fig. 2. Photo of the 300W PEMFC stack

To use the PEMFC stack as the backup and emergency power source, there is one polarization curve given in the operator's manual [5] that will be used in later analysis of the proposed model. Fig. 3 indicates the current-voltage and power characteristics of the PEMFC stack.

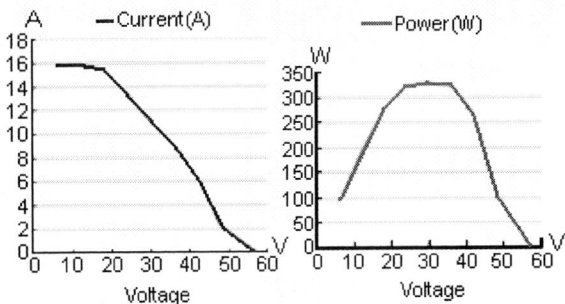

Fig. 3. Current-voltage and power characteristics of the 300W PEMFC

B. UPS System

In the PEMFC/battery hybrid UPS system, a boost active power factor corrector (PFC) with 160-275 VAC input voltage and fixed output voltage (±380 VDC) is analyzed and designed based on a high power factor pre-regular UC3854, which can make the input power factor (PF) of the AC/DC boost PWM rectifier be close to 1, restrict the input current's THD less than 5%, adopt the average current control and constant frequency control, and allow the frequency band of its current amplifier to be wide. Fig. 4 shows the single-phase active PFC AC/DC rectifier and its working pattern. The UC3854 working frequency is 100 kHz.

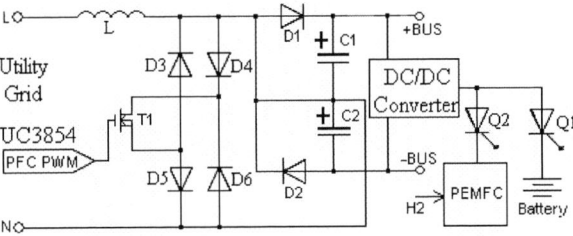

Fig. 4. Schematic diagram of an active PFC AC/DC rectifier with backup PEMFC/battery power supply

Using a TMS320F240 DSP, a DC/AC inverter has been designed to supply the load with a pure sine wave, as shown in Fig. 5, where the half-bridge inverter, LC filter and load are considered as the plant to be controlled. Since the switching frequency (the designed working frequency is 20 kHz) is much higher than the natural frequency and modulation frequency, the dynamics of the DC/AC inverter are mainly determined by its LC filter. Dead-time effect and inevitable loss in every part of the DC/AC inverter offer little damping. The damping effect can be considered as a small resistor connected in series with the filter inductor [6].

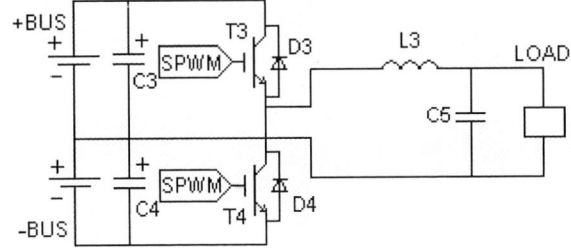

Fig. 5. Schematic diagram of the DC/AC inverter

A basic switch power system with universal input and adjustable output voltage is designed as the charger of the battery based on a high performance current mode PWM controller UC3845. Fig. 6 shows the schematic circuit model of the AC/DC charger.

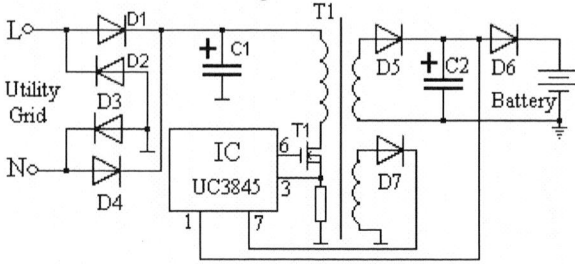

Fig. 6. Schematic diagram of AC/DC charger

A general and practical DC/DC converter for single-phase high frequency UPS is designed based on a regulating pulse width modulator UC3525. The PEMFC and battery are two kinds of low-voltage and high-current power source, so that their rated output voltage (36 VDC) should be boosted up to about ±380 VDC before the UPS inverter converts them into a 240 V, 50 Hz AC source. This boosting action is performed by DC/DC converters. Moreover, in the UPS system, energy storage units, such as a battery and super-capacitor, are required in the start-up stage of PEMFC generator or during the external load sudden change. Fig. 7 illustrates the schematic model of the DC/DC converter. The operating frequency of the power switches T_1 and T_2 is 20 KHz.

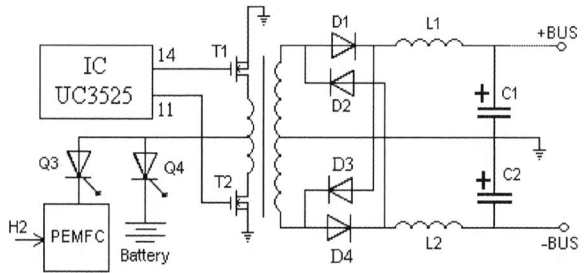

Fig. 7. Schematic diagram of DC/DC converter

C. Batteries

Generally speaking, in the hybrid PEMFC/battery UPS system, the PEMFC plays the role of main power supply under normal conditions and the battery provides the rest of the power when overload happens or when the PEMFC starts up. To realize this, the PEMFC needs a higher electrical potential than the battery. The battery type is LC-R127R2CH (PANASONIC™) with specifications of 12 V and 7.2 Ah.

III. PEMFC/BATTERY HYBRID EXPERIMENT

In the PEMFC/battery hybrid UPS system, when the utility grid power source is interrupted, the hydrogen will be supplied to the PEMFC stack that will supply the UPS system as the backup power source. The UPS system should ensure that there is enough fuel and battery capacity for providing the power needed by the external load. When the external load has a sudden change,

however, the hydrogen cannot be fed fast enough to the PEMFC stack. Therefore, the UPS system also uses the battery to protect the PEMFC from excessive use and to feed power smoothly to the external load. The PEMFC/battery hybrid power source in parallel structure is employed.

For theoretical analysis and experimental study, a passive connection diagram is designed, similar to the actual one, by implementing a device to connect the PEMFC and battery [7, 8], as shown in Fig. 8. In the experiment setup, it is examined and found out that the passive electronic circuits can work well for the UPS load.

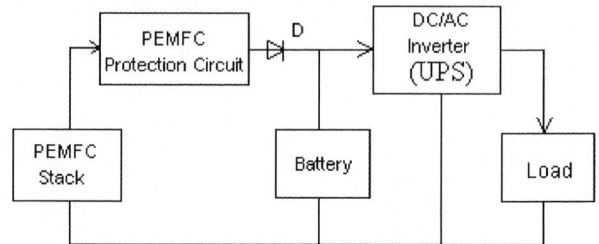

Fig. 8. Schematic diagram of connection between PEMFC and battery

In the PEMFC/battery hybrid UPS system used for the experiment, the PEMFC has series connection with its protection circuit, and parallel connection with the battery. The PEMFC in the UPS system is configured to output a constant power at all times, except when the load of the UPS has a sudden rise. The hybrid PEMFC/battery power source can also work when the external load exceeds the maximum power output of the PEMFC stack. In other words, the PEMFC/battery hybrid UPS system can supply power using the supplementary power stored in the battery. As revealed in Fig. 9, although the load suddenly changes from 60 W to 210 W, the UPS can meet the power output requirement. After the transient process, the PEMFC stack would gradually increase its output to meet the load and the battery is cut off or at the recharging condition.

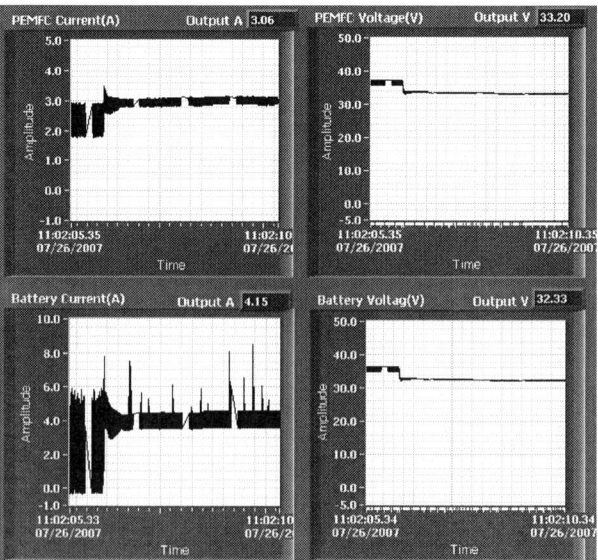

Fig. 9. Currents and voltages of the PEMFC and battery when the load of UPS changes from 60 W to 210 W

1664

IV. RESULTS AND DISCUSSIONS

A. Experimental Setup

Fig. 10 shows the experimental setup of the proposed PEMFC and battery hybrid UPS for backup and emergency power applications. The experimental setup consists of an UPS system, PEMFC generating and test system, lead-acid batteries and a data-acquisition system (a multifunction I/O device NI6036E), an analog voltage output device NI6713, a parallel digital I/O interface PCI-6503, and an analog multiplexer with temperature sensor AMUX-64T (National Instruments Australia Corporation). The UPS system with backup PEMFC and battery provides the AC power source to the linear loads (Lamp box) and nonlinear loads (PC computer), while the data-acquisition system measures and records the required information. In the PEMFC generating and test system, the hydrogen is regulated by mass flow controller (Type F-201C-GAS-22V, Bronkhorst). The temperature and humidity of hydrogen and air can be measured at the inlet by the hydro-transmitter (Type HD2008TV1, Delta OHM) as well as the pressure transmitter (Type AUS EX 1354X, Burkert) between the inlet of PEMFC stack. The output terminals of the UPS are connected to a lamp box and a PC computer that are used in constant voltage mode. All physical parameters like the currents and voltages of the UPS, the PEMFC stack and battery, the mass flow of the hydrogen, the pressure drop in the flow fields, the relative humidity and temperature of hydrogen are recorded with the data-acquisition system.

Fig. 10. Photo of experimental setup of the UPS with hybrid PEMFC and battery source

B. Performance of the Hybrid PEMFC/Battery Source

From Fig. 3, it can be seen that the voltage of the PEMFC decreases when the current increases. In other words, a load change causes a change in the output voltage. In the PEMFC/battery hybrid UPS system, during a load change or the startup of the PEMFC stack, the battery is needed as a supplementary power supply to keep the system voltage constant.

Fig. 11 shows the measured currents and powers of battery and PEMFC when the load changes. In the normal state, the load rise of the UPS keeps the current from the PEMFC higher than that from the battery over time. In other words, the PEMFC takes the majority of the load under normal conditions, and the battery provides the rest of the power.

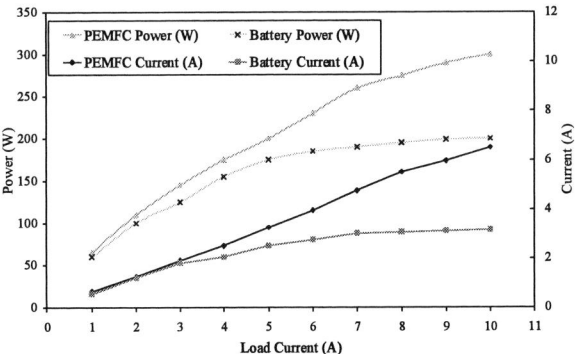

Fig. 11. Currents and powers of PEMFC and battery when the load current changes

C. Performance of the UPS System

The performances of the proposed hybrid UPS system are tested with the following specifications: input voltage of utility grid of 160-275 VAC, output voltage frequency of $50 \pm 5\%$ Hz, PEMFC/battery rated voltage of 36 VDC, and output power of 286 W. The experimental load includes a DELL™ computer (Model: HP-U2106F3) with the maximum input power of 213 W, and a monitor (Model: E772p) with a input power of 73W. Moreover, a lamp box is used as the supplementary load.

Figs. 12 and 13 illustrate the input voltage and output voltage of the UPS for utility grid input AC voltage failure and recovery. Both figures show that the uninterrupted output voltage has no overshoots and undershoots; confirming that a high quality output voltage of the designed UPS system with the PEMFC/battery hybrid power source is achieved. It can also be seen that very fast dynamic response has been achieved. Therefore, low sensitivity of output voltage against large load changes is obtained in the designed UPS system, as the good design concepts with the mature technology are included in the presented UPS system.

The performances of the designed UPS system with the PEMFC/battery hybrid power source are verified as follows: the output voltage of the UPS is 240±3% VAC, the output voltage frequency is 50±0.5% Hz, the input power factor is larger than 0.95, the output power factor is larger than 0.7, the input current's THD is less than 5%, and the transfer time is zero.

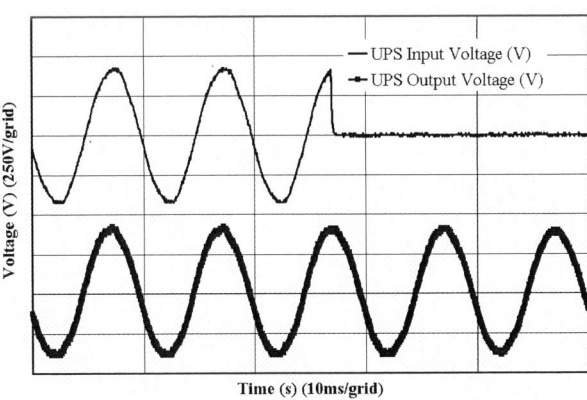

Fig. 12. Transitional waveform while the AC line is interrupted

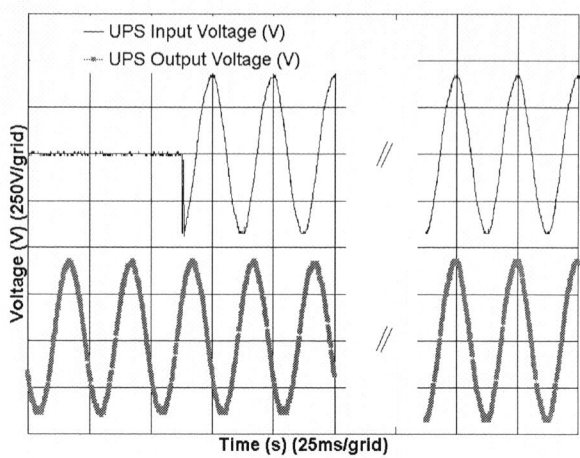

Fig. 13. Transitional waveform while the AC line recovers from backup status to the normal

V. CONCLUSIONS

A PEMFC/battery hybrid UPS system has been developed for backup and emergency power applications, which is mainly composed of a 36V/300W PEMFC stack, 3 cells 12V/7.2Ah lead-acid battery, and a UPS system, including an active PFC AC/DC rectifier, AC/DC charger, DC/AC inverter, DC/DC converter to boost 36VDC to $\pm$ 380VDC, and their data-acquisition and control devices. The performance tests on both the PEMFC/battery hybrid power source and the UPS system have been conducted at the normal conditions. The hybrid UPS system has been applied to a computer to evaluate the operation ability as the AC power source. When the UPS system supplies the rated current for the computer, the PEMFC plays the role of main power source taking the majority of the load under normal operating conditions, and the battery provides the rest of the power when the AC line is interrupted. Although this PEMFC/battery hybrid UPS system still requires improvements for the practical use and the product development in backup and emergency power applications, this study has demonstrated its potential as the next-generation backup power unit for electronic devices.

REFERENCES

[1] S. B. Bekiarov and A. Emadi, "Unintermptible power supplies: classification, operation, dynamics and control," in *Proc. IEEE Applied Power Electronics Conf. and Exposition*, Dallas, Texas, USA, 2002, pp. 597-604.

[2] K. Fok, "Metal hydride fuel cells: a new and practical approach for backup and emergency power applications," in *Proc. 28th Int. Telecom. Energy Conf.*, Rhode Island, USA, 2006, pp. 1-6.

[3] K. Tüber, M. Zobel, H. Schmidt, and C. Hebling, "A polymer electrolyte membrane fuel cell system for powering backup computers," *Journal of Power Sources*, vol. 122, pp. 1-8, 2003.

[4] J. S. Han and E. S. Park, "Direct methanol fuel-cell combined with a small backup battery," *Journal of Power Sources* vol. 112, pp. 477-483, 2002.

[5] Horizon Technology, "300W fuel cell stack operating instruments," available at www.horizonfuelcell.com.

[6] B. Liu, S. Duan, Y. Kang, and J. Chen, "Genetic algorithm optimized fuzzy repetitive controller for low cost UPS inverter application," in *Proc. IEEE Int. Conf. on Electrical Machines and Drives*, 2005, pp. 840-845.

[7] B. D. Lee, D. H. Jung, and Y. H. Ko, "Analysis of DMFC/battery hybrid power system for backup applications," *Journal of Power Sources*, vol. 131, pp. 207-212, 2004.

[8] L. Gao, Z. H. Jiang, and R. A. Dougal, "An actively controlled fuel cell/battery hybrid to meet pulsed power demands," *Journal of Power Sources,* vol. 130, pp. 202-207, 2004.

[9] B. J. Holland and J. G. Zhu, "Design of a 500W PEM fuel cell test system," in *Proc. Australasian Universities Power Engineering Conf.*, Melbourne, Australia, Sept. 29 - Oct. 2, 2002.

[10] Y. D. Zhan, J. G. Zhu, Y. G. Guo, and A. Rodriguez, "An intelligent controller for PEM fuel cell power system based on double closed-loop control," in *Proc. Australasian Universities Power Engineering Conf.*, Hobart, Australia, Sept. 25-28, 2005, pp. 174-179.

[11] Y. D. Zhan, J. G. Zhu, and Y. G. Guo, "Development of advanced hardware and software of proton exchange membrane fuel cell test systems," accepted for publication in *Australian Journal of Electrical & Electronic Engineering*, in press.

Design of the Two Parallel Inverter Modules by Circular Chain Control Technique

K. Piboonwattanakit[1], W. Khan-ngern[1]

[1]King Mongkut's Institute of Technology Ladkrabang (KMITL),
Research Center for Communications and Information Technology (ReCCIT),
Faculty of Engineering, Bangkok, Thailand, E-mail: kkveerac@kmitl.ac.th

Abstract-This paper presents the design technique of two parallel inverters. The parallel inverter can be shared the power with efficiently. The proposed technique is presented using the circular chain control (3C) to manage two parallel inverters operation. The inverter control is used by dsPIC 30F4011. The two set of 50 watt inverter are prototyped. The flowchart of operating condition is proposed to maintain the power sharing and the system efficiency. The simulated result using Simulink program confirms the agreement with the experimental results.

I. CIRCULAR CHAIN CONTROL (3C)

The parallel inverter is proposed that these are parallel with circular chain control [1], which makes the parallel inverter to share power with properly, equal current distribution. Additional, the parallel inverter is very stability. The last one module is tracked by the first module to form with circular chain connection. For the first, it is controlled by the last module too. The connections of current loop are circle loop. With 3C technique, the output voltage and out current of each inverter are controlled by an outer voltage loop and an inner current loop, respectively. The proposed control method with the inner loop is a fast dynamic response. Hence, the frequency amplitude and phase can be synchronized, and the current from parallel inverter can be distributed among inverter with properly [1]. The regulation with circular chain control (3C) is related each inverter, which makes each inverter to control its current that the current of each inverter equal with the current last module.

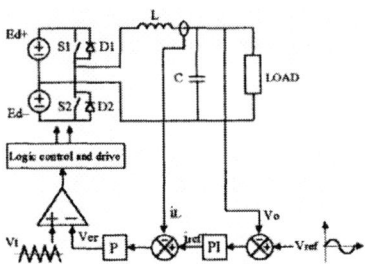

Fig. 1 Control of inverter one module.

In design of parameters in outer loop, It can be determined all parameter. But the design must be carefully, so, the conditions are the slop of feedback signal from inductor which must less than slop of a triangle waveform for comparing.

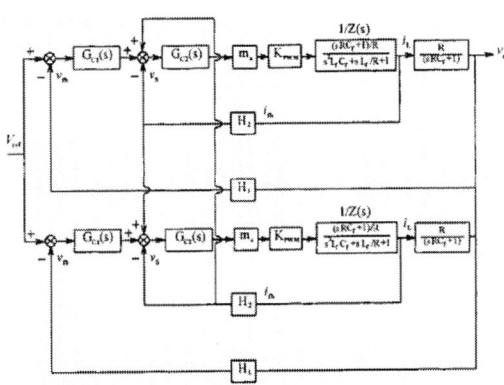

Fig. 4 Block diagram of parallel inverter.

The block diagram of the parallel inverter is represented. Further, the control is added to the controller which can be measured the load condition. Then can be decision to switch ON/OFF inverter that the power is enough for the load. For the two inverters, the capacitor of each inverter is 50 watt, therefore, it can be determined for start on the two inverters at the same time. Because the load with connected at the circuit can be known. At the same time, the microprocessor is received the voltage and current from sensor system for evaluating. If the load not exceed 50 watt, only one inverter is on. The load is between 50 to 100 watts, two inverters are on. The loads above 100 watt, two inverters are close for safety. These are can be represented the procedures and conditions as show in Fig. 5.

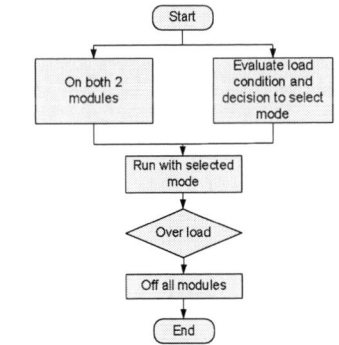

Fig. 5 The procedure of parallel inverter.

If the conditions load,
- Load   0-50   watts   module 1   on
- Load   51-100 watts   module 1, 2 on
- Load   > 100   watt   module 1, 2 off

II. SIMULATED RESULTS

The simulation is used MATLAB Simulink 7.2, which has the details as follow:

- The control of parallel inverter two modules with circular chain control (3C).
- The capability of each parallel inverter is 50 watt, bipolar.
- Voltage 48 VAC, $M_a = 0.8$
- IGBT, switching frequency 20 kHz
- The low-pass filter, $L = 820\ \mu H$, $C = 12$ pF

The simulation with MATLAB Simulink 7.2 shows as Fig. 6, and Fig. 7.

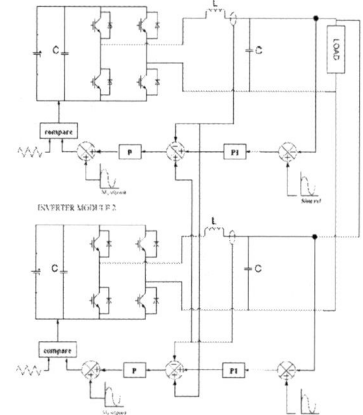

Fig. 6 Circular Chain Control (3C).

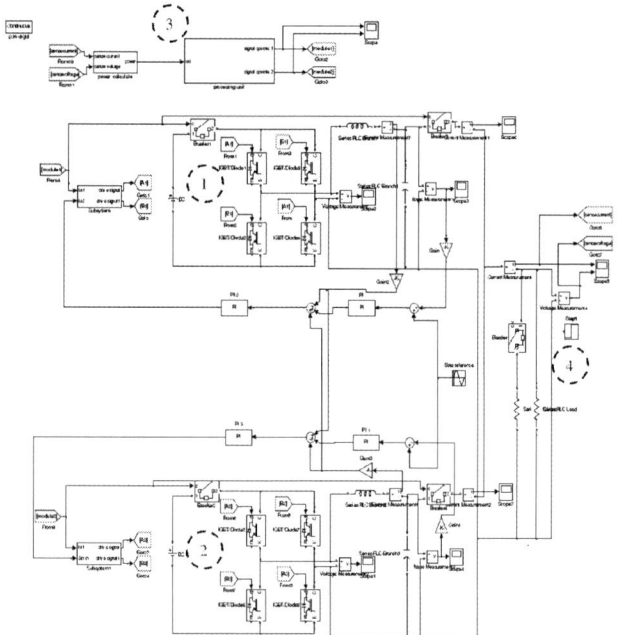

Fig. 7 The simulation of the parallel inverter.

Fig. 7 shows the simulation of the parallel inverter as follow: Part 1 is inverter 1, Part 2 is inverter 2, Part 3 is control system, and Part 4 is the load with 40 watt, 80 watt

- Test load at 40 % of the rated (40 watt)

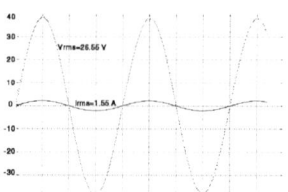

Fig. 8 The voltage and current at the load 40 watt.

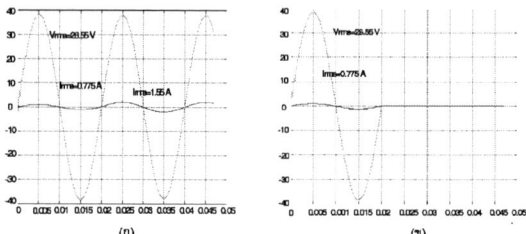

Fig. 9 The voltage and current of inverter (a) inverter 1. (b) inverter 2.

- Test load at 40-80 watts

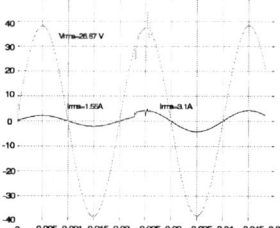

Fig. 10 The voltage and current of inverter at 40-80 watts.

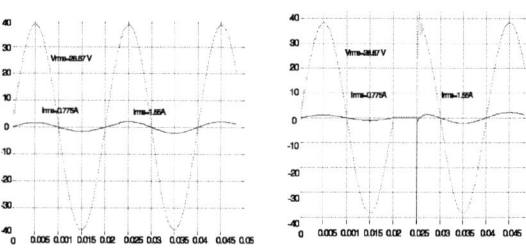

Fig. 11 The voltage and current of inverter at 40-80 watts (a) inverter1. (b) inverter 2.

- Test condition: The load changes the current at last inverter, which follows first inverter (the old current is $0.6\ A_{peak}$). It is change to $1.2\ A_{peak}$ at inverter 1. So, inverter 2 has fast response that it can be change the output current follows the current inverter 1 as shown in Fig. 12.

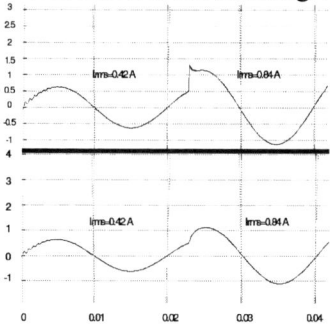

Fig. 12 The current of inverter 1 is changed, which makes the inverter 2, change follow the inverter 1.

The simulation with MATLAB Simulaink 7.2 can be showed that the stability of this parallel inverter, which is analysis only one inverter. Additional, the condition of load is verified, the parallel inverter can be regulated the current with properly.

III. PRACTICAL EXPERIMENTS

The proposed parallel inverter consists 6 parts.

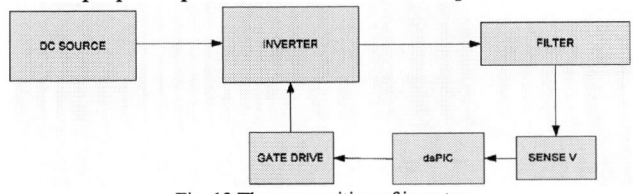

Fig. 13 The composition of inverter

A. DC source voltage

The DC source voltage is used from variac. Then, the rectifier is used for convert to Vdc, 48 V, which used diode bridge module KBPC35-08, and the capacitor 4000 mF with electrolytic and the resistor 5 ohm, 1 kilowatt with ceramic are connected to the circuit.

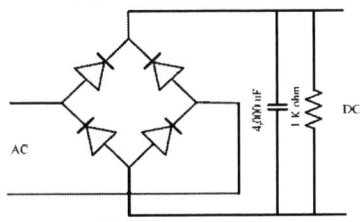

Fig. 14 The rectifier

B. Inverter module

The inverter is single phase (full bridge), used the IGBT HGTB30N60C3D

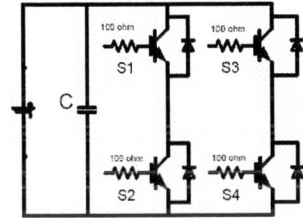

Fig 15 The inverter one module.

The detailed schematics of each module are shown as
- The input dc voltage source $\left(V_d\right)$ is average 48-60 V (normally 48 V).
- The input current $\left(I_d\right)$ is 1.3 A.
- The input power $\left(P_d\right)$ is 62.5 W
- The output voltage $\left(V_{o1}\right)$ is 27.15 V
- The output current $\left(I_{o1}\right)$ is 2.3 A
- The output power $\left(P_o\right)$ is 50 W
- The efficiency of system $\left(\eta\right)$ is 80 %

- The power factor (PF) is 0.8

C. Low pass filter

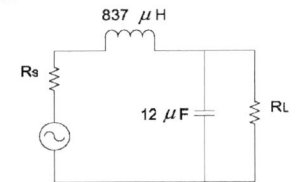

Fig. 16 The parameters of low pass filter.

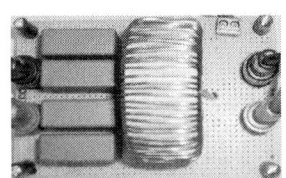

Fig. 17 The low pass filter connected with circuit.

D. Voltage probe and Current probe

The signal detector has 3 parts, and can be classified to 2 groups, which are the current probes for each inverter and the voltage probe for system.
- The voltage prove is used that it used potential transformer for detecting the voltage. Then, the signal from voltage detector is send to amplifier for amplifying. Next, the offset of signal up the level to 2 V., then, the signal is send to adjust the signal form for level voltage to proper with port of the controller, and the signal is only positive which used the diode to adjust the signal, finally, the signal from output diode is send to microcontroller for evaluating.

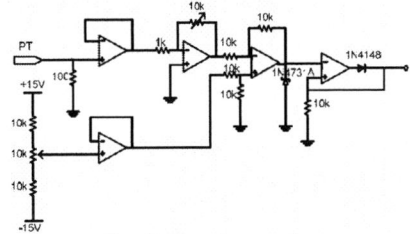

Fig. 18 The voltage probe.

The signal is detected by current probe at only output of system. The signal from detector is then to calculate the output power and the load condition. The current at rated is not exceed 25 A. So, the current probe is used, (LEM number LA25-NP), then, the ratio of input current with output current is selected as 1000:1. Afterwards, the signal from the detector is send to amplifier for amplifying and adjusts the offset.

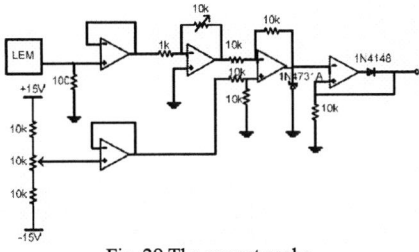

Fig. 29 The current probe.

E. Pulse modulator

The pulse modulator has received signal for regulating the switch from microcontroller, which used SPWM (Sinusoidal Pulse-Width Modulation), then, the pulse is amplified by IR2110, which is shown as Fig. 30. Afterwards, the pulse is send to gate of IGBT.

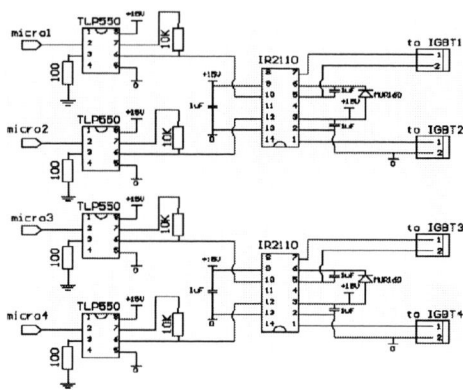

Fig. 30 The pulse modulator.

F. Microcontroller (dsPIC)

The control of the parallel inverter is use dsPIC 30F4011, which used the C language for writing the program, it is to control whole PWM modulator and the current probe and the voltage probe for calculate the load and decision the work mode, as shown in Fig. 31.

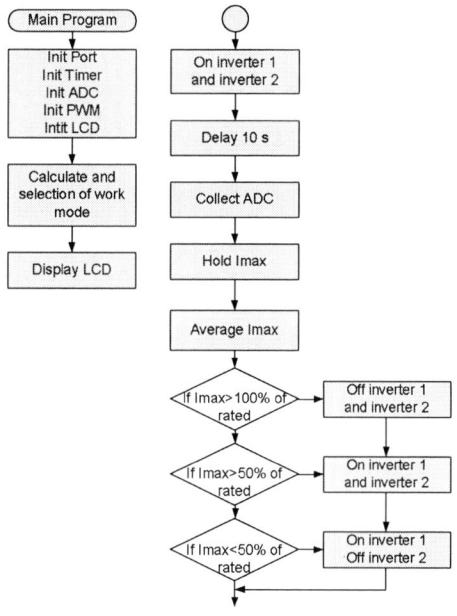

Fig. 31 The flowchart for evaluate of the main controller.

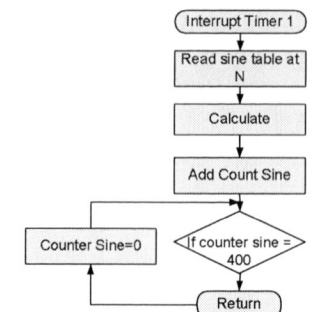

Fig. 32 The flowchart of the interrupt program.

IV. EXPERIMENTED RESULTS

- First experiment: Tests of each inverter while connecting with no load and load 30, 55, 100 and 145 watts ($m_a = 0.8$).

Inverter 1

No load

m_a	Output Power						
	P (W)	S (VA)	Q (VAR)	PF	DPF	THD_i (%)	THD_v (%)
0.8	0	0	0	0.99	0	95.7	0.9

Load 30, 55, 100, 145 watts ($m_a = 0.8$)

Load (W)	Input			Output				η
	$V_d(V)$	$I_d(A)$	$P_d(W)$	$V_L(V)$	$I_L(A)$	$cos\theta$	$P_o(W)$	(%)
30	48.2	0.5	24.1	21.01	0.94	0.99	19	78.84
55	48	0.9	43.2	20.42	1.52	1	30	69.44
100	47.8	1.5	71.7	19.72	2.62	1	51	71.13
145	47.9	2	95.8	19.35	3.73	1	71	74.11

Load 30, 55, 100, 145 watts

Load (W)	Output Power						
	P (W)	S (VA)	Q (VAR)	PF	DPF	THD_i (%)	THD_v (%)
30	19	19	1	0.99	1	6.2	6.8
55	31	30	2	1	1	8.2	8.7
100	51	51	2	1	1	10.6	10.7
145	71	71	3	1	1	11.7	11.9

Inverter 2

No load

m_a	Output Power						
	P (W)	S (VA)	Q (VAR)	PF	DPF	THD_i (%)	THD_v (%)
0.8	0	0	0	0.29	0.5	99.2	1.2

Load 30, 55, 100, 145 watts ($m_a = 0.8$)

Load (W)	Input			Output				η
	$V_d(V)$	$I_d(A)$	$P_d(W)$	$V_L(V)$	$I_L(A)$	$cos\theta$	$P_o(W)$	(%)
30	48.0	0.5	24	20.88	0.94	0.99	19	79.17
55	47.9	0.9	43.11	20.61	1.57	1	30	69.59
100	47.7	1.5	71.55	19.66	2.66	1	51	71.27
145	47.5	2	95	19.29	3.78	1	71	74.73

Load 30, 55, 100, 145 watts

Load (W)	Output Power						
	P (W)	S (VA)	Q (VAR)	PF	DPF	THD_i (%)	THD_v (%)
30	19	19	1	0.99	1	6.5	6.8
55	31	30	2	1	1	8.8	8.7
100	51	52	3	1	1	10.4	10.7
145	71	71	4	1	1	11.6	11.8

- Second experiment: Test at same load condition of one module and two modules at load 30, 55, 100, 145 watts ($m_a = 0.8$).

One module

Load (W)	Input			Output				η
	$V_d(V)$	$I_d(A)$	$P_d(W)$	$V_L(V)$	$I_L(A)$	$cos\theta$	$P_o(W)$	(%)
30	48.2	0.5	24.1	21.01	0.94	0.99	19	78.84
55	48	0.9	43.2	20.42	1.52	1	30	69.44
100	47.8	1.5	71.7	19.72	2.62	1	51	71.13
145	47.9	2	95.8	19.35	3.73	1	71	74.11

Load (W)	Output Power						
	P (W)	S (VA)	Q (VAR)	PF	DPF	THD_i (%)	THD_v (%)
30	19	19	1	0.99	1	6.2	6.8
55	31	30	2	1	1	8.2	8.7
100	51	51	2	1	1	10.6	10.7
145	71	71	3	1	1	11.7	11.9

Two modules

Load (W)	Input			Output				η
	$V_d(V)$	$I_d(A)$	$P_d(W)$	$V_L(V)$	$I_L(A)$	$cos\theta$	$P_o(W)$	(%)
30	48.3	0.65	28.86	21.71	1	0.99	22	76.23
55	48.1	1	48.2	21.10	1.62	1	34	70.54
100	47.5	1.5	74.24	20.6	2.82	1	57	76.78
145	46.7	2	97.78	19.55	3.87	1	77	78.75

Load (W)	Output Power						
	P (W)	S (VA)	Q (VAR)	PF	DPF	THD_i (%)	THD_v (%)
30	22	22	2	0.99	1	3.7	3.8
55	34	34	3	1	1	6	5.8
100	57	57	4	1	1	7.8	8
145	77	77	5	1	1	9.3	9.3

- Third experiment: Test of the parallel inverter at load 30, 55, 100, 145 watts.

The results of the parallel inverter while the load changing.

Load (W)	Input			Output				Work Mode
	$V_d(V)$	$I_d(A)$	$P_d(W)$	$V_L(V)$	$I_L(A)$	$cos\theta$	$P_L(W)$	
30	48.3	0.55	24	21.02	1	0.94	19	Inverter 1
55	47.9	0.9	38.72	20.44	1.56	1	31	Inverter 1
100	46.5	1.5	72.3	19.68	2.7	1	52	Inverter 1 Inverter 2
145	0	0	0	0	0	0	0	Overload (Stop)

Load (W)	Inverter 1				Inverter 2			
	$V_1(V)$	$I_1(A)$	$cos\theta_1$	$P_1(W)$	$V_2(V)$	$I_2(A)$	$cos\theta_2$	$P_2(W)$
30	21.08	0.94	0.99	19	0	0	0	0
55	20.44	1.56	1	31	0	0	0	0
100	19.68	1.28	0.99	23	19.68	1.32	0.99	24
145	0	0	0	0	0	0	0	0

Load (W)	Output Power						
	P (W)	S (VA)	Q (VAR)	PF	DPF	THD_i (%)	THD_v (%)
30	19	19	1	0.99	1	6.2	6.7
55	31	31	2	1	1	9.3	8.8
100	52	52	3	1	1	7.8	8.1
145	0	0	0	0	0	0	0

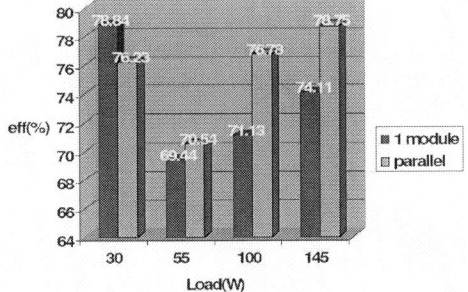

Fig. 33 The efficiency between one module and parallel.

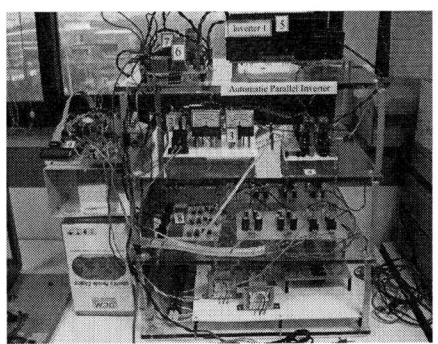

Fig. 34 The automatic parallel inverter.

The practical results are compared between one module, two modules and the parallel inverter. The one module and two modules have best efficiency when the load is not exceeded the rated of one inverter, but the parallel inverter has efficiency better than one module and two modules when the load exceeded the rated of one inverter.

V. CONCLUSIONS

The parallel inverter is proposed that each inverter can be shared the current with properly, fast dynamic response, high stability. The controller of parallel inverter can be controlled to distribute the power with load conditions. The simulation results is simulated when compared with the practical experiment, the performance of prototype are small different with the simulation result.

VI. ACKNOWLEDGEMENT

I am deeply indebted to my Assoc. Prof. Dr. W. Khanngern whose help, stimulating suggestions and encouragement helped me in all the time of research for and writing of this paper. I would like to express my gratitude to all those who gave me the possibility to complete this paper. Thank you for N. Wongvasupongsa, P. Madro, P. Srilaong, P. Munmai.

VII. REFERENCES

[1]. Tsai-Fu Wu, Yu-Kai Chen and Yong-Heh Huang "3C Strategy for Inverters in Parallel Operation Achieving an Equal Current Distribution," IEEE TRANSACTIONS ON INDUSTRIAL ELECTRONICS, VOL. 47, APRIL 2000.

[2]. วีระเชษฐ์ ขันเงิน, วุฒิพล ธาราธีรเศรษฐ์ "อิเล็กทรอนิกส์กำลัง (Power Electronic)", พิมพ์ครั้งที่ 1, 2547.

[3]. H. Van Der Broeck, U. Boeke, "A simple method for parallel operation of inverters," 1998, IEEE, pp. 143-150.

[4]. Ai Emadi , Abdolhosein Nasiri , Storan B.Bekiarov "Uninterruptiable Power Supply and Active Filter," first edition , CRC PRESS.

[5]. Hongying Wu, Dong Lin, Dehua Zhang, Kaiwei Yao, Jinfa Zhang "A Current-Mode Control Techniquewith Instantaneous Inductor-Current Feedback for UPS Inverters" IEEE Trans., On Power Electronic, 1999,pp 951-957.

[6]. Duan Shanxu, Meng Yu, Xiong Jian, Kang Yong, Chen Jian " Parallel Operation Control Technique of Voltage Source Inverters in UPS ," IEEE Trans., On Power Electronic , July 1999, pp.883-887.

Investigation of Topologies of Low Voltage Multilevel Inverters

Yanli Zhang, Wanmin Fei, and Shoufang Wang

School of Electrical and Automation Engineering, Nanjing Normal University, 78 Bancang Street, Nanjing, P.R. China

Abstract—This paper research on the topologies of main circuit of low-voltage high-performance power inverters. Two new topologies of multilevel inverter are proposed. Inverters with the proposal topologies share the advantages of simplicity of circuit and control method, fewer controllable semiconductor devices involved, high quality of output waveforms, high efficiency and so on. Simulations using PSIM 4.0 of the proposed adding-bidirectional-switch multilevel inverter based on multi-carrier SPWM method are carried out which proved the validity and practicability of the proposed topology.

Index Terms--low-voltage, topology, Main circuit, multilevel converter.

I. INTRODUCTION

Multilevel converter has become increasingly popular in recent years mainly due its high-voltage capacity achieved by employing devices with limited voltage rate [1]–[6]. The advantages of high quality waveforms, low electromagnetic compatibility (EMC) concerns, less switching losses are also great merits in low voltage power conversion situations. Audio amplifiers with great power are widely used in transmitter of television and radio stations, in acoustics facility and so on. If technology of multilevel power conversion is employed in audio amplifiers, power amplification of very high quality can be achieved, which will result in a completely new kind of power amplifier. Multilevel power conversion techniques used in low voltage ac motor driving systems can reduce the torque pulsation at low frequencies, increase the reliability of the bearing coil, decrease the size of output filter, all in all, can greatly enhance the performance of low voltage ac driving systems.

Topologies of multilevel converters employed in medium-high voltage systems, such as diode-clamping multilevel converters, flying capacitor multilevel converters, cascade multilevel converters and hybrid topologies composed of the three basic topologies, have some drawbacks when being used in low voltage power conversion systems. In order to achieve high-voltage capacity with devices of limited voltage rate, several devices in block state must be connected in series at any time, which results in the great number of devices employed. On the other hand, in the topology listed above, at least four devices in every phase are in conduction state at any time of energy output, this will

result in low efficiency of power conversion, especially in low voltage situation.

Aim to research on the applying of multilevel converter to low voltage circumstance, this paper proposed two topologies of multilevel voltage inverter, which has the advantages of high efficiency and simplicity in circuit and control method. The work mode of the proposed appending-bidirectional-switch multilevel converter is analyzed, Simulations using PSIM 4.0 based on multi-carrier SPWM method are carried out which proved the validity and practicability of the proposed topology.

II. TOPOLOGIES OF LOW VOLTAGE MULTILEVEL VOLTAGE INVERTERS

The basic topology of a multilevel voltage inverter is shown by fig 1, where the voltage levels can be obtained by several storage capacitors connected in series. Two or more than two switches in on state will cause short-circuit fault, so it should be assured that no more than one switch be in on state at any time. Different voltage level can be obtained on the load by turning on different switch. Fig 2 is the realization of Fig 1 according to the directions of the currents in the switches. Current in switches such as S1, S2, S3, S5, S6, S7 is in a single direction. Current in S4 is bidirectional, so a composite structure of bidirectional switch is adopted. Diode D1, D2 is employed to clamp the voltage across the devices. Let C1=C2=C3=C4=C5=C6, then voltages across the storage capacitors will be equal, that is Ud. Turn on S1 and turn off all the others, the voltage on the load will be 3Ud. Turn on S7 and turn off all the others, a voltage of -3Ud can be obtained. Other cases can be analyzed similarly.

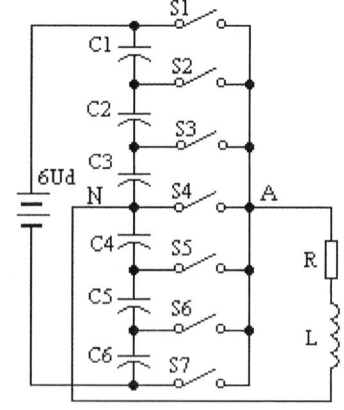

Fig. 1. Principle of the basic multilevel voltage inverter for low voltage power conversion .

This work was supported by natural science foundation of education department of Jiangsu province(No.2005111TSJB154)

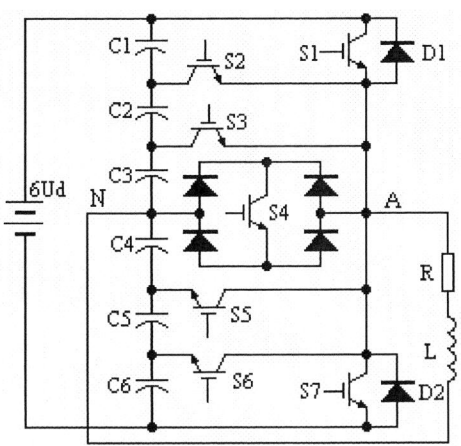

Fig. 2. Realization of the basic multilevel voltage inverters for low voltage power conversion

The proposed topology of multilevel voltage inverter is called switching-selecting multilevel inverter according to its principle. Because only one switch needed to be turned on at one time, the efficiency of proposed switch-selecting multilevel inverter is very high compared to inverters of other topologies.

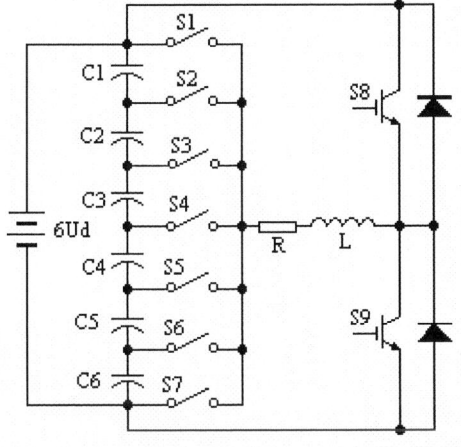

Fig. 3. Improved topology of the proposed low voltage multilevel voltage inverters

Circuit and control method of Switch-selecting multilevel inverter are very simple. Turn on one switch of

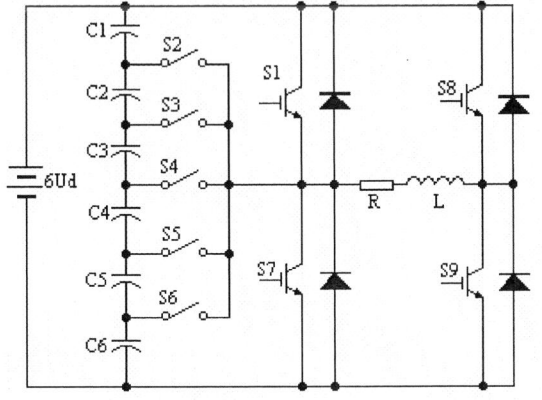

Fig. 4. Topology of the adding-bidirectional-switch multilevel voltage inverter

S1, S2 and S3 and turn off all the other switches, a positive voltage will be obtained on the load. Similarly, turn on one switch of S5, S6 and S7 and turn off all the other switches, a negative voltage will be obtained. Turn on S4 and turn off all the other switches zero voltage will be obtained. So peak value of the output voltage is 3Ud, half of the DC side voltage. If switch N point in Fig 1 between positive and negative poles of DC supply by a pair of switches, the peak value of the output voltage and the number of voltage levels will be increased greatly, topology obtained thus is shown by Fig.3. Considering the directions of the currents flowing in the switches S1~S7, switch S1 and S7 are single-directional, switches S2~S6 are bidirectional, so the topology shown by Fig 3 can be embodied by Fig 4, which can be considered as a bridge circuit with several bidirectional-switch and storage capacitors appended and be called adding-bidirectional-switch multilevel voltage inverter. For a adding-bidirectional-switch multilevel voltage inverter with N output voltage levels, 4+(N-3)/2 controllable electronic devices are needed, which is much less than that of other topologies.

Compared to switch-selecting multilevel voltage inverter, adding-bidirectional-switch multilevel inverter has more voltage levels and higher magnitude of output voltage with the same number of controllable electronic devices employed. The quality of output voltage waveform of adding-bidirectional-switch multilevel voltage inverter is higher, and it can make full use of devices and DC side voltage. Because two controllable electronic devices are in on state at anytime of power output, the efficiency of adding-bidirectional-switch multilevel voltage inverter is slightly lower than that of switch-selecting multilevel voltage inverter. The topology shown by Fig 2 and Fig 4 has its own merits and demerits and can be selected according to the demand in practice.

Fig 5 is one of the realizations of adding-bidirectional-switch multilevel voltage inverter.

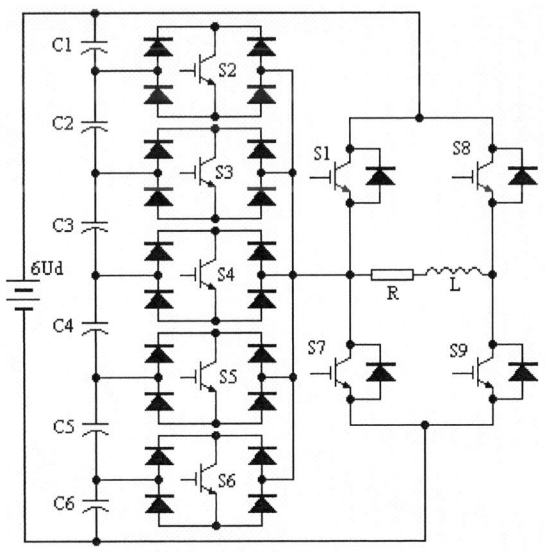

Fig. 5. Realization of the proposed add-bidirectional-switch multilevel voltage inverter

1673

III. CONTROL STRATEGY OF THE PROPOSED MULTILEVEL VOLTAGE INVERTER

Control method of switch-selecting multilevel voltage inverter is very simple: if we want a voltage of NUd, we can simply turn on switch S4-N and turn off all the other switches. Only the control method of adding-bidirectional-switch multilevel inverter is studied in this section.

A. Maximum value output and zero output modes

Turn off switches S2, S3, S4, S5, S6, adding-bidirectional-switch multilevel voltage inverter shown by Fig 4 or Fig 5 is simply a traditional two-level single phase inverter bridge. Turn on switch S1, S9 and turn off S7, S8, we can get 6Ud; Turn on switch S7, S8 and turn off S1, S9, we can get -6Ud; Turn on S1, S8 and turn off S7, S9, or turn on S7, S9 and turn off S1, S8, then zero voltage will be obtained.

B. Bidirectional switches work modes

Keep S9 in on state and S8 in off state, turn on SN (N=1, 2, 3, 4, 5, 6, 7) and turn off all the other switches, then a positive voltage of (7-N)Ud will be obtained, take N=3 for instance as shown in Fig 7 (a).

Keep S8 in on state and S9 in off state, turn on SN (N=1, 2, 3, 4, 5, 6, 7) and turn off all the other switches, then a negative voltage of (1-N)Ud will be obtained, take N=5 for instance as shown in Fig 7.

To sum up, for control of the adding-bidirectional-switch multilevel voltage inverter, it must be ensured that only one switch of S1, S2, S3, S4, S5, S6, S7 and only one switch of S8, S9 be in on state at anytime to avoid short-circuit fault. The states of the switches corresponding to different output voltage value are listed in table 1. From table 1 it can be seen that there are two switches in on state at any time, so the efficiency of adding-bidirectional-switch multilevel voltage inverter is higher than that of multilevel inverter employed in medium-high voltage systems. The number of controllable electronic devices is also much smaller and the main circuit is much simpler.

TABLE I
SWITCH STATE CORRESPONDING TO OUTPUT VOLTAGE

Output voltage	States of switches								
	S1	S2	S3	S4	S5	S6	S7	S8	S9
6Ud	1	0	0	0	0	0	0	0	1
5Ud	0	1	0	0	0	0	0	0	1
4Ud	0	0	1	0	0	0	0	0	1
3Ud	0	0	0	1	0	0	0	0	1
2Ud	0	0	0	0	1	0	0	0	1
Ud	0	0	0	0	0	1	0	0	1
0	0	0	0	0	0	0	1	0	1
0	1	0	0	0	0	0	0	1	0
-Ud	0	1	0	0	0	0	0	1	0
-2Ud	0	0	1	0	0	0	0	1	0
-3Ud	0	0	0	1	0	0	0	1	0
-4Ud	0	0	0	0	1	0	0	1	0
-5Ud	0	0	0	0	0	1	0	1	0
-6Ud	0	0	0	0	0	0	1	1	0

IV. SIMULATIONS

Simulations using PSIM 4.0 of the proposed adding-bidirectional-switch multilevel inverter based on multi-carrier SPWM method are carried out. SPWM mode is shown by Fig 7. Let Ud=1 and the amplitude of the reference sinusoidal voltage equal to 6 for convenience. Control signals of S8, S9 can be determined by the sign of reference sinusoidal voltage, turn on S9 and turn off S8 when the reference voltage is positive, turn on S8 and turn off S9 when the reference voltage is negative. According to the comparison of reference sinusoidal

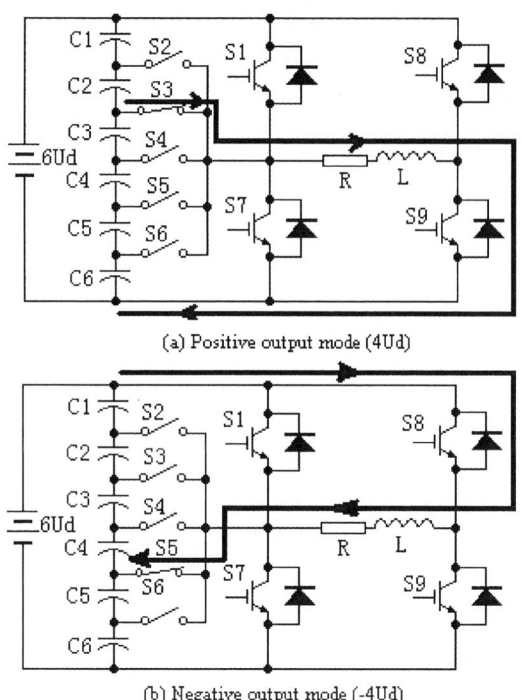

(a) Positive output mode (4Ud)

(b) Negative output mode (-4Ud)

Fig. 6. Bidirectional switch work mode

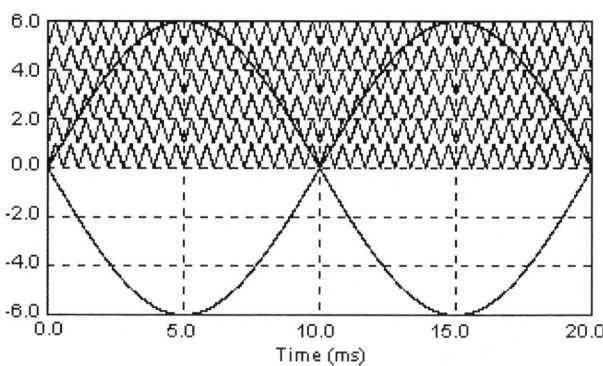

Fig. 7. SPWM based on multi-triangular carrier signals

1674

signal with triangular carriers, when the reference signal

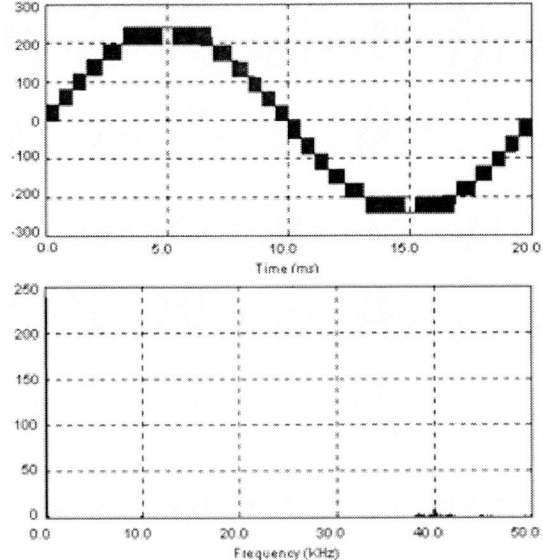

Fig. 8. Output voltage and its frequency spectrum

is between 0 and 1, turn on and turn off S7, S6 alternatively, when the reference signal is between 1 and 2, turn on and turn off S6, S5 alternatively, and so on. In the negative half cycle of the reference signal, when the absolute value of the reference signal is between 0 and 1, turn on and turn off S1, S2 alternatively, when the absolute value is between 1 and 2, turn on and turn off S2, S3 alternatively, and so on.

Let C1=C2=C3=C4=C5=C6=3300 μf, Ud=40V , R=24 ohm, L=1mH, fT=40KHz, then simulation waveform and its frequency spectrum are shown by Fig 8. Harmonics are concentrated around 40 kHz, the magnitudes of fundamental voltage and all harmonics is listed in table 2. The harmonic contents of output voltage are listed as follows:

TABLE II
LIST OF HARMONICS CONTENTS

Order	1st	761th	768th	770th	775th	777th
magnitude	239	0.8	1.5	2.4	0.8	1.8
Order	783th	787th	789th	793th	795th	797th
magnitude	2.1	1.0	1.76	3.0	1.0	4.4
Order	799th	801th	803th	805th	807th	811th
magnitude	10.0	8.7	4.0	0.75	3.0	1.56
Order	813th	817th	823th	825th	831th	833th
magnitude	0.92	2.19	1.56	1.0	2.29	1.73

Voltage waveforms on the controllable electronic devices are shown by Fig 9.

It can be seen that the output voltage waveform is excellent when the frequency of the carrier wave is 40KHz. Because the voltage varies only one level (here is 40V) every time, the frequency of the controllable electronic device can be higher. Variation of the voltage of the storage capacitors is also a problem as in diode-clamping multilevel inverter. The extent of unbalance of capacitor voltage has close relation with the capacitance of the storage capacitor and load current. In great power

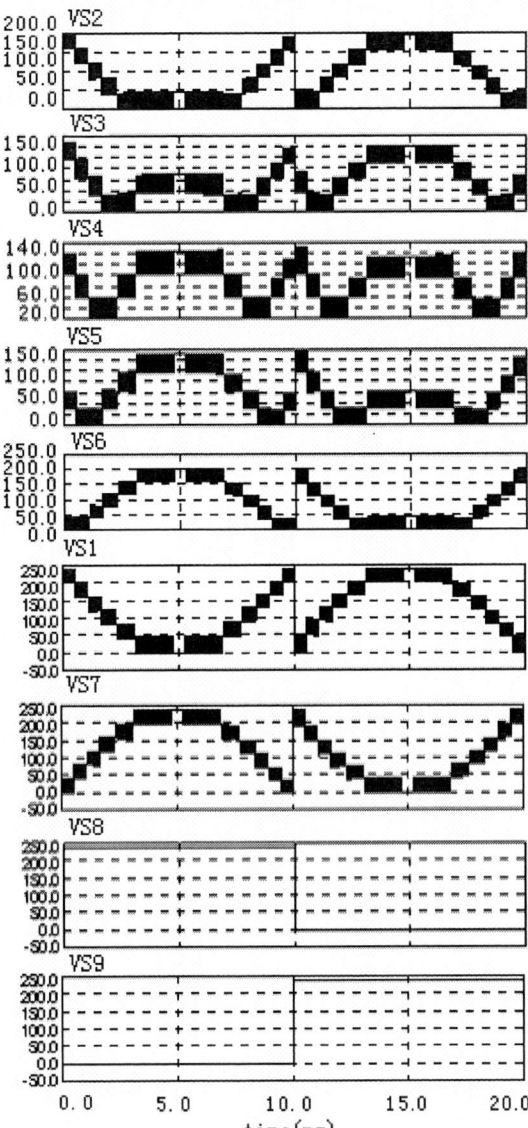

Fig. 9. Voltage across the controllable semiconductor devices

audio amplifiers, unbalance of capacitor voltages can be decreased by increase the capacitance of the storage capacitors as low voltage electrolytic capacitor is inexpensive. For three-phase application such as high performance variable frequency ac driving systems, besides increase the capacitance of the storage capacitors, the existing method for the control of the balance of capacitor voltages employed in medium-high voltage multilevel voltage inverters can also be adopted. In a word, the topologies of multilevel voltage inverters applicable to low voltage situations proposed in this paper are available.

V. CONCLUSION

Two topologies of multilevel voltage inverter applicable to low voltage power conversion is proposed in this paper. Advantages of Inverters with the proposed topology over existing ones include high efficiency, fewer controllable semiconductor devices involved, simplicity of circuit and control method and so on. Simulations using PSIM 4.0 of the proposed adding-bidirectional-switch multilevel inverter based on multi-carrier SPWM method are carried out which proved the validity and practicability of the proposed topology.

REFERENCES

[1] Leon M. Tolbert, Fang Z. Peng. Multilevel Converters for Large Electric Drives. Apecs 2000, Pages: 530~536

[2] Jih-Sheng Lai , Fang Zheng Peng. Multilevel Converters- A New Breed of Power Converters. IEEE Transactions on Industry Applications. Vol.32, No.3, MAY/JUNE 1996, Pages: 509~517

[3] Lai, Y.S.; Shyu, F.S. New topology for hybrid multilevel inverter. International Conference on Power Electronics, Machines and Drives, 2002. Pages: 211 – 216

[4] José Rodríguez, Jih-Sheng Lai, and Fang Zheng Peng. Multilevel Inverters: A Survey of Topologies, Controls, and Applications. IEEE Transactions on Industrial Electronics, Vol. 49, No. 4, AUGUST 2002, Pages: 724-738

[5] Rech, C, Grundling, H.A at el. A generalized design methodology for hybrid multilevel inverters. IEEE 2002 28th Annual Conference of the Industrial Electronics Society, Volume: 1, 5-8 Nov. 2002, Pages: 834 – 839

[6] Keith Corzine, Yakov Familiant. A New Cascaded Multilevel H-Bridge Drive. IEEE Trans. on Power Electronics, Vol.17, No. 1, January 2002, Pages: 125~131

Solution for PWM converter switching for Voltage Source Inverter using Non-Traditional Method

V. Jegathesan*, Dr. Jovitha Jerome**, *Member, IEEE*
*Karunya University, ** PSG College of Technology
Coimbatore, Tamilnadu, India

Abstract—This paper presents an efficient and reliable evolutionary-programming-based algorithm for specific harmonic elimination (SHE) switching pattern. This method eliminates the considerable amount of lower order line current harmonics in Pulse Width Modulation (PWM) inverter. Determination of pulse pattern for the elimination of some lower order harmonics of a PWM inverter necessitates solving a system of nonlinear transcendental equations. Evolutionary Programming is used to solve nonlinear transcendental function for PWM-SHE. In this proposed method 11th and 13th harmonics are eliminated using Evolutionary Programming. Simulation results using Matlab/Simulink are carried out to validate the solution.

Index Terms— Converter, Harmonics, Evolutionary Programming (EP), Pulse Width Modulation (PWM), Selective Harmonic Elimination (SHE).

I. INTRODUCTION

Pulse width modulation has been the subject of intensive research during the last few decades. Their implications in the design of AC drive systems depend on the machine type, the power level and the semiconductor used in the power converters. Different types of feed forward and feed backward pulse width modulation schemes having relevance for industrial application have been widely discussed [1].

The usage of power electronic equipments has been increased in recent years in industrial and consumer applications. Such loads draw the nonlinear sinusoidal current and voltage from the source and results in the harmonic in the networks [2]. They occurs frequently in variable frequency drives or any electronic devices using solid state switching to convert including AC or DC.

The characteristic harmonics (*h*) are based on the number of the rectifiers (pulse number) used in the circuit and can be determined by the equation (1).

$$h = (n \times p) \pm 1 \qquad (1)$$

where, *n* is an integer
　　　　p is pulse of rectifier.
For 12 pulse rectifier, the characteristic harmonics will be 11th & 13th and 23rd & 25th.

The undesirable lower order harmonics of a square wave can be eliminated and then fundamental voltage is controlled which is known as specific harmonic elimination (SHE). In SHE method, notches are created on the square wave at predetermined angles to eliminate the significant harmonic components and control the fundamental voltage.

Programmed PWM eliminating lower-order harmonics [3] generates high quality output spectra, which in turn results in minimum current ripples, thereby satisfying several performance criteria and contributes to overall improved performance. Performance characteristic of a rectifier/inverter power conversion scheme largely depends on the choice of the particular Pulse width modulation strategy employed. Programmed PWM techniques optimize a particular objective function, such as obtain selective elimination of harmonics and therefore are the most effective means of obtaining high performance.

An optimized PWM technique is proposed in [4] to reduce the harmonic distortion and to spread the harmonic energy.

Optimization algorithms are becoming increasingly popular in engineering design activities, where the emphasis is on the maximizing or minimizing a certain goal primarily because of the availability and affordability of high speed computers. They are extensively used in engineering design problems

The minimization of objective function used for the objective function used for SHE was done using traditional mathematical techniques such as Conjugate Gradient method (CGD) [5] and Newton's method (NR) [6]. These methods needs initial values to obtain the objective function and are based differential information, so they may produce local minimum solution which leads to undesirable pattern.

GA provides solution to nonlinear mathematical problems. GAs is inspired by the mechanism of natural selection, in which stronger individuals are likely to survive in a competing environment. GAs uses a direct analogy of such selection. In [7], GA is applied to eliminate the lower order harmonic in power converter. The 5th and 7th harmonic can be eliminated by using dual transformer. The 3rd and other triple harmonics can be ignored if the machine has an isolated neutral.

GA and CGD method are used to find the switching pattern for SHE to eliminate rectifier low input harmonics without having any initial guess for the switching pattern. The GA is used to provide the initial values [8].

978-1-4244-0644-9/07/$25.00 ©2007 IEEE

NR and GA are adopted to reduce the lower order line current harmonic (up to 15th) by developing the 'n' number of pulse per half cycle [9].

EP is a technique in the field of evolutionary computation. It is a powerful and general global optimization method which does not depend on the first and second differentials of the objective function of the problem to be optimized. The EP technique is based on the mechanism of natural selection [10].

The power circuit for Voltage Source Converter Drive system is given in Fig. 1. DC voltage is obtained by using 12 pulse Voltage Source Rectifier. The rectifier is connected to Voltage Source Inverter through DC link capacitor and AC source terminal exhibits current source characteristic by synchronous link inductor.

This paper presents a new method of line current harmonic reduction in PWM ac/dc converter using EP. The 11th and 13th harmonics are the characteristic harmonics are required to be eliminated. The objective is achieved by determining the switching pattern for the rectifier using EP. Simulation results are carried out and validated using MATLAB/SIMULINK.

Equation (2) has N variables $\left(\alpha_1 \, to \, \alpha_N \right)$ and a set of solution are obtained by equating N-1 harmonics to zero and assigning a specific value of the fundamental amplitude α_1, through the equation (3).

$$f_1(\alpha) = \frac{4}{\pi}[-1 - 2\sum_{k=1}^{n}(-1)^k \cos(\alpha_k)] - M = \varepsilon_1$$

$$f_2(\alpha) = \frac{4}{5\pi}[-1 - 2\sum_{k=1}^{n}(-1)^k \cos(5\alpha_k)] = \varepsilon_2 \qquad (3)$$

$$\cdots\cdots \qquad\qquad \cdots\cdots \qquad \cdots\cdots$$

$$f_N(\alpha) = \frac{4}{N\pi}[-1 - 2\sum_{k=1}^{n}(-1)^k \cos(n\alpha_k)] = \varepsilon_N$$

Where, the variables $\varepsilon_1 - \varepsilon_N$ are the normalized amplitude of the harmonics to be eliminated. The objective function of PWM-SHE technique is to minimize the harmonic content in the inverter line current and it is given in equation (4).

$$F(\alpha_1, \alpha_2, \alpha_3, \ldots, \alpha_N) = \varepsilon_1^2 + \varepsilon_2^2 + \ldots + \varepsilon_N^2 \qquad (4)$$

Subjected to the constraint equation (5),

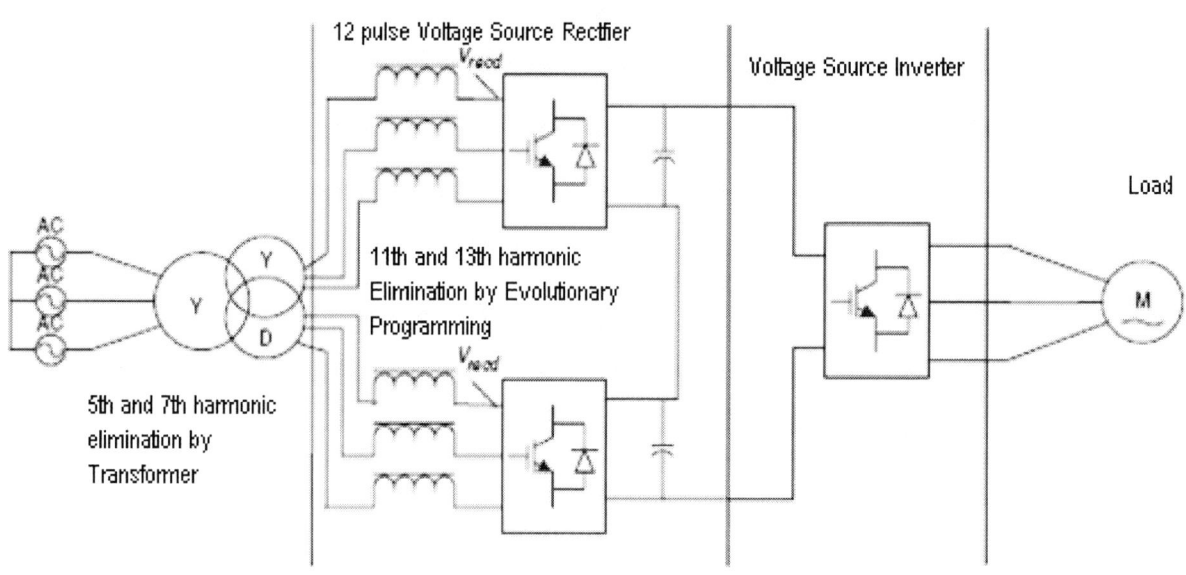

Fig. 1 Power circuit of Voltage Source Converter Drive System

II. PWM-SHE SWITCHING TECHNIQUES

The Fourier coefficients of the PWM-SHE switching pattern for a three phase line to neutral are given by the equation (2).

$$a_n = \frac{4}{n\pi}[-1 - 2\sum_{k=1}^{N}(-1)^k \cos(n\alpha_k)] \qquad (2)$$

$$b_n = 0$$

$$0 < \alpha_1 < \alpha_2 < \alpha_3 < \alpha_4 < \alpha_5 \ldots \alpha_N < \frac{\pi}{2} \qquad (5)$$

for Quarter-wave symmetric pulse pattern. In the proposed method $\alpha_1, \alpha_2, \alpha_3, \alpha_4 \, and \, \alpha_5$ solutions are expected with elimination of 5th, 7th, 11th, 13th and 17th harmonics.

III. EVOLUTIONARY PROGRAMMING METHOD TO SOLVE THE PROPOSED PWM SWITCHING PATTERN

Evolutionary Programming (EP) is powerful global optimization technique. EP has proved its effectiveness to handle to handle complex optimization problem. EP starts with a population of randomly generated candidate solution that evolves towards the better solution over a number of iterations. It uses the probabilistic rules to explore the complex search space.

Evolutionary programming is a probabilistic search technique, which generates the initial parent vectors distributed uniformly in intervals within the limits and obtains global optimum solution over number of iterations. The main stages of this technique are initialization, creation of offspring vectors by mutation, competition and selection of best vectors to evaluate best fitness solution. In this paper, an attempt has been made to determine the most optimal switching pattern to eliminate the lower order low line current harmonic in the Voltage Source Inverter. The implementation of EP algorithm is given below.

A. Initialization

The initial population (P_i) is generated after satisfying the equation (5) with randomly selected initial individual switching angle. The generated switching angles are distributed uniformly between their minimum and maximum limits by satisfying the equation (5).

B. Fitness of the candidate solutions

The Fitness Factor (FF) in this case attempts to minimize the error between the actual angles to the exact angles of the same, which is assumed to be the alpha values of the corresponding objective function and it is given in equation (6).

$$FF = \cfrac{K}{\int_0^1 (\alpha_{ref} - \alpha_{cal})^2 \, dt}$$

(6)

Where, K is an integer

α_{ref} is the average switching angle of the ith generation

α_{cal} is the specific switching angle of the jth individual

C. Producing new solutions by Mutation

An off-spring vector P_i' using the equation (7) is created from each parent vector by adding Gaussian random variable with zero mean and standard deviation σ_i, denoted as $N(0, \sigma_i^2)$.

$$P_i' = P_i + N(0, \sigma_i^2) \quad for \quad i = 1, 2, \ldots, Np-1 \quad (7)$$

$$\text{Where,} \quad \sigma_i = \beta \frac{f_i}{f_{min}} (P_{i-max} - P_{i-min})$$

and $N(0, 1)$ represents a Gaussian random variable with mean 0 and standard deviation 1.
Where, N_p is the number of population
β is the scaling factor
f_i is the fitness of the ith individual.
f_{min} is the minimum fitness of the entire population.

The created offspring vector must satisfy the minimum and maximum generation limits of the units. After adding a Gaussian random number to the parents, the element of offspring may violate the constraint equation given by (5). These violations are dealt with the equation (8).

$$P_i' = \begin{cases} P_{i-min} & if \quad P_i' < P_{i-min}' \\ P_{i-max} & if \quad P_i' > P_{i-max}' \end{cases}$$

(8)

D. Selection of individual by competition

In the competition stage, a selection mechanism is used to produce a new population from the existing population and the population is created by mutation. The selection technique used is the stochastic tournament method which is described in the following.

The parent solution P_i along with their corresponding off-springs formed by mutation P_i', each under goes a series of N_t tournaments with randomly selected opponents. Each individual is assigned a score w_s according to equation (9).

$$w_s = \sum_{i=1}^{N_t} w_i$$

$$w_t = \begin{cases} 1 & if \quad u_1 > \cfrac{\Delta f_s}{\Delta f_s + \Delta f_r} \\ 0 & otherwise \end{cases}$$

(9)

Where,

$$\Delta f_s = f_s - f_{min}$$

$$\Delta f_r = f_r - f_{min}$$

f_s is the objective function of the individual under consideration, f_r is the objective function of randomly selected opponent individual and f_{min} is the minimum objective function of an individual within the two population. The opponent is chosen at random from $2 \times Np$ individuals based on r=[$2 \times N_p \times u_2$+1]. u_1, u_2 are uniform random numbers in the interval [0, 1]. Individuals are ranked in descending order of their corresponding w_s score. The first N_p individuals are selected and transcribed along with their corresponding fitness values to be the parents of the next generation.

E. Termination criterion for EP

The above procedure is repeated until the maximum iteration count is reached.

The main stages and operations of the proposed Evolutionary Programming technique including initialization, mutation and competitions are shown in the flowchart of Fig.2.

START

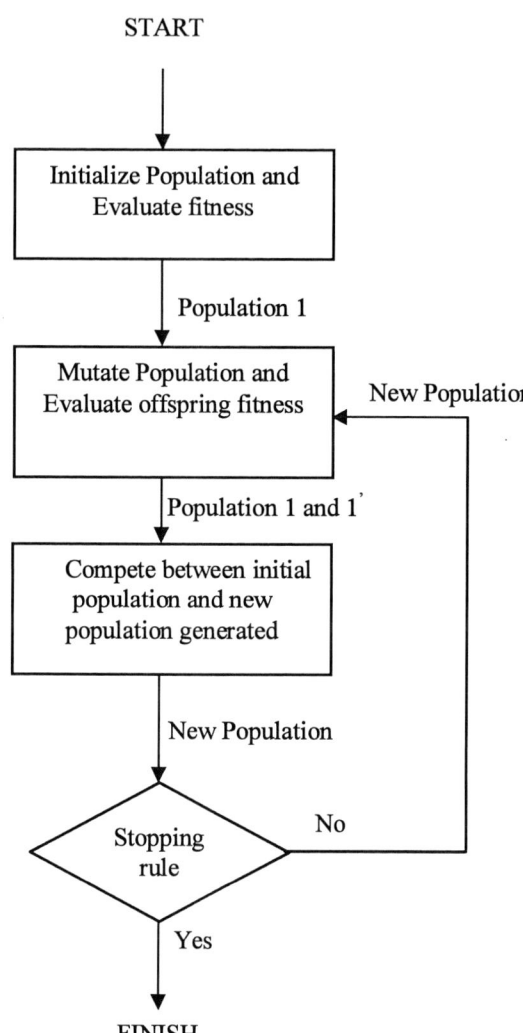

Fig. 2. Operation of Proposed Evolutionary Programming method

IV. OPTIMIZATION RESULTS

To obtain the best and optimum solution for the given non-linear transcendental equations, the following EP parameters are used.

Population size: 50
Number of generation: 200
Scaling factor: 0.01
Competition index: 30

After solving the five nonlinear functions of equations (3) simultaneously using MATLAB optimization toolbox, five angles are obtained. This process is repeated for the various modulation indexes from 0.05 to 1.25.

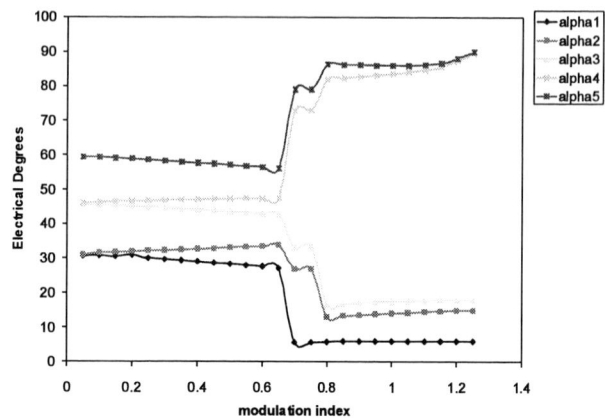

Fig. 3. Trajectory of calculated switching angles of proposed PWM-SHE switching pattern using EP.

Fig. 3 shows the Trajectory of calculated switching angles of proposed PWM-SHE switching pattern using EP. In this proposed approach, the trajectory is almost smooth except $0.7 < M < 0.75$. This proves that the switching angles obtained through Evolutionary programming method is better than obtaining switching angles using Genetic Algorithm method.

V. SIMULATION RESULTS

After obtaining the switching angles through the MATLAB using Evolutionary programming, the proposed system is developed using SIMULINK. The circuit uses 230V single phase AC supply sources (3nos) which is connected to the star connected primary winding of the dual transformer. The 12 pulse voltage source rectifier is developed by two six pulse rectifiers connected in series. This rectifier is being connected to Voltage Source Inverter through the Capacitor which is acting as a DC link between the rectifier and Inverter. Simulations were carried out on a Pentium III 933-MHz, 256–MB RAM processor. The Coding is written using MATLAB. The rating of the proposed AC drive system is

1. Power supply: 400 (line to line) 50 Hz
2. AC input impedance: 1 mH
3. DC link capacitance: 2x470µF

The harmonics are to be observed in the input line to line voltage and input current. With the five switching angles calculated, the whole switching pattern is constructed as shown in the Fig. 4 using quarter wave symmetry method. The line voltage and line current waveforms for the modulation index, M=0.5 is shown in Fig. 5.

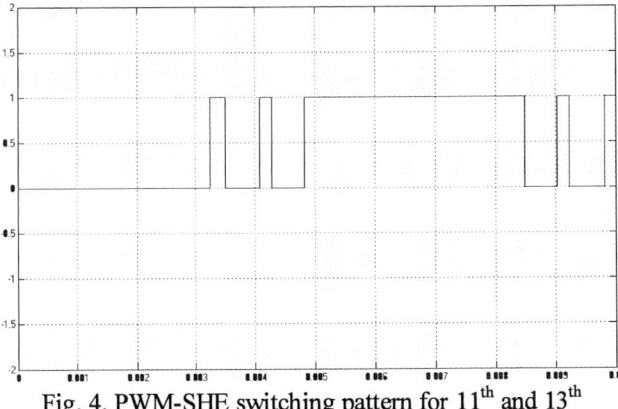

Fig. 4. PWM-SHE switching pattern for 11th and 13th harmonic elimination (M=0.5)

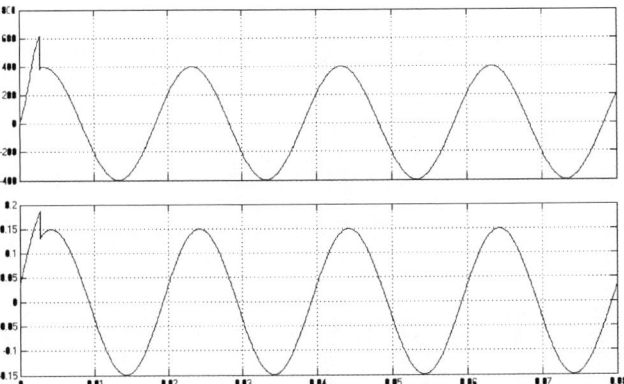

Fig. 5. Rectifier input Voltage and Current

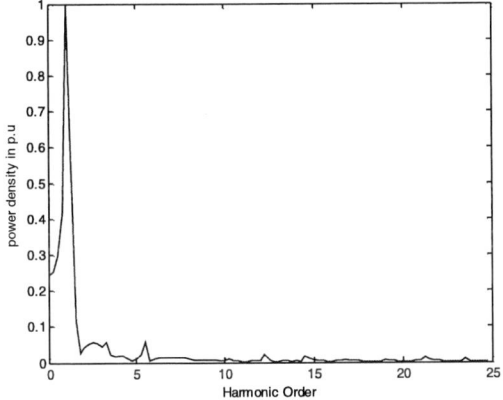

Fig. 6. Frequency Spectrum for input current (after 11th and 13th harmonics are eliminated using EP) at M=0.5

The frequency spectrum in Fig. 6 shows that almost 11th and 13th harmonics are eliminated for the modulation index value 0.5. In Fig. 7, we can see that 11th and 13th harmonics are not being eliminated for the modulation index value 0.7. From fig. 3 and Fig .6, we can find that the 11th and 13th harmonics are eliminated except for M=0.7 and M=0.75. Hence to avoid unpredictable inverter and its erroneous operation, it is beneficial to keep the modulation index in the range, from 0.05 to 0.65 and from 0.8 to 1.25.

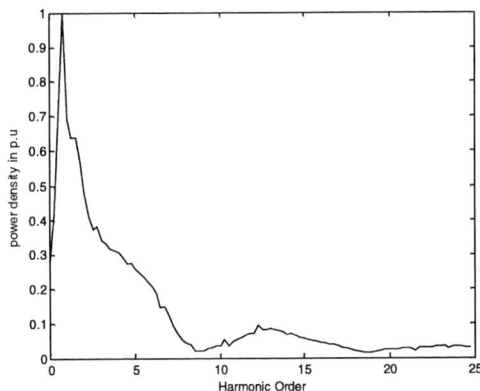

Fig. 7. Frequency Spectrum for input current at M=0.7

VI. CONCLUSION

An efficient technique of calculating switching angles through the EP method is illustrated. The proposed method avoids the traditional complex calculations. Analyzing the frequency spectrum states that except for few distortions, almost in all the modulation index, 11th and 13th harmonics are eliminated through the proposed method by using Evolutionary Programming except for M=0.7 and 0.75. The results shows that the operation range of modulation index in EP is larger compared to GA which operates in the range of 0.4 and 0.9. It remains to be a topic for further investigations to design similar real-time PWM system.

REFERENCES

[1] Joachim Holtz, "Pulse width modulation - A Survey", IEEE Trans. Industrial Electronics, Vol.39, No. 5, pp.410-420, Dec1992.

[2] V. E. Wagner, "Effect of harmonics on equipment," IEEE Trans. Power Delivery, Vol. 8, no. 2, pp. 672–680, Apr. 1993

[3] Jian Sun, Horst Grotsotollen, "Solving Nonlinear equations for Selective harmonic eliminated PWM using Predicted Initial values", IEEE Conf. Proc. IECON'92, pp. 259-264, 1992.

[4] K.L. Shi and Hui Li, "Optimized PWM strategy based on genetic algorithms", IEEE Trans. Industrial Electronics, Vol.52, No. 5, pp.1458-1461, Oct2005.

[5] Maswood, A.I, Wei,S., and Rahman, M.A, "A flexible way to generate PWM-SHE switching pattern using genetic algorithm", IEEE Applied Power Electronics (APEC) Conf. Proc., Anheim, californiea, USA, Vol.2, pp 1130-1134, 2001.

[6] J. Sun, S. Beineke, H. Grotsllen, "DSP based Real-time harmonic elimination of PWM Inverters" IEEE Power Electronics Specialists Conference PESC, Taipei, Taiwan. Vol. 2, pp. 679-685, 1994

[7] A.I. Maswood and S.Wei, "Genetic algorithm based solution in PWM converter switching", IEE Proc. Elect. Power Appl., Vol.152, No.3, pp. 473-478, May 2005.

[8] Shen, W., and Maswood, A.I., "A novel current source PWM drive topology with specific harmonic elimination switching", Proc. Canadian Conf. on Electrical and computer Engineering(CCECE), Halifax, Canada, No. 12, pp.53-55, 2000.

[9] K.Sundareswaran and Mullangi chanda, "Evolutionary Approach for Line Current harmonic Reduction in AC/DC converters", IEEE Trans. Industrial Electronics, Vol.49, No. 3, pp. 716-719,June 2002.

[10] Nidul Sinha,R., chakrabarti, and P.K.Chattopadhyay, "Evolutionary programming techniques for Economic load Dispatch", IEEE Trans. On Evolutionary Computation Computation, Vol.7, No.1, pp. 301-306, Feb 2003.

Piecewise Linear Control Surface for Single Input Nonlinear PI-Fuzzy Controller

S. M. Ayob, Z. Salam, and N. A. Azli

Department of Energy Conversion, Faculty of Electrical Engineering
Universiti Teknologi Malaysia, 81310 UTM Skudai, Johor, Malaysia
Email: shahrin@fke.utm.my

Abstract–This paper presents an analysis on piecewise linear control surface for a single-input nonlinear PI Fuzzy controller. The analysis is carried out to determine the conditions that should be met in order for the piecewise linear control surface approximation to be validated. An equivalent two-input PI-Fuzzy with conventional fuzzification, inference and deffuzification process is built for verification purpose and both controllers are then applied to a single phase inverter system. The dynamic response of both controllers are examined and compared. From the simulation results, it is shown that the approximation is valid.

Index Terms—PI-Fuzzy controller, control surface, piecewise linear approximation

I. INTRODUCTION

The main feature of a fuzzy controller is that it can control complex and ill-defined systems by translating the linguistic control strategy of experts' knowledge into an automatic control without knowing the detail mathematical model of the systems. For a UPS inverter system, its linear model can be obtained via a state-space averaging method. This linear model has been popularly used over the years in designing its linear theory based controllers [1,2]. These controllers basically provide an excellent steady state performance with low THD percentage but poor and slower dynamic response with nonlinear loads or large load disturbances. This is due to the fact that the model itself is developed based on a single operating point. Therefore when the operating point changes, the designed controller is no longer valid. Moreover, there are several parameters uncertainties that could not be modeled very well such as switching non-idealities and delays which can affect the overall controller performance [3]. Therefore, a non-model based controller such as a PI-Fuzzy controller is preferable for application purposes. However, since PI-Fuzzy controllers are not mathematical-based, heuristic design approaches could not be avoided and popularly known to be the main drawback of this type of controller. The controller also lacks on the aspect of stability theory and therefore impossible to be assessed. However, a lot of works have been done over the past few years that are devoted to minimise the approach and as a result, several design guidelines and stability theories have been proposed [4,5,6]. Previous researches have shown that the nonlinearity characteristic of this controller is strongly dependent on its input-output mapping or known as control surface. The PI-Fuzzy controller can be a linear PI controller if the control surface exhibits a linear surface as depicted in Fig. 1. This type of controller can also be known as a linear PI-Fuzzy controller while the nonlinear PI-Fuzzy is referred to a PI-Fuzzy controller with control surface as shown in Fig. 2. These surfaces are generated by applying several conditions on the selection of membership function shape, percentage of overlap between fuzzy sets, inference method, fuzzification and defuzzification method. Different conditions yield different control surfaces.

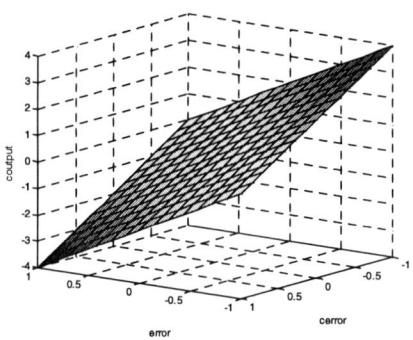

Fig. 1. Linear control surface

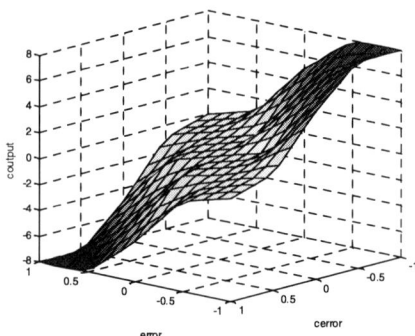

Fig. 2. Nonlinear control surface

The single input PI-Fuzzy based on the signed-distance method offers less number of rules, less tuning parameters and yet capable of providing identical performance of its equivalent two-input controller [7]. Instead of using two input variables, this controller uses a single input variable called "distance (d)" and is defined by (1).

$$d = \frac{\dot{e}(k) + \lambda e(k)}{\sqrt{1 + \lambda^2}} \qquad (1)$$

Where $e(k)$ is the error, $\dot{e}(k)$ is the change of error and λ is a constant.

It is a well known fact that linear PI controller suffer from poor large-signal performance, and in order to improve that, the control surface for the PI-Fuzzy should be made nonlinear as in Fig. 2. In the case of single input PI-Fuzzy controller, the nonlinear control surface can be approximated as a piecewise linear which can be realised by means of a look up table. This is possible, since the control surface for the sole input controller is a one-dimension mapping. However, several conditions should be met in order to realise the approximation. Hence, this paper extensively analyses the conditions that should be applied for the approximation to be valid.

II. CONTROL SURFACE APPROXIMATION ANALYSIS

Fuzzy controllers are very well known as controllers with complex algorithm and therefore they always demand a fast digital processor board in order to run the program. For power converters control, which associate with high frequency switching, the real-time implementation of fuzzy control is quite a challenging task. The targeted processor should be fast enough to execute fuzzy algorithm within one switching period.

To alleviate the problem, a signed-distance method has been proposed [7]. The method has significantly reduced the complexity of fuzzy algorithm by reducing the number of rules and inputs. By using this method, the input is reduced to a single input and the rule table can be constructed in one-dimension control surface mapping. To preserve the nonlinearity characteristic of the control surface and to lessen the on-line computation burden, an approximation of control surface can be done. For a one-dimension mapping, its approximation can be done via a look-up table. However, in order to do that, several conditions should be followed.

Let us consider the "distance" input has five fuzzy sets and five fuzzy sets for the output with each input and output sets labeled as Negative Big (NB), Negative Small (NS), Zero (Z), Positive Small (PS) and Positive Big (PB). The rules for this example are as follows:

IF d is Negative Big THEN output is Negative Big
IF d is Negative Small THEN output is Negative Small
IF d is Zero THEN output is Zero
IF d is Positive Small THEN output is Positive Small
IF d is Positive Big THEN output is Positive Big

For the inference process, the activation operator is set to be a PRODUCT (*), MAX for the accumulation operator and Center of Gravity (CoG) as the defuzzifier operator. For discussion purposes, six different control surfaces are considered as depicted in Figs. 4(a)-(f). The red line is the control surface generated from the inference process while the blue line is a linear unity slope line which represents a reference.

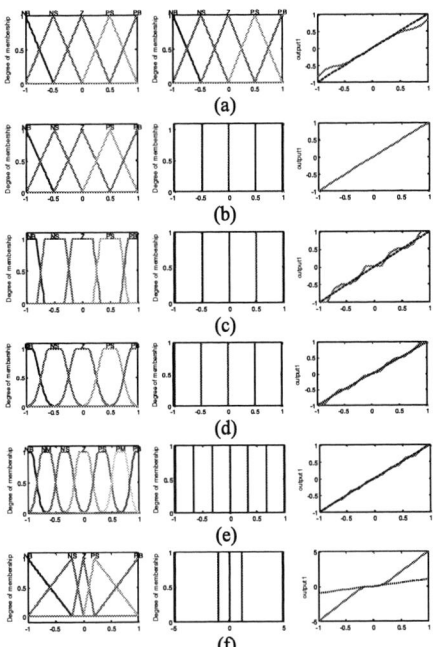

Fig. 4(a)-(f). Six different control surfaces for discussion purposes

From the figures, several conclusions on the generated control surface approximating the linear line can be made as follows,

a) The linear line approximation can only be realised with the singleton output sets.
b) The triangular input set must be used and should overlap at least 50% with the adjacent sets.
c) Either Centre of Gravity (CoG) or Centre of Gravity Singleton (CoGS) defuzzifier can be used if the output set is a singleton.

By setting fuzzy conditions as above and considering the generalised output equation of the single-input PI-Fuzzy as in (2), then the control surface equation of the single input PI-Fuzzy can be expressed as in (4).

$$\Delta u = \frac{\sum_{i}^{n} \mu_i S_i}{\sum_{i}^{n} \mu_i} \qquad (2)$$

where μ_i is the membership degree for i^{th} rule and S_i is the singleton membership output.

$$\mu_i = \frac{d_i - d}{d_i - d_{i-1}} \qquad (3)$$

where d_i and d_{i-1} are the peak of each fuzzy membership input sets where measured input is defined.

Substituting (3) into (2) gives,

$$\Delta u = \left(\frac{(S_i - S_{i-1})}{(d_i - d_{i-1})} d + \frac{(d_i S_{i-1} - d_{i-1} S_i)}{(d_i - d_{i-1})} \right) \qquad (4)$$

Equation (4) demonstrates that the output equation of the controller is actually a linear equation and can be written as (5).

$$\Delta u = \alpha d + \gamma \qquad (5)$$

The first term in (5), α, represents the ratio of the peak's location difference between output and input sets where the exact input value is defined. The equation actually represents the slope of the line. Therefore, different slope values can be obtained by changing the peak locations of input and output sets. Furthermore the second term in the equation, γ, will disappear when all the input and output sets are equally spaced as in Fig. 4(b). This will result in a straight linear line with a single slope. To obtain a better dynamic response compared to linear PI controllers, the surface should be made nonlinear and in order to obtain that, the input and output sets are placed unequally as can be depicted in Fig. 4(f). The nonlinear mapping as in Fig. 4(f) can be approximated as a piecewise linear mapping which can be realised using a look-up table. Fig. 5 shows the comparison of the control surfaces generated using two different methods. The solid red line is the control surface generated using the fuzzy conventional process (the same as in Fig. 4(f)) whiles the dash blue line is constructed using (4). As can be clearly seen, the dash blue line can approximate the line generated by the conventional process very well.

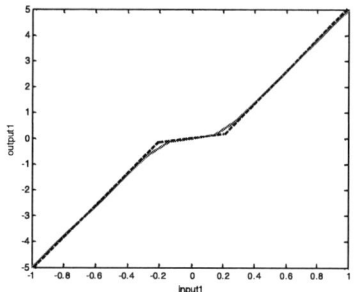

Fig. 5 Comparison of control surfaces generated using conventional process and using (4).

III. SIMULATIONS

Fig. 6 and Fig. 7 show the structure of the single-input PI-Fuzzy controller with the approximation and with conventional process, respectively. The Ψ is a piecewise linear mapping which represents the nonlinear control surface for the single input PI-Fuzzy controller. The variable, λ, has been defined in order for the controller to exhibit a similar linear PI's performance for input $|d| \leq 20$ units. The approximated control surface used in this work has unity gain for input range of $|d| \leq 20$ and 3.1875 unit gain for $|d|$ beyond that. The surface is set to saturate at input over 100 units. Table I shows the rule table used for the system in Fig. 7, while Fig. 8 shows the input and output sets used for the same system.

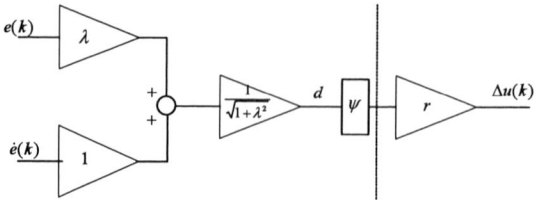

Fig. 6. Block diagram of the single input controller with approximation control surface.

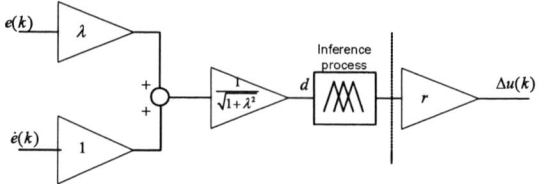

Fig. 7. Block diagram of the single input controller with conventional fuzzy process.

TABLE I
RULE TABLE FOR SINGLE INPUT PI-FUZZY CONTROLLER

d	LNB	LNS	LZ	LPS	LPB
ΔU	-275	-20	0	20	275

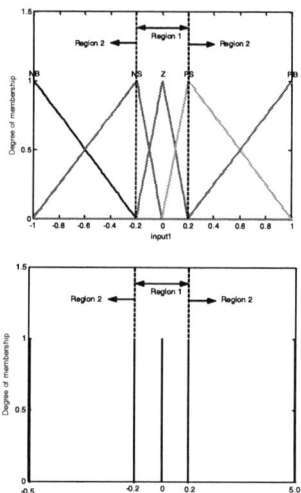

Fig. 8. Input and output fuzzy sets used for system in Fig. 7

Both controllers are built using MATLAB-Simulink. MATLAB's Fuzzy Toolbox is used to build the controller with the conventionl fuzzy process. Both controllers are then applied to control a typical single phase inverter system. To examine the excellent performance of the controllers, a small load (partial load disturbance) and large load disturbance (no load to full load) are imposed on the system. Both controllers reveal fast dynamic response that is identical to each other as shown in Fig. 9 and 10 which verifies the approximation.

1684

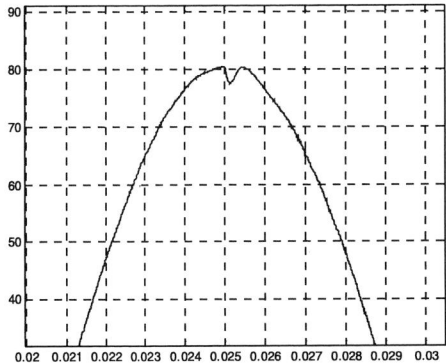

Fig. 9. Voltage response comparison between system in Fig. 6 and Fig. 7 for small load disturbance.

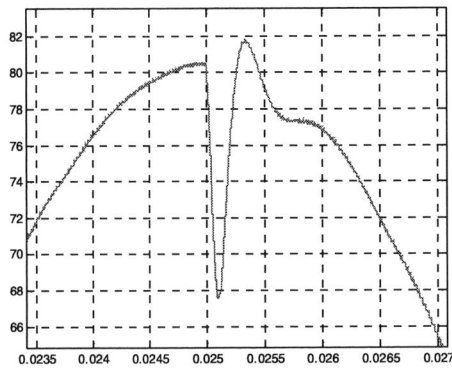

Fig. 10. Voltage response comparison between system in Fig. 6 and Fig. 7 for large load disturbance.

IV. CONCLUSION

In this paper, an analysis on a piecewise linear approximation control surface for a single input PI-Fuzzy controller has been presented. The approximation can be made by applying several soft conditions to the fuzzy parameters as discussed in detail earlier. For verification purposes, the single input PI-Fuzzy with conventional inference process and the single-input controller with the control surface approximation have been developed using MATLAB-Simulink. Both controllers have been applied to control the output voltage of a single phase inverter. The dynamic response of both controllers are found to be identical hence verifying the approximation.

REFERENCES

[1] M. J. Ryan, W. E. Brumsickle and R.D. Lorenz, "Control Topology Options for Single-Phase UPS Inverters", *IEEE Transaction on Industry Applications*, Vol. 33, No. 2, pp 493-501, March/April 1997.

[2] H. Deng, R.Oruganti and D. Srinivasan, "Modelling and Control of Single-Phase UPS Invter: A Survey", *The 6th International Conference on Power Electronics and Drive Systems (PEDS 2005)*, pp848-853, December 2005.

[3] A. Kawamura, T. Haneyoshi and R.G. Hoft, "Deadbeat Controlled PWM Inverter with Parameter Estimation Using Only Voltage Sensor", *IEEE Transaction on Power Electronics*, Vol. 3 (2), pp118 – 125, April 1988.

[4] "Fuzzy Control: Synthesis and Analysis", Edited by S. S. Farinwata, D. Filev and R. Langari, *John Wiley & Sons*, 2000.

[5] A. G. Perry, Y.-F. Liu and P. C. Sen, "A New Design Method for PI-like Fuzzy Logic Controllers for DC-to-DC Converters", *The 35th Annual IEEE Power Electronics Specialists Conference (PESC 2004)*, Vol.5, pp3751-3757, June 2004.

[6] A. Kandel,Y. Luo and Y.-Q. Zhang, "Stability Analysis of Fuzzy Control Systems", *Fuzzy Sets and Systems*, Vol.105, Issue 1, pp33-48, July 1999.

[7] B. J Choi, S.W. Kwak and B. K.Kim, " Design and Stability Analysis of Single-Input Fuzzy Logic Controller", *IEEE Transaction on Systems, Man and Cybernetics-Part B: Cybernetics*, Vol.30, No. 2, pp303-309, April 2000.

Open-Loop Control of a Stepping Motor through IP Network

K. Matsuo*, T. Miura*, and T. Taniguchi*

*Department of Electrical and Electronic Engineering, Akita University,
1-1 Tegata Gakuen-machi, Akita-shi, Akita 010-8502, JAPAN

Abstract–In this work, IP (internet protocol) network-based control system of a stepping motor is constructed and the control performance is examined. The network has transmission time delays, jitters and packet loss. These give disadvantages to the control system. Therefore, it is important to examine the performance degradations when a motor is driven under the system through IP networks. The position control experiments are done in open-loop using a stepping motor, which is one of the most popular one and an appropriate one for a position control. Then the control ability is examined against various network conditions.

Index Terms—Network-based control system, IP network, Stepping motor, Time delay

I. INTRODUCTION

Information and communication technologies have been growing up. Various services using IP (Internet Protocol) networks are recently provided owing to development of them, for example, e-mail, WWW (World Wide Web), etc. Moreover, the demands of the services requiring real time like IP telephony also increase. IP networks have infiltrated many aspects.

Therefore, constructing IP networks like LANs (local area networks) is easier. Moreover speed and capacity of the network are expanded, and the cost is also lower.

Introducing them begins by sending and receiving of a signal through IP network instead of RS-232C or GP-IB. Thus, it becomes possible to communicate between long distance through the Internet.

Due to these backgrounds, introducing IP networks in various control system will be also more important. Actually, there are many works about control systems through the network, i.e., network-based control system (e.g., [1] [2]).

The advantage is that multiple users share a line because IP networks are packet communication. On the other hand, the disadvantage is that transmission under high traffic load causes time delays, the jitters and packet losses.

The authors consider control systems of motors through IP networks. On one hand, the papers studying the system using DC motors which are the most basic one are increasing [1]-[3]. On the other hand, to the author's knowledge, there are not papers studying the system through the network using stepping motors. Hence, in this work, the authors want to study it.

A stepping motor can be position-controlled without sensors, that is, in open-loop. It differs from a control system of a DC motor having a feedback.

In this work, the authors actually construct an IP-network based position control system of a stepping motor. When the position control examinations through various IP network are done, the authors study how the performance is.

II. CONTROL SYSTEM

The position control system of a stepping motor used in this experiment is shown in Fig.1. The system is controlled in open loop. Hence the system has no feedback. PC1 is a function as a controller. PC2 is a function as a driver for the motor.

The stepping motor used in this work is two phases hybrid type one (PX244-02B, Oriental Motor) and the rated voltage and current are 6 V and 0.8 A respectively. The rated value of the fundamental step angle is 1.8 deg.

A stepping motor is rotated at each fundamental step angle. The motor is rotated and moved to next position, that is, in 1.8 deg, when the driver receives a pulsed signal. A rotary encoder detects an angle of the motor. The resolution is 20000 pulses/rev, where it is four multiple of 5000 pulses/rev.

In the system in Section III, when the rotor is moved to a next position, pulsed signal is sent through an IP network from PC1 to PC2, where the signal "1" is sent, and PC2 received the one. PC2 has the function of the driver. Exciting commands are given to the driving circuit according to each inputted pulses. Thus the motor is driven. Here the motor is without a load, and is driven using a single-phase excitation

In the system in Section IV, a signal sent from PC1 to PC2 is a one having a directly exciting phase instead of a pulse. Then PC2 receive the signal, and gives an exciting command to a drive circuit according to the phase ordered from PC1.

The detail will be described in Section III and IV, respectively.

As mentioned above, because the system is controlled in open loop, angle data by a rotary encoder is used for only experimental measurements. Both sampling time in PC1 and PC2 are 1 ms, and the OS used is Vine Linux.

PC1 and PC2 are equipped with NICs (Network Interface Cards), which is 100Base-T, and connected to IP networks through the ones. In the system, the receiving and sending of data between the network use UDP (User Diagram Protocol). The response time of

978-1-4244-0644-9/07/$25.00 ©2007 IEEE

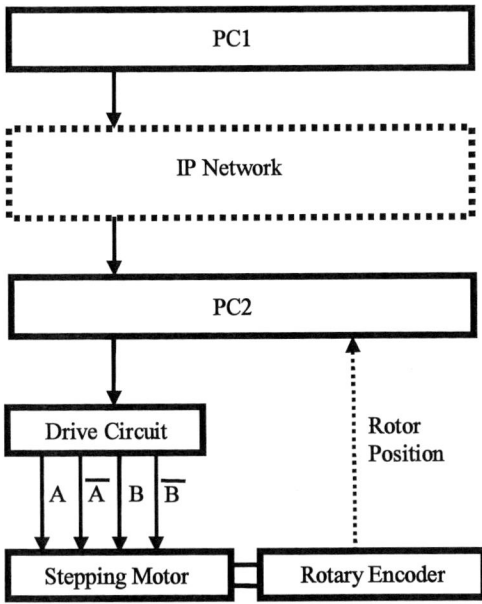

Fig.1 A position control system for a stepping motor through IP network.

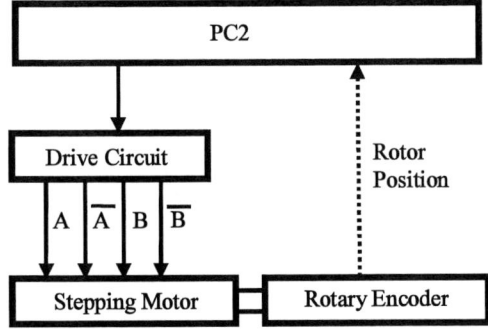

Fig.2 A control system in the case without IP network.

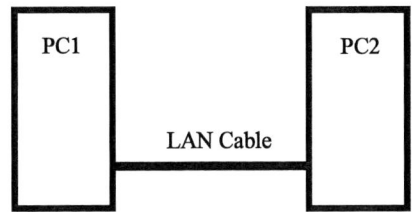

Fig.3 In case of IP network connected directly by a LAN cable between PC1 and PC2.

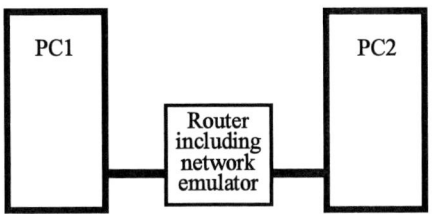

Fig.4 In case of IP network through the router including a network emulator between PC1 and PC2.

UDP is faster than of TCP (Transmission Control Protocol), but the reliability is lower owing to no confirmation. An IP network induces time delays which fluctuate and make packet losses, depending on the amount of a one's traffic. In Fig.1, for example, when data are sent from PC1 to PC2, there are cases that the sending order differs from the receiving order. That is, there are cases that a data sent later is previously received.

In the system, when the motor is moved to next position, the command is sent from PC1 to PC2. Then the interval between the commands is associated with the speed, and the number of the ones inputted to the driver means the rotating angle, which can be obtained by the rated value of a step angle multiplied by the inputted number. In this work, the authors examine how performance changes under various conditions of the network. The conditions will be shown in the next sections.

III. THE EXPERIMENTS AND ITS RESULTS

In this section, the results of position control experiments under various conditions of an IP network are shown when signals sent from PC1 to PC2 are pulses.

First, the experiments are done under the three conditions of the network described below. Here each experiment is done under the cases that the reference angles is 54 deg and each speed is 10 pps, 100 pps and 500 pps, where pps is pulses per second.

The conditions of the network are as follows:

(0)   the case having no IP network,

(1)   the case connected directly between PC1 and PC2,

(2)   the case connected through the router, in which network emulator, Nist net (described later), is installed, between PC1 and PC2. Then the constant time delay of packets through Router is 0ms, and 5 % of it are lost, (the setting by NIST Net).

The case (0) is shown in Fig.2. The case (0) is a normal control system of stepping motor having no network. Then the experiments are done for observing how the performance changes in cases with other actual networks. Here, in case (1) and (2) with IP networks, PC1 is the controller, and PC2 has only a function for sending exciting commands to the drive circuit according to a pulse sequence sent from PC1. In case of (0) without a network, PC1 is not used, and PC2 has a function as the controller. PC2 also commands to the driving circuit and makes the motor driven. Fig.3 shows the construction of the case (1), and Fig.4 shows of case (2). In case (2), a network emulator called Nist Net [4][5] is installed in Router. The emulator emulates a transmission delay time,

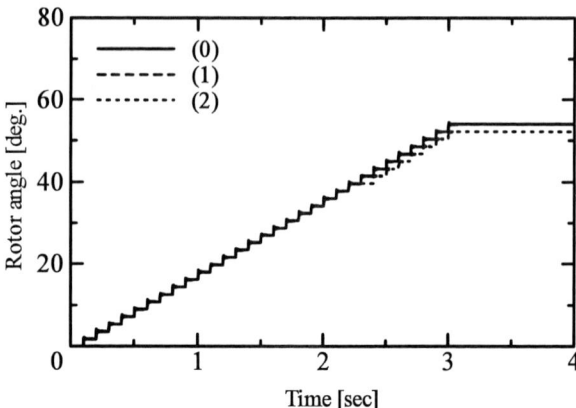

Fig.5 The results of the experiment (0)-(2) when the motor speed is 10pps.

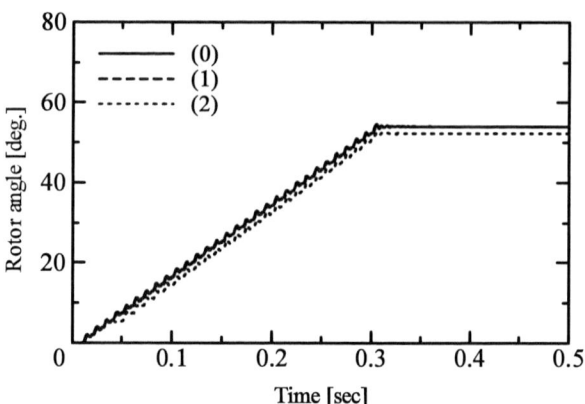

Fig.6 The results of the experiment (0)-(2) when the motor speed is 100pps.

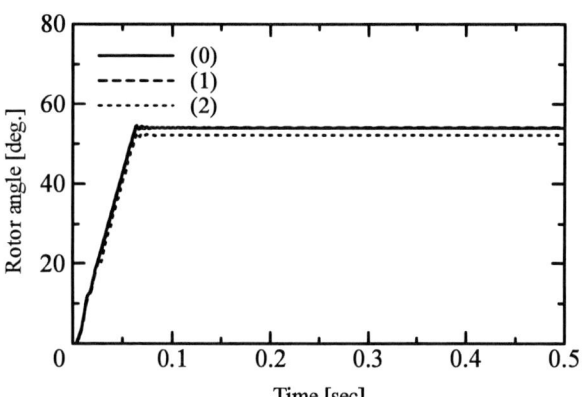

Fig.7 The results of the experiment (0)-(2) when the motor speed is 500pps.

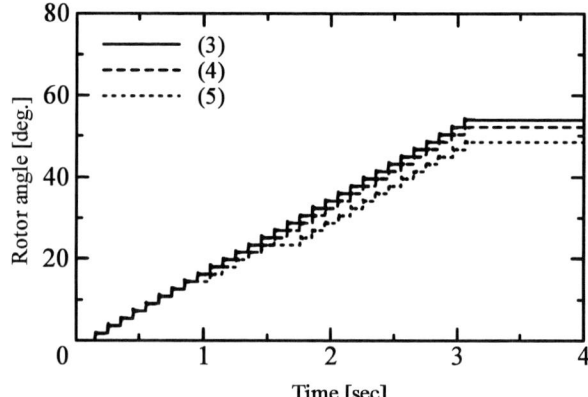

Fig.8 The results of the experiment (3)-(5) when the motor speed is 10pps.

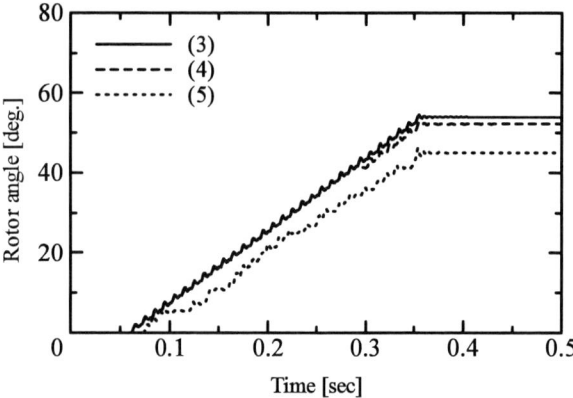

Fig.9 The results of the experiment (3)-(5) when the motor speed is 100pps.

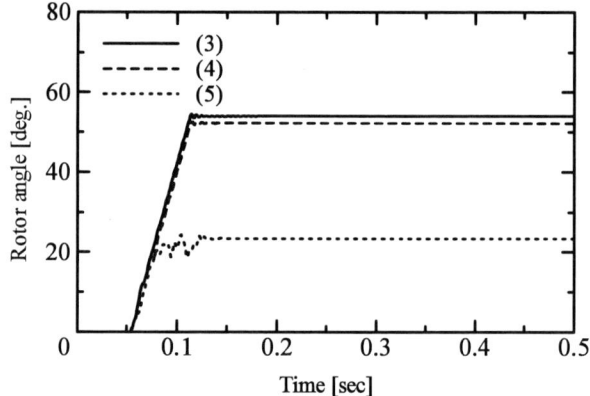

Fig.10 The results of the experiment (3)-(5) when the motor speed is 500pps.

the jitter and the rate of packet loss. The experiment is done under the condition of the network that the transmission time is set to a constant time of 0[ms] (no delay time). The rate of packets loss is set to 5%.

Then the experiments are done under the case (0)-(2). The results are shown in Fig.5, Fig.6 and Fig.7,

respectively. A difference between (0) and (1) is not observed as shown in the figures. In case (2), the rotor angle in the steady state is less than the reference, because of lost packets of the IP network. Moreover, the exciting command do not reached PC2 because pulses are lost owing to packet losses.

1688

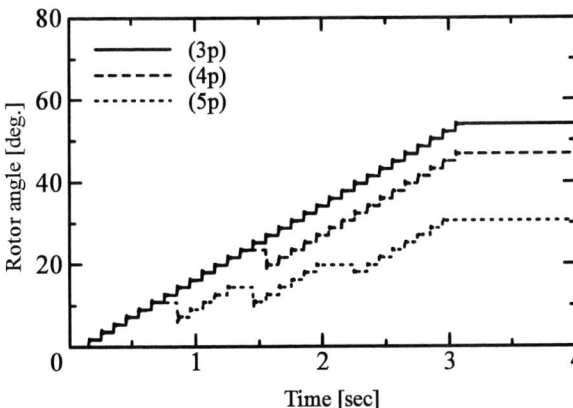

Fig.11 The results of the experiment (3d)-(5d) when the motor speed is 10pps.

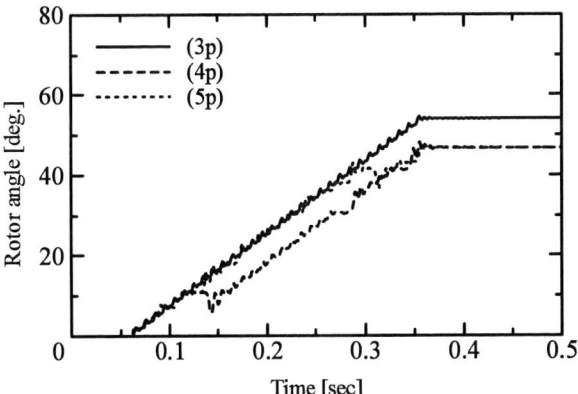

Fig.12 The results of the experiment (3d)-(5d) when the motor speed is 100pps.

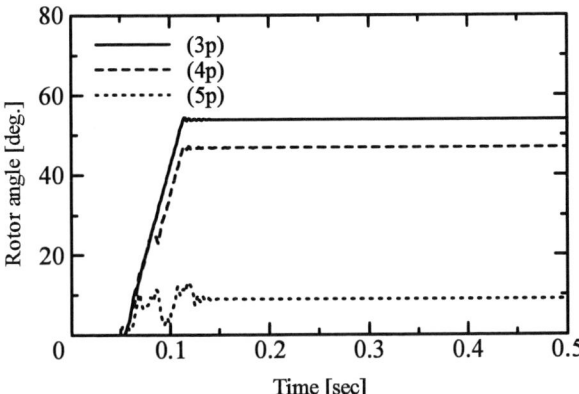

Fig.13 The results of the experiment (3d)-(5d) when the motor speed is 500pps.

Second, the network conditions are changed by the calculation ones of network emulator, also using NIST Net shown in Fig.4. The conditions are shown in (3)-(5) as follows:

(3) The constant time delay of packets through Router is 50 ms, (the setting by NIST Net),

(4) The constant time delay of packets through Router is 50 ms, and 5 % of it are lost, (the setting by NIST Net),

(5) Time delay of packets through Router is varied under heavy-tailed distribution with mean 50ms and variant 5ms, which is default one of NIST Net.

Then the reference angle is 54 deg, and each speed is 10 pps, 100 pps and 500 pps, and the experiments are done under the case (3)-(5). The results are shown in Fig.8, Fig.9 and Fig.10, respectively.

Fig.8-10 show that the motor is rotated with the delay of 50 ms, comparing to Fig.5-8. The figures also show that the displacement angle does not reach the reference 54 deg owing to the packet loss in the case (4) as well as in the case (2).

The displacement of the rotor angle in case (5) is more far to the reference than in case (4). This is probably due to the packet loss because the packet cannot arrive within

the waiting time owing to the condition that the communication time of the network is fluctuated, while the constant sampling time of the receiving side PC2 is divided into the constant one waiting for receiving and the control processing time of command for motor and the others.

IV. THE EXPERIMENTS IN CASE OF OTHER EXCITATION COMMAND AND ITS RESULTS

In this section, PC1 sends a packet including an actual phase command, that is phase A, $\overline{A}$, B or $\overline{B}$, against each changed interval, and PC2 actually changes an exciting winding after receiving the packet. Here, the difference from Section III is that the exciting command in this section is a specific phase instead of a pluses.

Then the reference angle of the motor is 54 deg. The conditions of experiment is same as in Section III. Those are refered as (3p), (4p) and (5p) to distinguish from (3), (4) and (5) in Section III. The driving expriment is performed under 10 pps, 100 pps and 500 pps. The results are shown in Fig.11 to Fig.13, and are similar to in Section III. However, after PC2 cannot receive the exciting command, the received one may not excite the valid phase demanded to rotate, since the exciting command specifies the phase. That is, it is found that a missed step is appeared. This is different from the case that the roter position is freezed when the command is not able to receive as the experiment in Section III. The rotor does not reach the reference position even in the previous experiment of (4d) and (5d) as well as (4) and (5).

In previous exiperiments, the pluse or the command for the exciting phase is sent from PC1 to PC2 to move the rotor to the next angle in the experiment performed until now. Therefore, the phase to excite at present is specified at each sampling time of PC1 in the next experiment. The quantity of the data is more than the previous method of command because the packet is sent continuously at each sampling time. However, it is guessed that this is more robust since the command for the exciting phase is immediately arrived even in the case that PC2 cannot send the packet normally. The experiment is done in 10

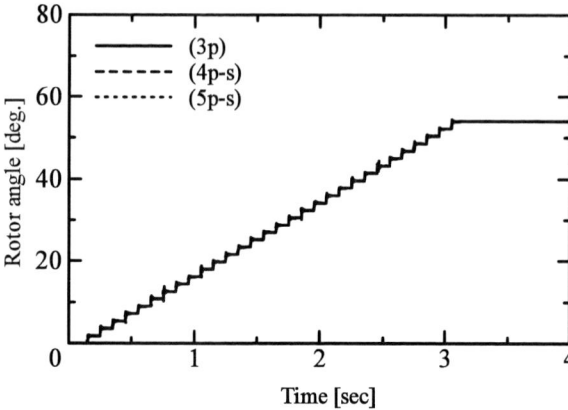

Fig.13 The results of the experiment (3p), (4p-s) and (5p-s) when the motor speed is 10pps.

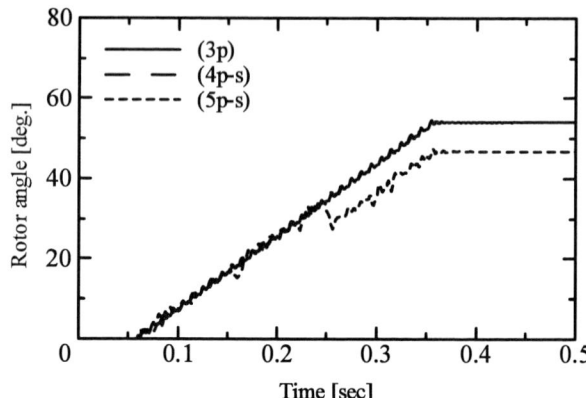

Fig.15 The results of the experiment (3p), (4p-s) and (5p-s) when the motor speed is 100pps.

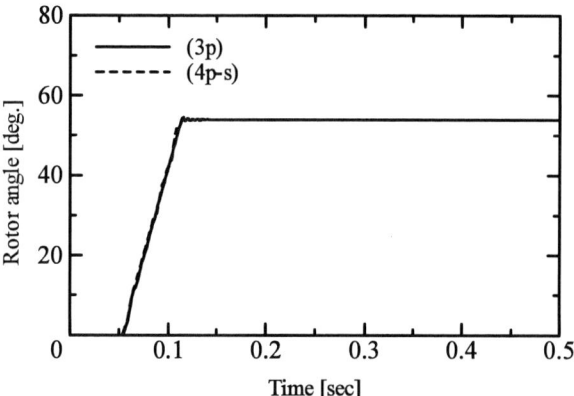

Fig.14 The results of the experiment (3p) and (4p-s) when the motor speed is 500pps.

fluctuated. Therefore, the experiment has been performed by the method that the signal of the phase to excite is sent continuously. The results have shown that this method has been effective since the rotor has reached the reference in the only case that the packet loss is happened.

The future work is such that an effective method is developed also against fluctuating time delays.

ACKNOWLEDGEMENT

The authors would like to thank Y. Akiyama, Department of Electrical and Electrinic Engineering, Akita University, for technical support.

REFERENCES

[1] Y. Tipsuwan and M.-Y. Chow, "Gain Scheduler Middleware: A Methodology to Enable Existing Controllers for Networked Control and Teleoperation - Part I: Networked Control," *IEEE Transactions Industrial Electronics*, Vol.51, No.6, pp.1218-1227, 2004.

[2] N. B. Loden and J. Y. Hung, "An Adaptive PID Controller for Network Based Control Systems," *Proc. of 31st Annual Conference of the IEEE Industrial Electronics Society (IECON 2005)*, North Carolina (USA), Nov. 2005, pp.2445-2450.

[3] K. Matsuo, T. Miura and T. Taniguchi, "Speed Control of a DC Motor System through Delay Time Variant Network," *Proc. of SICE-ICASE International Joint Conference 2006 (SICE-ICCAS 2006)*, Busan (South Korea), Oct. 2006, TA15-3, pp.399-404.

[4] NIST Net, http://www-x.antd.nist.gov/nistnet/

[5] M. Carson and D. Santay, "NIST Net: A Linux-based Network Emulation Tool," *ACM SIGCOMM Computer Communication Review*, Vol.33, No.3, pp.111-126, 2003.

pps as well as (4p) and (5p). Those are refered as (4p-s) and (5p-s). The results of (4p) are also shown in Fig.13 to compare. The results in 100 pps and 500 pps are also shown in Fig.14 and Fig.15, respectively. However, the results of (5p-s) are not shown in the figure as the motor cannot be driven. These results show that this method is effective since the rotor reaches the reference angle in the case (4p-s) of the packet loss. However, this method is not effective expect in 10 [pps] in the case that delay time is fluctuated.

V. CONCLUSIONS

An IP Network-based control system for a stepping motor has been constructed, and position control experiments have been done. Then the motor driving experiments have been done under the network environment produced by the several simulators. In this work, when it has been needed that the motor is moved by one step, the experiments have been done by the methods that the motor has been moved by sending one pulse command and that the next exciting phase has been directly specified. In both cases, the response has been disturbed and the rotor has not reached the reference angle when the transmission delay time has been

Fuzzy Logic Controller for Electric Vehicle Braking Strategy

Xixi. Wang[1], K.W.Eric Cheng[1], Xiaozhong Liao[2], Norbert C. Cheung[1] and Lei Dong[2]

[1]Department of Electrical Engineering, The Hong Kong Polytechnic University
[2] Department of Automatic Control , Beijing Institute of Technology, China

*Abstract--*In this paper, the braking strategy is developed for electric vehicle which includes regenerative braking and anti-lock brake system (ABS). Induction motor is used in EV to regenerate kinetic energy. This paper describes the preliminary research and implementation of a fuzzy logic controller to control wheel slip for ABS. The controller is used to enhance the vehicle stability and safety under highly nonlinear and time variant braking conditions. Simulation was used to derive a rule base. The robustness of the fuzzy-logic slip regulator is further tested by applying the resulting controller over a wide range controller over a wide range of operating conditions.

Index Terms-- Electrcial vehicle, ABS, Antilock braking system

I. INTRODUCTION

When braking electric vehicle whose wheels are driven by induction motor, the total braking torque is the sum of the motor regenerative braking torque and the hydraulic frictional torque. The distribution of the two kinds of torque is mainly determined by the braking mode. There are different braking modes which correspond to different control strategies: Emergency braking, Moderate braking and Downhill continuous braking

During low brake pedal force, only the regenerative braking torque is applied on the driving wheels, and is proportional to the pedal pressing force. When the pedal force is beyond a certain limit, the maximum regenerative braking torque is applied on the driving wheels, and the hydraulic braking torque is simultaneously applied on the driving wheels to top up the desired braking torque. Hence, the maximum regenerative braking torque can be kept constant to fully recover the kinetic energy. [1] But if sufficient braking force is applied, the wheel will "lock up," that is, slide without turning at all. It's very dangerous because a locked wheel has no lateral stability. The goal of ABS is to generate the largest possible brake force in real time while keeping the vehicle maneuverable and preventing the lock-up of car wheels.

In emergency braking, the car often slips when the road condition is not good. It is very dangerous for the car which will lose maneuverability. So this paper focuses on the first mode to study how regenerative

This work was supported by The Hong Kong Polytechnic University under the project number 1-BB86.

braking and anti-lock brake system work in emergency braking.

From Fig.1, where μ is tire-road adhesion coefficient, s is slip ratio, if the slip ratio is kept at the peak adhesion coefficient slip ratio, i.e. s_p point, the longitudinal force is the biggest, and the lateral force is well situated, the braking distance will be as short as possible while the vehicle is still controlled. If $s > s_p$, the lateral force will come to zero very soon. So ABS is to control the slip ratio in stable region ($0 < s < s_p$), and to approach s_p as near (Near s_p) as possible. Most control strategies define their performance goal as maintaining slip near a value of 0.2 throughout the braking trajectory. This represents a compromise between lateral stability, which is the best at zero slip, and maximum deceleration, which usually peaks for some value of slip between 0.1 and 0.3. The goal of ABS control then becomes the regulation of slip to a known and desired level [2] [3].

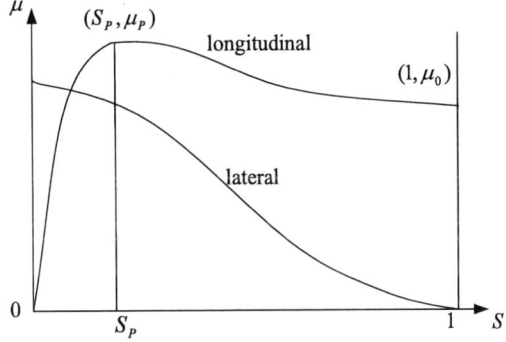

Fig.1 The $\mu - s$ characteristic curve

Recent research on braking system has been concentrated on Fuzzy control, for antilock braking systems[4], all wheel drives [5], antilock braking system using brake-by-wire [6], sensorless control [7]. In this paper, we intend to develop a fuzzy controller for the anti-lock brake system. Fuzzy logic, or the use of fuzzy sets, allows uncertain or inexact concepts to be represented. Fuzzy logic derives from a grouping of elements into classes that do not possess sharply defined boundaries. Fuzzy controllers differ from model-based controllers in that they encode heuristic knowledge. Despite the absence of analytical modeling information, systems governed by fuzzy controllers are often highly robust [8] [9]. In

978-1-4244-0644-9/07/$25.00 ©2007 IEEE

Section 2, the mathematical models for the braking dynamics are introduced, which include the model of the vehicle dynamic, induction motor and the brake-hydraulic system. In Section 3, a fuzzy controller for the ABS is developed. In Section 4, the software Simulink is used to build the simulation model for the full control system and provide simulation results. Finally, the conclusions are given.

II. THE PLANT

The system to be controlled is represented as four wheels of a vehicle in straightforward motion. In straight line braking, lateral tire forces do not exist. The model presented includes the effects of road roughness on the braking process. The model of the longitudinal vehicle dynamic [4] can be described as follows:

v_x − longitudinal speed of the vehicle, C_0 − wind block coefficient, A − wind area, f − roll resistance coefficient

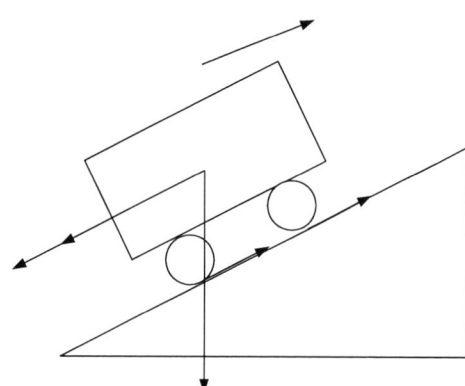

Fig.2 Road load

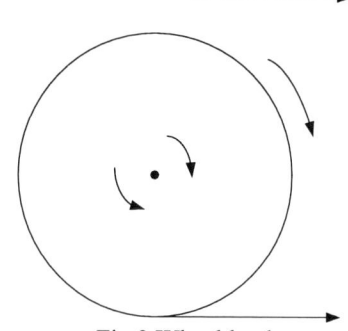

Fig.3 Wheel load

A. The wheel equations

Torque equilibrium equations of the four wheels :

$$J_\omega \cdot \overset{\bullet}{\omega}(1,1) = T_d(1,1) - F_x(1,1) \cdot R - M_f(1,1) \qquad (1)$$

$$J_\omega \cdot \overset{\bullet}{\omega}(1,2) = T_d(1,2) - F_x(1,2) \cdot R - M_f(1,2) \qquad (2)$$

$$J_\omega \cdot \overset{\bullet}{\omega}(2,1) = T_d(2,1) - F_x(2,1) \cdot R - M_f(2,1) \qquad (3)$$

$$J_\omega \cdot \overset{\bullet}{\omega}(2,2) = T_d(2,2) - F_x(2,2) \cdot R - M_f(2,2) \qquad (4)$$

$$M_f(i,j) = F_z(i,j) \cdot f \cdot R \qquad (5)$$

M_f − roll resistance moment, J_w − moment of inertia of wheels, R − radius of wheels.

B. Longitudinal tire force

The model will also consider the Longitudinal tire force :

$$F_x = \mu * F_z * sign(s_x) \qquad (6)$$

and also normal tire forces F_{z1} and F_{z2} .

This is the equation about the tire-road force which is the relationship between tyre braking force and wheel slip. μ is tire-road adhesion coefficient, s_x is slip ratio, and F_z is the normal force of the wheel.
The slip ratio model is:

$$s_x = \begin{cases} \dfrac{\omega R - v_x}{\omega R} & driving \\ \dfrac{\omega R - v_x}{v_x} & braking \end{cases} \qquad (7)$$

μ is tire-road adhesion coefficient, we can use linear model to express it, as shown in Fig.4 :

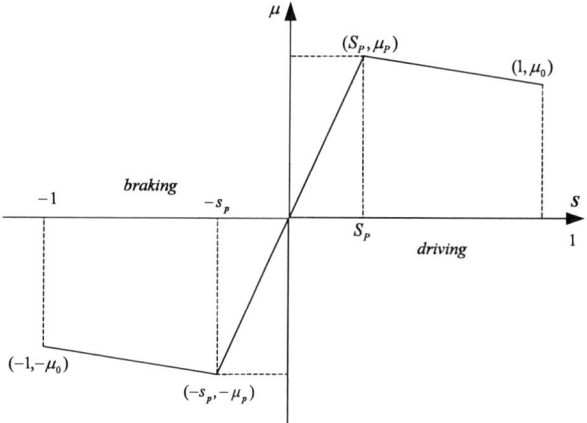

Fig.4 The linear model

Where μ_p is peak adhesion coefficient ; s_p is peak adhesion coefficient slip ratio ; μ_0 is the adhesion coefficient when the wheel locking, the corresponding equation is:

$$\mu = \begin{cases} \mu_p \dfrac{s_x}{s_p} & s_x \le s_p \\ \mu_p - \dfrac{(\mu_p - \mu_0)(s_x - s_p)}{1 - s_p} & s_x > s_p \end{cases} \qquad (8)$$

C. Tire forces

Neglecting aerodynamic drag force and taking moments about the contact point of the rear tire in Figure 5. when driving in horizontal road, the normal tire forces are:

$$F_z(1,1) = \frac{m(bg - \dot{v}_x h)}{2(a+b)}, \quad F_z(1,2) = \frac{m(bg - \dot{v}_x h)}{2(a+b)}$$

$$F_z(2,1) = \frac{m(ag + \dot{v}_x h)}{2(a+b)}, \quad F_z(2,2) = \frac{m(ag + \dot{v}_x h)}{2(a+b)}$$

$$(9)$$

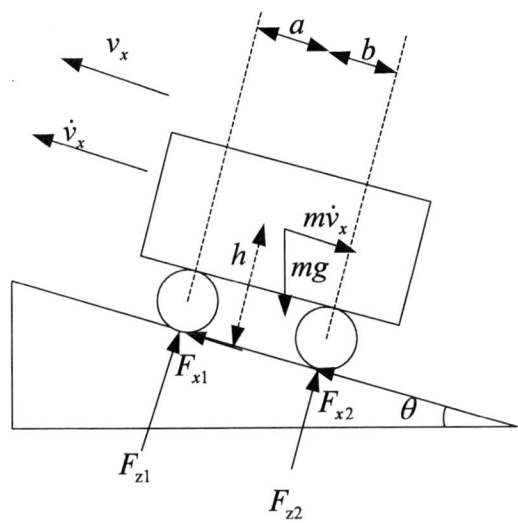

Fig.5 Calculation of normal tire forces

D. Model of the brake hydraulic dynamics

The essence of the ABS system is to control the oil flow of the build valve and the dump valve, thus to control the brake force. The sketch of ABS hydraulic system is shown in Fig. 5.

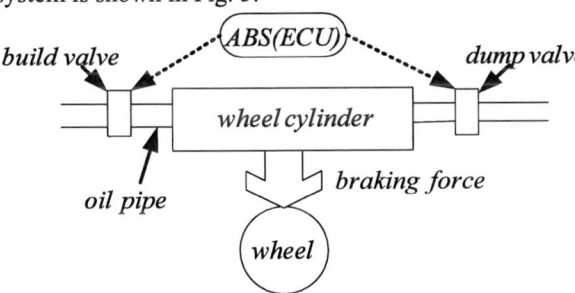

Fig.6 ABS hydraulic system

The valve opening of the electromagnetic valve is determined by the duty ratio of its current which can be controlled by PWM. There is only one valve open for each wheel at any time. When increasing pressure, build valve opens, and dump valve closes. When maintaining pressure, both valves close. And when decreasing pressure, build valve closes, and dump valve opens. The process is then represented in digital format as shown in Table 1. The whole process is correspond to (-1,1), the

minus for decreasing, zero for maintaining, and the positive for increasing. The absolute value represents the opening degree of the valve, which can be controlled by the duty ratio of the current. For example, -1 shows the state that the dump valve fully opens, and the system is decreasing pressure at its greatest degree.

TABLE I
THE ACTION OF VALVE WHEN CONTROL IS GIVEN

Valve control signal D	Input valve action	Output valve action
1	Fully open	Fully closed
(0,1)	Open at the percent of $|D|$	Fully closed
0	Fully closed	Fully closed
(-1,0)	Fully closed	Open at the percent of $|D|$
-1	Fully closed	Fully open

For the liquid flow of the valve, the simplified mathematic mode is

$$Q = C_v A \sqrt{\frac{2}{\rho}} \sqrt{|\Delta p|} * D \qquad (10)$$

Where D belongs to $(-1, \ 1)$, A is the maximum open area of the valve, C_v is a constant, ρ is the density of the liquid, Δp is the pressure difference of the valve.

To wheel cylinder, the differential equation is

$$\dot{P} = \frac{E}{v}(Q_{in} - Q_{out}) \qquad (11)$$

Hence,

$$P(t) - P(0) = \int \frac{E}{V} Q dt \qquad (12)$$

It's an integral process, where P is the oil pressure, E is a constant, and V is the volume of the wheel cylinder.

From the equations, it's shown that controlling the valve opening will finally affect the pressure of the oil in wheel cylinder. And the ABS system is controlling the value D using fuzzy logic controller.

III. CONTROLLER CONCEPT

Combining the hydraulic braking system and induction motor system, the induction motor ABS is shown in Fig 4. The system consists of the energy storage device, inverter drive for the motor, ABS and foot brake. The inverter is driven by rotor flux control.

1693

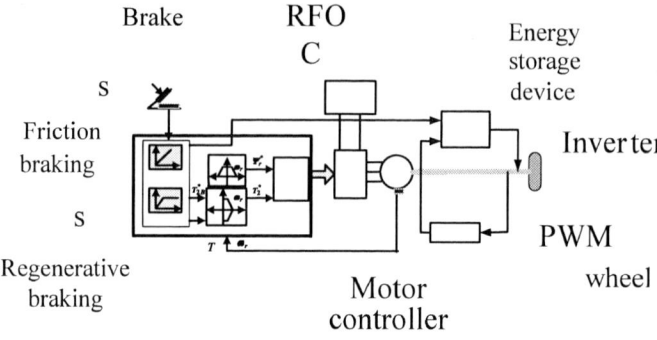

Fig4. Induction motor ABS

The induction motor is regenerative when braking. The Regenerative braking system utilizes the electric motor, providing negative torque to the driven wheels and converting kinetic energy to electrical energy for recharging the battery. The maximum regenerative power can not beyond rating power, and so the maximum regenerative torque. In order to decelerate the motor as soon as possible, and to regenerate more energy, the braking torque is equal to the rating torque. Because in this experiment, when the car reaches its highest speed, the motor magnetic field weakening is invalidate. The braking torque is not limited by the motor power and the vehicle speed.

If the braking force is low, the motors can provide the whole braking torque, and the deceleration is small, the wheel is impossible to be locked, so the ABS system will not work. However, in emergency braking, the process is very different. The friction brake becomes very crucial and the electrical control unit (ECU) of the car monitors the slip ratio of each wheel at real time.

There are many control methods for ABS, the most commonly used strategy is PID control. PID controller technology is very mature and the most widely used control strategy in continuous system. From the mathematic models, it's clearly that the tire model is nonlinear, which leads to the whole ABS nonlinear. The braking condition changes a lot at each braking time, so the crucial parameters such as tire-road adhesion coefficient are different at each time. PID controller is very sensitive to parameter variation. The proportional and integral parameters must be altered when condition changing, otherwise the result will be worse. So PID controller should self-adjust its parameters on line, which is complicated to be realized.

A fuzzy controller and its design principles have been developed for the anti-lock brake system. The fuzzy logic controller consists of three parts: the preprocessing part, the fuzzy logic control part, and the postprocessing part. The preprocessing part handles the hardware interface and calculates the necessary variables for the fuzzy logic control part. The fuzzy logic control part fuzzifies the inputs and maps them to a rule base to determine the control output. The fuzzy logic rules utilize standard max-min fuzzy inference. The center of area method is used to generate the control output. Output defuzzification is performed by computing the centroid of

all minima according to Mamdami's method. Each variable is defined in terms of seven fuzzy classes ranging from "negative large" to "positive large." Fig 5 shows the Fuzzy rule for the fuzzy controller.

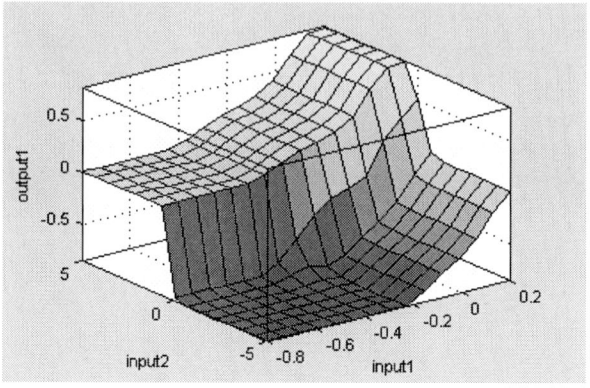

Fig.5 The rule for the fuzzy controller

(Input1 is the slip ratio difference, input2 is the rate of the difference and output is the valve opening)

IV. COMPUTER SIMULATION

In this research, we use the software Simulink to build the whole anti-lock braking system and its fuzzy controller. The mathematical models including the tire force model, the vehicle dynamics, motors and the model for the brake-hydraulic system discussed above are converted into Simulink models.

The main design intent for this braking system is preventing the wheels from locking and decreasing the stopping distance. That is maintaining slip near a value of 0.2 throughout the braking trajectory. The effect of the anti-lock braking system is often examined by braking on different surfaces, including dry road, wet road and icy road

The simulation results are shown below. Fuzzy controller ABS simulation has two parts due to different reference speeds as follows:

The reference speed is shown in the follwing figure. Brake each wheel with 50Nm at 20 seconds. It can be seen that the fuzzy scheme automatically adapts the slip control algorithm, identifies the unstable region of the slip curve and reduces the slip. Eventually the slip stabilizes.

From Fig 6, ABS doesn't work and the wheels keep regenerative braking. From Fig 7, it is clearly that the rear wheel is more apt to slip than the front wheel for its normal tire force is smaller than the front when braking. So ABS firstly takes effect on the rear wheels while the front wheels are regenerative braking. When the road condition is bad such as icy road, ABS becomes operative very soon as shown in Fig 8.

induction motor ABS has the same anti-lock effect as the mechanical ABS which is commonly in used, but can enhance the driving range. Furthermore, the motor response is much quicker than mechanical system, leading to faster response and shorter brake distance.

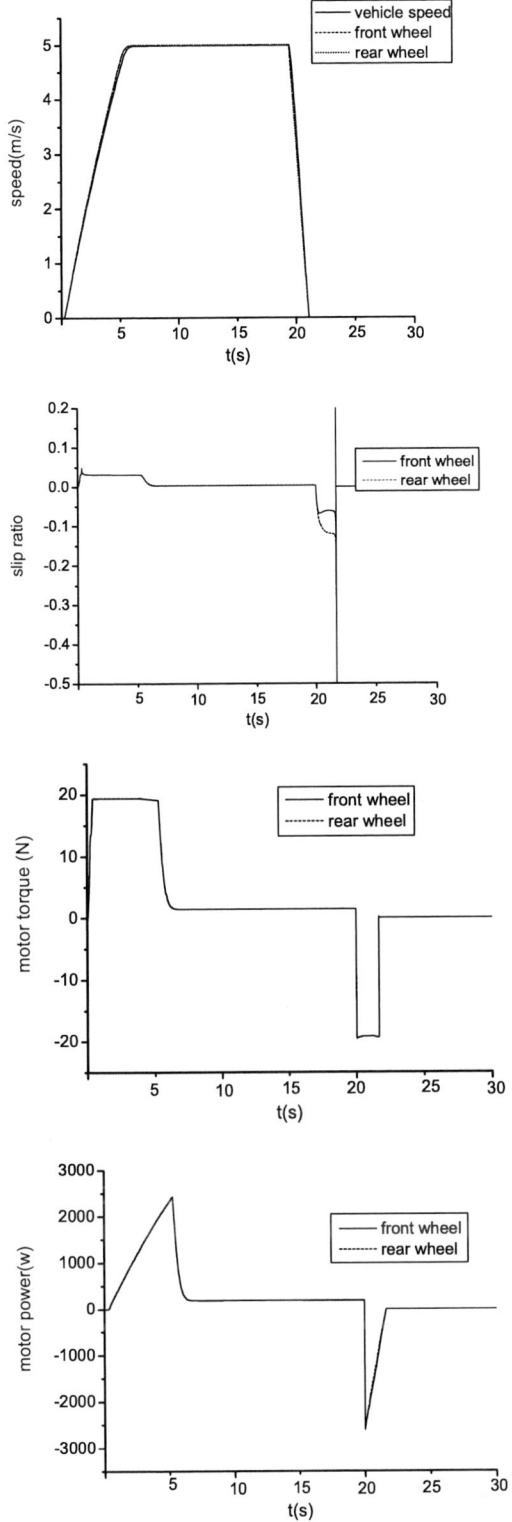

Fig 6: .Dry road curve for 5m/s, $\mu_p = 1, \mu_0 = 0.8$

The figures show three road conditions. Every condition includes four figures describing speed of the vehicle and the wheels, wheel slip ratio, motor torque and motor power. From this figures, it is shown that the electric vehicle regenerates power when braking no matter what the road condition is. Obviously, when the road condition better, more power can be regenerated. So

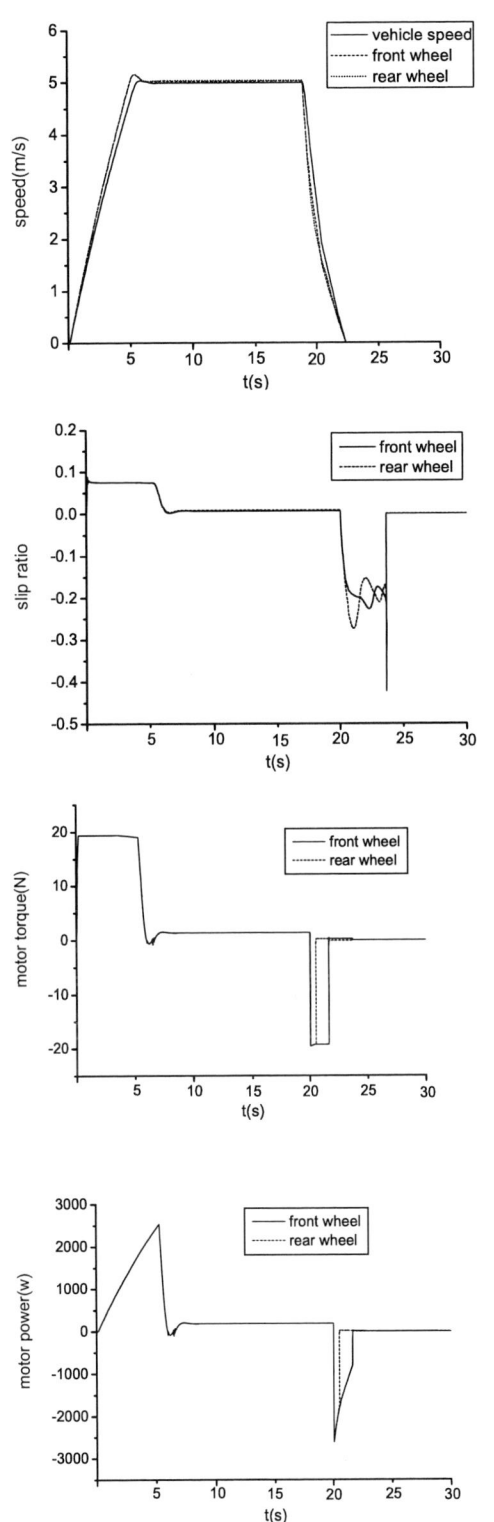

Fig 7: Wet road curve for 5m/s, $\mu_p = 0.4, \mu_0 = 0.2$

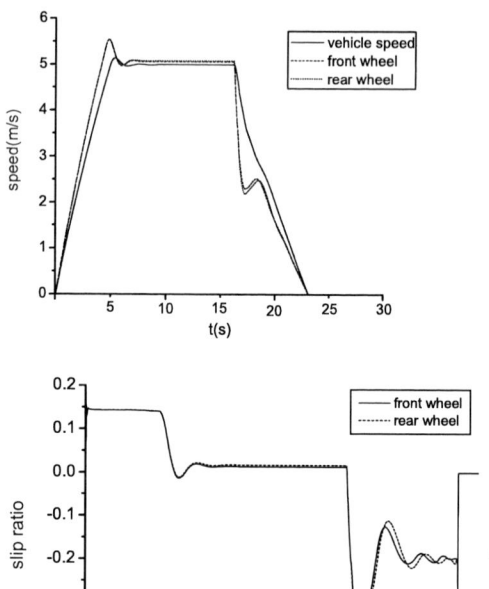

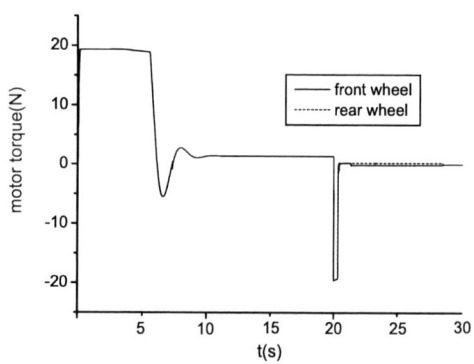

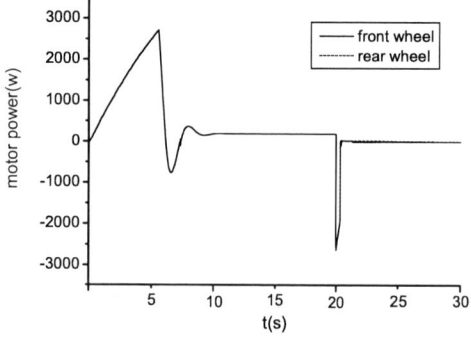

Fig8: Icy road curve for 5m/s, $\mu_p = 0.2, \mu_0 = 0.15$

From all the figures above, it is shown that fuzzy controller ABS has strong robustness and can minimize the brake distance while keeping the wheel unlocked.

V. CONCLUSION

In this paper, the braking system for electric vehicle has been developed. Motors provide maximum braking torque, the rest force is provided by the hydraulic braking system. If the braking torque is so large that the slip ratio is unstable, antilock braking system will take effect. The system mathematical models, such as vehicle dynamics, tire force ,the brake hydraulic system and motor control strategy, have been discussed and the fuzzy controller for the antilock braking system has been established. The simulations showed that a fuzzy controller is an effective means to provide braking torque control over operating conditions ranging from dry road to icy road. The robustness of the fuzzy controller has been shown with the ability to adjust to varying road surfaces. The method discussed in this paper could be used and more practical experiments should be done for improvement in the future.

REFERENCES

[1] Modern electric vehicle technology, Chan, C. C., New York ; Oxford : Oxford University Press, 2001

[2] Chih-Keng Chen, Yu-Chi Wang, "Fuzzy Control for the Anti-lock Brake System", *The 1996 Asian Fuzzy Systems Symposium.* 11-14 Dec. pp.:67 - 72

[3] Madau, D.P., Yuan, F., Davis. L.I., Jr. Feldkamp, L.A. "Fuzzy Logic Anti-Lock Brake System for a Limited Range Coefficient of Friction Surface", *IEEE Int.Conf. on Fuzzy Systems*, 1993. pp.:883 – 888.

[4] Khatun, P., Bingham, C.M., Schofield, N., Mellor, P.H., "Application of Fuzzy Control Algorithms for Electric Vehicle Antilock Braking/Traction Control Systems", *IEEE Transactions on Vehicular Technology*, Vol. 52, Issue 5, Sept. 2003. pp.:1356 – 1364

[5] Tahami, F., Kazemi, R., Farhanghi, S., "A Novel Driver Assist Stability System for All-Wheel-Drive Electric Vehicles", IEEE Transactions on Vehicular Technology, Vol. 52, Issue 3, May 2003. pp.:683 - 692

[6] Anwar, S., "An Anti-Lock Braking Control System for a Hybrid Electromagnetic/ Electro hydraulic Brake-By-Wire System", *American Control Conference*, Vol. 3, 30 June-2 July 2004. pp.:2699-2704.

[7] Jezernik, K , "Speed sensorless torque control of induction motor for EV's.", *7th International Workshop on Advanced Motion Control.* 3-5 July 2002. pp.:236 – 241.

[8] Mauer, G.F., "A Fuzzy Logic Controller for an ABS Braking System", *IEEE Transactions on Fuzzy Systems*, Vol. 3, Issue 4, Nov. 1995 . pp.381 - 388.

[9] Khatun, P., Bingham, C.M., Schofield, N., Mellor, P.H., "Application of Fuzzy Control Algorithms for Electric Vehicle Antilock Braking/Traction Control Systems", *IEEE Transactions on Vehicular Technology*, Vol. 52, Issue 5, Sept. 2003. pp.1356 – 1364

[10] Vehicle dynamics and control, Rajamani, Rajesh., New York : Springer, c2006.

Skid Steering in 4-Wheel-Drive Electric Vehicle

Gao Shuang[12], Norbert C. Cheung[1], K. W. Eric Cheng[1], Dong Lei[2], Liao Xiaozhong[2]

[1]Power Electronics Research Center, Dept. of EE, The Hong Kong Polytechnic University

[2]Dept. of Automatic Control, Beijing Institute of Technology

Abstract--This paper discusses skid steering applied to four wheel drive electric vehicles. In such vehicles, steering is achieved by differentially varying the speeds of the lines of wheels on different sides of the vehicle in order to induce yaw. Skid steer wheeled vehicles require elaborate tire model, so I choose the unite semi-empirical tire model. From this model, longitudinal and lateral tire force can be calculated by slip ratio directly. The vehicle model has 3-DOF, longitudinal, lateral and yaw direction, irrespective of suspension. Induction motor is chosen as the driven motor, and the control method is rotor flux field oriented vector control. To satisfy the requirement of the turn radius, the longitudinal slip must be controlled, so a method of slip limitation feedback is used in the simulation. When the vehicle is turning on a slippery surface, because of the drop at the coefficient of road adhesion, the drive wheels may slip. The traction control system reduces the engine torque and brings the slipping wheels into the desirable skid range. Some simulation results about the steering accuracy and maneuverability are given in the paper.

Index Terms--Skid steer, electric vehicle, induction motor, vector control

I. INTRODUCTION

AS is known to all, for skid steering vehicle, all of the wheels are non-steerable, and lateral slippage must appear [1]. Skid steering can be compact, light, require few parts, and exhibit agility from point turning to line driving using only the motions, components, and swept volume needed for straight driving. However, as the turn radius decreases from straight driving to a point turn, greater power and torque are required as a greater sideslip angle is encountered. For all turns skid steering requires greater power and torque than for explicit turning because sideslip angles are greater in all cases. However, it provides alternative method of steering and have certain advantage and disadvantage as compared with the conventional steering method. The lateral forces of skid steering are greater [2][3][4]. Fig. 1 shows the skid steer and explicit steer.

Skid steering is accomplished by creating a differential velocity between the inner and outer wheels.

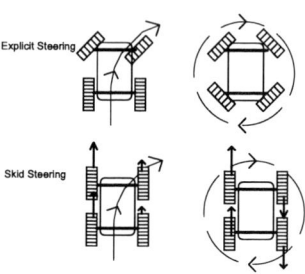

Fig. 1. Skid steer and explicit steer

There are three kinds of control strategies: with the inner wheels braked, by simultaneously speeding up the outer wheels and slowing down inner wheels, and by only speeding up the outer wheels [5]. It is found that the second method minimizes transients.

The increasing prospect of electric drive vehicles and the development of individual wheel traction motors, opens up the possibility of designing higher performance skid steer systems [5]. The use of separate traction motors at each wheel implies that torque to each drive wheel can be controlled independently. This kind of wheels are known as motorized wheels, which permit packaging flexibility by eliminating the central drive motor and the associated transmission and driveline components, including the transmission, the differential, the universal joints and the drive shaft [6]. So it provides a system that is regenerative by nature, by using motor braking, the power absorbed at one wheel is available for application at an opposing wheel.

By controlling the traction of each wheel individually, the vehicle yaw moment can be controlled directly [7]-[12]. However, perfect control of those independent motors will be a critical issue in this configuration.

In [6], a stability system in all-wheel-drive Electric Vehicle is introduced. The system comprises a fuzzy logic system that independently controls wheel torques to prevent vehicle spin. It also points out that yaw rate control of a vehicle by utilizing skid steering method is usually addressed as Direct Yaw-moment Control (DYC). It has been proved that DYC is more effective in enhancing vehicle stability than four wheel steering. Actually, the yaw moment resulting from difference in longitudinal tire force of left and right wheels is insignificantly influenced by lateral acceleration. On the contrary, the yaw moment generated by four-wheel steering decreases as the lateral acceleration increases.

The steering performance of a 4-wheel-drive (4WD) skid steer vehicle is presented in [5], the author states that the actual turn radius will be larger than the desired value because of the wheel slip.

This work was supported by The Hong Kong Polytechnic University under the project number 1-BB86.

Gao Shuang is with both the Beijing Institute of Technology and The Hong Kong Polytechnic University (e-mail:gao19820808@163.com).

Norbert C. Cheung and K.W.E.Cheng are also with the Power Electronics Research Center, The Hong Kong Polytechnic University, Kowloon, Hong Kong (e-mail: norbert.cheung@polyu.edu.hk, eeecheng@polyu,.edu.hk).

Dong Lei and Liao Xiaozhong are with the Beijing Institute of Technology (e-mail: pemc.bit@163.com).

978-1-4244-0644-9/07/$25.00 ©2007 IEEE

In this paper, we develop a 4WD skid steering electric vehicle, which is driven by four separate induction motors in each wheel, Fig. 2 shows the configuration of the motorized wheels on one side of the vehicle. The united semi-empirical tire model is developed here which is different from most papers in tire modeling where linear model is prevalent.

To distribute the torque of each wheel well enough, the slip ratio is fed back to adjust the given speed of the driven motors, then the torque is distributed again in order to satisfy the driver command. The objective is to analyze the behavior of skid steering applied to 4x4 electric vehicles when the inner and the outer wheel are given different speeds, the main characteristics are based on maneuverability, stability and the accuracy.

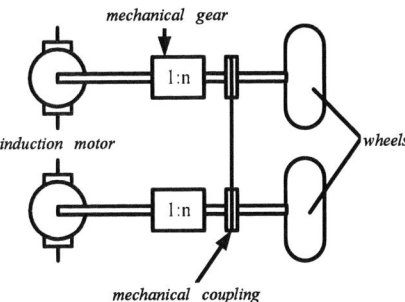

Fig. 2. Configuration of motorized wheel

II. SKID STEERING SYSTEM

The entire system consists of several models, including tire model, vehicle model, induction motor drive model and the slip limitation feedback controller.

A. Entire System

The entire system is shown in Fig 3. Skid steering electric vehicle model is shown in Fig. 4.

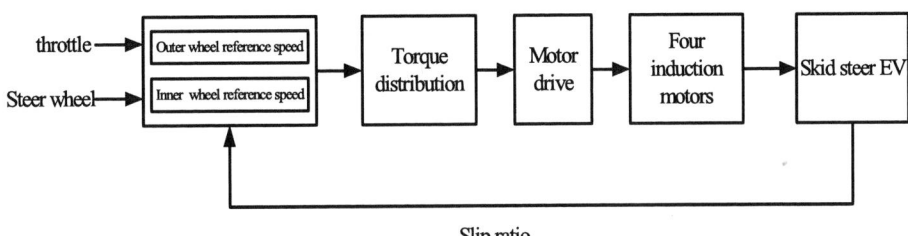

Fig. 3. Entire system

The input of the simulation system is the desired speed and the turn radius, then the inner and outer speed can be calculated from (1) and (2).

B. Tire Model

Regardless of the inability of the linear model to provide the wheel tractive force, which is the most important factor is that the linear model could not properly describe the overall vehicle dynamic behavior in some driving conditions. It is well known that the most accurate tire model is the magic formula model, developed by Bakker et al. But it is also the one requiring the longest computation time for the determination for the coefficients. The tire model used in this simulation study is relatively more complex than in a simple linear model, which is called unite semi-empirical tire model. From this model, longitudinal and lateral tire force can be calculated by slip ratio directly.

List of symbols:
μ —tire-road total friction coefficient

V_x —longitudinal tire speed

V_y —lateral tire speed

ω —wheel angular speed

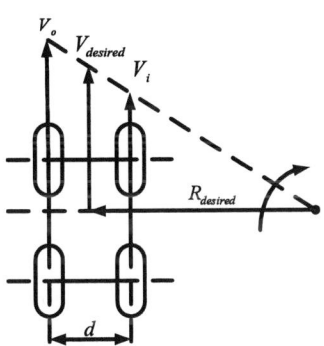

Fig. 4. Skid steering electric vehicle model

The radius of the turn can be calculated from similar triangles

$$R_{desired} = \frac{V_o + V_i}{V_o - V_i} \cdot \frac{d}{2} \qquad (1)$$

$$V_{desired} = \frac{V_o + V_i}{2} \qquad (2)$$

Where $R_{desired}$ and $V_{desired}$ are the desired speed and the turn radius, V_o and V_i denote the desired outer wheel speed and the inner wheel speed respectively, d represents the distance of the left and the right wheel.

1698

F_z —normal force

R —tire radius

μ_0 —tire-road static friction coefficient

V_0 —speed constant

k_x —tire initial infinitude relative longitudinal stiffness

k_y —tire initial infinitude relative lateral stiffness

E_y —tire force characteristic parameter

F_f — roll friction

F_w — wind friction

F_x — longitudinal tire force

F_y — lateral tire force

m — mass of vehicle

I_z — moments of inertia about Z axis

lf — distance from front axle to CG

lr — distance from rear axle to CG

d — distance between the left and the right wheel

θ — yaw angle

M_f — roll resistance moment

J_w — moment of inertia

g — acceleration of gravity

h — height of CG

C_0 —wind block coefficient

A — wind area

γ — yaw rate

f — roll resistance coefficient

u — longitudinal speed of CG in fixed reference frame

v — lateral speed of CG in fixed reference frame

T_d —electromagnetic torque of induction motor

n_p —number of poles

The tire-road relative slip velocity V_s and friction coefficient μ can be calculated by longitudinal speed V_x and lateral vehicle speed V_y and the wheel angular speed ω which are illustrated in Fig. 5.

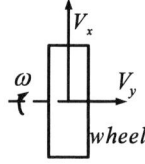

Fig. 5. Wheel speed

$$V_s = \sqrt{(V_x - \omega \cdot R)^2 + V_y^2} \tag{3}$$

$$\mu = \mu_0 \cdot e^{\frac{(V_x - \omega R)^2 + V_y^2}{V_0}} \tag{4}$$

longitudinal slip

$$s_x = \frac{V_x - \omega R}{\omega R} \tag{5}$$

lateral slip

$$s_y = \frac{V_y}{\omega R} \tag{6}$$

infinitude relative longitudinal slip

$$\phi_x = \frac{k_x \cdot s_x}{\mu} \tag{7}$$

infinitude relative lateral slip

$$\phi_y = \frac{k_y \cdot s_y}{\mu} \tag{8}$$

infinitude total slip

$$\phi = \sqrt{\phi_x^2 + \phi_y^2} \tag{9}$$

With (3)-(9), we get the longitudinal tire force and the lateral tire force (10), (11).

$$F_x = -\frac{\phi_x}{\phi} \cdot F_z \cdot \mu \cdot [1 - e^{-\phi - E_y \cdot \phi^2 - (E_y^2 + \frac{1}{12}) \cdot \phi^3}] \tag{10}$$

$$F_y = -\frac{\phi_y}{\phi} \cdot F_z \cdot \mu \cdot [1 - e^{-\phi - E_y \cdot \phi^2 - (E_y^2 + \frac{1}{12}) \cdot \phi^3}] \tag{11}$$

C. Vehicle Model

The coordinate system of the car shown in Fig. 6 is based at the bottom of the wheel where the X coordinate is in the direction of wheel travel. The Y coordinate is parallel to the axis of the wheel's rotation and the Z coordinate is perpendicular to the ground.

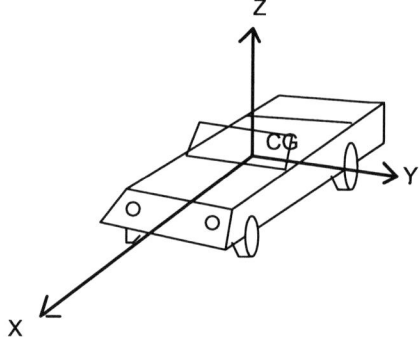

Fig. 6. Car coordinate system

We develop a vehicle dynamic model by neglecting some effects introduced by suspension and tire deformation.

The following assumptions are made:

1) Vehicle is moving on the horizontal plane

2) Vehicle speed is very low

3) Vehicle has no vertical motion

4) Vehicle has no pitch motion about Y axis or roll motion about X axis

Referring to Fig. 7, $O(X, Y)$ defines a fixed reference frame and $CG(x, y)$ is a moving frame attached to the vehicle body with origin at the center of mass CG. The center of mass is located at distances lf and lr from front and rear wheels respectively. The distance of left and right wheel is d.

The weight transfers are taken into consideration in the vehicle model. The lateral tire forces are thought of as causing resistance to turning, which is overcome by the differential longitudinal (or tractive) forces between the two sides of the vehicle. The lateral and longitudinal tire force induced yaw moments are balanced to provide the indicated yaw acceleration.

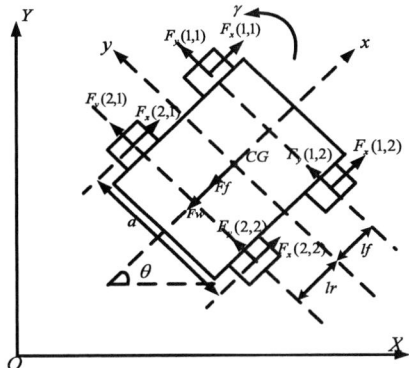

Fig. 7. Motion frame of skid steer EV

The 3-DOF motion equations in the moving frame can be written in (12)-(14), where the subscript (1,1), (1,2), (2,1) and (2,2) indicate the front left (FL) wheel, the front right (FR) wheel, the rear left (RL) wheel and the rear right (RR) wheel separately.

Longitudinal:

$$m(\dot{u} - \gamma \cdot v) = F_x(1,1) + F_x(1,2) + F_x(2,1) + F_x(2,2) - F_f - F_w \quad (12)$$

Lateral:

$$m(\dot{v} + \gamma \cdot u) = F_y(1,1) + F_y(1,2) + F_y(2,1) + F_y(2,2) \quad (13)$$

Yaw:

$$I_z \cdot \dot{\gamma} = [F_y(1,1) + F_y(1,2)] \cdot lf - [F_y(2,1) + F_y(2,2)] \cdot lr \quad (14)$$
$$- \frac{d}{2}[F_x(1,1) - F_x(1,2) + F_x(2,1) - F_x(2,2)]$$

We can also get the torque equilibrium equations (15) of the four wheels from moments on a wheel in Fig. 8.

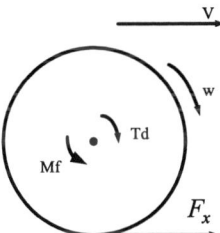

Fig. 8. Moments on a wheel

$$J_w \cdot \dot{\omega}(1,1) = T_d(1,1) - F_x(1,1) \cdot R - M_f(1,1)$$

$$J_w \cdot \dot{\omega}(1,2) = T_d(1,2) - F_x(1,2) \cdot R - M_f(1,2) \quad (15)$$

$$J_w \cdot \dot{\omega}(2,1) = T_d(2,1) - F_x(2,1) \cdot R - M_f(2,1)$$

$$J_w \cdot \dot{\omega}(2,2) = T_d(2,2) - F_x(2,2) \cdot R - M_f(2,2)$$

The longitudinal speed and lateral speed of each wheel are given by (16).

$$\begin{array}{ll} V_x(1,1) = u - \gamma \cdot \dfrac{d}{2} & V_y(1,1) = v + \gamma \cdot lf \\[2mm] V_x(1,2) = u + \gamma \cdot \dfrac{d}{2} & V_y(1,2) = v + \gamma \cdot lf \\[2mm] V_x(2,1) = u - \gamma \cdot \dfrac{d}{2} & V_y(2,1) = v - \gamma \cdot lr \\[2mm] V_x(2,2) = u + \gamma \cdot \dfrac{d}{2} & V_y(2,2) = v - \gamma \cdot lr \end{array} \quad (16)$$

The equations (17)-(23) can be derived from the vehicle locomotion

$$M_f(i.j) = F_z(i,j) \cdot f \cdot (1 + \frac{u^2}{1500}) \cdot R \quad (17)$$

$$F_z(1,1) = \frac{m}{lf + lr}[\frac{1}{2}g \cdot lr - \frac{1}{2}(\dot{u} - \gamma \cdot v) \cdot h + \frac{(\dot{v} + \gamma \cdot u) \cdot h \cdot lr}{d}]$$

$$F_z(1,2) = \frac{m}{lf + lr}[\frac{1}{2}g \cdot lr - \frac{1}{2}(\dot{u} - \gamma \cdot v) \cdot h - \frac{(\dot{v} + \gamma \cdot u) \cdot h \cdot lr}{d}] \quad (18$$

$$F_z(2,1) = \frac{m}{lf + lr}[\frac{1}{2}g \cdot lf + \frac{1}{2}(\dot{u} - \gamma \cdot v) \cdot h + \frac{(\dot{v} + \gamma \cdot u) \cdot h \cdot lr}{d}]$$

$$F_z(2,2) = \frac{m}{lf + lr}[\frac{1}{2}g \cdot lf + \frac{1}{2}(\dot{u} - \gamma \cdot v) \cdot h - \frac{(\dot{v} + \gamma \cdot u) \cdot h \cdot lr}{d}]$$

$$)$$

$$F_w = \frac{1}{21.15} \cdot C_0 \cdot A \cdot u^2 \quad (19)$$

$$F_f = m \cdot g \cdot (1 + \frac{u^2}{1500}) \quad (20)$$

The heading angle

$$\theta = \int \gamma \cdot dt \quad (21)$$

The longitudinal displacement in fixed reference frame

$$X = \int (u \cdot \cos\theta - v \cdot \sin\theta)dt \quad (22)$$

The lateral displacement in fixed reference frame

$$Y = \int (-u \cdot \sin\theta - v \cdot \cos\theta)dt \quad (23)$$

D. Induction Motor Drive

Vector control of induction motors has been widely used in high performance drive system. Field oriented induction motor drive systems offer high performance as well as independent control on torque and flux. The rotor flux field oriented vector control has been used as the induction motor drive method.

E. Slip Limitation Feedback Controller

Desired turn radius will only be achieved if no slippage occurs between the wheel and ground [5], to satisfy the requirement of the turn radius, the longitudinal slip must be controlled, so a method of slip limitation feedback is used in the simulation.

The desired turn radius is given in (1). The desired yaw rate

$$\gamma_{desired} = \frac{V_o - V_i}{d} \qquad (24)$$

where V_o and V_i denote the desired outer wheel speed and the inner wheel speed respectively, d represents the distance of the left and the right wheel.

Therefore

$$V_o = \omega_o \cdot R \qquad (25)$$
$$V_i = \omega_i \cdot R \qquad (26)$$

where ω_o and ω_i represent outer and inner wheel angular speed.

As is mentioned before, the longitudinal slip of the wheel is

$$s_x = \frac{V_x - \omega R}{\omega R} \qquad (27)$$

That is to say,

$$V_x = (s_x + 1) \cdot \omega R \qquad (28)$$

The actual turn radius will be

$$R_{actual} = \frac{V_o(1+s_o) + V_i(1+s_i)}{V_o(1+s_o) - V_i(1+s_i)} \cdot \frac{d}{2} \qquad (29)$$

The actual yaw rate

$$\gamma_{actual} = \frac{V_o(1+s_o) - V_i(1+s_i)}{d} \qquad (30)$$

When the vehicle turns, the outer wheel accelerates and the inner wheel decelerates, that is to say, for the outer wheel, the wheel rotate speed is greater than the wheel line speed, and for the inner wheel, the case is just reverse. So the longitudinal slip of the outer wheel s_o is less than zero, and the inner wheel slip s_i is greater than zero, from the equation of the actual turn radius, we can conclude that actual difference between the outer and inner speed will be less than the desired value, so the actual turn radius and yaw rate will be smaller than the desired values.

To satisfy the requirements of the turn radius and the yaw rate, the difference between the outer and inner speed must be controlled.

To increase the reference speed of the outer wheel and the inner wheel, the method of slip limitation feedback is used here. Firstly, the longitudinal slips of the four wheels are calculated and fed back to the reference speed controller, in the controller, the slips are limited to a range, then the reference speed are revised to a new value, and the output torque of the induction motor will be revised until the turn radius and the yaw rate reach to the desired values.

The reference speed are revised to

$$V_{oreference} = \frac{V_o}{1 + s_o} \qquad (31)$$

$$V_{ireference} = \frac{V_i}{1 + s_i} \qquad (32)$$

Replacing V_o and V_i in (1) and (24) by $V_{oreference}$ and $V_{ireference}$ respectively in the equations of actual turn radius and the yaw rate, then the actual turn radius and the yaw rate will be equal to the desired value.

III. SIMULATION RESULTS

Fig. 10 shows locus of the EV for different turn radius with tire-road friction coefficient is 0.8, and Fig. 11 to 14 illuminate the torque requirement for different turn radius for each wheel. We can see that the actual turn radius is well accordant with desired command from the driver, and as the turn radius becomes smaller the more torque is needed for each wheel at some constant speed.

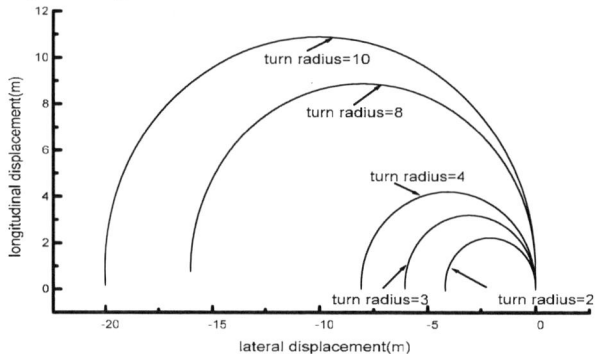

Fig. 10. Locus of the EV for different turn radius

1701

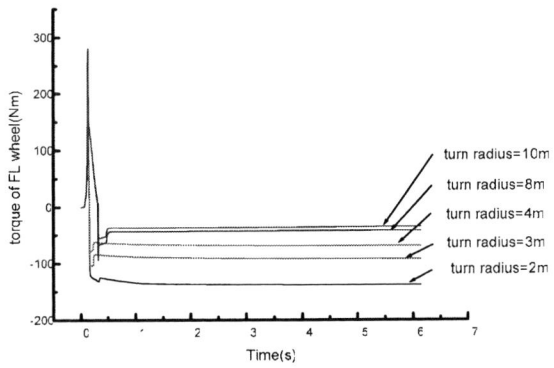

Fig. 11. Torque requirement of the FL wheel for different turn radius

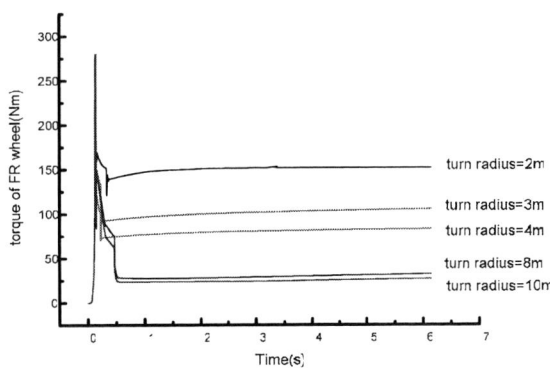

Fig. 12. Torque requirement of the FR wheel for different turn radius

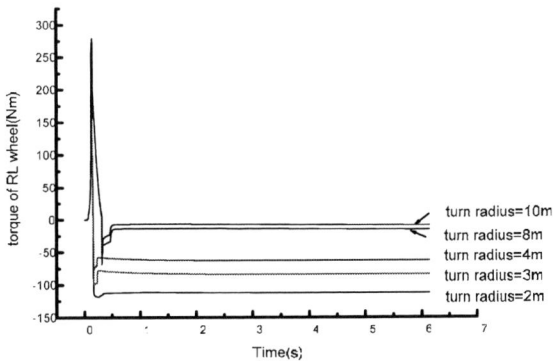

Fig. 13. Torque requirement of the RL wheel for different turn radius

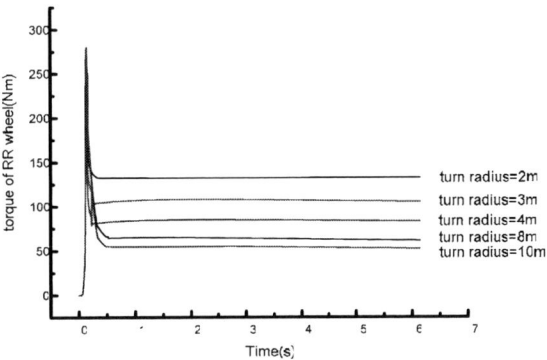

Fig. 14. Torque requirement of the RR wheel for different turn radius

IV. CONCLUSIONS

The paper describes a model for predicting the steering performance of a 4WD skid steer wheeled vehicle. The unite half-experience tire model is used here, which is more elaborate and the longitudinal and lateral tire force can be calculated by slip ratio directly. Induction motor is chosen as the driven motor, and the control method is rotor flux field oriented vector control. Slip limitation feedback controller is used to satisfy the requirement of the turn radius and the yaw rate. The simulation results show that the steering accuracy and maneuverability can be achieved, and the traction control system can adjust the motor torque to satisfy different road conditions.

V. REFERENCES

[1] D. Pazderski, K. Kozlowski, M. Lawniczak, "Practical stabilization of 4WD skid steering mobile robot," Proc. of the Fourth International Workshop on Robot Motion and Control, Puszczykowo, pp. 175-180, 2004.

[2] Benjamin Shamah, "Experimental Comparison of Skid Steering Vs. Explicit Steering for a Wheeled Mobile Robot," master's thesis, tech. report CMU-RI-TR-99-06, Robotics Institute, Carnegie Mellon University, March, 1999.

[3] E. Faruk Kececi, Gang Tao, "Adaptive vehicle skid control," Mechatronics 16 (2006) 291–301.

[4] Krzysztof, Kozowski, Dariusz, Pazderski, "Modeling and Control of 4wheel skid-steering mobile robot, " Int. J. Appl. Math. Comput. Sci., 2004, Vol. 14, No. 4, 477–496.

[5] Villiam R. Meldrum, Francis B.Hoogterp, Alexander R.Kovnat, "Modeling and Simulation of a Differential Torque Steered Wheeled Vehicle, " Technical Report 13777 July 1999.

[6] Farzad Tahami, Reza Kazemi and Shahrokh Farhangh, "A novel driver-assist stability system for all-wheel-drive Electric Vehicles," IEEE Transactions on Vehicular Technology, Vol. 52, No. 3, MAY 2003.

[7] R.E. Colyer, J.T. Economou, "Soft modeling and fuzzy logic control of wheeled skid-steer electric vehicles with steering prioritization, " International Journal of Approximate Reasoning 22 (1999) 31-52.

[8] Guilin Tao, Zhiyum Ma, Libing Zhou, Langru Li., "A novel driving and control system for direct-wheel-driven electric vehicle, " IEEE Transactions on MAgnetics, Jan. 2005 Vol.41, Issue: 1, Part 2 pp.497 – 500.

[9] U-Sok Chong, Eok Namgoong, Seung Ki Sul, "Torque steering control of 4-wheel drive electric vehicle" Power Electronics in Transportation, 1996. IEEE. pp. 159 – 164.

[10] Sakai S, Sado H, Hori Y, "Motion control in an electric vehicle with four independently driven in-wheel motors," IEEE/ASME Transactions on Mechatronics, March 1999 Vol. 4, Issue: 1, pp. 9 – 16.

[11] Pusca, R. Ait-Amirat, Y. Berthon, A. Kauffmann, J.M. CREEBEL, "Modeling and simulation of a traction control algorithm for an electric vehicle with four separate wheel drives," Vehicular Technology Conference, 2002. Proceedings. VTC 2002-Fall. 2002 IEEE 56th, Vol.3, pp. 1671 – 1675.

[12] Jalili-Kharaajoo, M. Besharati, F., "Sliding mode traction control of an electric vehicle with four separate wheel drives," IEEE Conference Emerging Technologies and Factory Automation, 2003. Proceedings. ETFA '03. Sept. 2003 Vol. 2 , pp. 291 - 296.

A Flexible Multi-Pulse Control Strategy for Universal Nail Collator

Chien-Lung Cheng, *Member, IEEE*, Shyi-Ching Chern, Jim-Chwen Yeh*, Ming-Yi Wu
Department of Electrical Engineering
* Power Mechanical Engineering
National Formosa University
Huwei, Yunlin, 632, Taiwan
clcheng@nfu.edu.tw

Abstract

In this paper, a new control method is proposed to improve nail collator. Developed nail collator can meet various nails. In this paper, VisSim DSP development system is used to speed the improvement of electrical control system and examine control strategy. The proposed method is proved effective by experiment. Control strategy of nail collator is flexible and meets various nails. The design and implementation of nail collator controller with high performances is reached.
Keywords: Nail Collator, Digital Signal Processor, VisSim

1. Introduction

The nail collator is designed to produce wire collated nails. The nail and wire are melted together by electric power to form a wire collated nails. The wire collated nails are used by nail gun. The wire collated nails are widely applied in the world. In recent years, power electronics technology has great progress. Improved controller in nail collator can reach the smaller transformer and high performance circuit. The conventional nail collator often only suits for specified nails. A universal nail collator that widely suits for various nails is very convenient and useful. Therefore, in this paper, an advanced controller based on multi-pulse is developed for universal nail collator.

In this paper, a digital signal process (DSP) TMS320LF2407A is used to set up controller. A VisSim DSP development system VS-ECD2407 is used to speed the controller development. VisSim DSP development system is dynamic system simulation software for Windows. An integrated design system environment based on VS-ECD2407 is set up. Only vision block graph needs to be constructed. It is very helpful to reduce lots of developed time.

2. The Proposed Method

2.1 System Overview

The nail collator is shown in Fig. 1. A programmable logic controller (PLC) controls the operation of nail collator. Fig. 2 shows a front panel to input required data. For example, sensor on or off, welding yes or no etc. should be selected properly. The flow chart of overall operation of nail collator is shown in Fig. 3.

Fig. 1. Nail collator.

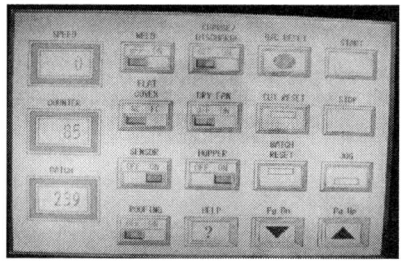

Fig. 2. Front panel of nail collator.

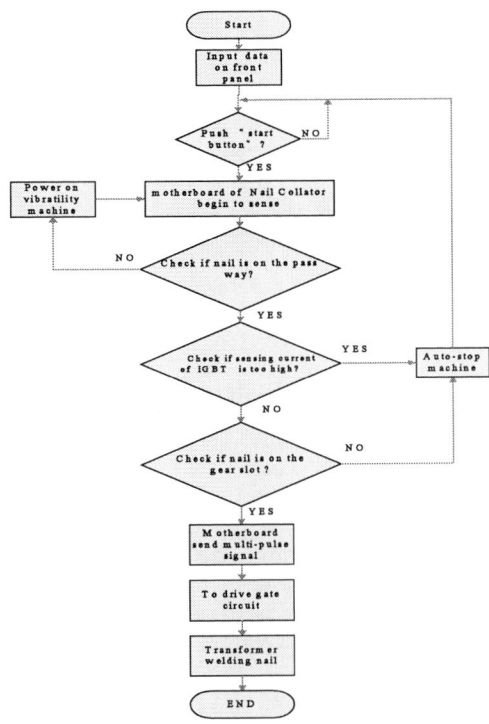

Fig.3. The flow chart of overall operation of nail collator

2.2 Welding Circuit

Fig. 4 shows a welding circuit of nail collator. Welding circuit contains transformer、IGBT、capacity and diode components. The IGBT is droved by multi-pulse controller. When IGBT is closed, the welding circuit is shown in Fig. 5. The energy can flow to secondary winding of transformer. When IGBT is opened, the welding circuit is shown in Fig. 6. The energy is blocked to secondary winding of transformer. The energy of leakage inductance flows into C1 and snubber circuit.

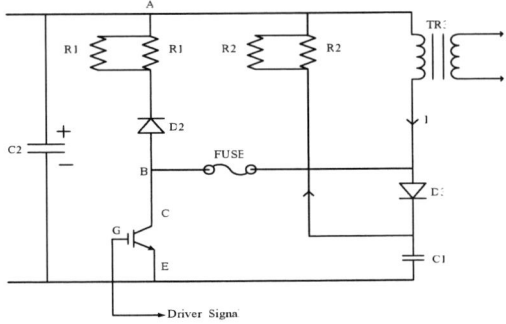

Fig. 4.   Welding circuit of nail collator.

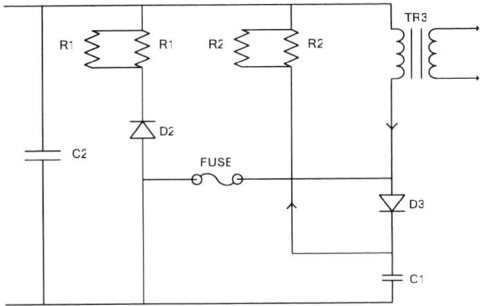

Fig. 5. Welding circuit under IGBT on state

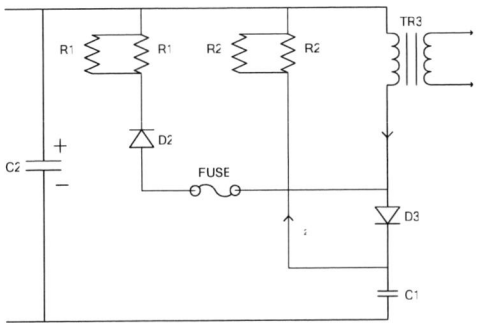

Fig.6. Welding circuit under IGBT off state

2.3 Controller

In this paper, VisSim DSP development system is used to set up controller of nail collator. The overall procedure of applying VisSim DSP development system is shown in Fig. 7.

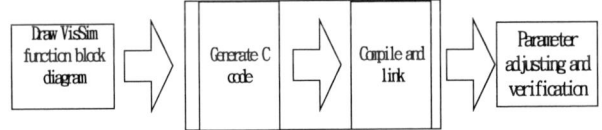

Fig. 7. VisSim DSP development procedure

2.3.1 Fixed Three-pulse Controller

Fig. 8 shows a developed block diagram algorithm based on VisSim DSP development system for fixed three-pulse controller. It is easily developed. A fixed three-pulse signal is sent to gate to drive IGBT. The fixed three-pulse signal is shown in Fig. 9. A darlington circuit is connected to drive IGBT effectively. The nail collator circuit with darlington circuit is shown

1704

in Fig. 10.

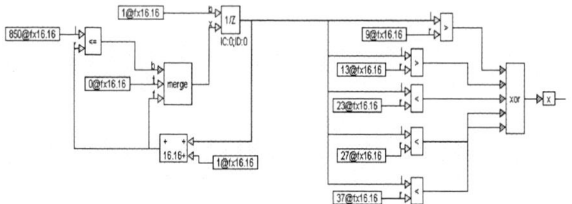

Fig. 8. Block diagram algorithm of fixed three-pulse controller

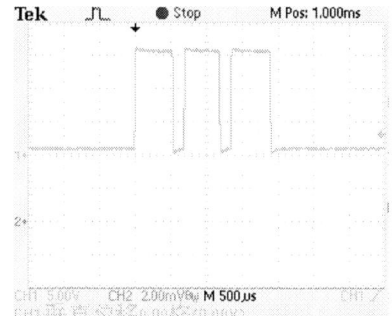

Fig.9. Fixed three-pulse control signal

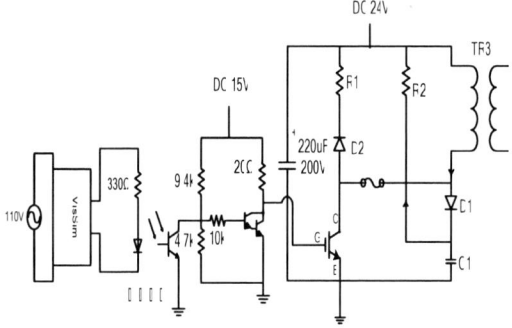

Fig.10. Nail collator circuit with darlington circuit.

2.3.2 Flexible Multi-pulse Controller

In order to develop universal nail collator suited for various nails, a flexible multi-pulse controller is developed. Fig. 11 shows a developed block diagram algorithm based on VisSim DSP development system for flexible multi-pulse controller. User can easily adjust the number and width of pulses. Instead of several hundred lines of assemble program codes, only some tens of VisSim block diagrams are required.

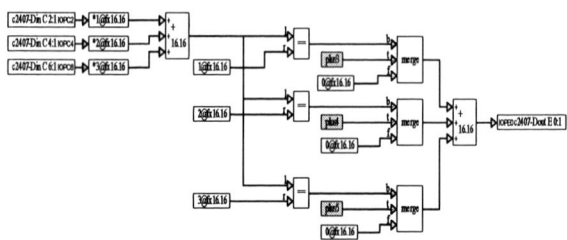

Fig. 11. Block diagram algorithm of flexible multi-pulse controller

3. Result and Discussion

Fig.12 shows experimental results of transformer primary winding in nail collator under fixed 3 pulses. CH1 is primary winding voltage and CH2 is primary winding current. Fig. 13 shows experimental results of transformer secondary winding in nail collator under fixed 3 pulses. CH1 is secondary winding voltage and CH2 is secondary winding current.

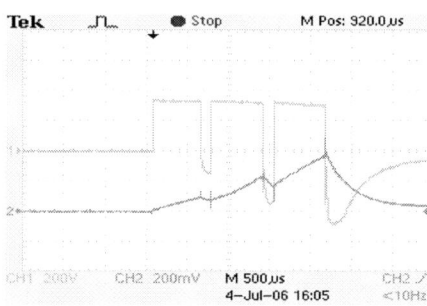

Fig.12. Experimental results of transformer primary winding in nail collator under fixed 3 pulses. CH1: primary winding voltage, CH2: primary winding current.

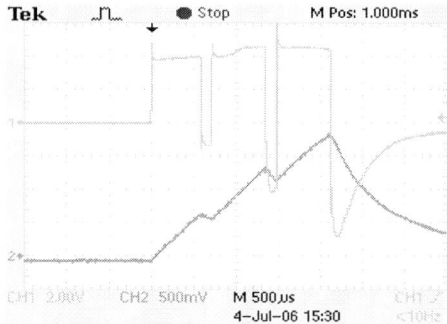

Fig.13. Experimental results of transformer secondary winding in nail collator under fixed 3 pulses. CH1: secondary winding voltage, CH2: secondary winding current.

Observing Figs. 12 and 13, when IGBT is turn-on, the primary and secondary currents obviously increases. The larger current caused the following problem:
(1) Easily make the component too hot and damage.
(2) The larger current need thicker wire, then the cost rises.

Observing Fig. 13, the current becomes very high after 3 pulses. It has very adverse influence for nail collator. The main reason is that accumulated V-t results in flux saturation. Therefore, enough off time is required. The on and off time of pulses in Figs. 12,13 are fixed. Correspondingly, the flux variation of transformer is shown in Fig. 14.

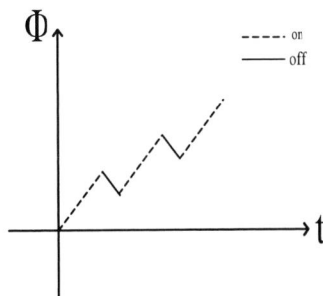

Fig.14. Flux variation of transformer.

To decrease flux saturation, ON-OFF time of pulses and pulses number are adjusted in this paper. By precisely calculating, flexible multi-pulses controller was developed by our group. Fig.15 shows an experimental result of transformer secondary winding in nail collator under flexible 3 pulses. The ON-OFF time is 1.5ms (on) -100us(off)- 200us (on) -100us(off) -200us(on). Fig.16 shows another experimental result of transformer secondary winding in nail collator under flexible 4 pulses. The ON-OFF time is 1.5ms (on) -100us(off)- 100us (on) -100us(off) -100us(on) -100us(off)- 100us (on).

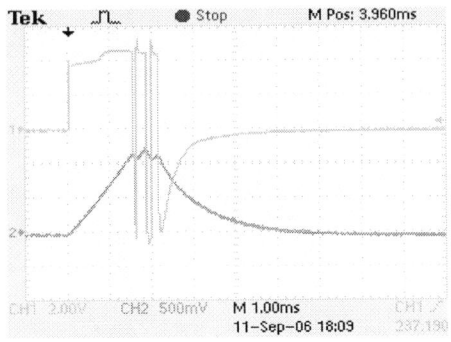

Fig.15 Experimental results of transformer secondary winding in nail collator under flexible 3 pulses. CH1: secondary winding voltage, CH2: secondary winding current. ON-OFF time: 1.5ms (on) -100us(off)- 200us (on) -100us(off) -200us(on)

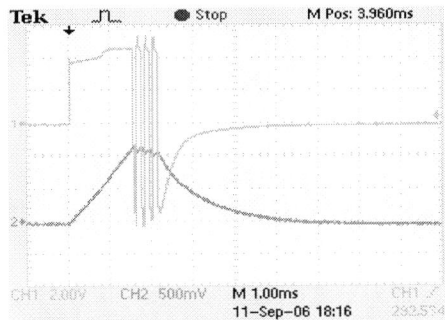

Fig.16 Experimental results of transformer secondary winding in nail collator under flexible 4 pulses. CH1: secondary winding voltage, CH2: secondary winding current. ON-OFF time: 1.5ms (on) -100us(off)- 100us (on) -100us(off) -100us(on) -100us(off)- 100us (on).

Comparing Figs.15, 16 with Fig.13, the current is equally distributed on each pulse. The peak value of current in Figs. 15, 16 obviously decreases to approximately 65% ~ 70% in Figs. 13. It is very useful to reduce device temperature.

4. Conclusion

This paper purposed a flexible multi-pulse controller for nail collator. The controller is easily designed and implemented by VisSim DSP development system. The flexible multi-pulse controller is examined with good performance by experiments. The welding current peak value for flexible multi-pulse controller is obviously decreases to 65% ~ 70% for fixed 3 pulses controller. It is helpful to protect device and reduce device cost. Besides, by controlling flux saturation, more power can output to nail welding. A large range of power output can satisfy various nail welding. A universal nail collator can be reached by using flexible multi-pulse controller.

References

[1] C. L. Cheng, S. W. Yang, J. P. Hong, C. C. HSU, S.C. Chern, C.Y.Yao, S.Y. Wang, 2005, "Design and Implementation of Nail Collator with High Performance", Pro. Of the 26th Symp. On Elec. Power Eng., pp.1881-1886.
[2] C. L. Cheng, S. C. Chern, B. W. Tang, J. C. Huang, Y. C. Hsu, J. C. Yeh, 2004, "A Simple Digital Power Converter Based on VisSim DSP Development System", Pro. Of the 25th Symp. On Electrical Power Engineering., pp.2053-2057.

[3] C. L. Cheng, C. Z. Hsu, Y. C. Hsu, B.W. Tang, J. C. Huang, C. Z. Hsu, "Digital and Implementation of A Simple Digital Controller Based On DSP for Boost Converter", Taiwan Power Electronics Conference, Conference Proceedings, 2004, pp.578-583.

[4] W. T. Chen, T. R. Chen, F. P. Chuo, T. T. Wu, T. P. Lin, L. S. Chen, 2005, "Study of Constant Control for Inverter-type Resistance Spot Welding", Taiwan Power Electronics Conference & Exhibition., pp.418-423.

[5] A.E. Fitzgerald, Charles Kingsley, Jr. Stephen D.U mans, "Electric Machinery", GAU LIH book corporation, 1995.

Cycloconverter Based Three Phase Induction Motor to Replace Flywheel of the Process Machine

M.V. Palandurkar* M. A. Chaudhari**   J. P. Modak***   S. G. Tarnekar*

* Deptt. of EE, GHRCE, Nagpur, M.S, India
** Deptt. of EE, YCCE, Nagpur, M.S, India
*** Deptt. of ME, PCE&A, Nagpur, M.S, India

Abstract-- **This paper details with the logic of developing an appropriate drive to eliminate flywheel of the process machine including provision for variation in load. This is achieved by using three phase cycloconverter based variable voltage variable frequency (VVVF) drive. To realize this, a demand torque characteristic of a specific process machine is studied. The cycle duration of demand torque characteristics is divided into suitable number of time intervals. The subdivisions of time intervals in form of frequencies are tabulated to simplify design procedure. Change in frequency on particular subdivision results in demand torque of the induction motor. To meet this new frequency and voltage, three phase cycloconverter is designed with required frequency output. Finally to support theoretical analysis, simulation results are presented.**

Keywords—Cycloconverter, Flywheel, Induction Motor, Process Machine.

I. INTRODUCTION

Generally there are two types of process machines, Type I and Type II. In Type I process machine, input motion is completely rotary and output motion is also rotary at uniform speed. In such process machine, demand torque does not vary cyclically, but varies with the load. In Type II, process machine makes use of link mechanism or cam mechanism or combination of linkage, cam and gears. For such process unit, at every instant, demand torque changes with respect to time. The arbitrary demand torque characteristic of any process machine can be estimated based on cycle time of operation, process resistance and inertia resistance [1, 2]. Hence, this variation is cyclic and cycle time is commensurate with rpm of process unit. But usually Induction motor cannot generate similar torque characteristics. Hence, the flywheel is required to make up for the difference of the torque in different sections of time axis. The flywheel decelerates its speed when torque is more than average torque and gains its speed when torque is increasing in nature or less than average torque. Presence of flywheel in the process machine increases torsional vibration and fatigue in the component of power transmission system. Therefore, it is necessary to eliminate flywheel from the design of any process machine in general. It is felt that by proper interfacing of power electronic devices this may be possible. Hence, in present paper among different control technique, variable voltage variable frequencies (VV-VF) method is chosen to design three phase cycloconverter to drive three phase Induction motor to get desired frequencies varying with different time intervals that generates supply torque characteristics matching with demand torque characteristics of the process machine.

Fig. 1 describes an arbitrary demand torque characteristic of any process machine, where T_d is demand torque and T_s is supply torque. This can be estimated based on cycle time of operation, process resistance and inertia resistance. These can be detailed based on intended operation and proposed details of partial mechanical design [1, 2]. It is evident from Fig. 1 that crank speed of input shaft of the process machine says it is 30 rpm. Complete cycle of operation of the process unit should be 2 sec. getting completed in one rotation of the input link of the process machine and speed of the input crank must be 30 rpm. The Fig. 1 shows that demand torque varies with time, which the motor cannot generate similar. Supply torque characteristic. Hence the flywheel is required to make up for the difference of the torque in sections AB & CD of time axis. It is known that flywheel will decelerate during intervals AB & CD whereas it will gain speed during intervals OA, BC, DE sections of time axis.

Fig. 2 describes the schematics of an arbitrary process unit, *P* along with usual mechanical power transmission system for torque amplification and speed reduction. Pulley *D2* is a flywheel cum power transmission pulley, pulley *D1* is driven pulley and *M* is the Induction motor. The portion of the system between *D2* and process unit is subjected to severe torsional vibrations and associated fatigue damage.

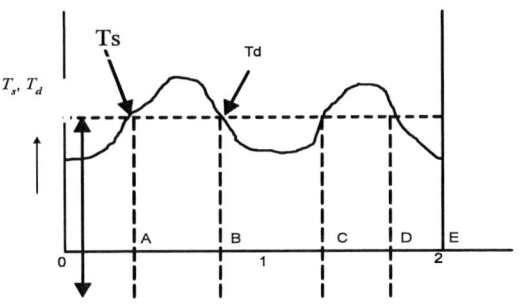

Fig.1. Arbitrary Torque- Time characteristic

978-1-4244-0644-9/07/$25.00 ©2007 IEEE   1708

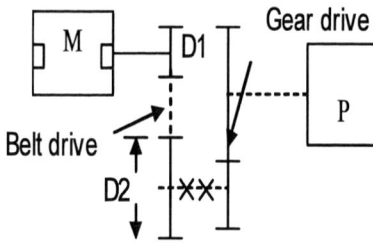

Fig. 2. Schematics of an Arbitrary Process Unit, Mechanical Power Transmission & 3 phase Induction Motor

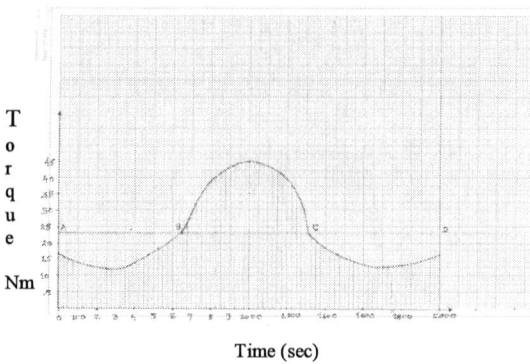

Time (sec)

Fig. 3. Demand Torque characteristic of a specified machine

II. ANALYSIS OF THE PROPOSED SOLUTION

With the advent of electric drives and power electronics circuitry [3], now it is possible to design power supply to induction motor to generate supply torque closely matching with demand torque resulting in elimination of flywheel, with only difference of instantaneous generated torque values would be 1/30 of demand torque of process unit input shaft. The expected supply torque characteristics would be like as shown in Fig. 3

As supply frequency is 50 cycles per second, the cycle time of one cycle of 3-phase supply is 20 ms and complete cycle of operation of the process unit should be 2 sec or 2000 ms (shown in Fig. 3). Now each linear part of the demand torque characteristics is divided in suitable time intervals. These time intervals are further divided into some equal subdivision such that frequency will be same for each division. Now for each division, average torque is calculated. Calculated magnitude of average torque will be different (depends upon nature of torque characteristics) with respect to same frequency of each division of specified time intervals. Division of demand torque characteristics is based on linearizing characteristics and nature of characteristics, hence frequency of all time intervals may not be same. But frequency is same in each subdivision of time intervals. Therefore, it is necessary to design a system using three phase cycloconverter to get desired frequencies varying with different time intervals. With these frequencies and calculated average torque (from Fig. 3) in each sub division in specified time intervals, require variable voltage can be calculated with given induction motor formula.

III. APPLICATION TO CASE STUDY

Proposed logic is described below for the procedure to be adopted for implementing this concept to a specific case study in which this application is demonstrated.
A process machine is so selected which comprises of some linkage mechanism as a main processor on account of non linear kinematics of the hardware, demand torque characteristics of the process machine is time variant shown in Fig. 3. The demand torque characteristics shows that the average supply torque is 23.8 N-m as generated by motor. The total cycle time of the process machine is

2000 ms. Let the assumed motor be three phase 415 V induction motor with synchronous speed as 1500 rpm. In this case, the average angular velocity of the input crank of the process unit must be 30 rpm. This gives torque amplification from motor shaft to the process unit input shaft of the order of $1500/30 = 50$. Thus the supply torque at the process unit input shaft is 23.8 X 50 = 1190 N-m. Hence the hp demand of the process unit with a given formula is,

$$
\begin{aligned}
hp &= \frac{2 \cdot \pi \cdot N \cdot T}{4500} \\
&= \frac{2 \cdot \pi \cdot 30 \cdot (23.8 \cdot 50)}{4500} \\
&= 5
\end{aligned}
\tag{1}
$$

Now for a given set of average torque and frequency of each sub division of specified time intervals, supply voltages are calculated with the help of induction motor torque formula, which is as under

$$
V_1^2 = \frac{S \cdot \omega_s \cdot (R_1 + R_2/s)^2 + (X_1 + X_2)^2 \cdot T}{0.3 \cdot R_2}
\tag{2}
$$

Where V_1 is impressed voltage per phase, R_1 is stator resistance per phase, R_2 is rotor resistance per phase referred to the stator, X_1 is stator reactance per phase, X_2 is rotor reactance referred to the stator referred to the stator, ω_s is synchronous angular velocity, T is generated torque or average torque of each division (calculated from demand torque characteristics) and S is slip.

IV. DESIGNING OF CYCLOCONVERTER CIRCUIT FOR DIFFERENT FREQUENCIES

Table 1 shows the selection of time interval by considering linearity of demand- torque characteristics. Change in frequency in different time intervals are 40Hz, 20 Hz, 33.3 Hz, and 13.33 Hz. Therefore it is necessary to design a three phase cycloconverter circuit to get desired frequencies [3, 4]. At the same time control circuit is designed to control the operation of cycloconverter so that for first 10 divisions three phase 12 pulse cycloconverter circuit of output frequency of 40 Hz will drive the motor and for next division, 20 Hz output of cycloconverter is required to drive the same

1709

TABLE I
SLECTION OF TIME INTERVAL BY CONSIDERING LINEARITY
OF DEMAND TORQUE CHARACTERISTICS

SR. NO.	TIME INTERVALS (Ms)	SUB DIVISION	FREQUENCY IN Hz (EACH DIV.)
1	0 -250	10	40
2	250-300	1	20
3	300 -325	1	40
4	325 – 375	1	20
5	375 – 500	5	40
6	500 – 600	5	50
7	600 – 660	2	33.3
8	660 – 675	2	13.33
9	675 – 1275	22	40
10	1275 – 1290	2	13.33
11	1290 - 1650	18	50
12	1650 - 1750	4	40
13	1750 - 2000	10	40

motor at that rated speed and so on. The block diagram of cycloconverter circuit is shown in Fig. 4. This is a new technique in which with the help of Matlab / PSIM simulation software, it is possible to simulate a cycloconverter circuit to get desire frequencies and to design a control scheme [5, 6] to operate a cycloconverter circuit at different frequencies to drive an Induction motor with given parameters to generate the supply torque matching with demand torque characteristics (shown in Fig. 3) of the process machine replacing flywheel, and keeping the speed constant.

V. RESULTS

A control scheme is designed and simulated to control the operation of cycloconverter that generate different frequencies, as per Table 1 in each specified intervals to drive a three phase induction motor at variable frequencies at variable supply voltage that generate the supply torque characteristics matching with demand torque characteristics of the specific process machine.

Some of the basic simulation results of three phase cycloconverter in PSIM software of frequencies 40 Hz and 20 Hz are simulated and given to three phase induction motor. Fig. 5 shows the three phase supply voltage of induction motor at 20 Hz frequency where V_{1im}, V_{2im} and V_{3im} are line voltages given to R phase, Y phase and B phase respectively and Fig. 6 shows the three phase currents of induction motor at same frequency where i_{1im}, i_{2im} and i_{3im} are currents in R phase, Y phase and B phase respectively. Similarly Fig. 7 shows the three phase supply voltage of induction motor at 40 Hz frequency where V'_{1im}, V'_{2im} and V'_{3im} are line voltages given to R phase, Y phase and B phase respectively and Fig. 8 shows the three phase currents of induction motor at same frequency where i'_{1im}, i'_{2im} and i'_{3im} are currents in R phase, Y phase and B phase respectively. Similarly three phase cycloconverter with 33.3 Hz frequency can be simulated and given to three phase induction motor. The designing and simulation of control scheme controlling the operation of three phase cycloconverter with different frequencies given to

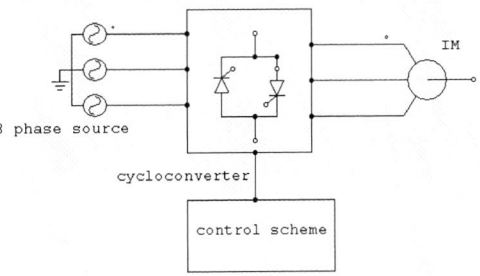

Fig. 4. Schematic diagram of Cycloconverter Based Induction Motor Drive

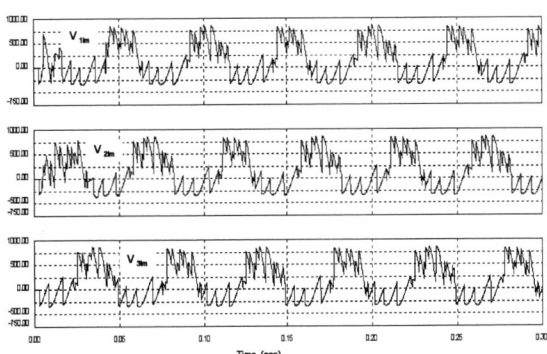

Fig. 5. Three phase voltages of Induction motor running with 20 Hz frequency

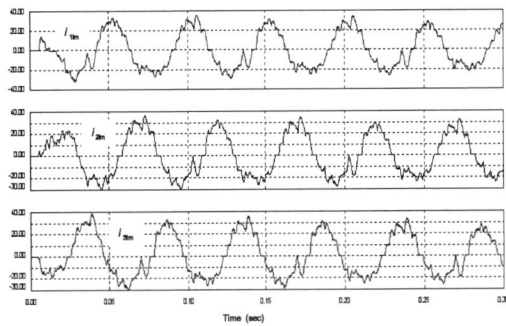

Fig. 6. Three phase currents in Induction motor running with 20 Hz frequency

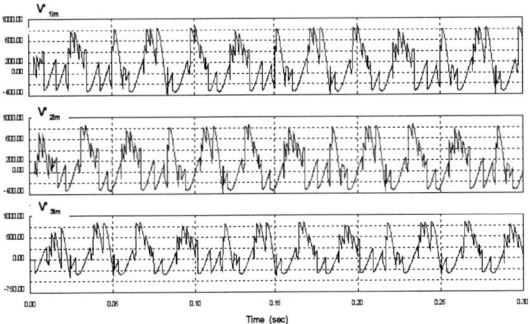

Fig. 7. Three phase voltages of Induction motor running with 40 Hz frequency

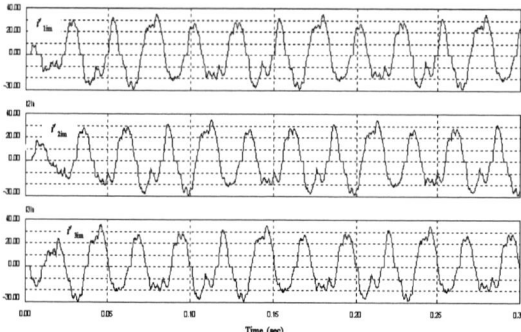

Fig. 8 Three phase currents in Induction motor running with 40 Hz frequency

[7] Thyristorised Power Controller by G K Dubey, S R Doralda, A Joshi, New Age International Publisher

[8] Deodatta Shingare, "Industrial and Power Electronics", Electrotech Publication Engineering Series, Ed. Third, 2006

induction motor in specified time interval as per Table 1, is in process that generates supply torque characteristics matching with demand toque characteristics of the process machine.

VI. CONCLUSION

The cycloconverter can be used to drive a three phase motor that generates supply torque characteristics matching with demand torque characteristics of certain process machine to replace flywheel, the latter causing torsional vibration and fatigue damage. This is a new technique in which the cycle duration of demand torque characteristics of specific process machine is divided into suitable number of time intervals and subdivisions of each time intervals in form of frequencies are tabulated to simplify design procedure. Finally control scheme is designed to control the operation of cycloconverter to drive the motor at different frequencies at variable voltage.

REFERENCES

[1] Joseph Edward Shigley, Charles R. Mischke, "Mechanical Engineering Design", Tata McGraw Hill, Ed. Second.

[2] Joseph Edward Shigley, John Joseph Vicker Jr., Theory of Machines & Mechanisms, Ed. Second 1995, McGraw Hill International

[3] E. Chiess, A. Monti, M. Matuonto, "Computer Simulation of a Cycloconverter Drive and Development of a Full Digital Field Oriented Control", 1993, The European Power Electronics Association, pp. 121-127.

[4] J. Zhang, G.P. Hunter, V.S. Ramsden, "A Single Phase Input Cycloconverter Driving a Three Phase Motor", 1993, The European Power Electronics Association, pp. 128-132.

[5] Songchun Zhang, Fenglin Wu, Shmin Shen, Shuchun Yang, "A Digital Controller Based Cycloconverter – Fed Drive", 1997 IEEE Transaction, pp. 637-641

[6] Stephen F. Gorman, Jimmie J. Cathey, Joseph a. weimer, "A Multi-Microprocessor Controller for a VV-VF Cycloconverter-Link Brushless DC Motor Drive", IEEE Transactions on Industrial electronics, Vol. 35, pp. 278-283, No. 2, May 1988

A Novel Zero-Voltage-Switching Single-Stage High-Power-Factor Electronic Ballast

Chien-Ming Wang*, Ching-Hung Su**, Chang-Hua Lin***, Maw-Yang Liu*, and Kuo-Lun Fang*

* Department of Electrical Engineering, National Ilan University, I-Lan, Taiwan
** Department of Electronic Engineering, Lunghwa University of Science and Technology, Taoyuan, Taiwan
*** Department of Electrical Engineering, Tatung University, Taipei, Taiwan

Abstract-- This paper proposes novel zero-voltage-switching (ZVS) single-stage high-power-factor electronic ballast for fluorescent lamps. In the proposed electronic ballast, all switching devices operate at ZVS turn on and turn off. And the proposed electronic ballast only uses a symmetrical half-bridge topology to procure the functions of a boost power-factor-correction converter and a half-bridge series parallel-loaded inverter. In spite of its simplicity, an excellent performance concerning load and supply is achieved, ensuring a sinusoidal and in phase supply current. The design equations are derived from the analyzed results based on fundamental approximation, and then an easy-to-use design tool is provided accordingly under considerations of filament heating and ignition. A prototype circuit designed for one 40-W fluorescent lamps operating at 40-kHz switching frequency and 110-V line voltage is built and tested to verify the analytical predictions.

Index Terms—electronic ballast, fluorescent lamps

I. INTRODUCTION

In resent years, the high frequency electronic ballasts have played a very important role for fluorescent lamps due to the benefits of light weight, small size, high luminous efficiency, and long lamp life. Most electronic ballasts are realized with load resonant inverters since they can provide an appropriate ignition voltage and then a stable arc current with a low crest factor for fluorescent. The peak detection rectifier is traditionally used by the resonant inverter to get the input dc voltage source. Nevertheless, this circuit will cause a large and sharp input current when the input ac source voltage reaches its peak. The harmonics included in the input current is harmful for the other electrical appliances, such as personal computers, and radios. In order to improve this drawback, a power factor correction (PFC) circuit must be attached to the electronic ballast to reduce the input line current harmonics. However, the two-stage solution requires more circuit components, resulting in higher cost and lower efficiency. For simplify the circuit of electronic ballast and reduced its cost, some single-stage electronic ballasts have been proposed by integrating PFC circuit into the inverter stage to perform both functions of the PFC and resonant inverter [1]-[6]. By sharing the active power switch and the control circuit, the component count can be effectively reduced. But this solution will loss the soft-switching function, the switching losses and electromagnetic interference (EMI) will be increased.

For improving these drawbacks and holding the advantages of single-stage electronic ballast, a novel

zero-voltage-switching single-stage high-power-factor electronic ballast is proposed in this paper. In the proposed electronic ballast, all switching devices operate at ZVS turn on and turn off. And the proposed electronic ballast only uses a symmetrical half-bridge topology to procure the functions of a boost power-factor-correction converter and a half-bridge series parallel-loaded inverter. In this electronic ballast, a high-power factor can be achieved by operating the converter at discontinuous conduction mode (DCM) at a fixed frequency with a constant duty cycle. In this way, the control circuit is simple, and low-cost commercial ICs are available. The proposed ballast circuit in analyzed and the design rules are listed, and the experimental results verify the analysis.

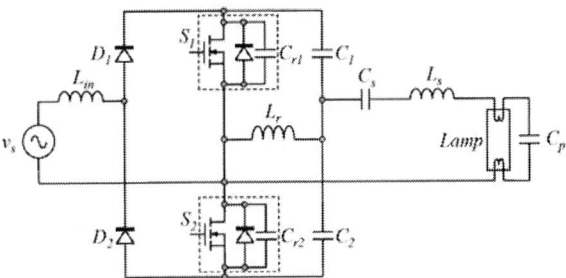

Fig. 1 Proposed novel ZVS-PWM single-stage high-power-factor electronic ballast

II. CIRCUIT OPERATION

The proposed zero-voltage-switching single-stage high-power-factor electronic ballast circuit is shown in Fig. 1. It includes a single-phase voltage supply, a boost PFC circuit composed of L_{in}, S_1, S_2, C_1, C_2, D_1, D_2, a half-bridge inverter for high-frequency supply to the lamp composed of S_1, S_2, C_1, C_2, C_s, L_s, C_P, and a fluorescent lamp, and a zero-voltage-switching cell to provide zero-voltage-switching function in switching devices composed of L_r, C_{r1}, C_{r2}. The power switches S_1 and S_2 are operated at fixed frequency with a constant duty cycle about 50%. The boost PFC circuit is operated in discontinuous conduction mode, the input current naturally follows the sinusoidal waveform of the input voltage, achieving unity power factor to the utility line. A series-resonant tank circuit L_s and C_s in series with the lamp network forms the load resonant circuit of the inverter. In the lamp network, only the capacitor C_P provides the current path for filament heating. Because the power switches are operated at a constant duty cycle about 50% and v_s is symmetrical voltage waveform. There should be no dc voltage in the load resonant circuit. In practice, however, a dc voltage may be presented in

978-1-4244-0644-9/07/$25.00 ©2007 IEEE   1712

the resonant circuit due to discrepant duty cycle and winding. This unwanted dc voltage can be easily blocked the capacitor C_s. Such a circuit topology has the advantages of simplicity and high efficiency. Since the circuit operates symmetrically, the circuit operation for the negative half-cycle of the line voltage is identical to that of the positive half-cycle except that the diodes. Thus, the circuit operation in the positive half-cycle is only described in this paper. And, to simplify the analysis, the following assumptions are made during one switching cycle.

1. All components and devices are ideal.
2. The input inductor L_{in} is large enough to assume that the input current I_{in} is constant during one switching cycle.

3. The filter capacitors C_1 and C_2 are large enough to assume that the voltages across capacitors C_1 and C_2 are constant during one switching cycle.

Based on these assumptions, circuit operations can be divided into five modes in accordance with the conducting power switch during one high-frequency cycle. Fig. 2 shows the operation modes for the positive and negative half-cycle of the line voltage. The input filter is omitted for simplicity. Fig. 3 illustrates the theoretical waveforms for each mode. To achieve a high power factor, the boost PFC circuit is operated in DCM. The resonant inverter is operated at a switching frequency above its resonance. The circuit operation is described as follows.

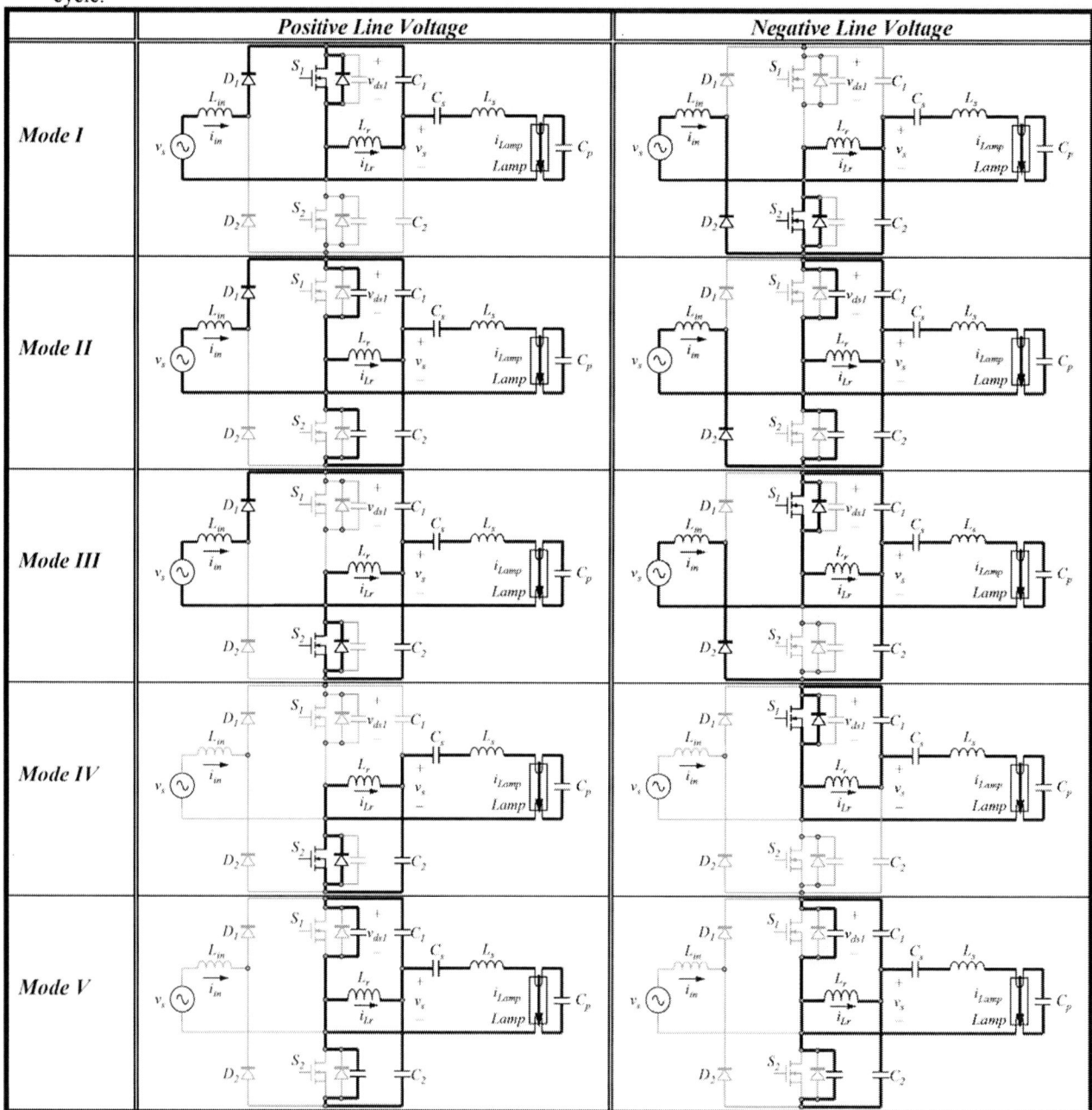

Fig. 2 Operation modes

Mode I $[t_0 < t < t_1]$:

1713

Before $t=t_0$, the switch S_1 and S_2 are turn-off state. The energy stored in input inductor L_{in} is delivered to the filter capacitors C_1 and C_2. The resonant inductor L_r and capacitors C_{r1}, C_{r2} perform the resonance behavior. The resonant voltage v_{Cr1} decreases and resonant voltage v_{Cr2} increases. The resonant current i_{Lr} increases. When the resonant voltage v_{Cr1} drop to zero at $t=t_0$. At this instant, the switch S_1 is activated by its gate signal $v_{gs1}(t)$ and turned on under zero voltage switching. Also, this mode begins. The input inductor L_{in} is charged by input voltage v_{in}. The energy stored in the filter capacitor C_1 supplies the output resonant circuit by negative value. Since the load circuit is designed to be inductive, the load resonant current i_{Lamp} is positive at this instant. It decreases and then increases when it reaches its peak value via the resonance of output resonant circuit. Because the boost PFC is operated DCM, the input current $i_{in}(t)$ increases linearly from zero. Simultaneously, the resonant inductor L_r is charged by voltage V_{C1} and the resonant current i_{Lr} linearly increases. When the duty cycle of switch S_1 is complete, this mode is finished.

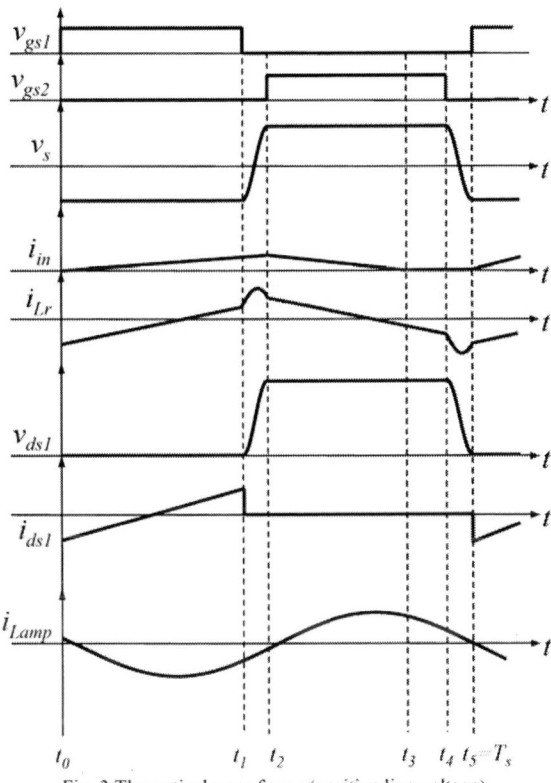

Fig. 3 Theoretical waveforms (positive line voltage).

Mode II $[t_1 < t < t_2]$:

At $t=t_1$, the switch S_1 is turned off and the switch S_2 maintains turned-off, and this mode begins. The input current $i_{in}(t)$ and the load resonant current i_{Lamp} continuously increase. The energy stored in the filter capacitor C_1 and C_2 simultaneously supplies the output resonant circuit. Also, the resonant inductor L_r and capacitors C_{r1}, C_{r2} perform the resonance behavior. The resonant voltage v_{Cr1} increases and resonant voltage v_{Cr2} decreases. The resonant current i_{Lr} decreases. When the resonant voltage v_{Cr2} drop to zero at $t=t_2$, this mode is finished.

Mode III $[t_2 < t < t_3]$:

At $t=t_2$, the resonant voltage v_{Cr2} drop to zero, the switch S_2 is activated by its gate signal $v_{gs2}(t)$ and turned on under zero voltage switching. Also, this mode begins. The energy stored in the input inductor L_{in} is delivered to the filter capacitors C_1 and C_2. Thus, the input current $i_{in}(t)$ deceases linearly from it peak value. The energy stored in the filter capacitor C_2 supplies the output resonant circuit by positive value. Since the load circuit is designed to be inductive, the load current i_{Lamp} is negative at this instant. It increases and then decreases when it reaches its peak value via the resonance of output resonant circuit. Simultaneously, the resonant inductor L_r is discharged by voltage V_{C2} and the resonant current i_{Lr} linearly decreases. When the input current $i_{in}(t)$ drops to zero, the diode D_1 is naturally turned off and this mode ends.

Mode IV $[t_3 < t < t_4]$:

During this mode, the voltage V_{C2} across the capacitor C_2 by positive value supplies continuously the output resonant circuit. The lamp current i_{Lamp} decreases continuously. Simultaneously, the resonant inductor L_r is continuously discharged by voltage V_{C2} and the resonant current i_{Lr} continuously and linearly decreases. When the duty cycle of switch S_2 is complete, this mode is finished.

Mode V $[t_4 < t < t_5]$:

At $t=t_4$, the switch S_2 is turned off and the switch S_1 maintains turned-off, and this mode begins. The load resonant current i_{Lamp} continuously increase. The energy stored in the filter capacitor C_1 and C_2 simultaneously supplies the output resonant circuit. Also, the resonant inductor L_r and capacitors C_{r1}, C_{r2} perform the resonance behavior. The resonant voltage v_{Cr1} decreases and resonant voltage v_{Cr2} increases. The resonant current i_{Lr} increases. When the resonant voltage v_{Cr1} drop to zero at $t=t_5$, this mode is finished.

After Mode V, the circuit operation is returned to the Mode I of the next cycle. The input current $i_{in}(t)$ returns to zero. The lamp current $i_{Lamp}(t)$ and resonant current $i_{Lr}(t)$ also return to their initial value, respectively. Therefore, the assumption previously made is proven to be valid. Through the analysis presented above, key waveforms of the proposed novel ZVS-PWM single-stage high-power-factor electronic ballast can be plotted as shown in Fig. 3.

III. REALIZATION AND EXPERIMENTATION

An electronic ballast for driving an fluorescent lamp (FL-40D) is illustrated as a design example. The circuit parameters are designed to operate the boost PFC circuit at DCM so that to a high input power factor can be achieved. The controller circuit uses the frequency modulation to ignite the lamp for getting the soft start characteristic so that the life of the lamp will be extended. The design procedure is outlined as follows.

Step 1—Circuit specification.

1714

Input Voltage v_{in}	$110\sqrt{2}\sin[2\pi(60)]t$
Switching Frequency f_s	50kHz
Lamp Power P_{Lamp}	38W
Lamp Voltage $V_{Lamp,max}$	150V
Lamp Equivalent Resistance R_{Lamp}	675Ω
Lamp Filament R_f	50Ω
Input Power Factor PF	>0.95

Step 2—Determining of the boost PFC circuit inductor and filter capacitors C_1 and C_2.

For satisfying the circuit requirement $PF>0.95$, the value of $a=V_C/V_{in,max}$ must be more than 1.25 from the reference [7]. Fortunately, the boost PFC circuit of proposed electronic ballast is operated at fixed frequency with a constant duty cycle of 50% and is operated in discontinuous conduction mode. Thus, the value of $a=V_C/V_{in,max}$ approximates 2. Thus, it satisfies the requirement. The boost inductor L_{in} can be found as follow.

$$y = \int_0^\pi \frac{\sin^2 \omega t}{a - \sin \omega t} d\omega t$$
$$= -2 - a\pi + \frac{2a^2}{\sqrt{a^2-1}}\left[\frac{\pi}{2} - \tan^{-1}(\frac{-1}{\sqrt{a^2-1}})\right]$$
$$= 1.42 \tag{1}$$

$$L_{in} = \frac{T_s V_{in,max}^2 ay}{8 P_{in} \pi} = \frac{T_s V_{in,max}^2 ay}{8 P_{Lamp} \pi} = 1.429 \times 10^{-3} \tag{2}$$

We use L_{in}=1.4mH. For minimize the ripple voltages of the voltages V_{C1} and V_{C2} we use $C_1=C_2$=470μF.

Step 3—Determining the resonant inductor L_r and capacitors C_{r1} and C_{r2}.

Because the power factor correction circuit is operated at discontinuous conduction mode (DCM) and at a fixed frequency with a constant duty cycle, the maximum input current $i_{in,max}$ can be obtained as follow.

$$i_{in,max} = \frac{0.5 T_s V_{in,max}}{L_{in}} = 1.11A \tag{3}$$

In order to achieve soft commutation at zero-voltage switching for both active switches, the following requirement should be satisfied.

$$i_{Lr,max} > i_{in,max} = 1.11A \tag{4}$$

Thus,

$$i_{Lr,max} = \frac{V_C T_s}{8 L_r} = \frac{a V_{in,max} T_s}{8 L_r} > 1.11A \tag{5}$$

$$\Rightarrow L_r < \frac{a V_{in,max} T_s}{8 \times 1.11} = 0.698mH \tag{6}$$

In this design, L_r=0.5mH is selected. For minimizing the influence of the resonant parameters and achieving easily soft commutation at zero-current-switching for both active switches, the resonant frequency f_r is selected as f_r=5f_s. Thus,

$$\omega_r = \frac{1}{\sqrt{L_r C_r}} = 2\pi f_r = 2\pi \times 5 f_s = 1.57 \times 10^6 \text{ rad/s} \tag{7}$$

$$\Rightarrow C_r = 8.11 \times 10^{-10} F$$

In this design, C_r=820pH is selected.

Step 4—Determining the resonant inductor L_s and capacitor C_p.

Assuming the starting peak voltage of the fluorescent lamp is 400V. The relative equations can be obtained from the reference [7]. Thus,

$$\frac{1}{\sqrt{(1-\omega^2 L_s C_p)^2 + (\omega R_f C_p)^2}} = \frac{400}{\frac{4}{\pi} \times 310} = 1.0134 \tag{8}$$

and

$$\frac{1}{\sqrt{(1-\omega^2 L_s C_p)^2 + (\frac{\omega L_s}{R_{Lamp}})^2}} = \frac{150}{\frac{4}{\pi} \times 310} = 0.38 \tag{9}$$

The resonant inductor L_S and capacitor C_P can be calculated as L_S=5.2mH and C_P=3.896nF by using the equations (8) and (9). L_S=5mH and C_P=3.9nF are selected in this design example.

In hardware realization, MOSFET's IRF840 and S3L60 are as the power switches and diodes, respectively. This proposed electronic ballast is controlled by the IC TL494 and some logic gate, which drive the active switch at suitable high switching frequencies from starting to steady-state operation. The experimental results are shown in Fig. 4, 5 and 6, respectively. Fig. 4 shows the waveforms of the input voltage and current, in which the waveforms of the input voltage and current are almost in phase. The measured power factor is greater than 0.95 and total current harmonic distortion (THD) is less than 8%. Fig. 5 shows the lamp voltage and current waveforms. The measured crest factor of the lamp current is below 1.5. The commutation phenomenon in the switch S_1 is measured in Fig. 6. The experimental result shown in Fig. 6 demonstrates that ZVS are achieved at constant frequency for switch S_1. Therefore, the switching losses for the main switches in proposed electronic ballast are practically zero. The efficiency of proposed electronic ballast is measured and it is equal 89%.

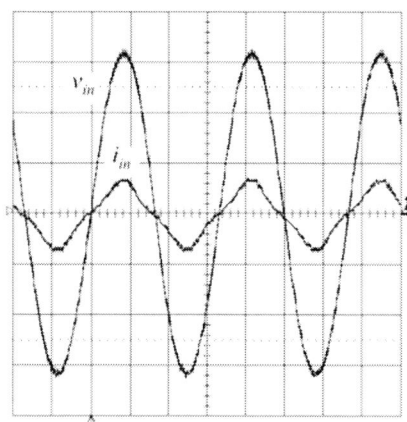

Fig. 4 Input voltage and current after EMI filter. (v_{in}: 50V/div; i_{in}:2A/div, time:5ms/div)

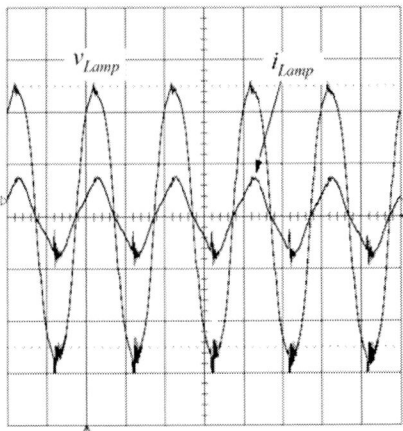

Fig. 5 Lamp voltage and current waveforms (v_{Lamp}: 50V/div; i_{Lamp}:2A/div, $time$:10µs/div)

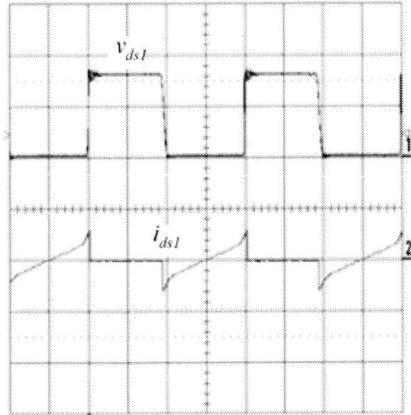

Fig. 6 Voltage and current waveforms of switch S_1. (v_{ds}: 200V/div; i_{ds}:2A/div, $time$:5µs/div).

IV. CONCLUSIONS

A novel zero-voltage-switching (ZVS) single-stage high-power-factor electronic ballast for fluorescent lamps has proposed in this paper with simple and compact configuration. In the proposed electronic ballast, all switching devices operate at ZVS turn on and turn off. The circuit operation was described and the design equations were derived. A prototype circuit of a design example was built and measured to verify the analytical predictions. In spite of the simplicity of circuit configuration, experimental results show that satisfactory performance can be achieved. The power factor is nearly unity and the THD is less than 8%. In addition, the PFC power flow path of the proposed circuit has two conduction drop, the conduction losses can also be reduced. Therefore, it can provide lower conduction loss than the conventional one. Compared with the other electronic ballasts, the performance of the input power factor and lamp crest factor are the same.

REFERENCE

[1] J. M. Alnonso, A. J. Calleja, F. J. Ferrero, E. Lopez, J. Ribas, and M. Rico-Secades, "Single-stage constant-wattage high-power-factor electronic ballast with dimming capability," in *IEEE PESC Conf. Rec.* 1998, pp. 2021-2027.

[2] R. de Oliveira Brioschi and J. L. F. Vieira, M "High-power-factor electronic ballast with constant DC-link volatge," in *IEEE Trans. Power Electron.* vol. 13, pp. 1030-1037, 1998.

[3] A. J. Calleja, J. M. Alonso, E. Lopez, J. Ribas, J. A. Martinez, and M. Rico-Secades, " Analysis and experimental results of a single-stage high-power-factor electronic ballast based on flyback converter," in *IEEE Trans. Power Electron.* vol. 14, pp. 998-1006, 1998.

[4] R. N. do Pardo, M. F. da Silva, M. Jungbeck, and A. R. Seidel, "Low cost high-power-factor electronic ballast for compact fluorescent lamps," in *Conf. Rec. IEEE-IAS Annu. Meeting*, 1999, pp. 256-261.

[5] C. S. Lin, and C. L. Chen, "A novel single-stage push-pull electronic ballast with high input power factor," in *IEEE Trans. Industrial Electronics.* vol. 48, no. 4, pp. 770-776, 2001.

[6] H. L. Cheng, C. S. Moo, and W. M. Chen, "A novel single-stage high-power-factor electronic ballast with symmetrical topology." in *IEEE Trans. Industrial Electronics.* vol. 50, no. 4, pp. 759-766, 2001.

[7] C. M. Wang, "A Novel Single-Switch Single-Stage Electronic Ballast With High Input Power Factor," in *IEEE Trans. Power Electron.* vol. 22, no. 3, pp. 797-803, 2007.

Opto-Mechatronic System Design of the LED Projector by Using Brushless DC Motor

Jian-Long KUO and Tzu-Hsuan FANG
Department of Mechanical and Automation Engineering,
National Kaohsiung First University of Science & Technology
Nantze, Kaohsiung 811, TAIWAN.
Tel: 886-7-6011000#2291
e-mail: JLKUO@ccms.nkfust.edu.tw

Abstract- This paper will propose a possible mechanism design of LED projector. The opto-mechatronic system will be studied. The analytical scheme of optical path will be described in detail. As compared with the other light sources, laser light usually has many advantages: higher brightness, higher directionality, pure monochromatism, and higher coherence. However, the LED device does not have these advantages. Therefore, this paper will develop the mechanism to project LED light to target screen. Biconvex lens and hexahedral mirror are used to focus the LED light into one spot on the target screen. Brushless DC Motor is used to rotate the mirror to provide the required projection.

Keywords: LED projector, opto-mechatronics, micro-controller , BLDCM, hexa-hedral mirror.

I. INTRODUCTION

This paper mainly focuses on the image projection method of the projector by means of one-dimension light emitting diode (LED) array. The one-dimension LED array is the light source of the projector as shown in Fig. 1.

Recently, the lighting technology is now rapidly developed to meet the requirements of high efficiency, long life cycle, and strong luminance. The next generation projection system integrates the optical, mechanism, electrical, and control aspects. Therefore, the integrated opto-mechatronic projector can further provide the more excellent visual images.

In the light source design of the projector, the microprocessor with 16-bit and its peripheral circuits are employed to actuate the one-dimension LED light source. Meanwhile, the microprocessor also sends the control signal to the inverter to drive the brushless DC motor (BLDCM). In the mechanism, the hexa-hedral mirror rotated by the BLDCM are assembled together. Therefore, the hexa-hedral mirror can rotate with the BLDCM. The hexa-hedral mirror provides six mirrors around its circumference. Moreover, the one-dimension light emitting from the one-dimension LED array will be expanded into 2-D image by the hexa-hedral mirror.

Therefore, when the hexa-hedral mirror is rotating, the proposed method will provide a novel projecting method for converting the one-dimensional light control signal into two-dimension image so as to display the required 2-D image, such that the cost of the projector system can be further reduced. [1]-[19]

II. SYSTEM DESIGN OF LED PROJECTOR

Usually, the LED projector can not transmit the light to the target projecting screen directly. The scattering phenomena make the projection light unable to focus in one perfect small spotlight. Even though the spotlight can reach the target screen, the spotlight is still not clearer. The improvement on the scattering problem is to use the convex lens controlled by the brushless DC motor.

The LED array is located at the two times of the focal length of the convex lens. Then the light through the lens will have image located at two times of the focal length and on the other side of the convex lens. The LEDs can have clearer spotlights by using the conven lens. The spotlights is regarded as pixels for the proposed LED display.

The hexa-hederal mirror is used to rotate and reflect the light to the target screen. There are six mirrors for the rotator. The incident light and the reflected light are in the same plane with the normal vector. Also, the angle of incidence is equal to the angle of the reflection.

The Snell's Law can be used describe the angle relation between incidence and refraction. The light impinging on a interface between two media with two different indices of refraction. The following relation can be satisfied.

978-1-4244-0644-9/07/$25.00 ©2007 IEEE

$$n_1 \sin \theta_1 = n_2 \sin \theta_2 \qquad (1)$$
or
$$\frac{\sin \theta_1}{\sin \theta_2} = \frac{n_2}{n_1} \qquad (2)$$

According to the theory of geometric optics, the LEDs are located around two times of the focal length. Then the light image is transmitted to the side of the convex lens. When the lights impinge on the rotator, the light begins to reflect and project onto the target screen to show the testing word *HI* on the screen.

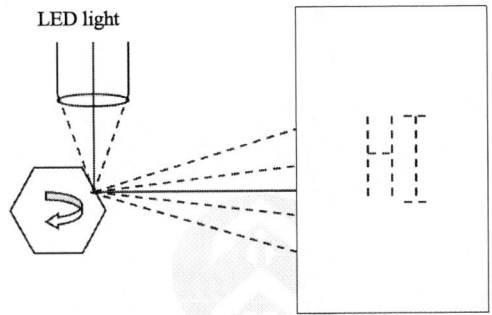

Fig. 1. Design of LED projector System

The LED projector is developed here to provide the long distance projection for the target screen. Usually, the Laser projector can directly provide the straight line to the target screen. However, the Laser diode is more expensive than LED diode. Again, the LED does not provide the straight line directly to the target screen.

There are two major problems for the long-distance projection:

Problem 1: The positioning of the projection

The initial position of the scan line is located at the left hand side of the screen. The end position of the scan line is located at the right hand side of the screen. The positioning of the scan line should be determined first. The brushless DC motor needs to be controlled to obtain the precise position control.

Problem 2: Scan rate servo control for the scan line

The scan rate is the scan times per second. For the test words *HI CHILIN*, the contour of the left character *H* and the right character *N* are clearer. However, the contour of the middle characters *C* and *H* are a little blurred. The scan rate should be higher to avoid the twinkling phenomena for the test words. To ensure the synchronous scan for the screen, scan rate servo control is required to provide the rotation. The rotator is rotating under the specific speed to keep the constant scan rate for the screen.

The AC servo motor can provide more precise position and speed control. The optical encoder has 2000 pulses per revolution. The AC servo motor with Hall sensors provides the electronic commutation. The word *HI* is shown on the screen to illustrate the projection function for the display.

III. DESIGN OF OPTO-MECHATRONIC SYSTEM

The mechanism of the rotator is composed of six mirrors so-called hexa-hedral mirror as shown in Fig. 2 ~ Fig. 4. In order to avoid the severe fluctuation, a bearing base is designed to keep the balanced rotation. The rotor of the brushless DC motor is connected to the rotator. This paper will focus on the servo control of the brushless DC motor. The six planes on the rotator is responsible for the scan rate of the target display. For one cycle of mechanical rotation, there will be six times scan rate to reveal the texts on the screen.

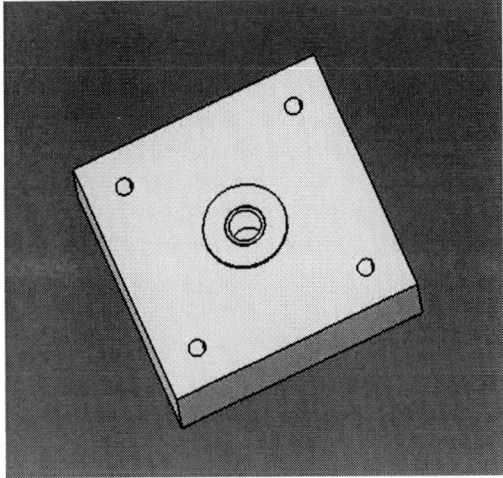

Fig. 2. 3-D view of bearing base for the rotator with hexa-hedral mirror.

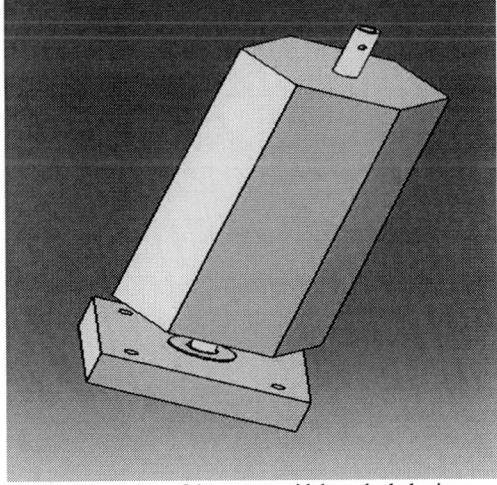

Fig. 3. 3-D view of the rotator with hexa-hedral mirror.

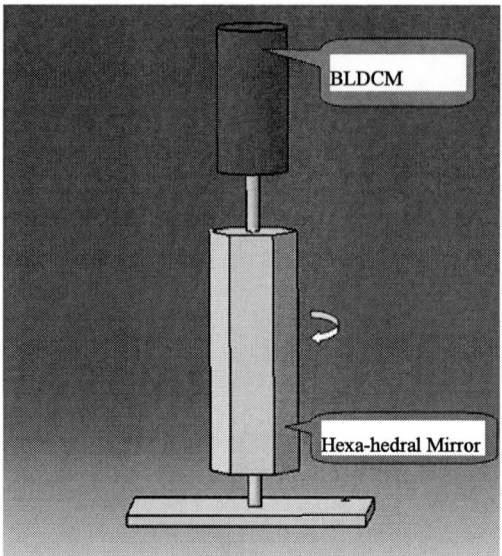

Fig. 4. Mechanism of the proposed LED projector.

IV. OPERATION OF BRUSHLESS DC MOTOR

As shown in Fig. 5 ~ Fig. 7 and TABLE 1, The driving method of the brushless DC motor is to energize the three phase windings by using switching logic obtained from the Hall sensors. The position of the rotor can be detected by using Hall sensors. 120-degree conducting pattern is used to control the motor current.

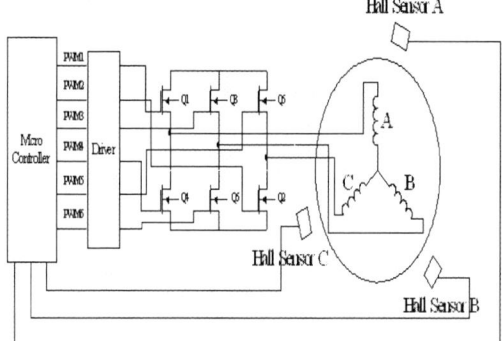

Fig. 5. Driver circuit for the brushless DC motor connected with the rotator with hexa-hedral mirror.

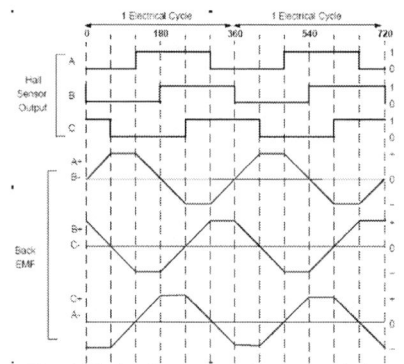

Fig. 6. Timing Sequence and voltage relation.

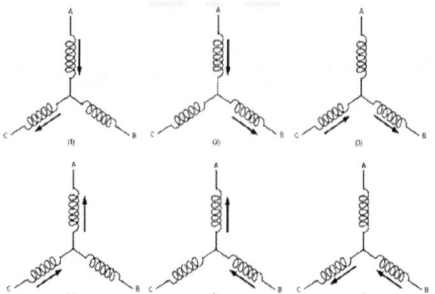

Fig. 7. Current control path for the brushless DC motor.

Table 1   Relation for the Hall sensor, PWM signals and phase current.

Sequence	Hall Sensor			Active PWMs	Phase Current		
	A	B	C		A	B	C
1	0	0	1	PWM1 PWM2	+	x	·
2	0	0	0	PWM1 PWM6	+	·	x
3	1	0	0	PWM5 PWM6	x	·	+
4	1	1	0	PWM5 PWM4	·	x	+
5	1	1	1	PWM3 PWM4	·	+	x
6	0	1	1	PWM3 PWM2	x	+	·

V. VERIFICATION ON THE CONSTRUCTION OF THE PROPOSED LED PROJECTOR

The LED projector is designed to provide basic verification for the projection for the display technology. The proposed system setup is shown in Fig. 8 ~ Fig. 10. The biconvex lens with focal length 24 cm is used to focus the LED light for a 48 cm long-distance transmission. The radius of the curvature of the biconvex lens is 48 cm which is two times of the focal length. The hexa-hedral mirror is rotated by the brushless DC motor. The mirror is used to reflect the LED light onto the target screen.

The proposed motor driver is shown in Fig. 8. The circuit board is developed to drive the brushless DC motor. PWM control signals come from the Microchip dsPIC micro-controller. The position and speed feedback control is programmed into the micro-controller.

The LED light is unlike the Laser light which has long-distance transmission. Therefore, bi-convex lens is used to provide the long-distance light transmission for the LED light. The LED light source is located at 48 cm which is two times of the focal length 24 cm. Then we will have an image located at the other side of the lens. At the distance 10 cm, the hexa-hedral mirror is located to reflect the image into another direction. The light is reflected by the mirror and then goes 38 cm to project the image onto the target screen. After the experimentation, the LED projector is successfully developed, the testing word HI is shown on the target screen.

VI. CONCLUSION

This paper has successfully developed a LED projector with opto-mechatronic mechanism. The LED light source in 1-D array can be expanded into 2-D screen by using a rotating hexa-hedral mirror. The mirror is controlled by a brushless DC motor. Results show that the proposed mechanism is reliable and achievable for the LED projection.

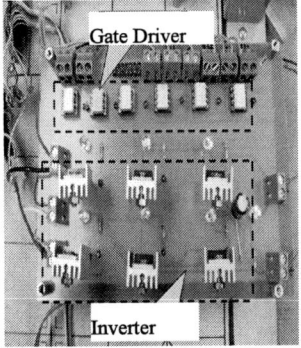

Fig. 8. Developed control circuit board for the LED projector.

Fig. 9. The testing word HI shown on the target screen.

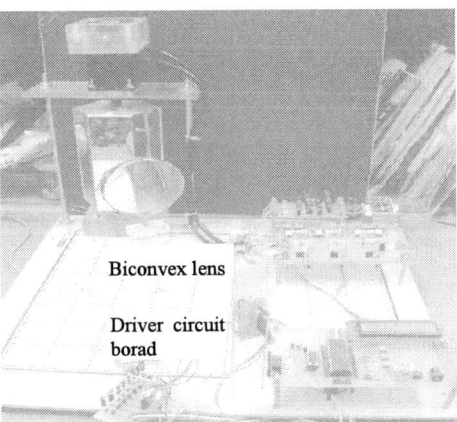

Fig. 10. The experimental setup for the proposed LED projector.

ACKNOWLEDGEMENTS

The authors would like thank the CHI-LIN Tech. Inc. for supporting the associated equipments.

REFERENCES

[1] Reber E. Fischer, Biljana Tadic-Galeh, "Optical System Design" , McGraw-Hill Companies, Inc, 2000.

[2] J. Warren Blaker, Peter Schaeffer, "Optics ： An Introduction for Technicians and Technologists", Prentice-Hall, Inc, 2000.

[3] T.Flowery and D. W. Petro, "Application considerations for PWM inverter-fed low-voltage induction motors," IEEE Transaction on Industry Applications, vol. 30, pp. 286-293, 1994.

[4] Masaharu Deguchi, Takesuke Maruyama, Hisao Inage,Takashi Kakuda, Futoshi Y amasaki, and Seiya Asai, "Development of High- Brightness Compact LC Projector", IEEE Transactions on Consumer Electronics, Vol. 41, No. 3, AUGUST 1995.

[5] Larry J. Hornbeck, "Digital Light Processing^{TM} ： A New MEMS-Based Display Technology", Texas Instruments, 09/01/1995.

[6] S. G. Johnson and J. A. Simmons, "Materials for Solid State light", Proc.Materials research Society Spring Meeting, April 1-5, 2002 in San Francisco,California.

[7] S. K. Tewksbury, "Semiconductor Materials", Microelectronic Systems Research Center, Dept. of Electrical and Computer Engineering, West Virginia University, Sept. 21, 1995.

[8] Hisashi Yajima and Hiroyuki Wakivaka, "Consideration on High-response of a linear DC Motor," 1997 IEEE, 0-7803-3862-6.

[9] Nasar, S. A. and Boldea, I., "Linear Electric Motors Theory, Design and Practical Applications, New Jersey:Prentice-Hall Inc., 1987.

[10] J.U. Chu, and I.H. Moon, G.W. Choi, J.C.Ryu, and M.S.Mun, "Design of BLDC motor controller for electric power wheelchair," in IEEE Mechatronics Conf. Rec., June., 2004, pp 92-97.

[11] K.H. Park, T.S. Kim, S.C.Ahn, and D.S. Hyun, "Speed Control of High-Performance Brushless DC Motor Drives by Load Torque Estimation," in IEEE Power Electronics Specialist Conf. Rec., Vol.4, June., 2003, pp 1677-1681.

[12] B. K. Lee, T. H. Kim, and M. Ehasani, "On the Feasibility of Four-Switch Three-phase BLDC Motor Drives for Low CostCommercial Applications: Topology and Control," IEEE Trans. On Power

Electronics, Vol.18, Jan., 2003, pp. 164-172.

[13] B.H. Kang, C.J.Kim, H.S. Mok, and G.H.Choe, "Analysis of torque ripple in BLDC motor with commutation time," in Proc. Industrial Electronics, Vol.2, June., 2001, pp. 1044-1408.

[14] Y.S. Jeon, H.S. Mok, G.H. Choe, D.K.Kim, and J.S. Ryu, "A New Simulation Model of BLDC Motor With Real Back EMF Waveform" Workshop on Computer in Power Electronics, July., 2000, pp. 217-220.

[15] Brushless DC (BLDC) Motor Fundamentals, Microchip, 2003.

[16] N. Matsui, T. Takashita, and K. Yasuda, "A new sensorless drive of brushless DC motor," IEEE IECON`92, pp.430-435, 1992.

[17] R.C. Becerra, T.M. Jahns, and M. Ehsani, "Four Quadrant sensorless Brushless ECM Drive," in IEEE Applied Power Electronics Conf. Rec., March., 1991, pp 202-209.

[18] K. Iizaka, H. Uzuhashi, "Microcomputer control for sensorless brushless DC motor," IEEE Trans. On Ind. Applicat, Vol.21, no. 4, May/June , 1985., pp. 595-601.

[19] "dsPIC30F4011/12 Data Sheet,"1/2005, Microchip.

The Color Measurement System of PWM-Controlled LCD by Using Back-Propagation Neural Network

Jian-Long Kuo and Xian-Lin Liu
Department of Mechanical and Automation Engineering,
Natinal Kaohsiung First University of Science & Technology,
Nantze, Kaohsiung 811, TAIWAN.
Tel: 886-7-6011000#2291
E-mail: JLKUO@ccms.nkfust.edu.tw

Abstract- This paper will develop a color measurement system that can describe the relation between device-dependent color space and device-independent color space. From the color theory, the relation might have some nonlinear phenomena. The nonlinear relation can be described by using the neural network with learning process. The back propagation algorithm is used for the learning process. The output of the neural network will provide color coordinate transformation. Therefore, the color chromaticity of the LCD can then be determined.

Keywords—Neural network, Color coordinates, Back-propagation Network, PWM-controlled gray scale.

I. INTRODUCTION

The LCD has becoming more and more popular display recently. However, its color space still has weakness as compare to the conventional CRT. Therefore, the systematic management for the color space of the LCD is required. The color management includes three steps: (a) Color Calibration (b) The characterization for the color space (c) The conversion for the color space. This paper will propose back propagation neural network method to deal with the nonlinear phenomena for the color measurement system. [1]-[21]

II. DEFINITION OF TRI-STUMULUS AND CIE STANDARD

The color signal for the LCD is controlled by the PWM method. The gray-scale signal is provided by the PWM-controlled signal which is widely used in power electronics. Duty based control signal is used to mix up the color space in three *RGB* colors. As shown in Fig. 1, the tri-stimulus values of CIE-XYZ can be given by:

$$X = k \int_{vis} R(\lambda) \bullet P(\lambda) \bullet \overline{x}(\lambda) d\lambda$$

$$Y = k \int_{vis} R(\lambda) \bullet P(\lambda) \bullet \overline{y}(\lambda) d\lambda \qquad (1)$$

$$Z = k \int_{vis} R(\lambda) \bullet P(\lambda) \bullet \overline{z}(\lambda) d\lambda$$

$$k = \frac{100}{\int_{vis} P(\lambda) \bullet \overline{y}(\lambda) d\lambda}$$

$R(\lambda)$: reflectance.

$P(\lambda)$: energy distribution.

$\overline{x}(\lambda), \overline{y}(\lambda), \overline{z}(\lambda)$: color matching function

k : normalization factor.

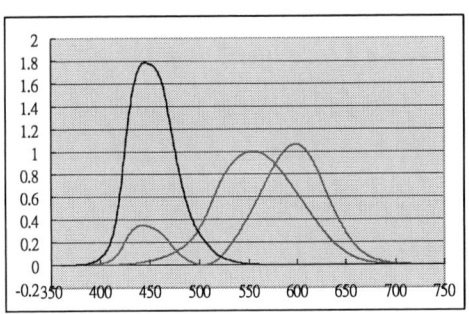

Fig. 1. The tristimulus of the CIE 1931.

By the definition of chromaticity coordinate:

$$x = X / [X + Y + Z]$$
$$y = Y / [X + Y + Z] \qquad (2)$$
$$Y = Z / [X + Y + Z]$$

For the CIE1960-UCS, The u is the x-coordinate and v is the y-coordinate. The relation between CIE1960-UCS and CIE1931-XYZ can be described as:

$$u = \frac{4X}{X + 15Y + 3Z}, v = \frac{6Y}{X + 15Y + 3Z} \qquad (3)$$

The transformation between CIE 1976-LUV and CIE 1960-UCS is given by
u'=u; v'=1.5*v,   (4)

III. COLOR MEASUREMENT

The original gamma curve of the LCD is usually not as expected cue to its physical property. The color reproduction is not natural enough. Different LCD also has different gamma curve and color space. Therefore, it

978-1-4244-0644-9/07/$25.00 ©2007 IEEE

is necessary to provide some method to adjust the color mismatch. The solution to this problem is to modify the input signal through the gamma table and then output the modified signal to the display.

A. Gamma curve measurement

As shown in Fig. 2, the gamma curve indicates the relation for the gray level and brightness. If the x defines 0~255 for different original gray level, then the y=f(x) means the practical gray level. Since the eyes have special sensitivity for different gray level, the gamma curve is usually not a straight line. The gamma correction formula can be expressed as:

$$L_{corrected} = L_{origin}^{\frac{1}{\gamma}}$$

(5)

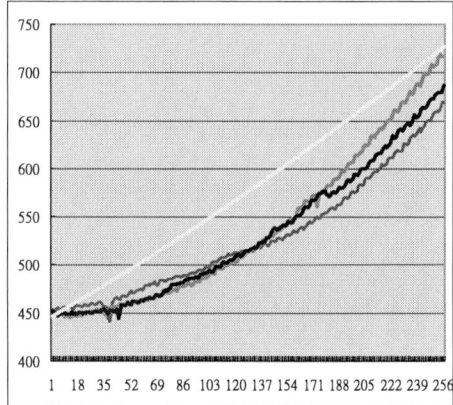

Fig. 2. The RGB measurement versus the respective gray scale.

B. Color Temperature Measurement

The color temperature indicates the white color for the display. The white color is defined from the theory of black-body radiation. The LCD display usually has two correlated color temperature: 6500K and 9300K. 6500K means the display has warm color tone and 9300K means the display has cold color tone.

The color temperature parameter in the OSD can be regulated from 6500~9300K. The all-white pattern is displayed on the screen. The color temperature of the display can be measured and then compared with standard color temperature. The color calibration is assessed the difference value between measured value and standard value.

McCamy proposed the calculation of color temperature by using the CIE XYZ color space as follows:

$$T_{cp} = -473n^3 + 3601n^2 - 6861n + 5514.31$$
$$n = (x - 0.332)/(y - 0.1858)$$

(6)

C. Color Saturation Measurement

The color saturation is defined to reveal the color reproduction ability of the display. The area of the color gamut for the LCD divided by the area of the color gamut defined by the NTSC. For example, the color saturation of one display is 69% of the color space defined by the NTSC. The conventional color saturation is 71% for one CRT display. The LCD usually does not have such a high color saturation like the CRT.

The area of the color space sometimes does not mean that there is the same image quality due to the RGB individual color reproduction. The three points of the RGB color do not always locate at the same coordinate. The color shift problem will lead to the ability of the color reproduction.

IV. BACK PROPAGATION NEURAL NETWORK

As shown in Fig. 3, the neural network is a parallel algorithm that has learning function like human beings. Multi-layer structure provide the solution to the nonlinear problem. The algorithms such as back-propagation and multi-layer perceptron belong to the supervisory learning network. It is very suitable for the application such as diagnosis an d prediction.

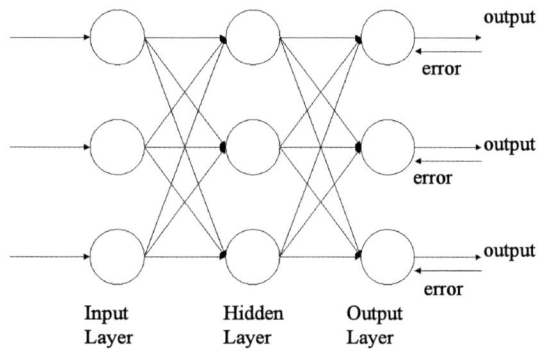

Fig. 3. The proposed back-propagation neural network.

The leraning process for the back propagation algorithm is as follows:
Step 1: clarify the network configuration: determine the layer number and the number of the neurons.
Step 2: assign the initial value for the weighted number.
Step 3: select one set of learning pattern.
Step 4: modify the weighted number.
Step 5: go back to the step 3 until the network converges.

V. COLOR MEASUREMENT SYSTEM

As shown in Fig. 4, the experimental setup is built up to verify the color measurement system.

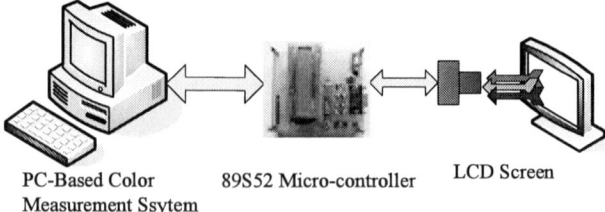

PC-Based Color   89S52 Micro-controller   LCD Screen
Measurement Ssytem

Fig. 4. The developed color measurement system.

1723

A. Color Sensor TCS230

TCS230 is a color sensor with 64 optical diodes. These diodes have four types. There are 16 photo diodes with red filter. There are 16 photo diodes with green filter. There are 16 photo diodes with blue filter. There are 16 photo diodes with without any filter. The photo diode without any filter can pass all the optical information.

These photo diodes are placed in interlaced manner to provide more uniform signal for the incident light. The ability of identification can be enhanced. The position error of color measurement can be eliminated.

There are two pins proving to select the required color filter. The operating frequency spread over the wide range from 2Hz to 500KHz. There are also two pins to select the scaling factor 100%, 20%, and 2%. The scaling factor is provided for scaling the intensity of the color signals. Wider measurement range can be achieved.

B. 89S52 Microcontroller

The microcontroller 89S52 is used to calculate the frequency of the color signal. Then the color signals are transmitted to PC based computer by RS232.

The specification of the control board with 89C52 is list as follows:
1. The power source of USB port is used.
2. The programmer mode is used though printer port.
3. 16*2 LCD module can be connected to show the basic information related to the color sensor.
4. RS232 circuit included.
5. There are six push-buttons. Four buttons are for general purpose, one is for reset function and one is for power source.

C. PC-Based Color Measurement System

VC++ MFC language is used to write the required color measurement program. The color diagram for the CIE-1931 and CIE-1964 can be shown on the computer screen.

When the software is starting, the control signal is transmitted to the 89S52. The RGB color value and brightness are measured by the color sensor. The Gamma value, brightness and color temperature can be calculated. The ICC profile can be generated thereafter. The device-dependent color difference can be compensated further. The gamma value, brightness and color temperature can be calibrated to adjust the color space reveal by the display.

As shown in Fig. 5, The color characterization can be described as the transformation from RGB color space to the XYZ color space. The measured XYZ value is used to compare with standard value. Since the XYZ is non-uniform color space. It is not easy assess the L value. Further transformation is required to obtain the uniform Lab color space. The CIE L*a*b* transformation formula can be expressed as follows:

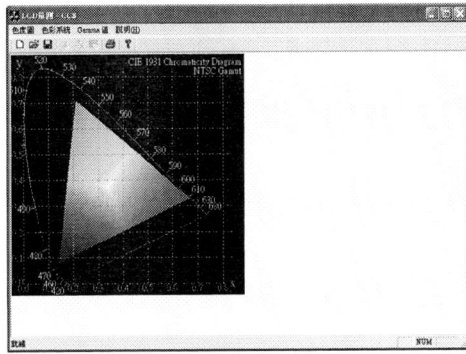

Fig. 5. The generated color gamut of the LCD by the developed color measurement system.

$$L* = 116 f(y) - 16$$
$$a* = 500[f(x) - f(y)]$$
$$b* = 200[f(y) - f(z)]$$
$$\Delta E*ab = \sqrt{(L_1 - L_2)^2 + (a_1 - a_2)^2 + (b_1 - b_2)^2}$$
where $Y/Y_n \leq 0.008856, L* = 903.3(Y/Y_n)$

$$f(x) \begin{cases} (X/Xn)^{1/3} & \text{if } X/Xn > 0.008856 \\ 7.787(X/Xn) + 16/116 & \text{if } X/Xn \leq 0.008856 \end{cases}$$

$$f(y) \begin{cases} (Y/Yn)^{1/3} & \text{if } Y/Yn > 0.008856 \\ 7.787(Y/Yn) + 16/116 & \text{if } Y/Yn \leq 0.008856 \end{cases}$$

$$f(z) \begin{cases} (Z/Zn)^{1/3} & \text{if } Z/Zn > 0.008856 \\ 7.787(Z/Zn) + 16/116 & \text{if } Z/Zn \leq 0.008856 \end{cases}$$

The white color for the above Xn, Yn and Zn is obtained from the RGB(255,255,255). The ICC based color management is produced an ICC profile by using the color calibration equipment. The color management software is used adjust the color space between the device-dependent color space and device-independent color space.

The ICC profile includes the input equipment and output equipment. The information revealed by the profile is the color space of the equipment, the color characterization, profile connection color space. That means it includes not only the type of the equipment but also the white point, the profile generator, the image information for the equipment. This can provide the color management for one specific equipment.

The International Color Consortium promote the ICC profile to provide the color management function. The color characterization for one equipment is recorded in the profile in advance. The color space is transformed into intermediate device-independent color space such as CIE-XYZ, CIE-LAB. Finally, the color space is transformed into output data format to generate the required colorimetric color reproduction.

As shown in Fig. 6 and Fig. 7, there is flexibility for the ICC profile when the color management system is manipulated. When the size of the data in the profile is increasing, higher resolution of color reproduction can be expected. The size of the data depends on different machine. For example, the size of the data for the CRT or LCD is about 1K bytes. However, the machine such as color printer required more data size.

The XYZ color space can provide the device-independent color (DIG). However, there is no unified standard suitable for different color equipment. ICC profile is a profile connection space based the color characterization. There is unified data format that is suitable for calibration among many different color equipments.

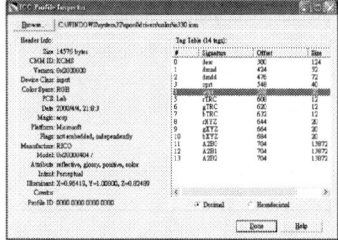

Fig. 6. The content of the ICC profile.

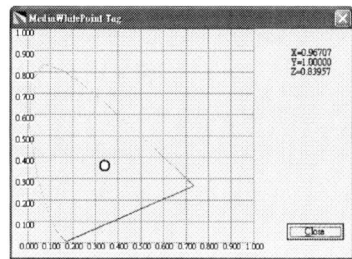

Fig. 7. The white-point defined in the ICC profile.

VI. DISCUSSION

There are two methods to describe the characterization of the LCD. One is by lookup table. The other is by mathematical method such as neural network. The look up table is built up based the testing experience. The difference between device-dependent color space and device-independent color space is described by the lookup table. The other method is by neural network method. The learning process can replace the conventional lookup table technique with learning algorithm. The nonlinear mapping can be described by using appropriate back-propagation neural network.

VII. CONCLUSION

In recent years, the LCD has made great progress in the production technology. The government emphasize the key technology for the future national technology development. The increasing performance and decreasing cost-down lead to the growing of the display industrial. However, in many vigorous application such as desktop publishing or image processing, the CRT is still popular because of its better color space. The LCD requires more specific color space management to provide better display quality. The color signal is provided by the PWM signal.

Duty based signal to mix up the color space for testing.

ACKNOWLEDGEMENTS

The authors would like thank the CHI-LIN Tech. Inc. for supporting the associated equipments.

REFERENCES

[1] TAOS, Inc., 2006, TCS230-E17, TAOS Corporation, Texas.

[2] Atmel Inc.,2006, AT89S52Datasheet，Atmel Corporation, USA.

[3] Jeff Prosise,1999, Programming Windows with MFC,Second Edition, Microsoft Press.

[4] M.D. Fairchild, 1998, Color Appearance Models, Addision Wesley, Reading, Massachusetts.

[5] Alma E. F. Taylor, 1990, Illumination Fundamentals ,Lighting Research Center, p.9.

[6] TAOS, Inc., 2006, The Science of Color, Todd Bishop and Glenn Lee, Texas.

[7] McCamy C. S., "Simulation of Daylight for Viewing and Measuring Color," Color Res. Appl., Vol.19, 1994, pp.437-445.

[8] TAOS, Inc., 2003, BASICS OF LIGHT AND COLOR, Plano, Texas.

[9] http://www.cie.co.at/cie/

[10] John Wiley & Sons, G. Wyszecki and W. S. Stiles, 1982, Color Science: Concepts and Methods, Quantitative Data and Formulae, 2nd edition, New York.

[11] Bentham Instruments, Colorimetry of Displays, Issue 1.00-January 1997.

[12] F. Rosenblatt, 1958, "The perception: a probabilistic model for information storage and organization in the brain," Psychological Review, Vol. 65, pp. 386-408.

[13] M. Minsky and S. Papert, Perceptrons, Cambridge, MA, MIT Press,1969.

[14] J. J. Hopfield and D. W. Tank, 1985,"Neural' computation of decisions in optimization problems, Biological Cybernetics, vol. 52, pp. 141 --152.

[15] D.E. Rumelhart and J. L. McClelland, eds., 1986 , Parallel Distributed Processing: Exploration in the Microstructure of Cognition, Vol. 1,Cambridge, MA, MIT Press.

[16] L. H.Tsoukalas, R. E. Uhrig, 1997, Fuzzy and neural approaches in engineering, John Wiley & Sons, New York.

[17] Robert J. Schalkoff, 1997,Artificial Neural Networks, McGRAW-HILL.

[18] J.M. Ortiz- Rodríguez, M.R.Martínez- Blanco, H.R. Vega-Carrillo, 2006 ,Robust Design of Artificial Neural Networks Applying the Taguchi methodology and DoE, cerma, pp. 131-136, Electronics, Robotics and Automotive Mechanics Conference (CERMA'06).

[19] Giovani, M.. 1983 .Response surface methodology and product optimization. Food Technol.37(11):41-45,83.

[20] G.E.P. Box and K.B. Wilson,1951, "On the Experimental Attainment of Optimum Conditions," *J. Royal Statistical Soc., Series B,* pp. 1-45.

[21] W.J. Hill and W.G. Hunter, 1966, A review of response surface methodology: a literature survey, Technometrics 8, pp. 571–590.

Gapped Air-cored Power Converter for Intelligent Clothing Power Transfer

Y. Lu[1], K.W.E.Cheng[1], Y. L. Kwok[2], K. W. Kwok[3], K.W. Chan[1] and N.C.Cheung[1]

[1] Department of Electrical Engineering, The Hong Kong Polytechnic University
[2] Institute of Textiles and Clothing, The Hong Kong Polytechnic University
[3] Department of Applied Physics, The Hong Kong Polytechnic University

Abstract-- The contactless power converter is examined here for power transfer in intelligent clothing which has a number of electronic devices that requires power. Using the contactless power converter that is based on magnetic coupling, the power can be transferred. Magnetic and circuit design, analysis and test will be presented in the paper. A gapped design is used for the magnetic device and a compensation method is implemented in the transformer primary and secondary sides for improvement for power transfer.

Index Terms-- Transformer, air-gap, contactless, power conversion, compensation, resonant converter, magnetics.

I. INTRODUCTION

Intelligent clothing is now a major development for protection, entertainment, security and military and other serious application. Missing person searching is needed and has been installed in intelligent clothing [1]. Temperature control is examined for used in very cold environment in clothing [2]. Medical application has also been found in clothing [3]. Protection clothing is also worn for robot [4]..The use of lighting controlled clothing for advertising or appearance has also been developing [5]. Therefore it is obvious that intelligent clothing is now a trent of development an d for clothing fashion as well. However, all the electronics and system installed in clothing requires electrical power. The power must be stored in the clothing through rechargeable battery or other energy storage devices. For rechargeable battery, a mechanism must be developed for the power transfer from external sources to the clothing. Metal connection is not feasible as it will affect the appearance, safety or other electronics in the clothing, therefore a contactless power transformer is developed that provide neat distance power

---

This work was supported by the Research Office, Hong Kong Polytechnic University under funding of Development of Intelligent Clothing Using Power Electronics Techniques of [project number G-YE17

K.W.E.Cheng, Y.Lu, and N.C.Cheung are also with Power Electronics Research Center , The Hong Kong Polytechnic University (e-mail: eeecheng@polyu.edu.hk).

transfer without direct contact.

A transformer may be used to provide voltage stepping for electrical energy supply to a load and, at same time, provide galvanic isolation. However, there is requirement recently to develop the transformer with large air-gap for the applications including contactless charger for electrical vehicles, robots and linear movable systems. By means of contactless transmission technology, some advantages are achieved such as no contact resistance, no wear and tear on the electrical contacts and trailing cable can be saved for movable systems. Contactless transformers are characterized by a small magnetizing inductance and large leakage inductance compared with traditional well coupling systems. Large primary current may be generated due to the small magnetizing inductance and high strain on switching devices may occur during transient process for the sake of large leakage inductance. Furthermore, this would also result in low power transfer ability and poor efficiency. Suitable compensation for contactless transformer plus resonant technology is a better way to solve these problems.

Contactless power transmission technology has been researched in recent years. Ref [6-7] analyzed compensations at both primary and secondary windings of a gapped transformer. The calculation of parameters of a gapped transformer has been presented in Ref [8]. The applications of the large air-gap have been also found in wireless power supplies [9], flyback and forward converter [10].

To enhance power transfer ability and improve efficiency, the compensation on both primary side and secondary side are investigated using the contactless transformer model which is based on mutual inductance. It is found that the relationship between primary current and secondary current of some contactless transformers can be quite different from that of traditional transformers due to the capacitive property of the secondary circuit. Because of the coupling between the primary and secondary is affected under the contactless manner, the leakage field increase and it is necessary to examine this phenomena.

A resonant technology is used to improve the power transfer through the air-gap. Finally, a PLL controller is implemented on the converter to ensure it operating at optimum switching frequency though the length of separate can be variable. The converter has been used in an intelligent system as a power supply.

II. STRUCTURE

The configuration of an air-gapped transformer is shown in Fig 1. Using high frequency operation, the size of the transformer can be reduced to very small. The primary size is energized by a high frequency switching circuit to provide the switching power signals. The high frequency AC is applied to the primary side of the transformer. The magnetic field generated is coupled to thye secondary side. As there is a large air-gap, it is expected that the coupling factor is lower as the air gap increases and the a portion of the field is not linking the secondary side coil and becomes leakage field. The selection of the operating frequency is also a concern. The frequency must be high in order to reduce the size of the converter and the transformer. However, the drawback is the high frequency loss including the loss in the switching components and other passive components.

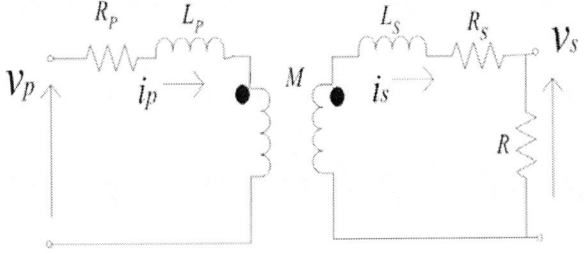

Fig. 1 The concept of the contactless converter

III. CONVENTIONAL MODELING OF GAPPED TRANSFORMERS

The contactless transformer is usually with a small magnetizing inductance and it is variable with respect to the length of air-gap. For a given primary voltage, the MMF is not constant in a contactless transformer. Hence the T equivalent circuit for analyzing a traditional transformer is not valid anymore.

The mutual inductance model can be employed to analyze the coupling between primary side and secondary side of the transformer.

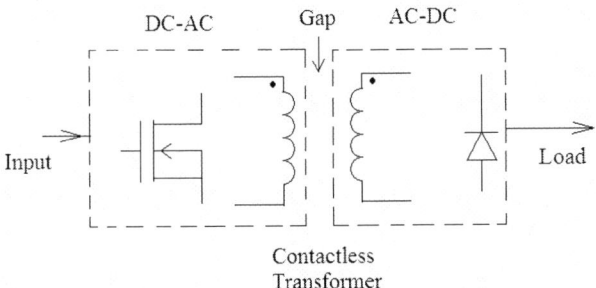

Fig. 2 Transformer as coupled windings

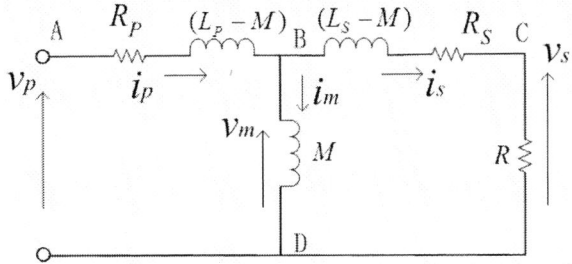

Fig. 3 Equivalent circuit of coupled windings

The coupling circuit of a gapped transformer is shown in Fig. 2 where M represents the mutual inductance. The electrical parameters of both sides are R_P, L_P, R_S and L_S. R is load resistance. The primary winding is driven by voltage v_p. Equations (1-2) can be obtained from Fig. 2.

$$i_p R_p + L_P \frac{di_p}{dt} - M \frac{di_s}{dt} = v_p \qquad (1)$$

$$-i_s R_s - L_S \frac{di_s}{dt} + M \frac{di_p}{dt} = v_s \qquad (2)$$

Based on these equations, the equivalent circuit can be achieved as shown in Fig. 3 for analysis. M is usually small due to loosely coupling. It may result in great current if the primary voltage is applied. The ability of power transfer of the transformer may also be limited and the efficiency would be decreased as well. Hence the power transfer ability and efficiency are both key issues need to be solved for this kind of systems.

IV. Gapped modeling of Transformer

The above model is conventional and must be re-examined for the gapped condition, As the air-gap is large, the coupling between the primary and secondary windings is low, the mutual inductance between two windings is also reduced significantly with distance between the primary and secondary windings.

The equations (1-2) is therefore re-written as:

$$i_p R_p + L_P \frac{di_p}{dt} - M_1 \frac{di_s}{dt} = v_p \qquad (1)$$

$$-i_s R_s - L_S \frac{di_s}{dt} + M_2 \frac{di_p}{dt} = v_s \qquad (2)$$

The flux linking between the windings are not the same and the coupling coefficient is reduced with the gap:

$$e_1 = \frac{d}{dt} \lambda_1 = \frac{dN_1 \phi_1}{dt} \qquad (3)$$

$$e_2 = \frac{d}{dt} \lambda_2 = \frac{dN_2 \phi_2}{dt} \qquad (4)$$

where e_1 and e_2 are the primary and secondary voltage of the transformer without including the resistive loss. Assume that the coupling coefficient between the two fluxes as:

$$\phi_2 = k\phi_1 \qquad (5)$$

It follows that

$$\frac{e_2}{e_1} = k\frac{N_2}{N_1} \qquad (6)$$

V. RESONANT COMPENSATION

Self-resonant technique can be used to solve the problems existed in a gapped transformer if it is properly compensated. The compensation mode can be series, parallel or series-parallel for primary or secondary winding [1, 11]. A resonant techniques that a parallel capacitor has been used to excited with the magnetizing inductance. The resultant current in the magnetizing circuit is reduced. The method allows air-core transformer to be practical. In summary, the resonant compensation modes can be considered in different configurations as shown in Fig. 4.

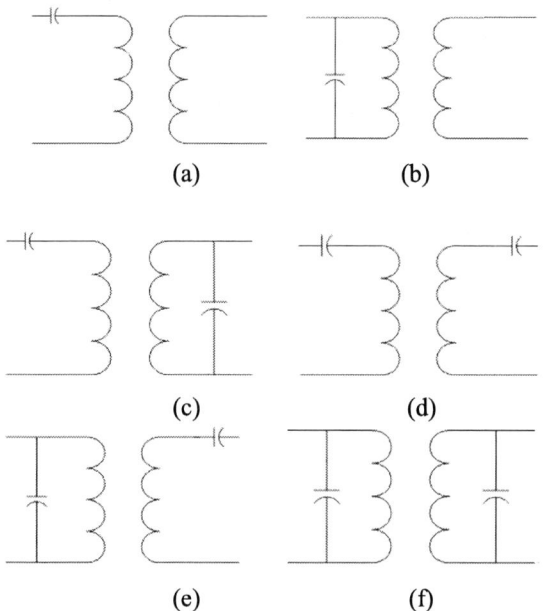

Fig.4 Topology variation for resonant capacitor compensation. (a) Series compensation on primary;)=(b) parallel compensation on Primary (c) Series compensated primary and parallel compensated secondary; (d) Series compensated primary and secondary; (e) Parallel compensated primary and series compensated secondary; (f) parallel compensated primary and secondary.

A. Efficiency of a contactless power converter

A contactless transformer is characterized by its large leakage inductance and low magnetizing inductance. It can be seen from Fig. 3 that the higher input current is generated for a gapped transformer due to its smaller magnetizing inductance, and, higher input voltage is required for the sake of high level of leakage. Hence efficiency of a contactless power converter may be poor. The lost power is mainly consumed in the gapped transformer and the driving circuit. Therefore reducing both the input current and the input voltage of the gapped transformer is the key issue to improve the efficiency and enhance the power transfer ability of a contactless power converter

B. Primary compensation

The primary current of a gapped transformer is shunted through the very small magnetizing inductance as shown in Fig. 3. It results in the poor efficiency. If the impedance of the branch circuit (A to C) can be reduced or the impedance of the branch circuit (B to D) can be increased through compensation, the efficiency would be improved. Series compensation on primary side can be achieved by connecting a capacitor in series with the primary winding. The impedance of the branch circuit (A to B) can be decreased by selecting suitable value of the compensation capacitor and frequency of the input voltage. The voltage dropped across the 'leakage' inductance is reduced and the power transfer ability is modified.

Half bridge circuit is widely used in resonant converters. Series compensation is a better choice for this situation. Thus the capacitor can be used as resonant component. Meanwhile, it also provides the voltage source for the lower switching device of the half bridge. In addition, the transformer would not saturate for the sake of the series capacitor. At resonance, the voltage across the resonant capacitor can be very high due to the large leakage on primary side and the high Q factor. But the high resonant voltage doesn't affect the voltage rating of the switching devices because the voltage across the capacitor and the voltage across the inductor cancel each other at resonance. It has been proved that the required primary compensation capacitance is independent of the load if the primary winding is series compensated [1].

Generally a gapped transformer can be considered as an inductive component. If the primary winding of it is parallel compensated using a capacitor, the leading current through the capacitor can compensate the inductive component of the primary current. The algorithm is similar to power factor improvement in alternative circuits. Therefore the input current of the compensated transformer can be reduced. The loss in the driving circuit is decreased for the sake of smaller output current of it. However, the primary current of the gapped transformer is not changed and the efficiency of it is not improved. Therefore parallel compensation cannot enhance power transfer ability of a gapped transformer.

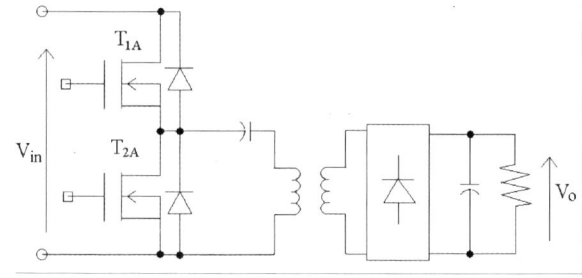

Fig. 5 Series resonant compensation of primary side in a half bridge converter

C. Secondary compensation

At resonance, I_m, which is the amplitude of the magnetizing current, is considered to be constant as the voltages across the resonant components almost cancel each other if series compensation is applied on primary side. Following analysis is based on the assumption. As shown in Fig. 6, a capacitor is connected in parallel with the load resistor to achieve parallel compensation. For the magnetizing voltage, the compensation circuit provides a leading current. If some secondary parameters such as (Ls-M) and Rs are not considered to simply the analysis, the compensation algorithm can be explained using a phasor diagram as shown in Fig. 7. The current through the compensation capacitor can compensate the magnetizing current and therefore the primary current is reduced. It also can be considered as parallel compensation can increase the equivalent magnetizing impedance (branch circuit B to D) and decrease the primary current. Thus the efficiency is improved. The other advantage of parallel compensation is that compensation function is always valid even the load is removed.

Fig. 8 shows the series compensation circuit where the capacitor is connected in series with the secondary winding. Same as parallel compensation in secondary side, series compensation can also make the secondary current to be leading the magnetizing voltage to compensate the magnetizing current. Therefore the primary current is reduced and the efficiency is improved. The compensation principle can be exhibited through a phasor diagram as shown in Fig. 9. It is assumed i_p, i_m and i_s represent the original primary current, magnetizing current and the secondary current. To simply the issue, it is suppose the current i_s is with the same phase with the magnetizing voltage. Suppose the secondary current supplying for load is not varied. It can be seen from the phasor diagram that the amplitude of the primary current would be reduced after compensation if the load current is maintained to be constant. The compensation effect is more obvious with respect to smaller compensation capacitance. This compensation capacitance can not be too small, or it would affect the loading ability of the gapped transformer. One drawback of the series compensation is that compensation function is affected by the load resistance and it would vanish if the load resistor is removed.

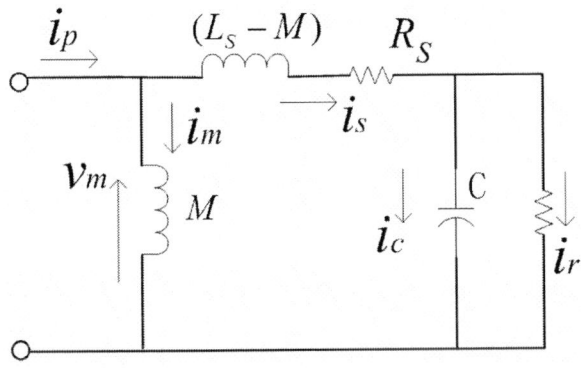

Fig. 6 Circuit of the parallel compensation in secondary side

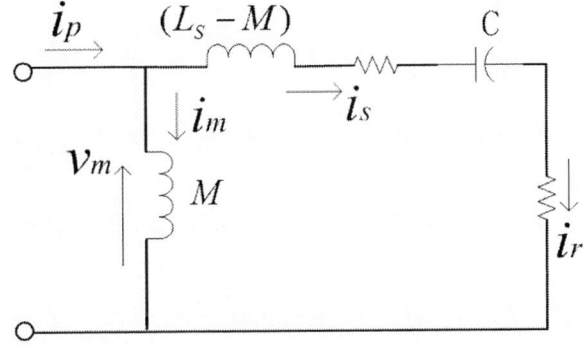

Fig. 8 Circuit of the series compensation in secondary side

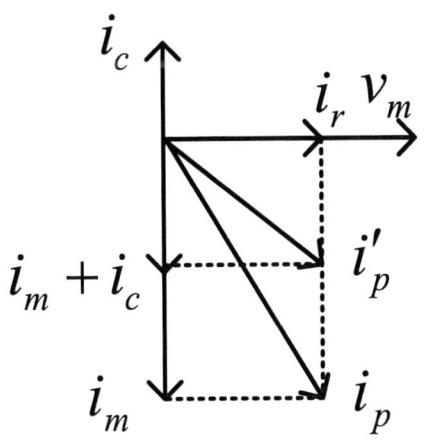

Fig. 7 Phasor diagram for secondary parallel compensation

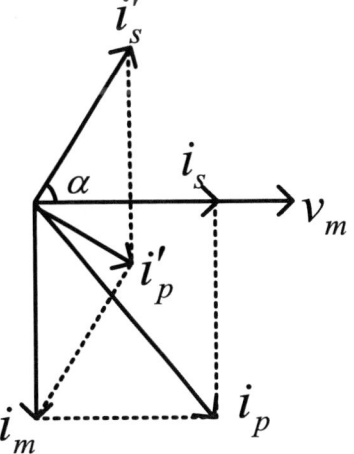

Fig. 9 Phasor diagram for secondary series compensation

1730

VI. MEASUREMENT OF A HEAVILY GAPPED TRANSFORMER

Experiments were carried out based on a gapped transformer constructed by two U-cores. The dimension of the U-core is that the limb cross-sectional diameter 14mm. The half-core is of base-wide 25mm and height 15mm. The relative permeability μr is around 1000. The numbers of turns of the primary/secondary windings of the transformer are 60 and 40, respectively. The separate length of the U-cores is 12mm.

An input voltage with square waveform is applied to the primary side of the gapped transformer through a series compensation capacitor as shown in Fig. 10. The voltage is generated by a half bridge circuit and the amplitude and frequency are 30V and 74kHz, respectively. Under this circumstance the current through windings of the transformer are quite sinusoidal.

In the first set of experiments, no compensation on the secondary side is applied. The measurement results are shown in Fig. 11(a). It can be seen that the load current is very small and the power transfer ability is weak. With primary current increase increases, the secondary current decreases. It means the special phenomenon occurs for the poor coupling of the transformer. Hence efficiency is low.

Series compensation in secondary side is applied in the second set of experiments. The compensation capacitor can be used for tunning α, which is the phase angle between the magnetizing voltage and the secondary current. By changing the load resistance, the current on both primary and secondary side can be regulated. For the following tests, the value of the compensation capacitance is 0.022uF. It can be seen from the measurement results shown in Fig. 11(b) that the secondary current is quite large than that in the first set of experiment, thus the power transfer ability is modified. The upper curve in Fig. 11(b) is consisted of different segments with both positive slop and negative slop. After the load resistance increases to certain value, the primary current increases while the secondary current decreases. That means power transfer ability and efficiency are both weaken within the range of load. The phenomenon is quite different from that in traditional transformers. For series compensation on secondary side, the compensation function is weakened when load resistance increases.

Same as series compensation in secondary side of the gapped transformer, parallel compensation can also make the secondary current to be leading the magnetizing voltage, thus the secondary current can compensate the magnetizing current to enhance power transfer ability and improve efficiency. In addition, this kind of compensation would not be affected by the load. It is still effective even though the load resistance to be removed. Hence parallel compensation is a better choice for secondary winding. It is suitable for those needed to be compensated within widely range of load.

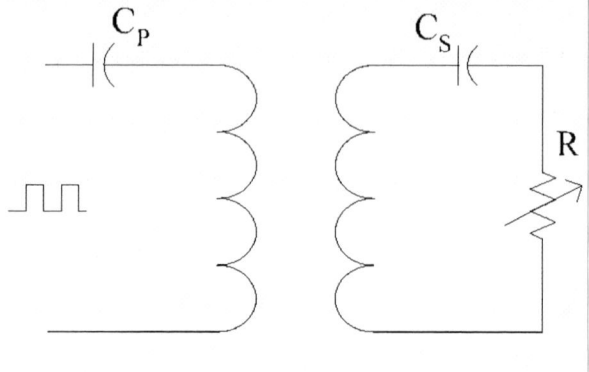

Fig. 10 Tested transformer with compensation

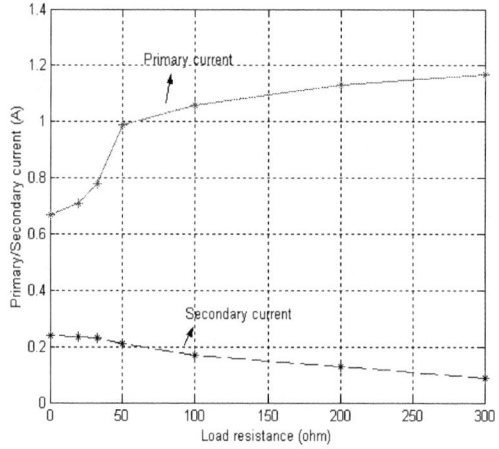

(a)

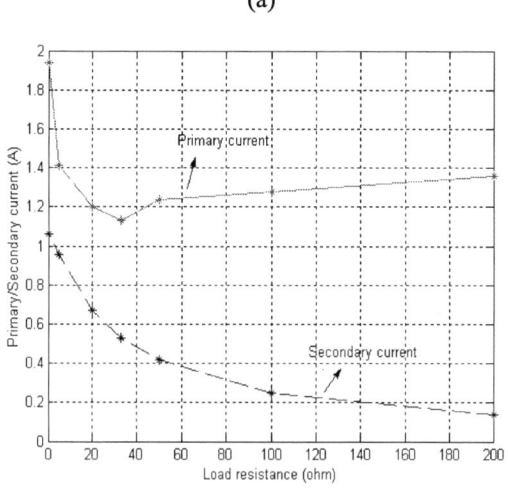

(b)

Fig. 11 Primary/secondary current vs load resistance for a gapped transformer. (a) No compensation on secondary side; (b) Series compensation being applied on secondary side.

Table1 shows the components use din the converter. Fog 12 shows then circuit design. The circuit is again a half-bridge circuit with resonant capacitor connected in the series-primary and parallel secondary manner.

TABLE I
ELECTRICAL SPECIFICATION OF THE CONVERTER

Operation	Specification
Input voltage E	30V
Output voltage Vo	12-28V
Output power Po	0-30W
Switching frequency	60-100 kHz
Separation length	5-12mm
Primary compensation capacitance	0.033µF
Secondary compensation capacitance	0.022µF

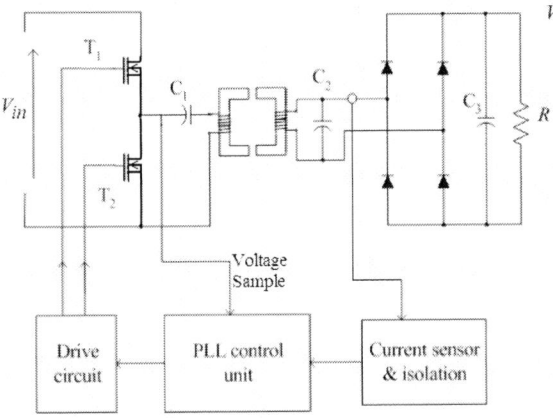

Fig. 12 Resonant contactless power converter using PLL control

Fig 13 shows the Finite element analysis results of the air-gapped transformer. It can be seen that the main flux linking between the primary and secondary sides are still strong but they are not equal. It is expected that the coupling coefficient for primary to secondary and secondary to primary are not equal.

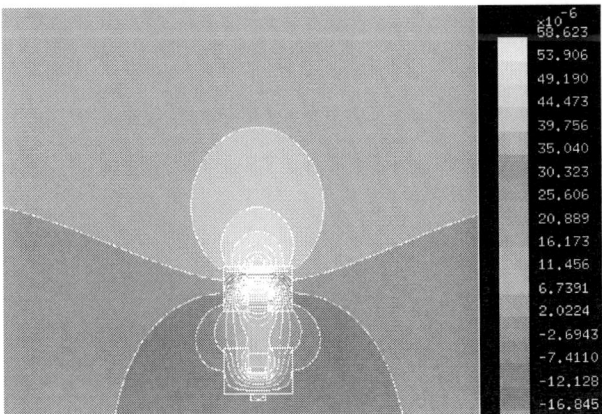

Fig 13: Finite element simulation of the contactless transformer.

VII. CONCLUSIONS

This paper focuses on investigating various compensation strategies of gapped transformers to enhance power transfer ability and improve efficiency of contactless power converters.

For compensation on primary side of a gapped transformer, parallel compensation can only reduce the output current of the driving circuit and the power transfer ability and the efficiency of the gapped transformer are not modified. Series compensation together with a half bridge driving circuit can achieve better effect.

Both series compensation and parallel compensation can achieve better effect by selecting suitable compensation capacitance and frequency of the driving voltage. However, series compensation effect is related with the load resistance. The compensation function is weakened while the load resistance increases. Parallel compensation is independent of the load resistance and therefore it is a better choice for secondary compensation.

A special phenomenon happening upon a transformer with large air-gap is presented. In this transformer, the primary current increases with the secondary current decreases in certain condition. This condition has been derived. Finally, PLL control is implemented on a contactless power converter where the used gapped transformer is primary series compensated and secondary parallel compensated. Both the ability of power transfer and the efficiency are modified through resonant technology.

VIII.   REFERENCES

[1]   Khosla, R.; Francionne, D.; Chu, D., "Intelligent Online Web Based Interactive Missing Person Clothing Identification System", *IEEE International Conference on Systems, Man and Cybernetics, ICSMC '06*, Vol. 5, 8-11 Oct. 2006, pp. 4022 – 4027.

[2]   Rantanen, J.; Alfthan, N.; Impio, J.; Karinsalo, T.; Malmivaara, M.; Matala, R.; Makinen, M.; Reho, A.; Talvenmaa, P.; Tasanen, M.; Vanhala, J., "Smart clothing for the arctic environment", T*he Fourth International Symposium on Wearable Computers*, 2000, 16-17 Oct, pp. 15 – 23

[3]   Dittmar, A.; Meffre, R.; De Oliveira, F.; Gehin, C.; Delhomme, G., "Wearable Medical Devices Using Textile and Flexible Technologies for Ambulatory Monitoring", 27th Annual IEEE-EMBS International Conference of the Engineering in Medicine and Biology Society, 2005, pp. 7161 - 7164

[4]   Yokoi, K.; Nakashima, K.; Kobayashi, M.; Mihune, H.; Hasunuma, H.; Yanagihara, Y.; Ueno, T.; Gokyuu, T.; Endou, K., "A tele-operated humanoid robot drives a backhoe in the open air", *IEEE/RSJ International Conference on Intelligent Robots and Systems (IROS 2003)*, Vol. 2, 27-31 Oct. 2003, pp. 1117 – 1122.

[5]   Akbari, M.A.; Takahashi, H.; Nakajima, M., "Discerning advisor: an intelligent advertising system for clothes

considering skin color", *International Conference on Cyberworlds,* 2005, 23-25 Nov., pp.8.

[6] C.S. Wang, G.A. Covic and O.H. Stielau, "Power transfer capability and bifurcation phenomena of loosely coupled inductive power transfer systems", IEEE Transactions on Industrial Electronics, Vol. 51, No. 1 February 2004, pp148-157.

[7] H. Abe, H. Sakamoto and K. Harada, "A noncontact charger using a resonant converter with parallel capacitor of the secondary coil", IEEE Trans. Ind. Applicat, Vol. 36, pp. 444-451, Mar/Apr. 2000.

[8] Y.Lu, K.W.E.Cheng and S.L.Ho, "Investigation of the leakage inductances and energy conversion of air-gapped transformers", IEE APSCOM, Nov 2003, Hong Kong, pp. 70-75

[9] O'Brien, K.; Scheible, G.; Gueldner, H., "Design of large air-gap transformers for wireless power supplies", IEEE Power Electronics Specialist, PESC. Vol.4 , June 15-19, 2003, pp. 1557 –1562.

[10] K.W.E.Cheng and Y.Lu, "Development of a Contactless power converter", IEEE International conference on Industrial Technology, ICIT'02, Bangkok, Dec 2002, Vol. 2, pp. 786-791.

[11] Cheng K.W.Eric, Chan H.L. and Sutanto D., "Development of a superconducting transformer using self-resonant techniques for DC/DC power conversion", *IEE Proceedings – Electric Power Appl.,* Volume 151, Issue 03. May 2004, pp. 296-302.

[12] K.W.E.Cheng, "Classical Switched-mode and resonant power converters, The Hong Kong Polytechnic University, ISBN: 962-367-364-7, Sep 2002.

Simulation Program for Switching Converters Using Numerical Fourier Transform

Yoshihiro Tomihisa*, Hirotaka Nakanishi*, Terukazu Sato*, Takashi Nabeshima*,

Kimihiro Nishijima*, Tadao Nakano*

* Oita University, 700 Dannoharu, Japan

Abstract--This paper presents a simulation method for the analysis of switching power converters. In the method, Numerical Fourier Transform is employed to obtain fast results. A circuit simulator using the proposed simulation method is implemented and demonstrated with a ripple regulator. The simulation results have good agreements with the experiments. The proposed method is applicable to any circuits including switching operation such as ripple regulators.

Index Terms—Simulation, Software, Frequency analysis, DC power supply

I. INTRODUCTION

Recently, for saving time and cost in designing switching converters, circuit simulators have been widely used. So far, many circuit simulators have been introduced [1-3], and they are classified into two categories, that is SPICE based and non-based. Generally, SPICE based simulators cannot perform frequency analysis (AC sweep) including switching operation. However it is possible to obtain frequency response of switching converters by processing the results of the SPICE simulator as follows:

Using SPICE:
(1) For a given measuring frequency, excite the input signal by sinusoidal waveform and find the steady state.
(2) Obtain the output waveform.
Using additional software:
(3) Calculate the gain and phase for the given measuring point.
(4) Repeat above for all measuring points.

As far as the authors know, SCAT [2] is non-SPICE based simulator and it seems that the above procedure is employed. It takes extremely long time to obtain the whole results because it must find the steady state for all given measuring points.

In this paper, new simulation algorithm for the frequency analysis is proposed. The input signal is exited by step function, and the output signal is transformed to frequency domain by using Numerical Fourier Transform. Since we have only to calculate the transient response for the lowest measuring frequency, it is rather fast to obtain whole results than the method using sinusoidal waveforms. The proposed method is applicable to any circuits including switching operations such as ripple

regulators. A circuit simulator using the proposed method is implemented and demonstrated with a ripple regulator.

II. NEWLY DEVELOPED CIRCUIT SIMULATOR

The Authors have been developed a simulation program named "Simple Circuit" or "SimSC" and it's features are introduced first. It has a circuit editor using GUI as shown in Fig. 1. Graphical circuits are converted to the State Space Equation and the voltage and current waveforms of each element are obtained on the Scope Window by solving it. Fig. 2. shows one of the examples of results. Since the Scope Window is linked together with the Circuit Editor, the waveforms of the element selected on it are displayed immediately. Simple Circuit provides the following basic circuit elements at present; Resistor, Capacitor, Inductor, Independent Voltage (Current) Source, Voltage Controlled Voltage (Current) Source, Current Controlled Voltage (Current) Source, Square (Sine, Triangular) Wave Generator, PWM Modulator, Comparator, Voltage Controlled Switch, Diode, Ground, and Port. And more, a sub-circuit can be packed as a new unit, and we can use it as a new element like 2 to 16 ports IC.

As for the analysis, the transient analysis is the fundamental and essential to the switched mode circuit analysis. The steady state and frequency analysis is based on it. The State Space Averaging Method [4] with the Newton Method is applied to Simple Circuit for the fast

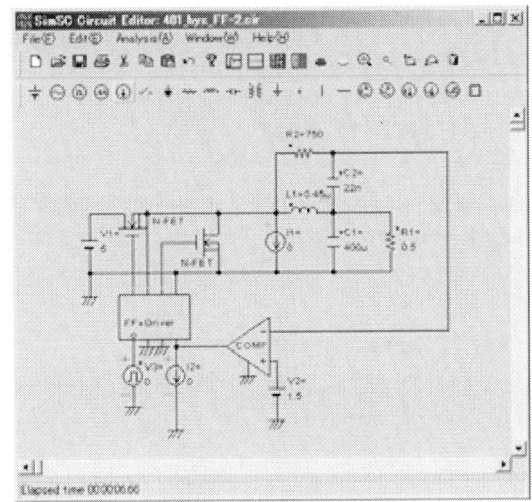

Fig. 1. Circuit Editor.

Financial support should be acknowledged here. Example: This work was supported by Japanese Ministry of Research.

978-1-4244-0644-9/07/$25.00 ©2007 IEEE

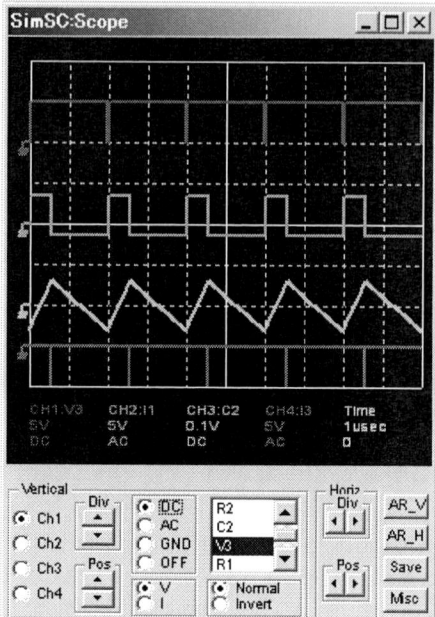

Fig. 2. Scope Window.

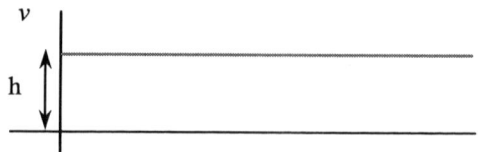

(a) Input voltage waveform.

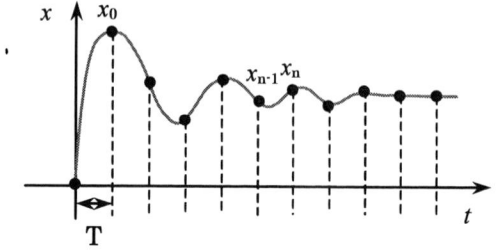

(b) Output voltage waveform.
Fig. 3. Input and output voltage waveform.

steady state analysis in which the state space equation is calculated numerically by small time step. For the frequency analysis, Simple Circuit has two methods, State Space Equation Method and Numerical Fourier Transform Method. The former is numerical calculation based on State Space Averaging Method. It can obtain fast results but cannot yield proper results for the circuits with variable duty cycle or with feedback loop such as hysteretic PWM converter. The Numerical Fourier Transform Method is the method that is developed for the proper analysis of switched mode power supplies with the feedback loop.

III. FREQUENCY ANALYSIS USING NUMERICAL FOURIER TRANSFORM

To obtain the frequency response of the transfer function of the input voltage v(s) to the output voltage x(s), the input signal v is changed by step function and the output voltage x changes as shown in Fig. 3.
The input signal v, Δv, is expressed by unit step function u(t) as follow.

$$\Delta v = hu(t) \tag{1}$$

and applying Raplace Transform, we have

$$\Delta v(s) = \frac{h}{s} \tag{2}$$

Taking the differential of x,

$$\Delta y(t) = \frac{d}{dt}\Delta x(t) \tag{3}$$

and applying Raplace Transform, we have

$$\Delta y(t) = s\Delta x(s) \tag{4}$$

From (2) and (4), we have

$$\frac{\Delta x(s)}{\Delta v(s)} = \frac{1}{h}\Delta y(s) \tag{5}$$

From (5), we can see that the transfer function of $\Delta x(s)/\Delta v(s)$ can be obtain by taking the Raplace Transform of the differential of the output signal x. If y is sampled periodically at the constant interval T, and it is expressed by sum of finite delta function as:

$$\Delta y = \Sigma y_n T \delta(t - nT) \tag{6}$$

where y can be approximated as

$$y_n = \frac{x_n - x_{n-1}}{T} \tag{7}$$

and we have

$$\frac{\Delta x(s)}{\Delta v(s)} = \frac{1}{h}\Sigma(x_n - x_{n-1})e^{-nT_s} \tag{8}$$

let s=jω, and applying the Eurer Equation,

$$\frac{\Delta x(s)}{\Delta v(s)} = \frac{1}{h}\Sigma(x_n - x_{n-1})\cos n\omega T$$

$$- j\frac{1}{h}\Sigma(x_n - x_{n-1})\sin n\omega T \tag{9}$$

The Equation (9) is the basic expression of the proposed algorithm, and the simulation program for the frequency analysis is implemented based on this equation.

IV. SIMULATION RESULTS

To demonstrate the performance of Simple Circuit, dynamic performances of four types of dc-to-dc converters are simulated bellow.

A. Buck Converter

Fig. 4. (a) shows a buck converter topology. In the figure, the symbol M1 is a PWM Modulator and a switch S1 is controlled by M1. The symbols +, - and arrows correspond voltage and current waveforms of same color in Scope Window, respectively. Fig 4. (b) is an example of transient analysis at start up.

1735

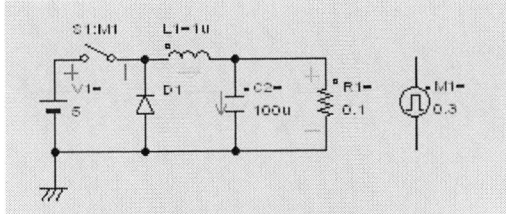

(a) Circuit configuration

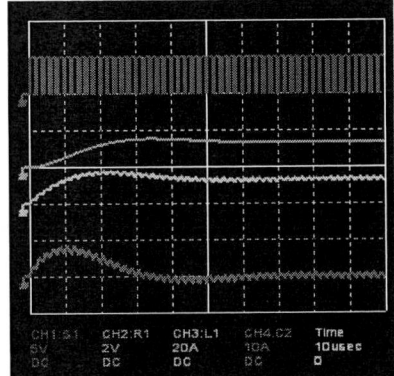

(b) Transient waveforms

Fig. 4. Buck Converter.

B. Frequency Characteristics

Bode plots of the transfer function $\Delta Vo(s)/\Delta D(s)$ are shown in Fig. 5. Where D denotes duty cycle of M1 and Vo is the voltage across R1. In the figure, (a) is in the case of the continuous conduction mode (CCM) and (b) is of the discontinuous conduction mode (DCM). These two modes are determined automatically even though the simulator was not programmed deliberately. From the figure, the well known characteristics of two pole transfer function in CCM and that of one pole in DCM are observed.

Fig. 6. shows the bode plots of $\Delta Vo(s)/\Delta V1(s)$. Fig. 6. (a) is in the case without feedback and (b) is with feedback. With the feedback, the DC gain is suppressed and the corner frequency is shifted at higher frequency. This is also well known characteristics. From the above results, the simulation algorithm of Simple Circuit may have the validity.

C. Boost and Buck-boost Converter

Fig. 7. and Fig. 8. show a boost and a buck-boost converter. As well known, these two converters have the zero in the right half plane, phase-shift in high frequency reaches to –270 degrees. Simple Circuit can derive this characteristic.

D. Ripple regulator

The circuit diagram of the ripple regulator is shown in Fig. 9. A flip-flop is included in this regulator to work at the fixed frequency. In The figure, the symbol FF is a flip-flop, and it synchronizes with the external clock V3. Fig. 10. is a simulation result of the transient analysis.

Bode plots of the transfer function $\Delta D(s)/\Delta Vo(s)$ are

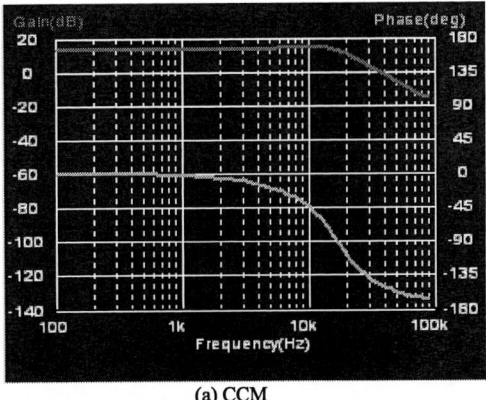

(a) CCM

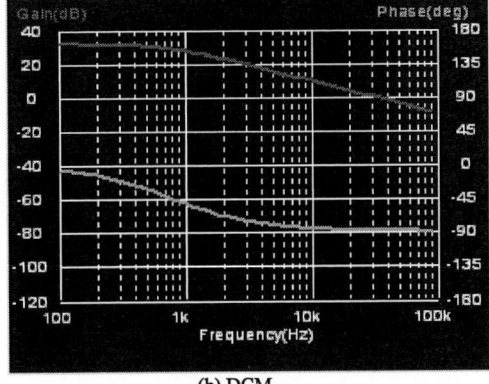

(b) DCM

Fig. 5. Frequency Characteristics of $\Delta Vo(s)/\Delta D(s)$.

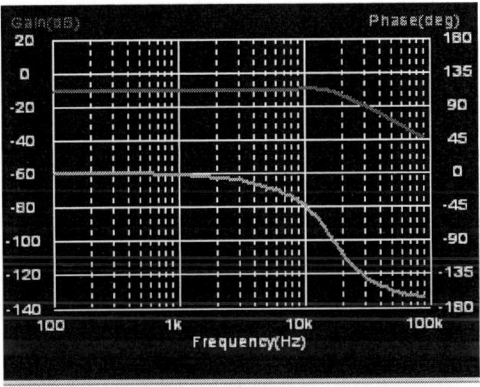

(a)Without feedback

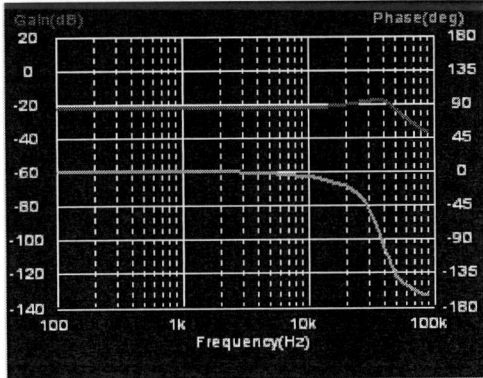

(b)With feedback

Fig. 6. Frequency Characteristics of $\Delta Vo(s)/\Delta V1(s)$.

1736

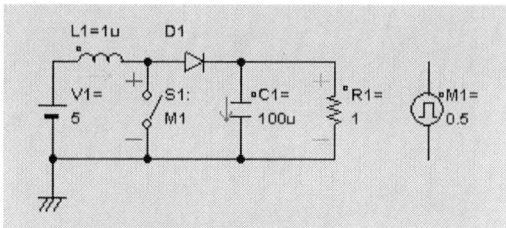

(a) Circuit configuration

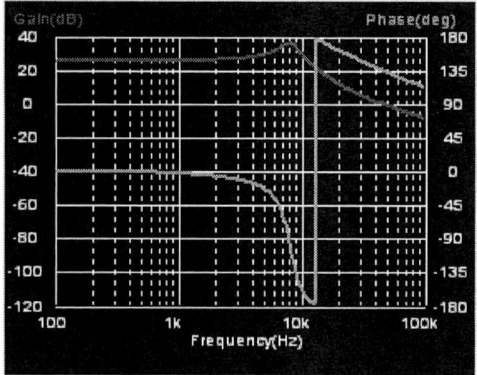

(b) ΔVo(s)/ΔD(s)
Fig. 7. Boost Converter.

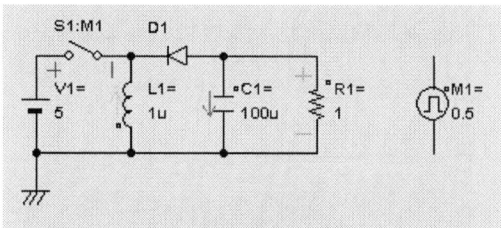

(a) Circuit configuration

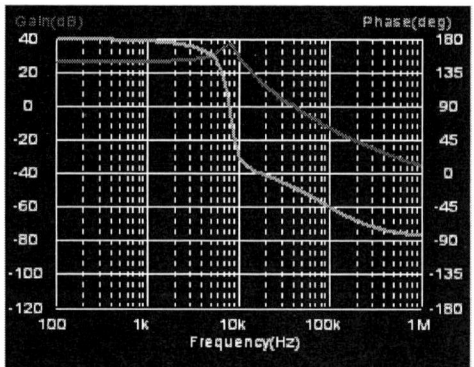

(b) ΔVo(s)/ΔD(s)
Fig. 8. Buck-boost Converter

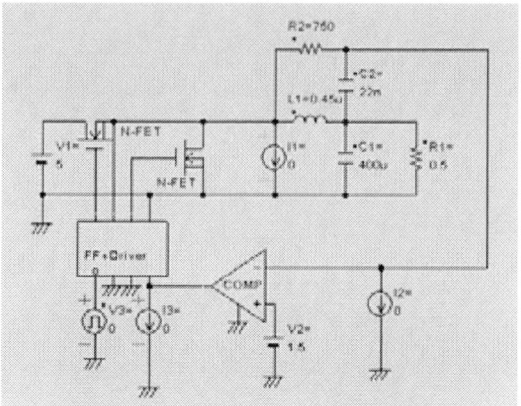

Fig. 9. Circuit diagram of a Ripple regulator.

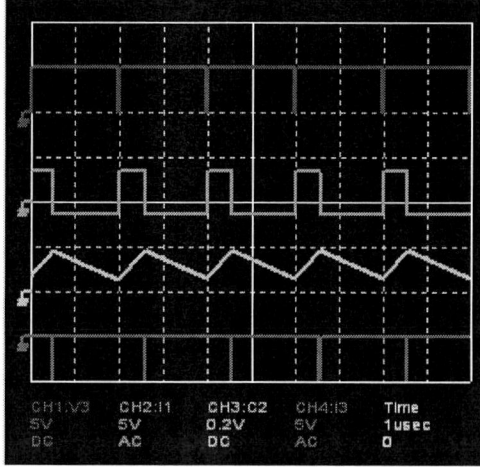

Fig. 10. Transient waveform of Ripple regulator.

shown in Fig. 11. Where D denotes duty cycle and Vo is the voltage across R1. In the figure, (a) is the simulated result and (b) is the experimental result, and we can see that ΔD(s)/ΔVo(s) shows the exact differential characteristic.

Fig.12. shows bode plots of the transfer function ΔD(s)/ΔV2(s), where V2 is the reference voltage. In the figure, we can see that ΔD(s)/ΔV2(s) has the inexact differential characteristic.

Bode plots of the loop transfer function are shown in Fig. 13. In the figure, the loop transfer function shows that the gain has a peak around 10kHz, and phase-shift in high frequency reaches to –270 degrees.

Finally Fig.14. shows bode plots of the transfer function ΔVo(s)/ΔV2(s).

In these figures, we can see that simulated results by Simple Circuit have good agreement with the measurement and it is confirmed that the proposed method is valid.

1737

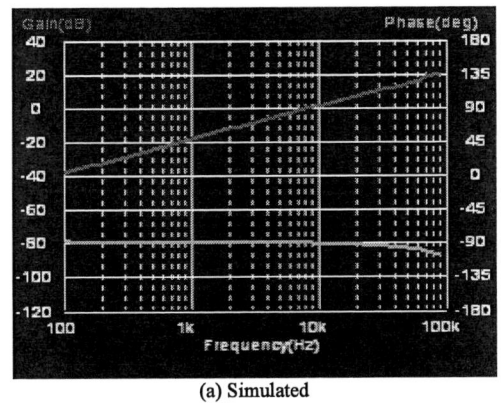

(a) Simulated

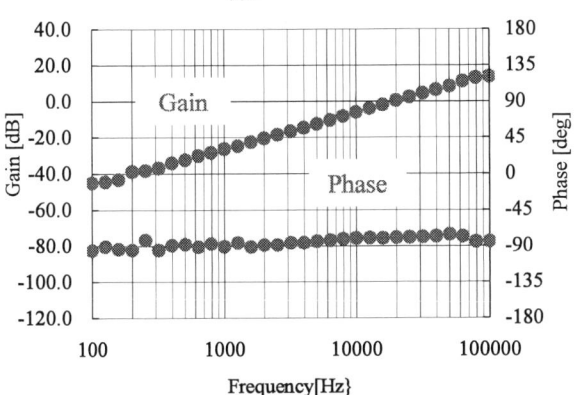

(b) Experimental

Fig. 11. Bode plot of $\Delta D(s)/\Delta Vo(s)$ of Ripple

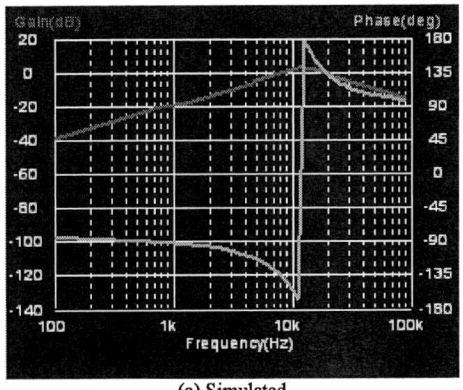

(a) Simulated

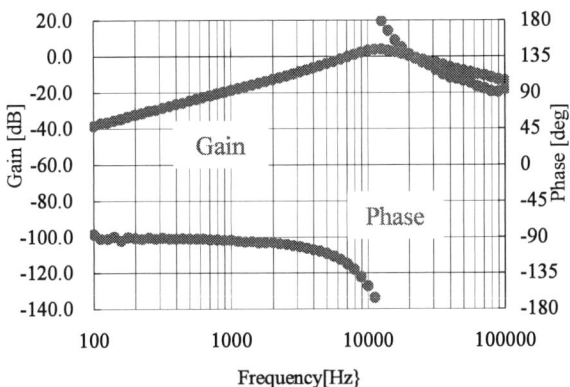

(b) Experimental

Fig. 13. Bode plot of Loop transfer function.

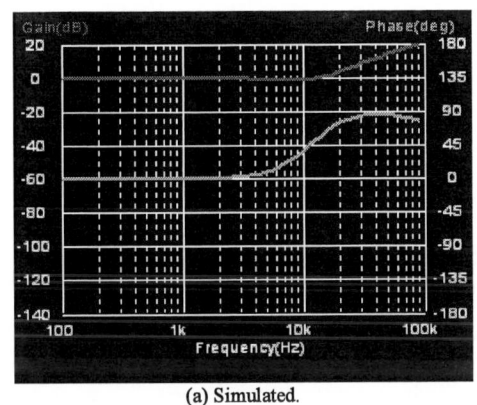

(a) Simulated.

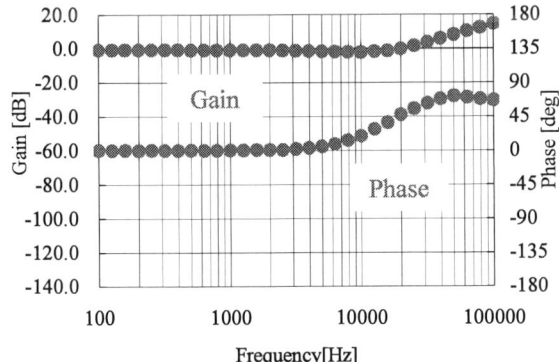

(b) Experimental.

Fig. 12. Bode plot of $\Delta D(s)/\Delta V2(s)$ of Ripple regulator.

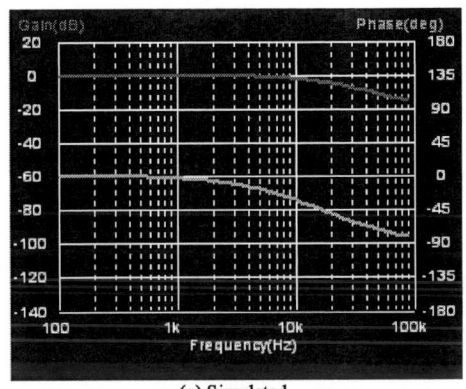

(a) Simulated.

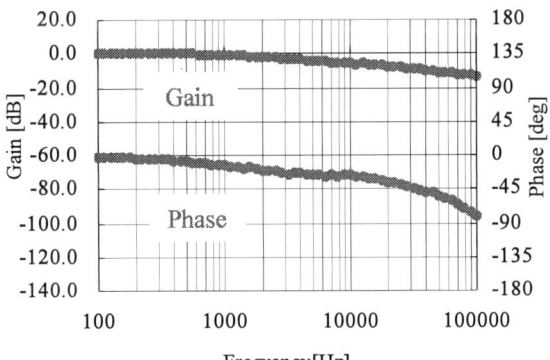

(b) Experimental.

Fig. 14. Bode plot of $\Delta Vo(s)/\Delta V2(s)$ of Ripple regulator.

1738

V. Conclusions

Newly developed simulation program for switching power converters is presented. Numerical Fourier Transform is employed to obtain fast results of frequency response. Simulations and experiments were performed for the three basic PWM converters and a constant frequency ripple regulator: and it is verified that proposed method is valid.

References

[1] M.H Rashid, "SPICE for Circuits and Electronics Using Pspice, "Prentice-Hall International Editions, 1990.

[2] M.Nakahara, "A fast computer algorithm for switching converters, "IEEE Trans. on PE, Vol12, No. 1, pp.180-186, Jan. 1997.

[3] T.Sato et.al, "A Novel Switching Converter Simulation Algorithm by Averaged State Transition Matrices, "Proceedings of Intelec'98

[4] R.D. Middlebrook et.al, "Ageneral unified approach to modeling switching-converter power stages, "IEEE PESC' 76 Record, pp. 18-34, June 1976

[5] T. Nabeshima, et.al, "Analysis and Design Considerations of a Buck Converter with a Hysteretic PWM controller, "IEEE PESC'04 CD-ROM, pp. 1711-1716

The Most Suitable Application of SiC Diode

Tomoaki Makino*, Atsushi Hirota**, and Satoshi Nagai*

* Department of Electrical and Electronic Engineering, Tsuyama National College of Technology,
624-1 Numa, Tsuyama-City, Okayama-Pref., 708-8509 JAPAN

** Department of Electrical and Computer Engineering, Akashi National College of Technology,
679-3 Nishioka, Uozumi-cho, Akashi-City, Hyogo-Pref., 674-8501 JAPAN

Abstract--In recent years, next generation silicon carvide(SiC) diodes were put to practical use. This SiC diodes have many advantages compared with fast recovery diode(Si-FRD) from the aspect of small recovery charge. To make use of this advantage, the most suitable operating conditions is to use at current lead mode of high frequency inverter circuit. In this paper, the authors evaluated SiC diode + IGBT inverter at current lead mode and Power MOSFET inverter at current lag mode. From the experimental results, switching loss of SiC + IGBT inverter over 100kHz frequency operation range is sufficiently small. As a result, high efficiency inverter using IGBTs over 100kHz frequency range can be realized for practical use.

Index Terms--SiC device, switching loss, inverter application

I. INTRODUCTION

Nowadays, silicon PN junction diodes are widely used for power conversion applications. Only for low voltage applications, Schottkey barrier diode(Si-SBD) are used. And GaAs-SBD[1] was put to practical use, but is not widespread. Comparing these devices, SiC-SBD[2] has following advantages; high voltage ratings, low leakage current, high temperature operation, no thermal runaway, small reverse recovery charge. But it has following disadvantages; low surge current rating, relatively expensive.

II. MEASUREMENT OF RECOVERY LOSSES

The authors have evaluated recovery losses of following diodes;
SiC diode : Infineon technologies SDT 10S30[3]
Si-SBD : International rectifier MBR20100CT[4]
Si-FRD : International rectifier 60EPU04[5].

Table I show specifications of these diodes. Fig. 1. shows switching loss measurement circuit and ideal operating waveforms are shown in Fig. 2. In Fig. 2., during the IGBT operating, the inductor current increases

TABLE I

SPECIFICATIONS OF DIODES

	V_{RM}(V)	I_F(A)	t_{rr}(ns)	Q_{rr}(nC)
SDT10S30	300	10	-	23
MBR20100	100	10	-	-
60EPU04	400	60	50	375

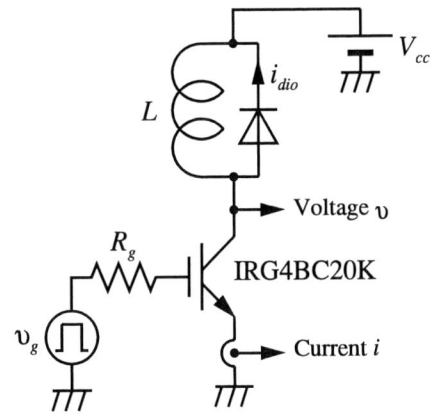

Fig. 1. Mesurement circuit diagram.

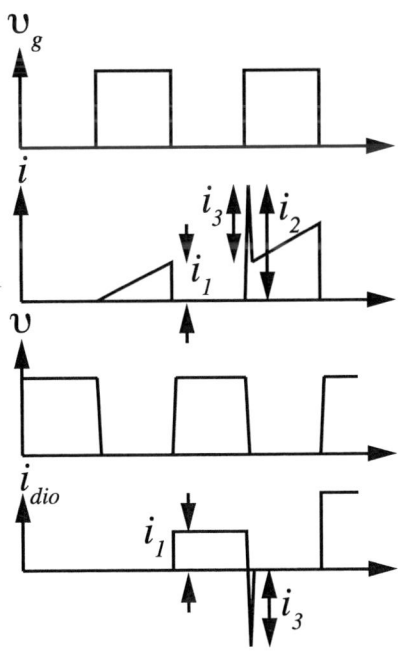

Fig. 2. Mesurement circuit waveforms.

TABLE II
MEASURED RECOVERY LOSSES (J)

SiC Diode	Si-SBD	Si-FRD
5.916×10^{-7}	1.489×10^{-6}	9.407×10^{-6}

linearly. About 10μsec later, the IGBT turns off and inductor current i_1 flows diode from the IGBT. During the IGBT turn-off operation, inductor current i_1 can be through as constant value. When the IGBT turns on again, the recovery current of diode and sum of inductor current flows through the IGBT. i_3 corresponds to recovery current and calculated i_1 subtracted from i_2.

Table II shows measured recovery losses of diodes at $Vcc = 100V$. From this table, recovery loss of SiC diode is much smaller than those of Si-FRD or Si-SBD.

III. EVALUATIONS OF SWITCHING LOSSES USING PRACTICAL INVERTER CIRCUIT

In this paper, the authors evaluated practical switching losses using conventional Single Ended Push Pull(SEPP) inverter circuit shown in Fig. 3. The inverter using SiC diodes and IGBTs[6], the operating frequency is set lower than the load resonant frequency. In this case, the load current phase leads to voltage and turn-off loss of IGBTs is minimized. But turn-on loss occurs by reverse recovery charge of diodes. The inverter using power MOSFETs[7], operating frequency is set higher than load resonant frequency. In this operating mode, turn-on loss becomes minimized but turn-off losses depend on switching characteristics of the power MOSFET. Usually, converter operating frequency using IGBTs is limited about 20kHz by its tail current. But using current lead mode with SiC diodes, it is expected that switching losses become to be much reduced.

The load resonant circuit parameters are as follows;
$R = 12.5\Omega$
$L = 100\mu H$

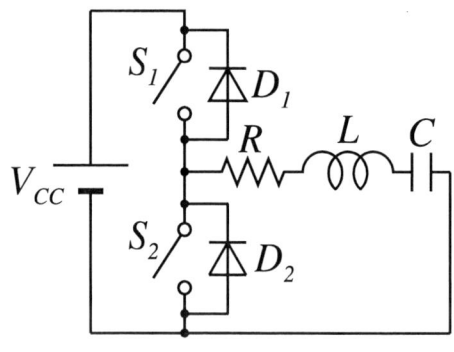

Fig. 3. SEPP inverter circuit.

$C = 0.0188\mu F$
Resonant frequency = 120kHz.

Fig. 4. shows voltage and current waveforms of the power MOSFET inverter operating at 130kHz, and DC input is 100V 0.89A. Fig. 5. shows voltage and current waveforms of SiC diode + IGBT inverter operating at 110kHz, and DC input is 100V 0.89A. Fig. 6. shows expanded turn-off waveforms of the MOSFET inverter. And Fig. 7. shows expanded turn-on waveforms of the SiC diode + IGBT inverter. Table III, table IV and Fig. 8. show measured switching losses of Fig. 6. and Fig. 7. Fig. 8. shows that the switching loss of the SiC diode + IGBT inverter is greater than that of the MOSFET inverter, but absolute values are small enough.

IV. EVALUATIONS OF EXPERIMENTAL RESULTS

From the experimental results of section III, the

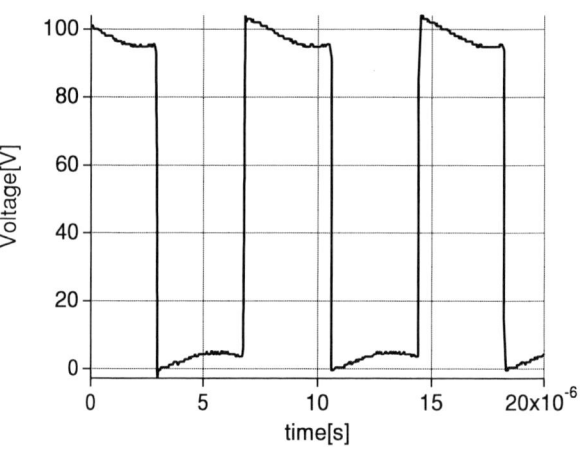

(a) Voltage waveforms.

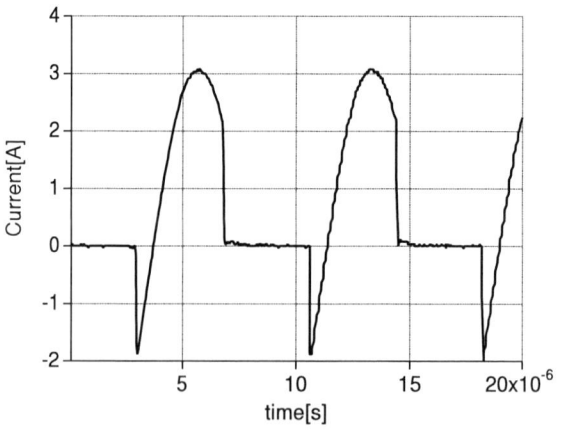

(b) Currnet waveforms.

Fig. 4. Operating waveforms of the MOSFET inverter.

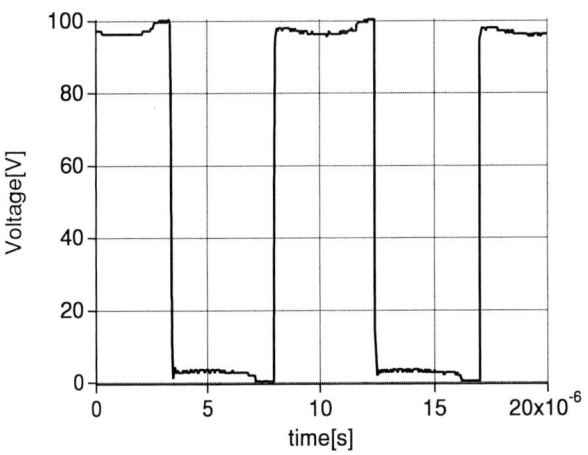

(a) Voltage waveforms.

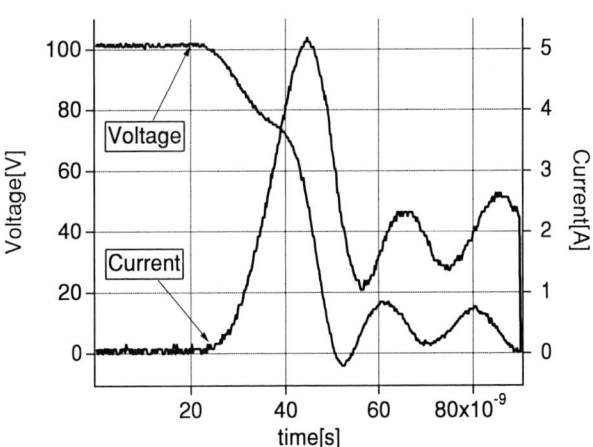

Fig. 7. Trun-on waveforms of IGBT.

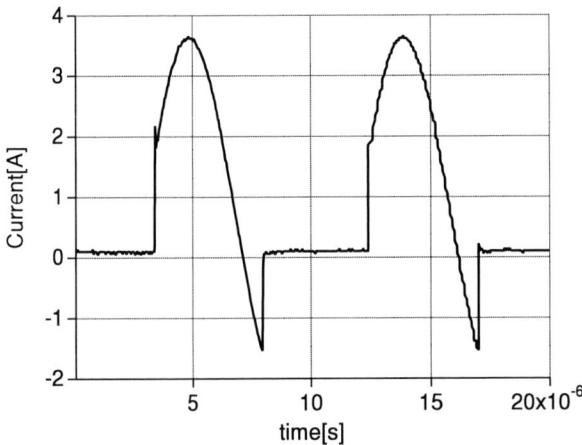

(b) Currnet waveforms.

Fig. 5. Operating waveforms of the SiC diode + IGBT inverter.

TABLE III

MEASURED TURN-OFF LOSSES OF MOSFET

DC Voltage[V]	DC Current[A]	Switching Loss[J]
50	0.45	6.199×10^{-7}
100	0.85	2.341×10^{-6}
150	1.22	5.509×10^{-6}

TABLE IV

MEASURED TURN-ON LOSSES OF SiC+IGBT

DC Voltage[V]	DC Current[A]	Switching Loss[J]
50	0.45	1.050×10^{-6}
100	0.85	3.905×10^{-6}
150	1.33	9.902×10^{-6}

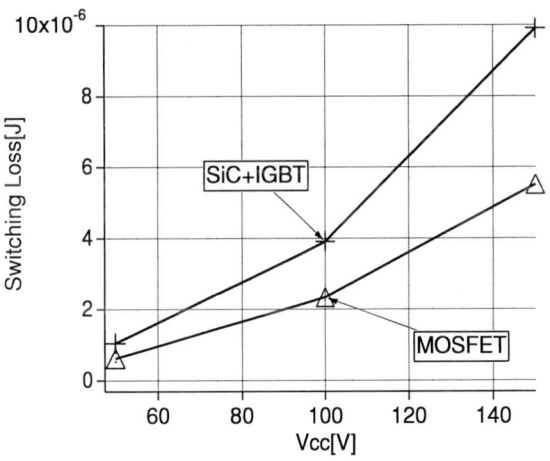

Fig. 8. Measured switching losses.

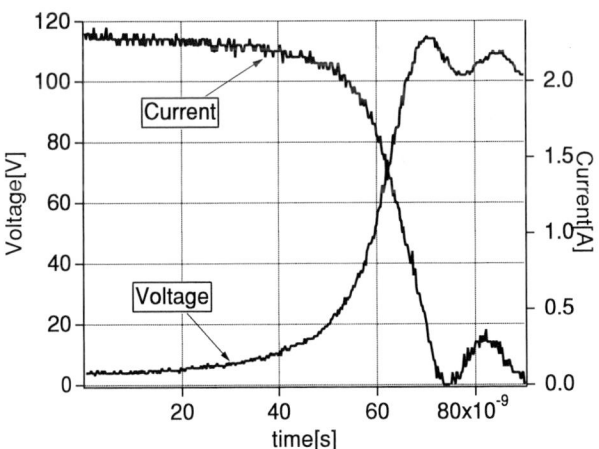

Fig. 6. Trun-off waveforms of MOFSFET.

switching losses of SiC diode + IGBT inverter are significantly reduced to small value over 100kHz operating frequency range. For example, calculated switching loss at 110kHz is 1.089W, when the DC input is 150V 1.33A. These results realized by that tail current losses are reduced to almost zero and reduced recovery loss of the SiC diode.

1742

V. Conclusions

In this paper, it has become clear that the SiC diode + IGBT inverter operating at current lead mode, the switching losses were significantly reduced. From experimental result, high efficiency power converter using IGBTs over 100kHz operating frequency range can be realized. As a result, high frequency high power converter with low cost and small volume will be realized.

References

[1] XIS homepage, http://www.ixys.com/Appasp/pdpgsd01.asp

[2] "thinQ!TM Silicon Carbide Schottky Diodes: An SMPS Circuit Designer's Dream Comes True!", Christian Miesner, Roland Rupp, Holger Kapels, Michael Krach and Ilia Zverev

[3] SDT10S30 Datasheet, Infenion Technologies

[4] MBR20100CT Datasheet, International Rectifier

[5] 60EPU04 Datasheet, International Rectifier

[6] 2SK2996 Datasheet, Toshiba

[7] IRG4BC30F Datasheet, International Rectifier

Multi-Domain System Simulation and Rapid Prototyping of Digital Control Algorithms using VHDL-AMS

P.J. Randewijk

Department of Electrical & Electronic Engineering, Stellenbosch University, Stellenbosch, South Africa

Abstract—This paper will discuss the applicability of VHDL-AMS as a research tool for multi-domain system (e.g. an electrical drive system) simulation as well as the rapid prototyping and testing of digital control algorithms for these systems.

Index Terms—multi-domain, system simulation, VHDL-AMS

I. HISTORY OF VHDL-AMS

VHDL was originally developed by the US Department of Defence (DoD) to simulate the behaviour of ASIC devices. Eventually the DoD granted all language definition rights to the IEEE and so the IEEE Standard 1076-1987 (IEEE Standard VHDL Language Reference Manual) was born. This sparked industrial interest and extensive investment in VHDL. Although initially intended as a simulation language only, VHDL quickly developed into a language not only capable to simulate and design ASICs, but also to implement complex logic designs into programmable logic (e.g. CPLD and FPGA) type devices. Some companies, e.g. http://www.altera.com/Altera and http://www.xilinx.com/Xilinx, provide free VHDL compilers to program their products with. The VHDL language have since been revised twice. The current standard is called the IEEE Standard 1076-2002.

In the early 1990s the need to simulate systems consisting of a combination of analog and digital signals led to the establishment of the IEEE Working Group 1076.1. This Working Group started to develop a set of extensions to the VHDL language to facilitate the simulation of analog and mixed signal systems. These extensions were published as the IEEE Standard 1076.1-1999, (IEEE standard VHDL analog and mixed-signal extensions). IEEE Standard 1076.1 was coined VHDL-AMS by industry to distinguish it from the original VHDL.

Due to VHDL-AMS's powerful extended instruction set, industry was (and still is some what) reluctant to try and develop analog and mixed signal hardware that could be programmed using the full extend of VHDL-AMS as this seems to be an almost impossible task. VHDL-AMS was deemed more applicable to the simulation of mixed-signal systems (i.e. containing both analog and digital elements, and the boundary between them) with the focus on IC design. A number of companies therefore started to develop simulation software based on VHDL-AMS for this purpose, e.g. Ansoft and MentorGraphics.

The strength of VHDL-AMS however became apparent in the multi-disciplinary modelling and design of electro-mechanical, mechatronic and micro electro-mechanical systems (MEMS). To cater for multi-disciplinary modelling, a further set of extensions to VHDL-AMS was proposed in 2001 and was finally adopted by the IEEE in 2004, as the IEEE Standard 1076.1.1-2004 (IEEE Standard VHDL Analog and Mixed-Signal Extensions-Packages for Multiple Energy Domain Support). These extensions defined a standard set of *terminals*, *natures*, *quantities*, *units*, *symbols*, *constants* and *tolerances* for a common standard interface across the various domains that could co-exist within a *"multiple energy domain"* VHDL-AMS simulation.

II. A BRIEF INTRODUCTION TO VHDL-AMS

In order to simulate the analog behaviour of an analog `entity`, certain extensions to VHDL's external interface (i.e. ports) were introduced. In VHDL, a digital `entity` is usually declared with only `signal` ports, [1]. In VHDL-AMS, an analog `entity` can now be declared with `terminal` or `quantity` ports as well, [2].

This is accomplished by the following additional keywords:

- `terminal`
- `nature`
  - `through`
  - `across`
- `quantity`

A. A simple example

Our first example will be that of a simple resistor, modeled in VHDL-AMS, as shown in Listing 1 and Listing 2.

```
library ieee;

use ieee.electrical_systems.all;

entity resistor is
  generic(
    R : resistance := 1.0 -- [Ohm]
      );
  port(
    terminal t1, t2 : electrical
      );
end entity resistor;
```

Listing 1.   A VHDL-AMS "interface description" for a resistor.

```
14  architecture simple of resistor is
15    quantity v across t1 to t2;
16    quantity i through t1 to t2;
17  begin
18    v==i*R;
19  end architecture simple;
```

Listing 2. A VHDL-AMS "behavioural model" for a resistor.

We start in line 1 by telling the VHDL-AMS simulator that we want to use the IEEE library[1] and in line 3 that from this library we want to use all of the definitions as defined in the IEEE's electrical_systems package.

Next, starting in line 5, we describe how the resistor's entity (i.e. the interface description) should look like. First we need to list all of the entity's parameter or properties. For our simple resistor, we have only one generic parameter, its resistance[2]. We can assign a default value for each generic parameter, if required. For our resistor, a default value was chosen as $1\,\Omega$, line 7. Secondly, in line 9, we need to describe how our resistor can be connected to other entities. We define our simple resistor as a two terminal device, and that these *terminals* are electrical[3] by nature. This permits us to connect our resistor to *terminals* of other models provided that they are also of the electrical nature.

Starting in line 14, the architecture (i.e. the behavioural model) of our resistor is described. Here, the first step is to define the across and the through *branch quantities* associated with the electrical *terminals*[4] in order to write down the "differential equations" for our simple resistor model in terms of these *branch quantities*. This is done in lines 15 and 16. The next step is to write the *simultaneous statement* ('==' notation) or "deferential equation" for our simple resistor which is a mere one line of code (line 18) and is nothing else but "Ohm's Law" for a resistor,

$$v_R(t) = i_R(t) \cdot R \ . \tag{1}$$

What the *simultaneous statement* does, is that it instructs the analog solver (e.g. Simplorer®) to find solution at each analog time step for the *quantities* on both sides of the '==' so that they are equal.

B. A more complex example

Let us consider a more complex example, and that of an inductor, as shown in Listing 3. For our inductor's entity description, two generic parameters were chosen, the inductance value 'L' (line 7) and the parasitic series resistance value, 'rL' (line 8). For the behavioural model, we decided

---

[1]As defined by the IEEE Standard 1076.1.1-2004.

[2]The IEEE's Electrical Systems Package defines resistance as a subtype of real with a unit attribute value of "Ohm" and a symbol attribute value of "Ohm".

[3]In the IEEE's Electrical Systems Package electrical and magnetic *natures* are defined.

[4]In the IEEE's Electrical Systems Package, for electrical *terminals*, the predefined across type is voltage, with a unit attribute value of "Volt" and a symbol attribute value of "V", and the predefined through type of current, with a unit attribute value of "Ampere" and a symbol attribute value of "A", both are of the subtype real.

```
1   library ieee;
2
3   use ieee.electrical_systems.all;
4
5   entity inductor is
6     generic(
7       L  : inductance := 1.0E-3;  -- [H]
8       rL : resistance := 1.0E-3   -- [Ohm]
9         );
10    port(
11      terminal t1, t2 : electrical
12        );
13  end entity inductor;
14
15  architecture simple of inductor is
16    quantity vL across iL through t1 to t2;
17  begin
18    vL==L*iL'dot;
19  end architecture simple;
20
21  architecture more_complex of inductor is
22    quantity vt across iL through t1 to t2;
23    quantity vL : voltage;
24  begin
25    vL==vt-rL*iL;
26    vL==L*iL'dot;
27  end architecture more_complex;
```

Listing 3. A VHDL-AMS inductor model.

that we want to have a 'simple' model for our inductor that ignores the parasitic series resistance, line 15–19, as well as a 'more_complex' model that takes the parasitic series resistance into account, line 21–27. The through and across *branch quantities* for our 'simple' inductor model are defined in a concise manner on line 16 as appose to the "two line" declaration used for our resistor model.

A new attribute, the 'dot' or *derivative of time* attribute is introduced in line 18 in order to write the *simultaneous statement*, i.e. "differential equation", for our simple inductor,

$$v_L(t) = L\frac{d}{dt}i_L(t) \ . \tag{2}$$

An alternative approach could have been to write the "differential equation" in integral format, using the 'integ' attribute and replacing line 18 with,

```
iL==vL'integ/L;
```

to realise,

$$i_L(t) = \frac{1}{L}\int_0^t v_L(t) \ . \tag{3}$$

For the 'more_complex' simulation model of our inductor, the parasitic resistor value, is taken into account. Once again the across and through *branch quantities* needs to be defined, line 22. The names chosen for the *branch quantities* need not be the same as for the 'simple' model, as the scope of the names will only be limited to the architectural model in which it was defined.

In line 23 an additional *free quantity* is defined. The allows (actually forces) us to define another *simultaneous statement* to cater for the voltage drop across the parasitic series resistor as shown in line 25.

1745

TABLE I
MULTI DOMAIN NATURES AVAILABLE IN VHDL-AMS

nature	across	through
electrical	voltage	current
magnetic	mmf	flux
translational	displacement	force
translational_velocity	velocity	force
rotational	angle	torque
rotational_velocity	angular_velocity	torque
fluidic	pressure	vflow
thermal	temperature	heat_flow

Because our inductor now has two architectures, we can choose which one to use in our simulation, as long as the `entity` description is *generic* to both.

III. MULTI DOMAIN MODELLING

A. Introduction

So far only *quantities* of the `electrical` nature has been presented.

In this section an example of a simulation with a `electrical` and a `magnetic` nature will be shown, followed by a simulation with a `electrical` and a `rotational_velocity` nature. All the different *natures* that can be used to describe a terminal(s) in VHLD-AMS together with their corresponding `across` and `through` *branch quantities*, as defined by the IEEE Standard 1076.1.1, are listed in Table I.

B. An electromagnetic VHDL-AMS example

A transformer winding is a good example of an `entity` with two types of *natures*, an `electrical` nature, with *branch quantities* of `voltage` across and `current` `through` as well as a `magnetic` nature, with *branch quantities* of `mmf` across and `flux` through. A VHDL-AMS example of a transformer core winding is shown in Listing 4 in order to describing how both of these *natures* can co-exist within a single VHDL-AMS model.

```
1  library ieee;
2
3  use ieee.electrical_systems.all;
4
5  entity winding is
6    generic(
7      N : in real := 1.0;        -- [turns]
8      r : in resistance := 0.0 -- [Ohm]
9          );
10   port(
11     terminal e1, e2 : electrical;
12     terminal m1, m2 : magnetic
13         );
14  end entity winding;
15
16  architecture simple_linear of winding is
17    quantity vt across i through e1 to e2;
18    quantity f  across phi through m1 to m2;
19  begin
20    vt==-N*phi'dot+i*r;
21    f ==N*i;
22  end architecture simple_linear;
```

Listing 4. The entity description of a simple winding.

The `generic` parameters for our winding model is the number of turns, 'N', and the parasitic resistance of the winding, 'r'. This is similar to that of the inductor discussed previously. The major difference between the winding model and that of the inductor, is that the winding has two sets of *terminals*, a set of `electrical` and a set of `magnetic` *terminals*, line 17 and 18 respectively.

In the architectural description of the winding, names for the *branch quantities* associated with each `terminal`'s `across` and `through` *quantity* are assigned. For each set of `branch` quantities, a *simultaneous statement* (i.e. "differential equation") is required to describe the behaviour of the winding. The two "differential equations" necessary to model the behaviour of the winding, are shown in (4) and (5), with the VHDL-AMS implementation given in lines 20 and 21 respectively.

$$v_t(t) = -N\frac{d}{dt}\Phi + iR \quad (4)$$

$$\mathcal{F} = Ni \quad (5)$$

Because VHDL-AMS is a strongly typed language, only *terminals* of the same `nature` can be connected together. To illustrate this, an Ansoft Simplorer® SV (Student Version) [3] schematic of a push-pull switch-mode power supply is shown in Fig. 1. In the schematic it can be seen that the circuit to the left of windings **n01** and **n01** and that to the right of windings **n011** and **n012** are `electrical` circuits. Where as in between the winding, a `magnetic` circuit exists. Simplorer® automatically assigns different colours to different "types" of circuits. The defaults are 'black' for `electrical` circuits and 'orange' for `magnetic` circuits.

Also shown (for interest sake), is Simplorer®'s built-in Jiles-Atherton model for a 3F3 Ferroxcube ferrite core. The only parameter that needs to be supplied are the effective length (l_e) and the effective area (A_e). For an EFD10 core,

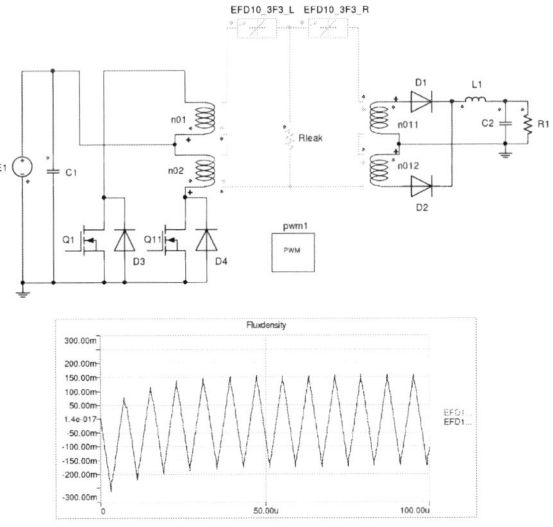

Fig. 1. Simplorer Push-Pull Converter Simulation

1746

$l_e = 23.1\,\text{mm}$ and $A_e = 7.2\,\text{mm}^2$. For this Simplorer® simulation, the core is split in two, and a leakage reluctance, **Rleak**, is added to simulate the effect of non ideal coupling between the primary and secondary due to (say) the glass tape used for isolation purposes.

Simplorer® has built-in models for a large range of ferrite cores for the following manufacturers, AVX, Epcos, Ferroxcube, Magnetics, Micrometals, Steward and TDK.

C. An electromechanical VHDL-AMS example

An permanent magnet DC motor is a good example of a typical electromechanical system [4], that has both an `electrical` and a "mechanical" nature. A VHDL-AMS model for the electromechanical behaviour of a permanent magnet DC motor is shown in Listing 5.

Once again the `generic` parameters are listed in the `entity` declaration (line 9–18). For the permanent magnet DC motor, they are:

- armature resistance ('R_a')
- armature inductance ('L_a')
- motor torque constant ('K_T')
- armature moment of inertia ('J_a')
- armature damping ('B_a')

```
1  library ieee;
2
3  use ieee.math_real.all;
4  use ieee.electrical_systems.all;
5  use ieee.mechanical_systems.all;
6
7  entity pmdcm is
8    generic(
9      -- Armature Resistance
10     R_a : resistance := 1.0;
11     -- Armature Inductance
12     L_a : inductance := 1.0E-3;
13     -- Motor Torque Constant
14     K_T : real := 1.0;
15     -- Armature's Moment of Inertia
16     J_a : moment_inertia := 1.0E3;
17     -- Armature's Damping
18     B_a : damping := 1.0E3
19         );
20    port(
21      -- Electrical Terminals
22      terminal t1, t2   : electrical;
23      -- Rotational Velocity Terminal(s)
24      terminal m : rotational_v;
25      -- Speed of the Motor in rpm
26      quantity n : out real
27         );
28  end entity pmdcm;
29
30  architecture simple_linear of pmdcm is
31    constant o2n : real := 60.0/(2.0*math_pi);
32    quantity v_t across i_a through t1 to t2;
33    quantity omega_m across tau_m through m to
34      rotational_v_ref;
35  begin
36    v_t  ==i_a*R_a+L_a*i_a'dot+K_T*omega_m;
37    tau_m==K_T*i_a-J_a*omega_m'dot-B_a*omega_m;
38    n    ==o2n*omega_m;
39  end architecture simple_linear;
```

Listing 5. An Electromechanical VHDL-AMS model of a Permanent Magnet DC Motor.

The next step is to define the set of `port` definitions for our VHDL-AMS model. The first `port` definition, line 22, are the (now familiar) `electrical` *terminals*. For the second set of *terminals* however, we have a choice between `rotational` or `rotation_velocity` *terminals*.

For servo applications, the `rotational` nature would be a good choice with `angle` and `torque` as the respective `across` - and `through` - *branch quantities*. For a speed control application however, the `rotational_velocity` nature would be a better choice with `rotational_velocity` and `torque` as the respective `across` and `through` *branch quantities*. Our example choose the latter, see line 24.

What is interesting to note, is that only one `terminal` 'm' was specified. We will come to that later.

Another interesting thing to note, is the declaration in line 26. Here a new type of `port` is defined, a *port quantity*, 'n'. This *port quantity* (the speed of the motor in rpm) is also defined as a **out** quantity. *Port quantities* can either be declared as **in** or **out** (as appose to a *signal* that can be **in**, **out** or **inout**). This *port quantity* can now be used for plotting - or control purposes (see the next section).

In the course of this paper, we have now discussed all types of quantities defined in VHDL-AMS:

- *branch quantities*
- *free quantities*
- *port quantities*

In the `architecture` declaration, `across` and `through` *branch quantities* are once again assigned for each set of *terminals*. For the `electrical` *terminals*, this is not a problem. But what about our "single" `rotational_velocity` terminal?

When we measure the speed of a motor's shaft, we usually measure it with respect to standstill. This is similar when measuring a line voltage with respect to ground potential. In VHDL-AMS a reference (i.e. ground) is defined for each type of `nature`. For `rotational_velocity` this is the `rotational_velocity_ref` and for `electrical` it is the `electrical_ref`. Thus we can define our *branch quantities* as shown in line 33 and 34 were `rotational_v_ref` is an `alias` for `rotational_velocity_ref`.

For each set of *branch quantities* a *simultaneous statement* is once again required, as well as for the *port quantity* defined. The differential equations necessary to describe the `electrical` and `rotational_velocity` nature of the permanent magnet DC motor is given in (6) and (7) and implemented in lines 36 and 37 respectively.

$$v_t = i_a R_a + L \frac{d}{dt} i_a + e_a$$
$$= i_a R_a + L \frac{d}{dt} i_a + K_T \omega_m \quad (6)$$

$$\tau_m = \tau_e - J_a \frac{d}{dt} \omega_m - B_a \omega_m$$
$$= K_T i_a - J_a \frac{d}{dt} \omega_m - B_a \omega_m \quad (7)$$

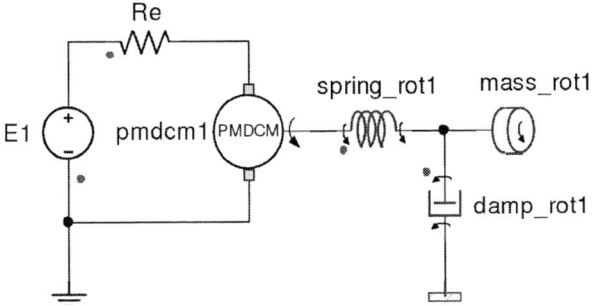

Fig. 2. Electromechanical system using a VHDL-AMS model of an permanent magnet DC machine.

The *simultaneous statement* to convert the speed of motor for rad/s to rpm for *port quantity* 'n', is shown in line 38.

This VHDL-AMS model was imported into a custom library inside Simplorer® SV to create a new VHDL-AMS element (i.e component). A custom symbol for the new element was created and is shown in Fig. 2 together with other predefined "electrical" and "mechanical" components to simulate a coupled electromechanical systems.

The figure clearly shows how we can connect a `rotational_velocity` "circuit" to our DC motor to simulate the behaviour of a mechanical load...

IV. CONTROL SYSTEM BLOCK DIAGRAM MODELLING IN VHDL-AMS

The examples shown up till now, focused only on the modelling of an existing "plant", but did not address the modelling of a control system to be used. Analogous to the *derivative of time*, `dot` attribute, VHDL-AMS provides two attribute ideally suited to simulate transfer functions in either the continuous time domain or the discrete time domain, by either performing a Laplace transform (using the `ltf`

```
 1 entity lead_lag is
 2   generic(
 3     K   : real := 1.0; -- Gain
 4     z_1 : real := 1.0; -- 1st zero
 5     z_2 : real := 1.0; -- 2nd zero
 6     p_1 : real := 1.0; -- 1st pole
 7     p_2 : real := 1.0  -- 2nd pole
 8         );
 9   port(
10         quantity input  : in real;
11         quantity output : out real
12         );
13 end entity lead_lag;
14
15 architecture implementation of lead_lag is
16   constant num : real_vector :=
17     (K*z_1*z_2, K*(z_1+z_2), K);
18   constant den : real_vector :=
19     (p_1*p_2, p_1+p_2, 1.0);
20 begin
21   output==input'ltf(num,den);
22 end architecture implementation;
```

Listing 6. A VHDL-AMS model for a Lead-Lag Compensator, in the Continuous Time Domain

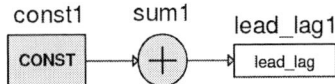

Fig. 3. Hybrid control system using VHDL-AMS blocks together with predefined Simplorer blocks.

attribute) or a Z-transform (using the `ztf` attribute) on a *port quantity*.

It it therefore possible to simulate a typical lead-lag compensator (8), in VHDL-AMS as shown in Listing 6. On thing to note however, is that the 'num' and 'den' constants used for the numerator and denominator coefficients of the transfer function, must be in ascending powers of *s*. It is now possible to import our Lead-Lag compensator as an VHDL-AMS "element block" into Simplorer® and use it together with Simplorer®'s own "blocks" for "hybrid" control purposes, as shown in Fig. 3.

$$
\begin{aligned}
G_c(s) &= K \frac{(s + z_1)(s + z_2)}{(s + p_1)(s + p_2)} \\
       &= K \frac{s^2 + (z_1 + z_2)s + z_1 z_2}{s^2 + (p_1 + p_2) + p_1 p_2}
\end{aligned} \tag{8}
$$

To simulate a discrete transfer function, the 'ztf', Z-Domain Transform, attribute is used analogous to the 'ltf' attribute, but with an additional sampling time parameter that also needs to be supplied.

V. DIGITAL CONTROL ALGORITHM SIMULATION

Most of the examples discussed so far could just as easily have been simulated in Simplorer® without using any VHDL-AMS models, using the predefined circuit- or block elements instead.

However, when trying to simulate somewhat more complex electrical drive or power electronic systems, block element does not suffice to simulate the control system when trying to simulate digital or numerical control algorithms. Simplorer® does have the ability to simulate these type of control systems using graphical state machines, but these state machines simulations are difficult to debug and are not portable. Also these graphical state machine implementation differs completely from the PC, microcontroller, DSP and/or FPGA code that will be used to implement these control algorithms.

This is where VHDL-AMS really comes to the fore. With the standard 'if-then-else' or 'case-when' conditional statements, the 'for-loop' or 'while-loop' loop statements, the ability to define custom *functions* and *procedures* combined with the IEEE's `math_real` package[5], Simplorer® can be used to simulate highlevel language code with minimum amount of porting (i.e. from C or even Fortran), using VHDL-AMS.

[5]This includes all the mathematical functions, e.g. 'sin', 'cos', 'tan', 'sqrt', etc.

1748

```vhdl
function spline(x, y : real_vector; yp1, ypn : real)
    return real_vector is
  variable n : integer := x'high;
  variable p, qn, sig, un : real;
  variable u, y2 : real_vector(x'low to x'high);
begin
  if (yp1 > 0.99E30) then
    y2(1):=0.0;
    u(1):=0.0;
  else
    y2(1):=-0.5;
    u(1):=(3.0/(x(2)-x(1)))*((y(2)-y(1))/(x(2)-
         x(1))-yp1);
  end if;
  for i in (x'low+1) to (n-1) loop
    sig:=(x(i)-x(i-1))/(x(i+1)-x(i-1));
    p:=sig*y2(i-1)+2.0;
    y2(i):=(sig-1.0)/p;
    u(i):=(6.0*((y(i+1)-y(i))/(x(i+1)-x(i))-(y(i)-
         y(i-1))/(x(i)-x(i-1)))/(x(i+1)-x(i-1))-
         sig*u(i-1))/p;
  end loop;
  if (ypn > 0.99E30) then
    qn:=0.0;
    un:=0.0;
  else
    qn:=0.5;
    un:=(3.0/(x(n)-x(n-1)))*(ypn-(y(n)-y(n-1))/
         (x(n)-x(n-1)));
  end if;
  y2(n):=(un-qn*u(n-1))/(qn*y2(n-1)+1.0);
  for k in n-1 downto x'low loop
    y2(k):=y2(k)*y2(k+1)+u(k);
  end loop;
  return y2;
end function spline;
```

Listing 7.   A VHDL-AMS spline second derivative calculation function.

A. Spline interpolation of wind turbine power curves

When trying to simulate a wind generator system, the non-linear transfer function of the wind turbine needs to be taken into account. Usually the turbine manufacturer can provide

```vhdl
function splint(xa, ya, y2a : real_vector; x : real)
    return real is
  variable k, khi, klo  : integer;
  variable a, b, h      : real;
begin
  klo:=xa'low;
  khi:=xa'high;
  while (khi-klo) > 1 loop
    k:=(khi+klo)/2;
    if xa(k)>x then
      khi:=k;
    else
      klo:=k;
    end if;
  end loop;
  h:=xa(khi)-xa(klo);
  if h=0.0 then
    report "Bad xa input to splint"
    severity note;
  end if;
  a:=(xa(khi)-x)/h;
  b:=(x-xa(klo))/h;
  return a*ya(klo)+b*ya(khi)+((a**3-a)*y2a(klo)+
       (b**3-b)*y2a(khi))*(h**2)/6.0;
end function splint;
```

Listing 8.   A VHDL-AMS spline interpolation function

```vhdl
...
package numerical_recipes is
  ...
  function spline(x, y : real_vector; yp1, ypn : real)
    return real_vector;
  function splint(xa, ya, y2a : real_vector; x : real)
    return real;
  ...
end numerical_recipes;

package body numerical_recipes is
...
end numerical_recipes;
```

Listing 9.   A custom VHDL-AMS numerical package.

a set of power curves for the wind turbine as measured at different wind speeds. To use these curves with in a simulation program, may provide a problem. These curves can however easily be simulated with a combination of a number of spline functions together with normal linear interpolation for a complete two-dimensional curve fitting algorithm (e.g. power or torque as a function of both wind speed and tip speed).

The easiest way to implement the spline function, was to use the Fortran 77 spline and splint functions as defined in [5] and port them to VHDL-AMS. The ported spline and splint functions is listed in Listing 7 and Listing 8 respectively.

VHDL-AMS allows one to group function together into a package. In Listing 9 a skeleton layout of the numerical_recipes *package* is shown that was written for this example. In the package *declaration* section only the headers or interface description of the functions are listed were as the function description itself is placed in the package body section. This is similar to the entity and architecture parts discussed previously to allow for

```vhdl
library peds;

use peds.numerical_recipes.all;

entity wind_turbine is
    port(
        quantity wind_speed : in  real;
        quantity rpm        : in  real;
        quantity power      : out real
        );
end wind_turbine;

architecture simple of wind_turbine is
begin
  turbine: procedural is
    -- 11 m/s Wind speed curve --
    constant rpm_11  : real_vector(1 to 8)
      := (0.0,60.0,100.0,200.0,300.0,400.0,500.0,600.0);
    constant pwr_11  : real_vector(1 to 8)
      := (0.0,0.12,0.36,2.4,2.9,2.76,2.48,2.08);
    constant d2pwr_11 : real_vector(1 to 8)
      := spline(rpm_11, pwr_11, 0.0, -0.12/20.0);

  begin
    power:= splint(rpm_11, pwr_11, d2pwr_11, rpm);
  end procedural turbine;
end architecture simple;
```

Listing 10.   A VHDL-AMS spline interpolation function

"information hiding" by encoding only the `architecture` or `package` body parts of the VHDL-AMS element models or *packages*.

In Listing 10 the VHDL-AMS implementation of a wind turbine's power curve as measured at 11 m/s is shown utilising the `spline` and `splint` functions from the `numerical_recipes` *package* which, for this example, was placed in the `peds` *library*. Data for other wind speeds still need to be added to this model. For wind speeds values other than the few for which data is available, linear interpolation will have be used to obtain the power value from the two adjacent wind speed power curves at the required tip speed. This will allow for the power or torque produced by the wind turbine to be estimated given the wind speed and tip speed.

B. Space vector pulse width modulation for an active power filter

The follow example will briefly discuss how to simulate space vector pulse width modulation (SVPWM) as well as a simple active power filtering control algorith, using VHDL-AMS. This will be illustrated using a Simplorer® SV simulation of an active power filter (APF) with SVPWM to compensate for the harmonics produced by a three-phase six-

```
83  -- Calculate new duty cycle value every 1/(fs*2)
84  if update = clk_hi_val then
85      v_s_abc:=(v_s_a, v_s_b, v_s_c);
86      v_s_zab:=abc_to_zab(v_s_abc);
87
88      i_s_abc:=(i_s_a, i_s_b, i_s_c);
89      i_s_zab:=abc_to_zab(i_s_abc);
90
91      i_l_abc:=(i_l_a, i_l_b, i_l_c);
92      i_l_zab:=abc_to_zab(i_l_abc);
93
94      p_l     :=dotp(v_s_zab,i_l_zab);
95
96      g_l     :=p_l/(v_s_zab(zero) **2+
97                     v_s_zab(alpha) **2+
98                     v_s_zab(beta)  **2);
99
100     i_c_zab:=(i_s_zab(zero)  -g_l*v_s_zab(zero),
101               i_s_zab(alpha) -g_l*v_s_zab(alpha),
102               i_s_zab(beta)  -g_l*v_s_zab(beta));
103
104     v_r_zab:=(L_f*fs*i_c_zab(zero)  +v_s_zab(zero),
105               L_f*fs*i_c_zab(alpha) +v_s_zab(alpha),
106               L_f*fs*i_c_zab(beta)  +v_s_zab(beta));
107
108     D       :=space_vector_dcc(v_r_zab,V_dc);
109  end if;
```

Listing 11. The "main" section of the APF's Control Algorithm.

pulse rectifier. The Simplorer® schematic is shown in Fig. 4.

Due to the limited space available, only parts of the VHDL-AMS code will be presented however. The "main" section of the APF's control algorithm is shown in Listing 11.

In order to calculate the reference voltage for the VSI, the supply voltage, the supply current and the load current needs to be "measured" by our VHDL-AMS control block. For each value that needs to be "measured" a *port quantity* needs to be defined in our VHDL-AMS model, e.g. `v_s_a`, for phase a of the supply voltage, etc.

As soon as an instance of our VHDL-AMS model is used, (**apf1** in this case), the *output name* of the *output property* of the Simplorer® *element* that is required to be "measured" must to be typed next to the appropriate *port quantity* of **apf1**.[6]

In order to calculate the voltage reference for the VSI in space vector format, the supply voltage, supply current and load current needs to be converted to space vector format. This is done in lines 85 − 92 by making use of a custom VHDL-AMS function `abc_to_zab`. The instantaneous power delivered to the load can now be calculated using the dot product of the supply voltage and the load current [6] as shown in (9) and implemented in line 94 using a custom VHDL-AMS function `dotp`.

$$p_l(t) = \mathbf{v}_s(t) \cdot \mathbf{i}_l(t) \tag{9}$$

The next step is to calculate the instantaneous conductance of the load, as shown in (10) and implemented in lines 96 − 98.

$$g_l(t) = \frac{p_l(t)}{||\mathbf{v}_s(t)||^2} \tag{10}$$

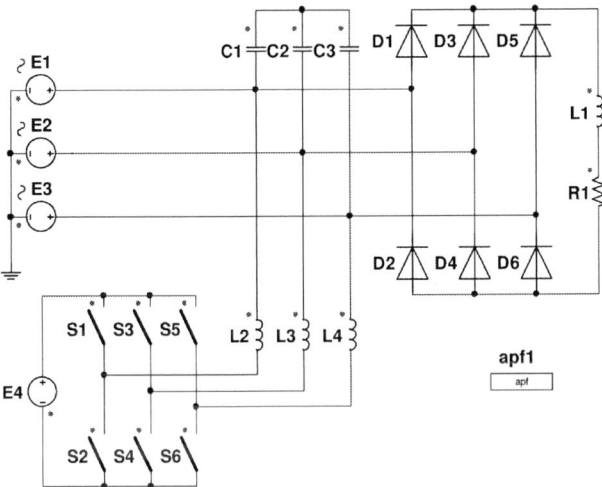

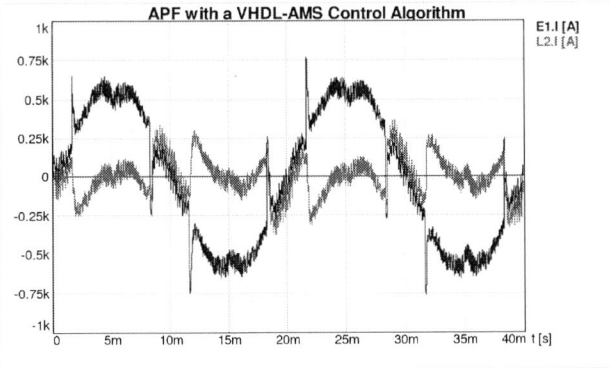

Fig. 4.   An Active Power Filter utilising Space-Vector Pulse Width Modulation with all the controls implemented in VHDL-AMS.

[6]With **E1** the AC voltage source used for phase a, **E1.EMF** needs to be typed next to the `v_s_a` parameter name in the **apf1** block.

```
164 -- Space Vector Duty Cycle Calculation --
165 function space_vector_dcc
166     (v_ref_zab : in space_vector;
167      Ud        : in real)
168      return real_vector is
169
170   variable v_ref_sec : real;
171   variable d0, d1, d2, d3, d4, d5, d6 : real;
172   variable DA, DB, DC : real;
173
174 begin
175   v_ref_sec := sec_det_06(v_ref_zab);
176
177   case (v_ref_sec) is
178
179     when 1.0=>d1:=(v_ref_zab(alpha)-
180                   v_ref_zab(beta)/tan60)/(C*Ud);
181               d2:=v_ref_zab(beta)/(sin60*C*Ud);
182               d0:=d1+d2;
183               if d0 > 1.0 then
184                  d1:=d1/d0;
185                  d2:=d2/d0;
186               end if;
187               d0:=1.0-d1-d2;
188               DA:=d1+d2+d0/2.0;
189               DB:=d2+d0/2.0;
190                  DC:=d0/2.0;
191 ...
```

Listing 12.   Space Vector Duty Cycle Calculation function.

The compensation current that needs to be injected by the APF, can now be calculated as shown in (11) and is implemented in lines 100 − 102.

$$\mathbf{i}_c(t) = \mathbf{i}_l(t) - g_l\mathbf{v}_s(t) \qquad (11)$$

With the switching frequency (f_s) and the filter induc-

```
82  -- Sector Detector --
83  function sec_det_06
84      (s : in space_vector)
85      return real is
86
87    variable sector : real;
88
89  begin
90    if s(beta) > 0.0 then
91      if s(beta) > abs(s(alpha))*tan60 then
92        -- Sector II --
93        sector := 2.0;
94      elsif s(alpha) > 0.0 then
95        -- Sector I --
96        sector := 1.0;
97      else
98        -- Sector III --
99        sector := 3.0;
100     end if;
101   elsif s(beta) < -abs(s(alpha))*tan60 then
102     -- Sector V --
103     sector := 5.0;
104   elsif s(alpha) > 0.0 then
105     -- Sector VI --
106     sector := 6.0;
107   else
108     -- Sector IV --
109     sector := 4.0;
110   end if;
111   return sector;
112 end function sec_det_06;
```

Listing 13.   A function to calculate in which sector a space vector is situated.

tance (L_f=**L2.L**=**L3.L**=**L4.L**) known, the reference voltage for the VSI can be calculated as shown in (12) and implemented in lines 104 − 106.

$$\mathbf{v}_r(t) = L_f \frac{\mathbf{i}_c(t)}{\frac{1}{f_s}} + \mathbf{v}_s(t) \qquad (12)$$

The duty cycle for the VSI is now calculated using the space_vector_dcc VHDL-AMS custom function, in line 108. A partial listing of the space_vector_dcc function is given in Listing 12. A description of space vector PWM is left to the reader [7] .

As a final example of how DSP-type algorithms can be implemented in VHDL-AMS, the listing of yet another custom VHDL-AMS function, sec_det_06, is given in Listing 13.

This function calculates in which sector the reference voltage is situated, and is called by space_vector_dcc in line 174. This custom VHDL-AMS function together with the abc_to_zab and dotp VHDL-AMS functions, forms part of a VHDL-AMS space_vector *package* that was developed by the author to specifically deal with space vector calculations for complex Simplorer® simulations of advanced power electronic topologies.

VI. CONCLUSION

In this paper the most common "extensions" to VHDL that became later became known as VHDL-AMS were presented. The suitability of VHDL-AMS for multi domain system modelling was discussed. Examples of normal electrical, electromagnetic, electromechanical, continuous transfer functions and typical digital control algorithm as used for advanced power electronic drive system modelling was shown. The power of VHDL-AMS allows researchers to rapid prototype and test complicated digital control algorithms for multi-domain systems

All of the Simplorer® simulations that was presented, was done in the Student Version of Simplorer®, Simplorer® SV further illustrating VHDL-AMS and Simplorer® SV's suitability even for post graduate research work.

REFERENCES

[1] P. J. Ashenden, *The Designer's Guide to VHDL*, 2nd ed. Morgan Kaufmann Publishers, 2002.

[2] ——, *The System Designer's Guide to VHDL-AMS*. Morgan Kaufmann Publishers, 2003.

[3] Ansoft's simplorer website. [Online]. Available: http://www.ansoft.com/products/em/simplorer/

[4] C. M. Close, D. H. Frederick, and J. C. Newell, *Modeling and Analysis of Dynamic Systems*, 3rd ed. John Wiley & Sons, Inc., 2002.

[5] W. Press, B. Flannery, S. Teukolsky, and W. Vetterling, *Numerical Recipes in FORTRAN: The Art of Scientific Computing*, 2nd ed. Cambridge University Press, Sep. 1992.

[6] F. Z. Peng, J. Ott, G.W., and D. Adams, "Harmonic and reactive power compensation based on the generalized instantaneous reactive power theory for three-phase four-wire systems," *IEEE Trans. Power Electron.*, vol. 13, no. 6, pp. 1174–1181, Nov. 1998.

[7] H. van der Broeck, H.-C. Skudelny, and G. Stanke, "Analysis and realization of a pulsewidth modulator based on voltage space vectors," *IEEE Trans. Ind. Applicat.*, vol. 24, no. 1, pp. 142–150, 1988.

Reforming Power Electronics Laboratory

Xiaohan Guan, Weiping Zhang, Xusen Zhao, and Yuanchao Liu

Green Power & Energy System Laboratory, North China University of Technology, Beijing 100041, China
Tel. (Fax): 86-010-88802880 Email: gxh@ncut.edu.cn

Abstract– This Paper introduces the reforming for power electronics laboratory under the guide of the principle of "Four Combinations" in NCUT. All the experiments are divided into four levels: conventional basic experiments, comprehensive experiments, innovative research program and advanced demo experiments. Some new experiment equipments are developed, including a set of network-type experiment system, the platforms for innovative research programs and a series of advanced demo experiment apparatus. The new experiment contents are arranged around the controllable switching devices and the digital analysis and design technology based on simulation has been concluded. Teachers' research results have been turned into experiment contents and students are encouraged to participate in the research. Not only have the new experiment system cultivated the students' experimental skills and innovative ability, but also enhanced their knowledge level and professional qualities.

Index Terms-- Experiment Reforming, Power electronics, Power electronics experiment apparatus.

I. INTRODUCTION

Today with the development of information technology, aviation & aerospace industry and renewable energy industry, some new requirements have been put forward to the power electronics technology. The development of the novel circuits and the related techniques focusing on the controllable switching devices are greatly promoted. Colleges all over the world have been carrying on the reforms to the traditional course of power electronics in succession and a series of new methods and new thinking have been proposed. Because power electronics is a course of very strong practicality, the experiment teaching plays an importance role. Developing some reasonable power electronics experiments and some corresponding equipments to match the course reform has become a significant issue.

Many famous universities have reorganized or innovated in the power electronics experiments, combining with the course contents reformation. Many characteristic experimental programs have been put forward. For example, the University of Minnesota has developed a building block (power pole) methodology. The supporting power pole circuit board consists of the modules of power pole, on-board isolated drive circuits, PWM generator, fault protection, output filter, and switched load. The power pole can be reconfigured to be DC/DC (including isolated DC/DC) converters and DC/AC inverters. Digital control has been integrated. It

has the novel features of tightly coupling with lectures, use of low voltages (<50V) for enhanced safety, easy to use and low cost [1][2].

In Zhejiang University in China, a remote experiment system for power electronics has been built up, which is an important part of the NETLAB (Electrical and Electronic Network Laboratory). Some experiments such as SCR rectifier, single-phase inverter and DC/DC converter, have been set up. The real experiment instruments can be controlled through network. For part of the experiments, the whole process can be watched via video surveillance. The students can do the experiments through the network whenever and wherever possible, which improves the utilization ratio of experimental facilities [3][4].

Huazhong University of Science and Technology in China has set up an independent experiment course for power electronics, into which the relative contents concerned the courses of signals and systems, control theory and detection technology are combined. All the experiment items are divided into four levels: basic, comprehensive, design and innovative experiments. The items are expanded to 23, 3 compulsory and 20 elective, which leave a free space for students to develop their own interests. Some following practical exercises, such as curriculum design, extra curriculum innovative practice and electric competition, are arranged to strengthen the students' ability [5].

At present universities all over the world have paid attention to the cultivation of the students' ability to design and innovate through the experiment exercises, to help the students lay foundation for research and their future carriers. Consulting the reforms carried on by the domestic and international universities, we have developed a series of updated power electronics experiments to reform the conventional experiments and the corresponding laboratories have been built up. This paper will introduce the reformation and some corresponding equipments briefly and their characteristics will be summarized afterwards.

II. REFORMING POWER ELECTRONICS LABORATORY

Not only is the experiment the verification of the theory but also the rapid development of the research and industry should be followed tightly. So our reformation has combined with the course contents, the teachers and the graduates' research results as well as the projects of companies. Its main contributions include: (1) Four kinds of experiments, including conventional basic experiments, comprehensive experiments, innovative research program and advanced demo experiments; (2)

This work was supported by 2005 Beijing Municipal College Education & Teaching Reform Project: Reform and Exploration of Course System for Power Electronics.

978-1-4244-0644-9/07/$25.00 ©2007 IEEE

Four combinations. Combination of the above four kinds of experiments; Combination of the cultivation of students' ability with teachers' research; Combination of the university's intramural laboratory with collaborative companies; Combination of scientific research and cultivation of the graduate students with socialization services. Under the guide of the four combinations, a series of the power electronics experiments have been developed for each level. Each will be introduced briefly in the following.

A. Conventional basic experiments

Conventional basic experiments give priority to the verification experiments and are carried on with the theoretic course synchronously. It is the purposes of this kind of experiment to provide an opportunity for students to train their basic experimental skills and to master the usage of the conventional experimental instruments, thereby to develop their ability to analyze, think and solve problems independently. The conventional basic experiment items [6][7], taking DC/DC converters as examples, include: (a) The testing of the electric characteristics of power devices; (b) The study on Flyback DC/DC converter or double-switch Forward DC/DC converter; (c) The study on half-bridge or full-bridge DC/DC converter; (d) The study on Boost PFC (Power factor correction) circuit.

This kind of experiments is centering on the controllable switching devices, such as MOSFET, IGBT, and their application in power conversion, which follows the trend of the current power electronics technology closely. The circuits selected in this kind of experiments are all mature circuits. In this phase, our emphases are not put on cultivating the students' innovative ability, but on giving them some general impression on power electronics.

B. Comprehensive experiments

In order to improve the students' ability step by step, the comprehensive experiments are arranged just after the conventional basic experiments. During the comprehensive experiments, the prospective functions, specifications and even experimental procedures are provided, and the students are required to build up a small-size converter system innovatively to meet the requirements.. To keep the balance of inhering and innovating, most parameters of the experimental circuits have been given, only a few parameters or functions are required to design. Students should verify the known parameters and components through theoretical analysis first, then design and optimize the unknown parameters.

The comprehensive experimental items already set up include [6][7][8]: (a) Single-stage active PFC-AC/DC converter, (b) Zero-voltage quasi-resonant converter, (c) Series-parallel LLC resonant converter, (d) Flyback active clamped converter, (e) EMI filter. These experimental items almost cover each aspect of power electronics and even some new technology.

Computer simulation is used as a supporting method during the process. To decrease the blindness of experiments and ensure the design reasonable students

are asked to employ simulation tools to verify the parameters beforehand, which helps them grasp the digital analysis and design method based on simulation.

After the design and analysis of some complete power electronics systems, the students would have ability to establish independently a complete system composed of switching devices, main circuit, control and driving circuit, protecting circuit and so on, all indispensable. During the experiments the knowledge learned in the course will be verified synthetically and the opinion of system be built up. As well the experimental skills and the experiences accumulated in the first kind of experiments will be exercised more expertly.

To carry on the above two kinds of experiments, a network type experiment platform has been customized, shown in Figure 1. All the experimental platforms in a laboratory can be connected with the teacher computer through laboratory inner network. The teacher computer can communicate with each platform through photoelectric-isolate data collector system to confirm the equipments to operate safely. On the one hand the teacher may keep watch on the experimental procedure of all the platforms by collecting experimental data; on the other hand the teacher may answer students' question or help to solve the problems met with during the experiments in time.

The experimental platforms adopt interface structure, which is convenient for replacing and enlarging experiments. Various power supplies are provided and its safety is good.

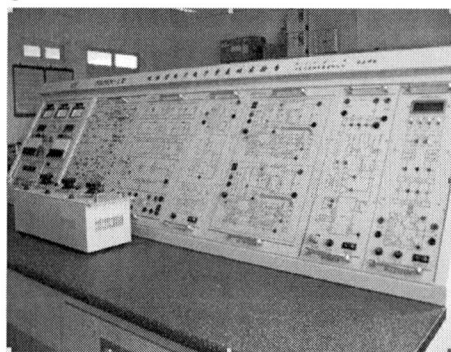

Fig.1 Network type power electronics experiment platform

C. Innovative research program

Innovative research program is carried on in the form of elective courses, extra-curriculum activities, teachers' research projects as well as the training for the National Electronic Design Competition for College students. This program, to the great extent, aims at encouraging the excellent senior students and the graduates to participate in the researches facing the industry, scientific research and the development of practical products. Students should be taught in accordance with their aptitude, inspired their interests and improved their personal abilities during the whole process of designing, producing, and adjusting.

In the program, the experiment projects and the corresponding platforms are developed synchronously. Some of the projects and their corresponding platforms are listed in Table 1 [6][7][8].

TABLE I EXPERIMENT PROJECTS AND PLATFORMS FOR INNOVATIVE RESEARCH PROGRAM

No.	Experiment projects	Experiment platforms
1	Design of the inductance for 575W electronic power supply	Testing platform for power magnetic elements
2	Study on the small signal model of the conversion system	Measuring system of small signal characteristics
3	Active/passive EMI test for 300W computer power supply	Testing platform for active/passive EMI filters
4	Study on the control theory of resonant converter	Measuring platform for closed-loop frequency characteristics
	Measuring of the closed-loop frequency characteristics of resonant network	
5	Design of electronic ballasts for 4~6kW HID lamp	Ballast of high power HID (high intensity discharge) lamp
6	Design of matching networks for PT converter	PT (piezoelectricity transformer) converter
7	Analysis and design of 4~6kW PFC circuit	PFC experiment platform
8	Study on 2500W PWM-DC/DC converter	Medium power PWM-DC/DC converter
9	Study on 1000W high frequency resonant inverter	Medium power high frequency resonant inverter
10	Study on 500W DSP controlled high-performance low-frequency inverter	DSP controlled power converter

The implementation of the innovative research program has made some progresses in cultivating the students' practical ability and broadening their knowledge insight. We also encourage students to publish their research results by paper. Until now there are 16 papers having been published. The students have also got good results in National Electronic Design Competition for College students. For example, in 2005, the students won the second-national prize and the first prize in Beijing district with the project "Three-phase sine waves inverter".

D. Advanced demonstrate experiments

The high level demo experiments are to illustrate the up to date research achievements to students.

Students always have the questions such as "where is the frontier of power electronics?", "what should I study now and what can I do in the future?" or "in which areas has our university got the prominent achievements?" To answer these questions, we turn the typical research achievements into the demo experiments. At the same time, we have collaborated with the companies to develop series of advanced demo experiment apparatus which can be utilized not only by the students but also by the teachers in their researches. The developed advanced demo experiments and their corresponding apparatus are listed in Table 2. Several of the apparatus are shown in Figure 2~4.

TABLE II ADVANCED DEMO EXPERIMENT PROJECTS AND APPARATUS

No.	Experiment projects	Apparatus
1	Study on the switching power supply based on TOP Switch	Switching power supply designed by TOP Switch
2	Measurement of 300W switching power supply for computer	300W switching power supply for computer
3	Study on the performances of fluorescent lamp with electronic ballast	Demonstrators of electronic ballast for fluorescent lamp
4	Study on the performances of 1k~12kW HID lamps	~12kW HID lamps
5	Study on the solar and wind complementary power generation system	Solar and wind complementary power generation system
6	Study on the air condition system	Air condition system utilizing terrestrial heat
7	Study on 50~150W power PT converter	High power PT converter system
8	Study on the arc light of HID lamps	Imaging system of arc light
9	Measurement of the closed-loop characteristics of Buck converter	Closed-loop measuring system for switched converter
10	Study on the power supply for program-controlled switch	Power supply for program-controlled switch
11	Study on 600W integrated power system	600W integrated power system
12	Measurement of the anti-jamming ability of 575W switching power supply	EMC measurement system (IEC standard measurement)

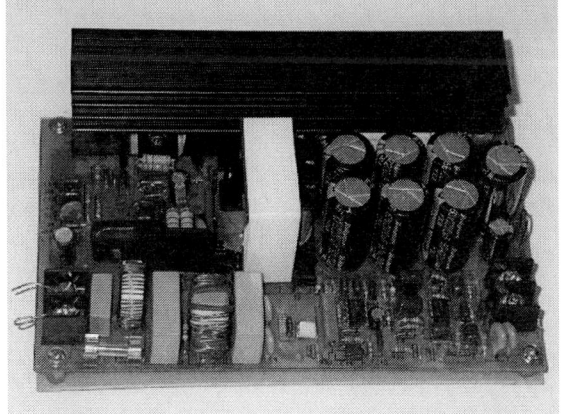

Fig.2 Demonstrator of electronic ballast for Hid lamp

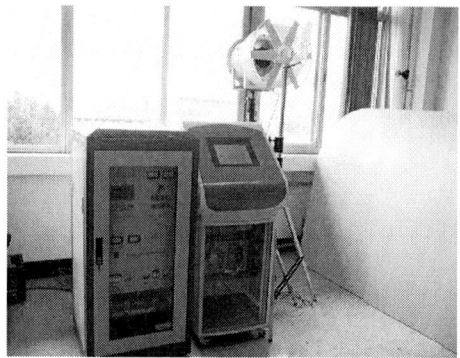

Fig.3 Complementary system of solar and wind power

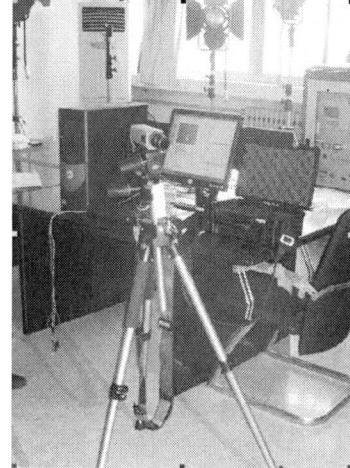

Fig.4 Imaging system of arc light of HID lamps

Through the advanced demo experiments, students can know about the engineering backgrounds, basic principles and practical values and find out the relationship between the experiments and the course knowledge. It greatly broadens the students' horizon, inspires their passion and points out clearly the direction of their professional development.

The demo experiments' another purpose is to stimulate the students' interest in learning power electronics. Just at the beginning of the course, students are led to visit the laboratories and watch some demo experiments. Through the visiting and watching, the students may draw such intuitive conclusion that the power electronics is a very interesting and useful technology.

Although all the experiments are divided into 4 kinds, they form an interactive system intrinsically. The conventional basic experiments train the basic experimental skills and lay the foundation for the later 3 kinds of experiments. From the comprehensive experiments the students' independent design ability are cultivated step by step, also their experimental skills are improved at the same time. In the innovative research program, all the abilities to be required in the industry will be trained comprehensively, including design, simulating, installation, adjusting, etc.

III. THE FEATURES OF THE REFORMING FOR POWER ELECTRONICS LABORATORIES

The on-going reformation for the power electronics laboratories has the following novel features.

A. Taking the "Four Combinations" as the basic principle

This feature helps to combine the course teaching and the experiment teaching together, to combine the improvement of experiments with the scientific research and the service for industry, and to improve the teachers' and the students' academic level. The latest research results can be turned into students' experiment contents in time to make the experiment develop synchronously with the new research results.

B. Setting up the experiment content in layers

The power electronics experiments are divided into 4 kinds, with the conventional basic experiments as the foundation to train experimental skills, with the comprehensive experiments as the important method to improve students' practical ability, with the innovative research program and the advanced demo experiments as the carrier to catch up with research achievements. In such appropriate sequence, students' experimental and creative ability will be improved continuously and their knowledge level and expertise quality will be enhanced step by step.

C. Arranging the experiment contents around the controllable switching devices and their applications

In order to meet the requirements of high frequency, high efficiency and high performances in the field of information, aviation and the use of renewable energy, etc, and matching the course reforming carried on at the same time, the new experiment contents are focusing on the controllable switching devices and their application in power conversion. MOSFET, IGBT and their related circuits are the main research objects.

D. Modernizing the experiment methods and contents

Simulation technology has been introduced into the experiment as an assistant method just in the undergraduates' teaching. Digital analysis and design technology based on the simulation results has been generalized. The mature industrial technology and the research achievements, which can be viewed as the source of renewal, have been turned into students' experiment contents.

E. Developing experiment apparatus with foresight

Our teachers and the collaborated companies have perfectly cooperated to develop a few of suitable apparatus for all levels of experiments. They can meet the requirements of teaching and research, students' extra-curriculum activities, cultivation of graduate students and some public services.

IV. CONCLUSIONS

The results of the reforming for power electronics laboratory in our universities are introduced in this paper. Under the guide of the basic principle of "Four Combination", series of experiment platforms and equipments have been developed in four levels. The experiment content is arranged around the controllable switching devices and its application in power conversion. Digital analysis and design technology based on simulation has been introduced. Teachers' research results have been turned into experiment contents and students are encouraged to participate in the research. Not only have the new experiment system cultivated the students' experimental skills and helped them verify the theoretical and simulation results, but also fostered effectively the students' ability to discover, analyze and solve problems independently. Both the students' original creative ability and their expertise qualities have been enhanced.

REFERENCES

[1] Ned Mohan, William P. Robbins, Paul Imbertson, et.al., "Restructuring of First Courses in Power Electronics and Electric Drives That Integrates Digital Control", *IEEE Transactions on Power Electronics, Vol.18, No.1*, Jan. 2003, pp.429-437.

[2] William Robbins, Ned Mohan, et.al., "A Building-Block-Based Power Electronics Instructional Laboratory", *Power Electronics Specialists Conference*, 2002. pesc02. 2002 IEEE 33rd Annual, pp.467-472.

[3] Dehong Xu, Hao Ma, Yasuyuki Nishida, "Education on Power Electronics in China", *The 2005 International Power Electronics Conference*, pp.521-525.

[4] Lin Qun, Ma Hao, Zhu Shan'an, "Reforming for Power Electronic Experiment Based on network", *2003 National Power Electronics and Drives Teaching Seminar*, Hangzhou, China, pp.75-79.

[5] Huazhong University of Science and Technology, http://202.114.4.28/2005/C47/kcms-2.htm

[6] Ying Jianping, "Fundamentals of Power Electronic Technology", *China Machine Press*, Beijing, 2003.

[7] Lin Weixun, "Modern Power Electronic Circuits", *Zhejiang University Press*, Hangzhou, 2002.7.

[8] Zhang Weiping, "Modeling and Control for DC-DC Converters", *China Electric Power Press*, Beijing, 2006.1.

Online performance monitoring and testing of electrical equipment using Virtual Instrumentation

S.S. Murthy, Raghu K. Mittal, Avneesh Dwivedi, G. Pavitra, and Sonika Choudhary

Dept. of Electrical Engg., IIT Delhi, India

Abstract—With the advancement in signal processing technology and the advent of new sophisticated sensors, there is a need to change the way experiments are conducted in laboratories where conventional methods are still used. This paper proposes a method of experimentation where a programmable computer acts like a virtual instrument and can be programmed to do experiments without having to use old and conventional equipment like ammeters, voltmeters, wattmeters, power analyzers and even oscilloscopes. Online performance monitoring and testing of an induction motor and a transformer have been carried out through Virtual Instrumentation to show the benefits of such a method.

Index Terms--Induction Motor, LabVIEW, Transformer, Virtual Instrumentation.

I. INTRODUCTION

Virtual Instrumentation (VI) has been growing popular since its inception in the late 1970s, mainly because of its versatility and the cost advantages offered. Its underlying principle is the use of customizable software and modular measurement hardware to create user-defined measurement systems, called virtual instruments.

The primary difference between 'natural' instrumentation and virtual instrumentation is the software component of a virtual instrument. The software enables complex and expensive equipment to be replaced by simpler and less expensive hardware; for example, analog to digital converter can act as a hardware complement of a virtual oscilloscope. Software packages like National Instruments' LabVIEW and other graphical programming languages make it easier for non-programmers to develop virtual systems. LabVIEW has been used in Vibration monitoring and analysis [1]. It can also be used in higher education for teaching of instrumentation, instrument control, signal processing, programming, or any discipline that requires data acquisition, control and data analysis such as electronics and chemistry [2].

Online performance monitoring of a motor and electrical equipment testing have been done using LabVIEW to show the benefits of such virtual systems [3]. Basically, this system consists of voltage, current and speed sensors that can be hooked to any electrical equipment and then attached to a Personal Computer through a Universal Serial Bus (USB) port, a standard port that is available in most computers of today. The

Data Acquisition Board acquires various signals of interest. Then using G programming of LabVIEW, the computer can be programmed to monitor voltages, currents and the speed signal as well as calculate various parameters of the equipment such as instantaneous power, efficiency, harmonics etc as well as for calculating equivalent circuit parameters of the motor and transformer [4].

II. INDUCTION MOTOR

Induction motors are the "work horses" in industry, consuming about 70% of the electricity produced in the world. Their applications range from manifold- domestic, commercial, heavy machinery, drives, traction, agriculture, defense etc. Needless to say, their proper working is essential. Hence online performance monitoring of Induction motors is important.

Performance monitoring of the motor has been done using LabVIEW environment in order to prove the ease of use and utility of virtual instrumentation. Torque, power, efficiency and losses of the three-phase induction motor have been chosen as the performance parameters which have been calculated by building VIs.

A. Hardware Implementation

On a Y-connected three phase induction motor, three voltage sensors (Fig.1) and two current sensors (Fig.2) were connected.

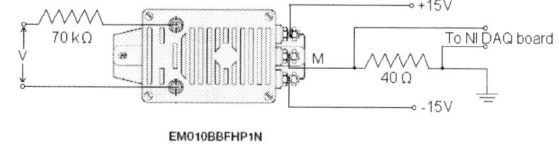

Fig. 1. Voltage Sensor Connections

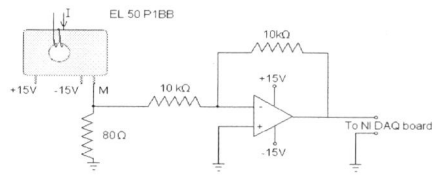

Fig. 2. Current Sensor Connections

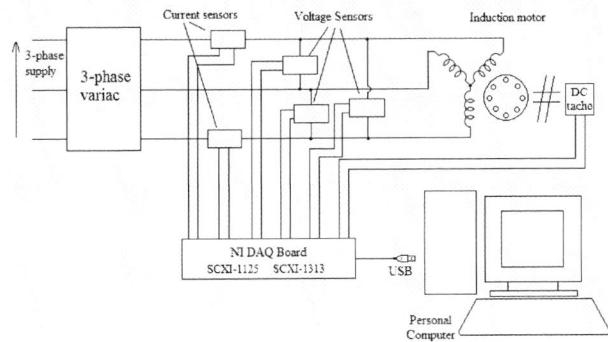

Fig. 3. Block diagram of the connections done

Only two voltage sensors and two current sensors are needed since the sum of the three voltages and currents is zero. Additionally a speed sensor is also required for power calculations.

The complete set-up used for experimentation has been shown in Fig.3.

B. Parameter Determination

Using the IEEE standard procedure for Polyphase Induction Machines [5], the No-load and blocked rotor tests were conducted to determine various parameters of an induction motor. The stator resistance R_s was measured separately.

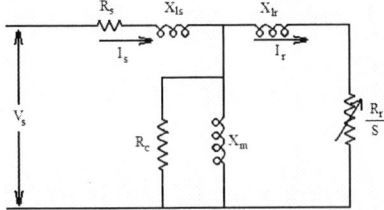

Fig. 4. Per phase equivalent circuit parameters of a 3-phase Induction Motor

C. Calculation of Three Phase Power

The instantaneous power for the induction motor can be calculated as:

$$p = v_a i_a + v_b i_b + v_c i_c$$
$$= v_{ac} i_a + v_{bc} i_b$$

where v_{ac} and v_{bc} are the line voltage and i_a and i_b are the line currents.

D. Torque Determination

In order to calculate the motor torque, there is a need for effecting a d-q transformation in order to formulate the necessary equations. The transformation is done to d-q axis fixed to the stator frame of reference.

With usual notations, the flux linkages in stator reference frame are given by,

$$\lambda_{sd} = L_s i_{sd} + L_m i_{rd}$$
$$\lambda_{sq} = L_s i_{sq} + L_m i_{rq}$$

$$\lambda_{rd} = L_r i_{rd} + L_m i_{sd}$$
$$\lambda_{rq} = L_r i_{rq} + L_m i_{sq}$$

Using the above flux linkages d-q voltages can be written as:

$$v_{sd} = R_s i_{sd} + p \lambda_{sd}$$
$$v_{sq} = R_s i_{sq} + p \lambda_{sq}$$
$$v_{rd} = R_r i_{rd} + p \lambda_{rd} + \omega_r \lambda_{rq}$$
$$v_{rq} = R_r i_{rq} + p \lambda_{rq} - \omega_r \lambda_{rd}$$

From these equations, flux linkage can be calculated as:

$$v_{sd} = R_s i_{sd} + p \lambda_{sd}$$
$$\rightarrow \lambda_{sd} = \int (v_{sd} - R_s i_{sd}) \, dt$$
$$v_{sq} = R_s i_{sq} + p \lambda_{sq}$$
$$\rightarrow \lambda_{sq} = \int (v_{sq} - R_s i_{sq}) \, dt$$

Thus torque can be calculated by the following formula:

$$T_e = (3/2)(P/2)(\lambda_{sd} i_{sq} - \lambda_{sq} i_{sd})$$

where
$$v_{sd} = \sqrt{(2/3)}[v_a - v_b/2 - v_c/2] = \sqrt{(2/3)}[v_{ab}/2 + v_{ac}/2]$$
$$v_{sq} = (1/\sqrt{2}) v_{bc}$$
$$i_{sd} = \sqrt{(2/3)}(i_a - i_b/2 - i_c/2) = \sqrt{(3/2)} i_a$$
$$i_{sq} = \sqrt{(2/3)}[i_b \sqrt{3}/2 - i_c \sqrt{3}/2] = (1/\sqrt{2})(2i_b + i_a)$$

The d-q axis voltages and currents can also be seen experimentally. As can be seen, there is no requirement of speed signal for torque determination.

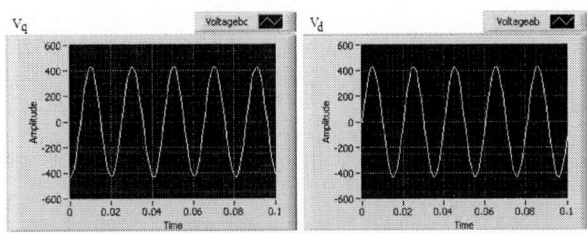

Fig. 5. d-q axis voltages

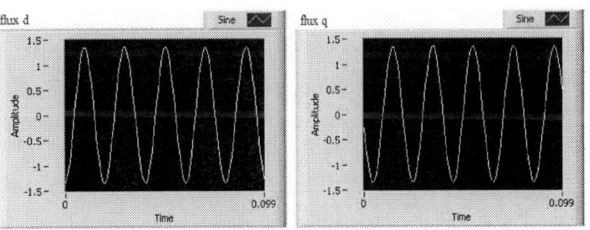

Fig. 6. Flux plots

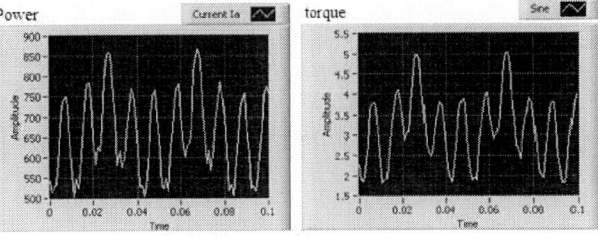

Fig. 7. Power and torque curves

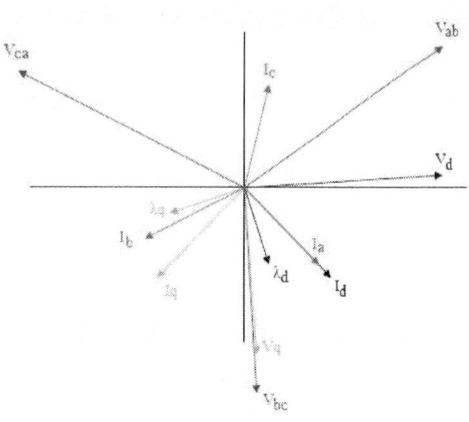

Fig. 8. Phasor diagram showing all voltage, current and flux linkage phasors

E. Efficiency and Loss Calculation

The efficiency calculation does require a speed signal in order to calculate the output mechanical power.

Efficiency = Power Output / Power Input * 100
where
Power output = $T * \omega_m$
Power input = $v_{ac}i_A + v_{bc}i_B$

Now losses = Power output - Power input

Losses include rotor and stator copper loss, core loss, friction and windage loss. Stator and rotor copper losses are determined using the currents and various machine parameters. Thus losses due to friction and windage can be calculated.

F. Load Test of Induction Motor

The load test on the induction motor was conducted by loading the coupled DC motor and the armature voltage, armature current and speed were observed. The loading on the motor was changed to find out how its power factor, instantaneous power, torque were changing with respect to the input power. All the observations are in Table I.

TABLE I
TEST RESULTS OF INDUCTION MOTOR USING VI

Sr No	Iarm (A)	PF	Torque (N-m)	Input Power (W)	Loss (W)	Efficiency (%)
1	0	0.1528	0.0048	318.76	318.02	0.23
2	1	0.2704	0.822	506.05	380.61	24.79
3	1.9	0.3513	1.575	663.78	424.59	36.03
4	3.1	0.4538	2.5425	896.19	510.61	43.02
5	4	0.5104	3.2665	1062.4	568.04	46.53
6	5.1	0.5691	4.023	1232.6	626.71	49.15
7	6.3	0.6251	4.838	1412.9	689.84	51.17
8	7	0.6520	5.195	1496.9	722.66	51.72
9	7.8	0.6806	5.635	1595.4	760.30	52.34

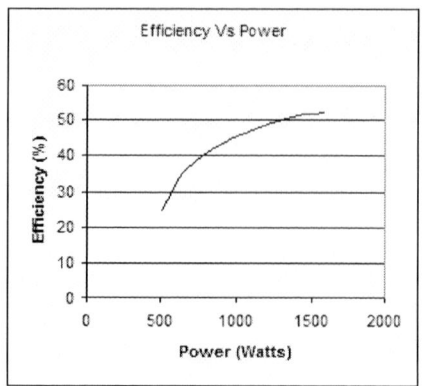

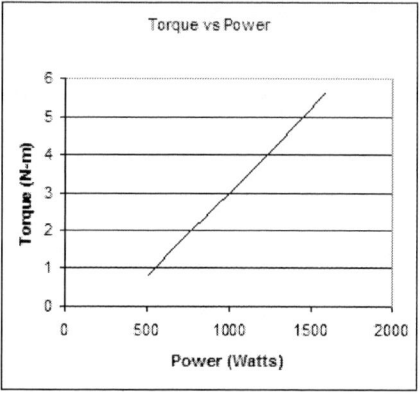

Fig. 9. Performance curves

From Table I, the graphs between Efficiency and Torque Vs Input Power were plotted (Fig.9.) to find out how the performance of the Induction motor varied with load.

III. TRANSFORMER

Here the case of a single phase transformer has been considered and standard tests such as Open Circuit, Short circuit etc. were conducted.

The experimentation (Fig.10) was done on a transformer of 2 kVA rating with a turns ratio of 2:1. One current sensor and a voltage sensor were connected on the high voltage primary side and output of the sensors given to the daughter board or the I/O interfacing board SCXI-1125. Data Acquisition Tool (DAQ) was used to acquire the waveform through the interfacing board to show the instantaneous voltage and current. The signals were appropriately calibrated to show actual values in magnitude in output waveform.

VIs were created to calculate power, power factor, harmonics in input current, flux, and to obtain the B-H loop along with the calculation of the transformer equivalent circuit parameters.

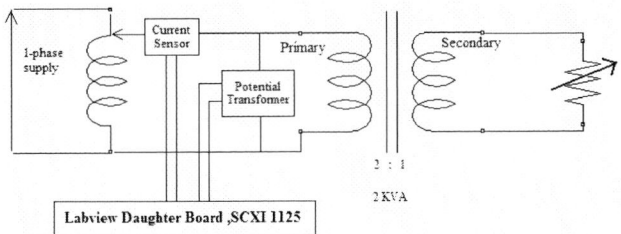

Fig. 10. Setup for testing transformer

The power factor was calculated using two different methods:

1. By using the phase of individual signals i.e. by calculating the difference between the zero crossings of current and voltage.
2. By calculating it as the ratio of real power and the apparent power.

Both methods were found to be in close agreement with each other.

A. No Load Test and Short Circuit Test

The results from No-Load Test are as follows:

Rms Voltage = 206.37 V, Rms current = 1.017 A
Power factor = 0.15, Average Power = 29.94 Watts
R_0 = 1319.09 Ω, X_0 = 205.25 Ω

The results from Short-Circuit Test are as follows:

Rms Voltage = 25.09 V, Rms current = 10.17 A
Power factor = 0.91, Average Power = 222.98 Watts
R_1 = 2.24 Ω, X_1 = 1.017 Ω

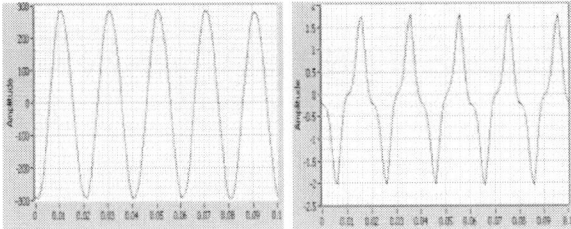

Fig. 11. Primary Voltage (V) and Current (A) under No-load test

B. B-H loop at rated voltage

Nature of BH loop of core material is a very useful information on a transformer. This paper explains a simple way to obtain the same using VI. Flux signal was obtained by integrating the voltage signal with respect to time. Further, B-H loop was plotted using an XY graph giving current signal to X-axis and ∫Vdt to Y-axis.

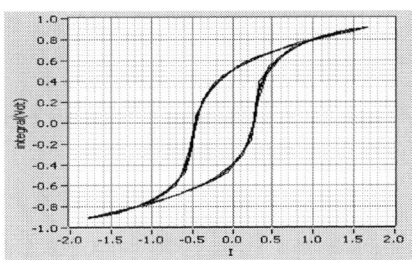

Fig. 12. B-H curve for the transformer

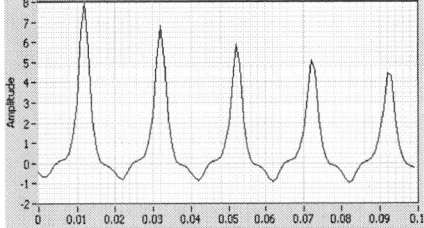

Fig. 13. Magnetic inrush current

C. Inrush Current

The magnetic inrush current of the transformer was obtained and the harmonics in the inrush current analyzed. The percentage of second harmonic was found to be predominant (52%).

All the above results were verified using conventional measurement equipment like Power Analyzers and Digital Multimeters.

IV. CONCLUSION

From the various experiments conducted using Virtual Instrumentation, the accuracy and reliability of the results were found to be quite satisfactory. This further strengthens the future of use of virtual instrumentation for conducting tests using voltage and current sensors and is beneficial in doing harmonic analysis, monitoring starting and stopping transients in an induction motor, power factor determination, and online power and torque calculation. Virtual Instrumentation also has wider applications in condition monitoring, effect of unbalances, thermal effects, noise and vibration, energy audit/conservation, Direct torque control, Field oriented control and transient and dynamic studies.

There is huge scope for adoption of these methods in various laboratories as well as industries. It can also be used in universities for conducting experiments on electrical machines and to teach the students various topics like transients in voltages and currents, harmonics, concepts of real and apparent power etc. which are not easily comprehended because of the inability of older equipment to make these quantities visible to the experimenter.

Thus a package consisting of various voltage, current sensors, a Data Acquisition Board and a bouquet of LabVIEW programs can be used for conducting complete testing of electrical equipment like a three-phase induction motor and a transformer.

REFERENCES

[1] Asan Gani M.J.E. Salami: "A LabVIEW based Data Acquisition System for Vibration Monitoring and Analysis." *IEEE 2002, Student Conference on Research and Development Proceedings, Shah Alam, Malayasia.* Pg 62-65

[2] J.Anthony Vento : "Application of Labview in Higher Education Laboratories" Session 25B4, IEEE 1988, Frontiers in Education Conference Proceedings. Pages 444-447

[3] Halit Eren and Chun Che Fung: "A Virtual Instrumentation of Electric Drive Systems for Automation and Testing." *IEEE 2000,* Pages 1500-1505

[4] F. Filippetti. S. Pirani. L. Tommasini G . Franceschini : "A LabVIEW based virtual instrument for on-line induction motor parameters identification." *Industrial Electronics, 1995. ISIE '95., Proceedings of the IEEE International Symposium* on Volume 2, 10-14 July 1995 Page(s):648 - 653 vol.2

[5] IEEE Std 112™-2004 : "IEEE Standard Test Procedure for Polyphase Induction Motors and Generators". *IEEE Power Engineering Society.*

[6] NI documentation and help files

A Balancing Strategy and Implementation of Current Equalizer for High Power LED Backlighting

Chang-Hua Lin*, Tsung-You Hung**, Chien-Ming Wang***, and Kai-Jun Pai****

*Tatung University, Taipei, R.O.C., ***National Ilan University, Ilan, R.O.C.
**St. John's University, Taipei, R.O.C., ****National Taiwan University of Science & Technology, Taipei, R.O.C.

Abstract– This paper presents a current equalizer to balance the driving currents for high power LEDs (light emitting diode). The proposed strategy combined PWM strategy and digital dimming control to achieve the same illumination for each driven diodes. To obtain the same average current, the duty cycle of LEDs is modulated according to its current amplitude instead of traditional current amplitude adjusting. Complete mathematical analysis and design considerations are detailed. The experimental results are close to the theoretical predictions, and the measured tolerance errors are all below 4%.

Index Terms-- High Power LED, Balancing Strategy, Current Equalizer, Backlighting.

I. INTRODUCTION

Liquid crystal displays (LCDs) have became the master stream on the market of flat display panels (FDPs). A backlight module is essential and its performance is vital to the display quality of the FDP [1-6]. At present, there are two major backlight sources available, namely a white LED and a cold cathode fluorescent lamp (CCFL). CCFL is usually more suitable for larger display panels due to its cost advantage over a white LED. Recently, LED has gradually replaced with CCFL as backlight sources, especially in smaller panel applications, due to improvement on luminous efficiency, long life, and wide chromaticity. Currently, the combination of an LCD and a LED seems to have attracted a constant, or even an increasing, global demand for a thinner FDP with fine performance and decent efficiency [1-6]. In the past, some driving methods, such as constant voltage driving and constant current driving, or control circuit [7] are proposed to achieve constant-luminance control. However, it is not easy to keep uniformity in each driving currents at different loops for large-scale backlighting. In this paper, a balancing strategy combined from PWM control and digital dimming [8] is proposed and implemented to act as a current equalizer for driving multi-LCDs modular.

II. THE BASIC CONFIGURATION OF LED DRIVER

In general, there are two types of connection for LED driving, parallel-connection and series-connection, as shown in Fig. 1. In Fig. 1a, the driving circuit with low voltage is easy to be implemented; however, both the electrical characteristics variation between the LEDs and the resistance derivations of limiting resistor resulted in unbalanced driving current. In addition, Fig. 1b shows a most popular circuit configuration, which are widely used in many applications. The series-connection driving configuration with a constant current source can ensure same forward current for each LED to obtain same luminance and to expand life span. However, general constant current source possess complicated control and higher cost, and it still cannot guarantee that constant luminance control is available in multi-loop configurations. Moreover, in order to avoid over voltage or broken loop, Fig. 2 shows a series-connection driving circuit with a protection circuit to avoid occurring of device broken, but its cost is still too high.

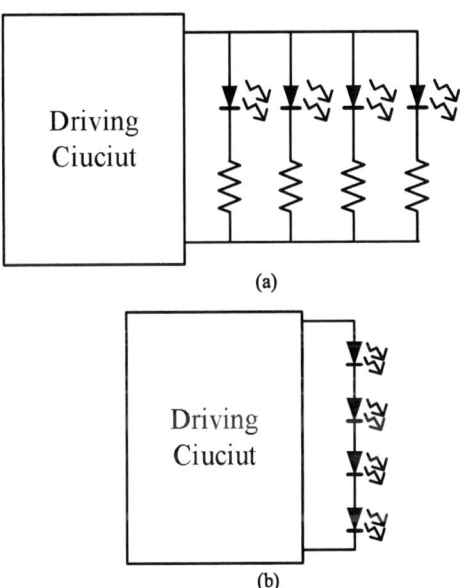

(a)

(b)

Fig. 1 The configuration of LED driving (a) a parallel-connection (b) a series-connection.

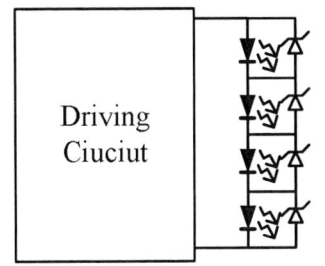

Fig. 2 A series-connection with protection device.

978-1-4244-0644-9/07/$25.00 ©2007 IEEE   1762

III. ANALYSIS OF THE BALANCING STRATEGY

With the increasing demand in large scale FDP market and rapid improvement in LED technologies, large scale backlighting by using LED has became a technical choke point. In this paper, the concept of balancing strategy to accomplish average current driving is illustrated as Fig. 3. Suppose the product of the current amplitude i_{LED} and its duty cycle D can be defined as a constant value ρ, then the average current can be expressed as follows:

$$i_{Lavg} = i_{LED} \times D = \rho \qquad (1)$$

The realized current equalizer will modulate the duty cycles of LEDs according the current amplitudes of each loop to result in identical average driving current for all sets of LEDs.

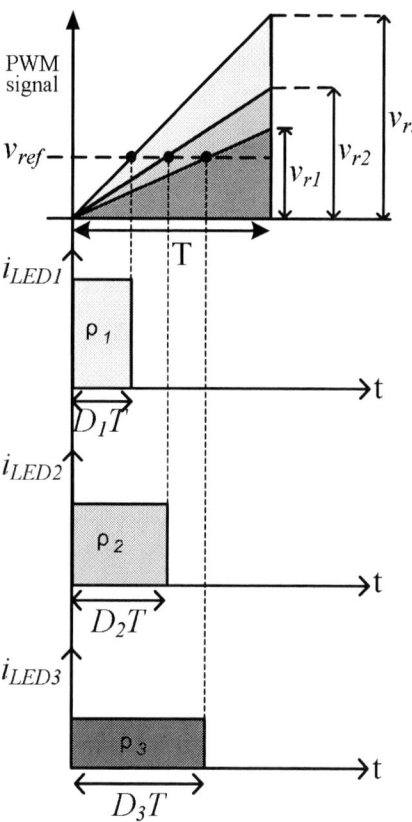

Fig. 3 A conceptual diagram of the proposed balancing strategy.

In this paper, the objective of the proposed balancing controller is to maintain the average current at a preset value instead of the digital dimming control. In general, the digital dimming control has being applied to backlight module for adjusting lamp current as shown in Fig. 4, where a low frequency dimming (LFD) signal modulates the average lamp current according to its duty cycle. In addition, the amplitude of the modulated lamp current just has two levels like the traditional PWM control. However, in the proposed strategy, the saw-tooth waveform, whose amplitude is proportional to the forward current, is utilized to compare with a dc level to produce a corresponding duty cycle instead of changing current amplitude directly. Namely, we modulate the duty cycle of LED by sampling its current amplitude to

produce a corresponding saw-tooth wave whose amplitude is proportional to the sensed current amplitude. Therefore, the purpose of this work is to avoid nonlinear feedback control to lead an identical average forward current for each LEDs. Fig. 5 depicted the relationship between the PWM signal and the duty cycle of the forward current, and it can be expressed as follows:

$$\frac{D}{1} = \frac{v_{c2-ref}}{v_{c2}} \qquad (2)$$

where v_{c2-ref} is the preset dc level, and v_{c2} is the amplitude of the saw-tooth waveform. i.e,

$$D \times v_{c2} = v_{c2-ref} \qquad (3)$$

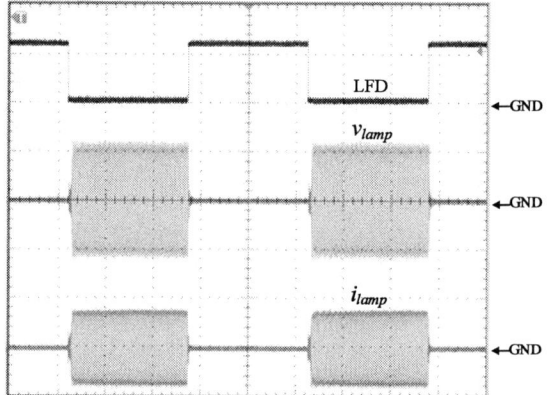

Fig. 4 A typical waveforms of backlight module with digital dimming control.

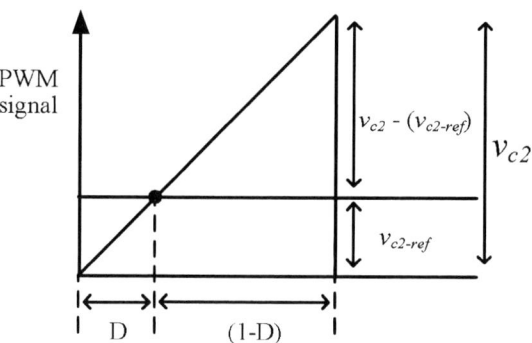

Fig. 5 The relationship between PWM signal and duty cycle.

By comparing the mentioned saw-tooth wave with a dc level, so that pulse width modulation is generated. From (2) and (3), it is easy to realize that the product of current amplitude and duty cycle can be kept as a constant value in the proposed strategy. The duty cycle will vary as long as the current amplitude changes. The multiple LED driving system with the proposed current equalizers are shown in Fig. 6. In this configuration, although each series-connection set is independent, the average driving currents through each loop are the same for each other.

IV. IMPLEMENTATION OF THE CURRENT EQUALIZER

In Fig. 6, a resistor R_s is utilized to sample the forward current and convert to a sampling signal v_{pt} as

$$v_{pt} = i_{LED} \times R_s \qquad (4)$$

Next, the sampling signal v_{pt} is passed through a low pass filter as shown as Fig. 7 to obtain the signal v_b as follows:

$$v_b \approx \left| v_{pt} \right| e^{\frac{-T}{RC}} = \left| v_{pt} \right| e^{\frac{-1}{RCf}} \qquad (5)$$

This equation can be rewritten as

$$v_b = \left(i_{LED} \times R_s \times e^{\frac{-1}{RCf}} \right) \qquad (6)$$

To obtain a higher sampling signal, a noninverting amplifier is cascaded between the low pass filter and error amplifier as shown in Fig. 6. The voltage gain α of this amplifier can be given as

$$v_{b2} = v_{b2-ref} \qquad (7)$$

$$\frac{v_{b2}}{v_b} = \alpha = 1 + \frac{R_2}{R_1} \qquad (8)$$

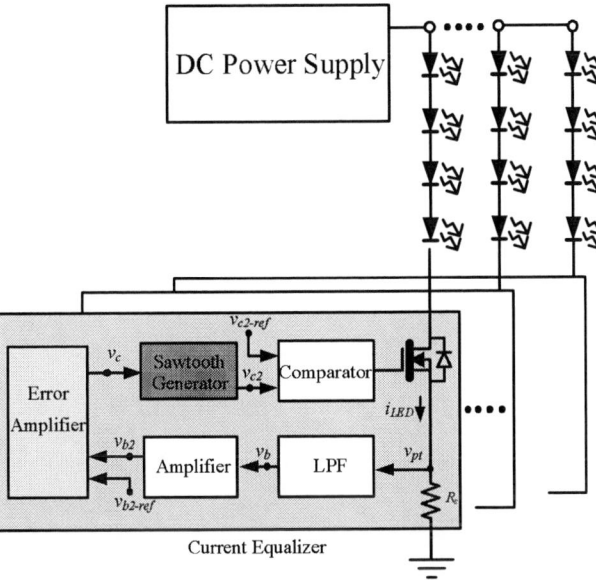

Fig. 6 The configuration of the proposed LED driving system with current equalizers.

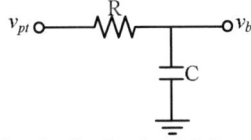

Fig. 7 A low pass filter for filtering the high frequency components.

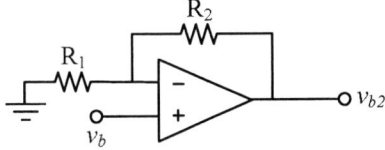

Fig. 8 The sampling signal is enlarged by a noninverting amplifier.

And then the amplified signal v_{b2} is entered the error amplifier as shown in Fig. 9 to complete feedback

compensation to generate a control signal v_c. Moreover, the control signal v_c will determine the amplitude of the saw-tooth waveform. By comparing the reference signal v_{c2-ref}, a low frequency control signal LFD is produced to maintain the identical average forward current.

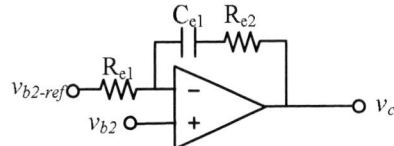

Fig. 9 A typical error amplifier.

V. DESIGN CONSIDERATION AND EXPERIMENTS

In this section, we will implement a current equalizer for driving the high power LEDs and demonstrate the proposed balancing performance. The design considerations of these parameters are discussed below:

1) Obtain the electrical specifications of LED

The adopted high power LED, which is characterized by its material, driving type and temperature, is inherently a nonlinear load, and hence influences the design of the driving circuit. The LED used in this paper is EHP-A08B/UT01-P01, which has rated power of 1W, forward current of 500mA(max), forward voltage of 4V(max), optical efficiency of 31lm/W, and color temperature of 5600K. Among the mentioned parameters above, the most important one in the proposed circuit structure is the forward current.

2) Determine the average current and tolerance error

The selected driving current should be operated within the maximum forward current of 500mA. In practical design, a reasonable operating range of 0.1A~0.4A is chosen. Suppose the average current i_{Lavg}=0.1A and its tolerance error≦ 5% are preset, respectively.

3) Estimate the related parameters of the controller

First, we can use a lower value sampling resistor R_s of 2Ω, and then construct a low pass filter as shown in Fig. 7, which is composed of R and C and its cut off frequency should much lower than the operating frequency of LFD to filter the high frequency components. In this paper, we select the operating frequency of LFD at 100Hz. And then R=300kΩ, C=1μF, whose cut off frequency satisfy the mentioned requirement, are arranged respectively. Moreover, the filtered signal v_b can be enlarged by an amplifier according (8). The voltage gain α is calculated about 10 when the v_{b2-ref}=2V. Hence, R_2=9kΩ and R_1=1kΩ are used in practice. Finally, the error amplifier shown in Fig. 9 is closed related to the system stability, R_{e1}=10kΩ, R_{e2}=1.5kΩ, and C_{e1}=100nF are chosen in practical design.

Fig. 10 shows the LED currents driven by four levels constant dc voltage sources at various current levels. As

1764

shown in Fig. 10, the driving current of LED fluctuates according to the dc voltage source. Namely, the illumination of panel will flash as long as voltage source is unstable. To verify the validity of the proposed control strategy, Fig. 11 illustrates the comparison of the measured LED current under various conditions with the proposed control mechanism. From the experimental results, the duty cycles are all inverse proportional to the corresponding current amplitude, which are all close to the derived equations in (2) and (3). Finally, Fig. 12 displays a prototype of a LED driving system with the proposed current equalizer.

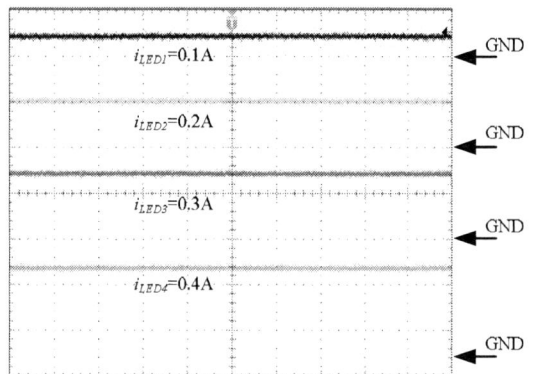

Fig. 10 Measured waveforms of LED current at various driving level (Ver: 200mA/div; Hor: 100us/div)

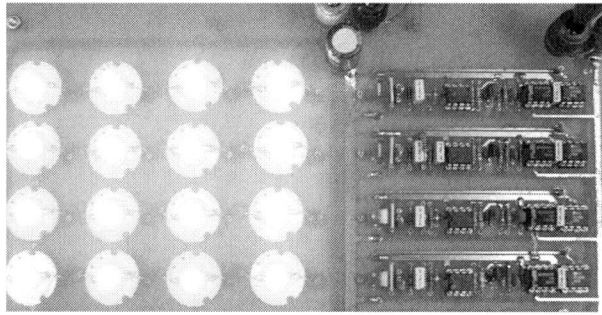

Fig. 12 The prototype of the proposed LED driving system with current equalizer.

VI. CONCLUSION

In this paper, in order to obtain identical luminance for driving high power LEDs, a current equalizer is implemented by using the balancing strategy, which combines PWM strategy with digital dimming control. With the proposed control strategy, the product of the forward current amplitude and duty cycle always maintains a preset fixed value to obtain a constant average driving current. Therefore, the proposed control strategy and the implemented current equalizer not only provides the capability to balance each average driving current in each LCD module to achieve constant-luminance control, but also offers linear and easy design rules to avoid nonlinear feedback relationship. The measured maximum current variations in experiments are all below 4%.

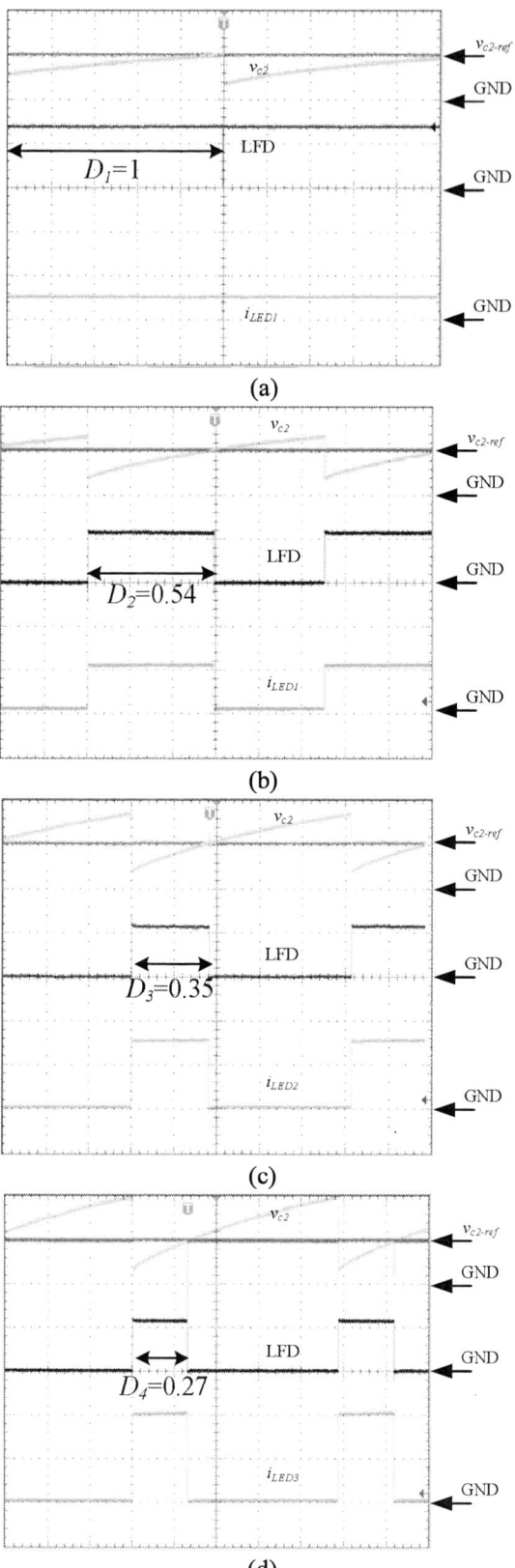

Fig. 11 Measured waveforms of v_{c2}, v_{c2-ref}, LFD, i_{LED} with the proposed current equalizer. (a)i_{LED1}=0.1A (b)i_{LED2}=0.2A (c)i_{LED3}=0.3A (d)i_{LED4}=0.4A (Ver: 2V/div for v_{c2} ; Ver: 2V/div for v_{c2-ref} ; Ver: 10V/div for LFD ; Ver: 200mA/div for i_{LED} ; Hor: 2ms/div)

ACKNOWLEDGEMENT

This work was sponsored by the National Science Council, Taiwan, R.O.C., Project number: NSC 95-2221-E-129-024.

REFERENCES

[1] P. Narra, and D. S. Zinger, "An Effective LED Dimming Approach," in *Proc. IEEE IAS'04 Conf.*, pp. 1671-1676, 2004.

[2] I. H. Oh, "A Single-Stage Power Converter for a Large Screen LCD Back-Lighting," in *Proc. IEEE APEC'06 Conf.*, pp. 1058-1063, 2006.

[3] M. Nishikawa, Y. Ishizuka, H. Matsuo, and K. Shigematsu, "An LED Drive Circuit with Constant-Output-Current Control and Constant-Luminance Control," in *Proc. IEEE INTELEC'06 Conf.*, sec. 8-2, 2006.

[4] P. L. M. Catala, and P. G.oyhenetche, "An Integrated digital PFM DC/DC Boost converter for a power management application: A RGB backlight LED system driver," in *Proc. IEEE IECON'02 Conf.*, pp. 37-42, 2002.

[5] C. C. Chen, C. Y. Wu, P. C. Lu, Y. M. Chen, and T. F. Wu, "Sequential Color LED Backlight Driving System for LCD Panel," in *Proc. IEEE IPEMC'06 Conf.*, pp. 237-241, 2006.

[6] C. C. Chen, C. Y. Wu, and T. F. Wu, "Fast Transition Current-Type Burst Mode Dimming Control for the LED Backlight Driving System of LCD TV," in *Proc. IEEE PESC'06 Conf.*, pp. 432-438, 2006.

[7] F. Kazuo, and F. Tsutomu, "LED Device ," *Japan Patent 2006120860*, May 11, 2006.

[8] C. H. Lin, Y. Lu, and K. J. Pai "Digital-Dimming Controller with Current Spikes Elimination Technique for LCD Backlight Electronic Ballast," in *Proc. IEEE APEC'04 Conf.*, vol. 1, pp. 153-158, 2004.

Modeling of the Parasitical Capacitance Effect in LCD Panel and Corresponding Elimination Strategy

Chang-Hua Lin*, Tsung-You Hung**, Chien-Ming Wang***, and Kai-Jun Pai****

*Tatung University, Taipei, R.O.C., **St. John's University, Taipei, R.O.C.,
***National Ilan University, Ilan, R.O.C. ****National Taiwan University of Science & Technology, Taipei, R.O.C.

Abstract—A mathematical model of parasitical capacitance in LCD panel is conducted to explore the influence of leakage current effect on electric characteristics. First, a class D resonant backlight inverter is employed to act as the main circuit. Next, the phase angle variations caused by the parasitic capacitances is considered as a reference parameter in the proposed control strategy. By using the primary-side control and incorporating the DPLL technique to form a feedback mechanism to track the optimal operating frequency. And then the influence of parasitic capacitance can be reduced so as to eliminate the leakage current effect, hence, the system efficiency and stability will be improved. Complete mathematical analysis and design considerations are detailed. Experimental results agree with the theoretical predictions.

Index Terms—Parasitical Capacitance, Leakage Current Effect, Backlight Inverter, DPLL.

I. INTRODUCTION

Inside a flat display panel (FDP), a backlight module is essential and its performance is vital to the display quality of the FDP. Nowadays, there are two major backlight sources available, namely a white LED and a cold cathode fluorescent lamp (CCFL). Usually a CCFL is more suitable for larger display panels due to its cost advantage over a white LED. Currently, the combination of an LCD and a CCFL seems to have attracted a constant, or even an increasing, global demand for a thinner FDP with fine performance and decent efficiency [1-10]. In the past, in order to suppress EMI, the LCD is always covered with conductive coatings. These conductive coatings are connected to the system grounds so that there are significant parasitic capacitances occurring between the cables and the system grounds [11]. In general, the primary ground represents the system ground, so the parasitic capacitances cause severe significant leakage currents along the display housing. Consequently, the lamp currents will be absorbed by these parasitic currents flowing to the system ground. Therefore, the conventional backlight systems cannot provide preset driving current for CCFL, and thus obtain a lower efficiency. In this paper, a mathematical model of the parasitic capacitance is derived and a new control scheme is proposed to deal with the said shortcomings for backlight module. The proposed control scheme integrates the primary-side control and the digital phase-locked loop (DPLL) technique to reduce the influence of parasitic capacitance and to track an optimal operating frequency so that the leakage current effect is removed.

II. THE LEAKAGE CURRENT EFFECT AND ANALYSIS OF RESONANT INVERTER

Fig. 1 illustrates the leakage current effect resulted from the parasitic capacitances in LCD housing. Generally, a backlight system is located behind the LCD panel and with a metal rear plate. There are many parasitic paths between cables and metal construction due to the induced parasitic capacitances. All parasitic paths will cause that the backlight system wastes more energy to maintain desired current in the lamp. In order to reduce the leakage current phenomenon and to promote the system efficiency, there are two key strategies are adopted in this paper. First, the secondary-side of boost transformer is closed without connecting with the system ground, i.e., the used CCFL is floated to remove all of the leakage current paths. Therefore, the lamp current can be regulated and monitored by the primary-side control. Next, a DPLL controller is essential to track the optimal operating frequency when the parasitic capacitances influence the preset resonant characteristics.

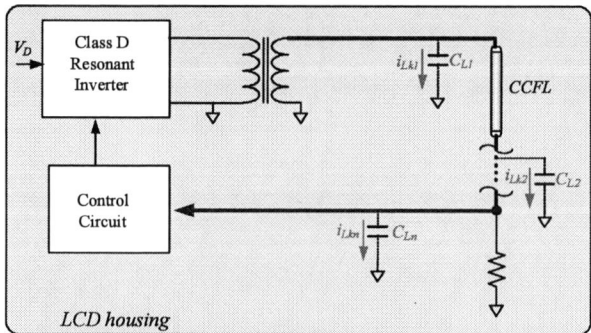

Fig. 1 The leakage current effect hides between the LCD housing and the backlight system with conductive coating.

The schematic of the employed class D resonant inverter is shown in Fig. 2, which is composed of a choke inductor L_r, a resonant capacitor C_r, a ballast capacitor C_B, and two alternative switching power switches S_1 and S_2. When the operating frequency is close to the resonant frequency, the load quality factor Q_L is high enough to filter the dc current component and current harmonics of the main circuit [4]. The main circuit is also accompanied with the boost transformer T_1 to generate sinusoidal

978-1-4244-0644-9/07/$25.00 ©2007 IEEE

voltage and current for driving CCFL. The equivalent circuit of Fig. 2 is shown in Fig. 3a, in which the magnetizing inductance, the external resonant components, parasitic capacitances, and the secondary reflected impedance are all merged to form a resonant tank [1], [6].

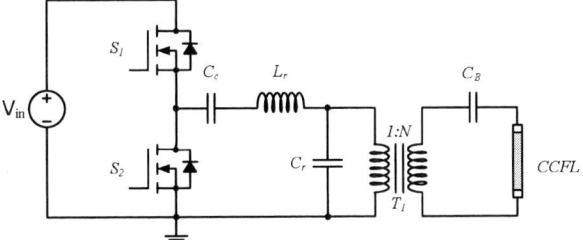

Fig. 2 The schematic of the employed class D resonant inverter.

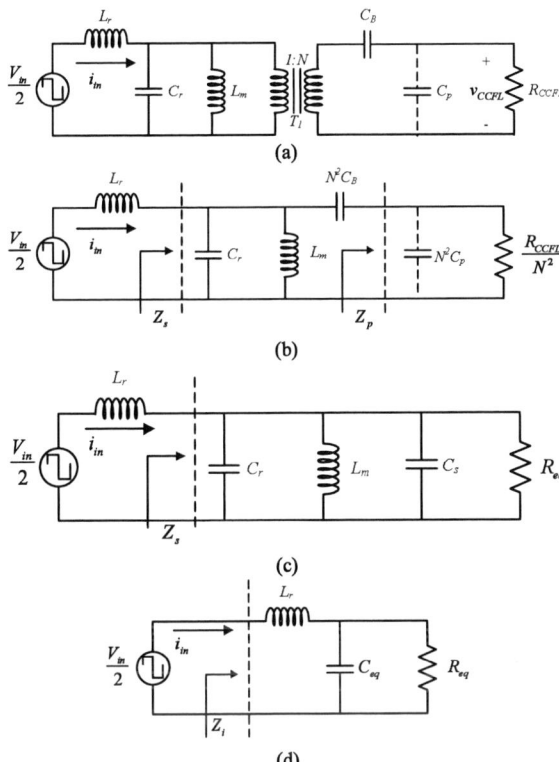

(a)

(b)

(c)

(d)

Fig. 3 (a) The equivalent circuit of the class D resonant inverter, (b) the reflected equivalent circuit of Fig. 3a, (c) the equivalent circuit of Fig. 3b, (d) the simplified circuit of 3c.

To simplify the circuit model, the CCFL is replaced with R_{CCFL}, and reflects all the secondary components to the primary-side as shown in Fig. 3b-3d, respectively. From Fig. 3d, the input impedance can be given as

$$Z_i = R_{eq} + j\left(\omega L_r - \frac{1}{\omega C_{eq}}\right) = \frac{R_{eq}\left[1 - \left(\frac{\omega}{\omega_o}\right)^2 + j\frac{1}{Q_L}\left(\frac{\omega}{\omega_o}\right)\right]}{1 + j\frac{1}{Q_L}\left(\frac{\omega}{\omega_o}\right)} \quad (1)$$

$$= |Z_i|\angle\phi$$

where

$$|Z_i| = \sqrt{\frac{\left(Q_L^2\omega^4 - 2\omega_o^2 Q_L^2\omega^2 + \omega^2\omega_o^2 + Q_L^2\omega_o^4\right)R_{eq}}{\omega_o^2 Q_L^2\left(\omega^2 + Q_L^2\omega^2\right)}} \quad (2)$$

$$\phi = \tan^{-1}\left\{Q_L\left(\frac{\omega}{\omega_o}\right)\left[\left(\frac{\omega}{\omega_o}\right)^2 + \frac{1}{Q_L^2} - 1\right]\right\} \quad (3)$$

$$\omega_o = \frac{1}{\sqrt{L_r C_{eq}}} \quad (4)$$

$$Q_L = \omega_o C_{eq} R_{CCFL} \quad (5)$$

If the quality factor Q_L is high enough, the corner frequency f_o is near the resonant frequency f_r [13], i.e.

$$f_o \cong f_r \cong \frac{1}{2\pi\sqrt{L_r C_{eq}}} \quad (6)$$

According (6), the relationship of the resonant frequency and parasitic capacitance can be depicted as Fig. 4. From Fig. 4, it is apparent that the resonant frequency f_r is inverse proportional to the parasitic capacitance C_p. Namely, the operating point of the resonant tank will depart from the preset one as long as the parasitic capacitance effect occurs. Thus, the traditional feedback mechanism is incapable of keeping optimal operating frequency when environmental parameters change.

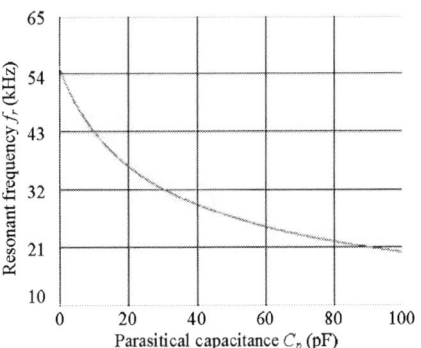

Fig. 4 The relationship between the resonant frequency and the parasitic capacitance.

From Fig. 3b, when the parasitic capacitance is considered, we can derive the practical lamp current as follows:

$$i_{CCFL} = i_{in}\left|\frac{\frac{X_{Cr} \cdot X_{Lm}}{X_{Cr} + X_{Lm}}}{\frac{X_{Cr} \cdot X_{Lm}}{X_{Cr} + X_{Lm}} + Z_p + X_{CB}} \cdot \frac{X_{Cp}}{\frac{R_{CCFL}}{N^2} + X_{Cp}}\right| \quad (7)$$

where

$$I_m = \frac{2V_{in}}{\pi|Z_i|} \quad (8)$$

The relationship curve of the parasitic capacitance versus the lamp current can be depicted by Matlab simulation according to (7). In the simulated results, it is noted that leakage current effect is considerable when the parasitic capacitance cannot be neglected. Combing (3), (7), and (8), the mathematical model of the parasitic capacitance can be obtained.

$$C_p\left(i_{CCFL}, \phi_{Zi}\right) = \frac{i_{in}\pi\left(A_1 - A_2\right) - A_3 A_4}{\omega N^2 R_{CCFL} \cdot \left\{i_{in}\pi A_4 + [A_3 A_5]\right\}} \quad (9)$$

where

$$A_1 = R_{CCFL}\left[X_{Cr}\left(X_{Lm}+X_{Lr}\right)+X_{Lm}X_{Lr}\right] \quad (10)$$

$$A_2 = N^2\left\{X_{Lm}\left[X_{Cr}\left(X_{CB}+X_{Lr}\right)+X_{CB}X_{Lr}\right]\right\} \quad (11)$$

$$A_3 = -2V_{in}\sin\left(\omega t-\phi_{Zi}\right) \quad (12)$$

$$A_4 = R_{CCFL}\left(X_{Cr}+X_{Lm}\right)+N^2\left[X_{Cr}\left(X_{CB}+X_{Lm}\right)+X_{CB}X_{Lm}\right] \quad (13)$$

$$A_5 = X_{Cr}\left(X_{CB}+X_{Lm}\right)+X_{CB}X_{Lm} \quad (14)$$

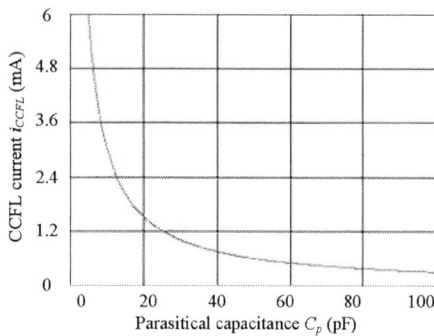

Fig. 5 The relationship between the lamp current and the parasitic capacitance.

To ensure that the CCFL can be started in normal conditions, the minimal turn ratio of a boost transformer T_t can be determined using the lowest input voltage V_D and the starting voltage $v_{start(rms)}$ by sinusoidal approximation.

$$N \geq \dfrac{\sqrt{2}v_{start\,(rms)}}{\dfrac{4}{\pi}\times\dfrac{V_D}{2}} \quad (15)$$

Since CCFLs possess inherently negative dynamic impedance characteristics after igniting, like other gas discharge lamps, the ballast capacitor C_{B} is required to stabilize the lamp current [1], [6], [7]. For reducing the nonlinear behavior of the CCFL, the reactance of C_{B} is generally arranged to be greater than the CCFL impedance. The transfer function of lamp voltage v_d versus v_{in} can be calculated from Fig. 3d by sinusoidal approximation as follows:

$$\frac{v_{CCFL}}{V_{in}} = \frac{N^3 C_B R_{CCFL} L_m\left(B_1+jB_2\right)}{B_3+jB_4} \quad (16)$$

The magnitude of (16) is obtained as

$$\left|\frac{v_{CCFL}(j\omega)}{V_{in}(j\omega)}\right| = \sqrt{\frac{\left(N^2\omega^3 R_{CCFL}L_m\right)^2\left[\left(B_1\right)^2+\left(B_2\right)^2\right]}{\left(B_3\right)^2+\left(B_4\right)^2}} \quad (17)$$

Where

$$B_1 = N^2\omega^3 R_{CCFL}L_m\left[C_r\left(C_p+C_r\right)+N^2 C_B C_p L_r\right]+\omega R_{CCFL}\left(C_p-C_B\right)\left(N^2 L_r+L_m\right) \quad (18)$$

$$B_2 = L_m - N^2 L_r\left[1+\omega^2 L_m\left(C_r+N^2 C_B\right)\right] \quad (19)$$

$$B_3 = \omega^3 R_{CCFL}L_m\left[C_p\left(C_B+C_r\right)+C_B C_r\right]-\omega R_{CCFL}\left(C_p+C_B\right) \quad (20)$$

$$B_4 = 1+\omega^2 L_m\left(C_r+N^2 C_B\right) \quad (21)$$

From (17), we see that the lamp voltage v_{CCFL} is closely related to the input voltage, operating frequency, lamp impedance, transformer characteristics (L_{m} and N), resonant tank components (L_r and C_{eq}), and parasitic capacitance C_p.

III. DIGITAL PHASE-LOCKED LOOP CONTROLLER

As mentioned previously, the electrical characteristics of CCFL are highly sensitive to the parasitic capacitances, which will result in the resonant frequency variation and thus significantly influence the preset operating point and system efficiency. Therefore, a new controller based on the DPLL technique is proposed to track the optimal operating frequency, so as to raise the system efficiency. The block diagram of the DPLL controller proposed in this paper is shown in Fig. 6, where the DPLL module SN74LS297 is adopted, and the L counter needs to be provided additionally [8-10]. The basic structure of a DPLL consists of a phase detector, a divide-by-K counter, an Increment/Decrement (Inc/Dec) circuit, and a divide-by-L counter.

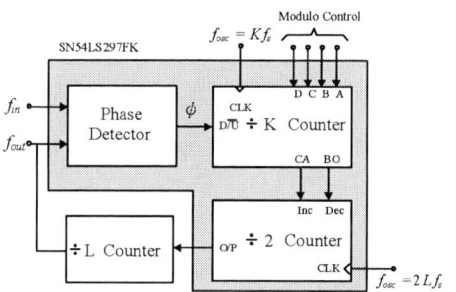

Fig. 6 The typical structure of the employed DPLL module.

According to Fig. 6, if the K counter and the Inc/Dec circuit both adopt the same timing clock f_{osc}, then their output frequencies, f_K and f_{ID}, can be expressed, respectively, as follows:

$$f_K = \frac{K_d\ \phi\ f_{osc}}{K} \quad (22)$$

$$f_{ID} = \frac{f_{osc}+f_K}{2} \quad (23)$$

where K_d is the gain of the phase detector, ϕ is the phase error, and K is the modulus of the K counter.

$$K = \log_2\left(\frac{f_{osc}}{f_s}\right) \quad (24)$$

$$L = \log_2\left(\frac{f_{osc}}{2f_s}\right) \quad (25)$$

Because of $K_d \times \phi = +/-1$ at the boundary of the lock-in range, the DPLL's output frequency f_{out} and lock-in range f_L can be derived as

$$f_{out} = \frac{f_{ID}}{L} = f_s \pm f_L \quad (26)$$

$$f_L = \frac{f_{osc}}{2^{(K+L)}} \quad (27)$$

where f_s is the selected operating frequency of the backlight module. The appropriate values of above-mentioned parameters of the DPLL can be determined from (22) and (27) as well as the specifications of the employed system.

Fig. 7 illustrates the block diagram of the complete system configuration, where a sensing resistor R_s incorporates the DPLL module to form a primary-side controller as the control mechanism. In this scheme, we regulate the lamp current by feeding a phase angle of the equivalent input impedance Z_s on the primary-side of the

1769

main circuit, so the CCFL can be arranged in a floating form. Consequently, the secondary ground is separated from the primary-side ground to cut off the leakage current paths. As shown in this figure, the phase detector captures the driving signal of resonant inverter, v_{comp}, and the sampled signal of input current, v_{squ}, from the shaping circuit so as to sense the phase difference of these two signals. The sensed phase difference which equals the phase angle of the input impedance and vary with the change of environmental conditions, and, based on the phase discrepancy, outputs the signal ϕ to the K counter, which in turn generates periodic pulses with carry or borrow indication to regulate the output frequency of Inc/Dec circuit. And then the output frequency of L counter is adjusted immediately to synchronize the output frequency of DPLL and hence govern the operating frequency of the inverter. Therefore, the v_{comp} and v_{squ} remain synchronized with a fixed phase difference, resulting in the fast tracking of the optimal operating frequency of system, and thus reducing the leakage current phenomenon as well as achieving the maximum output efficiency. Therefore, the phase difference ϕ can be fixed to a certain value by using the DPLL technique even under the perturbation of environmental parameters variation. Moreover, according to Fig. 3d, we can utilize the derived mathematical model (9) to verify and analyze the influence of the parasitic capacitance.

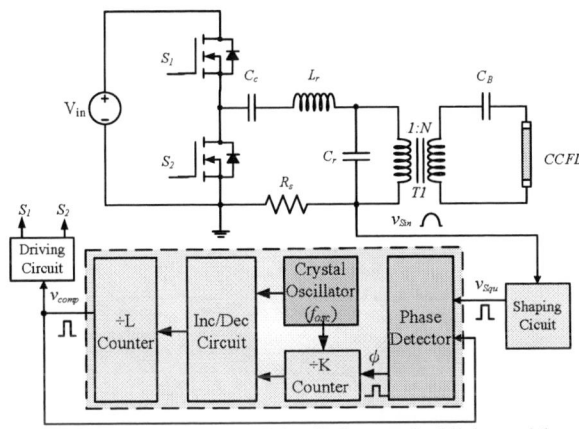

Fig. 7 The block diagram of the single-stage backlight module incorporating the primary-side control with the DPLL technique.

IV. DESIGN CONSIDERATIONS AND EXPERIMENTS

The CCFL used in this paper is FL-24315, which has rated power of 3.6W, operating voltage v_{CCFL} of 610V$_{rms}$, operating current i_{CCFL} of 6mA$_{rms}$, and starting voltage v_{start} of 720V$_{rms}$. The operating frequency of CCFL is generally selected within the range of 20~80kHz. In this paper, we select the operating frequency f_s of 56kHz and input voltage V$_D$ = 12V$_{DC}$. The most important parameters in the proposed circuit structure are the turns ratio N of the boost transformer, the magnetizing inductor L_m, the resonant inductor L_r, the resonant capacitor C_r, the ballast capacitor C_B as well as the DPLL parameters, such as f_{out}, N, M and f_L. The design considerations of these parameters are discussed below

1) Estimate the turns ratio N

The boost transformer should provide enough turns ratio to ignite the CCFL. By substituting mentioned above parameters v_{start} and V_{in} into (15), we obtain

$$N \geq \frac{\sqrt{2}v_{start\,(rms)}}{\frac{4}{\pi} \times \frac{V_{in}}{2}} = \frac{720\sqrt{2}}{\frac{4}{\pi} \times \frac{12}{2}} = 133 \tag{28}$$

2) Calculate L_r and C_r

From (28), we get the actual value of turns ratio with 150. In addition, we have the magnetizing inductance L_m=29.64µH, the ballast capacitor C_B=30pF, the lamp resistor R_{CCFL}=v_{CCFL}/i_{CCFL}=102kΩ, and the operating frequency f_s=56kHz. By substituting these parameters into (15) and (17), we have

$$\left| \frac{v_{CCFL}(j\omega)}{V_{in}(j\omega)} \right| = \frac{620\sqrt{2}}{\frac{4}{\pi} \times \frac{12}{2}} = 114 \tag{29}$$

From the preset conditions and (6), we have

$$f_r = \frac{1}{2\pi\sqrt{L_r C_{eq}}} = 56kHz \tag{30}$$

By combining (29) and (30), the parameters for this design are given by L_R=34.36µH (use 34.92µH), C_r=445nF (use 470nF).

3) Determine the DPLL parameters

The operating frequency f_s=56kHz is selected in this paper, hence the center frequency is preset at 56kHz, the lock-in frequency f_L=1.95kHz, and the lock-in range Δf_L=3.9kHz. From Fig. 6, (24), and (25), we can estimate L=5.16 and K=6.16 when the crystal frequency f_{osc}≅4MHz is adopted. In practical design, the modulus of the counter L=5 and K=6 are used.

4) Evaluate the parasitic capacitance C_p

Substituting the measured lamp current i_{CCFL} and phase angle ϕ into (3), (7), and (9), we can simulate the 3D relationship, so as to evaluate the corresponding parasitic capacitance C_p with respect to the lamp current and phase angle as shown in Fig. 8.

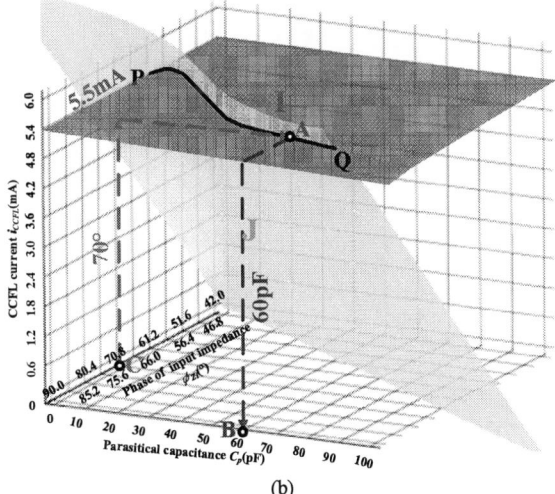

(b)

Fig. 8 The 3D simulation relationship among C_p, i_{CCFL}, and ϕ.

Fig. 8 clearly exhibits the 3D relation among the lamp current i_{CCFL}, phase angle ϕ, and the parasitic capacitance C_p. Assuming the preset lamp current i_{CCFL} =5.5mA as the plane I, which intercross the surface J derived from (9) along the curve $\overline{PQ}$, then we can obtain the corresponding parasitic capacitance C_p from this curve if the input phase angle ϕ is given.

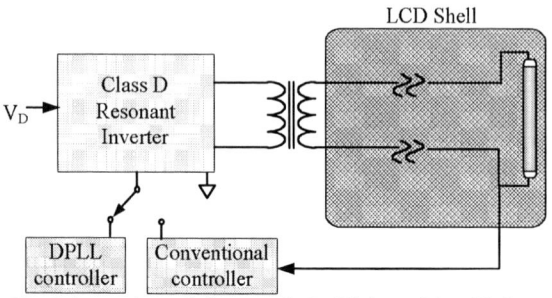

Fig. 9 A experimental diagram of a backlight module with the LCD shell.

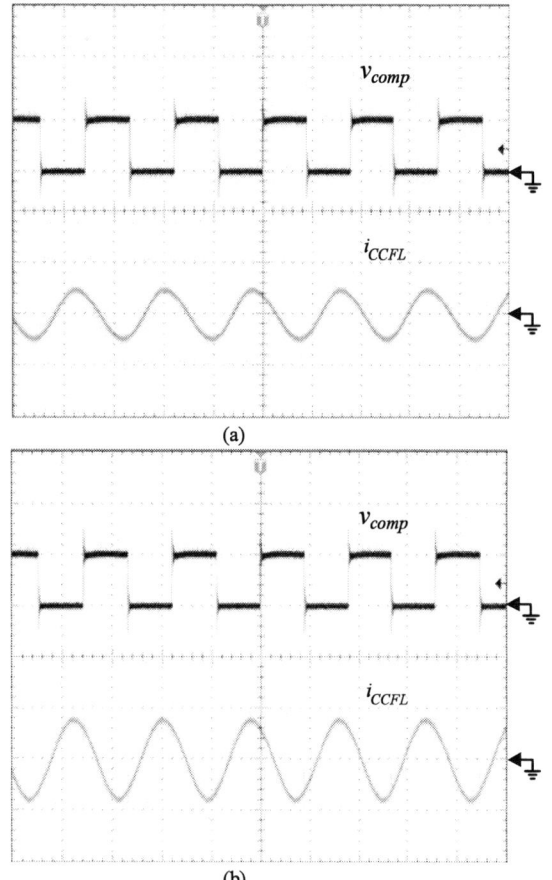

Fig. 10 Measured waveforms of v_{comp}, and i_{CCFL} with LCD shell at various approaches (a) with conventional feedback (b) with the proposed method (Ver: 10V/div for v_{comp}; Ver: 10mA/div for i_{out}; Hor: 10μs/div)

Fig. 9 displays an experimental diagram of a backlight module with LCD shell. To verify the validity of the proposed control strategy, the cable of CCFL is connected to the secondary-side of boost transformer and fixed on the LCD shell to exhibit the leakage current

effect. Fig. 10 shows the comparison of the measured lamp current i_{CCFL} under the parasitic effect with and without the proposed control mechanism, respectively. As Fig. 10a indicates, the measured lamp current, which is suffering from the parasitic capacitance effect, is merely 2mA. It is obvious to note that the lamp current is maintained at normal operating value of 6mA shown in Fig. 10b when the proposed strategy is embedded in the system.

Next, the input phase angles ϕ of the resonant tank under different test conditions are shown in Fig. 11, respectively. Fig. 11a displays the phase difference of 66 degrees between v_{comp} and the input current i_{in} by using traditional control. Obviously, with the proposed controller, the phase difference is down to 50 degrees as shown in Fig. 11b.

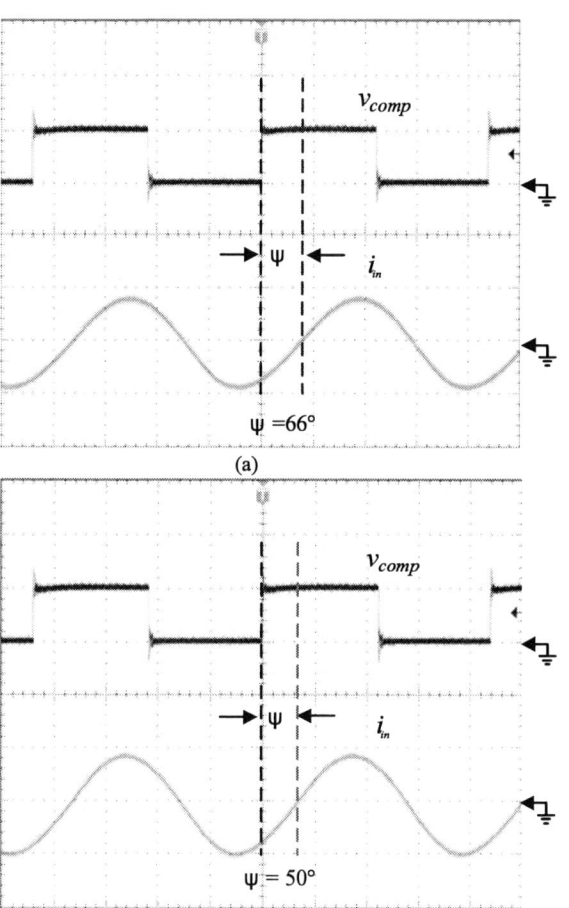

Fig. 11 Measured waveforms of v_{comp}, and i_{in} with LCD shell at various approaches (a) with conventional feedback (b) with the proposed method (Ver:5V/div for v_{comp}; Ver:200mA/div for i_{in}; Hor:4μ s/div)

In the final experiments, the lamp voltage v_{CCFL}, lamp current i_{CCFL}, and lamp power p_{CCFL} are measured under various test conditions as shown in Fig. 12. With the traditional control, Fig. 12a shows the lamp voltage v_{CCFL}=520V$_{rms}$, lamp current i_{CCFL}=2.5mA$_{rms}$, and lamp power p_{CCFL}=1.3W. However, the lamp voltage v_{CCFL}=560V$_{rms}$, lamp current i_{CCFL}=5.5mA$_{rms}$, and lamp power p_{CCFL}=3.08W when the proposed controller is incorporated.

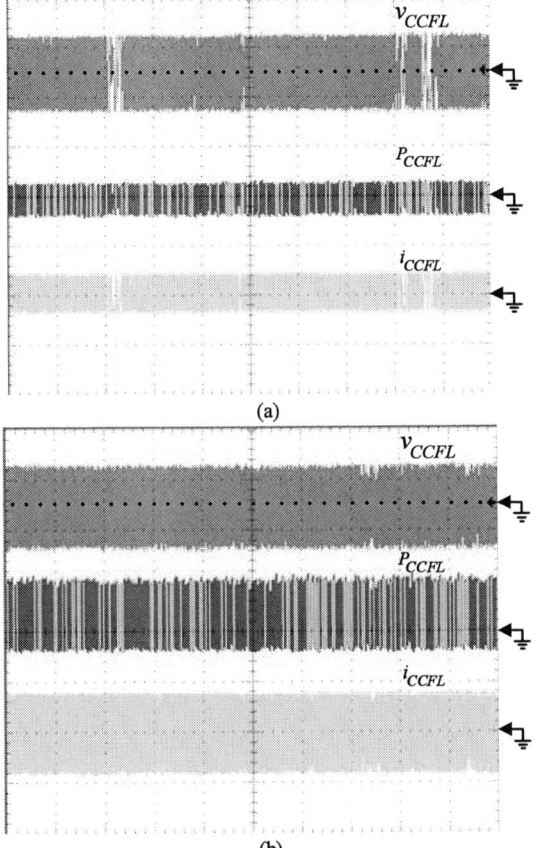

(a)

(b)

Fig. 12 Measured waveforms of v_{CCFL}, i_{CCFL}, and p_{CCFL}, with LCD shell at various approaches (a) with conventional feedback (b) with the proposed method (Ver:1kV/div for v_{CCFL}; Ver:10mA/div for i_{CCFL}; Ver:5W/div for P_{CCFL}; Hor:10s/div)

V. CONCLUSION

In this paper, a class D resonant inverter is employed to verify the leakage current effect resulted from the parasitic capacitances. A merged model and a model simplification are both used to derive the optimal parameters for the backlight system. In addition, the primary-side control and DPLL technique are integrated to eliminate the said leakage current, thus, to enhance the system stability and efficiency. Complete analysis and design considerations are conducted. Finally, some experimentation is used to demonstrate the derived mathematical model and the proposed strategy.

ACKNOWLEDGEMENT

This work was sponsored by the National Science Council, Taiwan, R.O.C., Project number: NSC 95-2221-E-129-024.

REFERENCES

[1] M.S. Lin, W. J. Ho, F. Y. Shih, D. Y. Chen and Y. P. Wu, "A cold-cathode fluorescent lamp driver circuit with synchronous primary-side dimming control, " *IEEE Trans. Ind. Electr.*, Vol. 45, No. 2, Apr., 1998.

[2] M. Jordan and J. A. O'Connor, " Resonant fluorescent lamp converter provides efficient and compact solution, " *Proc. IEEE-APEC'93 Conf. Record*, pp. 424-431, Mar. 1993.

[3] G. H. Kweon, Y. C. Lim, and S. H. Yang, " An analysis of the backlight inverter by topology, " *Proc. IEEE ISIE'01 Conf.*, Vol. 2, pp. 896- 900, June 2001.

[4] C. S. Moo, W. M. Chen and H. K. Hsien " Electronic ballast with piezoelectric transformer for cold cathode fluorescent lamps, " *Proc. IEE Proceedings-Electric Power Applications*, Vol. 150, pp. 278-282, May 2003.

[5] R. L. Lin and Y. T. Chen, " Electronic ballast for fluorescent lamps with phase-locked loop control scheme , " *IEEE Trans. Power Electronics*, Vol. 21, No. 1, pp. 254-262, January 2006.

[6] C. S. Moo, W. M. Chen, and H. K. Hsien, " Electronic ballast with piezoelectric transformer for cold cathode fluorescent lamps, " *Proc. IEE Proceedings-Electric Power Applications*, Vol. 150, pp. 278-282, May 2003.

[7] R. L. Lin and Y. T. Chen, " Phase-locked-loop-control-based electronic basllast for fluorescent lamps, " *IEE Proc.-Electr. Power Appl.*, Vol. 152, NO. 3, pp. 669-676, May 2005.

[8] R. L. Lin and C. H. Wen , " PLL control scheme for the electronic ballast with a current-equalization network, " *Journal Of Display Technology*, Vol. 2, No. 2, pp. 160-169, June 2006 .

[9] G. H. Kweon, Y. C. Lim and S. H. Yang, " An analysis of the backlight inverter by topologies , " *IEEE International Symposium Industrial Electronics, 2001. Proceedings. ISIE 2001. Conf.* , Vol. 2, 12-16, pp. 896-900 June 2001.

[10] C. H. Lin, Y. C. Chen, Y. Lu and F. L. Wen, " DPLL Technique Applied to Backlight Module for Eliminating Temperature Effect in Piezoelectric transformer," *Proc. IEEE PEDS 2005 International Conference*, vol. 2, pp. 1040-1045, Nov. 2005.

[11] S. K., O. O., H. H., S. T., A. F., T. I. and S. H., "Third Order Longitudinal Mode Piezoelectric Ceramic Transformer and Its Application to High-Voltage Power Inverter," *IEEE Ultrasonics Symposium, 1994. Proceedings.*, Vol. 1, pp. 525-530, 1-4 Nov. 1994.

[12] C. H. Lin, T. Y. Hung, and Y. Lu, "The Elimination of the Leakage Current Effect for LCD Backlight Module Based on the Primary-side Control and the DPLL Technique," in *Conf. Rec. IEEE ICIEA'07*, pp. 1950-1955, Harbin, China, May 23-25, 2007.

[13] M. K. Kazimierczuk & D. Czarkowski, *Resonant Power Converters*. Wliey Interscience, 1995, pp. 201-239.

[14] Application Hand "SN54LS297FK," Teax Instrument Semiconductor Inc.

On-line SOC Estimation of Battery for Wireless Tram Car

Hiroyuki Miyamoto*, Masayuki Morimoto*, Katsuaki Morita**

* Department of Electrical and Electronic Engineering, Tokai University,
1117 Kita-Kaname Hiratsuka city Kanagawa prefecture 259-1292, Japan
** Mitsubishi Heavy Industries, 1-1 Wada-Oki Mihara city Hiroshima prefecture 723-0042, Japan

Abstract— **In recent years, the battery driven tram car (wireless tram) is researched and developed. The important information for the system is the battery state-of-charge (SOC) during the tram operation. SOC can be estimated by calculating battery open-circuit-voltage (OCV). In this paper, on-line SOC estimation by equivalent circuit in the current fluctuating condition is described. The experimental result will be also shown.**

Index Terms— **equivalent circuit, Lithium-ion battery, open-circuit-voltage, state-of-charge, wireless tram car**

NOMENCLATURE

C	Capacitance of equivalent circuit
I	Battery current
$I_{(n-1)}$	Battery current just before current change
$I_{(n)}$	Battery current just after current change
OCV	Battery open-circuit-voltage
R_1, R_2	Internal resistance of equivalent circuit
SOC	Battery state-of-charge
t	Time
$t_{(n)}$	Time at current change
V	Battery terminal voltage

I. INTRODUCTION

APM (Automated People Mover) and LRT (Light Rail Transit) are introducing in many cities, because of solving the problems of traffic jam, noise and environmental pollution. Recently, the battery driven tram car is developed, because the technology of energy storage device is progressing [1], [2]. Such a tram car system is called as the wireless tram car system. Wireless tram car is a new transportation system, which is driven by on-board battery. In this system, there is no catenary wire in the street. In this tram car, lithium-ion battery for electric vehicle (EV) is used, because of the high energy density, high power density, and its low weight.

The system configuration of the wireless tram car is shown in Fig.1. The energy of the traction is supplied by on-board battery. So that pantograph is not required to feed electricity. The motor and the inverter of wireless tram car are the same ones of conventional tram with catenary system.

The capability of mileage of tram car depends on capacity of the battery. Larger battery enables the longer distance. However, the weight of the car increases. Also, the mileage decreases by the weight of the battery. Therefore, in order to decrease weight of the vehicle, the

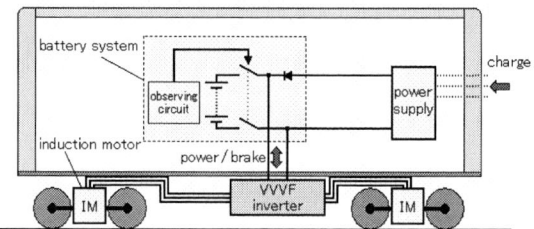

Fig. 1. System configuration of wireless tram car

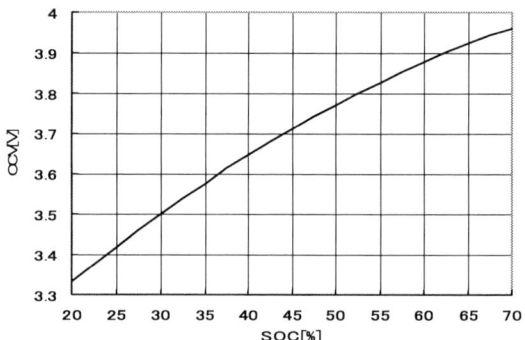

Fig. 2. The relation between SOC and OCV

battery should be as small as possible. The small battery requires frequent charge at the station. In this energy management concept, the state-of-charge (SOC) of the battery is important information to manage the energy of the vehicle. But it is difficult to know the SOC directly during operation.

In this paper, we propose the method of SOC estimation which uses only battery current and terminal voltage during the tram operation.

II. ESTIMATION OF SOC

A. Method of SOC Estimation

There are several methods of battery SOC estimation. They are shown as follows.

1) *Current Integration*: The SOC is estimated from the integration of the input and output energy of the battery. The tram car repeats acceleration and deceleration at every station. So the estimation error also integrates and increases.

2) *Open-Circuit-Voltage (OCV) Measurement*: OCV is the battery terminal voltage when the battery current is zero. OCV has simple relation to SOC as shown in Fig.2. So if we know the battery OCV,

SOC can be estimated easily. However, when the tram is running, the battery current is not zero. Therefore, it is difficult to measure directly OCV during the tram operation.

3) *Equivalent Circuit*: This method is to estimate SOC using an equivalent circuit of the battery. Therefore, the accurate determinating the constants of equivalent circuit is required. With the equivalent circuit, OCV can be calculated by using the measurement value of battery current and terminal voltage. So this is a possible method for the system during the operation.

The current integration method tends to increase the error. The OCV can not be measured during the tram operation. Therefore, we propose the equivalent circuit method as the best method of SOC estimation for the wireless tram car system. The detail of this method will be described in the next section.

B. Proposed Method

Our proposed method is the estimation of SOC by using the equivalent circuit. The equivalent circuit of lithium-ion battery is shown in Fig.3 [3]. In Fig.3, "I" express a battery current, "V" express a battery terminal voltage, and "OCV" express a battery real voltage. In this circuit, the battery internal resistance is divided into two parts, "R_1" and "R_2". "R_1" is the ohmic resistance that is dominated the rapid change of current. "R_2" is the polarization resistance that is connected to capacitance "C" in parallel. R_2C circuit has a long transient response time.

From Fig.3, the OCV at time "t" can be represented as follows [4].

$$OCV = V + (R_1 + R_2)I - R_2 I \exp\left(-\frac{t}{R_2 C}\right) \quad (1)$$

OCV can be calculated by equation (1) during the constant current charge or discharge. The accuracy of equation (1) is assured by the experiment. The experimental result of constant current charge/discharge is shown in Fig.4 and Fig.5. The error of SOC estimated by the equivalent circuit is less than 1.5%.

However, the current of the tram car is not constant, but varies by the running condition. Equation (1) should be extended to the current variation condition. The OCV at step-wise current variation can be expressed in equation (2) [4].

$$OCV = V + (R_1 + R_2)I - (R_2 I_{(n)} - R_2 I_{(n-1)}) \exp\left(-\frac{t - t_{(n)}}{R_2 C}\right) \quad (2)$$

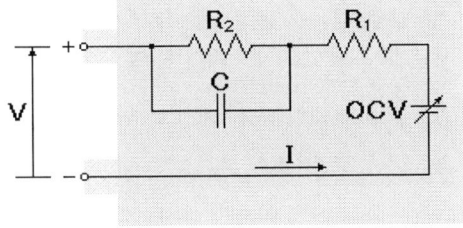

Fig. 3. Equivalent circuit of lithium-ion battery

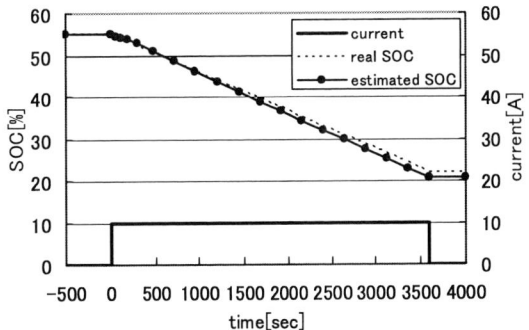

Fig. 4. The result of SOC estimation (10A discharge)

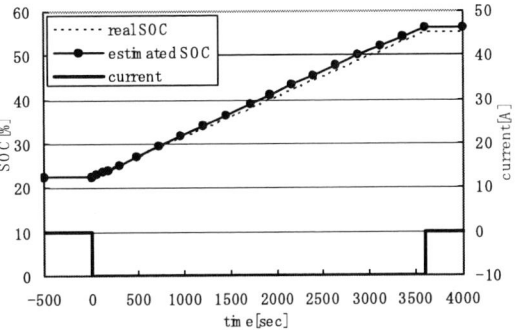

Fig. 5. The result of SOC estimation (10A charge)

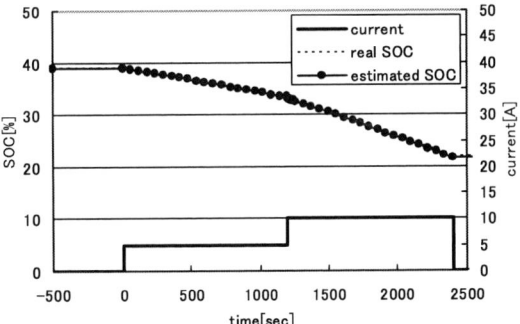

Fig. 6. The result of SOC estimation (5A→10A discharge)

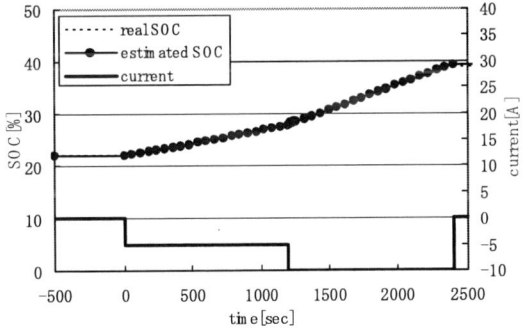

Fig. 7. The result of SOC estimation (5A→10A charge)

Where, "$I_{(n-1)}$" express the current just before current change, "$I_{(n)}$" express the current just after current change, and "$t_{(n)}$" express the time at current change. The Fig.6 and Fig.7 show the result of 5A-10A step-wise variation of the current. The error of SOC is less than 0.5% even at the instant of current change. This result shows that the equivalent circuit as shown in Fig.3 can be applied to current variation condition.

1774

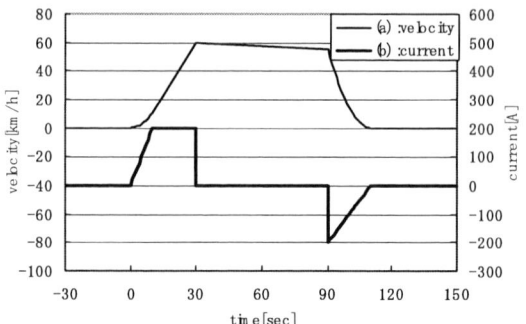

Fig. 8. Test data of actual wireless tram car

Fig. 9. The battery used in the experiment

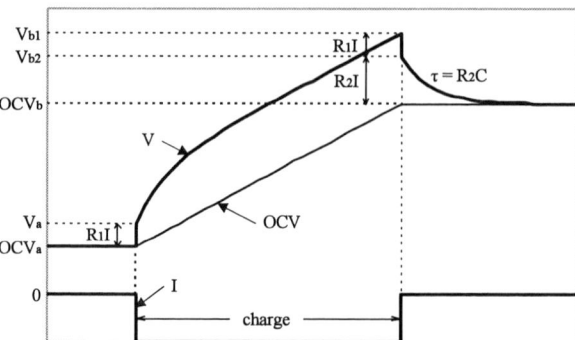

Fig. 10. Voltage variation at charging

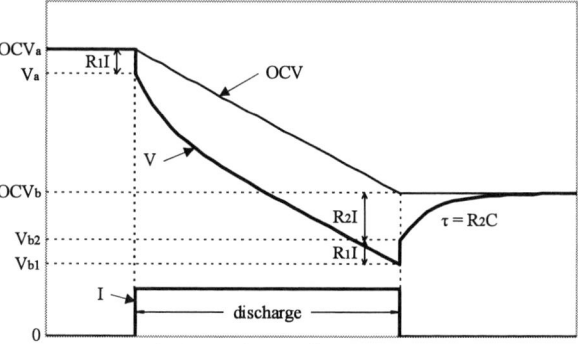

Fig. 11. Voltage variation at discharging

TABLE I
CONSTANTS OF EQUIVALENT CIRCUIT

$R_1 = 1.2\,[\text{m}\Omega]$
$R_2 = 2.3\exp(0.08I) + 0.8\,[\text{m}\Omega]$ (charge)
$\quad = 2.0\exp(-0.04I) + 1.1\,[\text{m}\Omega]$ (discharge)
$C = 55\,[\text{kF}]$

The actual operation of tram is represented by run-curve of the tram car as shown in Fig.8. The run-curve is time variation of velocity, shown by line (a) in Fig.8. The current at this time is shown by line (b). The complicated variation of current can be expressed by the summation of step-wise variation. So we verify the proposed method in the run-curve experiment.

III. EXPERIMENT

The battery used in the experiment is the lithium-ion battery as shown in Fig.9. This battery is made for electric vehicle (EV) and hybrid electric vehicle (HEV). The nominal voltage is 3.8V, and nominal capacity is 30Ah. The actual tram car uses the several modules of this battery in series.

Firstly, in order to determine the equivalent circuit, the battery charges and discharges at the constant current condition. The voltage variation at charging is shown in Fig.10. R_1 is determined by the voltage drop at current change. R_1 can calculate by equation (3).

$$R_1 = \frac{OCV_a - V_a}{I} = \frac{V_{b2} - V_{b1}}{I} \tag{3}$$

R_1 can be also calculated at the end of charging. R_2 is determined by the voltage difference of the transient phenomena has finished. R_2 can calculate by equation (4).

$$R_2 = \frac{OCV_b - V_{b2}}{I} \tag{4}$$

C is determined by the time constant at transient phenomena. The time constant is expressed by R_2C. C can calculate by equation (5).

$$C = \frac{\tau}{R_2} \tag{5}$$

The equation (3)-(5) can also calculate in the case of discharge, as shown in Fig.11. The constants of equivalent circuit of this battery are shown in Table. I. The value of R_2 depends on the battery current.

Next, the battery current variation is calculated from the actual run-curve of the tram, as shown in Fig.12, which simulates acceleration, coasting, deceleration, and stopping. In the experiment, the current is 1/10 of actual tram shown in Fig.8. Also, the time is 10/1 of actual tram. Because, there is the limitation of the experimental utensils.

The battery voltage variation by the current change shown in Fig.12 is measured. The experimental result of voltage variation is shown in Fig.13. In Fig.13, the line (a) is the measurement value of battery terminal voltage. The line (b) is the real OCV that is determined by the measurement of OCV before and after the experiment. The line (c) is the estimated OCV that is calculated by equation (2). The result show that estimated OCV traces real OCV.

The result of SOC estimation during the operation is shown in Fig.14. The maximum error between real SOC and estimated SOC is 1.0%. SOC can be estimated the error of less than 5% even at the transitional part.

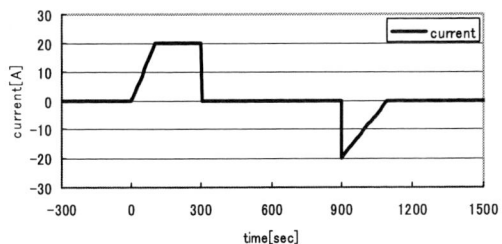

Fig. 12. Current simulation

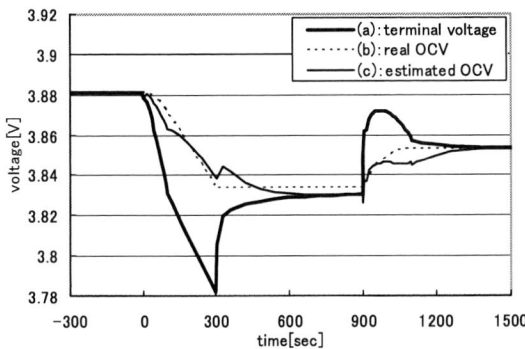

Fig. 13. Experimental data (voltage profile)

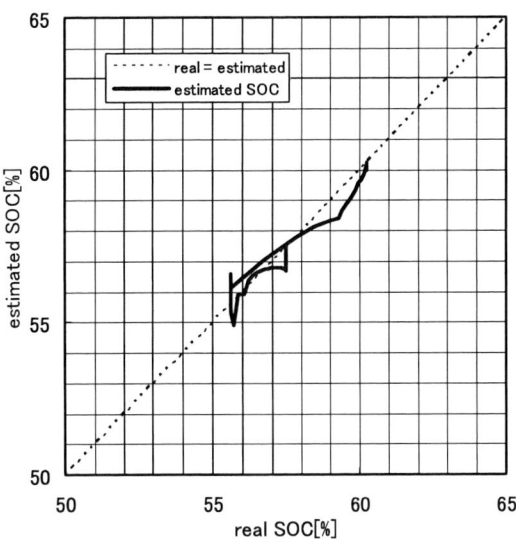

Fig. 14. Experimental data (SOC estimation)

IV. CONCLUSIONS

This paper presented the method of on-line SOC estimation for wireless tram car during the operation. As a result, the estimation error of SOC is less than 5% in the running condition. This result shows that the proposed method can be applied to actual wireless tram car.

REFERENCES

[1] "Battery-powered Tram Developed by RTRI", Japan Railway & Transport Review No.38, www.jrtr.net
[2] "World's First Hybrid Railcar", Japan Railway & Transport Review No.36, www.jrtr.net
[3] L. Gao, S. Liu, R. A. Dougal, "Dynamic lithium-ion battery model for system simulation", *IEEE Transactions on Components and Packaging Technologies*, Vol.25, No.3, pp.495-505, 2002.
[4] Hiroyuki Miyamoto, Hitoshi Takakura, Taku Aisaka, Yukio Sato, Masayuki Morimoto, "On-line SOC Estimation for the Wireless Tram System", *IEEJ National Convention 2007*, 5-154, 2007. (In Japanese)

Narrow- control-bandwidth Operation of Piezoelectric-transformer Converter

Weiping Zhang, Xiaoqiang Zhang, Yuzhou Lei, and Yuanchao Liu

College of Information Engineering, North China Univ. of Tech., Beijing, 100041, P.R.China
Tel. (Fax): 86-010-88802880, Email: zwp@ncut.edu.cn

Abstract-- According to the operating and efficiency properties of piezoelectric-transformer, an input matching network (abbrev. IMN) attached to it is proposed in this paper. The main contributions are as follows: (1) The operating principles of Piezoelectric-transformer are analyzed and its IMN is put forward. This enable Piezoelectric-transformer converter operating area close to the resonant frequency as well as the control bandwidth narrow enough to improve the efficiency. (2)Analyses the electric property of IMN, and draw a conclusion, in agreement with analyze. (3)The simulation is achieved by PSPICE, and the results show that operating bandwidth of the entire resonant network, including IMN and PT, are much narrower than that without IMN but only Piezoelectric-transformer.

Index Terms--Input matching networks , narrow-control-bandwidth, piezoelectric-transformer converter.

I. INTRODUCTION

Piezoelectric-transformer (abbrev PT) has been developed and been used in many applications in the past few decades [1~3]. The equivalent circuit of PT is similar as an LCC resonate tank. It has high voltage gain, and little circulating current when the operational frequency is closed to the resonate frequency. The property of PT's voltage gain vs. frequency is shown in Fig.1. If the range of the operational frequency is very narrow and the operational frequency is very close to the resonate frequency, the slop of the voltage gain is higher. That is to say, the output voltage can be easily regulated through variable frequency. This converter, being sensitive to line and load variation, usually require a wide –frequency range to achieve output regulation. This wide –frequency range results in complicated EMI filter design and poor utilization of magnetic components (or PT). The curve of PT's efficiency vs. frequency is shown in Fig. 2. To alleviate the design difficulty induced by the variable-frequency control and to optimize the utilization of PT, new circuit topologies and control concept, enabling PT converter operation at a narrow-control- bandwidth, has been proposed in this paper. If the operational frequency works in a narrow bandwidth, the EMI filter can be easily designed to meet the requirement of IEC standard.

According to above analysis, there are a lot of advantages if the operational frequency works in a narrow bandwidth for PT converter. However, when the line voltage changes from high (260Vrms) to low (90Vrams) and the load changes varies from full (100%) to light (1%) load, the very extent range of operational frequency is required to regulate the output voltage. In this paper, we are going to propose a new topology, which can regulate the output voltage during narrow-bandwidth.

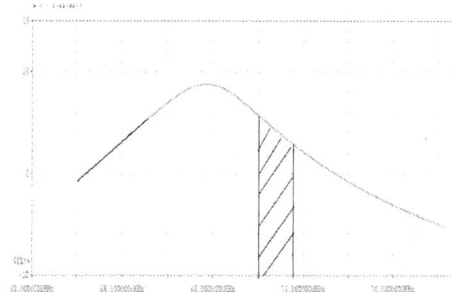

Fig.1 Voltage gain vs. frequency of PT

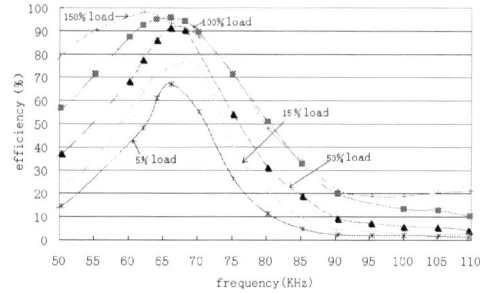

Fig.2 The efficiency vs. frequency of PT

II. THE ELECTRIC PROPERTY OF TYPICAL PT-DC/DC CONVERTER

In order to study the electric properties of PT, we chose a typical PT (output power 50W), which is produced by some Chinese company. The equivalent circuit [4] has been shown in Fig.3, and its parameters are shown in Table I.

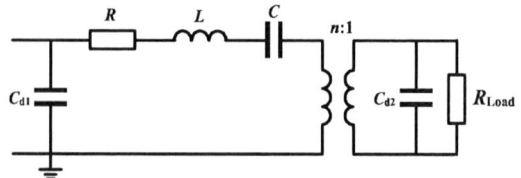

Fig.3 The equivalent circuit of typical PT

TABLE I PARAMETERS OF PT

R(Ω)	L(mH)	C(nF)	Cd2(uF)	n	Cd1(nF)
4.3	3.12	2.145	0.226	5	10.94

Project supported by Natural Science foundation of China(N0.50477054)
Project supported by Beijing Natural Science foundation of China (No. 4052011)

978-1-4244-0644-9/07/$25.00 ©2007 IEEE

A. The voltage gain of PT

The equivalent circuit of PT can be considered as a resonant tank, which has a similar electric performance to LCC—series-parallel resonant tank. The electric property of a LCC resonant tank can be represented by the peak values of voltage gain, resonant frequency and the characteristic factor.

To simplify the analysis, an analytic model without transformer shown in Fig.4, is used to replace the equivalent circuit of PT. The analytic model can be simplified a series equivalent circuit of PT shown in Fig.5. Its parameters, shown in Table I, were applied to simulate the electric properties by PSPICE, while the load of the PT is that $R_L=370$ohms (Ro=3Ω), 24.67kohms (Ro=200Ω). The simulation results are shown in Fig.6.

To simplify the analysis of PT's electrical characteristics, PT has no internal loss because $R<<R_L$, and the input capacitor has useless for PT's voltage gain M_{PT}. Under the above assumption, the equivalent circuit for PT becomes a LCC resonant network, and the voltage gain is given by equation (1)

$$M_{PT} = \frac{\omega_n Q_p (1+1/C_n)}{\sqrt{\left(1-\omega_n^2 \frac{1+C_n}{C_n}\right)^2 + \omega_n^2 Q_p^2 \frac{(1+C_n)^4}{C_n^2}(1-\omega_n^2)^2}} \quad (1)$$

$C_{eq} = C \cdot C_p / C + C_p$, $C_n = C_p / C$, $\omega_n = \omega_s / \omega_o$

$\omega_o = 1/\sqrt{L \cdot C_{eq}}$, $Z_o = \sqrt{L/C_{eq}}$, $Q_p = R_L / Z_o$

If PT is operated in a narrow bandwidth and the center operating frequency ω_S is very close to the resonant frequency ω_O, the equation (1) can be rewritten as

$$M_{PT} \approx \omega_n Q_p (1+1/C_n) / \left| \left(1-\omega_n^2 \frac{1+C_n}{C_n}\right)\right| \quad (2)$$

If selection of the operational frequency ω_S is great than the resonant frequency ω_O, the slope of voltage gain is as

$$\frac{dM_{PT}}{d\omega_n} = \frac{-Q_p \cdot (1+C_n) \cdot \left(C_n + \omega_n^2(1+C_n)\right)}{\left(\omega_n^2(1+C_n)-C_n\right)^2} \quad (3)$$

It can be seen from above equation that the minimum slope occurs in the case of full load.

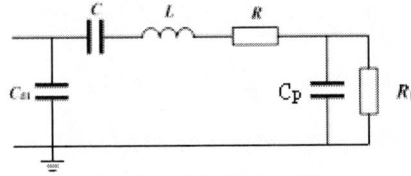

Fig.4 Analytic Model of PT

Fig.5 Series equivalent circuit of PT

When ω_n is greater than unit and very closed to unit, the voltage gain and the slop become as

$$M_{PT} \approx Q_p(1+C_n)$$

$$dM_{PT}/d\omega_n \approx -Q_p \cdot (1+C_n) \cdot (1+2C_n) = -M_{PT}(1+2C_n)$$

It can be seen that the input resistance R_{L1} is in inverse proportion to PT's ac load R_{LOAD}. For example, when $f_s=68$kHz, $C_{d2}=0.226\mu F$, (DC load $R_o=3$ohms, 6ohms, 200ohms), R_L is 370ohms, 740ohms and 24.67 kiloohms, the voltage gain is 1.436,2.873, and 95.754 respectively, the slope is –13.543, -27.086, and –902.854.

B. The input impedance of PT

The input impedance of PT play important role in design of input match network. Changing the parallel branch consisted of Cp and R_L (shown in Fig.4) to a series branch, a series equivalent circuit is shown in Fig.5

The series equivalent resistance,

$$R_L' = n^2 \cdot R_{Load} / (1+q_p^2) \quad (4)$$

Where, $q_p = \omega_s R_{Load} C_{d2}$, If $q_p^2 >> 1$, the equation (4) can be simplified as

$$R_L' \approx n^2 / \omega_s^2 C_{d2}^2 R_{Load} \quad (5)$$

If the operational frequency ωS is very close to the resonant frequency ωo, the input resistance is the load of input match network R_{L1},

$$R_{L1} = n^2 / R_{Load} \cdot \omega_s^2 \cdot C_{d2}^2 + R \quad (6)$$

It can be seen that the input resistance RL1 is in inverse proportion to PT's ac load R_{Load}. For example, when $f_s=68$ kHz, $C_{d2}=0.226\mu F$, (DC load Ro=3ohms, 6ohms, 200ohms), R_{Load} is 14.804ohms, 29.609ohms and 986.96 ohms, the input impedance $R_{L1}$is 125.906ohms, 84.986ohms, 7.016ohms.

The series equivalent capacitor,

$$C_p' = \frac{1+q_p^2}{q_p^2} \cdot \frac{C_{d2}}{n^2} = \frac{1+q_p^2}{q_p^2} C_p \quad (7)$$

If $q_p^2 >> 1$, the equation (7) can be simplified as

$$C_p' \approx C_p$$

C. The simulation results

The analytic model without transformer, shown in Fig.4, is used to replace the equivalent circuit of PT. Its parameters shown in Table I were applied to simulate the electric properties by PSPICE, while the load of the PT is that RL=370ohms (Ro=3Ω)、24.67kohms (Ro=200Ω). The simulative results are shown in Fig.6.

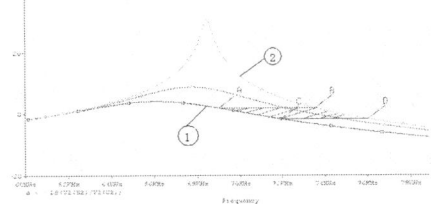

Fig.6 Voltage gain of PT

In Fig.6, the curve ① and curve ② represent respectively the voltage gain with heavy load and light load. For a DC/DC converter, the output voltage Uo is fixed, the range of input voltage is from Uimin to Uimax for the line voltage fluctuated range. In order to regulate the output voltage, the voltage gain should be required to change from maximum voltage gain M (max) to minimum voltage gain M (min). In Fig.6, the straight line AB corresponds to M (max) and the straight line CD corresponds to M (min). The operational area, shown by the shadow part, is made of the curve ① , curve ②, the straight line AB and straight line CD.

The maximum voltage gain M (max) to minimum voltage gain M (min) can be calculated by the following formulas,

$$M(\max) = U_o / U_{i\min} \qquad M(\min) = U_o / U_{i\max}$$

If PT works in the operational area (shadow part) shown in Fig.6, like as LCC resonate converter, there are following intrinsic problems of PT converter operating in the frequency control:

(1).At full-load and low-line voltage (A point in Fig.6), the control bandwidth drops significantly. This phenomenon can be explained by the fact that the low-frequency gain of the control-to-output transfer function, which is determined by line voltage and slope of the curve (1), is much lower at low-line full load case than that is under other operating conditions. This gain variation range can be reduced if the operating region is selected far away from the resonant peak, but that selection will cause more reactive power flowing in and out of the resonant tank, and will increase the loss of PT's internal resistor R, shown in Fig. 3. That is not beneficial for PT's efficiency.

The voltage gain characteristic will be very flat, to regulate the output voltage, the switching frequency will be very high, which is not practical, because the range of the operational frequency is very wide.

(2).This operational area maybe are far away from the resonate frequency, which will hurt PT efficiency shown in Fig.2 very much.

(3).This operational area shown in Fig.6 cannot cover the line variation range and the load variation range.

In addition, PT is a passive device and a DC/AC converter is usually required to run it. In the equivalent circuit of PT, shown in Fig.3, there is large capacitor in its input port, which may cause many problems when applying it to implement a converter or an inverter.

Based on the above discussion, it is obvious that some additional network is required so that the selection operating area is very close to the resonate frequency and the control bandwidth is very narrow to improve the efficiency.

III. THE PURPOSES OF INPUT MATCHING NETWORK

An input matching networks (abbrev. IMN) is defined as the interface circuit between the output port of a DC/AC converter and the input port of PT. The purposes of using IMN are as the follows:

(1).The operating area of PT is closed to PT's resonate frequency to reduce the reactive power flow in the PT and to improve the entire efficiency to make sure that PT runs in the whole operating area in higher efficiency.

(2).The control bandwidth will be narrow in order to alleviate the design difficulty induced by the wide variable-frequency control, to optimal utilization of PT and magnetic components in IMN and to attenuate the high-order harmonics to eliminate the noise among its load.

(3).The operating area of whole resonate tank, including IMN and PT, should cover the line fluctuating range and the load variation range.

(4).The input impendence of IMN is inductive to make sure ZVS to be achieved among the interest rang of the load variation to decrease the switching loss and noise in DC/DC converter.

Based on above requirements, a LCC resonant network is selected as IMN, shown in Fig.7. In this circuit, C_{d1} is the input capacitor; R_{L1} is the input resistance that is determined by equation (6) and the input resistance R_{L1} is in inverse proportion to PT's AC load R_{LOAD}. If PT's AC loads R_{LOAD} is 14.804ohms, 29.609ohms and 986.96 ohms, the input impedance R_{L1} is 125.906ohms, 84.986ohms, and 7.016ohms. When PT has the full load and the operating frequency is much closed to the resonant frequency of PT, the minimum voltage gain and the slop occur, however, the maximum voltage gain and the slop for IMN is achieved because its load R_{L1} is the maximum, vice versa. So, the voltage gains and slops for IMN and PT complement each other. It is verified that LCC-IMN is reasonable. If IMN is designed properly, for full load, the entire slope of voltage gain, including the IMN and PT, could be increased and for light load, the entire slope of voltage gain could be decreased. So the operating frequency range could be decreased to implement narrow bandwidth operating for PT converter.

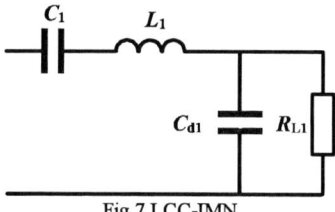

Fig.7 LCC-IMN

IV. ANALYSIS OF ENTIRE RESONANT NETWORK

Fig.8 shows entire resonant network, including IMN and PT. Analysis is carried out under the following assumptions:

(1).The IMN and PT have almost same resonant frequency, $\omega_{01} = \omega_{02}$.

(2).The entire resonant is operated in a narrow bandwidth and the center operating frequency ω_S is much closed to the resonant frequency.

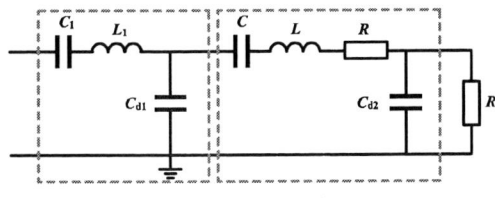

Fig. 8 The entire resonant network

Based on above assumptions, each of voltage gain and slop could be calculated by the formulas in section II when $\omega_S = \omega_0$.

According to equation (1), we have the following formulas:

The voltage gain of IMN,

$$M_{IMN} \approx Q_{p1}(1+C_{n1}) \qquad (8)$$

$$C_{eq1} = C_1 \cdot C_{d1}/(C_1 + C_{d1}) \quad , \quad C_n = C_{d1}/C_1 ,$$

$$\omega_{o1} = 1/\sqrt{L_1 \cdot C_{eq1}} \quad , \qquad Z_{o1} = \sqrt{L_1/C_{eq1}} ,$$

$$, Q_{p1} = R_{L1}/Z_{o1} \quad R_{L1} = R_L/(1+R_L^2 \cdot \omega_s^2 \cdot C_{d2}^2) + R$$

The voltage gain of PT,

$$M_{PT} \approx Q_{p2}(1+C_{n2}) \qquad (9)$$

$$C_{eq2} = C \cdot C_p/(C+C_p) \quad , \qquad C_n = C_p/C \quad ,$$

$$\omega_{o2} = 1/\sqrt{L \cdot C_{eq2}} \quad , \qquad Z_{o2} = \sqrt{L/C_{eq2}} \quad ,$$

$$Q_{p2} = R_L/Z_{o2}$$

Entire voltage gain,

$$M = M_{IMN} \cdot M_{PT} = Q_{p1}Q_{p2}(1+C_{n1})(1+C_{n2})$$

$$= \left(\frac{R_L^2}{1+\omega_s^2 R_L^2 C_p^2} + RR_L\right) \cdot \frac{(1+C_{n1})(1+C_{n2})}{Z_{o1}Z_{o2}} \qquad (10)$$

If $\dfrac{R_L^2}{1+\omega_s^2 R_L^2 C_p^2} \gg RR_L$ and $\omega_s^2 R_L^2 C_p^2 \gg 1$, then

$$M \approx \frac{1}{\omega_s^2 C_p^2} \cdot \frac{(1+C_{n1})(1+C_{n2})}{Z_{o1}Z_{o2}} \qquad (11)$$

The entire voltage gain has no relationship with load R_L. If the following condition is satisfied,

$$\frac{R_{Lmin}}{(1+\omega_s^2 R_{Lmin}^2 C_p^2)} > \frac{Z_{01}}{(1+C_{n1})} > \frac{R_{Lmax}}{(1+\omega_s^2 R_{Lmax}^2 C_p^2)} \qquad (12)$$

Then, for full load, $R_L = R_{Lmin}$, $M_{IMN} > 1$, $M > M_{PT}$, the entire voltage gain could be increased; for light load, $R_L = R_{Lmax}$, $M_{IMN} < 1$, $M < M_{PT}$, the entire voltage gain could be decreased. Therefore, adding an IMN can reduce the operational frequency range to regulate output voltage.

According to equation (3), the slop of IMN's voltage gain is

$$\left.\frac{dM_{IMN}}{d\omega_n}\right|_{\omega_n \approx 1} = -M_{IMN}(1+2C_{n1}) \qquad (13)$$

The slop of PT's voltage gain,

$$\left.\frac{dM_{PT}}{d\omega_n}\right|_{\omega_n \approx 1} = -M_{PT}(1+2C_{n2}) \qquad (14)$$

The slop of entire voltage gain,

$$\left.\frac{dM}{d\omega_n}\right|_{\omega_n \approx 1} = M_{IMN} \cdot \left(1+\frac{1+2C_{n1}}{1+2C_{n2}}\right) \cdot \frac{dM_{PT}}{d\omega_n} \qquad (15)$$

For full load, $M_{IMN} > 1$ then $\left|\dfrac{dM}{d\omega_n}\right| > \left|\dfrac{dM_{PT}}{d\omega_n}\right|$.

In the case of full load, compared with the slop of PT's voltage gain, the slop of entire voltage gain becomes steeper, which is useful to reduce the operational frequency range.

If the operational frequency bandwidth of PT is $\Delta\omega_{PT}$, the bandwidth of entire resonant network is given by

$$\Delta\omega = \Delta\omega_{PT} / \left\{ M_{IMN} \cdot \left(1+\frac{(1+2C_{n1})}{(1+2C_{n2})}\right) \right\} \qquad (16)$$

The operating bandwidth will become narrow.

V. SIMULATION RESULTS

Fig.8 shows entire resonant network, including IMN and PT. The simulation results are shown in Fig.9, and the load of the PT is that R_L=370ohms, 24.67kohms. In Fig.9 the curve ② represents the voltage gain of entire resonant network with heavy load and light load, and the curve ① is the voltage gain of PT without IMN. The simulation results shows the operating area of whole resonate tank, including IMN and PT becomes A'B'C'D', and operating area of PT without IMN is ABCD. So operating bandwidth of PT reduces about 40%, which is from 67.5 kHz to 71 kHz instead of 67 kHz to 73 kHz.

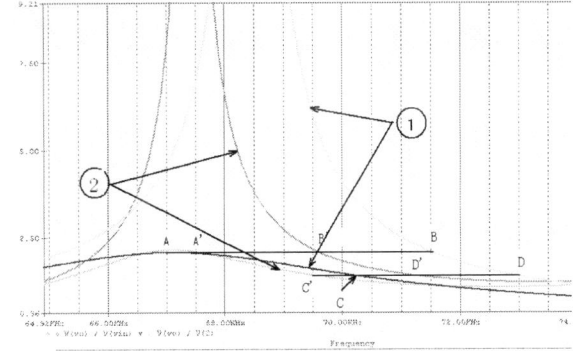

Fig. 9 Voltage gain of entire resonant network

VI. CONCLUSIONS

This paper analyses the electric property of typical PT-DC/DC converter and proposes a new circuit topologies and control methods to resolve the problems of PT converter operating in the frequency control. Based on the theoretical analysis, simulation results are provided. Some conclusions can be made as follows:(1) It is important to add an input matching network for PT converter to operate at a narrow-control-bandwidth.(2) The formulas of calculating $\Delta\omega$ has been proved, given out in the formula [16].(3) The simulation results show that the operating bandwidth of the entire resonant network, including IMN and PT, is much narrow than that without IMN but only with PT converter. This is consistent with the theoretic results.

REFERENCES

[1] Takashi Yamane, Sunao Hamamura, Toshiyuki Zaitsu, "Efficiency Improvement of Piezoelectric-transformer DC/DC Converter," *PESC' 98, 29th Annual, Vol.2*, pp. 1504~1510

[2] Ray L. Lin, Fred C. Lee, Eric M. Baker, and Dan Y. Chen, "Inductorless Piezoelectric Transformer Ballast for linear Fluorescent Lamps," *Proceedings of CPES Power Electronics Seminar , Sept.17~19, 2000*,pp.309~314.

[3] Eric M. Baker, Weixing Huang, Dan Y. Chen, and Fred C. Lee, "Radial Mode Piezoelectric Transformer Design for Fluorescent lamp Ballast Applications," *Proceedings of CPES Power Electronics Seminar, April.23~25, 2001*, pp.105~112.

[4] C.Y.Lin, "Design and analysis of Piezoelectric Transformer Converter", Ph.D. dissertation, Virginia Tech., July 1997.

Modified Map of Variable Active Passive Reactance for Stability Evaluation with Consideration of Capacitor Mode

S. Mohammad Shariatmadar*, Jalal Nazarzadeh**

*Lecturer of Azad University -Branch of Naragh, Naragh, Iran
PhD Student of Azad University-Branch of Science & Research, Tehran, Iran,
Shariatmadar@iau-naragh.ac.ir
** PhD of Electrical Engineering, Shahed University, Tehran, Iran,
Nazarzadeh@shahed.ac.ir

Abstract--Variable active passive reactance (VAPAR) can generate virtual variable inductance in power circuits .There are two poincare maps for studying of stability in the switching systems. One of them is first order poincare map. It is suitable for the systems with one fixed point on every switching period however; if the system has low periodic response in steady state the first order poincare map can not analyze behavior of the systems. In this case modified poincare map can determine stability of the systems. In this paper VAPAR stability with ac source include capacitor mode by modified poincare map is investigated. The stability region for period of time varying system can be found by eigenvalues loci of modified poincare map. For computing of jacobian matrix, the fixed point on each period is calculated by averaging theory.

Index Terms-- Modified Poincare Map, Second Order Poincare Map, VAPAR, Variable Reactance.

I. INTRODUCTION

Nowadays converters are widely used for traction drivers, telecommunications and etc. Inverters are one of the most important converters that used in many applications such as an active filter, static var compensator and etc. A variable active passive reactance (VAPAR) is a power circuit that generates virtual variable inductance including negative values and consists of inverter [1].

It is very important to examine the stability of VAPAR when it is employed in RL configuration. Two main techniques are available for more accurate approach: averaging and poincare map modeling [2].The state space averaging method formulates the dynamic equations in state space form for each of topological model.The low frequency properties retained i.e. it will only be able to predict low frequency behavior. More complete information can be achieved by sample data modeling using poincare map [2]. It is able to predict standard bifurcation and instability and will be used in this study.

Unlike a dc-dc converter, modified poincare map may be necessary for the stability analysis of VAPAR. In the steady-state condition the output of first order poincare map of VAPAR with ac source is time varying. However a modified poincare map converts the problem

from one of analyzing the stability of an orbit to that of analyzing the stability of a fixed point [3]. In [4] stability analysis of VAPAR is investigated in case of dc source and battery for inverter instead of capacitor. In this case, the system had only one fixed point and first order poincare map is constructed.

In this paper, first the configuration and operation of VAPAR is reviewed and after that modified poincare map is developed for VAPAR system. Also jacobian matrix belonging to its fixed trajectory is determined. Finally, characteristic multipliers for stability region are calculated.

II. SYSTEM DESCRIPTION

Schematic model of VAPAR is shown in Fig.1. In practical, if the terminal voltage V_t and current i_t are defined by the following:

$$V_t = L_v \frac{di_t}{dt} \tag{1}$$

then VAPAR behaves as a virtual inductance L_v (see Fig.1.b. VAPAR consists of an inductor and a controlled voltage source. Generally, a PI controller is used for adjusting of virtual inductance to L_v . The adjusting error of the system is (see Fig.1.b)

$$e = \phi_{ref} - \phi \tag{2}$$

$$\phi_{ref} = \int V_t dt \tag{3}$$

$$\phi = L_v . i_t \tag{4}$$

where ϕ, L_v and ϕ_{ref} are the virtual flux, virtual inductance and reference flux, respectively [5]. Behavior of VAPAR equals to virtual inductance L_v if error in equation (2) follow to zero. It can provide not only positive virtual inductance values, but also negative ones, that used to cancel existing series undesired inductances (note that in this case, i.e., for $L_v < 0$, the reference flux ϕ_{ref} will change its sign, so a negative feedback is maintained).

The dc-bus voltage of an inverter has great influence on its operation . In the case when a capacitor is

978-1-4244-0644-9/07/$25.00 ©2007 IEEE

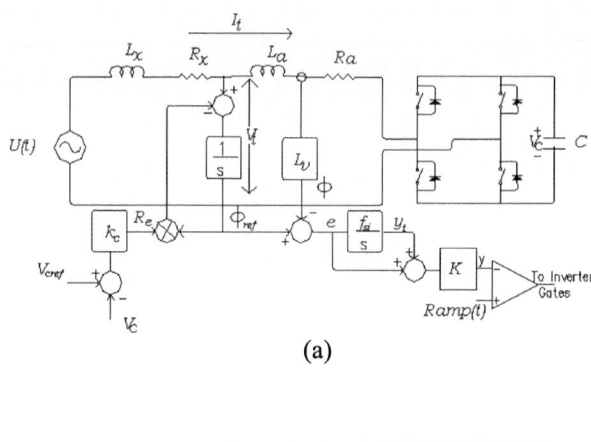

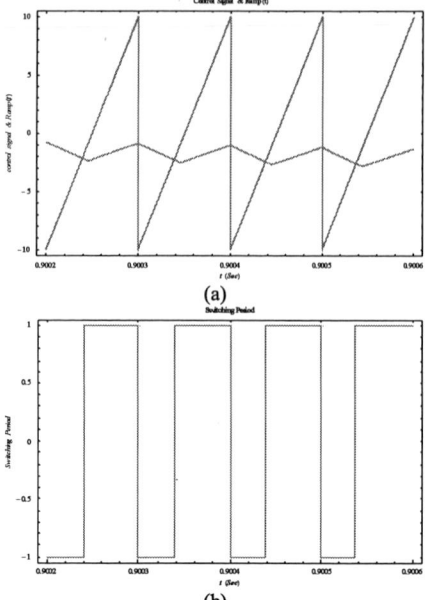

(a)

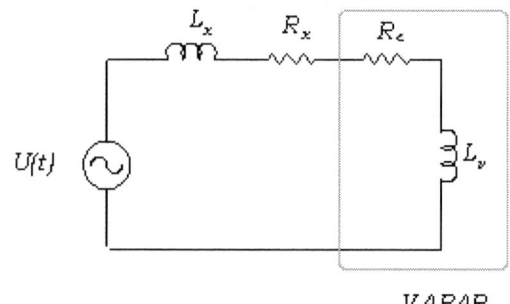

(b)

Fig.1. (a). VAPAR close loop system topology (b). Principle of voltage control.

connected to the inverter dc bus, as shown in Fig. 1(a), the capacitor voltage should first be established to the reference voltage and second be stabilized. The capacitor voltage control with the following specifications, is required for establishment and stabilization of the capacitor voltage.

1) At establishment of the capacitor voltage.

a) The capacitor voltage must be established from zero to the reference voltage.

2) After establishment.

a) In a practical circuit, losses always exist such as an inductor-conduction loss, device-switching loss, etc. Therefore, the capacitor energy is consumed in these loss components and the capacitor voltage decreases gradually. The capacitor voltage must be maintained to be over the minimum level by the control.

b) The capacitor-voltage control must not degrade the inherent characteristics of VAPAR as an energy storage element.

In order to regulate the capacitor voltage, VAPAR is controlled to generate an RL series impedance as shown in Fig.1.b Inverter loss component can be represented with an equivalent resistance R_e. In the proposed method R_e is controlled in proportion to error between reference voltage capacitor V_{cref} and actual capacitor voltage V_c

$$R_e = k_c(V_{cref} - V_c) \qquad (5)$$

where k_c and f_{si} are a proportional capacitor voltage

Fig.2 (a)PWM frequency , control signal (b) switching period

loop gain and main integrating controller gain, respectively [6].

Present study is concerned with the stability of this virtual inductance when employed in a series connection with an external inductance L_x and ac source supplied VAPAR with capacitor mode.

III. ANALYTICAL MAP OF VAPAR

The system investigated is variable-structure piecewise linear. The structure of the active circuit varies during the operation and is depicted in Fig.1.a (the state variable y_t of the PI controller is also pointed out). Under the constant frequency PWM switching illustrated in Fig.2, the switching function $q(t)$ represents the sign of the difference between the ramp waveform and the signal $y(t)$. The duty ratio $0 < dn < 1$ of the nth cycle is determined by the switching instant t_s when they cross within that cycle.

Consequently, two periodically toggling linear circuits can be identified during each cycle (Fig.2). The Poincaré map is determined by solving their state equations:

There are two switching state for VAPAR system, because all of switches in the inverter are on or off.

So, the dynamical equation of the VAPAR can be described as

$$X'(t) = A_1 . X(t) + B.U(t)$$
$$X'(t) = A_2 . X(t) + B.U(t) \qquad (6)$$

where A_1, A_2 and B are state matrices and input matrix

$$A_1 = \begin{pmatrix} -\dfrac{R_a + R_x}{L_a + L_x} & 0 & 0 & \dfrac{L_v}{L_a + L_x} \\ \dfrac{L_x R_a - L_a R_x}{L_v(L_a + L_x)} & -(V_{cref} - V_c)k_c & 0 & -\dfrac{L_x}{L_a + L_x} \\ -fsi & fsi & 0 & 0 \\ \dfrac{-1}{L_v C} & 0 & 0 & 0 \end{pmatrix}$$

$$A_2 = \begin{pmatrix} -\dfrac{R_a + R_x}{L_a + L_x} & 0 & 0 & -\dfrac{L_v}{L_a + L_x} \\ \dfrac{L_x R_a - L_a R_x}{L_v(L_a + L_x)} & -(V_{cref} - V_c)k_c & 0 & \dfrac{L_x}{L_a + L_x} \\ -fsi & fsi & 0 & 0 \\ \dfrac{1}{L_v C} & 0 & 0 & 0 \end{pmatrix}$$

$$\qquad (7)$$

$$B = \left(\dfrac{V_m L_v}{L_a + L_x} \quad \dfrac{V_m L_v}{L_a + L_x} \quad 0 \quad 0 \right)^T \qquad (8)$$

where $R_a, R_x, L_a, L_x, V_{cref}, V_c, k_c$ and f_{si} are VAPAR resistance, load resistance, series inductance of VAPAR , inductance of load , capacitor reference voltage, capacitor voltage , capacitor control loop feedback and gain of integrator controller respectively.

Also $X(t)$ and $U(t)$ represent the state of the system and time varying input of system

$$X(t) = (\phi \quad \phi_{ref} \quad y_t \quad V_c)^T \qquad (9)$$

$$U(t) = V_m \sin(\omega t) \qquad (10)$$

where y_t and V_m are output of integrating controller and amplitude of ac source, respectively.

For finding the first order poincare map in the nth switching cycle solution of equation (6) for state vector are

$$X(n+dn) = e^{\int_{nT}^{nT+dnT} A_1 d\tau} X(n) + \int_{nT}^{nT+dnT} e^{\int_{nT}^{nT+dnT} A_1 d\tau} B.U(\tau)d\tau \qquad (11)$$

$$X(n+1) = e^{\int_{nT+dnT}^{nT+T} A_2 d\tau} X(n+dn) + \int_{nT+dnT}^{nT+T} e^{\int_{nT+dnT}^{nT+T} A2 d\tau} B.U(\tau)d\tau \qquad (12)$$

For simplicity assume $e^{At} \simeq I + A\,t$. The poincare map is obtained by solutions

$$X(n+1) = P(X(n), dn) \qquad n=1,2,...,M \qquad (13)$$

if n varied from 1 to M all of source period can consider in the map as (14)

Dc input systems naturally have only one fixed point (X_{ss}) and corresponding duty ratio (d_{ssn}). But in the system with sinusoidal input in every switching X_{ss} and d_{ssn} varied. Depending on the ratio of fs (switching frequency) and f_e (source frequency) the modified poincare map can obtained by cascading f_s/f_e solutions of first order poincare map [3].In this case f_s/f_e is considered an integer.

$$P(X(n), dn) = (I + A_2(1 - dn)T)[(I + A_1 dnT)X(n)$$
$$+ B(\dfrac{1}{\omega}\cos(\omega nT) - \dfrac{1}{\omega}\cos(\omega(n + dn)T))$$
$$+ A_1 B(\dfrac{1}{\omega}(dnT \cos(\omega nT)) - \dfrac{1}{\omega^2}(\sin(\omega(nT + dnT))$$
$$- \sin(\omega nT)] + B(\dfrac{1}{\omega}\cos(\omega(n + dn)T)$$
$$- \dfrac{1}{\omega}\cos(\omega(n + 1)T))$$
$$+ A_2 B(\dfrac{1}{\omega}((1 - dn)T \cos(\omega(n + dn)T))$$
$$- \dfrac{1}{\omega^2}(\sin(\omega(nT + T) - \sin(\omega(n + dn)T)) \qquad (14)$$

For finding VAPAR modified map cascade (14) for $f_s/f_e = M$ and modified map is obtained as complicated equation (15).This equation is a function of $X(n)$ and d_n, d_{n+1}, d_{n+2},.... and d_{n+m-1}.

$$X(n + m) = \prod_{i=m}^{1}(I + A_2(1 - d_{n+i-1})T)(I + A_1 d_{n+i-1}T)X(n)$$
$$+ \sum_{i=1}^{m}(\prod_{j=m-1}^{i}[(I + A_2(1 - d_{n+j})T)(I + A_1 d_{n+j}T)$$
$$((I + A_2(1 - d_{n+j-1})T)[\dfrac{B}{\omega}(\cos(\omega(n + i - 1)T)$$
$$- \cos(\omega(n + i - 1 + d_{n+i-1})T)$$
$$+ \dfrac{A_1 B}{\omega}((d_{n+i-1}T)\cos(\omega(n + i - 1)T)$$
$$- \dfrac{1}{\omega}(\sin(\omega(n + i - 1 + d_{n+i-1})T)) -$$
$$\sin(\omega(n + i - 1)T)] + \dfrac{B}{\omega}(\cos(\omega(n + i - 1)T +$$
$$d_{n+i-1}T) - \cos(\omega(n + i)T) + \dfrac{1}{\omega^2}\cos(\omega(n + 1)T)) +$$
$$\dfrac{A_1 B}{\omega}(d_{n+1}T \cos(\omega(n + 1)T) -$$
$$\dfrac{1}{\omega}(\sin(\omega(n + i + d_{n+i})T)) + \sin(\omega(n + i)T)] \qquad (15)$$

Relation between duty ratio d_n and the state vector $X(n)$ at the beginning of that cycle in Fig.1 can be seen as

$$g(X(n), dn) = y_t - l - (H - l)dn = 0 \qquad (16)$$

$$y_t = cX(n + dn) \qquad (17)$$

where $c = K(-1 \quad 1 \quad 1 \quad 0)$ and K, H and l are gain of PI controller, upper and lower limit of ramp function, respectively.

There are many parametric matrices in equation (15) that its linearization is very complicated. Then other methods must be introduced.

IV. LINEARIZATION
Equation (13) can be written for any point with related fixed point. At a fixed point, we get

$$X(n+i) = P(X(n+i-1), d_{n+i-1}) \qquad (18)$$

For small deviation jacobian is

$$\delta X(n+i) = \left[\frac{\partial P}{\partial X(n+i-1)} + \frac{\partial P}{\partial d_{n+i-1}}\frac{\partial d_{n+i-1}}{\partial X(n+i-1)}\right]\delta X(n+i-1) \qquad (19)$$

Equation (18) must be computed for its fixed point (related X_{ss} and d_{ssn}); because of time varying source, fixed points are changed in every switching period. The equation (18) could not be suitable for presenting of the system behavior because, for $f_s/f_e = M$ switching period equation (18) must be computed (with substituting related X_{ss} and d_{ssn} in each point).Then M produced matrices multiple to each other for finding of modified poincare map of the system.

For computing fixed point in each switching period, averaging system must be solved. Eigenvalues of modified poincare map display stability of system around of average trajectory in each period. In fact modified poincare map describes system for whole period. But first order poincare map depends on starting point of map. If system has disturbance in a part of route and first order poincare map starts in this region characteristic multipliers are out of unit circle. And if system in the region of route has normal condition characteristic multipliers of first order poincare map are in the unit circle. For this reason modified poincare map must be used for stability investigation in whole period.

This kind of maps describes all of system manner. In equation (18) we have

$$\frac{\partial P}{\partial X} = (I + A_2(1 - d_{ssn}(n))T)(I + A_1 d_{ssn}(n)T)$$

$$\frac{\partial P}{\partial dn} = (A_1 T - A_2 T + A_1 A_2 T^2 + 2A_1 A_2 d_{ssn}(n)T^2)X_{ss}(n)$$

$$+(-A_2 T)(B(\frac{1}{\omega}\cos(\omega nT) - \frac{1}{\omega}\cos(\omega(n+dn)T)) + \qquad (20)$$

$$A_1 B(\frac{1}{\omega}(dnT\cos(\omega nT)) - \frac{1}{\omega^2}(\sin(\omega(nT+dnT) - $$

$$\sin(\omega nT) + (I + A_2(1 - d_{ssn}(n))T)BT\sin(\omega(nT+dnT) + $$

$$A_1 B(\frac{1}{\omega}\cos(\omega nT) - \frac{T}{\omega}\cos(\omega(n+dn)T)) - BT\sin(\omega(nT+dnT)$$

$$+(\frac{A_2 BT}{\omega}\cos(\omega(n+dn)T) - A_2 BT^2(1 - d_{ssn}(n))\sin(\omega(nT+dnT)$$

$$+\frac{A_2 BT}{\omega}\cos(\omega(n+dn)T))$$

And from algebraic equation (16) for small changes can be derived

$$\frac{\partial dn}{\partial X(n)} = \frac{A}{B}$$

$$if \qquad (21$$

$$A = -c(I + A_1 dssn(n).T$$

$$and$$

$$B = c(A_1 TX_{ss}(n) + TB\sin(\omega(n + dssn(n))T)) +$$

$$A_1 B(\frac{T}{\omega}Cos(\omega nT) - \frac{T}{\omega}Cos(\omega(n+dssn(n)T)) - 20$$

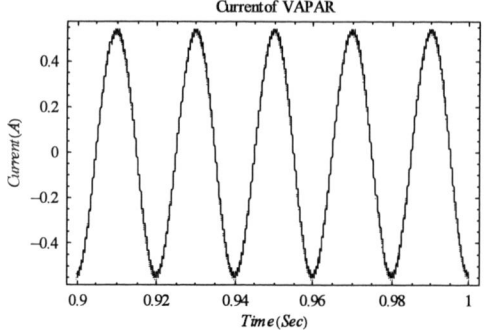

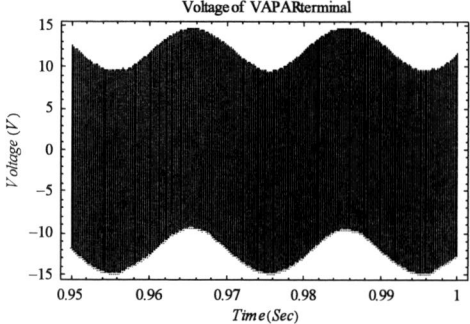

Fig.3.Current and voltage of VAPAR for L_v=0.08 H

For calculation of linear approximation modified poincare map first equations (19), (20) and (21) are substituted into (18) for related M fixed point. After that M resulted map multiple to sequential each other. Finally eigenvalues (characteristic multipliers) of linear approximation of modified poincare map can be investigate VAPAR stability.

V. SIMULATION RESULTS

Robust stability when the virtual inductance (L_v) is varied , can be checked by system eigenvalues of modified poincare map (parameters of VAPAR listed in Table. I).According to the stability condition as long as the characteristic multipliers lie inside the unit circle the periodic steady state solution is asymptotically stable. As the L_v is increased a real characteristic multiplier moves out of unit circle and system is unstable.

In the case study if L_v varied until 91 mH, the system is stable and when this value increases from 91 mH characteristic multipliers moves out of unit circle and system is unstable. In Fig.3 and 4 current and voltage of VAPAR are shown for three cases, 80 mH, 91 mH and 110mH.

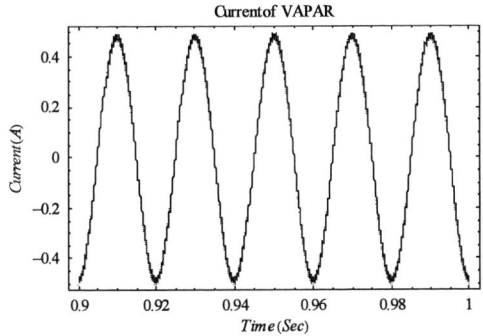

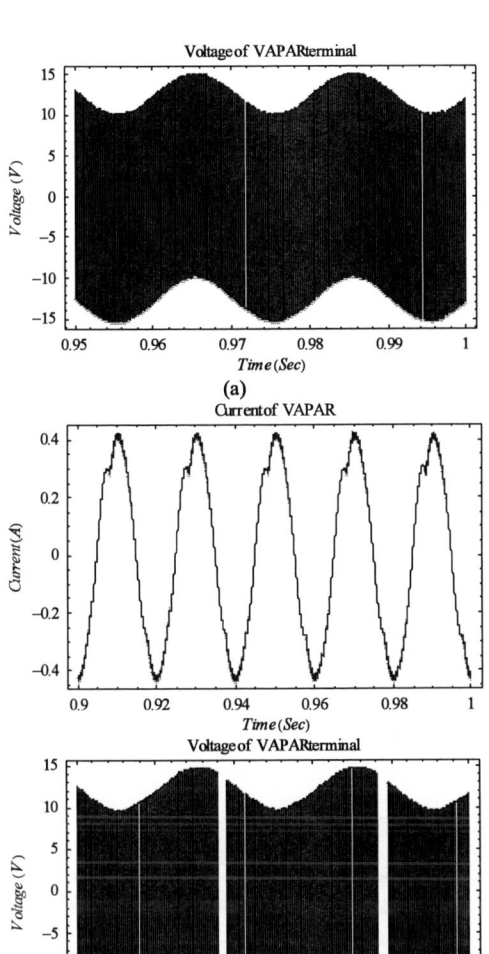

Fig.4.Current and voltage of VAPAR for(a) L_v=0.091 H (b) L_v=0.110 H

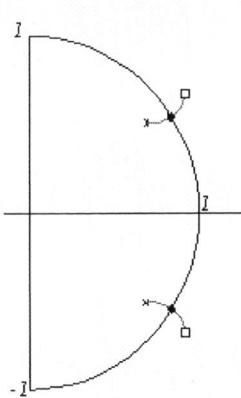

Fig.5. Loci of characteristic multipliers for increasing virtual inductance (×)for L_v=80 mH, (•)for L_v=91 mH and (□) for L_v=110 mH

It can be seen if L_v increase from 91 mH, system was unstable. For inductance 110 mH current is non symmetric and voltage is disturbed. In the critical eigenvalues of linear approximation modified poincare map for L_v = 91 mH is equal 1. In Fig.5 loci for first characteristic multiplier can be seen. Eigenvalues for this problem calculated for n=0 to n=200(from t=0 to t=2π). If calculation started from another point eigenvalues changed and this is because of time varying source. Chaos and bifurcation can be traced with changing n from zero to M in VAPAR. If eigenvalues equal -1, system will be bifurcated. And period-doubling will be seen in VAPAR terminals.

VI. CONCLUSIONS

This paper presents stability investigation of a VAPAR with ac source and capacitor in its inverter. VAPAR can generate virtual impedance in the steady state operation. For stability investigation, modified poincare map model contributed. Average modeling for finding fixed points is considered.

TABLE.I.
PARAMETERS AND VARIABLES VALUES

$U(t)$	$20\ sin(\omega t)$	R_a	0.8 ohm
L_x	30 mH	V_{cref}	18 v
R_x	5 ohm	K	1000
L_a	4.51 mH	T	0.01 μsec
ω	314	f_{si}	1000
K_c	3	H,l	10,-10

REFERENCES

[1] H. Funato and A. Kawamura, "Realization of Negative Inductance Using Variable Active Passive Reactance", *IEEE Transaction On Power Electronics*,Vol. 12, Issue 4,pp.589 - 596, July 1997.

[2] S. Banerjee and G. Verghese, "Nonlinear Phenomena in Power Electronics", *IEEE Press*, New York, April 2001.

[3] S. K. Mazumder, "A Novel Approach to Predict the Instability and Analyze The Dynamics of A Single Phase Bidirectional Boost Converter", *Power Electronics Specialists Conference*, *33rd Annual*, vol.. 4, pp.1711 – 1716, June 2002.

[4] O. Dranga, H. Funato and C. K. Tse, "Stability Analysis of Power Comprising Virtual Inductance", *ISCAS2004*, IEEE 2004.

[5] H. Funato and A. Kawamura, "Proposal of Variable Active Passive Reactance", *IEEE International Conference on Industrial Electronics, Control, Instrumentation, and Automation*,vol.1,pp.381-384,Nov 1994.

[6] H. Funato and A. Kawamura, "Control of Variable Active Passive Reactance and Negative Inductance", *Proc. IEEE PESC '94, Taipei, Taiwan*, pp. 189–196, June 1994.

Design of the Longitudinal Mode Piezoelectric Transformer

Shine-Tzong Ho*

* Mechanical Engineering Department, National Kaohsiung University of Applied Sciences,
415 Chien-Kung road, San-min district, Kaohsiung, Taiwan

Abstract– **The purpose of the present paper is to establish a method of design for a longitudinal mode piezoelectric transformer (PT). This method is based on the electromechanical model of the PT, which is derived by the Rayleigh-Ritz assumed mode energy method and equivalent circuit method. The assumed mode energy method is used to model the distributed piezoceramic structure for estimating the parameters of the equivalent circuit model of the PT. Moreover, the equivalent circuit model is used to estimate the electrical characteristics of the PT. Such as voltage step-up ratio, output power and efficiency can be derived by the equivalent circuit model. Therefore, the performance of the PT can be predicted by the geometrical dimensions and material parameters. This method can be used to make a comparison between various kinds of the design of the longitudinal mode PT. The design principle of a longitudinal mode PT will be discussed based the proposed method.**

Index Terms– **Piezoelectric transformer, Longitudinal mode, Design, Electromechanical model.**

I. INTRODUCTION

The idea of a piezoelectric transformer was first implemented by Rosen in 1956 [1], as shown in Fig.1(a). It used the coupling effect between electrical and mechanical energy of piezoelectric materials. A sinusoidal signal is used to excite mechanical vibrations by the inverse piezoelectric effect via the driver section. Due to the direct piezoelectric effect, an output voltage can be induced in the generator part. The PT offers many advantages over the conventional electromagnetic transformer such as high power-to-volume ratio, electromagnetic field immunity, and nonflammable.

In the literature [2-5], many piezoelectric transformers have been proposed and a few of them found practical applications. For example, a Rosen-type PT has been adopted in cold cathode fluorescent lamp inverters for liquid-crystal display. Sakurai proposed two types of the longitudinal mode PT and studied their power transmission characteristics [2]. The one is a second-order longitudinal mode PT, as shown in Fig.1(c). The other one is a third-order longitudinal mode PT, as shown in Fig.1(d). Kanayama developed a longitudinal mode PT which has an alternately poled structure [3], as shown in Fig.1(e). This type of PT is designed to achieve lower input voltage for specific output power. Also, the PT with multilayer structure to provide high-output power

was studied by Kanayama, which may be used in various kinds of power supply units [4]. In reference [4], the PTs shown in Fig.1(b) and Fig.(f) can also be seen. For convenience, the six types of the longitudinal mode PT, as shown in Fig.1(a)-(f), are called as Type A, B, C, D, E, F, respectively. Furthermore, these types of PT are analyzed and made a comparison in order to understand the design principle for the longitudinal mode PT for various purposes. However, because of the mode coupling effect and the complexity of vibration modes at high frequency, the conventional lumped-equivalent circuit method may not accurately predict the dynamic behaviors of the PT.

In this paper, the electromechanical models for various kinds of the longitudinal mode PT are analyzed. Especially, Type C has been verified by experiments and a very good agreement is obtained [5]. In order to establish the model, vibration characteristics of the

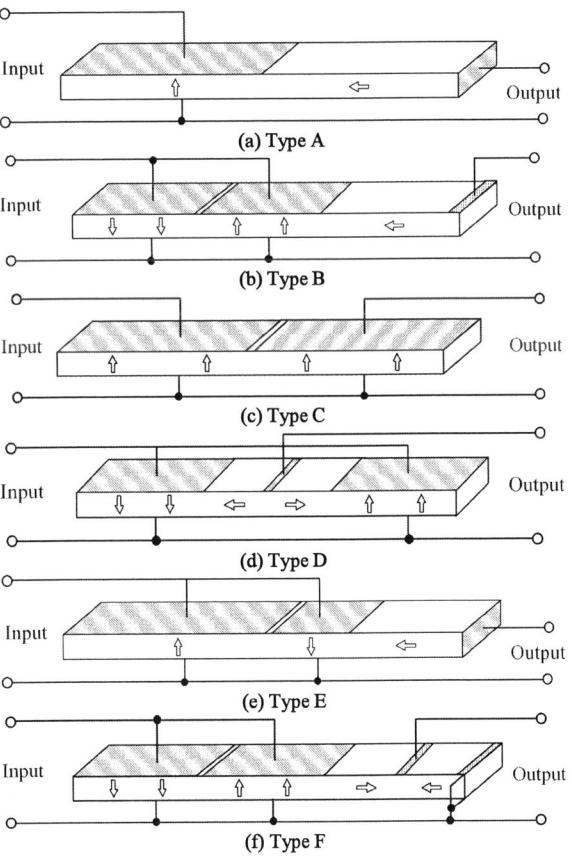

Fig.1 Structure of longitudinal mode piezoelectric transformers.

piezoelectric plate with free boundary conditions are analyzed in advance, and the natural frequencies and mode shapes are obtained. In addition, an equivalent circuit model of the PT is obtained based on the equations of the motion for the coupling electromechanical system. Furthermore, the voltage step-up ratio, output power, and efficiency for various types of PTs will be conducted and compared. Also, the optimal load resistance and the maximum efficiency for the PTs are calculated in this paper.

II. THEORETICAL ANALYSIS

A. Longitudinal Vibration Modes

The longitudinal mode type of PTs, as shown in Fig.1(a)(b)(c)(d)(e)(f), consists of two or three poled piezoceramic beams with equal cross sections rigidly boned together, or a single piezoceramic beam with both ends poled separately. A sinusoidal voltage is supplied to excite mechanical vibrations by the inverse piezoelectric effect via the input part. An output voltage can be induced in the output part due to the direct piezoelectric effect. A mode is excited if the supply frequency coincides with a particular vibration mode, and the frequency of these modes depends on the dimensions of the PT and its material properties. Considering a piezoelectric beam of length L with constant cross-sectional area A, its free vibration equation can be obtained as the following.

$$\frac{\partial^2 u(x,t)}{\partial x^2} = \beta^2 u(x,t) \qquad (1)$$

where $\beta^2 = \rho\omega^2/E$. ρ, ω and E represent density, frequency and Young's modulus, respectively. The solution of Eq.(1) can be written as $u(x,t)=U(x)G(t)$ and the nth longitudinal modes $U_n(x)$ are determined by the free-free ends of boundary conditions, as the following.

$$U_n(x) = \cos\beta_n x \qquad (2)$$

The resonant frequencies are given by

$$\sin\beta_n L = 0, \quad \beta_n = n\pi/L, \quad n=1,2,3\ldots \qquad (3)$$

where n represents the order of the mode. Fig.2 shows the stress distributions of the first three modes, which can be calculated by the differentiation of Eq.(2).

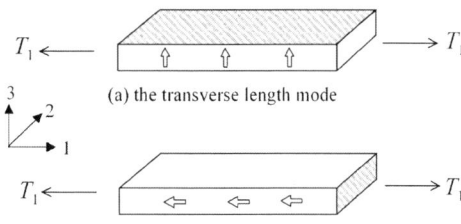

(a) the transverse length mode

(b) the parallel longitudinal mode

Fig.3 Structure of longitudinal mode piezoelectric transformers.

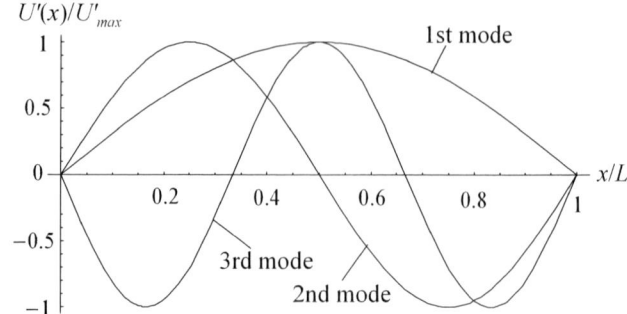

Fig.2 Normalized stress of the mode shapes for the PT.

B. The Constitutive Equations

The constitutive equations of a piezoelectric material can be written in the following forms.

$$S_i = s_{ij}^E T_j + d_{mi}^t E_m, \qquad (4)$$

$$D_n = d_{nj} T_j + \varepsilon_{nm}^T E_m$$

where the subscripts m, n=1,2,3 and i, j=1,2,3,4,5,6. T_j and S_i represent the stress vector and the strain vector respectively. E_m is the electric field vector, and D_n is the electric displacement vector. d_{nj} is a matrix of piezoelectric constants, ε_{nm}^T is a matrix of dielectric permitivities (at constant stress), and s_{ij}^E is the elastic compliance matrix (at constant electric field).

Considering the poling and vibrating direction in each section of the piezoelectric beam, the structure of the beam can be divided into two types of sections: the parallel longitudinal mode section and the transverse length mode section, as shown in Fig.3(a)(b).

For the transverse length mode section, alternating T_1 is applied symmetrically to the ends of a piezoelectric plate, as shown in Fig.3(a). The magnitude of T_1 can be assumed to be mush larger than T_2 and T_3. Also, we can assume that only E_3, D_3 will be applied or detected. Therefore, this system can be simplified to one dimensional constitutive equations of a piezoelectric material, so equations (4) can be rewritten as

$$S_1 = s_{11}^E T_1 + d_{31} E_3, \qquad (5)$$

$$D_3 = d_{31} T_1 + \varepsilon_{33}^T E_3,$$

where S_1 is the strain of x direction, T_1 is the stress of x direction, E_3 is the electric field of z direction, and D_3 is the electric displacement of z direction.

For the parallel longitudinal mode section of Fig.3(b), the magnitude of T_1 can also be assumed to be mush larger than T_2 and T_3, but only E_1, D_1 will be applied or detected in this section. Therefore, this system can be simplified to one dimensional constitutive equations, so equations (4) can be rewritten as

$$S_1 = s_{33}^E T_1 + d_{33} E_1, \tag{6}$$

$$D_1 = d_{33} T_1 + \varepsilon_{33}^T E_1. \tag{}$$

III. ELECTROMECHANICAL MODEL OF THE PT

A. Electromechanical Model of the PT

To model the dynamic behavior of a longitudinal mode PT, equations of motion for the coupling system of the PT can be obtained by using Hamilton's principle. From Hagood's paper [6], we have a generalized form as the following.

$$\int_{t_1}^{t_2} \partial[T - U + W_e + W_i + W_o]dt = 0 \tag{7}$$

where T is the kinetic energy, U is the potential energy of the system, W_e is the electrical energy stored within the piezoceramic beam, W_i is the applied electric energy in the driving portion, and W_o is the applied electric energy in the receiving portion. T, U, W_i, W_o can be written as

$$T = \frac{1}{2} \int_V \rho \dot{u}^2(x,t)dV \tag{8}$$

$$U = \frac{1}{2} \int_V S^T T dV \tag{9}$$

$$W_e = \frac{1}{2} \int_V E^T D dV \tag{10}$$

$$\partial W_i = -\partial \varphi_i \cdot q_i, \quad \partial W_o = -\partial \varphi_o \cdot q_o \tag{11}$$

where φ_i and q_i are the electric potential and the applied charge in the input part, respectively. φ_o and q_o are the electric potential and the applied charge in the output part. V means the whole volume of the PT. By substituting Eqs.(8)-(11) into Eq.(7), the equations of motion for the PT can be written in Laplace transform as

$$(m_n s^2 + d_n s + k_{nt} + k_{np})X_n + A_o V_o = A_i V_i \tag{12}$$

$$A_i X_n + C_i V_i = q_i \tag{13}$$

$$A_o X_n = C_o V_o + q_o \tag{14}$$

where V_i and V_o represent the input and output voltage, q_i and q_o represent the input and output charge respectively. It is noted that the input current I_i and the output current I_o can be calculated by the differentiation of the input and output charge, respectively. X_n represent mechanical displacement of the PT. The mass m_n for the PT and the stiffness k_{nt}, the input turn ratio A_{it}, the output turn ratio A_{ot}, the input capacitance C_{it} and the output capacitance C_{ot} for the tranverse length mode section of the PT can be obtained from the following.

$$m_n = \int_0^L \rho b h U^2 dx \tag{15}$$

$$k_{nt} = \int_0^{l_t} \frac{bh}{s_{11}^E} \left(\frac{\partial U}{\partial x}\right)^2 dx \tag{16}$$

$$A_{it} = \frac{1}{h} \int_{V_T} \frac{d_{31}}{s_{11}^E} \frac{\partial U}{\partial x} dV_T \tag{17}$$

$$A_{ot} = \frac{1}{h} \int_{V_T} \frac{d_{31}}{s_{11}^E} \frac{\partial U}{\partial x} dV_T \tag{18}$$

$$C_{it} = \int_{V_T} \frac{\varepsilon_{33}^T}{h^2} dV_T \tag{19}$$

$$C_{ot} = \int_{V_T} \frac{\varepsilon_{33}^T}{h^2} dV_T \tag{20}$$

where V_T and l_t represent the volume and length in the transverse length mode section, respectively. The stiffness k_n, the input turn ratio A_{ip}, output turn ratio A_{op}, the input capacitance C_{ip} and the output capacitance C_{op} for the parallel longitudinal mode section of the PT can be obtained from the following.

$$k_{np} = \int_0^{l_p} \frac{bh}{s_{33}^E} \left(\frac{\partial U}{\partial x}\right)^2 dx \tag{21}$$

$$A_{ip} = \frac{1}{l_p} \int_{V_P} \frac{d_{33}}{s_{33}^E} \frac{\partial U}{\partial x} dV_P \tag{22}$$

$$A_{op} = \frac{1}{l_p} \int_{V_P} \frac{d_{33}}{s_{33}^E} \frac{\partial U}{\partial x} dV_P \tag{23}$$

$$C_{ip} = \int_{V_P} \frac{\varepsilon_{33}^T}{l_p^2} dV_P \tag{24}$$

$$C_{op} = \int_{V_P} \frac{\varepsilon_{33}^T}{l_p^2} dV_P \tag{25}$$

where V_P and l_p represent the volume and length in the parallel length mode section, respectively. For the whole system of PT, the input turn ratio A_i may be A_{it} or A_{ip}, that is depended on the design of the poling and vibrating direction in the input part. Also, the output turn ratio A_o, the input capacitor C_i and the output capacitor C_o can be decided in this manner. The damping coefficient d_n for the nth vibration mode can be calculated by

$$d_n = \frac{k_n}{\omega_c Q_m}. \tag{26}$$

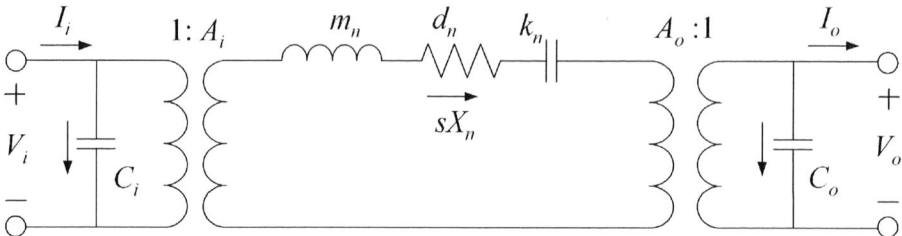

Fig. 4. Equivalent circuit of piezoelectric transformer.

where $k_n=k_{nt}+k_{np}$, ω_c is the resonant frequency and Q_m is the quality factor of the piezoelectric material. On the other hand, the equivalent circuit model of the PT can be established based on Eqs.(12)-(14) and shown in Fig.4. From the equivalent circuit model, we can see that Eq.(12) satisfies Kirchhoff's voltage law equation, which shows that the input voltage A_iV_i is the sum of the output voltage A_oV_o and the voltage difference $(m_ns^2+d_ns+k_n)X_n$. Eq.(13) satisfies Kirchhoff's current law equation in the input part, which shows that the input current I_i is the sum of the current flowing through mechanical impedance $(m_ns+d_n+k_n/s)$ and the current flowing through input capacitor C_i. Eq.(14) satisfies Kirchhoff's current law equation in the output part, which shows that the current flowing through impedance $(m_ns+d_n+k_n/s)$ is the sum of the current flowing through output capacitor C_o and the output current I_o.

B. Characteristics of the PT

Based the coupling equations of motion and the equivalent circuit model, as shown in Fig.4, the characteristics of the PT can be calculated. In order to compare the performance of various kinds of the longitudinal mode PTs, the voltage step-up ratio, output power, and efficiency of the PT will be derived as follows.

The voltage step-up ratio for the PT with a load resistance R_L in the output part can be obtained based on Eqs.(12) and (14) as the following.

$$\frac{V_o(s)}{V_i(s)}=\frac{sA_iA_oR_L}{(m_ns^2+d_ns+k_n)(sC_oR_L+1)+sA_o^2R_L} \quad (27)$$

If the electrodes in the output part of the PT is short-circuited, the voltage step-up ratio for the PT can be obtained as zero by substituting $R_L=0$ into Eq.(27). In addition, Eq.(27) shows that the higher the load resistance R_L, the higher the voltage step-up ratio at the narrow band near its resonance frequency. The maximum voltage step-up ratio as a function of frequency can be obtained as Eq.(28) when the load resistance R_L approach infinite.

$$\frac{V_o(s)}{V_i(s)}=\frac{A_oA_i}{(m_ns^2+d_ns+k_n)C_o+A_o^2} \quad (28)$$

On the other hand, if the natural frequency ω_c is chosen as the operating frequency in the PT, then the voltage step-up ratio of Eq.(27) can be simplified as

$$\frac{V_o}{V_i}=\frac{A_iA_o}{j\omega_cd_nC_o+d_n/R_L+A_o^2}. \quad (29)$$

The output power of the PT can be calculated by the power consumption of the load resistance R_L as the following.

$$P_o=|V_o|^2/R_L=\frac{A_i^2A_o^2V_i^2}{R_L[(\omega_c^2d_n^2C_o^2+(d_n/R_L+A_o^2)^2]} \quad (30)$$

According to the equivalent circuit of the PT, the input power of the PT can be calculated by the sum of the power consumption of the damping d_n and that of the load resistance R_L. Eq.(14) shows that the current flowing through d_n is $(sC_oV_o+I_o)/A_o$, thus the input power of the PT can be obtained as

$$P_i=V_o^2[d_n(\omega_cC_o/A_o)^2+d_n/(R_LA_o)^2+1/R_L] \quad (31)$$

Therefore, the efficiency of the PT can be obtained by Eqs.(30) and (31).

$$\eta=\frac{P_o}{P_i}=\frac{A_o^2}{d_nR_L\omega_c^2C_o^2+d_n/R_L+A_o^2} \quad (32)$$

The maximum efficiency can be calculated by the differentiation of Eq.(32). Thus, the maximum efficiency can be obtained when the optimal load resistance R_L is

$$R_L=1/(\omega_cC_o). \quad (33)$$

Substituting Eq.(33) into Eq.(32) gives the maximum efficiency

$$\eta_{max}=\frac{A_o^2}{2d_n\omega_cC_o+A_o^2} \quad (34)$$

It is noted that the smaller the damping coefficient d_n, the higher the maximum efficiency. On the other hand, a high efficiency of PT can be designed by raising its output turn ratio A_o. Depend on Eq.(18) and (23), we can conclude that it is necessary to choose a large value of d_{31} or d_{33}, and consider the relation between the output electrode and the stress distribution of the vibration mode in order to obtain a large value of A_o.

IV. NUMERICAL RESULTS AND DISCUSSION

A. Verification of the Electromechanical Mode

To verify the electromechanical model, a longitudinal mode of PT with 10 mm in width, 2 mm in thickness and 35 mm in length has been verified [5]. The PT has silver electrodes on two opposite surfaces and is poled along its thickness direction. The transformer structures are considered to be fabricated using the piezoelectric material APC840 by APC International, USA. The material properties provided by the supplier are listed in Table I. According to Table I, parameters of the equivalent circuit of the PTs can be calculated by Eqs.(15)-(26) and shown in Table II. Also, the resonance frequencies, the optimal resistances and the efficiencies are calculated and listed in Table II.

TABLE I   Properties of piezoelectric material

Piezoelectric coefficient d_{31}	-132.55×10^{-12} C/N
Piezoelectric coefficient d_{33}	300×10^{-12} C/N
Coupling factor k_{31}	0.33
Coupling factor k_{33}	0.68
Young's modulus Y_{11}^E	7.6×10^{10} N/m^2
Young's modulus Y_{33}^E	6.3×10^{10} N/m^2
Dielectric constant $\varepsilon_{33}^T / \varepsilon_o$	1385
Density ρ	7600 g/cm^3
Mechanical quality factor Q_m	1400

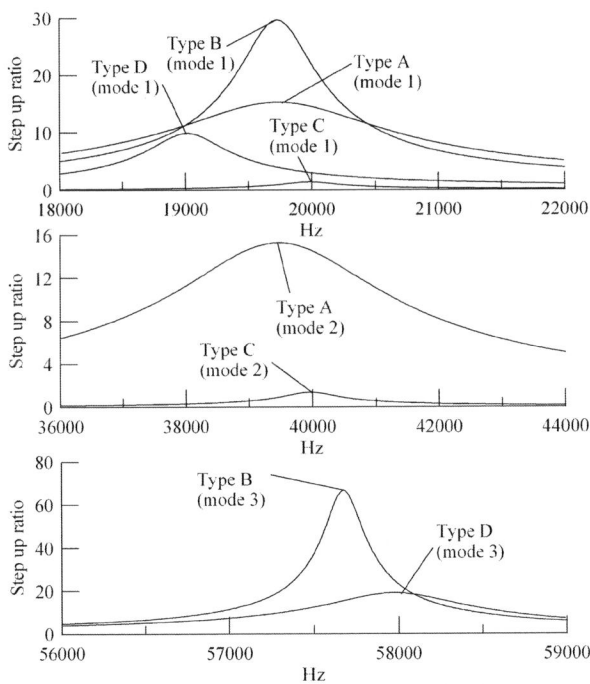

Fig.5 Voltage step-up ratio for the optimal load resistance.

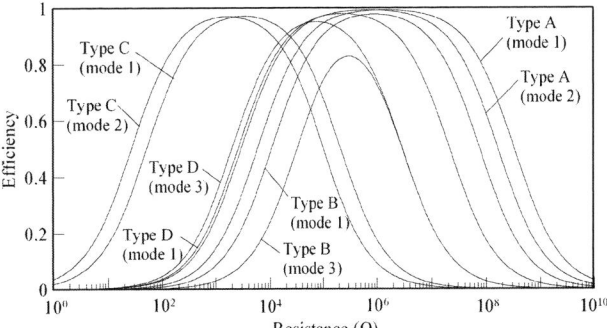

Fig.6 Efficiency as a function of load resistance.

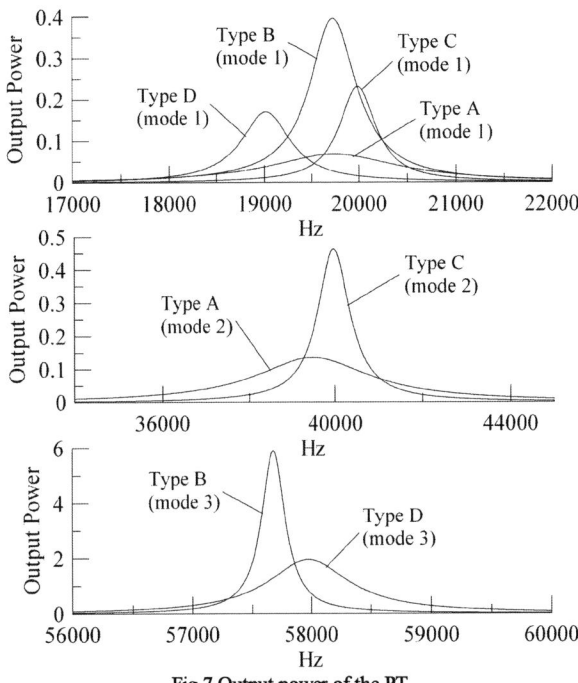

Fig.7 Output power of the PT.

B. Voltage Step-up Ratio, Output Power, and Efficiency

The voltage step-up ratios as a function of frequency at the optimal load resistances are calculated and shown in Fig.5. It shows that Type B operating in the third order longitudinal vibration mode has the highest step-up ratio for the optimal load resistance connected with its output electrodes due to its high ratio of the input capacitance to the output capacitance. In contrast to the result, Type C operating in mode 1 has the lowest step-up ratio for the optimal load resistance due to its low ratio of the input capacitance to the output capacitance. The exact values of the voltage step-up ratio can be calculated by Eq.(27).

The efficiencies as a function of load resistance are calculated and shown in Fig.6. It shows that Type B operating at mode 3 has the lowest efficiency in comparison with the other types due to its low output turn ratio A_o. The exact values of the voltage step-up ratio can be calculated by Eq.(34). On the other hands, the output powers as a function of frequency at the optimal load resistances when the input voltage is 20V are calculated and shown in Fig.7. It shows that Type B operating in mode 3 has the highest output power in comparison with

1792

the other types due to its high input turn ratio A_i and output turn ratio A_o.

V. CONCLUSION

In this paper, the electromechanical model for various kinds of the longitudinal mode PT is presented. Based on the electromechanical model, an equivalent circuit of the PT is shown. Also, the voltage step-up ratio, output power and efficiency of the PT are calculated, and the optimal load resistance and the maximum efficiency for the PT have been obtained. The model presented here serves as a useful design tool for optimizing the configuration of the PT. We can conclude the design principle of the longitudinal mode PT as in the following. Basically, the mechanical quality factor Q_m should be chose as high as possibility, thus the higher efficiency will be obtained in the PT. Also, the higher output turn ratio A_o should be designed in order to obtain a high efficiency of the PT. For a higher voltage step-up ratio, the PT should be designed to having the high ratio of C_i/C_o. In the last, a higher input turn ratio A_i should also be designed in order to obtain a high output power of the PT.

REFERENCES

[1] C. A. Rosen, "Ceramic Transformers and Filters," *Proceedings of Electronic Comp. Symp.*, pp.205-211, 1956.

[2] K. Sakurai, K. Ohnishi and Y. Tomikawa, "Presentation of a New Equivalent Circuit of a Piezoelectric Transformer under High-Power Operation," *Jpn. J. Appl. Phys.*, vol.38, pp.5592-5597, 1999.

[3] K. Kanayama and N. Maruko, "Properties of Alternately Poled Piezoelectric Transformers," *Jpn. J. Appl. Phys.*, vol. 36, pp. 3048-3049, 1997.

[4] K. Kanayama, N. Maruko and H. Saigoh, "Development of the Multilayer Alternately Poled Piezoelectric Transformers," *Jpn. J. Appl. Phys.*, vol. 36, pp. 2891-2895, 1998.

[5] S. T. Ho, "Electromechanical Model of a Longitudinal Mode Piezoelectric Transformer," *The Seventh International Conference on Power Electronics and Drive Systems*, Bangkok, 2007.

[6] N. W. Hagood, W. H. Chung and A. V. Flotow, "Modeling of Piezoelectric Actuator Dynamics for Active Structural Control," *J. of Intell. Mater. Syst. And Struct.*, vol. 1, pp. 327-354, 1990.

Table II. Parameters of the equivalent circuit

Structure	Type A		Type B		Type C		Type D	
Mode n	1	2	1	3	1	2	1	3
Effective mass m_m	0.00608 kg	0.00608 kg	0.00608 kg	0.00608 kg	0.00608 kg	0.00608 kg	0.00608 kg	0.00608 kg
Effective damping d_n	0.5157 N-s/m	1.0315 N-s/m	0.5502 N-s/m	1.5711 N-s/m	0.5157 N-s/m	1.0786 N-s/m	0.5104 N-s/m	1.5711 N-s/m
Effective stiffness k_n	8.574×10^7 N/m	3.45×10^8 N/m	9.063×10^7 N/m	7.957×10^8 N/m	9.376×10^7 N/m	3.75×10^8 N/m	8.399×10^7 N/m	7.957×10^8 N/m
Input capacitance C_i	2.4523nF		3.2698nF		2.4523nF		3.2698nF	
Output capacitance C_o	6.1308pF		9.1962pF		2.4523nF		36.785pF	
Input turn ratio A_i	0.10074	0.20148	0.15110	-0.40295	0.10074	0.20148	0.10074	0.40295
Output turn ratio A_o	-0.00945	0.0189	-0.007088	-0.007088	0.100738	-0.201476	-0.014175	0.02835
Resonance frequency	18900Hz	37800Hz	19431Hz	57577Hz	1976Hz	3952Hz	18707Hz	57578Hz
The optimal Resistance R_{opt}	1.3735MΩ	686767Ω	890675Ω	300580kΩ	3284Ω	164Ω	231290kΩ	7514Ω
Efficiency η_{max}	99.166%	99.166%	97.685%	82.774%	97%	96.875%	97.851%	95.055%

The Comparison of Conducted EMI Emission and Electrical Performances of Lamps

C. Uyaisom and W. Khan-ngern

Department of Electrical Engineering, Faculty of Engineering
and Research Center for Communications and Information Technology
King Mongkut's Institute of Technology Ladkrabang, Bangkok, Thailand
Phone/Fax: +(662) 7373000 Ext. 3322
E-mail: chutipon@eau.ac.th, kkveerac@kmitl.ac.th

Abstract--This paper presents the comparison the conducted electromagnetic interference (EMI) emission of lamps. The test set has been gotten public favor that consists of typical lamps. The types of lamp, such as incandescent lamp (40 W), new compact fluorescent lamp or energy saving lamp or Lord-Ta-Keab in Thai (8 W), compact fluorescent lamp with magnetic core ballast (36 W), and compact fluorescent lamp with electronic ballast (36 W). The effect of lamps on the conducted EMI emission are compared to verify the effectiveness of electrical performance. The test setup is referred to the CISPR-16 standard.

Index Terms-- conducted EMI emission, and lamp

I. INTRODUCTION

The lamps are one of the most popular light sources used in the house, industry, department store, hospital, university, and so on. The electrical engineer who design the lighting system should consider on the level of luminous flux, luminous efficacy, life time, high power factor, high-efficiency, energy saving, and low cost. However, the some type of the lamps can generate the conducted electromagnetic interference (EMI) due to their high switching frequency operations [1-4].

II. EXPERIMENTS

The test set was found from public favor and consisted of typical lamp to test conducted EMI emissions and electrical performance. The types of lamp, such as incandescent lamp (40 W), new compact fluorescent lamp or energy saving lamp or Lord-Ta-Keab in Thai (8 W), compact fluorescent lamp with magnetic core ballast (36 W), and compact fluorescent lamp with electronic ballast (36 W). Four case studies are summarized in table 1.

TABLE I
MEASURED DIAGRAM

Case	Type of lamp	Power rate
1.	Incandescent lamp	40 W
2.	* New compact fluorescents lamp or energy saving lamp or Lord-Ta-Keab in Thai	8 W
3.	Compact fluorescent lamp with magnetic core ballast	36 W
4.	Compact fluorescent lamp with electronic ballast	36 W

* The new compact fluorescents lamp or energy lamp saving or Lord-Ta-Keab in Thai (8 W) can be up to 80% energy saving over standard incandescent bulbs (40 W).

III. TEST SET UP FOR MEASURING

Test setup is referred to the CISPR-16 standard and the conducted EMI is measured with Line Impedance Stabilization Network (LISN) which has the frequency range between 9 kHz and 30 MHz. LISN was used to provide standard impedance for the measurement and provide the constant 50 ohm impedance for the measurement stability is shown in figure 1.

The instrument details of EMI test consists of EMI Analyzer (Agilent model E7401A 9 kHz-1.5 GHz) and LISN (EMCO model 3810/2 50Ω 50/250μH 9 kHz-30 MHz). The diagram EMI test is shown in figure 2.

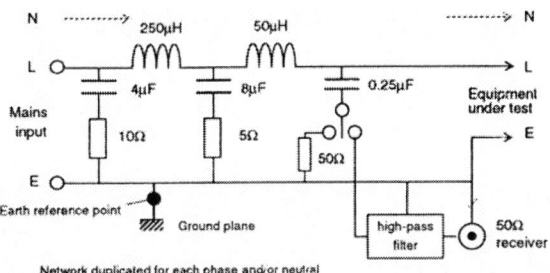

Fig. 1. Line Impedance Stabilization Network (LISN) circuit

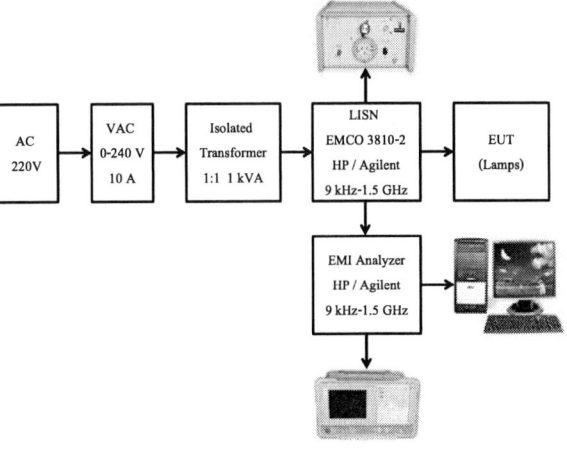

Fig. 2. Diagram for EMI testing

IV. Experimental results and analysis

A. Voltage, current waveform and Electrical performance

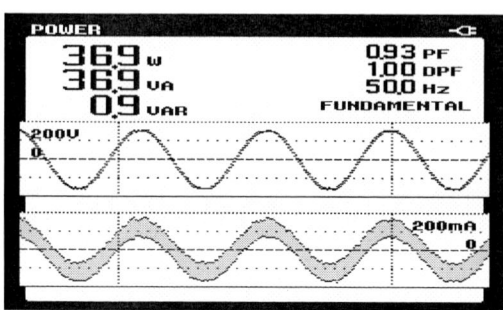

Fig. (3a).

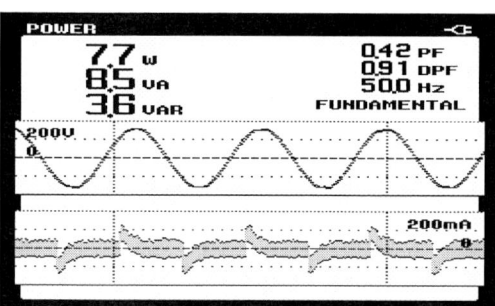

Fig. (3b).

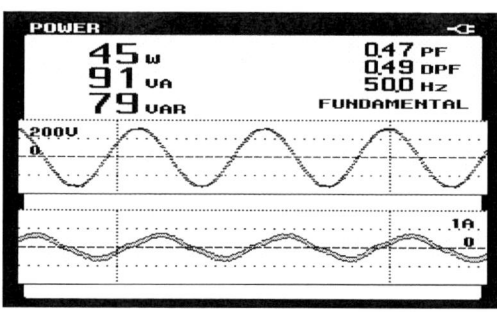

Fig. (3c).

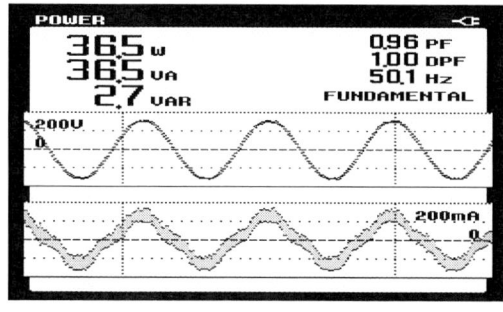

Fig. (3d).

Fig. 3. Voltage, current waveform, power and power factor of
(3a) incandescent lamp (40 W)
(3b) energy saving lamp (8 W)
(3c) compact fluorescent lamp with magnetic core ballast (36 W)
(3d) compact fluorescent lamp with electronic ballast (36 W)

B. Spectra of total harmonic distortion

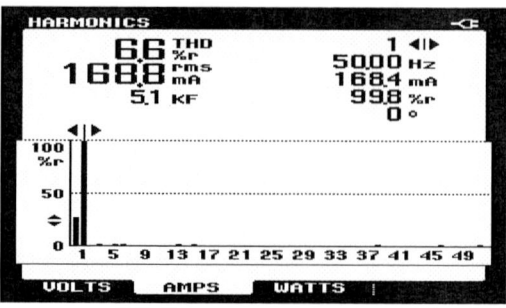

Fig. (4a).

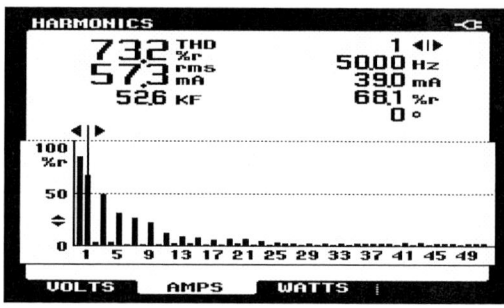

Fig. (4b).

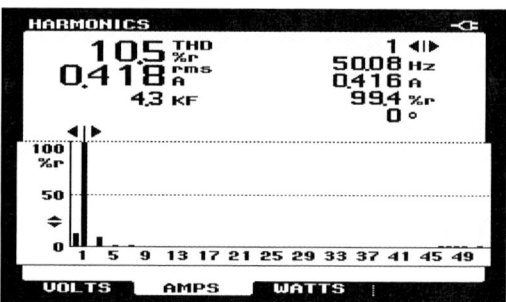

Fig. (4c).

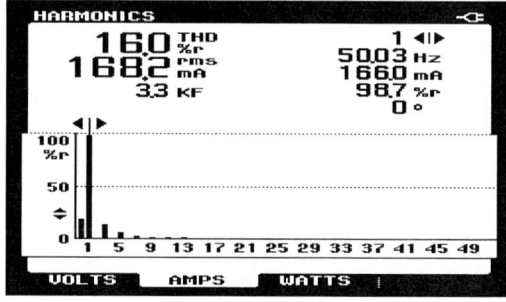

Fig. (4d).

Fig. 4. Spectra of total harmonic distortion (% THD_i) of
(4a) incandescent lamp (40 W)
(4b) energy saving lamp (8 W)
(4c) compact fluorescent lamp with magnetic core ballast (36 W)
(4d) compact fluorescent lamp with electronic ballast (36 W)

1795

C. Experimental EMI noise

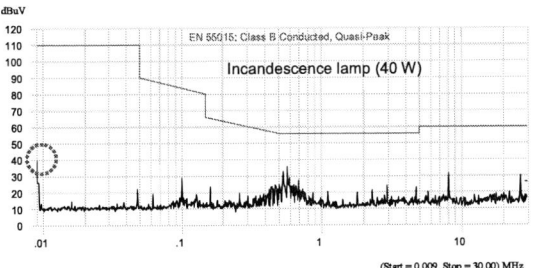

Fig. (5a). Max EMI noise = 40 dBμV at 9 kHz

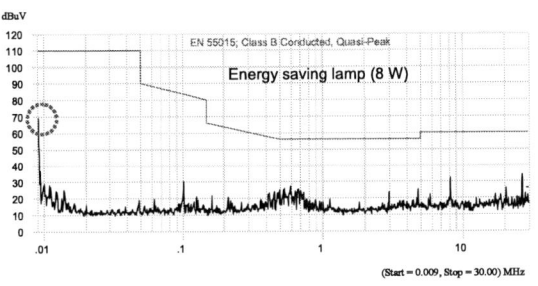

Fig. (5b). (Max EMI noise = 69 dBμV at 9 kHz)

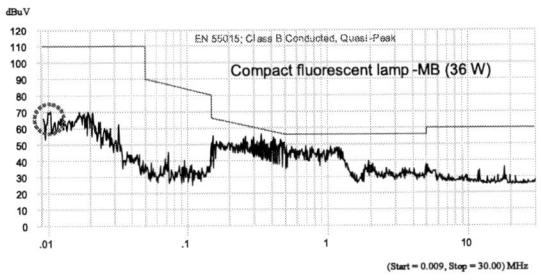

Fig. (5c). (Max EMI noise = 70 dBμV at 10 kHz)

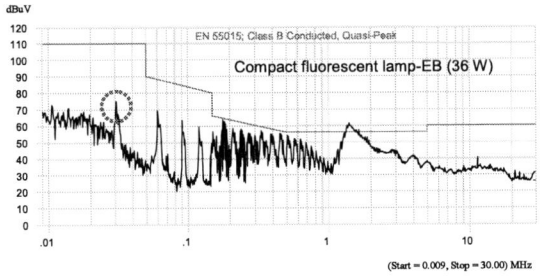

Fig. (5d). (Max EMI noise = 75 dBμV at 31 kHz)

Fig. 5. EMI noise of lamps
(5a) incandescent lamp (40 W)
(5b) energy saving lamp (8 W)
(5c) compact fluorescent lamp with magnetic core ballast (36 W)
(5d) compact fluorescent lamp with electronic ballast (36 W)

D. Comparison of spectra conducted EMI noise

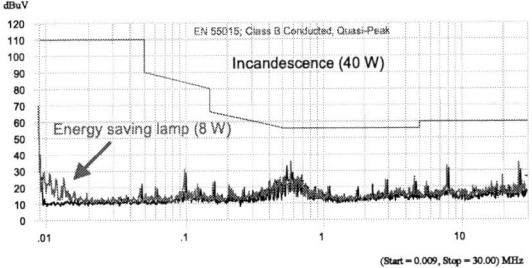

Fig. (6a).

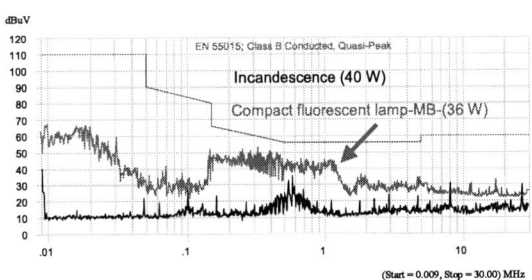

Fig. (6b).

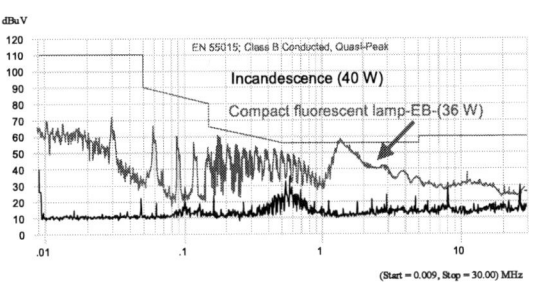

Fig. (6c).

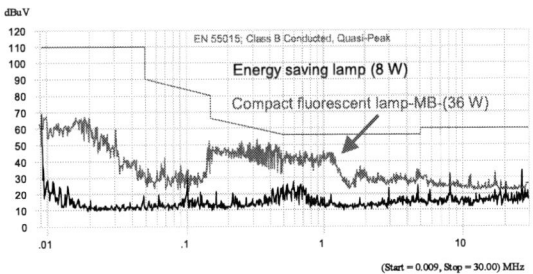

Fig. (6d).

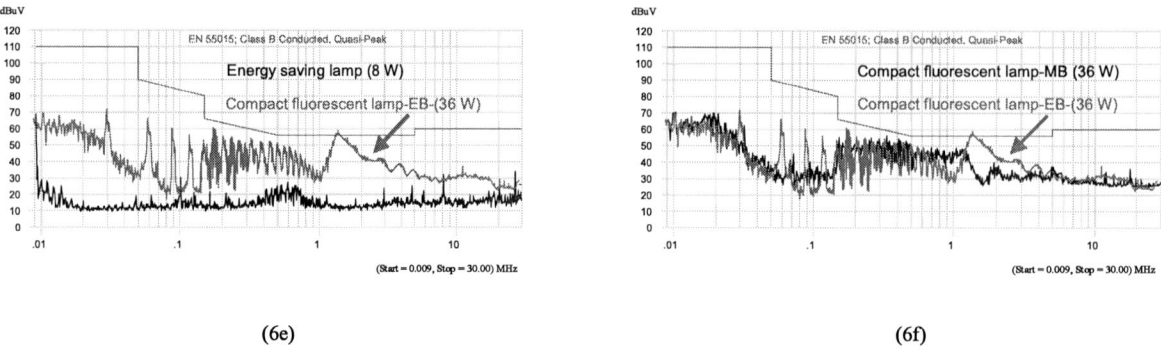

(6e)   (6f)

Fig. 6. Comparison of EMI noise
(6a) incandescent lamp (40 W) and energy saving lamp (8 W)
(6b) incandescent lamp (40 W) and compact fluorescent lamp with magnetic core ballast (36 W)
(6c) incandescent lamp (40 W) and compact fluorescent lamp with electronic ballast 36 (W)
(6d) energy saving lamps (8 W) and compact fluorescent with magnetic core ballast (36 W)
(6e) energy saving lamps (8 W) and compact fluorescent lamp with electronic ballast (36 W)
(6f) compact fluorescent lamp with magnetic core ballast (36 W) and compact fluorescent lamp with electronic ballast (36 W)

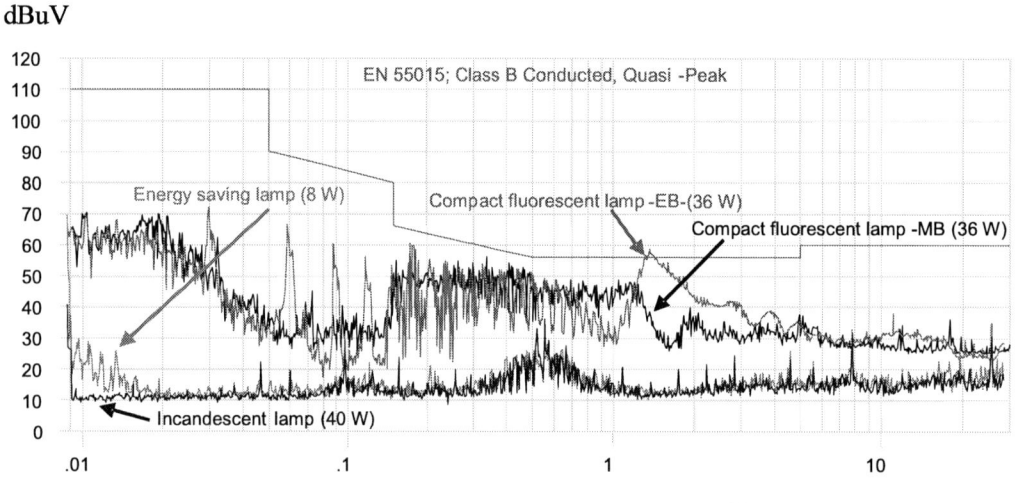

(Start = 0.009, Stop = 30.00) MHz

Fig. 7. Comparison of spectra conducted EMI emission of lamps

TABLE II
EXPERIMENTAL RESULTS OF ELECTRICAL PERFERMANCE (constant input voltage at 220 V)

Type of lamp	Power rate (W)	luminous flux (lumen)	luminous efficacy (lm/W)	Price (Bath) / set	V_{in} (V)	I_{in} (mA)	P_{in} (W)	PF	% THD		EMI	
									V	I	Max (dBuV)	Freq. (kHz)
1. Incandescent lamp	40	430	11	28	220	180.9	36.9	0.93	1.1	6.6	40	9
2. Energy saving lamp	8	400	49	128	220	86.4	7.7	0.42	1.1	73.2	69	9
3. Compact fluorescent lamp with magnetic core ballast	36	2,600	73	355	220	424	45	0.47	1.3	10.5	70	10
4. Compact fluorescent lamp with electronic ballast	36	2,600	73	535	220	171.9	36.5	0.96	1.4	16.0	75	31

From the experimental results, the input voltage waveform, current waveform, power, power factor and percentage of total voltage and current harmonic distortion were measured and analyzed via a Power Meter and Analyzer; Fluke 43B Power Quality Analyzer.

Figure 3 shows the voltage, current waveforms, power and power factor with each lamp. The maximum input power of the lamp is compact fluorescent lamp with magnetic core ballast (36 W) is 45 watts, and the minimum input power of the lamp is energy saving lamp (8 W) is 7.7 watts. The maximum power factor of the lamp is incandescent lamp (40 W) is 0.93, and the minimum power factor of the lamp is energy saving lamp (8 W) is 0.42 .

Figure 4 shows the spectra and percentage of total harmonic distortion (voltage and current). The maximum %THD_i of the lamp is energy saving lamp (8 W) is 73.2 percentage and the minimum %THD_i of the lamp is incandescent lamp (40 W) is 6.6 percentage.

Figure 5 shows the experimental results of EMI noise due to each lamp. Figure 6-7 shows the experimental results and comparison of experimental results of the conducted EMI emission due to each lamp. The Y-axis shows magnitude of EMI noise voltage in dBμV and X-axis is switching frequency in log scale unit with starting frequency of 9 kHz and stopping frequency of 30 MHz with referred to the CISPR-15 standard.

For the example, It can be seen that the peaks of the noise levels are about 40, 60, 70, and 75 dBμV at incandescent lamp (40 W), energy saving lamp (8 W), compact fluorescent lamp with magnetic core ballast (36 W), and compact fluorescent lamp with electronic ballast (36 W), respectively.

The table 2 shows the comparison of electrical performance and conducted EMI emission between incandescent lamp (40 W), energy saving lamp (8 W), compact fluorescent lamp with magnetic core ballast (36 W), and compact fluorescent lamp with electronic ballast (36 W), respectively. For the example, the conducted EMI emissions form the compact fluorescent lamp with electronic ballast (36 W) are very high level due to their high switching frequency operations and the conducted EMI emissions from incandescent lamp are very low level due to it's not used switching device operations.

From the table 2, It can be considered in two comments for selection types of lamp to reduce the EMI noise.

- Low input power and low EMI should be used in energy saving lamp.
- High power factor and low EMI should be used in incandescent lamp.

However, the energy saving lamp and incandescent lamp are low luminous flux and luminous efficacy.

Finally, the electrical engineer who design the lighting system should be careful in selection types of lamp, power rate, luminous flux, luminous efficacy and electrical performance of lamps to reduce the EMI noise.

V. CONCLUSIONS

This paper has proposed the comparison of the conducted electromagnetic interference (EMI) emission and electrical performance of lamps. The test set is found from public favor and consists of typical incandescent lamp (40 W), energy saving lamp (8 W), compact fluorescent lamp with magnetic core ballast (36 W), and compact fluorescent lamp with electronic ballast (36 W), respectively. The measured results of the conducted EMI emission from each lamp are compared to the effectiveness of electrical performance.

ACKNOWLEDGMENT

The author would like to thank Assoc. Prof. Dr. Sombat Teekasup, Faculty of Engineering Eastern Asia University, Patumtanee, Thailand and Mr. Komkrit Karanun, Electrical and Electronics Institute Bangkok Thailand for their advice and fully support.

REFERENCES

[1] W. Khan-ngern and V. Tarateeraseth, " Power Electronics", King Mongkut's Institute of Technology Ladkrabang (KMITL) Bangkok Thailand, September 2004, pp. 787-808.

[2] P. Prasit, "Lighting System Design", Bangkok Thailand, pp. 61-160.

[3] S.Y.R. Hui, W. Yan, H. Chung, P.W. Tam and G. Ho, "Energy Efficiency Comparison of Dimmable Electromagnetic and Electronic Ballast Systems", IEEE IAS 2005, pp. 2775-2781.

[4] Antonio J. Calleja, J. Marcos Alonso, Emilio L´opez, Javier Ribas, Juan Angel Mart´ynez, and Manuel Rico-Secades, "Analysis and Experimental Results of a Single-Stage High-Power Factor Electronic Ballast Based on Flyback Converter", IEEE Transactions on Power Electronics, Vol. 14 No. 6 November 1999. pp. 998-1006.

Mr. Chutipon Uyaisom, He received the degree of B.Eng. from Mahanakron University of Technology (MUT) in 1996 and master degree of M.Eng. from King Mongkut's Institute of Technology Ladkrabang (KMITL) in 2003. His research in the area of EMI & EMC on power electronic system and illumination system.

Assoc. Prof. Dr. Werachet Khan-ngern, He received the degree of B.Eng. and M.Eng from King Mongkut's Institute of Technology Ladkrabang (KMITL) in 1982 and 1988, repectively. He received his Ph.D and DIC from Imperial Collage of Science Technology and Medicine, the University of London in 1997 in the area of power electronicd. He is also a technical committee no 890: in EMC for the Ministry of Industry. Dr.Khan-ngern continues his research in the area of EMI & EMC and power electronic system.

Neural Identification of Average Model of STATCOM using DNN and MLP

M. Tavakoli Bina* and S. Rahimzadeh*

* Faculty of Electrical Engineering, K. N. Toosi University of Technology, P. O. Box 16315–1355, Tehran 16314, Iran,
E-mail: tavakoli@ieee.org

Abstract—Modeling of STATCOM is conventionally performed in the *time-domain*. Amongst them, dq-theory is well-known in which state-space equations are used for the analysis. Power systems, however, use the *frequency-domain* information in phasor-related studies such as load flow analysis. Because *time-domain* models of FACTS controllers cannot be directly applied to the power system analysis, an intelligent model can usefully bridge the time-domain information to the corresponding frequency-domain data. This paper proposes two neural network identifiers based on the existing time-domain average model of STATCOM. Extended resultant bridge presents an average-neural model of STATCOM, which can be analytically applied to power systems. To this extent, design and development of two neural network identifiers are performed using the dynamic neural network (DNN) and the multi-layer perceptron (MLP). To verify the developed models, the exact solutions obtained from the average model of STATCOM are compared with the outcomes of the DNN and the MLP identifiers. Moreover performance of the two identifiers is accordingly compared as well.

Index Terms-- DNN, FACTS controllers, MLP, modeling, neural-averaging, STATCOM.

I. INTRODUCTION

DEREGULATED systems consider transmission lines as the principal components of the electricity market for both producers and consumers of energy. At the same time, an optimal power flow (OPF) determines the amount of energy to be transferred through each transmission line. This further helps the market to achieve a competitive pricing tool. Therefore, it is necessary to develop accurate models in order to establish a fair pricing system. In fact, if the operation of the modelled equipments is close to that of their exact devices, the energy pricing will be more accurate. In particular, this would be crucial when FACTS devices are engaged in the OPF for mitigation of congestion of transmission systems (CTS).

Typically, in [1]–[5], FACTS devices are suggested to alleviate and/or regulate the CTS. Additionally, the FACTS devices are modelled as either pure reactive elements (e.g. inductors and capacitors) or independent voltage/current sources. However, power losses of FACTS devices are *not* included in the analysis by the introduced models, assuming negligible *energy consumption* by the device itself. When the number and capacity of the employed FACTS devices increases,

considerable energy losses is cancelled in power flow analysis (i.e. part of the network load is cancelled). This undermines the correctness of the process of energy pricing management.

Meanwhile, the principal objective is to bridge the instantaneous models to the power system single-frequency requirement. For example, power flow analysis is widely used in power systems in order to control active and reactive power as well as protection systems and pre-fault calculations. Moreover, deregulated power systems and the CTS control are additional applications in electricity market. This paper develops a bridging intelligent identifier that includes power losses in the analysis. Moreover, we assume an existing average model of STATCOM that is based on the well-known stat-space averaging technique [6]-[8]. The average model appropriately takes into account the low-frequency variations of the DC–link of the converter as well as the power losses related to the AC-side. It should be noted that the switches are treated as ideal.

However, one issue concerned with this model is that for each switching period a considerable number of differential equations have to be solved. This depends on the switching frequency, and takes long to process the OPF. To remedy this issue, the neural network modelling technique is employed to link the instantaneous outcomes of the STATCOM to the single–frequency power system analysis. The developed model produces power losses as well as angles and magnitudes that are suitable for phasor analysis in steady-states. Here it is examined two identifiers; the dynamic neural network (DNN) and the multi-layer perceptron (MLP). The objective is to compare the accuracy and reliability of the two identifiers. In this paper the developed model is called *average–neutral (AN)* model of STATCOM.

II. AVERAGE MODEL

Averaging technique is a common approach to the modeling of power converters. Switch-mode converters have a discontinuous behavior, which is very complex in analysis. Average modeling approximates the behavior of the converter from a periodic discontinuous system to a periodic continuous one, maintaining smooth waveforms by removing high order harmonics. Average model of STATCOM is presented in [6], shown here by Fig. 1(a).

978-1-4244-0644-9/07/$25.00 ©2007 IEEE

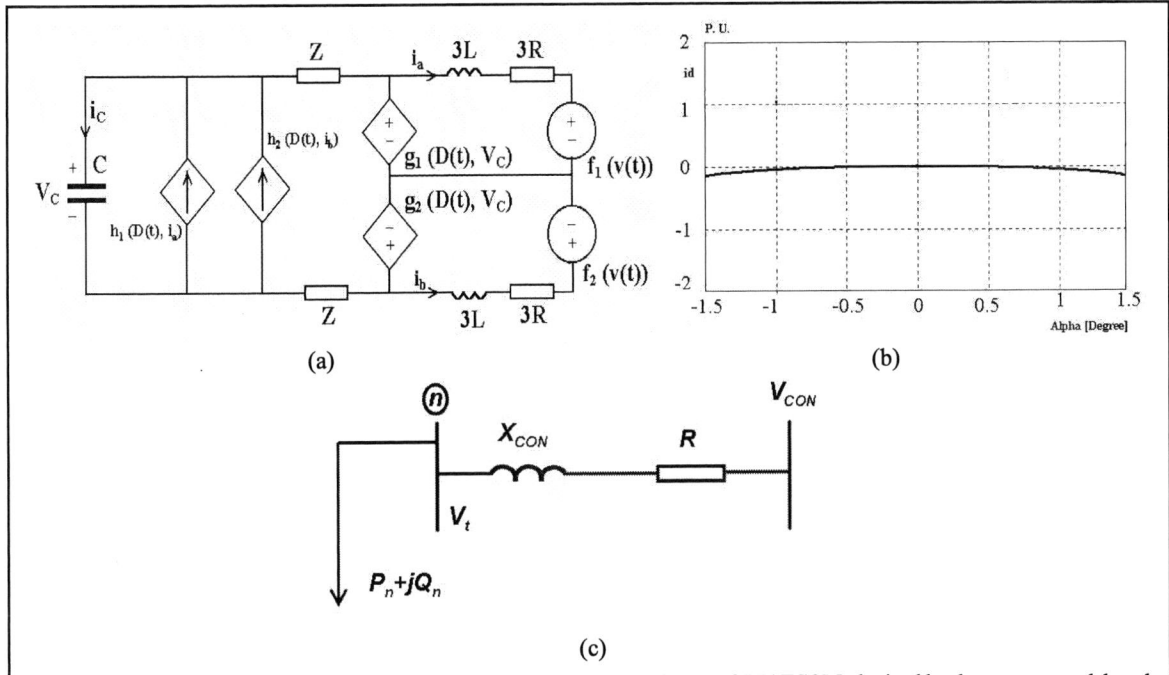

Figure 1: (a) Average circuit model of STATCOM, (b) typical internal power losses of STATCOM obtained by the average model, and (c) adaptation of the average model connected to the power system by adding a bus for STATCOM.

In this model, L introduces the equivalent coupling inductance between the converter and the power system. The resistance R is part of the compensator losses related to the interconnection of the converter to the power system. The other part of the power losses corresponds to the converter losses that are absorbed by the proper modulation of the converter switches. Fig. 1(b) shows typical STATCOM power losses in P.U. against the phase shift between the converter output and the power system voltage (α) that is obtained by the average model.

While the average model presents a time-dependent circuit, a PQ or PV model is essential for the power flow analysis. Hence, here it is performed adaptive analysis to get the supplied active and reactive powers of STATCOM (P_{CON} and Q_{CON}). A new bus is added for every STATCOM as the converter AC voltage, which is connected to an existing bus n through the commutation reactance (X_{CON}) and the AC resistance (R).

III. IDENTIFICATION OF STATCOM MODEL USING NEURAL NETWORK

Average model of Fig. 1(a) describes a state-space model in a circuit format, which solving differential equations will lead eventually to a steady-state solution. Meanwhile, moving from one steady state to another takes time to complete the transient regime that is not suitable for the OPF. An OPF program seeks among the feasible region for a desirable solution. Thus, it is necessary for the OPF to be performed as fast as possible. One approach to achieve a fast OPF is the identification of the average model of STATCOM using the neural network. The average model is analyzed as a reference to generate required training data for the *average-neural model (AN)*.

Training data can be produced in two steps. First, Fig. 1(a) is assumed as the exact model suggested in [6]. Then, to cover operating range of STATCOM, magnitude of the terminal voltage is varied within $|V_t| \in [0.7, 1.2]$ P.U. by small steps (e.g. 0.01 P.U. (see Fig. 1(c))). Also, the phase angle between the converter voltage V_{CON} and the terminal voltage V_t ($\angle(V_{CON}, V_t)$) is varied within $\alpha \in [-1.5°, 1.5°]$ by small steps (e.g. 0.01°). For the given small steps, total number of operating points sums up to 15000 set of steady-state training data for the STATCOM.

Second, for every operating point, the *time–domain* model of Fig. 1(a) is solved. Then, the steady-state phasors are obtained from the instantaneous solution. This is used to calculate and store the absorbing active as well as generating/absorbing reactive powers of STATCOM delivered to bus n. Gathering all the calculated data leads to formation of a database for single-frequency operation of STATCOM in power system.

The next step will be the study and selection of a suitable neural network identifier for the average model

1800

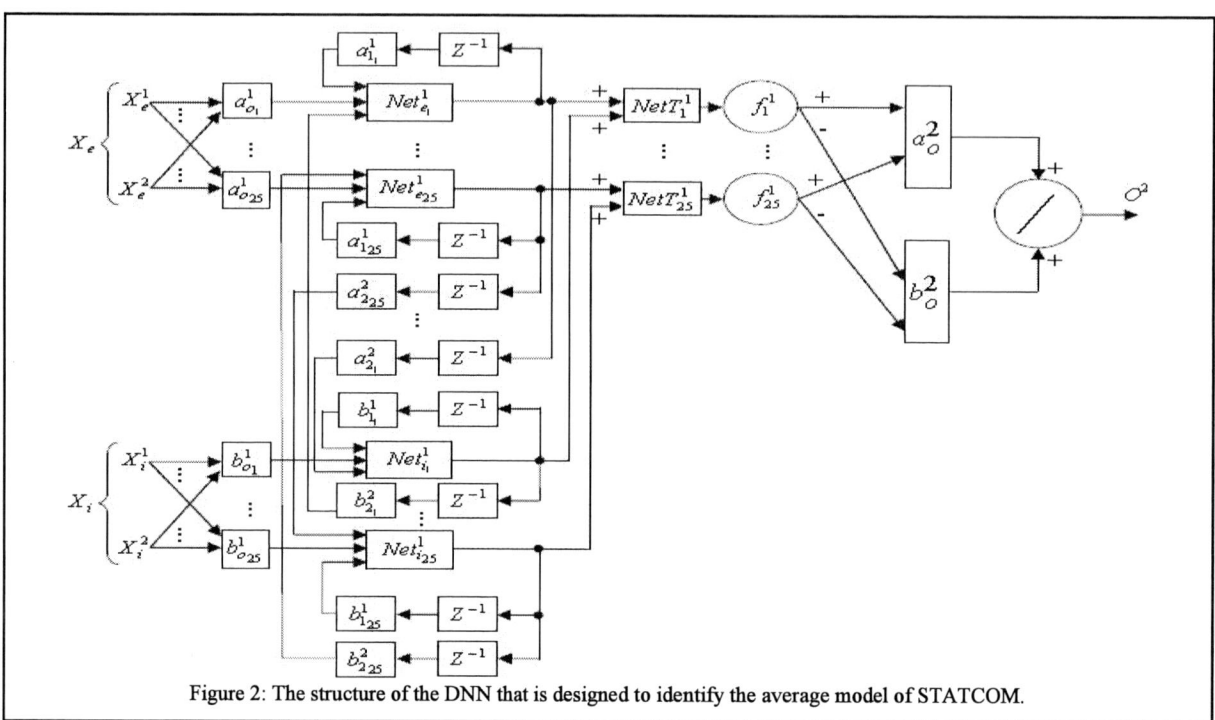

Figure 2: The structure of the DNN that is designed to identify the average model of STATCOM.

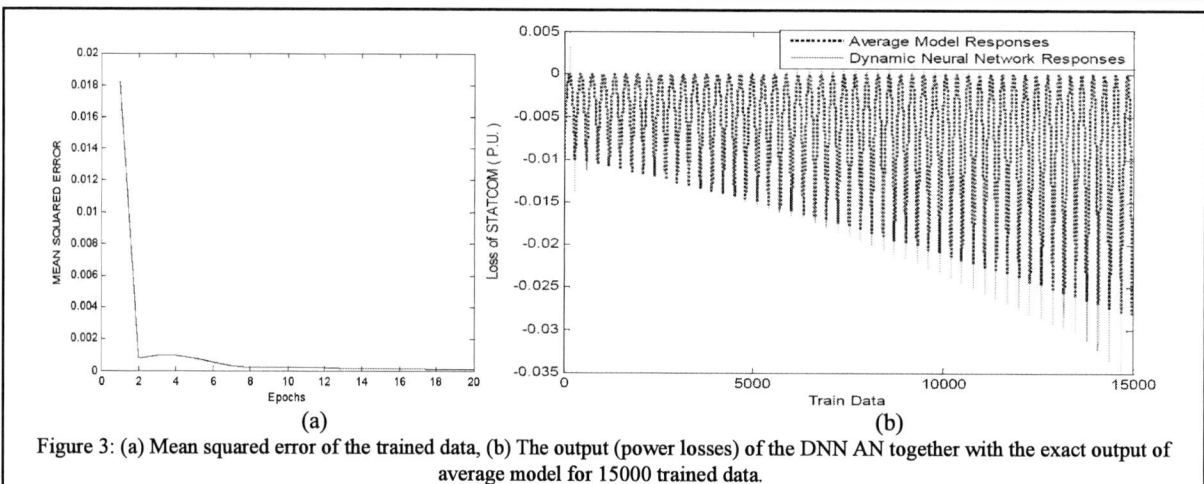

Figure 3: (a) Mean squared error of the trained data, (b) The output (power losses) of the DNN AN together with the exact output of average model for 15000 trained data.

of Fig. 1(a). Here we identify the average model of STATCOM by two well-known neural network identifiers, dynamic neural network (*DNN*) and multi-layer perceptron (*MLP*). It is noticeable that other identifiers should also be investigated that is left for future studies. Outcomes of the *AN* identifiers of STATCOM are then correspondingly compared.

A. The DNN neural identifier

There are various structures for the dynamic neural network. The employed structure for identifying the AN model is given by Fig. 2 in which the DNN includes two layers. The hidden layer has delay blocks taken from the neurons, which take the data history of the network into account for the progressing output. Index e corresponds to the exhibitory positive classes (e.g. positive input

vector $\mathbf{X_e}$), and index i relates to inhibitory negative classes (e.g. negative input vector $\mathbf{X_i}$). Each neuron from the exhibitory class has a delayed input from its corresponding neuron in inhibitory class and vice versa. Output of each neuron is applied to a non-linear neuron activation function to be able to model non-linear systems.

Then, the outcomes of the activation functions are weighted by $\mathbf{a_0^2}$ for exhibitory classes, and by $\mathbf{b_0^2}$ for inhibitory classes. These weighted productions are eventually applied to the linear activation function of the output layer to get the output of the DNN. The advantages of the DNN are non-linear modelling capability as well as the fast network convergence

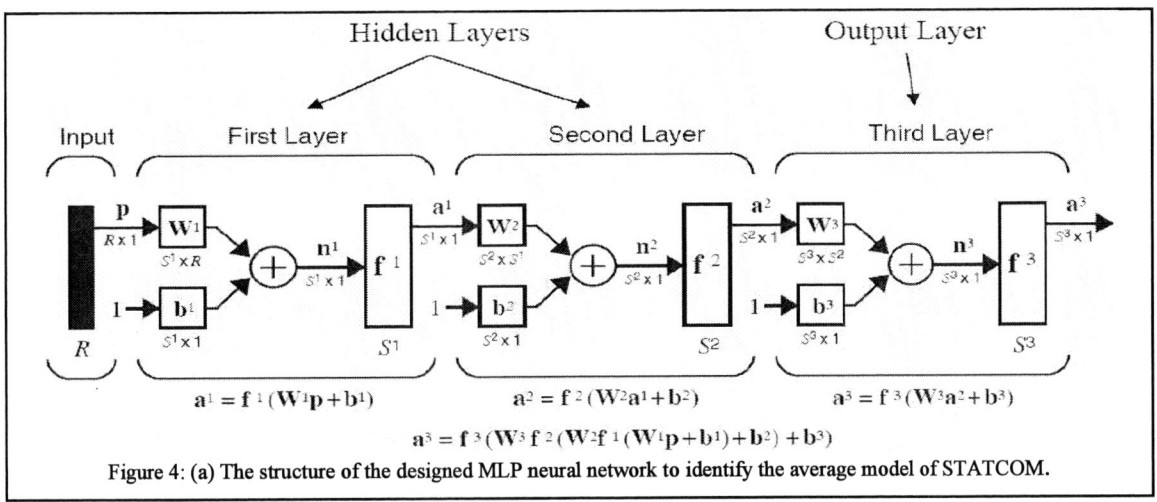

Figure 4: (a) The structure of the designed MLP neural network to identify the average model of STATCOM.

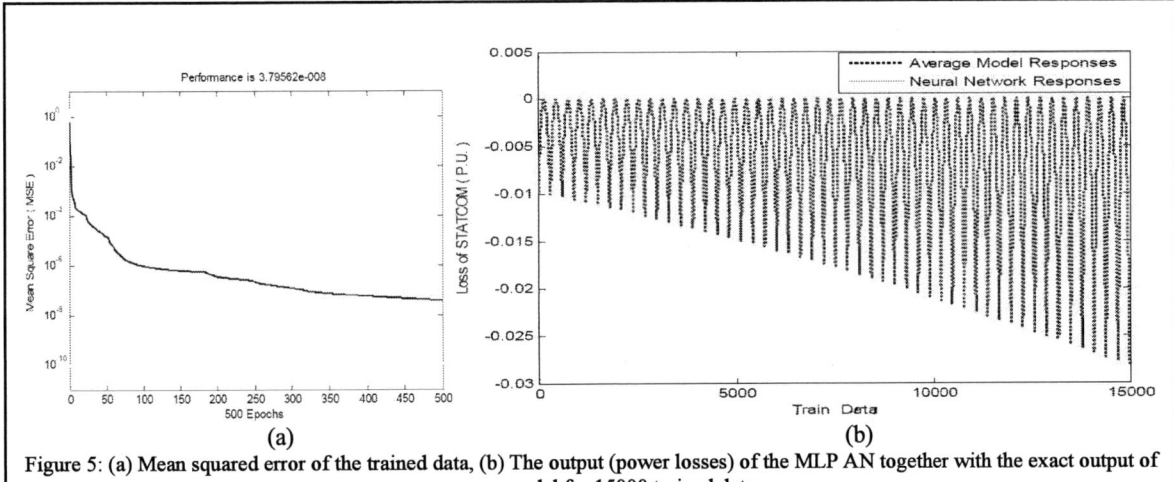

(a)   (b)

Figure 5: (a) Mean squared error of the trained data, (b) The output (power losses) of the MLP AN together with the exact output of average model for 15000 trained data

because of the inherit delays of the structure. The following relationships describe the designed feed forward state of the network:

$$\begin{cases} \mathbf{X_e} = [X_e^1, X_e^2]^t \\ \mathbf{X_i} = [X_i^1, X_i^2]^t \end{cases} \quad (1)$$

$$Net_{e_j}^1(t) = a_{o_j}^1 X_e(t) + a_{1_j}^1 Net_{e_j}^1(t-1) - b_{2_j}^1 Net_{i_j}^1(t-1) \quad (2)$$

$$Net_{i_j}^1(t) = b_{o_j}^1 X_i(t) + b_{1_j}^1 Net_{i_j}^1(t-1) - a_{2_j}^1 Net_{e_j}^1(t-1) \quad (3)$$

$$NetT_j^1(t) = Net_{e_j}^1(t) + Net_{i_j}^1(t) \quad (4)$$

$$O_j^1(t) = f_j^1(t) = \frac{1 - e^{-NetT_j^1(t).k}}{1 + e^{-NetT_j^1(t).k}} \quad (5)$$

All parameters of the network are trained using the back propagation method, and the structure is designed

by trial end error. In this research, different structures were examined in which 25 neurons is considered for the hidden layer alongside with one neuron for the output layer.

Designed structure of Fig. 2 is simulated with MATLAB, where Fig. 3(a) shows the reduction in mean squared error of the training data, Fig. 3(b) depicts the output of the DNN as well as the power losses analysed by the average model, and details of zooming Fig. 3(b) for various terminal voltages over $\alpha \in [-1.5°, 1.5°]$ will be presented in the full paper. Simulation results indicate that the AN model of STATCOM, developed by the DNN, introduces the mean squared error of 0.12% and mean error of 0.3775% for the trained data. This situation confirms that the DNN is unable to develop successfully the *AN* model of STATCOM because the DNN uses the history of the response to identify the progressing output.

B. The AN model using the MLP

The MLP neural network is a global estimator, which an initial processing on training data is taken place followed by designing its structure for the AN model. Structure of this design is shown by Fig. 4, where three hidden layers are proposed for the MLP neural network. By examining and performing various tests, eight neurons are considered for the first and the second layers and one neuron for the third layer. Neuron functions of the first and the second layers are considered identical as follows:

$$f^1(n) = f^2(n) = \frac{e^n - e^{-n}}{e^n + e^{-n}} \qquad (6)$$

Also, weighting matrices of the three layers are $\mathbf{W}^1$, $\mathbf{W}^2$ and $\mathbf{W}^3$, and the 2×1 input vector $\mathbf{P}$ includes the terminal voltage and injected reactive power of STATCOM to bus n as shown by Fig. 1(c). Three vectors $\mathbf{b}^1$, $\mathbf{b}^2$ and $\mathbf{b}^3$ are the threshold values of the neurons, and three outputs of the three layers are $\mathbf{a}^1$, $\mathbf{a}^2$ and $\mathbf{a}^3$.

To train this network, again 15000 operating states are used. Figure 9(a) demonstrates the reduction of the mean squared error during the training process, and Fig. 5(b) gives the response of the MLP *AN* model to these trained data alongside with the exact power losses of the average model. Simulation results show that when the AN model of STATCOM is identified by the MLP, then the error between the average and the AN models is considerably low. The mean error is equal to 0.0147%, and the mean squared error is 0.000003795%. Hence, the MLP identifies the average model with an acceptable error, much lower compared to the DNN.

IV. CONCLUSION

Average model of STATCOM describe its exact operation, while high frequency ripples are ignored. This model can be used for steady state analysis of power systems such as load flow program. However, an identifier is needed to bridge the time domain average model to the frequency domain power system analysis.

This paper aims at doing this objective by designing two neural network identifiers; the dynamic neural network (DNN) and the multi-layer perceptron (MLP). These neural networks are trained using up to 15000 steady state operating data that are obtained by simulating the average model of STATCOM. Outcomes of the two identifiers are compared with the exact solutions of the average model. The results show that the MLP provides much more accurate outcomes compared to the DNN. Therefore, the MLP identifier can be applied to the power system planning and analysis purposes.

ACKNOWLEDGEMENT

The authors would like to thank the support of the Research Laboratory of power quality and reactive power control in K. N. Toosi University of Technology.

REFERENCES

[1] A. Berizzi and et. al., "Enhanced Security-Constrained OPF with FACTS Devices", *IEEE Transactions on Power Systems*, vol. 20, no. 3, August 2005, pp. 1597-1605.

[2] W. Feng and G. B. Shrestha , "Allocation of TCSC Device to Optimize Total Transmission Capacity in a Competitive Power Market ", *IEEE PES Winter Meeting 2001*, vol. 2, Issue 2, pp. 587-593.

[3] G. M. Huang and P. yan, "Establishing Pricing Schemes for FACTS Device in Congestion Management", *IEEE PES General Meeting 2003*, vol. 2, Issue 13-14, July 2003, pp. 1025-1030.

[4] S. C. Srivastava and P.Kumar, "Optimal Power Dispatch in Deregulated Market Considering Congestion Management", *IEEE, International Conf. on Electric Utility Deregulation and Restructuring and Power Technologies Proceedings*, DRPT 2000, vol. 1, pp. 53-59.

[5] S. Verma et al, "FACTS Devices Location for Enhancement of Total Transfer Capability", *IEEE PES Winter Meeting 2001*, vol. 2, 2001, pp. 522-527.

[6] M. Tavakoli Bina and D. C. Hamil, "Average Circuit Model for Angle-Controlled STATCOM", IEE Proceedings.-Electric Power Applications, vol. 152, no.3, May 2005, pp. 653-659.

[7] P. T. Krien and et al, "On the use of averaging for the analysis of the power electronic systems", *IEEE Transactions on Power Electronics*, vol. 5 , no. 2 , pp. 182-190, April 1990.

[8] *M. Tavakoli Bina and M. Panahlou*, "Application of Averaging Technique to the Power System Optimum Placement and sizing of Static Compensators" *International Power Engineering Conference, 2005, IPEC 2005*, 29 Nov.-2 Dec. 2005, pp.1-6.

Hybrid Simulation of Power Systems with Dynamic Phasor SVC Transient Model

Zhijun E *, K. W. Chan **, and D. Z. Fang *

* School of Electrical Engineering and Automation, Tianjin University, Tianjin 300072, P. R. China
** Department of Electrical Engineering, The Hong Kong Polytechnic University, Hong Kong SAR

Abstract--This paper proposes a new hybrid simulator which interfaces a dynamic phasor (DP) model of a static VAR compensator (SVC) into the conventional transient stability program (TSP). In this hybrid simulator, the electromagnetic transient simulation of SVC is implemented using dynamic phasor model while the transient stability simulation of the external system is carried out using fundamental phasor model. By taking the fundamental and fifth harmonics into consideration, the dynamic phasor model of SVC presented in this paper is a large-signal, time-variant model which can simulate the nonlinear characteristics of SVC with fast execution speed and high accuracy. The use of DP modeling eliminates the problem of phase discontinuity and dc-offset found in conventional TSP-EMTP hybrid simulation. The effectiveness of this approach is validated using the benchmark results obtained by the DCG EMTP software on the 9-bus and New England 39-bus power systems.

Index Terms--hybrid simulation, dynamic phasor, transient stability simulation, static VAR compensator.

I. INTRODUCTION

Digital simulators using modern computational techniques are powerful tools for power system stability and security analysis. Though transient stability program (TSP) based on fundamental frequency phasor model can quickly simulate the behavior of system after disturbance, it cannot provide detailed dynamic responses of highly nonlinear components such as HVDC links and FACTS devices especially under asymmetrical faults. On the other hand, electromagnetic transients program (EMTP) can provide accurate and detail resolution of all types of transients; but it is not an ideal tool to study dynamics and security of large size power system network because of its high demands on computational resources.

Hybrid simulators, which integrate TSP and EMTP into one integrated tool with capability of simulating lager power systems, can provide accurate dynamics of nonlinear components. In hybrid simulation [1-4], a power system is split into a transient stability (TS) subsystem and an electromagnetic transient (EMT) subsystem along the connection buses called as the interface buses. The EMT subsystem, which includes nonlinear components, is simulated using EMT models and produces voltage and current waveforms; whereas, the remaining TS subsystem is represented using a TSP. Comparing with pure TSP, fast electromagnetic transient

responses of HVDC or FACTS devices can be obtained by the EMTP simulation. Since TSP adopts much larger integration time-step than that of EMTP, hybrid simulation has higher computational efficiency and requires less computer storage than the pure EMTP. However, the phase discontinuity in TSP equivalent and the effect of dc-offset in EMTP equivalent are the main problems which reduce the accuracy of this type of TSP-EMTP hybrid simulation [5].

Based on the time-varying Fourier coefficient series of the system variables, dynamic phasor [6-9] model (DP) can catch the dynamic behavior of the original detailed model. So far dynamic phasor modeling has been successfully applied in power systems for the modeling of nonlinear components. In recent years, a number of dynamic phasor models of nonlinear components, such as HVDC [10], TCSC [11], STATCOM [12] and UPFC [13], have been proposed and satisfactory performances were obtained. Dynamic phasor modeling has also been successfully applied in dynamic simulation and analysis of the power system with nonlinear devices.

In this paper, a new hybrid simulator which interfaces dynamic phasor model to conventional transient stability simulator is presented. In the new simulator, the whole power network consists of two subsystems: the dynamic phasor subsystem such as nonlinear components and the external subsystem. The two subsystems are simulated in different model types and time steps separately. By using relatively larger simulation time step in electromagnetic transient simulation, higher speed is obtained in new hybrid simulator based on dynamic phasor model than TSP-EMTP hybrid simulation. And less error is produced in the interface because both TSP and DP are based on phasor domain.

II. DYNAMIC PHASOR MODEL OF SVC

A. Basic concept of dynamic phasor

The approach of dynamic phasor is firstly called as the method of state-space averaging and based on the time-varying Fourier coefficients. A complex time domain waveform $x(\tau)$ can be represented on the interval $\tau \in (t-T, t]$ using a Fourier series of the form:

$$x(\tau) = \sum_{k=-\infty}^{\infty} X_k(t) e^{jk\omega_s \tau} \quad (1)$$

where, $\omega_s = 2\pi/T$ and $X_k(t)$ are the complex time-varying Fourier coefficients, which called as dynamic

---

The authors gratefully acknowledge the support of the Hong Kong Polytechnic University under Project A-PA2L.

phasor. The k th dynamic phasor at time t is determined by the following expression:

$$X_k(t) = \frac{1}{T}\int_{t-T}^{t} x(\tau)e^{-jk\omega_s\tau}d\tau = \langle x \rangle_k(t) \quad (2)$$

The dynamic phasor method is based on the idea of frequency decomposition, and focus on the dynamics of the significant Fourier coefficient. There are two key and useful properties of the dynamic phasors:

1. k-phasor differentiaj characteristic:

For the k th Fourier coefficient, the differential with time satisfy the following formula:

$$\frac{dX_k}{dt}(t) = \left\langle \frac{dx}{dt} \right\rangle_k(t) - jk\omega_s X_k(t) \quad (3)$$

2. Product of dynamic phasors:

The k th phasor of a product of two time-domain waveform $x(\tau)$ and $y(\tau)$ can be obtained by the following operation:

$$\langle xy \rangle_k = \sum_{i=-\infty}^{\infty} \langle x \rangle_{k-i} \langle y \rangle_i \quad (4)$$

Also, the time domain waveform $x(\tau)$ can be transformed back from its dynamic phasors by the following equation:

$$x(\tau) = \mathrm{Re}(X_k(t)e^{jk\omega_0\tau})$$
$$= X_{-k}(t)e^{-jk\omega_0\tau} + X_{-(k-1)}(t)e^{-j(k-1)\omega_0\tau} + \cdots$$
$$+ X_{k-1}(t)e^{j(k-1)\omega_0\tau} + X_k(t)e^{jk\omega_0\tau} \quad (5)$$

Moreover, since $x(\tau)$ is real,

$$X_{-k} = X_k^*$$

where the operator * means the conjugate of a complex number.

B. Dynamic phasor model of SVC

A TCR is one of the most important building blocks of thyristor-based SVC. Although it can be used alone, it is more often employed in conjunction with fixed or thyristor-switched capacitors to provide rapid, continuous control of reactive power over the entire selected lagging-to-leading range. In this paper, a three-phase SVC is interfaced into TSP to implement the hybrid simulation. The SVC consists of three delta-connection single-phase TCRs, fixed capacitor and filter circuit. The DP model of every component is described as follow.

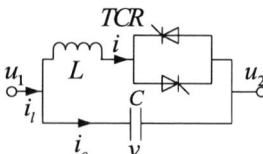

Fig. 1: A single phase circuit of SVC

1. Dynamic phasor model of TCR

A single phase TCR is shown in Fig. 2. If the two thyristor valves are fired symmetrically in the positive and negative half-cycles of supply voltage, only odd-order harmonics would be produced. In addition, the delta connection of the three single phase TCRs could prevent the triple harmonics from percolating into the transmission lines. However, since the 5th harmonics can not be cancelled out in the lines, it has to be taken into consideration in the dynamic phasor model of SVC for realistic modelling of its dynamics.

From the circuit, its time-domain model can be obtained as:

$$\begin{cases} C\dfrac{dv}{dt} = i_l - i \\[2mm] L\dfrac{di}{dt} = sv \end{cases} \quad (6)$$

where s is switch function, $s = 1$ when one thyristor is full conducting, and $s = 0$ when both thyristors are shut. With $<x>_k$ rewritten as X_k, the dynamic phasor model of SVC is obtained by the differential characteristic as the following formula:

$$\begin{cases} C\dfrac{dV_k}{dt} = -jk\omega_s CV_k + I_{lk} - I_k \\[2mm] L\dfrac{dI_k}{dt} = -jk\omega_s LI_k + <sv>_k \end{cases} \quad k=1,5 \quad (7)$$

where, $\langle sv \rangle_k$ can be calculated with

$$\langle sv \rangle_k = \sum_{l=-5,-1,1,5} S_{k-l}V_l \quad (8)$$

The complete TCR dynamic phasor model is as follow:

$$C\frac{dV_k^R}{dt} - k\omega_s CV_k^I = I_{lk}^R - I_k^R$$

$$C\frac{dV_k^I}{dt} + k\omega_s CV_k^R = I_{lk}^I - I_k^I$$

$$L\frac{dI_k^R}{dt} - k\omega_s LI_k^I = \sum_{m+n=k}[S_m^R V_n^R - S_m^I V_n^I]$$
$$+ \sum_{-m+n=k}[S_m^R V_n^R + S_m^I V_n^I]$$
$$+ \sum_{m-n=k}[S_m^R V_n^R + S_m^I V_n^I]$$

$$L\frac{dI_k^I}{dt} + k\omega_s LI_k^R = \sum_{m+n=k}[S_m^I V_n^R - S_m^R V_n^I]$$
$$+ \sum_{-m+n=k}[-S_m^I V_n^R + S_m^R V_n^I]$$
$$+ \sum_{m-n=k}[S_m^I V_n^R - S_m^R V_n^I] \quad (9)$$

where the superscript R and I denote the real and imaginary parts of the defined quantities.

2. Dynamic phasor model of switch function s:

The dynamic model of the switching function which simulates the nonlinear operation of TCR[13] is a key element of TCR DP model.

$$S_0 = \frac{1}{T}\int_{t-T}^{t} s(\tau)\cdot d\tau = \frac{\sigma}{\pi} \quad (10)$$

$$S_m = \frac{1}{T}\int_{t-T}^{t} s(\tau)\cdot e^{-jm\omega_s\tau}d\tau = \frac{j}{m\pi}[e^{-jm\tau} - e^{-jm\alpha}]$$
$$= \frac{1}{m\pi}[\sin m(\alpha + \sigma) - \sin m\alpha]$$
$$+ \frac{j}{m\pi}[\cos m(\alpha + \sigma) - \cos m\alpha] \quad (m \neq 0) \quad (11)$$

1805

The delay angle α and conduction angle σ depend on the closed-loop control of SVC.. For each simulation time step, α and τ are calculated from control loop.

For phase B and C, the models of switch function are different with phase A because of the angle difference between phases. The switch function models of phase B and C can be calculated by the following equations.

$$\begin{cases} S_{k,B} = e^{k*(-2\pi/3)} S_{k,A} \\ S_{k,C} = e^{k*(2\pi/3)} S_{k,A} \end{cases} \tag{12}$$

3. *Dynamic phasor model of control circuit* :

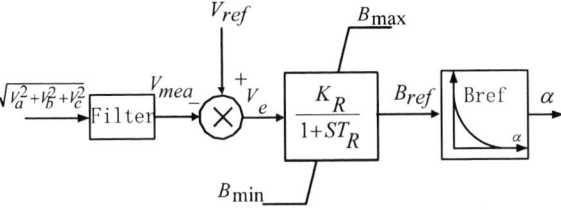

Fig. 2: Block diagram of SVC control circuit

A reliable control system model is absolutely necessary for SVC practical operation. The time-domain model of SVC control circuit is shown in Fig.2. In the dynamic model of SVC, the same voltage control unit is used. The difference is the value of V_{mea} is calculated from the dynamic phasors of voltage instead of the instantaneous value. Note that several steps simulation with smaller step-time should be taken to get correct firing angle at the beginning of the whole simulation.

4. *Dynamic phasor model of filter cuicirts*

The filter circuit is also necessary to model SVC accurately. Actually it is a RLC circuit which consists of one RL circuit and one capacitor circuit shown as Fig.3. The dynamic phasor model of RL circuit and capacitor circuit can be deduced similarly.

For RL circuit, we have the time-domain model:

$$v(t) = L\frac{di(t)}{dt} + Ri(t) \tag{13}$$

and the dynamic phasors are

$$L\frac{dI_k}{dt} = V_k - jk\omega_s L I_k - R I_k \tag{14}$$

For capacitor circuit, we have the time-domain model:

$$i(t) = C\frac{dv(t)}{dt} \tag{15}$$

and the dynamic phasors are

$$\frac{dV_k}{dt} = \frac{1}{C}I_k - jk\omega_s V_k \tag{16}$$

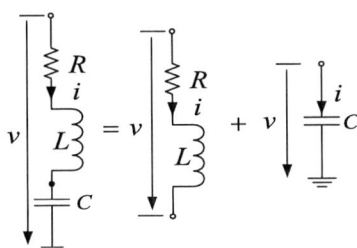

Fig. 3: Single-phase filter circuit

III. HYBRID SULATION BASED ON DYNAMIC PHASORS

A hybrid simulator takes advantage of the computational inexpensive dynamic representation of the main network in a TSP with the accurate dynamic modeling of the remaining components. In order to enable TSP and EMTP to interface and run as a single package, the following issues have to be addressed.

A. Network partitioning

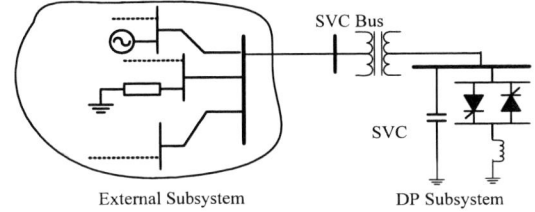

Fig. 4: The system decomposition

In the new simulator, the whole power network is partitioned into two parts along interface bus: the dynamic phasor subsystem and the external subsystem as shown in Fig. 4. The dynamic phasor subsystem comprises of nonlinear components such as HVDC links and FACTS devices which need to be simulate in detail. Compared to conventional TSP-DP hybrid simulator, the nonlinear components are simulated by dynamic phasors model instead of full time-domain model. Similarly the external subsystem is simulated with conventional fundamental frequency phasor model. In a regular fixed time interval, the TSP and DP exchange the necessary data to form and coordinate the hybrid simulation.

B. Equivalent between TS and DP subsystem

Based on the interface parameters, the external subsystem is viewed as a Norton equivalent current source with shunt impedance. The dynamic phasors of current and impedance can be easily transformed from the results of transient stability simulation. And only the fundamental dynamic phasor of current source can be transformed to the DP subsystem simulation. In TSP, the nonlinear component is considered as dynamic load injection at interface bus which is transmitted from dynamic phasor simulator. Since the results obtained by the DP simulation are also phasors, the equivalent load can be easily calculated easily by the dynamic phasors of voltage and current.

In TSP-EMTP hybrid simulation, the fundamental frequency phasors of voltage and current are extracted from the waveforms obtained by the EMTP simulation. Because of the effect of dc-offset, the error of extraction reduced the accuracy of the hybrid simulator by incorrect power injection [5]. However, such reduction will be avoided in TSP-DP hybrid simulator because the results of DP simulation are also phasors.

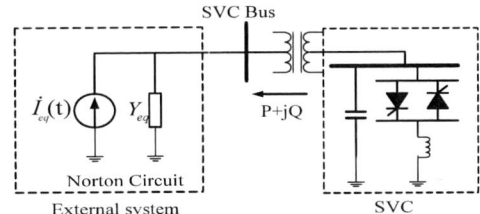

Fig. 5: The equivalent between external subsystem and DP subsystems

1806

C. interface protocol

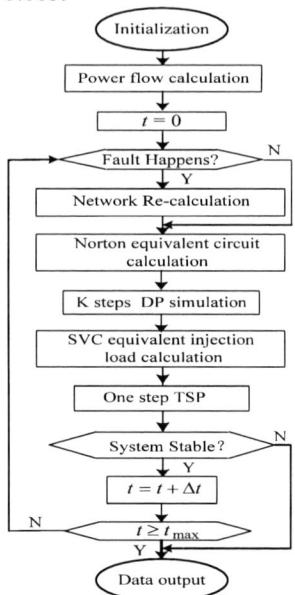

Fig. 6: The flowchart of hybrid simulation

The procedure of the parameters exchanging process is similar to the one of conventional hybrid simulation [1-2][14]. The parameters exchanging accomplished at every TSP step time-point. The TSP simulates one step and then DP simulator runs several steps. The difference is the number of DP subsystem simulation steps between two parameters exchanging point. Because of larger step being taken, less steps DP simulation needs to be implemented in the new hybrid simulator. Such interface is also easy to implement because the TSP and DP simulators are both based on phasors domain. The flowchart of the hybrid simulation is shown in Fig 6.

IV. CASE STUDY

Case studies are performed on the WSCC 9-bus system and New England 10-generator test system. Generators are represented by two-axis generator model without exciter and all loads are modeled by constant impedance. The time step of TSP is 20ms while the one for DP is 500μs. The convergence precision is ε=10⁻⁶. For the comparison of the accuracy, DCG EMTP simulations of the two test systems are also carried out to produce the benchmark results.

A. Case 1

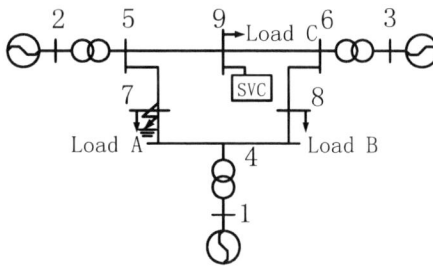

Fig. 7: The IEEE 9-bus test system

The first case study is performed on the 9-bus system including a SVC shown as Fig. 7. A three-phase SVC, whose electromagnetic model [14] is installed at bus 9.

Simulation starts from steady state. A three-phase short circuit fault occurs in bus 7 at 0.5s and disappears at 0.8s.

1. Electromagnetic transient assessment

The emphasis of the EMT assessment is placed on the waveforms of voltage of the SVC bus. Both the curves from the new hybrid simulator and DCG EMTP software are presented in same figure for comparison. The results of comparison are shown in Fig. 8.

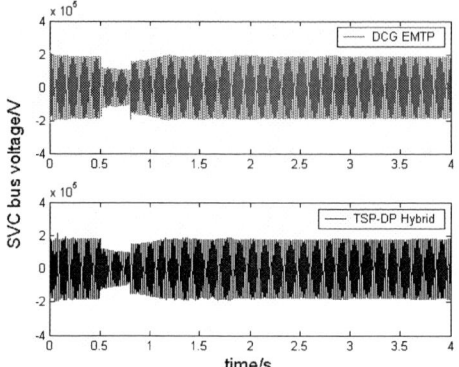

Fig. 8: Waveform of Phase A voltage of SVC bus

2. Ttransient stability assessment

The emphasis of the TS assessment is placed on the waveforms of voltage at the generator bus. The results of comparison are shown in Fig. 9-11. It is clearly show that the dynamic phasors from the new hybrid simulator trace the envelope of the instantaneous voltage curves from the benchmark closely as shown in Fig. 8. The RMS value of voltage in bus 1, 2 and 3 from TSP can also predict the system transient stability correctly from Fig. 9-11. As shown above, the TSP-DP hybrid simulator can simulate the dynamic response of whole system to disturbance precisely.

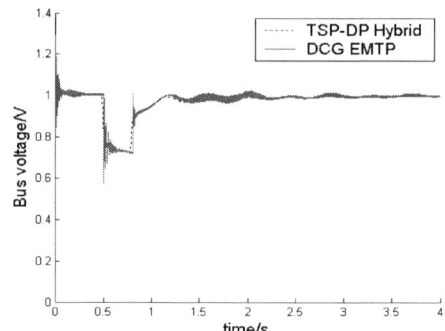

Fig. 9: RMS Value of voltage at bus1

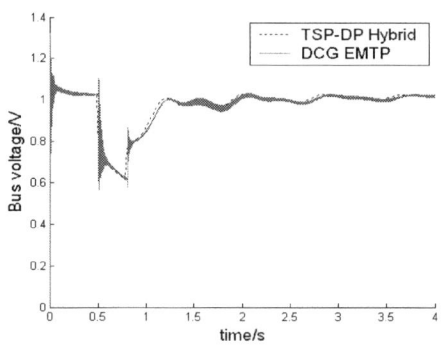

Fig. 10: RMS Value of voltage at bus2

1807

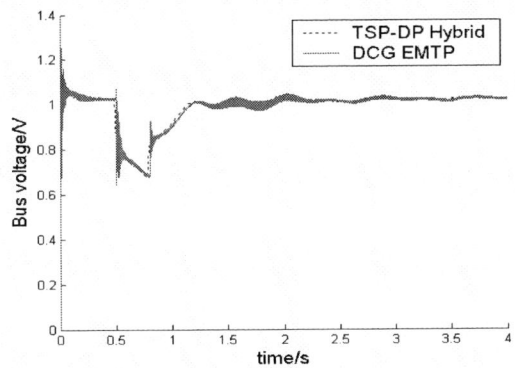

Fig. 11: RMS Value of voltage at bus3

B. Case 2

The 10-generater New England test system is also applied to study the validity of the new approach. The SVC is installed at bus 12. Simulation begins in steady state. A 3-phase short circuit fault occurs in bus 1 at 1.2 s and disappears at 1.3 s.

1. Electromagnetic transient assessment

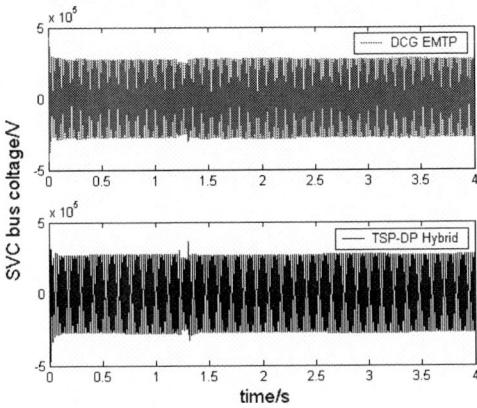

Fig. 12: Waveform of Phase A voltage of SVC bus

The emphasis of the EMT assessment is still focused on the waveforms of voltage of the SVC bus. Both the curves from the new hybrid simulator and DCG EMTP software are presented in same figure for comparison. The results of comparison are shown in Fig. 12.

2. Ttransient stability assessment

The emphasis of the TS assessment is placed on rotor angle and the waveforms of voltage at the generator bus. The results of comparison are shown in Fig. 13-15.

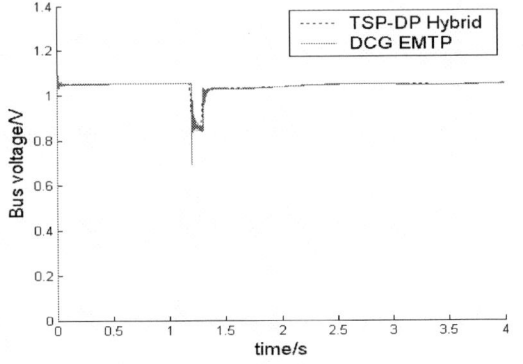

Fig. 13: RMS Value of voltage at bus30

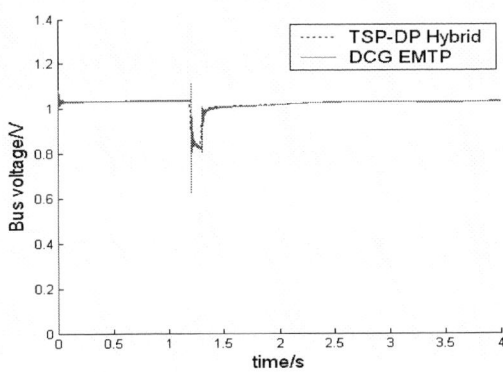

Fig. 14: RMS Value of voltage at bus 31

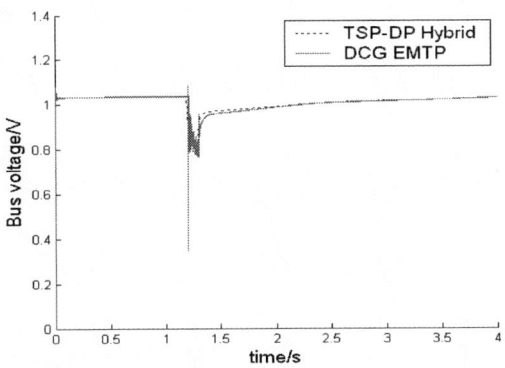

Fig. 15: RMS Value of voltage at bus 39

Waveforms in Fig. 12 show that the curves of voltage obtained by the new hybrid simulator match the benchmark closely. Fig. 13-15 demonstrate the RMS value of voltage in bus 30, 31 and bus 39 obtained by TSP are also credible. As shown above, the hybrid simulation approach based on dynamic phasors can simulate the dynamic response of SVC to disturbance precisely

V. Conclusion

A new hybrid simulator interfacing dynamic phasor into transient stability program is proposed in this paper. In this simulator, FACTS devices are simulated by DP model and the remaining network is simulated by TSP model. The 9-bus system and New England system including a SVC are used to evaluate the new simulator. By comparing with the benchmark results of DCG EMTP software, it is clearly shown that the TSP-DP hybrid simulator can simulate the whole system quickly and effectively.

References

[1] M. D. Heffernan, K. S. Turner, J. Arrillaga, C. P. Arnold, "Computation of AC-DC System Disturbance, Part I, II and III," IEEE Trans. on Power Apparatus and Systems, vol. PAS-100, Nov. 1981, pp. 4341-4348.

[2] J. Reeve, R. Adapa. "A New Approach To Dynamic Analysis of AC Networks Incorporating Detailed Modeling of DC Systems Part I: Principles and Implementations," IEEE Transactions on Power Delivery, vol.3, Oct. 1988, pp. 2005-2011.

[3] G. W. J. Anderson, N. R. Watson, C. P. Arnold, J. Arrillaga, "A new hybrid algorithm for analysis of HVDC and FACTS systems," in Proc. 1995 IEEE Energy Management and Power Delivery Conf., pp. 462-467

[4] M. Sultan, J. Reeve, R. Adapa, "Combined transient and dynamic analysis of HVDC and FACTS systems," IEEE Trans. on PD, vol.13, pp. 1271-1277, Oct. 1998.

[5] H. T. Su, K. W. Chan, L. A. Snider. "Evaluation study for the integration of electromagnetic transients simulator and transient stability simulator," Electric Power Systems Research, 75, 2005, pp. 67-78.

[6] S. R. Sanders, J. M. Noworolski, X. Z. Liu, G. C. Verghese, "Generalized averaging method for power conversion circuits", IEEE Trans. on Power Electronics, vol. 6, Apr. 1991, pp. 251-259.

[7] C. L. Demarco, G. C. Verghese, "Bring phasor dynamics into the power system load flow", 25th North America Power Symposium,1993.

[8] V. Venkatasubramanian, "Tools for dynamic analysis of the general large power system using time-varyingphasors", International Journal on Electric Power and Energy Systems, December 1994,pp. 365-376.

[9] J. Mahdavi, A. Emaadi, M. D. Bellar, et al. "Analysis of power electronic converters using the generalized state2space averaging approach". IEEE Trans. on Circuits System, vol. 48, Aug.1997, pp.767-770.

[10] H. J. Zhu, Z. X. Cai, H. M. Liu, Y. X. Ni, "Multi-infeed HVDC/AC power system modeling and analysis with dynamic phasor application," in Proc. 2005 IEEE/PES Transmission and Distribution Conference and Exhibition: Asia and Pacific, pp.1-67.

[11] P. Mattavelli, A. M. Stankovic, G. C. Verghese, "SSR analysis with dynamic phasor model of thyristor-controlled series capacitor," IEEE Transactions on Power Systems, vol.14, Feb. 1999, pp. 200-208.

[12] Q. G. Qi, K. W. Chan, S. Y. R. Hui, "Dynamic phasors applied to a 3-level STATCOM: modeling, simulation and experimental verification," in Proc. 2004 IEEE-PES/CSEE International Conference on Power System Technology.

[13] P. C. Stefanov, A. M. Stankovic, "Modeling of UPFC operation under unbalanced conditions with dynamic phasors," IEEE Transactions on Power Systems, vol. 17, May. 2002, pp. 395-403.

[14] H. T. Su, K. W. Chan, L. A. Snider, "Interfacing an electromagnetic SVC model into the transient stability simulation," in Proc. 2002 IEEE PowerCon International Conf., pp. 1568- 1572.

CONTROL OF CURRENT- SOURCE ACTIVE POWER FILTER USING UNIT VECTOR TEMPLATE IN THREE PHASE FOUR WIRE UNBALNCED SYSTEM

Vadirajacharya.K[1],   Pramod Agarwal[2] and   H.O.Gupta[3]
Student Member IEEE   *Member IEEE*   *Member IEEE*
[1] Research scholar, [2,3] Professor in Electrical Engineering
Department of Electrical Engineering, Indian Institute of Technology, Roorkee. (Uttarakhand). India

Abstract- In recent years there has been considerable interest in the development and application of active power filters using current source active power filters for harmonic filtering. The unbalanced load resulting zero sequence current can affect the performance of shunt active filter adversely. In this paper a novel active, filter control for a three-phase four-wire unbalance system has been designed. This method allows the calculation of the reference current under non-sinusoidal and unbalanced three-phase supply voltage conditions. The proposed filter is based on Akagi's Instantaneous Power Theory (IPT), which are organized in different independent blocks. We introduce a Unit Vector Template to derive fundamental voltage signal from an unbalanced source. The reference currents are then compared with actual source current in a hystersis band controller. Extensive simulation results obtained using MATLAB/SIMULINK under R-L non- linear loading conditions are presented and discussed.

Key words: Active filtering, CSI, Harmonic filtering, IPT, Shunt filters, Unbalance system

I. INTRODUCTION

In recent years there has been considerable increase in the use of power electronics equipment in domestic, commercial and industries at low voltage distribution systems. These devices utilize switched-mode power converters, which typically draw excessive harmonic currents resulting in alarming levels of harmonic distortion. One of the suggested methods to neutralize the current harmonics produced by these nonlinear loads is to use active power filters [1,2]. These active filters are divided in to two types: voltage source active filters (VSAF) and current source active filters (CSAF). With the availability of new IGBTs with reverse blocking capability, the use of current source active filter is increasing due to its inbuilt short circuit protection capability, higher efficiency at low power loads, simple open loop current control and effective filtering of harmonics[3]. Fig. 1 shows the most

commonly used main circuit structure of the current-source active power filter [4]. For proper operation of APF, it is necessary to maintain a DC link current constant, which is slightly greater than the peak of the filter current demand.

Harmonic filtering by active power filters, based on injection method, is performed by detecting the harmonic distortion in the power line current and injecting a current equal-but opposite to the distortion into the power line to compensate for the harmonics. There are different control methods that can be used to generate the compensating current that cancels the harmonics in the load current [5]. It is reported in many literature that Akagi's instantaneous reactive power theory is one among commonly used technique to detect reactive and harmonic current components. It has been proved with effective operation and good performance under balanced voltage source conditions. This method has been well developed for the purpose of line current compensation. However, when the voltage source is significantly unbalanced, great errors may result from calculation formula by direct application of the instantaneous power theory [6-8] Reviewing system performance using the instantaneous power theory, it is difficult to operate efficiently under unbalanced three-phase voltage sources to achieve reactive and harmonic current compensation. In this paper, it is verified that the Akagi's IPT can be extended to unbalance system using fundamental voltage template derived from unbalance source.

This paper presents a simple and robust control strategy for three-phase current source active filter used for current compensation in three-phase four-wire unbalance system. It uses a unit vector template to derive fundamental signal from the unbalanced distorted signal, which is then used to generate reference signal for derivation of compensating signal. The performance of the system is verified by extensive simulation on SIMPOWERSYSTEM (SPS) of MATLAB/ SIMULINK environment.

The system configuration under consideration is presented in section II. The proposed control technique based on unit vector template generation

978-1-4244-0644-9/07/$25.00 ©2007 IEEE   1810

is given in section III. Simulation results are discussed in section IV.

II. SYSTEM CONFIGURISATION

Fig. 2.1 shows a system configurisation of the current source active power filter, installed for compensating both reactive power and harmonic current generated by the three-phase full wave uncontrolled bridge rectifier. The active power filter system consists of a current-source PWM bridge built with six IGBT switches.

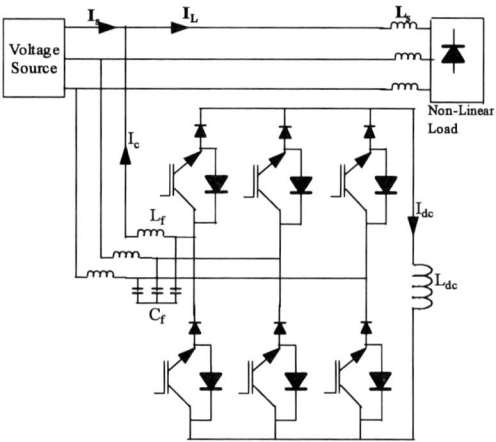

Fig. 2.1. Configurisation of current source active power filter

They have to stand the active filter dc link current i_{dc}. The semiconductor devices are under bidirectional voltage stresses and the maximum values of these are the peak value of the supply line-to-line voltage. Because of these and the very low reverse voltage, blocking capability of the IGBT's additional diodes have to be connected in series with the transistors. Series diode can be eliminated by use of RB-IGBT. The PWM Bridge of the current-source APF is connected to mains through the second order filter (L_fC_f) and as the energy storage of the current-source APF there is an inductor L_{dc} with a dc current i_{dc} flowing through it. The theoretical minimum dc link current is zero, but the current should be at least as high as the peak value of the compensating current.

III. CONTROL STRATEGY

The control structure of an APF can be decomposed in the following blocks: a first block, which allows recovering an equilibrated and balanced voltage system. A second block identifies and filters the harmonics in order to specify the required reference currents for the control algorithms. The third block, reinjects the currents via the APF's power circuit,

composed of an inverter and an output filter. The APF's first control block is achieved through unit vector template, the harmonic identification and filtering is done using Akagi's Instantaneous Power Theory (IPT)[9]. This technique computes the instantaneous active and reactive powers from the measured currents under sinusoidal balanced conditions. Instantaneous active and reactive powers can be decomposed into dc components related to the fundamental frequency and into AC components related the harmonic distortions. A filtering operation is then used to separate the terms produced by the harmonic distortions from the DC components related to the fundamental frequency. The IPT instantaneously identifies the power properties of equilibrated and balanced systems and specifies the required reference currents for the control strategy. Hysteresis band controller is used for generating gating signals for APF.

A. Extraction of Unit Vector Template

The shunt filter performs satisfactorily under balanced and stiff voltage source. If the supply voltage is unbalanced, the compensation signals derived are not accurate. For proper operation of active filter, instantaneous power is derived from unit vector template and source current. The schematic diagram of unit vector template generation is as given below in fig. 3.1[10].

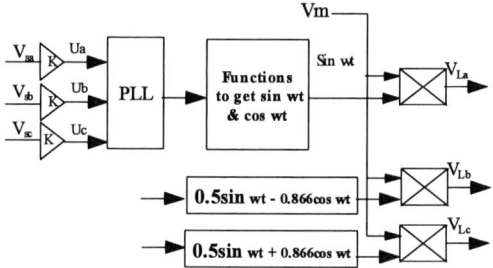

Fig. 3.1. Extraction of Unit Vector Template

The 3-φ distorted input source voltage at PCC contains fundamental component and distorted component. To get unit input voltage vectors Vs, the input voltage is sensed and multiplied by gain K equal to $1/V_m$, where V_m is peak amplitude of fundamental input voltage. These unit input voltage vectors are then taken to phase locked loop (PLL). With proper phase delay, the unit vector templates for different phase are generated as follows

$$U_a = Sin\,\omega t$$
$$U_b = Sin(\omega t - 120^0)$$
$$U_c = Sin(\omega t + 120^0)$$

(1)

1811

The block diagram for the control of shunt active filter is as shown in fig. 3.2.

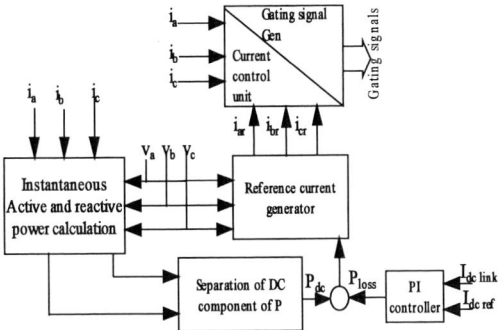

Fig. 3.2. Control block diagram of Shunt controller

The source voltage and load current measured are transformed to α-β co-ordinates using expression 2-3

$$\begin{bmatrix} v_\alpha \\ v_\beta \end{bmatrix} = \begin{bmatrix} 1 & -1/2 & -1/2 \\ 0 & \sqrt{3}/2 & -\sqrt{3}/2 \end{bmatrix} \begin{bmatrix} v_a \\ v_b \\ v_c \end{bmatrix} \quad (2)$$

$$\begin{bmatrix} i_\alpha \\ i_\beta \end{bmatrix} = \begin{bmatrix} 1 & -1/2 & -1/2 \\ 0 & \sqrt{3}/2 & -\sqrt{3}/2 \end{bmatrix} \begin{bmatrix} i_a \\ i_b \\ i_c \end{bmatrix} \quad (3)$$

Assuming the effects by LC filters are negligible, the instantaneous active and reactive power are calculated using equation 6.

$$\begin{bmatrix} p \\ q \end{bmatrix} = \begin{bmatrix} v_\alpha & v_\beta \\ -v_\beta & v_{d\alpha} \end{bmatrix} \begin{bmatrix} i_\alpha \\ i_\beta \end{bmatrix} \quad (4)$$

To compute harmonic free unity power factor, three-phase currents, compensating powers p_c and q_c are selected as

$$p_c = p_{ldc} + p_{loss} \quad (5)$$
$$q_c = 0 \quad (6)$$

Where p_{loss} is the instantaneous active power corresponding to switching loss and resistive loss of APF. Total instantaneous active power is calculated by adding real power loss due to switching as shown in fig.3.2. The orthogonal components of the fundamental current are then obtained by using total instantaneous active power and voltage vector as shown in equation (7).

$$\begin{bmatrix} i_\alpha \\ i_\beta \end{bmatrix} = \begin{bmatrix} v_\alpha & v_\beta \\ -v_\beta & v_\alpha \end{bmatrix}^{-1} \begin{bmatrix} p_c \\ q_c \end{bmatrix} \quad (7)$$

The a-b-c components of fundamental reference current are obtained as given in equation (8).

$$\begin{bmatrix} i^*_{sa} \\ i^*_{sb} \\ i^*_{sc} \end{bmatrix} = \begin{bmatrix} 2/3 & 0 \\ -1/3 & 1/\sqrt{3} \\ -1/3 & -1/\sqrt{3} \end{bmatrix} \begin{bmatrix} i_\alpha \\ i_\beta \end{bmatrix} \quad (8)$$

These reference currents are then compared with actual source current in a hystersis controller band to derive the switching signals to shunt filter.

IV. SIMULATION RESULTS

To verify the performance of APF the system was first simulated with balanced source voltage ,using Power System Blockset in MATLAB/SIMULINK environment. All the compensators are implemented using equivalent discrete blocks. The sampling time is selected as $1e^{-6}$s. Fig. 4.1 to 4.6 shows the simulation results of the proposed system with RL load. The Non-linear load current shown in fig 4.1 is having magnitude of 3.85 Amp and THD of 27.71%. The APF is switched on at 0.01 sec. The variation of source current with respect to time is as shown in fig. 4.2. The THD of source current reduces from 27.71% to 0.45% in almost 0.05 sec. The dc link current gradually increases from an initial value of zero amperes and fairly stabilizes at 5.4 ampere in 0.1 sec, as shown in fig. 4.4.

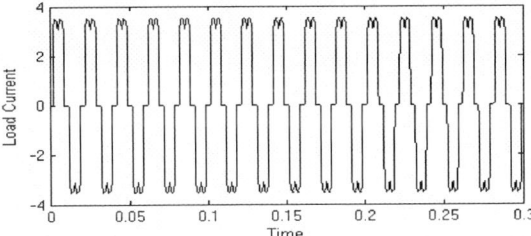

Fig. 4.1. Load current

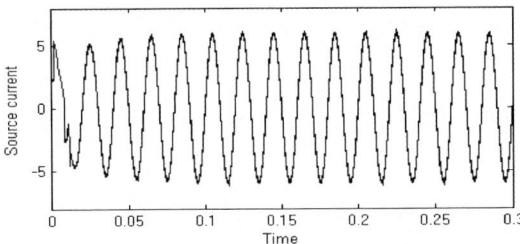

Fig. 4.2. Source current with balanced voltage source

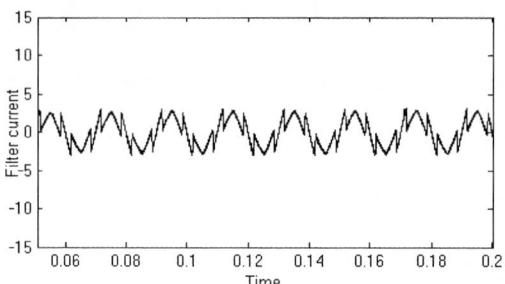

Fig. 4.3. Compensation current with balanced voltage source

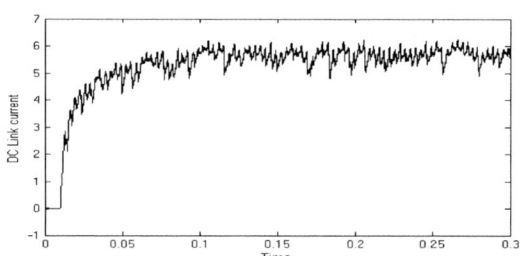

Fig. 4.4. D.C. Link current with balanced voltage source

The variation source current in all other phases is same with THD of 0.45% as shown in fig 4.5.

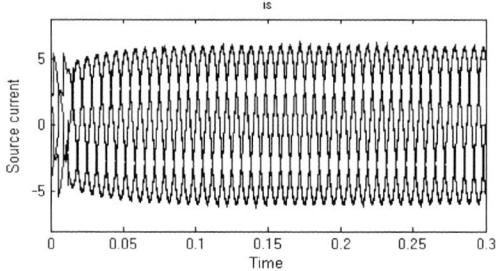

Fig. 4.5. Source Current in all the three phases

The APF also compensates for reactive power. Fig. 4.6 show the performance of APF with respect to reactive power compensation.

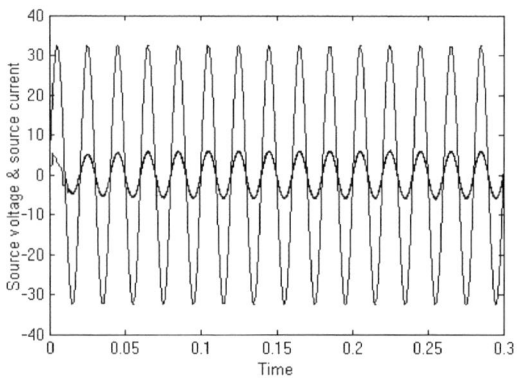

Fig. 4.6. Source Voltage and Source current

If the supply voltage is unbalanced due to unequal loading or any other reason, the shunt filter fails to perform effectively. An unbalance of ±5% is introduced in supply voltage to verify its performance. The performance of shunt filter during such condition is as shown in fig. 4.7-9. The THD of source current is reduced from 27.71% to 3.84% instead of 0.45% as with balanced supply. The dc link current oscillation also increases which effects on stability of filter operation.

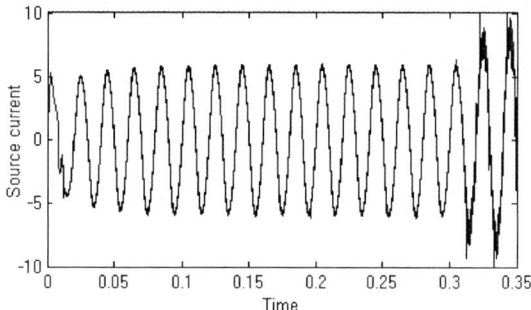

Fig. 4.7. Source current with unbalanced voltage source

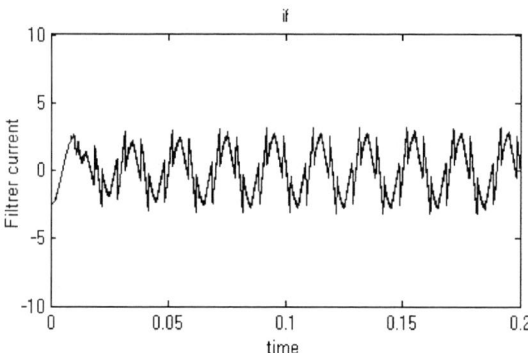

Fig.4.8. Compensation current with unbalanced voltage source

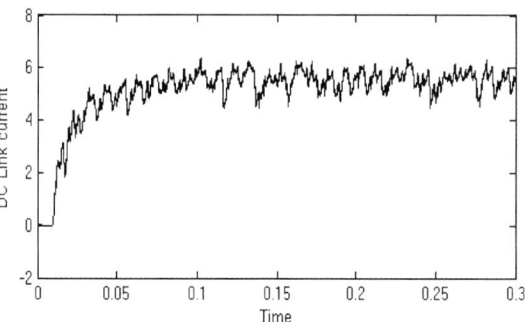

Fig.4.9. D.C. Link current with unbalanced voltage source

The variation source current in phase a is 3.84%, in b is 3.52% while in phase c is 4.01% .The variation of source current in all the three-phases is as shown in fig 4.10.

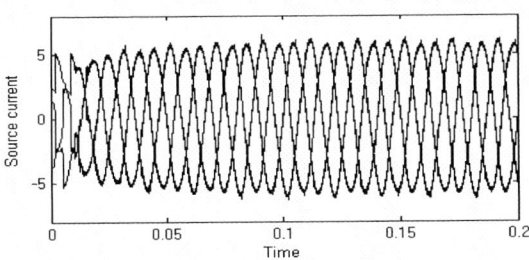

Fig 4.10. Source Current in all the three -phases with unbalance voltage source.

If this unbalance voltage is decomposed in to balanced one, the performance of shunt filter will regain as in balanced system. Fig.4.11 shows the unit vector template obtained from unbalance voltage source.

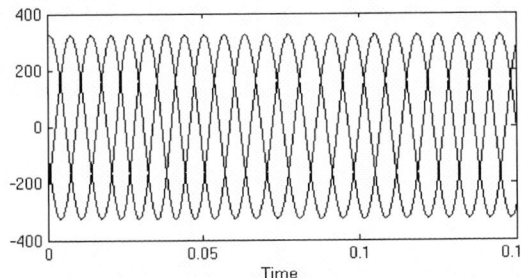

Fig. 4.11. Unit vector template

.Fig. 4.12-15 shows the performance of shunt active filter for unbalanced source voltage, but with unit vector template. The performance of shunt active filter regains its efficiency even source voltage is unbalance, the source current is sinusoidal and in phase with voltage as in balance system. The THD of source current is again reduces from 15.55% to 0.45%. The dc link current also stables at 5.4 amp as shown in fig. 4.14

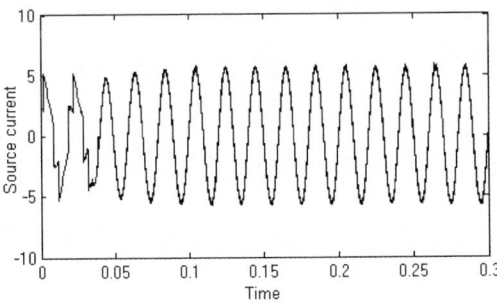

Fig. 4.12. Source current with unit vector template

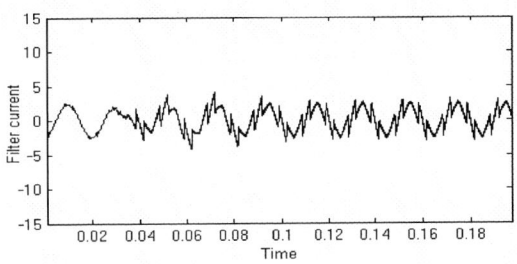

Fig. 4.13. Compensation current with unit vector template

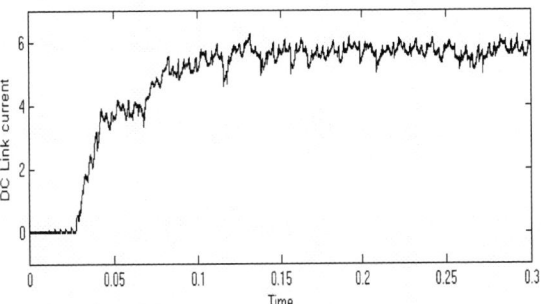

Fig. 4.14. D.C. Link current with unit vector template

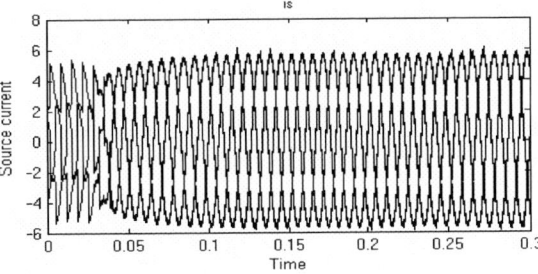

Fig. 4.15. Source Voltage and Source current

V. CONCLUSIONS

In this paper performance of current source, Active Power Filter is analyzed for different loading conditions. The shunt active filter performs efficiently at balanced condition.

However, in three-phase four wire system where unbalance is very common in such cases it is required to derive unit vector template of supply voltage before deriving instantaneous active and reactive power for compensation. The control strategy proposed is very simple, robust and accurate. The system is simulated using MATLAB/ Simulink for R-L load using uncontrolled bridge rectifier. The simulation result validates the proposed scheme for elimination of current harmonics effectively even under unbalanced condition using Akagi's IRP theory.

1814

VI. REFERENCES

[1] V.B. Bhavaraju and Prasad N. Enjeti, "Analysis and design of an active power filter for balancing unbalanced loads," IEEE Trans. Power Electronics, Vol. 8, No. 4 October 1993,pp 640-645

[2] H.Akagi, " New trends in active power filter for power conditioning", IEEE Transactions on Industrial Applications ,Volume.32, No. 6, Nov/Dec 1996,pp1312-1322

[3] Routimo, M. Salo, M. & Tuusa, H. " Comparison of voltage-source and current-source shunt active power filters. IEEE 36th Annual Power Electronics Specialists Conference, PESC05, June 12-16, 2005, Recife, Brazil. pp. 2571-2577.

[4] M. Salo and H. Tuusa, H., "A novel open-loop control method for a current-source active power filter", *IEEE Trans. Ind. Electr.*, Vol. 50, No. 2, pp. 313-321, 2003.

[5] T.C.Green and J.H.Marks, "Control techniques for active power filters", IEE proceedings of Electric power Applications, Vol. 152, No. 2, March 2005, pp 369-381.

[6] H. Akagi, Y. Kanazawa, and A. Nabae, "Instantaneous Reactive Power Comprising Switching Devices Without Energy Storage Components",IEEE Trans. on IA, Vol. IA-20, No. 3, May /June,1984, pp. 625-630.

[7] T. Furuhashi, S. Okuma, and Y. Uchikawa,"A Study on the Theory of Instantaneous Reactive Power", IEEE Trans. on IE, Vol. IE-37, No. 1, February 1990, pp. 86-90

[8] C. E. Lin, C. L. Chen, C. L. Huang, "Reactive and Harmonic Current Compensation for Unbalanced Three-phase Systems", IASTED Power High Tech Conference, March 1991, Tainan, Taiwan, p. 317-321.

[9] Djaffar Ould Abdeslam, Patrice Wira, Jean Mercklé, Damien Flieller, and Yves-André Chapuis, "A Unified Artificial Neural Network Architecture for Active Power Filters", IEEE transactions on industrial Electronics, Vol.54, No-1, Feb2007,p-p 61-76

[10] V. Khadkikar, P. Aganval, A. Chandra, A.O. Bany and T.D. Nguyen, "A Simple New Control Technique For Unified Power Quality Conditioner (UPQC)", 2004 11th International Conference on Harmonics and Quality of Power, Lake Placid, New York, USA, Sept 12-15 2004, pp 289-293.

VII. BIOGRAPHIES

Vadirajacharya graduated in Electrical Engineering from PDA college of Engineering Gulabarga, India in 1984. He completed his PG in Electrical Engineering from SGSITS, Indore, India in 1998. Currently he is pursuing research in the field of power quality converters at Electrical Engineering Dept Indian Institute of Technology Roorkee, India. His fields of interest include Power quality, and Energy auditing.

Pramod Agarwal obtained his Bachelor's degree in Electrical Engineering from University of Roorkee, now Indian Institute of Technology Roorkee (IITR), India. He received his PG and completed his PhD in Electrical Engineering from IITR in 1985 & 1995 respectively. Currently he is Professor in the department of Electrical Engineering, IITR. His fields of interest include Electrical Machines, Power Electronics, Power quality, Microprocessors and microprocessor-controlled Drives, Active power filters, High power factor Converters, Multilevel Converters, and application of dSPACE for the control of Power Converters.

H.O.Guta obtained his Bachelor's degree in Electrical Engineering from University of Jabalpur, India. He received his PG and completed his PhD in Electrical Engineering from University of Roorkee, India in 1975 and 1980 respectively. Currently he is Professor in the department of Electrical Engineering, Indian Institute of Technology Roorkee, India. His fields of interest include Database Management, Transformer, Power system analysis and optimization, System Engineering.

Improved Control of Three Phase Active Filters Using Genetic Algorithms

Bhim Singh, *Senior Member IEEE*, and Varun Singhal

Abstract – This paper deals with a new control technique of system parameter evaluation for a parallel three phase active filters (AF) to eliminate harmonics and to compensate reactive power of non-linear loads. The work is motivated by the need to minimize the burden on the dc-link capacitor voltage under dynamic load conditions. Genetic Algorithm is used to obtain an optimized control of dc-link voltage. A three phase diode rectifier with R-L loading is used as a non-linear load. The system dynamic performance is analyzed based on its transient response to load change and load unbalancing. Moreover, the robustness of the system to filter parameter variation is also studied. The AF is found effective to meet IEEE-519 standard recommendations on harmonics level.

Index Terms – Active Filter, Harmonic Compensation, Power Factor Correction, Genetic Algorithm

I. INTRODUCTION

Solid State control of ac power is in extensive use in a number of applications such as adjustable speed drives, furnaces, and computer power supplies. These power converters behave as non-linear loads to ac supply system and cause harmonic injection and low power-factor problems. Sometime due to use of traction and furnaces some unbalancing may also be present in three phase loads. Active filters have appeared to be a promising solution [1-7] to negate these effects.

Several control schemes for active filters can be found in the literature including direct control [1], instantaneous reactive power theory [2], and sliding mode control [3] to name a few. However, as reported in [4] most of these control algorithms need a number of transformations and are difficult to implement. The improved control algorithm proposed in [4] takes care of all the problems encountered in previous algorithms. However, this algorithm faces the problem of rigorous hit and trial analysis to evaluate the controller parameters of the system.

Genetic Algorithms are evolving as an efficient optimization algorithm to optimize any objective function. A number of researchers have tried with different set of objective functions to obtain system controller parameters using genetic algorithm (GA) [5-7]. However, the objective functions being realized are fairly complex. The power of GA's to evaluate system

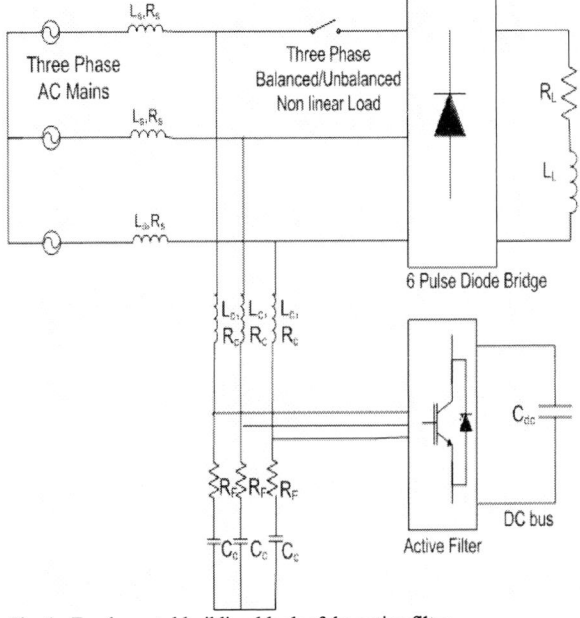

Fig. 1. Fundamental building block of the active filter

controller parameters from a random set of population make it ideal for offline optimizations.

This paper presents an indirect current control of a three phase active filter [4]. The three wire system(Fig.1) is made up of standard three phase IGBT based VSI bridge with the input ac inductors L_c R_c, R-C filter R_F Cc; and a dc bus capacitor C_{dc} to obtain a self supporting dc bus for an effective current control. The load is a 6 pulse diode rectifier fed R-L load. The controller system parameters are evaluated using an offline evaluation of objective function using genetic algorithm. The system is simulated in Matlab Simulink environment using Power Block Set to study its dynamic response under load perturbations and load unbalancing. The performance of APF is evaluated to system parameter perturbations.

II. INDIRECT CURRENT CONTROL

Fig 2 shows the control scheme of the active filter. The band pass filtered voltages (v_{sa}, v_{sb}, v_{sc}) from point of common coupling are used to evaluate unit current vectors in phase with supply voltages as (1).

$$u_{sa} = v_{sa}/V_{sp}; \ u_{sb} = v_{sb}/V_{sp} \text{ and } u_{sc} = v_{sc}/V_{sp} \qquad (1)$$

where, V_{sp} is the amplitude of supply voltage computed

Bhim Singh and Varun Singhal are with Department of Electrical Engineering at Indian Institute of Technology, (IIT) New Delhi (bsingh@ee.iitd.ac.in and varun.singhal2@gmail.com)

978-1-4244-0644-9/07/$25.00 ©2007 IEEE

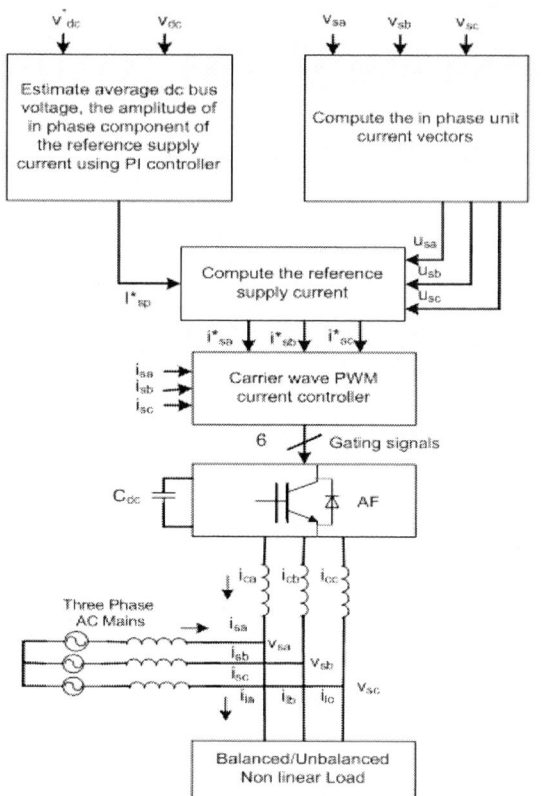

Fig.2 Control Scheme of the Active Filter

as:

$$V_{sp} = \left\{ 2/3 (v_{sa}^2 + v_{sb}^2 + v_{sc}^2) \right\}^{\frac{1}{2}} \qquad (2)$$

The self supporting dc bus of the AF is realized using a PI controller over the sensed (v_{dc}) and reference ($v_{dc}*$) values of dc bus voltage of the AF. The PI voltage controller on the dc bus voltage of the AF provides the amplitude (I^*_{sp}) of the in phase components of reference supply currents obtained by multiplying the amplitude with unit current vectors (1). This reference source current is compared with sensed source current and fed to a 15 kHz PWM current controller to generate the gating signals of the IGBT's used in the VSI bridge working as an AF.

The controller parameters to be tuned for the system are the proportional and integral gain constants on the error in dc voltages. For tuning of these parameters a new offline genetic algorithm (GA) approach is employed.

III. GENETIC ALGORITHM FOR DESIGN OF CONTROLLER PARAMETERS

As explained in detail in [5-7], GA involves selection, mixing and mutation of components. Like a biological system, the selection is driven by an organism's ability to survive in the environment. GA performs a search for a multidimensional space based on the control parameters containing a hyper surface known as the fitness surface.

In GA, a "population" of problem solutions is maintained in the form of "chromosomes". These chromosomes are formed from the combination of individual solution called "genes". Based on the fitness functions values obtained from the first set of population, the parents are evaluated on fitness scaling criteria. A more prominent parent would contribute more to the next children and vice versa. These selected parents then produce next set of children based on three operations:

Elite Count: The number of parents to be retained who have the minimum value of fitness function (the best in the pool). This is to ensure that the best solution is not lost.

Crossover: Two parents combine segments of their genes together to generate children. This ensures that all the solutions existing nearby parents are checked for their survival.

Mutation: A gene can be mutated from its parent by introducing unprecedented information as a part of parent's gene. This ensures that sufficient random set of solutions are tried for.

Once the children (of next generation) are obtained the next population is reevaluated and this process goes on and on until any of the stopping criteria is reached. These criteria could be reaching the maximum number of generations, or reaching stall generation limit (reaching a limit on number of generations for which the best controller parameters remain the same). Hence, the final controller parameters are obtained to the subject of minimum value of pre set objective function.

Fig. 3 shows the flowchart of genetic algorithm employed for calculating the controller parameters K_p and K_i. The GA parameters are initialized as:

Number of Population (N_p):	12
Range for initial population ($0, n_r$):	(0, 10)
Maximum number of Generation (N_g):	25
Stall Generation limit (N_s):	15
Elite Count (e_n):	2
Crossover fraction (u_c):	0.7
Crossover ratio (u_r):	0.8

After the initialization process the first set of population is generated through a uniform distribution over ($0, n_r$). Each chromosome of this population then runs for the complete simulation to evaluate the objective function, J. The motive for the design of the objective function is to bring down the peak overshoot and undershoot characteristics of the dc link voltage under dynamic and unbalanced load conditions. Hence, the objective function is calculated as the integral of square of error of v_{dc} reference and actual v_{dc} (3).

$$J = \int_0^T \left(v_{dc}^* - v_{dc} \right)^2 dt \qquad (3)$$

Integration of the square of error is realized numerically by Simpson's rule:

$$\int_a^b f(x)dx \approx \frac{h}{3}(f(x_0) + 4f(x_1) + 2f(x_2) + ... + f(x_n)) \quad (4)$$

After the objective function J is evaluated for all set of population, 'e_n' number of elite children corresponding to the

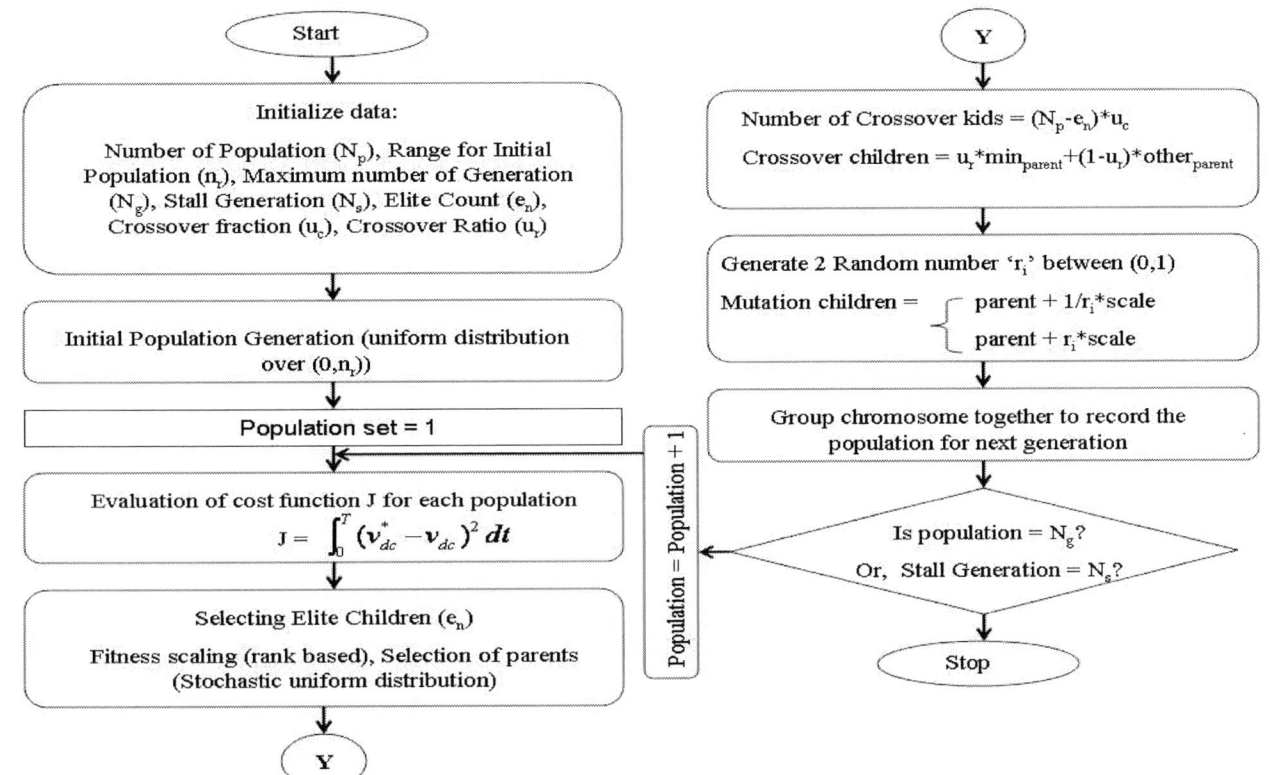

Fig.3 Flowchart of the Genetic Algorithm

best set of K_p and K_i are retained. Then fitness scaling is performed to give ranks to all the members of the population. Based on their ranks the parents are selected using a stochastic uniform distribution function. These parents are the one with a better survival rate and would be responsible for the generation of children for next set of population.

The numbers of crossover children are given by $(N_p-e_n)^*u_c$. For our case this would be equal to 7 and accordingly number of mutation children would be 3. Crossover children are formed by taking two parents according to their ranks and if J(parent1) < J(parent2) then crossover children = u_r*parent1+ $(1-u_r)$*parent2 and vice versa, where, u_r is the crossover ratio. Once the crossover children are obtained, the next step would be to generate mutation children so that there is enough diversity in the population.

The mutation is an essential tool of genetic algorithm as it ensures that the K_p K_i combinations are tried over a varied range and none of the points are left over which could have global minima. To perform mutation, two different random numbers are generated between 0 and 1 and then they or their inverse are added to the parent combinations of K_p and K_i as shown in Fig. 3. The scale is a naturally decaying parameter which decays proportional to inverse of current generation (starting from 1 when Population =1) after every generation. This is essential for the system to converge as the number of generation reaches close to N_g. As the new sets of children are generated, the iteration is checked for the stopping criteria. If the criterion is met the final results are reported otherwise the complete iteration is performed again.

IV. SIMULATION RESULTS

A three phase AF system is modeled in Matlab Simulink environment using Power Block Sets. The three phase AF system (Fig.1) is made up of standard three phase IGBT based VSI bridge with the input ac inductors, R-C filter and a dc bus capacitor to obtain a self supporting dc bus for an effective current control. The load is a 6-pulse diode rectifier fed R-L load. The system parameters are as follows (refer to Fig.1 for nomenclature):

V_s = 415 V (line rms); R_s = 0.1 Ω ; L_s = 0.2 mH;

R_c = 0.1 Ω ; L_c = 1.8 mH, C_c = 3.4 μF; R_f = 4 Ω ;

C_{dc} = 1500 μF; R_L = 5 Ω ; L_L = 15 mH; V_{dc} (ref) = 650 V

Switching frequency: 15 kHz

A. Genetic Algorithm Results

Fig.4 shows the results obtained with the set objective function and crossover and mutation techniques discussed in much detail above. As it can be inferred from Fig.4 (d) the GA is terminated due to algorithm reaching its stopping criterion of maximum number of generation i.e. 25 whereas only 40% of stall generations limit was reached.

Fig.4 (a - b) show the value of objective function and trend of K_p and K_i as a function of number of generations. The first best score corresponds to a K_p K_i value of (3.5, 3.23) with a score of 31.367; finally terminating at K_p, K_i value of (3.7974, 4.0999) with a score of 12.254. This implies a reduction in initial score upto 155.9%. With the same final set of K_p and K_i stalling for 8 generations it is concluded that the obtained set of K_p and K_i is optimal for the

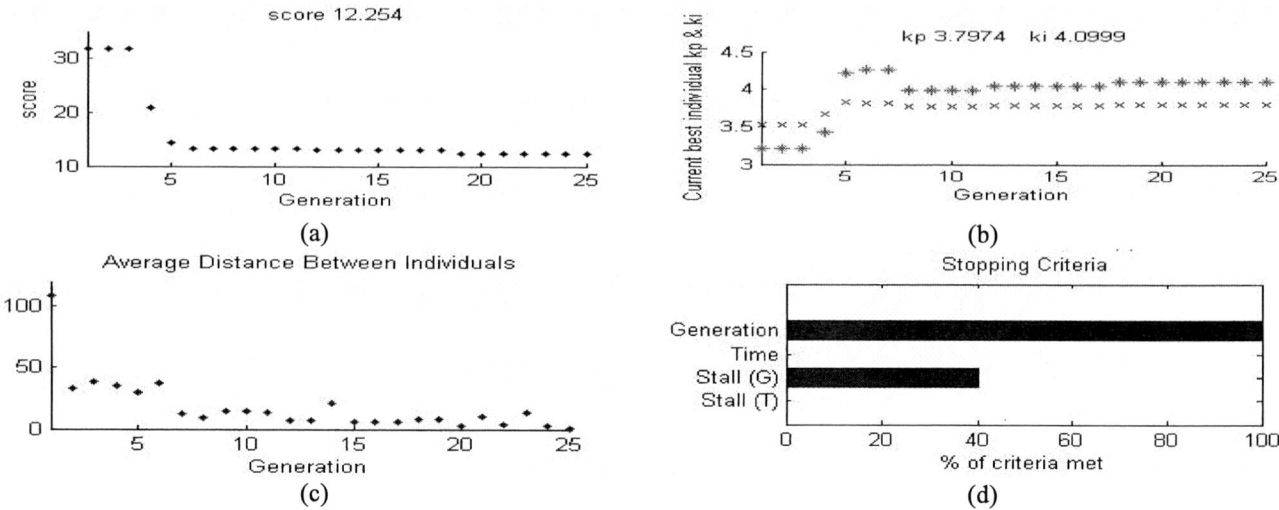

Fig.4 (a) Objective function, b) (Kp, Ki), and (c) Average Distance between Individuals vs. Number of Generations (d) Stopping Criterion for the Genetic Algorithm.

set objective function. Fig.4 (c) shows the average distance between the individuals for each generation. This graph is necessary to depict the diversity and the convergence of the genetic algorithm. As it can be seen during the initial generations the average distance between the individual has been of the order of 50's. This implies enough diverse population is tested for its survival in the initial stage. But as the number of generations increased the mutation scale is gradually reduced and the average distance between the individuals also reduced accordingly.

B. Dynamic Performance of AF

Some of the system parameters required for simulation study are specified in the beginning. With the control parameters obtained from Genetic Algorithm the performance of the AF is analyzed considering the following cases:

Case 1) Transient Response to Load Change:

To get a measure of transient response system load is changed from 5 Ω to 3 Ω and then to 2 Ω as shown in Fig.5 (a). The first variation of V_{dc} graph is for control parameters obtained through GA and the second is for $K_p = 0.47$ and $K_i = 8.3$.

Table 1 draws a comparison between the performances for the two. As it can be seen clearly the overall stress on the dc-link capacitor in term of voltage dip and swell is 50 % less then the latter case though the % overshoot is minimal in latter.

TABLE I
DYNAMIC PERFORMANCE OF APF FOR LOAD CHANGE

K_p	K_i	% undershoot	% overshoot	Settling time (cycles)
3.7974	4.0999	4.6	3	2.5
0.47	8.3	9.2	0.9	3.5

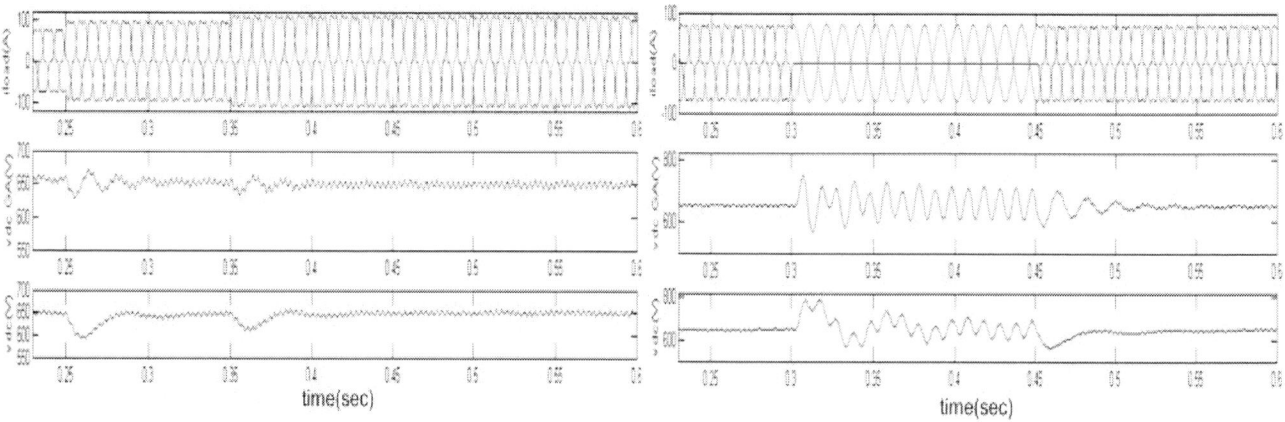

Fig. 5 Dynamic Response of V_{dc} for (a) load variation from 5 Ω to 3 Ω at 0.25 sec and then to 2 Ω at 0.35 sec (b) Three phase to single phase load at 0.3 sec and back to three phase load at 0.45 sec

1819

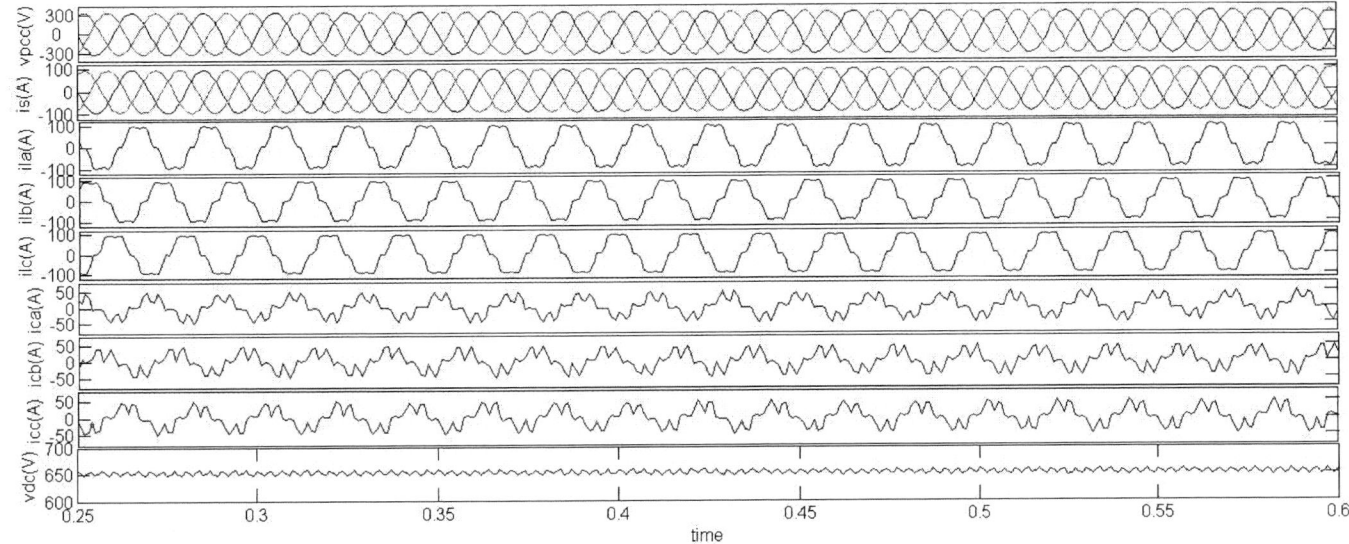

Fig. 6 Steady State Response of APF

Case 2) Transient Response to Load Unbalancing:

A three phase load is changed to a single phase load by opening phase a of the three phase system at 0.3 sec and then put back at 0.45 sec to study the transient response (Fig.5 (b)). As it can be seen from Table II, again the stress on dc-link capacitor in term of dip and swell from 650 V is much less then latter case. The high oscillations at steady state for single phase system are due to presence of 2nd harmonic component on the dc side. Moreover, the settling time with the genetically obtained parameters is far better then latter case.

TABLE II
DYNAMIC PERFORMANCE OF APF FOR UNBALANCING

Kp	Ki	% undershoot	% overshoot	Settling time (cycles)
3.7974	4.0999	10	15.3	3.5
0.47	8.3	12.3	23	5

Case 3) Robustness under Filter Parameter Variations:

Tables III and IV show the value of settling time of dc voltage and THDs in source current for the variation in values of filter parameters L_c and C_{dc}, respectively during the load perturbations. It can be noticed that the variations in values of filter inductance and dc capacitor have no significant effect on the settling time whatsoever. Moreover, the THDs in source current always remain in normal range as recommended by IEEE-519 limits.

TABLE III
PERFORMANCE OF CONTROLLER FOR VARIATION IN FILTER INDUCTANCE

L_c (mH)	THD (%)	Settling time (cycles)
2.5	4.06	2.5
1.8	1.47	2.5
1	1.14	2.5

TABLE IV
PERFORMANCE OF CONTROLLER FOR VARIATION IN DC LINK CAPACITOR

C_{dc} (uF)	THD (%)	Settling time (cycles)
1500	1.47	2.5
2000	1.29	2.5
2500	1.11	2.5

C. Unity Power Factor Operation

Fig.7 shows the source current and voltage for phase 'a' vs. the time highlighting unity power factor operation of the system. The voltage is scaled by a factor of (100/340) to stress on the mapping of two waveforms. Fig.6 shows the steady state waveform for the system under discussion with genetically obtained controller parameters.

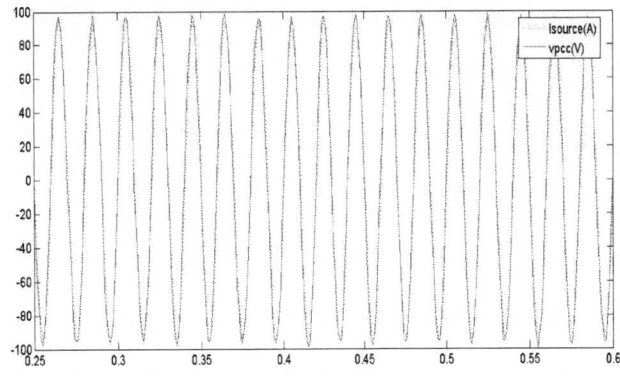

Fig.7 I source and Vpcc for phase 'a' vs. time

V. CONCLUSION

A technique to evaluate AF system controller parameters has been given for an active filter using genetic algorithm. The detail of control scheme and genetic algorithm is provided to achieve optimum gain. Finally, the performance of the filter to various loads perturbations and unbalancing has been shown in terms of reduction in stress on the dc link capacitor voltage.

VI. REFERENCES

[1] B. Singh, K. Al-Haddad, and A. Chandra, "A new control approach to three phase active filter for harmonics and reactive power compensation," IEEE Trans. Power Syst., vol. 13, pp. 133-138, Feb 1998.

[2] H. Akagi, Y. Kanazawa, and A. Nabae, "Instantaneous reactive power compensators comprising switching devices without energy storage components," IEEE Trans. Ind. Applicat., vol. IA-20, pp. 625–630, May/June 1984.

[3] B. Singh, K. Al-Haddad, and A. Chandra, "Active power filter with sliding mode control", Proc. of Inst. Elect. Eng., Generation, Transm., Distrib., vol. 144, pp. 564–568, Nov. 1997.

[4] A. Chandra, B. Singh, B.N. Singh, and Kamal Al-Haddad, "An Improved Control Algorithm of Shunt Active Filter for Voltage Regulation, Harmonic Elimination, Power Factor Correction, and Balancing of Nonlinear Loads", IEEE Trans on Power Elec., vol. 15, no. 3, May 2000.

[5] M El-Habrouk; M.K. Darwish; "A new control technique for active power filters using a combined genetic algorithm/conventional analysis", IEEE Trans on Industrial Electronics, vol. 49, no. 1, pp. 58-65, Feb. 2002.

[6] S. Mishra, and C.N. Bhende, "Bacterial Foraging Technique-Based Optimized Active Power Filter for Load Compensation", IEEE Trans on Power Delivery, vol. 22, no. 1, pp. 457-465, Jan 2007.

[7] M. Marinelli, A. Dell'Aquilla; "Improved current control of active power filters using genetic algorithms"; in Proc. of IEEE International Symposium on Ind. Elec. 2004, vol. 2, pp. 1447-1451, May 2004.

VII. BIOGRAPHIES

Bhim Singh (SM'99) was born in Rahamapur, India, in 1956. He received the B.E. electrical degree from the University of Roorkee, Roorkee, India, in 1977, and the M.Tech. and Ph.D. degrees from the Indian Institute of Technology (IIT), New Delhi, India, in 1979 and 1983, respectively. In 1983, he joined the Department of Electrical Engineering, University of Roorkee, as a Lecturer. He became a Reader there in 1988. In 1990, he joined the Department of Electrical Engineering, IIT, as an Assistant Professor. He became an Associate Professor in 1994 and a Full Professor in 1997. His fields of interest include power electronics, electrical machines and drives, active filters, static var compensators, analysis, and digital control of electrical machines. Dr. Singh is a Fellow of the Indian National Academy of Engineering (INAE), Institution of Engineers (India), and Institution of Electronics and Telecommunication Engineers (IETE) and a Life Member of the Indian Society for Technical Education (ISTE), System Society of India (SSI), and the National Institution of Quality and Reliability (NIQR).

Varun Singhal was born in Sangroor, India in 1986. He has received the B. Tech degree in electrical engineering in power from Indian Institute of Technology (IIT), New Delhi in 2007. His fields of interest include power electronics, active filters, and control of electrical machines.

A Fuzzy Adaptive Detecting Approach of Harmonic Currents for Active Power Filter

Yilong Qu, Weipu Tan, and Yihan Yang

Abstract– **Successful control of active power filter requires an accurate current reference. Base on adaptive interference canceling theory, this paper presents an adaptive least mean squares (LMS) algorithm employing fuzzy step size for the fast estimation of harmonic current signals in power networks. The learning parameter of the proposed algorithm is constrained by two variable parameters. By using a fuzzy logic based step size update method which causes an automatic suitable adjustment of the step size, the present method can provide fast convergence and noise rejection for the tracking of fundamental and harmonic components from distorted signals. The results of computer simulations and physical experiments have proved the proposed algorithm is practicable.**

Index Terms-- **Active power filter, harmonic current detection, adaptive filter, fuzzy rule based**

I. INTRODUCTION

In the present ac power networks, the effect of harmonic currents has become a considerable problem due to the increased introduction of power electronic devices. The active power filter (APF), relatively well-researched in recent years, is a quit promising approach for bring harmonic pollution under control[1]. And accurately estimation of the harmonic components in a load current is a prerequisite for evolving suitable control strategies for APF. This greatly affects the performance of the APF for eliminating and reducing the effects of harmonics in power system. Then many algorithms are available to detect harmonic currents and generate the reference signal for the APF. Conventional detecting methods for harmonic currents include the digital low-pass or high-pass filters, the Fast Fourier Transform (FFT)[2] and the method based on instantaneous reactive power theory[3]. However they all have some limitations. The digital low-pass filters or high-pass filters have the disadvantage that they are not possible to obtain a fast transient response which changes remarkably as a function of frequency. Further, a small change in frequency may cause a significant phase shift. FFT is the most widely used algorithm for detection and extraction of harmonics. However, it is not a flexible method that once the sampling frequency is determined, its processing speed and accuracy is determined. And its performance is adversely affected by decaying d.c. components or a low signal-to –

Y. Qu, W. Tan and Y. Yang are with the Department of Electrical & Electronic Engineering, North China Electric Power University, Beijing, 102206, China (e-mail: quyilong142@163.com).

This work was supported by Beijing Natural Science Foundation of China (No. 3062018)

noise ratio. The instantaneous reactive power theory is based on complex voltage and current transforms and their inverse transforms. Although this method provides very precise solution, it is known to suffer from some shortcoming (the distorted voltage or the nonlinear load), which make it difficult to provide an accurate basis for active power filters.

Since the above detecting methods for harmonic currents have various limitations. To provide a fast convergence speed and more robust response with various inputs, some of the known signal processing techniques is introduced to use for the harmonic detection. Based on the adaptive noise canceling theory, some adaptive detecting methods for harmonic are introduced for APF application [4,5]. They had been proved feasible, not only its configuration is simple, but also its algorithm is easy to realize by analog hardware circuit.

II. BASIC PRINCIPLE OF ADAPTIVE DETECTING METHOD FOR HARMONIC CURRENT

The adaptive noise canceling technique based on the Wiener theory has been widely used in many signal processing applications. It can maintain the system in the best operating state by continuously self-studying and self-adjusting from start to end [6]. According to the adaptive noise canceling theory, the principle of adaptive detecting method for harmonic current can be illustrated in Fig. 1.

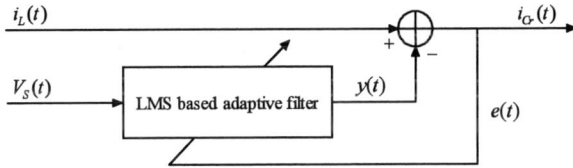

Fig.1. the principle of adaptive detecting method for harmonic current

With the decomposition of non-sinusoidal periodic currents in time domain, assuming the mains voltage is a pure sinusoidal wave, it is represented as

$$V_s(t) = V_m \sin(\omega t) \qquad (1)$$

The nonlinear load current can be represented as

$$i_L(t) = i_1 \cos \varphi_1 \sin(\omega t) + i_1 \sin \varphi_1 \cos(\omega t)$$
$$+ \sum_{n=2}^{\infty} i_n \sin(n\omega t + \varphi_n) \qquad (2)$$

It also can be represented as

$$i_L(t) = i_p(t) + i_q(t) + i_h(t) \qquad (3)$$

Where, $i_p(t)$ represents the active current, $i_q(t)$ represents the reactive current and $i_h(t)$ represents the sum of all

978-1-4244-0644-9/07/$25.00 ©2007 IEEE

harmonic. If the active power filter can generate $i_C(t)$, which is equivalent to $i_h(t)$, then we obtain the APF control propose. And we also note that the $i_1 = i_p(t) + i_q(t)$ represents the fundamental component of the nonlinear load current, and also correlate with the main voltage $V_s(t)$.

$$i_C(t) = i_L(t) - (i_p(t) + i_q(t))$$
$$= i_L(t) - [w_1, w_2] \times V_s(t) \qquad (4)$$

Then according to the adaptive noise canceling theory, if we obtain $W = [w_1, w_2]$, the current reference of the APF is easy to calculate. From Fig. 1, $y(t)$ is the output of the adaptive filter and the output of the system is $i_{Cr}(t)$, as follows:

$$i_{Cr} = i_L - y = i_C + (i_1 - y) \qquad (5)$$

Squaring, one obtains

$$i_{Cr}^2 = i_C^2 + (i_1 - y)^2 + 2i_C(i_1 - y) \qquad (6)$$

Taking expectation of both sides of (6), and realizing that i_C is uncorrelated with i_1 and y because of the orthogonality of trigonometric functions, yields

$$E[i_{Cr}^2] = E[i_C^2] + E[(i_1 - y)^2] + 2E[i_C(i_1 - y)]$$
$$= E[i_C^2] + E[(i_1 - y)^2] \qquad (7)$$

Note that, $E[i_{Cr}^2]$ will be unaffected when the adaptive filter network is adjusted to minimize $E[i_{Cr}^2]$. accordingly,

$$\min E[i_{Cr}^2] = E[i_C^2] + \min E[(i_1 - y)^2] \qquad (8)$$

So, the output y is then a best least squares estimate of i_1. It is seen that the smallest one is

$$E[i_{Cr}^2] = E[i_C^2] \qquad (9)$$

When this is achievable, the current reference is generated successfully.

Therefore, the conventional adaptive filtering using the least mean square (LMS) algorithm as follows shown

$$W(k) = W(k-1) + \mu i_C(k-1) V_s(k-1) \qquad (10)$$

$$i_C(k) = i_L(k) - W(k) V_s(k) \qquad (11)$$

Where μ is called convergence coefficient (or step size) controls stability and speed of convergence.

III. FUZZY LMS ADAPTIVE FILTER STRUCTURE

Since a key parameter to the design of the LMS-based adaptive filters is the step size, which governs the steady-state performance and the convergence characteristic [7]. In general, a large step size leads to rapid convergence but results in large steady-state misadjustment error, the results are opposite when a small step size is used. With the actual APF application, we object is to provide fast convergence and noise rejection for the tracking of fundamental and harmonic components from distorted signals. However the conventional adaptive filtering

with a constant step size can not rapid respond to the changing environment, and is more sensitive to the noise disturbance.

A fuzzy logic system is used to perform the nonlinear mapping of the input data vector into a scalar output value [8]. To obtain fast response characteristic of harmonic detection, it most commonly approach is to utilize a large step size during the transient state and shifts to a smaller step size during the steady state. By using a fuzzy logic based step size update method which causes an automatic suitable adjustment of the step size, this paper presents a new LMS-based adaptive filter, just as Fig.2. shown: the fuzzy-logic controlled step size adaptive filter network is employed to perform the nonlinear mapping of the squared error e^2 and the squared error variation Δe^2 into the step size μ in response to rapid statistical variations of the i_L.

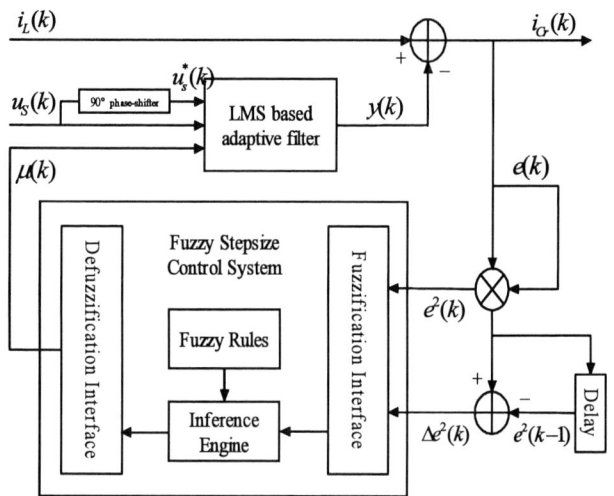

Fig. 2. Fuzzy LMS Adaptive Filter Structure

The fuzzy step size control system (FSS) is based on the principle of fuzzy logic developed by Zadeh [8], which is used to handle the linguistic concepts. It consists of three main processors: Fuzzification interface, Fuzzy rule inference engine, and Defuzzification interface. Then the general format for the adaptive harmonic currents detection with fuzzy step size algorithm controller can be written as

$$i_{Cr}(k) = i_L(k) - y(k) = i_L(k) - W(k) u_s^*(k) \qquad (12)$$

$$\mu(k) = FSS(e^2(k), \Delta e^2(k)) \qquad (13)$$

$$W(k+1) = W(k) + \mu i_C(k) u_s^*(k) \qquad (14)$$

The FSS in Fig. 2 uses the squared error ($e^2(k)$) and the squared error variation ($\Delta e^2(k)$) as the input variables at the kth iteration. The input variables to the FSS are transformed to the respective degree via membership functions (MBFs), with centroids of the large (L), medium (M), and small (S), are selected to cover the universe of the input and output variables, as illustrated in Fig. 3. The fuzzy-logic adaptive filters are constructed from a set of changeable fuzzy IF-THEN rules. The fuzzy rules come either from human experts or by accumulating past experience in the practical applications.

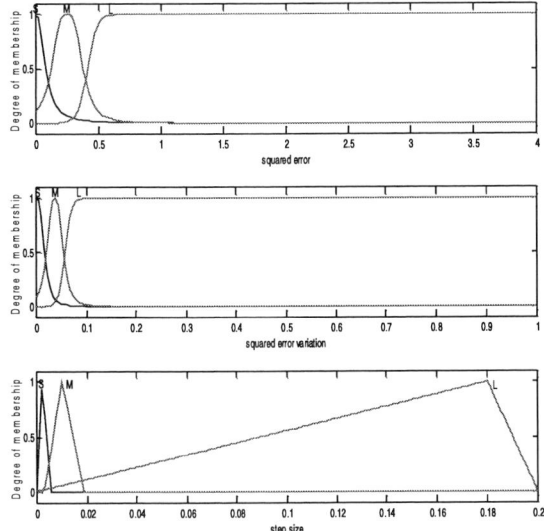

Fig.3 MBFs of the fuzzy-inference-based algorithms spread over their respective universe.

This paper uses the past hundred computer simulation results to design the fuzzy rules, as illustrated in Table. 1. Here, the symbol " μ " is used to represent time-varying variables step size. For instance, IF the squared error is "M" and the squared error variation is "L", THEN the value of step size (μ) is "M". Here the minimum limit for the step size μ_{min} is set to zero, and the maximum limit for the step size μ_{max} can be determined as the step size bound proposed by Widrow [6]:

$$0 < \mu < \frac{1}{NP_x} = \frac{1}{tr(R)} \quad (15)$$

where $tr(R)$ is the trace of the input autocorrelation matrix R.

TABLE I
PREDICATE BOXES FOR THE FUZZY LMS ALGORITHM

		e^2		
Δe^2	μ	S	M	L
	S	S	S	S
	M	S	S	M
	L	M	M	L

In what follows, the fuzzified input variables, which contain the degrees of the antecedents (IF-part) of a fuzzy rule, are combined using the fuzzy "OR " operator, which selects the maximum value of the two, to obtain a single value. Table I. also illustrated the relations between the MBFs and the fuzzy rules in the FSS. Subsequently, this is followed by the implication process, which defines the reshaping task of the consequent (THEN-part) of the fuzzy rule based on the antecedent. The input for the implication process is a single number given by the antecedent, and the output is a fuzzy set. A minimum operation is generally employed to truncate the output fuzzy set for each rule. Since decisions are based on the testing of all of the rules in the FSS, the rules need to be combined in some manner in order to make a decision.

Aggregation is the process by which the fuzzy sets that represent the outputs of each rule are combined into a single fuzzy set. The process for obtaining a crisp output value from the resulting fuzzy set is called defuzzification. To derive the variable μ from the FSS, the fuzzy centroid-defuzzification calculation of μ is employed, which return the center of area under the aggregate MBF's curve, as follows:

$$\mu(k) = \sum_{i=1}^{n} \mu^i(k) m_B(\mu^i(k)) \Big/ \sum_{i=1}^{n} m_B(\mu^i(k)) \quad (16)$$

where n is the number of sections used for approximating the area under the aggregated MBFs. Then the desired output obtain from the fuzzy rules in Table. 1.

IV. SIMULATION AND EXPERIMENT RESULTS

A. Computer simulation

To verify the performance of the proposed fuzzy LMS adaptive filter, one computer simulations were developed using MATLAB, as illustrated in Fig. 4 and Fig. 5. An input nonlinear load current signal consisting of a unity fundamental component, the 3rd harmonic content of 0.3, the 5th harmonic content of 0.1, and the 11th harmonic content of 0.1 is applied to the fuzzy adaptive filter. Fig. 4 shows performance of the filter in extracting the harmonic and the fundamental components. The results indicate the present algorithm can rapid respond with the change of the nonlinear load current less than quarter period of the fundamental period (20 ms). Here, the data sampling frequency is 10kHz.

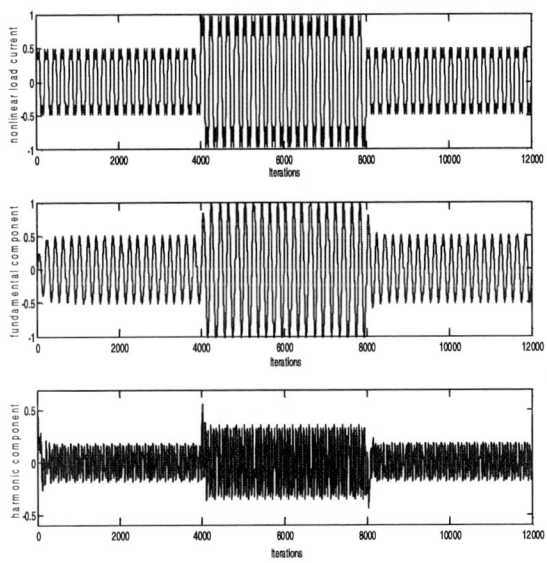

Fig. 4 Simulation results with the fuzzy LMS algorithm (a)

The fundamental component estimation square error of the desired and actual filtered, and the varied step sizes in each iteration operation are also shown in Fig. 5. It is observed that the present filter is not capable of providing a fast transient response while having negligible steady-state error in cases where highly polluted signals or amplitude variations are involved. A number of computer simulations with different operating conditions also show that it is desired to have a transient response of not more than half cycle and a steady-state error of not more than 1% with respect to a polluted input signal whose amplitude may suddenly change up or down to 50%.

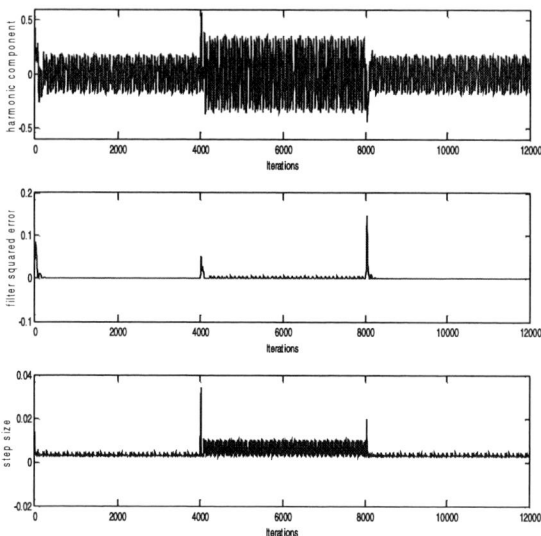

Fig. 5 Simulation results with the fuzzy LMS algorithm (b)

B. Physical experiment

In this paper, to verify the present fuzzy adaptive filter algorithm, a single-phase active power filter physical development system is also setup. As shown in Fig. 6, the whole system is simple consist with a commercial power supply, a nonlinear load controlled by thyristor-controlled switch, a full-bridge VSI unit and a control system based on DSP (TMS320F2812). The zero-crossing detection unit generates synchronous signal for DSP in order to generate the referenced $u_s^*(k)$, and the A/D unit samples all currents for 100 data in one grid cycle.

In practice, to obtain fast respond for control requirement, we made a 100*3 point preliminary definition table for the squared error, the squared error variation, and the step size. And with each two sample cycle, we look-up table for the step size. A 32-bit fixed-point DSP is used for generated PWM impulse with programmable dead-beat time.

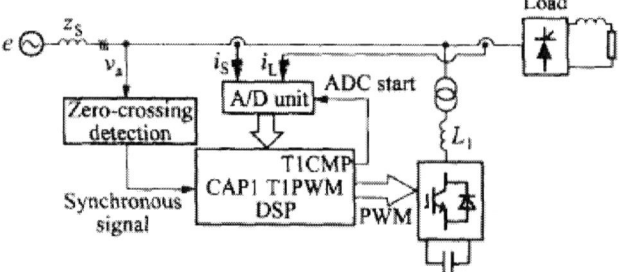

Fig.6 single-phase shunt APF system controlled by DSP

Performance of the APF (Fig. 6), with respect to harmonic elimination is evaluated by means of time-domain simulation in a typical nonlinear environment. The nonlinear load is controlled by thyristor-controlled switch which can generate

rapid variable load current for algorithm test. Parameters of the simulated system are summarized in Table. II.

TABLE II
PREDICATE BOXES FOR THE FUZZY LMS ALGORITHM

Ac source	Load	APF	LC filter
Vs = 110v	L = 7.8mH	L = 5mH	L = 0.4 mH
f = 50Hz	R = 10 Ω	V_{dc} = 500v	C = 5 µF
		C = 4000 µF	f = 5 kHz
		f_{sw} = 10kHz	

Fig. 7 shows the performance of the APF to a highly distorted load current with rapid change controlled fire-angle of the nonlinear load. The main current after compensation is depicted in the up figure of Fig. 7, and the extracted harmonic current (reference current for APF) is depicted in the low figure of Fig. 7.

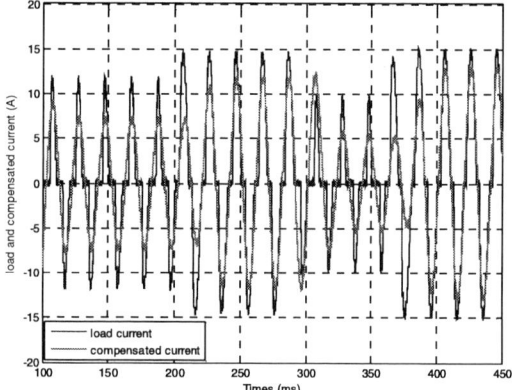

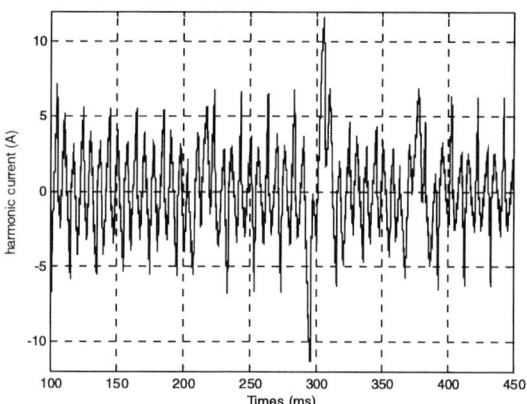

Fig. 7 Physical simulation

As shown in Fig. 7, for difference firing angle of the thyristor, not only the amplitude but also the harmonic components of the nonlinear load current are changed rapidly. However the inject current of APF also traced rapidly enable the supply current to be normalize within one grid cycle. The transient time of almost one cycle is observed in the response before the stead-state is achieved. This result is in agreement with the simulation results above.

It proved that the new fuzzy-logic controlled adaptive filter algorithm can provide fast convergence and noise rejection for the tracking of fundamental and harmonic components from distorted signals.

V. CONCLUSION

Accurately estimation of the harmonic components in a load current is a prerequisite for evolving suitable control strategies for APF, and greatly affects the performance of the APF. However the conventional adaptive filter for APF has some shortcomings. Based on adaptive noise canceling technique, the adaptive filter for APF generates reference current has been proved feasible. For accurately and synchronous detection and extraction of harmonics, this paper presents a novel adaptive least mean squares (LMS) algorithm employing fuzzy step size. The learning parameter of the proposed fuzzy based algorithm is constrained by two variable parameters: the filter system output squared error and squared error variation. By using a fuzzy logic based step size update method which causes an automatic suitable adjustment of the step size, the present method can provide fast convergence and noise rejection for the tracking of fundamental and harmonic components from distorted signals.

The results of the verification of proposed algorithm by computer simulations and physical experiments show that the proposed method is practicable.

VI. REFERENCES

[1] H. Akagi, "New trends in active filters for power conditioning," IEEE Trans. Ind. Applicat. ,vol. 32, pp. 1312-1322, Nov. 1996.

[2] Oran. Brigham. E, "The Fast Fourier Transform and its applications", Prentice-Hall International, 1988.

[3] T. Tanaka and H. Akagi, "A new method of harmonic power detection based on the instantaneous active power in three-phase circuits," IEEE Tans. Power Delivery, vol. 10, pp. 1737-1742, Oct. 1995.

[4] Shiguo Luo and Zhencheng Hou, "An adaptive detecting method for harmonic and reactive currents," IEEE Trans. Ind. Electron, vol.42 pp. 85-89, Feb 1995.

[5] Qun Wang, Ning Wu, and Zhaoan Wang, "A neuron adaptive detecting approach of harmonic current for APF and its realization of analog circuit," IEEE Trans. Ins. Measure, vol. 50, pp. 77-84, Feb 2001.

[6] A. H. Sayed, "Fundamentals of Adaptive Filtering," New York: Wiley, 2003.

[7] W. S. Gan, "Designing a fuzzy step size LMS algorithm," IEE Pro.-Vis. Image Signal Process, vol. 144, pp. 261-266, Oct 1997.

[8] C. Von Altrock, "Fuzzy Logic and Neurofuzzy Applications Explained." Englewood Cliffs, NJ: Prentice-Hall, 1995.

[9] R. H. Kwong and E. W. Johnston, "A variable step size LMS ," algorithm IEEE Trans. Signal Process., vol. 40,no.7, pp. 1633-1642, Jul. 1992.

VII. BIOGRAPHIES

Yilong Qu was born in Nong'an County, Jilin Province, China in 1979. He graduated from North China Electric Power University in 2001. And then He has studied for the doctor's degree up to now. The main aspect he researches is the power electric and its application in power network.

Weipu Tan was born in Harbin City, China in 1963. He received his Ph.D. degree from Harbin Institute of Technology in 1999, now working in North China Electric Power University as an associate professor. His researches mainly focus on the power system analysis and automation.

Yang Yihan was born in Tieling City, Liaoning Province in 1927. He is a famous professor in China. He graduated from Harbin Institute of Technology as a graduate student in 1952. Then, he has taught in Northeast Institute of Technology, Harbin Institute of Technology and North China Electric power University. He has been a director of CSEE, and staff room director and departmental director.

Comparative Evaluation of Harmonic Extraction Techniques for Three-Phase Three-Wire Active Power Filter

R. Chudamani*, Krishna Vasudevan*, and C.S. Ramalingam*

* Indian Institute of Technology Madras, Chennai-600036, India
Corresponding author: Krishna Vasudevan Email: Krishna@ee.iitm.ac.in

Abstract-- **In this paper the performance of four harmonic extraction algorithms for active power filters are evaluated based on their steady state and dynamic performance. The techniques which are considered for comparative study are (i) filtering approach using the conventional low pass filter, (ii) instantaneous real and imaginary power (*pq*) theory, (iii) instantaneous active and reactive currents method (*dq*), and (iv) nonlinear least squares (NLS) estimation method. Comparison is based on simulation studies carried out in SABER. It is shown through simulation results that the dynamic behaviour of the harmonic extraction approach has an effect on the active power filter ratings. The simulation study reveals the superiority of the nonlinear least squares approach. This method provides good dynamic response and also extracts the load current harmonics explicitly.**

Index Terms-- **Nonlinear least squares, Harmonic elimination, Compensation lag.**

I. INTRODUCTION

The distortion of Utility supply arising out of widespread use of nonlinear loads is a cause of great concern. Adjustable speed drives, electric arc furnaces, switched mode power supplies, and power converters which employ thyristors, are typical cases. Conventionally, these problems were addressed by using conventional passive LC filters. The main drawback of these filters is that they provide fixed compensation. Moreover, the values of these components require fine-tuning, and they also suffer from the problems of large size and ageing. An effective solution to these problems is to use active power filters. These equipments use power electronics to inject suitable anti-phase harmonics in a manner that the utility sees an effective linear load. Determining the harmonics is a very important part of the Active Power filter controller. This involves extraction of harmonic components of load current and a suitable control strategy for the power electronic interface that enable the utility to see a linear load. In this paper, we confine ourselves with the methods used for harmonic extraction and their influence on the filter behaviour.

Several methods have been proposed for extracting the harmonic content [1]–[9]. In [1] the authors use a low pass filter to obtain the harmonics. While simple, the method suffers from the problem of filtering delay during load dynamics and introduces phase shift in the fundamental it extracts during frequency fluctuations. In

[2] the harmonics are extracted based on the instantaneous real and reactive power (*pq*) theory. This method works well under steady state, balanced and sinusoidal conditions of supply voltage. The instantaneous active and reactive current i_d-i_q method reported in [3] is identical to the *pq* method under sinusoidal and balanced voltage conditions; but it introduces error in the harmonic components under non-sinusoidal voltage conditions, as it requires the system frequency information. Further, both the above methods make use of a low-pass filter, which suffers from the problem of inherent lag, which causes problems when sudden load changes occur in the system. Due to delay in tracking the reference currents, the duration for which active power filter has to source/sink the fundamental current during the load transients, increases. This increases the power loss taking place in the inverter switches. The neural network approach [4]-[6] tracks the references currents without delay, but it requires large number of data for training. The probability of getting erroneous results when the network sees an unknown waveform can be high. In the adaptive filtering technique [7], the determination of current amplitude and the frequency is essential for the filter to track the fundamental component accurately. In the Transformed PI control strategy [8], accurate system frequency information and several parallel control loops are required. In [9], the authors propose a nonlinear least squares (NLS) based approach for system frequency determination and detection of current harmonics. The performance of these algorithms under steady state conditions has been well-studied in the literature, whereas the performance under load dynamics are yet to be addressed.

This paper presents a comparative evaluation of the harmonic detection algorithms based on the performance of the active power filter under dynamic load conditions. The algorithms that are considered for comparative study are (i) filtering approach using the conventional low pass filter [1], (ii) instantaneous real and imaginary power (*pq*) theory [2], (iii) instantaneous active and reactive currents theory (*dq* method) [3], and (iv) the NLS method [9]. A brief description of each algorithm with necessary equations is provided in section II. In section III the simulation results are presented and a comparative analysis is made based on the simulation results. A shunt active power filter (three phase, three wire) is used for

978-1-4244-0644-9/07/$25.00 ©2007 IEEE

testing these algorithms. An active power filter consists of four essential parts namely, (i) the signal conditioning circuit, (ii) the reference current generation circuit, (iii) the control circuit, and (iv) the power converter (see Fig.1). The signal conditioning circuit acquires the essential voltage and current signals to provide accurate system information. The reference current generation circuit generates the required harmonic currents to be amplified and injected into the lines, at the point where the load is connected. The control circuit compares the reference currents (i_a^*, i_b^*, i_c^*) and the injected currents (i_a, i_b, i_c) and generates the gating signals for the devices used in the power converter. The simulation study is carried out using SABER.

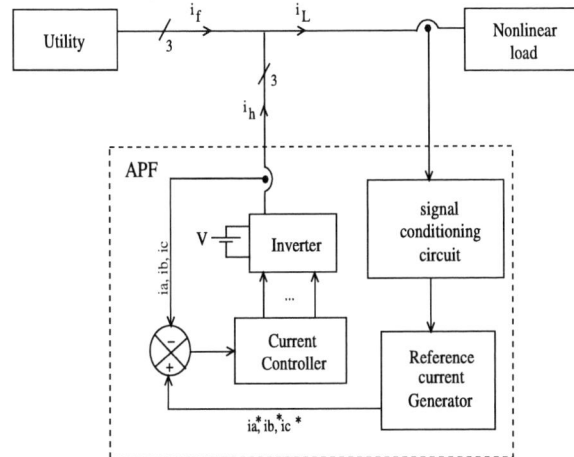

Fig.1 Major functional blocks of a shunt active power filter

II. DESCRIPTION OF DIFFERENT HARMONIC EXTRACTION ALGORITHMS

A. Filtering Approach using a Conventional Low Pass Filter

In this approach, a low pass filter is first used for extracting the fundamental component of the load current. When this fundamental component is subtracted from the original current waveform, the harmonic part of the current is obtained [1]. The operations are shown in Fig. 2

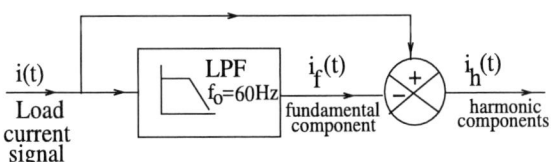

Fig.2 Block diagram of the conventional LPF scheme

These are then used as the references for the current controller in Fig. 1. In order to test the efficacy of this scheme, a fourth order Butterworth filter with a cut-off frequency of 60 Hz was used. Its transfer function is given by,

$$H(s) = \frac{2 \times 10^{10}}{s^4 + 9.85 \times 10^2 s^3 + 4.85 \times 10^5 s^2 + 1.4 \times 10^8 s + 2 \times 10^{10}} \quad (1)$$

The magnitude and the phase response of the filter are shown in Fig. 3 and Fig. 4 respectively. The gain at the fundamental frequency is -1.08 dB and the roll-off is -80 dB/decade. The gain at $f=5f_0$ (5th harmonic component) is -49.7 dB and at $f=7f_0$ it is -61.3 dB. Being a three wire system, triplen harmonics are not considered.

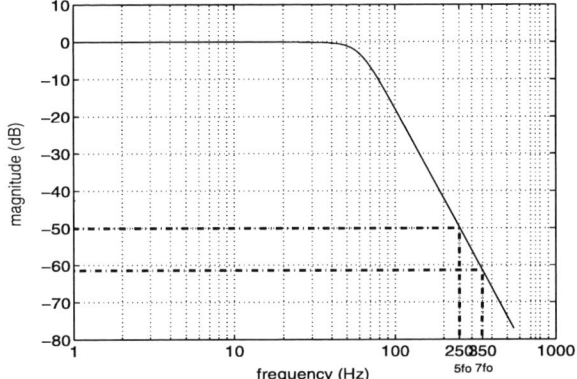

Fig.3 Magnitude response of the fourth order LPF

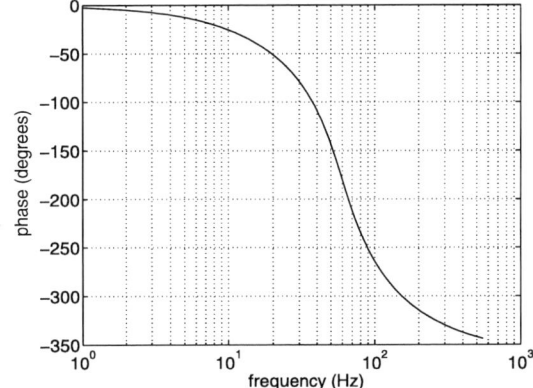

Fig.4 Phase response of the fourth order LPF

Thus, the low pass filter is used to obtain the fundamental component of the load current while attenuating all harmonics as best as it can. Therefore the design of low pass filter represents a trade off between the lag at fundamental frequency and the attenuation it provides to the dominant harmonic.

B. Instantaneous Real and Reactive Power Theory (pq Method)

In this method the cancellation references i_a^*, i_b^*, and i_c^* are obtained from the instantaneous active power p and reactive power q [2]. For the computation of p and q, the three phase voltages at the point of common coupling (PCC) and load currents must first be transformed to the stationary two axis ($\alpha\beta$) co-ordinates. The instantaneous real and reactive power p and q are determined through the following equations (2)–(6).

$$C = \sqrt{\frac{2}{3}} \begin{bmatrix} 1 & -\frac{1}{2} & -\frac{1}{2} \\ 0 & \sqrt{\frac{3}{2}} & -\sqrt{\frac{3}{2}} \end{bmatrix} \quad (2)$$

$$V_{\alpha\beta} = C \times V_{abc} \tag{3}$$

$$I_{\alpha\beta} = C \times I_{abc} \tag{4}$$

$$p = v_\alpha \times i_\alpha + v_\beta \times i_\beta \tag{5}$$

$$q = v_\alpha \times i_\beta - v_\beta \times i_\alpha \tag{6}$$

Both power quantities p and q consists of dc and ac components. While the dc components $\overline{p}$ and $\overline{q}$, arise due to the fundamental, the ac components $\widetilde{p}$ and $\widetilde{q}$ are a result of harmonic components. The ac components of p and q, when obtained, may be transformed back to get the harmonic currents using (7) and (8). These form the references to the current controller. $\widetilde{p}$, $\widetilde{q}$ are determined by first extracting $\overline{p}$ and $\overline{q}$, using a very low cut-off low pass filter and then subtracting them from p and q respectively. A low pass filter is thus embedded into the system. The signal flow in this scheme is shown in Fig. 5.

$$\begin{bmatrix} i_\alpha^* \\ i_\beta^* \end{bmatrix} = \begin{bmatrix} v_\alpha & v_\beta \\ -v_\beta & v_\alpha \end{bmatrix} \tag{7}$$

$$\begin{bmatrix} ia^* \\ ib^* \\ ic^* \end{bmatrix} = [C]^T \begin{bmatrix} i_\alpha^* \\ i_\beta^* \end{bmatrix} \tag{8}$$

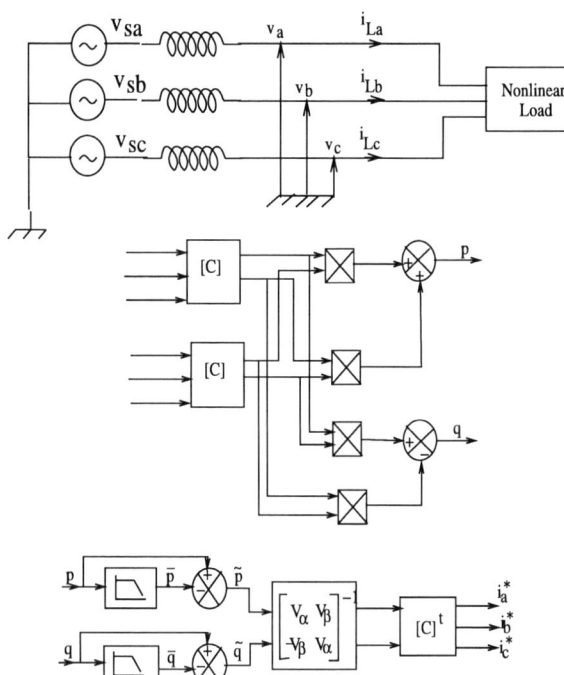

Fig.5 Block diagram of the 'pq method' scheme

C. Instantaneous Active and Reactive Current Theory (dq method)

In this method the current references i_a^*, i_b^*, and i_c^* are obtained from the instantaneous active and reactive current components i_{Lq} and i_{Ld} of the nonlinear load, in the synchronous reference frame [3]. These are obtained from the stationary frame currents determined in (4) by applying a further transformation as in (9).

$$\begin{bmatrix} i_{Ld} \\ i_{Lq} \end{bmatrix} = \begin{bmatrix} \cos\theta & \sin\theta \\ -\sin\theta & \cos\theta \end{bmatrix} \begin{bmatrix} i_{L\alpha} \\ i_{L\beta} \end{bmatrix} \tag{9}$$

where

$$\theta = \tan^{-1}\left(\frac{v_\beta}{v_\alpha}\right) \tag{10}$$

The real and reactive currents consist of dc and ac components, and as before, the dc components correspond to the fundamental component. The ac component of i_{Ld} ($\widetilde{i}_{Ld}$) and i_{Lq}($\widetilde{i}_{Lq}$) arise due to the harmonics of the load current. They may be obtained from i_{Ld} and i_{Lq} using a high pass filter. Alternatively, they may be obtained by using a low pass filter as in the earlier method. The actual reference currents are obtained by transforming these ac components using (11) into the stationary frame ($\alpha\beta$) and then using (8) to get the three phase quantities. The signal flow is depicted in Fig. 6.

$$\begin{bmatrix} i_\alpha^* \\ i_\beta^* \end{bmatrix} = \begin{bmatrix} \cos\theta & \sin\theta \\ -\sin\theta & \cos\theta \end{bmatrix}^{-1} \begin{bmatrix} i_{L\alpha} \\ i_{L\beta} \end{bmatrix} \tag{11}$$

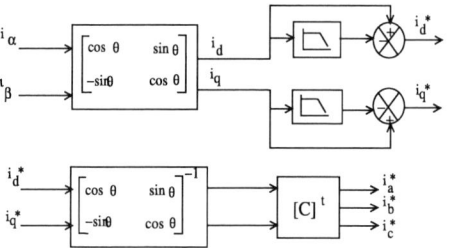

Fig. 6 Block diagram for dq method

D. Nonlinear Least Squares Approach (NLS Method)

This method proposed in [9], is summarized here. Any periodic signal can be expressed as a Fourier series, i.e., as a linear combination of harmonically related sinusoids. The current drawn by a nonlinear load can therefore be written as

$$i(t) = a_0 + \sum_{n=1}^{\infty} [a_n \cos(n\omega_0 t) + b_n \sin(n\omega_0 t)] \tag{12}$$

where $\sqrt{a_n^2 + b_n^2}$ is the amplitude of the nth harmonic component and ω_0 is the fundamental frequency in radians/sec. Since $i(t)$ does not usually contain a dc component, $a_0=0$. In practice, we limit the number of harmonics to be estimated to a finite number N. This converts (12) to an approximation. This is particularly

1829

true if measurement errors and/or additive noise are present. In order to determine the harmonic contents of the load current we have to estimate a_n and b_n for $n=1,2,...,N$. In this expression, we have $2N$ unknown parameters; if ω_0 were also not known precisely, we have one additional parameter. The procedure to obtain these unknowns may be explained as follows. Let us say that ω_0 is known. To solve for a_n and b_n, we assume that $i(t)$ is known at M uniformly sampled points, i.e., at $i(t_k)$ for $k=0,1,...,M-1$. This leads to the following set of M equations:

$$i(t_k) = a_0 + \sum_{n=1}^{N} [a_n \cos(n\omega_0 t_k) + b_n \sin(n\omega_0 t_k)] \qquad (13)$$

where $k=0,1,...,M-1$.
In vector notation (13) can be written as
$$\mathbf{Hx} = \mathbf{y} \qquad (14)$$
where $\mathbf{y}$ is the vector of samples $i(t_k)$ and $\mathbf{x}$ is the vector of unknowns a_n and b_n. The number of samples M required to solve these simultaneous equations is equal to $2N$. But in the presence of measurement and/or additive noise an overdetermined system of equations i.e., $M \geq 2N$ will give a better solution. Since the system is overdetermined, the least squares solution $\mathbf{x}$ is given by the following equation.
$$\mathbf{x} \approx (\mathbf{H^T H})^{-1} \mathbf{H^T y} \qquad (15)$$

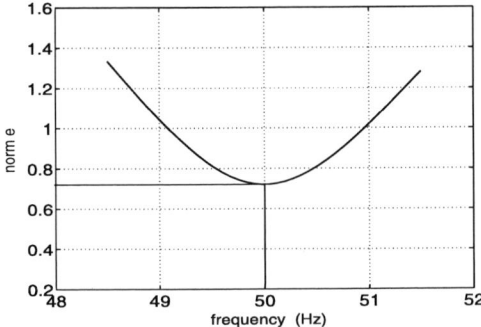

Fig. 7 Variation of the squared error with frequency

If the system frequency ω_0 is known, then the problem is a linear least squares problem. If ω_0 is not known precisely, the linear least squares problem is transformed into a nonlinear one [10]. Even though $2N+1$ unknowns are present, the linearly entering $2N$ amplitudes variables can be eliminated, resulting in a one-dimensional nonlinear least squares problem. The standard trick is to eliminate the linear variables by substituting (15) into (14), resulting in (16).

$$(\mathbf{H^T H})^{-1} \mathbf{H^T y} \approx \mathbf{y} \qquad (16)$$
The error vector e is given by
$$e = (\mathbf{I} - (\mathbf{H^T H})^{-1} \mathbf{H^T}) \mathbf{y} \qquad (17)$$
The system frequency ω_0 is considered to be the one that minimizes e. Since we have only one parameter, a simple grid search is enough to locate the minimum. The typical behaviour of the error norm, for a particular set of sampled waveform values, as the frequency search is done, is shown in Fig. 7. It is seen that the norm reaches a

minimum at a particular frequency, which is taken as the estimated frequency. Once ω_0 is estimated, (15) can be used to estimate $\mathbf{x}$ and hence a_n and b_n. From the solution vector $\mathbf{x}$, a_1 and b_1 are known and hence the fundamental. Fig. 8 shows the signal flow when this algorithm is used for harmonic extraction.. This is then subtracted from the load current to get the harmonic current reference for the current controller.

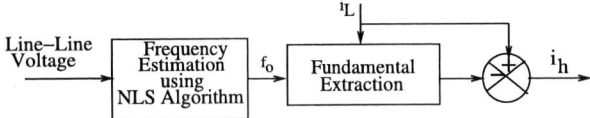

Fig.8 Block diagram for NLS scheme

III. SIMULATION RESULTS AND DISCUSSION

Using SABER we studied the behaviour of the shunt Active Power Filter shown in Fig. 9 for the four algorithms the results are presented. The following conditions of supply voltage were considered: (a) balanced sinusoidal voltage conditions with step load change (b) with non-sinusoidal supply voltage. We also studied the system behaviour when the load was time varying.

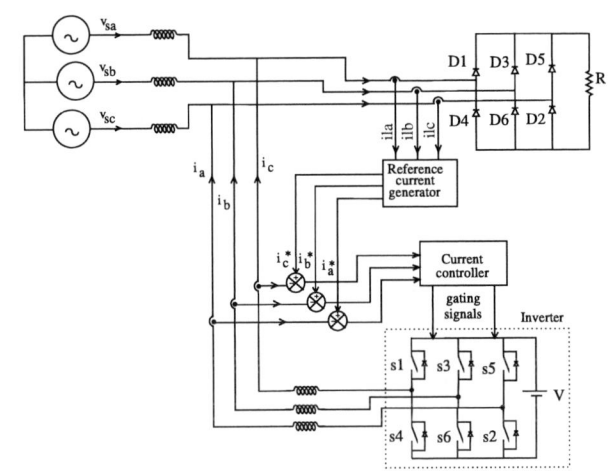

Fig. 9 Schematic diagram of a shunt active power filter used for simulation study

A. Estimation delay and its effects

a. Compensation lag

It can be seen from the earlier section that all methods considered other than NLS, incorporate a low pass filter. Fig. 10 compares the delay in compensation caused by these algorithms for a step load change. The source voltage is balanced and sinusoidal. Step load change increase and decrease, occur at 60 ms and 140 ms respectively. It can be seen from the figure that a delay in compensation exists in all algorithms. In the case of the NLS algorithm (Fig. 10(f)), the delay is essentially due to the time required to acquire quarter cycle information while the delay in others is the low pass filter. The LPF used is a fourth order Butterworth filter which has a

1830

settling time of 25 ms. The delay marked in the plot of Fig. 10(c) corresponds to the waveform reaching 98% of the fundamental amplitude. The low pass filters used in the *pq* and *dq* methods implementation in this paper are identical (second order Butterworth, with a 10 Hz cut-off), and with a settling time of 100 ms. The delay caused by this filter is higher because of the lower cut-off frequency even though its order is lower than the filter used in the LPF approach. The performance of *pq* and *dq* algorithms is therefore similar on this context.

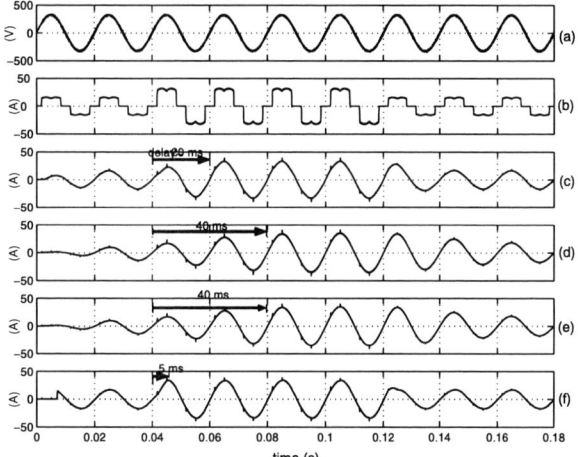

Fig. 10 Simulation results for transient performance of the APF (a) Voltage at PCC (b) Source current before Compensation (c) Source current after compensation with LPF method (d) with *pq* method (e) with *dq* method (f) with NLS method

b. Active power flow through the APF

The delay in the fundamental extraction is a common feature of all methods, but its value differs with the harmonic extraction algorithm used. This delay causes active power to flow through the APF during sudden load change. as shown in Fig. 11(a) and (b). Fig. (a) shows the results obtained with the LPF method, whereas Fig. (b) corresponds to the NLS method.

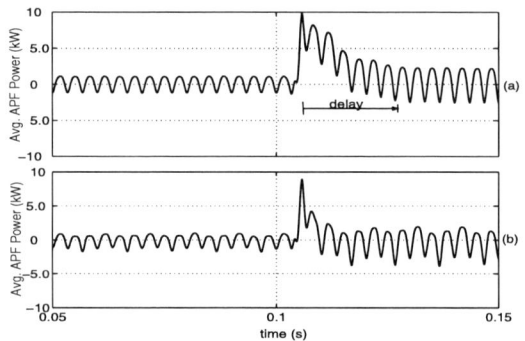

Fig. 11 Average power flow through the APF (a) Using LPF method (b) Using NLS method

Step load increase occurs at 0.1 s. This causes the average power flow through the active filter to increase temporarily. It takes a longer time for this power to reach zero when an LPF is used for generating the reference currents. The delay in compensation is due to the delay in extracting the fundamental component of the load

current. Simulation results for only the LPF method are given as it is representative of *pq* and *dq*, as they also employ a low pass filter for obtaining the harmonic components of the load current. The simulation results presented in this section are shown in Table 1. From the tabulated values it is clear that the Total Harmonic Distortion in the source current after compensation, when a sinusoidal supply is given to the nonlinear load, differs only marginally. Delay in compensation is the predominant effect.

TABLE I
STEADY STATE PERFORMANCE ANALYSIS

Parameter	Method			
	LPF	PQ	Dq	NLS
THD in the source current with sinusoidal supply	2.6%	2.72%	2.55%	2.37%
Response to step load change	≥20 ms	≥40 ms	≥40 ms	5 ms

B. Performance with time varying loads

Performance of the above methods with continuously varying loads is also studied and presented below. For the purpose of simulation, a time varying load is considered whose variation with time is given in Fig. 12(a). The load power varies between 3.1 kW and 5.2 kW. The average power flow from the active power filter is shown in Fig. 12(b). Table 2 shows that the power loss per load cycle is the least in the NLS method compared to all other methods. It is obvious from the tabulated values that in *pq* and *dq* methods the power dissipation in the inverter switches is considerably high. This is due to the large delay in tracking the fundamental component.

TABLE II
COMPARATIVE ANALYSIS WITH VARYING LOAD FOR A LOAD SWING OF 4 KW TO 8 KW

Parameter	Method			
	LPF	PQ	Dq	NLS
Peak-peak power in Watts	1025	2695	2630	345
APF loss in Watts and as a % of Avg. load power	137 (2.28%)	190 (3.2%)	185 (3.08%)	83 (1.38%)

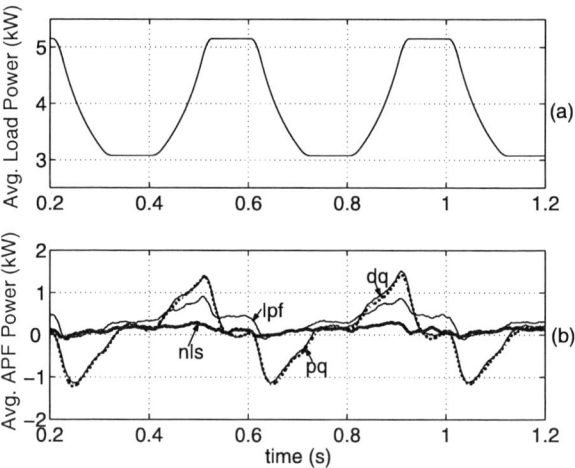

Fig. 12 (a) Load power variation with time. (b) Average power flow through the APF for (i) LPF, (ii) pq (iii) dq, and (iv) NLS methods

C. Effect of voltage distortion at PCC

Fig. 13 shows the essential waveforms for *pq* approach when the voltage at PCC is distorted. A THD of 10% is considered for the voltage waveform having 5th, 7th and *11*th harmonics considered. It can be seen that in this case the compensation is incomplete and the source current after compensation as shown in Fig. 13(b) is still significantly distorted. The reason for this can be understood by a simple example.

The following analysis shows how an nth harmonic component in the load current manifests itself in the reference current when the supply voltage contains harmonics. Consider the case where the voltage at PCC contains a 5th harmonic component as given by

$$v_a = V_1 \sin \omega t + V_5 \sin 5\omega t \qquad (18)$$

For simplicity, let us look at a nonlinear load that draws a pure 5th harmonic current alone, which can be modeled as

$$i_a = I_5 \sin 5\omega t \qquad (19)$$

The active filter is then required to generate the current required by the load (a pure 5th harmonic). We apply the pq method for extracting the harmonics as follows.

When the three phase voltages and currents are converted to the stationary($\alpha\beta$) coordinates we get

$$v_\alpha = \sqrt{\frac{3}{2}}(V_1 \sin \omega t + V_5 \sin 5\omega t) \qquad (20)$$

$$v_\beta = -\sqrt{\frac{3}{2}}(V_1 \cos \omega t + V_5 \cos 5\omega t) \qquad (21)$$

$$i_\alpha = \sqrt{\frac{3}{2}}(I_5 \sin 5\omega t) \qquad (22)$$

$$i_\beta = -\sqrt{\frac{3}{2}}(I_5 \cos 5\omega t) \qquad (23)$$

The instantaneous active and reactive power p and q can be obtained using (5) and (6) as given by (24) and (25).

$$p = \frac{3}{2}(V_1 I_5 \cos 4\omega t) \qquad (24)$$

$$q = \frac{3}{2}(V_1 I_5 \sin 4\omega t) \qquad (25)$$

From (24) and (25) it is clear that the ac components of *p* and *q* are $p = \frac{3}{2}(V_1 I_5 \cos 4\omega t)$ and $q = \frac{3}{2}(V_1 I_5 \sin 4\omega t)$ respectively. Using these quantities and (7)–(8), the reference current of phase '*a*' in three phase coordinates obtained is given by

$$i_a^* = \frac{V_1^2 I_5 \sin(5\omega t) + V_1 V_5 I_5 \sin(9\omega t)}{V_1^2 + V_5^2 + 2V_1 V_5 \cos(4\omega t)} \qquad (26)$$

It can be shown that i_a^* contains a 5th harmonic component which is not in phase with the 5th harmonic component of the load current. This explains why the '*pq* algorithm' is unable to compensate harmonics in the source current completely. This is evident form the value of THD in the source current waveform with distorted voltage, which is computed to be 10.77%. This is substantial compared to the result obtained with balanced sinusoidal voltage. However, it may be noted that the NLS method is insensitive to voltage distortion. The LPF method on the other hand, is insensitive only when the harmonic currents caused by the distortion are attenuated sufficiently by the filter. The *dq* method is insensitive to distortion only if the angle of the fundamental voltage component can be estimated accurately. If the supply voltage is nonsinusoidal, the expression used for computing θ is not valid and it requires the estimation of θ through some other means. Usually phase locked loops (PLL) are used, but they suffer from problems of delay in generating the synchronizing signal, due to fluctuations in supply voltage and frequency [11]–[13].

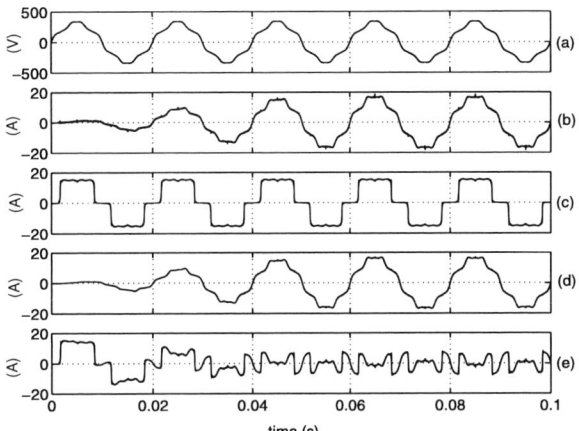

Fig. 13 Simulation results with pq method under nonsinusoidal supply voltage conditions (a) Voltage at PCC (b) Source current after compensation (c) Source current before compensation (d) Fundamental component (d) Harmonic components

IV. CONCLUSIONS

This paper compared four different algorithms for harmonic extraction for a shunt active power filter. Our conclusions are valid for other topologies as well. It has been demonstrated that when a low pass filter is used the algorithms not only cause a delay in compensation, but also require active power flow through the filter. When load is time varying, the delay manifests as increased power dissipation in the filter. It can be seen that the NLS algorithm, in which the delay is minimal, minimizes these effects.

The paper also investigated the effect of voltage distortion at PCC. The NLS method is superior in performance, which comes at no extra cost. The method is computation intensive and requires fast processing to minimize the computational delay. We are working on simplifying the computational cost of the NLS method. Further, availability of high speed DSP and FPGA makes the real time implementation feasible and is being pursued.

REFERENCES

[1] Luis A. Moran, Juan W. Dixon and Aogel R. Wallace, "A Three-phase Active Power Filter Operating with Fixed Switching Frequency for Reactive and Current Harmonic Compensation", *IEEE Transactions on Industrial Electronics*, Vol. 42, No. 4, August 1995.

[2] Hirofumi Akagi, Yoshihira Kanazawa and Akira Nabae, "Instantaneous Reactive Power Compensators Comprising Switching Devices without Energy Storage Components," *IEEE Transaction on Industry Applications*, Vol. IA-20, No.3, May/June 1984, pp.625-631.

[3] Vasco Soares, Pedro Verdelho and Gil Marques, "Active Power Filter Control Circuit based on the Instantaneous Active and Reactive Current i_q-i_d method," *Power Electronics Specialists Conference*, 1997, PESC'97 Record, 28th annual IEEE Vol. 2, 22-27 June 1997, pp.1096-1101.

[4] L.L. Lai, C.T. Tse, W.L. Chan and A.T.P. So, "Real-time Frequency and Harmonic Evaluation using Artificial Neural Networks," *IEEE Transactions on Power Delivery*, Vol.14, No. 1,January 1999, pp.52-59.

[5] M. Rukonuzzaman, Katsumi Nishida and Mutsuo Nakaoda, "Adaptive Neural Network," *Industry Applications Conference*, 2003, 38th IAS Annual meeting, Conference record of the Vol. 2, 12-16 Oct. 2003, pp.1215-1221.

[6] John H. Marks and Tim. C. Green, "Predictive Transient Following Control of Shunt and Series Active Power Filters," *IEEE Transactions on Power Electronics*, Vol.17, No.4, July 2002, pp.574-584.

[7] Sami Valiviita and Seppo J. Ovaska, "Delay less Method to Generate Current Reference for Active Filters," *IEEE Transactions on Industrial Electronics*, Vol. 45, No. 4, August 1998, pp.559-569.

[8] Raghavan N, Ellinger T, Grossmann U, Buettner A, Pedzoldt J, Vasudevan and K Swarup KS, "Simulation based Comparison of PQ Theory and Transformed Series Hybrid Active Power Filter," 13th *National Power System Conference India*, Dec. 27-30,2004, Vol. II, pp.920-924.

[9] R. Chudamani, Krishna Vasudevan and C.S. Ramalingam, "Nonlinear Least Squares Current Estimator for Three Phase Loads," *IEEE Conference on Industrial Technology*, Dec.15-17, Mumbai, India, pp.1581-1585.

[10] S.M. Kay, "Fundamentals of Statistical Signal Processing: Estimation Theory," Prentice-Hall, Englewood Cliff, NJ, 1993.

[11] Dragan Jovcic, "Phase Locked Loop System for FACTS," *IEEE Transactions on Power Systems*, Vol. 18, No.3, August 2003, pp.1116-1124.

[12] Joonsuk Lee and Beomsup Kim, "A Low-Noise Fast-Lock Phase-Locked Loop with adaptive Bandwidth control," *IEEE Journal of Solid-State Circuits*, Vol.35, No.8, August 2000, pp.1137-1145.

[13] Masoud Karimi-Ghartemani and M. Reza Iravani, "A Signal Processing Module for Power System Applications," *IEEE Transactions on Power Delivery*, Vol.18, No.4, October 2003, pp.1118-1124.

Hybrid Passive Filter Design for Distribution Systems with Adjustable Speed Drives

M.A.S. Masoum, *SM IEEE*, A. Ulinuha, S. Islam, *SM IEEE,* and K. Tan, *M IEEE*

Department of Electrical and Computer Engineering, Curtin University of Technology, Perth, WA, Australia

Abstract–Drive systems inject significant low order harmonics currents into distribution system and deteriorate the quality of electric power. This paper models variable frequency and PWM adjustable speed drives as harmonic current sources and performs power flow analysis before filtering to identify highly distorted buses and the spectrum of harmonic frequencies with unacceptable THD levels as specified by IEEE-519 standard. The IEEE 30-bus system with penetration of nonlinear adjustable speed drive loads is used to study the effectiveness of passive filters and to investigate the impact of their location and tuning frequencies on the quality of voltage and current waveforms. Simulation results before and after the installations of filters are presented, compared and analyzed for different nonlinear loading and filter configurations. It is shown that the number, locations and tuning frequencies of filters have major impacts on the overall quality of the distribution system.

Index Terms– Variable frequency, PWM, drives systems, power quality, THD, passive filter and harmonic power flow

I. INTRODUCTION

DRIVE systems are nonlinear devices with non-sinusoidal waveforms that inject low order harmonic currents into distribution system. Electric utilities are very concern about the fast growth of large electric drives in the industrial sectors of distribution system. Variable frequency and PWM drives are considered as one of the biggest contributors to power quality problems due to their high-power ratings [1-3]. The nonlinear v-i characteristics of drive systems may result in triplen harmonic currents, neutral conductor problems (and increased losses), transformer saturation (and overheating), power-factor capacitor failures, unsatisfactory performance of fuses, circuit breakers and relays (e.g., longer or shorter tripping times depending on the harmonic magnitude and spectrum), etc. Consideration of these problems requires investigating their impacts on power system devices, loads and equipments and, if necessary, suppression or prevention of their generation. Consequently, harmonic filters are often utilized to eliminate or limit the injected harmonics [4-6]. The common approach is to place power filters at the terminal of the drive and tune them to attenuate the injected harmonic currents without considering their effects on the rest of the power system.

For the power system consisting of a number of nonlinear loads (e.g., variable speed drives), a vast variety of harmonic currents with different orders, magnitudes and phase angles are injected at different locations (buses). Therefore, for accurate filter modeling, the propagation of harmonics as well as the impact of nonlinear drive systems on the rest of the power system needs to be considered.

There are many solution approaches to compensate for harmonic injections of nonlinear loads [5], including (passive, active and hybride) filters, unified power quality conditioners (UPQCs) and active power line conditioners (APLCs).

Passive filters are the easiest and most cost effective method for harmonic compensation. There are a large number of low-power nonlinear loads in single-phase power system, such as ovens, air conditioners, fluorescent lamps, TVs, computers, power supplies, printers, copiers and battery chargers. Low-cost compensation of these residential nonlinear loads can be achieved using hybrids of passive filters (e.g., shunt or series combinations of a group of passive filters). Three-phase power system is highly polluted by a large number of small rating to reasonable power level nonlinear loads such as adjustable speed drives and HVDC transmission systems in high power rating. These loads can be compensated using either hybrids of passive filters or hybrids of active and passive filters depending on the nature of the AC system. Compensation of single-phase high-rating traction systems are effectively performed with a hybrid of active filters. Vastly distributed single-phase nonlinear loads in three-phase four-wire systems may be compensated using a number of hybrids of passive filters, active filters or hybrids of passive and active filters.

This paper performs harmonic flow analysis in the presence of nonlinear drive systems to determine highly distorted buses and the most detrimental harmonic frequencies. A hybrid of shunt passive filters (tuned at the dominated harmonic frequencies) is used to compensate for the injected harmonics. The power quality of the system (including the overall THD and individual THD levels of each bus) are examined before and after the placement of the filters. The effects of filter location as well as the number of filter units are investigated. Variable speed drives are modelled as harmonic current sources and decouple harmonic power flow (DHPF) algorithm is performed to model the distorted power system. The IEEE 30-bus system with a number of variable-frequency and PWM adjustable speed drives is used for the analysis and simulations.

II. HARMONIC MODELING OF AC DRIVES

The literature is rich with documents on classification, modeling and analyses of nonlinear drive systems [3-9].

Modern electric drives utilize rectifier circuits and are classified as dc and ac drives.

DC drives employ controlled rectifiers to realize variable dc voltages while ac drives usually have PWM inverters with variable voltage and variable frequency technology. In ac drives, a dc capacitor is normally applied between the rectifier and the PWM inverter to limit the low-ripple dc voltage. Many industrial loads such as the paper industry use controlled ac drives with voltage-source inverter (VSI) that operate based on PWM switching. However, the dc capacitor magnifies line harmonics and may cause power quality problems. DC drives are directly connected to dc motors without a dc capacitor. Therefore, dc drives have large equivalent inductances on the dc bus and cause less harmonic distortions on the line-side, as compared to their counterpart ac drives. Furthermore, harmonics distortions of ac drives become more severe under light load conditions. One approach to improve the performance of ac drives is to place an inductor on the dc bus in addition to the dc capacitor [3]. This will reduce reflection of harmonics into lines from the VSI unit; however, there are cost and implementation issues.

A counterpart to the PWM-VSI electric drive is the current-source inverter (CSI)-based PWM-operated drive system; the dc bus includes a dc inductor (with no dc capacitor) and the load side is connected to a three-phase ac capacitor in parallel with the motor [3]. With this improved configuration, the capacitor and dc inductor act as a filter unit while the ac inductor serves as an energy storage element. Comparison between PWM-CSI and PWM-VSI electric drives can be performed based on output voltage regulation, line current waveform, load-side voltage/current waveforms, and cost [7-9]:

Voltage regulation- the line-to-line output voltages of PWM-CSI and PWM-VSI drives are:

$$V_{rms}^{PWM-CSI} = \sqrt{3}\,\frac{V_g}{D} = 1.732\,\frac{V_g}{D}$$

$$V_{rms}^{PWM-VSI} = \frac{\sqrt{3}}{2\sqrt{2}}\,DV_g = 0.612DV_g$$

(1)

where V_g is the equivalent dc source voltage and D is the duty ratio of the inverter which is less than one. According to Eq.1, the CSI configuration has better voltage regulation capabilities due to the substantial energy stored in the inductor as well as its boost characteristics.

Line current harmonics- PWM-CSI drives inherently have low line harmonics. On the other hand, due to the reflection of the dc capacitor in a VSI-based drive, the line current waveforms are highly distorted and become discontinuous in many cases, particularly under light load conditions. The total harmonic distortion THD of VSI-based drives may be as high as 60% with very high penetrations of 5th and 7th harmonics.

Output (load side) voltage and current waveforms- typical voltage waveforms of the VSI-based drives have high THD values of about 40% (with lowest harmonic order equal to the switching frequency of about 5-10 kHz), as compared with acceptable THD levels experienced with CSI-based configurations (about 6%). Due to the high-frequency switching, the current waveforms of both types of drives are acceptable and have low THD levels (e.g., less than 2%).

Cost considerations- PWM-CSI drives have certain limitations that increase their overall cost. The dc inductor and the large ac capacitor demand extra space and increase the cost. In addition, the ac capacitor may cause self excitation with magnetizing inductor of the rotating machine. The self-excitation problem may be solved by coordinating the dc inductor, ac capacitor and motor operating conditions, such that the resonant frequency does not coincide with any harmonic frequencies of the system and/or nonlinear loads.

For analysis and modelling of small to moderate size power systems, detailed and accurate models of dc and ac drives can be utilized to generate accurate results. However, due to memory storage and convergence problems of harmonic power flow algorithms, detailed nonlinear models of drive systems should not be used for the simulation of large distorted distribution systems with high penetration of nonlinear loads. This paper uses harmonic current sources to model drive systems and employs a decoupled harmonic power flow algorithm [10-13] to simulate the distorted distribution system that usually includes other nonlinear devices and/or loads. An inherent advantage of this simple approach (in addition to its simplicity, fast convergence and acceptable accuracy), is the possibility of using measured waveform current (or voltage) of the drive to estimate the harmonic model without the need for detailed models. This is particularly attractive for most electric utilities that have limited information about the configuration, types and number of drive systems for their industrial customers. Table 1 shows the typical harmonic spectrum of variable-frequency and PWM adjustable speed drives used in this paper [8-9].

TABLE 1
HARMONIC SPECTRUM OF TYPICAL VARIABLE-FREQUENCY AND PWM ADJUSTABLE SPEED DRIVES

h	variable frequency drive (675 kW, 439 kVAr)		PWM-adjustable speed drive (350 kW, 175 kVAr)	
	magnitude [%]	phase angle [degree]	magnitude [%]	phase angle [degree]
1	100	0	100	0
5	23.52	111	82.8	-135
7	6.08	109	77.5	69
11	4.57	-158	46.3	-62
13	4.2	-178	41.2	139
17	1.8	-94	14.2	9
19	1.37	-92	9.7	-155
23	0.75	-70	1.5	-158
25	0.56	-70	2.5	98
29	0.49	-20	0	0
31	0.54	7	0	0

III. DECOUPLED HARMONIC POWER FLOW

At the fundamental frequency, system is modeled using the conventional (sinusoidal) power flow approach. The magnitudes and phase angles of bus voltages are calculated using the following mismatch equations [10-13]:

$$P_i - \sum_{j=i-1}^{i+1} \left|Y_{ji}^1\right|\left|V_j^1\right|\left|V_i^1\right|\cos(\delta_i^1 - \delta_j^1 - \theta_{ji}^1) = 0$$

$$(2)$$

$$Q_i - \sum_{j=i-1}^{i+1} \left|Y_{ji}^1\right|\left|V_j^1\right|\left|V_i^1\right|\sin(\delta_i^1 - \delta_j^1 - \theta_{ji}^1) = 0$$

where

$$Y_{ji}^1 = \left|Y_{ji}^1\right|\angle\theta_{ji}^1 = \begin{cases} -y_{ji}^1, & \text{if } j \neq i \\ y_{i-1,i}^1 + y_{i+1,i}^1 + y_{ci}^1, & \text{if } j = i \end{cases} \quad (3)$$

while P_i, Q_i, V_i^1 and y_{ci}^1 are the total active power, reactive power, fundamental voltage and admittance of shunt capacitor at bus i, and $y_{i,i+1} = 1/(R_{i,i+1} + jX_{i,i+1})$ is the admittance of line section between bus i and bus $i+1$.

At harmonic frequencies, power system is modeled as combination of passive elements and current sources. The system can then be considered as a passive element with multiple harmonic injection currents. Linear loads are modeled with a resistance in parallel with an inductance to account for the respective active and reactive loads at fundamental frequency. Nonlinear loads are considered as ideal harmonic current sources that generate harmonic currents and inject them into the system. The admittance-matrix-based harmonic power flow is the most widely used method as it is based on the frequency-scan process. In this approach, admittance of system components will vary with the harmonic order. If skin effect is ignored at higher frequencies, the resulting h^{th} harmonic frequency load admittance, shunt capacitor admittance and feeder admittance are respectively given by the following equations [13]:

$$y_{li}^h = \frac{P_{li}}{\left|V_i^1\right|^2} - j\frac{Q_{li}}{h\left|V_i^1\right|^2}$$

$$y_{ci}^h = hy_{ci}^1 \quad (4)$$

$$y_{i,i+1}^h = \frac{1}{R_{i,i+1} + jhX_{i,i+1}}$$

where P_{li} and Q_{li} are the respective active and reactive linear loads at bus i.

The nonlinear load is treated as harmonic current sources and the hth harmonic current injected at bus i introduced by the nonlinear load with real power P_n and reactive power Q_n is:

$$I_i^1 = [(P_{ni} + jQ_{ni})/V_i^1]^* $$

$$I_i^h = C(h)I_i^1 \quad (5)$$

where I_i^1 is the fundamental current and I_i^h is the hth harmonic current determined by $C(h)$, which is the ratio of the hth harmonic to the fundamental current. $C(h)$ can be obtained by field test and Fourier analysis for all customers along the distribution feeder [13].

For decouple harmonic power flow calculation, loop equations are written at each harmonic frequency of interest. Each loop is formed including the source nodes. After modifying admittance matrix and the associated harmonic currents, the harmonic load flow problem can then be calculated using the following equation [13]:

$$Y^h V^h = I^h \quad (6)$$

At any bus i, the rms voltage is defined as:

$$\left|V_i\right| = \left(\sum_{h=1}^{H} \left|V_i^h\right|^2\right)^{1/2} \quad (7)$$

where H is the maximum harmonic order considered. After solving load flow for different harmonic orders, the total harmonic distortion of voltage at bus i (THD_{vi}) is computed as:

$$THD_{vi}(\%) = \frac{\left(\sum_{n\neq1}^{H} \left|V_i^h\right|^2\right)^{1/2}}{\left|V_i^1\right|} \times 100\% \quad (8)$$

IV. HYBRID OF PASSIVE FILTERS

The configuration and complexity of the filter depends on harmonic spectrum and nature of the distortion. If a nonlinear load is locally causing significant harmonic distortion, passive filters may be installed to prevent the harmonic currents from being injected into the system. These filters are inexpensive compared with most other mitigating devices. Passive filters are composed of only passive elements (inductance, capacitance, and resistance) tuned to the harmonic frequencies. In practice, passive filters are added to the system starting with the lowest trouble harmonic (e.g., installing a seventh-harmonic filter usually requires that a fifth-harmonic filter also be included).

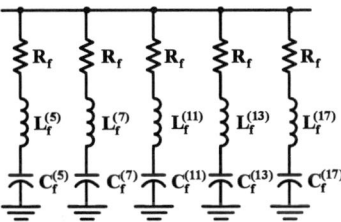

Fig. 1. Hybrid of passive filters employing five series-resonance filters tuned at 5th, 7th, 11th, 13th and 17th harmonics

Since drive systems mainly inject low order harmonics, a hybrid filter block consisting of a number of band-pass filters tuned to dominating harmonic frequencies [4] will be used (Fig. 1). The resonant frequency of each shunt branch is:

$$\omega_h = 2\pi h f = 1/\sqrt{L_f^h C_f^h} \quad (9)$$

where h is the order of harmonic that is being attenuated, f is the fundamental frequency, L_f^h and C_f^h are filter inductor and capacitor, respectively. For variable-frequency drives, two filter branches tuned at the 5th and the 7th harmonic are used (Fig. 1). For PWM adjustable speed drives (Table 1) all filter branches are activated.

V. RESULT AND DISCUSSION

The 23kV distribution system [14-15] is used in this paper (Fig. 2). A PWM-adjustable speed drive and a variable frequency drive (Table 1) are placed at buses 15 and 18, respective. At the fundamental frequency, all loads including nonlinear drives are considered as constant power loads. Three cases will be considered.

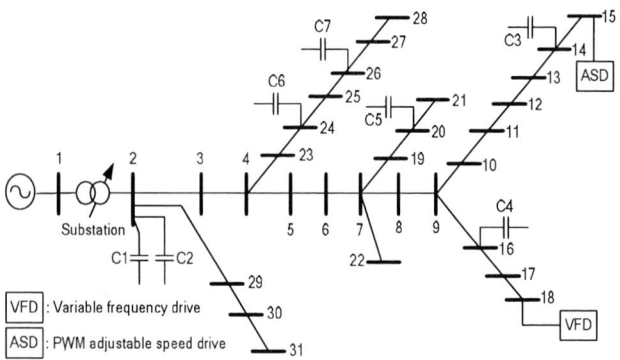

Fig. 2. The IEEE 30-bus system used for simulations [14-15]

Case A. Harmonic Compensation with Two Hybrid Flirters

Harmonic analysis before the installation of passive filters indicate very high distortion of both voltage and current waveforms (Fig. 3 and Table 2) due to the presence of the two ac drives with relatively high power ratings (Table 1). According to the IEEE-519 standard [7], the individual harmonic current injections at bus 15 (for 5th, 7th, 11th, 13th, 17th and 19th harmonics) and bus 18 (for 5th, 7th and 11th harmonics) are higher than the allowable level of 4% and the electric utility has the right to reject (disconnect) these nonlinear loads. In addition, propagation of injected harmonic currents has resulted in unacceptable THDv levels (e.g., larger that 5%) at buses 12 to 15. Buses 11 and 16 to 18 are also experiencing relatively high THDv of more than 4% (Fig. 6).

The common procedure is to require the consumers of nonlinear loads (at buses 15 and 18) to installation passive filter banks at the terminals of the AC drives. Figure 3 and Table 2 show the simulation results after the installation of the following filter banks, tuned at the dominated frequencies:

- **Filter Bank 1-** A hybrid passive filter with five shunt branches tuned to the 5th, 7th, 11th, 13th and 17th harmonic frequencies (Eq. 9 with f_1=50Hz), installed at bus 15 with
  $R_f = 100\Omega$, $L_f = 100mH$ (for all branches), $C_f^{(5)} = 4.05\mu F$,
  $C_f^{(7)} = 2.07\mu F$, $C_f^{(11)} = 0.84\mu F$, $C_f^{(13)} = 0.6\mu F$, $C_f^{(17)} = 0.35\mu F$.

- **Filter Bank 2-** A hybrid passive filter with only two shunt branches tuned to the 5th and 7th harmonic frequencies (Eq. 9 with f_1=50Hz), installed at bus 18 with
  $R_f = 100\Omega$, $L_f = 100mH$ (for all branches),
  $C_f^{(5)} = 4.05\mu F$, $C_f^{(7)} = 2.07\mu F$.

According to Table 2, the performance of these filters are satisfactory; the maximum level of THDv has dropped from 6.51% (at bus 15) to 2.17% (at bus 18) while the average system THDv is within the permissible limits of the IEEE-519 standard [7].

Case B. Harmonic Compensation with One Hybrid Flirter

This case examines the possibility of meeting IEEE-519 limits with only one filter bank. Simulations results before and

after the installation of Filter Bank 1 at bus 15 are shown in Figs. 4 and 6 and Table 2. Therefore, by performing harmonic load flow analysis, it may be possible to eliminate unnecessary filter banks and reduce the overall cost of harmonic compensation.

Case C. Impact of Filter Location on Power Quality

To show the impact of filter locations on the overall power quality of system, Filter Bank 1 is installed between the two drive systems at bus 9. Based on the results (Fig. 5, Table 2), poor placement of power filters will not only deteriorate their performances, but might also hearten harmonic propagation and cause additional power quality problems in other sectors of the system (e.g., in buses 10-15 and 16-18 for Case C).

TABLE 2
SUMMARY OF SIMULATION RESULTS FOR THE IEEE 30 BUS SYSTEM (FIG. 2)
BEFORE AND AFTER THE INSTALLATION OF FILTER(S)

	before filtering		after filtering (Case A)	
	minimum	maximum	minimum	maximum
rms voltage	0.8382 pu at bus 15	1.0003 pu at bus 1	0.8295 pu at bus 15	1.0003 pu at bus 2
THDv	1.8492 % at bus 5	6.5089 pu at bus 15	1.0548 % at bus 11	2.1691 % at bus 18
system THDv	3.1820%		1.4818%	
	after filtering (Case B)		after filtering (Case C)	
	minimum	maximum	minimum	maximum
rms voltage	0.8337 pu at bus 5	1.0001 pu at bus 2	0.8469 pu at bus 15	1.0002 pu at bus 2
THDv	1.2520 % at bus 22	1.8629 % at bus 18	1.0938 % at bus 6	5.8453 % at bus 15
system THDv	1.4950 %		1.8707 %	

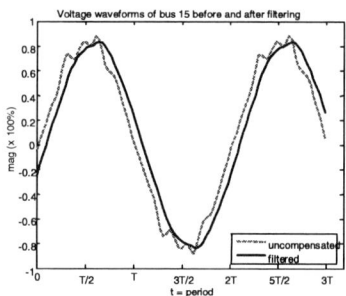

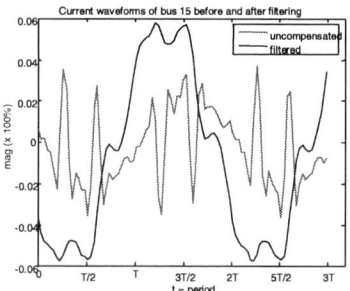

Fig. 3. Voltage and current waveforms at bus 15 before and after the installation of two filter banks at buses 15 and 18 (Case A)

1837

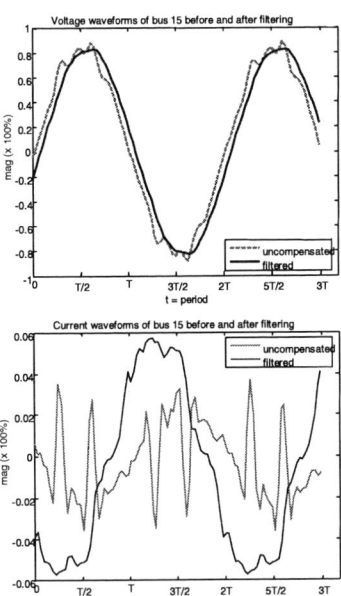

Fig. 4. Voltage and current waveforms at bus 15 before and after the installation of one filter bank at bus 15 (Case B)

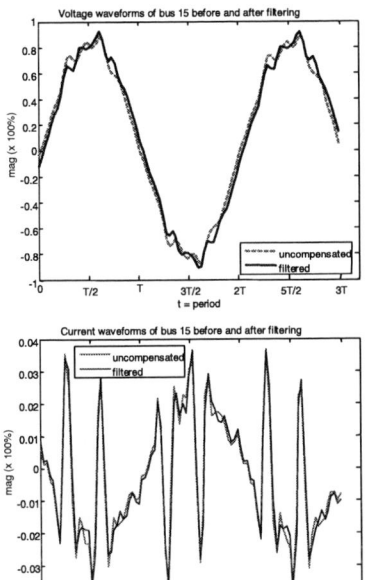

Fig. 5. Voltage and current waveforms at bus 15 before and after the installation of one filter bank at bus 9 (Case C)

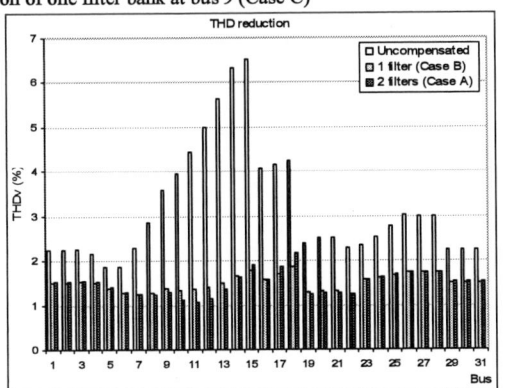

Fig. 6. Comparison of THDv levels of the IEEE 30-bus system (Fig. 2) before and after the installation of one (Case B) and two (Case A) filter banks

VI. CONCLUSION

Distorted distribution systems with nonlinear ac drive loads are modeled and the impact of numbers, locations and tuning frequencies of hybrid filters (consisting of shunt passive branches tuned at the dominating harmonic frequencies) are investigated. A decoupled harmonic power flow algorithm is used to simulate the distribution system and nonlinear ac drives are modeled with harmonic current sources. This simple approach is very practical and convenient for the analysis of large distorted industrial systems with inadequate information about the nonlinear loads (e.g., parameters and ratings of ac drives). Simulation results indicate that the common approach of placing filter banks at the terminals of each drive system is not always the best and most economical solution. Simulation results indicate that the number, locations and tuning frequencies of filters have major impacts on the overall quality of the distribution system. It might be possible to control and limit the overall system distortion, as well as the individual bus THDv levels with fewer filter banks if system conditions before and after harmonic compensation are carefully studied and analyzed. This required fast and relatively accurate algorithms and solution approaches, as presented in this paper.

VII. REFERENCES

[1] M. Villablanca, W. Flores, C. Cuevas and P. Armijo, "Harmonic reduction in adjustable-speed synchronous motors", *IEEE Transactions on Energy Conversion*, vol. 16, no. 3, pp. 239-245, 2001.

[2] J. Faiz, H. Barati and E. Akpinar, "Harmonic analysis and performance improvement of slip energy recovery induction motor drives", *IEEE Transactions on Power Electronics*, vol. 16, no. 3, pp. 410-417, 2001.

[3] Y. Yin and A.Y. Wu, "A Low-Harmonic Electric Drive System

[4] Based on Current-Source Inverter", *IEEE Transactions on Power Industry Applications*, vol. 34, no. 1, pp. 227-235, 1998.

[5] J.C. Das, "Passive Filters- Potentialities and Limitations", *IEEE Transactions on Power Industry Applications*, vol. 40, no. 1, pp. 232-241, 2004.

[6] B. Singh, V. Verma, A. Chandra and K. Al-Hadded, "Hybrid Filters for Power Quality Improvement", *IEE Proceedings-Generation, Transmission and Distribution*, vol. 152, no. 3, pp. 365-378, 2005.

[7] Y.M. Chen, "Passive Filter Design using Genetic Algorithms", *IEEE Transactions on Industrial Electronics*, vol. 50, no. 1, pp. 202-208, 2003.

[8] *IEEE Recommended Practice and Requirements for Harmonic Control in Electric Power Systems*, IEEE Standard 519, 1992.

[9] N. Mohan, T. M. Undeland, and W. P. Robbins, *Power Electronics:Converters, Applications, and Design*. New York: Wiley, 1989.

[10] M. H. Rashid, *Power Electronics, Circuits, Devices, and Applications*, 2nd ed. Englewood Cliffs, NJ: Prentice-Hall, 1993.

[11] D. Xia and G. T. Heydt, "Harmonic power flow studies, part I - Formulation and Solution," *IEEE Trans. on Power Apparatus and System*, vol. 101, no. 6, pp. 1257-1265, 1982.

[12] T. S. Chung and H. C. Leung, "A genetic algorithm approach in optimal capacitor selection with harmonic distortion considerations," *International Journal of Electrical Power & Energy Systems*, vol. 21, no. 8, pp. 561-569, 1999.

[13] H.-C. Chin, "Optimal shunt capacitor allocation by fuzzy dynamic programming," *Electric Power Systems Research*, vol. 35, no. 2, pp. 133-139, 1995.

[14] Y. Baghzouz and S. Ertem, "Shunt capacitor sizing for radial distribution feeders with distorted substation voltage," *IEEE Trans. on Power Delivery*, vol. 5, no. 2, pp. 650-657, 1990.

[15] J. J. Grainger and S. Civanlar, "Volt/var control on distribution systems with lateral branches using shunt capacitors and voltage regulators. Part III: The numerical result," *IEEE Transactions on Power Apparatus and Systems*, vol. 104, no. 11, pp. 3291-3297, 1985.

[16] Z. Hu, X. Wang, H. Chen, and G. A. Taylor, "Volt/VAr control in distribution systems using a time-interval based approach," *IEE Proceedings-Generation, Transmission and Distribution*, vol. 150, no. 5, pp. 548-554, 2003.

A Graphic User Interface-based Program for Voltage Sag Calculation

T. Tayjasanant, K. Yossombut, and P. Sawatpipat

Department of Electrical Engineering, Faculty of Engineering, Chulalongkorn University 10330, Thailand

Abstract–This paper presents a graphic user interface (GUI)-based program for voltage sag calculation. The GUI feature provides the user with visual interaction and facilitates the analysis. The analysis is based on the bus impedance matrix and the symmetrical-components transformation. Voltages during a fault at various buses and percentages of sag are calculated. Simulation results on 13-bus balanced industrial distribution system are summarized.

Index Terms-- Voltage Sag, fault calculation, bus impedance matrix.

I. INTRODUCTION

The main source of voltage sag—a decrease in rms voltage to between 0.1 and 0.9 p.u. for duration of 0.5 cycle to 1 minute [1]—is a fault in the power system. Most of the interest in voltage sag is directed to sags due to short circuit faults because these sags are the most severe and cause major equipment to trip. During a fault, the high fault current is passing through impedances along the fault path to the fault location causing voltages to sag or dip. Voltage increases as soon as a fault clearing device interrupts the flow of fault current. Knowing voltage magnitudes during a fault can help to assess the system performance. One of the methods is a stochastic prediction method [2]. Because a fault is a main source of sag so fault calculation technique can be used to determine sag magnitudes during a fault. The fault calculation in this paper is based on the bus impedance matrix (Z_{BUS}) and symmetrical-components transformation [3]-[6]. The objective of this paper is to present a developed GUI-based program that can facilitate the calculation of voltage sag during a fault.

In order to facilitate the calculation of harmonic impedance, a program with Graphical User Interface (GUI) feature is proposed. The GUI feature provides the user with visual interactive capability with less tedious and repetitive work on parameter adjustments and calculations. The user with limited knowledge of the fault analysis can easily simulate the program. The design environment is based on the MATLAB Graphical User Interface Development Environment (GUIDE) [7].

This paper is organized as follows. Section II presents the background on voltage sag calculation. The bus impedance matrix and symmetrical–components transformation are utilized for the sag calculation. Four types of faults on buses and lines are considered. In Section III, the developed GUI-based program is described. Results from a simulation on 13-bus balanced industrial distribution system are shown on execution displays of the program. Lastly, Section IV summarizes the conclusion.

II. VOLTAGE SAG CALCULATION

Voltages during the fault (sag magnitudes) in the network can be computed using an analytical approach from the fault calculation. One of short-circuit calculations is based on the bus impedance matrix and the symmetrical-components transformation. Faults can be categorized into two groups depending on the location of fault. They are faults at buses and faults on lines.

A. Faults at Buses

Consider voltages at bus m when there is a fault at bus i, the voltages at bus m can be derived from pre-fault voltages (subscript -pf) at bus m, transfer bus impedances between buses m and i and fault current at bus i.

$$\begin{bmatrix} V_m^{(0)} \\ V_m^{(1)} \\ V_m^{(2)} \end{bmatrix} = \begin{bmatrix} 0 \\ V_{m-pf}^{(1)} \\ 0 \end{bmatrix} - \begin{bmatrix} Z_{mi}^{(0)} & 0 & 0 \\ 0 & Z_{mi}^{(1)} & 0 \\ 0 & 0 & Z_{mi}^{(2)} \end{bmatrix} \begin{bmatrix} I_i^{(0)} \\ I_i^{(1)} \\ I_i^{(2)} \end{bmatrix} \quad (1)$$

where superscripts 0, 1 and 2 are zero-, positive- and negative-sequence components, respectively. Phase voltages can be obtained from using the transformation matrix [A] as given in (2).

$$\begin{bmatrix} V_a \\ V_b \\ V_c \end{bmatrix} = [A] \begin{bmatrix} V_m^{(0)} \\ V_m^{(1)} \\ V_m^{(2)} \end{bmatrix} \quad (2)$$

where the transformation matrix [A] is defined as

$$[A] = \begin{bmatrix} 1 & 1 & 1 \\ 1 & 1\angle -120° & 1\angle 120° \\ 1 & 1\angle 120° & 1\angle -120° \end{bmatrix} \quad (3)$$

Each type of fault possesses different form of sequence fault currents ($I_i^{(0)}$, $I_i^{(1)}$ and $I_i^{(2)}$) which will be provided later.

This work was supported by the Grant for Development of New Faculty Staff, Chulalongkorn University.

978-1-4244-0644-9/07/$25.00 ©2007 IEEE

B. Faults on Lines

The concept of fault at a bus in the previous section can be extended to a fault on lines by viewing the fault location on the line as a fictitious bus. Consider a fault on the line between buses k and j, voltages at bus m can be derived from pre-fault voltages at bus m, transfer bus impedances between buses m and fictitious bus p and fault current at bus p.

$$[V_m^{(012)}] = [V_{m-pf}^{(012)}] - [Z_{mp}^{(012)}][I_p^{(012)}] \qquad (4)$$

It can be seen from (4) that the form is similar to (1) by replacing bus i with a fictitious bus p. The position of p is determined by the value of ϕ on the line which is between 0 and 1. The parameter ϕ is defined as the ratio of length between bus k and location p (L_{kp}) to the length of the line k and j (L_{kj}) or L_{kp}/L_{kj}. Values of $[Z_{mp}^{(012)}]$ and $[Z_{pp}^{(012)}]$ can be calculated from (5) and (6) [6].

$$[Z_{mp}^{(012)}] = (1-\phi)[Z_{mk}^{(012)}] + \phi[Z_{mj}^{(012)}] \qquad (5)$$

$$[Z_{mp}^{(012)}] = (1-\phi)^2[Z_{mk}^{(012)}] + \phi^2[Z_{mj}^{(012)}]$$
$$+ 2\phi(1-\phi)[Z_{kj}^{(012)}] + \phi(1-\phi)[z_{kj}^{(012)}] \qquad (6)$$

where $[Z_{xy}^{(012)}]$ is a diagonal 3x3 transfer bus impedance matrix between bus x and bus y whose diagonal elements are zero-, positive- and negative-sequence impedances and $[z_{kj}^{(012)}]$ is a diagonal 3x3 matrix whose diagonal elements are zero-, positive- and negative-sequence impedances of line between buses k and j.

Equations of sequence currents at the fault location p ($I_p^{(012)}$) for four types of faults are provided in (7) to (10) where Z_f is a fault impedance. For a three-phase fault, $I_p^{(012)}$ is

$$\begin{bmatrix} I_p^{(0)} \\ I_p^{(1)} \\ I_p^{(2)} \end{bmatrix} = \begin{bmatrix} 0 \\ \dfrac{V_{p-pf}^{(1)}}{Z_{pp}^{(1)} + Z_f} \\ 0 \end{bmatrix} \qquad (7)$$

For a single line-to-ground fault, $I_p^{(012)}$ is

$$\begin{bmatrix} I_p^{(0)} \\ I_p^{(1)} \\ I_p^{(2)} \end{bmatrix} = \begin{bmatrix} I_p^{(1)} \\ \dfrac{V_{p-pf}^{(1)}}{Z_{pp}^{(0)} + Z_{pp}^{(1)} + Z_{pp}^{(2)} + 3Z_f} \\ I_p^{(1)} \end{bmatrix} \qquad (8)$$

For a line-to-line fault, $I_p^{(012)}$ is

$$\begin{bmatrix} I_p^{(0)} \\ I_p^{(1)} \\ I_p^{(2)} \end{bmatrix} = \begin{bmatrix} 0 \\ \dfrac{V_{p-pf}^{(1)}}{Z_{pp}^{(1)} + Z_{pp}^{(2)} + Z_f} \\ -I_p^{(1)} \end{bmatrix} \qquad (9)$$

For a double line-to-ground fault, $I_p^{(012)}$ is

$$\begin{bmatrix} I_p^{(0)} \\ I_p^{(1)} \\ I_p^{(2)} \end{bmatrix} = \begin{bmatrix} -I_p^{(1)} \dfrac{Z_{pp}^{(2)}}{Z_{pp}^{(0)} + Z_{pp}^{(2)} + 3Z_f} \\ \dfrac{V_{p-pf}^{(1)}}{Z_{pp}^{(1)} + \dfrac{Z_{pp}^{(2)}(Z_{pp}^{(0)} + 3Z_f)}{Z_{pp}^{(0)} + Z_{pp}^{(2)} + 3Z_f}} \\ -I_p^{(1)} \dfrac{Z_{pp}^{(0)} + 3Z_f}{Z_{pp}^{(0)} + Z_{pp}^{(2)} + 3Z_f} \end{bmatrix} \qquad (10)$$

Phase voltages at the fault can be calculated from sequence components as in (2).

III. DEVELOPED PROGRAM AND SIMULATION RESULTS

A. Developed GUI-based Program

The program is divided into three sections: the input, the processing section and the output. The detail of each section is summarized as follows:

1) The Input

The data can be input to the program from two different methods. The first method is by loading the system information (in per unit) from an Excel file. Fig. 1 shows the format of the already-prepared .xls file. Data in 11 columns are required for the analysis. The first two columns (Left bus and Right bus) specify the connection of the bus. The next six columns are components' positive-, negative- and zero-sequence impedances. The ninth column is the reactive power from a capacitor. The tenth column determines the code of the element:

1 = Generator, 2 = Line, 3 = Transformer and *4 = Load.*

The eleventh column specifies the connection of a transformer:

0 = YG-yg,   1 = YG-y,   2 = Y-yg,   3 = Y-y,   4 = Δ-Δ,
5 = YG-Δ,   6 = Y-Δ,   7 = Δ-yg,   8 = Δ-y

The second method is by entering the system information by the user. Four components (Generator, Line impedance, Transformer and Load) can be chosen. The user can also add or delete the component.

1841

2) The Processing Section

The processing part will calculate voltages at various buses based on the fault conditions from the input section using equations from Section II.

3) The Output

After the processing part is finished, the user can select the bus of interest to view the voltages. The output shows the voltage magnitudes and phases for pre- and post-fault conditions. Moreover, the presentation using phasor plots is also displayed. The user can visualize the change in magnitudes and phases form the phasor plots easily.

Procedures for running the program are in the following orders.

1. Browse the data of system components by clicking "**BROWSE**" button. Or the user inputs the data directly. After all system components are entered, the user clicks "**FINISH**" button. The details of system components will be displayed on the listbox.
2. Select the fault information such as fault type, fault impedance and the location of fault.
3. Select the bus of interest to view the voltage sag magnitudes and phases.
4. The user can choose to use the pre-fault voltages from load-flow results by clicking "**LOAD PRE-FAULT VOLTAGE**" button. If this information is not available, the user can tick the checkbox "Pre-fault voltage = 1.0 pu". The assumption of pre-fault voltages equal to 1.0 pu will be used. The format of the load-flow result starts with bus number, voltage magnitude and voltage phase angle, all in the column format.
5. Click "**Calculate**" button to start the simulation.
6. The user can click "**CLEAR**" button to clear the results and start a new case.

Fig. 2 summarizes the steps for using the voltage sag calculation program.

B. Simulation Results

A 13-bus balanced industrial distribution system in [8] is used for the simulation of the developed GUI-based program. Fig. 3 shows the diagram of the studied system. The plant is fed from a utility supply at 69 kV and the local plant distribution system operates at 13.8 kV. Due to the balanced nature of this example, only positive sequence data is provided. Capacitance of the short overhead line and all cables are neglected. It is assumed that all transformers are YG-yg connections (connection code = "0").

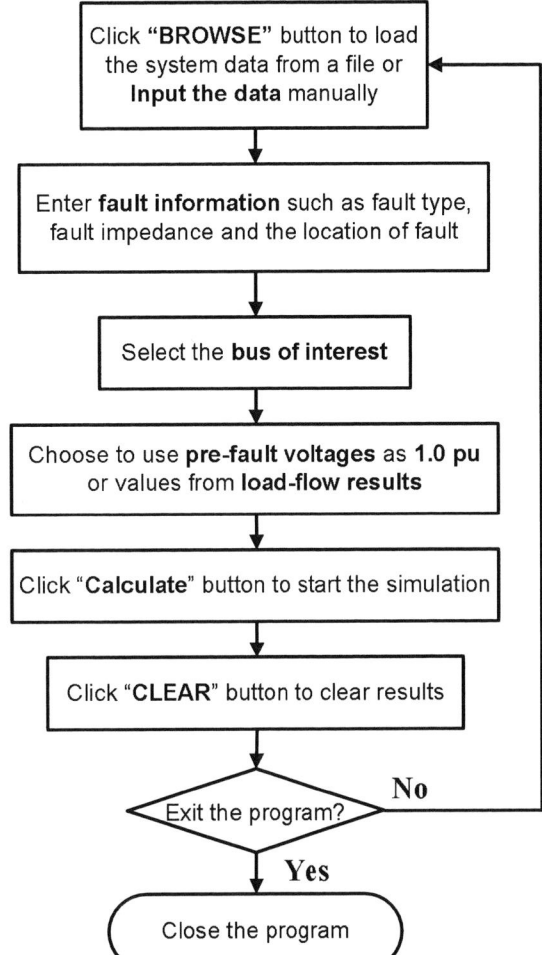

Fig. 2. Steps for using the developed GUI-based program.

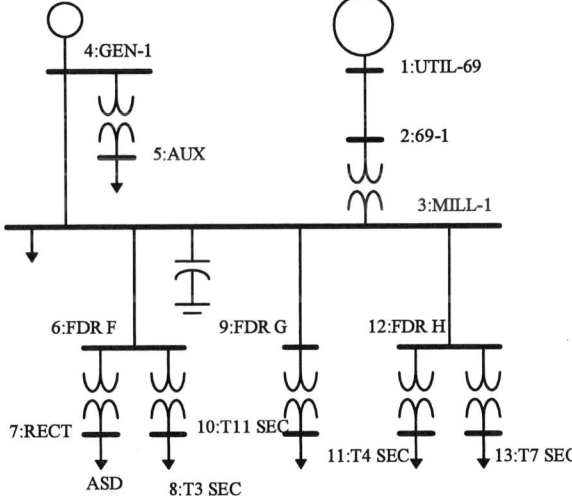

Fig. 3. System diagram for the simulation.

The system is described by the following data and data in Tables I and II.

1. The supply system equivalent impedance (transient and sub-transient) is $0.05 + j1$ per unit based on 100 MVA. The zero-sequence impedance is 1/3 of this value.

2. The local (in-plant) generator has an internal impedance of $X = 0.25$ per-unit based on the generator rated kVA which is 2,000 kVA.

3. The plant power factor correction capacitors are rated at 5,000 kVar.

TABLE I
PER-UNIT LINE AND CABLE IMPEDANCE DATA (10 MVA BASE)

From	To	R (p.u.)	X (p.u.)
UTIL-69	69-1	0.00139	0.00296
MILL-1	GEN-1	0.00122	0.00243
MILL-1	FDR F	0.00075	0.00063
MILL-1	FDR G	0.00157	0.00131
MILL-1	FDR H	0.00109	0.00091

TABLE II
TRANSFORMER DATA

From	To	Voltage	% R	% X
69-1	MILL-1	69:13.8	0.4698	7.9862
GEN1	AUX	13.8:0.48	0.9593	5.6694
FDR F	RECT	13.8:0.48	0.7398	4.4388
FDR F	T3 SEC	13.8:4.16	0.7442	5.9537
FDR G	T11 SEC	13.8:0.48	0.8743	5.6831
FDR H	T4 SEC	13.8:0.48	0.8363	5.4360
FDR H	T7 SEC	13.8:2.4	0.4568	5.4810

Table III tabulates load-flow results and the load data. The model of load is assumed to be a constant impedance load ($Z_L = |V_L|^2/(P_L - jQ_L)$) in order to include the effect of load currents. Bold numbers in Table III are obtained from the conventional load-flow analysis.

TABLE III
LOAD-FLOW RESULTS

Bus	V_{mag} (p.u.)	θ (deg)	P_{gen} kW	Q_{gen} kVar	P_{load} kW	Q_{load} kVar
UTIL-69	1.000	0.00	**7364**	**579**		
69-1	**0.9988**	**−0.12**				
MILL-1	**0.9943**	**−2.37**			2240	2000
GEN1	0.995	**−2.37**	2000	**2853**		
Aux	0.9702	**−3.51**			600	530
FDR F	**0.9940**	**−2.37**				
RECT	0.9756	**−4.68**			1150	290
T3 SEC	0.9459	**−4.83**			1310	1130
FDR G	**0.9940**	**−2.37**				
FDR H	**0.9936**	**−2.37**				
T4 SEC	0.9792	**−3.05**			370	330
T7 SEC	0.9508	**−4.67**			2800	2500
T11 SEC	**0.9571**	**−3.94**			810	800

A solid three-phase fault ($Z_f = 0$) at bus **12** is simulated. Tables IV and V summarize voltage magnitudes for cases where 1.0 pu and load-flow results are used as pre-fault voltages, respectively. The percentage of the sag is denoted as ΔV value. It can be calculated from

$$\Delta V = \frac{V_{pre-fault} - V_{post-fault}}{V_{pre-fault}} \times 100\% \qquad (9)$$

TABLE IV
VOLTAGE MAGNITUDES FROM USING PRE-FAULT VOLTAGES = 1.0 P.U.

Bus No.	Pre-fault Voltages		During-fault Voltages		ΔV (%)
	Magnitude (per unit)	Angle (deg.)	Magnitude (per unit)	Angle (deg)	
1	1.0	0.0	0.36533	-1.77	63.47
2	1.0	0.0	0.34585	-0.62	65.42
3	1.0	0.0	0.01022	-41.74	98.98
4	1.0	0.0	0.01242	-38.16	98.76
5	1.0	0.0	0.03646	18.90	96.35
6	1.0	0.0	0.01041	-40.52	98.96
7	1.0	0.0	0.04221	50.68	95.78
8	1.0	0.0	0.06601	31.08	93.40
9	1.0	0.0	0.01042	-40.96	98.96
10	1.0	0.0	0.04902	23.52	95.10
11	1.0	0.0	0.01868	38.84	98.13
12	1.0	0.0	0.00000	0.00	100.00
13	1.0	0.0	0.05827	41.18	94.17

TABLE V
VOLTAGE MAGNITUDES FROM USING VOLTAGES FROM LOAD-FLOW RESULTS AS PRE-FAULT VOLTAGES

Bus No.	Pre-fault Voltages		During-fault Voltages		ΔV (%)
	Magnitude (per unit)	Angle (deg.)	Magnitude (per unit)	Angle (deg)	
1	1.0000	0.00	0.36959	2.31	63.04
2	0.9988	-0.12	0.34985	3.45	64.97
3	0.9943	-2.37	0.01069	-41.61	98.93
4	0.9950	-2.37	0.01347	-36.85	98.65
5	0.9702	-3.51	0.01309	-37.69	98.65
6	0.9940	-2.37	0.01065	-41.49	98.93
7	0.9756	-4.68	0.01039	-43.85	98.94
8	0.9459	-4.83	0.01016	-43.96	98.93
9	0.9940	-2.37	0.01066	-41.92	98.93
10	0.9571	-3.94	0.01034	-43.41	98.92
11	0.9792	-3.05	0.00002	110.27	100.00
12	0.9936	-2.37	0.00000	0.00	100.00
13	0.9508	-4.67	0.00004	-69.35	100.00

Comparing Tables IV and V, post-fault voltages obtained from using load-flow results as pre-fault voltages are more accurate than ones from using 1.0 pu as pre-fault voltages. For example, the voltage at bus 13 is supposed to be close to 0 pu due to a solid three-phase fault at bus 12. The post-fault voltage from using load-flow results as pre-fault voltages is 0.00004 pu, while the value from using 1.0 pu as pre-fault voltages is 0.05827 pu. Therefore, if the load-flow results are available and used as pre-fault voltages, post-fault voltages would be more accurate than assuming pre-fault magnitudes as 1.0 pu.

Fig. 4 illustrates the execution display of the simulation when there is a three-phase fault at bus 12. Voltages at bus 3 are chosen as the bus of interest. The output section displays voltage magnitudes and phases and percentages of sag. Phasor diagrams showing pre-fault and post-fault voltage phasors are also provided. The phase angle shift—the phase angle difference between pre- and post-fault conditions—in voltage phasors can be visualized from the phasor diagrams section. The program does not take into account of the phase shift due to Δ-Y transformer connections. These features of the program will be included for the future work.

IV. CONCLUSION

This paper presents a program for voltage sag calculation developed under MATLAB GUI environment. The GUI feature provides the visual interactive capability and facilitates the analysis such as changing the fault type, fault impedance and fault location and checking the sensitivity of results. The user with limited knowledge of the voltage sage calculation can easily simulate the program. The calculation of voltage sag is based on the bus impedance matrix and the symmetrical-components transformation. The developed program has been tested using simulation results from a 13-bus balanced industrial distribution system. Pre-fault voltages from load-flow results should be used for more accurate sag calculation.

REFERENCES

[1] *IEEE Recommended Practice for Monitoring Electric Power Quality,* IEEE Standard. 1159, 1995.

[2] M. H. J. Bollen, *Understanding Power Quality Problems: Voltage Sags and Interruptions*, IEEE Press Series on Power Engineering, 2000.

[3] G. W. Stagg and A. H. El-Abiad, *Computer Methods in Power System Analysis*, McGraw-Hill International Editions, 1968.

[4] P. M. Anderson, Analysis of Faulted Power Systems, IEEE Press, New York, 1995.

[5] J. J. Grainger and W. D. Stevenson, Jr., *Power System Analysis*, McGraw-Hill International Editions, 1994

[6] E. E. Juárez and A. Hernández, "An Analytical Approach for Stochastic Assessment of Balanced and Unbalanced Voltage Sags in Large Systems," *IEEE Trans. Power Delivery*, Vol.21, No.3, July 2006, pp. 1493-1500.

[7] *Creating Graphical User Interface*, MATLAB 7, Mathworks, 2007, Available: http://www.mathworks.com.

[8] Task Force on Harmonics Modeling and Simulation, "Test Systems for Harmonics Modeling and Simulation," *IEEE Trans. Power Delivery*, Vol. 14, No. 2, April 1999, pp. 579-587.

	A	B	C	D	E	F	G	H	I	J	K
1	Left bus	Right bus	R+	X+	R-	X-	R0	X0	Cap	Element	Xfmr conn
2	1	1	0.05	1	0.05	1	0.01663	0.3325	0	1	
3	1	2	0.0139	0.0296	0.0139	0.0296	0.0139	0.0296	0	2	
4	2	3	0.03132	0.53241	0.03132	0.53241	0.03132	0.53241	0	3	0
5	3	3	24.5579	21.9267	24.5579	21.9267	24.5579	21.9267	0.05	4	
6	3	4	0.0122	0.0243	0.0122	0.0243	0.0122	0.0243	0	2	
7	3	6	0.0075	0.0063	0.0075	0.0063	0.0075	0.0063	0	2	
8	3	9	0.0157	0.0131	0.0157	0.0131	0.0157	0.0131	0	2	
9	3	12	0.0109	0.0091	0.0109	0.0091	0.0109	0.0091	0	2	
10	4	4	0	12.5	0	12.5	0	12.5	0	1	
11	4	5	0.63953	3.7796	0.63953	3.7796	0.63953	3.7796	0	3	0
12	5	5	88.1218	77.841	88.1218	77.841	88.1218	77.841	0	4	
13	6	7	0.59184	3.55104	0.59184	3.55104	0.59184	3.55104	0	3	0
14	7	7	77.8163	19.6233	77.8163	19.6233	77.8163	19.6233	0	4	
15	6	8	0.43142	3.45142	0.43142	3.45142	0.43142	3.45142	0	3	0
16	8	8	39.1611	33.7802	39.1611	33.7802	39.1611	33.7802	0	4	
17	9	10	0.58287	3.78873	0.58287	3.78873	0.58287	3.78873	0	3	0
18	10	10	57.2481	56.5413	57.2481	56.5413	57.2481	56.5413	0	4	
19	12	11	0.55753	3.624	0.55753	3.624	0.55753	3.624	0	3	0
20	11	11	144.332	128.729	144.332	128.729	144.332	128.729	0	4	
21	12	13	0.12181	1.4616	0.12181	1.4616	0.12181	1.4616	0	3	0
22	13	13	17.9649	16.0401	17.9649	16.0401	17.9649	16.0401	0	4	

Fig. 1. Format of the system data file.

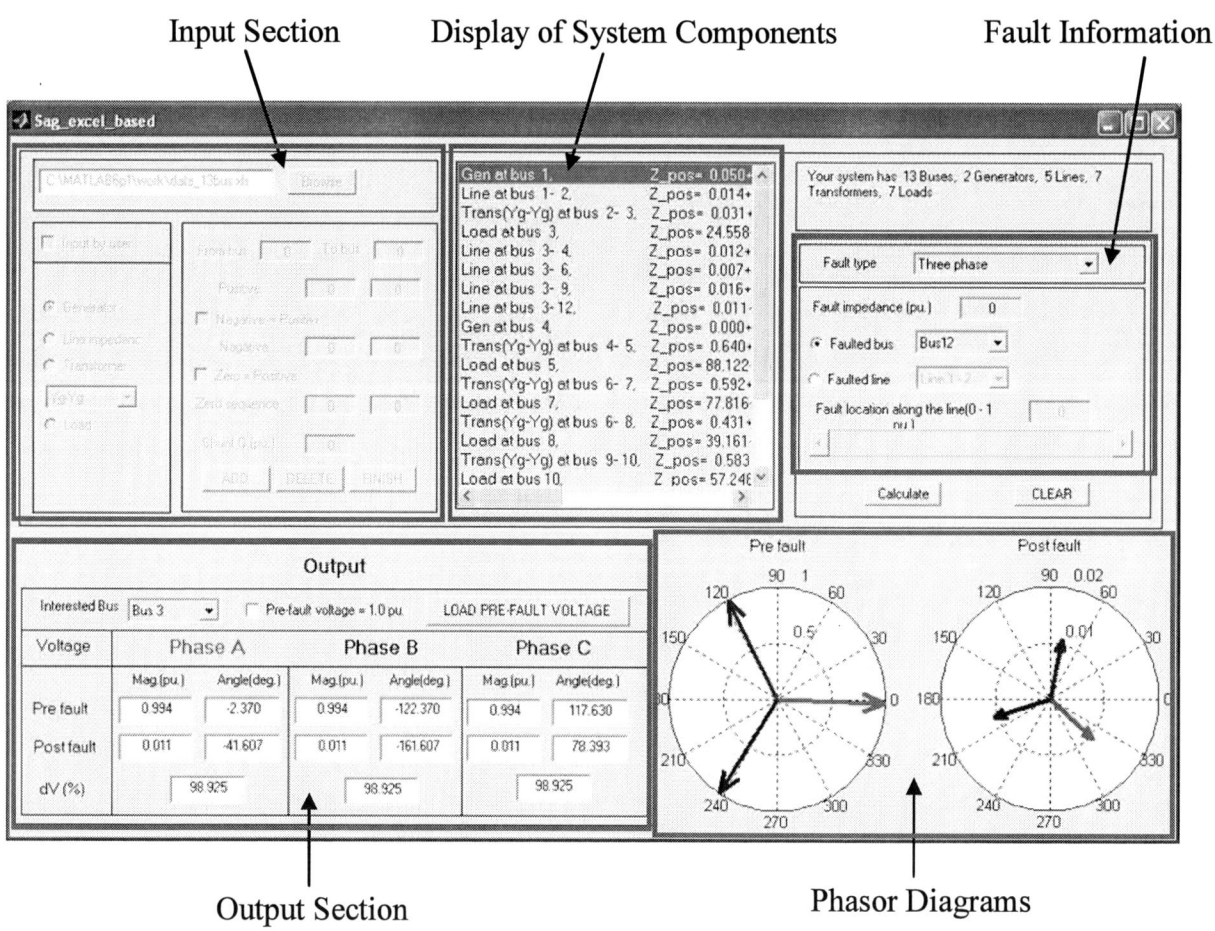

Fig. 4. Execution display for the simulation.

Operational Characteristics of Fault Current Limiting Reactor Combined with Multi-Functional Inverter

S. H. Ko*, S. H. Lim**, S. R. Lee***, S. W. Lee****, I. C. Kim*****, S. H. Ko*****, and H. S. Kim *****

* Advanced Graduate Education Center of Jeonbuk for Electronics & Information Tech., Chonbuk National Uni., Korea
** School of Electrical Eng., Soongsil Uni., Seoul, Korea
*** School of Electronic & Information Eng., Kunsan National Uni., Kunsan, Korea
**** Institute of TMS Information Tech., Yonsei Uni., Seoul, Korea
***** Division of Electronics & Information Eng., Chonbuk National Uni., Korea

Abstract--This paper deals with the fault current limiting reactor (FCLR) combined with a multi-functional inverter system, which aims at the integration of power quality improvement, uninterruptible power supply (UPS) and fault current limitation. The system consists of a FCLR with a switch and the single-phase voltage source inverter (VSI). The FCLR, which has coil 1 and coil 2 wound in parallel through pure resistance with switch, can perform the fault current limiting operation during a fault period. The multi-functional inverter is connected between coil 3 and grid, which can be operated to compensate the reactive power demanded by nonlinear and variation loads, and be controlled to provide a sinusoidal voltage at the fundamental value (220V/60Hz) for load during a grid fails. It is shown through the simulation for the its operation that the proposed system could both protect the system from over-current by a line-ground fault and outages by a grid fail (abnormal utility power condition).

Index Terms--High Power quality, Multi-functional inverter, UPS, Fault current limiting reactor.

I. INTRODUCTION

With the growth of the industry and the augmentation of hi-technology power electronic application (e.g., information system, internet data center, life support system), the power electronics industry have been researched and developed in order to improve power quality demand by nonlinear load and to protect critical equipment cause by sudden accident such as line-ground happen(short circuit) and grid fail(abnormal utility power condition). Recently, the parallel processing topologies has become more popular. As it is inherently efficient, compact and economical, it offers numerous functions requiring a minimum number of power conversions [1-7].

The parallel processing system mainly controls the power flow and quality by the bi-directional voltage source inverter (VSI) since it has to control the power conversion between the dc voltage source and the grid for battery energy storage (BES) systems or/and renewable energy sources (RES) applications. VSIs are intrinsically efficient, compact and economical devices to control

power flow and provide quality supply [8-9]. The VSIs can be further classified into voltage-controlled VSI(VCVSI) and current-controlled VSI(CCVSI), depending on their control mechanism. There are advantage and limitations associated with each control mechanism. For instance, VCVSIs can provide voltage support to the load, while CCVSIs can provide current support. The CCVSI is faster in response compared to the VCVSI as its power flow is controlled by the switching instant, whereas in the VCVSI the power flow is controlled by adjusting the voltage across the decoupling inductor (fundamental frequency). Generally, the advantages of one type of VSI are considered as a limitation of the other type [9]. However, the increase of the fault current has imposed a severe burden on the related machinery in the grid. To protect the system from the short circuit, most of topologies have been dependent on the circuit breaker or series reactor [10-12].

In this paper, a fault current limiting reactor (FCLR) combined with a multi-functional inverter system is presented. The operation of this system could be divided into three modes (normal, UPS and FCL). In normal mode, the inverter based on the algorithm of the current control can compensate the reactive power demanded by nonlinear or variation in loads. In UPS mode, the proposed system can provide a sinusoidal voltage at the fundamental value (220V/60Hz) for the load when the grid fails. In FCL mode, the fault current can be limited by opening the switch and the limiting impedance of the system can be changed by adjusting the resistance connected with switch. Fault current limiting reactor (FCLR), as one of solutions to decrease the fault current. The advantageous functions of FCLR such as automatic fault-current sensing, automatic recovering and faster fault current-limiting. The main purpose of this system are to compensate the current harmonic and reactive power demand by nonlinear load, to supply clean uninterruptible power during grid fail, and to limit over current cause by line-ground accident at the point of installation on power distribution for critical load. The operation of the proposed system is confirmed through the simulation and its usefulness is discussed.

This work was supported in part by the New & Renewable Energy Development project (2007-N-PV08-03-0) of the KEMCO.

978-1-4244-0644-9/07/$25.00 ©2007 IEEE

II. FCLR COMBINED WITH MULTI-FUNCTIONAL INVERTER

The schematic diagram of the FCLR combined with multi-functional inverter is shown in Fig. 1.

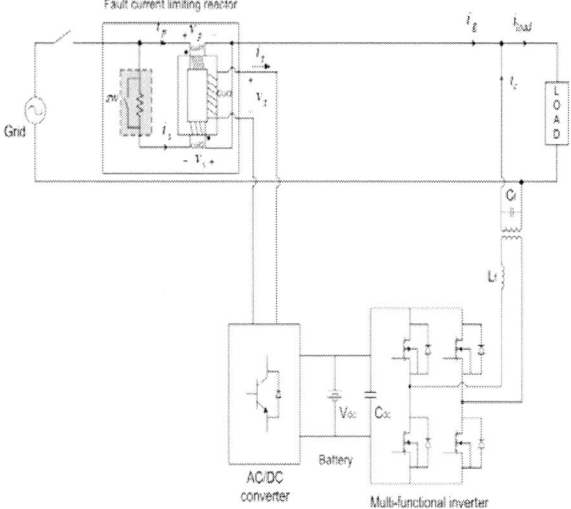

Fig. 1 Schematic diagram of the FCLR combined with multi-function inverter.

This system consists of a FCLR with a switch using a transformer and the single-phase VSI. The FCLR consists of the primary winding(L_p, coil1), the secondary winding(L_s, coil2) with the power switch and the third winding(L_t, coil3), which are wound in parallel through the same iron core. The multi–functional inverter is connected in series on the DC side(battery and coil 3) and in parallel on the AC side(grid).

A. Operation principle of the multi-functional inverter

As shown the Fig. 1, i_p, i_s and i_t are the primary, the secondary and the third current, respectively. The primary and the secondary windings are wound to counteract each other's flux, which the flux generated from the primary winding is cancelled out by one from the secondary winding if the power switch is closed. This means that the voltage across the primary winding (V_1) and the secondary (V_2) one are zeros. Therefore, the simplified equivalent schematic diagram of the system at normal mode (Fig. 2(a)) and UPS mode (Fig. 2(b)) can be presented as Fig. 2.

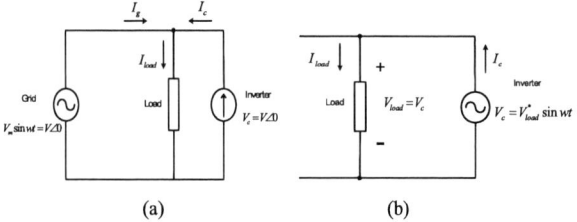

(a)   (b)

Fig. 2 The equivalent circuit diagram of the proposed system at each operation mode. (a) Nor mode. (b) UPS mode.

In normal mode, as a multi-functional inverter controls the current flow using the VSI switching instants, it is modeled as a current source. As the output voltage of the

inverter is filtered, this current can be assumed to have only a fundamental frequency depending upon the grid (60 Hz). As the current generated from the inverter can be controlled independently from the voltage, the active and reactive power controls are decoupled. Hence, unity power factor operation for the entire range of the load is possible.

From Fig. 2, the load current (I_{load}) is continuously supplied by the current of the grid (I_g) and inverter (I_c), and it can be expressed as (1);

$$I_{load} = I_g + I_c \qquad (1)$$

In the normal mode, the active power of the load (P_{load}) demand should be equal to the grid power. Hence the desired grid current can be rewritten as follows;

$$I_g^* = \text{Re}[I_{load}] = \frac{\text{Re}[P_{load}]}{V_g} \qquad (2)$$

To manage the power flow of the entire system, the output current of the inverter should be controlled to compensate the reactive current of the grid depending on the load condition. Therefore, the inverter output current must be controlled to meet the reactive current of the load demand as presented in (3);

$$I_c = I_{load} - I_g^* \qquad (3)$$

When an abnormal utility power condition (grid fail), the multi-functional inverter is forced to operate in the UPS mode to supply a sinusoidal voltage at the fundamental value (220V/60Hz) to the load. The phase locked loop (PLL) is responsible for synchronizing the inverter output voltage with the grid voltage. The sampling from the load voltage and current, inverter current and battery bank voltage is also used to generate the required inverter voltage (V_c^*) (for UPS operation). After comparing the required/reference values and the actual variables, and error signal is generated to feed a PI controller. After generating the desired reference signal, it is given to the PWM generator block to generate the required switching signals.

B. Operation principle of the FCLR

In the normal mode, the voltages across coils 1,2, and 3 are zero. It means that the impedance in FCLR is zero. In short circuit accident such as line-ground fault, if the switch open, the voltage induced in the primary winding is not equal to the voltage in the secondary one any more, which the fluxes generated by each winding are not cancelled out each other and the fault current can be limited by the impedance of the FCLR. Fig. 3 shows the simulation waveforms of FCLR using the switch at line-ground happens. Fig. 3(a) shows the voltage of switch, coil1, coil2 and coil3 (V_{sw}, V_p, V_s and V_t) in each winding kept zeros by closing the switch during a normal mode. However, the voltages in each winding were

1847

generated by opening the power switch when the line-ground fault happened. It was also observed from Fig. 3(b) that the grid current, equal to the sum of currents in the primary (I_p), the secondary (I_s) and third (I_t) windings, was limited by generation of voltages in each winding.

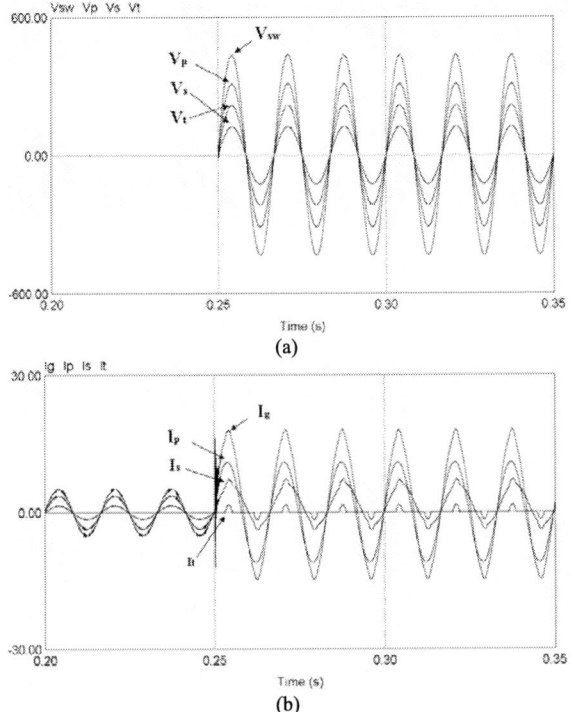

Fig. 3 Simulation waveforms of the FCLR using the switch at line-ground fault happens. (a) Voltage of switch, coil1, coil2 and coil3 (V_{sw}, V_p, V_s, V_t). (b) Current of grid, coil1, coil2 and coil3 (I_{sw}, I_p, I_s and I_t).

III. SIMULATION RESULTS

A PSim simulator was used for simulation in order to investigate the operational characteristics of the FCLR combined with multi-functional inverter. The operation characteristics were analyzed into three modes (the normal mode, the UPS mode and the mode). Nonlinear and variation load conditions for the simulation were determined to be the 1KVA load capacity. Table 1 illustrates the simulation condition and parameter values.

TABLE I
SIMULATION CONDITION AND PARAMETERS

Parameter	Value	Parameter	Value
Vac	220[Vrms]	Fsw	10[kHz]
Fs	60[Hz]	Lp	60[mH]
Lf	3[mH]	Ls	24[mH]
Cf	2[uF]	Lt	42[mH]

Where, F_s is the fundamental frequency, F_{sw} is the switching frequency, and L_p, L_s and L_t are the inductance values of primary, secondary and third winding , respectively.

The simulation was conducted to evaluate the performance of the proposed system in the presence of different operation modes. Where V_g and V_{load} are the voltage waveforms of the grid and load, and I_g, I_c and I_{load} are current waveform of the grid, inverter and load, respectively. Also P_g, P_c and P_{load} are the active power waveforms of the grid, inverter and load, and I_p, I_s and I_t

are the current waveforms of the primary(coil1), secondary(coil2) and third(coil3) in the FCLR, respectively.

Fig. 4 shows that the simulation results in the normal mode where the load current changes from peak value of 8.1A$_{peak}$ to 11.9A$_{peak}$ for the sixth period and then, returns to 8.1A$_{peak}$ after sixth period. The system can provide all the reactive power demanded by the nonlinear load and hence the grid supplies only the remaining active power. In this case, the proposed system prevents a low order harmonics from being injected into the grid. The frequency analysis for grid voltage, grid current, inverter current and load current obtain by simulation is shown in Fig. 5.

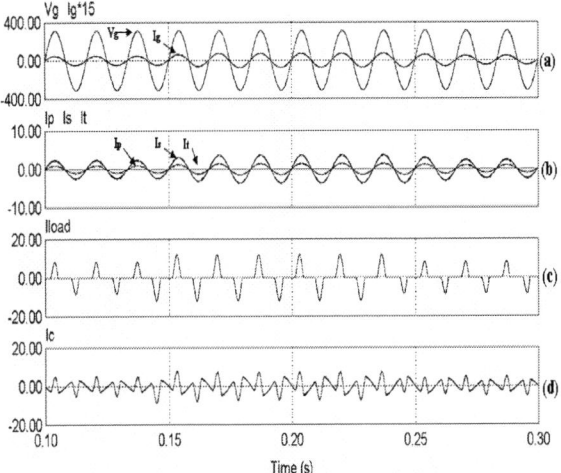

Fig. 4 Simulation waveforms results of normal mode for state and transient response. (a) Grid voltage and current. (b) Current of coil 1, coil 2 and coil3. (c) Load current. (d) Inverter current.

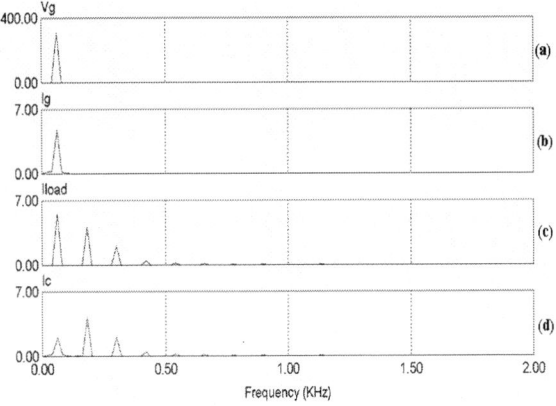

Fig. 5 Frequency analysis. (a) Grid voltage. (b) Grid current. (c) Load current. (d) Inverter current.

As shown Fig. 5, all the reactive power associated with low order harmonics from the nonlinear load(load ITHD: 62.7%) could be supplied by the inverter. It is shown that the proposed system can achieve unity power factor and satisfies THD(grid ITHD: 1.1%) requirements of voltage and current for the full range of the load.

Fig. 6 shows that the simulation results in the UPS mode at nonlinear load condition. It assumed that at

1848

0.25ms the grid fails and the inverter had to supply the load. As it shown, before grid failure the system supplied the reactive power demand by the nonlinear load. And the proposed system picked up the load rapidly after grid failed.

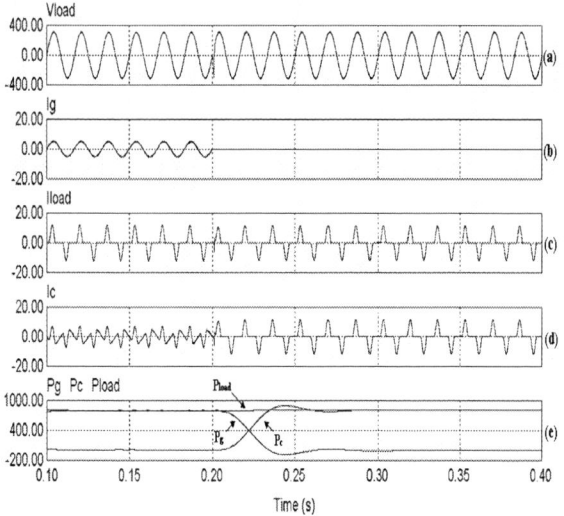

Fig. 6 Simulation waveforms results of UPS mode for state and transient response. (a) Load voltage. (b) Grid current. (c) Load current. (d) Inverter current. (e) Active power flow of grid, inverter and load.

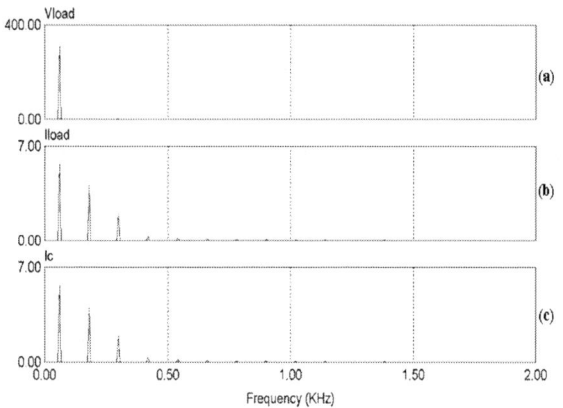

Fig. 7 Frequency analysis. (a) Load voltage. (b) Load current. (c) Inverter current.

Fig. 7 shows that the spectrums of the load voltage(VTHD: 1.2%), the load current and the inverter current. It is shown that the proposed system can compensate low order harmonic of the load voltage in order to meet IEEE standards in the presence of nonlinear loads. Fig 8 shows that simulation results in the FCL mode when the line-ground fault happens. Where V_p, V_s and V_t are the voltage waveforms of the primary (coil1), secondary(coil2) and third(coil3) in the FCLR, respectively.

As shown Fig. 8, the system compensated the reactive power demand by the nonlinear load before a fault happens. With the turning-off operation of power switch in the FCLR in case of the line-ground fault, the voltage across the primary and the secondary winding can be induced, which leads to fault current limiting operation as

shown in Fig. 8(a). During a FCL mode, the inverter operation was useless because the reactive power of the grid did not exist. On the other hand, the fault current level was expected to be adjusted through the operation of AC/DC converter connected between the battery and the FCLR`s coil 3, whose current was induced by the generation of its voltage due to the magnetic flux linkage between coil 3 and other two coil during a fault time.

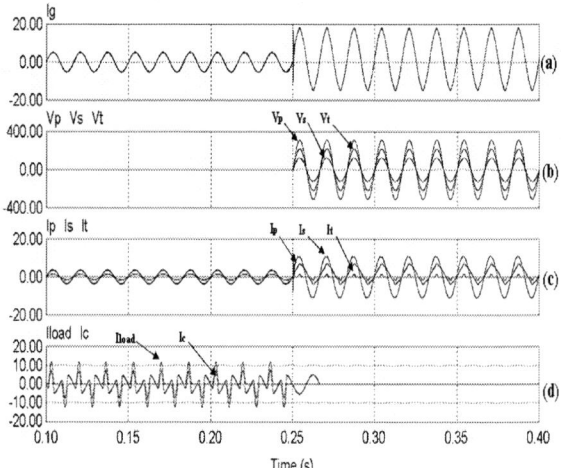

Fig. 8 Simulation waveforms results of FCL mode for state and transient response. (a) Grid current. (b) Voltage of coil 1, coil 2 and coil 3. (c) Current of coil 1, coil 2 and coil 3. (d) Current of load and inverter.

IV. CONCLUSIONS

This paper presents a fault current limiting reactor (FCLR) combined with a multi-functional inverter system. The suggested topology consisted of the FCLR with a switch and the multi-functional inverter. The operation of this topology could be divided into the normal mode, UPS mode, and FCL mode. In the normal mode, the inverter based on the algorithm of the current control can compensate the reactive power demanded by nonlinear or variation in loads. In UPS mode, the proposed system can provide a sinusoidal voltage at the fundamental value (220V/60Hz) for the load when the grid fails. In FCL mode, the fault current can be limited by opening the switch and the limiting impedance of the system can be changed by adjusting the resistance connected with switch.

We will research the operational characteristics of this FCLR through the control both the current of coil3 ant the voltage of dc capacitor using AC/DC converter.

REFERENCES

[1] M. El-Harbrouk, M.K. Darwish, and P. Mehta, "A Survey of Active Filters and Reactive Power Compensation Techniques," *International conference on Power Electronics and Variable Speed Drivers*, pp. 7-12, Sept. 2000.

[2] B.H. Kwon, J.H. Choi, and T.W. Kim, "Improved Single-Phase Line-Interactive UPS," *IEEE Trans. Ind. Electron.*, vol. 48, no. 4, Aug. 2001.

[3] S.H. Lim, S.H. Ko, S.R. Lee, S.U. Lee, C.H. Jeon, and B.S. Han, "Analysis of Fault Current Limiting and Reactive

Power Compensating Operations of New Transformer Topology with Current Controlled Inverter," *in proc 35th Annu. Power Electronics Specialists Conference*, vol. 2, pp. 980-983, June 2004.

[4] M. Dai, M.N. Marwail, J.W. Jung and A. Keyhani, "Power flow control of a single distributed generation unit with nonlinear local load", *Conf. Rec. of IEEE PES 2004,* vol. 1, pp. 398-403, Oct. 2004.

[5] T. Kawabata, N. Sashida, Y. Yamamoto, K. Ogasawara and Y. Yamasaki, "Parallel Processing Inverter System", *IEEE Trans. power Electronics*, vol. 6, no. 3, pp. 442-450, July 1991.

[6] S. Rathmann, and H.A. Warner, "New Generation UPS Technology, The Delta conversion Principle," *in Proc. 33th Annu. Industry Application Conference*, vol. 4, pp. 2389-2395, Oct. 1996.

[7] C.V. Nayar, M. Ashari, and W.W.L. Keerthipala, "A Grid-Interactive Photovoltaic Uninterruptible Power Supply System Using Battery Storage and a Back Up Diesel Generator," *IEEE Tran. Energy Conver.*, vol. 15, no. 3, Sept. 2000.

[8] A. J. Baronian and S. B. Dewan, "An adaptive digital control of current source inverter suitable for parallel processing inverter systems", *Conf. Rec. of IEEE IAS 1995-Annual Meeting,* vol. 3, pp. 2670-2671, Oct. 1995.

[9] S.H. Ko, S.R. Lee, H. Dehbonei, and C.V. Nayar, "Application of Voltage and Current Controlled Voltage Source Inverters for Distributed Generation Systems," *IEEE. Trans. Energy Conver.*, vol. 21, no. 3, pp. 782-792, Sept. 2006.

[10] C.A. Falcone, J.E. Beehler, W.E. Mekolites, and J. Grazen, "Curent limiting device-A utility`s need," *IEEE Trans. Power App. Syst.*, vol. PAS-93, pp. 1768-1775, Nov. 1974.

[11] M. Steurer, K. Frohlich, W. Holaus, and K. kaltenegger, "A novel hybrid current-limiting circuit breaker for medium voltage: principle and test results," *IEEE Trans. Power Del.*, vol. 18. pp. 460-467, Apr. 2003.

[12] S.H. Lim, S.R. Lee, H.S. C, and B.S. Han, "Analysis of Operational Characteristics of Flux-Lock Type SFCL Combined With Power Compensator," *IEEE Trans. Applied Superconductivity*, vol. 15, no. 2, pp. 2043-2046, June. 2005.

Low Cost AC Solid State Circuit Breaker

W. Pusorn[1], W. Srisongkram[1,2], W. Subsingha[1], S. Deng-em[1] and P. N. Boonchiam[1]
[1]Department of Electrical Engineering, RMUTT, Thailand   [2]Department of Electrical Engineering, RMUTSB, Thailand

***Abstract*--This paper presents a Solid State Circuit Breaker
(SSCB) for a low distribution level. When voltage disturbances
occurred, the SSCB operates for keeping the system limitations.
Power semiconductor devices are compared and selected by
using the performances of their devices. SSCB is provided with
circuitry for permitting off/on/reset operations actuated by a
mechanical toggle control switch for close simulation of magnetic
circuit breaker operation. A power continuity feature is provided
for load power by locating the control switch so that it opens all
of the circuit, except the static power switch branch, between the
source and the load, to provide a high degree of fail-safe
protection for the apparatus. The synchronous reference frame is
used to detect the current disturbances and control the switches.
Simulation results are shown that SSCB offers advantages when
compared to present solutions. A switching circuit that is
particularly applicable to high voltage 3 phases solid state circuit
breakers (SSCB). The circuit comprises a solid state switch; an
energy absorbing device connected in parallel with said switch to
form a parallel combination; an inductor connected in series with
said parallel combination; and a control circuit connected to said
solid state switch and adapted to turn off said switch in response
to a predetermined voltage/current condition.**

Index terms--**Solid State Circuit Breaker**

I. INTRODUCTION

Advanced current interruption technology, utilizing high
power solid-state circuit breaker (SSCB), offers a viable
solution to the transmission and distribution system problems
caused by high available fault current. Although the power
industry has been interested in this concept for decades, it
appears that the time has now arrived when the selling price
can be low enough to justify significant sales. By providing
almost instantaneous voltage/current limiting, the SSCB
alleviates the short circuit condition in both downstream and
upstream devices by limiting fault currents coming from the
sources of high short circuit capacity. The advantages of
added functions that a conventional circuit breaker cannot
offer help to justify the higher cost associated with a solid
state breaker.

To interrupt the current, the SSCB must rapidly insert an
energy-absorbing element (e.g. resistor) into the circuit to
limit the fault current. In addition to limiting the fault current,
the SSCB can also limit the inrush current (soft start
capability), even for capacitive loads, by gradually phasing in
the switching device rather than making an abrupt transition
from an open to a closed position. A solid state breaker can
offer the following advantages: limited fault current, limited
inrush current (soft start), even for capacitive loads, repeated
operations with high reliability and without wear-out, reduced
switching surges, improved power quality for unfaulted lines.

As presented in [1], the present solutions dealing with short
circuit protection are mechanical circuit breakers. After having
detected a short circuit or over-load situation, some time
elapses prior to open the switches mechanically. Subsequently,
an arc occurs, which initially has little impact on the current.
As a result, turning off a short-circuit will take at least 100 ms
(without detection time). So, solid state circuit breakers based
on power semiconductors potentially offer enormous
advantages when compared to conventional solutions, since a
solid state breaker is able to switch in a few micro second.

The rest of this paper is organized as follows; Section 2
presents the comparison of power semiconductors,
considering the requirements of solid state breaker application.
The topology is proposed in Section 3. Section 4 gives the
design criteria of the breaker. Simulation results are shown in
Section 5. Finally, the discussion and conclusion are given in
Section 6.

II. BASIC OPERATION OF SSCB

To analysis the fundamental behavior of the proposed
semiconductor switch, a single-phase equivalent circuit, as
shown in Fig. 2, is used. The grid is represented by voltage
source and line impedance. In this example, a pure resistive
load is shorted by an ideal short-circuit with zero resistance.
The typical waveform of current in and voltage across the
switch are depicted in Fig. 3. At approximately 2 ms, a short-
circuit occurs and the current rises very fast. After a small
delay time, the semiconductors switch opens. The energy
stored in the line inductance is dissipated in high-energy
varistors, which are connected in parallel with the
semiconductors. Consequently, the current decreases. When
the current reaches zero, the switch has to block the line to
neutral voltage of the grid.

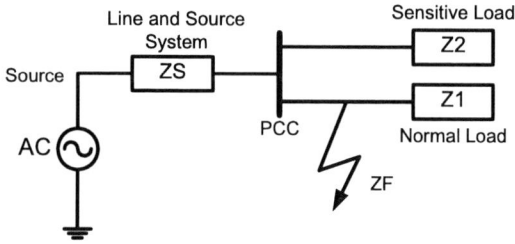

Fig. 1 Equivalent circuit of simple distribution system.

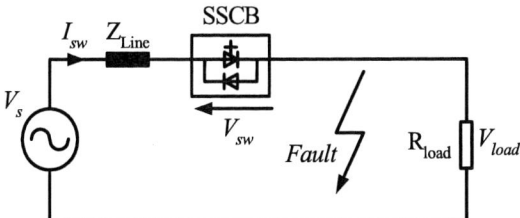

Fig. 2 Single phase equivalent circuit of SSCB protection.

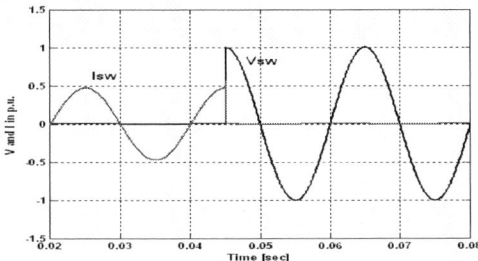

Fig. 3 Voltage across and current in ideal circuit breaker.

$$Z_L = Z_1 // Z_2 \qquad (1)$$

The voltage at PCC is:

$$\overline{V}_{PCC} = \frac{\overline{Z}_N}{\overline{Z}_S + \overline{Z}_N} \qquad (2)$$

$$V_{PCC} = \left| \frac{Z_N}{Z_S + Z_N} \right| \qquad (3)$$

$$\phi_{PCC} = \tan^{-1}\left(\frac{X_N}{R_N} \right) - \tan^{-1}\left(\frac{X_N + X_S}{R_N + R_S} \right) \qquad (4)$$

where

$Z_1 = R_1 + jX_1$ Normal Load

$Z_2 = R_2 + jX_2$ Sensitive Load

$\overline{V}_{PCC}$: Voltage at PCC

V_{PCC} : Voltage amplitude at PCC

ϕ_{PCC} : Phase angle at PCC

The voltage and phase angle is changed when fault occurred as follows:

$$Z_F = Z_2 // Z_F \qquad (5)$$

$$\overline{V}_{PCC(F)} = \frac{\overline{Z}_F}{\overline{Z}_S + \overline{Z}_F} \qquad (6)$$

$$V_{PCC(F)} = \left| \frac{\overline{Z}_F}{\overline{Z}_S + \overline{Z}_F} \right| \qquad (7)$$

$$\phi_{PCC(F)} = \tan^{-1}\left(\frac{X_F}{R_F} \right) - \tan^{-1}\left(\frac{X_F + X_S}{R_F + X_S} \right) \qquad (8)$$

where

$Z_f = R_f + jX_f$ Fault impedance

$Z_2 = R_2 + jX_2$ Sensitive load

$\overline{V}_{PCC(F)}$: Voltage at PCC when fault occurred

$V_{PCC(F)}$: Voltage amplitude at PCC when fault occurred

$\phi_{PCC(F)}$: Phase angle at PCC when fault occurred

The normal voltage in low voltage system is typically much higher than the maximum blocking voltage of today's semiconductors although their ratings have increased significantly [3]. Therefore, series connection of devices is necessary.

III. POWER ELECTRONICS DEVICES

Power electronics devices are solid-state devices or transistors capable of modulating or converting electrical power. A power electronic device enables the production of power management modules that can handle all of the electric power control and conversion functions required to move power from the generating and storage sources to the ultimate loads. Electric energy will play an increasingly important role in future energy system because of its controllability, safety, and high efficiency.

There are different ways to implement the SSCB. Since we want to provide sub-half cycle current limiting, we need to either use semiconductor devices with turn-off capability such as GTO, IGBT or IGCT or to use an SCR switch together with a forced commutation circuit. The former option offers the advantage of using a simple power circuit and very high speed operation. The current can be switched into an energy absorber in a few micro seconds. In contrast to inverter applications, the switching losses of these devices turns out to be a minor issue in this application. Here, the conduction behavior and conduction losses are essential. As s result, the IGBT has the advantage that, as a transistor, it limits the current automatically. Hence, current cannot exceed a certain value. The GTO and IGCT, the current is not limited. Thus, the detection time has to be short enough to assure a safe turn-off [4].

IGBT 1200 V 54 A is selected in this simulation study. The reason to select this device is and based on the current limiting and simple driver circuit. Fig. 4 shows two different types of IGBT.

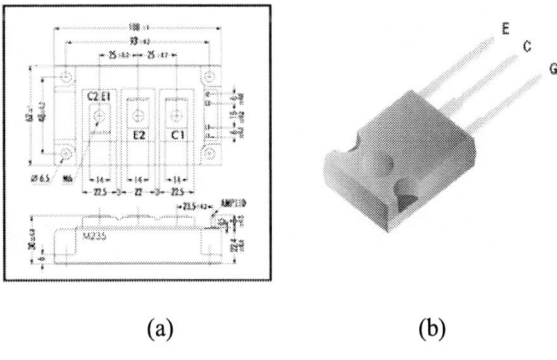

(a)   (b)

Fig. 4 Outline drawing and equivalent circuit of IGBT.

IV. CIRCUIT DESIGN

First, the designer define the fast simulation capability allows to propose an equivalent electrical circuit based on the basic cells of a library (a depleted PMOS transistor and two IGBT depletion mode IGBT) and providing the functionality required.

The designer can highlight the electrical parameter influences of the basic cell on the electrical characteristics of the "circuit breaker" function. In this case the most important parameters are the threshold voltages and the linear resistance value of the IGBT transistor.

Since each phase of the breaker must withstand a high peak voltage, it is necessary to connect semiconductor devices in series. This has been done in many cases for applications such as ac to dc converter stations and the necessary precautions are well understood.

A modular approach offers important advantages during design, testing, manufacturing and service stages. Some of the more important are:

o Design is simplified because within the module we are dealing with comparatively low voltages and the individual module is much smaller than the whole breaker.

o Testing is also simplified because of the reduced voltage level. This will translate into substantial time and cost savings as prototype tests are iterated.

o During manufacturing we will save time and money by mass producing small modules, testing them and finally stacking them up to build breakers (like Liberty ships). The same modules can be applied to many different voltage ratings.

o During service we can have single modules as spares and then replace the failed module rather than repairing the valve itself by replacing a failed IGBT. The module can then be sent to the shop or the factory for repairs.

o We can use the same modules for different voltage breakers by stacking the appropriate number of modules depending on the voltage level of the breaker. This way we reap the benefits of mass production even further by:

o Having the same building block for different voltage breakers, thus simplifying manufacturing, testing and repair costs.

o Reducing maintenance expenses by needing only one type of spare module, thus reducing the cost of spare parts and simplifying training of field personnel.

V. PROPOSED TOPOLOGY

The circuit breaker should consist of several modules to achieve a flexible adaptation for different current and voltage levels. For example, in Thailand there is 380 V low distribution voltage system. It has already been mentioned that the blocking that the blocking voltage of the breaker must be higher than the maximum grid voltage. Fig. 5 shows the proposed topology that consists of four IGBTs with snubbers and two diodes.

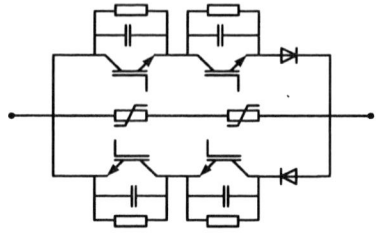

Fig. 5 proposed topology of SSCB

Fig. 5 shows the simplified block diagram of SSCB configuration. The switch operation has only two possible values: zero and one. The required local control functions are relatively simple once the system controller has made the decision for line connection/disconnection and reflected in a reduced set of functions at that level. The current space vector is required to detect the amplitude of the load current. The transformed currents are converted into the synchronous rotating frame (SRF), producing I_d and I_q, by using the output angle of the software phase lock loop (SPLL). Therefore, a means of phase tracking to obtain the positive sequence current component is required.

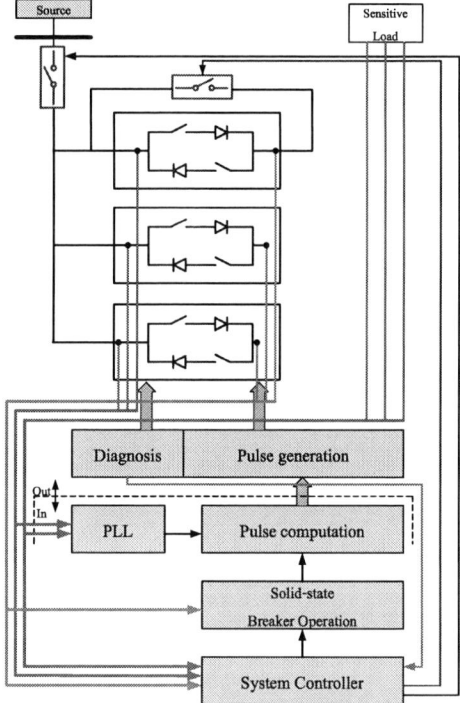

Fig. 6 SSCB control system and interface

The SPLL was implemented by sampling the instantaneous phase utility voltages, I_{sa}, I_{sb}, I_{sc} and converting them into voltages, $V_{s\alpha}$ and $V_{s\beta}$.

$$\begin{bmatrix} I_{L\alpha} \\ I_{L\beta} \end{bmatrix} = \sqrt{\frac{2}{3}} \begin{bmatrix} 1 & -1/2 & -1/2 \\ 0 & \sqrt{3}/2 & -\sqrt{3}/2 \end{bmatrix} \begin{bmatrix} I_{La} \\ I_{Lb} \\ I_{Lc} \end{bmatrix} \qquad (9)$$

These transformed voltages are converted into a SRF, producing I_{sd} ans I_{sq}, by rotating through the output angle of the SPLL, θ.

$$\begin{bmatrix} I_{Ld} \\ I_{Lq} \end{bmatrix} = \begin{bmatrix} \cos\theta & \sin\theta \\ -\sin\theta & \cos\theta \end{bmatrix} \begin{bmatrix} I_{L\alpha} \\ I_{L\beta} \end{bmatrix} \tag{10}$$

The I_{Ld} and I_{Lq} component was normalized using:

$$I_{Ldn} = \frac{I_{Ld}}{\sqrt{I_{sd}^2 + I_{sq}^2}} \tag{11}$$

$$I_{Lqn} = \frac{I_{Lq}}{\sqrt{I_{Ld}^2 + I_{Lq}^2}} \tag{12}$$

And the amplitude of the current detection:

$$I_{amp} = \sqrt{I_{Ldn}^2 + I_{Lqn}^2} \tag{13}$$

Whenever, the overcurrent or fault occurred, the detected operator will send the command to the gate pin of SSCB. Then the SSCB is turned off automatically. Anther way the SSCB will receive the turn on signal when no disturbance. Therefore the controller has two functions as detecting the current amplitude and sending the gate signal to SSCB. To ensure the system design, the simulation results are given in next section.

VI. TOPOLOGY OF IGBT

This article identifies the key parametric considerations for comparing IGBT performance in specific SMPS (switch mode power supply) applications. Parameters such as switching losses are investigated in both hard-switched and soft-switched ZVS (zero-voltage-switching) topologies. The three main power switch losses: turn-on, conduction and turn-off are described relative to both circuit and device characteristics. The impact of diode recovery performance on hard-switched topologies is also discussed illustrating that diode recovery is the dominant factor determining MOSFET or IGBT turn-on switching

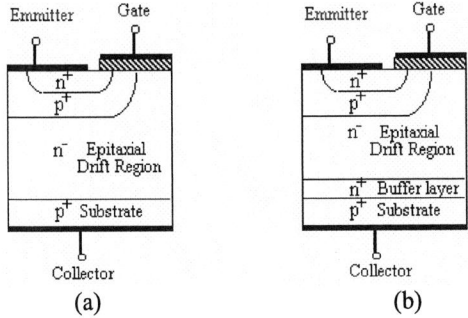

Fig.7 (a) Non Punch Through (NPT) IGBT (b) Punch Through (PT) IGBT.

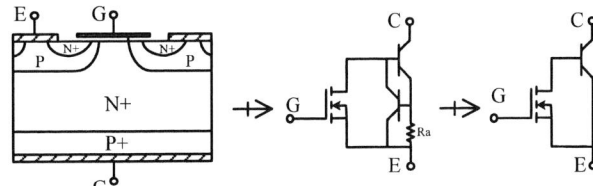

Figure.8 Cross Section and Equivalent Circuit for IGBT

Figure.8 shows a common symbol of the IGBT and its equivalent circuit. The IGBT is designed such that the turn-on and turn-off times of the device can be affected by the gate-to emitter source impedance. Its equivalent input capacitance is lower than a Power MOSFET with a comparable current and voltage rating. The device is turned on by applying a positive voltage between the gate and the emitter, VGE. In switching applications the device operates in the saturation region.

The IGBT turns on when the collector-emitter voltage is positive and greater than Vf and a positive signal is applied at the gate input (g > 0). It turns off when the collector-emitter voltage is positive and a 0 signal is applied at the gate input (g = 0). The IGBT device is in the off state when the collector-emitter voltage is negative. Note that many commercial IGBTs do not have the reverse blocking capability. Therefore, they are usually used with an antiparallel diode.

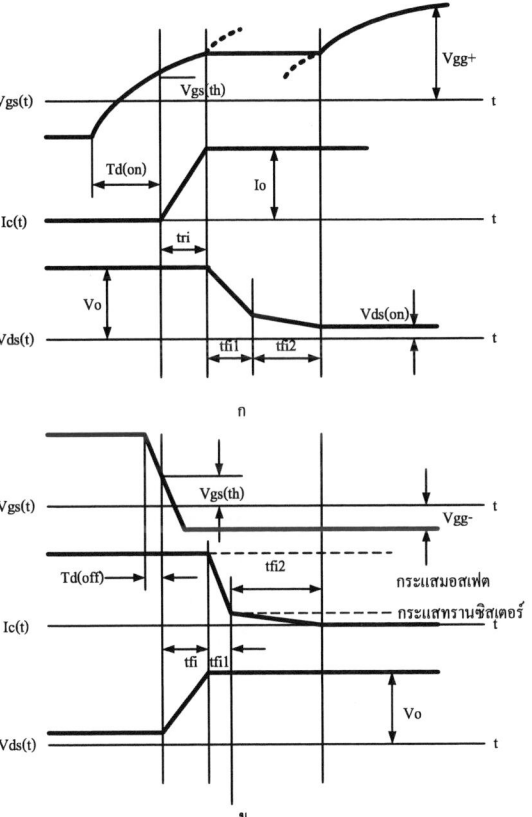

Fig.9 (a) IGBT turn-on switching transient with inductive load
(b) IGBT turn-off switching transient with inductive load

(a) Turn-on Transients

The turn-on switching transient of an IGBT with an inductive load are shown in Fig. 10(a). The turn-on switching transients of IGBTs are very similar to MOSFETs since the IGBT is essentially acting as a MOSFET during most of the turn-on interval. With gate voltage applied across the gate to emitter terminals of the IGBT, the gate to emitter voltage rises up in an exponential fashion from zero to VGE(th) due to the circuit gate resistance (RG) and the gate to emitter capacitance (Cge). The Miller effect capacitance (Cgc) effect is very small due to the high voltage across the device terminals. Beyond VGE(th), the gate to emitter voltage continues to rise as before and the drain current begins to increase linearly as shown above. Due to the clamp diode, the collector to emitter voltage remains at Vdc as the IGBT current is less than Io. Once the IGBT is carrying the full load current but is still in the active region, the gate to emitter voltage becomes temporarily clamped to VGE,Io, which is the voltage required to maintain the IGBT current at Io. At this stage, the collector to emitter voltage starts decreasing in two distinctive intervals tfv1 and tfv2. The first time interval corresponds to the traverse through the active region while the second time interval corresponds to the completion of the transient in the ohmic region. During these intervals, the Miller capacitance becomes significant where it discharges to maintain the gate to source voltage constant. When the Miller capacitance is fully discharged, the gate to emitter voltage is allowed to charge up to VG and the IGBT goes into deep saturation. The resultant turn on switching losses are shown in the above figure. The on energy loss is approximately estimated via,

$$E_{on} = \frac{V_{dc}I_o}{2} \times t_{on} \qquad (14)$$

The above switching waveforms are ideal in the since that the clamp diode reverse recovery effects are neglected. If these effects are included, an additional spike in the current waveform results as shown in the previous figure. As a result, additional energy losses will be incurred within the device.

(b) Turn-off Transients

The turn-off switching transient of an IGBT with an inductive load are shown in Fig. 10(b). When a negative gate signal is applied across the gate to emitter junction, the gate to emitter voltage starts decreasing in a linear fashion. Once the gate to emitter voltage drops below the threshold voltage (VGE(th)), the collector to emitter voltage starts increasing linearly. The IGBT current remains constant during this mode since the clamp diode is off. When the collector to emitter voltage reaches the dc input voltage, the clamp diode starts conducting and the IGBT current falls down linearly. The rapid drop in the IGBT current occurs during the time interval tfi1 which corresponds to the turn-off of the MOSFET part of the IGBT (Fig. 9). The tailing of the collector current during the second interval tfi2 is due to the stored charge in the n-drift region of the device. This is due to the fact that the MOSFET is off and there is no reverse voltage applied to the

IGBT terminals that could generate a negative drain current so as to remove the stored charge. The only way for stored charge removal is by recombination within the n- drift region. Since it is desirable that the excess carriers lifetime be large so as to reduce the on-state voltage drop, the duration of the tail current becomes long. This will result in additional switching losses within the device. This time increases also with temperature similar to the tailing effect in BJTs. Hence, a trade off between the on-state voltage drop and faster turn-off times must be made.

$$E_{off} = \frac{V_{dc}I_o}{2} \times t_{off} \qquad (15)$$

VII. SIMULATION RESULTS

To show the performance of SSCB, we use the model of IGBT in power blockset in Matlab/simulink. The parameters of IGBT are in Table 1. The switching waveforms of the test model of IGBT is shown in Fig. 5 which the response of turn-on and turn off IGBT. The rating of and time delay are very important because we

TABLE I: Parameters of IGBT

Description	Rated
Rated voltage	600 V
Rated current	63 A
Gate to Emitter Threshold Voltage	6.0 V
Switching SOA	60 A
Leakage current	± 100 μA
Current rise time	45 ns
Current Turn-on Delay Time	40 ns
Current fall time	275 ns
Current Turn-Off Delay Time	400 ns
Turn On Energy	1050 μJ
Turn Off Energy	2500 μJ

Fig. 10 to 13 show the simulation results of performance of SSCB that used in low voltage level. The current waveform without on/off of SSCB when the overcurrent occurred is shown in Fig. 10. The voltage at this point is reduced the amplitude of voltage as shown in Fig. 11.

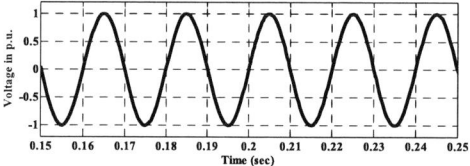

Fig.10 Phase A voltage when normal case

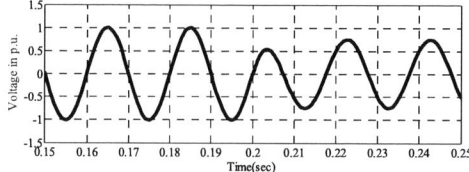

Fig. 11 Phase A voltage when fault occurred at 200 ms

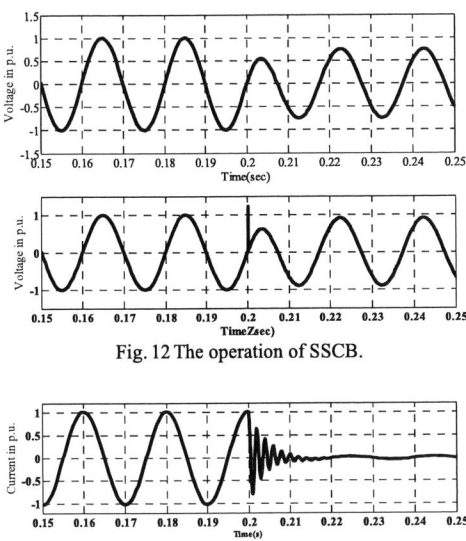

Fig. 12 The operation of SSCB.

Fig.13 Phase A current.

Fig. 12 shows the voltage and current waveform when SSCB operates and the current did not pass to the load. The oscillation of the current in Fig. 13 is depended on the snubber circuit of power semiconductor devices that uses for reducing the voltage and current stress. However, if we consider in the ideal case this oscillation does not appear in the system.

VIII. CONCLUSION

This paper shows the performance of IGBT when is used to be SSCB. The selected topology presents the situation topology when compared each other. The fast action of SSCB is represented in ideal for avoiding drawbacks concerning state-of-the-art mechanical breaker. The solid-state breaker turned off only for 100 micro seconds in comparison to conventional circuit breaker 100 ms today. Simulation results are shown that solid state circuit breakers offers advantages when compared to present solutions.

IX. REFERENCES

[1] Ekstrom, A.; Bennich, P.; De Oliveira, M.; Wikstrom, A.: "Design and Control of a Current-Controlled Current Limiting Device, EPE, Graz 2001

[2] Klingbeil, L.; Kalkner, W.; Heinrich, Ch.: "Fast Acting Solid-State Circuit Breaker using state-of-the-art power electronic devices", EPE, Graz 2001

[3] De Doncker, R. W.: "Recent Power Electronics Developments for FACTS and Customed Power", Korea Germany Advanced Power Electronics Symposium, 1998

[4] Tosato, F.: "Voltage Sags Mitigation on Distribution Utilities", ETEP Vol. 1 1, No. I , January/Febntary 2001

[5] Russ, M.; Sommer, R.; Zaiser, G.: "Spannungszwischenkreisumrichter im Mittelspannungsbereich", ETG, Bad Nauheim , 2002

[6] "Power Electronics: Converters, Applications and Design", Mohan, Undeland and Robbins, Wiley, 1989.

A Variable Gain Control Scheme of Digital Automatic Voltage Regulator for AC Generator

Dong-Hee Lee, Jin-Woo Ahn, and Tae-Won Chun*

Dept. of EE & Mechatronics, Kyungsung University, Busan, Korea
* Dept. of Electrical Engineering, Ulsan University, Ulsan, Korea

Abstract—The voltage control characteristic of synchronous generation system in power plant is dependent on the performance of the excitation control. AVR (Automatic Voltage Regulator) adjusts the excitation voltage and current for the control of synchronous generation system. The AVR controller must have capabilities to acquire the demanded transient response that is able to over come the time-delay of exciter and to damp out the inherent instability of generator and prime mover.

This paper presents a variable gain control scheme of DAVR (Digital Automatic Voltage Regulator) for AC brushless generator. In order to improve dynamic response and steady state performance, a PID controller with a simple gain adjustor is used. And the gains of PID controller are updated by the terminal voltage and load current in order to improve dynamic response and stability. The structure of the proposed gain adjustor is designed by a a simple linear function and the gain is increased in-proportional to load current and terminal voltage. The proposed control scheme is verified by experiments.

Index Terms—DAVR, AC brushless generator, variable gain control scheme.

I. Introduction

Automatic voltage regulators for brushless excited synchronous generators have traditionally utilized analog electronics circuits[1-3]. In recent, DSP(Digital Signal Processor) or micro-processor are much used for AVR(Automatic Voltage Regulator) systems, due to the power control performance, flexibility, and cost advantage[4-8]. DAVR(Digital AVR) based on DSP or micro-processor regulates various output signals of a AC generator by controlling the exciter field voltage and current. So, the control performance of AC generator system is dependent on the excitation system with AVR control scheme.

Although, the various control method are studied for excitation system, the general PID control method is much used in practical application due to the easy approach[9-10].

The control performance of the general PID control method is dependent on the control gain. And the proper gain determination is very difficult for a wide operating range of AC generator excitation system. The higher gain is proper for fast dynamic response from load variation, but the output voltage of AC generator may has oscillation from high control gain due to the time-delay

of rms voltage detection. In contrary, the low control gain is good for smooth output voltage in steady state, but dynamic response is not good in load variation.

This paper presents a variable gain control scheme of DAVR(Digital Automatic Voltage Regulator) for AC brushless generator. In order to improve dynamic response and steady state performance, a PID controller with a simple gain adjustor is used. And the gains of PID controller are updated by the terminal voltage and load current in order to improve dynamic response and stability. The structure of the proposed gain adjustor is designed by a a simple linear function and the gain is increased in-proportional to load current and terminal voltage. In the steady state, the small control gain is selected for generator voltage control. And the control gain is decreased according to terminal voltage and load current increasing. So, it can get fast dynamic response with high stability without any oscillation of terminal voltage. The calculation of control gain is very simple, and has linear characteristics according to terminal voltage error.

The proposed control scheme of DAVR system is verified by experimental results.

II. AC Brushless Generator and Exciter System

DAVR controlled AC brushless generator and exciter system are shown in Fig. 1. This type of power system is much used in ship and emergency power supply system due to the reliability, maintenance and easy user interface.

The rotating exciter is installed on the shaft of AC generator, and the field winding of exciter is connected to AVR controller. The induced voltage of exciter armature winding from exciter field current is rectified by the 3-phase diode module, and supplied to field winding of AC generator. Actually, the terminal voltage of AC generator can be controlled by the exciter field current connected to AVR controller.

DAVR controller has main digital processor, generator terminal voltage, grid voltage and load current detecting part, power electronics part for the control of exciter. The basic structure of DAVR system is very simple, but detail care is required in the design of system due to the control resolution, self-starting and timing-delay of the generator system. Furthermore, in the brushless AC generator, field windings of generator is

978-1-4244-0644-9/07/$25.00 ©2007 IEEE

located on the shaft, so the field voltage and current can not be directly detected. The information used for the generator control is exciter field voltage and current. The power electronics of a conventional AVR controller using SCR is shown in Fig. 1(a), which uses fire angle control method. PWM controlled IGBT chopper in Fig. 1(b) and digital control technique are used for exciter field control for a high dynamics and high performance nowadays.

In the Fig. 1, AVR controller produces exciter current reference according to terminal voltage error. And the internal current controller adjusts the field current in order to keep the reference by changing PWM duty ratio.

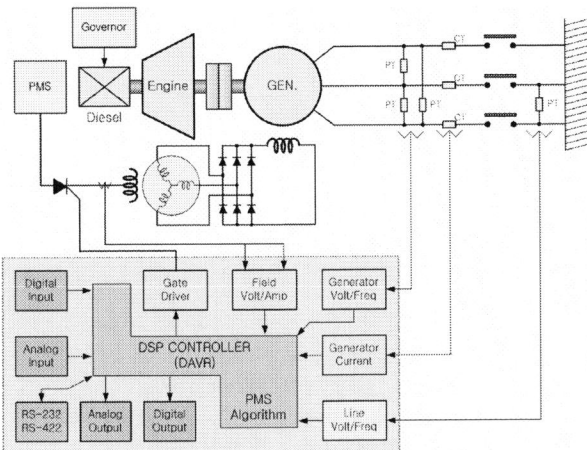

(a) SCR controlled AC brushless generator

(b) PWM controlled AC brushless generator
Fig. 1 Excitation control system for AC brushless generator system

Fig. 2 shows a simple block diagram of AC generator and PID controlled exciter system.

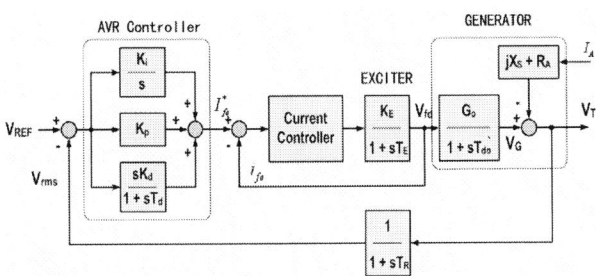

Fig. 2 Block diagram of AC generator and PID controller

Here, K_E and T_E are the gain and time-constant of the exciter, G_o and T'_{do} are the incremental gain and time-constant of the generator, X_S and R_A are the synchronous reactance and armature resistance respectively. T_R denotes sensing time-delay of terminal rms voltage.

The internal generation voltage can be produced by field voltage as follow.

$$V_G(s) = \frac{G_o}{1 + sT'_{do}} \cdot V_{fd}(s) \tag{1}$$

where G_o is the incremental gain of generator.

And the terminal output voltage is determined by load current and impedance.

$$V_T(s) = V_G(s) - sX_S \cdot I_A(s) - R_A \cdot I_A(s) \tag{2}$$

The mathematical model of AC generator terminal voltage can be simply expressed in d-q axis vector model[2].

$$\begin{aligned}
V_d(s) &= X_q \cdot I_q(s) - R_A \cdot I_d(s) \\
V_q(s) &= V_G(s) - X_D(s) \cdot I_d(s) - R_A \cdot I_q(s)
\end{aligned} \tag{3}$$

$$X_D(s) = \frac{X_d + sX'_d T'_{do}}{1 + sT'_{do}} \tag{4}$$

where R_A : armature resistance

$I_d(s)$, $I_q(s)$: d-q axis generator currents

X_d, X_q : d-q axis synchronous reactance

X'_d : d-axis transient reactance

T'_{do} : open-loop field time-constant

In order to control the generator voltage, AVR controller adjusts the exciter current, and the control performance is changed by the AVR controller gain. The transfer function of AVR controller can be obtained as follows.

$$G(s) = K_p + \frac{K_i}{s} + K_d \cdot s \tag{5}$$

where, K_p, K_i and K_d are proportional, integral and derivative gain.

If controller gain is high, the dynamic response of the system is fast, but it may have some over shoot and oscillation due to the time-delay of the system. But, the system is stable in low control gain with slow dynamic response. To improve the dynamic response and system stability, adaptive and fuzzy controller are investigated. Well designed adaptive or fuzzy controller has good performance, but designed controller is not proper in other generator system, and the design process is somewhat difficult.

In this paper, a simple variable gain control scheme is proposed to improve control performance. The variable gain controller is practically used in AC servo and variable speed motor drive. The control gain of motor drive is changed according to motor speed or torque. In the light load and low speed, the higher gain can improve the control performance, and the lower gain can insure the system stability in high speed and heavy load. A similar idea is introduced in AC generator system.

III. THE PROPOSED CONTROL SCHEME

As state, AC generator system has time-delay in measuring the actual rms terminal voltage and controlling exciter current for voltage control. If the control gain is high, as the dynamics of the PID controller is faster than the generator system, generator system will be unstable. On the other hand, generator system is stable but low dynamic performance with low control gain.

Fig. 3 shows the proposed variable gain control scheme for AC brushless generator system.

The basic structure is same as conventional PID controller, but a simple gain adjustor is added. The D-term of PID controller is fixed due to the complex of gain adjustor and strong effects to system stability. In Fig. 3, K_{p1}, K_{p2} and K_{i1}, K_{i2} are the gains of PI controller. Input factors of gain adjustor are selected based on valued rms terminal voltage V_{rms} and load current I_{rms}. Two input factors are normalized and multiple by weight factors w_f and $1-w_f$, respectively. In this paper, the weight factor w_f is 0.5. According to two inputs each gains are simply determined.

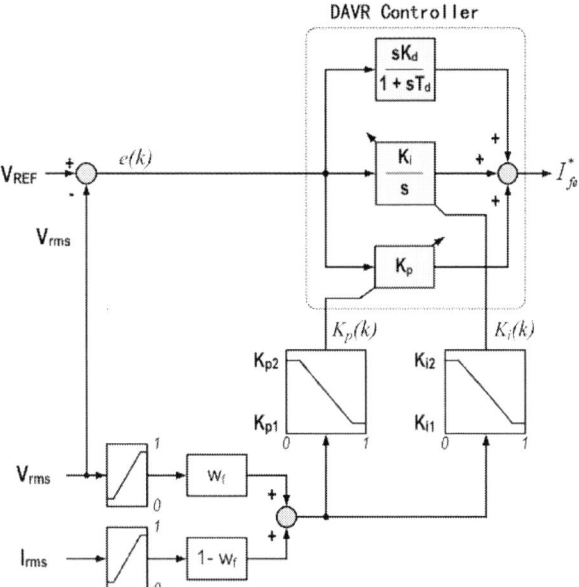

Fig. 3 Block diagram of the proposed variable gain control scheme

The gain table is designed that high gain is selected in light load and high voltage error.

The reference exciter current can be obtained as follows.

$$e(k) = V_{REF}(k) - V_{rms}(k) \tag{6}$$

$$I_{fe}^*(k) = K_p(k) \cdot e(k) + K_i(k) \cdot \sum_{n=0}^{k} e(k) \cdot T_S + K_d \cdot \frac{e(k) - e(k-1)}{T_d} \tag{7}$$

And the gain of controller can be obtained by the generator voltage, current and pre-set look-up table.

$$k_f = w_f \cdot norm(V_{rms}) + (1-w_f) \cdot norm(I_{rms}) \tag{8}$$

$$K_p(k) = func(K_{p2}, K_{p1}, k_f) \tag{9}$$

$$K_i(k) = func(K_{i2}, K_{i1}, k_f) \tag{10}$$

IV. EXPERIMENTAL RESULTS

In order to verify the proposed control scheme, practical experiments are implemented. 50kW-380V 3-phase emergence AC generator with heat load is .

The main controller of digital AVR is designed by TMS320F2811. Embedded 12bit 8-channel ADC modules are used for detecting exciter field current, rms terminal voltage, grid voltage, load current and temperature. Basic parameters of control system and generator are saved in the serial connected EEPROM.

Self-starting of generator is implemented by the resident flux of generator. The initial voltage produced by the resident flux of generator is very small. So the digital controller can not be operated due to the low power.

Fig. 4 shows the experimental configuration. The 50[kW], 3-phase emergency AC brushless generator and designed DAVR controller is used for the experiment.

Fig. 5 shows the experimental result of self-starting. When the diesel-engine is starting, generator voltage is smoothly increased from residual flux. And the exciter current is controlled as a fixed self-starting value during initialization of DAVR controller. After the initialization of DAVR controller, the reference is smoothly increased to target voltage. In the designed system, the slope of the soft-starting can be adjusted.

Fig. 6 shows the experimental results of reference voltage changing. The reference is changed 110% value of rated voltage. In the load and no-load case, the experimental results show the stable and fast dynamic response with the proposed control scheme.

Fig. 7 shows the experimental results of sudden load change. In the rated voltage control, full-load is suddenly changed. But the generator voltage is fast recovered in the designed DAVR. The recovery time is about 0.3[sec] in experimental results.

(a) Diesel engine and AC brushless generator

(b) Digital AVR controller

Fig. 4 Experimental configurations

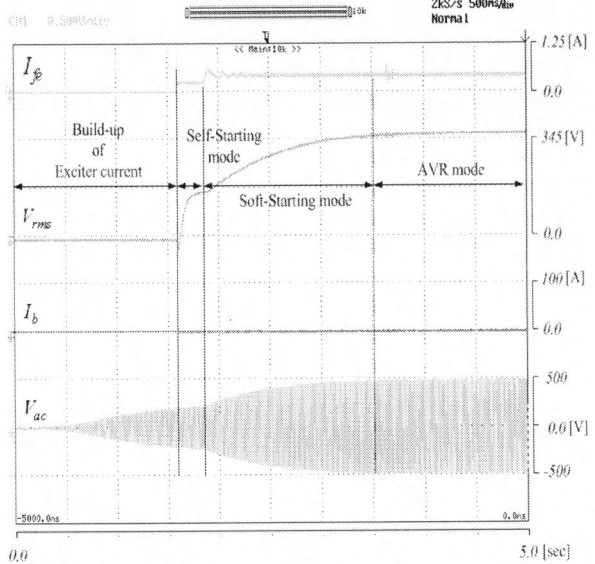

Fig. 5 Starting of the generator

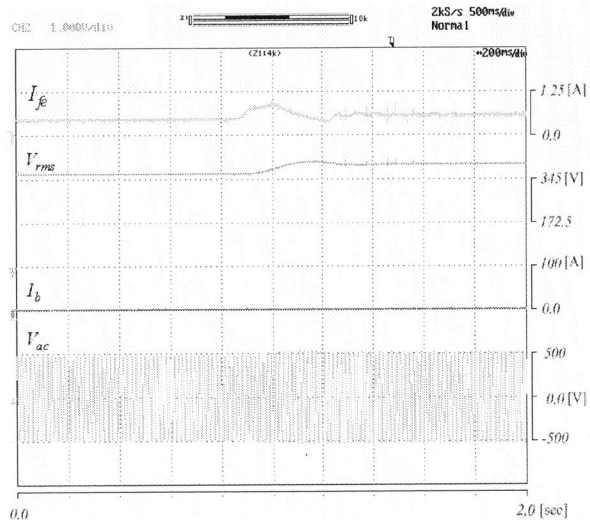

(a) Reference voltage change(no-load)

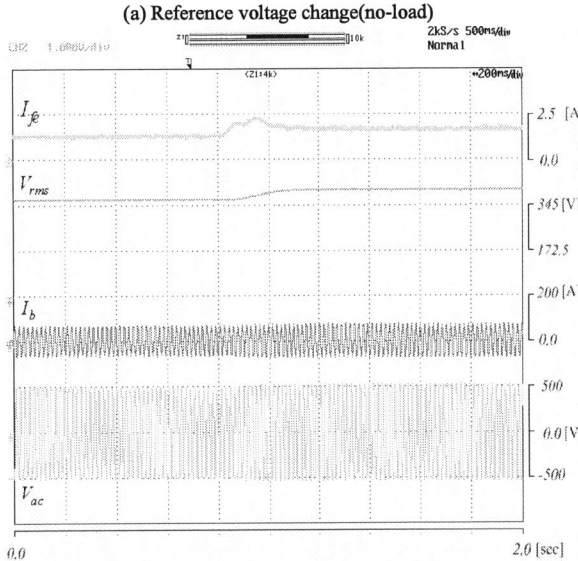

(b) Reference voltage change(full-load)

Fig. 6 Experimental result of reference voltage change(380→418V)

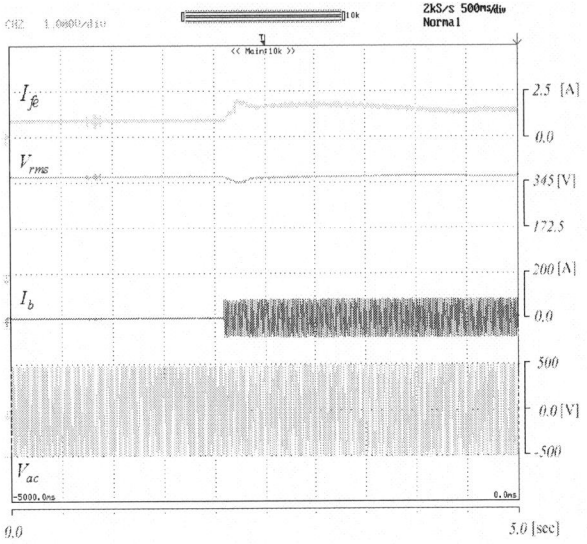

(a) Sudden load injection(full-load)

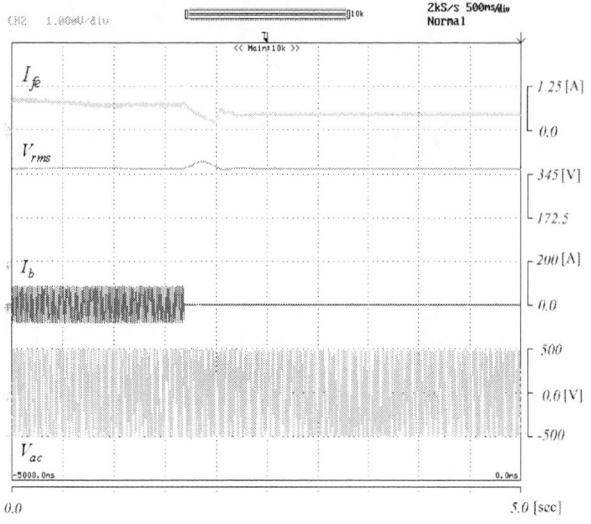

(b) Sudden load rejection(full-load)

Fig. 7 Experimental result in load change

V. CONCLUSIONS

In this paper, a simple variable gain control scheme for AC brushless generator system is presented. The proposed variable gain control scheme consists of conventional PID controller and a simple gain adjustor. The designed gain adjustor updates proportional and integral gain of PID controller according to generator voltage and load current. Gain tables of gain adjustor are designed by simple linear function with upper and lower limiter.

The computer simulation and experimental results show the good performance in a dynamic response and a steady state.

ACKNOWLEDGEMENT

This research was supported by the program for the training of Graduate Students in the Regional Innovation which was conducted by the Ministry of Commerce, Industry and Energy, Korea.

REFERENCES

[1] "An American National Standard IEEE Standard Definitions Excitation System for Synchronous Machines", pp. 8-14, 1986.

[2] L. W.`Matsch, J. D. Morgan, „Electromagnetic and Electromagnetical Machines", John wiley & Sons, pp. 214-215, 1987.

[3] P. Kundur, M. Klin, G. J. Rogers, and M. S. Zywno, "Application of Power System Stabilizer for Enhancement of Overall System Stability", IEEE Trans. on Power Systems, Vol. 4, No. 2. pp. 614-626, May, 1989.

[4] K. A. Riddle, "Renovation of a paper mill steam driven turbine-generator", presented at IEEE Pulp and Paper Technical Conference, Rome, GA, 1995.

[5] IEEE Guide for Specification for Excitation Systems", IEEE Std. 421. 4. 1990

[6] Gao Feng, Qin Yihong, Xu Guoyu. A method for automatic design of fuzzy voltage regulator[J]. Automation of Electric Power Systems, 1995, 19(8): 5-9.

[7] Tan Ximei, Zhou Shuangxi. Analysis of new fuzzy logic excitation controller[J]. Automation of Electric Power Systems, 1998, 22(8): 8-11.

[8] J.H.Anderson. The Control of Synchronous Machine Using Optimal Control Theory. Proceedings of IEEE Transactions on PAS. 1971, 59: 25-35

[9] A. Godhwani, M. J. Basler, and T. W. Eberly, "Commissioning and operational experience with a modern digital excitation system", IEEE Trans. Energy Conversiton, Vol. 13, pp. 183-187, June, 1998.

[10] A. Godhwani and M. J. Basler, "Design, test and simulation results of a var/power factor controller implemented in a modern digital excitation system", IEEE Power Engineering Society Summer Meeting 1998, SanDiego, CA.

A Graphic User Interface-based Program for Harmonic Impedance Calculation

T. Tayjasanant

Department of Electrical Engineering, Faculty of Engineering, Chulalongkorn University 10330, Thailand

Abstract–This paper presents a graphic user interface (GUI)-based program for harmonic impedance calculation. The GUI feature provides the user with visual interaction and facilitates the analysis. The harmonic impedance calculation is based on two steady-state conditions of voltage and current waveforms. Modal transformations and weighting average method are presented and proposed. The program has been tested using two actual field measurements.

Index Terms–Harmonic impedance, harmonic measurement, harmonics.

I. INTRODUCTION

Harmonic impedance of the network is an important information for harmonic filter design, harmonic limit compliance and harmonic resonance prediction. Reference [1] classifies measurement methods for harmonic impedance into two types: the steady-state-based and transient-based methods. The steady-state-based method has been used by references [1]-[7]. Increments in harmonic voltages and currents are used to calculate the harmonic impedance. Typical disturbances are injections of harmonic currents from external harmonic sources or switching of a network component such as a shunt capacitor or a transformer.

In order to facilitate the calculation of harmonic impedance, a program with Graphical User Interface (GUI) feature is proposed. The GUI feature provides the user with visual interactive capability with less tedious and repetitive work on parameter adjustments and calculations. The user with limited knowledge of the harmonic impedance calculation can easily simulate the program. The design environment is based on the MATLAB Graphical User Interface Development Environment (GUIDE) [8].

The objective of this paper is to present a developed GUI-based program that can facilitate the calculation of harmonic impedance. The impedance calculation is derived from two steady-state conditions, i.e. prior and after the disturbance. The application of modal transformations and weighting average method for the final impedance value are also discussed.

This paper is organized as follows. Section II presents the background on the harmonic impedance calculation. The symmetrical-components and Clarke (α–β–0)

transformations are utilized for the calculation. The weighting average method is proposed for the calculation of the final value of harmonic impedance from two modal transformations. In Section III, the developed GUI-based program is described. Results from running two actual field measurements are shown on execution displays of the program. Lastly, Section IV summarizes the conclusion.

II. HARMONIC IMPEDANCE CALCULATION

A. Conversion of Line Voltages to Phase Voltages

Voltage and current signals are normally taken from the low-voltage side of potential and current transformers. In some cases, the voltage signal is available only from two line-to-line connected potential transformers. For the harmonic impedance calculation used in this paper, three-phase voltages are required for modal transformations.

Three phase voltages can be derived from two line-to-line voltages by assuming that the system is balanced, i.e. $V_a + V_b + V_c = 0$. Combinations of line-to-line voltages can be categorized into three major groups: Group 1 consists of V_{ab} and V_{bc}; Group 2 consists of V_{bc} and V_{ca}; and Group 3 consists of V_{ca} and V_{ab} as shown in Table I.

TABLE I
THE DERIVATION OF THREE PHASE VOLTAGES FROM TWO LINE VOLTAGES

Group	V_1	V_2	V_a	V_b	V_c
1	V_{ab}	V_{bc}	$\dfrac{2V_1 + V_2}{3}$	$\dfrac{-V_1 + V_2}{3}$	$\dfrac{-V_1 - 2V_2}{3}$
2	V_{bc}	V_{ca}	$\dfrac{-V_1 - 2V_2}{3}$	$\dfrac{2V_1 + V_2}{3}$	$\dfrac{-V_1 + V_2}{3}$
3	V_{ca}	V_{ab}	$\dfrac{-V_1 + V_2}{3}$	$\dfrac{-V_1 - 2V_2}{3}$	$\dfrac{2V_1 + V_2}{3}$

The conversion of Group 1 for V_a is provided for an example. Let $V_{ab} = V_1$ and $V_{bc} = V_2$,

$$V_a + V_b + V_c = 3V_a - 2V_a + V_b + V_c = 0$$

$$3V_a = 2V_a - V_b - V_c = (2V_a - 2V_b) + (V_b - V_c)$$

$$3V_a = 2V_{ab} + V_{bc}$$

$$V_a = \frac{2V_{ab} + V_{bc}}{3} = \frac{2V_1 + V_2}{3}$$

This work was supported by the Grant for Development of New Faculty Staff, Chulalongkorn University.

Note that $V_{cb} = -V_{bc}$ so if V_{ab} and V_{cb} are available instead of V_{ab} and V_{bc} then $V_a = (2V_1-V_2)/3$. Similar derivations can be applied for other equations. All equations are summarized in Table I.

B. Method for Harmonic Impedance Calculation

Assuming that the internal system voltage and system impedance are constant during the measurement period, the impedance at the location of the switching capacitor can be computed from [1], [7]

$$Z_h = \frac{V_{post-h} - V_{pre-h}}{I_{post-h} - I_{pre-h}} = \frac{\Delta V_h}{\Delta I_h} \tag{1}$$

where V_{post-h} and I_{post-h} are post-disturbance h harmonic voltage and current phasors, and V_{pre-h} and I_{pre-h} are pre-disturbance h harmonic voltage and current phasors.

When there is a frequency variation is the system, a phase shift exists between pre- and post-disturbance data. It is recommended to compensate this phase shift for harmonic impedance calculation [7]. The phase shift per cycle (θ) can be calculated from the (fundamental-frequency) phase difference between two selected cycles in pre- or post-disturbance *voltages* divided by the number of cycles separating the two selected voltage waveforms. Fig. 1 shows plots of voltage waveforms and fundamental RMS voltages from one field measurement (Case 2). The disturbance starts at the beginning of the 7th cycle, the pre-disturbance period is between cycles 1 and 6 and the post-disturbance period is between cycles 10 to 14. If the 1st and 5th cycles are selected, the phase angle shift per cycle can be calculated from the fundamental-frequency phase difference between these two cycles and then divided by four. The post-disturbance voltage and current phasors are then correct.

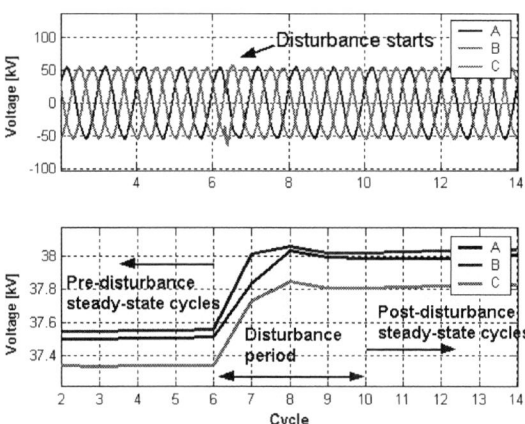

Fig. 1. Voltage waveform and magnitude plots for the calculation of phase shift per cycle.

Therefore, the new equation for harmonic impedance calculation after the frequency variation is compensated is given in (2).

$$Z_h = \frac{V_{post-h}e^{j\theta hn} - V_{pre-h}}{I_{post-h}e^{j\theta hn} - I_{pre-h}} \tag{2}$$

where n is the number of cycles separating the selected pre- and post-disturbance cycles and h is the harmonic order of interest.

It should be noted that the computed impedance is sensitive to the selections of pre- and post-disturbance cycles [9]. The selection of pre-disturbance cycle should not be problematic because it can be easily selected from fundamental-frequency voltage and current plots. However, the post-disturbance cycle has to be selected carefully. The post-disturbance cycles in the steady-state condition should be chosen, not during the transient condition. In addition, (2) assumes that the phase shift at harmonic is linearly proportional to the harmonic order.

C. The Application of Modal Transformations

Two modal transformations—Symmetrical-components and Clarke (α–β–0) transformations—are considered in this paper.

1) Symmetrical-Components Transformation

This transformation has been widely applied for the analysis of three-phase unbalanced power system. This transformation decouples three-phase voltages into positive-, negative- and zero-sequence components as given in (3).

$$\begin{bmatrix} V_0 \\ V_+ \\ V_- \end{bmatrix} = \frac{1}{3} \begin{bmatrix} 1 & 1 & 1 \\ 1 & 1\angle 120° & 1\angle -120° \\ 1 & 1\angle -120° & 1\angle 120° \end{bmatrix} \begin{bmatrix} V_a \\ V_b \\ V_c \end{bmatrix} \tag{3}$$

This transformation deals with complex number operation and it can only be applied to steady-state voltage and current phasors. The operation of complex numbers tends to introduce errors to phase angles of voltages and currents. The harmonic impedance can be calculated by substituting sequence components of voltages and currents into (2). The calculation considers the sequence characteristic of harmonics, e.g. the 5th harmonic is predominantly in negative sequence and 7th harmonic is predominantly in positive sequence. In practice, it is possible that at a certain harmonic, the harmonic current excitation is insufficient, i.e. the harmonic current change (ΔI_h) is small. This results in unreliable harmonic impedance values.

2) Clarke or α–β–0 Transformation

This transformation decouples three-phase voltages to alpha, beta and 0 components as given in (4).

$$\begin{bmatrix} V_\alpha \\ V_\beta \\ V_0 \end{bmatrix} = \frac{1}{\sqrt{6}} \begin{bmatrix} 2 & -1 & -1 \\ 0 & \sqrt{3} & -\sqrt{3} \\ \sqrt{2} & \sqrt{2} & \sqrt{2} \end{bmatrix} \begin{bmatrix} V_a \\ V_b \\ V_c \end{bmatrix} \tag{4}$$

This transformation is a real transformation because it can be applied to instantaneous voltages and currents. There are no complex number operations to be concerned with. Moreover, there are two impedance values (Z_α and Z_β) available to cross-check each other. Generally, both alpha and beta components of harmonic current are numerically comparable. There is no sequence characteristic of harmonics to be considered as in the symmetrical-components transformation. Consequently, the Clarke transformation is more attractive than the symmetrical-components transformation for the calculation of harmonic impedance [10].

If the system impedance is balanced or symmetrical, $Z_\alpha = Z_\beta = Z_+ = Z_-$. It should be bear in mind that both symmetrical-components and α–β–0 transformations only work for balanced systems. In order to deal with unbalanced system, three independent capacitor switchings are required for the calculation of harmonic impedance [11].

D. Weighting Average Method

Unreliable harmonic impedances should be filtered out when the percentage of harmonic current change is less than a certain threshold. This paper proposes a weighting average method for the harmonic impedance calculation using harmonic current changes. The final harmonic impedance value is weighted by the harmonic current change associated with each component. The harmonic current changes (in percentage) from symmetrical-components ($\Delta I_{h\text{-}sym}$) and α–β–0 transformations ($\Delta I_{h\text{-}\alpha}$ and $\Delta I_{h\text{-}\beta}$) are defined as follows:

$$\Delta I_{h\text{-}sym} = \frac{|\Delta I_{h\text{-}sequence}|}{|I_{1+}|} \times 100\ \% \tag{5}$$

$$\Delta I_{h\text{-}\alpha} = \frac{|\Delta I_{h\text{-}\alpha}|}{|I_{1\alpha}|} \times 100\ \% \tag{6.a}$$

$$\Delta I_{h\text{-}\beta} = \frac{|\Delta I_{h\text{-}\beta}|}{|I_{1\beta}|} \times 100\ \% \tag{6.b}$$

where $|I_{1+}|$, $|I_{1\alpha}|$ and $|I_{1\beta}|$ are fundamental-frequency current magnitudes of the positive-sequence, alpha- and beta-components, respectively.

There are one Z_h value from the symmetrical-components transformation and two Z_h values from the α–β–0 transformation. Therefore, these three Z_h values

will be weighted with corresponding harmonic current changes to yield the single Z_h value as provided in (7).

$$Z_h = \frac{Z_{h\text{-}sym}\Delta I_{h\text{-}sym} + Z_{h\text{-}\alpha}\Delta I_{h\text{-}\alpha} + Z_{h\text{-}\beta}\Delta I_{h\text{-}\beta}}{\Delta I_{h\text{-}sym} + \Delta I_{h\text{-}\alpha} + \Delta I_{h\text{-}\beta}} \tag{7}$$

Fig. 2 summarizes the algorithm for the calculation of harmonic impedance.

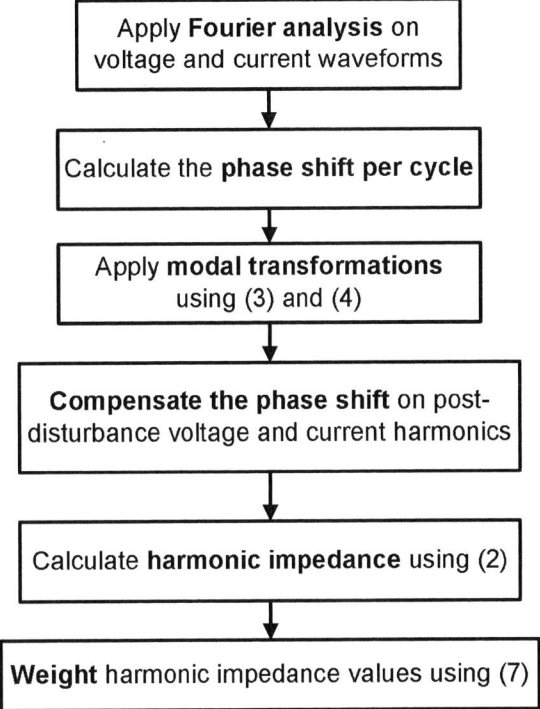

Fig. 2. The diagram of the harmonic impedance calculation.

III. DEVELOPED PROGRAM AND VERIFICATION STUDIES

A. Developed GUI-based Program

Procedures for running the program are in the following orders.

1. Enter number of samples per cycle and click the **"Load Waveform Data and Plot Fundamental Components"** button.
2. Browse the data file which contains three-phase voltages and three-phase currents in the column format. The conversion in Table I can be used to prepare the data file if original voltage waveforms are two line-to-line voltages.
3. The program then produces six plots: voltage and current waveforms, fundamental-frequency RMS voltage and current and fundamental-frequency active and reactive powers.
4. From the fundamental-frequency RMS voltage and current plots, the user selects and enters the pre- and post-disturbance cycles. Typically, the disturbance starts at the cycle where there is a change in the voltage or current magnitudes. The post-disturbance cycle is suggested to be the one

when the steady-state values are rather constant (see cases in Section *III.B*).

5. Click "**Compare Spectra**" button in order to view the spectra comparison of voltage and current between pre- and post-disturbance cycles in the bar chart. Only values from phase A are shown in Figs 4 and 5. The greater the difference, the better harmonic impedances will be.

6. Enter the harmonic order for the harmonic impedance calculation and then click "**Calculate Zh**" button. The harmonic impedance value based on the weighting average method in (7) will be shown with the harmonic current change equals to $(\Delta I_{h-sym} + \Delta I_{h-\alpha} + \Delta I_{h-\beta})/3$.

7. The user can click "**CLEAR**" button to clear plots and results. The user can start a new simulation by repeating Steps 1 to 6.

8. After finishing the simulation, the user can click "**EXIT**" button to exit the program. The three operations of loading data and plotting components, clearing results and closing the program can also be performed via the menu "**File**" on the top left of the main display.

Fig. 3 summarizes the steps for using the harmonic impedance calculation program.

B. Field Measurement Results

Two field measurements were used to test the developed GUI-based program. Two capacitor switching operations were performed at the customer side in order to find the utility-side harmonic impedance. Switching waveforms have been recorded and processed by the harmonic impedance calculation presented in this paper. Figs. 4 and 5 show the execution displays of these two cases. Neglecting the negative sign, case 1 gives $Z_5 = 80.5 + j227.5\ \Omega$ with $\Delta I = 7.8\ \%$ and case 2 gives $Z_5 = 72.9 + j206.8\ \Omega$ with $\Delta I = 8.1\ \%$. Thus, the average of the 5th harmonic impedance of the utility side at the capacitor switching location is about $76 + j217\ \Omega$ [12]. This 5th harmonic impedance was found to be very close to the model used for frequency scan studies.

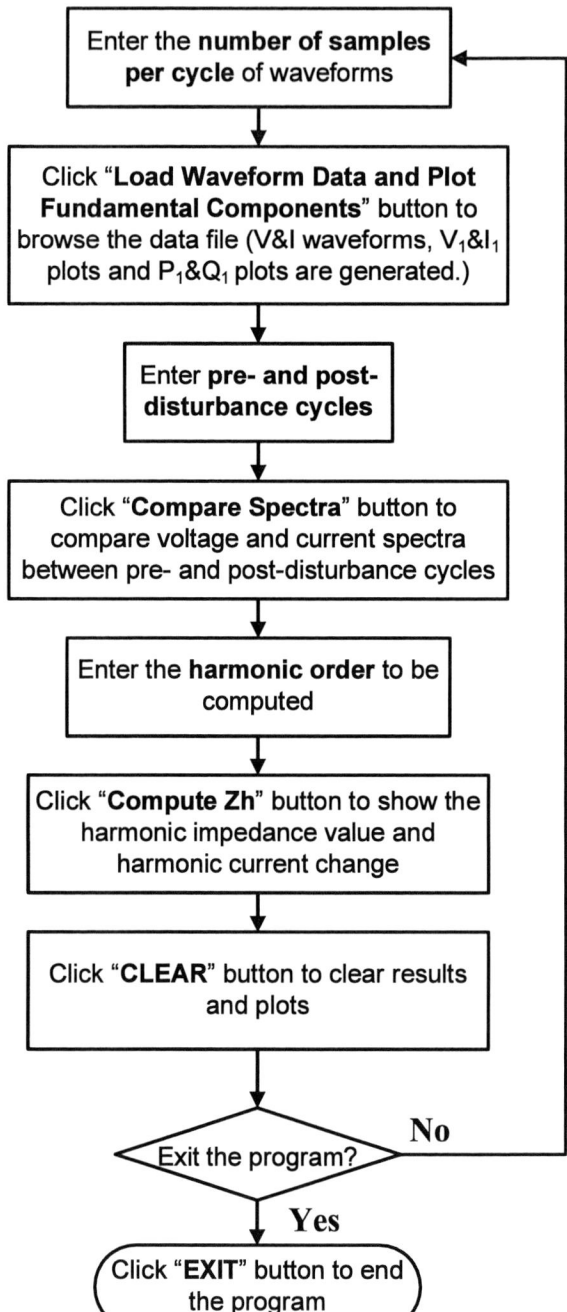

Fig. 3. Steps for using the developed GUI-based program.

IV. CONCLUSION

This paper presents a program for harmonic impedance calculation developed under MATLAB GUI environment. The GUI feature provides the visual interactive capability and facilitates the analysis such as changing the pre- or post-disturbance cycles and checking the sensitivity of results. The user with limited knowledge of the harmonic impedance calculation can easily simulate the program. The calculation of harmonic impedance is based on two steady-state conditions. The application of modal transformations and weighting average has been discussed. The developed program has been tested using two actual field measurements.

REFERENCES

[1] A. Robert and T. Deflandre, "Guide for Assessing the Network Harmonic Impedances," CIGRE 36.05, Working Group CC02 Rep., Mar. 1993.

[2] J. Bergeal and L. Moller, "Analysis of the spectrum impedance of a network, use of digital methods," in *Proc. Int. Conf. Electricity Distribution (CIRED)*, 1983, Paper c.05.

[3] R. Gretsch and R. Weber, "Disturbances on a medium-voltage supply system caused by harmonics, measurement, computer simulation and remedial measures," in *Proc. Int. Wroclaw Symp. Electromagnetic Compatibility (EMC'88)*, 1988, pp. 715–719.

[4] E. Duggan and R. E. Morrison, "A noninvasive technique for the measurement of power system harmonic impedance," in *Proc. Int. Conf. Electricity Distribution (CIRED)*, 1991, Paper 2.02.

[5] A. de Oliveira, J. C. de Oliveira, J. W. Resende, and M. S. Miskulin, "Practical approaches for AC system harmonic impedance measurements," *IEEE Trans. Power Delivery*, vol. 6, pp. 1721–1726, Oct. 1991.

[6] Y. Xiao, J.C. Maun, H. B. Mahmoud, T. Detroz and S. Do, "Harmonic Impedance Measurement using Voltage and Current Increments from Disturbing Loads," *9th IEEE International Conference on Harmonics and Quality of Power (ICHQP)*, Oct. 1-4, 2000, Florida, USA, pp. 220-225.

[7] W. Xu, E. E. Ahmed, X. Zhang and X. Liu, "Measurement of Network Harmonic Impedances: Practical Implementation Issues and Their Solutions," *IEEE Trans. Power Delivery*, Vol. 17, No. 1, pp. 210-216, Jan. 2002.

[8] *Creating Graphical User Interface*, MATLAB 7, Mathworks, 2007, Available: http://www.mathworks.com.

[9] K. S. Prakash, O. P. Malik, and G. S. Hope, "Amplitude Comparator Based Algorithm for Directional Comparison Protection of Transmission Lines," *IEEE Trans. Power Delivery*, Vol. 4, No. 4, Oct. 1989, pp. 2032-2041.

[10] E. Ahmed, W. Xu, and X. Liu, "Application of Modal Transformations for Power System Harmonic Impedance Measurement," *Electrical Power and Energy Systems*, Vol. 23, Issue 2, Feb. 2001, pp. 147-154.

[11] M. Nagpal, W. Xu, and J. H. Sawada, "Harmonic Impedance Measurement using Three-phase Transients," *IEEE Trans. Power Delivery*, Vol. 13, No.1, Jan 1998, pp. 272-277..

[12] C. Li, W. Xu and T. Tayjasanant, "A Critical Impedance Based Method for Identifying Harmonic Sources," *IEEE Trans. Power Delivery*, Vol. 19, No.2, April 2004, pp. 671-678.

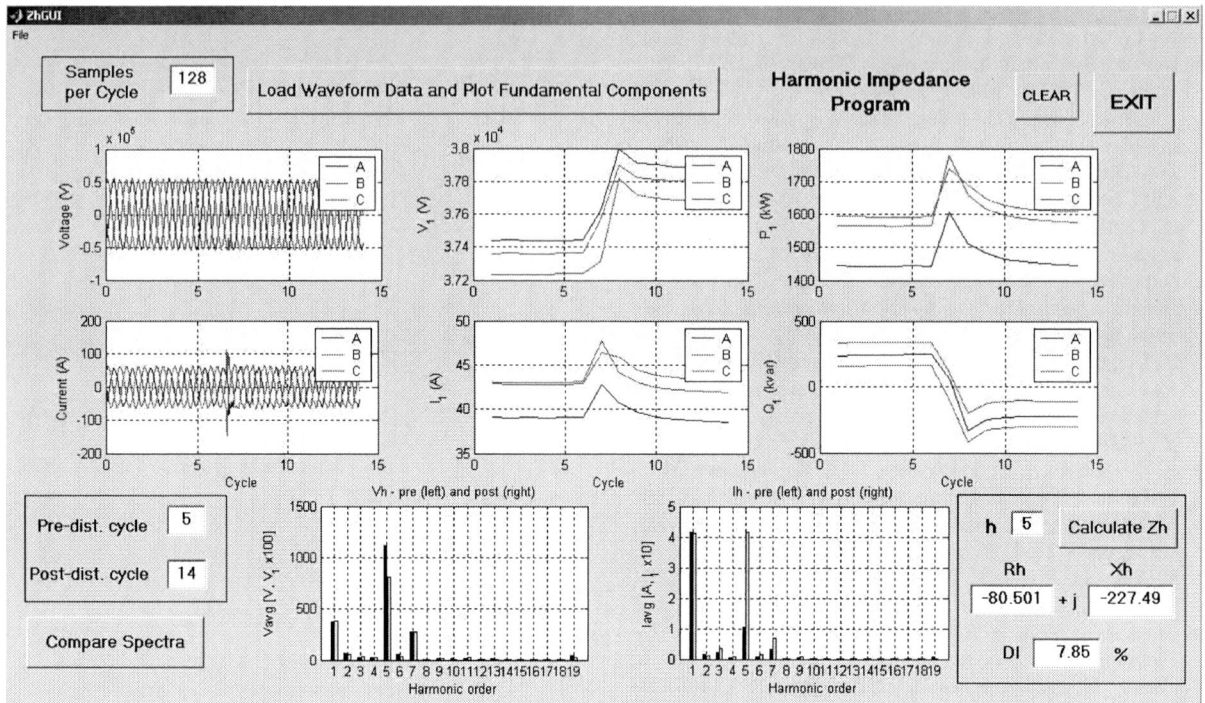

Fig. 4. The execution display for field measurement Case 1.

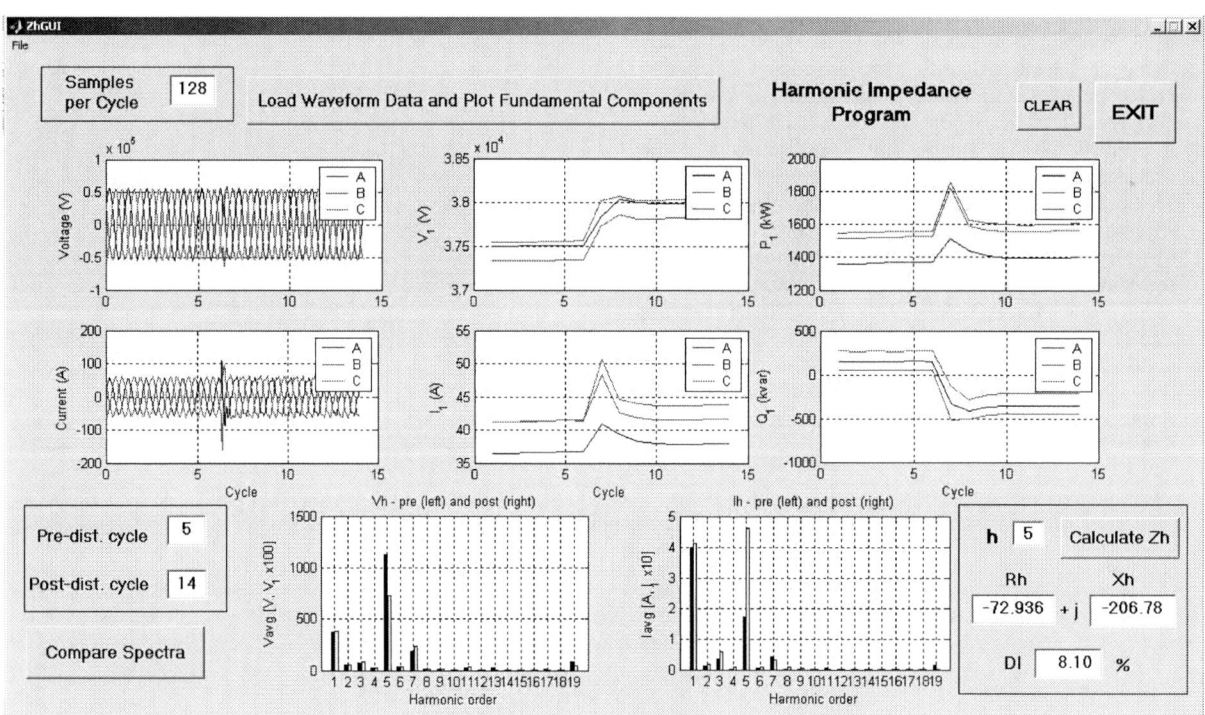

Fig. 5. The execution display for field measurement Case 2.

The analysis and simulation of power circuits for AC high-voltage converters

Y.Y. Skorokhod*, S.I. Volskiy**

* Transconverter LTD, Moscow, Russia
** Moscow Aviation Institute (State Technical University), Moscow, Russia

Abstract — In this paper different AC high-voltage auxiliary converter for railway transport power wire schemes and functioning diagrams are presented. Also here have been carried out estimations of power losses in AC high-voltage auxiliary converters according for the different power modes. Here is presented the problem of losses influence in IGBT-transistors on commutation frequency of AC high-voltage auxiliary converter. Comparative analysis of power losses in AC high-voltage auxiliary converters is executed according to different power diagrams and IGBT-transistors of different classes is carried out.

So being guided by above mention factors optimum AC high-voltage auxiliary converters power diagrams have been obtained taking in consideration power semiconductors elements modern state.

Index Terms — high-voltage auxiliary converter, IGBT-transistors, power losses, step-up converters.

I. INTRODUCTION

In Russia and Europe railway transport traction power nets main voltage parameters are DC current voltage at nominal value 3000 V deviating in the range from 2200 V to 4000 V and at the same time AC current single-phase voltage at nominal 25 kV, 50 Hz deviating in the range from 19 kV to 29 kV [1, 2]. And it should be noted that for main AC power net feeding on board of rolling-stock is realized by using extension lead of traction transformer at voltage value 2200 – 3600 V or 1150 – 1885 V. But for the sake of feeding auxiliary equipment at rolling-stock is used as a rule AC three-phase voltages of 380 V, 220 V, 50 Hz value and DC voltage of 110 V value. Hence for the sake of converting traction voltage into auxiliary voltage it is necessary to employ high-voltage auxiliary converters.

Following requirements have been referred to as most important:
- wide range of high input voltage RMS value (within 1150-1885 V);
- reduction of power converter performance at current harmonic composition and power factor of the input power network;
- high robustness and electromagnetic compatibility;
- high efficiency and reliability, simplicity of electrical and mechanical design, modular principle of assembly;
- acceptable mass and dimensions indices.

It is generally known that utilizing principle of electrical energy high frequency conversion in power converters is effective and attractive way. It make possible improving of the consumed current harmonic composition and power factor of the input power network increasing. Besides due to this principle it becomes possible to reduce mass and dimensions of converters. This is why the principles proposed are widely used in the power converters of different types.

II. AC HIGH-VOLTAGE AUXILIARY CONVERTER POWER WIRE SCHEME

AC high-voltage auxiliary converter power charts synthesis can be realized by different means.

One of them presupposes including single link high-frequency stabilizer of input power net into AC high-voltage auxiliary converter structure. Such approach was realized in the AC high-voltage auxiliary converter represented at Fig. 1. It consists of the following components:
- input choke (L1);
- single link high-frequency stabilizer of input power net (A1);
- intermediate step-down high-frequency transformer (TV1);
- intermediate voltage rectifier (A2);
- three-phase voltage source inverter (A3);
- output sinus filter (A4).

Input choke L1 in combination with high-voltage step-up converter of high-frequency stabilizer of input power net provides power net current filtering at AC high-voltage auxiliary converter input.

High-frequency stabilizer of input power net A1 transforms AC single-phase voltage of 1150 – 1885 V value, 50 Hz into stabilized high-frequency intermedeate voltage 4200 V.

At the same time high-frequency stabilizer of input power net minimizes undesirable input current harmonic components it also can fulfills functions of power factor corrector.

Intermediate voltage rectifier A2 provides rectification of high-frequency intermediate voltage.

Three-phase voltage source inverter A3 transforms DC voltage from rectifier A2 output into three-phase stabilized PWM-modulated AC voltage which has RMS of basic harmonic value equal to 380 V and basic frequency 50 Hz.

978-1-4244-0644-9/07/$25.00 ©2007 IEEE

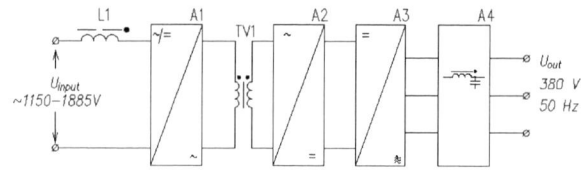

Fig. 1. Single link AC high-voltage auxiliary converter block scheme.

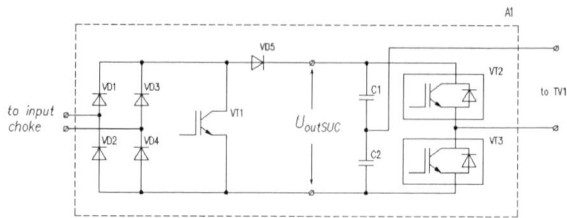

Fig. 2. Single link high-frequency stabilizer of input power net power wire scheme.

Output sinus filter A4 smoothes down PWM-voltage value of three-phase voltage source inverter. Filter output is connected directly to the load.

High-frequency stabilizer of input power net power wire scheme synthesis can also be realized by different means. First of all in the course of choosing high-frequency stabilizer of input power net power wire scheme it is necessary to take into account performance parameters of modern semiconductors devices. And it is necessary to keep in mind that one of the principal criteria for choosing chart configuration is losses level in semiconductor devices.

Single link high-frequency stabilizer of input power net power wire scheme version based at IGBT-transistors rated at collector-emitter voltage value 6500 V [3] is represented at Fig. 2. This chart includes following components: rectifier composed of diodes VD1…VD4, step-up voltage converter composed semiconductor devices VT1, VD5, capacitors C1, C2 and high-frequency inverter composed of IGBT-transistors VT2, VT3.

Four link AC high-voltage auxiliary converter version including high-voltage step-up converter based at IGBT-transistors rated at collector-emitter voltage value 1700 V [2] is represented at Fig. 3.

High-frequency stabilizer of input power net in the chart shown consists of four connected in series similar units analogous in the design to single link unit A1 represented at Fig. 2.

Decision concerning quantity of links has been taken depending on AC high-voltage auxiliary converter input voltage, collector-emitter voltage of IGBT-transistors and principle of connected in series transistors choice:

$$N_{links} > \frac{U_{outSUC}}{V_{CESVT} \cdot k}$$

where N_{links} - quantity of connected in series units (whole number);

U_{outSUC} - output voltage of step-up voltage converter;

V_{CESVT} - maximum IGBT-transistor collector-emitter voltage in turned off state;

k - factor, which determines IGBT-transistor functioning with required reliability at U_{outSUC} value set (usually assumed not more then 0,67).

For the sake of comparative estimation of the AC high-voltage auxiliary converter structural block schemes (Fig. 1 and Fig. 3) analysis of losses in semiconductor devices of single link and four links high-frequency stabilizer of input power net has been completed at the mode of power supply by input AC voltage in the range of 1150 – 1885 V, 50 Hz.

III. LOSSES LEVEL IN HIGH-VOLTAGE STEP-UP VOLTAGE CONVERTER SEMICONDUCTOR DEVICES

A. Mathematical techniques of losses level calculation in semiconductor devices

Modern power IGBT-transistors and diodes has comparatively small static losses in turned off condition. At the same time power diodes has small dynamic losses in the course of turning off the device and especially in the course of turning it in. Taking in account these facts in the course of analysis it should be assumed that static losses of IGBT-transistors and diodes in the turned off condition are equal to zero.

According to above stated assumptions static losses in IGBT-transistor and diodes are determined by the following expressions:

$$S_{VT} = f_{VT} \cdot \int_{0}^{t_{pVT}} i_{VT} \cdot v_{CE(sat)VT}(i_{VT})dt \qquad (1)$$

$$S_{VD} = f_{VD} \cdot \int_{0}^{t_{pVD}} i_{VD} \cdot v_{FVD}(i_{VD})dt \qquad (2)$$

where f_{VT} and f_{VD} - transistors and diodes commutations frequency;

i_{VT} - current flowing through transistor when it is in turned on condition;

i_{VD} - current flowing through diode when it is in open condition;

$v_{CE(sat)VT}(i_{VT})$ - transistor saturation voltage at i_{VT} current value;

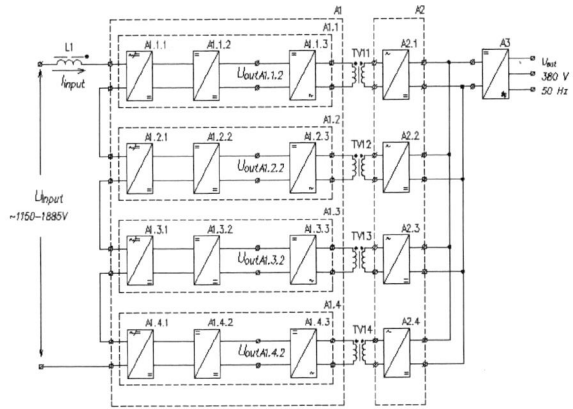

Fig. 3. Four links AC high-voltage auxiliary converter structural block scheme.

$v_{FVD}(i_{VD})$ - turned on diode condition at i_{VD} current value;

t_{pVT} and t_{pVD} - turned on IGBT-transistor and diode condition duration.

IGBT-transistor dynamic losses are determined by the following equation [3, 4]:

$$P_{VT} = \left(\frac{E_{onVT} \cdot U_{onVT}}{V_{CCVT}} + \frac{E_{offVT} \cdot U_{offVT}}{V_{CCVT}} \right) \cdot f_{VT} \qquad (3)$$

where E_{onVT}, E_{offVT} - energy of losses at IGBT-transistor turning on and turning off measured at the currents I_{onVT} and I_{offVT} value and voltage V_{CCVT} value;

U_{onVT}, U_{offVT} - values of voltages at which transistor turns on and turns off;

I_{onVT}, I_{offVT} - values of currents at which transistor turns on and turns off;

V_{CCVT} - transistor voltage value at which power losses values E_{onVT}, E_{offVT} are presented in reference sources.

B. Single link high-frequency stabilizer of input power net losses in semiconductor devices

For analysis execution the next assumption are assumed:
- load is presented by pure active resistance;
- AC high-voltage auxiliary converter RMS input current value during period of output voltage at frequency 50 Hz does not change;
- step-up voltage converter and high-frequency inverter input current and output voltage fluctuations are negligible small;
- calculation is carried out in uninterruptable current mode;

- losses in reverse diodes of IGBT-transistors in high-frequency inverter were not taken in account.

In the course of losses calculation in single link high-voltage step-up converter unit A1 it is necessary to take into account that at AC current mode input current I_{in} during half period of input voltage leaks only through diodes VD1,VD4 or VD2, VD3. And it also should be mentioned that calculation of the total losses in diodes VD1 and VD4 (VD2 and VD3) is fulfiled according expression (4) which was deduced on the base of expression (2):

$$S_{VD1,4} = 2 \cdot \frac{Q \cdot V_F(I_{in})_{VD1}}{\eta \cdot U_{in}} \qquad (4)$$

where Q - AC high-voltage auxiliary converter output power;

$V_F(I_{in})_{VD1}$ - voltage drop at turned on diodes VD1, VD4 average current value I_{in} in accordance to AC high-voltage auxiliary converter output voltage period;

η - AC high-voltage auxiliary converter efficiency factor;

U_{in} - AC high-voltage auxiliary converter input voltage half-period average value.

Half-period average value I_{in} of the AC high-voltage auxiliary converter input current is determined by the expression:

$$I_{in} = \frac{Q}{\eta \cdot U_{in}} \qquad (5)$$

Principle of single link step-up converter functioning is discussed in detail in the work [5].

Dependence of step-up voltage converter output voltage of the input voltage is determined by expression:

$$U_{outSUC} = \frac{U_{in}}{1 - \gamma_{VT1}} \qquad (6)$$

where $\gamma_{VT1} = \dfrac{t_{pVT1}}{T_{VT1}}$ - duty factor of transistor VT1;

t_{pVT1} - transistor VT1 turned on condition duration;

$T_{VT1} = \dfrac{1}{f_{VT1}}$ - transistor VT1 commutation period;

f_{VT1} - transistor VT1 commutation frequency.

For static losses calculation in the step-up voltage converter the following expressions where deduced taking in account expressions (1), (2), (6):

$$S_{VT1} = \frac{Q \cdot V_{CE(sat)VT1}(I_{VT1})}{U_{in} \cdot \eta} \cdot \gamma_{VT1} \qquad (7)$$

$$S_{VD5} = \frac{Q \cdot V_{FVD5}(I_{VD5})}{U_{in} \cdot \eta} \cdot (1 - \gamma_{VT1}) \qquad (8)$$

where $V_{CE(sat)VT1}(I_{VT1})$ - transistor VT1 saturation voltage at mean current value I_{VT1} in the duration of turned on condition;

$V_{FVD5}(I_{VD5})$ - turned on diode VD5 voltage drop at mean current value I_{VD5} in the duration of turned on condition.

In the step up voltage converter transistor VT1 and diode VD5 middle value current in the duration of turned on condition is equal to input current middle value:

$$I_{VT1} = I_{VD5} = I_{in} = \frac{Q}{U_{in} \cdot \eta} \qquad (9)$$

In order to find out transistor VT1 dynamic losses there was obtained expression (10) according to expression (3):

$$P_{VT1} = \left(E_{onVT1} + E_{offVT1} \right) \cdot \frac{U_{outSUC} \cdot f_{VT1}}{V_{CCVT1}} \qquad (10)$$

where f_{VT1} - transistor VT1 functioning frequency;

E_{onVT1}, E_{offVT1} - transistor VT1 energy of losses during turning on and turning off and measured at I_{VT1} and V_{CCVT1} value;

V_{CCVT1} - transistor VT1 voltage value at which E_{onVT1}, E_{offVT1} is given in reference data books.

High frequency inverter transistors losses calculation is possible to perform using expressions (11) and (12) and taking in account expressions (2) and (3):

$$S_{VT2,3} = 2 \cdot S_{VT2} = \frac{2 \cdot Q \cdot V_{CE(sat)VT2}(I_{VT2})}{U_{outSUC} \cdot \eta} \qquad (11)$$

$$P_{VT2,3} = 2 \cdot P_{VT2} =$$
$$= \left(E_{onVT2} + E_{offVT2} \right) \cdot \frac{U_{outSUC} \cdot f_{VT2}}{V_{CCVT2}} \qquad (12)$$

where $V_{CE(sat)VT2}(I_{VT2}) = V_{CE(sat)VT3}(I_{VT3})$ - transistors VT2 and VT3 saturation voltage in the course

of duration of turned on condition at the current $I_{VT2} = I_{VT3}$ mean value;

$E_{onVT2} = E_{onVT3}$ and $E_{offVT2} = E_{offVT3}$ - power of losses at transistors VT2 and VT3 turning on and turning off and measured at the current $I_{VT2} = I_{VT3}$ value and the voltage $V_{CCVT2} = V_{CCVT3}$ value;

$V_{CCVT2} = V_{CCVT3}$ - transistor VT1 voltage value at which $E_{onVT2} = E_{onVT3}$ and $E_{offVT2} = E_{offVT3}$ is given in reference data books;

$f_{VT2} = f_{VT3}$ - transistors VT2 and VT3 commutation frequency.

Mean I_{VT2} current value during turned on condition is determined by expression:

$$I_{VT2} = I_{VT3} = \frac{2 \cdot Q}{U_{outSUC} \cdot \eta} \qquad (13)$$

C. Four links high-frequency stabilizer of input power net losses in semiconductor devices

High-frequency stabilizer of input power net power chart (Fig. 3) make it possible to realize two IGBT-transistor control algorithms asynchronous and synchronous (Fig. 4).

Asynchronous control mode is most perspective because it permits to reduce four links AC high-voltage auxiliary converter IGBT-transistors frequency of functioning at least four times. At the same time values of input choke and input current fluctuation amplitude does not change. This principle is thoroughly examined in the papers [6, 7].

Taking these circumstances in account only asynchronous control mode should be considered.

Power losses in A1.1.1...A1.4.1 units taking in account simultaneous current flow through eight diodes may be determined using following expression:

$$S_{A1.1.1...4} = 8 \cdot \frac{Q \cdot V_{FVD}(I_{in})}{\eta \cdot U_{in}} \qquad (14)$$

where $V_{FVD}(I_{in})$ - voltage drop at turned on diode when mean input current I_{in} flows through it. Value of I_{in} current is determined by expression (5).

Four links high-frequency stabilizer of input power net A1.1.2...A1.4.2 control characteristic concerning asynchronous mode is determined by the following expression:

$$U_{outA1.1.2} = ... = U_{outA1.4.2} = \frac{U_{in}}{4} \cdot \frac{1}{1 - \gamma_{VT1}} \qquad (15)$$

1871

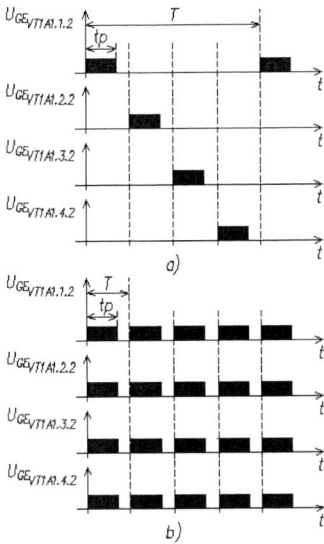

Fig. 4. Step-up voltage converter transistors asynchronous a) and synchronous b) control algorithms.

where $\gamma_{VT1} = \gamma_{VT1A1.1.2} = \ldots = \gamma_{VT1A1.4.2} = \dfrac{t_{pVT1}}{T_{VT1}}$ - duty factor identical for all A1.1.2...A1.4.2 units transistors.

For finding out static losses in AC high-voltage auxiliary converter IGBT-transistors and diodes there were deduced expressions:

$$S_{VT1A1.1.2...A1.4.2} = 4 \cdot S_{VT1A1.1.2} =$$
$$= \frac{4 \cdot Q \cdot V_{CE(sat)VT1}(I_{VT1})}{U_{in} \cdot \eta} \cdot \gamma_{VT1} \qquad (16)$$

$$S_{VD5A1.1.2...A1.4.2} = 4 \cdot S_{VD5A1.1.2} =$$
$$= \frac{4 \cdot Q \cdot V_{FVD5}(I_{VD5})}{U_{in} \cdot \eta} \cdot (1 - \gamma_{VT1}) \qquad (17)$$

where $V_{CE(sat)VT1}(I_{VT1})$ - transistor VT1 saturation voltage value at the current I_{VT1} during duration of turned on condition;

$V_{FVD5}(I_{VD5})$ - voltage drop at diode VD5 at turned on condition with current average value I_{VD1} which exist during turned on condition duration.

In this case currents I_{VT1} and I_{VD5} values are equal to mean AC high-voltage auxiliary converter current $I_{VT1} = I_{VD5} = I_{in}$ which is determined according to expression (5).

Dynamic losses in step-up voltage converter IGBT-

transistors and diodes are determined by expression:

$$P_{VTA1.1.2...A1.4.2} = 4 \cdot P_{VT1A1.1.2} =$$
$$= 4 \cdot \left(E_{onVT1} + E_{offVT1} \right) \cdot \frac{U_{outA1.1.2} \cdot f_{VT1}}{V_{CCVT1}} \qquad (18)$$

where $U_{outA1.1.2}$ - high-voltage step-up converter unit A1.1.2 output voltage.

Static and dynamic losses in high-frequency inverter four blocks A1.1.3...A1.4.3 can be calculated using expressions (19) and (20) analogous to expressions (11) and (12):

$$S_{VT2,3A1.1.3...A1.4.3} = 4 \cdot S_{VT2,3A1.1.3} =$$
$$= \frac{2 \cdot Q \cdot V_{CE(sat)VT2}(I_{VT2})}{U_{outA1.1.2} \cdot \eta} \qquad (19)$$

$$P_{VT2,3A1.1.3...1.4.3} = 4 \cdot P_{VT2A1.1.3} =$$
$$= \left(E_{onVT2} + E_{offVT2} \right) \cdot \frac{4 \cdot U_{outA1.1.2} \cdot f_{VT2}}{V_{CCVT2}} \qquad (20)$$

where $V_{CE(sat)VT2}(I_{VT2})$ - transistor VT2 saturation voltage at mean current I_{VT2} during turned on condition.

Transistor VT2, VT3 mean current I_{VT2}, I_{VT3} during turned on condition is determined by expression:

$$I_{VT2} = I_{VT3} = \frac{Q}{2 \cdot \eta \cdot U_{outA1.1.2}} \qquad (21)$$

IV. ANALYSIS OF SINGLE LINK AND FOUR LINKS HIGH-FREQUENCY STABILIZER OF INPUT POWER NET LOSSES

Below are represented results of power losses calculation in the high-frequency stabilizer of input power net having output power 80 kVA in case it is supplied by 1150 – 1885 V AC voltage source.

During calculation of losses in case high-frequency stabilizer of input power net asynchronous control mode transistor VT1 commutation frequency f_{VT1} (at collector-emitter voltage of units A1.1.2...A1.4.2 at 1700 V value in the four links high-voltage step-up converter) was taken four times lower than commutation frequency of transistor VT1 of A1 unit in single chain high-frequency stabilizer of input power net (collector-emitter voltage 6500 V). This circumstance made it possible to equalize both charts concerning the inductance of the input choke value and input current fluctuation value.

Dependence of static losses in one link and four links

high-frequency stabilizer of input power net on AC high-voltage auxiliary converter input voltage is presented at Fig. 5. At these figures can be seen that static power losses in four links high-frequency stabilizer of input power net is practically three times higher than in single link high-frequency stabilizer of input power net.

Dependence of dynamic losses power in semiconductor devices of single link and four links high-frequency stabilizer of input power net on transistors commutation frequency at input voltage values 1150 V and 1885 V is presented it Fig. 6.

It is possible to see at these charts that dynamic power losses in single link high-frequency stabilizer of input power net is much higher (2 - 10 times depending on commutation frequency) then power losses in four links high-frequency stabilizer of input power net taking in account that input choke inductor and input current fluctuation amplitude are of the same value.

It depends on higher (more then 12 times higher) power losses at 65-class IGBT-transistors then 17-class transistors losses at the process of commutation.

We can also see at the Fig. 6 that highest dynamic losses take place at AC input voltage value 1150 V. This happens because increment of the current value in the process of commutation has higher influence at increment of dynamic losses than increment of commutated voltage value has.

And it should be taken into account that for the IGBT-transistors designed at 6500 voltage value load this phenomena is peculiar in higher degree than to IGBT-transistors designed at 1700 voltage value load.

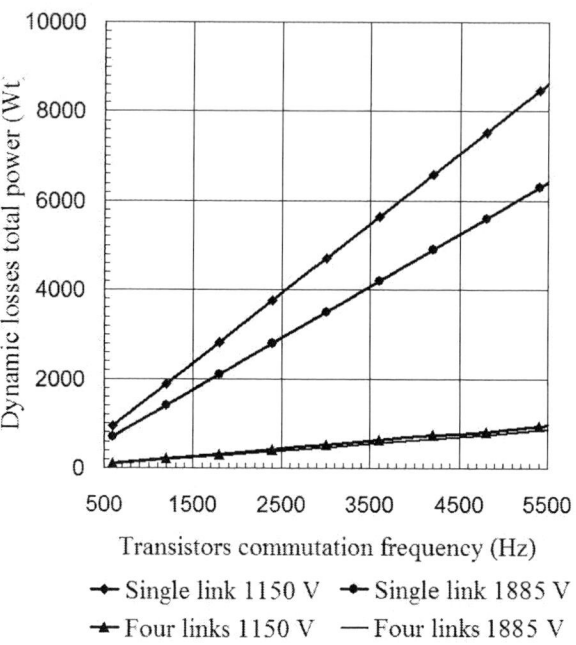

Fig. 6. Dependence of dynamic power losses in single link and four links high-frequency stabilizer of input power net on IGBT-transistors commutation frequency.

Total static and dynamic losses in single link and four links high-frequency stabilizer of input power net depending on commutation frequency of IGBT-transistors at AC input voltage 1150 V are shown at Fig.7. At this picture can be clearly seen that at the commutation frequency about 700 Hz and higher an obvious advantage of four links high-frequency stabilizer of input power net over single link high-frequency stabilizer of input power net is displayed in the area of total losses in semiconductor devices.

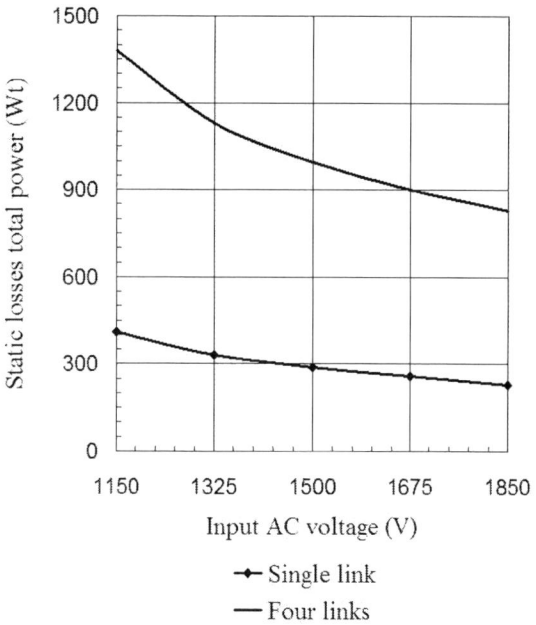

Fig. 5. Dependence of the total power static losses in the high-frequency stabilizer of input power net semiconductor devices of single link and four links on AC high-voltage auxiliary converter.

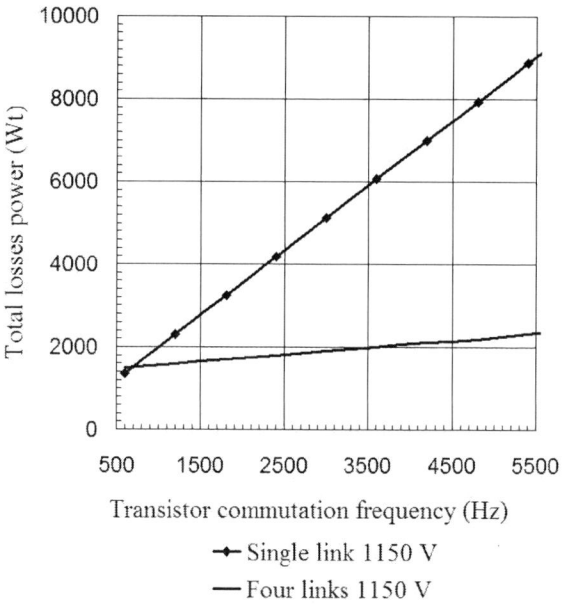

Fig. 7. Dependence of total losses in semiconductor devices of single link and four links high-frequency stabilizer of input power net of IGBT-transistors commutation frequency.

This phenomenon can be explained by more extensive growth of dynamic losses power along IGBT-transistors commutation frequency increase in single link high-frequency stabilizer of input power net in comparison to high-frequency stabilizer of input power net.

V. CONCLUSIONS

Comparative analysis of power losses in semiconductor devices between single link and four links high-frequency stabilizer of input power net revealed:

- single link high-frequency stabilizer of input power net has approximately three times lower static semiconductor losses in comparison with four links high-frequency stabilizer of input power net along the whole range of input voltage value due to smaller number connected in series semiconductor devices;

- single link high-frequency stabilizer of input power net has much more higher dynamic losses of semiconductor devices in comparison with four links high-frequency stabilizer of input power net due to much more higher IGBT-transistors at the collector-emitter voltage value 6500 V commutation power than IGBT-transistors at the collector-emitter voltage value 1700 V.

Power losses ratio of single link and four links high-frequency stabilizer of input power net increases practically proportionally to transistor commutation frequency increase.

Use of four links high-frequency stabilizer of input power net is much more preferable in comparison to single link high-frequency stabilizer of input power net at transistors commutation frequency higher than 700 Hz due to much more lower dynamic power losses of low voltage IGBT-transistors in comparison to high-voltage IGBT-transistors.

REFERENCES

[1] Sleptsov M.A., Savina T.I. Electric transport power supply. Moscow, *MPEI edition*, 2001 year, 48 pages.

[2] GOST 6962-75 Electric powered transport supplied by contact net, 1975.

[3] http://www.infineon.com/. 1999 - 2007 *Infineon Technologies AG.*

[4] Power Electronics. Modules. Drivers. Systems. *SEMIKRON International GmbH.* 05/2005.

[5] B.W. Williams. Power Electronics. Devices, drivers and applications. *Macmillan, London,* 1987.

[6] V.Y. Shergin, Y.Y. Skorokhod, S.I. Volsky. The analysis and simulation of power circuits for high voltage converter. *Conference Proceedings PCIM'2007.*

[7] S.I. Volsky, Y.Y. Skorokhod, V.Y. Shergin. The analysis and simulation of power circuits for high voltage converter. *Conference Proceedings IPEMC'2006.*

A Single Stage Flyback PFC Converter for Testing Distance Relay Systems

V. Fernão Pires [*+], J. F. Martins [*+], and J. Fernando Silva [#]

* Escola Superior Tecnologia Setúbal / Instituto Politécnico Setúbal, Setúbal , Portugal
[#] Instituto Superior Técnico / Universidade Técnica de Lisboa, Lisboa, Portugal
[+] LabSEI – Laboratório de Sistemas Eléctricos Industriais, Setúbal, Portugal

Abstract–In this paper, a single stage flyback pfc converter is used as a power supply of a testing distance relay system. The flyback pfc converter topology, operation and control are presented. A fast and robust sliding mode control of the input current is also presented. The sliding mode controller enables also high power factor and nearly sinusoidal input current. The dc output voltage of the converter is controlled by a proportional integral (PI) controller. Simulation and experimental results from a laboratory prototype are presented and discussed.

Index Terms—High power factor, Sliding mode controller, Single stage flyback PFC Converter, Testing Distance Relay Systems.

I. INTRODUCTION

High-power-factor rectifiers are receiving vast attention during the last years. These power converters can be divided in two groups [1]. The first group employs a two-stage power factor corrector (pfc) topology, where an ac/dc front end converter that is used to shape the line current is followed by a dc/dc stage to regulate the dc output voltage. The drawbacks of the two-stage approach are more components count and low overall conversion efficiency and reliability. The second group, single-stage PFCs, is advantageous since it uses only a single-stage converter to perform current shaping and output voltage control [2-6].

Most pfc converters requiring isolation between the ac and the dc sides have pointed out the flyback structure. This structure is convenient for building up switch-mode power supplies with good harmonic regulation for small and medium power levels [7,8], owing to the simplicity of its structure, to the existence of a high-frequency isolation between the input and output stages, and to the ability of multiple-output stages and to the variety of the output DC voltage values that can be obtained.

In this work, a single-stage pfc converter for testing distance relay systems is proposed. This test system requires isolation between the ac and the dc sides, as well as two different ac output voltages.

This paper is organized into four sections. The first one is this introduction. In section II it is presented the proposed single stage flyback pfc converter as power supply of a testing distance relay system. The system controllers of the proposed power supply are described in section III. In section IV are presented several simulation and experimental results. The results are obtained using a Matlab/Simulink-based simulator. Finally, in section IV the conclusions of the work are synthecized.

II. PROPOSED SINGLE-STAGE PFC FLYBACK CONVERTER

The testing system of distance relay ensures that this relay types will have a proper operation. So, their operating characteristics should be tested The developed testing system is composed by a computer equipment, controller I/O interface and power amplifiers. In the computer equipment there is a simulator that provides several choices for the fault type. According the fault type it is generated the test signals that are converted in current and voltages references for the power amplifiers controllers.

The power amplifier system is composed by two four wire three-phase DC/AC converters (Fig. 1). One of the converters is used as a voltage amplifier, while the other is used as a current amplifier. The DC power supply voltages of the voltage and current amplifiers are different, since the three-phase voltage converter requires high voltages at very low output currents, while the current converter requires very low voltages at high output currents [9].

For the power supply of the amplifier system it was required isolation between the ac and the dc sides, as well as two different ac output voltages. So, considering the required characteristics of the power supply, a single-stage pfc flyback converter providing two output voltages was adopted (Fig. 2).

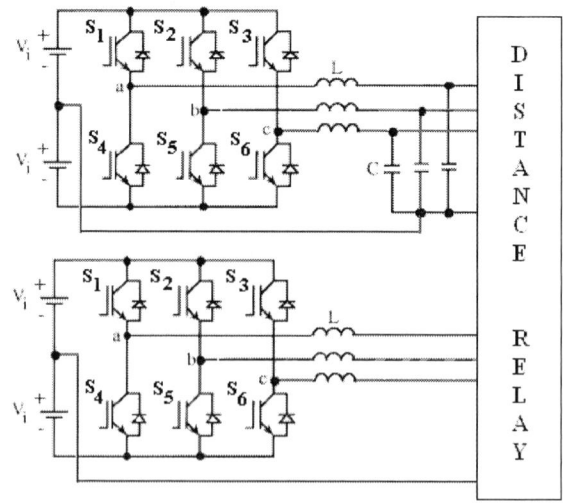

Fig. 1. Testing distance relay system.

978-1-4244-0644-9/07/$25.00 ©2007 IEEE

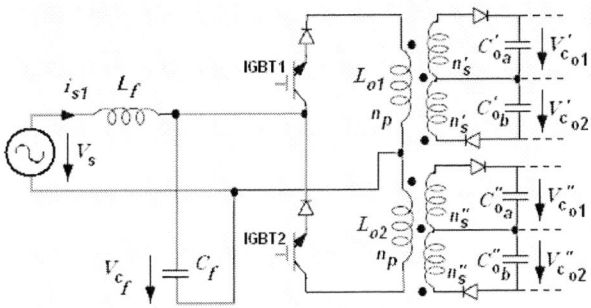

Fig. 2. Proposed single-stage pfc flyback converter.

One of the advantages of this topology is that the rectifier equilibrates the output capacitor voltages due to the magnetic coupling between the transformer windings.

III. CONTROL OF THE PFC FLYBACK CONVERTER

To obtain a suitable dynamic model, let us assume zero losses in the inductors, capacitors and power semiconductors. The states of the two IGBT plus diode switches can be represented by the time dependent variables α_1 and α_2, defined as:

$$\alpha_i = \begin{cases} 1 & , Switch\ i\ is\ on \\ 0 & , Switch\ i\ is\ of \end{cases} , \quad i = 1,2 \tag{1}$$

By circuit analysis of Fig. 2 (using the displayed state variables) a simplified switched state-space model of the rectifier can be obtained:

$$
\begin{cases}
\dfrac{di_s}{dt} = -\dfrac{R_f}{L_f} i_s - \dfrac{1}{L_f} v_{C_f} + \dfrac{1}{L_f} v_s \\[2mm]
\dfrac{dv_{C_f}}{dt} = \dfrac{1}{C_f} i_s - \dfrac{\alpha_1 \beta_1}{C_f} i_{L_{o1}} + \dfrac{\alpha_2 \beta_2}{C_f} i_{L_{o2}} \\[2mm]
\dfrac{d\psi_o}{dt} = (\alpha_1 \beta_1 - \alpha_2 \beta_2) v_{C_f} - \\[2mm]
\qquad \alpha_1 \beta_1 \dfrac{n_p}{n_s'} v_{C_{o1}}' - \alpha_2 \beta_2 \dfrac{n_p}{n_s'} v_{C_{o2}}' \\[2mm]
\dfrac{dv_{C_{o1}}}{dt} = \dfrac{1}{C_{o_a}'} i_{L_{o1}} - \dfrac{n_p}{n_s' R_o C_{o_a}'} v_{C_{o1}}' \\[2mm]
\dfrac{dv_{C_{o2}}}{dt} = \dfrac{1}{C_{o_b}'} i_{L_{o2}} - \dfrac{n_p}{n_s' R_o C_{o_b}'} v_{C_{o2}}'
\end{cases}
\tag{2}
$$

Using the dynamics of (2), it is possible to obtain de expressions for the input current and output voltage controller. To design a robust current controller, which responds to the output voltage regulation and sinusoidal supply current waveshaping, a sliding mode control technique will be used. Considering the i_s input line current as the controlled output, from (2) the state space equations in the controllability canonical form can be obtained. In fact, making:

$$\theta = \frac{v_s - R_f\, i_s - v_{C_f}}{L_f} \tag{3}$$

The state space equations in the controllability canonical form are.

$$\frac{d}{dt}\begin{bmatrix} i_s \\ \theta \end{bmatrix} = \begin{bmatrix} \theta \\[2mm] -\dfrac{R_f}{L_f}\theta - \dfrac{1}{L_f C_f} i_s + \\[2mm] +\dfrac{\omega}{L_f} V_{s\max} \cos(\omega t) - \dfrac{\alpha}{L_f C_f} i_{Lo} \end{bmatrix} \tag{4}$$

From these new state space equations, it can be concluded that is current has a strong relative degree of two [10] (since only its second time derivative contains the control variable). Therefore, it can be concluded that the sliding surface ensuring the robustness of the closed loop system [11], is:

$$S\left(e_{i_s}, e_\theta\right) = \left(i_{sref} - i_s\right) + k\left(\theta_{ref} - \theta\right) \tag{5}$$

where k is the parameter related to the time constant of the desired first order response of input source current is ($k{>}0$). From (3) and (5), the sliding surface is:

$$
S\left(e_{i_s}, e_\theta\right) = \left(i_{sref} - i_s\right) + k\,\frac{di_{sref}}{dt} - \\
- \frac{k}{L_f}\left(v_s - R_f i_s - v_{C_f}\right)
\tag{6}
$$

The switching strategy implementation is accomplished with a three level hysteretic comparator in order to limit the maximum switching frequency (Fig. 2).

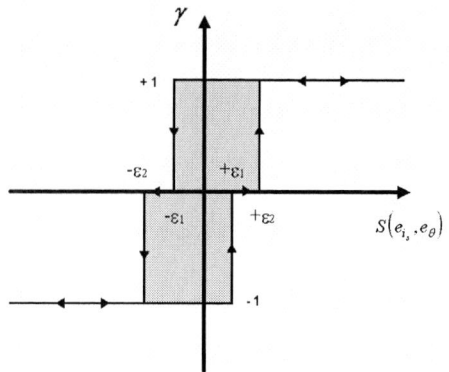

Fig. 2. Three level hysteretic comparator for the implementation of the switching strategy.

Thus, the strategy related to the switching law (6) with variables α1 and α2 (gating signals of the switches IGBT1 and IGBT2), is given by:

$$\alpha_1 = \begin{cases} 1 & , \gamma = 1 \\ 0 & , \gamma \neq 1 \end{cases} \quad (7)$$

$$\alpha_2 = \begin{cases} 1 & , \gamma = -1 \\ 0 & , \gamma \neq -1 \end{cases} \quad (8)$$

To obtain the expressions for the voltage controller, model (2) can be used. In fact, using a current mode modulator to enforce the reference value for the current is (that usually exhibit a fast dynamic compared with the dynamics of V_{C_o}), a quasi first order linear model (9) of the rectifier can be obtained from (1):

$$\frac{dV_{C_o}}{dt} \approx \frac{(1-\alpha)}{C_o} i_{L_{oref}} - \frac{1}{R_o\,C_o} V_o \quad (9)$$

A linear PI regulator, sampling the error between the output voltage reference $V_{C_{oref}}$ and the actual output, can be designed to provide the reference value for the current i_{sref}, since there is a direct relationship between the currents i_{L_o} and i_s according equations (1). The PI regulator constants (10 and 11) will depend on the load parameters (R_o), on the output capacitor value (C_o), on the delay of the current controller (small delay t_d of the i_{L_o} current) and on the required damping factor ζ.

$$K_P \approx \frac{C_o}{4\,\zeta^2\,t_d} \quad (10)$$

$$K_I \approx \frac{1}{4\,\zeta^2\,t_d\,R_o} \quad (11)$$

To ensure inrush and short-circuit-proof operation, a limiter was used at the output of the PI compensator.

IV. SIMULATION AND EXPERIMENTAL RESULTS

In table I it is presented the main parameters of the practical implementation of the proposed single stage flyback pfc converter.

The simulation results are obtained using a Matlab/Simulink-based simulator.

Fig. 3 shows the obtained simulation results of the input voltage and line current. From this result it is possible to confirm, that the proposed rectifier provide almost sinusoidal input current and high power factor. The input source current is dominated by the mains frequency sinusoid and the switching frequency components are greatly attenuated.

TABLE I
CONVERTER CHARACTERISTICS

$V_{s\,max}$	150 V
f	50 Hz
L_f	1 mH
R_f	0.5 Ω
C_f	5 μF
$L_{o1} = L_{o2}$	0.5 mH
$C_{o1} = C_{o2}$	470 μF

Fig. 4 shows the experimental results of the input voltage and line current. From this result it is possible to verify that this experimental result is similar to the correspondent simulation result. In this way it is confirmed that this converter provide almost sinusoidal input current and high power factor.

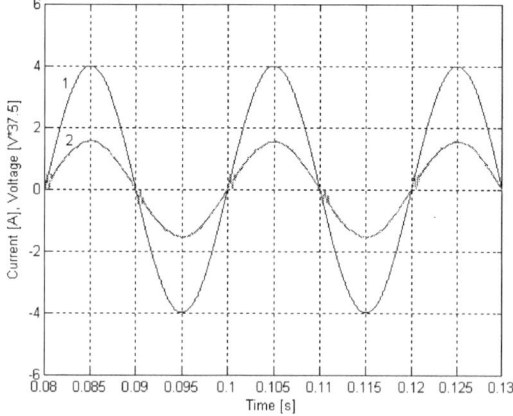

Fig. 3. 1 – Simulation results of 1 - Input source voltage (V_s), 2 - Input line current (i_s).

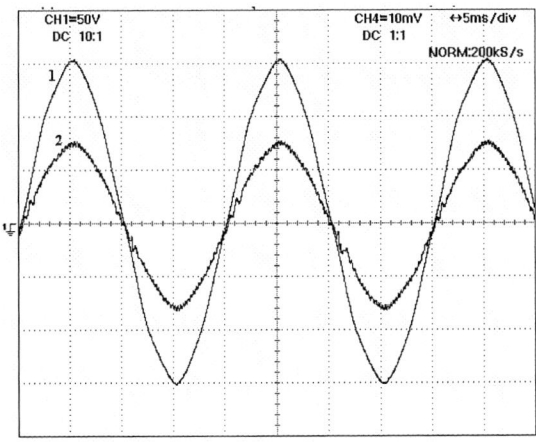

Fig. 4. 1 - Experimental results of 1 - Input source voltage
(V_s -20V/Div), 2 - Input line current (i_s -2A/Div).

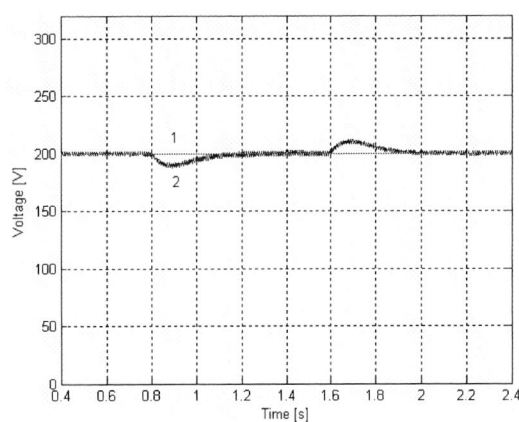

Fig. 6. 1 – Simulation results of 1 - Input source voltage (V_s), 2 - Input
line current (i_s).

The total harmonic distortion (THD) of the input line
current for a hysteresis of ε=0.5 is 6.4 %. Fig. 5 shows the
frequency spectrum of the input line current. As can be
seen by this figure the third, fifth, seventh, ninth, eleventh,
thirteenth and fifteen order harmonics have values of
1.1%, 1.5%, 1.3%, 0.4%, 0.3%, 0.3% and 0.4%,
respectively of the fundamental line current.

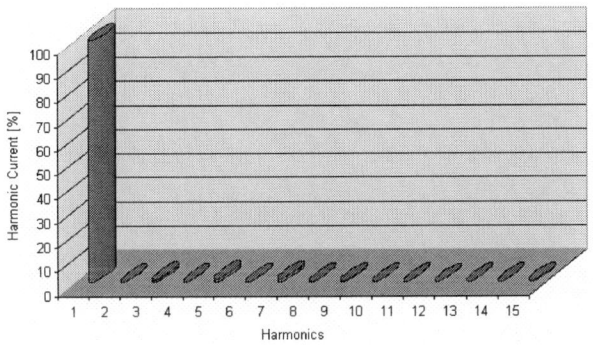

Fig. 5. Input line current harmonic of the experimental rectifier.

The transient characteristics are investigated with a load
step response, in order to analyze the closed loop behavior
of the power supply. Figure 6 shows the simulation
transient responses with a PI controller for a 30% load
change. This figure shows that it is possible to achieve a
sufficiently fast voltage regulation.

Fig. 7 shows the experimental result of the rectifier
output voltage under a step change in the load. From this
result it is possible to verify that this experimental result is
similar to the correspondent simulation result. In this way
it is confirmed that using the PI controller it is possible to
achieve a sufficiently fast voltage regulation.

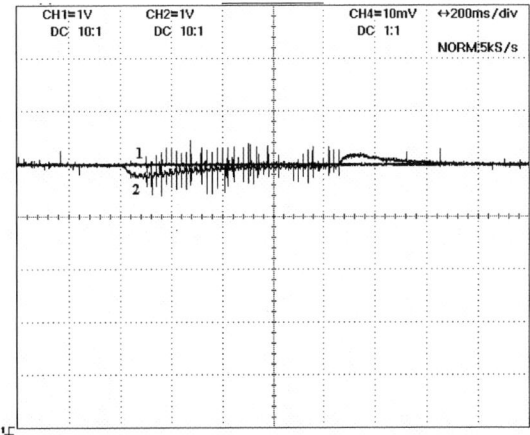

Fig. 7. 1 - Experimental results of 1 - Output voltage
reference (V_{oref} – 40V/Div), 2 - Rectifier output
voltage(V_o-40V/Div).

So, from the obtained results it is possible to confirm
that this topology is a valuable solution as a power supply
for this application.

V. CONCLUSIONS

a single stage flyback pfc converter is used as a power
amplifier system for testing the operating characteristics
of distance relay systems

In this paper it was presented a power supply for a
testing distance relay system. For this power supply it was
required isolation between the ac and the dc sides, as well
as two different ac output voltages. So, considering the
required characteristics of the power supply, a single-stage
pfc flyback converter providing two output voltages was
adopted. To control the input line current of this power
supply a fast and robust sliding mode control was used.
With this system controller it was possible to obtain a high
power factor and nearly sinusoidal input current. To
control the dc output voltage of the power supply was
adopted a proportional integral controller.

Simulation and experimental results from a laboratory
prototype were presented. From these results it was

1878

possible to confirm that this power supply presents good quality ac current and widely controllable output voltages.

REFERENCES

[1] L. H. Dixon, "High power factor preregulators for offline power supply, " Unitrode Switching Regulated Power Supply Design Seminar Manual. Unitrode Corporation, Lexington, MA.

[2] S. Buso, G. Spiazzi, D. Tagliavia, "Simplified Control Technique for High-Power-Factor Flyback Cuk and Sepic Rectifiers Operating in CCM" *IEEE Transactions on Industrial Aplications*, vol. 36, no. 5, pp. 1413-1418, September/October 2000

[3] R. Redl. L. Balogh, N. 0. Sokal. "A new family of single-stage isolated power-factor correctors with fast regulation of the output voltage", *Proc. of IEEE Power Electronics Specialist Conference (PESC'94)*, pp. 1137-1144, 1994.

[4] Se-Hwan Chun; Nam-Ju Park; Dong-Yun Lee; Dong-Seok Hyun, "A new AC/DC rectifier using one power circuit for unity power factor and dual output application", *Proc of Annual Conference of the IEEE Industrial Electronics Society*, vol. 2, pp. 1596 - 1600, 2003.

[5] R. Erickson, M. Madigan, S. Singer, "Design of a simple high-power-factor rectifier based on the flyback converter", *Proc. of the Applied Power Electronics Conference and Exposition (APEC '90)*, pp. 792 - 801, 1990.

[6] W. Huai, I. Batarseh, Z. Guangyong, P. Kornetzky, "A single-switch AC-DC converter with power factor correction" *IEEE Transactions on Power Electronics*, vol. 15, no. 3, pp. 421-430, May 2000.

[7] L. Rossetto, G. Spiazzi, P. Tenti, "Control Techniques for power Factor Correction Converters", *Proc. of the International Conference on Power Electronics & Motion Control (PEMC94)*, pp. 1310-1318, Warsaw, Poland, 1994.

[8] W. Tang, F.C. Lee, R.B. Ridley, I. Cohen, "Charge Control: Modeling, Analysis and Design", *IEEE Transactions on Power Electronics*, vol. 8, no. 4, pp. 396-403, October 1993.

[9] R. Marçal, R Rodrigues, C. Fortunato, L. Sousa Martins, V. Fernão Pires, "A Power Amplifier System for Testing of Distance Relay Operating Characteristic", *Proc. Of the 12th International Conference on Power Electronics & Motion Control (EPE-PEMC 2006)*, pp 1757-1761, 2006, Portoroz, Slovenia.

[10] J. Hung, W. Gao, "Variable structure control: A Survey" *IEEE Transactions on Industrial Electronics*, vol. 40, no. 1, pp. 2 22, February 1993.

[11] A. Groef, P. Bosh, H. Visser, "Multi-input variable structure controllers for electronic converters" *Proc. of the European Conference on Power Electronics and Applications (EPE'91)*, pp. 001 006, Firenze, September 1991.

H-Infinity Control Theory Apply to New Type Arc-suppression Coil System

Yilong QU, Weipu Tan, and Yihan Yang

Abstract– In the paper, the advanced H-Infinity optimization control theory is applied to a new type Principal-auxiliary arc-suppression coil (ASC) based on active filter technology. The H-Infinity standard control block diagram and the state space expression of model matching system are given. The auxiliary ASC composes of a single-phase shunt voltage source inverter (VSI), which produces little harmonics and automatically compensates the grounding capacity current with rapid response in a wide range. By injecting the active current, it realizes the full component of the earth-fault current. In this paper, the work principle of the new device, and a combinational compensation mode of presetting mode and following-setting mode are analyzed in detail. A continuous time H-Infinity controller is also designed which achieves all performances and robustness properties required. The results of the computer simulation and 10kV physical simulation experiments prove the validity of our approach.

Index Terms—H-Infinity control theory, arc-suppression coil, principal-auxiliary

I. INTRODUCTION

IN the context of electricity market, the users are paying increasingly more attention to the reliability of MV power system. More and more MV systems become resonant grounded systems to suppress grounded arcs more effectively. With the increase of the capacitive current, more and more grounded arcs are hard to self-extinguish. In the resonant grounded systems, most grounded arcs can self-extinguish with the fault spot current compensated by the inductive current coming from the arc-suppression coil. So it greatly improves the security and reliability of the power networks and is widely used in the world wide [1].

The conventional design of ASC controller is PI controller but it is designed by linear control system, which has strong nonlinearity, but may not be deal with operating point changes. The concept of H-Infinity control design has been applied widely in power system, in order to obtain high performance and robust stability even under uncertainty in plant parameters. Also its concept can reduce influence of disturbance to the

---

Yilong Qu, Weipu Tan, Yihan Yang is with the Department of Electrical & Electronic Engineering, North China Electric Power University, Beijing 102206 China (e-mail: quyilong142@163.com).

This work is supported by Beijing Natural Science Foundation (No. 3062018)

output of a system. In the paper, the advanced H-Infinity optimization ASC control design is applied to a new type Principal-auxiliary arc-suppression coil (ASC) based on active filter technology. This control scheme is known to provide an efficient solution for the compensation of current harmonics generated by distorting earth-fault current. [3,4]

We first present a new design of principal-auxiliary arc-suppression coil. Then based on it we establish the system model and control objectives in Section 3. By using standard algorithm by MATLAB robust toolbox we obtain a regulator K(s) of order 17th which gives very satisfactory simulation results. 10kV physical simulation experiments prove our approach in Section 4.

II. DESIGN OF PRINCIPAL-AUXILIARY ARC-SUPPRESSION COIL

The classical passive compensation of the earth fault current by an arc-suppression coil (ASC), which is tuned to the system capacitance, is only able to minimize the capacitive reactive part of the fault current, and to eliminate this part in the ideal case of exact compensation. The remaining part, which will not be compensated, consists of three components [2]:

-The remain reactive component owing to the ASC not accurate turned;

-The active component owing to the resistance loss of the ASC, feeders, and so on;

-The harmonic component owing to the non-linear characteristic of the power network itself;

Due to various studies and many years of experience it is well known that, despite ideal tuning of the ASC, the non compensable earth-fault current can reach up to 10% of the whole capacitive extinction current of the system. And if the earth-fault cannot be eliminated by the classical passive ASC, a so-called restrik of earth-fault arc will occur. This has badly threatened the system safety and operation.

As mentioned above, the traditional ASC couldn't fully compensate the ground current, and this is the uppermost defection of it.

A. The new type Principal-auxiliary Arc-suppression Coil System

To solve these problems, in the paper, a new type Principal-auxiliary arc-suppression coil based on active filter technology is designed [3~5]. See Fig. 1.

978-1-4244-0644-9/07/$25.00 ©2007 IEEE

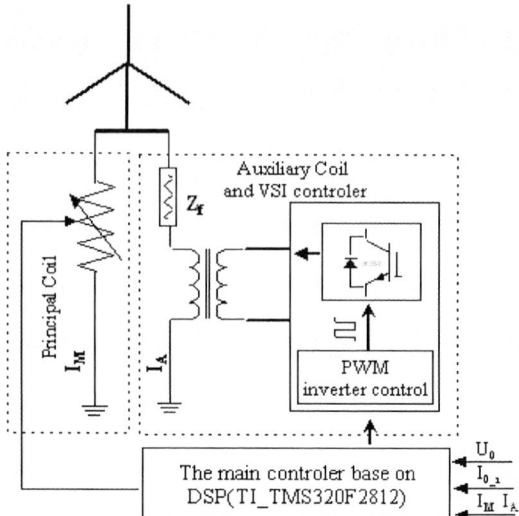

Fig.1. The new type Principal-auxiliary Arc-suppression Coil System.

We take some automatic tap-adjusting ASC as the principal part which would compensate most earth-fault current and significant decrease the capacity of auxiliary coil. This make the auxiliary coil has fast-response characteristic.

In shunt with the principal part is the auxiliary coil consists of a single-phase shunt voltage source inverter (VSI) connected to a dc capacitor, which produces no harmonics and could quickly and automatically compensates the grounding capacity current with rapid response in a wide range. The VSI composes of two Mitsubishi IPM (Intelligent Power Module): PM75DSA75. And the main controller is based on a DSP (TI_TMS320F2812).

The adjustment of the auxiliary coil is to exactly compensate earth-fault current. To enable the compensation precision, it should cover it should cover all the taps of the principal arc-suppression coil. For example, in some 10kV power network, the capacity of the principal arc-suppression coil is 315kVA, with a current-adjustment range of 20A-50A. Here we choose the auxiliary coil is 45kVA about 15% capacity of the principal coil, with a current-adjustment range of 3A-10A.

B. Basic working principle of the auxiliary coil

As it shown in Fig.1, there is a shunt voltage source inverter based on PWM control in auxiliary coil, and it can be an equivalent controllable voltage source U_{LA}. The Z_f consists of two parts: the output filter resistance which mainly filters the high order harmonic component in output voltage, and the output reactance L_A. According to [11], the basic working principle of the auxiliary coil is: as soon as the earth-fault occurs, there is a voltage U_0 between neutral and ground. Then with interaction of the U_0 and U_{LA} on the output reactance L_A, we will obtain the following equation:

$$L\frac{di_{LA}}{dt} = (U_{LA}(t) - U_0(t)) \quad (1)$$

Then with the controllable U_{LA} which differs to the U_0 both in amplitude and phase angle, we can obtain arbitrary output auxiliary coil current I_{LA} at will.

Now with the active injecting current I_{LA}, the new type arc-suppression coil can not only compensate reactive component of the earth-fault current, but also compensate the active component and harmonic with the active full compensation apparatus. And with the application of IGBT, it also can respond to an earth-fault in millisecond class.

C. The new compensation mode of ASC operation

For effectively reducing earth-fault current and strictly limiting the bias voltage in neutral point of the network, there are two compensation modes for the application of the arc-suppression coil, which are presetting compensation mode and following-setting compensation mode [6-8]. The two modes have their unique advantage, but they all have insufficiencies. The compensation precision of the presetting mode is restricted by the adjusting precision of the tap-adjusting arc-suppression coil, and the fault spot residual current is in closely related to the short velocity of the damping resistor. The following-setting mode improves the compensation precision and it can avoid the occurrence of series resonance. However when the system is in this compensation mode, the fault spot residual current is rather large on the fault moment, which does no good to the self-extinction of the instantaneous grounded arcs.

Based on the above analysis, this paper proposed a combination compensation mode of presetting mode and following-setting mode. In this mode, the arc-suppression coil is preset to some place that is about 15% over-compensation when the system is healthy, and the auxiliary arc-suppression coil is quickly following-set to compensate the earth-fault residual current when the single-phase-to-ground fault occurs. This mode makes the best use of the two above compensation modes, and have two obvious advantages described as follows:

1) This compensation mode make the principal arc-suppression coil act on the ground fault moment in no time, and it can suppress almost 70~80% grounded arcs in the MV power networks.

2) The auxiliary arc-suppression coil can then quickly inject the active current to fully compensate fault current when the grounded arc doesn't self-extinguish, which is sufficient to extinguish the predominant grounded arcs in the MV power networks.

On the fault line selection, the methods such as active power method, fifth harmonic content method and wavelet method are applied. Moreover with the new type arc-suppression coil and the new compensated mode, a method called abrupt change of zero-sequence current is also introduced. The principle of new method is that the abrupt change of zero-sequence current in the fault line is the largest before and after the adjustment of the auxiliary arc-suppression coil. All the above methods are modified to output some last fault line result through a whole set of fault measures and evidence theory , which is high reliable to realize the auto-turning for the new type arc-suppression, the accurate measure of the network capacitive current and the injecting current control method for the auxiliary arc-suppression coil are the key problems. The fast-response characteristic and the accurate PWM switching control make it possible for earth-fault exact compensation.

III. SYSTEM MODELIZATION AND H-INFINITY CONTROL DESIGN

From the Fig.1.and reference [1.2], a simplified electrical circuit representation for the system (Fig. 2.) during Phase-A single-phase earth-fault can be derived.

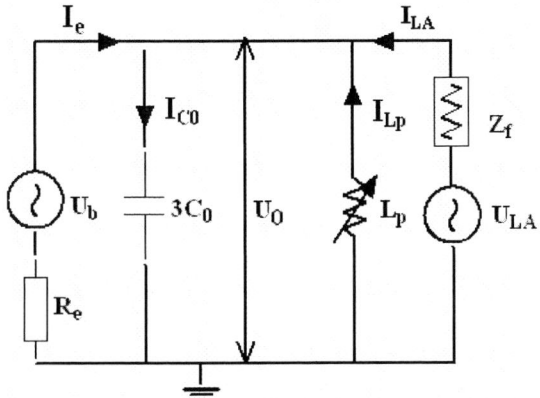

Fig.2.Simplified electrical circuit representation for the system during Phase-A single-phase earth-fault

The equivalent circuit was based on the following parameter:

-The Fault spot current Ie; The zero sequence capacitance current ICO; The current in principal coil ILP; The output current of the auxiliary coil ILA;

-The synthetic voltage of the 3-Phase power source during earth-fault (phase-B) Ub; The voltage on the neutral point U0; The continuous average voltage output of the auxiliary coil ULA, with results of the PWM control;

- The Fault spot resistance Re; The total 3-Phase zero sequence capacitance 3C0; The filter resistance of the auxiliary coil Zf;

-The resistance earth fault happens in phase-B

Our control object is to minimizing the Fault spot current Ie, as: $I_e = I_{LA} - (I_{c0} - I_{LP})$

For the principal can only compensate the reactive fault spot current component t, the remains components had to been compensated by the output current of the auxiliary coil I_{LA} inside the harmonic component.

Now we could deduced the following transfer function model [1]:

$$I_e(s) = \underbrace{\frac{N_1(s)}{D(s)}(U_{LA}(s) - U_b(s))}_{G(s)} + \underbrace{\frac{N_2(s)}{D(s)}(I_{C0}(s) - I_{LP}(s))}_{d(s)} \quad (2)$$

With:
$$N_1(s) = -1, N_2(s) = Z_f(s)$$
$$D(s) = R_e + Z_f(s)$$

It should be remarked that this rather "crude" model will be precious for the control design of the controller.

Modern automatic system theories provide the method to define the control problem in a standardized form (see Fig.3), by distinguishing control input u, external disturbances inputs v, outputs to be controlled z and measurement outputs y.

According to equation 1, G(s) represents the transfer between control input ($u=U_{LA}-U_b$) and the current I_e. d(s) is an output additive disturbance. I_{C0} and I_{LP} could be measured from CT in each feed lines. The control outputs are I_{LA}, which has to be filtered by Z_f, the I_e has to be minimizing. Then finally we obtain the following closed loop scheme (see Fig.4).

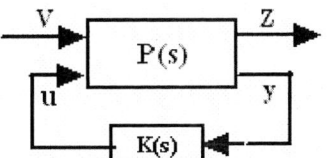

Fig.3. Standard control scheme

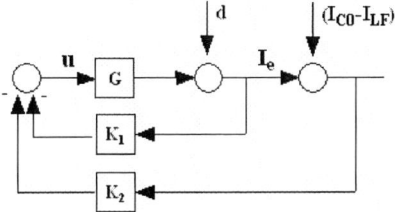

Fig.4. Closed loop structure

The closed loop transfer function is given by:

$$\begin{pmatrix} I_e - (I_{C0} - I_{LP}) \\ I_e \end{pmatrix} = \begin{pmatrix} \dfrac{1}{1+G(K_1+K_2)} & \dfrac{1+GK_1}{1+G(K_1+K_2)} \\ \dfrac{1}{1+G(K_1+K_2)} & \dfrac{1+GK_2}{1+G(K_1+K_2)} \end{pmatrix} \times \begin{pmatrix} d \\ (I_{C0}-I_{LP}) \end{pmatrix} \quad (3)$$

With: $T = \dfrac{1+GK_1}{1+G(K_1+K_2)} = \dfrac{I_e - (I_{C0} - I_{LP})}{I_e}$

Robustness specification can be taken into account by imposing: $\underbrace{\sup_{\omega}}|T(j\omega)| << \varepsilon$ ε take as infinitesimal positive real number.

Using standard algorithm by MATLAB robust toolbox we obtain a regulator K(s) of order 17th which gives very satisfactory simulation results. A balanced method based on Gram matrix is applied to reduce the controller order, finally we get a regulator K(s) of 7th which be used for DSP main controller.

IV. COMPUTER SIMULATION RESULTS

Corresponding to above analysis and design, we simulated the whole system with PSCAD/EMTDC. The simulation system parameters are similar with our 10kVA HV simulation system as Part V shown below.

The results on this computer simulation system are showing in Fig. 5.6. The waveform of neutral point voltage, main coil current, fault spot current, auxiliary coil current are shown when the auxiliary coil quit operation or put into operation .In Fig. 5, the fault happened at 400ms when the auxiliary coil quit operation, the compensate current from the main coil is about 10A. As the whole system capacitive current is about 8A, the fault spot current is about 2A. In Fig. 6, when the auxiliary coil put into operation at 830ms, which is controlled by timer-logic controller in PSCAD. We can see when the auxiliary coil put into operation, the spot current quickly reduces to about 0.2A. This result has proved performance of the new principal-auxiliary arc-suppression coil proposed in the paper, it can quickly compensate the fault spot current to a significant small current about zero.

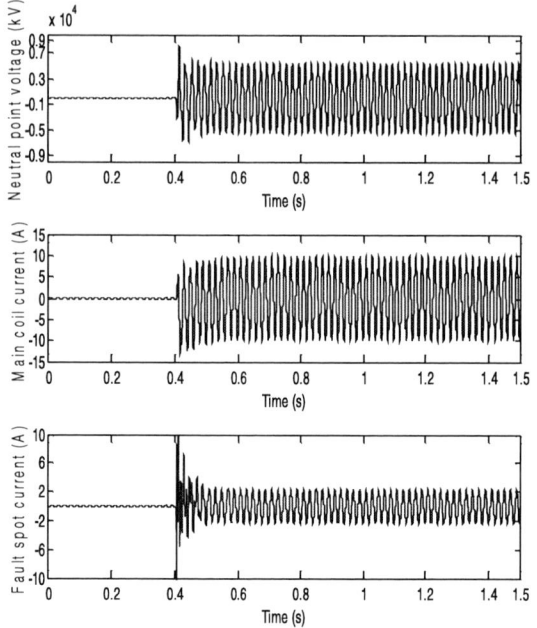

Fig. 5. Fault spot and main coil current
without the auxiliary coil put into operation

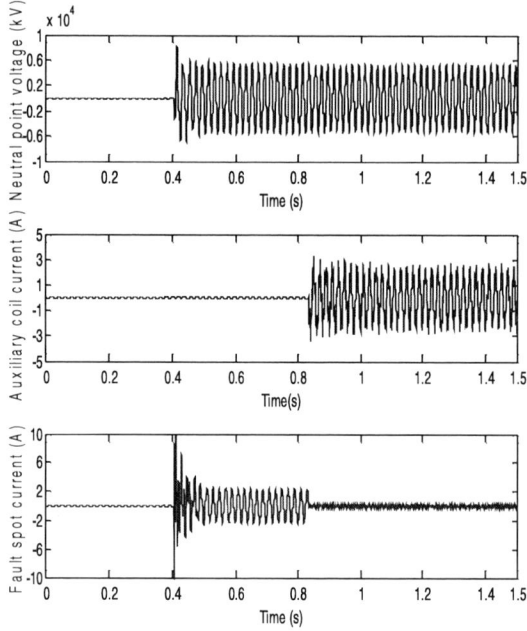

Fig. 6. Fault spot and auxiliary coil current
with the auxiliary coil put into operation

V. 10KV HIGH VOLTAGE PHYSICAL SIMULATING TEST

In order to verify the feasibility of this compensation mode and the control theory, a large amount of experiments also have been performed in the 10kV high voltage physical simulation system. The central device of the physical

simulation system is the 10kV power network model, which consists of a 380/10kV transformer, a new type Principal-auxiliary Arc-suppression Coil System on the neutral point and 6 lines on the 10kV bus bar. The line model adopts the π model. The total capacitive current is 8.4A under the maximum operating mode of the 10kV system. All the devices in the system (including the CT, PT, and circuit breaker etc.) are the same as those used in the field, which can well simulate the single-phase earth-faults in the power network.

In the experiments, the system is operating on the maximum operation mode, and the principal coil is operating on the over-compensation degree is 18%, and the auxiliary coil quit operation. Many experiments are performed, in which arcing faults with all gaps are simulated to verify the acting performance of suppressing all above arcs. It is found out that the arcs will not self-extinguish when the gaps are small. Then we put the auxiliary coil into operation with the new compensation mode proposed in the paper.

The waves of this typical test on this system are showing in Fig.7.8.9.the waves of fault spot current I_f, voltage on the neutral point U_0 and voltage on the fault phase U_B after the auxiliary coil put into operation.

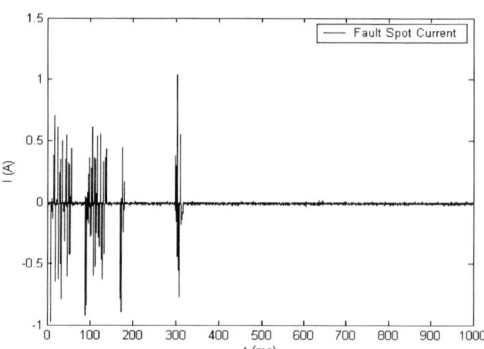

Fig. 7. Fault spot current
after the auxiliary coil put into operation

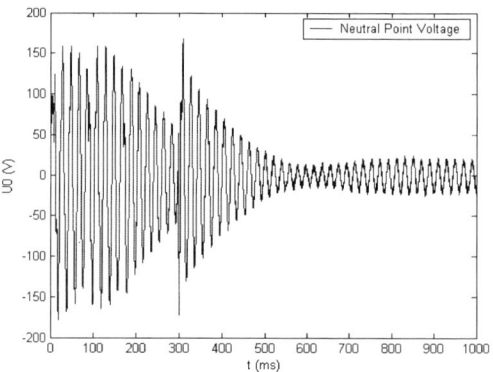

Fig. 8. Voltage on neutral point
after the auxiliary coil put into operation

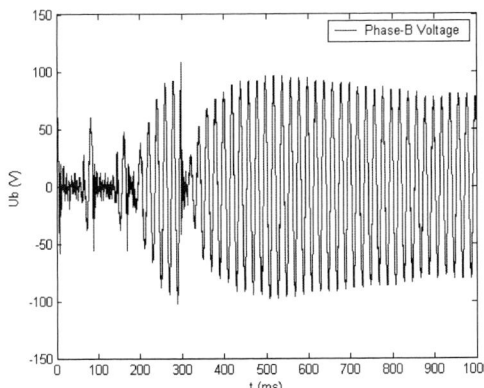

Fig. 9. Voltage on fault phase (phase-B)
after the auxiliary coil put into operation

VI. CONCLUSION

1. This paper presents a new type Principal-auxiliary arc-suppression coil (ASC) based on active filter technology. It can make the arcs more easily to self-extinguish and improve the security and reliability of the power networks.

3. A continuous time H-Infinity controller is designed which achieves all performances and robustness properties required.

2. The results of the 10kV physical simulation experiments verify the reliability of this compensation mode and the new principal-auxiliary arc-suppression coil proposed in this paper.

VII. REFERENCES

[1] Griffel .D, Leitloff .V, Harmand .Y, Bergeal .J, "A New Deal for Safety and Quality on MV Networks", Power Delivery, IEEE Transactions on, Volume: 12, Issue: 4, Oct. 1997, Pages: 1428 – 1433

[2] Yao Wannian, Cao Meiyue, "Resonance to Ground in Power System" Beijing: China Electric Power Press, 2000,8

[3] Chevrel. P, Pina. F, "H∞ control design methodology for active filtering problems", Control Applications, 1997., Proceedings of the 1997 IEEE International Conference on 5-7 Oct. 1997 Page(s):755 – 760

[4] Chevrel. P, Auger. F, Machmoum. M, "H∞ control for a single-phase active power filter: a systematic approach", Power Electronics Specialists Conference, 1996. PESC '96 Record., 27th Annual IEEE Volume 2, 23-27 June 1996 Page(s):1112 - 1118 vol.2

[5] C. Tuttas, "Compensation of capacitive loads by a voltage source active filter", ETEP, Vol 2, n1, pp15-19, 1992

[6] Cai Xu, Liu Jie, "Dynamic Resonance Adjustment Device of Arc-suppression Coil with Magnetic Bias", Automation of Electric Power Systems, 2002, 26(15)

[7] Xu Yuqin, Chen Zhiye, "The Method for Automatic Compensation and Detection of Earth Faults in Distribution Network", Power System Technology, 2002. Proceedings. PowerCon 2002. International Conference on, Volume: 3, 13-17 Oct. 2002, Pages: 1753 - 1757 vol.3

[8] Li Fushou, "Operation of Power Network with Neutral Point non-valid grounded" Beijing: Irrigation & Electric Power Press, 1993,10

[9] Cai Xu, Liu Yong, Hu Chunqiang, Zuo Hongyan. "New Resonance Earth System with Magnetic Bias and Its Protection", Proceeding of the CSEE, 2004, 24 (6); 44-49

[10] Sheng Jianke, Chen Qiaofu, Xiong Yali, Zhang Yu, Jia Zhengchun. "A New Type Automatic Resonant Arc-Suppressing Coil Based on Controllable Magnetic Flux", Transactions of China Electrotechnical Society, 2005, 20 (2); 88-93

[11] Jiang Qirong, Zhao Dongyuan, Chen Jianye, "Active power filter: structure, principles and control", Beijing : Scientific Press, 2005.

VIII. BIOGRAPHIES

Yilong Qu was born in Nong an County, Jinlin Province, China in 1979. He graduated from North China Electric Power University as an undergraduate in 2002. And then He has studied for the doctor's degree in North China Electric Power University up to now. The main aspect he researches is the automation and protection of distributing power network.

Weipu Tan was born in Harbin City, China in 1963. He received his Ph.D. degree from Harbin Institute of Technology in 1999. He is now working in North China Electric Power University as an associate professor. His researches mainly focus on the power electronics technology & its application in power systems.

Yihan Yang was born in Tieling County, Liaoning Province on Feb. 7, 1927. He is a famous professor in China. He received Bachelor's degree in 1949 in Northeast University, and received Master's degree in 1952 in Harbin Institute of Technology. After that, he has been teaching in Northeast Institute of Technology, Harbin Institute of Technology and North China Electric power University.

Characteristics of a novel topology of a DC-AC Converter for Fuel Cells

K. Fukushima*, T. Ninomiya*, I. Norigoe**, Y. Harada**, K. Tsukakoshi**, and Z. Dai**

* Kyushu University, 744, Moto-oka, Nishi-ku, Fukuoka, Japan

** EBARA DENSAN .LTD, 11-1, Ota-ku, Tokyo, Japan

Abstract— This paper proposes a novel DC-AC converter topology for fuel cells. The proposed circuit topology has no need smoothing circuit because boosted voltage from first step provides directly to PWM inverter. The notice point is that the output voltage from isolating step (third step) is pulsed waveform, and pulse voltage is also made from DC at the converting step. As the result, the smoothing step will be able to be cut if the demanded voltage in one switching period provides to converting AC step within pulse voltage is output. Furthermore, this topology is expected to achieve ZVS of the PWM switches because pulsed voltage has certainly zero voltage periods in the switching period. As the results, it will be possible to reduce the size of this component.

Moreover, input current-ripple is damaged to the life span of fuel cells. Using the topology, the method of input current-ripple reduction is shown.

This paper shows the advantage the converter by the experimental results.

Key words—DC-AC Converter, Fuel Cells, Pulse-link, Current-ripple Reduction

I. INTRODUCTION

Today, environmental issues such as global heating and air pollution have become international crisis. And new clean energy system is strongly demanded. Therefore, several clean energy system have been researched and developing, recently. Among them, the generating systems used by fuel cells have come to attention. Fuel cells convert chemical energy directly to electricity and thermal energy. As the result, fuel cells are high efficiency about energy conversion. And now, the home co-generation system using both electricity and thermal energy from fuel cells are researched actively [1]. In order to be in use, the cost of this system is important [2][3]. And also, the small size of this component is demanded.

When fuel cells are used as home co-generation system, DC-AC converter is needed. The specifications for the DC-AC converter for fuel cells are shown below:

1. It can be isolated between fuel cells and load stage.
2. It boosts input voltage that is from fuel cells to a level of commercial voltage.
3. It does not make the input current ripple.

First thing is needed for safety. Second and third things are original specifications to fuel cells. The input voltage provided from fuel cells is much lower than commercial voltage. So, the converter is necessary to boost input voltage more than commercial voltage. Fuel cells are generated electricity by chemical reaction. The reaction time is much slower than commercial frequency. So, if current ripple is large, the power from fuel cells will not be able to provide for loads well. As the result, the current-ripple not only affects the fuel cell capacity, but also damages the fuel consumption and life-span [4][5][6].

The conventional DC-AC converter for fuel cells is shown Fig. 1. The conventional DC-AC converter component for fuel cells has two stages. First stage is isolated boost DC-DC converter, and second stage is PWM inverter. In this component, there are 4 steps - boosting, isolating, smoothing, and converting AC. In this topology, the large capacitor is inserted between first stage and second stage because the capacitor absorbs the current ripple from commercial frequency by maintaining the boosted DC voltage. As the result, the conventional topology is difficult to reduce the size of the component. And, the cost becomes expensive because the large capacitor is usually used electrical double layer capacitor.

To overcome these problems, this paper proposes a novel DC-AC converter topology shown Fig. 2. The proposed circuit topology needs no smoothing step because boosted voltage from first step provides directly to PWM inverter. The notice point is that the output voltage from isolating step is pulsed waveform, and also pulse voltage is made from DC at the converting AC step. As the result, the smoothing step will be able to be cut if the demanded voltage in one switching period provides to converting AC step within pulse voltage is output. This concept has known as AC link or pulse link [7][8]. Here, we call this topology as Pulse-link DC-AC converter. Furthermore, it is clear that this topology can achieve zero-voltage-switching (ZVS) of the PWM inverter switches. As these results, this topology is expected to reduce the size of this component.

This paper analyzes the steady-state characteristics of Pulse-link inverter topology. And it makes clear the advantage of this topology by the experimental results.

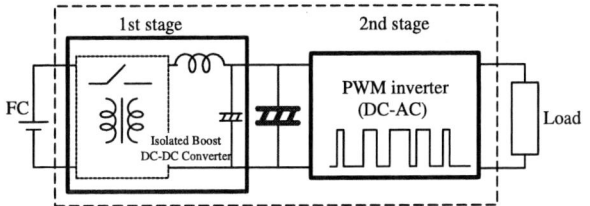

Fig. 1: The conventional DC-AC converter component for fuel cells.

II. OPERATING STATES

Fig. 2 shows the proposed circuit topology. As mentioned above, this topology has two stages, and this converter provides boosted pulse-voltage directly to PWM inverter. And between two stages, series LC circuit is connected in parallel in order to reduce current-ripple. The value of the capacitor using this LC circuit is much less than the conventional one.

This topology provides boosted voltage pulse that is controlled by switch Q_1 directly to PWM inverter. The switching sequence of this converter is shown as Fig. 3. This converter has 5 switches. As mentioned before, switch Q_1 controls the boost pulse from input voltage. And, from S_1 to S_4 are PWM switches. S_1 and S_4 are controlled to make output voltage sinusoidal waveform, while S_2 and S_3 are decided the plus/minus of output voltage. And control combination of S_1, S_3 and S_2, S_4 is a pair. Q_1 and S_1/S_4 are synchronous at rising time. Here, it is analyzed when the output voltage is in the positive semicircle period. So, there are 3 states in one switching period shown Fig. 3.

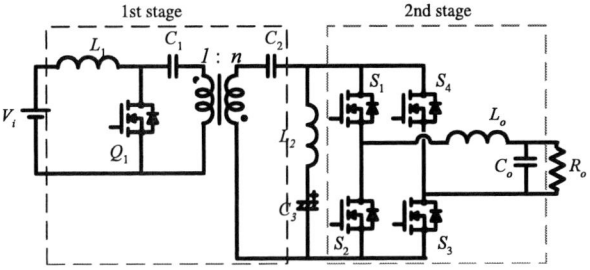

Fig. 2: The proposed circuit topology.

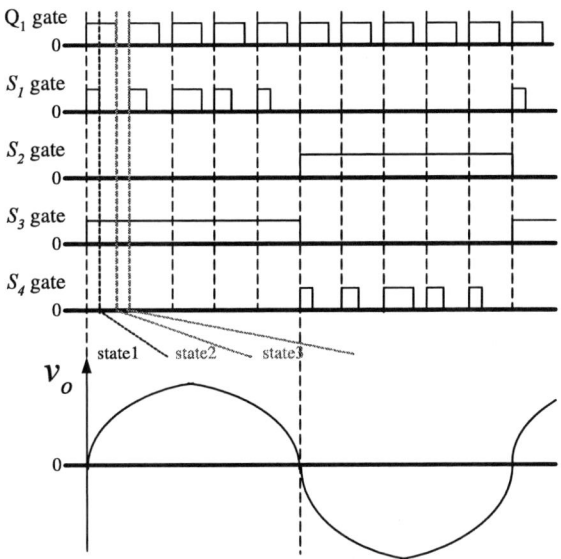

Fig. 3: The proposed circuit topology.

A. State 1 (Q_1: ON, S_1: ON, S_3: ON)

Fig. 4(a) shows the equivalent circuit of state 1. State equation on state 1 is written as below:

$$
\begin{cases}
v_{L_1} = V_i - r_{Q_1}\hat{i}_{L_1} - nr_{Q_1}\hat{i}_{L_2} - nr_{Q_1}\hat{i}_{L_o} \\
v_{L_2} = \hat{v}_{C'} - \hat{v}_{C_3} - nr_{Q_1}\hat{i}_{L_1} - n^2 r_{Q_1}\hat{i}_{L_2} - n^2 r_{Q_1}\hat{i}_{L_o} \\
v_{L_o} = \hat{v}_{C'} - \hat{v}_{C_o} - nr_{Q_1}\hat{i}_{L_1} + (r_{s1} + r_{s4} - n^2 r_{Q_1})\hat{i}_{L_2} - (r_{s1} + r_{s4})\hat{i}_{L_o} \\
i_{C'} = -\hat{i}_{L_2} - \hat{i}_{L_o} \\
i_{C_3} = \hat{i}_{L_2} \\
i_{C_o} = \hat{i}_{L_o} - \dfrac{\hat{v}_{C_o}}{R_o}
\end{cases}
$$

(1)

, where, $C' = \dfrac{C_1 C_2}{C_1 + n^2 C_2}$

B. State 2 (Q_1: ON, S_1: OFF, S_3: ON)

Fig. 4(b) shows the equivalent circuit of state 2. State equation on state 2 is written as below:

$$
\begin{cases}
v_{L_1} = V_i - r_{Q_1}\hat{i}_{L_1} - nr_{Q_1}\hat{i}_{L_2} \\
v_{L_2} = \hat{v}_{C'} - \hat{v}_{C_3} - nr_{Q_1}\hat{i}_{L_1} - n^2 r_{Q_1}\hat{i}_{L_2} \\
v_{L_o} = -\hat{v}_{C_o} - nr_{Q1} - (r_{s4} + r_{D3})\hat{i}_{L_o} \\
i_{C'} = -\hat{i}_{L_2} \\
i_{C_3} = \hat{i}_{L_2} \\
i_{C_o} = \hat{i}_{L_o} - \dfrac{\hat{v}_{C_o}}{R_o}
\end{cases}
$$

(2)

C. State 3 (Q_1: OFF, S_1: OFF, S_3: ON)

Fig. 4(c) shows the equivalent circuit of state 3. State equation on state 3 is written as below:

$$
\begin{cases}
v_{L_1} = V_i - \dfrac{1}{n}\hat{v}_{C'} - \dfrac{1}{n}\left(r_{D1'}+r_{D2'}\right)\hat{i}_{L1} - \left(r_{D1'}+r_{D2'}\right)\hat{i}_{L_2} \\[2mm]
v_{L_2} = -\hat{v}_{C_3} - \dfrac{1}{n}\left(r_{D1'}+r_{D2'}\right)\hat{i}_{L_1} - \left(r_{D1'}+r_{D2'}\right)\hat{i}_{L_2} - r_{D2'}\hat{i}_{L_o} \\[2mm]
v_{L_o} = -\hat{v}_{C_o} - \dfrac{1}{n}r_{D1}\hat{i}_{L_1} - r_{D2}\hat{i}_{L_2} - \left(r_{D2'}+r_{D3'}\right)\hat{i}_{L_o} \\[2mm]
i_{C'} = \dfrac{1}{n}\hat{i}_{L_1} - \hat{i}_{L_2} \\[2mm]
i_{C_3} = \hat{i}_{L_2} \\[2mm]
i_{C_o} = \hat{i}_{L_o} - \dfrac{\hat{v}_{C_o}}{R_o}
\end{cases}
\tag{3}
$$

D. Steady state

From above equations, the state-averaging vector is written below by using state space averaging method. Here, on-resistance and conduction losses are ignored.

$$
\frac{dX}{dt} =
\begin{bmatrix}
0 & 0 & 0 & -\dfrac{\frac{1}{n}(1-D_{Q_1})}{L_1} & 0 & 0 \\[2mm]
0 & 0 & 0 & \dfrac{D_{Q_1}}{L_2} & -\dfrac{1}{L_2} & 0 \\[2mm]
0 & 0 & 0 & \dfrac{d_{s1}}{L_o} & 0 & -\dfrac{1}{L_o} \\[2mm]
\dfrac{\frac{1}{n}(1-D_{Q_1})}{C'} & -\dfrac{1}{C'} & -\dfrac{d_{s1}}{C'} & 0 & 0 & 0 \\[2mm]
0 & \dfrac{1}{C_3} & 0 & 0 & 0 & 0 \\[2mm]
0 & 0 & \dfrac{1}{C_o} & 0 & 0 & -\dfrac{1}{C_o R_o}
\end{bmatrix}
X +
\begin{bmatrix}
\frac{1}{L_1} \\ 0 \\ 0 \\ 0 \\ 0 \\ 0
\end{bmatrix} V_i
\tag{4}
$$

, where $X = \begin{bmatrix} \hat{i}_{L1} & \hat{i}_{L2} & \hat{i}_{Lo} & \hat{v}_{C'} & \hat{v}_{C3} & \hat{v}_{Co} \end{bmatrix}^T$.

And from (4), the steady-state characteristics are shown below:

$$
V_{c'} = \frac{n}{1-D_{Q_1}}V_i
\tag{5}
$$

$$
V_o = \frac{n d_s(t)}{1-D_{Q_1}}V_i
\tag{6}
$$

Furthermore, from (5), the peak voltage pulse that is input to PWM inverter (v_{inv_in}) is written below

$$
v_{inv\_in} = \frac{n}{1-D_{Q_1}}V_i
\tag{7}
$$

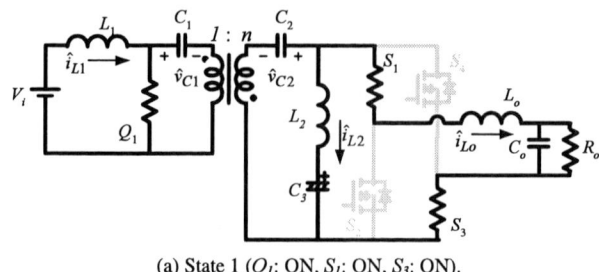

(a) State 1 (Q_1: ON, S_1: ON, S_3: ON).

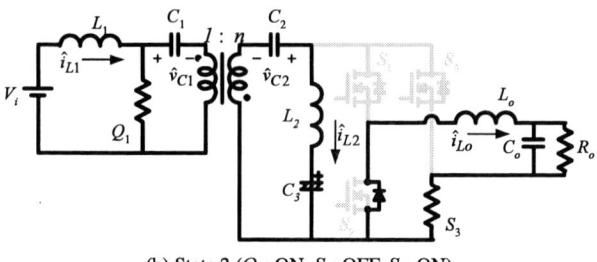

(b) State 2 (Q_1: ON, S_1: OFF, S_3: ON).

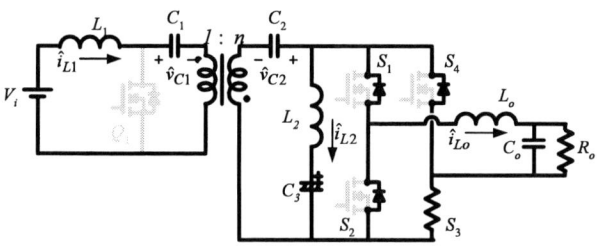

(c) State 3 (Q_1: OFF, S_1: OFF, S_3: ON).

Fig. 4: Equivalent circuit of each state.

Here, D_{Q_1} is duty ratio of switch Q_1. And, $d_{s1}(t)$ is duty ratio of PWM inverter switch of S_1/S_4. $d_{s1}(t)$ is changed shown as (8) in order to make output voltage to be sinusoidal waveforms.

$$
ds(t) = d_{s1\_max} \cdot \sin\left(2\pi \cdot 50t\right)
\tag{8}
$$

Moreover, the relationship of D_{Q_1} and d_{s1_max} is limited (9), because PWM inverter is provided voltage only when Q_1 is ON.

$$
D_{Q_1} \geq d_{s1\_max}
\tag{9}
$$

III. EXPERIMENTAL RESULTS

A. Setting of parameters

In order to evaluate the performance of the proposed circuit, the experimental circuit is implemented by using the specifications and parameters in Table 1. From table 1, C_1 is 3[mF], and it is aluminum electrolytic capacitor. C_1 is decided from the allowable current. Primary-side is flown large current, so capacitance of C_1 becomes large value. If C_1 is film capacitor or conductive polymer electrolytic capacitor, the value will become less. However, the capacitor these types are expensive, so the cost of this unit becomes higher. Furthermore, primary-side is low voltage, so the size of aluminum electrolytic capacitor is not so large even if the value is large because withstand-voltage is low. Therefore, large value of aluminum electrolytic capacitor is used at C_1 in this experiment.

B. Output voltage waveform and input current waveform

Fig. 5 shows the experimental waveforms of output voltage (v_o), input current (i_i), and inductor current of L_2 (i_{L2}). And table II shows the experimental measurement. From those results, it is considered that output voltage is achieved to output commercial voltage. Furthermore, it is considered that inductor current of i_{L2} oscillates low frequently with zero crossing.

Fig. 6 shows the waveforms of input voltage to PWM inverter (v_{inv_in}), drain-source voltage of switch Q_1 (v_{ds}), and inductor current of L_2 (i_{L2}). From the waveforms of v_{inv_in} and v_{ds}, it is considered that boosted voltage pulse v_{inv_in} is generated when switch Q_1 is ON. Furthermore, the peak voltage of the voltage pulse v_{inv_in} is shown from equation (7). In this experiment, duty ratio of Q_1 is 0.7, so the peak voltage of v_{inv_in} is 200[V] by equation (7). And the peak voltage of output voltage is 140[V] calculated from equation (6). This peak voltage is corresponding to the peak voltage of commercial alternating voltage in 100[V].

Furthermore, it is noticed that i_{L2} is always crossing zero in one switching period. When i_{L2} is positive, i_{L2} makes charge of C_3. On the other hand, when i_{L2} is negative, i_{L2} provides to PWM inverter and charge to C_2. And, i_{L2} is oscillated at low frequency by shown Fig. 5. This means that in one switching period, when the power provided to load is high it is provided from series L_2 and C_3 circuit in addition to input power. While the power provided to load in one switching period is low, the input power is stored at series L_2 and C_3 circuit. As the results, this topology is reduced the input current ripple.

C. Reduction of surge of v_{inv_in}

Furthermore, the waveform of v_{inv_in} is occurred surge. This surge is occurred by charging parasitic capacitance of MOSFET for PWM inverter. To overcome this surge, clamp capacitor (C_{clamp}) and auxiliary MOSFET (Q_2) are inserted shown fig. 7. Q_2 is turned ON just before Q_1 is turned ON, and duty ratio of Q_2 is short.

Figure 8 shows the experimental waveforms of v_o, i_{L2} and i_i. Figure 9 shows the experimental waveforms of v_{inv_in}, v_{dsQ1} and i_{dQ1} when clamp-circuit is inserted, and table III is the experimental measurement. From table III, it is considered that efficiency is as same as the

Table I:: Experimental circuit values.

Symbol	Description	value
Vi	Input voltage	20[V]
$L1$	Input inductance	400[uH]
$L2$	Middle inductance	1[mH]
LM	Magnetizing inductance	38.8[uH]
$C1$	primary-side capacitance	3[mF]
$C2$	Secondary-side capacitance	330[uF]
$C3$	Middle capacitance	330[uF]
n	Turn ratio	3
Lo	Output inductance	3[mH]
Co	Outout capacitance	9.4[uF]
fs	Switching frequency	30[kHz]
$DQ1$	Duty ratio of $Q1$	0.7
$ds\ max$	Maximimum duty ratio of $S1$ and $S4$	0.7
Ro	Output resistance	100[Ω]

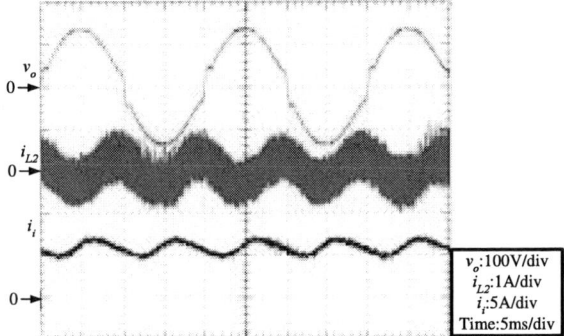

Fig. 5: Experimental waveforms of v_o, i_{L2}, and i_i.

Table II:: Experimental measurement.

Symbol	Description	value
Vi	Input voltage	20[V]
Ii	Input current	6.11[A]
Vo	Output voltage	99.9[V(rms)]
R_o	Output resistance	100[Ω]
η	Efficiency	82[%]

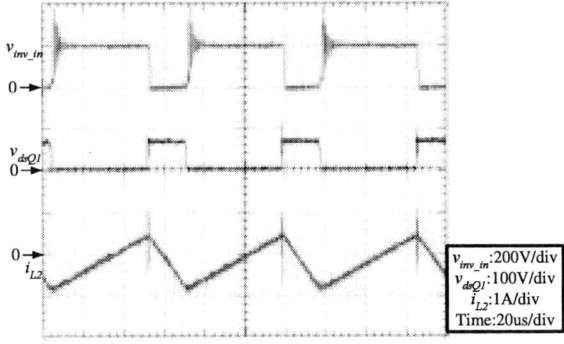

Fig. 6: Experimental waveforms of v_{inv_in}, v_{ds}, and i_{L2}.

previous topology. From fig. 9, v_{inv_in} is not occurred surge. As the result, improvement circuit is able to use the MOSFET which is lower withstanding voltage than first topology, so it is expected to improve the efficiency.

IV. IMPROVEMENT METHOD OF INPUT CURENT-RIPPLE

From fig. 5, it is observed that input current ripple is input current ripple is 3[Ap_p]. It is not enough to reduce input current ripple. So, here is mentioned a control method to reduce input current ripple.

A. Control method

In order to control input current, duty ratio of switch Q_1 (D_{Q1}) is controlled. From fig. 6, it is observed that when output voltage is around zero crossing, input current is minimum point. This means that input power is not needed, so extra power is about to go back to input stage.

To avoid the above thing, it is controlled D_{Q1} by detecting input current. Fig. 8 shows the experimental component of current detecting. In the experiment, FPGA is used and it is controlled by digital-control with A/D converter.

In the experiment, current transformer detects input current. 1[A] is converted to 0.125[V] used by current transformer. And the converted voltage is input to A/D converter. 0.01[V] is corresponding to 1 binary data at A/D converter.

By using the binary data, the duty ratio signal of switch Q_1 ($Q_{1signal}$) is calculated by below equation:

$$Q_{1signal} = Q_{1signal\_ref} + k\left(I_{in} - I_{ref}\right) \cdots(10)$$

, where $Q_{1signal\_ref}$ is corresponding to the binary data that duty ratio of switch Q_1 (D_{Q1}) is 0.7, and I_{ref} is converted reference input current to binary data.

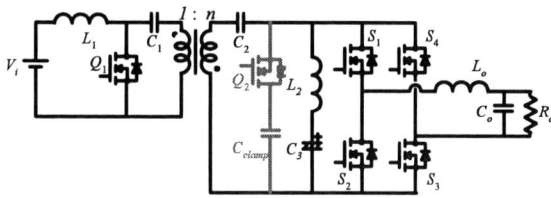

Fig. 7: Improvement of circuit topology.

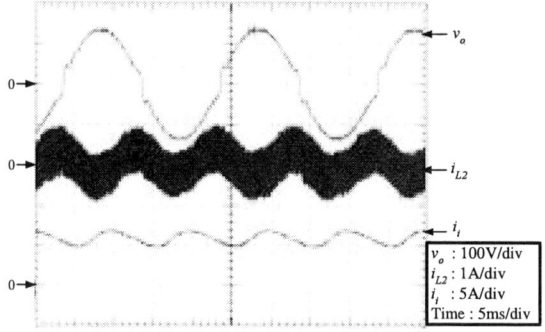

Fig. 8: Experimental waveforms of v_o, i_{L2}, and i_i.

B. Experimental result

Fig. 11 shows the experimental waveforms of v_o, and i_i.. From the figure, it is considered that output voltage is output commercial voltage, and input current has fewer ripples. This means that the input current regulation is achieved.

Furthermore, it is observed that i_{L2} is oscillated much actively. This means that series L_2 and C_3 circuit is regulated in order to both current ripple is canceled and provide to PWM inverter. From the results, it is considered that the proposed circuit topology is achieved to reduce input current ripple. This topology is expected to be used for fuel cells.

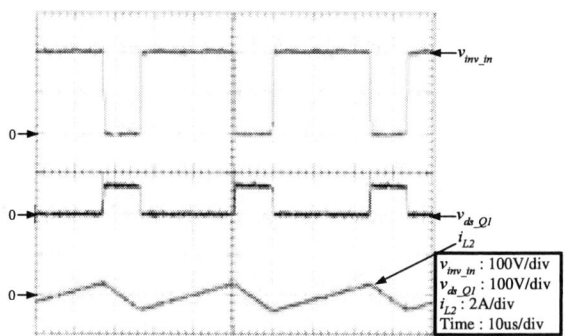

Fig. 9: Experimental waveforms of v_{inv_in}, v_{ds}, and i_{L2}.

Table III:: Experimental measurement.

Symbol	Description	value
Vi	Input voltage	20[V]
Ii	Input current	6.0[A]
Vo	Output voltage	99[V(rms)]
R_o	Output resistance	100[Ω]
η	Efficiency	82[%]

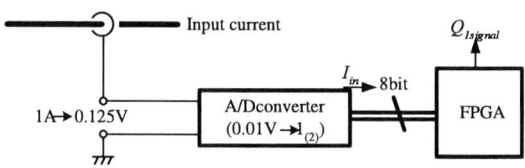

Fig. 9: Experimental component of current sensing.

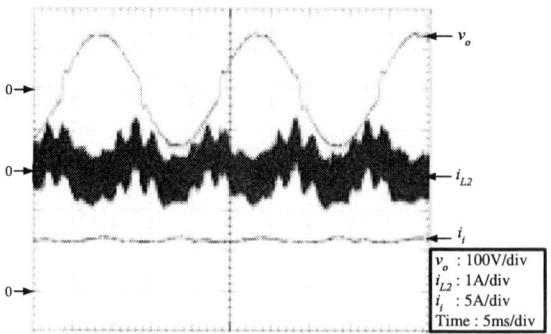

Fig. 11: Experimental waveforms of v_o, i_{L2}, and i_i.

V. CONCLUSION

This paper proposes a novel DC-AC converter for fuel cells. This converter provides the boosted voltage pulse directly to PWM inverter. We call this topology as Pulse-link inverter. And this paper is shown that this topology is achieved to convert DC to AC. Furthermore, the surge reduction circuit is inserted, improvement circuit topology is shown.

Moreover, in order to reduce input current ripple, the control method of this topology is mentioned. As the result, the input current is improved to be fewer ripples. The optimum parameters of this proposed topology and improvement of the efficiency is future work.

REFERENCES

[1] S. Sumiyoshi, H. Omori, and Y. Nishida, "Power Conditioner Consisting of Utility Interactive Inverter and Soft-Switching DC-DC Converter for Fuel-Cell Cogeneration System," PCC-Nagoya 2007, pp.455-462, Apr. 2007.

[2] Wang, F. Z. Peng, J. Anderson, A. Joseph, and R. Buffenbarger, "Low Cost Fuel Cell Converter System for Residential Power Generation," IEEE transactions on Power Electronics, Vol.19, No.5, Sep. 2004.

[3] R. Gopinath, S. Kim, J. Hahn, P. N. Enjeti, M. B. Yeary, and J. W. Howze, "Development of a Low Cost Fuel Cell Inverter System With DSP Control," IEEE transactions on Power Electronics, Vol.19, No.5, Sep. 2004.

[4] S. Moon, J. Lai, S. Park and C. Liu, "Impact of SOFC Fuel Cell Source Impedance on Low Frequency AC Ripple," Power Electronics Specialists Conference, Proc. of IEEE PESC 2006, pp.2037-2042, Jun. 2006.

[5] W. Choi, P.N. Enjeti and J.W. Howze,"Development of an Equivalent Circuit Model of a Fuel Cell to Evaluate the Effects of Inverter Ripple Current," Proc. of IEEE APEC 2004, pp. 255-361, Feb. 2004.

[6] G. Fontes, C. Turpin, R. Saiset, T. Meynard, and S. Astier, "Interactions between fuel cells and power converters Influence of current harmonics on a fuel cell stack," Proc. of PESC 2004, pp. 4729-4735, 2004.

[7] P. T. Kerin, R. S. Balog, and X. Geng, "High-Frequency Link Inverter for Fuel Cells Based on Multiple-Carrier PWM," IEEE Transaction on PE, Vol. 19, No. 5, pp. 1279-1288, Sep. 2004.

[8] D. Chen and L. Li, "Novel Static Inverters With High Frequency Pulse DC Link," IEEE Transaction on PE, Vol. 19, No. 4, pp. 971-978, Jul. 2004.

A Comparative Study of PWM Schemes for Grid Connected PV Cell

Vineeta Agarwal *, *Member IEEE,* and Alok Vishwakarma **
* MNNIT, Allahabad, India
** NCPS DADRI, NTPC Ltd., India,

Abstract – A photovoltaic (PV) cell connected to the grid system has been modeled in MATLAB environment. Various characteristics have been plotted with changing atmospheric condition. An operating point at which photovoltaic cell exhibits maximum efficiency and generates maximum power is tracked by, a method called voltage Based Maximum Power Point tracking (VMPPT). To inject a sinusoidal current into the grid system, two modulations techniques namely; Sine PWM and State Vector PWM (SVPWM) have been implemented, and their performance is evaluated based on different parameters. Simulation results demonstrate the fast and effective response.

Index Terms—Distortion factor, Harmonics, Modulation Index, Maximum Power Point (MPP), Photovoltaic (PV) cell, Space Vector,

I. INTRODUCTION

Photovoltaic (PV) generation is becoming increasingly important as a renewable source since it offers many advantages such as incurring no fuel costs, not being polluting, requiring little maintenance, and emitting no noise, among others [1]. The motivation for using AC drives in solar powered drives is the fact that 3-phase induction motor is a very rugged motor and requires no maintenance [2]. Power generated by PV generators, and injected into the grid, is gaining more and more visibility in the area of PV applications. This is mainly because the world's power demand is steadily increasing. The amount of power generated by a PV depends on the operating voltage of the array. It's voltage-current and voltage-power characteristic curves specify a unique operating point at which maximum possible power is delivered [3]. At the maximum power point the PV cell operates at its highest efficiency. The maximum power point (MPP) varies with solar insulation and temperature [4].

A number of methods have been developed to determine MPP [5] – [7]. PV systems connected to the utility lines have strict regulations, which relate to harmonic distortion and power factor. With growing use of power electronics, there is a tendency of the harmonic distortion levels to increase. The line current at the input of the diode bridge rectifier deviates significantly from a sinusoidal waveform and this distorted current can also lead to distortion in the line voltage. Moreover, many modern instruments use microprocessor based digital controllers, which are sensitive to variations in the voltage and current waveforms [8]. Therefore, to increase the PV system utilization the power conversion must be designed to provide harmonic and reactive power compensation also.

This paper presents a typical application of photovoltaic cell connected to a grid consisting of a dc-dc converter and an inverter. The dc-dc converter is controlled to track the maximum power point of the photovoltaic array and the inverter is to control the grid voltage in such a way that the grid current has low total harmonic distortion (THD) and it is in phase with the grid voltage. Two modulations techniques namely; Sine PWM and State Vector PWM (SVPWM) have been implemented to inject a sinusoidal current into the grid and their performance is evaluated based on different parameters.

II. GRID CONNECTED PV SYSTEMS

A PV system can be connected to a grid in two ways: 1) PV directly to the Grid wherein inverter has to perform MPPT function, and 2) PV to the Grid via DC-DC converter wherein dc-dc converter performs MPPT as shown in Fig. 1. The dc–dc converter is responsible to ensure that the PV module(s) is operated at the maximum power point [9] and dc–ac inverter controls the grid current. The photovoltaic modules are made up of silicon solar cell consisting of a p-n junction fabricated in a thin wafer or layer of semiconductor. It converts sunlight into electricity. The simplest equivalent circuit [10] of a solar cell is a current source in parallel with a shunt diode and a. series resistance R_S shown in Figure 2. The output of the current source is directly proportional to the light falling on the cell.

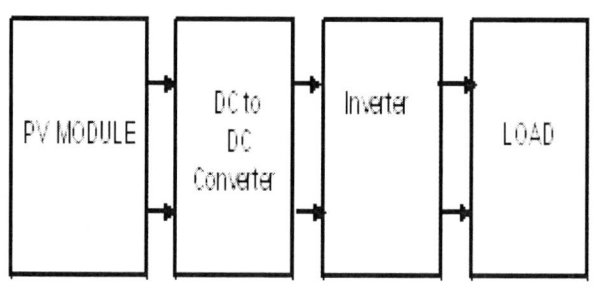

Fig. 1. Block diagram of grid connected PV system

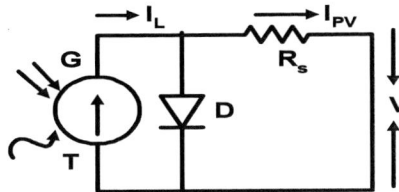

Fig. 2. Circuit diagram of the PV model

The current - voltage output characteristic of a solar cell has an exponential characteristic similar to that of a diode and given by (1)

$$I_{PV} = I_L - I_0 \left[e^{\frac{q}{\eta KT}(v+iR_s)} - 1 \right] \quad (1)$$

Where, I_L = light generated current
I_0 = reverse saturation current

Fig. 3 shows the current - voltage and power - voltage characteristics of the solar cell. It can be noted that the characteristics is highly non linear. In general, the output current of a solar array is the function of the insolation and temperature. The various important parameters of the cell are the short circuit current I_{SC}, which is almost equal to I_L. It can be noted that the cell is safe with a short circuit. The magnitude of this current depends on the area of cell and is directly proportional to solar insolation.

The open circuit voltage, V_{OC}, varies logarithmically with solar insolation, hence does not vary very much. It is sensitive to temperature. In general, a silicon solar cell, gives an output voltage of around 0.7V under open circuit condition. When many such cells are connected in series a solar PV module is obtained. Normally in a module there are 36 cells which amount for a open circuit voltage of about 20V. For obtaining higher power output the solar PV modules are connected in series and parallel combinations forming solar PV arrays [11]. The nonlinear V-I characteristics of M parallel strings with N series cells per string is given by (2)

$$V_{PV} = \frac{N}{\lambda} \ln \left(\frac{I_{SC} - I_{PV} + MI_0}{MI_0} \right) - \frac{N}{M} \times R_s \times I_{PV} \quad (2)$$

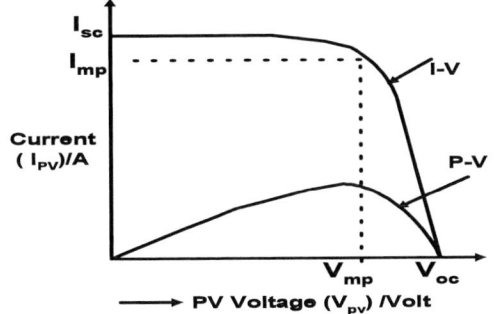

Fig. 3 V-I and P-V Characteristics of Solar Cell

where λ is a constant coefficient and depends on the cell material.

It can be noted that maximum power is available at a certain voltage V_{mp} and current I_{mp} only. So in order to obtain the maximum efficiency of PV cell, it is necessary to find automatically the voltage V_{mp} or current I_{mp} at which a PV array should operate to obtain the maximum power output P_{mp} under a given temperature and irradiance.

III. MAXIMUM POWER TRACKING

As the solar radiation varies throughout the day, the power output also varies. The principle of maximum power tracking can be explained with the help of Fig. 4 where the line having slope $1/R_o$ represents a constant load R_o. If this load is connected directly across PV cell, it will operate a power P_a in spite of the fact that maximum power P_b is available from the array. Thus, a power conditioner or DC –DC Converter is introduced between the inverter and the solar PV module which adapts the load to the array so that load characteristics is transformed along locus of maximum points and maximum power is transformed from the array. The duty cycle, D, of this converter is changed till the peak power point is obtained.

A method called, **voltage based maximum power point tracking** (VMPPT), **[12], has been used** to vary the duty cycle of the chopper. **In order to** determine the operating point corresponding to maximum power for different insolation levels equation (2) is used for computing the partial derivative of power. At P_{max}, it is zero. So value of V_{mp} is obtained for different insolation levels. A plot between V_{oc} and V_{mp} is plotted as shown in Fig. 5.

It can be seen that there exists almost a linear relationship between V_{oc} and V_{mp}. Thus,

$$V_{mp} = M_v \times V_{oc}. \quad (3)$$

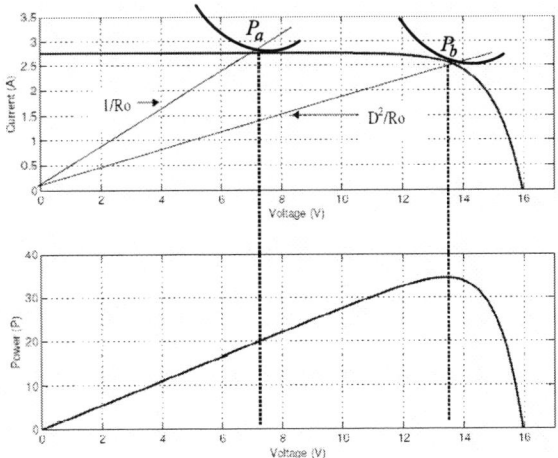

Fig. 4 Load Adapter for Resistive load

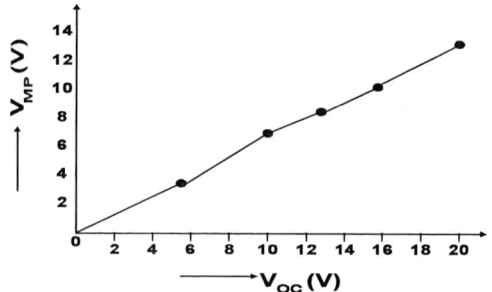

Fig. 5. Plot between Voc and Vmp

M_v is called as voltage factor and it is the slope of the linear relationship. It generally lies in the range of 0.65 to 0.80. From this curve, M_v is obtained as 0.69.

Considering a step down converter with output voltage V_o, and input voltage V_i,

$$V_o = D \times V_i \qquad (4)$$

Solving for the Impedance transfer ratio D^2

$$R_o = D^2 \times R_i \qquad (5)$$

Where R_o is output impedance and R_i is input impedance as seen by the source

$$R_i = R_o / D^2 \qquad (6)$$

Thus output resistance R_o remains constant and by changing the duty cycle, the input resistance R_i seen by the source changes. So the resistance corresponding to the peak power point is obtained by changing the duty cycle.

IV. DIFFERENT MODULATION TECHNIQUES

The output of PV cell drives a three-phase inverter shown in Fig. 6. In order to optimize the harmonics, the option of operating the pair of switching elements many times during each fundamental ac cycle, is exercised. Two schemes 1) sine PWM and 2) Space vector PWM are reviewed and implemented on simulink which woks in MATLAB environment in the following section.

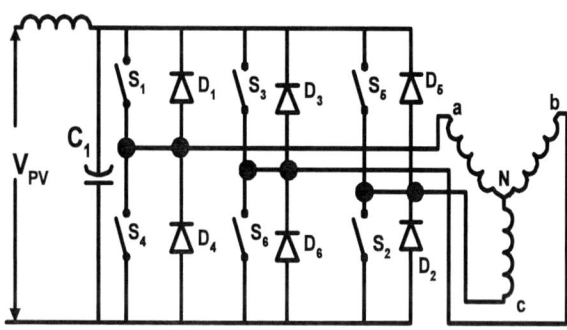

Fig. 6 Three -phase Inverter

A. Sinusoidal pulse width Modulation

In sine PWM the switching instants of the inverter legs are decided by the points of intersection of a high frequency, f_c, anywhere from 450 hertz to a few kilohertz) triangular wave, called carrier, having amplitude A_C and a sine as the reference wave of amplitude A at the frequency, f, the required fundamental frequency of output voltage. The pulses will be generated at the instants when sinusoidal output is high than triangular wave as shown in Fig. 7 for phase "a". The maximum value of the fundamental component is only 0.866 V_{PV}.

For modulation index, m_f less than one, the largest harmonic amplitude in the output voltage are associated with harmonics of order $(f_c/f) \pm 1$. Thus by increasing the number of pulses per half cycle the order of dominant harmonic frequency can be raised which can then be filtered out easily. For modulation index greater than one low order harmonics appear since for $m_f > 1$, pulse width is no longer a sine function of angular position of the pulse.

The limitation with SPWM is that the amplitude of the sine wave has to be increased in promotional to its frequency in order to keep the motor flux constant. Therefore at higher motor speeds a situation arises where the amplitude of the sine wave becomes equal to that of the triangle. If the sine wave amplitude were to increase any further the linearity between the sine and the fundamental component of the inverter output voltage is lost. This situation is referred to as over-modulation. A further consequence of over modulation is that frequency components with low harmonic numbers (5th. 7th etc.) begin to appear in the output voltage. From the implementation point of view [13], also sine triangle PWM requires some care, where the triangle frequency is limited by the switching speed of the power devices. Analog realizations give the best results in tens of output waveform quality. However, analog realizations require considerable ingenuity especially to achieve synchronization

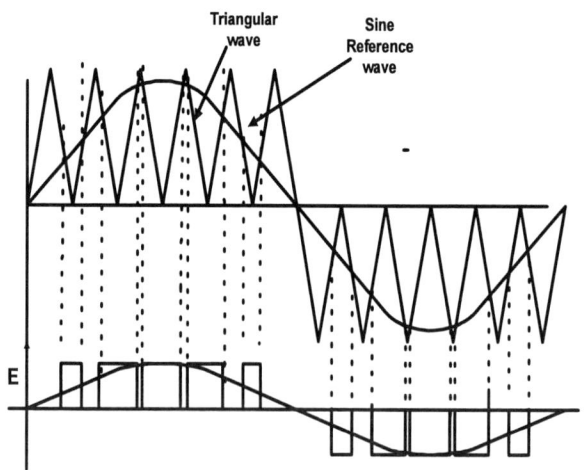

Fig. 7. Sinusoidal pulse width modulation technique

B. Space Vector PWM (SVPWM)

The space vector PWM method is based on the space vector representation of the voltages in the α, β planes. The α, β components are found by park transform, where the total power, as well as the impedances, remains unchanged [14]. In Fig. 6, S1 to S_6 are the six power switches that shape the output, which are controlled by the switching variables a, a', b, b', and c c'. When upper switches is turned on, i.e., when a, b or c is 1, the corresponding lower transistor is switched off, i.e., the corresponding a', b' or c' is 0. Therefore, the on and off states of the upper transistors S_1, S_3 and S_5 can be used to determine the output voltage. The relationship between the switching variable vector [a, b, c] 'and the line-to-line voltage vector $[V_{ab}\ V_{bc}\ V_{ca}]$ is given by (7):

$$\begin{bmatrix} V_{ab} \\ V_{bc} \\ V_{ca} \end{bmatrix} = V_{PV} \begin{bmatrix} 1 & -1 & 0 \\ 0 & 1 & -1 \\ -1 & 0 & 1 \end{bmatrix} \begin{bmatrix} a \\ b \\ c \end{bmatrix} \quad (7)$$

Also, the relationship between the switching variable vector [a, b, c] ' and the phase voltage vector $[V_{an}\ V_{bn}\ V_{cn}]$ ' can be expressed below

$$\begin{bmatrix} V_{an} \\ V_{bn} \\ V_{cn} \end{bmatrix} = \frac{V_{PV}}{3} \begin{bmatrix} 2 & -1 & -1 \\ -1 & 2 & -1 \\ -1 & -1 & 2 \end{bmatrix} \begin{bmatrix} a \\ b \\ c \end{bmatrix} \quad (8)$$

There are eight possible combinations of on and off patterns for the three upper power switches in Fig. 6. The on and off states of the lower power devices are opposite to the upper one and so are easily determined once the states of the upper power transistors are determined. According to equations (7) and (8), the eight switching vectors, output line to neutral voltage (phase voltage), and output line-to-line voltages in terms of DC-link V_{PV}, are given in Table I. Fig. 8 shows the eight inverter voltage vectors (V_0 to V_7). To implement the space vector PWM, the voltage equations in the *abc* reference frame can be transformed into the stationary *dq* reference frame that consists of the horizontal (d) and vertical (q) axes as depicted in Fig. 9.

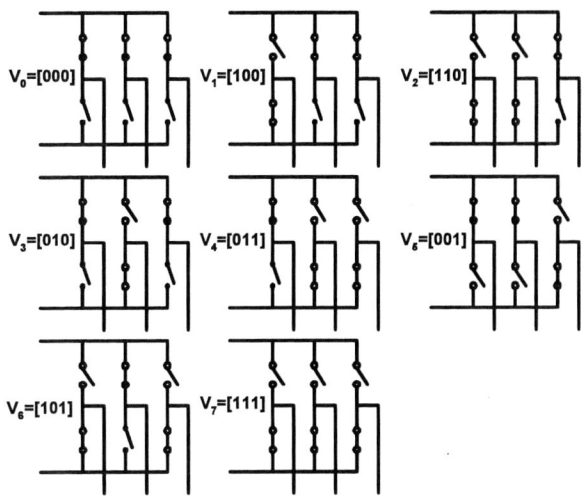

Fig. 8. Eight-inverter voltage vectors (V_0 to V_7)

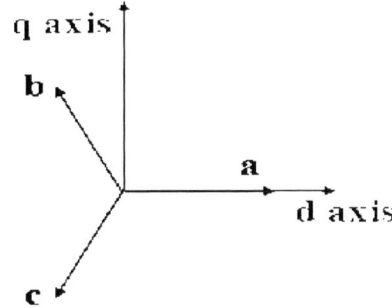

Fig. 9. Relationship between abc Reference Frame & Stationary dq reference frame

The relation between two reference frames is given by (9)

$$f\ dq0 = Ks\ \mathbf{f}\ abc \quad (9)$$

$$\text{where}\quad Ks = \frac{2}{3} \begin{bmatrix} 1 & -1/2 & -1/2 \\ 0 & \sqrt{3}/2 & -\sqrt{3}/2 \\ 1/2 & 1/2 & 1/2 \end{bmatrix}, \quad (10)$$

$\mathbf{f}\ dq\ 0 = [fd\ fq\ f0]$ ', $\mathbf{f}\ abc = [fa\ fb\ fc]$ ', and f denotes either a voltage or a current variable.

As described in Fig. 9, this transformation is equivalent to an orthogonal projection of [a, b, c] t onto the two-dimensional perpendicular to the vector [1, 1, 1] t (the equivalent d-q plane) in a three-dimensional coordinate system. As a result, six non-zero vectors and two zero vectors are possible. Six nonzero vectors (V_1 - V_6) shape the axes of a hexagonal as depicted in Fig. 10, and feed electric power to the load. The angle between any adjacent two non-zero vectors is 60 degrees. Meanwhile, two zero vectors (V_0 and V_7) are at the origin and apply zero voltage to the load. The eight vectors are called the basic space vectors and are denoted by V_0, V_1, V_2, V_3, V_4, V_5, V_6, and V_7. The same transformation can be applied to the desired output voltage to get the desired reference voltage vector V_{ref} in the d-q plane.

TABLE I INSTANTANEOUS BASIC VOLTAGE VECTORS

Voltage Vectors	Switching Vectors			Line to neutral			Line to Line		
	a	b	c	V_{an}	V_{bn}	V_{cn}	V_{ab}	V_{bc}	V_{ca}
V_0	0	0	0	0	0	0	0	0	0
V_1	1	0	0	2/3	-1/3	-1/3	1	0	-1
V_2	1	1	0	1/3	1/3	-2/3	0	1	-1
V_3	0	1	0	-1/3	2/3	-1/3	-1	1	0
V_4	0	1	1	-2/3	1/3	1/3	-1	0	1
V_5	0	0	1	-1/3	-1/3	2/3	0	-1	1
V_6	1	0	1	1/3	-2/3	1/3	1	-1	0
V_7	1	1	1	0	0	0	0	0	0

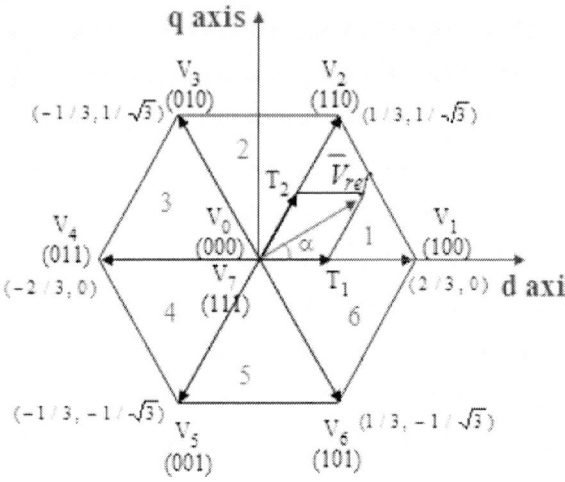

Fig. 10. Basic Switching vectors and sectors

$$T1 = T_s \times a.\sin(60 - \alpha)/\sin(\pi/3)$$
$$T2 = T_s \times a.\sin\alpha/\sin(\pi/3) \qquad (15)$$
$$T0 = T_s - T1 - T2$$

As the reference vector moves in other sectors the corresponding boundary vectors of the sector should be created during the intervals *T1* and *T2*. In order to minimize the number of switching in the inverter it is desirable that switching should take place in one phase of the inverter only for transition from one state to another. This can be achieved if the following switching sequence is used   0-1-2-7 | 7-2-1-0 | 0-1-2. Therefore the zero intervals are divided into two equal halves of length T0/2. These half intervals are placed at the beginning and end of every sampling interval. If the half at the beginning is realized as V_0 then at the end is realized as V_7 and vice versa. The switching sequence during successive sampling intervals will then be as shown in Fig. 11.

It can be seen that the chopping frequency f_c of each phase of the inverter is given by $f_c = f_s/2$ It should be pointed out that during the sampling interval the desired reference vector is approximated in the average sense, since the volt-seconds arc equated. However, instantaneously the actual vectors produced by the inverter are different from the reference vector and therefore instantaneous voltage deviations or voltage 'ripple' exists. As a result harmonic current will also flow. By following the above-mentioned switching sequence harmonics are reduced to some extent.

The objective of space vector PWM technique is to approximate the reference voltage vector V using the eight switching patterns. One simple method of approximation is to generate the average output of the inverter in a small period, T to be the same as that of V_{ref} in the same period. Consider that at a particular sampling instant the reference V_{ref} is situated in sector 1 as shown in Fig. 11. It is intuitively clear that the reference vector can be reproduced best during the period till the next sample by switching the inverter to create the vectors V_1, V_2, V_0 and V_7 in some sequence. Selecting any of the other vectors would result in a greater deviation of the actual vector from the desired reference and would thus contribute the harmonics.

V.   MATLAB SIMULATION

SUMILINK software and its facilities are used to model a grid-connected solar system with a VMPPT characteristics simulated by employing the cell short circuit current (Isc) as a measure of insolation level. For voltage-based, a block "VMPPT" computes PV cell

The switching pattern can be calculated as follows. Assume that the sampling period T_S is divided into three subintervals T1, T2 and T0. The inverter is switched so as to produce the vector V_I for T1 seconds, V_2 for T2 seconds and zero (either V_0 or V_7) for To seconds, The subintervals have to be calculated so that Ibe volt - seconds produced by these vectors along the a and b axes are the same as those produced by the desired reference vector V_{ref} i.e.

$$V_{PV} \times T1 + V_{PV}\cos 60 \times T2 = \left| V_{ref} \right|\cos\alpha \times T_s \qquad (11)$$

and $\qquad V_{PV}\sin 60 \times T2 = \left| V_{ref} \right|\sin\alpha \times T_s \qquad (12)$

where $\left| V_{ref} \right|$ is tbe amplitude or the length of the reference vector. Defining amplitude ratio a = $\left| V_{ref} \right|/V_{dc}$, Then (11) and (12) can be rewritten as.

$$T1 + T2\cos 60 \times T2 = a \times T_s\cos\alpha , \qquad (13)$$

$$T2 \times \sin 60 = a \times T_s\sin\alpha \qquad (14)$$

Solving for T1 and T2

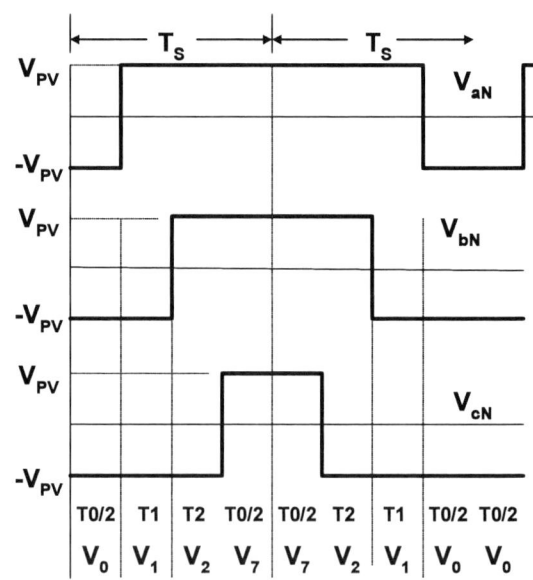

Fig. 11 Inverter phase to neutral voltages

1895

voltage using Isc and compares it with the PV output voltage to calculate the firing commands to the switch the MOSFET.

Performance is obtained for both PWM schemes and comparison has been made for suitability in grid connected photovoltaic system with following identical parameters. A 400V DC bus can be obtained by cascading several PV modules. Load: R=20 ohm, L = 20 mH, Vdc = 400V, Switching frequency = 21 KHz

Fig. 12 shows the line voltage V_{ab} along with its harmonic spectrum for Sine PWM and Fig. 13 shows the same plots for SVPWM.

Figures 14 and 15 show the phase voltages V_{an} along with their harmonic spectrum for Sine PWM and SVPWM respectively.

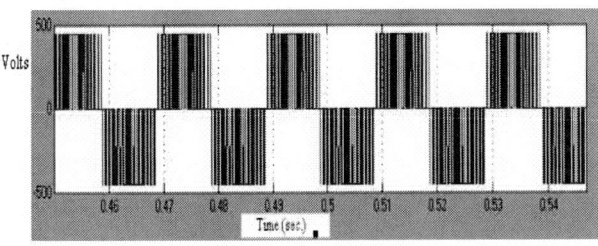

(a) Line Voltage for SPWM

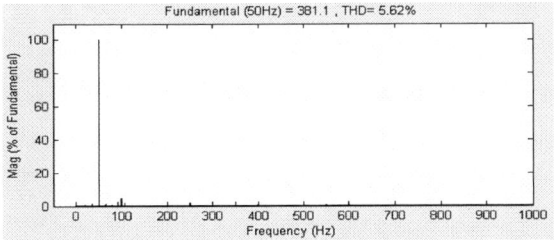

(b) Line Voltage FFT for SPWM

Fig. 12. Simulation results for SPWM

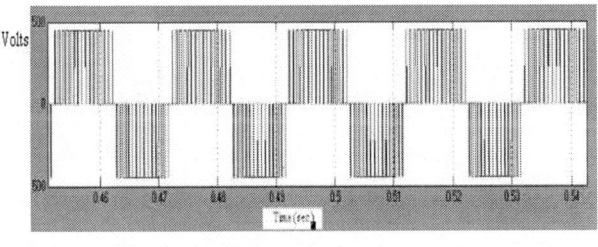

(a) Line Voltage

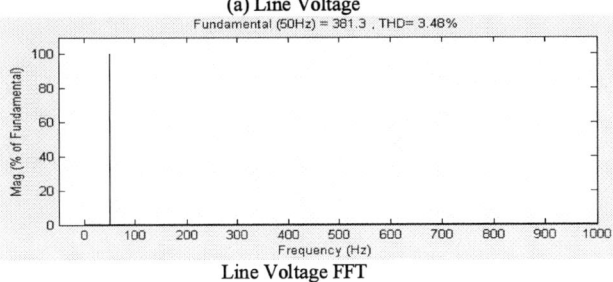

Line Voltage FFT

Fig. 13. Simulation results for SVPWM

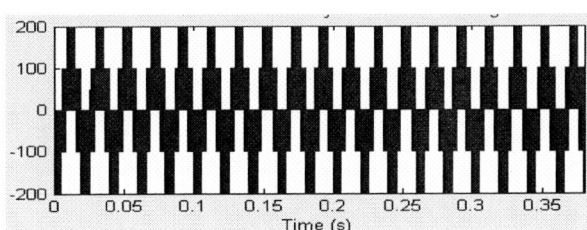

(a) Phase Voltage for SPWM

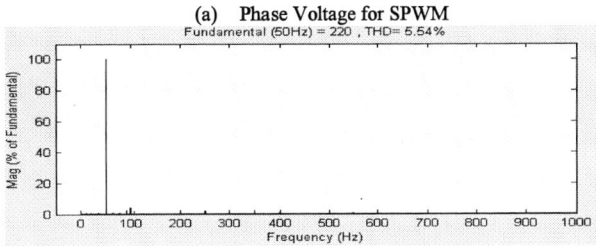

(b) FFT Analysis for SPWM

Fig 14   Simulation results for SPWM

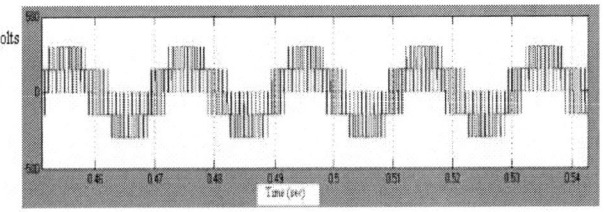

(a)   Phase Voltage

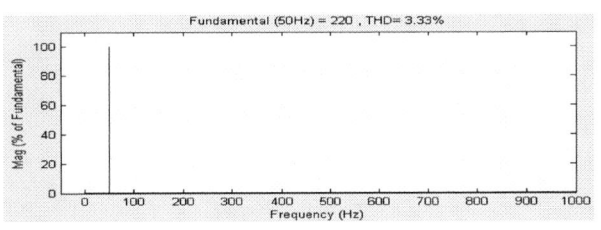

FFT of phase Voltage

Fig. 15 Simulation results for SVPWM

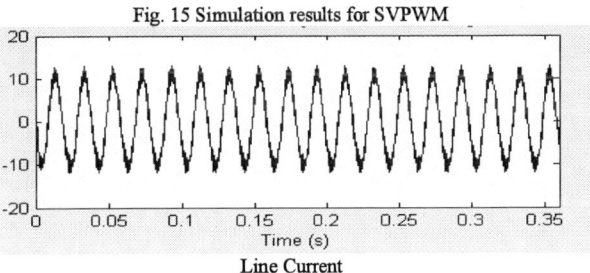

Line Current

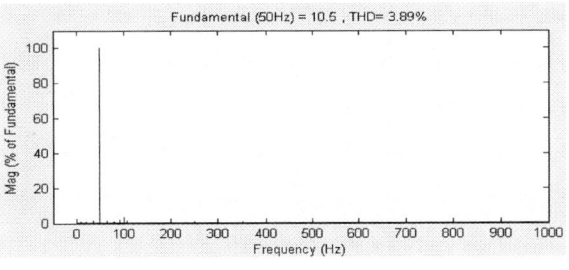

(b) FFT analysis of Line current

Fig 16   Simulation Results for SPWM

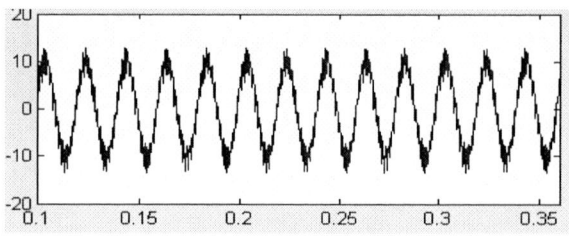

(a) Line Current

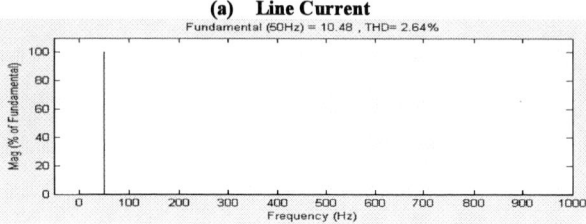

(b) FFT of Line Current

Fig 17   Simulation results for SVPWM

Table II shows the comparison between sine PWM inverter and Space Vector PWM inverter. The variation in total harmonic distortion in inverter output voltage waveform for Sine PWM technique (SPWM) and Space vector based PWM technique (SVPWM) for different modulation index is shown in Fig. 18. It can be seen that SVPWM gives lower THD in voltage as well as current. So for Grid connected PV system, SVPWM is better choice to comply with the grid standards.

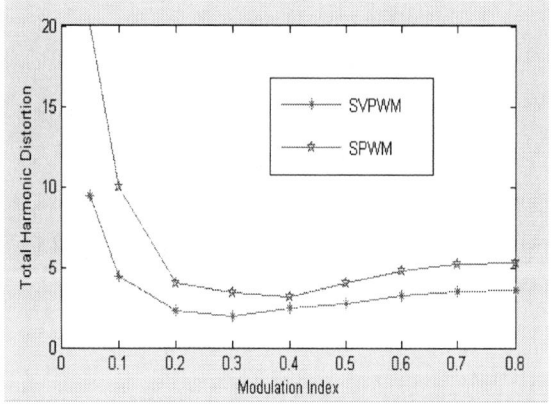

Fig. 18 Harmonic analysis of inverter output voltage

TABLE II COMPARATIVE RESULTS OF SPWM AND SVPWM

	SPWM		SVPWM	
	Fundamental	THD (in % Of fundamental)	Fundamental	THD (in % Of fundamental)
Line Voltage	381.1 V	5.62%	381.3V	3.48%
Phase Voltage	220V	5.54%	220V	3.33%
Current	10.5 A	3.89%	10.48 A	2.64%

VI. CONCLUSIONS

Modelling of solar cell has been done based on single diode model. A MATLAB script file has been written to obtain the various characteristics with changing atmospheric condition. It is seen that there exists a maximum point, which changes with changing atmospheric condition. To track the maximum power point, voltage based method has been implemented. Simulation of SVPWM inverter has been done and it is shown that it is better choice than SPWM for connecting PV system to the grid to comply with grid standards. Simulation results demonstrate the fast and effective response under changing atmospheric condition.

REFERENCES

[1] C. Hua, C. Shen, " Comparative study of Peak Power Tracking Techniques for Solar Storage Systems, " IEEE Applied Power Electronics Conference and Exposition Proceedings, Vol. 2, Feb. 1998.

[2] Arbind kumar, B.G. Fernandes, K chatterjee, "DTC of Open –End Winding Induction Motor Drive Using Space Vector Modulation With reduced switching frequency", *IEEE Power Electronics Specialists Conference, Germany, 2004*

[3] E.V. Solodovnik, S. Liu, and R. A. Dougal, "Power controller design for maximum power tracking in solar installations", *IEEE Transactions on Power Electronics, Vol. 19, Sept. 2004*

[4] C. Hua and C. Shen, "Study of Maximum Power Tracking Techniques and Control of DC/DC Converters for Photovoltaic Power System," *29th Annual JEEE PESC*, IEEE Computer Soc. Press, New York, USA, 1998

[5] Nicola Femia, Giovanni Spagnuolo, "Optimization of Perturb and Observe maximum Power Point Tracking Method", IEEE transactions on Power Electronics ,Vol. 20, No. 4,July 2005.

[6] G. Walker, "Evaluating MPPT converter topologies using a MATLAB PV model," *Journal off Electrical & Electronics Engineering, Australia*, vol. 21, 2001

[7] E.Koutroulis, K.Kalaitzakis and N.C.Voulgaris., "Development of a microcontroller- based, photovoltaic maximum power point tracking control system", *IEEE Trans. Power Electronics., vol. 16, Jan. 2001.*

[8] Johan H. R. Enslin, and Peter J. M. Heskes, "Harmonic Interaction between a Large Number of Distributed Power Inverters and the Distribution Network'', *IEEE Transactions on Power Electronics, Vol. 19, No. 6, NOV. 2004*

[9] Sachin Jain, & Vivek Agrawal, "A new algorithm for rapid tracking of approximate maximum power point in photovoltaic system," IEEE Power Electronics Letters,Vol.2,No.1, March 2004

[10] J.A. Gow ,C.D. Manning, "Development of a photovoltaic array model for use III power electronics simulation studies," IEE Proceedings on Electric Power Applications ,Vol. 146 ,no.2 ,March 1999

[11] F. Lasnier, "Solar Photovoltaic Handbook, etal. AIT Bangkok 1988

[12] Vineeta Agarwal & Alok Vishwakarma "Maximum Power Point Tracking Using VMPPT, " Proceedings of First Power Engineering and Optimization Conference" PEOCO 2007 6th June, 2007, Shah Alam, Malaysia, pp. 121-126,

[13] Du Yonggui and Xie Keming, " *The SPWM inversion mode of suppressing harmonic waves based on genetic algorithm*", Proc. of the 5th International Conf. On Electrical Machines and Systems 2001, (ICEMS 2001), vol. 2, 18-20 Aug. 2001, pp. 1304 – 1307.

[14] D.C.Lee ad G.N .Lee, "A Novel Over modulation Technique for Space Vector PWM Inverters", IEEE Trans. On Power Electronics, vol. 13, Nov. 1998, pp. 1144-1151.

This page intentionally left blank.

Author Index

A

Abdi, Ehsan 1096
Abe, Seiya..1388
Abjadi, N. R...1442
Achara, P. 394
Adélaide, L. 569
Adya, A. 731
Afjei, E. 722
Agarwal, Pramod...1810
Agarwal, Vineeta...1891
Ahmadian, H. Molla 1147
Ahn, Jin-Woo..1857
Almardy, M. 677
Alonge, F. 959
Amaral, Acácio M. R..587, 643
Amirudin, Dessy...534
Amrane, F. 569
An, Young-Joo..1527
Andersen, P. Scavenius...886
Ang, Y. 382
Ang, Yong-Ann...376
Aodsup, K. 937
Arab, G.R. 1449
Aree, P. 703
Arvindan, A. N...480
Ashaibi, Ahmed Ali...1242
Ataei, S. 722
Athab, Hussain S. 869, 874
Attaviriyanupap, Pathom...1102
Auger, Francois...1368
Ayob, S. M...1274, 1363, 1682
Azli, N. A...475, 1041, 1274, 1363, 1682

B

Bac, Nguyen Xuan..1501
Baharom, R. 1626
Baiju, M.R. 1047
Banerjee, Subrata..812
Bartholet, M.T. 257
Batzies, Ekkehard...1316
Bhat, A.K.S. 677
Bhuvaneswari, G..310
Bi, C. 1082
Bina, M. Tavakoli..465, 1060, 1065, 1799
Binder, A. 249
Bingham, C. M. 382
Bingham, Chris...376
Binh, Tran Cong..1195
Biswas, S. K. 1352, 1605
Blaabjerg, Frede...226, 541, 1247, 1376
Boonchiam, P. N...937, 1851
Boonyaroonate, Itsda..1078, 1383
Bosing, M. 1160
Branco, P.J. Costa...917
Brauer, Helge J. 716
Buatti, Gustavo M...643

Bunlaksananusorn, C. 977, 1575

C

Cangemi, T. 959
Cardoso, A. J. Marques...587, 643
Carstensen, Christian...912
Chalermyanont, Kusumal...295
Champa, P. 703
Chan, K. W. 1394, 1727, 1804
Chan, Shun-Yu...305
Chang, David..305
Chang, Tsin-Yuan..581
Chang, Y. D. 1321
Chao, Ma Xian...665
Chatratana, S. 1495
Chaudhari, M. A. 1708
Chen, Jiaxin 510
Chen, L. 1538
Chen, Sufen 1262
Chen, W. C. 440
Chen, Wei 1636
Chen, Y. H. 456, 1278
Chen, Y. M. 1321
Chenfeng, Yang..745
Cheng, Chien-Lung..749, 1703
Cheng, K. W. Eric...1691, 1697
Cheng, K.W.E...1727
Cheng, Qiang...1330
Chengfeng, Yang...270, 427
Chereau, Vinciane..1368
Chern, Shyi-Ching..749, 1703
Cheung, N.C...1727
Cheung, Norbert C...1691, 1697
Chi, Chien-An..388
Chiang, Wen-Jung..824
Chien, F.T. 660
Chin, Li-Yuan..305
Chivite-Zabalza, F. Javier.......................................788, 796, 804
Cho, B. H. 401
Cho, Kyu Min..665, 1142
Choi, S. J. 401
Choudhary, Sonika..1757
Chrin, P. 1575
Chudamani, R. 1827
Chudoung, Nakharet..1213
Chun, Tae-Won...1857
Chunkag, V..863
Ciobotaru, M. 226
Colak, Baris..1507
Corradini, L. 600
Cosic, A. 1301
Cruden, A. 1182

D

Dahlan, N.Y...527
Dahono, P. A. 1267
Dahono, Pekik Argo..534

Author Index

Dai, Z. 1885
Dananjayan, P. 626
Davat, Bernard 1
Deb, N. K. 1352, 1605
Dehbonei, H. 1657
Deleroi, W. 1495
Deng-Em, S. 1851
Deni, ... 534, 1267
Densei-Lambda, K.K. 280
Desai, Hardik P. 829
Dhomane, G. A. 1590
Dick, Christian P. 448
D'ippolito, F. 959
Ditmanson, C. 556
Doki, Shinji 999
Doncker, R. W. De 907, 912, 1160
Doncker, Rik W. De 213, 327, 333, 448, 710, 716
Dong, Lei 1691
Dong, Ming-Chui 607, 614
Dong, Yang 1340
Dorkmai, Pramoch 697
Dorrell, David G 886, 922, 1167, 1174
Duan, S.X. 551
Duan, Shanxu 836, 842
Dwivedi, Avneesh 1757
Dzung, Phan Quoc 1195, 1202, 1501

E

Ekkaravarodome, Chainarin 1383
Ertan, H. Bulent 1507
Eskandari, B. 1060, 1065

F

Fang, D. Z. 1394, 1804
Fang, Kuo-Lun 1610, 1712
Fang, Tzu-Hsuan 1717
Fei, Wanmin 350, 354, 1672
Ferraz, Antonio 817
Ferreira, O.C. 1017
Fidler, Peter 327
Fingerhuth, S. 1160
Finney, S.J. 299, 1242
Finney, Steve J. 1255
Foroosh, S. Chini 1465
Forsyth, Andrew J. 788, 796, 804
Foster, M. P. 382
Foster, Martin 376
Fuengwarodsakul, Nisai H. 710
Fukuda, Shoji 1070
Fukushima, K. 1885

G

Gairola, Sanjay 738, 899
Gao, F. 1247, 1376
Garg, Vipin 310
Geethalakshmi, B. 626

Goel, P.K. 941
Gonthier, L. 322
Gopinath, Anish 1047
Goyal, Devendra 1520
Grant, D M 368
Grantham, Colin 1284
Gruber , W. 574
Guan, Xiaohan 994, 1752
Gueldner, H. 1006
Guldner, H. 556
Guo, Youguang 275, 510, 1662
Gupta, H.O. 1810
Gupta, J.R.P. 731

H

Hai, Quach Thanh 1033
Hajian, M. 1449
Hamzah, M.K. 527, 1626
Hamzah, N.R. 527, 1626
Han, Ying-Duo 607, 614
Hansen, P. E. 886
Haque, M. Tarafdar 620
Harada, Y. 1885
Hasegawa, Masaru 1543
Hellinger, R. 1006
Hennen, Martin D. 716, 907
Hew, W.P. 1514
Heyun, Lin 270, 427, 745
Higuchi, Kohji 280
Hinkkanen, Marko 406
Hirokawa, Masahiko 1388
Hirota, Atsushi 1740
Ho, Shine-Tzong 498, 1788
Hoang, Nguyen Minh 1195, 1501
Hofmann, W. 781, 1538
Hotait, Hadi A 299, 1255
Hothongkham, Prasopchok 1236
Hsieh, C. T. 1567
Hsu, Chih-Jen 286
Huang, P. L. 1321
Hung, Tsung-You 1762, 1767
Hwu, K. I. 338, 456, 692, 1278

I

Idris, Z. 527
Iov, F. 226
Ishitobi, Manabu 504
Islam, S. 1834
Iso, Osamu 1102

J

Jang, B.H. 1657
Jangjaempradit, Saksit 1641
Jangwanitlert, A. 989, 1412
Janjornmanit, Suchart 1327
Jayashree, E. 1555

Author Index

Jeevananthan, S. 1221
Jegathesan, V. 1677
Jerome, Jovitha 1677
Jeung, Giwoo 1092
Jian, Guo 270, 427, 745
Jou, Hurng-Liahng 493, 824
Jovanovic, Milutin G 922
Junge, Christian 1533
Jwo, W. S. 1560

K

Kadir, M. N. Abdul 1514
Kaewsingha, Aswin 1327
Kamnarn, U. 863
Kamper, M.J. 420, 1017, 1295
Kando, M. 488
Kang, Yong...551, 836
Kano, Masaru..414
Kanthaphayao, Y. 863
Kanzi, K. 465
Karunakar, K 1620
Karutz, P. 574
Kasal, Gaurav Kumar 357
Kasper, K. A. 1160
Kavitha, A. 595
Kazimierczuk, Marian K.................................1136
Kennel, R.M. 1017
Kerz, O. 363
Khaehintung, Noppadol 847, 1429
Khajeh, A. 1455
Khalil, Ahmed G. Abo- 1471
Khan, P. K. Shadhu. 869, 874
Khan-Ngern, Werachet 460, 1335
Khomfoi, Surin 1055, 1228
Khun, C. 488
Kim, Dong-Hun 1092
Kim, H. S. 1846
Kim, Hee Jun 665, 1142
Kim, Heung-Geun 1092
Kim, I. C. 1846
Kim, In Dong 1092
Kinnaraes, V. 1483
Kinnares, V. 1356, 1489
Kinnares, Vijit 1236
Kittiratsatcha, S. 977
Ko, S. H. 1657, 1846, 1846
Ko, T.K. 1657
Ko, Yi-Pin 388
Kobayashi, Takayuki 265
Kock, H.W. De 1017
Koenig, Andreas 327
Kohama, Teruhiko 1417
Kok, W. Sae- 368
Kolar, J.W. 257, 574
Kongsuk, P. 937
Kongthawornwattana, P. 977
Konig, Andreas 448

Krein, Philip T. 221
Krismadinata, ...1290
Kubota, Hisao 265
Kulvitit, Youthana 342, 697
Kumar, S. Ganesh 1632
Kumar, S. Krishna 1632
Kumchaiyo, Ruthapong 1078
Kunakorn, Anantawat 1429
Kuo, J. S. 440
Kuo, Jian-Long 1717, 1722
Kurokawa, F. 968, 1398
Kusuhara, Yoshito 954
Kwok, K. W. 1727
Kwok, Y. L. 1727
Kwon, Soon Kurl 504

L

Laczynski, T. 1645
Lafzi, A. 620
Lai, Y. M. 1262
Lai, Yen-Shin 1586
Lakhdari, Z. 569
Lan, Yi-Hung 749
Lee, Chien-Min 1586
Lee, Dong-Choon 1471
Lee, Dong-Hee 1527, 1857
Lee, Hong Hee 1027, 1033
Lee, S. R. 1657, 1846
Lee, S. W. 1657, 1846
Lee, Yuang-Shung 286, 388
Lei, Dong 949, 1340, 1697
Lei, Yuzhou 291, 1777
Leibfried, T. 363, 726
Lenke, Robert U. 213
Lenwari, W. 470
Leou, Rong Ceng 546
Lerdudomsak, Smith......................................999
Li, X. 551
Li, Y. J. 440
Liang, C. 1376
Liao, C.N. 660
Liao, Xiaozhong 1691
Lijie, Wang 949
Lim, P. Y. 475, 1041
Lim, S. H. 1846
Lim, T.C. 299
Lin, Chang-Hua 1610, 1712, 1762, 1767
Lin, Chih-Hong 1549
Lin, H. C. 1567
Lin, Hung-Chih 581
Lin, Min 1423
Lipo, Thomas A. 1308
Liu, B.Y. 551
Liu, Bangyin 836, 842, 1636
Liu, Dikai 275
Liu, Fei 842
Liu, Maw-Yang 1610, 1712

Author Index

Liu, Xian-Lin 1722
Liu, Yi-Hwa 546
Liu, Yuanchao 291, 1615, 1752, 1777
Liu, Z. 551
Loh, P. C. 1247, 1376, 1620
Loh, Poh Chiang 541
Loron, Luc 1368
Lu, Haiyan 275
Lu, Y. 1727
Lu, Zhengyu 354, 1088
Luomi, Jorma 406

M

Ma, Yu 632, 1088, 1601
Macheiner, P. 880
Madawala, U. K. 648, 654
Makany, Ph. 569
Makino, Tomoaki 1740
Manmek, Thip 773
Mao, Peng 1615
Markadeh, Gh. R. Arab 1442
Marques, Gil D. 636
Martin, F. 363
Martins, J. F. 894, 1875
Masoum, Amir S. 767
Masoum, M.A.S. 767, 1834
Massoud, A.M. 299
Massoud, Ahmed M. 1255
Massoud, Ahmed 1242
Matsui, Keiju 1543
Matsui, N. 1398
Matsui, Y. 1109
Matsui, Yasuaki 1102
Matsuo, K. 1686
Matsuse, K. 394, 685
Matsuse, Kouki 521, 1460
Mattavelli, P. 600
Mattavelli, Paolo 760
Mcmahon, Richard 1096
Medagam, Peda V 1477
Mekhilef, S. 1514
Meng, Peipei 632
Mertens, A. 1645
Meyer, Christoph 213
Milani, A. Roshan 620
Miri, A. M. 726
Mirmousa, H. 1404
Mishima, Tomokazu 563
Mithulananthan, N. 937
Mittal, A.P. 731
Mittal, Raghu K. 1757
Miura, T. 1686
Miyamoto, Hiroyuki 1773
Moallem, Ali 983, 1147
Modak, J. P. 1708
Moghani, J. S. 1455
Mondal, N. 1352, 1605

Moon, Y.H. 1657
Morimoto, Masayuki 1641, 1773
Morita, Katsuaki 1773
Moses, Paul S. 767
Mossner, K. 363
Mudannayake, Chathura P. 773
Mun, Sang Pil 504, 563
Mura, Florian 213
Muraoka, Hidekazu 563
Murthy, S.S. 941, 1123, 1757

N

Nabeshima, Takashi 858, 1423, 1734
Naetiladdanon, Sumate 755
Nagai, Satoshi 1740
Nakagawa, Shin 954
Nakanishi, Hirotaka 858, 1734
Nakano, Kazushi 280
Nakano, Tadao 858, 1734
Nakaoka, Mutsuo 504, 563
Nakayama, Asahi 954
Nandhakumar, R. 1221
Nathakaranakule, Adisak 1383
Navi, K. 722
Nazarzadeh, Jalal 1782
Neammanee, B. 1129
Neuhaus, Christoph R. 710
Ngern, W. Khan- 488, 515, 1667, 1794
Ngoc, Ha Pham 1102
Nguyen, Binhminh 1434
Nho, Eui-Chel 1527
Nho, Nguyen Van 1027, 1033
Nia, S.Hosein 1449
Ninomiya, T. 1885
Ninomiya, Tamotsu 954, 1388, 1417
Nishijima, Kimihiro 858, 1423, 1734
Nishimura, Jun. 1460
Noguchi, Toshihiko 414, 1595, 1651
Noor, S.Z. Mohammad 1626
Norigoe, I. 1885
Nussbaumer, T. 257, 574

O

Obata, S. 671
Ogura, K. 1109
Oh, Won Seok 1142
Oka, Kazuo 521, 1460
Okuma, Shigeru 999
Omori, Hideki 563
Opanuruk, Puckapon 342
Oranpiroj, Kosol 318
Owatchaiphong, Satit 912
Ozdemir, Engin 1055
Ozdemir, Sule 1055

Author Index

P

Pai, Kai-Jun 1762, 1767
Pal, Jayanta 812
Palandurkar, M.V. 1708
Panda, Sanjib K. 852
Park, Hong-Geuk 1471
Park, J. H. 401
Pashajavid, E. 465
Passal, A. 322
Patel, H. K. 829
Pavitra, G. 1757
Peng, S.T. 930
Phuong, Le Minh 1195, 1202, 1501
Piboonwattanakit, K. 1667
Piippo, Antti..406
Pinto, A.J.P. 1123
Pires, A. J. 894, 917
Pires, V. Fernao 894, 1875
Plum, Thomas 327, 333
Pothana, Aravind 1152
Pothi, N. 1208
Pourboghrat, Farzad 1477
Prasad, Dinkar...812
Prasertsit, Anuwat 295
Premrudeepreechacharn, Suttichai.....318, 1208
Pusorn, W. 1851

Q

Qian, Zhaoming 632, 1088, 1601
Qu, Yilong 1822, 1880

R

Rafael, Silviano.......................................917
Rahim, Nasrudin Abd 1290
Rahimzadeh, S. 1799
Rahman, Muhammed Fazlur......................1284
Rakpenthai, C. 1208
Ramalingam, C.S. 1827
Ramli, M. Z. 1274
Randewijk, P.J. 420, 1744
Rentzsch, M. 556
Ribeiro, Antonio C...................................636
Ribeiro, Hugo 643
Ritchie, E. 1167
Rizqiawan, Arwindra 534
Rockhill, Andrew A. 1308
Rong, Runjie 541
Rossouw, F.G. 1295
Rost, J. 1006

S

Sadarangani, Chandur.....................1012, 1301
Saggini, S. 600
Saha, Bishwajit 504, 563
Saito, Y. 671

Sakulhirirak, D. 515
Salam, Z. 1274, 1363, 1682
Sanajit, N. 1412
Sangampai, Pairote 295
Sangwongwanich, Somboon 1213
Sankar, S. Siva..1632
Sano, Kohji 1595
Saparon, A. 527, 1626
Saritsiri, Kritsada 460
Sato, Akira 1651
Sato, S. 1109
Sato, Terukazu.................858, 1423, 1734
Sawatpipat, P. 1840
Sawetsakulanond, B. 1356, 1483, 1489
Schmidt, I. 1115
Schneider, T. 249
Scholler, Tobias 1316
Schroder, D. 1187
Schuster, H. 1187
Sebastiao, Pedro J...................................636
Sekine, T. 671
Sekiya, Hiroo 1136
Selvaraj, Jeyraj 1290
Senicar, Florian 1533
Sera, D. 226
Sezgin, Volkan 1507
Shah, Laxman 1182
Shahbazi, M. 1455
Shao, Shiyi 1096
Shariatmadar, S. Mohammad 1782
Sharma, Deepen......................................973
Sharma, V. K. 480
Shen, C.L. 930
Shi, Hu 949
Shiang, J. Z. 433, 1560
Shibano, Yusuke......................................265
Shisha, Samer 1012
Shuang, Gao 949, 1697
Shuhua, Fang 270, 427, 745
Silber, S. 257
Silva, J. Fernando 1875
Sim, J. M. 401
Sing, Bhim 941
Singer, A. 781
Singh, Bhim...........58, 310, 357, 731, 738, 899, 1520, 1816
Singhal, Varun.......................................1816
Sinha, S. 1352, 1605
Sirisuk, Phaophak 847, 1429
Sirisumrannukul, S. 1495
Skorokhod, Y.Y. 1868
Sode-Yome, A. 937
Soh, C.S. 1082
Soltani, J. 1442, 1449
Somsiri, P. 703
Son, Kwang-Myoung 1471
Songboonkaew, J. 989
Soter, Stefan 1533
Soulard, J. 1301

Author Index

Sousa, Duarte M. 636, 817
Srisongkram, W. 1851
Stone, D. A. 382
Stone, David 376
Stumberger, R.H. 1346
Su, Ching-Hung 1610, 1712
Su, Y.-H. 433
Subsingha, W. 1851
Sudhakar, S. Bala..1581
Sudmee, W. 1129
Suetsugu, Tadashi 1136
Sugawara, A. 1109
Sugimura, Hisayuki 504, 563
Sukita, S. 968
Sumner, M. 470
Sun, J. Q. 1394
Sun, Yu-Hua 493
Supriatna, E. G. 1267
Suryawanshi, H. M. 1590
Svechkarenko, D. 1301

T

Ta, Minh C. 1434
Tahami, F. 1147, 1465
Tai, Sio-Un 607, 614
Takeda, T. 1109
Takegami, Eiji 280
Tan, K. 1834
Tan, Siew-Chong 1262
Tan, Weipu 1822, 1880
Taniguchi, T. 1686
Tansatit, Tanvaa............................342, 697
Tarateeraseth, V. 515
Tarnekar, S. G. 1708
Tayjasanant, T. 1840, 1862
Tedeschi, Elisabetta 760
Tenca, Pierluigi 1308
Teng, Jen-Hao...305
Teng, L. Y. 1041
Tenti, P. 600
Tenti, Paolo...760
Teo, K.K. 1082
Teodorescu, R. 226, 1247
Teshnizi, Hesameddin Mirzaee.....................983
Theinmontri, Surapon...................................295
Thrimawithana, D. J. 648, 654
Tiwari, S.K. 941
To, Huu-Phuc...1284
Tolbert, Leon M. 1055, 1228
Tomihisa, Yoshihiro 858, 1734
Tomioka, Satoshi 280
Tomita, H. 671
Trevisan, D. 600
Tsai, Y.T. 660
Tse, Chi K. 1262
Tseng, S. Y. 440, 433, 1321, 1560, 1567
Tsukakoshi, K. 1885

Tsunesada, Ryota..1417
Tungpimonrut, K. 703

U

Ueda, Shigeta...1070
Ulinuha, A. 1834
Uma, G. 595, 1555, 1632
Uyaisom, C. 1794

V

Vadirajacharya, K..1810
Vaigundamoorthi, M. 1555
Vargas, Ismael Araujo-...................788, 796
Vasudevan, Krishna..............1152, 1827
Veerachary, M. 973, 1581
Veszpremi, K. 1115
Vilathgamuwa, D M 1247, 1620
Vinh, Pham Quang 1195, 1202, 1501
Viriya, P. 394, 685
Vishwakarma, Alok 1891
Vogelsberger, M.A. 1346
Volskiy, S.I. 1868
Vorlander, M. 1160

W

Walker, J. A. 1167
Wang, Chengzhi 1636
Wang, Chien-Ming...........1610, 1712, 1762, 1767
Wang, Hua..1662
Wang, Peng 541
Wang, Qi 350
Wang, R-J. 420
Wang, Shoufang 1672
Wang, Shuhong..275
Wang, Shun-Chung 546
Wang, Xixi...1691
Wangsathitwong, S. 1495
Watanabe, Kazushi 280
Watanabe, Takayuki 1136
Wegener, Ralf 1533
Weihrauch, N. C. 886
Welker, Volkmar 1316
Weller, A. 1006
Westermaier, C. 1187
Williams, Barry W..............299, 1182, 1242, 1255
Wipasuramonton, P. 703
Wolbank, T.M. 880, 1346
Wong, Man-Chung 607, 614
Wu, Jinn-Chang.....................................493, 824
Wu, Ming-Yi 1703
Wu, T. F 1321
Wu, Xinhui 852

X

Xiaozhong, Liao 949, 1340, 1697

A-6

Author Index

Xie, Xiaogao 1601
Xiping, Liu 270, 427, 745
Xu, Hai 1142
Xu, Jianxin 852
Xu, Pengwei 842
Xu, Yun 1636

Y

Yachiangkam, Samart 1327
Yang, C. M. 1560
Yang, Yihan 1822, 1880
Yau, Y. T. 338, 692
Yeh, Jim-Chwen 749, 1703
Yeon, Jae Eul 665
Yingkayun, Krisda 318
Yongyuth, N. 685
Yoothanom, N. 515
Yoshida, Takatsugu 1070
Yoshimura, S. 671
Yoshioka, Satoshi 1543
Yossombut, K. 1840
Yun, S. T. 401

Z

Zanchetta, P. 470
Zhan, Yuedong 1662
Zhang, Dongyan 994
Zhang, H.B. 299
Zhang, Junming 632
Zhang, Weiping 291, 994, 1330, 1615, 1752, 1777
Zhang, Xiaofeng 1088
Zhang, Xiaoqiang 291, 1777
Zhang, Yanli 350, 354, 1672
Zhao, Xusen 1752
Zhijun, E. 1804
Zhu, G.R. 551
Zhu, Jianguo 275, 510, 1662
Zirn, Oliver 1316
Zolghadri, M.R. 1404
Zolghadri, Mohammadreza 983
Zoller, T. 726
Zou, Yunping 1636